Key Papers in
The Development of Information Theory

D1567113

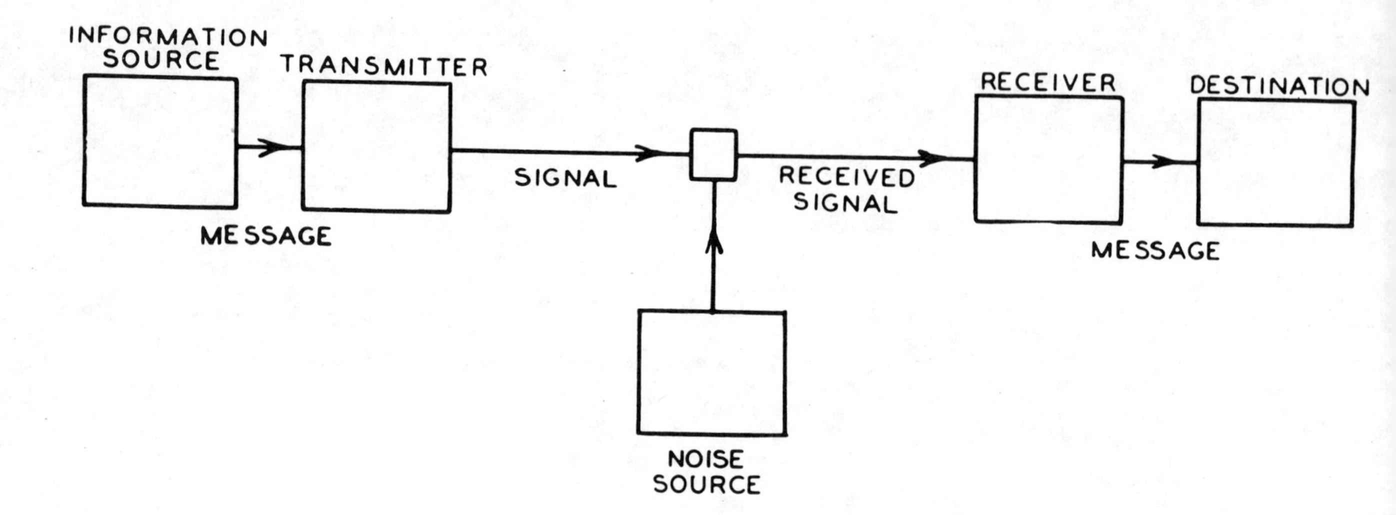

The Information Theorist's Coat of Arms

Shannon's canonical block diagram of the one-way communication system. (Reproduced with permission from "A Mathematical Theory of Communication," C. E. Shannon, *Bell System Technical Journal*, October 1948.)

Key Papers in
The Development of Information Theory

Edited by
David Slepian
Member, Technical Staff
Bell Laboratories
and
Professor of Electrical Engineering
University of Hawaii

A volume in the IEEE PRESS Selected Reprint Series, prepared under the sponsorship of the IEEE Information Theory Group.

The Institute of Electrical and Electronics Engineers, Inc. New York

IEEE PRESS SELECTED REPRINT SERIES

Clearing the Air: The Impact of the Clean Air Act on Technology
Edited by John C. Redmond, John C. Cook, and A. A. J. Hoffman
Active Inductorless Filters
Edited by Sanjit K. Mitra
A Practical Guide to Minicomputer Applications
Edited by Fred F. Coury
Power Semiconductor Applications, Volume I: General Considerations
Edited by John D. Harnden, Jr. and Forest B. Golden
Power Semiconductor Applications, Volume II: Equipment and Systems
Edited by John D. Harnden, Jr. and Forest B. Golden
Semiconductor Memories
Edited by David A. Hodges
Minicomputers: Hardware, Software, and Applications
Edited by James D. Schoeffler and Ronald H. Temple
Digital Signal Processing
Edited by Lawrence R. Rabiner and Charles M. Rader
Laser Theory
Edited by Frank Barnes
Integrated Optics
Edited by Dietrich Marcuse
Literature in Digital Signal Processing: Terminology and Permuted Title Index
Edited by Howard D. Helms and Lawrence R. Rabiner
Laser Devices and Applications
Edited by Ivan P. Kaminow and A. E. Siegman
Computer-Aided Filter Design
Edited by George Szentirmai
Key Papers in the Development of Coding Theory
Edited by E. R. Berlekamp

1973 IEEE Press
Editorial Board
E. E. Grazda, *Editor*

Robert Adler
Walter Beam
M. K. Enns
R. C. Hansen
R. K. Hellmann
E. W. Herold
Y. C. Ho
W. G. Howard
Thomas Kailath
Dietrich Marcuse
D. T. Michael
Sanjit Mitra
J. H. Pomerene
Irving Reingold
A. E. Seigman
J. B. Singleton

W. R. Crone, *Managing Editor*

Copyright © 1974 by
THE INSTITUTE OF ELECTRICAL AND ELECTRONICS ENGINEERS, INC.
345 East 47 Street, New York, N. Y. 10017
All rights reserved.

International Standard Book Numbers
Clothbound: 0-87942-027-8
Paperbound: 0-87942-028-6

Library of Congress Catalog Card Number 73-77997

PRINTED IN THE UNITED STATES OF AMERICA

Contents

Part III: Many Terminal Channels

Introduction

Probably no single work in this century has more profoundly altered man's understanding of communication than C. E. Shannon's *A Mathematical Theory of Communication*, first published in 1948. Shannon observed that a great many communication systems can be represented by the block diagram shown in the frontispiece. He constructed simple mathematical models that describe the macroscopic functioning of these blocks, and studied how these models fit together. There resulted theorems of great power, elegance, generality, and beauty. They have shed much understanding on the elusive true nature of the communication process and have delineated its inherent limitations.

The ideas in Shannon's paper were soon picked up by communication engineers and mathematicians around the world. They were elaborated upon, extended, and complemented with new related ideas. The subject thrived and grew to become a well-rounded and exciting chapter in the annals of science.

Twenty-five years have now elapsed, and it is a fitting moment to take stock. The present volume, together with a companion one—*Key Papers in the Development of Coding Theory*, edited by E. R. Berlekamp—is intended both to commemorate this special anniversary of Shannon's extraordinary contribution, and to provide the reader with a detailed reference guide to the many developments that followed.

The collection of papers that follows should not be mistaken for a textbook. A novice to the subject would be ill-advised to start his education here; any of the texts listed in the Reference section will better serve his needs. Rather, this volume is for the advanced student in information theory, for the active research worker in the field, and for the student of its history. It presents in one convenient collection much of the basic source material of the subject.

In choosing papers to include within these covers, many arbitrary decisions had to be made. First of all, it was decided that information theory would mean Shannon theory—ideas derived from coding, entropy, channel capacity, etc.—and not the wider discipline of statistical communication theory that is also sometimes called information theory. It was clear from the start that the subject of error-correcting codes had grown so as to warrant a book of its own. It was at first decided that all matters concerned with explicit code construction would appear in the second volume, whereas bounds on codes and other theoretical matters closer to the heart of Shannon's original papers would appear here. Of course, this separation turned out to be difficult to define clearly, and borderline papers were ultimately assigned rather arbitrarily to one book or the other. As matters developed, it turned out to be convenient to keep *all* papers dealing with non-block codes with the code construction papers, and so no material on sequential or convolutional codes is to be found within the present volume.

The task of selecting a small set of "key" papers from the large literature of 25 years of active research in information theory has been most difficult. I have tried to limit my choices to contributions that in some way advanced the central theory or provided new ideas. Many excellent and important papers concerned with applications or the working out of difficult details have consequently been omitted.

Again, it was quite beyond my ability to deal adequately with the vast Soviet literature: I have included a few well-known shorter papers as representative samples to whet the reader's appetite. An excellent detailed guide to this larger literature is provided by Dobrushin's recent survey article listed in Section III of the References.

The list of criteria, rationalizations, and apologies for what is here included and what is here omitted could go on at length. In the last analysis, I must assume responsibility for the choices; these are papers that I have liked; these are papers that I have found important. Nor are they the only ones in their category. Space limitations have unhappily prevented inclusion of them all.

Concerning the papers themselves, they fall rather naturally by subject matter into three groups: The Classical Source and Channel; Rate Distortion Theory; and Many Terminal Channels. Within each category, the papers have been listed chronologically. I feel that they speak for themselves and need no further introduction. Each has contributed in some important way to the body of knowledge now known as information theory.

References—A Mini-List

I. Introductory Material

A. Elementary

[1] J. R. Pierce, *Symbols, Signals and Noise*. New York: Harper and Row, 1961.

[2] D. Slepian, "Information theory," *IRE Stud. Quart. EE Dig.*, pp. 10-18, Nov. 1962.

[3] B. McMillan and D. Slepian, "Information theory," *Proc. IRE*, vol. 50, pp. 1151-1157, May 1962.

[4] C. E Shannon, "Information theory," *Encyclopedia Britannica*, vol. 12. Chicago: Encyclopedia Britannica, Inc., 1970, pp. 246-249.

[5] E. N. Gilbert, "Information theory after 18 years," *Science*, vol. 152, pp. 320-326, Apr. 15, 1966.

B. Advanced

[1] R. Gallager, "Information theory," in *The Mathematics of Physics and Chemistry,* vol. 2, H. Margenau and G. Murphy, Ed. Princeton, N.J.: Van Nostrand, 1964, pp. 190–248.

[2] A. D. Wyner, "Another look at the coding theorem of information theory," *Proc. IEEE,* vol. 58, pp. 894–913, June 1970.

II. Textbooks—Listed Roughly in Order of Increasing Difficulty and Comprehensiveness

A. Elementary

[1] J. F. Young, *Information Theory.* New York: Interscience, 1951.

[2] G. Raisbeck, *Information Theory.* Cambridge, Mass.: M.I.T. Press, 1964.

[3] F. M. Ingels, *Information and Coding Theory.* San Francisco: Intext Educational Publishers, 1971.

[4] D. A. Bell, *Information Theory and its Engineering Applications.* New York: Pitman, 1962.

[5] N. Abramson, *Information Theory and Coding.* New York: McGraw-Hill, 1963.

B. Intermediate

[1] F. M. Reza, *Introduction to Information Theory.* New York: McGraw-Hill, 1961.

[2] W. Meyer-Eppler, *Grundlagen und Anwendungen der Informationstheorie.* Berlin: Springer, 1959.

C. Advanced

[1] A. Ya. Khinchin, *Mathematical Foundations of Information Theory.* New York: Dover, 1957.

[2] A. Feinstein, *Foundations of Information Theory.* New York: McGraw-Hill, 1958

[3] R. M. Fano, *Transmission of Information.* New York: Wiley, M.I.T. Press, 1961.

[4] R. Ash, *Information Theory.* New York: Interscience, 1965.

[5] J. Wolfowitz, *Coding Theorems of Information Theory.* Berlin: Springer, 1964.

[6] F. Jelinek, *Probabilistic Information Theory.* New York: McGraw-Hill, 1968.

[7] T. Berger, *Rate Distortion Theory.* Englewood Cliffs, N.J.: Prentice-Hall, 1971.

[8] R. G. Gallager, *Information Theory and Reliable Communication.* New York: Wiley, 1968.

[9] M. S. Pinsker, *Information and Information Stability.* San Francisco: Holden-Day, 1964.

III. Bibliographies and History

[1] S. Kotz, *Recent Results in Information Theory.* London: Methuen, 1966.

[2] R. L. Dobrushin, "A survey of Soviet research in information theory," *IEEE Trans. Inform. Theory,* vol. IT-18, pp. 703–724, Nov. 1972.

[3] J. R. Pierce, "The early days of information theory," *IEEE Trans. Inform. Theory,* vol. IT-19, pp. 3–8, Jan. 1973.

[4] D. Slepian, "Information theory in the fifties," *IEEE Trans. Inform. Theory,* vol. IT-19, pp. 145–147, Mar. 1973.

[5] A. J. Viterbi, "Information theory in the sixties," *IEEE Trans. Inform. Theory,* vol. IT-19, pp. 257–262, May 1973.

Part I

The Classical Source and Channel

A Mathematical Theory of Communication

By C. E. SHANNON

INTRODUCTION

THE recent development of various methods of modulation such as PCM and PPM which exchange bandwidth for signal-to-noise ratio has intensified the interest in a general theory of communication. A basis for such a theory is contained in the important papers of Nyquist[1] and Hartley[2] on this subject. In the present paper we will extend the theory to include a number of new factors, in particular the effect of noise in the channel, and the savings possible due to the statistical structure of the original message and due to the nature of the final destination of the information.

The fundamental problem of communication is that of reproducing at one point either exactly or approximately a message selected at another point. Frequently the messages have *meaning*; that is they refer to or are correlated according to some system with certain physical or conceptual entities. These semantic aspects of communication are irrelevant to the engineering problem. The significant aspect is that the actual message is one *selected from a set* of possible messages. The system must be designed to operate for each possible selection, not just the one which will actually be chosen since this is unknown at the time of design.

If the number of messages in the set is finite then this number or any monotonic function of this number can be regarded as a measure of the information produced when one message is chosen from the set, all choices being equally likely. As was pointed out by Hartley the most natural choice is the logarithmic function. Although this definition must be generalized considerably when we consider the influence of the statistics of the message and when we have a continuous range of messages, we will in all cases use an essentially logarithmic measure.

The logarithmic measure is more convenient for various reasons:

1. It is practically more useful. Parameters of engineering importance such as time, bandwidth, number of relays, etc., tend to vary linearly with the logarithm of the number of possibilities. For example, adding one relay to a group doubles the number of possible states of the relays. It adds 1 to the base 2 logarithm of this number. Doubling the time roughly squares the number of possible messages, or doubles the logarithm, etc.

2. It is nearer to our intuitive feeling as to the proper measure. This is closely related to (1) since we intuitively measure entities by linear comparison with common standards. One feels, for example, that two punched cards should have twice the capacity of one for information storage, and two identical channels twice the capacity of one for transmitting information.

3. It is mathematically more suitable. Many of the limiting operations are simple in terms of the logarithm but would require clumsy restatement in terms of the number of possibilities.

The choice of a logarithmic base corresponds to the choice of a unit for measuring information. If the base 2 is used the resulting units may be called binary digits, or more briefly *bits*, a word suggested by J. W. Tukey. A device with two stable positions, such as a relay or a flip-flop circuit, can store one bit of information. N such devices can store N bits, since the total number of possible states is 2^N and $\log_2 2^N = N$. If the base 10 is used the units may be called decimal digits. Since

$$\begin{aligned} \log_2 M &= \log_{10} M/\log_{10} 2 \\ &= 3.32 \log_{10} M, \end{aligned}$$

a decimal digit is about $3\frac{1}{3}$ bits. A digit wheel on a desk computing machine has ten stable positions and therefore has a storage capacity of one decimal digit. In analytical work where integration and differentiation are involved the base e is sometimes useful. The resulting units of information will be called natural units. Change from the base a to base b merely requires multiplication by $\log_b a$.

By a communication system we will mean a system of the type indicated schematically in Fig. 1. It consists of essentially five parts:

1. An *information source* which produces a message or sequence of messages to be communicated to the receiving terminal. The message may be of various types: e.g. (a) A sequence of letters as in a telegraph or teletype system; (b) A single function of time $f(t)$ as in radio or telephony; (c) A function of time and other variables as in black and white television—here the message may be thought of as a function $f(x, y, t)$ of two space coordinates and time, the light intensity at point (x, y) and time t on a pickup tube plate; (d) Two or more functions of time, say $f(t)$, $g(t)$, $h(t)$—this is the case in "three dimensional" sound transmission or if the system is intended to service several individual channels in multiplex; (e) Several functions of several variables—in color television the message consists of three functions $f(x, y, t)$, $g(x, y, t)$, $h(x, y, t)$ defined in a three-dimensional continuum—we may also think of these three functions as components of a vector field defined in the region—similarly, several black and white television sources would produce "messages" consisting of a number of functions of three variables; (f) Various combinations also occur, for example in television with an associated audio channel.

2. A *transmitter* which operates on the message in some way to produce a signal suitable for transmission over the channel. In telephony this operation consists merely of changing sound pressure into a proportional electrical current. In telegraphy we have an encoding operation which produces a sequence of dots, dashes and spaces on the channel corresponding to the message. In a multiplex PCM system the different speech functions must be sampled, compressed, quantized and encoded, and finally interleaved properly to construct the signal. Vocoder systems, television, and frequency modulation are other examples of complex operations applied to the message to obtain the signal.

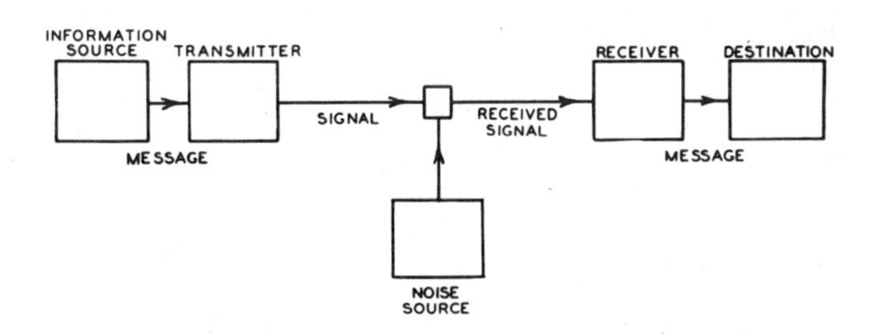

Fig. 1—Schematic diagram of a general communication system.

3. The *channel* is merely the medium used to transmit the signal from transmitter to receiver. It may be a pair of wires, a coaxial cable, a band of radio frequencies, a beam of light, etc.

4. The *receiver* ordinarily performs the inverse operation of that done by the transmitter, reconstructing the message from the signal.

5. The *destination* is the person (or thing) for whom the message is intended.

We wish to consider certain general problems involving communication systems. To do this it is first necessary to represent the various elements involved as mathematical entities, suitably idealized from their physical counterparts. We may roughly classify communication systems into three main categories: discrete, continuous and mixed. By a discrete system we will mean one in which both the message and the signal are a sequence of

[1] Nyquist, H., "Certain Factors Affecting Telegraph Speed," *Bell System Technical Journal*, April 1924, p. 324; "Certain Topics in Telegraph Transmission Theory," *A. I. E. E. Trans.*, v. 47, April 1928, p. 617.
[2] Hartley, R. V. L., "Transmission of Information," *Bell System Technical Journal*, July 1928, p. 535.

Reprinted with permission from *Bell Syst. Tech. J.*, vol. 27, pp. 379–423, July 1948. Copyright © 1948, The American Telephone and Telegraph Company.

discrete symbols. A typical case is telegraphy where the message is a sequence of letters and the signal a sequence of dots, dashes and spaces. A continuous system is one in which the message and signal are both treated as continuous functions, e.g. radio or television. A mixed system is one in which both discrete and continuous variables appear, e.g., PCM transmission of speech.

We first consider the discrete case. This case has applications not only in communication theory, but also in the theory of computing machines, the design of telephone exchanges and other fields. In addition the discrete case forms a foundation for the continuous and mixed cases which will be treated in the second half of the paper.

PART I: DISCRETE NOISELESS SYSTEMS

1. The Discrete Noiseless Channel

Teletype and telegraphy are two simple examples of a discrete channel for transmitting information. Generally, a discrete channel will mean a system whereby a sequence of choices from a finite set of elementary symbols $S_1 \cdots S_n$ can be transmitted from one point to another. Each of the symbols S_i is assumed to have a certain duration in time t_i seconds (not necessarily the same for different S_i, for example the dots and dashes in telegraphy). It is not required that all possible sequences of the S_i be capable of transmission on the system; certain sequences only may be allowed. These will be possible signals for the channel. Thus in telegraphy suppose the symbols are: (1) A dot, consisting of line closure for a unit of time and then line open for a unit of time; (2) A dash, consisting of three time units of closure and one unit open; (3) A letter space consisting of, say, three units of line open; (4) A word space of six units of line open. We might place the restriction on allowable sequences that no spaces follow each other (for if two letter spaces are adjacent, it is identical with a word space). The question we now consider is how one can measure the capacity of such a channel to transmit information.

In the teletype case where all symbols are of the same duration, and any sequence of the 32 symbols is allowed the answer is easy. Each symbol represents five bits of information. If the system transmits n symbols per second it is natural to say that the channel has a capacity of $5n$ bits per second. This does not mean that the teletype channel will always be transmitting information at this rate—this is the maximum possible rate and whether or not the actual rate reaches this maximum depends on the source of information which feeds the channel, as will appear later.

In the more general case with different lengths of symbols and constraints on the allowed sequences, we make the following definition:
Definition: The capacity C of a discrete channel is given by

$$C = \lim_{T \to \infty} \frac{\log N(T)}{T}$$

where $N(T)$ is the number of allowed signals of duration T.

It is easily seen that in the teletype case this reduces to the previous result. It can be shown that the limit in question will exist as a finite number in most cases of interest. Suppose all sequences of the symbols $S_1, \cdots, S_n$ are allowed and these symbols have durations $t_1, \cdots, t_n$. What is the channel capacity? If $N(t)$ represents the number of sequences of duration t we have

$$N(t) = N(t - t_1) + N(t - t_2) + \cdots + N(t - t_n)$$

The total number is equal to the sum of the numbers of sequences ending in $S_1, S_2, \cdots, S_n$ and these are $N(t - t_1), N(t - t_2), \cdots, N(t - t_n)$, respectively. According to a well known result in finite differences, $N(t)$ is then asymptotic for large t to X_0^t where X_0 is the largest real solution of the characteristic equation:

$$X^{-t_1} + X^{-t_2} + \cdots + X^{-t_n} = 1$$

and therefore

$$C = \log X_0$$

In case there are restrictions on allowed sequences we may still often obtain a difference equation of this type and find C from the characteristic equation. In the telegraphy case mentioned above

$$N(t) = N(t-2) + N(t-4) + N(t-5) + N(t-7) + N(t-8) + N(t-10)$$

as we see by counting sequences of symbols according to the last or next to the last symbol occurring. Hence C is $-\log \mu_0$ where μ_0 is the positive root of $1 = \mu^2 + \mu^4 + \mu^5 + \mu^7 + \mu^8 + \mu^{10}$. Solving this we find $C = 0.539$.

A very general type of restriction which may be placed on allowed sequences is the following: We imagine a number of possible states $a_1, a_2, \cdots, a_m$. For each state only certain symbols from the set $S_1, \cdots, S_n$ can be transmitted (different subsets for the different states). When one of these has been transmitted the state changes to a new state depending both on the old state and the particular symbol transmitted. The telegraph case is a simple example of this. There are two states depending on whether or not a space was the last symbol transmitted. If so then only a dot or a dash can be sent next and the state always changes. If not, any symbol can be transmitted and the state changes if a space is sent, otherwise it remains the same. The conditions can be indicated in a linear graph as shown in Fig. 2. The junction points correspond to the states and the lines indicate the symbols possible in a state and the resulting state. In Appendix I it is shown that if the conditions on allowed sequences can be described in this form C will exist and can be calculated in accordance with the following result:

Theorem 1: Let $b_{ij}^{(s)}$ be the duration of the sth symbol which is allowable in state i and leads to state j. Then the channel capacity C is equal to $\log W$ where W is the largest real root of the determinant equation:

$$\left| \sum_s W^{-b_{ij}^{(s)}} - \delta_{ij} \right| = 0.$$

where $\delta_{ij} = 1$ if $i = j$ and is zero otherwise.

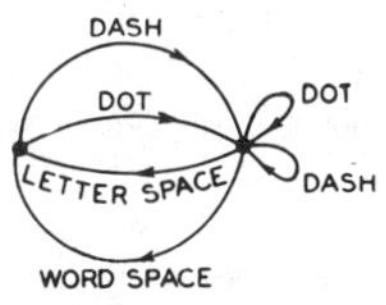

Fig. 2—Graphical representation of the constraints on telegraph symbols.

For example, in the telegraph case (Fig. 2) the determinant is:

$$\begin{vmatrix} -1 & (W^{-2} + W^{-4}) \\ (W^{-3} + W^{-6}) & (W^{-2} + W^{-4} - 1) \end{vmatrix} = 0$$

On expansion this leads to the equation given above for this case.

2. The Discrete Source of Information

We have seen that under very general conditions the logarithm of the number of possible signals in a discrete channel increases linearly with time. The capacity to transmit information can be specified by giving this rate of increase, the number of bits per second required to specify the particular signal used.

We now consider the information source. How is an information source to be described mathematically, and how much information in bits per second is produced in a given source? The main point at issue is the effect of statistical knowledge about the source in reducing the required capacity of the channel, by the use of proper encoding of the information. In telegraphy, for example, the messages to be transmitted consist of sequences of letters. These sequences, however, are not completely random. In general, they form sentences and have the statistical structure of, say, English. The letter E occurs more frequently than Q, the sequence TH more frequently than XP, etc. The existence of this structure allows one to make a saving in time (or channel capacity) by properly encoding the message sequences into signal sequences. This is already done to a limited extent in telegraphy by using the shortest channel symbol, a dot, for the most common English letter E; while the infrequent letters, Q, X, Z are represented by longer sequences of dots and dashes. This idea is carried still further in certain commercial codes where common words and phrases are represented by four- or five-letter code groups with a considerable saving in average time. The standardized greeting and anniversary telegrams now in use extend this to the point of encoding a sentence or two into a relatively short sequence of numbers.

We can think of a discrete source as generating the message, symbol by symbol. It will choose successive symbols according to certain probabilities depending, in general, on preceding choices as well as the particular symbols in question. A physical system, or a mathematical model of a system which produces such a sequence of symbols governed by a set of probabilities is known as a stochastic process.[3] We may consider a discrete source, therefore, to be represented by a stochastic process. Conversely, any stochastic process which produces a discrete sequence of symbols chosen from a finite set may be considered a discrete source. This will include such cases as:

1. Natural written languages such as English, German, Chinese.
2. Continuous information sources that have been rendered discrete by some quantizing process. For example, the quantized speech from a PCM transmitter, or a quantized television signal.
3. Mathematical cases where we merely define abstractly a stochastic process which generates a sequence of symbols. The following are examples of this last type of source.

(A) Suppose we have five letters A, B, C, D, E which are chosen each with probability .2, successive choices being independent. This would lead to a sequence of which the following is a typical example.

B D C B C E C C C A D C B D D A A E C E E A
A B B D A E E C A C E E B A E E C B C E A D

This was constructed with the use of a table of random numbers.[4]

(B) Using the same five letters let the probabilities be .4, .1, .2, .2, .1 respectively, with successive choices independent. A typical message from this source is then:

A A A C D C B D C E A A D A D A C E D A
E A D C A B E D A D D C E C A A A A A D

(C) A more complicated structure is obtained if successive symbols are not chosen independently but their probabilities depend on preceding letters. In the simplest case of this type a choice depends only on the preceding letter and not on ones before that. The statistical structure can then be described by a set of transition probabilities $p_i(j)$, the probability that letter i is followed by letter j. The indices i and j range over all the possible symbols. A second equivalent way of specifying the structure is to give the "digram" probabilities $p(i, j)$, i.e., the relative frequency of the digram $i\ j$. The letter frequencies $p(i)$, (the probability of letter i), the transition probabilities $p_i(j)$ and the digram probabilities $p(i, j)$ are related by the following formulas.

$$p(i) = \sum_j p(i, j) = \sum_j p(j, i) = \sum_j p(j) p_j(i)$$

$$p(i, j) = p(i) p_i(j)$$

$$\sum_j p_i(j) = \sum_i p(i) = \sum_{i,j} p(i, j) = 1.$$

As a specific example suppose there are three letters A, B, C with the probability tables:

$p_i(j)$	j: A	B	C
i: A	0	$\frac{4}{5}$	$\frac{1}{5}$
B	$\frac{1}{2}$	$\frac{1}{2}$	0
C	$\frac{1}{2}$	$\frac{2}{5}$	$\frac{1}{10}$

i	$p(i)$
A	$\frac{9}{27}$
B	$\frac{16}{27}$
C	$\frac{2}{27}$

$p(i, j)$	j: A	B	C
i: A	0	$\frac{4}{15}$	$\frac{1}{15}$
B	$\frac{8}{27}$	$\frac{8}{27}$	0
C	$\frac{1}{27}$	$\frac{4}{135}$	$\frac{1}{135}$

A typical message from this source is the following:

A B B A B A B A B A B A B A B B B A B B B B B A B
A B A B A B A B B B A C A C A B B A B B B B A B B
A B A C B B B A B A

The next increase in complexity would involve trigram frequencies but no more. The choice of a letter would depend on the preceding two letters but not on the message before that point. A set of trigram frequencies $p(i, j, k)$ or equivalently a set of transition probabilities $p_{ij}(k)$ would be required. Continuing in this way one obtains successively more complicated stochastic processes. In the general n-gram case a set of n-gram probabilities $p(i_1, i_2, \cdots, i_n)$ or of transition probabilities $p_{i_1, i_2, \cdots, i_{n-1}}(i_n)$ is required to specify the statistical structure.

(D) Stochastic processes can also be defined which produce a text consisting of a sequence of "words." Suppose there are five letters A, B, C, D, E and 16 "words" in the language with associated probabilities:

.10 A	.16 BEBE	.11 CABED	.04 DEB
.04 ADEB	.04 BED	.05 CEED	.15 DEED
.05 ADEE	.02 BEED	.08 DAB	.01 EAB
.01 BADD	.05 CA	.04 DAD	.05 EE

Suppose successive "words" are chosen independently and are separated by a space. A typical message might be:

DAB EE A BEBE DEED DEB ADEE ADEE EE DEB BEBE BEBE BEBE ADEE BED DEED DEED CEED ADEE A DEED DEED BEBE CABED BEBE BED DAB DEED ADEB

If all the words are of finite length this process is equivalent to one of the preceding type, but the description may be simpler in terms of the word structure and probabilities. We may also generalize here and introduce transition probabilities between words, etc.

These artificial languages are useful in constructing simple problems and examples to illustrate various possibilities. We can also approximate to a natural language by means of a series of simple artificial languages. The zero-order approximation is obtained by choosing all letters with the same probability and independently. The first-order approximation is obtained by choosing successive letters independently but each letter having the same probability that it does in the natural language.[5] Thus, in the first-order approximation to English, E is chosen with probability .12 (its frequency in normal English) and W with probability .02, but there is no influence between adjacent letters and no tendency to form the preferred digrams such as *TH*, *ED*, etc. In the second-order approximation, digram structure is introduced. After a letter is chosen, the next one is chosen in accordance with the frequencies with which the various letters follow the first one. This requires a table of digram frequencies $p_i(j)$. In the third-order approximation, trigram structure is introduced. Each letter is chosen with probabilities which depend on the preceding two letters.

3. The Series of Approximations to English

To give a visual idea of how this series of processes approaches a language, typical sequences in the approximations to English have been constructed and are given below. In all cases we have assumed a 27-symbol "alphabet," the 26 letters and a space.

1. Zero-order approximation (symbols independent and equi-probable).

   XFOML RXKHRJFFJUJ ZLPWCFWKCYJ FFJEYVKCQSGXYD QPAAMKBZAACIBZLHJQD

2. First-order approximation (symbols independent but with frequencies of English text).

   OCRO HLI RGWR NMIELWIS EU LL NBNESEBYA TH EEI ALHENHTTPA OOBTTVA NAH BRL

3. Second-order approximation (digram structure as in English).

   ON IE ANTSOUTINYS ARE T INCTORE ST BE S DEAMY ACHIN D ILONASIVE TUCOOWE AT TEASONARE FUSO TIZIN ANDY TOBE SEACE CTISBE

4. Third-order approximation (trigram structure as in English).

   IN NO IST LAT WHEY CRATICT FROURE BIRS GROCID PONDENOME OF DEMONSTURES OF THE REPTAGIN IS REGOACTIONA OF CRE

5. First-Order Word Approximation. Rather than continue with tetragram, $\cdots$, n-gram structure it is easier and better to jump at this point to word units. Here words are chosen independently but with their appropriate frequencies.

   REPRESENTING AND SPEEDILY IS AN GOOD APT OR COME CAN DIFFERENT NATURAL HERE HE THE A IN CAME THE TO OF TO EXPERT GRAY COME TO FURNISHES THE LINE MESSAGE HAD BE THESE.

6. Second-Order Word Approximation. The word transition probabilities are correct but no further structure is included.

[3] See, for example, S. Chandrasekhar, "Stochastic Problems in Physics and Astronomy," *Reviews of Modern Physics*, v. 15, No. 1, January 1943, p. 1.

[4] Kendall and Smith, "Tables of Random Sampling Numbers," Cambridge, 1939.

[5] Letter, digram and trigram frequencies are given in "Secret and Urgent" by Fletcher Pratt, Blue Ribbon Books 1939. Word frequencies are tabulated in "Relative Frequency of English Speech Sounds," G. Dewey, Harvard University Press, 1923.

THE HEAD AND IN FRONTAL ATTACK ON AN ENGLISH WRITER THAT THE CHARACTER OF THIS POINT IS THEREFORE ANOTHER METHOD FOR THE LETTERS THAT THE TIME OF WHO EVER TOLD THE PROBLEM FOR AN UNEXPECTED

The resemblance to ordinary English text increases quite noticeably at each of the above steps. Note that these samples have reasonably good structure out to about twice the range that is taken into account in their construction. Thus in (3) the statistical process insures reasonable text for two-letter sequence, but four-letter sequences from the sample can usually be fitted into good sentences. In (6) sequences of four or more words can easily be placed in sentences without unusual or strained constructions. The particular sequence of ten words "attack on an English writer that the character of this" is not at all unreasonable. It appears then that a sufficiently complex stochastic process will give a satisfactory representation of a discrete source.

The first two samples were constructed by the use of a book of random numbers in conjunction with (for example 2) a table of letter frequencies. This method might have been continued for (3), (4), and (5), since digram, trigram, and word frequency tables are available, but a simpler equivalent method was used. To construct (3) for example, one opens a book at random and selects a letter at random on the page. This letter is recorded. The book is then opened to another page and one reads until this letter is encountered. The succeeding letter is then recorded. Turning to another page this second letter is searched for and the succeeding letter recorded, etc. A similar process was used for (4), (5), and (6). It would be interesting if further approximations could be constructed, but the labor involved becomes enormous at the next stage.

4. Graphical Representation of a Markoff Process

Stochastic processes of the type described above are known mathematically as discrete Markoff processes and have been extensively studied in the literature.[6] The general case can be described as follows: There exist a finite number of possible "states" of a system; $S_1, S_2, \cdots, S_n$. In addition there is a set of transition probabilities; $p_i(j)$ the probability that if the system is in state S_i it will next go to state S_j. To make this Markoff process into an information source we need only assume that a letter is produced for each transition from one state to another. The states will correspond to the "residue of influence" from preceding letters.

The situation can be represented graphically as shown in Figs. 3, 4 and 5. The "states" are the junction points in the graph and the probabilities and letters produced for a transition are given beside the corresponding line. Figure 3 is for the example B in Section 2, while Fig. 4 corresponds to the example C. In Fig. 3 there is only one state since successive letters are independent. In Fig. 4 there are as many states as letters. If a trigram example were constructed there would be at most n^2 states corresponding to the possible pairs of letters preceding the one being chosen. Figure 5 is a graph for the case of word structure in example D. Here S corresponds to the "space" symbol.

5. Ergodic and Mixed Sources

As we have indicated above a discrete source for our purposes can be considered to be represented by a Markoff process. Among the possible discrete Markoff processes there is a group with special properties of significance in communication theory. This special class consists of the "ergodic" processes and we shall call the corresponding sources ergodic sources. Although a rigorous definition of an ergodic process is somewhat involved, the general idea is simple. In an ergodic process every sequence produced by the process is the same in statistical properties. Thus the letter frequencies, digram frequencies, etc., obtained from particular sequences will, as the lengths of the sequences increase, approach definite limits independent of the particular sequence. Actually this is not true of every sequence but the set for which it is false has probability zero. Roughly the ergodic property means statistical homogeneity.

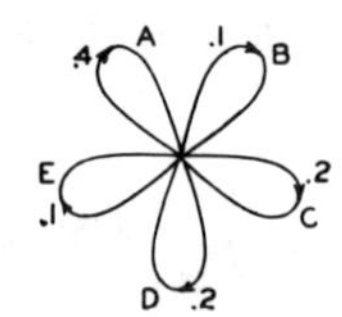

Fig. 3—A graph corresponding to the source in example B.

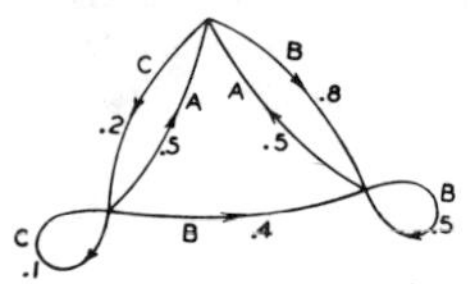

Fig. 4—A graph corresponding to the source in example C.

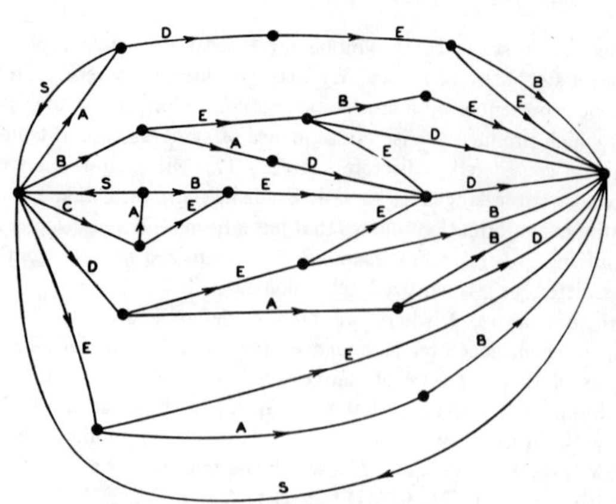
Fig. 5—A graph corresponding to the source in example D.

All the examples of artificial languages given above are ergodic. This property is related to the structure of the corresponding graph. If the graph has the following two properties[7] the corresponding process will be ergodic:

1. The graph does not consist of two isolated parts A and B such that it is impossible to go from junction points in part A to junction points in part B along lines of the graph in the direction of arrows and also impossible to go from junctions in part B to junctions in part A.
2. A closed series of lines in the graph with all arrows on the lines pointing in the same orientation will be called a "circuit." The "length" of a circuit is the number of lines in it. Thus in Fig. 5 the series BEBES is a circuit of length 5. The second property required is that the greatest common divisor of the lengths of all circuits in the graph be one.

If the first condition is satisfied but the second one violated by having the greatest common divisor equal to $d > 1$, the sequences have a certain type of periodic structure. The various sequences fall into d different classes which are statistically the same apart from a shift of the origin (i.e., which letter in the sequence is called letter 1). By a shift of from 0 up to $d - 1$ any sequence can be made statistically equivalent to any other. A simple example with $d = 2$ is the following: There are three possible letters a, b, c. Letter a is followed with either b or c with probabilities $\frac{1}{3}$ and $\frac{2}{3}$ respectively. Either b or c is always followed by letter a. Thus a typical sequence is

a b a c a c a c a b a c a b a b a c a c

This type of situation is not of much importance for our work.

If the first condition is violated the graph may be separated into a set of subgraphs each of which satisfies the first condition. We will assume that the second condition is also satisfied for each subgraph. We have in this case what may be called a "mixed" source made up of a number of pure components. The components correspond to the various subgraphs. If $L_1, L_2, L_3, \cdots$ are the component sources we may write

$$L = p_1L_1 + p_2L_2 + p_3L_3 + \cdots$$

where p_i is the probability of the component source L_i.

Physically the situation represented is this: There are several different sources $L_1, L_2, L_3, \cdots$ which are each of homogeneous statistical structure (i.e., they are ergodic). We do not know *a priori* which is to be used, but once the sequence starts in a given pure component L_i it continues indefinitely according to the statistical structure of that component.

[6] For a detailed treatment see M. Frechet, "Methods des fonctions arbitraires. Theorie des énénements en chaine dans le cas d'un nombre fini d'états possibles." Paris, Gauthier-Villars, 1938.

[7] These are restatements in terms of the graph of conditions given in Frechet.

As an example one may take two of the processes defined above and assume $p_1 = .2$ and $p_2 = .8$. A sequence from the mixed source

$$L = .2\,L_1 + .8\,L_2$$

would be obtained by choosing first L_1 or L_2 with probabilities .2 and .8 and after this choice generating a sequence from whichever was chosen.

Except when the contrary is stated we shall assume a source to be ergodic. This assumption enables one to identify averages along a sequence with averages over the ensemble of possible sequences (the probability of a discrepancy being zero). For example the relative frequency of the letter A in a particular infinite sequence will be, with probability one, equal to its relative frequency in the ensemble of sequences.

If P_i is the probability of state i and $p_i(j)$ the transition probability to state j, then for the process to be stationary it is clear that the P_i must satisfy equilibrium conditions:

$$P_j = \sum_i P_i p_i(j).$$

In the ergodic case it can be shown that with any starting conditions the probabilities $P_j(N)$ of being in state j after N symbols, approach the equilibrium values as $N \to \infty$.

6. Choice, Uncertainty and Entropy

We have represented a discrete information source as a Markoff process. Can we define a quantity which will measure, in some sense, how much information is "produced" by such a process, or better, at what rate information is produced?

Suppose we have a set of possible events whose probabilities of occurrence are $p_1, p_2, \cdots, p_n$. These probabilities are known but that is all we know concerning which event will occur. Can we find a measure of how much "choice" is involved in the selection of the event or of how uncertain we are of the outcome?

If there is such a measure, say $H(p_1, p_2, \cdots, p_n)$, it is reasonable to require of it the following properties:

1. H should be continuous in the p_i.
2. If all the p_i are equal, $p_i = \frac{1}{n}$, then H should be a monotonic increasing function of n. With equally likely events there is more choice, or uncertainty, when there are more possible events.
3. If a choice be broken down into two successive choices, the original H should be the weighted sum of the individual values of H. The meaning of this is illustrated in Fig. 6. At the left we have three possibilities $p_1 = \frac{1}{2}, p_2 = \frac{1}{3}, p_3 = \frac{1}{6}$. On the right we first choose between two possibilities each with probability $\frac{1}{2}$, and if the second occurs make another choice with probabilities $\frac{2}{3}, \frac{1}{3}$. The final results have the same probabilities as before. We require, in this special case, that

$$H(\tfrac{1}{2}, \tfrac{1}{3}, \tfrac{1}{6}) = H(\tfrac{1}{2}, \tfrac{1}{2}) + \tfrac{1}{2}H(\tfrac{2}{3}, \tfrac{1}{3})$$

The coefficient $\frac{1}{2}$ is because this second choice only occurs half the time.

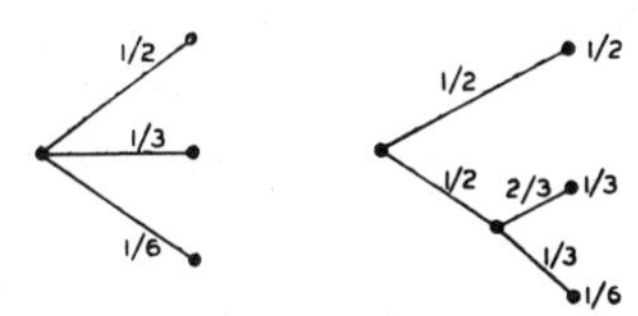

Fig. 6—Decomposition of a choice from three possibilities.

In Appendix II, the following result is established:

Theorem 2: The only H satisfying the three above assumptions is of the form:

$$H = -K \sum_{i=1}^{n} p_i \log p_i$$

where K is a positive constant.

This theorem, and the assumptions required for its proof, are in no way necessary for the present theory. It is given chiefly to lend a certain plausibility to some of our later definitions. The real justification of these definitions, however, will reside in their implications.

Quantities of the form $H = -\Sigma\, p_i \log p_i$ (the constant K merely amounts to a choice of a unit of measure) play a central role in information theory as measures of information, choice and uncertainty. The form of H will be recognized as that of entropy as defined in certain formulations of statistical mechanics[8] where p_i is the probability of a system being in cell i of its phase space. H is then, for example, the H in Boltzmann's famous H theorem. We shall call $H = -\Sigma\, p_i \log p_i$ the entropy of the set of probabilities $p_1, \cdots, p_n$. If x is a chance variable we will write $H(x)$ for its entropy; thus x is not an argument of a function but a label for a number, to differentiate it from $H(y)$ say, the entropy of the chance variable y.

The entropy in the case of two possibilities with probabilities p and $q = 1 - p$, namely

$$H = -(p \log p + q \log q)$$

is plotted in Fig. 7 as a function of p.

The quantity H has a number of interesting properties which further substantiate it as a reasonable measure of choice or information.

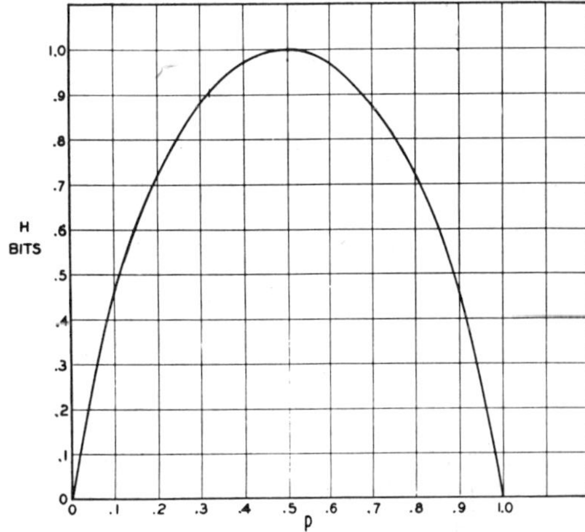

Fig. 7—Entropy in the case of two possibilities with probabilities p and $(1 - p)$.

1. $H = 0$ if and only if all the p_i but one are zero, this one having the value unity. Thus only when we are certain of the outcome does H vanish. Otherwise H is positive.

2. For a given n, H is a maximum and equal to $\log n$ when all the p_i are equal $\left(\text{i.e., } \frac{1}{n}\right)$. This is also intuitively the most uncertain situation.

3. Suppose there are two events, x and y, in question with m possibilities for the first and n for the second. Let $p(i, j)$ be the probability of the joint occurrence of i for the first and j for the second. The entropy of the joint event is

$$H(x, y) = -\sum_{i,j} p(i, j) \log p(i, j)$$

while

$$H(x) = -\sum_{i,j} p(i, j) \log \sum_j p(i, j)$$

$$H(y) = -\sum_{i,j} p(i, j) \log \sum_i p(i, j).$$

It is easily shown that

$$H(x, y) \le H(x) + H(y)$$

with equality only if the events are independent (i.e., $p(i, j) = p(i)\, p(j)$). The uncertainty of a joint event is less than or equal to the sum of the individual uncertainties.

4. Any change toward equalization of the probabilities $p_1, p_2, \cdots, p_n$ increases H. Thus if $p_1 < p_2$ and we increase p_1, decreasing p_2 an equal amount so that p_1 and p_2 are more nearly equal, then H increases. More generally, if we perform any "averaging" operation on the p_i of the form

$$p_i' = \sum_j a_{ij}\, p_j$$

where $\sum_i a_{ij} = \sum_j a_{ij} = 1$, and all $a_{ij} \ge 0$, then H increases (except in the special case where this transformation amounts to no more than a permutation of the p_j with H of course remaining the same).

[8] See, for example, R. C. Tolman, "Principles of Statistical Mechanics," Oxford, Clarendon, 1938.

5. Suppose there are two chance events x and y as in 3, not necessarily independent. For any particular value i that x can assume there is a conditional probability $p_i(j)$ that y has the value j. This is given by

$$p_i(j) = \frac{p(i, j)}{\sum_j p(i, j)}.$$

We define the *conditional entropy* of y, $H_x(y)$ as the average of the entropy of y for each value of x, weighted according to the probability of getting that particular x. That is

$$H_x(y) = -\sum_{i,j} p(i, j) \log p_i(j).$$

This quantity measures how uncertain we are of y on the average when we know x. Substituting the value of $p_i(j)$ we obtain

$$H_x(y) = -\sum_{ij} p(i, j) \log p(i, j) + \sum_{ij} p(i, j) \log \sum_j p(i, j)$$

$$= H(x, y) - H(x)$$

or

$$H(x, y) = H(x) + H_x(y)$$

The uncertainty (or entropy) of the joint event x, y is the uncertainty of x plus the uncertainty of y when x is known.

6. From 3 and 5 we have

$$H(x) + H(y) \geq H(x, y) = H(x) + H_x(y)$$

Hence

$$H(y) \geq H_x(y)$$

The uncertainty of y is never increased by knowledge of x. It will be decreased unless x and y are independent events, in which case it is not changed.

7. The Entropy of an Information Source

Consider a discrete source of the finite state type considered above. For each possible state i there will be a set of probabilities $p_i(j)$ of producing the various possible symbols j. Thus there is an entropy H_i for each state. The entropy of the source will be defined as the average of these H_i weighted in accordance with the probability of occurrence of the states in question:

$$H = \sum_i P_i H_i$$

$$= -\sum_{i,j} P_i p_i(j) \log p_i(j)$$

This is the entropy of the source per symbol of text. If the Markoff process is proceeding at a definite time rate there is also an entropy per second

$$H' = \sum_i f_i H_i$$

where f_i is the average frequency (occurrences per second) of state i. Clearly

$$H' = mH$$

where m is the average number of symbols produced per second. H or H' measures the amount of information generated by the source per symbol or per second. If the logarithmic base is 2, they will represent bits per symbol or per second.

If successive symbols are independent then H is simply $-\Sigma\, p_i \log p_i$ where p_i is the probability of symbol i. Suppose in this case we consider a long message of N symbols. It will contain with high probability about p_1N occurrences of the first symbol, p_2N occurrences of the second, etc. Hence the probability of this particular message will be roughly

$$p = p_1^{p_1N} p_2^{p_2N} \cdots p_n^{p_nN}$$

or

$$\log p \doteq N \sum_i p_i \log p_i$$

$$\log p \doteq -NH$$

$$H \doteq \frac{\log 1/p}{N}.$$

H is thus approximately the logarithm of the reciprocal probability of a typical long sequence divided by the number of symbols in the sequence. The same result holds for any source. Stated more precisely we have (see Appendix III):

Theorem 3: Given any $\epsilon > 0$ and $\delta > 0$, we can find an N_0 such that the sequences of any length $N \geq N_0$ fall into two classes:

1. A set whose total probability is less than ϵ.
2. The remainder, all of whose members have probabilities satisfying the inequality

$$\left| \frac{\log p^{-1}}{N} - H \right| < \delta$$

In other words we are almost certain to have $\frac{\log p^{-1}}{N}$ very close to H when N is large.

A closely related result deals with the number of sequences of various probabilities. Consider again the sequences of length N and let them be arranged in order of decreasing probability. We define $n(q)$ to be the number we must take from this set starting with the most probable one in order to accumulate a total probability q for those taken.

Theorem 4:

$$\lim_{N\to\infty} \frac{\log n(q)}{N} = H$$

when q does not equal 0 or 1.

We may interpret $\log n(q)$ as the number of bits required to specify the sequence when we consider only the most probable sequences with a total probability q. Then $\frac{\log n(q)}{N}$ is the number of bits per symbol for the specification. The theorem says that for large N this will be independent of q and equal to H. The rate of growth of the logarithm of the number of reasonably probable sequences is given by H, regardless of our interpretation of "reasonably probable." Due to these results, which are proved in appendix III, it is possible for most purposes to treat the long sequences as though there were just 2^{HN} of them, each with a probability 2^{-HN}.

The next two theorems show that H and H' can be determined by limiting operations directly from the statistics of the message sequences, without reference to the states and transition probabilities between states.

Theorem 5: Let $p(B_i)$ be the probability of a sequence B_i of symbols from the source. Let

$$G_N = -\frac{1}{N} \sum_i p(B_i) \log p(B_i)$$

where the sum is over all sequences B_i containing N symbols. Then G_N is a monotonic decreasing function of N and

$$\lim_{N\to\infty} G_N = H.$$

Theorem 6: Let $p(B_i, S_j)$ be the probability of sequence B_i followed by symbol S_j and $p_{B_i}(S_j) = p(B_i, S_j)/p(B_i)$ be the conditional probability of S_j after B_i. Let

$$F_N = -\sum_{i,j} p(B_i, S_j) \log p_{B_i}(S_j)$$

where the sum is over all blocks B_i of $N - 1$ symbols and over all symbols S_j. Then F_N is a monotonic decreasing function of N,

$$F_N = NG_N - (N - 1)\, G_{N-1},$$

$$G_N = \frac{1}{N} \sum_1^n F_N,$$

$$F_N \leq G_N,$$

and $\lim_{N\to\infty} F_N = H$.

These results are derived in appendix III. They show that a series of approximations to H can be obtained by considering only the statistical structure of the sequences extending over $1, 2, \cdots N$ symbols. F_N is the better approximation. In fact F_N is the entropy of the Nth order approximation to the source of the type discussed above. If there are no statistical influences extending over more than N symbols, that is if the conditional probability of the next symbol knowing the preceding $(N - 1)$ is not changed by a knowledge of any before that, then $F_N = H$. F_N of course is the conditional entropy of the next symbol when the $(N - 1)$ preceding ones are known, while G_N is the entropy per symbol of blocks of N symbols.

The ratio of the entropy of a source to the maximum value it could have while still restricted to the same symbols will be called its *relative entropy*. This is the maximum compression possible when we encode into the same alphabet. One minus the relative entropy is the *redundancy*. The redundancy of ordinary English, not considering statistical structure over greater distances than about eight letters is roughly 50%. This means that when we write English half of what we write is determined by the structure of the language and half is chosen freely. The figure 50% was found by several independent methods which all gave results in this neighborhood. One is by calculation of the entropy of the approximations to English. A second method is to delete a certain fraction of the letters from a sample of English text and then let someone attempt to restore them. If they can be restored when 50% are deleted the redundancy must be greater than 50%. A third method depends on certain known results in cryptography.

Two extremes of redundancy in English prose are represented by Basic English and by James Joyces' book "Finigans Wake." The Basic English vocabulary is limited to 850 words and the redundancy is very high. This is reflected in the expansion that occurs when a passage is translated into Basic English. Joyce on the other hand enlarges the vocabulary and is alleged to achieve a compression of semantic content.

The redundancy of a language is related to the existence of crossword puzzles. If the redundancy is zero any sequence of letters is a reasonable text in the language and any two dimensional array of letters forms a crossword puzzle. If the redundancy is too high the language imposes too many constraints for large crossword puzzles to be possible. A more detailed analysis shows that if we assume the constraints imposed by the language are of a rather chaotic and random nature, large crossword puzzles are just possible when the redundancy is 50%. If the redundancy is 33%, three dimensional crossword puzzles should be possible, etc.

8. Representation of the Encoding and Decoding Operations

We have yet to represent mathematically the operations performed by the transmitter and receiver in encoding and decoding the information. Either of these will be called a discrete transducer. The input to the transducer is a sequence of input symbols and its output a sequence of output symbols. The transducer may have an internal memory so that its output depends not only on the present input symbol but also on the past history. We assume that the internal memory is finite, i.e. there exists a finite number m of possible states of the transducer and that its output is a function of the present state and the present input symbol. The next state will be a second function of these two quantities. Thus a transducer can be described by two functions:

$$y_n = f(x_n, \alpha_n)$$

$$\alpha_{n+1} = g(x_n, \alpha_n)$$

where: x_n is the nth input symbol,

α_n is the state of the transducer when the nth input symbol is introduced,

y_n is the output symbol (or sequence of output symbols) produced when x_n is introduced if the state is α_n.

If the output symbols of one transducer can be identified with the input symbols of a second, they can be connected in tandem and the result is also a transducer. If there exists a second transducer which operates on the output of the first and recovers the original input, the first transducer will be called non-singular and the second will be called its inverse.

Theorem 7: The output of a finite state transducer driven by a finite state statistical source is a finite state statistical source, with entropy (per unit time) less than or equal to that of the input. If the transducer is non-singular they are equal.

Let α represent the state of the source, which produces a sequence of symbols x_i; and let β be the state of the transducer, which produces, in its output, blocks of symbols y_j. The combined system can be represented by the "product state space" of pairs (α, β). Two points in the space, (α_1, β_1) and $(\alpha_2 \beta_2)$, are connected by a line if α_1 can produce an x which changes β_1 to β_2, and this line is given the probability of that x in this case. The line is labeled with the block of y_j symbols produced by the transducer. The entropy of the output can be calculated as the weighted sum over the states. If we sum first on β each resulting term is less than or equal to the corresponding term for α, hence the entropy is not increased. If the transducer is non-singular let its output be connected to the inverse transducer. If H_1', H_2' and H_3' are the output entropies of the source, the first and second transducers respectively, then $H_1' \geq H_2' \geq H_3' = H_1'$ and therefore $H_1' = H_2'$.

Suppose we have a system of constraints on possible sequences of the type which can be represented by a linear graph as in Fig. 2. If probabilities $p_{ij}^{(s)}$ were assigned to the various lines connecting state i to state j this would become a source. There is one particular assignment which maximizes the resulting entropy (see Appendix IV).

Theorem 8: Let the system of constraints considered as a channel have a capacity C. If we assign

$$p_{ij}^{(s)} = \frac{B_j}{B_i} C^{-\ell_{ij}^{(s)}}$$

where $\ell_{ij}^{(s)}$ is the duration of the sth symbol leading from state i to state j and the B_i satisfy

$$B_i = \sum_{s,j} B_j C^{-\ell_{ij}^{(s)}}$$

then H is maximized and equal to C.

By proper assignment of the transition probabilities the entropy of symbols on a channel can be maximized at the channel capacity.

9. The Fundamental Theorem for a Noiseless Channel

We will now justify our interpretation of H as the rate of generating information by proving that H determines the channel capacity required with most efficient coding.

Theorem 9: Let a source have entropy H (bits per symbol) and a channel have a capacity C (bits per second). Then it is possible to encode the output of the source in such a way as to transmit at the average rate $\frac{C}{H} - \epsilon$ symbols per second over the channel where ϵ is arbitrarily small. It is not possible to transmit at an average rate greater than $\frac{C}{H}$.

The converse part of the theorem, that $\frac{C}{H}$ cannot be exceeded, may be proved by noting that the entropy of the channel input per second is equal to that of the source, since the transmitter must be non-singular, and also this entropy cannot exceed the channel capacity. Hence $H' \leq C$ and the number of symbols per second $= H'/H \leq C/H$.

The first part of the theorem will be proved in two different ways. The first method is to consider the set of all sequences of N symbols produced by the source. For N large we can divide these into two groups, one containing less than $2^{(H+\eta)N}$ members and the second containing less than 2^{RN} members (where R is the logarithm of the number of different symbols) and having a total probability less than μ. As N increases η and μ approach zero. The number of signals of duration T in the channel is greater than $2^{(C-\theta)T}$ with θ small when T is large. If we choose

$$T = \left(\frac{H}{C} + \lambda\right) N$$

then there will be a sufficient number of sequences of channel symbols for the high probability group when N and T are sufficiently large (however small λ) and also some additional ones. The high probability group is coded in an arbitrary one to one way into this set. The remaining sequences are represented by larger sequences, starting and ending with one of the sequences not used for the high probability group. This special sequence acts as a start and stop signal for a different code. In between a sufficient time is allowed to give enough different sequences for all the low probability messages. This will require

$$T_1 = \left(\frac{R}{C} + \varphi\right) N$$

where φ is small. The mean rate of transmission in message symbols per second will then be greater than

$$\left[(1 - \delta)\frac{T}{N} + \delta\frac{T_1}{N}\right]^{-1} = \left[(1 - \delta)\left(\frac{H}{C} + \lambda\right) + \delta\left(\frac{R}{C} + \varphi\right)\right]^{-1}$$

As N increases δ, λ and φ approach zero and the rate approaches $\frac{C}{H}$.

Another method of performing this coding and proving the theorem can be described as follows: Arrange the messages of length N in order of decreasing probability and suppose their probabilities are $p_1 \geq p_2 \geq p_3 \ldots \geq p_n$. Let $P_s = \sum_1^{s-1} p_i$; that is P_s is the cumulative probability up to, but not including, p_s. We first encode into a binary system. The binary code for message s is obtained by expanding P_s as a binary number. The expansion is carried out to m_s places, where m_s is the integer satisfying:

$$\log_2 \frac{1}{p_s} \leq m_s < 1 + \log_2 \frac{1}{p_s}$$

Thus the messages of high probability are represented by short codes and those of low probability by long codes. From these inequalities we have

$$\frac{1}{2^{m_s}} \leq p_s < \frac{1}{2^{m_s-1}}.$$

The code for P_s will differ from all succeeding ones in one or more of its m_s places, since all the remaining P_i are at least $\frac{1}{2^{m_s}}$ larger and their binary expansions therefore differ in the first m_s places. Consequently all the codes are different and it is possible to recover the message from its code. If the channel sequences are not already sequences of binary digits, they can be ascribed binary numbers in an arbitrary fashion and the binary code thus translated into signals suitable for the channel.

The average number H' of binary digits used per symbol of original message is easily estimated. We have

$$H' = \frac{1}{N} \Sigma m_s p_s$$

But,

$$\frac{1}{N} \Sigma \left(\log_2 \frac{1}{p_s}\right) p_s \leq \frac{1}{N} \Sigma m_s p_s < \frac{1}{N} \Sigma \left(1 + \log_2 \frac{1}{p_s}\right) p_s$$

and therefore,

$$-\Sigma p_s \log p_s \leq H' < \frac{1}{N} - \Sigma p_s \log p_s$$

As N increases $-\Sigma p_s \log p_s$ approaches H, the entropy of the source and H' approaches H.

We see from this that the inefficiency in coding, when only a finite delay of N symbols is used, need not be greater than $\frac{1}{N}$ plus the difference between the true entropy H and the entropy G_N calculated for sequences of length N. The per cent excess time needed over the ideal is therefore less than

$$\frac{G_N}{H} + \frac{1}{HN} - 1.$$

This method of encoding is substantially the same as one found independently by R. M. Fano.[9] His method is to arrange the messages of length N in order of decreasing probability. Divide this series into two groups of as nearly equal probability as possible. If the message is in the first group its first binary digit will be 0, otherwise 1. The groups are similarly divided into subsets of nearly equal probability and the particular subset determines the second binary digit. This process is continued until each subset contains only one message. It is easily seen that apart from minor differences (generally in the last digit) this amounts to the same thing as the arithmetic process described above.

10. Discussion

In order to obtain the maximum power transfer from a generator to a load a transformer must in general be introduced so that the generator as seen from the load has the load resistance. The situation here is roughly analogous. The transducer which does the encoding should match the source to the channel in a statistical sense. The source as seen from the channel through the transducer should have the same statistical structure as the source which maximizes the entropy in the channel. The content of Theorem 9 is that, although an exact match is not in general possible, we can approximate it as closely as desired. The ratio of the actual rate of transmission to the capacity C may be called the efficiency of the coding system. This is of course equal to the ratio of the actual entropy of the channel symbols to the maximum possible entropy.

In general, ideal or nearly ideal encoding requires a long delay in the transmitter and receiver. In the noiseless case which we have been considering, the main function of this delay is to allow reasonably good matching of probabilities to corresponding lengths of sequences. With a good code the logarithm of the reciprocal probability of a long message must be proportional to the duration of the corresponding signal, in fact

$$\left|\frac{\log p^{-1}}{T} - C\right|$$

must be small for all but a small fraction of the long messages.

If a source can produce only one particular message its entropy is zero, and no channel is required. For example, a computing machine set up to calculate the successive digits of π produces a definite sequence with no chance element. No channel is required to "transmit" this to another point. One could construct a second machine to compute the same sequence at the point. However, this may be impractical. In such a case we can choose to ignore some or all of the statistical knowledge we have of the source. We might consider the digits of π to be a random sequence in that we construct a system capable of sending any sequence of digits. In a similar way we may choose to use some of our statistical knowledge of English in constructing a code, but not all of it. In such a case we consider the source with the maximum entropy subject to the statistical conditions we wish to retain. The entropy of this source determines the channel capacity which is necessary and sufficient. In the π example the only information retained is that all the digits are chosen from the set 0, 1, ..., 9. In the case of English one might wish to use the statistical saving possible due to letter frequencies, but nothing else. The maximum entropy source is then the first approximation to English and its entropy determines the required channel capacity.

11. Examples

As a simple example of some of these results consider a source which produces a sequence of letters chosen from among A, B, C, D with probabilities $\frac{1}{2}$, $\frac{1}{4}$, $\frac{1}{8}$, $\frac{1}{8}$, successive symbols being chosen independently. We have

$$H = -\left(\tfrac{1}{2}\log\tfrac{1}{2} + \tfrac{1}{4}\log\tfrac{1}{4} + \tfrac{2}{8}\log\tfrac{1}{8}\right)$$
$$= \tfrac{7}{4} \text{ bits per symbol.}$$

Thus we can approximate a coding system to encode messages from this source into binary digits with an average of $\frac{7}{4}$ binary digit per symbol. In this case we can actually achieve the limiting value by the following code (obtained by the method of the second proof of Theorem 9):

A	0
B	10
C	110
D	111

The average number of binary digits used in encoding a sequence of N symbols will be

$$N\left(\tfrac{1}{2} \times 1 + \tfrac{1}{4} \times 2 + \tfrac{2}{8} \times 3\right) = \tfrac{7}{4}N$$

It is easily seen that the binary digits 0, 1 have probabilities $\frac{1}{2}$, $\frac{1}{2}$ so the H for the coded sequences is one bit per symbol. Since, on the average, we have $\frac{7}{4}$ binary symbols per original letter, the entropies on a time basis are the same. The maximum possible entropy for the original set is $\log 4 = 2$, occurring when A, B, C, D have probabilities $\frac{1}{4}$, $\frac{1}{4}$, $\frac{1}{4}$, $\frac{1}{4}$. Hence the relative entropy is $\frac{7}{8}$. We can translate the binary sequences into the original set of symbols on a two-to-one basis by the following table:

00	A'
01	B'
10	C'
11	D'

This double process then encodes the original message into the same symbols but with an average compression ratio $\frac{7}{8}$.

As a second example consider a source which produces a sequence of A's and B's with probability p for A and q for B. If $p << q$ we have

$$H = -\log p^p(1-p)^{1-p}$$
$$= -p \log p\,(1-p)^{(1-p)/p}$$
$$\doteq p \log \frac{e}{p}$$

In such a case one can construct a fairly good coding of the message on a 0, 1 channel by sending a special sequence, say 0000, for the infrequent symbol A and then a sequence indicating the *number* of B's following it.

[9] Technical Report No. 65, The Research Laboratory of Electronics, M. I. T.

This could be indicated by the binary representation with all numbers containing the special sequence deleted. All numbers up to 16 are represented as usual; 16 is represented by the next binary number after 16 which does not contain four zeros, namely 17 = 10001, etc.

It can be shown that as $p \rightarrow 0$ the coding approaches ideal provided the length of the special sequence is properly adjusted.

PART II: THE DISCRETE CHANNEL WITH NOISE

11. Representation of a Noisy Discrete Channel

We now consider the case where the signal is perturbed by noise during transmission or at one or the other of the terminals. This means that the received signal is not necessarily the same as that sent out by the transmitter. Two cases may be distinguished. If a particular transmitted signal always produces the same received signal, i.e. the received signal is a definite function of the transmitted signal, then the effect may be called distortion. If this function has an inverse—no two transmitted signals producing the same received signal—distortion may be corrected, at least in principle, by merely performing the inverse functional operation on the received signal.

The case of interest here is that in which the signal does not always undergo the same change in transmission. In this case we may assume the received signal E to be a function of the transmitted signal S and a second variable, the noise N.

$$E = f(S, N)$$

The noise is considered to be a chance variable just as the message was above. In general it may be represented by a suitable stochastic process. The most general type of noisy discrete channel we shall consider is a generalization of the finite state noise free channel described previously. We assume a finite number of states and a set of probabilities

$$p_{\alpha, i}(\beta, j).$$

This is the probability, if the channel is in state α and symbol i is transmitted, that symbol j will be received and the channel left in state β. Thus α and β range over the possible states, i over the possible transmitted signals and j over the possible received signals. In the case where successive symbols are independently perturbed by the noise there is only one state, and the channel is described by the set of transition probabilities $p_i(j)$, the probability of transmitted symbol i being received as j.

If a noisy channel is fed by a source there are two statistical processes at work: the source and the noise. Thus there are a number of entropies that can be calculated. First there is the entropy $H(x)$ of the source or of the input to the channel (these will be equal if the transmitter is non-singular). The entropy of the output of the channel, i.e. the received signal, will be denoted by $H(y)$. In the noiseless case $H(y) = H(x)$. The joint entropy of input and output will be $H(xy)$. Finally there are two conditional entropies $H_x(y)$ and $H_y(x)$, the entropy of the output when the input is known and conversely. Among these quantities we have the relations

$$H(x, y) = H(x) + H_x(y) = H(y) + H_y(x)$$

All of these entropies can be measured on a per-second or a per-symbol basis.

12. Equivocation and Channel Capacity

If the channel is noisy it is not in general possible to reconstruct the original message or the transmitted signal with *certainty* by any operation on the received signal E. There are, however, ways of transmitting the information which are optimal in combating noise. This is the problem which we now consider.

Suppose there are two possible symbols 0 and 1, and we are transmitting at a rate of 1000 symbols per second with probabilities $p_0 = p_1 = \frac{1}{2}$. Thus our source is producing information at the rate of 1000 bits per second. During transmission the noise introduces errors so that, on the average, 1 in 100 is received incorrectly (a 0 as 1, or 1 as 0). What is the rate of transmission of information? Certainly less than 1000 bits per second since about 1% of the received symbols are incorrect. Our first impulse might be to say the rate is 990 bits per second, merely subtracting the expected number of errors. This is not satisfactory since it fails to take into account the recipient's lack of knowledge of where the errors occur. We may carry it to an extreme case and suppose the noise so great that the received symbols are entirely independent of the transmitted symbols. The probability of receiving 1 is $\frac{1}{2}$ whatever was transmitted and similarly for 0. Then about half of the received symbols are correct due to chance alone, and we would be giving the system credit for transmitting 500 bits per second while actually no information is being transmitted at all. Equally "good" transmission would be obtained by dispensing with the channel entirely and flipping a coin at the receiving point.

Evidently the proper correction to apply to the amount of information transmitted is the amount of this information which is missing in the received signal, or alternatively the uncertainty when we have received a signal of what was actually sent. From our previous discussion of entropy as a measure of uncertainty it seems reasonable to use the conditional entropy of the message, knowing the received signal, as a measure of this missing information. This is indeed the proper definition, as we shall see later. Following this idea the rate of actual transmission, R, would be obtained by subtracting from the rate of production (i.e., the entropy of the source) the average rate of conditional entropy.

$$R = H(x) - H_y(x)$$

The conditional entropy $H_y(x)$ will, for convenience, be called the equivocation. It measures the average ambiguity of the received signal.

In the example considered above, if a 0 is received the *a posteriori* probability that a 0 was transmitted is .99, and that a 1 was transmitted is .01. These figures are reversed if a 1 is received. Hence

$$\begin{aligned} H_y(x) &= -[.99 \log .99 + 0.01 \log 0.01] \\ &= .081 \text{ bits/symbol} \end{aligned}$$

or 81 bits per second. We may say that the system is transmitting at a rate 1000 − 81 = 919 bits per second. In the extreme case where a 0 is equally likely to be received as a 0 or 1 and similarly for 1, the a posteriori probabilities are $\frac{1}{2}$, $\frac{1}{2}$ and

$$\begin{aligned} H_y(x) &= -[\tfrac{1}{2} \log \tfrac{1}{2} + \tfrac{1}{2} \log \tfrac{1}{2}] \\ &= 1 \text{ bit per symbol} \end{aligned}$$

or 1000 bits per second. The rate of transmission is then 0 as it should be.

The following theorem gives a direct intuitive interpretation of the equivocation and also serves to justify it as the unique appropriate measure. We consider a communication system and an observer (or auxiliary device) who can see both what is sent and what is recovered (with errors due to noise). This observer notes the errors in the recovered message and transmits data to the receiving point over a "correction channel" to enable the receiver to correct the errors. The situation is indicated schematically in Fig. 8.

Theorem 10: If the correction channel has a capacity equal to $H_y(x)$ it is possible to so encode the correction data as to send it over this channel and correct all but an arbitrarily small fraction ϵ of the errors. This is not possible if the channel capacity is less than $H_y(x)$.

Roughly then, $H_y(x)$ is the amount of additional information that must be supplied per second at the receiving point to correct the received message.

To prove the first part, consider long sequences of received message M' and corresponding original message M. There will be logarithmically $TH_y(x)$ of the M's which could reasonably have produced each M'. Thus we have $TH_y(x)$ binary digits to send each T seconds. This can be done with ϵ frequency of errors on a channel of capacity $H_y(x)$.

The second part can be proved by noting, first, that for any discrete chance variables x, y, z

$$H_y(x, z) \geq H_y(x)$$

The left-hand side can be expanded to give

$$H_y(z) + H_{yz}(x) \geq H_y(x)$$

$$H_{yz}(x) \geq H_y(x) - H_y(z) \geq H_y(x) - H(z)$$

If we identify x as the output of the source, y as the received signal and z as the signal sent over the correction channel, then the right-hand side is the equivocation less the rate of transmission over the correction channel. If the capacity of this channel is less than the equivocation the right-hand side will be greater than zero and $H_{yz}(x) \geq 0$. But this is the uncertainty of what was sent, knowing both the received signal and the correction signal. If this is greater than zero the frequency of errors cannot be arbitrarily small.

Example:
Suppose the errors occur at random in a sequence of binary digits: probability p that a digit is wrong and $q = 1 - p$ that it is right. These errors can be corrected if their position is known. Thus the correction channel need only send information as to these positions. This amounts to transmitting from a source which produces binary digits with probability p for 1 (correct) and q for 0 (incorrect). This requires a channel of capacity

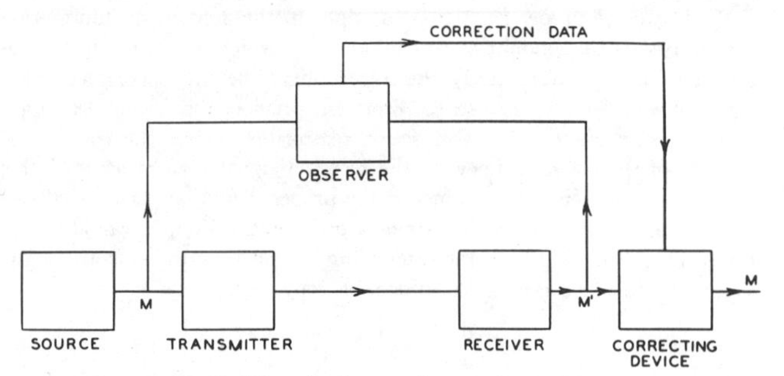

Fig. 8—Schematic diagram of a correction system.

$$-[p \log p + q \log q]$$

which is the equivocation of the original system.

The rate of transmission R can be written in two other forms due to the identities noted above. We have

$$R = H(x) - H_y(x)$$
$$= H(y) - H_x(y)$$
$$= H(x) + H(y) - H(x, y).$$

The first defining expression has already been interpreted as the amount of information sent less the uncertainty of what was sent. The second measures the amount received less the part of this which is due to noise. The third is the sum of the two amounts less the joint entropy and therefore in a sense is the number of bits per second common to the two. Thus all three expressions have a certain intuitive significance.

The capacity C of a noisy channel should be the maximum possible rate of transmission, i.e., the rate when the source is properly matched to the channel. We therefore define the channel capacity by

$$C = \text{Max}\,(H(x) - H_y(x))$$

where the maximum is with respect to all possible information sources used as input to the channel. If the channel is noiseless, $H_y(x) = 0$. The definition is then equivalent to that already given for a noiseless channel since the maximum entropy for the channel is its capacity.

13. The Fundamental Theorem for a Discrete Channel with Noise

It may seem surprising that we should define a definite capacity C for a noisy channel since we can never send certain information in such a case. It is clear, however, that by sending the information in a redundant form the probability of errors can be reduced. For example, by repeating the message many times and by a statistical study of the different received versions of the message the probability of errors could be made very small. One would expect, however, that to make this probability of errors approach zero, the redundancy of the encoding must increase indefinitely, and the rate of transmission therefore approach zero. This is by no means true. If it were, there would not be a very well defined capacity, but only a capacity for a given frequency of errors, or a given equivocation; the capacity going down as the error requirements are made more stringent. Actually the capacity C defined above has a very definite significance. It is possible to send information at the rate C through the channel *with as small a frequency of errors or equivocation as desired* by proper encoding. This statement is not true for any rate greater than C. If an attempt is made to transmit at a higher rate than C, say $C + R_1$, then there will necessarily be an equivocation equal to a greater than the excess R_1. Nature takes payment by requiring just that much uncertainty, so that we are not actually getting any more than C through correctly.

The situation is indicated in Fig. 9. The rate of information into the channel is plotted horizontally and the equivocation vertically. Any point above the heavy line in the shaded region can be attained and those below cannot. The points on the line cannot in general be attained, but there will usually be two points on the line that can.

These results are the main justification for the definition of C and will now be proved.

Theorem 11. Let a discrete channel have the capacity C and a discrete source the entropy per second H. If $H \leq C$ there exists a coding system such that the output of the source can be transmitted over the channel with an arbitrarily small frequency of errors (or an arbitrarily small equivocation). If $H > C$ it is possible to encode the source so that the equivocation is less than $H - C + \epsilon$ where ϵ is arbitrarily small. There is no method of encoding which gives an equivocation less than $H - C$.

The method of proving the first part of this theorem is not by exhibiting a coding method having the desired properties, but by showing that such a code must exist in a certain group of codes. In fact we will average the frequency of errors over this group and show that this average can be made less than ϵ. If the average of a set of numbers is less than ϵ there must exist at least one in the set which is less than ϵ. This will establish the desired result.

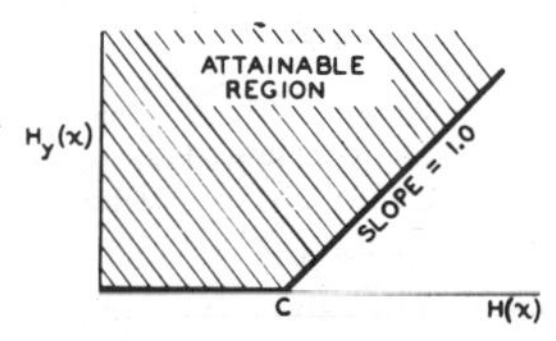

Fig. 9—The equivocation possible for a given input entropy to a channel.

The capacity C of a noisy channel has been defined as

$$C = \text{Max}\,(H(x) - H_y(x))$$

where x is the input and y the output. The maximization is over all sources which might be used as input to the channel.

Let S_0 be a source which achieves the maximum capacity C. If this maximum is not actually achieved by any source let S_0 be a source which approximates to giving the maximum rate. Suppose S_0 is used as input to the channel. We consider the possible transmitted and received sequences of a long duration T. The following will be true:

1. The transmitted sequences fall into two classes, a high probability group with about $2^{TH(x)}$ members and the remaining sequences of small total probability.
2. Similarly the received sequences have a high probability set of about $2^{TH(y)}$ members and a low probability set of remaining sequences.
3. Each high probability output could be produced by about $2^{TH_y(x)}$ inputs. The probability of all other cases has a small total probability.

All the ϵ's and δ's implied by the words "small" and "about" in these statements approach zero as we allow T to increase and S_0 to approach the maximizing source.

The situation is summarized in Fig. 10 where the input sequences are points on the left and output sequences points on the right. The fan of cross lines represents the range of possible causes for a typical output.

Now suppose we have another source producing information at rate R with $R < C$. In the period T this source will have 2^{TR} high probability outputs. We wish to associate these with a selection of the possible channel inputs in such a way as to get a small frequency of errors. We will set up this association in all possible ways (using, however, only the high probability group of inputs as determined by the source S_0) and average the frequency of errors for this large class of possible coding systems. This is the same as calculating the frequency of errors for a random association of the messages and channel inputs of duration T. Suppose a particular output y_1 is observed. What is the probability of more than one message in the set of possible causes of y_1? There are 2^{TR} messages distributed at random in $2^{TH(x)}$ points. The probability of a particular point being a message is thus

$$2^{T(R-H(x))}$$

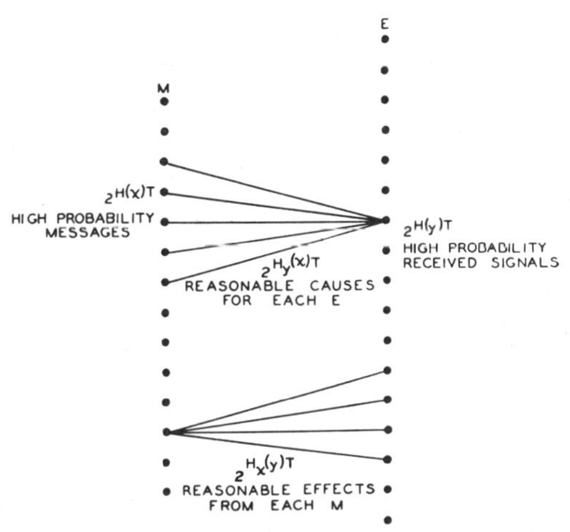

Fig. 10—Schematic representation of the relations between inputs and outputs in a channel.

The probability that none of the points in the fan is a message (apart from the actual originating message) is

$$P = [1 - 2^{T(R - H(x))}]^{2^{TH_y(x)}}$$

Now $R < H(x) - H_y(x)$ so $R - H(x) = -H_y(x) - \eta$ with η positive. Consequently

$$P = [1 - 2^{-TH_y(x) - T\eta}]^{2^{TH_y(x)}}$$

approaches (as $T \to \infty$)

$$1 - 2^{-T\eta}.$$

Hence the probability of an error approaches zero and the first part of the theorem is proved.

The second part of the theorem is easily shown by noting that we could merely send C bits per second from the source, completely neglecting the remainder of the information generated. At the receiver the neglected part gives an equivocation $H(x) - C$ and the part transmitted need only add ϵ. This limit can also be attained in many other ways, as will be shown when we consider the continuous case.

The last statement of the theorem is a simple consequence of our definition of C. Suppose we can encode a source with $R = C + a$ in such a way as to obtain an equivocation $H_y(x) = a - \epsilon$ with ϵ positive. Then $R = H(x) = C + a$ and

$$H(x) - H_y(x) = C + \epsilon$$

with ϵ positive. This contradicts the definition of C as the maximum of $H(x) - H_y(x)$.

Actually more has been proved than was stated in the theorem. If the average of a set of numbers is within ϵ of their maximum, a fraction of at most $\sqrt{\epsilon}$ can be more than $\sqrt{\epsilon}$ below the maximum. Since ϵ is arbitrarily small we can say that almost all the systems are arbitrarily close to the ideal.

14. Discussion

The demonstration of theorem 11, while not a pure existence proof, has some of the deficiencies of such proofs. An attempt to obtain a good approximation to ideal coding by following the method of the proof is generally impractical. In fact, apart from some rather trivial cases and certain limiting situations, no explicit description of a series of approximation to the ideal has been found. Probably this is no accident but is related to the difficulty of giving an explicit construction for a good approximation to a random sequence.

An approximation to the ideal would have the property that if the signal is altered in a reasonable way by the noise, the original can still be recovered. In other words the alteration will not in general bring it closer to another reasonable signal than the original. This is accomplished at the cost of a certain amount of redundancy in the coding. The redundancy must be introduced in the proper way to combat the particular noise structure involved. However, any redundancy in the source will usually help if it is utilized at the receiving point. In particular, if the source already has a certain redundancy and no attempt is made to eliminate it in matching to the channel, this redundancy will help combat noise. For example, in a noiseless telegraph channel one could save about 50% in time by proper encoding of the messages. This is not done and most of the redundnacy of English remains in the channel symbols. This has the advantage, however, of allowing considerable noise in the channel. A sizable fraction of the letters can be received incorrectly and still reconstructed by the context. In fact this is probably not a bad approximation to the ideal in many cases, since the statistical structure of English is rather involved and the reasonable English sequences are not too far (in the sense required for theorem) from a random selection.

As in the noiseless case a delay is generally required to approach the ideal encoding. It now has the additional function of allowing a large sample of noise to affect the signal before any judgment is made at the receiving point as to the original message. Increasing the sample size always sharpens the possible statistical assertions.

The content of theorem 11 and its proof can be formulated in a somewhat different way which exhibits the connection with the noiseless case more clearly. Consider the possible signals of duration T and suppose a subset of them is selected to be used. Let those in the subset all be used with equal probability, and suppose the receiver is constructed to select, as the original signal, the most probable cause from the subset, when a perturbed signal is received. We define $N(T, q)$ to be the maximum number of signals we can choose for the subset such that the probability of an incorrect interpretation is less than or equal to q.

Theorem 12: $\lim_{T\to\infty} \frac{\log N(T, q)}{T} = C$, where C is the channel capacity, provided that q does not equal 0 or 1.

In other words, no matter how we set our limits of reliability, we can distinguish reliably in time T enough messages to correspond to about CT bits, when T is sufficiently large. Theorem 12 can be compared with the definition of the capacity of a noiseless channel given in section 1.

15. Example of a Discrete Channel and Its Capacity

A simple example of a discrete channel is indicated in Fig. 11. There are three possible symbols. The first is never affected by noise. The second and third each have probability p of coming through undisturbed, and q of being changed into the other of the pair. We have (letting $\alpha = -[p \log$

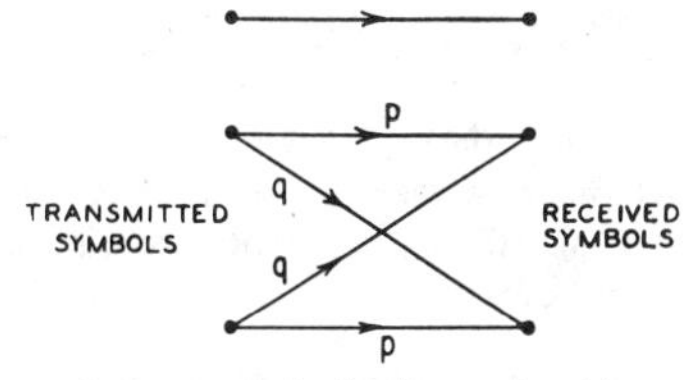

Fig. 11—Example of a discrete channel.

$p + q \log q]$ and P and Q be the probabilities of using the first or second symbols)

$$H(x) = -P \log P - 2Q \log Q$$

$$H_y(x) = 2Q\alpha$$

We wish to choose P and Q in such a way as to maximize $H(x) - H_y(x)$, subject to the constraint $P + 2Q = 1$. Hence we consider

$$U = -P \log P - 2Q \log Q - 2Q\alpha + \lambda(P + 2Q)$$

$$\frac{\partial U}{\partial P} = -1 - \log P + \lambda = 0$$

$$\frac{\partial U}{\partial Q} = -2 - 2 \log Q - 2\alpha + 2\lambda = 0.$$

Eliminating λ

$$\log P = \log Q + \alpha$$

$$P = Qe^{\alpha} = Q\beta$$

$$P = \frac{\beta}{\beta + 2} \qquad Q = \frac{1}{\beta + 2}.$$

The channel capacity is then

$$C = \log \frac{\beta + 2}{\beta}.$$

Note how this checks the obvious values in the cases $p = 1$ and $p = \frac{1}{2}$. In the first, $\beta = 1$ and $C = \log 3$, which is correct since the channel is then noiseless with three possible symbols. If $p = \frac{1}{2}$, $\beta = 2$ and $C = \log 2$. Here the second and third symbols cannot be distinguished at all and act together like one symbol. The first symbol is used with probability $P = \frac{1}{2}$ and the second and third together with probability $\frac{1}{2}$. This may be distributed in any desired way and still achieve the maximum capacity.

For intermediate values of p the channel capacity will lie between log 2 and log 3. The distinction between the second and third symbols conveys some information but not as much as in the noiseless case. The first symbol is used somewhat more frequently than the other two because of its freedom from noise.

16. The Channel Capacity in Certain Special Cases

If the noise affects successive channel symbols independently it can be described by a set of transition probabilities p_{ij}. This is the probability, if symbol i is sent, that j will be received. The maximum channel rate is then given by the maximum of

$$\sum_{i,j} P_i p_{ij} \log \sum_i P_i p_{ij} - \sum_{i,j} P_i p_{ij} \log p_{ij}$$

where we vary the P_i subject to $\Sigma P_i = 1$. This leads by the method of Lagrange to the equations,

$$\sum_j p_{sj} \log \frac{p_{sj}}{\sum_i P_i p_{ij}} = \mu \qquad s = 1, 2, \cdots.$$

Multiplying by P_s and summing on s shows that $\mu = -C$. Let the inverse of p_{sj} (if it exists) be h_{st} so that $\sum_s h_{st} p_{sj} = \delta_{tj}$. Then:

$$\sum_{s,j} h_{st} p_{sj} \log p_{sj} - \log \sum_i P_i p_{it} = -C \sum_s h_{st}.$$

Hence:

$$\sum_i P_i p_{it} = \exp \left[C \sum_s h_{st} + \sum_{s,j} h_{st} p_{sj} \log p_{sj}\right]$$

or,

$$P_i = \sum_t h_{it} \exp \left[C \sum_s h_{st} + \sum_{s,j} h_{st} p_{sj} \log p_{sj}\right].$$

This is the system of equations for determining the maximizing values of P_i, with C to be determined so that $\Sigma P_i = 1$. When this is done C will be the channel capacity, and the P_i the proper probabilities for the channel symbols to achieve this capacity.

If each input symbol has the same set of probabilities on the lines emerging from it, and the same is true of each output symbol, the capacity can be easily calculated. Examples are shown in Fig. 12. In such a case $H_x(y)$ is independent of the distribution of probabilities on the input symbols, and is given by $-\Sigma\, p_i \log p_i$ where the p_i are the values of the transition probabilities from any input symbol. The channel capacity is

$$\text{Max } [H(y) - H_x(y)]$$
$$= \text{Max } H(y) + \Sigma\, p_i \log p_i.$$

The maximum of $H(y)$ is clearly $\log m$ where m is the number of output

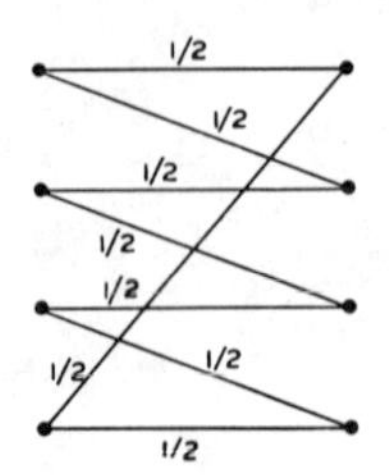

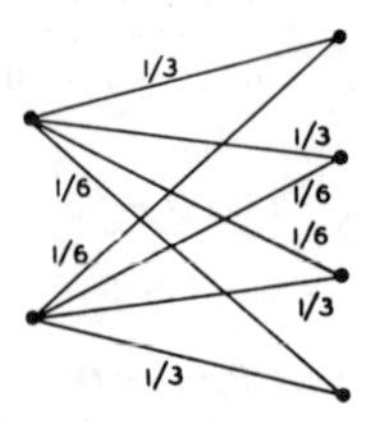

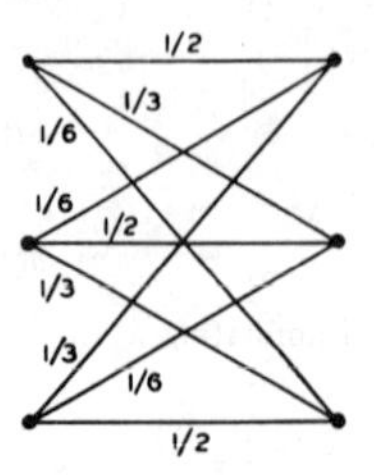

Fig. 12—Examples of discrete channels with the same transition probabilities for each input and for each output.

symbols, since it is possible to make them all equally probable by making the input symbols equally probable. The channel capacity is therefore

$$C = \log m + \Sigma\, p_i \log p_i.$$

In Fig. 12a it would be

$$C = \log 4 - \log 2 = \log 2.$$

This could be achieved by using only the 1st and 3d symbols. In Fig. 12b

$$C = \log 4 - \tfrac{2}{3} \log 3 - \tfrac{1}{3} \log 6$$
$$= \log 4 - \log 3 - \tfrac{1}{3} \log 2$$
$$= \log \tfrac{1}{3}\, 2^{\frac{5}{3}}.$$

In Fig. 12c we have

$$C = \log 3 - \tfrac{1}{2} \log 2 - \tfrac{1}{3} \log 3 - \tfrac{1}{6} \log 6$$
$$= \log \frac{3}{2^{\frac{1}{2}} 3^{\frac{1}{3}} 6^{\frac{1}{6}}}.$$

Suppose the symbols fall into several groups such that the noise never causes a symbol in one group to be mistaken for a symbol in another group. Let the capacity for the nth group be C_n when we use only the symbols in this group. Then it is easily shown that, for best use of the entire set, the total probability P_n of all symbols in the nth group should be

$$P_n = \frac{2^{C_n}}{\Sigma 2^{C_n}}.$$

Within a group the probability is distributed just as it would be if these were the only symbols being used. The channel capacity is

$$C = \log \Sigma 2^{C_n}.$$

17. An Example of Efficient Coding

The following example, although somewhat unrealistic, is a case in which exact matching to a noisy channel is possible. There are two channel symbols, 0 and 1, and the noise affects them in blocks of seven symbols. A block of seven is either transmitted without error, or exactly one symbol of the seven is incorrect. These eight possibilities are equally likely. We have

$$C = \text{Max } [H(y) - H_x(y)]$$
$$= \tfrac{1}{7} [7 + \tfrac{8}{8} \log \tfrac{1}{8}]$$
$$= \tfrac{4}{7} \text{ bits/symbol}.$$

An efficient code, allowing complete correction of errors and transmitting at the rate C, is the following (found by a method due to R. Hamming):

Let a block of seven symbols be $X_1, X_2, \ldots X_7$. Of these X_3, X_5, X_6 and X_7 are message symbols and chosen arbitrarily by the source. The other three are redundant and calculated as follows:

X_4 is chosen to make $\alpha = X_4 + X_5 + X_6 + X_7$ even

X_2 " " " " $\beta = X_2 + X_3 + X_6 + X_7$ "

X_1 " " " " $\gamma = X_1 + X_3 + X_5 + X_7$ "

When a block of seven is received α, β and γ are calculated and if even called zero, if odd called one. The binary number $\alpha\,\beta\,\gamma$ then gives the subscript of the X_i that is incorrect (if 0 there was no error).

APPENDIX 1

The Growth of the Number of Blocks of Symbols With A Finite State Condition

Let $N_i(L)$ be the number of blocks of symbols of length L ending in state i. Then we have

$$N_j(L) = \sum_{is} N_i(L - b_{ij}^{(s)})$$

where $b_{ij}^1, b_{ij}^2, \ldots b_{ij}^m$ are the length of the symbols which may be chosen in state i and lead to state j. These are linear difference equations and the behavior as $L \to \infty$ must be of the type

$$N_j = A_j W^L$$

Substituting in the difference equation

$$A_j W^L = \sum_{i,s} A_i W^{L - b_{ij}^{(s)}}$$

or

$$A_j = \sum_{is} A_i W^{-b_{ij}^{(s)}}$$

$$\sum_i \left(\sum_s W^{-b_{ij}^{(s)}} - \delta_{ij}\right) A_i = 0.$$

For this to be possible the determinant

$$D(W) = |a_{ij}| = \left|\sum_s W^{-b_{ij}^{(s)}} - \delta_{ij}\right|$$

must vanish and this determines W, which is, of course, the largest real root of $D = 0$.

The quantity C is then given by

$$C = \operatorname*{Lim}_{L \to \infty} \frac{\log \Sigma A_j W^L}{L} = \log W$$

and we also note that the same growth properties result if we require that all blocks start in the same (arbitrarily chosen) state.

APPENDIX 2

Derivation of $H = -\Sigma p_i \log p_i$

Let $H\left(\frac{1}{n}, \frac{1}{n}, \cdots, \frac{1}{n}\right) = A(n)$. From condition (3) we can decompose a choice from s^m equally likely possibilities into a series of m choices each from s equally likely possibilities and obtain

$$A(s^m) = m\,A(s)$$

Similarly

$$A(t^n) = n\,A(t)$$

We can choose n arbitrarily large and find an m to satisfy

$$s^m \leq t^n < s^{(m+1)}$$

Thus, taking logarithms and dividing by $n \log s$,

$$\frac{m}{n} \leq \frac{\log t}{\log s} \leq \frac{m}{n} + \frac{1}{n} \quad \text{or} \quad \left|\frac{m}{n} - \frac{\log t}{\log s}\right| < \epsilon$$

where ϵ is arbitrarily small.

Now from the monotonic property of $A(n)$

$$A(s^m) \leq A(t^n) \leq A(s^{m+1})$$

$$m\,A(s) \leq nA(t) \leq (m+1)\,A(s)$$

Hence, dividing by $nA(s)$,

$$\frac{m}{n} \leq \frac{A(t)}{A(s)} \leq \frac{m}{n} + \frac{1}{n} \quad \text{or} \quad \left|\frac{m}{n} - \frac{A(t)}{A(s)}\right| < \epsilon$$

$$\left|\frac{A(t)}{A(s)} - \frac{\log t}{\log s}\right| \leq 2\epsilon \qquad A(t) = -K \log t$$

where K must be positive to satisfy (2).

Now suppose we have a choice from n possibilities with commeasurable probabilities $p_i = \frac{n_i}{\Sigma n_i}$ where the n_i are integers. We can break down a choice from Σn_i possibilities into a choice from n possibilities with probabilities $p_i \ldots p_n$ and then, if the ith was chosen, a choice from n_i with equal probabilities. Using condition 3 again, we equate the total choice from Σn_i as computed by two methods

$$K \log \Sigma n_i = H(p_1, \ldots, p_n) + K\Sigma\, p_i \log n_i$$

Hence

$$H = K\,[\Sigma\, p_i \log \Sigma\, n_i - \Sigma\, p_i \log n_i]$$

$$= -K\,\Sigma\, p_i \log \frac{n_i}{\Sigma n_i} = -K\,\Sigma\, p_i \log p_i .$$

If the p_i are incommeasurable, they may be approximated by rationals and the same expression must hold by our continuity assumption. Thus the expression holds in general. The choice of coefficient K is a matter of convenience and amounts to the choice of a unit of measure.

APPENDIX 3

Theorems on Ergodic Sources

If it is possible to go from any state with $P > 0$ to any other along a path of probability $p > 0$, the system is ergodic and the strong law of large numbers can be applied. Thus the number of times a given path p_{ij} in the network is traversed in a long sequence of length N is about proportional to the probability of being at i and then choosing this path, $P_i p_{ij} N$. If N is large enough the probability of percentage error $\pm\,\delta$ in this is less than ϵ so that for all but a set of small probability the actual numbers lie within the limits

$$(P_i p_{ij} \pm \delta)N$$

Hence nearly all sequences have a probability p given by

$$p = \Pi\, p_{ij}^{(P_i p_{ij} \pm \delta)N}$$

and $\frac{\log p}{N}$ is limited by

$$\frac{\log p}{N} = \Sigma(P_i p_{ij} \pm \delta) \log p_{ij}$$

or

$$\left|\frac{\log p}{N} - \Sigma P_i p_{ij} \log p_{ij}\right| < \eta.$$

This proves theorem 3.

Theorem 4 follows immediately from this on calculating upper and lower bounds for $n(q)$ based on the possible range of values of p in Theorem 3.

In the mixed (not ergodic) case if

$$L = \Sigma\, p_i L_i$$

and the entropies of the components are $H_1 \geq H_2 \geq \ldots \geq H_n$ we have the

Theorem: $\operatorname*{Lim}_{N \to \infty} \frac{\log n(q)}{N} = \varphi(q)$ is a decreasing step function,

$$\varphi(q) = H_s \quad \text{in the interval} \quad \sum_1^{s-1} \alpha_i < q < \sum_1^{s} \alpha_i .$$

To prove theorems 5 and 6 first note that F_N is monotonic decreasing because increasing N adds a subscript to a conditional entropy. A simple substitution for $p_{B_i}(S_j)$ in the definition of F_N shows that

$$F_N = N\,G_N - (N-1)\,G_{N-1}$$

and summing this for all N gives $G_N = \frac{1}{N} \Sigma\, F_N$. Hence $G_N \geq F_N$ and G_N monotonic decreasing. Also they must approach the same limit. By using theorem 3 we see that $\operatorname*{Lim}_{N \to \infty} G_N = H$.

APPENDIX 4

Maximizing the Rate for a System of Constraints

Suppose we have a set of constraints on sequences of symbols that is of the finite state type and can be represented therefore by a linear graph.

Let $\ell_{ij}^{(s)}$ be the lengths of the various symbols that can occur in passing from state i to state j. What distribution of probabilities P_i for the different states and $p_{ij}^{(s)}$ for choosing symbol s in state i and going to state j maximizes the rate of generating information under these constraints? The constraints define a discrete channel and the maximum rate must be less than or equal to the capacity C of this channel, since if all blocks of large length were equally likely, this rate would result, and if possible this would be best. We will show that this rate can be achieved by proper choice of the P_i and $p_{ij}^{(s)}$.

The rate in question is

$$\frac{-\Sigma P_i p_{ij}^{(s)} \log p_{ij}^{(s)}}{\Sigma P_{(i)} p_{ij}^{(s)} \ell_{ij}^{(s)}} = \frac{N}{M}.$$

Let $\ell_{ij} = \sum_s \ell_{ij}^{(s)}$. Evidently for a maximum $p_{ij}^{(s)} = k \exp \ell_{ij}^{(s)}$. The constraints on maximization are $\Sigma P_i = 1$, $\sum_j p_{ij} = 1$, $\Sigma\, P_i(p_{ij} - \delta_{ij}) = 0$.

Hence we maximize

$$U = \frac{-\Sigma P_i p_{ij} \log p_{ij}}{\Sigma P_i p_{ij} \ell_{ij}} + \lambda \sum_i P_i + \Sigma \mu_i p_{ij} + \Sigma \eta_j P_i(p_{ij} - \delta_{ij})$$

$$\frac{\partial U}{\partial p_{ij}} = -\frac{MP_i(1 + \log p_{ij}) + NP_i \ell_{ij}}{M^2} + \lambda + \mu_i + \eta_i P_i = 0.$$

Solving for p_{ij}

$$p_{ij} = A_i B_j D^{-\ell_{ij}}.$$

Since

$$\sum_j p_{ij} = 1, \qquad A_i^{-1} = \sum_j B_j D^{-\ell_{ij}}$$

$$p_{ij} = \frac{B_j D^{-\ell_{ij}}}{\sum_s B_s D^{-\ell_{is}}}.$$

The correct value of D is the capacity C and the B_j are solutions of

$$B_i = \Sigma B_j C^{-\ell_{ij}}$$

for then

$$p_{ij} = \frac{B_j}{B_i} C^{-\ell_{ij}}$$

$$\Sigma P_i \frac{B_j}{B_i} C^{-\ell_{ij}} = P_j$$

or

$$\Sigma \frac{P_i}{B_i} C^{-\ell_{ij}} = \frac{P_j}{B_j}$$

So that if λ_i satisfy

$$\Sigma \gamma_i C^{-\ell_{ij}} = \gamma_j$$

$$P_i = B_i \gamma_i$$

Both of the sets of equations for B_i and γ_i can be satisfied since C is such that

$$| C^{-\ell_{ij}} - \delta_{ij} | = 0$$

In this case the rate is

$$-\frac{\Sigma P_i p_{ij} \log \frac{B_j}{B_i} C^{-\ell_{ij}}}{\Sigma P_i p_{ij} \ell_{ij}}$$

$$= C - \frac{\Sigma P_i p_{ij} \log \frac{B_j}{B_i}}{\Sigma P_i p_{ij} \ell_{ij}}$$

but

$$\Sigma P_i p_{ij} (\log B_j - \log B_i) = \sum_j P_j \log B_j - \Sigma P_i \log B_i = 0$$

Hence the rate is C and as this could never be exceeded this is the maximum, justifying the assumed solution.

(To be continued)

A Mathematical Theory of Communication

By C. E. SHANNON

(Concluded from July 1948 issue)

PART III: MATHEMATICAL PRELIMINARIES

In this final installment of the paper we consider the case where the signals or the messages or both are continuously variable, in contrast with the discrete nature assumed until now. To a considerable extent the continuous case can be obtained through a limiting process from the discrete case by dividing the continuum of messages and signals into a large but finite number of small regions and calculating the various parameters involved on a discrete basis. As the size of the regions is decreased these parameters in general approach as limits the proper values for the continuous case. There are, however, a few new effects that appear and also a general change of emphasis in the direction of specialization of the general results to particular cases.

We will not attempt, in the continuous case, to obtain our results with the greatest generality, or with the extreme rigor of pure mathematics, since this would involve a great deal of abstract measure theory and would obscure the main thread of the analysis. A preliminary study, however, indicates that the theory can be formulated in a completely axiomatic and rigorous manner which includes both the continuous and discrete cases and many others. The occasional liberties taken with limiting processes in the present analysis can be justified in all cases of practical interest.

18. Sets and Ensembles of Functions

We shall have to deal in the continuous case with sets of functions and ensembles of functions. A set of functions, as the name implies, is merely a class or collection of functions, generally of one variable, time. It can be specified by giving an explicit representation of the various functions in the set, or implicitly by giving a property which functions in the set possess and others do not. Some examples are:

1. The set of functions:

$$f_\theta(t) = \sin (t + \theta).$$

Each particular value of θ determines a particular function in the set

2. The set of all functions of time containing no frequencies over W cycles per second.
3. The set of all functions limited in band to W and in amplitude to A.
4. The set of all English speech signals as functions of time.

An *ensemble* of functions is a set of functions together with a probability measure whereby we may determine the probability of a function in the set having certain properties.[1] For example with the set,

$$f_\theta(t) = \sin (t + \theta),$$

we may give a probability distribution for θ, $P(\theta)$. The set then becomes an ensemble.

Some further examples of ensembles of functions are:

1. A finite set of functions $f_k(t)$ $(k = 1, 2, \cdots, n)$ with the probability of f_k being p_k.
2. A finite dimensional family of functions

$$f(\alpha_1, \alpha_2, \cdots, \alpha_n; t)$$

with a probability distribution for the parameters α_i:

$$p(\alpha_1, \cdots, \alpha_n)$$

For example we could consider the ensemble defined by

$$f(a_1, \cdots, a_n, \theta_1, \cdots, \theta_n; t) = \sum_{n=1}^{n} a_n \sin n(\omega t + \theta_n)$$

with the amplitudes a_i distributed normally and independently, and the phrases θ_i distributed uniformly (from 0 to 2π) and independently.

3. The ensemble

$$f(a_i, t) = \sum_{n=-\infty}^{+\infty} a_n \frac{\sin \pi(2Wt - n)}{\pi(2Wt - n)}$$

with the a_i normal and independent all with the same standard deviation $\sqrt{N}$. This is a representation of "white" noise, band-limited to the band from 0 to W cycles per second and with average power N.[2]

4. Let points be distributed on the t axis according to a Poisson distribution. At each selected point the function $f(t)$ is placed and the different functions added, giving the ensemble

$$\sum_{k=-\infty}^{\infty} f(t + t_k)$$

where the t_k are the points of the Poisson distribution. This ensemble can be considered as a type of impulse or shot noise where all the impulses are identical.

5. The set of English speech functions with the probability measure given by the frequency of occurrence in ordinary use.

An ensemble of functions $f_\alpha(t)$ is *stationary* if the same ensemble results when all functions are shifted any fixed amount in time. The ensemble

$$f_\theta(t) = \sin (t + \theta)$$

is stationary if θ distributed uniformly from 0 to 2π. If we shift each function by t_1 we obtain

$$\begin{aligned} f_\theta(t + t_1) &= \sin (t + t_1 + \theta) \\ &= \sin (t + \varphi) \end{aligned}$$

[1] In mathematical terminology the functions belong to a measure space whose total measure is unity.

[2] This representation can be used as a definition of band limited white noise. It has certain advantages in that it involves fewer limiting operations than do definitions that have been used in the past. The name "white noise," already firmly intrenched in the literature, is perhaps somewhat unfortunate. In optics white light means either any continuous spectrum as contrasted with a point spectrum, or a spectrum which is flat with *wavelength* (which is not the same as a spectrum flat with frequency).

Reprinted with permission from *Bell Syst. Tech. J.*, vol. 27, pp. 623–656, Oct. 1948. Copyright © 1948, The American Telephone and Telegraph Company.

with φ distributed uniformly from 0 to 2π. Each function has changed but the ensemble as a whole is invariant under the translation. The other examples given above are also stationary.

An ensemble is *ergodic* if it is stationary, and there is no subset of the functions in the set with a probability different from 0 and 1 which is stationary. The ensemble

$$\sin (t + \theta)$$

is ergodic. No subset of these functions of probability $\neq 0, 1$ is transformed into itself under all time translations. On the other hand the ensemble

$$a \sin (t + \theta)$$

with a distributed normally and θ uniform is stationary but not ergodic. The subset of these functions with a between 0 and 1 for example is stationary.

Of the examples given, 3 and 4 are ergodic, and 5 may perhaps be considered so. If an ensemble is ergodic we may say roughly that each function in the set is typical of the ensemble. More precisely it is known that with an ergodic ensemble an average of any statistic over the ensemble is equal (with probability 1) to an average over all the time translations of a particular function in the set.[3] Roughly speaking, each function can be expected, as time progresses, to go through, with the proper frequency, all the convolutions of any of the functions in the set.

Just as we may perform various operations on numbers or functions to obtain new numbers or functions, we can perform operations on ensembles to obtain new ensembles. Suppose, for example, we have an ensemble of functions $f_\alpha(t)$ and an operator T which gives for each function $f_\alpha(t)$ a result $g_\alpha(t)$:

$$g_\alpha(t) = Tf_\alpha(t)$$

Probability measure is defined for the set $g_\alpha(t)$ by means of that for the set $f_\alpha(t)$. The probability of a certain subset of the $g_\alpha(t)$ functions is equal to that of the subset of the $f_\alpha(t)$ functions which produce members of the given subset of g functions under the operation T. Physically this corresponds to passing the ensemble through some device, for example, a filter, a rectifier or a modulator. The output functions of the device form the ensemble $g_\alpha(t)$.

A device or operator T will be called invariant if shifting the input merely shifts the output, i.e., if

$$g_\alpha(t) = Tf_\alpha(t)$$

implies

$$g_\alpha(t + t_1) = Tf_\alpha(t + t_1)$$

for all $f_\alpha(t)$ and all t_1. It is easily shown (see appendix 1) that if T is invariant and the input ensemble is stationary then the output ensemble is stationary. Likewise if the input is ergodic the output will also be ergodic.

A filter or a rectifier is invariant under all time translations. The operation of modulation is not since the carrier phase gives a certain time structure. However, modulation is invariant under all translations which are multiples of the period of the carrier.

Wiener has pointed out the intimate relation between the invariance of physical devices under time translations and Fourier theory.[4] He has shown, in fact, that if a device is linear as well as invariant Fourier analysis is then the appropriate mathematical tool for dealing with the problem.

An ensemble of functions is the appropriate mathematical representation of the messages produced by a continuous source (for example speech), of the signals produced by a transmitter, and of the perturbing noise. Communication theory is properly concerned, as has been emphasized by Wiener, not with operations on particular functions, but with operations on ensembles of functions. A communication system is designed not for a particular speech function and still less for a sine wave, but for the ensemble of speech functions.

[3] This is the famous ergodic theorem or rather one aspect of this theorem which was proved is somewhat different formulations by Birkhoff, von Neumann, and Koopman, and subsequently generalized by Wiener, Hopf, Hurewicz and others. The literature on ergodic theory is quite extensive and the reader is referred to the papers of these writers for precise and general formulations; e.g., E. Hopf "Ergodentheorie" Ergebnisse der Mathematic und ihrer Grenzgebiete, Vol. 5, "On Causality Statistics and Probability" Journal of Mathematics and Physics, Vol. XIII, No. 1, 1934; N. Weiner "The Ergodic Theorem" Duke Mathematical Journal, Vol. 5, 1939.

[4] Communication theory is heavily indebted to Wiener for much of its basic philosophy and theory. His classic NDRC report "The Interpolation, Extrapolation, and Smoothing of Stationary Time Series," to appear soon in book form, contains the first clear-cut formulation of communication theory as a statistical problem, the study of operations on time series. This work, although chiefly concerned with the linear prediction and filtering problem, is an important collateral reference in connection with the present paper. We may also refer here to Wiener's forthcoming book "Cybernetics" dealing with the general problems of communication and control.

19. Band Limited Ensembles of Functions

If a function of time $f(t)$ is limited to the band from 0 to W cycles per second it is completely determined by giving its ordinates at a series of discrete points spaced $\frac{1}{2W}$ seconds apart in the manner indicated by the following result.[5]

Theorem 13: Let $f(t)$ contain no frequencies over W.
Then

$$f(t) = \sum_{-\infty}^{\infty} X_n \frac{\sin \pi(2Wt - n)}{\pi(2Wt - n)}$$

where

$$X_n = f\left(\frac{n}{2W}\right).$$

In this expansion $f(t)$ is represented as a sum of orthogonal functions. The coefficients X_n of the various terms can be considered as coordinates in an infinite dimensional "function space." In this space each function corresponds to precisely one point and each point to one function.

A function can be considered to be substantially limited to a time T if all the ordinates X_n outside this interval of time are zero. In this case all but $2TW$ of the coordinates will be zero. Thus functions limited to a band W and duration T correspond to points in a space of $2TW$ dimensions.

A subset of the functions of band W and duration T corresponds to a region in this space. For example, the functions whose total energy is less than or equal to E correspond to points in a $2TW$ dimensional sphere with radius $r = \sqrt{2WE}$.

An *ensemble* of functions of limited duration and band will be represented by a probability distribution $p(x_1 \cdots x_n)$ in the corresponding n dimensional space. If the ensemble is not limited in time we can consider the $2TW$ coordinates in a given interval T to represent substantially the part of the function in the interval T and the probability distribution $p(x_1, \cdots, x_n)$ to give the statistical structure of the ensemble for intervals of that duration.

20. Entropy of a Continuous Distribution

The entropy of a discrete set of probabilities $p_1, \cdots p_n$ has been defined as:

$$H = -\sum p_i \log p_i .$$

In an analogous manner we define the entropy of a continuous distribution with the density distribution function $p(x)$ by:

$$H = -\int_{-\infty}^{\infty} p(x) \log p(x)\, dx$$

With an n dimensional distribution $p(x_1, \cdots, x_n)$ we have

$$H = -\int \cdots \int p(x_1 \cdots x_n) \log p(x_1, \cdots, x_n)\, dx_1 \cdots dx_n .$$

If we have two arguments x and y (which may themselves be multi-dimensional) the joint and conditional entropies of $p(x, y)$ are given by

$$H(x, y) = -\iint p(x, y) \log p(x, y)\, dx\, dy$$

and

$$H_x(y) = -\iint p(x, y) \log \frac{p(x, y)}{p(x)}\, dx\, dy$$

$$H_y(x) = -\iint p(x, y) \log \frac{p(x, y)}{p(y)}\, dx\, dy$$

where

$$p(x) = \int p(x, y)\, dy$$

$$p(y) = \int p(x, y)\, dx.$$

The entropy of continuous distributions have most (but not all) of the properties of the discrete case. In particular we have the following:

[5] For a proof of this theorem and further discussion see the author's paper "Communication in the Presence of Noise" to be published in the *Proceedings of the Institute of Radio Engineers.*

1. If x is limited to a certain volume v in its space, then $H(x)$ is a maximum and equal to $\log v$ when $p(x)$ is constant $\left(\frac{1}{v}\right)$ in the volume.
2. With any two variables x, y we have
$$H(x, y) \leq H(x) + H(y)$$
with equality if (and only if) x and y are independent, i.e., $p(x, y) = p(x)\, p(y)$ (apart possibly from a set of points of probability zero).
3. Consider a generalized averaging operation of the following type:
$$p'(y) = \int a(x, y) p(x)\, dx$$
with
$$\int a(x, y)\, dx = \int a(x, y)\, dy = 1, \qquad a(x, y) \geq 0.$$
Then the entropy of the averaged distribution $p'(y)$ is equal to or greater than that of the original distribution $p(x)$.
4. We have
$$H(x, y) = H(x) + H_x(y) = H(y) + H_y(x)$$
and
$$H_x(y) \leq H(y).$$
5. Let $p(x)$ be a one-dimensional distribution. The form of $p(x)$ giving a maximum entropy subject to the condition that the standard deviation of x be fixed at σ is gaussian. To show this we must maximize
$$H(x) = -\int p(x) \log p(x)\, dx$$
with
$$\sigma^2 = \int p(x) x^2\, dx \quad \text{and} \quad 1 = \int p(x)\, dx$$
as constraints. This requires, by the calculus of variations, maximizing
$$\int [-p(x) \log p(x) + \lambda p(x) x^2 + \mu p(x)]\, dx.$$
The condition for this is
$$-1 - \log p(x) + \lambda x^2 + \mu = 0$$
and consequently (adjusting the constants to satisfy the constraints)
$$p(x) = \frac{1}{\sqrt{2\pi}\,\sigma} e^{-(x^2/2\sigma^2)}.$$

Similarly in n dimensions, suppose the second order moments of $p(x_1, \cdots, x_n)$ are fixed at A_{ij}:
$$A_{ij} = \int \cdots \int x_i x_j p(x_1, \cdots, x_n)\, dx_1 \cdots dx_n.$$
Then the maximum entropy occurs (by a similar calculation) when $p(x_1, \cdots, x_n)$ is the n dimensional gaussian distribution with the second order moments A_{ij}.
6. The entropy of a one-dimensional gaussian distribution whose standard deviation is σ is given by
$$H(x) = \log \sqrt{2\pi e}\,\sigma.$$
This is calculated as follows:
$$p(x) = \frac{1}{\sqrt{2\pi}\,\sigma} e^{-(x^2/2\sigma^2)}$$
$$-\log p(x) = \log \sqrt{2\pi}\,\sigma + \frac{x^2}{2\sigma^2}$$
$$H(x) = -\int p(x) \log p(x)\, dx$$
$$= \int p(x) \log \sqrt{2\pi}\,\sigma\, dx + \int p(x) \frac{x^2}{2\sigma^2}\, dx$$
$$= \log \sqrt{2\pi}\,\sigma + \frac{\sigma^2}{2\sigma^2}$$
$$= \log \sqrt{2\pi}\,\sigma + \log \sqrt{e}$$
$$= \log \sqrt{2\pi e}\,\sigma.$$

Similarly the n dimensional gaussian distribution with associated quadratic form a_{ij} is given by
$$p(x_1, \cdots, x_n) = \frac{|a_{ij}|^{\frac{1}{2}}}{(2\pi)^{n/2}} \exp\left(-\tfrac{1}{2}\Sigma a_{ij} X_i X_j\right)$$
and the entropy can be calculated as
$$H = \log (2\pi e)^{n/2} |a_{ij}|^{\frac{1}{2}}$$
where $|a_{ij}|$ is the determinant whose elements are a_{ij}.
7. If x is limited to a half line ($p(x) = 0$ for $x \leq 0$) and the first moment of x is fixed at a:
$$a = \int_0^\infty p(x) x\, dx,$$
then the maximum entropy occurs when
$$p(x) = \frac{1}{a} e^{-(x/a)}$$
and is equal to $\log ea$.
8. There is one important difference between the continuous and discrete entropies. In the discrete case the entropy measures in an ***absolute*** way the randomness of the chance variable. In the continuous case the measurement is *relative to the coordinate system*. If we change coordinates the entropy will in general change. In fact if we change to coordinates $y_1 \cdots y_n$ the new entropy is given by
$$H(y) = \int \cdots \int p(x_1 \cdots x_n) J\left(\frac{x}{y}\right) \log p(x_1 \cdots x_n) J\left(\frac{x}{y}\right) dy_1 \cdots dy_n$$
where $J\left(\frac{x}{y}\right)$ is the Jacobian of the coordinate transformation. On expanding the logarithm and changing variables to $x_1 \cdots x_n$, we obtain:
$$H(y) = H(x) - \int \cdots \int p(x_1, \cdots, x_n) \log J\left(\frac{x}{y}\right) dx_1 \cdots dx_n.$$
Thus the new entropy is the old entropy less the expected logarithm of the Jacobian. In the continuous case the entropy can be considered a measure of randomness *relative to an assumed standard*, namely the coordinate system chosen with each small volume element $dx_1 \cdots dx_n$ given equal weight. When we change the coordinate system the entropy in the new system measures the randomness when equal volume elements $dy_1 \cdots dy_n$ in the new system are given equal weight.
In spite of this dependence on the coordinate system the entropy concept is as important in the continuous case as the discrete case. This is due to the fact that the derived concepts of information rate and channel capacity depend on the *difference* of two entropies and this difference *does not* depend on the coordinate frame, each of the two terms being changed by the same amount.
The entropy of a continuous distribution can be negative. The scale of measurements sets an arbitrary zero corresponding to a uniform distribution over a unit volume. A distribution which is more confined than this has less entropy and will be negative. The rates and capacities will, however, always be non-negative.
9. A particular case of changing coordinates is the linear transformation
$$y_j = \sum_i a_{ij} x_i.$$
In this case the Jacobian is simply the determinant $|a_{ij}|^{-1}$ and
$$H(y) = H(x) + \log |a_{ij}|.$$
In the case of a rotation of coordinates (or any measure preserving transformation) $J = 1$ and $H(y) = H(x)$.

21. Entropy of an Ensemble of Functions

Consider an ergodic ensemble of functions limited to a certain band of width W cycles per second. Let
$$p(x_1 \cdots x_n)$$
be the density distribution function for amplitudes $x_1 \cdots x_n$ at n successive sample points. We define the entropy of the ensemble per degree of freedom by

$$H' = -\lim_{n\to\infty} \frac{1}{n} \int \cdots \int p(x_1 \cdots x_n) \log p(x_1, \cdots, x_n)\, dx_1 \cdots dx_n .$$

We may also define an entropy H per second by dividing, not by n, but by the time T in seconds for n samples. Since $n = 2TW$, $H' = 2WH$.

With white thermal noise p is gaussian and we have

$$H = \log \sqrt{2\pi eN},$$

$$H' = W \log 2\pi eN.$$

For a given average power N, white noise has the maximum possible entropy. This follows from the maximizing properties of the Gaussian distribution noted above.

The entropy for a continuous stochastic process has many properties analogous to that for discrete processes. In the discrete case the entropy was related to the logarithm of the *probability* of long sequences, and to the *number* of reasonably probable sequences of long length. In the continuous case it is related in a similar fashion to the logarithm of the *probability density* for a long series of samples, and the *volume* of reasonably high probability in the function space.

More precisely, if we assume $p(x_1 \cdots x_n)$ continuous in all the x_i for all n, then for sufficiently large n

$$\left| \frac{\log p}{n} - H' \right| < \epsilon$$

for all choices of $(x_1, \cdots, x_n)$ apart from a set whose total probability is less than δ, with δ and ϵ arbitrarily small. This follows from the ergodic property if we divide the space into a large number of small cells.

The relation of H to volume can be stated as follows: Under the same assumptions consider the n dimensional space corresponding to $p(x_1, \cdots, x_n)$. Let $V_n(q)$ be the smallest volume in this space which includes in its interior a total probability q. Then

$$\lim_{n\to\infty} \frac{\log V_n(q)}{n} = H'$$

provided q does not equal 0 or 1.

These results show that for large n there is a rather well-defined volume (at least in the logarithmic sense) of high probability, and that within this volume the probability density is relatively uniform (again in the logarithmic sense).

In the white noise case the distribution function is given by

$$p(x_1 \cdots x_n) = \frac{1}{(2\pi N)^{n/2}} \exp - \frac{1}{2N} \Sigma x_i^2 .$$

Since this depends only on Σx_i^2 the surfaces of equal probability density are spheres and the entire distribution has spherical symmetry. The region of high probability is a sphere of radius $\sqrt{nN}$. As $n \to \infty$ the probability of being outside a sphere of radius $\sqrt{n(N + \epsilon)}$ approaches zero and $\frac{1}{n}$ times the logarithm of the volume of the sphere approaches $\log \sqrt{2\pi eN}$.

In the continuous case it is convenient to work not with the entropy H of an ensemble but with a derived quantity which we will call the entropy power. This is defined as the power in a white noise limited to the same band as the original ensemble and having the same entropy. In other words if H' is the entropy of an ensemble its entropy power is

$$N_1 = \frac{1}{2\pi e} \exp 2H'.$$

In the geometrical picture this amounts to measuring the high probability volume by the squared radius of a sphere having the same volume. Since white noise has the maximum entropy for a given power, the entropy power of any noise is less than or equal to its actual power.

21. Entropy Loss in Linear Filters

Theorem 14: If an ensemble having an entropy H_1 per degree of freedom in band W is passed through a filter with characteristic $Y(f)$ the output ensemble has an entropy

$$H_2 = H_1 + \frac{1}{W} \int_W \log |Y(f)|^2\, df.$$

The operation of the filter is essentially a linear transformation of coordinates. If we think of the different frequency components as the original coordinate system, the new frequency components are merely the old ones multiplied by factors. The coordinate transformation matrix is thus essentially diagonalized in terms of these coordinates. The Jacobian of the transformation is (for n sine and n cosine components)

TABLE I

GAIN	ENTROPY POWER FACTOR	ENTROPY POWER GAIN IN DECIBELS	IMPULSE RESPONSE
$1-\omega$	$\frac{1}{e^2}$	-8.68	$\frac{\sin^2 \pi t}{(\pi t)^2}$
$1-\omega^2$	$\left(\frac{2}{e}\right)^4$	-5.32	$2\left[\frac{\sin t}{t^3} - \frac{\cos t}{t^2}\right]$
$1-\omega^3$	0.384	-4.15	$6\left[\frac{\cos t - 1}{t^4} - \frac{\cos t}{2t^2} + \frac{\sin t}{t^3}\right]$
$\sqrt{1-\omega^2}$	$\left(\frac{2}{e}\right)^2$	-2.66	$\frac{\pi}{2}\frac{J_1(t)}{t}$
(gain 1, falling linearly to 0 over the last α before $\omega = 1$)	$\frac{1}{e^{2\alpha}}$	$-8.68\,\alpha$	$\frac{1}{\alpha t^2}\left[\cos(1-\alpha)t - \cos t\right]$

$$J = \prod_{i=1}^{n} |Y(f_i)|^2$$

where the f_i are equally spaced through the band W. This becomes in the limit

$$\exp \frac{1}{W} \int_W \log |Y(f)|^2\, df.$$

Since J is constant its average value is this same quantity and applying the theorem on the change of entropy with a change of coordinates, the result follows. We may also phrase it in terms of the entropy power. Thus if the entropy power of the first ensemble is N_1 that of the second is

$$N_1 \exp \frac{1}{W} \int_W \log |Y(f)|^2\, df.$$

The final entropy power is the initial entropy power multiplied by the geometric mean gain of the filter. If the gain is measured in *db*, then the output entropy power will be increased by the arithmetic mean *db* gain over W.

In Table I the entropy power loss has been calculated (and also expressed in *db*) for a number of ideal gain characteristics. The impulsive responses of these filters are also given for $W = 2\pi$, with phase assumed to be 0.

The entropy loss for many other cases can be obtained from these results. For example the entropy power factor $\frac{1}{e^2}$ for the first case also applies to any gain characteristic obtained from $1 - \omega$ by a measure preserving transformation of the ω axis. In particular a linearly increasing gain $G(\omega) = \omega$, or a "saw tooth" characteristic between 0 and 1 have the same entropy loss. The reciprocal gain has the reciprocal factor. Thus $\frac{1}{\omega}$ has the factor e^2. Raising the gain to any power raises the factor to this power.

22. Entropy of the Sum of Two Ensembles

If we have two ensembles of functions $f_\alpha(t)$ and $g_\beta(t)$ we can form a new ensemble by "addition." Suppose the first ensemble has the probability density function $p(x_1, \cdots, x_n)$ and the second $q(x_1, \cdots, x_n)$. Then the density function for the sum is given by the convolution:

$$r(x_1, \cdots, x_n) = \int \cdots \int p(y_1, \cdots, y_n) \cdot q(x_1 - y_1, \cdots, x_n - y_n)\, dy_1, dy_2, \cdots, dy_n .$$

Physically this corresponds to adding the noises or signals represented by the original ensembles of functions.

The following result is derived in Appendix 6.

Theorem 15: Let the average power of two ensembles be N_1 and N_2 and let their entropy powers be $\bar{N}_1$ and $\bar{N}_2$. Then the entropy power of the sum, $\bar{N}_3$, is bounded by

$$\bar{N}_1 + \bar{N}_2 \leq \bar{N}_3 \leq N_1 + N_2 .$$

White Gaussian noise has the peculiar property that it can absorb any other noise or signal ensemble which may be added to it with a resultant entropy power approximately equal to the sum of the white noise power and the signal power (measured from the average signal value, which is normally zero), provided the signal power is small, in a certain sense, compared to the noise.

Consider the function space associated with these ensembles having n dimensions. The white noise corresponds to a spherical Gaussian distribution in this space. The signal ensemble corresponds to another probability distribution, not necessarily Gaussian or spherical. Let the second moments of this distribution about its center of gravity be a_{ij}. That is, if $p(x_1, \cdots, x_n)$ is the density distribution function

$$a_{ij} = \int \cdots \int p(x_i - \alpha_i)(x_j - \alpha_j)\, dx_1, \cdots, dx_n$$

where the α_i are the coordinates of the center of gravity. Now a_{ij} is a positive definite quadratic form, and we can rotate our coordinate system to align it with the principal directions of this form. a_{ij} is then reduced to diagonal form b_{ii}. We require that each b_{ii} be small compared to N, the squared radius of the spherical distribution.

In this case the convolution of the noise and signal produce a Gaussian distribution whose corresponding quadratic form is

$$N + b_{ii} .$$

The entropy power of this distribution is

$$[\Pi(N + b_{ii})]^{1/n}$$

or approximately

$$= [(N)^n + \Sigma b_{ii}(N)^{n-1}]^{1/n}$$

$$\doteq N + \frac{1}{n}\Sigma b_{ii} .$$

The last term is the signal power, while the first is the noise power.

PART IV: THE CONTINUOUS CHANNEL

23. The Capacity of a Continuous Channel

In a continuous channel the input or transmitted signals will be continuous functions of time $f(t)$ belonging to a certain set, and the output or received signals will be perturbed versions of these. We will consider only the case where both transmitted and received signals are limited to a certain band W. They can then be specified, for a time T, by $2TW$ numbers, and their statistical structure by finite dimensional distribution functions. Thus the statistics of the transmitted signal will be determined by

$$P(x_1, \cdots, x_n) = P(x)$$

and those of the noise by the conditional probability distribution

$$P_{x_1, \cdots, x_n}(y_1, \cdots, y_n) = P_x(y).$$

The rate of transmission of information for a continuous channel is defined in a way analogous to that for a discrete channel, namely

$$R = H(x) - H_y(x)$$

where $H(x)$ is the entropy of the input and $H_y(x)$ the equivocation. The channel capacity C is defined as the maximum of R when we vary the input over all possible ensembles. This means that in a finite dimensional approximation we must vary $P(x) = P(x_1, \cdots, x_n)$ and maximize

$$-\int P(x) \log P(x)\, dx + \iint P(x, y) \log \frac{P(x, y)}{P(y)}\, dx\, dy.$$

This can be written

$$\iint P(x, y) \log \frac{P(x, y)}{P(x)P(y)}\, dx\, dy$$

using the fact that $\iint P(x, y) \log P(x)\, dx\, dy = \int P(x) \log P(x)\, dx$. The channel capacity is thus expressed

$$C = \lim_{T \to \infty} \max_{P(x)} \frac{1}{T} \iint P(x, y) \log \frac{P(x, y)}{P(x)P(y)}\, dx\, dy.$$

It is obvious in this form that R and C are independent of the coordinate system since the numerator and denominator in $\log \frac{P(x, y)}{P(x)P(y)}$ will be multiplied by the same factors when x and y are transformed in any one to one way. This integral expression for C is more general than $H(x) - H_y(x)$. Properly interpreted (see Appendix 7) it will always exist while $H(x) - H_y(x)$ may assume an indeterminate form $\infty - \infty$ in some cases. This occurs, for example, if x is limited to a surface of fewer dimensions than n in its n dimensional approximation.

If the logarithmic base used in computing $H(x)$ and $H_y(x)$ is two then C is the maximum number of binary digits that can be sent per second over the channel with arbitrarily small equivocation, just as in the discrete case. This can be seen physically by dividing the space of signals into a large number of small cells, sufficiently small so that the probability density $P_x(y)$ of signal x being perturbed to point y is substantially constant over a cell (either of x or y). If the cells are considered as distinct points the situation is essentially the same as a discrete channel and the proofs used there will apply. But it is clear physically that this quantizing of the volume into individual points cannot in any practical situation alter the final answer significantly, provided the regions are sufficiently small. Thus the capacity will be the limit of the capacities for the discrete subdivisions and this is just the continuous capacity defined above.

On the mathematical side it can be shown first (see Appendix 7) that if u is the message, x is the signal, y is the received signal (perturbed by noise) and v the recovered message then

$$H(x) - H_y(x) \geq H(u) - H_v(u)$$

regardless of what operations are performed on u to obtain x or on y to obtain v. Thus no matter how we encode the binary digits to obtain the signal, or how we decode the received signal to recover the message, the discrete rate for the binary digits does not exceed the channel capacity we have defined. On the other hand, it is possible under very general conditions to find a coding system for transmitting binary digits at the rate C with as small an equivocation or frequency of errors as desired. This is true, for example, if, when we take a finite dimensional approximating space for the signal functions, $P(x, y)$ is continuous in both x and y except at a set of points of probability zero.

An important special case occurs when the noise is added to the signal and is independent of it (in the probability sense). Then $P_x(y)$ is a function only of the difference $n = (y - x)$,

$$P_x(y) = Q(y - x)$$

and we can assign a definite entropy to the noise (independent of the statistics of the signal), namely the entropy of the distribution $Q(n)$. This entropy will be denoted by $H(n)$.

Theorem 16: If the signal and noise are independent and the received signal is the sum of the transmitted signal and the noise then the rate of transmission is

$$R = H(y) - H(n)$$

i.e., the entropy of the received signal less the entropy of the noise. The channel capacity is

$$C = \operatorname*{Max}_{P(x)} H(y) - H(n).$$

We have, since $y = x + n$:

$$H(x, y) = H(x, n).$$

Expanding the left side and using the fact that x and n are independent

$$H(y) + H_y(x) = H(x) + H(n).$$

Hence

$$R = H(x) - H_y(x) = H(y) - H(n).$$

Since $H(n)$ is independent of $P(x)$, maximizing R requires maximizing $H(y)$, the entropy of the received signal. If there are certain constraints on the ensemble of transmitted signals, the entropy of the received signal must be maximized subject to these constraints.

24. Channel Capacity with an Average Power Limitation

A simple application of Theorem 16 is the case where the noise is a white thermal noise and the transmitted signals are limited to a certain average power P. Then the received signals have an average power $P + N$ where N is the average noise power. The maximum entropy for the received signals occurs when they also form a white noise ensemble since this is the greatest possible entropy for a power $P + N$ and can be obtained by a suitable choice of the ensemble of transmitted signals, namely if they form a white noise ensemble of power P. The entropy (per second) of the received ensemble is then

$$H(y) = W \log 2\pi e(P + N),$$

and the noise entropy is

$$H(n) = W \log 2\pi e N.$$

The channel capacity is

$$C = H(y) - H(n) = W \log \frac{P + N}{N}.$$

Summarizing we have the following:

Theorem 17: The capacity of a channel of band W perturbed by white thermal noise of power N when the average transmitter power is P is given by

$$C = W \log \frac{P + N}{N}.$$

This means of course that by sufficiently involved encoding systems we can transmit binary digits at the rate $W \log_2 \frac{P + N}{N}$ bits per second, with arbitrarily small frequency of errors. It is not possible to transmit at a higher rate by any encoding system without a definite positive frequency of errors.

To approximate this limiting rate of transmission the transmitted signals must approximate, in statistical properties, a white noise.[6] A system which approaches the ideal rate may be described as follows: Let $M = 2^s$ samples of white noise be constructed each of duration T. These are assigned binary numbers from 0 to $(M - 1)$. At the transmitter the message sequences are broken up into groups of s and for each group the corresponding noise sample is transmitted as the signal. At the receiver the M samples are known and the actual received signal (perturbed by noise) is compared with each of them. The sample which has the least R.M.S. discrepancy from the received signal is chosen as the transmitted signal and the corresponding binary number reconstructed. This process amounts to choosing the most probable (*a posteriori*) signal. The number M of noise samples used will depend on the tolerable frequency ϵ of errors, but for almost all selections of samples we have

$$\lim_{\epsilon \to 0} \lim_{T \to \infty} \frac{\log M(\epsilon, T)}{T} = W \log \frac{P + N}{N},$$

so that no matter how small ϵ is chosen, we can, by taking T sufficiently large, transmit as near as we wish to $TW \log \frac{P + N}{N}$ binary digits in the time T.

Formulas similar to $C = W \log \frac{P + N}{N}$ for the white noise case have been developed independently by several other writers, although with somewhat different interpretations. We may mention the work of N. Wiener,[7] W. G. Tuller,[8] and H. Sullivan in this connection.

In the case of an arbitrary perturbing noise (not necessarily white thermal noise) it does not appear that the maximizing problem involved in determining the channel capacity C can be solved explicitly. However, upper and lower bounds can be set for C in terms of the average noise power N and the noise entropy power N_1. These bounds are sufficiently close together in most practical cases to furnish a satisfactory solution to the problem.

Theorem 18: The capacity of a channel of band W perturbed by an arbitrary noise is bounded by the inequalities

$$W \log \frac{P + N_1}{N_1} \le C \le W \log \frac{P + N}{N_1}$$

where

P = average transmitter power
N = average noise power
N_1 = entropy power of the noise.

Here again the average power of the perturbed signals will be $P + N$. The maximum entropy for this power would occur if the received signal were white noise and would be $W \log 2\pi e(P + N)$. It may not be possible to achieve this; i.e. there may not be any ensemble of transmitted signals which, added to the perturbing noise, produce a white thermal noise at the receiver, but at least this sets an upper bound to $H(y)$. We have, therefore

$$C = \max H(y) - H(n)$$
$$\le W \log 2\pi e(P + N) - W \log 2\pi e N_1.$$

This is the upper limit given in the theorem. The lower limit can be obtained by considering the rate if we make the transmitted signal a white noise, of power P. In this case the entropy power of the received signal must be at least as great as that of a white noise of power $P + N_1$ since we have shown in a previous theorem that the entropy power of the sum of two ensembles is greater than or equal to the sum of the individual entropy powers. Hence

$$\max H(y) \ge W \log 2\pi e(P + N_1)$$

and

$$C \ge W \log 2\pi e(P + N_1) - W \log 2\pi e N_1$$
$$= W \log \frac{P + N_1}{N_1}.$$

As P increases, the upper and lower bounds approach each other, so we have as an asymptotic rate

$$W \log \frac{P + N}{N_1}$$

If the noise is itself white, $N = N_1$ and the result reduces to the formula proved previously:

$$C = W \log \left(1 + \frac{P}{N}\right).$$

If the noise is Gaussian but with a spectrum which is not necessarily flat, N_1 is the geometric mean of the noise power over the various frequencies in the band W. Thus

$$N_1 = \exp \frac{1}{W} \int_W \log N(f)\, df$$

where $N(f)$ is the noise power at frequency f.

[6] This and other properties of the white noise case are discussed from the geometrical point of view in "Communication in the Presence of Noise," loc. cit.
[7] "Cybernetics," loc. cit.
[8] Sc. D. thesis, Department of Electrical Engineering, M.I.T., 1948.

Theorem 19: If we set the capacity for a given transmitter power P equal to

$$C = W \log \frac{P + N - \eta}{N_1}$$

then η is monotonic decreasing as P increases and approaches 0 as a limit.

Suppose that for a given power P_1 the channel capacity is

$$W \log \frac{P_1 + N - \eta_1}{N_1}$$

This means that the best signal distribution, say $p(x)$, when added to the noise distribution $q(x)$, gives a received distribution $r(y)$ whose entropy power is $(P_1 + N - \eta_1)$. Let us increase the power to $P_1 + \Delta P$ by adding a white noise of power ΔP to the signal. The entropy of the received signal is now at least

$$H(y) = W \log 2\pi e(P_1 + N - \eta_1 + \Delta P)$$

by application of the theorem on the minimum entropy power of a sum. Hence, since we can attain the H indicated, the entropy of the maximizing distribution must be at least as great and η must be monotonic decreasing. To show that $\eta \to 0$ as $P \to \infty$ consider a signal which is a white noise with a large P. Whatever the perturbing noise, the received signal will be approximately a white noise, if P is sufficiently large, in the sense of having an entropy power approaching $P + N$.

25. The Channel Capacity with a Peak Power Limitation

In some applications the transmitter is limited not by the average power output but by the peak instantaneous power. The problem of calculating the channel capacity is then that of maximizing (by variation of the ensemble of transmitted symbols)

$$H(y) - H(n)$$

subject to the constraint that all the functions $f(t)$ in the ensemble be less than or equal to $\sqrt{S}$, say, for all t. A constraint of this type does not work out as well mathematically as the average power limitation. The most we have obtained for this case is a lower bound valid for all $\frac{S}{N}$, an "asymptotic" upper band $\left(\text{valid for large } \frac{S}{N}\right)$ and an asymptotic value of C for $\frac{S}{N}$ small.

Theorem 20: The channel capacity C for a band W perturbed by white thermal noise of power N is bounded by

$$C \geq W \log \frac{2}{\pi e^3} \frac{S}{N},$$

where S is the peak allowed transmitter power. For sufficiently large $\frac{S}{N}$

$$C \leq W \log \frac{\frac{2}{\pi e} S + N}{N} (1 + \epsilon)$$

where ϵ is arbitrarily small. As $\frac{S}{N} \to 0$ (and provided the band W starts at 0)

$$C \to W \log \left(1 + \frac{S}{N}\right).$$

We wish to maximize the entropy of the received signal. If $\frac{S}{N}$ is large this will occur very nearly when we maximize the entropy of the transmitted ensemble.

The asymptotic upper bound is obtained by relaxing the conditions on the ensemble. Let us suppose that the power is limited to S not at every instant of time, but only at the sample points. The maximum entropy of the transmitted ensemble under these weakened conditions is certainly greater than or equal to that under the original conditions. This altered problem can be solved easily. The maximum entropy occurs if the different samples are independent and have a distribution function which is constant from $-\sqrt{S}$ to $+\sqrt{S}$. The entropy can be calculated as

$$W \log 4S.$$

The received signal will then have an entropy less than

$$W \log (4S + 2\pi eN)(1 + \epsilon)$$

with $\epsilon \to 0$ as $\frac{S}{N} \to \infty$ and the channel capacity is obtained by subtracting the entropy of the white noise, $W \log 2\pi eN$

$$W \log (4S + 2\pi eN)(1 + \epsilon) - W \log (2\pi eN) = W \log \frac{\frac{2}{\pi e} S + N}{N} (1 + \epsilon).$$

This is the desired upper bound to the channel capacity.

To obtain a lower bound consider the same ensemble of functions. Let these functions be passed through an ideal filter with a triangular transfer characteristic. The gain is to be unity at frequency 0 and decline linearly down to gain 0 at frequency W. We first show that the output functions of the filter have a peak power limitation S at all times (not just the sample points). First we note that a pulse $\frac{\sin 2\pi Wt}{2\pi Wt}$ going into the filter produces

$$\frac{1}{2} \frac{\sin^2 \pi Wt}{(\pi Wt)^2}$$

in the output. This function is never negative. The input function (in the general case) can be thought of as the sum of a series of shifted functions

$$a \frac{\sin 2\pi Wt}{2\pi Wt}$$

where a, the amplitude of the sample, is not greater than $\sqrt{S}$. Hence the output is the sum of shifted functions of the non-negative form above with the same coefficients. These functions being non-negative, the greatest positive value for any t is obtained when all the coefficients a have their maximum positive values, i.e. $\sqrt{S}$. In this case the input function was a constant of amplitude $\sqrt{S}$ and since the filter has unit gain for D.C., the output is the same. Hence the output ensemble has a peak power S.

The entropy of the output ensemble can be calculated from that of the input ensemble by using the theorem dealing with such a situation. The output entropy is equal to the input entropy plus the geometrical mean gain of the filter;

$$\int_0^W \log G^2 \, df = \int_0^W \log \left(\frac{W - f}{W}\right)^2 df = -2W$$

Hence the output entropy is

$$W \log 4S - 2W = W \log \frac{4S}{e^2}$$

and the channel capacity is greater than

$$W \log \frac{2}{\pi e^3} \frac{S}{N}.$$

We now wish to show that, for small $\frac{S}{N}$ (peak signal power over average white noise power), the channel capacity is approximately

$$C = W \log \left(1 + \frac{S}{N}\right).$$

More precisely $C/W \log \left(1 + \frac{S}{N}\right) \to 1$ as $\frac{S}{N} \to 0$. Since the average signal power P is less than or equal to the peak S, it follows that for all $\frac{S}{N}$

$$C \leq W \log \left(1 + \frac{P}{N}\right) \leq W \log \left(1 + \frac{S}{N}\right).$$

Therefore, if we can find an ensemble of functions such that they correspond to a rate nearly $W \log \left(1 + \frac{S}{N}\right)$ and are limited to band W and peak S the result will be proved. Consider the ensemble of functions of the following type. A series of t samples have the same value, either $+\sqrt{S}$ or $-\sqrt{S}$, then the next t samples have the same value, etc. The value for a series is chosen at random, probability $\frac{1}{2}$ for $+\sqrt{S}$ and $\frac{1}{2}$ for $-\sqrt{S}$. If this ensemble be passed through a filter with triangular gain characteristic (unit gain at D.C.), the output is peak limited to $\pm S$. Furthermore the average power is nearly S and can be made to approach this by taking t sufficiently large. The entropy of the sum of this and the thermal noise can be found by applying the theorem on the sum of a noise and a small signal. This theorem will apply if

$$\sqrt{t}\,\frac{S}{N}$$

is sufficiently small. This can be insured by taking $\frac{S}{N}$ small enough (after t is chosen). The entropy power will be $S + N$ to as close an approximation as desired, and hence the rate of transmission as near as we wish to

$$W \log \left(\frac{S+N}{N}\right).$$

PART V: THE RATE FOR A CONTINUOUS SOURCE

26. Fidelity Evaluation Functions

In the case of a discrete source of information we were able to determine a definite rate of generating information, namely the entropy of the underlying stochastic process. With a continuous source the situation is considerably more involved. In the first place a continuously variable quantity can assume an infinite number of values and requires, therefore, an infinite number of binary digits for exact specification. This means that to transmit the output of a continuous source with *exact recovery* at the receiving point requires, in general, a channel of infinite capacity (in bits per second). Since, ordinarily, channels have a certain amount of noise, and therefore a finite capacity, exact transmission is impossible.

This, however, evades the real issue. Practically, we are not interested in exact transmission when we have a continuous source, but only in transmission to within a certain tolerance. The question is, can we assign a definite rate to a continuous source when we require only a certain fidelity of recovery, measured in a suitable way. Of course, as the fidelity requirements are increased the rate will increase. It will be shown that we can, in very general cases, define such a rate, having the property that it is possible, by properly encoding the information, to transmit it over a channel whose capacity is equal to the rate in question, and satisfy the fidelity requirements. A channel of smaller capacity is insufficient.

It is first necessary to give a general mathematical formulation of the idea of fidelity of transmission. Consider the set of messages of a long duration, say T seconds. The source is described by giving the probability density, in the associated space, that the source will select the message in question $P(x)$. A given communication system is described (from the external point of view) by giving the conditional probability $P_x(y)$ that if message x is produced by the source the recovered message at the receiving point will be y. The system as a whole (including source and transmission system) is described by the probability function $P(x, y)$ of having message x and final output y. If this function is known, the complete characteristics of the system from the point of view of fidelity are known. Any evaluation of fidelity must correspond mathematically to an operation applied to $P(x, y)$. This operation must at least have the properties of a simple ordering of systems; i.e. it must be possible to say of two systems represented by $P_1(x, y)$ and $P_2(x, y)$ that, according to our fidelity criterion, either (1) the first has higher fidelity, (2) the second has higher fidelity, or (3) they have equal fidelity. This means that a criterion of fidelity can be represented by a numerically valued function:

$$v(P(x, y))$$

whose argument ranges over possible probability functions $P(x, y)$.

We will now show that under very general and reasonable assumptions the function $v(P(x, y))$ can be written in a seemingly much more specialized form, namely as an average of a function $\rho(x, y)$ over the set of possible values of x and y:

$$v(P(x, y)) = \iint P(x, y)\, \rho(x, y)\, dx\, dy$$

To obtain this we need only assume (1) that the source and system are ergodic so that a very long sample will be, with probability nearly 1, typical of the ensemble, and (2) that the evaluation is "reasonable" in the sense that it is possible, by observing a typical input and output x_1 and y_1, to form a tentative evaluation on the basis of these samples; and if these samples are increased in duration the tentative evaluation will, with probability 1, approach the exact evaluation based on a full knowledge of $P(x, y)$. Let the tentative evaluation be $\rho(x, y)$. Then the function $\rho(x, y)$ approaches (as $T \to \infty$) a constant for almost all (x, y) which are in the high probability region corresponding to the system:

$$\rho(x, y) \to v(P(x, y))$$

and we may also write

$$\rho(x, y) \to \iint P(x, y) \rho(x, y)\, dx, dy$$

since

$$\iint P(x, y)\, dx\, dy = 1$$

This establishes the desired result.

The function $\rho(x, y)$ has the general nature of a "distance" between x and y.[9] It measures how bad it is (according to our fidelity criterion) to receive y when x is transmitted. The general result given above can be restated as follows: Any reasonable evaluation can be represented as an average of a distance function over the set of messages and recovered messages x and y weighted according to the probability $P(x, y)$ of getting the pair in question, provided the duration T of the messages be taken sufficiently large.

The following are simple examples of evaluation functions:

1. R.M.S. Criterion.

$$v = \overline{(x(t) - y(t))^2}$$

In this very commonly used criterion of fidelity the distance function $\rho(x, y)$ is (apart from a constant factor) the square of the ordinary euclidean distance between the points x and y in the associated function space.

$$\rho(x, y) = \frac{1}{T}\int_0^T [x(t) - y(t)]^2\, dt$$

2. Frequency weighted R.M.S. criterion. More generally one can apply different weights to the different frequency components before using an R.M.S. measure of fidelity. This is equivalent to passing the difference $x(t) - y(t)$ through a shaping filter and then determining the average power in the output. Thus let

$$e(t) = x(t) - y(t)$$

and

$$f(t) = \int_{-\infty}^{\infty} e(\tau) k(t - \tau)\, dt$$

then

$$\rho(x, y) = \frac{1}{T}\int_0^T f(t)^2\, dt.$$

3. Absolute error criterion.

$$\rho(x, y) = \frac{1}{T}\int_0^T |\, x(t) - y(t)\, |\, dt$$

4. The structure of the ear and brain determine implicitly an evaluation, or rather a number of evaluations, appropriate in the case of speech or music transmission. There is, for example, an "intelligibility" criterion in which $\rho(x, y)$ is equal to the relative frequency of incorrectly interpreted words when message $x(t)$ is received as $y(t)$. Although we cannot give an explicit representation of $\rho(x, y)$ in these cases it could, in principle, be determined by sufficient experimentation. Some of its properties follow from well-known experimental results in hearing, e.g., the ear is relatively insensitive to phase and the sensitivity to amplitude and frequency is roughly logarithmic.
5. The discrete case can be considered as a specialization in which we have tacitly assumed an evaluation based on the frequency of errors. The function $\rho(x, y)$ is then defined as the number of symbols in the sequence y differing from the corresponding symbols in x divided by the total number of symbols in x.

[9] It is not a "metric" in the strict sense, however, since in general it does not satisfy either $\rho(x, y) = \rho(y, x)$ or $\rho(x, y) + \rho(y, z) \geq \rho(x, z)$.

27. The Rate for a Source Relative to a Fidelity Evaluation

We are now in a position to define a rate of generating information for a continuous source. We are given $P(x)$ for the source and an evaluation v determined by a distance function $\rho(x, y)$ which will be assumed continuous in both x and y. With a particular system $P(x, y)$ the quality is measured by

$$v = \iint \rho(x, y)\, P(x, y)\, dx\, dy$$

Furthermore the rate of flow of binary digits corresponding to $P(x, y)$ is

$$R = \iint P(x, y) \log \frac{P(x, y)}{P(x)P(y)}\, dx\, dy.$$

We define the rate R_1 of generating information for a given quality v_1 of reproduction to be the minimum of R when we keep v fixed at v_1 and vary $P_x(y)$. That is:

$$R_1 = \underset{P_x(y)}{\text{Min}} \iint P(x, y) \log \frac{P(x, y)}{P(x)P(y)}\, dx\, dy$$

subject to the constraint:

$$v_1 = \iint P(x, y)\rho(x, y)\, dx\, dy.$$

This means that we consider, in effect, all the communication systems that might be used and that transmit with the required fidelity. The rate of transmission in bits per second is calculated for each one and we choose that having the least rate. This latter rate is the rate we assign the source for the fidelity in question.

The justification of this definition lies in the following result:

Theorem 21: If a source has a rate R_1 for a valuation v_1 it is possible to encode the output of the source and transmit it over a channel of capacity C with fidelity as near v_1 as desired provided $R_1 \leq C$. This is not possible if $R_1 > C$.

The last statement in the theorem follows immediately from the definition of R_1 and previous results. If it were not true we could transmit more than C bits per second over a channel of capacity C. The first part of the theorem is proved by a method analogous to that used for Theorem 11. We may, in the first place, divide the (x, y) space into a large number of small cells and represent the situation as a discrete case. This will not change the evaluation function by more than an arbitrarily small amount (when the cells are very small) because of the continuity assumed for $\rho(x, y)$. Suppose that $P_1(x, y)$ is the particular system which minimizes the rate and gives R_1. We choose from the high probability y's a set at random containing

$$2^{(R_1 + \epsilon)T}$$

members where $\epsilon \to 0$ as $T \to \infty$. With large T each chosen point will be connected by a high probability line (as in Fig. 10) to a set of x's. A calculation similar to that used in proving Theorem 11 shows that with large T almost all x's are covered by the fans from the chosen y points for almost all choices of the y's. The communication system to be used operates as follows: The selected points are assigned binary numbers. When a message x is originated it will (with probability approaching 1 as $T \to \infty$) lie within one at least of the fans. The corresponding binary number is transmitted (or one of them chosen arbitrarily if there are several) over the channel by suitable coding means to give a small probability of error. Since $R_1 \leq C$ this is possible. At the receiving point the corresponding y is reconstructed and used as the recovered message.

The evaluation v_1' for this system can be made arbitrarily close to v_1 by taking T sufficiently large. This is due to the fact that for each long sample of message $x(t)$ and recovered message $y(t)$ the evaluation approaches v_1 (with probability 1).

It is interesting to note that, in this system, the noise in the recovered message is actually produced by a kind of general quantizing at the transmitter and is not produced by the noise in the channel. It is more or less analogous to the quantizing noise in P.C.M.

28. The Calculation of Rates

The definition of the rate is similar in many respects to the definition of channel capacity. In the former

$$R = \underset{P_x(y)}{\text{Max}} \iint P(x, y) \log \frac{P(x, y)}{P(x)P(y)}\, dx\, dy$$

with $P(x)$ and $v_1 = \iint P(x, y)\rho(x, y)\, dx\, dy$ fixed. In the latter

$$C = \underset{P(x)}{\text{Min}} \iint P(x, y) \log \frac{P(x, y)}{P(x)P(y)}\, dx\, dy$$

with $P_x(y)$ fixed and possibly one or more other constraints (e.g., an average power limitation) of the form $K = \iint P(x, y)\, \lambda(x, y)\, dx\, dy$.

A partial solution of the general maximizing problem for determining the rate of a source can be given. Using Lagrange's method we consider

$$\iint \left[P(x, y) \log \frac{P(x, y)}{P(x)P(y)} + \mu\, P(x, y)\rho(x, y) + \nu(x)P(x, y) \right] dx\, dy$$

The variational equation (when we take the first variation on $P(x, y)$) leads to

$$P_y(x) = B(x)\, e^{-\lambda\rho(x,y)}$$

where λ is determined to give the required fidelity and $B(x)$ is chosen to satisfy

$$\int B(x) e^{-\lambda\rho(x,y)}\, dx = 1$$

This shows that, with best encoding, the conditional probability of a certain cause for various received y, $P_y(x)$ will decline exponentially with the distance function $\rho(x, y)$ between the x and y is question.

In the special case where the distance function $\rho(x, y)$ depends only on the (vector) difference between x and y,

$$\rho(x, y) = \rho(x - y)$$

we have

$$\int B(x) e^{-\lambda\rho(x-y)}\, dx = 1.$$

Hence $B(x)$ is constant, say α, and

$$P_y(x) = \alpha e^{-\lambda\rho(x-y)}$$

Unfortunately these formal solutions are difficult to evaluate in particular cases and seem to be of little value. In fact, the actual calculation of rates has been carried out in only a few very simple cases.

If the distance function $\rho(x, y)$ is the mean square discrepancy between x and y and the message ensemble is white noise, the rate can be determined. In that case we have

$$R = \text{Min}\, [H(x) - H_y(x)] = H(x) - \text{Max}\, H_y(x)$$

with $N = \overline{(x - y)^2}$. But the Max $H_y(x)$ occurs when $y - x$ is a white noise, and is equal to $W_1 \log 2\pi e\, N$ where W_1 is the bandwidth of the message ensemble. Therefore

$$R = W_1 \log 2\pi e Q - W_1 \log 2\pi e N$$
$$= W_1 \log \frac{Q}{N}$$

where Q is the average message power. This proves the following:

Theorem 22: The rate for a white noise source of power Q and band W_1 relative to an R.M.S. measure of fidelity is

$$R = W_1 \log \frac{Q}{N}$$

where N is the allowed mean square error between original and recovered messages.

More generally with any message source we can obtain inequalities bounding the rate relative to a mean square error criterion.

Theorem 23: The rate for any source of band W_1 is bounded by

$$W_1 \log \frac{Q_1}{N} \leq R \leq W_1 \log \frac{Q}{N}$$

where Q is the average power of the source, Q_1 its entropy power and N the allowed mean square error.

The lower bound follows from the fact that the max $H_y(x)$ for a given $\overline{(x - y)^2} = N$ occurs in the white noise case. The upper bound results if we place the points (used in the proof of Theorem 21) not in the best way but at random in a sphere of radius $\sqrt{Q - N}$.

Acknowledgments

The writer is indebted to his colleagues at the Laboratories, particularly to Dr. H. W. Bode, Dr. J. R. Pierce, Dr. B. McMillan, and Dr. B. M. Oliver for many helpful suggestions and criticisms during the course of this work. Credit should also be given to Professor N. Wiener, whose elegant solution of the problems of filtering and prediction of stationary ensembles has considerably influenced the writer's thinking in this field.

APPENDIX 5

Let S_1 be any measurable subset of the g ensemble, and S_2 the subset of the f ensemble which gives S_1 under the operation T. Then

$$S_1 = TS_2.$$

Let H^λ be the operator which shifts all functions in a set by the time λ. Then

$$H^\lambda S_1 = H^\lambda TS_2 = TH^\lambda S_2$$

since T is invariant and therefore commutes with H^λ. Hence if $m[S]$ is the probability measure of the set S

$$m[H^\lambda S_1] = m[TH^\lambda S_2] = m[H^\lambda S_2]$$

$$= m[S_2] = m[S_1]$$

where the second equality is by definition of measure in the g space the third since the f ensemble is stationary, and the last by definition of g measure again.

To prove that the ergodic property is preserved under invariant operations, let S_1 be a subset of the g ensemble which is invariant under H^λ, and let S_2 be the set of all functions f which transform into S_1. Then

$$H^\lambda S_1 = H^\lambda TS_2 = TH^\lambda S_2 = S_1$$

so that $H^\lambda S_1$ is included in S_1 for all λ. Now, since

$$m[H^\lambda S_2] = m[S_1]$$

this implies

$$H^\lambda S_2 = S_2$$

for all λ with $m[S_2] \neq 0, 1$. This contradiction shows that S_1 does not exist.

APPENDIX 6

The upper bound, $\bar{N}_3 \leq N_1 + N_2$, is due to the fact that the maximum possible entropy for a power $N_1 + N_2$ occurs when we have a white noise of this power. In this case the entropy power is $N_1 + N_2$.

To obtain the lower bound, suppose we have two distributions in n dimensions $p(x_i)$ and $q(x_i)$ with entropy powers $\bar{N}_1$ and $\bar{N}_2$. What form should p and q have to minimize the entropy power $\bar{N}_3$ of their convolution $r(x_i)$:

$$r(x_i) = \int p(y_i)q(x_i - y_i)\,dy_i.$$

The entropy H_3 of r is given by

$$H_3 = -\int r(x_i)\log r(x_i)\,dx_i.$$

We wish to minimize this subject to the constraints

$$H_1 = -\int p(x_i)\log p(x_i)\,dx_i$$

$$H_2 = -\int q(x_i)\log q(x_i)\,dx_i.$$

We consider then

$$U = -\int [r(x)\log r(x) + \lambda p(x)\log p(x) + \mu q(x)\log q(x)]\,dx$$

$$\delta U = -\int [[1 + \log r(x)]\delta r(x) + \lambda[1 + \log p(x)]\delta p(x)$$
$$+ \mu[1 + \log q(x)\delta q(x)]]\,dx.$$

If $p(x)$ is varied at a particular argument $x_i = s_i$, the variation in $r(x)$ is

$$\delta r(x) = q(x_i - s_i)$$

and

$$\delta U = -\int q(x_i - s_i)\log r(x_i)\,dx_i - \lambda\log p(s_i) = 0$$

and similarly when q is varied. Hence the conditions for a minimum are

$$\int q(x_i - s_i)\log r(x_i) = -\lambda\log p(s_i)$$

$$\int p(x_i - s_i)\log r(x_i) = -\mu\log q(s_i).$$

If we multiply the first by $p(s_i)$ and the second by $q(s_i)$ and integrate with respect to s we obtain

$$H_3 = -\lambda H_1$$

$$H_3 = -\mu H_2$$

or solving for λ and μ and replacing in the equations

$$H_1\int q(x_i - s_i)\log r(x_i)\,dx_i = -H_3\log p(s_i)$$

$$H_2\int p(x_i - s_i)\log r(x_i)\,dx_i = -H_3\log p(s_i).$$

Now suppose $p(x_i)$ and $q(x_i)$ are normal

$$p(x_i) = \frac{|A_{ij}|^{n/2}}{(2\pi)^{n/2}}\exp -\tfrac{1}{2}\Sigma A_{ij}x_ix_j$$

$$q(x_i) = \frac{|B_{ij}|^{n/2}}{(2\pi)^{n/2}}\exp -\tfrac{1}{2}\Sigma B_{ij}x_ix_j.$$

Then $r(x_i)$ will also be normal with quadratic form C_{ij}. If the inverses of these forms are a_{ij}, b_{ij}, c_{ij} then

$$c_{ij} = a_{ij} + b_{ij}.$$

We wish to show that these functions satisfy the minimizing conditions if and only if $a_{ij} = Kb_{ij}$ and thus give the minimum H_3 under the constraints. First we have

$$\log r(x_i) = \frac{n}{2}\log\frac{1}{2\pi}|C_{ij}| - \tfrac{1}{2}\Sigma C_{ij}x_ix_j$$

$$\int q(x_i - s_i)\log r(x_i) = \frac{n}{2}\log\frac{1}{2\pi}|C_{ij}| - \tfrac{1}{2}\Sigma C_{ij}s_is_j - \tfrac{1}{2}\Sigma C_{ij}b_{ij}.$$

This should equal

$$\frac{H_3}{H_1}\left[\frac{n}{2}\log\frac{1}{2\pi}|A_{ij}| - \tfrac{1}{2}\Sigma A_{ij}s_is_j\right]$$

which requires $A_{ij} = \dfrac{H_1}{H_3}C_{ij}$.

In this case $A_{ij} = \dfrac{H_1}{H_2}B_{ij}$ and both equations reduce to identities.

APPENDIX 7

The following will indicate a more general and more rigorous approach to the central definitions of communication theory. Consider a probability measure space whose elements are ordered pairs (x, y). The variables x, y are to be identified as the possible transmitted and received signals of some long duration T. Let us call the set of all points whose x belongs to a subset S_1 of x points the strip over S_1, and similarly the set whose y belongs to S_2 the strip over S_2. We divide x and y into a collection of non-overlapping measurable subsets X_i and Y_i approximate to the rate of transmission R by

$$R_1 = \frac{1}{T}\sum_i P(X_i, Y_i)\log\frac{P(X_i, Y_i)}{P(X_i)P(Y_i)}$$

where

$P(X_i)$ is the probability measure of the strip over X_i
$P(Y_i)$ is the probability measure of the strip over Y_i
$P(X_i, Y_i)$ is the probability measure of the intersection of the strips.

A further subdivision can never decrease R_1. For let X_1 be divided into $X_1 = X_1' + X_1''$ and let

$$P(Y_1) = a \qquad P(X_1) = b + c$$
$$P(X_1') = b \qquad P(X_1', Y_1) = d$$
$$P(X_1'') = c \qquad P(X_1'', Y_1) = e$$
$$P(X_1, Y_1) = d + e$$

Then in the sum we have replaced (for the X_1, Y_1 intersection)

$$(d + e)\log\frac{d + e}{a(b + c)} \quad \text{by} \quad d\log\frac{d}{ab} + e\log\frac{e}{ac}.$$

It is easily shown that with the limitation we have on b, c, d, e,

$$\left[\frac{d+e}{b+c}\right]^{d+e} \leq \frac{d^d e^e}{b^d c^e}$$

and consequently the sum is increased. Thus the various possible subdivisions form a directed set, with R monotonic increasing with refinement of the subdivision. We may define R unambiguously as the least upper bound for the R_1 and write it

$$R = \frac{1}{T} \iint P(x, y) \log \frac{P(x, y)}{P(x)P(y)} \, dx \, dy.$$

This integral, understood in the above sense, includes both the continuous and discrete cases and of course many others which cannot be represented in either form. It is trivial in this formulation that if x and u are in one-to-one correspondence, the rate from u to y is equal to that from x to y. If v is any function of y (not necessarily with an inverse) then the rate from x to y is greater than or equal to that from x to v since, in the calculation of the approximations, the subdivisions of y are essentially a finer subdivision of those for v. More generally if y and v are related not functionally but statistically, i.e., we have a probability measure space (y, v), then $R(x, v) \leq R(x, y)$. This means that any operation applied to the received signal, even though it involves statistical elements, does not increase R.

Another notion which should be defined precisely in an abstract formulation of the theory is that of "dimension rate," that is the average number of dimensions required per second to specify a member of an ensemble. In the band limited case 2W numbers per second are sufficient. A general definition can be framed as follows. Let $f_\alpha(t)$ be an ensemble of functions and let $\rho_T[f_\alpha(t), f_\beta(t)]$ be a metric measuring the "distance" from f_α to f_β over the time T (for example the R.M.S. discrepancy over this interval.) Let $N(\epsilon, \delta, T)$ be the least number of elements f which can be chosen such that all elements of the ensemble apart from a set of measure δ are within the distance ϵ of at least one of those chosen. Thus we are covering the space to within ϵ apart from a set of small measure δ. We define the dimension rate λ for the ensemble by the triple limit

$$\lambda = \lim_{\delta \to 0} \lim_{\epsilon \to 0} \lim_{T \to \infty} \frac{\log N(\epsilon, \delta, T)}{T \log \epsilon}.$$

This is a generalization of the measure type definitions of dimension in topology, and agrees with the intuitive dimension rate for simple ensembles where the desired result is obvious.

Communication in the Presence of Noise*

CLAUDE E. SHANNON†, MEMBER, IRE

Summary—A method is developed for representing any communication system geometrically. Messages and the corresponding signals are points in two "function spaces," and the modulation process is a mapping of one space into the other. Using this representation, a number of results in communication theory are deduced concerning expansion and compression of bandwidth and the threshold effect. Formulas are found for the maximum rate of transmission of binary digits over a system when the signal is perturbed by various types of noise. Some of the properties of "ideal" systems which transmit at this maximum rate are discussed. The equivalent number of binary digits per second for certain information sources is calculated.

* Decimal classification: 621.38. Original manuscript received by the Institute, July 23, 1940. Presented, 1948 IRE National Convention, New York, N. Y., March 24, 1948; and IRE New York Section, New York, N. Y., November 12, 1947.

† Bell Telephone Laboratories, Murray Hill, N. J.

I. Introduction

A GENERAL COMMUNICATIONS system is shown schematically in Fig. 1. It consists essentially of five elements.

1. *An information source.* The source selects one message from a set of possible messages to be transmitted to the receiving terminal. The message may be of various types; for example, a sequence of letters or numbers, as in telegraphy or teletype, or a continuous function of time $f(t)$, as in radio or telephony.

2. *The transmitter.* This operates on the message in some way and produces a signal suitable for transmission to the receiving point over the channel. In teleph-

Reprinted from *Proc. IRE*, vol. 37, pp. 10–21, Jan. 1949.

ony, this operation consists of merely changing sound pressure into a proportional electrical current. In teleg-

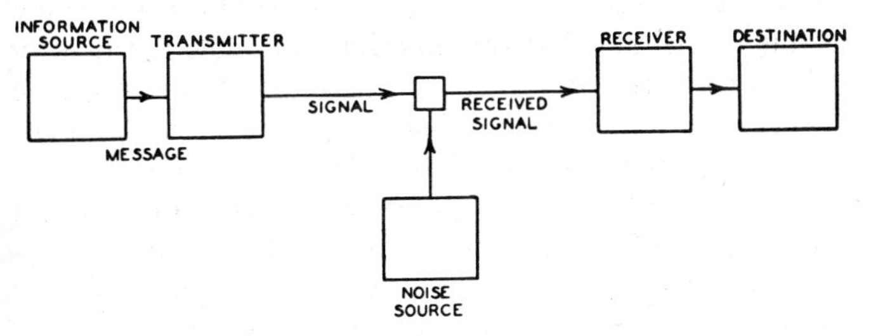

Fig. 1—General communications system.

raphy, we have an encoding operation which produces a sequence of dots, dashes, and spaces corresponding to the letters of the message. To take a more complex example, in the case of multiplex PCM telephony the different speech functions must be sampled, compressed, quantized and encoded, and finally interleaved properly to construct the signal.

3. *The channel.* This is merely the medium used to transmit the signal from the transmitting to the receiving point. It may be a pair of wires, a coaxial cable, a band of radio frequencies, etc. During transmission, or at the receiving terminal, the signal may be perturbed by noise or distortion. Noise and distortion may be differentiated on the basis that distortion is a fixed operation applied to the signal, while noise involves statistical and unpredictable perturbations. Distortion can, in principle, be corrected by applying the inverse operation, while a perturbation due to noise cannot always be removed, since the signal does not always undergo the same change during transmission.

4. *The receiver.* This operates on the received signal and attempts to reproduce, from it, the original message. Ordinarily it will perform approximately the mathematical inverse of the operations of the transmitter, although they may differ somewhat with best design in order to combat noise.

5. *The destination.* This is the person or thing for whom the message is intended.

Following Nyquist[1] and Hartley,[2] it is convenient to use a logarithmic measure of information. If a device has n possible positions it can, by definition, store $\log_b n$ units of information. The choice of the base b amounts to a choice of unit, since $\log_b n = \log_b c \log_c n$. We will use the base 2 and call the resulting units binary digits or bits. A group of m relays or flip-flop circuits has 2^m possible sets of positions, and can therefore store $\log_2 2^m = m$ bits.

If it is possible to distinguish reliably M different signal functions of duration T on a channel, we can say that the channel can transmit $\log_2 M$ bits in time T. The *rate* of transmission is then $\log_2 M/T$. More precisely, the *channel capacity* may be defined as

$$C = \lim_{T\to\infty} \frac{\log_2 M}{T}. \tag{1}$$

A precise meaning will be given later to the requirement of reliable resolution of the M signals.

II. The Sampling Theorem

Let us suppose that the channel has a certain bandwidth W in cps starting at zero frequency, and that we are allowed to use this channel for a certain period of time T. Without any further restrictions this would mean that we can use as signal functions any functions of time whose spectra lie entirely within the band W, and whose time functions lie within the interval T. Although it is not possible to fulfill both of these conditions exactly, it is possible to keep the spectrum within the band W, and to have the time function very small outside the interval T. Can we describe in a more useful way the functions which satisfy these conditions? One answer is the following:

Theorem 1: *If a function $f(t)$ contains no frequencies higher than W cps, it is completely determined by giving its ordinates at a series of points spaced $1/2W$ seconds apart.*

This is a fact which is common knowledge in the communication art. The intuitive justification is that, if $f(t)$ contains no frequencies higher than W, it cannot change to a substantially new value in a time less than one-half cycle of the highest frequency, that is, $1/2W$. A mathematical proof showing that this is not only approximately, but exactly, true can be given as follows. Let $F(\omega)$ be the spectrum of $f(t)$. Then

$$f(t) = \frac{1}{2\pi}\int_{-\infty}^{\infty} F(\omega)e^{i\omega t}d\omega \tag{2}$$

$$= \frac{1}{2\pi}\int_{-2\pi W}^{+2\pi W} F(\omega)e^{i\omega t}d\omega, \tag{3}$$

since $F(\omega)$ is assumed zero outside the band W. If we let

$$t = \frac{n}{2W} \tag{4}$$

where n is any positive or negative integer, we obtain

$$f\left(\frac{n}{2W}\right) = \frac{1}{2\pi}\int_{-2\pi W}^{+2\pi W} F(\omega)e^{i\omega \frac{n}{2W}}\, d\omega. \tag{5}$$

On the left are the values of $f(t)$ at the sampling points. The integral on the right will be recognized as essentially the nth coefficient in a Fourier-series expansion of the function $F(\omega)$, taking the interval $-W$ to $+W$ as a fundamental period. This means that the values of the samples $f(n/2W)$ determine the Fourier coefficients in the series expansion of $F(\omega)$. Thus they determine $F(\omega)$, since $F(\omega)$ is zero for frequencies greater than W, and for

[1] H. Nyquist, "Certain factors affecting telegraph speed," *Bell Syst. Tech. Jour.*, vol. 3, p. 324; April, 1924.
[2] R. V. L. Hartley, "The transmission of information," *Bell Sys. Tech. Jour.*, vol. 3, p. 535–564; July, 1928.

lower frequencies $F(\omega)$ is determined if its Fourier coefficients are determined. But $F(\omega)$ determines the original function $f(t)$ completely, since a function is determined if its spectrum is known. Therefore the original samples determine the function $f(t)$ completely. There is one and only one function whose spectrum is limited to a band W, and which passes through given values at sampling points separated $1/2W$ seconds apart. The function can be simply reconstructed from the samples by using a pulse of the type

$$\frac{\sin 2\pi Wt}{2\pi Wt}. \tag{6}$$

This function is unity at $t=0$ and zero at $t=n/2W$, i.e., at all other sample points. Furthermore, its spectrum is constant in the band W and zero outside. At each sample point a pulse of this type is placed whose amplitude is adjusted to equal that of the sample. The sum of these pulses is the required function, since it satisfies the conditions on the spectrum and passes through the sampled values.

Mathematically, this process can be described as follows. Let x_n be the nth sample. Then the function $f(t)$ is represented by

$$f(t) = \sum_{n=-\infty}^{\infty} x_n \frac{\sin \pi(2Wt - n)}{\pi(2Wt - n)}. \tag{7}$$

A similar result is true if the band W does not start at zero frequency but at some higher value, and can be proved by a linear translation (corresponding physically to single-sideband modulation) of the zero-frequency case. In this case the elementary pulse is obtained from $\sin x/x$ by single-side-band modulation.

If the function is limited to the time interval T and the samples are spaced $1/2W$ seconds apart, there will be a total of $2TW$ samples in the interval. All samples outside will be substantially zero. To be more precise, we can define a function to be limited to the time interval T if, and only if, all the samples outside this interval are exactly zero. Then we can say that any function limited to the bandwidth W and the time interval T can be specified by giving $2TW$ numbers.

Theorem 1 has been given previously in other forms by mathematicians[3] but in spite of its evident importance seems not to have appeared explicitly in the literature of communication theory. Nyquist,[4,5] however, and more recently Gabor,[6] have pointed out that approximately $2TW$ numbers are sufficient, basing their arguments on a Fourier series expansion of the function over the time interval T. This gives TW sine and $(TW+1)$ cosine terms up to frequency W. The slight discrepancy is due to the fact that the functions obtained in this way will not be strictly limited to the band W but, because of the sudden starting and stopping of the sine and cosine components, contain some frequency content outside the band. Nyquist pointed out the fundamental importance of the time interval $1/2W$ seconds in connection with telegraphy, and we will call this the Nyquist interval corresponding to the band W.

The $2TW$ numbers used to specify the function need not be the equally spaced samples used above. For example, the samples can be unevenly spaced, although, if there is considerable bunching, the samples must be known very accurately to give a good reconstruction of the function. The reconstruction process is also more involved with unequal spacing. One can further show that the value of the function and its derivative at every other sample point are sufficient. The value and first and second derivatives at every third sample point give a still different set of parameters which uniquely determine the function. Generally speaking, any set of $2TW$ independent numbers associated with the function can be used to describe it.

III. Geometrical Representation of the Signals

A set of three numbers x_1, x_2, x_3, regardless of their source, can always be thought of as co-ordinates of a point in three-dimensional space. Similarly, the $2TW$ evenly spaced samples of a signal can be thought of as co-ordinates of a point in a space of $2TW$ dimensions. Each particular selection of these numbers corresponds to a particular point in this space. Thus there is exactly one point corresponding to each signal in the band W and with duration T.

The number of dimensions $2TW$ will be, in general, very high. A 5-Mc television signal lasting for an hour would be represented by a point in a space with $2\times 5 \times 10^6 \times 60^2 = 3.6\times 10^{10}$ dimensions. Needless to say, such a space cannot be visualized. It is possible, however, to study analytically the properties of n-dimensional space. To a considerable extent, these properties are a simple generalization of the properties of two- and three-dimensional space, and can often be arrived at by inductive reasoning from these cases. The advantage of this geometrical representation of the signals is that we can use the vocabulary and the results of geometry in the communication problem. Essentially, we have replaced a complex entity (say, a television signal) in a simple environment (the signal requires only a plane for its representation as $f(t)$) by a simple entity (a point) in a complex environment ($2TW$ dimensional space).

If we imagine the $2TW$ co-ordinate axes to be at right angles to each other, then distances in the space have a simple interpretation. The distance from the origin to a

[3] J. M. Whittaker, "Interpolatory Function Theory," Cambridge Tracts in Mathematics and Mathematical Physics, No. 33, Cambridge University Press, Chapt. IV; 1935.

[4] H. Nyquist, "Certain topics in telegraph transmission theory," *A.I.E.E. Transactions*, p. 617; April, 1928.

[5] W. R. Bennett, "Time division multiplex systems," *Bell Sys. Tech. Jour.*, vol. 20, p. 199; April, 1941, where a result similar to Theorem 1 is established, but on a steady-state basis.

[6] D. Gabor, "Theory of communication," *Jour. I.E.E.* (London), vol. 93; part 3, no. 26, p. 429; 1946.

point is analogous to the two- and three-dimensional cases

$$d = \sqrt{\sum_{n=1}^{2TW} x_n^2} \tag{8}$$

where x_n is the nth sample. Now, since

$$f(t) = \sum_{n=1}^{2TW} x_n \frac{\sin \pi(2Wt - n)}{\pi(2Wt - n)}, \tag{9}$$

we have

$$\int_{-\infty}^{\infty} f(t)^2 dt = \frac{1}{2W} \sum x_n^2, \tag{10}$$

using the fact that

$$\int_{-\infty}^{\infty} \frac{\sin \pi(2Wt - m)}{\pi(2Wt - m)} \frac{\sin \pi(2Wt - n)}{\pi(Wt - n)} dt = \begin{cases} 0 & m \neq n \\ \frac{1}{2W} & m = n. \end{cases} \tag{11}$$

Hence, the square of the distance to a point is $2W$ times the energy (more precisely, the energy into a unit resistance) of the corresponding signal

$$\begin{aligned} d^2 &= 2WE \\ &= 2WTP \end{aligned} \tag{12}$$

where P is the average power over the time T. Similarly, the distance between two points is $\sqrt{2WT}$ times the rms discrepancy between the two corresponding signals.

If we consider only signals whose average power is less than P, these will correspond to points within a sphere of radius

$$r = \sqrt{2WTP}. \tag{13}$$

If noise is added to the signal in transmission, it means that the point corresponding to the signal has been moved a certain distance in the space proportional to the rms value of the noise. Thus noise produces a small region of uncertainty about each point in the space. A fixed distortion in the channel corresponds to a warping of the space, so that each point is moved, but in a definite fixed way.

In ordinary three-dimensional space it is possible to set up many different co-ordinate systems. This is also possible in the signal space of $2TW$ dimensions that we are considering. A different co-ordinate system corresponds to a different way of describing the same signal function. The various ways of specifying a function given above are special cases of this. One other way of particular importance in communication is in terms of frequency components. The function $f(t)$ can be expanded as a sum of sines and cosines of frequencies $1/T$ apart, and the coefficients used as a different set of co-ordinates. It can be shown that these co-ordinates are all perpendicular to each other and are obtained by what is essentially a rotation of the original co-ordinate system.

Passing a signal through an ideal filter corresponds to projecting the corresponding point onto a certain region in the space. In fact, in the frequency-co-ordinate system those components lying in the pass band of the filter are retained and those outside are eliminated, so that the projection is on one of the co-ordinate lines, planes, or hyperplanes. Any filter performs a linear operation on the vectors of the space, producing a new vector linearly related to the old one.

IV. Geometrical Representation of Messages

We have associated a space of $2TW$ dimensions with the set of possible signals. In a similar way one can associate a space with the set of possible messages. Suppose we are considering a speech system and that the messages consist of all possible sounds which contain no frequencies over a certain limit W_1 and last for a time T_1.

Just as for the case of the signals, these messages can be represented in a one-to-one way in a space of $2T_1W_1$ dimensions. There are several points to be noted, however. In the first place, various different points may represent the same message, insofar as the final destination is concerned. For example, in the case of speech, the ear is insensitive to a certain amount of phase distortion. Messages differing only in the phases of their components (to a limited extent) sound the same. This may have the effect of reducing the number of essential dimensions in the message space. All the points which are equivalent for the destination can be grouped together and treated as one point. It may then require fewer numbers to specify one of these "equivalence classes" than to specify an arbitrary point. For example, in Fig. 2 we have a two-dimensional space, the set of points in a square. If all points on a circle are regarded as equivalent, it reduces to a one-dimensional space—a point can now be

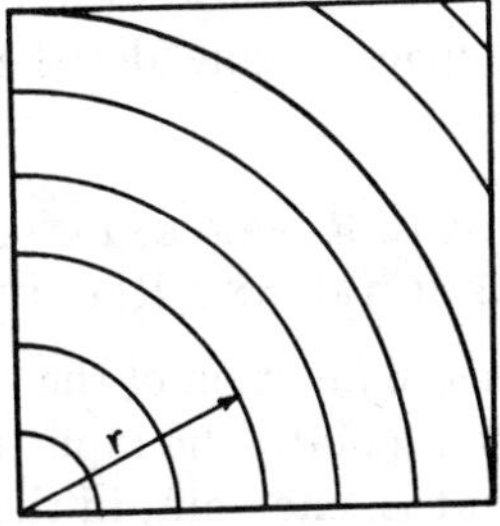

Fig. 2—Reduction of dimensionality through equivalence classes.

specified by one number, the radius of the circle. In the case of sounds, if the ear were completely insensitive to phase, then the number of dimensions would be reduced by one-half due to this cause alone. The sine and cosine components a_n and b_n for a given frequency would not need to be specified independently, but only $\sqrt{a_n^2+b_n^2}$; that is, the total amplitude for this frequency. The reduction in frequency discrimination of the ear as frequency increases indicates that a further reduction in dimensionality occurs. The vocoder makes use to a considerable extent of these equivalences among speech sounds, in the first place by eliminating, to a large degree, phase information, and in the second place by lumping groups of frequencies together, particularly at the higher frequencies.

In other types of communication there may not be any equivalence classes of this type. The final destination is sensitive to any change in the message within the full message space of $2T_1W_1$ dimensions. This appears to be the case in television transmission.

A second point to be noted is that the information source may put certain restrictions on the actual messages. The space of $2T_1W_1$ dimensions contains a point for *every* function of time $f(t)$ limited to the band W_1 and of duration T_1. The class of messages we wish to transmit may be only a small subset of these functions. For example, speech sounds must be produced by the human vocal system. If we are willing to forego the transmission of any other sounds, the effective dimensionality may be considerably decreased. A similar effect can occur through probability considerations. Certain messages may be possible, but so improbable relative to the others that we can, in a certain sense, neglect them. In a television image, for example, successive frames are likely to be very nearly identical. There is a fair probability of a particular picture element having the same light intensity in successive frames. If this is analyzed mathematically, it results in an effective reduction of dimensionality of the message space when T_1 is large.

We will not go further into these two effects at present, but let us suppose that, when they are taken into account, the resulting message space has a dimensionality D, which will, of course, be less than or equal to $2T_1W_1$. In many cases, even though the effects are present, their utilization involves too much complication in the way of equipment. The system is then designed on the basis that all functions are different and that there are no limitations on the information source. In this case, the message space is considered to have the full $2T_1W_1$ dimensions.

V. Geometrical Representation of the Transmitter and Receiver

We now consider the function of the transmitter from this geometrical standpoint. The input to the transmitter is a message; that is, one point in the message space. Its output is a signal—one point in the signal space. Whatever form of encoding or modulation is performed, the transmitter must establish some correspondence between the points in the two spaces. Every point in the message space must correspond to a point in the signal space, and no two messages can correspond to the same signal. If they did, there would be no way to determine at the receiver which of the two messages was intended. The geometrical name for such a correspondence is a mapping. The transmitter maps the message space into the signal space.

In a similar way, the receiver maps the signal space back into the message space. Here, however, it is possible to have more than one point mapped into the same point. This means that several different signals are demodulated or decoded into the same message. In AM, for example, the phase of the carrier is lost in demodulation. Different signals which differ only in the phase of the carrier are demodulated into the same message. In FM the shape of the signal wave above the limiting value of the limiter does not affect the recovered message. In PCM considerable distortion of the received pulses is possible, with no effect on the output of the receiver.

We have so far established a correspondence between a communication system and certain geometrical ideas. The correspondence is summarized in Table I.

TABLE I

Communication System	*Geometrical Entity*
The set of possible signals	A space of $2TW$ dimensions
A particular signal	A point in the space
Distortion in the channel	A warping of the space
Noise in the channel	A region of uncertainty about each point
The average power of the signal	$(2TW)^{-1}$ times the square of the distance from the origin to the point
The set of signals of power P	The set of points in a sphere of radius $\sqrt{2TW\,P}$
The set of possible messages	A space of $2T_1W_1$ dimensions
The set of actual messages distinguishable by the destination	A space of D dimensions obtained by regarding all equivalent messages as one point, and deleting messages which the source could not produce
A message	A point in this space
The transmitter	A mapping of the message space into the signal space
The receiver	A mapping of the signal space into the message space

VI. Mapping Considerations

It is possible to draw certain conclusions of a general nature regarding modulation methods from the geometrical picture alone. Mathematically, the simplest types of mappings are those in which the two spaces have the same number of dimensions. Single-sideband amplitude modulation is an example of this type and an especially simple one, since the co-ordinates in the signal space are proportional to the corresponding co-ordinates in the message space. In double-sideband transmission the signal space has twice the number of co-ordinates, but they occur in pairs with equal values. If there were only one dimension in the message space and two in the signal space, it would correspond to mapping a line onto a square so that the point x on the line is represented by (x, x) in the square. Thus no significant use is made of the extra dimensions. All the messages go into a subspace having only $2T_1W_1$ dimensions.

In frequency modulation the mapping is more involved. The signal space has a much larger dimensionality than the message space. The type of mapping can be suggested by Fig. 3, where a line is mapped into a three-dimensional space. The line starts at unit distance from

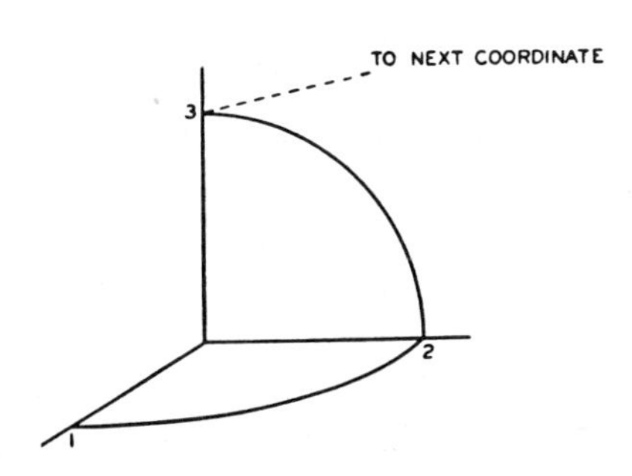

Fig. 3—Mapping similar to frequency modulation.

the origin on the first co-ordinate axis, stays at this distance from the origin on a circle to the next co-ordinate axis, and then goes to the third. It can be seen that the line is lengthened in this mapping in proportion to the total number of co-ordinates. It is not, however, nearly as long as it could be if it wound back and forth through the space, filling up the internal volume of the sphere it traverses.

This expansion of the line is related to the improved signal-to-noise ratio obtainable with increased bandwidth. Since the noise produces a small region of uncertainty about each point, the effect of this on the recov-

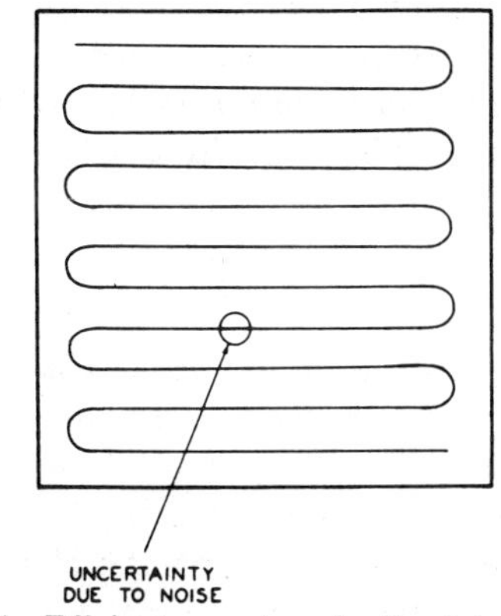

Fig. 4—Efficient mapping of a line into a square.

ered message will be less if the map is in a large scale. To obtain as large a scale as possible requires that the line wander back and forth through the higher-dimensional region as indicated in Fig. 4, where we have mapped a line into a square. It will be noticed that when this is done the effect of noise is small relative to the length of the line, provided the noise is less than a certain critical value. At this value it becomes uncertain at the receiver as to which portion of the line contains the message. This holds generally, and it shows that *any system which attempts to use the capacities of a wider band to the full extent possible will suffer from a threshold effect when there is noise.* If the noise is small, very little distortion will occur, but at some critical noise amplitude the message will become very badly distorted. This effect is well known in PCM.

Suppose, on the other hand, we wish to reduce dimensionality, i.e., to compress bandwidth or time or both. That is, we wish to send messages of band W_1 and duration T_1 over a channel with $TW < T_1W_1$. It has already been indicated that the effective dimensionality D of the message space may be less than $2T_1W_1$ due to the properties of the source and of the destination. Hence we certainly need no more than D dimension in the signal space for a good mapping. To make this saving it is necessary, of course, to isolate the effective co-ordinates in the message space, and to send these only. The reduced bandwidth transmission of speech by the vocoder is a case of this kind.

The question arises, however, as to whether further reduction is possible. In our geometrical analogy, is it possible to map a space of high dimensionality onto one of lower dimensionality? The answer is that it is pos i-ble, with certain reservations. For example, the points of a square can be described by their two co-ordinates which could be written in decimal notation

$$x = .a_1a_2a_3 \cdots$$
$$y = .b_1b_2b_3 \cdots . \tag{14}$$

From these two numbers we can construct one number by taking digits alternately from x and y:

$$z = .a_1b_1a_2b_2a_3b_3 \cdots . \tag{15}$$

A knowledge of x and y determines z, and z determines both x and y. Thus there is a one-to-one correspondence between the points of a square and the points of a line.

This type of mapping, due to the mathematician Cantor, can easily be extended as far as we wish in the direction of reducing dimensionality. A space of n dimensions can be mapped in a one-to-one way into a space of one dimension. Physically, this means that the frequency-time product can be reduced as far as we wish when there is no noise, with exact recovery of the original messages.

In a less exact sense, a mapping of the type shown in Fig. 4 maps a square into a line, provided we are not too particular about recovering exactly the starting point, but are satisfied with a near-by one. The sensitivity we noticed before when increasing dimensionality now takes a different form. In such a mapping, to reduce TW, there will be a certain threshold effect when we perturb the message. As we change the message a small amount, the corresponding signal will change a small amount, until some critical value is reached. At this point the signal will undergo a considerable change. In topology it is shown[7] that it is not possible to map a region of higher dimension into a region of lower dimension *continuously.* It is the necessary discontinuity which produces the threshold effects we have been describing for communication systems.

[7] W. Hurewitz and H. Wallman, "Dimension Theory," Princeton University Press, Princeton, N. J.; 1941.

This discussion is relevant to the well-known "Hartley Law," which states that "... an upper limit to the amount of information which may be transmitted is set by the sum for the various available lines of the product of the line-frequency range of each by the time during which it is available for use."[2] There is a sense in which this statement is true, and another sense in which it is false. It is not possible to map the message space into the signal space in a one-to-one, continuous manner (this is known mathematically as a *topological* mapping) unless the two spaces have the same dimensionality; i.e., unless $D=2TW$. Hence, if we limit the transmitter and receiver to continuous one-to-one operations, there is a lower bound to the product TW in the channel. This lower bound is determined, not by the product W_1T_1 of message bandwidth and time, but by the number of *essential* dimension D, as indicated in Section IV. There is, however, no good reason for limiting the transmitter and receiver to topological mappings. In fact, PCM and similar modulation systems are highly discontinuous and come very close to the type of mapping given by (14) and (15). It is desirable, then, to find limits for what can be done with no restrictions on the type of transmitter and receiver operations. These limits, which will be derived in the following sections, depend on the amount and nature of the noise in the channel, and on the transmitter power, as well as on the bandwidth-time product.

It is evident that any system, either to compress TW, or to expand it and make full use of the additional volume, must be highly nonlinear in character and fairly complex because of the peculiar nature of the mappings involved.

VII. The Capacity of a Channel in the Presence of White Thermal Noise

It is not difficult to set up certain quantitative relations that must hold when we change the product TW. Let us assume, for the present, that the noise in the system is a white thermal-noise band limited to the band W, and that it is added to the transmitted signal to produce the received signal. A white thermal noise has the property that each sample is perturbed independently of all the others, and the distribution of each amplitude is Gaussian with standard deviation $\sigma=\sqrt{N}$ where N is the average noise power. How many different signals can be distinguished at the receiving point in spite of the perturbations due to noise? A crude estimate can be obtained as follows. If the signal has a power P, then the perturbed signal will have a power $P+N$. The number of amplitudes that can be reasonably well distinguished is

$$K\sqrt{\frac{P+N}{N}} \tag{16}$$

where K is a small constant in the neighborhood of unity depending on how the phrase "reasonably well" is interpreted. If we require very good separation, K will be small, while toleration of occasional errors allows K to be larger. Since in time T there are $2TW$ independent amplitudes, the total number of reasonably distinct signals is

$$M=\left[K\sqrt{\frac{P+N}{N}}\right]^{2TW}. \tag{17}$$

The number of bits that can be sent in this time is $\log_2 M$, and the rate of transmission is

$$\frac{\log_2 M}{T}=W\log_2 K^2\frac{P+N}{N}\ \text{(bits per second)}. \tag{18}$$

The difficulty with this argument, apart from its general approximate character, lies in the tacit assumption that for two signals to be distinguishable they must differ at some sampling point by more than the expected noise. The argument presupposes that PCM, or something very similar to PCM, is the best method of encoding binary digits into signals. Actually, two signals can be reliably distinguished if they differ by only a small amount, provided this difference is sustained over a long period of time. Each sample of the received signal then gives a small amount of statistical information concerning the transmitted signal; in combination, these statistical indications result in near certainty. This possibility allows an improvement of about 8 db in power over (18) with a reasonable definition of reliable resolution of signals, as will appear later. We will now make use of the geometrical representation to determine the exact capacity of a noisy channel.

Theorem 2: *Let P be the average transmitter power, and suppose the noise is white thermal noise of power N in the band W. By sufficiently complicated encoding systems it is possible to transmit binary digits at a rate*

$$C=W\log_2\frac{P+N}{N} \tag{19}$$

with as small a frequency of errors as desired. It is not possible by any encoding method to send at a higher rate and have an arbitrarily low frequency of errors.

This shows that the rate $W\log(P+N)/N$ measures in a sharply defined way the capacity of the channel for transmitting information. It is a rather surprising result, since one would expect that reducing the frequency of errors would require reducing the rate of transmission, and that the rate must approach zero as the error frequency does. Actually, we can send at the rate C but reduce errors by using more involved encoding and longer delays at the transmitter and receiver. The transmitter will take long sequences of binary digits and represent this entire sequence by a particular signal function of long duration. The delay is required because the transmitter must wait for the full sequence before the signal is determined. Similarly, the receiver must wait for the full signal function before decoding into binary digits.

We now prove Theorem 2. In the geometrical representation each signal point is surrounded by a small region of uncertainty due to noise. With white thermal noise, the perturbations of the different samples (or coordinates) are all Gaussian and independent. Thus the probability of a perturbation having co-ordinates $x_1, x_2, \cdots, x_n$ (these are the differences between the original and received signal co-ordinates) is the product of the individual probabilities for the different co-ordinates:

$$\prod_{n=1}^{2TW} \frac{1}{\sqrt{2\pi 2TWN}} \exp - \frac{x_n^2}{2TWN}$$
$$= \frac{1}{(2\pi 2TWN)^{TW}} \exp \frac{-1}{2TW} \sum_1^{2TW} x_n^2.$$

Since this depends only on

$$\sum_1^{2TW} x_n^2,$$

the probability of a given perturbation depends only on the *distance* from the original signal and not on the direction. In other words, the region of uncertainty is spherical in nature. Although the limits of this region are not sharply defined for a small number of dimensions ($2TW$), the limits become more and more definite as the dimensionality increases. This is because the square of the distance a signal is perturbed is equal to $2TW$ times the average noise power during the time T. As T increases, this average noise power must approach N. Thus, for large T, the perturbation will almost certainly be to some point near the surface of a sphere of radius $\sqrt{2TWN}$ centered at the original signal point. More precisely, by taking T sufficiently large we can insure (with probability as near to 1 as we wish) that the perturbation will lie within a sphere of radius $\sqrt{2TW(N+\epsilon)}$ where ϵ is arbitrarily small. The noise regions can therefore be thought of roughly as sharply defined billiard balls, when $2TW$ is very large. The received signals have an average power $P+N$, and in the same sense must almost all lie on the surface of a sphere of radius $\sqrt{2TW(P+N)}$. How many different transmitted signals can be found which will be distinguishable? Certainly not more than the volume of the sphere of radius $\sqrt{2TW(P+N)}$ divided by the volume of a sphere of radius $\sqrt{2TWN}$, since overlap of the noise spheres results in confusion as to the message at the receiving point. The volume of an n-dimensional sphere[8] of radius r is

$$V = \frac{\pi^{n/2}}{\Gamma\left(\frac{n}{2}+1\right)} r^n. \tag{20}$$

Hence, an upper limit for the number M of distinguishable signals is

$$M \leq \left(\sqrt{\frac{P+N}{N}}\right)^{2TW}. \tag{21}$$

Consequently, the channel capacity is bounded by:

$$C = \frac{\log_2 M}{T} \leq W \log_2 \frac{P+N}{N}. \tag{22}$$

[8] D. M. Y. Sommerville, "An Introduction to the Geometry of N Dimensions," E. P. Dutton, Inc., New York, N. Y., 1929; p. 135.

This proves the last statement in the theorem.

To prove the first part of the theorem, we must show that there exists a system of encoding which transmits $W \log_2 (P+N)/N$ binary digits per second with a frequency of errors less than ϵ when ϵ is arbitrarily small. The system to be considered operates as follows. A long sequence of, say, m binary digits is taken in at the transmitter. There are 2^m such sequences, and each corresponds to a particular signal function of duration T. Thus there are $M=2^m$ different signal functions. When the sequence of m is completed, the transmitter starts sending the corresponding signal. At the receiver a perturbed signal is received. The receiver compares this signal with each of the M possible transmitted signals and selects the one which is nearest the perturbed signal (in the sense of rms error) as the one actually sent. The receiver then constructs, as its output, the corresponding sequence of binary digits. There will be, therefore, an over-all delay of $2T$ seconds.

To insure a frequency of errors less than ϵ, the M signal functions must be reasonably well separated from each other. In fact, we must choose them in such a way that, when a perturbed signal is received, the nearest signal point (in the geometrical representation) is, with probability greater than $1-\epsilon$, the actual original signal.

It turns out, rather surprisingly, that it is possible to choose our M signal functions at random from the points inside the sphere of radius $\sqrt{2TWP}$, and achieve the most that is possible. Physically, this corresponds very nearly to using M different samples of band-limited white noise with power P as signal functions.

A particular selection of M points in the sphere corresponds to a particular encoding system. The general scheme of the proof is to consider all such selections, and to show that the frequency of errors averaged over all the particular selections is less than ϵ. This will show that there are particular selections in the set with frequency of errors less than ϵ. Of course, there will be other particular selections with a high frequency of errors.

The geometry is shown in Fig. 5. This is a plane cross section through the high-dimensional sphere defined by a typical transmitted signal B, received signal A, and the origin 0. The transmitted signal will lie very

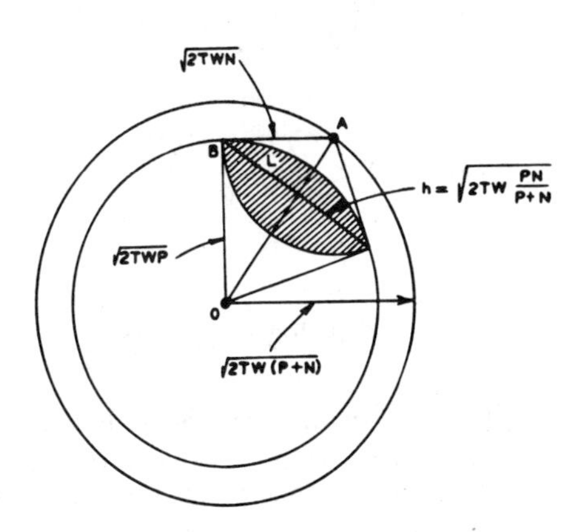

Fig. 5—The geometry involved in Theorem 2.

close to the surface of the sphere of radius $\sqrt{2TWP}$, since in a high-dimensional sphere nearly all the volume is very close to the surface. The received signal similarly will lie on the surface of the sphere of radius $\sqrt{2TW(P+N)}$. The high-dimensional lens-shaped region L is the region of possible signals that might have caused A, since the distance between the transmitted and received signal is almost certainly very close to $\sqrt{2TWN}$. L is of smaller volume than a sphere of radius h. We can determine h by equating the area of the triangle OAB, calculated two different ways:

$$\tfrac{1}{2}h\sqrt{2TW(P+N)} = \tfrac{1}{2}\sqrt{2TWP}\sqrt{2TWN}$$

$$h = \sqrt{2TW\frac{PN}{P+N}}.$$

The probability of any particular signal point (other than the actual cause of A) lying in L is, therefore, less than the ratio of the volumes of spheres of radii $\sqrt{2TW\,PN/P+N}$ and $\sqrt{2TWP}$, since in our ensemble of coding systems we chose the signal points at random from the points in the sphere of radius $\sqrt{2TWP}$. This ratio is

$$\left(\frac{\sqrt{2TW\frac{PN}{P+N}}}{\sqrt{2TWP}}\right)^{2TW} = \left(\frac{N}{P+N}\right)^{TW}. \tag{23}$$

We have M signal points. Hence the probability p that all except the actual cause of A are *outside* L is greater than

$$\left[1-\left(\frac{N}{P+N}\right)^{TW}\right]^{M-1}. \tag{24}$$

When these points are outside L, the signal is interpreted correctly. Therefore, if we make P greater than $1-\epsilon$, the frequency of errors will be less than ϵ. This will be true if

$$\left[1-\left(\frac{N}{P+N}\right)^{TW}\right]^{(M-1)} > 1-\epsilon. \tag{25}$$

Now $(1-x)^n$ is always greater than $1-nx$ when n is positive. Consequently, (25) will be true if

$$1-(M-1)\left(\frac{N}{P+N}\right)^{TW} > 1-\epsilon \tag{26}$$

or if

$$(M-1) < \epsilon\left(\frac{P+N}{N}\right)^{TW} \tag{27}$$

or

$$\frac{\log(M-1)}{T} < W\log\frac{P+N}{N} + \frac{\log\epsilon}{T}. \tag{28}$$

For any fixed ϵ, we can satisfy this by taking T sufficiently large, and also have $\log(M-1)/T$ or $\log M/T$ as close as desired to $W \log P+N/N$. This shows that, with a random selection of points for signals, we can obtain an arbitrarily small frequency of errors and transmit at a rate arbitrarily close to the rate C. We can also send *at* the rate C with arbitrarily small ϵ, since the extra binary digits need not be sent at all, but can be filled in at random at the receiver. This only adds another arbitrarily small quantity to ϵ. This completes the proof.

VIII. Discussion

We will call a system that transmits without errors at the rate C an ideal system. Such a system cannot be achieved with any finite encoding process but can be approximated as closely as desired. As we approximate more closely to the ideal, the following effects occur: (1) The rate of transmission of binary digits approaches $C=W\log_2(1+P/N)$. (2) The frequency of errors approaches zero. (3) The transmitted signal approaches a white noise in statistical properties. This is true, roughly speaking, because the various signal functions used must be distributed at random in the sphere of radius $\sqrt{2TWP}$. (4) The threshold effect becomes very sharp. If the noise is increased over the value for which the system was designed, the frequency of errors increases very rapidly. (5) The required delays at transmitter and receiver increase indefinitely. Of course, in a wide-band system a millisecond may be substantially an infinite delay.

In Fig. 6 the function $C/W=\log(1+P/N)$ is plotted with P/N in db horizontal and C/W the number of bits per cycle of band vertical. The circles represent PCM systems of the binary, ternary, etc., types, using positive and negative pulses and adjusted to give one error in about 10^5 binary digits. The dots are for a PPM system with two, three, etc., discrete positions for the pulse.[9]

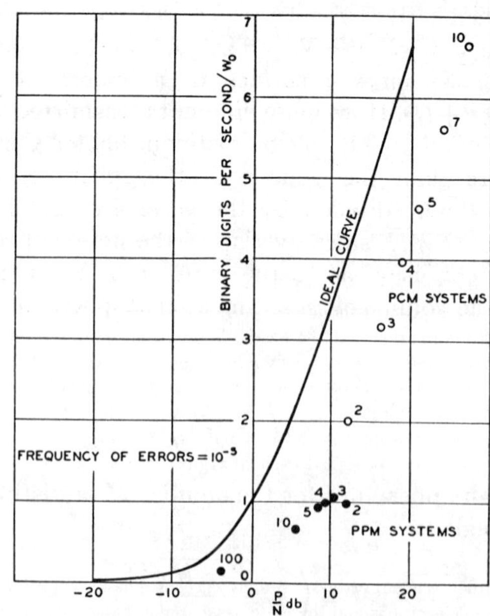

Fig. 6—Comparison of PCM and PPM with ideal performance.

[9] The PCM points are calculated from formulas given in "The philosophy of PCM," by B. M. Oliver, J. R. Pierce, and C. E. Shannon, Proc. I.R.E., vol. 36, pp. 1324–1332; November, 1948. The PPM points are from unpublished calculations of B. McMillan, who points out that, for very small P/N, the points approach to within 3 db of the ideal curve.

The difference between the series of points and the ideal curve corresponds to the gain that could be obtained by more involved coding systems. It amounts to about 8 db in power over most of the practical range. The series of points and circles is about the best that can be done without delay. Whether it is worth while to use more complex types of modulation to obtain some of this possible saving is, of course, a question of relative costs and valuations.

The quantity $TW \log(1+P/N)$ is, for large T, the number of bits that can be transmitted in time T. It can be regarded as an exchange relation between the different parameters. The individual quantities T, W, P, and N can be altered at will without changing the amount of information we can transmit, provided $TW \log(1+P/N)$ is held constant. If TW is reduced, P/N must be increased, etc.

Ordinarily, as we increase W, the noise power N in the band will increase proportionally; $N=N_0W$ where N_0 is the noise power per cycle. In this case, we have

$$C = W \log\left(1 + \frac{P}{N_0W}\right). \tag{29}$$

If we let $W_0=P/N_0$, i.e., W_0 is the band for which the noise power is equal to the signal power, this can be written

$$\frac{C}{W_0} = \frac{W}{W_0} \log\left(1 + \frac{W_0}{W}\right). \tag{30}$$

In Fig. 7, C/W_0 is plotted as a function of W/W_0. As we increase the band, the capacity increases rapidly until the total noise power accepted is about equal to the

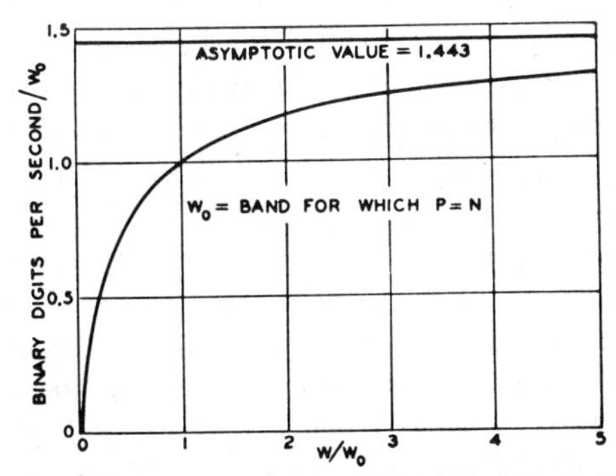

Fig. 7—Channel capacity as a function of bandwidth.

signal power; after this, the increase is slow, and it approaches an asymptotic value $\log_2 e$ times the capacity for $W=W_0$.

IX. Arbitrary Gaussian Noise

If a white thermal noise is passed through a filter whose transfer function is $Y(f)$, the resulting noise has a power spectrum $N(f)=K|Y(f)|^2$ and is known as Gaussian noise. We can calculate the capacity of a channel perturbed by any Gaussian noise from the white-noise result. Suppose our total transmitter power is P and it is distributed among the various frequencies according to $P(f)$. Then

$$\int_0^W P(f)df = P. \tag{31}$$

We can divide the band into a large number of small bands, with $N(f)$ approximately constant in each. The total capacity for a given distribution $P(f)$ will then be given by

$$C_1 = \int_0^W \log\left(1 + \frac{P(f)}{N(f)}\right)df, \tag{32}$$

since, for each elementary band, the white-noise result applies. The maximum rate of transmission will be found by maximizing C_1 subject to condition (31). This requires that we maximize

$$\int_0^W \left[\log\left(1 + \frac{P(f)}{N(f)}\right) + \lambda P(f)\right]df. \tag{33}$$

The condition for this is, by the calculus of variations, or merely from the convex nature of the curve $\log(1+x)$,

$$\frac{1}{N(f) + P(f)} + \lambda = 0, \tag{34}$$

or $N(f)+P(f)$ must be constant. The constant is adjusted to make the total signal power equal to P. For frequencies where the noise power is low, the signal power should be high, and vice versa, as we would expect.

The situation is shown graphically in Fig. 8. The curve is the assumed noise spectrum, and the three lines correspond to different choices of P. If P is small, we cannot make $P(f)+N(f)$ constant, since this would require negative power at some frequencies. It is easily shown, however, that in this case the best $P(f)$ is obtained by making $P(f)+N(f)$ constant whenever possible, and making $P(f)$ zero at other frequencies. With low values of P, some of the frequencies will not be used at all.

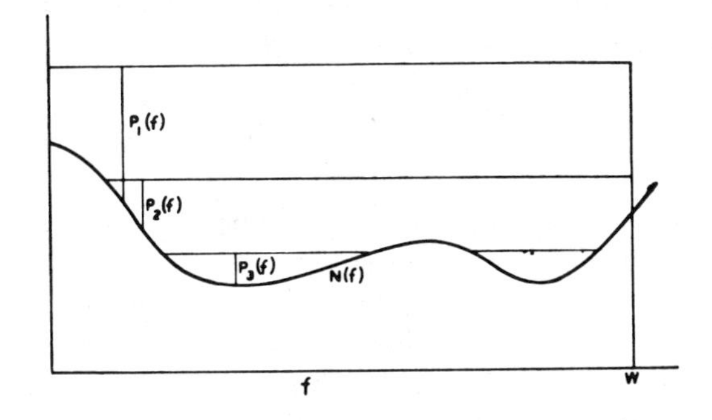

Fig. 8—Best distribution of transmitter power.

If we now vary the noise spectrum $N(f)$, keeping the total noise power constant and always adjusting the signal spectrum $P(f)$ to give the maximum transmission, we can determine the worst spectrum for the noise. This turns out to be the white-noise case. Although this only shows it to be worst among the Gaussian noises, it will be shown later to be the worst among all possible noises with the given power N in the band.

X. The Channel Capacity with an Arbitrary Type of Noise

Of course, there are many kinds of noise which are not Gaussian; for example, impulse noise, or white noise that has passed through a nonlinear device. If the signal is perturbed by one of these types of noise, there will still be a definite channel capacity C, the maximum rate of transmission of binary digits. We will merely outline the general theory here.[10]

Let $x_1, x_2, \cdots, x_n$ be the amplitudes of the noise at successive sample points, and let

$$p(x_1, x_2, \cdots, x_n)dx_1 \cdots dx_n \tag{35}$$

be the probability that these amplitudes lie between x_1 and x_1+dx_1, x_2 and x_2+dx_2, etc. Then the function p describes the statistical structure of the noise, insofar as n successive samples are concerned. The *entropy*, H, of the noise is defined as follows. Let

$$H_n = -\frac{1}{n}\int \cdots \int p(x_1, \cdots, x_n) \cdot \log_e p(x_1, \cdots, x_n)dx_1, \cdots, dx_n. \tag{36}$$

Then

$$H = \lim_{n\to\infty} H_n. \tag{37}$$

This limit exists in all cases of practical interest, and can be determined in many of them. H is a measure of the randomness of the noise. In the case of white Gaussian noise of power N, the entropy is

$$H = \log_e \sqrt{2\pi eN}. \tag{38}$$

It is convenient to measure the randomness of an arbitrary type of noise not directly by its entropy, but by comparison with white Gaussian noise. We can calculate the power in a white noise having the same entropy as the given noise. This power, namely,

$$\overline{N} = \frac{1}{2\pi e}\exp 2H \tag{39}$$

where H is the entropy of the given noise, will be called the *entropy power* of the noise.

A noise of entropy power $\overline{N}$ acts very much like a white noise of power $\overline{N}$, insofar as perturbing the message is concerned. It can be shown that the region of uncertainty about each signal point will have the same volume as the region associated with the white noise. Of course, it will no longer be a spherical region. In proving Theorem 1 this volume of uncertainty was the chief property of the noise used. Essentially the same argument may be applied for any kind of noise with minor modifications. The result is summarized in the following:

Theorem 3: *Let a noise limited to the band W have power N and entropy power N_1. The capacity C is then bounded by*

$$W \log_2 \frac{P+N_1}{N_1} \leq C \leq W \log_2 \frac{P+N}{N_1} \tag{40}$$

where P is the average signal power and W the bandwidth.

If the noise is a white Gaussian noise, $N_1=N$, and the two limits are equal. The result then reduces to the theorem in Section VII.

For any noise, $N_1<N$. This is why white Gaussian noise is the worst among all possible noises. If the noise is Gaussian with spectrum $N(f)$, then

$$N_1 = W \exp \frac{1}{W}\int_0^W \log N(f)df. \tag{41}$$

The upper limit in Theorem 3 is then reached when we are above the highest noise power in Fig. 8. This is easily verified by substitution.

In the cases of most interest, P/N is fairly large. The two limits are then nearly the same, and we can use $W \log (P+N)/N_1$ as the capacity. The upper limit is the best choice, since it can be shown that as P/N increases, C approaches the upper limit.

XI. Discrete Sources of Information

Up to now we have been chiefly concerned with the channel. The capacity C measures the maximum rate at which a random series of binary digits can be transmitted when they are encoded in the best possible way. In general, the information to be transmitted will not be in this form. It may, for example, be a sequence of letters as in telegraphy, a speech wave, or a television signal. Can we find an equivalent number of bits per second for information sources of this type? Consider first the discrete case; i.e., the message consists of a sequence of discrete symbols. In general, there may be correlation of various sorts between the different symbols. If the message is English text, the letter E is the most frequent, T is often followed by H, etc. These correlations allow a certain compression of the text by proper encoding. We may define the entropy of a discrete source in a way analogous to that for a noise; namely, let

$$H_n = -\frac{1}{n}\sum_{i,j,\cdots,s} p(i,j,\cdots,s)\log_2 p(i,j,\cdots,s) \tag{42}$$

where $p(i, j, \cdots, s)$ is the probability of the sequence of symbols $i, j, \cdots, s$, and the sum is over all sequences of n symbols. Then the entropy is

$$H = \lim_{n\to\infty} H_n. \tag{43}$$

[10] C. E. Shannon, "A mathematical theory of communication," *Bell Sys. Tech. Jour.*, vol. 27, pp. 379–424 and 623–657; July and October, 1948.

It turns out that H is the number of bits produced by the source for each symbol of message. In fact, the following result is proved in the Appendix.

THEOREM 4. *It is possible to encode all sequences of n message symbols into sequences of binary digits in such a way that the average number of binary digits per message symbol is approximately H, the approximation approaching equality as n increases.*

It follows that, if we have a channel of capacity C and a discrete source of entropy H, it is possible to encode the messages via binary digits into signals and transmit at the rate C/H of the original message symbols per second.

For example, if the source produces a sequence of letters A, B, or C with probabilities $p_A=0.6$, $p_B=0.3$, $p_C=0.1$, and successive letters are chosen independently, then $H_n=H_1=-[0.6 \log_2 0.6+0.3 \log_2 0.3 +0.1 \log_2 0.1]=1.294$ and the information produced is equivalent to 1.294 bits for each letter of the message. A channel with a capacity of 100 bits per second could transmit with best encoding $100/1.294=77.3$ message letters per second.

XII. CONTINUOUS SOURCES

If the source is producing a continuous function of time, then without further data we must ascribe it an infinite rate of generating information. In fact, merely to specify exactly one quantity which has a continuous range of possibilities requires an infinite number of binary digits. We cannot send continuous information *exactly* over a channel of finite capacity.

Fortunately, we do not need to send continuous messages exactly. A certain amount of discrepancy between the original and the recovered messages can always be tolerated. If a certain tolerance is allowed, then a definite finite rate in binary digits per second can be assigned to a continuous source. It must be remembered that this rate depends on the nature and magnitude of the allowed error between original and final messages. The rate may be described as the rate of generating information *relative to the criterion of fidelity*.

Suppose the criterion of fidelity is the rms discrepancy between the original and recovered signals, and that we can tolerate a value $\sqrt{N_1}$. Then each point in the message space is surrounded by a small sphere of radius $\sqrt{2T_1W_1N_1}$. If the system is such that the recovered message lies within this sphere, the transmission will be satisfactory. Hence, the number of different messages which must be capable of distinct transmission is of the order of the volume V_1 of the region of possible messages divided by the volume of the small spheres. Carrying out this argument in detail along lines similar to those used in Sections VII and IX leads to the following result:

THEOREM 5: *If the message source has power Q, entropy power $\bar{Q}$, and bandwidth W_1, the rate R of generating information in bits per second is bounded by*

$$W_1 \log_2 \frac{\bar{Q}}{N_1} \leqq R \leqq W_1 \log_2 \frac{Q}{N_1} \tag{44}$$

where N_1 is the maximum tolerable mean square error in reproduction. If we have a channel with capacity C and a source whose rate of generating information R is less than or equal to C, it is possible to encode the source in such a way as to transmit over this channel with the fidelity measured by N_1. If $R>C$, this is impossible.

In the case where the message source is producing white thermal noise, $\bar{Q}=Q$. Hence the two bounds are equal and $R=W_1 \log Q/N_1$. We can, therefore, transmit white noise of power Q and band W_1 over a channel of band W perturbed by a white noise of power N and recover the original message with mean square error N_1 if, and only if,

$$W_1 \log \frac{Q}{N_1} \leqq W \log \frac{P+N}{N} \cdot \tag{45}$$

APPENDIX

Consider the possible sequences of n symbols. Let them be arranged in order of decreasing probability, $p_1 \geqq p_2 \geqq p_3 \cdots \geqq p_s$. Let $P_i=\sum_1^{i-1} p_j$. The ith message is encoded by expanding P_i as a binary fraction and using only the first t_i places where t_i is determined from

$$\log_2 \frac{1}{p_i} \leqq t_i < 1+\log_2 \frac{1}{p_i} \cdot \tag{46}$$

Probable sequences have short codes and improbable ones long codes. We have

$$\frac{1}{2^{t_i}} \leqq p_i \leqq \frac{1}{2^{t_i-1}} \cdot \tag{47}$$

The codes for different sequences will all be different. P_{i+1}, for example, differs by p_i from P_i, and therefore its binary expansion will differ in one or more of the first t_i places, and similarly for all others. The average length of the encoded message will be $\sum p_i t_i$. Using (46),

$$-\sum p_i \log p_i \leqq \sum p_i t_i < \sum p_i(1-\log p_i) \tag{48}$$

or

$$nH_n \leqq \sum p_i t_i < 1+nH_n. \tag{49}$$

The average number of binary digits used per message symbol is $1/n \sum p_i t_i$ and

$$H_n \leqq \frac{1}{n} \sum p_i t_i < \frac{1}{n}+H_n. \tag{50}$$

As $n \to \infty$, $H_n \to H$ and $1/n \to 0$, so the average number of bits per message symbol approaches H.

Prediction and Entropy of Printed English

By C. E. SHANNON

(*Manuscript Received Sept. 15, 1950*)

A new method of estimating the entropy and redundancy of a language is described. This method exploits the knowledge of the language statistics possessed by those who speak the language, and depends on experimental results in prediction of the next letter when the preceding text is known. Results of experiments in prediction are given, and some properties of an ideal predictor are developed.

1. Introduction

In a previous paper[1] the entropy and redundancy of a language have been defined. The entropy is a statistical parameter which measures, in a certain sense, how much information is produced on the average for each letter of a text in the language. If the language is translated into binary digits (0 or 1) in the most efficient way, the entropy H is the average number of binary digits required per letter of the original language. The redundancy, on the other hand, measures the amount of constraint imposed on a text in the language due to its statistical structure, e.g., in English the high frequency of the letter E, the strong tendency of H to follow T or of U to follow Q. It was estimated that when statistical effects extending over not more than eight letters are considered the entropy is roughly 2.3 bits per letter, the redundancy about 50 per cent.

Since then a new method has been found for estimating these quantities, which is more sensitive and takes account of long range statistics, influences extending over phrases, sentences, etc. This method is based on a study of the predictability of English; how well can the next letter of a text be predicted when the preceding N letters are known. The results of some experiments in prediction will be given, and a theoretical analysis of some of the properties of ideal prediction. By combining the experimental and theoretical results it is possible to estimate upper and lower bounds for the entropy and redundancy. From this analysis it appears that, in ordinary literary English, the long range statistical effects (up to 100 letters) reduce the entropy to something of the order of one bit per letter, with a corresponding redundancy of roughly 75%. The redundancy may be still higher when structure extending over paragraphs, chapters, etc. is included. However, as the lengths involved are increased, the parameters in question become more erratic and uncertain, and they depend more critically on the type of text involved.

2. Entropy Calculation from the Statistics of English

One method of calculating the entropy H is by a series of approximations $F_0, F_1, F_2, \cdots$, which successively take more and more of the statistics of the language into account and approach H as a limit. F_N may be called the N-gram entropy; it measures the amount of information or entropy due to statistics extending over N adjacent letters of text. F_N is given by[1]

$$\begin{aligned} F_N &= -\sum_{i,j} p(b_i, j) \log_2 p_{b_i}(j) \\ &= -\sum_{i,j} p(b_i, j) \log_2 p(b_i, j) + \sum_i p(b_i) \log p(b_i) \end{aligned} \tag{1}$$

in which: b_i is a block of N-1 letters [(N-1)-gram]

j is an arbitrary letter following b_i

$p(b_i, j)$ is the probability of the N-gram b_i, j

$p_{b_i}(j)$ is the conditional probability of letter j after the block b_i, and is given by $p(b_i, j)/p(b_i)$.

The equation (1) can be interpreted as measuring the average uncertainty (conditional entropy) of the next letter j when the preceding N-1 letters are known. As N is increased, F_N includes longer and longer range statistics and the entropy, H, is given by the limiting value of F_N as $N \to \infty$:

$$H = \lim_{N\to\infty} F_N . \tag{2}$$

The N-gram entropies F_N for small values of N can be calculated from standard tables of letter, digram and trigram frequencies.[2] If spaces and punctuation are ignored we have a twenty-six letter alphabet and F_0 may be taken (by definition) to be $\log_2 26$, or 4.7 bits per letter. F_1 involves letter frequencies and is given by

$$F_1 = -\sum_{i=1}^{26} p(i) \log_2 p(i) = 4.14 \text{ bits per letter.} \tag{3}$$

The digram approximation F_2 gives the result

$$\begin{aligned} F_2 &= -\sum_{i,j} p(i,j) \log_2 p_i(j) \\ &= -\sum_{i,j} p(i,j) \log_2 p(i,j) + \sum_i p(i) \log_2 p(i) \\ &= 7.70 - 4.14 = 3.56 \text{ bits per letter.} \end{aligned} \tag{4}$$

The trigram entropy is given by

$$\begin{aligned} F_3 &= -\sum_{i,j,k} p(i,j,k) \log_2 p_{ij}(k) \\ &= -\sum_{i,j,k} p(i,j,k) \log_2 p(i,j,k) + \sum_{i,j} p(i,j) \log_2 p(i,j) \\ &\doteq 11.0 - 7.7 = 3.3 \end{aligned} \tag{5}$$

In this calculation the trigram table[2] used did not take into account trigrams bridging two words, such as WOW and OWO in TWO WORDS. To compensate partially for this omission, corrected trigram probabilities $p(i, j, k)$ were obtained from the probabilities $p'(i, j, k)$ of the table by the following rough formula:

$$p(i,j,k) = \frac{2.5}{4.5} p'(i,j,k) + \frac{1}{4.5} r(i)p(j,k) + \frac{1}{4.5} p(i,j)s(k)$$

where $r(i)$ is the probability of letter i as the terminal letter of a word and $s(k)$ is the probability of k as an initial letter. Thus the trigrams within words (an average of 2.5 per word) are counted according to the table; the bridging trigrams (one of each type per word) are counted approximately by assuming independence of the terminal letter of one word and the initial digram in the next or vice versa. Because of the approximations involved here, and also because of the fact that the sampling error in identifying probability with sample frequency is more serious, the value of F_3 is less reliable than the previous numbers.

Since tables of N-gram frequencies were not available for $N > 3$, F_4, F_5, etc. could not be calculated in the same way. However, word frequencies have been tabulated[3] and can be used to obtain a further approximation. Figure 1 is a plot on log-log paper of the probabilities of words against frequency rank. The most frequent English word "the" has a probability .071 and this is plotted against 1. The next most frequent word "of" has a probability of .034 and is plotted against 2, etc. Using logarithmic scales both for probability and rank, the curve is approximately a straight line with slope -1; thus, if p_n is the probability of the nth most frequent word, we have, roughly

$$p_n = \frac{.1}{n} . \tag{6}$$

Zipf[4] has pointed out that this type of formula, $p_n = k/n$, gives a rather good approximation to the word probabilities in many different languages. The formula (6) clearly cannot hold indefinitely since the total probability Σp_n must be unity, while $\sum_1^\infty .1/n$ is infinite. If we assume (in the absence of any better estimate) that the formula $p_n = .1/n$ holds out to the n at which the

[1] C. E. Shannon, "A Mathematical Theory of Communication," *Bell System Technical Journal*, v. 27, pp. 379–423, 623–656, July, October, 1948.

[2] Fletcher Pratt, "Secret and Urgent," Blue Ribbon Books, 1942.

[3] G. Dewey, "Relative Frequency of English Speech Sounds," Harvard University Press, 1923.

[4] G. K. Zipf, "Human Behavior and the Principle of Least Effort," Addison-Wesley Press, 1949.

Reprinted with permission from *Bell Syst. Tech. J.*, vol. 30, pp. 50-64, Jan. 1951. Copyright © 1951, The American Telephone and Telegraph Company.

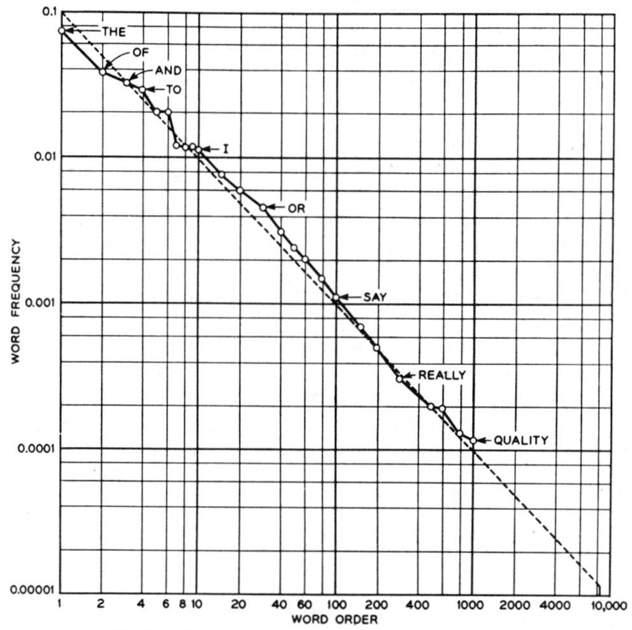

Fig. 1—Relative frequency against rank for English words.

total probability is unity, and that $p_n = 0$ for larger n, we find that the critical n is the word of rank 8,727. The entropy is then:

$$-\sum_{1}^{8727} p_n \log_2 p_n = 11.82 \text{ bits per word,} \tag{7}$$

or $11.82/4.5 = 2.62$ bits per letter since the average word length in English is 4.5 letters. One might be tempted to identify this value with $F_{4.5}$, but actually the ordinate of the F_N curve at $N = 4.5$ will be above this value. The reason is that F_4 or F_5 involves groups of four or five letters regardless of word division. A word is a cohesive group of letters with strong internal statistical influences, and consequently the N-grams within words are more restricted than those which bridge words. The effect of this is that we have obtained, in 2.62 bits per letter, an estimate which corresponds more nearly to, say, F_5 or F_6.

A similar set of calculations was carried out including the space as an additional letter, giving a 27 letter alphabet. The results of both 26- and 27-letter calculations are summarized below:

	F_0	F_1	F_2	F_3	F_{word}
26 letter.	4.70	4.14	3.56	3.3	2.62
27 letter.	4.76	4.03	3.32	3.1	2.14

The estimate of 2.3 for F_8, alluded to above, was found by several methods' one of which is the extrapolation of the 26-letter series above out to that point. Since the space symbol is almost completely redundant when sequences of one or more words are involved, the values of F_N in the 27-letter case will be $\frac{4.5}{5.5}$ or .818 of F_N for the 26-letter alphabet when N is reasonably large.

3. Prediction of English

The new method of estimating entropy exploits the fact that anyone speaking a language possesses, implicitly, an enormous knowledge of the statistics of the language. Familiarity with the words, idioms, clichés and grammar enables him to fill in missing or incorrect letters in proof-reading, or to complete an unfinished phrase in conversation. An experimental demonstration of the extent to which English is predictable can be given as follows: Select a short passage unfamiliar to the person who is to do the predicting. He is then asked to guess the first letter in the passage. If the guess is correct he is so informed, and proceeds to guess the second letter. If not, he is told the correct first letter and proceeds to his next guess. This is continued through the text. As the experiment progresses, the subject writes down the correct text up to the current point for use in predicting future letters. The result of a typical experiment of this type is given below. Spaces were included as an additional letter, making a 27 letter alphabet. The first line is the original text; the second line contains a dash for each letter correctly guessed. In the case of incorrect guesses the correct letter is copied in the second line.

```
(1) THE ROOM WAS NOT VERY LIGHT A SMALL OBLONG
(2) ----ROO------NOT-V-----I------SM----OBL----          (8)
(1) READING LAMP ON THE DESK SHED GLOW ON
(2) REA----------O------D----SHED-GLO--O--
(1) POLISHED WOOD BUT LESS ON THE SHABBY RED CARPET
(2) P-L-S-----O---BU--L-S--O------SH-----RE--C------
```

Of a total of 129 letters, 89 or 69% were guessed correctly. The errors, as would be expected, occur most frequently at the beginning of words and syllables where the line of thought has more possibility of branching out. It might be thought that the second line in (8), which we will call the *reduced text*, contains much less information than the first. Actually, both lines contain the same information in the sense that it is possible, at least in principle, to recover the first line from the second. To accomplish this we need an identical twin of the individual who produced the sequence. The twin (who must be mathematically, not just biologically identical) will respond in the same way when faced with the same problem. Suppose, now, we have only the reduced text of (8). We ask the twin to guess the passage. At each point we will know whether his guess is correct, since he is guessing the same as the first twin and the presence of a dash in the reduced text corresponds to a correct guess. The letters he guesses wrong are also available, so that at each stage he can be supplied with precisely the same information the first twin had available.

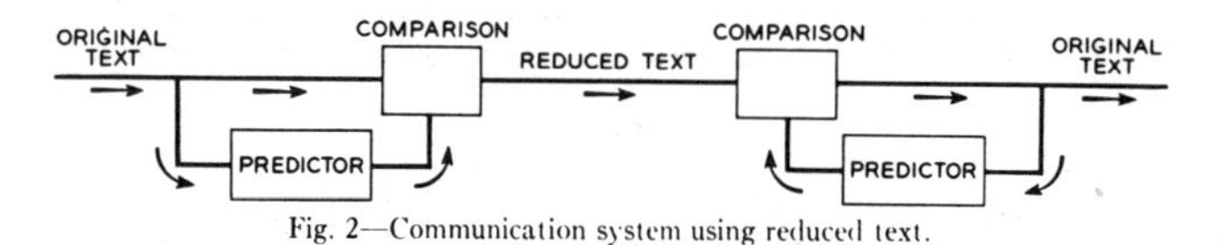

Fig. 2—Communication system using reduced text.

The need for an identical twin in this conceptual experiment can be eliminated as follows. In general, good prediction does not require knowledge of more than N preceding letters of text, with N fairly small. There are only a finite number of possible sequences of N letters. We could ask the subject to guess the next letter for each of these possible N-grams. The complete list of these predictions could then be used both for obtaining the reduced text from the original and for the inverse reconstruction process.

To put this another way, the reduced text can be considered to be an encoded form of the original, the result of passing the original text through a reversible transducer. In fact, a communication system could be constructed in which only the reduced text is transmitted from one point to the other. This could be set up as shown in Fig. 2, with two identical prediction devices.

An extension of the above experiment yields further information concerning the predictability of English. As before, the subject knows the text up to the current point and is asked to guess the next letter. If he is wrong, he is told so and asked to guess again. This is continued until he finds the correct letter. A typical result with this experiment is shown below. The first line is the original text and the numbers in the second line indicate the guess at which the correct letter was obtained.

```
(1) T H E R E   I S   N O   R E V E R S E   O N   A   M O T O R C Y C L E   A
(2) 1 1 1 5 1 1 2 1 1 2 1 1 15 1 17 1 1 1 2 1 3 2 1 2 2 7 1 1 1 1 4 1 1 1 1 1 3 1
(1) F R I E N D   O F   M I N E   F O U N D   T H I S   O U T
(2) 8 6 1 3 1 1 1 1 1 1 1 1 1 1 6 2 1 1 1 1 1 1 2 1 1 1 1 1 1
(1) R A T H E R   D R A M A T I C A L L Y   T H E   O T H E R   D A Y
(2) 4 1 1 1 1 1 1 11 5 1 1 1 1 1 1 1 1 1 1 1 6 1 1 1 1 1 1 1 1 1 1 1 1     (9)
```

Out of 102 symbols the subject guessed right on the first guess 79 times, on the second guess 8 times, on the third guess 3 times, the fourth and fifth guesses 2 each and only eight times required more than five guesses. Results of this order are typical of prediction by a good subject with ordinary literary English. Newspaper writing, scientific work and poetry generally lead to somewhat poorer scores.

The reduced text in this case also contains the same information as the original. Again utilizing the identical twin we ask him at each stage to guess as many times as the number given in the reduced text and recover in this way the original. To eliminate the human element here we must ask our subject, for each possible N-gram of text, to guess the most probable next letter, the second most probable next letter, etc. This set of data can then serve both for prediction and recovery.

Just as before, the reduced text can be considered an encoded version of the original. The original language, with an alphabet of 27 symbols, A,

$B, \cdots, Z$, space, has been translated into a new language with the alphabet $1, 2, \cdots, 27$. The translating has been such that the symbol 1 now has an extremely high frequency. The symbols 2, 3, 4 have successively smaller frequencies and the final symbols $20, 21, \cdots, 27$ occur very rarely. Thus the translating has simplified to a considerable extent the nature of the statistical structure involved. The redundancy which originally appeared in complicated constraints among groups of letters, has, by the translating process, been made explicit to a large extent in the very unequal probabilities of the new symbols. It is this, as will appear later, which enables one to estimate the entropy from these experiments.

In order to determine how predictability depends on the number N of preceding letters known to the subject, a more involved experiment was carried out. One hundred samples of English text were selected at random from a book, each fifteen letters in length. The subject was required to guess the text, letter by letter, for each sample as in the preceding experiment. Thus one hundred samples were obtained in which the subject had available $0, 1, 2, 3, \cdots, 14$ preceding letters. To aid in prediction the subject made such use as he wished of various statistical tables, letter, digram and trigram tables, a table of the frequencies of initial letters in words, a list of the frequencies of common words and a dictionary. The samples in this experiment were from "*Jefferson the Virginian*" by Dumas Malone. These results, together with a similar test in which 100 letters were known to the subject, are summarized in Table I. The column corresponds to the number of preceding letters known to the subject plus one; the row is the number of the guess. The entry in column N at row S is the number of times the subject guessed the right letter at the Sth guess when (N-1) letters were known. For example,

TABLE I

	1	2	3	4	5	6	7	8	9	10	11	12	13	14	15	100
1	18.2	29.2	36	47	51	58	48	66	66	67	62	58	66	72	60	80
2	10.7	14.8	20	18	13	19	17	15	13	10	9	14	9	6	18	7
3	8.6	10.0	12	14	8	5	3	5	9	4	7	7	4	9	5	
4	6.7	8.6	7	3	4	1	4	4	4	4	5	6	4	3	5	3
5	6.5	7.1	1	1	3	4	3	6	1	6	5	2	3			4
6	5.8	5.5	4	5	2	3	2			1	4	2	3	4	1	2
7	5.6	4.5	3	3	2	2	8		1	1	1	4	1		4	1
8	5.2	3.6	2	2	1	1	2	1	1	1	1		2	1	3	
9	5.0	3.0	4		5	1	4		2	1	1	2		1		1
10	4.3	2.6	2	1	3		3	1					2			
11	3.1	2.2	2	2	2	1			1	3		1	1	2	1	
12	2.8	1.9	4		2	1	1	1			2	1	1		1	1
13	2.4	1.5	1	1	1	1	1	1	1	1		1	1			
14	2.3	1.2		1			1					1				1
15	2.1	1.0	1	1							1	1	1			
16	2.0	.9					1			1					1	
17	1.6	.7	1		2	1	1				1		2	2		
18	1.6	.5													1	
19	1.6	.4			1	1			1		1					
20	1.3	.3		1		1	1									
21	1.2	.2														
22	.8	.1														
23	.3	.1														
24	.1	.0														
25	.1															
26	.1															
27	.1															

the entry 19 in column 6, row 2, means that with five letters known the correct letter was obtained on the second guess nineteen times out of the hundred. The first two columns of this table were not obtained by the experimental procedure outlined above but were calculated directly from the known letter and digram frequencies. Thus with no known letters the most probable symbol is the space (probability .182); the next guess, if this is wrong, should be E (probability .107), etc. These probabilities are the frequencies with which the right guess would occur at the first, second, etc., trials with best prediction. Similarly, a simple calculation from the digram table gives the entries in column 1 when the subject uses the table to best advantage. Since the frequency tables are determined from long samples of English, these two columns are subject to less sampling error than the others.

It will be seen that the prediction gradually improves, apart from some statistical fluctuation, with increasing knowledge of the past as indicated by the larger numbers of correct first guesses and the smaller numbers of high rank guesses.

One experiment was carried out with "reverse" prediction, in which the subject guessed the letter preceding those already known. Although the task is subjectively much more difficult, the scores were only slightly poorer. Thus, with two 101 letter samples from the same source, the subject obtained the following results:

No. of guess	1	2	3	4	5	6	7	8	>8
Forward	70	10	7	2	2	3	3	0	4
Reverse	66	7	4	4	6	2	1	2	9

Incidentally, the N-gram entropy F_N for a reversed language is equal to that for the forward language as may be seen from the second form in equation (1). Both terms have the same value in the forward and reversed cases.

4. Ideal N-Gram Prediction

The data of Table I can be used to obtain upper and lower bounds to the N-gram entropies F_N. In order to do this, it is necessary first to develop some general results concerning the best possible prediction of a language when the preceding N letters are known. There will be for the language a set of conditional probabilities $p_{i_1, i_2, \cdots, i_{N-1}}(j)$. This is the probability when the (N-1) gram $i_1, i_2, \cdots, i_{N-1}$ occurs that the next letter will be j. The best guess for the next letter, when this (N-1) gram is known to have occurred, will be that letter having the highest conditional probability. The second guess should be that with the second highest probability, etc. A machine or person guessing in the best way would guess letters in the order of decreasing conditional probability. Thus the process of reducing a text with such an ideal predictor consists of a mapping of the letters into the numbers from 1 to 27 in such a way that the most probable next letter [conditional on the known preceding (N-1) gram] is mapped into 1, etc. The frequency of 1's in the reduced text will then be given by

$$q_1^N = \Sigma p(i_1, i_2, \cdots, i_{N-1}, j) \tag{10}$$

where the sum is taken over all (N-1) grams $i_1, i_2, \cdots, i_{N-1}$ the j being the one which maximizes p for that particular (N-1) gram. Similarly, the frequency of 2's, q_2^N, is given by the same formula with j chosen to be that letter having the second highest value of p, etc.

On the basis of N-grams, a different set of probabilities for the symbols in the reduced text, $q_1^{N+1}, q_2^{N+1}, \ldots, q_{27}^{N+1}$, would normally result. Since this prediction is on the basis of a greater knowledge of the past, one would expect the probabilities of low numbers to be greater, and in fact one can prove the following inequalities:

$$\sum_{i=1}^{S} q_i^{N+1} \geq \sum_{i=1}^{S} q_i^N \qquad S = 1, 2, \cdots. \tag{11}$$

This means that the probability of being right in the first S guesses when the preceding N letters are known is greater than or equal to that when only (N-1) are known, for all S. To prove this, imagine the probabilities $p(i_1, i_2, \cdots, i_N, j)$ arranged in a table with j running horizontally and all the N-grams vertically. The table will therefore have 27 columns and 27^N rows. The term on the left of (11) is the sum of the S largest entries in each row, summed over all the rows. The right-hand member of (11) is also a sum of entries from this table in which S entries are taken from each row but not necessarily the S largest. This follows from the fact that the right-hand member would be calculated from a similar table with (N-1) grams rather than N-grams listed vertically. Each row in the N-1 gram table is the sum of 27 rows of the N-gram table, since:

$$p(i_2, i_3, \cdots, i_N, j) = \sum_{i_1=1}^{27} p(i_1, i_2, \cdots, i_N, j). \tag{12}$$

The sum of the S largest entries in a row of the N-1 gram table will equal the sum of the $27S$ selected entries from the corresponding 27 rows of the N-gram table only if the latter fall into S columns. For the equality in (11) to hold for a particular S, this must be true of every row of the N-1 gram table. In this case, the first letter of the N-gram does not affect the set of the S most probable choices for the next letter, although the ordering within the set may be affected. However, if the equality in (11) holds for all S, it follows that the ordering as well will be unaffected by the first letter of the N-gram. The reduced text obtained from an ideal N-1 gram predictor is then identical with that obtained from an ideal N-gram predictor.

Since the partial sums

$$Q_S^N = \sum_{i=1}^{S} q_i^N \qquad S = 1, 2, \cdots \tag{13}$$

are monotonic increasing functions of N, <1 for all N, they must all approach limits as $N \to \infty$. Their first differences must therefore approach limits as $N \to \infty$, i.e., the q_i^N approach limits, q_i^∞. These may be interpreted as the relative frequency of correct first, second, $\cdots$, guesses with knowledge of the entire (infinite) past history of the text.

The ideal N-gram predictor can be considered, as has been pointed out, to be a transducer which operates on the language translating it into a sequence of numbers running from 1 to 27. As such it has the following two properties:

1. The output symbol is a function of the present input (the predicted next letter when we think of it as a predicting device) and the preceding (N-1) letters.
2. It is *instantaneously* reversible. The original input can be recovered by a suitable operation on the reduced text without loss of time. In fact, the inverse operation also operates on only the (N-1) preceding symbols of the reduced text together with the present output.

The above proof that the frequencies of output symbols with an N-1 gram predictor satisfy the inequalities:

$$\sum_1^S q_i^N \geq \sum_1^S q_i^{N-1} \qquad S = 1, 2, \cdots, 27 \tag{14}$$

can be applied to any transducer having the two properties listed above. In fact we can imagine again an array with the various (N-1) grams listed vertically and the present input letter horizontally. Since the present output is a function of only these quantities there will be a definite output symbol which may be entered at the corresponding intersection of row and column. Furthermore, the instantaneous reversibility requires that no two entries in the same row be the same. Otherwise, there would be ambiguity between the two or more possible present input letters when reversing the translation. The total probability of the S most probable symbols in the output, say $\sum_1^S r_i$, will be the sum of the probabilities for S entries in each row, summed over the rows, and consequently is certainly not greater than the sum of the S largest entries in each row. Thus we will have

$$\sum_1^S q_i^N \geq \sum_1^S r_i \qquad S = 1, 2, \cdots, 27 \tag{15}$$

In other words ideal prediction as defined above enjoys a preferred position among all translating operations that may be applied to a language and which satisfy the two properties above. Roughly speaking, ideal prediction collapses the probabilities of various symbols to a small group more than any other translating operation involving the same number of letters which is instantaneously reversible.

Sets of numbers satisfying the inequalities (15) have been studied by Muirhead in connection with the theory of algebraic inequalities.[5] If (15) holds when the q_i^N and r_i are arranged in decreasing order of magnitude, and also $\sum_1^{27} q_i^N = \sum_1^{27} r_i$, (this is true here since the total probability in each case is 1), then the first set, q_i^N, is said to *majorize* the second set, r_i. It is known that the majorizing property is equivalent to either of the following properties:

1. The r_i can be obtained from the q_i^N by a finite series of "flows." By a flow is understood a transfer of probability from a larger q to a smaller one, as heat flows from hotter to cooler bodies but not in the reverse direction.
2. The r_i can be obtained from the q_i^N by a generalized "averaging" operation. There exists a set of non-negative real numbers, a_{ij}, with $\sum_j a_{ij} = \sum_i a_{ij} = 1$ and such that

$$r_i = \sum_j a_{ij}(q_j^N). \tag{16}$$

5. Entropy Bounds from Prediction Frequencies

If we know the frequencies of symbols in the reduced text with the ideal N-gram predictor, q_i^N, it is possible to set both upper and lower bounds to the N-gram entropy, F_N, of the original language. These bounds are as follows:

$$\sum_{i=1}^{27} i(q_i^N - q_{i+1}^N) \log i \leq F_N \leq -\sum_{i=1}^{27} q_i^N \log q_i^N. \tag{17}$$

The upper bound follows immediately from the fact that the maximum possible entropy in a language with letter frequencies q_i^N is $-\sum q_i^N \log q_i^N$. Thus the entropy per symbol of the reduced text is not greater than this. The N-gram entropy of the reduced text is equal to that for the original language, as may be seen by an inspection of the definition (1) of F_N. The sums involved will contain precisely the same terms although, perhaps, in a different order. This upper bound is clearly valid, whether or not the prediction is ideal.

[5] Hardy, Littlewood and Polya, "Inequalities," Cambridge University Press, 1934.

The lower bound is more difficult to establish. It is necessary to show that with any selection of N-gram probabilities $p(i_1, i_2, \ldots, i_N)$, we will have

$$\sum_{i=1}^{27} i(q_i^N - q_{i+1}^N) \log i \leq \sum_{i_1,\cdots,i_N} p(i_1 \cdots i_N) \log p_{i_1} \cdots {}_{i_{N-1}}(i_N) \tag{18}$$

The left-hand member of the inequality can be interpreted as follows: Imagine the q_i^N arranged as a sequence of lines of decreasing height (Fig. 3). The actual q_i^N can be considered as the sum of a set of rectangular distributions as shown. The left member of (18) is the entropy of this set of distributions. Thus, the ith rectangular distribution has a total probability of $i(q_i^N - q_{i+1}^N)$. The entropy of the distribution is $\log i$. The total entropy is then

$$\sum_{i=1}^{27} i(q_i^N - q_{i+1}^N) \log i.$$

The problem, then, is to show that any system of probabilities $p(i_1, \ldots, i_N)$, with best prediction frequencies q_i has an entropy F_N greater than or equal to that of this rectangular system, derived from the same set of q_i.

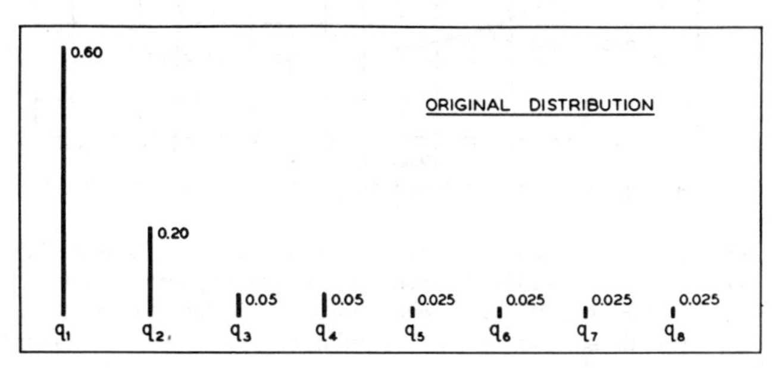

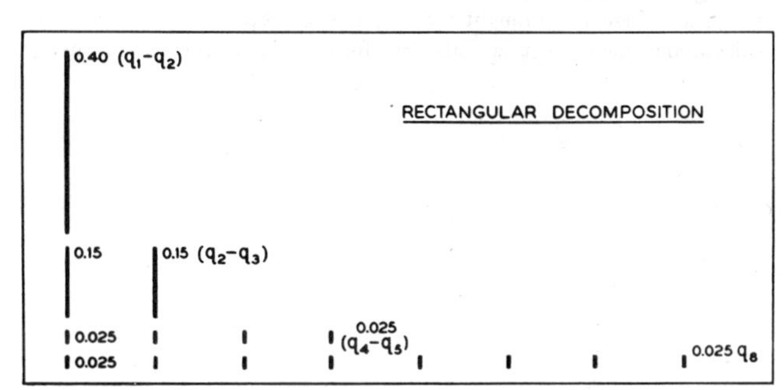

Fig. 3—Rectangular decomposition of a monotonic distribution.

The q_i as we have said are obtained from the $p(i_1, \ldots, i_N)$ by arranging each row of the table in decreasing order of magnitude and adding vertically. Thus the q_i are the sum of a set of monotonic decreasing distributions. Replace each of these distributions by its rectangular decomposition. Each one is replaced then (in general) by 27 rectangular distributions; the q_i are the sum of 27 x 27^N rectangular distributions, of from 1 to 27 elements, and all starting at the left column. The entropy for this set is less than or equal to that of the original set of distributions since a termwise addition of two or more distributions always increases entropy. This is actually an application of the general theorem that $H_y(x) \leq H(x)$ for any chance variables x and y. The equality holds only if the distributions being added are proportional. Now we may add the different components of the same width without changing the entropy (since in this case the distributions *are* proportional). The result is that we have arrived at the rectangular decomposition of the q_i, by a series of processes which decrease or leave constant the entropy, starting with the original N-gram probabilities. Consequently the entropy of the original system F_N is greater than or equal to that of the rectangular decomposition of the q_i. This proves the desired result.

It will be noted that the lower bound is definitely less than F_N unless each row of the table has a rectangular distribution. This requires that for each possible (N-1) gram there is a set of possible next letters each with equal probability, while all other next letters have zero probability.

It will now be shown that the upper and lower bounds for F_N given by (17) are monotonic decreasing functions of N. This is true of the upper bound since the q_i^{N+1} majorize the q_i^N and any equalizing flow in a set of probabilities increases the entropy. To prove that the lower bound is also monotonic decreasing we will show that the quantity

$$U = \sum_i i(q_i - q_{i+1}) \log i \tag{20}$$

is increased by an equalizing flow among the q_i. Suppose a flow occurs from q_i to q_{i+1}, the first decreased by Δq and the latter increased by the same amount. Then three terms in the sum change and the change in U is given by

$$\Delta U = [-(i-1)\log(i-1) + 2i \log i - (i+1)\log(i+1)]\Delta q \tag{21}$$

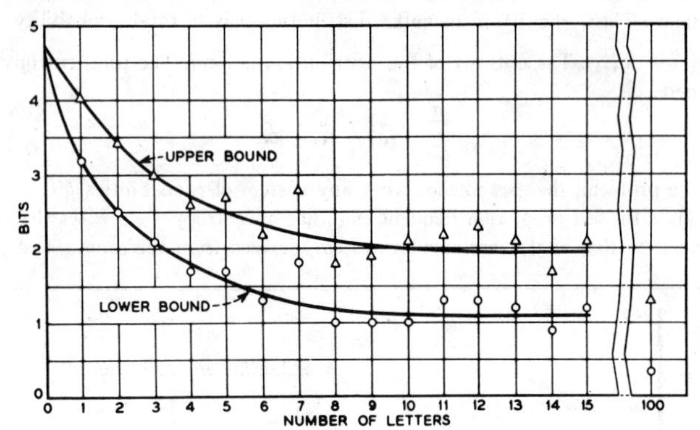

Fig. 4—Upper and lower experimental bounds for the entropy of 27-letter English.

The term in brackets has the form $-f(x-1) + 2f(x) - f(x+1)$ where $f(x) = x \log x$. Now $f(x)$ is a function which is concave upward for positive x, since $f''(x) = 1/x > 0$. The bracketed term is twice the difference between the ordinate of the curve at $x = i$ and the ordinate of the midpoint of the chord joining $i - 1$ and $i + 1$, and consequently is negative. Since Δq also is negative, the change in U brought about by the flow is positive. An even simpler calculation shows that this is also true for a flow from q_1 to q_2 or from q_{26} to q_{27} (where only two terms of the sum are affected). It follows that the lower bound based on the N-gram prediction frequencies q_i^N is greater than or equal to that calculated from the $N + 1$ gram frequencies q_i^{N+1}.

6. Experimental Bounds for English

Working from the data of Table I, the upper and lower bounds were calculated from relations (17). The data were first smoothed somewhat to overcome the worst sampling fluctuations. The low numbers in this table are the least reliable and these were averaged together in groups. Thus, in column 4, the 47, 18 and 14 were not changed but the remaining group totaling 21 was divided uniformly over the rows from 4 to 20. The upper and lower bounds given by (17) were then calculated for each column giving the following results:

Column	1	2	3	4	5	6	7	8	9	10	11	12	13	14	15	100
Upper.......	4.03	3.42	3.0	2.6	2.7	2.2	2.8	1.8	1.9	2.1	2.2	2.3	2.1	1.7	2.1	1.3
Lower.......	3.19	2.50	2.1	1.7	1.7	1.3	1.8	1.0	1.0	1.0	1.3	1.3	1.2	.9	1.2	.6

It is evident that there is still considerable sampling error in these figures due to identifying the observed sample frequencies with the prediction probabilities. It must also be remembered that the lower bound was proved only for the ideal predictor, while the frequencies used here are from human prediction. Some rough calculations, however, indicate that the discrepancy between the actual F_N and the lower bound with ideal prediction (due to the failure to have rectangular distributions of conditional probability) more than compensates for the failure of human subjects to predict in the ideal manner. Thus we feel reasonably confident of both bounds apart from sampling errors. The values given above are plotted against N in Fig. 4.

Acknowledgment

The writer is indebted to Mrs. Mary E. Shannon and to Dr. B. M. Oliver for help with the experimental work and for a number of suggestions and criticisms concerning the theoretical aspects of this paper.

A Method for the Construction of Minimum-Redundancy Codes*

DAVID A. HUFFMAN†, ASSOCIATE, IRE

Summary—**An optimum method of coding an ensemble of messages consisting of a finite number of members is developed. A minimum-redundancy code is one constructed in such a way that the average number of coding digits per message is minimized.**

INTRODUCTION

ONE IMPORTANT METHOD of transmitting messages is to transmit in their place sequences of symbols. If there are more messages which might be sent than there are kinds of symbols available, then some of the messages must use more than one symbol. If it is assumed that each symbol requires the same time for transmission, then the time for transmission (length) of a message is directly proportional to the number of symbols associated with it. In this paper, the symbol or sequence of symbols associated with a given message will be called the "message code." The entire number of messages which might be transmitted will be called the "message ensemble." The mutual agreement between the transmitter and the receiver about the meaning of the code for each message of the ensemble will be called the "ensemble code."

Probably the most familiar ensemble code was stated in the phrase "one if by land and two if by sea." In this case, the message ensemble consisted of the two individual messages "by land" and "by sea", and the message codes were "one" and "two."

In order to formalize the requirements of an ensemble code, the coding symbols will be represented by numbers. Thus, if there are D different types of symbols to be used in coding, they will be represented by the digits $0, 1, 2, \cdots, (D-1)$. For example, a ternary code will be constructed using the three digits 0, 1, and 2 as coding symbols.

The number of messages in the ensemble will be called N. Let $P(i)$ be the probability of the ith message. Then

$$\sum_{i=1}^{N} P(i) = 1. \tag{1}$$

The length of a message, $L(i)$, is the number of coding digits assigned to it. Therefore, the average message length is

$$L_{av} = \sum_{i=1}^{N} P(i)L(i). \tag{2}$$

The term "redundancy" has been defined by Shannon[1] as a property of codes. A "minimum-redundancy code" will be defined here as an ensemble code which, for a message ensemble consisting of a finite number of members, N, and for a given number of coding digits, D, yields the lowest possible average message length. In order to avoid the use of the lengthy term "minimum-redundancy," this term will be replaced here by "optimum." It will be understood then that, in this paper, "optimum code" means "minimum-redundancy code."

The following basic restrictions will be imposed on an ensemble code:

(a) No two messages will consist of identical arrangements of coding digits.

(b) The message codes will be constructed in such a way that no additional indication is necessary to specify where a message code begins and ends once the starting point of a sequence of messages is known.

Restriction (b) necessitates that no message be coded in such a way that its code appears, digit for digit, as the first part of any message code of greater length. Thus, 01, 102, 111, and 202 are valid message codes for an ensemble of four members. For instance, a sequence of these messages 1111022020101111102 can be broken up into the individual messages 111-102-202-01-01-111-102. All the receiver need know is the ensemble code. However, if the ensemble has individual message codes including 11, 111, 102, and 02, then when a message sequence starts with the digits 11, it is not immediately certain whether the message 11 has been received or whether it is only the first two digits of the message 111. Moreover, even if the sequence turns out to be 11102, it is still not certain whether 111-02 or 11-102 was transmitted. In this example, change of one of the two message codes 111 or 11 is indicated.

C. E. Shannon[1] and R. M. Fano[2] have developed ensemble coding procedures for the purpose of proving that the average number of binary digits required per message approaches from above the average amount of information per message. Their coding procedures are not optimum, but approach the optimum behavior when N approaches infinity. Some work has been done by Kraft[3] toward deriving a coding method which gives an average code length as close as possible to the ideal when the ensemble contains a finite number of members. However, up to the present time, no definite procedure has been suggested for the construction of such a code

* Decimal classification: R531.1. Original manuscript received by the Institute, December 6, 1951.

† Massachusetts Institute of Technology, Cambridge, Mass.

[1] C. E. Shannon, "A mathematical theory of communication," *Bell Sys. Tech. Jour.*, vol. 27, pp. 398–403; July, 1948.

[2] R. M. Fano, "The Transmission of Information," Technical Report No. 65, Research Laboratory of Electronics, M.I.T., Cambridge, Mass.; 1949.

[3] L. G. Kraft, "A Device for Quantizing, Grouping, and Coding Amplitude-modulated Pulses," Electrical Engineering Thesis, M.I.T., Cambridge, Mass.; 1949.

Reprinted from *Proc. IRE*, vol. 40, pp. 1098–1101, Sept. 1952.

to the knowledge of the author. It is the purpose of this paper to derive such a procedure.

Derived Coding Requirements

For an optimum code, the length of a given message code can never be less than the length of a more probable message code. If this requirement were not met, then a reduction in average message length could be obtained by interchanging the codes for the two messages in question in such a way that the shorter code becomes associated with the more probable message. Also, if there are several messages with the same probability, then it is possible that the codes for these messages may differ in length. However, the codes for these messages may be interchanged in any way without affecting the average code length for the message ensemble. Therefore, it may be assumed that the messages in the ensemble have been ordered in a fashion such that

$$P(1) \geqq P(2) \geqq \cdots \geqq P(N-1) \geqq P(N) \qquad (3)$$

and that, in addition, for an optimum code, the condition

$$L(1) \leqq L(2) \leqq \cdots \leqq L(N-1) \leqq L(N) \qquad (4)$$

holds. This requirement is assumed to be satisfied throughout the following discussion.

It might be imagined that an ensemble code could assign q more digits to the Nth message than to the $(N-1)$st message. However, the first $L(N-1)$ digits of the Nth message must not be used as the code for any other message. Thus the additional q digits would serve no useful purpose and would unnecessarily increase L_{av}. Therefore, for an optimum code it is necessary that $L(N)$ be equal to $L(N-1)$.

The kth prefix of a message code will be defined as the first k digits of that message code. Basic restriction (b) could then be restated as: No message shall be coded in such a way that its code is a prefix of any other message, or that any of its prefixes are used elsewhere as a message code.

Imagine an optimum code in which no two of the messages coded with length $L(N)$ have identical prefixes of order $L(N)-1$. Since an optimum code has been assumed, then none of these messages of length $L(N)$ can have codes or prefixes of any order which correspond to other codes. It would then be possible to drop the last digit of all of this group of messages and thereby reduce the value of L_{av}. Therefore, in an optimum code, it is necessary that at least two (and no more than D) of the codes with length $L(N)$ have identical prefixes of order $L(N)-1$.

One additional requirement can be made for an optimum code. Assume that there exists a combination of the D different types of coding digits which is less than $L(N)$ digits in length and which is not used as a message code or which is not a prefix of a message code. Then this combination of digits could be used to replace the code for the Nth message with a consequent reduction of L_{av}. Therefore, all possible sequences of $L(N)-1$ digits must be used either as message codes, or must have one of their prefixes used as message codes.

The derived restrictions for an optimum code are summarized in condensed form below and considered in addition to restrictions (a) and (b) given in the first part of this paper:

(c) $L(1) \leqq L(2) \leqq \cdots \leqq L(N-1) = L(N)$. (5)

(d) At least two and not more than D of the messages with code length $L(N)$ have codes which are alike except for their final digits.

(e) Each possible sequence of $L(N)-1$ digits must be used either as a message code or must have one of its prefixes used as a message code.

Optimum Binary Code

For ease of development of the optimum coding procedure, let us now restrict ourselves to the problem of binary coding. Later this procedure will be extended to the general case of D digits.

Restriction (c) makes it necessary that the two least probable messages have codes of equal length. Restriction (d) places the requirement that, for D equal to two, there be only two of the messages with coded length $L(N)$ which are identical except for their last digits. The final digits of these two codes will be one of the two binary digits, 0 and 1. It will be necessary to assign these two message codes to the Nth and the $(N-1)$st messages since at this point it is not known whether or not other codes of length $L(N)$ exist. Once this has been done, these two messages are equivalent to a single composite message. Its code (as yet undetermined) will be the common prefixes of order $L(N)-1$ of these two messages. Its probability will be the sum of the probabilities of the two messages from which it was created. The ensemble containing this composite message in the place of its two component messages will be called the first auxiliary message ensemble.

This newly created ensemble contains one less message than the original. Its members should be rearranged if necessary so that the messages are again ordered according to their probabilities. It may be considered exactly as the original ensemble was. The codes for each of the two least probable messages in this new ensemble are required to be identical except in their final digits; 0 and 1 are assigned as these digits, one for each of the two messages. Each new auxiliary ensemble contains one less message than the preceding ensemble. Each auxiliary ensemble represents the original ensemble with full use made of the accumulated necessary coding requirements.

The procedure is applied again and again until the number of members in the most recently formed auxiliary message ensemble is reduced to two. One of each of the binary digits is assigned to each of these two composite messages. These messages are then combined to form a single composite message with probability unity, and the coding is complete.

TABLE I

Optimum Binary Coding Procedure

Message Probabilities												
Original Message Ensemble	*Auxiliary Message Ensembles*											
	1	2	3	4	5	6	7	8	9	10	11	12
												→1.00
											→0.60	
										→0.40	0.40	
									→0.36	0.36		
								→0.24	0.24	0.24		
0.20	0.20	0.20	0.20	0.20	0.20	0.20	0.20	0.20	0.20			
							→0.20	0.20	0.20			
0.18	0.18	0.18	0.18	0.18	0.18	0.18	0.18	0.18				
						→0.18	0.18	0.18				
					→0.14	0.14	0.14					
0.10	0.10	0.10	0.10	0.10	0.10	0.10	0.10					
0.10	0.10	0.10	0.10	0.10	0.10	0.10						
0.10	0.10	0.10	0.10	0.10	0.10	0.10						
				→0.10	0.10							
		→0.08	0.08	0.08	0.08							
			→0.08	0.08								
0.06	0.06	0.06	0.06	0.06								
0.06	0.06	0.06	0.06									
0.04	0.04	0.04	0.04									
*0.04	0.04	0.04										
0.04	0.04	0.04										
0.04	0.04											
	→0.04											
0.03												
0.01												

Now let us examine Table I. The left-hand column contains the ordered message probabilities of the ensemble to be coded. N is equal to 13. Since each combination of two messages (indicated by a bracket) is accompanied by the assigning of a new digit to each, then the total number of digits which should be assigned to each original message is the same as the number of combinations indicated for that message. For example, the message marked *, or a composite of which it is a part, is combined with others five times, and therefore should be assigned a code length of five digits.

When there is no alternative in choosing the two least probable messages, then it is clear that the requirements, established as necessary, are also sufficient for deriving an optimum code. There may arise situations in which a choice may be made between two or more groupings of least likely messages. Such a case arises, for example, in the fourth auxiliary ensemble of Table I. Either of the messages of probability 0.08 could have been combined with that of probability 0.06. However, it is possible to rearrange codes in any manner among equally likely messages without affecting the average code length, and so a choice of either of the alternatives could have been made. Therefore, the procedure given is always sufficient to establish an optimum binary code.

The lengths of all the encoded messages derived from Table I are given in Table II.

Having now determined proper lengths of code for each message, the problem of specifying the actual digits remains. Many alternatives exist. Since the combining of messages into their composites is similar to the successive confluences of trickles, rivulets, brooks, and creeks into a final large river, the procedure thus far described might be considered analogous to the placing of signs by a water-borne insect at each of these junctions as he journeys downstream. It should be remembered that the code which we desire is that one which the insect must remember in order to work his way back upstream. Since the placing of the signs need not follow the same rule, such as "zero-right-returning," at each junction, it can be seen that there are at least 2^{12} different ways of assigning code digits for our example.

TABLE II

Results of Optimum Binary Coding Procedure

i	$P(i)$	$L(i)$	$P(i)L(i)$	*Code*
1	0.20	2	0.40	10
2	0.18	3	0.54	000
3	0.10	3	0.30	011
4	0.10	3	0.30	110
5	0.10	3	0.30	111
6	0.06	4	0.24	0101
7	0.06	5	0.30	00100
8	0.04	5	0.20	00101
9	0.04	5	0.20	01000
10	0.04	5	0.20	01001
11	0.04	5	0.20	00110
12	0.03	6	0.18	001110
13	0.01	6	0.06	001111
			$L_{av}=3.42$	

The code in Table II was obtained by using the digit 0 for the upper message and the digit 1 for the lower message of any bracket. It is important to note in Table I that coding restriction (e) is automatically met as long as two messages (and not one) are placed in each bracket.

Generalization of the Method

Optimum coding of an ensemble of messages using three or more types of digits is similar to the binary coding procedure. A table of auxiliary message ensembles similar to Table I will be used. Brackets indicating messages combined to form composite messages will be used in the same way as was done in Table I. However, in order to satisfy restriction (e), it will be required that all these brackets, with the possible exception of one combining the least probable messages of the original ensemble, always combine a number of messages equal to D.

It will be noted that the terminating auxiliary ensemble always has one unity probability message. Each preceding ensemble is increased in number by $D-1$ until the first auxiliary ensemble is reached. Therefore, if N_1 is the number of messages in the first auxiliary ensemble, then $(N_1-1)/(D-1)$ must be an integer. However $N_1=N-n_0+1$, where n_0 is the number of the least probable messages combined in a bracket in the original ensemble. Therefore, n_0 (which, of course, is at least two and no more than D) must be of such a value that $(N-n_0)/(D-1)$ is an integer.

In Table III an example is considered using an ensemble of eight messages which is to be coded with four digits; n_0 is found to be 2. The code listed in the table is obtained by assigning the four digits 0, 1, 2, and 3, in order, to each of the brackets.

TABLE III

Optimum Coding Procedure for $D=4$

Message Probabilities					
Original Message Ensemble	*Auxiliary Ensembles*			$L(i)$	*Code*
			→1.00		
		→0.40			
0.22	0.22	0.22		1	1
0.20	0.20	0.20		1	2
0.18	0.18	0.18		1	3
0.15	0.15			2	00
0.10	0.10			2	01
0.08	0.08			2	02
	→0.07				
0.05				3	030
0.02				3	031

Acknowledgments

The author is indebted to Dr. W. K. Linvill and Dr. R. M. Fano, both of the Massachusetts Institute of Technology, for their helpful criticism of this paper.

A Comparison of Signalling Alphabets

By E. N. GILBERT

(Manuscript received March 24, 1952)

Two channels are considered; a discrete channel which can transmit sequences of binary digits, and a continuous channel which can transmit band limited signals. The performance of a large number of simple signalling alphabets is computed and it is concluded that one cannot signal at rates near the channel capacity without using very complicated alphabets.

INTRODUCTION

C. E. Shannon's encoding theorems[1] associate with the channel of a communications system a capacity C. These theorems show that the output of a message source can be encoded for transmission over the channel in such a way that the rate at which errors are made at the receiving end of the system is arbitrarily small provided only that the message source produces information at a rate less than C bits per second. C is the largest rate with this property.

Although these theorems cover a wide class of channels there are two channels which can serve as models for most of the channels one meets in practice. These are:

1. The binary channel

This channel can transmit only sequences of binary digits 0 and 1 (which might represent hole and no hole in a punched tape; open-line and closed line; pulse and no pulse; etc.) at some definite rate, say one digit per second. There is a probability p (because of noise, or occasional equipment failure) that a transmitted 0 is received as 1 or that a transmitted 1 is received as 0. The noise is supposed to affect different digits independently. The cpacity of this channel is

$$C = 1 + p \log p + (1 - p) \log (1 - p) \tag{1}$$

bits per digit. The log appearing in Equation (1) is log to the base 2; this convention will be used throughout the rest of this paper.

2. The low-pass filter

The second channel is an ideal low-pass filter which attenuates completely all frequencies above a cutoff frequency W cycles per second and which passes frequencies below W without attenuation. The channel is supposed capable of handling only signals with average power P or less. Before the signal emerges from the channel, the channel adds to it a noise signal with average power N. The noise is supposed to be white Gaussian noise limited to the frequency band $|\nu| < W$. The capacity of this channel is

$$C = W \log\left(1 + \frac{P}{N}\right) \tag{2}$$

bits per second.

Shannon's theorems prove that encoding schemes exist for signalling at rates near C with arbitrarily small rates of errors without actually giving a constructive method for performing the encoding. It is of some interest to compare encoding systems which can easily be devised with these ideal systems. In Part I of this paper some schemes for signalling over the binary channel will be compared with ideal systems. In Part II the same will be done for the low-pass filter channel.

PART I

THE BINARY CHANNEL

1. Error-Correcting Alphabets

Imagine the message source to produce messages which are sequences of letters drawn from an alphabet containing K letters. We suppose that the letters are equally likely and that the letters which the source produces at different times are independent of one another. (If the source given is a finite state source which does not fit this simple description, it can be converted into one which approximately does by a preliminary encoding of the type described in Shannon's Theorem 9.) To transmit the message over the binary channel we construct a new alphabet of K letters in which the letters are different sequences of binary digits of some fixed length, say D digits. Then the new alphabet is used as an encoding of the old one suitable for transmission over the channel. For example, if the source produced sequences of letters from an alphabet of 3 letters, a typical encoding with $D = 5$ might convert the message into a binary sequence composed of repetitions of the three letters.

00000
11100
and 00111

If $K = 2^D$, the alphabet consists of all binary sequences of length D and hence if any of the digits of a letter is altered by noise the letter will be misinterpreted at the receiving end of the channel. If K is somewhat smaller than 2^D it is possible to choose the letters so that certain kinds of errors introduced by the noise do not cause a misinterpretation at the receiver. For example, in the three letter alphabet given above, if only one of the five digits is incorrect there will be just one letter (the correct one) which agrees with the received sequence in all but one place. More generally if the letters of the alphabet are selected so that each letter differs from every other in at least $2k + 1$ out of the D places, then when k or fewer errors are made the correct interpretation of the received sequence will be the (unique) letter of the alphabet which differs from the received sequence in no more than k places. An alphabet with this property will be called a *k error correcting alphabet*[2].

Error correcting alphabets have the advantage over the random alphabets which Shannon used to prove his encoding theorems that they are uniformly reliable whereas Shannon's alphabets are reliable only in an average sense. That is, Shannon proved that the probability that a letter *chosen at random* shall be received incorrectly can be made arbitrarily small. However, a certain small fraction of the letters of Shannon's alphabets are allowed a much higher probability of error than the average. This kind of alphabet would be undesirable in applications such as the signalling of telephone numbers; one would not want to give a few subscribers telephone numbers which are received incorrectly more often than most of the others. It is only conjectured that the rate C can be approached using error correcting alphabets. The alphabets which are to be considered here are all error correcting alphabets.

A geometric picture of an alphabet is obtained by regarding the D digits of a sequence as coordinates of a point in Euclidean D dimensional space. The possible received sequences are represented by vertices of the unit cube. A k error correcting alphabet is represented by a set of vertices, such that each pair of vertices is separated by a distance at least $\sqrt{2k+1}$

Let $K_0(D, k)$ be the largest number of letters which a D dimensional k error correcting alphabet can contain. Except when $k = 1$, there is no general method for constructing an alphabet with $K_0(D, k)$ letters, nor is $K_0(D, k)$ known as a function of D and k. Crude upper and lower bounds for $K_0(D, k)$ are given by the following theorem.

Theorem 1. The largest number of letters $K_0(D, k)$ satisfies

$$\frac{2^D}{N(D, 2k)} \le K_0(D, k) \le \frac{2^D}{N(D, k)} \tag{3}$$

where

$$N(D, k) = \sum_{r=0}^{k} C_{D,r}$$

is the number of sequences of D digits which differ from a given sequence in 0, 1, $\cdots$, or k places.

[1] C. E. Shannon, "A Mathematical Theory of Communication," *Bell System Tech. J.*, **27**, p. 379–423 and pp. 623–656, 1948, theorems 9, 11, and 16 in particular.

[2] R. W. Hamming, "Error Detecting and Error Correcting Codes," *Bell System Tech. J.*, **29**, pp. 147–160, 1950.

Reprinted with permission from *Bell Syst. Tech. J.*, vol. 31, pp. 504–522, May 1952. Copyright © 1952, The American Telephone and Telegraph Company.

Proof

The upper bound is due to R. W. Hamming and is proved by noting that for each letter S of a k error correcting alphabet there are $N(D, k)$ possible received sequences which will be interpreted as meaning S. Hence $N(D, k)\ K_0(D, k) \leq 2^D$, the total number of sequences.

The lower bound is proved by a random construction method. Pick any sequence S_1 for the first letter. There remain $2^D - N(D, 2k)$ sequences which differ from S_1 in $2k + 1$ or more places. Pick any one of these S_2 for the second letter. There remain at least $2^D - 2N(D, 2k)$ sequences which differ from both S_1 and S_2 in $2k + 1$ or more places. As the process is continued, there remain at least $2^D - rN(D, 2k)$ sequences, which differ in $2k + 1$ or more places from $S_1, \cdots, S_r$, from which S_{r+1} is chosen. If there are no choices available after choosing S_K, then $2^D - KN(D, 2k) \leq 0$ so the alphabet $(S_1, \cdots, S_K)$ has at least as many letters as the lower bound (3).

For all the simple cases (D and k not very large) investigated so far the upper bound is a better estimate of $K_0(D, k)$ than the lower bound. The upper and lower bounds differ greatly, as may be seen from a quick inspection of Table I. For example, in the case of a ten dimensional two error correcting alphabet, the bounds are 2.7 and 18.3.

2. *Efficiency Graph*

The first step in constructing an efficiency graph for comparing alphabets is to decide on what constitutes reliable transmission. The criterion used here is that on the average no more than one letter in 10^4 shall be misinterpreted.

TABLE I

TABLE OF $2^D/N(D, k)$

k = ...	1	2	3	4	5	6	7
$D =$ 3	2						
4	3.2						
5	5.3	2					
6	9.1	2.9					
7	16	4.4	2.9				
8	28.4	6.9	2.8				
9	51.2	11.1	3.9	2			
10	93.1	18.3	5.8	2.7			
11	170.7	30.6	8.8	3.6	2		
12	315.8	51.8	13.7	5.2	2.6		
13	585.2	89.0	21.6	7.5	3.4	2	
14	1092.3	154.4	34.9	11.1	4.7	2.5	
15	2048	270.8	56.8	16.8	6.6	3.3	2

Missing entries are numbers between 1 and 2.

This sort of criterion might be appropriate for a channel transmitting English text. For other messages it is not always appropriate. For example, if the messages are telephone numbers, one would naturally require that the probability of mistaking a telephone number be small, say less than 10^{-4}. If the telephone numbers are L decimal digits long, and if the alphabet has K different letters in it (so that it takes about $L \log 10/\log K$ letters to make up a telephone number) the probability of making a mistake in a single letter should be required to be less than about

$$\frac{10^{-4} \log K}{L \log 10}$$

which gives alphabets with large K an advantage over alphabets with small K.

Since the probability that exactly r binary digits out of D shall be received incorrectly is $C_{D,r}p^r(1 - p)^{D-r}$, we achieve the required reliability with a D-dimensional k-error correcting alphabet provided p satisfies

$$\sum_{r=k+1}^{D} C_{D,r}p^r(1 - p)^{D-r} \leq 10^{-4}. \tag{4}$$

The value of p which makes the inequality hold with the equals sign determines the noisiest channel over which the alphabet can be used safely.

Let K be the number of different letters in the alphabet. Then the rate in bits per digit at which information is being recieved is

$$R = \frac{\log K}{D}. \tag{5}$$

In Equation (5) we have neglected a term which takes account of the information lost due to channel noise. This is legitimate because all but 10^{-4} of the letters are received correctly.

The worst tolerable probability p of (4) and the rate R of Equation (5) determine the noise combating ability of an alphabet. To compare different alphabets one may represent them as points on an efficiency graph of R versus p. Fig. 1 is an efficiency graph on which the values (p, R) for a number of simple error correcting alphabets have been plotted. Each point on the graph is labelled with the two numbers k, D in that order. The alphabets represented were not found by any systematic process and are not all proved to be best possible (i.e., to have the largest K) for the stated values of k and D. Fortunately, R depends on K only logarithmically so that it is not likely the points representing the best possible alphabets lie far away from the plotted points.

The solid line represents the curve

$$R = C = 1 + p \log p + (1 - p) \log (1 - p).$$

According to Shannon's theorems, all alphabets are represented by points lying below this line.

The efficiency graph only partially orders the alphabets according to

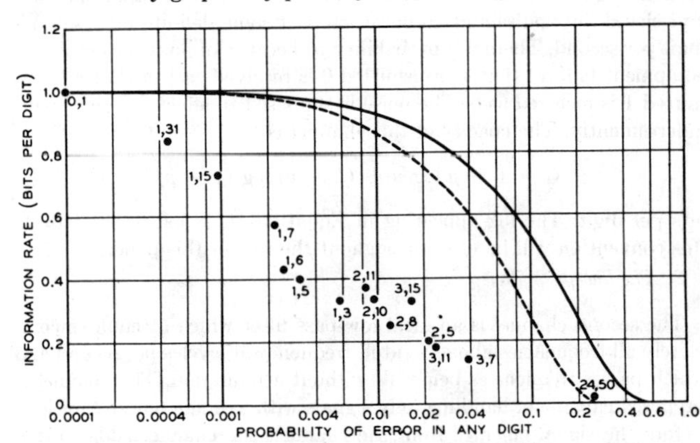

Fig. 1—Probability of error in a letter is 10^{-4}.

their invulnerability to noise. For example, it is clear that the alphabet 3, 15 is better than 2, 8. However, without further information about the channel, such as knowledge of p, there is no reasonable way of choosing between 3, 15 and 3, 7.

3. *Large Alphabets*

We have been unable to prove that there are error correcting alphabets which signal at rates arbitrarily close to C while maintaining an arbitrarily small probability of error for any letter. A result in this direction is the following theorem.

Theorem 2. Let any positive ϵ and δ be given. Given a channel with $p < \frac{1}{4}$ there exists an error correcting alphabet which can signal over the channel at a rate exceeding $R_0 - \epsilon$ where

$$R_0 = 1 + 2p \log 2p + (1 - 2p) \log (1 - 2p)$$

bits per digit and for which the probability of error in any letter is less than δ.

Proof

The probability of error in any letter is the sum on the left of (4). This is a sum of terms from a binomial distribution which, as is well known, tends to a Gaussian distribution with mean Dp and variance $Dp(1 - p)$ for large D. Hence there is a constant $A(\delta)$ such that all k error correcting alphabets with sufficiently large D have a letter error probability less than δ provided

$$k \geq Dp + A(\delta)\ (Dp(1 - p))^{1/2} \tag{6}$$

Let $k(D)$ be the smallest integer which satisfies (6) and consider an alphabet which corrects $k(D)$ errors and contains $K_0(D, k(D))$ letters.

By Equation (5) and the lower bound of Theorem 1, this alphabet signals at a rate $R(D)$ satisfying

$$1 - \frac{1}{D} \log N(D, 2k(D)) \le R(D).$$

Since $p < \frac{1}{4}$, $2k(D) < D/2$ for large D and hence

$$N(D, 2k(D)) < (2k(D) + 1)C_{D,2k(D)}.$$

Then an application of Stirling's approximation for factorials shows that as $D \to \infty$

$$1 - \frac{1}{D} \log N(D, 2k(D)) \to R_0.$$

Hence by taking D large enough one obtains an alphabet with rate exceeding $R_0 - \epsilon$ and letter error probability less than δ.

The rate R_0 appears on the efficiency graph as a dotted line.

It has not been shown that no error-correcting alphabet has a rate exceeding R_0. In fact, one alphabet which exceeds R_0 in rate is easy to construct. If the noise probability p is greater than $\frac{1}{4}$, then $R_0 = 0$. The alphabet with just two letters

0 0 0 0 . . . 0

and

1 1 1 1 . . . 1

will certainly transmit information at a (small) positive rate, and with a 10^{-4} probability of errors if D is large enough, as long as $p < \frac{1}{2}$.

Using a more refined lower bound for $K_0(D, k)$ it might be shown that there are error-correcting alphabets which signal with rates near C. If one repeats the calculation that led to R_0 using the upper bound (3) (which seems to be a better estimate of the true $K_0(D, k)$) instead of the lower bound (3), one is led to the rate C instead of R_0.

The condition (4) is more conservative than necessary. The structure of the alphabet may be such that a particular sequence of more than k errors may occur without causing any error in the final letter. This is illustrated by the following simple example due to Shannon: the alphabet with just two letters

0 0 0 0 0 0
1 1 1 0 0 0

corrects any single error but also corrects certain more serious errors such as receiving 0 0 1 1 1 1 for 0 0 0 0 0 0. An alphabet designed for practical use would make efficient enough use of the available sequences so that any sequence of much more than k errors causes an error in the final letter; the random alphabets constructed above probably do not. If this kind of error were properly accounted for, the rate R_0 could be improved, perhaps to C.

4. *Other Discrete Channels*

If instead of transmitting just 0's and 1's the channel can carry more digits

$$0, 1, 2, \cdots, n$$

a similar theory can be worked out. The simplest kind of noise in this channel changes a digit into any one of the n other possible numbers with probability p/n. Then the capacity of the channel is

$$C = \log (n + 1) + p \log \frac{p}{n} + (1 - p) \log (1 - p).$$

Error-correcting alphabets for this channel can also be constructed and the criterion (4) for good transmission remains unchanged. The proof of theorem 1 can be repeated with little change using

$$N(D, k) = \sum_{r=0}^{k} C_{D,r} n^r$$

as the number of sequences which can be reached after k or fewer errors [the terms 2^D in (1) and (3) are replaced by $(n + 1)^D$]. Once more, using the lower bound, one finds an expression for R_0 which is the same as the one for C but with p replaced by $2p$.

Part II

THE LOW PASS FILTER

1. *Encoding and Detection*

If $f(t)$ is a signal emerging from a low pass filter (so that its spectrum is confined to the frequency band $|\nu| < W$ cycles per second) then $f(t)$ has a special analytic form given by the sampling theorem[3]

$$f(t) = \sum_{m=-\infty}^{\infty} f\left(\frac{m}{2W}\right) \frac{\sin \pi (2Wt - m)}{\pi(2Wt - m)} \tag{7}$$

Thus the signal is completely determined by the sequence of sample values $f(m/2W)$. The average power of the signal $f(t)$ is measured by

$$P = \lim_{T\to\infty} \frac{1}{2T} \int_{-T}^{T} f^2(t)\, dt$$

which can be expressed in terms of the sample values as follows

$$P = \lim_{M\to\infty} \frac{1}{2M} \sum_{m=-M}^{M} f^2\left(\frac{m}{2W}\right). \tag{8}$$

As in Part I, consider a message source producing a sequence of letters from an alphabet of K equally likely letters. To transmit this information over the low pass filter we must encode the sequence into a function $f(t)$ of the form (7), or in other words into a sequence of sample values $f(m/2W)$. To do this, we construct a new alphabet containing K letters which are different sequences of real numbers of some fixed length, say D places. When we let the letters of the new alphabet correspond to letters of the old one the message is translated into a sequence of real numbers which we use for the sequence $f(m/2W)$.

If the K letters of the sequence alphabet are

$$\begin{aligned} S_1&: a_{11}, \cdots, a_{1D} \\ S_2&: a_{21}, \cdots, a_{2D} \\ &\;\;\vdots \\ S_K&: a_{K1}, \cdots, a_{KD}, \end{aligned}$$

the expression (8) for the average power of the function $f(t)$ becomes

$$P = \frac{1}{DK} (d_1^2 + d_2^2 + \cdots + d_K^2) \tag{9}$$

where

$$d_i^2 = \sum_{j=1}^{D} a_{ij}^2.$$

If the D numbers in the sequence S_i are regarded as coordinates of a point in Euclidean D dimensional space, d_i^2 represents the square of the distance from the point representing S_i to the origin.

When $f(t)$ is transmitted, the received signal will be $f(t) + n(t)$ where $n(t)$ is some (unknown) white Gaussian noise signal. The noise signals $n(t)$ are characterized by the fact that their sample values $n(m/2W)$ are independently distributed according to Gaussian laws. That is,

$$\text{Prob}\left(n\left(\frac{m}{2W}\right) \le X\right) = \frac{1}{\sqrt{2\pi}\sigma} \int_{-\infty}^{X} e^{-y^2/2\sigma^2}\, dy. \tag{10}$$

The variance σ^2 of the distribution of noise samples is, by an application of (8), the power of this ensemble of noise signals.

At the receiving end of the channel, there is a detector which observes each block of D sample values $f(m/2W) + n(m/2W)$ and tries to decide which one of the K letters $S_1, \cdots, S_K$ was sent. In terms of the geometric picture, the detector divides all of D dimensional space into K non-overlapping regions $U_1, \cdots, U_K$ with the property that, if the D received sample values are represented by a point in U_i, the detector

[3] C. E. Shannon, "*Communication in the Presence of Noise*," *Proc. I. R. E.*, **37**, pp. 10–21, Jan. 1949.

decides that S_i was sent. By Equation (10), the probability that the detector picks the wrong letter when S_i is sent is

$$p_i = \frac{1}{(2\pi)^{D/2}\sigma^D} \int \int_{\bar{U}_i} \cdots \int e^{-r_i^2/2\sigma^2}\, dy_1 \cdots dy_D \tag{11}$$

where $\bar{U}_i$ is the set of all points not in U_i and r_i is the distance from $(y_1, \cdots, y_D)$ to the point representing S_i.

For any given alphabet the best possible detector (in the sense that it minimizes the average probability of making an error in guesssing a letter) is called a *maximum likelihood detector*. The region U_i for a maximum likelihood detector consists of all points $(y_1, \cdots, y_D)$ which are closer to the point S_i than to any other letter point $S_j (r_i < r_j$ for all $j \neq i)$. To prove that this choice of U_i is best possible consider any other detector such that U_i contains a set V of points in which $r_i > r_j$. A direct calculation shows that the detector obtained by removing V from U_i and making V part of U_j has a smaller probability of error per letter. The set of points equidistant from two given points is a hyperplane. The region U_i of a maximum likelihood detector is a convex region bounded by segments of the hyperplanes

$$r_i = r_1, \qquad r_i = r_2, \cdots.$$

To compare signalling alphabets under the most favorable possible circumstances, we always compute letter error probabilities assuming that the detector is a maximum likelihood detector.

2. *Computation of error probabilities*

Exact evaluation of the letter error probability integral (11) is impossible except in a few special cases. Fortunately we are only interested in (11) when σ is small enough in comparison to the size of U_i to make the integral small. Then fairly accurate approximate formulas can be derived.

Theorem 3. Let R_{ij} be the distance between letter points S_i and S_j. Then

$$1 - \prod_{j \neq i} (1 - Q_{ij}) \leq p_i \leq \sum_{j \neq i} Q_{ij} \tag{12}$$

where

$$Q_{ij} = \frac{1}{\sqrt{2\pi}} \int_{R_{ij}/2\sigma}^{\infty} e^{-x^2/2}\, dx.$$

The proof of Theorem 3 follows from the fact that Q_{ij} is the probability that, when S_i is transmitted, the received sequence will be closer to S_j than to S_i.

In the cases to be computed Q_{ij} is a rapidly decreasing function of R_{ij} and the only terms worth keeping in (12) are the ones for which R_{ij} is the smallest of the numbers $R_{i1}, \cdots, R_{iK}$. Moreover since the Q_{ij} are all small enough so that the upper and lower bounds differ only by a few per cent, the upper bound is a good approximation to p_i. Then a simple approximate formula for the average letter error probability $p = (p_1 + \cdots + p_K)/K$ is

$$p = \frac{N}{\sqrt{2\pi}} \int_{r_0/\sigma}^{\infty} e^{-x^2/2}\, dx \tag{13}$$

where $2r_0$ is the smallest of the $K(K-1)/2$ distances R_{ij} and N is the average over all letters in the alphabet of the number of letter points which are a distance $2r_0$ away.

3. *Efficiency graph*

The efficiency graph to be described was constructed originally to compare alphabets for signalling telephone numbers of length equal to ten decimal digits. It was desired that on the average only one telephone number in 10^4 should be received incorrectly. As described in Part I section 2, if the telephone numbers are encoded into sequences of letters from an alphabet of K letters, we must require that the average probability of error in any letter be

$$p = 10^{-5} \log_{10} K \tag{14}$$

or smaller.

Given an alphabet, one can compute with the help of (13) and (14) and a table of the error integral the largest value of the noise power σ^2 which can be tolerated. The average power of the transmitted signal is P given by Equation (9). Hence we can compute the smallest signal to noise ratio

$$Y = P/\sigma^2 \tag{15}$$

which will be satisfactory.

A letter containing $\log K$ bits of information is transmitted during an interval of $D/2W$ seconds. Hence the rate at which information is received is

$$R = \frac{2W \log K}{D} \tag{16}$$

bits per second. Again Equation (16) ignores a term representing information lost due to channel noise which is negligible because the error probability is low.

The efficiency graph, Fig. 2, is a chart on which the signal to noise ratio Y in db [computed from Equation (15)] is plotted against the signalling rate per unit bandwidth $R/W = (2 \log K)/D$ for different alphabets. An alphabet is considered poor if its point on the efficiency graph lies far above the ideal curve $R/W = C/W = \log(1 + Y)$.

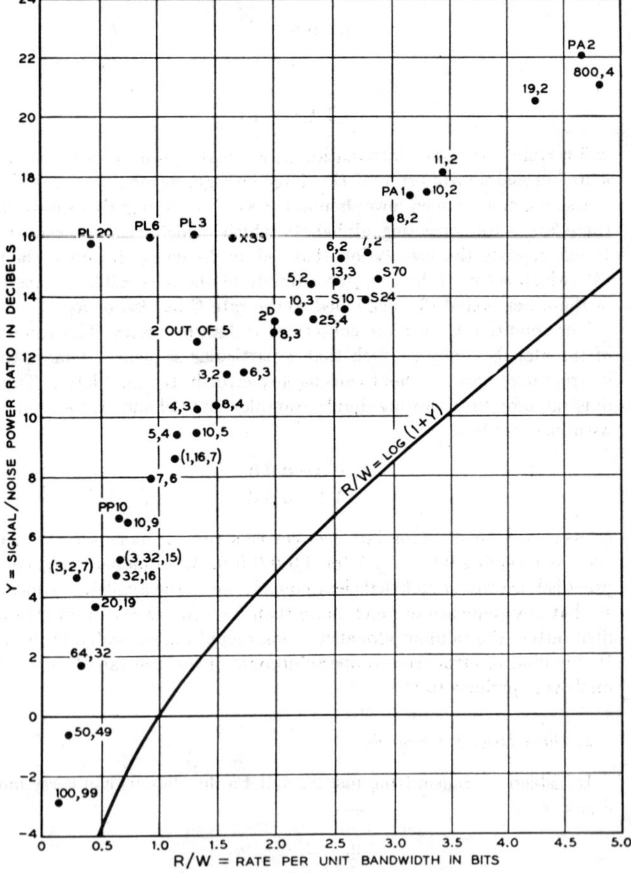

Fig. 2—Probability is 10^{-4} that an error is made in a 10 digit decimal number.

4. *The alphabets*

The alphabets which appear on the efficiency graph are the following:

excess three (*XS3*): the ten sequences of 4 binary digits which represent 3, 4, $\cdots$, and 12 in binary notation;

two out of five: the ten sequences of five binary digits which contain exactly two ones;

pulse position (*PP10*): the ten sequences of ten binary digits which contain exactly one one;

2^D *binary*: all of 2^D sequences of D binary digits.

pulse amplitude (*PAn*): the $2n + 1$ sequences of length 1 consisting

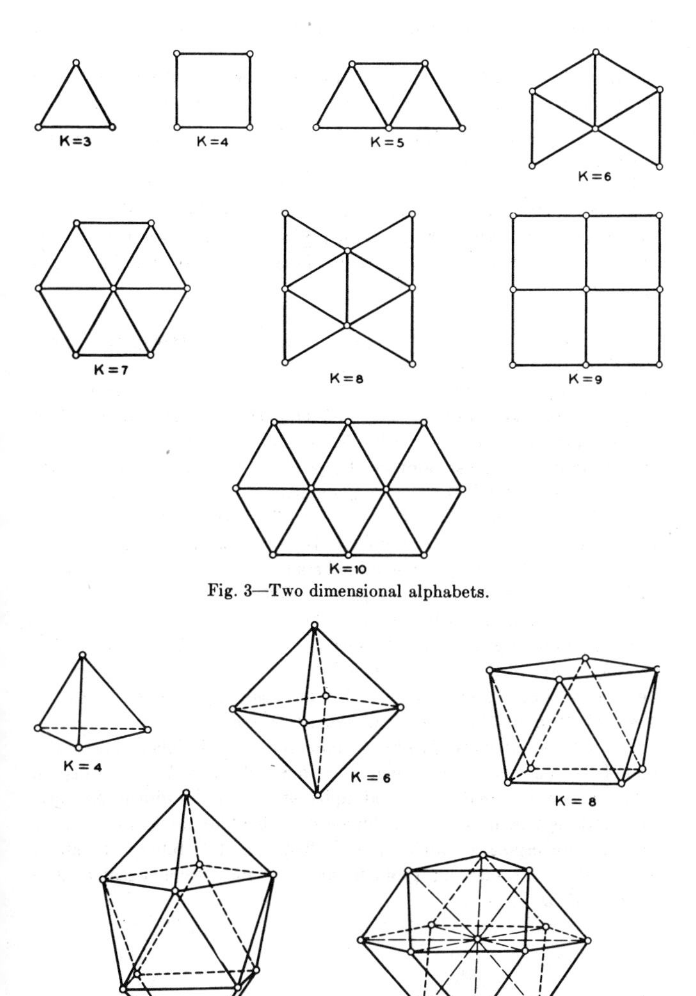

Fig. 3—Two dimensional alphabets.

Fig. 4—Three dimensional alphabets.

where r_1 is the rms distance from the origin to the points of a configuration A, R_0 is the distance from the origin to the centroid of A, and r_2 is the rms distance from the points of A to the centroid of A.

In plotting points on the efficiency graph the notation K, D is used for the best K-letter D-dimensional alphabet which has been found. The arrangement of points for various K, 2 and K, 3 alphabets is given in Figs. 3 and 4. In these figures two points are joined by a straight line if the distance between them is 1 (which is the value we have adopted for the minimum allowed separation $2r_0$). Although not shown, the origin is always at the centroid of the figure. To aid interpretation of these diagrams we have included Fig. 5 which demonstrates how all the signals of a typical alphabet can be generated. The functions of time shown in

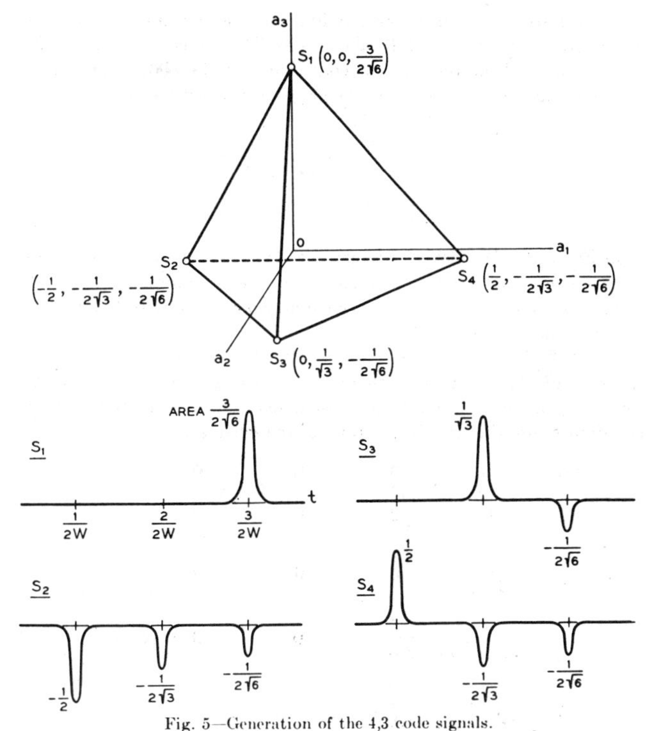

Fig. 5—Generation of the 4,3 code signals.

Fig. 5 are not the code signals themselves but impulse functions which are to be passed through a low pass filter with cutoff at W c.p.s. to form the code signals.

The best possible higher dimensional alphabets can be described more easily verbally than pictorially. In four dimensions we have found four alphabets.

The *25,4* alphabet consists of the origin and all 24 points in 4 dimensional space having two coordinates equal to zero and the remaining two equal to $1/\sqrt{2}$ or $-1/\sqrt{2}$. Each of the 24 points lies a unit distance away from the origin and its 10 other nearest neighbors; they are, in fact, the vertices of a regular solid. This alphabet has an advantage beyond its high efficiency. The code signals are composed entirely of positive and negative pulses of fixed energy and so should be easier to generate than most of the other codes which appear in this paper.

The *800, 4* alphabet is constructed in the following way: Consider a lattice of points throughout the entire 4-dimensional space formed by taking all the linear combinations with integer coefficients of a basic set of four vectors. That is, the lattice points are of the form $C_1v_1 + C_2v_2 + C_3v_3 + C_4v_4$ where $C_1, \cdots, C_4$ are integers and the v_i are the four given vectors. In connection with our problem it is of interest to know what lattice, (i.e. what choice of v_1, v_2, v_3, v_4) has all lattice points separated at least unit distance from one another and at the same time packs as many points as possible into the space per unit volume. When a solution to this "packing problem" is known, it is clear that a good alphabet can be obtained just by using all the lattice points which are contained inside a hypersphere about the origin as the letter points. Many of the two dimensional alphabets illustrated in the sketches are related in this way to the corresponding two dimensional packing prob-

of $-n, -n+1, \cdots, n$. This alphabet gives rise to a sort of quantized amplitude modulation.

pulse length (PLn): the $n + 1$ sequences of n binary digits of the form $11 \cdots 10 \cdots 0$, i.e., a run of ones followed by a run of zeros.

Minimizing alphabets (K, D): The above alphabets are taken from actual practice. They are convenient because, aside from PAn, they require a signal generator with only two amplitude levels. If we ignore ease of generating the signals as a factor, a great many geometric arrangements of points suggest themselves as possible good alphabets. The principle by which one arrives at good alphabets may be described as follows. When a D and K have been determined which give the desired information rate R [by Equation (16)] try to arrange the K letter points in D dimensional space in such a way that the distances between pairs of points are all greater than some fixed distance and that the average of the K squared distances to the origin is minimized. By Equations (9) and (13) it is seen that, apart from the small influence of the factor N, this process must minimize the signal to noise ratio Y required.

Ordinarily it is difficult to prove that a configuration is a minimizing one. Even to recognize a configuration which leads to a relative minimum (*i.e.* a minimum over all nearby configurations) is not always easy. The eight vertices of a cube, for example, do not give a relative minimum. Consequently, most of the alphabets to be described are only conjectured to be "best possible." Each of them satisfies one necessary requirement of minimizing alphabets that the centroid of the point configuration (assuming a unit mass at each letter point) lies at the origin. That this condition is necessary follows from the easily derived identity

$$r_2^2 = r_1^2 - R_0^2$$

lem (which is solved by letting v_1 and v_2 be a pair of unit vectors 60° apart). A solution to the four dimensional packing problem is afforded by

$$v_1 = \frac{1}{\sqrt{2}}, \frac{1}{\sqrt{2}}, 0, 0$$

$$v_2 = \frac{1}{\sqrt{2}}, 0, \frac{1}{\sqrt{2}}, 0$$

$$v_3 = \frac{1}{\sqrt{2}}, 0, 0, \frac{1}{\sqrt{2}}$$

$$v_4 = 0, \frac{1}{\sqrt{2}}, \frac{1}{\sqrt{2}}, 0.$$

This lattice contains two points per unit volume (twice as dense as the cubic lattice in which $v_1, \cdots, v_4$ are orthogonal to one another) and each point has 18 nearest neighbors. A hypersphere of radius 3 about the origin has a volume $(\pi^2/2)3^4$, about 400. Thus it contains about 800 lattice points. Take these as the code points of the 800, 4 code. Their average squared distances from the origin can be estimated as

$$\frac{\int_0^3 r^5\,dr}{\int_0^3 r^3\,dr} = \frac{2}{3}(3)^2 = 6.$$

N in Equation (13) may be estimated at 18; this is conservative because some lattice points outside the sphere are being counted.

The two remaining four dimensional alphabets belong to two families of D-dimensional alphabets.

The 4, 3; 5, 4; $\cdots$; $D + 1$, $D \cdots$ alphabets are the vertices of the simplest regular solid in D-dimensional space. For example, 4, 3 is a tetrahedron. Such a solid can be constructed from $D + 1$ vertices whose coordinates are the first $D + 1$ rows of the scheme

$$\begin{array}{cccccc}
0 & 0 & 0 & 0 & 0 & \cdots \\
1 & 0 & 0 & 0 & 0 & \cdots \\
\frac{1}{2} & \frac{3}{2\sqrt{3}} & 0 & 0 & 0 & \cdots \\
\frac{1}{2} & \frac{1}{2\sqrt{3}} & \frac{4}{2\sqrt{6}} & 0 & 0 & \cdots \\
\frac{1}{2} & \frac{1}{2\sqrt{3}} & \frac{1}{2\sqrt{6}} & \frac{5}{2\sqrt{10}} & 0 & \cdots \\
\frac{1}{2} & \frac{1}{2\sqrt{3}} & \frac{1}{2\sqrt{6}} & \frac{1}{2\sqrt{10}} & \frac{6}{2\sqrt{15}} & \cdots \\
\cdot & \cdot & \cdot & \cdot & \cdot & \cdots \\
\cdot & \cdot & \cdot & \cdot & \cdot & \cdots \\
\cdot & \cdot & \cdot & \cdot & \cdot & \cdots
\end{array}$$

The vertices all lie a distance $\sqrt{D/2(D+1)}$ from the centroid of the figure.

6, 3; 8, 4; $\cdots$; $2D$, D, $\cdots$ are obtained by placing a point wherever any positive or negative coordinate axis intersects the sphere of radius $1/\sqrt{2}$ about the origin. Thus it follows that 6, 3 consists of the vertices of an octohedron.

Error correcting alphabets $((k, K, D))$: The error correcting alphabets discussed in Part I can be converted into good alphabets for this channel by replacing all digits which equalled 0 by -1. Three error correcting alphabets appear on the chart; each is labelled by three numbers signifying (k, K, D).

Slepian alphabets (SD): Using group theoretic methods, D. Slepian has attempted to construct families of alphabets which signal at rates approaching C. Although this goal has not yet been reached, families of alphabets depending on the parameter D have been found which approach the ideal curve to within 6.2 db and then get worse as $D \to \infty$. In the simplest of these families of alphabets, $D = 2m$ is even and the letters consist of all the $2^m C_{2m,m}$ sequences containing m zeros, the remaining places being filled by ± 1. The best alphabet in this family is the one with $D = 24$. It lies 6.23 db away from the ideal curve and contains 1.1×10^{10} letters. The alphabets of this family for $D = 10$, 24, and 70 appear on the efficiency graph labelled $S10$, $S24$, and $S70$.

The conclusion to which one is forced as a result of this investigation is that one cannot signal over a channel with signal to noise level much less than 7 db above the ideal level of Equation (2) without using an unbelievably complicated alphabet. No ten digit alphabet tolerates less than 7.7 db more than the ideal signal to noise ratio.

It would be interesting to know more about good higher dimensional alphabets. They are very much more difficult to obtain. The regular solids, which provided some good alphabets in 3 and 4 dimensions, provide nothing new in 5 or more dimensions; there are only three of them and they correspond to our $D + 1$, D; $2D$, D, and 2^D binary alphabets. Worse still, the packing problem also becomes unmanageable after dimension 5.

ACKNOWLEDGMENT

The author wishes to thank R. W. Hamming, L. A. MacColl, B. McMillan, C. E. Shannon, and D. Slepian for many helpful suggestions during the investigation summarized by this paper.

THE BASIC THEOREMS OF INFORMATION THEORY

By Brockway McMillan

Bell Telephone Laboratories

Summary. This paper describes briefly the current mathematical models upon which communication theory is based, and presents in some detail an exposition and partial critique of C. E. Shannon's treatment of one such model. It then presents a general limit theorem in the theory of discrete stochastic processes, suggested by a result of Shannon's.

1. General models of the communication problem.

1.0. *Introduction.* For the purposes of this exposition, information theory is the body of statistical mathematics which has developed, largely over the last decade, out of efforts to understand and improve the communications art. We shall not attempt a history of this development, nor any detailed justification for its existence, since either of these efforts would take us further into the technics of communication than is desirable in a short essay.

It suffices to say here that this discipline has come specifically to the attention of mathematicians and mathematical statisticians almost exclusively through the book [1] of N. Wiener and the paper [2] of C. E. Shannon.

In the remainder of this section we shall describe very broadly the kind of problem to which these two works are addressed.

1.1. *A simple model.* The simplest mathematical model of the communication problem is like the problem of parameter estimation. A parameter θ, usually ranging over a fairly abstract or at least multi-dimensional domain, represents the transmitted message. A variable, y, also fairly abstract in general, represents the received message. In realistic situations the received message is seldom a mathematically exact copy, or even an exactly predictable mutilation, of the original transmitted message. Hence, y is represented as a random variable whose distribution depends upon the parameter θ. The communication problem then is: given a sample of one value of y, to estimate the unknown θ.

There are two reasons why this model may not seem at first look to be a good one for the communication problem. One is merely that our most usual media of communication, direct acoustic transmission of voice and the written or printed word, are ones in which essentially exact transmission is possible and we are not aware of the underlying statistical nature of the problem. This is clearly a matter of degree, however, and almost anyone can find in his own experience instances in which the statistical aspect of the problem was evident.

Another apparent failing of this model is in fact real, and has led to refinement of the model. There are communication problems, mostly in technical fields, where it is realistic to assume that the recipient of y has no a priori knowledge

Received 8/9/51.

Reprinted with permission from *Ann. Math. Stat.*, vol. 24, pp. 196–219, June 1953.

about the parameter θ. The usual situation in human experience, however, is one in which there is a great deal of a priori knowledge about the possible values of θ. There are simple experiments with mutilated text, spoken or written, which will convince one that he can, and often does, exploit his own a priori knowledge of language, speaker, and subject matter to assist in deciphering what he reads and hears. A realistic model must include this possibility.

1.2. *Stochastic transmitted message.* It was Wiener who first clearly pointed out that we may, and indeed often must, regard the transmitted message itself as a random variable drawn from a universe whose distribution function reflects our a priori knowledge of the situation. Cogent statements of this philosophy may be found both in [1] and [2]. This leads us to a model in which we have two abstract random variables, say x representing the transmitted message (replacing the parameter θ), and y the received message. There is then a joint distribution function for x and y which contains in it the complete mathematical description of the situation. One ordinarily thinks of this distribution function as being "factored" into an a priori distribution for x, representing the universe of possible messages, and a conditional distribution for y knowing x, representing for each x the universe of possible mutilations thereof.

In this second and more important model, one can still regard the communication problem as one of estimation: given the y value of a joint sample (x, y), to estimate the x value. This view is particularly appropriate in discussing the work of [1]. Here, the x and y are numerically valued time series and there is a natural numerical way to measure the deviation between the estimated and true values of x, namely, by the variance of estimate.

The statistician may alternatively wish to regard the communication problem (in either model) as one of testing hypotheses. The observed y has a distribution depending on the hypothesis "x;" the problem is to decide which x is obtained at the time of observation. This view is more appropriate to the work of [2], wherein the time series are abstract valued, and no natural measure of the "wrongness" of an incorrectly adopted hypothesis is available. In the second model, the a priori distribution for hypotheses x eliminates one kind of testing error, so that in this model there is a simple criterion of performance, namely, the total probability in the (x, y) universe of all events (x, y) in which the hypothesis adopted is correct. The reader will observe this particular criterion in sections 6 and 8.

The distinction between estimation, on the one hand, and testing among many hypotheses, on the other, is not sharp. We shall use "estimation" as a loose word to refer to the kind of model here set up for communication.

1.3. *Peculiarities of engineering applications.* Information theory is distinguished from a general study of models like these in two important respects. In the first place, as noted, the random quantities x and y of interest are, naturally, time series. Furthermore, the passage of time is explicitly recognized and the distinction between past events, which can be known, and future events which cannot be known, is carefully observed.

In the second place, the kind of question considered in information theory, particularly by Shannon, reflects the peculiar interests of communication engineers. To illustrate this, we might go back to the jointly distributed abstract variables x and y of 1.2 above and the estimation problem there stated: given a sample of one value of y, to estimate the corresponding x. Typically, a practicing statistician facing this kind of problem will find himself confronted with a given joint distribution function for the variables, or at least committed to choosing one which he thinks is representative, and his attention is directed toward such questions as the following.

a. By what criterion shall various estimates of x be compared?

b. Given the criterion, what is the best estimate of x which can be made, and how good is it?

c. How do competing methods of estimating x compare with the best?

These questions of course appear in a communication context, too. The entire effort of [1] is concentrated in this general area. It often happens, however, that the communication engineer has a freedom that the statistician seldom has, that of controlling, at least in part, the joint distribution with which he must deal. How this comes about will be discussed in a moment. We can see at once, however, that his interest in question (b) above will then extend to asking, in addition, how he can optimize his best estimate over the additional freedom he has.

1.4. *The additional freedom.* The additional freedom enjoyed by a communication engineer is like the freedom granted the designer of an experiment. Typically, technology provides the engineer with a communicating device or medium; a random variable y whose range Y represents, as above, the events which can take place at the receiving point, and a probability distribution for y which depends upon a parameter θ. As above, the range Θ of θ represents the possible events at the transmitting point. In addition, one is given a quite separate random variable x whose range X is the universe of possible messages with a probability measure appropriate thereto.

No relation is yet specified between the message x and the "stimulus" θ which is applied to the communication medium, and it is here that the extra freedom lies. Subject to limitations set by the necessary distinction between past and future, one is free to choose a mapping function $f(x)$ from X into Θ, $\theta = f(x)$. This corresponds to choosing some kind of encoding or modulation scheme transforming the original message into a form suitable for transmission.

To illustrate the effect of this, suppose that the distribution function of y has a density $\rho(\theta; y)$ with respect to some fixed underlying measure ν in the y universe, and that the distribution of x has a density $\sigma(x)$ with respect to some underlying measure μ in the x universe. Then if one fixes the relation above between x and θ, the function $\sigma(x)\rho(f(x); y)$ in $X \otimes Y$ represents the density of the resulting joint distribution of x and y relative to the product measure $\mu \otimes \nu$. It is this joint distribution with which the communication engineer works.

To the practising engineer, the most interesting theorems of Shannon's paper relate to what can be achieved by varying the encoding process represented by the function $f(x)$. The strong theorems now known are all of an asymptotic kind.

1.5. *Role of Fourier analysis.* Even the casual reader will observe in [1], and in the latter part of [2], a preoccupation with Fourier analysis. It may be well to point out that this is a kind of accident; it happens that most practical communication media are governed by linear time-invariant differential equations. Hence, the first applications of information theory have been to systems which are naturally best handled by the tools of Fourier or Laplace analysis.

2. Terminology and concepts.

2.0. *Limitation to discrete model.* We shall confine our attention to the first part of Shannon's paper [2]. This whole paper relates to the second model of the communication problem described above, with an emphasis on the kind of question discussed in 1.3. The first part of that paper is based on a fairly specific kind of model. The stochastic processes which it admits are all derived from Markov processes having finitely many states. The auxiliary devices, encoders, etc., which are admitted are defined by similar constructions. We adopt the term "finitary" to denote a restriction to these classes of objects without at this point repeating Shannon's definitions in detail. (There is a restriction, tacit in [2] but nowhere made explicitly, to devices whose graphs have the property that the terminal state of any transition is uniquely fixed when the initial state and the letter emitted are given. For the present, we take "finitary" to include this limitation.)

The central concepts of [2] may be introduced well enough here by a glossary of terms. At this purely descriptive level, we may be quite general and admit things which are not finitary.

2.1. *Sample space and measurable sets.* Let A be a finite set. We call such a set an alphabet and will have occasion to introduce further alphabets A_1, B, etc. These are all abstract finite sets. An element of A will be called a letter of A, or simply a letter when no ambiguity results.

Let I denote the set of integers: $I = (\cdots, -1, 0, 1, 2, \cdots)$.

Given an alphabet A, denote by A^I the class of infinite sequences

$$x = (\cdots, x_{-1}, x_0, x_1, x_2, \cdots)$$

where each $x_t \varepsilon A$, $t \varepsilon I$. Here x is an element of A^I, and we call x_t the letter of x at time t.

A basic set (in A^I) is a subset of A^I obtained by specifying

(i) an integer $n \geqq 1$,

(ii) a finite sequence $\alpha_0, \alpha_1, \cdots, \alpha_{n-1}$ of letters $\alpha_k \varepsilon A$.

(iii) an integer t, $-\infty < t < \infty$.

The basic set resulting from this specification consists of all sequences $x \varepsilon A^I$ such that

$$x_{t+k} = \alpha_k, \qquad 0 \leqq k \leqq n - 1.$$

Let F_A be the Borel field of subsets of A^I determined by the basic sets.

2.2. *Glossary.* Our glossary now reads:

2.21. *Information source.* If μ is a probability measure defined over the Borel field F_A, the ensemble or stochastic process $[A^I, F_A, \mu]$ is an information source. Since the space A^I is fixed by the alphabet, and the Borel field F_A is always that determined by the basic sets, we can specify a source by the pair of symbols $[A, \mu]$.

2.22. *Stationary and ergodic sources.* Consider a source $[A, \mu]$. Let T be the coordinate-shift transformation defined as follows. If $x = (\cdots, x_{-1}, x_0, x_1, \cdots)$ then $Tx = (\cdots, x'_{-1}, x'_0, x'_1, \cdots)$, where $x'_t = x_{t+1}$, $t \varepsilon I$. Then T preserves membership in F_A (measurability). The source will be called stationary if (i) below holds, and ergodic if (i) and (ii) both hold.

(i) If $S \varepsilon F_A$, then $\mu(S) = \mu(TS)$.

(ii) If $S = TS$, then either $\mu(S) = 0$ or $\mu(S) = 1$.

2.23. *Transducer.* A transducer is characterized by two alphabets, A and B, and a function τ from A^I to B^I: given $x \varepsilon A^I$, $\tau(x) \varepsilon B^I$. A transducer differs from a general functional relationship in that it cannot anticipate.

If $x^{(1)} \varepsilon A^I$ and $x^{(2)} \varepsilon A^I$ and t_0 is an integer such that

$$x_t^{(1)} = x_t^{(2)} \qquad \text{for } t \leqq t_0,$$

then

$$y_t^{(1)} = y_t^{(2)} \qquad \text{for } t \leqq t_0,$$

where

$$y^{(i)} = \tau(x^{(i)}), \qquad i = 1, 2.$$

We can specify a transducer by the symbol $[A, \tau, B]$.

2.24. *Channel or communication channel.* A channel is characterized by two alphabets A and B, and a list of probability measures ν_θ defined over F_B, one for each $\theta \varepsilon A^I$. Here we have used θ to denote the "parameter" in conformance with an earlier notation.

Like a transducer, a channel cannot anticipate. That is, informally, if

$$\theta_t^{(1)} = \theta_t^{(2)} \qquad \text{for } t \leqq t_0, \tag{1}$$

we must have

$$\nu_1(S) = \nu_2(S), \tag{2}$$

where $\nu_i(S)$ denotes the value of $\nu_\theta(S)$ when $\theta = \theta^{(i)}$, for any set $S \varepsilon F_B$ which depends only on letters occuring before $t_0 + 1$. More precisely stated, (1) must imply (2) for any set $S \varepsilon F_B$ such that "$y_t^{(1)} = y_t^{(2)}$ for $t \leqq t_0$, and $y^{(1)} \varepsilon S$" implies "$y^{(2)} \varepsilon S$."

A transducer is a special case of a channel; it is a channel in which the received signal y is determined exactly by the transmitted signal θ.

We can specify a channel by the symbol $[A, \nu_\theta, B]$.

2.25. *Stationarity*. The concept of stationarity extends to channels and transducers. It suffices to define a stationary channel, a stationary transducer is a special case. Referring to the definition of a channel, this channel will be called stationary if, for any $S \varepsilon F_B$, $\nu_\theta(S) = \nu_{T\theta}(TS)$, where T is the coordinate-shift transformation.

2.26. We have so worded the definitions above that all sources are "letter generators" producing one new letter for each unit of time, and channels and transducers accept and produce one letter for each unit of time. In a careful setting of the theory, one must account for the phenomena of compression and expansion which appear when languages are translated. For example, a long business message of fairly stereotyped form, when encoded for transmission by cable, may appear in a form having many fewer letters or words than the original. There are several ways of accommodating the mathematics to this situation, but these details are unimportant in a first look at the subject and will be ignored from here on. The fact of so ignoring them does not invalidate any theorem that will be stated. It merely leaves a gap between these theorems and certain useful interpretations of them.

3. Entropy.

3.0. *Entropy*. The terms defined in Section 2, suitably hedged, are the concepts with which [2] deals. (For purposes of exposition, we have defined channels and transducers quite differently from [2]. The disparity is largely but not entirely verbal. (Cf. 10.2.)) The principal tool for their quantitative study is the concept of entropy.

Let $p_1, p_2, \cdots, p_n$ be a finite and exhaustive list of probabilities: $p_i \geqq 0$, $1 \leqq i \leqq n$, $p_1 + p_2 + \cdots + p_n = 1$. The entropy of this list is defined to be

$$H(p_1, p_2, \cdots, p_n) = -\sum_{i=1}^{n} p_i \log p_i = \text{Expectation}\ (-\log p).$$

It is by now traditional to use logs to the base 2 in this definition, but the choice of base affects the value of H only by a constant factor. We shall use the base 2.

3.1. *Marginal entropies*. To change the notation slightly, suppose that α and β run over finite index sets (alphabets) A and B, and that $p(\alpha, \beta)$ is the probability of the joint event (α, β). That is $p(\alpha, \beta) \geqq 0$, $\sum_{\alpha \varepsilon A} \sum_{\beta \varepsilon B} p(\alpha, \beta) = 1$. The entropy of this list of probabilities is denoted by $H(\alpha, \beta)$:

$$H(\alpha, \beta) = -\sum_{\alpha} \sum_{\beta} p(\alpha, \beta) \log p(\alpha, \beta).$$

We can define also two marginal entropies

$$H(\alpha) = -\sum_{\alpha} \sum_{\beta_1} p(\alpha, \beta_1) \log \left(\sum_{\beta} p(\alpha, \beta)\right),$$

$$H(\beta) = -\sum_{\beta} \sum_{\alpha_1} p(\alpha_1, \beta) \log \left(\sum_{\alpha} p(\alpha, \beta)\right),$$

and two average conditional entropies

$$H_\beta(\alpha) = H(\alpha, \beta) - H(\beta),$$
$$H_\alpha(\beta) = H(\alpha, \beta) - H(\alpha).$$

3.2. *Average conditional entropy.* These latter are called average conditional entropies because of the following formula: fix β and consider the conditional probabilities for the various $\alpha \varepsilon A$. These are $q_\beta(\alpha) = p(\alpha, \beta)/\sum_{\alpha_1} p(\alpha_1, \beta)$. The entropy of this list is

$$-\sum_\alpha q_\beta(\alpha) \log q_\beta(\alpha) = \frac{1}{r(\beta)} \sum_\alpha p(\alpha, \beta) \log p(\alpha, \beta) + \frac{1}{r(\beta)} \sum_\alpha p(\alpha, \beta) \log \sum_{\alpha_2} p(\alpha_2, \beta) \tag{1}$$

where $r(\beta)$ is defined by (2) below.

This expression is the entropy of the conditional distribution of α when it is known that a particular β has occurred. The a priori probability of this β is

$$\sum_{\alpha_1} p(\alpha_1, \beta) = r(\beta). \tag{2}$$

To average (1) over all β, we multiply it by (2) and sum over β. The result is seen to be $H_\beta(\alpha)$. This last entropy, then, is the average over all β of the entropies of the conditional distribution of α when β is known.

3.3. *Properties.* Shannon [2] gives a fairly complete heuristic justification for regarding the entropy of a list of probabilities as a measure of one's a priori uncertainty as to which of the possible events will actually occur in a given trial. In the course of this demonstration, he introduces the most important mathematical properties of the H function. These are (i) its positivity, (ii) a kind of convexity property implied by the convexity of the function $-x \log x$, (iii) that composition law which permitted the identification above of the average value of (1) over β, with the earlier defined $H_\beta(\alpha)$, and (iv) $H = 0$ if and only if there is exactly one event of nonzero probability.

The convexity property (ii) mentioned above leads to the general inequality $H_\beta(\alpha) \leqq H(\alpha)$; that is, verbally, a condition (i.e. an a priori restriction on the "freedom of choice") never increases an entropy. This statement must however be taken only in the average sense in which it is stated: for any particular β, the entropy of the conditional distribution of α bears no provable relation to the marginal entropy $H(\alpha)$. It is only in the average over-all β that an inequality obtains.

4. The entropy rate of a source.

4.0. *Definition.* So far we have considered the entropy of a list of probabilities. The entropy rate of a stationary source $[A, \mu]$ is most easily defined as follows. Given $x \varepsilon A^I$, we use either of the bracket notations

(1) $$[x_t, x_{t+1}, \cdots, x_{t+n-1}], [t, t+n-1; x]$$

to denote that basic set $S \subseteq A^I$ which consists of all x' such that

$$x'_{t+h} = x_{t+h}, \qquad 0 \leqq h \leqq n-1.$$

The second notation will be used when it is to be emphasized that the basic set depends upon a particular infinite sequence x.

The possible basic sets (1), as x ranges over A^I, or, alternatively, as the x_{t+h}, $0 \leqq h \leqq n-1$, range independently over A, partition A^I into a^n measurable subsets, where a is the number of letters in the alphabet A. These subsets represent all the possible sequences of n consecutive letters. They have the respective probabilities

(2) $$\mu([x_t, x_{t+1}, \cdots, x_{t+n-1}]).$$

Our stationarity assumption makes this list of probabilities independent of t. There is then a unique number F_n, independent of t, which is the entropy of this list (2) of probabilities. We shall show presently that the limit

(3) $$\lim_{n\to\infty} \frac{1}{n} F_n$$

always exists. The value of this limit is defined to be the entropy rate of the source $[A, \mu]$.

4.1. *Interpretation.* One cannot escape the heuristic meaning of this rate; one considers the possible long sequences of text as his universe of events, and evaluates the uncertainty F_n of the outcome of a trial. This uncertainty is then prorated among the n letters. These letters represent interdependent but possibly not determinately related elementary events whose concatenation generates the universe. The result, F_n/n, represents in the limit the average uncertainty per letter generated by the source.

4.2. *Defining F_n as an integral.* We shall now prove the existence of the limit (3). The proof follows Shannon's in a different notation.

Given any $x \varepsilon A^I$, the basic set (1) defined by that x contains x. The probability (2) then may be regarded as a step function of x, equal for each x to the probability of that basic set containing x which is specified by letter values at times $t, t+1, \cdots, t+n-1$. In the same way, the definition

(4) $$f_n(x) = -\frac{1}{n} \log \mu([0, n-1; x])$$

defines a nonnegative step function of x. One verifies at once from the definition of F_n that

(5) $$\frac{1}{n} F_n = \int_{A^I} f_n(x)\, d\mu(x).$$

Regarding (2) and (4) as functions of x in this way permits us to phrase certain key problems in the language of integration theory.

4.3. *Another definition of H.* Consider now the special conditional probabilities

$$p_n(x) = \frac{\mu([x_{-n}, x_{-n+1}, \cdots, x_{-1}, x_0])}{\mu([x_{-n}, \cdots, x_{-1}])}, \qquad n \geqq 1. \tag{6}$$

Again we use the device of representing these as step functions of x. In words, $p_n(x)$ is the conditional probability of observing at time zero the letter x_0 of x, when it is known that the letters occurring at times $t = -n, -n+1, \cdots, -1$ are exactly those of x.

Define

$$\begin{aligned} g_0(x) &= f_1(x) \\ g_n(x) &= -\log p_n(x), \qquad n \geqq 1. \end{aligned} \tag{7}$$

Then $g_n(x) \geqq 0$.

One verifies by direct calculation from (6) and (7) that

$$G_n = \int_{A^I} g_n(x)\, d\mu(x) \tag{8}$$

is the average conditional entropy of the next letter when n preceding letters are known. The inequality stated earlier, that adjoining a condition cannot increase an entropy, can be used to show that the G_n form a monotone sequence:

$$G_0 \geqq G_1 \geqq G_2 \geqq \cdots \geqq 0.$$

Therefore

$$\lim_{n \to \infty} G_n = H \tag{9}$$

certainly exists. The verbal interpretation of G_n, the average conditional entropy of the next letter after a long segment of text is already known, suggests that the limit H in (9) is again the average uncertainty per letter generated by the source, that is, H is the entropy rate defined in (3). The proof in 4.4 below that this is indeed so, proves the existence of the limit (3).

4.4. *Identification of two definitions.* By a direct calculation from the definitions it is found that

$$f_N(x) = \frac{1}{N} \sum_{k=0}^{N-1} g_k(T^k x). \tag{10}$$

If one integrates this and uses the assumed stationarity of μ, he obtains

$$\frac{1}{N} F_N = \frac{1}{N}(G_0 + G_1 + \cdots + G_N). \tag{11}$$

Therefore F_N/N represents the first Cesaro mean of a monotonely convergent sequence. It follows that the limit (3) exists and indeed is approached monotonely. A further consequence of (11) is that $F_N/N \geqq G_N \geqq H$.

5. The capacity of a channel.

5.0. *Channel and source.* We wish now to examine a stationary channel "driven" by a stationary source. Consider a source $[A, \mu]$, and a channel $[A, \nu_\theta, B]$. Denote by $C = A \otimes B$ the alphabet of pairs (α, β), $\alpha \varepsilon A$, $\beta \varepsilon B$. Then C^I is the class of all infinite sequences

$$(\cdots, (x_{-1}, y_{-1}), (x_0, y_0), (x_1, y_1), \cdots)$$

where $x_t \varepsilon A$, $y_t \varepsilon B$, $t \varepsilon I$. In an obvious way we can also regard $C^I = A^I \otimes B^I$, that is, as the class of paired sequences (x, y), $x \varepsilon A^I$, $y \varepsilon B^I$. It is known that the Borel field F_C is determined by the sets $X \otimes Y$ where $X \varepsilon F_A$, $Y \varepsilon F_B$. We define a measure ω for sets in F_C by the formula

$$\int_{C^I} h(x, y)\, d\omega(x, y) = \int_{A^I} d\mu(x) \int_{B^I} h(x, y)\, d\nu_x(y)$$

valid for all positive measurable $h(x, y)$. Here ν_x is the measure over F_B which is induced by the channel when the input sequence is $x \varepsilon A^I$.

The stochastic process $[C^I, F_C, \omega]$ is now a source, which we denote by $[C, \omega]$. It is easily shown that if the original source $[A, \mu]$ and the channel are stationary, then the source $[C, \omega]$ is stationary.

5.1. *Marginal distributions.* The source $[C, \omega]$ represents the joint distribution of x and y, of input to and output from the channel. The source $[A, \mu]$ represents the marginal distribution of the input. The marginal distribution of the output is represented by the source $[B, \eta]$, where the measure η over F_B is defined by

$$\int_{B^I} k(y)\, d\eta(y) = \int_{A^I} d\mu(x) \int_{B^I} k(y)\, d\nu_x(y) .$$

This marginal source is staionary if $[A, \mu]$ and $[A, \nu_\theta, B]$ are.

5.2. *Causation.* It is worth noting that the implication of causation in our language here, as we speak of a channel driven by a source, results from the fact that we consider the channel $[A, \nu_\theta, B]$ as a pregiven thing, existing independently of any particular source $[A, \mu]$; this is the typical situation in the communications art. Actually, the joint process $[C, \omega]$ is a completely symmetrical concept, as to the roles of x and y, and one may consider, at will, the conditional probabilities $\nu_x(S)$, $x \varepsilon A^I$, $S \varepsilon F_B$, the conditional probabilities of y-events, knowing x, or the conditional probabilities, say, $\bar{\mu}_y(U)$, $y \varepsilon B^I$, $U \varepsilon F_A$, of x-events, knowing y. (Indeed, given the joint process, one will find that each of these conditional probabilities ν_x, respectively $\bar{\mu}_y$, are measures for, respectively, almost all $x(\mu)$, almost all $y(\eta)$.)

It happens that in most applications the ν_x are pregiven, and the $\bar{\mu}_y$ derivative.

5.3. *Channel capacity.* To use Shannon's notation, let $H(x, y)$ denote the entropy rate of the source $[C, \omega]$, $H(x)$ the entropy rate of the marginal source $[A, \mu]$, and $H(y)$ that of the marginal source $[B, \eta]$. The quantity $R = H(x) + H(y) - H(x, y)$ is defined to be the transmission rate achieved by the source

$[A, \mu]$ over the channel $[A, \nu_\theta, B]$. The supremum or least upper bound of these rates, as μ is allowed to vary, is defined to be the capacity of that channel.

5.4. *Interpretation.* An intuitive interpretation of the rate $H(x) + H(y) - H(x, y)$ can be obtained if we assume that the quantity $H_x(y) = H(x, y) - H(x)$ can be given the same verbal interpretation when the x and y are stochastic processes that it was given earlier when the random quantities involved were drawn from finite populations. That it can, in the same limiting sense that the entropy concept has been carried over to stochastic processes, is easy to show. Foregoing this demonstration, we observe that $R = H(y) - H_x(y)$; that is, the rate of transmission R is the marginal rate of the output, $H(y)$, diminished by that amount of uncertainty at the output which arises from the average uncertainty of y even when x is known, that is, by $H_x(y)$, the average conditional entropy of y when x is known. In this verbal way, at least, R represents that portion of the "randomness" or average uncertainty of each output letter which is not assignable to the randomness created by the channel itself.

Another observation here is also pertinent. Because of the symmetry of R in x and y (which is more than a mere consequence of the notation!) we also have $R = H(x) - H_y(x)$. This shows R as the rate of the original source diminished by the average uncertainty as to the input x when the output y is known.

6. The fundamental theorem.

6.0. *As justifying the theory.* So far, we have introduced a list of what is hoped are natural-seeming concepts, and have stated a few mathematical results to help justify the rather picturesque language used in introducing them. The concepts themselves can only be justified as objects worthy of mathematical attention by the existence of theorems relating them. There is one such theorem, the so-called fundamental theorem for a noisy channel ([2], Theorem 11), which in itself performs this task completely. We shall quote this theorem and sketch its proof. This will complete our general exposition and lead us to our general limit theorem.

6.1. THE THEOREM. The fundamental theorem relates to this question. Suppose we are given a stationary channel with input alphabet A, and a stationary ergodic source with alphabet A_1. We are permitted to insert a stationary transducer $[A_1, \tau, A]$ between the source and channel, to create in effect, a new stationary channel with input alphabet A_1. With this freedom, what is the optimum transmission rate which can be achieved between source and output?

For the class of finitary sources, channels, and transducers, admitted in the model used in [2], this question is answered by Shannon's theorem: Let the given channel have capacity C and the given source have rate H. Then if $H < C$, for any $\epsilon > 0$ there exists a transducer such that a rate $R > H - \epsilon$ can be achieved. If $H \geqq C$, there exists similarly a transducer such that $C \geqq R > C - \epsilon$. No rate greater than C can be achieved.

Actually, Shannon's proof of this theorem proves the following more complete result.

THEOREM. *Let the given channel have capacity C and the given source have rate H. If $H < C$, then, given any $\epsilon > 0$, there exists an integer $n(\epsilon)$ and a transducer (depending on ϵ) such that when $n(\epsilon)$ consecutive received letters are known, the corresponding n transmitted letters can be identified correctly with probability at least $1 - \epsilon$. If $H > C$ no such transducer exists.*

This statement is perhaps more satisfying to a statistician, in that the logarithmic quantities H and C appear only in the hypotheses. The conclusion is then given in terms of the criterion of performance suggested in 1.2.

6.2. *Interpretation.* In the vernacular, this theorem asserts that if a channel has adequate capacity C, an infinitesimal margin being mathematically adequate, then virtually perfect transmission of the material from the source can be achieved, but not otherwise. Here, of course, we have used "virtually perfect" to describe transmission at a rate

$$R = H - \epsilon_1 \geqq H - \epsilon. \tag{1}$$

The sense in which this is to be interpreted as virtually perfect transmission is, of course, an asymptotic one and refers to the rate at which certain probabilities decay as the amount of available received text increases.

Engineering experience has been that the presence in the channel of perturbations, noise, in the engineer's language, always degrades the exactitude of transmission. Our verbal interpretation above leads us to expect that this need not always be the case; that perfect transmission can sometimes be achieved in spite of noise. This practical conclusion runs so counter to naive experience that it has been publicly challenged on occasion. What is overlooked by the challengers is, of course, that "perfect transmission" is here defined quantitatively in terms of the capabilities of the channel or medium, perfection can be possible only when transmission proceeds at a slow enough rate. When it is pointed out that merely by repeating each message sufficiently often one can achieve virtually perfect transmission at a very slow rate, the challenger usually withdraws. In doing so, however, he is again misled, for in most cases the device of repeating messages for accuracy does not by any means exploit the actual capacity of the channel.

Historically, engineers have always faced the problem of *bulk* in their messages, that is, the problem of transmitting rapidly or efficiently in order to make a given facility as useful as possible. The problem of noise has also plagued them, and in many contexts it was realized that some kind of exchange was possible, for example, noise could be eliminated by slower or less "efficient" transmission. Shannon's theorem has given a general and precise statement of the asymptotic manner in which this exchange takes place.

The statistician will recognize the exchange between bulk and noise as akin to the more or less general exchange between sample size and validity or significance.

7. The asymptotic equipartition property.

7.0. *A Basic Lemma.* The theorem quoted in 6.1 is termed fundamental in

[2] because it answers a question which is clearly fundamental in the communications art, and because it defines the applicability of the central concept of channel capacity. Many of the later results in [2] then concern the calculation of capacities for practical or interesting channels.

The proof of this fundamental theorem rests directly on a lemma (Theorem 3 of [2]) which itself is a basic limit theorem in the theory of stochastic processes. As a mathematical theorem, this lemma requires very little of the specialized imagery of communication theory for its understanding. A mathematician, therefore, is likely to regard it as the more fundamental element. A generalization of it is the one contribution of the present paper.

7.1. *Shannon's form.* The basic limit theorem, as given in Theorem 3 of [2], asserts that the text from an ergodic finitary source possesses what we shall call an asymptotic equipartition property. The basic sets

$$[x_0, x_1, \cdots, x_{n-1}], \tag{1}$$

as x ranges over A^I describe a partition of A^I, as we noted earlier: a partition into a^n events, each one of which is the occurrence of a particular string of n letters. Shannon's Theorem 3 asserts that, if H is the rate of a finitary ergodic source, then, given $\epsilon > 0$ and $\delta > 0$, there exists an $n_0(\epsilon, \delta)$ such that, given any $n \geqq n_0$, the basic sets (1) above can be divided into two classes:

(i) a class whose union has μ-measure less than ϵ,

(ii) a class each member E of which has a measure $\mu(E)$ such that $| H + 1/n \log \mu(E) | < \delta$.

That is, this theorem asserts the possibility of dividing the long segments of text from a finitary source into a class of roughly equally probable segments plus a residual class of small total probability.

7.2. *Stronger form.* Let us introduce here the step functions $f_n(x)$ defined in (4) of Section 4: $f_n(x) = -1/n \log \mu([0, n-1; x])$. In terms of these, the possibility of dividing the long segments of text into the categories (i) and (ii) above is easily seen to be equivalent to the assertion that the sequence $(f_n(x))$ converges in probability to the constant H.

We shall say that a source $[A, \mu]$ has the asymptotic equipartition property, AEP, if the sequence $(f_n(x))$ converges in probability to a constant.

Shannon's Theorem 3 then asserts that a finitary ergodic source has the AEP. We shall improve this in Section 9 to read as follows.

THEOREM. *For any source $[A, \mu]$, the sequence $(f_n(x))$ converges in L^1 mean (μ). If $[A, \mu]$ is ergodic, and has rate H, this sequence converges in L^1 mean to the constant H.*

Since L^1 convergence here implies convergence in probability, (a fact easily proved,) we have the

COROLLARY. *Every ergodic source has the AEP.*

These are the limit theorems mentioned in the Summary. As we shall see in Section 8, they permit extending Shannon's fundamental theorem, 6.1, to other than finitary sources.

7.3. *Interpretation.* Returning to 7.1 and the description there of the AEP, we see that most of the probability must be accounted for by the aggregate (ii) of "likely" long sequences. That is, if the source has the AEP, there are, for large enough n, 2^{nH} likely basic sets $[x_0, \cdots, x_{n-1}]$, roughly equally probable, accounting in the aggregate for all but a small fraction of the total probability.

7.4. *Another corollary.* The proof in [2] of the fundamental theorem uses also a consequence of the AE property. We examine this consequence briefly.

Consider a stationary source $[A, \mu]$ and a stationary channel $[A, \nu_\theta, B]$. Suppose that both $[A, \mu]$, and the joint process $[C, \omega]$ which results when this source drives the channel, have the AEP. We can write

$$(2)\qquad \begin{aligned} &-\frac{1}{n}\log \omega([0, n-1; x] \otimes [0, n-1; y]) \\ &= -\frac{1}{n}\log \frac{\omega([0, n-1; x) \otimes [0, n-1; y)]}{\mu([0, n-1; x])} - \frac{1}{n}\log \mu([0, n-1; x]). \end{aligned}$$

Our hypothesis that the joint process has the AEP now implies that the left member of this equation converges in measure to a constant, namely the entropy rate of the joint process, $H(x, y)$. (Here the notation is misleading. In $H(x, y)$, the x and y are labels merely. Equation (2) involves x and y as specific variables.) Also by hypothesis the second term on the right converges in probability to $H(x)$, the entropy rate of $[A, \mu]$. It follows then that the first term on the right converges in probability also to a constant, which constant must then be $H_x(y)$, by 5.4.

Now the first term on the right of (2) is

$$(3)\qquad -\frac{1}{n}\log \bar{\mu}_{n,x}([0, n-1; y]),$$

where the argument of the logarithm is the conditional probability of $[0, n-1; y]$ knowing that $[0, n-1; x]$ has occurred. We have therefore proved the following.

Corollary. *If* $[A, \mu]$ *and* $[C, \omega]$ *have the AEP, then the functions* (3) *converge in probability to a constant.*

8. Proof of the fundamental theorem.

8.0. *Introduction.* For simplicity, we do not examine the question of ergodicity and consider only the most interesting of the cases cited in the statement of the theorem (6.1), that in which we are given a finitary source $[A_1, \mu]$ of entropy rate H and a finitary channel $[A, \nu_\theta, B]$ of capacity $C > H$, both stationary. Our problem is then, given $\epsilon > 0$, to exhibit an $n(\epsilon)$ and a finitary transducer $[A_1, \tau, A]$ such that, when the given source drives the channel through this transducer it is possible at the receiver, given $n(\epsilon)$ consecutive received letters, to identify the corresponding $n(\epsilon)$ transmitted letters correctly with a probability exceeding $1 - \epsilon$. Here the probability is not conditional (i.e., not given the received letters) but in the universe of joint events at transmitter and receiver.

We shall review Shannon's argument. He does not supply detailed epsilontics here, and we shall not either. Generally, the manner in which they could be supplied is evident enough, though at one point we must consider a detail. (My efforts to make them simple have so far failed, however.)

8.1. *The "likely" events.* The channel $[A, \nu_\theta, B]$ has capacity $C = H + 2\gamma$, say, where $\gamma > 0$. There, therefore, exists a source, say $[A, \mu^*]$, which achieves over this channel a rate

$$R^* \geqq C - \gamma = H + \gamma. \tag{1}$$

We will use asterisks to denote quantities referring to this source. Let H^* be the entropy rate of $[A, \mu^*]$, and let $[C, \omega^*]$ denote the joint process of input (x) and output (y) when $[A, \mu^*]$ drives the channel. For a simpler notation let $K^* = H_y^*(x)$, the average conditional entropy of input to $[C, \omega^*]$ when output is known. Then by definition (5.3)

$$R^* = H^* - K^*. \tag{2}$$

We now invoke the AEP for the processes $[A_1, \mu]$, $[A, \mu^*]$, and $[C, \omega^*]$. For large n there are roughly 2^{nH} equally likely basic sets $[0, n - 1; w]$ from $[A_1, \mu]$, call these the likely outputs of $[A_1, \mu]$. Similarly there are roughly 2^{nH^*} equally likely basic sets $[0, n - 1; x]$ from $[A, \mu^*]$, the likely outputs of $[A, \mu^*]$. Furthermore, consider the possible basic sets $[0, n - 1; y]$ at the output of the channel. With the exception of an aggregate of these of small total probability in $[C, \omega^*]$, the conditional probabilities in $[C, \omega^*]$ of the $[0, n - 1; x]$, knowing $[0, n - 1; y]$, are such that roughly, there are 2^{nK^*} equally likely $[0, n - 1; x]$ for each $[0, n - 1; y]$, call these the likely antecedents to $[0, n - 1; y]$.

In each of these definitions the "likely" objects in sum exhaust most of the probability. In particular, the likely antecedents of $[0, n - 1; y]$ exhaust most of the a posteriori probability in $[C, \omega^*]$ of the basic sets $[0, n - 1; x]$ when $[0, n - 1; y]$ is known. Let us use the word "package" to mean "the aggregate of likely antecedents to a given $[0, n - 1; y]$."

8.2. *Marked basic sets.* The nub of Shannon's proof lies in the fact that the packages are so small that it is easy to find 2^{nH} of them which are disjoint. Indeed, suppose one designates, "marks," 2^{nH} of the likely basic sets $[0, n - 1; x]$ from $[A, \mu^*]$, doing so at random. Then the probability that a particular $[0, n - 1; x]$ be marked in this process is $2^{n(H-H^*)}$. Consider the 2^{nK^*} likely antecedents of some $[0, n - 1; y]$. The conditional probability that two or more of these get marked, knowing that one of them is marked, is of the order of

$$2^{nK^*} \cdot 2^{n(H-H^*)} = 2^{n(H-R^*)} \leqq 2^{-n\gamma},$$

by (1) and (2). This probability may be made small by choosing a large n.

8.3. *Distinguishability a posteriori of marked inputs.* Conceptually, we now have this situation: some 2^{nH} basic sets $[0, n - 1; x]$ have been specially marked. Given a $[0, n - 1; y]$, the received message, and knowing in addition that a marked basic set $[0, n - 1; x]$ has been transmitted (has occurred) there is but a small conditional probability, in the joint universe of $[C, \omega^*]$ and of random markings, that either of the following events has occurred.

(i) The actual $[0, n - 1; x]$ which occurred is not a likely antecedent of $[0, n - 1; y]$;

(ii) The actual $[0, n - 1; x]$ which occurred is a likely antecedent of $[0, n - 1; y]$, but there are other marked $[0, n - 1; x]$ in the same package.

There is now virtual certainty in the joint universe of $[C, \omega^*]$ and random markings that the actual $[0, n - 1; x]$ is a unique marked likely antecedent of $[0, n - 1; y]$ when we know $[0, n - 1; y]$ a priori, and that a marked $[0, n - 1; x]$ is transmitted. That is, by making a marking at random, one is almost certain to have chosen a limited vocabulary of 2^{nH} basic sets $[0, n - 1; x]$ which are almost certain to be distinguishable a posteriori, knowing $[0, n - 1; y]$.

8.4. *The transducer*. The next step is deceptively simple. One shows easily that, given a marking, a finitary transducer can be described which maps the 2^{nH} likely $[0, n - 1; w]$ from $[A_1, \mu]$ on to the marked $[0, n - 1; x]$ from $[A, \mu^*]$. When one drives this transducer from $[A_1, \mu]$, the likely output basic sets $[0, n - 1; x]$ are just those which were marked. Therefore, when one operates the channel from $[A_1, \mu]$ through this transducer he has essentially only the vocabulary of marked basic sets appearing at the input to the channel. Let us call the resulting joint process of input x to, and output y from, the channel the source $[C, \omega]$. This source itself depends on the marking.

If the probabilities sketched in 8.3 can be relied on for this new situation, it is evident that we have described a transducer, depending on a random marking, which, when $[0, n - 1; y]$ is given, permits the correct identification of the $[0, n - 1; x]$ which occurred in all but a set of cases of small probability (a posteriori, knowing $[0, n - 1; y]$) in the joint universe of random markings and events in $[C, \omega]$. We can assume that for all likely $[0, n - 1; x]$ the input $[0, n - 1; w]$ which produced it is unique. Then the average, over the joint universe of markings and events in $[C, \omega]$, of the probability that the actual $[0, n - 1; w]$ which occurred is not the one determined by this procedure is small. By the Tchebycheff inequality, then, all but a small fraction of the markings will describe transducers which make the probability of misidentifying the actual $[0, n - 1; w]$ *simultaneously* small for all but a small fraction of the $[0, n - 1; y]$.

8.5. *Critique*. This argument shows that it is somehow easy to describe a transducer which will make the probability of error small. There is, however, a gap in the argument. The probabilities calculated in 8.3 were based on $[C, \omega^*]$. In 8.4 we used these as though they applied to any $[C, \omega]$ which might arise when a marking had been made. If they are both ergodic, and this we are tacitly assuming, ω and ω^* are either identical or else each assigns unit probability to a null set of the other. (This is almost trivial to prove. To my knowledge it was first explicity noted by G. W. Brown.) The probabilities in 8.3 are based on relations which hold only almost everywhere in $[C, \omega^*]$, and therefore, possibly, at most on a null set in $[C, \omega]$. This point is not touched on in [2].

In 10.1 we shall show that finitary channels have a kind of continuity which permits passage from $[C, \omega^*]$ to $[C, \omega]$, when n is large enough, without serious modification of the probabilities. Shannon's argument is then valid, though incomplete, for finitary channels. Indeed, it is valid for any channel having this kind of continuity, but I have not yet found a satisfying formulation of the property or isolation of the class.

9. The limit theorem.

9.0. *Introduction* This section is devoted principally to the proof of the theorem quoted in 7.2, which has as a corollary that every ergodic source has the AEP. We recall the definitions of 2.1 and 2.2, and use the following notation.

Given any fixed $x \in A^I$, The symbols $[t, t + n - 1; x]$, $[x_t, \cdots, x_{t+n-1}]$ denote that basic set which consists of all $x' \in A^I$ such that $x'_{t+k} = x_{t+k}$, $k = 0, 1, \cdots, n - 1$.

Given a source $[A, \mu]$, the symbols $\int f(x)\, d\mu(x)$, $\int f\, d\mu$, denote integration over the space A^I. Integration over a measurable subset $S \subseteqq A^I$ is denoted by one of $\int_S f(x)\, d\mu(x)$, $\int_S f\, d\mu$.

Following [3], we append "(μ)" to a statement which holds almost everywhere with respect to μ, or to a statement involving mean convergence relative to μ.

9.1. *The Theorem.* Given the source $[A, \mu]$ we define the following step functions of $x \in A^I$.

$$(1) \qquad \begin{aligned} p_n(x) &= \frac{\mu([-n, 0; x])}{\mu([-n, -1; x])}, && n \geqq 1, \\ p_0(x &= \mu([x_0]); && \\ g_n(x) &= -\log p_n(x), && n \geqq 0; \\ f_n(x) &= -\frac{1}{n} \log \mu(0, n - 1; x]), && n \geqq 1. \end{aligned}$$

The function $p_n(x)$ is the conditional probability that the letter which occurs at time $t = 0$ is x_0 when it is known that the letters between time $t = -n$ and $t = -1$ are also those of the infinite sequence x. The definitions of $g_n(x)$ and $f_n(x)$ need no comment. They are related by the important and easily verified formula

$$(2) \qquad f_N(x) = \frac{1}{N} \sum_{k=1}^{N-1} g_k(T^k x).$$

What is now to be proved is the

THEOREM. *For any source $[A, \mu]$, the sequence $(f_n(x))$ converges in L^1 mean (μ). If $[A, \mu]$ is ergodic, the limit of this sequence is almost everywhere constant and equal to H, the information rate of $[A, \mu]$.*

9.2. Proof. The proof of this theorem requires the following intermediate results.

(i) The sequence $(p_n(x))$ converges almost everywhere (μ).

(ii) Each $g_n(x) \varepsilon L^1(\mu)$, and the sequence $(g_n(x))$ converges in L^1 mean (μ).

These will be established in 9.3 and 9.4, respectively. Granted the second of them, the theorem to be proved follows easily, as we now show.

We have $g_n(x) \varepsilon L^1$, and $\lim_n \int | g_n - g | \, d\mu = 0$ for some function $g \varepsilon L^1$. Then the mean ergodic theorem (e.g., [4], equation 2.42) implies that $\sum_{k=0}^{N-1} g(T^k x)/N$ converge in L^1 mean to an invariant function $h(x) = h(Tx)$. When μ is ergodic $h(x) = H$, a constant, almost everywhere.

By (2) of 9.1

$$\int | f_N - h | \, d\mu \leqq \int \left| \frac{1}{N} \sum_{k=0}^{N-1} [g_k(T^k x) - g(T^k x)] \right| d\mu(x) + \int \left| \frac{1}{N} \sum_{k=0}^{N-1} g(T^k x) - h(x) \right| d\mu(x)$$

$$\leqq \frac{1}{N} \sum_{k=0}^{N-1} \int | g_k(x) - g(x) | \, d\mu(x) + \int \left| \frac{1}{N} \sum_{k=0}^{N-1} g(T^k x) - h(x) \right| d\mu(x).$$

In the second inequality we use the stationarity of μ to obtain the first term. This term represents the first Cesaro mean of a sequence which by hypothesis has zero as a limit, hence it has also the limit zero. The second term also goes to zero as $N \to \infty$, by the mean ergodic theorem. We conclude then that $f_n \to h$ in L^1 mean, and that $f_n \to H$ in L^1 mean when μ is ergodic. We identify this constant H with the entropy rate of $[A, \mu]$ in 9.4.

9.3. *First Lemma.* We now prove that the sequence $(p_n(x))$ converges almost everywhere (μ). For any given set $D \varepsilon F_A$ define, in analogy with 9.1, (1)

$$p_n(x, D) = \frac{\mu([-n, -1; x] \cap D)}{\mu([-n, -1; x])}, \qquad n \geqq 1;$$

this is the conditional probability of D knowing $x_{-n}, x_{-n+1}, \cdots, x_{-1}$. It is a result of Doob [5] that such a sequence of conditional probabilities is a martingale (positive and bounded) and converges almost everywhere.

Given $\alpha \varepsilon A$, let D_α denote the basic set of all $x \varepsilon A^I$ with $x_0 = \alpha$. Given any $x \varepsilon A^I$, the value of $p_n(x)$ is one of the numbers $p_n(x, D_\alpha)$ obtained as α ranges over the finite set A. Therefore

$$| p_n(x) - p_m(x) | \leqq \sum_{\alpha \varepsilon A} | p_n(x, D_\alpha) - p_m(x, D_\alpha) |, \tag{3}$$

because the left member is, for each x, one of the summands on the right.

Except for x in a certain null set (μ), each term on the right of (3) converges to zero as m and n go to infinity, by the result of Doob quoted above. By (3), then, the sequence $p_n(x)$ converges almost everywhere (μ), say to $p(x)$.

It follows at once that the sequence $g_n(x) = -\log p_n(x)$ converges almost everywhere to $g(x) = -\log p(x)$, if we admit convergence to $+\infty$.

9.4. *Second Lemma.* We must now show that $g_n(x) \varepsilon L^1$ and that the sequence $g_n(x)$ converges in mean (μ). The integrability of the $g_n(x)$ is simple to establish directly, but will follow automatically from stronger results which are needed later. We need a uniform bound for the contribution of the "unbounded part" of g_n to the value of $\int g_n \, d\mu$. We shall therefore show that

$$\int_{A_{n,L}} g_n \, du \leqq O(L2^{-L}) \tag{4}$$

uniformly in n, where $A_{n,L}$ is the set of x's such that $g_n(x) \geqq L$.

Let $E_{n,K}$ be the set of x's where

$$K \leqq g_n(x) < K + 1. \tag{5}$$

Let B denote a typical basic set $[-n, -1; x]$. Given $\alpha \varepsilon A$, let D_α, as before, be the basic set of all x such that $x_0 = \alpha$. By its definition, $g_n(x)$ is constant over each $B \Lambda D_\alpha$, in fact, it has there the value $-\log[\mu(B \Lambda D_\alpha)/\mu(B)]$. Hence $g_n(x)$ is measurable.

Let a be the number of letters in the alphabet A. There are altogether finitely many, namely a^{n+1}, sets $B \Lambda D_\alpha$ covering A^I. Since $g_n(x) \geqq 0$ everywhere, we have

$$A^I = \bigcup_B \bigcup_{K=0}^{\infty} B \Lambda E_{n,K}.$$

For fixed n, K, let D^K range over those D_α such that $B \Lambda D_{n,K} \neq \phi$. Then the step character of $g_n(x)$ implies that $B \Lambda E_{n,K} = \bigcup_{D^K} B \Lambda D^K$. Therefore

$$\int_{B \Lambda E_{n,K}} g_n \, d\mu = \sum_{D^K} \int_{B \Lambda D^K} g_n \, d\mu \tag{6}$$

and, furthermore, over any $B \Lambda D^K$, (5) holds. Therefore $-\log[\mu(B \Lambda D^K)/\mu(B)] \geqq K$, or $\mu(B \Lambda D^K) \leqq 2^{-K}\mu(B)$. From (6), then,

$$\int_{B \Lambda E_{n,K}} g_n \, d\mu < \sum_{D^K} (K + 1)2^{-K}\mu(B) < a(K + 1)2^{-K}\mu(B).$$

We have then that

$$\int_{E_{n,K}} g_n \, d\mu = \sum_B \int_{B \Lambda E_{n,K}} g_n \, d\mu < (K + 1)2^{-K}, \tag{7}$$

since $\sum \mu(B) = 1$. The right member of (7) is the Kth term of a convergent series and is independent of n. Since $A_{n,L} = \bigcup_{K \geqq L} E_{n,K}$, (4) follows at once.

That $g_n \varepsilon L^1$ follows by summing (7) over all $K \geqq 0$. This summation gives a uniform bound, say β, for $\int g_n \, d\mu$. Define $g_n^L(x) = \inf (g_n(x), L)$, $g^L(x) = \inf(g(x),$

L). Then $\lim_n g_n^L(x) = g^L(x), (\mu)$, and this convergence is dominated by the integrable function L. Hence

$$\text{(8)} \qquad \lim_n \int | g_n^L - g^L | \, d\mu = 0,$$

and $g^L \varepsilon L^1$. Furthermore

$$\text{(9)} \qquad \int g^L \, d\mu = \lim_n \int g_n^L \, d\mu \leqq \limsup_n \int g_n \, d\mu \leqq \beta.$$

By (9) and the definition of the left-hand side,

$$\int g \, d\mu = \lim_{L \to \infty} \int g^L \, d\mu \leqq \beta$$

whence $g \varepsilon L^1$. Furthermore

$$\text{(10)} \qquad \lim_{L \to \infty} \int | g - g^L | \, d\mu = \lim_L \int (g - g^L) \, d\mu = 0.$$

We have now

$$\int | g_n - g | \, d\mu \leqq \int | g_n - g_n^L | \, d\mu + \int | g_n^L - g^L | \, d\mu + \int | g^L - g | \, d\mu.$$

The first term on the right is dominated by

$$\int_{A_{n,L}} g_n \, d\mu$$

where $A_{n,L}$ is the set over which $g_n(x) \geqq L$. By (4) and (8) therefore,

$$0 \leqq \limsup_n \int | g_n - g | \, d\mu \leqq 0(L2^{-L}) + \int | g^L - g | \, d\mu.$$

We let $L \to \infty$ and use (10) to conclude that $\lim_n \int | g_n - g | \, d\mu = 0$.

This establishes the mean convergence of the sequence $(g_n(x))$.

It was shown in 4.3 that the entropy rate of $[A, \mu]$ is $\lim_{n \to \infty} \int g_n \, d\mu$. From what we have just shown, $\lim_{n \to \infty} \int g_n \, d\mu = \int g \, d\mu$. In 9.2, $h(x)$ is defined as the limit of $1/N \sum_{k=0}^{N-1} g(T^k x)$ and we know by the ergodic theorem then that $\int h \, d\mu = \int g \, d\mu$. When μ is ergodic $h(x) = H$, a constant, almost everywhere. Therefore

$$H = \int H \, d\mu = \int h \, d\mu = \int g \, d\mu = \lim \int g_n \, d\mu.$$

This identifies H with the entropy rate of $[A, \mu]$.

10. Finitary devices.

10.0. *Sources*. Shannon's Markov-like sources, which we have here called finitary, are defined by a construction equivalent (in a sense to be made precise later) to one now to be described.

Consider a Markov process with finitely many states, enumerated $1, 2, \cdots, S$. Let p_{ij} be the probability of the transition from state j to state i, and ξ_i the stationary probability of occupancy of state i, so that

$$\xi_i = \sum_{j=1}^{S} p_{ij}\xi_j, \qquad 1 \leqq i \leqq S.$$

Let B be the alphabet whose letters are the symbols $1, 2, \cdots, S$. We may suppose that the Markov process makes a transition at each time $\tau = t + \frac{1}{2}$, $t = 0, \pm 1, \pm 2, \cdots$. We define the stationary information source $[B, \nu]$ by the rule that the letter which occurs at time $t \varepsilon I$ is the name of the state in which the Markov process is at that time. The p_{ij} and the ξ_i are enough to define this source. A source defined in this way will be called a finite Markov source.

Let A be an arbitrary alphabet and let φ be a function from B to A: $\alpha = \varphi(\beta)$, $\beta \varepsilon B$, $\alpha \varepsilon A$. Given $y \varepsilon B^I$, say $y = (\cdots, y_{-1}, y_0, y_1, \cdots)$, we define $x = \Phi(y) \varepsilon A^I$ by $x = (\cdots, x_{-1}, x_0, x_1, \cdots)$, where $x_t = \varphi(y_t)$, $t \varepsilon I$. Let μ be the measure over F_A defined by this construction. In the notation of [3], $\mu = \nu\Phi^{-1}$. The source $[A, \mu]$ we will call a *projection* of $[B, \nu]$. The notion of projection clearly applies even when $[B, \nu]$ is not Markov.

An arbitrary source $[A, \mu]$ will be called finitary if it is a projection of some finite Markov source $[B, \nu]$. Shannon's sources are all of this kind in the sense that, given any source of his, there is a projection of a finite Markov source which produces the same ensemble of text, and conversely.

Consider now an $[A, \mu]$, a projection by φ of $[B, \nu]$. Given $\alpha \varepsilon A$, let $\varphi^{-1}(\alpha)$ denote that subset of B consisting of all β such that $\varphi(\beta) = \alpha$. We will call $[A, \mu]$ *unifilar* if for each $\alpha \varepsilon A$ and each state $i \varepsilon B$ there is at most one transition from i to $\varphi^{-1}(\alpha)$ which has nonzero probability.

The definition of [2], paragraph 7, and certain related results, are tacitly restricted to finitary and unifilar sources. The word "finitary" as we use it in discussing the proof of the fundamental theorem (Section 8) may, however, be interpreted in the wider sense defined above: being a projection of a finite Markov process.

10.1. *Channels*. We now frame a definition of "finitary channel" consonant with that just given for finitary sources. A finitary channel is specified by:

(i) An input alphabet A.

(ii) An output alphabet B.

(iii) A finite set $D = (1, 2, \cdots, K)$ of states. We treat D as an alphabet.

(iv) A set of Markov transition matrices, $\| q_{ij}(\alpha) \|$, one matrix for each $\alpha \varepsilon A$. Each element $q_{ij}(\alpha)$ represents a conditional probability of transition from state $j \varepsilon D$ to state $i \varepsilon D$ knowing that the input letter is α. We have $\Sigma_i\, q_{ij}(\alpha) = 1$ for each j and α.

(v) A function ψ from D to B. The output letter from the channel is $\psi(i)$ whenever the transition is to the state $i \varepsilon D$.

Consider a stationary source $[A, \mu]$ driving the channel so specified. Let λ_j be the stationary probability of finding the channel in state $j \varepsilon D$, (if such a probability exists). Then the probability that the letter α be presented to the channel and that the channel make a transition to state $i \varepsilon D$ is

$$\sum_j \mu([\alpha]) q_{ij}(\alpha) \lambda_j . \tag{1}$$

Stationarity of the system requires now that the sum of these numbers over all $\alpha \varepsilon A$ be λ_i. That is, the vector $(\lambda_i, i \varepsilon D)$ must be invariant under left multiplication by the Markov matrix

$$Q = \| \sum_{\alpha \varepsilon A} \mu([\alpha]) \, q_{ij}(\alpha) \| .$$

At least one such invariant probability vector exists. If, for example, each matrix $\| q_{ij}(\alpha) \|$ has a unique such invariant vector, then in general the λ_i will also be unique and they will be continuous functions of the letter frequencies of the source.

Given the λ_i above, the joint probability that letters $\alpha_1, \cdots, \alpha_n$ be presented to the channel and that the corresponding sequence of states of the channel be $i_1, i_2, \cdots, i_n$ is similar to the expression (1):

$$\mu([\alpha_1, \cdots, \alpha_n]) \sum_{j \varepsilon D} q_{i_n i_{n-1}}(\alpha_n) \cdots q_{i_2 i_1}(\alpha_2) q_{i_1 j}(\alpha_1) \lambda_j . \tag{2}$$

The joint probability of input letters $[\alpha_1, \cdots, \alpha_n]$ and output letters $[\beta_1, \cdots, \beta_n]$ is found by summing (2) for

(1) all i_1 in $\psi^{-1}(\beta_1)$,

(2) all i_2 in $\psi^{-1}(\beta_2)$,

.

.

.

(n) all i_n in $\psi^{-1}(\beta_n)$.

The conditional probability of $[\beta_1, \cdots, \beta_n]$ knowing $[\alpha_1, \cdots, \alpha_n]$ is then

$$\sum_n \cdots \sum_1 \sum_{j \varepsilon D} q_{i_n i_{n-1}}(\alpha_n) \cdots q_{i_1 j}(\alpha_1) \lambda_j , \tag{3}$$

where $\sum_k$ denotes the summation of i_k over $\psi^{-1}(\beta_k)$. The expression (3) depends on $[\alpha_1, \cdots, \alpha_n]$ and $[\beta_1, \cdots, \beta_n]$, and not otherwise upon past history. It is independent of the source except for the continuous dependence of the λ_j upon the letter frequencies. This continuity is sufficient for the proof in Section 8, since it is easy there to guarantee that the source $[A, \mu^*]$ and the source which results from putting $[A_1, \mu]$ through the transducer there defined have virtually the same letter frequencies.

10.2. *A discrepancy*. The purist will observe that in 2.24 we defined a channel as a set of conditional probability measures ν_θ over outputs, where θ represents the input sequence. The construction in 10.1 is not obviously of this kind, since the measures ν_θ there obtained might well depend not only on θ but also on the particular source-ensemble in mind at the moment. We will not clarify the point here. Some pedagogic license was used in 2.24, and it is simpler to enlarge the notion of channel beyond that defined there than to try to reconcile the two definitions.

11. A useful theorem. Let Ω be an abstract countable set of elements ω. Let A be a finite set, an alphabet. Let μ be a probability measure over a Borel field containing all sets $S \otimes W$, where $S \varepsilon F_A$ and $W \subsetneq \Omega$. Define the measures μ_ω over F_A by $\mu_\omega(S) = \mu(S \otimes \omega)/\mu(A^I \otimes \omega)$. This definition is valid for almost every ω. Define $\bar{\mu}$ over F_A by $\bar{\mu}(S) = \mu(S \otimes \Omega)$. Suppose that the source $[A, \bar{\mu}]$ is ergodic and has rate H. Suppose that

$$(1) \qquad \int [-\log \mu(A^I \otimes \omega)]\, d\mu(x \otimes \omega) < \infty.$$

Then the functions $f_n(x, \omega) = -(\log \mu\, ([0, n-1; x] \otimes \omega))/n$ converge in L^1 mean to H relative to μ. Considering ω as a parameter, for almost every ω the functions $f_n(x, \omega)$ converge in L^1 mean to H relative to μ_ω.

PROOF. Since $\mu([0, n-1; x] \otimes \omega) \leqq \bar{\mu}([0, n-1; x])$ we have

$$(2) \qquad f_n(x, \omega) \geqq -\frac{1}{n} \log \bar{\mu}([0, n-1; x]) = g_n(x),$$

where the second equality sign defines $g_n(x)$. Fix n and consider the countable list of events $[0, n-1; x] \otimes \omega$. By the composition law (3.2), extended to infinite sums, the entropy of this list of events is the sum of the entropy of the $[0, n-1; x]$ and the conditional entropy of ω knowing $[0, n-1; x]$:

$$(3) \qquad H([0, n-1; x] \otimes \omega) = H([0, n-1; x]) + H_{x,n}(\omega).$$

Now the convexity law (3.3) implies that the average conditional entropy $H_{x,n}(\omega)$ is always less than the unconditional entropy of ω, which latter is the integral asserted to be finite in (1). Hence there is a finite K such that for all n

$$(4) \qquad H_{x,n}(\omega) \leqq K.$$

From (2), (3), (4) and the definitions of $\bar{\mu}$ and the entropies,

$$\int |f_n(x, \omega) - g_n(x)|\, d\mu(x \otimes \omega) \int f_n(x, \omega)\, d\mu(x \otimes \omega) - \int g_n(x)\, d\bar{\mu}(x)$$

$$= \frac{1}{n} H([0, n-1; x] \otimes \omega - \frac{1}{n} H([0, n-1; x]) \leqq \frac{1}{n} K.$$

Therefore

$$\int | f_n - H | \, d\mu \leqq \int | f_n - g_n | \, d\mu + \int | g_n - H | \, d\bar{\mu} \leqq \frac{1}{n} K + \int | g_n - H | \, d\bar{\mu}.$$

Since by hypothesis and (9.1) g_n tends to H in L^1 mean ($\bar{\mu}$), the first conclusion of the theorem follows.

For the second conclusion of the theorem, we note that

$$\int | f_n(x, \omega) - H | \, d\mu(x \otimes \omega) = \sum_{\omega} \mu(A^I \otimes \omega) \int | f_n(x, \omega) - H | \, d\mu_\omega(x).$$

Since the left-hand side has limit zero, every term on the right for which $\mu(A^I \otimes \omega) \neq 0$ must have limit zero.

As an application of this theorem let $[B, \nu]$ be a stationary source. Let $[A, \bar{\mu}]$ be a projection by φ of $[B, \nu]$. Let Ω coincide with the alphabet B and for $S \varepsilon F_A$ define μ by $\mu(S \otimes \omega) = \nu(\Phi^{-1}(S) \cap D_\omega)$ where D_ω is the set of $y \varepsilon B^I$ such that $y_{-1} = \omega$. The theorem then implies (if $[A, \bar{\mu}]$ is ergodic) that the rate of $[A, \mu]$ may be calculated by considering conditional probabilities knowing that the letter ω occurred at time -1. When $[B, \nu]$ is finite and Markov, this often leads to simplified calculations.

As another application, consider a fixed countable partition of A^I into sets $S_\omega \varepsilon F_A$. Given an ergodic source $[A, \bar{\mu}]$, define $\mu(S \otimes \omega)$ by $\mu(S \otimes \omega) = \bar{\mu}(S \cap S_\omega)$. The theorem then implies that the entropy rate of $[A, \bar{\mu}]$ can be calculated using only partitions which refine the given one.

REFERENCES

[1] NORBERT WIENER, *Extrapolation, Interpolation, and Smoothing of Stationary Time Series*, John Wiley and Sons, 1949.

[2] C. E. SHANNON, "A mathematical theory of communication," *Bell System Technical Journal*, Vol. 27, pp. 379–423, pp. 623–656.

[3] P. R. HALMOS, *Measure Theory*, D. Van Nostrand, 1950.

[4] NORBERT WIENER, "The ergodic theorem," *Duke Mathematical Journal*, Vol. 5, pp. 1–18.

[5] J. L. DOOB, *Stochastic Processes*, John Wiley and Sons, 1953.

A NEW BASIC THEOREM OF INFORMATION THEORY *

Amiel Feinstein
Research Laboratory of Electronics, Massachusetts Institute of Technology
Cambridge, Massachusetts

INTRODUCTION

Information theory, in the restricted sense used in this paper, originated in the classical paper of C. E. Shannon, in which he gave a precise mathematical definition for the intuitive notion of information. In terms of this definition it was possible to define precisely the notion of a communication channel and its capacity. Like all definitions that purport to deal with intuitive concepts, the reasonability and usefulness of these definitions depend for the most part on theorems whose hypotheses are given in terms of the new definitions but whose conclusions are in terms of previously defined concepts. The theorems in question are called the fundamental theorems for noiseless and noisy channels. We shall deal exclusively with noisy channels.

By a communication channel we mean, in simplest terms, an apparatus for signaling from one point to another. The abstracted properties of a channel that will concern us are: (a) a finite set of signals that may be transmitted; (b) a set (not necessarily finite) of signals that may be received; (c) the probability (or probability density) of the reception of any particular signal when the signal transmitted is specified. A simple telegraph system is a concrete example. The transmitted signals are a long pulse, a short pulse, and a pause. If there is no noise in the wire, the possible received signals are identical with the transmitted signals. If there is noise in the wire, the received signals will be mutilations of the transmitted signals, and the conditional probability will depend on the statistical characteristics of the noise present.

We shall now sketch the definitions and theorems mentioned. Let X be a finite abstract set of elements x, and let p( ) be a probability distribution on X. We define the "information content" of X by the expression $-\sum_X p(x) \log_2 p(x)$, where the base 2 simply determines the fundamental unit of information, called "bit". One intuitive way of looking at this definition is to consider a machine that picks, in a purely random way but with the given probabilities, one x per second from X. Then $-\log_2 p(x_o)$ may be considered as the information or surprise associated with the event that x_o actually came up. If each event x consists of several events, that is, if $x = \{a, b, \ldots\}$, we have the following meaningful result: $H(X) \leq H(A) + H(B) + \ldots$ with equality if, and only if, the events a, b, . . . are mutually independent.

We are now in a position to discuss the first fundamental theorem. We set ourselves the following situation. We have the set X. Suppose, further, that we have some "alphabet" of D "letters" which we may take as 0, . . . , D-1. We wish to associate to each x a sequence of integers 0, . . . , D-1 in such a way that no sequence shall be an extension of some shorter sequence (for otherwise they would be distinguishable by virtue of their length, which amounts to introducing a $D+1^{th}$ "variable"). Now it is easy to show that a set of D elements has a maximum information content when each element has the same probability, namely 1/D. Suppose now that with each x we associate a sequence of length N_x. The maximum amount of information obtainable by "specifying" that sequence is $N_x \log_2 D$ bits. Suppose $N_x \log_2 D = -\log_2 p(x)$; then $\sum_X p(x) N_x = H(X)/\log_2 D$ is the average length of the sequence. The first

*This work was supported in part by the Signal Corps; the Office of Scientific Research, Air Research and Development Command; and the Office of Naval Research.

Reprinted from *IRE Trans. Inform. Theory*, vol. IT-4, pp. 2-22, Sept. 1954.

fundamental theorem now states that if we content ourselves with representing sequences of x's by sequences of integers 0, . . . , D-1, then if we choose our x-sequences sufficiently long, the sequences of integers representing them will have an average length as little greater then $H(X)/\log_2 D$ as desired, but that it is not possible to do any better than this.

To discuss the second fundamental theorem, we now take, as usual, X to be the set of transmitted messages and Y the set of received signals. For simplicity we take Y finite. The conditional probability mentioned above we denote by $p(y/x)$. Let $p(\ )$ be a probability distribution over X, whose meaning is the probability of each x being transmitted. Then the average amount of information being fed into the channel is $H(X) = -\sum_X p(x) \log_2 p(x)$. Since in general the reception of a y does not uniquely specify the x transmitted, we inquire how much information was lost in transmission. To determine this, we note that, inasmuch as the x was completely specified at the time of transmission, the amount of information lost is simply the amount of information necessary (on the average, of course) to specify the x. Having received y, our knowledge of the respective probability of each x having been the one transmitted is given by $p(x/y)$. The average information needed to specify x is now $-\sum_X p(x/y) \log_2 p(x/y)$. We must now average this expression over the set of possible y's. We obtain finally

$$\sum_Y p(y) \left[-\sum_X p(x/y) \log_2 p(x/y) \right] = -\sum_Y \sum_X p(x, y) \log_2 p(x/y) \equiv H(X/Y)$$

often called the equivocation of the channel. The rate at which information is received through the channel is therefore $R = H(X) - H(X/Y)$. A precise statement of the fundamental theorem for noisy channels is given in section II.

I. For the sake of definiteness we begin by stating a few definitions and subsequent lemmas, more or less familiar.

Let X and Y be abstract sets consisting of a finite number, α and β, of points x and y. Let $p(\ )$ be a probability distribution over X, and for each $x \in X$ let $p(\ /x)$ denote a probability distribution over Y. The totality of objects thus far defined will be called a communication channel.

The situation envisaged is that X represents a set of symbols to be transmitted and Y represents the set of possible received signals. Then $p(x)$ is the a priori probability of the transmission of a given symbol x, and $p(R/x)$ is the probability of the received signal lying in a subset R of Y, given that x has been transmitted. Clearly, $\sum_{x \in Q} p(x)\, p(R/x)$ represents the joint probability of R and a subset Q of X, and will be written as $p(Q, R)$. Further, $p(X, R) \equiv p(R)$ represents the absolute probability of the received signal lying in R. (The use of p for various different probabilities should not cause any confusion.)

The "information rate" of the channel "source" X is defined by $H(X) = -\sum_X p(x) \log p(x)$, where here and in the future the base of the logarithm is 2. The "reception rate" of the channel is defined by the expression

$$\sum_X \sum_Y p(x, y) \log \frac{p(x, y)}{p(x)\, p(y)} \geqslant 0$$

If we define the "equivocation" $H(X/Y) = -\sum_{X}\sum_{Y} p(x,y)\log p(x/y)$ then the reception rate is given by $H(X) - H(X/Y)$. The equivocation can be interpreted as the average amount of information, per symbol, lost in transmission. Indeed we see that $H(X/Y) = 0$ if and only if $p(x/y)$ is either 0 or 1, for any x, y, that is, if the reception of a y uniquely specifies the transmitted symbol. When $H(X/Y) = 0$ the channel is called noiseless. If we interpret $H(X)$ as the average amount of information, per symbol, required to specify a given symbol of the ensemble X, with $p(\ )$ as the only initial knowledge about X, then $H(X) - H(X/Y)$ can be considered as the average amount, per symbol transmitted, of the information obtained by the (in general) only partial specification of the transmitted symbol by the received signal.

Let now u(v) represent a sequence of length n (where n is arbitrary but fixed) of statistically independent symbols x(y), and let the space of all sequences be denoted by U(V). In the usual manner we can define the various "product" probabilities. The n will be suppressed throughout. It is now simple to verify the following relations:

$$\log p(u) = \sum_{i=1}^{n} \log p(x_i), \text{ where } u = \{x_1, \ldots, x_n\} \tag{1}$$

$$\log p(u/v) = \sum_{i=1}^{n} \log p(x_i/y_i), \text{ where } v = \{y_1, \ldots, y_n\} \tag{2}$$

$$H(X) = -\frac{1}{n}\sum_{U} p(u)\log p(u) \tag{3}$$

$$H(X/Y) = -\frac{1}{n}\sum_{U}\sum_{V} p(u,v)\log p(u/v) \tag{4}$$

The weak law of large numbers at once gives us the following lemma, which is fundamental for the proof of Shannon's theorem (see also section V).

LEMMA 1. For any ϵ, δ there is an $n(\epsilon,\delta)$ such that for any $n \geq n(\epsilon,\delta)$ the set of u for which the inequality $|H(X) + (1/n)\log p(u)| < \epsilon$ does not hold has $p(\ )$ probability less then δ. Similarly, but with a different $n(\epsilon,\delta)$, the set of pairs (u, v) for which the inequality $|H(X/Y) + (1/n)\log p(u/v)| < \epsilon$ does not hold has $p(\ ,\ )$ probability less than δ.

In what follows we shall need only the weaker inequalities $p(u) < 2^{-n(H(X)-\epsilon)}$ and $p(u/v) > 2^{-n(H(X/Y)+\epsilon)}$. The probability of these inequalities failing will be denoted by δ^- and δ^+, respectively.

The following lemma is required to patch up certain difficulties caused by the inequalities of lemma 1 failing to hold everywhere.

LEMMA 2. Let Z be a (u, v) set of $p(\ ,\ )$ probability greater than $1 - \delta_1$ and U_o a set of u with $p(U_o) > 1 - \delta_2$. For each $u \in U$ let A_u be the set of v's such that $(u, A_u) \in Z$. Let $U_1 \subset U_o$ be the set of $u \in U_o$ for which $p(A_u/u) \geq 1 - \alpha$. Then $p(U_1) > 1 - \delta_2 - (\delta_1/\alpha)$.

PROOF. Let U_2 be the set of u for which $p(A_u^c/u) > \alpha$, where A_u^c is the complement of A_u. Then $p(u, A_u^c) \geq \alpha p(u)$ for $u \in U_2$, and $\sum_{U_2} (u, A_u^c)$ is, by the definition of A_u, outside Z. Hence

$$\delta_1 \geq \sum_{U_2} p(u, A_u^c) \geq \alpha p(U_2), \quad \text{or } p(U_2) \leq \frac{\delta_1}{\alpha}$$

Thus $p(U_2 \cdot U_o) \leq \delta_1/\alpha$ and, using $U_1 = U_o - U_o \cdot U_2$, we have

$$p(U_1) = p(U_o) - p(U_o \cdot U_2) > 1 - \delta_2 - \frac{\delta_1}{\alpha}$$

II. We have seen that, by our definitions, the average amount of information received, per symbol transmitted, is $H(X) - H(X/Y)$. However, in the process of transmission an amount $H(X/Y)$ is lost, on the average. An obvious question is whether it is, in some way, possible to use the channel in such a manner that the average amount of information received, per symbol transmitted, is as near to $H(X) - H(X/Y)$ as we please, while the information lost per symbol is, on the average, as small as we please. Shannon's theorem asserts (1), essentially, that this is possible. More precisely, let there be given a channel with rate $H(X) - H(X/Y)$. Then for any $e > 0$ and $H < H(X) - H(X/Y)$ there is an $n(e,H)$ such that for each $n \geq n(e,H)$ there is a family $\{u_i\}$ of message sequences (of length n) of number at least 2^{nH}, and a probability distribution on the $\{u_i\}$ such that, if only the sequences $\{u_i\}$ are transmitted, and with the given probabilities, then they can be detected with average probability of error less than e. The method of detection is that of maximum conditional probability, hence the need for specifying the transmission probability of the $\{u_i\}$. By average probability of error less than e is meant that if e_i is the fraction of the time that when u_i is sent it is misinterpreted, and p_i is u_i's transmission probability, then $\sum_i e_i p_i < e$.

A sufficient condition (2) for the above-mentioned possibility is the following:

For any $e > 0$ and $H < H(X) - H(X/Y)$ there is an $n(e, H)$ of such value that among all sequences u of length $n \geq n(e, H)$ there is a set $\{u_i\}$, of number at least 2^{nH}, such that:

1. to each u_i there is a v-set B_i with $p(B_i/u_i) > 1 - e$
2. the B_i are disjoint.

What this says is simply that if we agree to send only the set $\{u_i\}$ and always assume that, when the received sequence lies in B_i, u_i was transmitted, then we shall misidentify the transmitted sequence less than a fraction e of the time. As it stands, however, the above is not quite complete; for, if C is the largest number such that for $H < C$ there is an $n(e, H)$ and a set of at least 2^{nH} sequences u_i satisfying 1 and 2, C is well defined in terms of $p(X/Y)$ alone. However, $H(X) - H(X/Y)$ involves $p(X)$ in addition to $p(X/Y)$. One might guess that C is equal to l.u.b. $(H(X) - H(X/Y))$ over all choices of $p(\ )$. This is indeed so, as the theorem below shows. Note the important fact that we have here a way of defining the channel capacity C without once mentioning information contents or rates. (Strictly speaking we should now consider the channel as being defined simply by $p(y/x)$.) These remarks evidently apply equally well to Shannon's theorem, as we have stated it. We go now to the main theorem.

THEOREM. For any $e > 0$ and $H < C$ there is an $n(e, H)$ such that among all sequences u of length $n \geq n(e, H)$ there is a set $\{u_i\}$, of number at least 2^{nH} such that:

1. to each u_i there is a v-set B_i, with $p(B_i/u_i) > 1 - e$.
2. the B_i are disjoint.

This is not possible for any $H > C$.

PROOF. Let us note here that if we transmit the u_i with equal probability and use a result of section III (namely $P_e \leq e$) we immediately obtain the positive assertion of Shannon's theorem. We shall first indicate only the proof that the theorem cannot hold for $H > C$, which is well known. Indeed if one could take $H > C$ then, as shown in section III one would have, for n sufficiently large, the result that the information rate per symbol would exceed C. But this cannot be (3). Q. E. D. In the following we will take $p(\ )$ as that for which the value C is actually attained (4). We shall see, however, that no use of this fact is actually made in what follows, other than, of course, $C = H(X) - H(X/Y)$.

For given $\epsilon_1, \delta_1^+, \epsilon_2, \delta_2^-$, let $n_1(\epsilon_1, \delta_1^+)$, $n_2(\epsilon_2, \delta_2^-)$ be as in lemma 1 for $p(u/v) > 2^{-n(H(X/Y)+\epsilon_1)}$ and $p(u) < 2^{-n(H(X)-\epsilon_2)}$, respectively. Let us henceforth consider n as fixed and $n \geq \max\left(n_1(\epsilon_1, \delta_1^+), n_2(\epsilon_2, \delta_2^-)\right)$. For Z and U_o in lemma 2 we take, respectively, the sets on which the first two inequalities stated above hold. Then for any $u \in U_1$ (with a as any fixed number $< e$) and v in the corresponding A_u we have:

$$\frac{p(u/v)}{p(u)} > \frac{2^{-n(H(X/Y)+\epsilon_1)}}{2^{-n(H(X)-\epsilon_2)}} = 2^{n(C-\epsilon_1-\epsilon_2)}, \text{ or}$$

$$\frac{p(u, v)}{p(u)} > 2^{n(C-\epsilon_1-\epsilon_2)} p(v)$$

Summing v over A_u we have

$$\frac{p(u, A_u)}{p(u)} > 2^{n(C-\epsilon_1-\epsilon_2)} p(A_u)$$

Since $1 \geq p(A_u/u)$ we have finally

$$p(A_u) < 2^{-n(C-\epsilon_1-\epsilon_2)}$$

Let $u_1, \ldots, u_N$ be a set M of members of U such that:

a. to each u_i there is a v-set B_i with $p(B_i/u_i) > 1 - e$

b. $p(B_i) < 2^{-n(C-\epsilon_1-\epsilon_2)}$ (See footnote 5.)

c. the B_i are disjoint

d. the set M is maximal, that is, we cannot find a u_{N+1} and a B_{N+1} such that the set $u_1, \ldots, u_{N+1}$ satisfies (a) to (c).

Now for any $u \in U_1$ there is by definition an A_u such that $p(A_u/u) \geq 1 - a > 1 - e$ and as we have seen above, $p(A_u) < 2^{-n(C-\epsilon_1-\epsilon_2)}$. Furthermore, for any $u \in U_1$, $A_u - A_u \cdot \sum_i B_i$ is disjoint from the B_i, and certainly

$$p\left(A_u - A_u \cdot \sum_i B_i\right) < 2^{-n(C-\epsilon_1-\epsilon_2)}$$

If u is not in M, we must therefore have

$$p\left(A_u - A_u \cdot \sum_i B_i/u\right) \leq 1 - e$$

In other words, $p\left(A_u \cdot \sum_i B_i/u\right) \geq e - \alpha$, or certainly

$$p\left(\sum_i B_i/u\right) \geq e - \alpha, \text{ for all } u \in U_1 - M \equiv U_1 - M \cdot U_1$$

Now

$$p\left(\sum_i B_i\right) = \sum_U p\left(\sum_i B_i/u\right) p(u) \geq \left\{\sum_{U_1 - M \cdot U_1} + \sum_{M \cdot U_1}\right\} p\left(\sum_i B_i/u\right) p(u)$$

$$\geq (e-\alpha)\left[1 - \beta_2^- - \frac{\delta_1^+}{\alpha} - p(M \cdot U_1)\right] + (1-e)\, p(M \cdot U_1) \geq (e-\alpha)\left[1 - \delta_2^- - \frac{\delta_1^+}{\alpha}\right]$$

if $e \leq 1/2$, since then $1 - e \geq e - \alpha$.

On the other hand, $p\left(\sum_i B_i\right) < N2^{-n(C-\epsilon_1-\epsilon_2)}$. Hence

$$N2^{-n(C-\epsilon_1-\epsilon_2)} > (e-\alpha)\left[1 - \delta_2^- - \frac{\delta_1^+}{\alpha}\right]$$

If $e > 1/2$ then, using $p(M \cdot U_1) < N2^{-n(H(X)-\epsilon_2)}$, we would obtain

$$N2^{-n(C-\epsilon_1-\epsilon_2)} > (e-\alpha)\left[1 - \delta_2^- - \frac{\delta_1^+}{\alpha} - N2^{-n(H(X)-\epsilon_2)}\right]$$

Since the treatment of both cases is identical, we will consider $e \leq 1/2$.

To complete the proof we must show that for any e and $H < C$ it is possible to choose ϵ_1, ϵ_2, δ_1^+, δ_2^-, $\alpha < e$, and $n \geq \max\left(n_1(\epsilon_1, \delta_1^+), n_2(\epsilon_2, \delta_2^-)\right)$ in such a way that the above inequality requires $N \geq 2^{nH}$. Now it is clear that, if, having chosen certain fixed values for the six quantities mentioned, the inequality fails upon the insertion of a given value (say N^*) for N, then the smallest N for which the inequality holds must be greater than N^*. Let us point out that N will in general depend upon the particular maximal set considered.

We take $N^* = 2^{nH}$ and $\alpha = e/2$. Then we can take δ_1^+, δ_2^-, and ϵ_2 so small and n so large that

$$\left[1 - \delta_2^- - \frac{\delta_1^+}{\alpha}\right] \text{ is } > \frac{2}{3} \text{ say.}$$

We obtain finally $e/3 < 2^{-n(C-H-\epsilon_2-\epsilon_1)}$. Choosing ϵ_2 and ϵ_1 sufficiently small so that $C - H - \epsilon_2 - \epsilon_1 > 0$ we see that for sufficiently large n the inequality $e/3 < 2^{-n(C-H-\epsilon_2-\epsilon_1)}$ fails. Hence for $\alpha = e/2$, for ϵ_1, ϵ_2, δ_1^+, δ_2^- sufficiently small the insertion of $N^* = 2^{nH}$ for N causes the inequality to fail for all n sufficiently large. Thus $N > N^* = 2^{nH}$ for such n. Q.E.D.

It is worthwhile to emphasize that the codes envisaged here, unlike those of Shannon, are uniformly good, i. e., the probability of error for the elements of a maximal set is uniformly $<e$. These codes are therefore error correcting, which answers in the affirmative the question as to whether the channel capacity can be approached using such codes (6).

If we wish to determine how e decreases as a function of n, for fixed H, we have (7):

$$e \leq a + \frac{A}{B - (\delta_1^+/a)}, \quad \text{where } A = 2^{-n(C-H-\epsilon_1-\epsilon_2)}, \quad B = 1 - \delta_2$$

To eliminate the "floating" variable a, we proceed as follows. For $a > 0$

$$a + \frac{A}{B - (\delta_1^+/a)} \quad \text{achieves its minimum value at } a = \frac{(A\delta_1^+)^{1/2} + \delta_1^+}{B}$$

and this value, namely, $\frac{1}{B}\left[A^{1/2} + (\delta_1^+)^{1/2}\right]^2$, is greater than $\frac{(A\delta_1^+)^{1/2} + \delta_1^+}{B}$

If we take

$$a = \frac{(A\delta_1^+)^{1/2} + \delta_1^+}{B} \quad \text{and } e = \frac{1}{B}\left[A^{1/2} + (\delta_1^+)^{1/2}\right]^2$$

then $a < e$. Hence $\frac{1}{B}\left[A^{1/2} + (\delta_1^+)^{1/2}\right]^2$ is an upper bound for the minimum value of e which is possible for a given H. This expression is still a function of ϵ_1 and ϵ_2. The best possible upper bound which can be obtained in the present framework is to minimize with respect to ϵ_1 and ϵ_2. This cannot be done generally and in closed form.

Let us remark, however, that at this point we cannot say anything concerning a lower bound for e. In particular, the relation $a < e$ is a condition that is required only if we wish to make use of the framework herein considered.

III. Let us consider a channel (i. e., (S, s), (R, r), p( ) and p(/s) where s is a transmitted and r a received symbol) such that to each s there is an r-set A_s such that $p(A_s/s) \geq 1 - e$ and the A_s are disjoint. For each r let $p_e(r) = 1 - p(s_r/r)$ where s_r is such that $p(s_r/r) \geq p(s/r)$ for all $s \neq s_r$. (Then $p_e(r)$ is simply the probability that when r is received an error will be made in identifying the symbol transmitted, assuming that whenever r is received s_r will be assumed to have been sent.) Now the inequality $\ln a \leq a - 1$ can be used to show that

$$H(S/R) \leq -P_e \log P_e - (1 - P_e) \log (1 - P_e) + P_e \log (N-1)$$

where $P_e = \sum_R p(r)\, p_e(r)$ and N is the number of symbols in S.

We now make use of the special properties of the channel considered. We have

$$P_e = \sum_R p(r)(1 - p(s_r/r)) = 1 - \sum_R p(r)\,p(s_r/r)$$

$$= 1 - \sum_S \sum_{A_s} p(r)\,p(s_r/r) - \sum_{R-\Sigma_S A_s} p(r)\,p(s_r/r)$$

$$\leq 1 - \sum_S \sum_{A_s} p(r)\,p(s/r) - \sum_{R-\Sigma_S A_s} p(r)\,p(s_o/r)$$

$$= 1 - \sum_{S-s_o} \sum_{A_s} p(r)\,p(s/r) - \sum_{R-\Sigma_{S-s_o} A_s} p(r)\,p(s_o/r)$$

$$= 1 - \sum_{S-s_o} p(s)\,p(A_s/s) - p(s_o)\,p\left(R - \sum_{S-s_o} A_s/s_o\right)$$

$$\leq 1 - \sum_{S-s_o} p(s)(1-e) - p(s_o)(1-e) = e \quad \text{where } s_o \text{ is any } s \text{ (8)}.$$

Then $H(S/R) \leq -e \log e - (1-e)\log(1-e) + e \log(N-1)$ since for $e < 1/2$ the left side of the above inequality is an increasing function of e. (We assume of course $e < 1/2$.)

Let us consider the elements $u_1, \ldots, u_N$ of some maximal set as the fundamental symbols of a channel. Then regardless of what $p(u_i)$ is, $i = 1, \ldots, N$, the channel is of the type considered above. Hence $P_e < e$ (where e is as in II) and

$$H(U/V) \leq -e \log e - (1-e)\log(1-e) + e \log(N-1)$$

Here $H(U/V)$ represents the average amount of information lost per sequence transmitted. The average amount lost per symbol is $1/n\, H(U/V)$. Now for $N = 2^{nH}$ and $H < C$, $e = e(n) \to 0$ as $n \to \infty$. Thus $1/n\, H(U/V) \to 0$ as $n \to \infty$. In particular if we take $p(u_i) = 2^{-nH}$, then $1/n\, [H(U) - H(U/V)] \to H$ as $n \to \infty$. (This is the proof mentioned in footnote 2.)

Actually, a much stronger result will be proven, namely, that for $N = 2^{nH}$, $H < C$ (and H fixed, of course) the equivocation per sequence $H(U/V)$, goes to zero as $n \to \infty$. Since $\log(N-1) \approx nH$, a sufficient condition that $H(U/V) \to 0$ as $n \to \infty$ is that $e(n)\,n \to 0$ as $n \to \infty$.

We saw that $e \leq \frac{1}{B}\left[A^{1/2} + \left(\delta_1^+\right)^{1/2}\right]^2$ where $B = 1 - \delta_2$ and $A = 2^{-n(C-H-\epsilon_1-\epsilon_2)}$.

Now if we take ϵ_1, ϵ_2 sufficiently small so that $C - H - \epsilon_1 - \epsilon_2 > 0$ and $H(X) - H - \epsilon_2 > 0$, then the behavior of δ_1^+ as $n \to \infty$ is the only unknown factor in the behavior of e. If the original X consists of only x_1, x_2, and Y consists of only y_1, y_2, and if $p(x_1/y_2) = p(x_2/y_1)$, then $\log p(x/y)$ is only two-valued. If we take $\epsilon_1 = \epsilon(n)$ as vanishing, for $n \to \infty$, faster than $n^{-1/6}$, then a theorem on large deviations (9) is applicable and shows that δ_1^+, and hence e, approaches zero considerably faster than $1/n$.

We omit the details inasmuch as a proof of the general case will be given in section V.

IV. Up till now we have considered the set Y of received signals as having a finite number of elements y. One can, however, easily think of real situations where this is not the case, and where the set Y is indeed nondenumerable. Our terminology and notation will follow the supplement of (10).

We define a channel by:

1. the usual set X and a probability distribution p( ) over X
2. a set Ω of points ω
3. a Borel field F of subsets Λ of Ω
4. for each $x \in X$, a probability measure p(/x) on F.

We define the joint probability $p(x, \Lambda) = p(x)\, p(\Lambda/x)$ and $p(\Lambda) = p(X, \Lambda) = \sum_X p(x, \Lambda)$. Since $p(x, \Lambda) \leq p(\ )$ for any x, Λ, we have by the Radon-Nikodym theorem

4.1 $$p(x, \Lambda) = \int_\Lambda p(x/\omega)\, p(d\omega) \text{ where } p(x/\omega) \text{ may be taken as } \leq 1 \text{ for all } x, \omega.$$

As the notation implies, $p(x/\omega)$ plays the role of a conditional probability.

We define $H(X) = -\sum_X p(x) \log p(x)$, as before. In analogy with the finite case we define

4.2 $$H(X/Y) = -\sum_X \int_\Omega \log p(x/\omega)\, p(x, d\omega)$$

To show that the integral is finite we see first, by section 4.1 that

$$p\left(x, \{p(x/\omega) = 0\}\right) = 0$$

Furthermore, putting

$$\Lambda_i = \left\{\frac{1}{2^{i+1}} < p(x/\omega) \leq \frac{1}{2^i}\right\}$$

we have, since $p(\Lambda_i) \leq p(\Omega) = 1$, that

$$\int_{\Lambda_i} p(x/\omega)\, p(d\omega) \leq \frac{p(\Lambda_i)}{2^i} \leq \frac{1}{2^i}$$

Hence

4.3 $$p\left(x, \left\{\frac{1}{2^{i+1}} < p(x/\omega) \leq \frac{1}{2^i}\right\}\right) \leq \frac{1}{2^i}$$

We therefore have

4.4 $$-\int_\Omega \log p(x/\omega)\, p(x, d\omega) < \sum_{i=0}^{\infty} \frac{i+1}{2^i} < \infty$$

by the ratio test.

Everything we have done in sections I, II, and III can now be carried over without change to the case defined above. A basic theorem in this connection is that we can find a finite number of disjoint sets Λ_j, $\sum_j \Lambda_j = \Omega$ such that $-\sum_X \sum_j p(x, \Lambda_j) \log p(x/\Lambda_j)$ approximates H(X/Y) as closely as desired. Since we make no use of it, we shall not prove it, though it follows easily from the results given above and from standard integral approximation theorems.

V. We shall now show that $e = e(n)$ goes to zero, as $n \to \infty$, faster than $1/n$, which will complete the proof that the equivocation goes to zero as the sequence length $n \to \infty$.

As previously mentioned, it is the behavior of δ_1^+, of lemma 1 that we must determine. The mathematical framework briefly is as follows.

We have the space $X \otimes \Omega$ of all pairs (x, ω) and a probability measure $p(\ ,\ )$ on the measurable sets of $X \otimes \Omega$. We consider the infinite product space $\prod_{i=1}^{\infty} \otimes (X \otimes \Omega)_i$ and the corresponding product measure

$$\prod_{i=1}^{\infty} \otimes\, p_i(\ ,\ ) \equiv p_\infty(\ ,\ ).$$

Let us denote a "point" of $\prod_{i=1}^{\infty} \otimes (X \otimes \Omega)_i$ by $(x_\infty, \omega_\infty) \equiv \{(x_1, \omega_1), (x_2, \omega_2), \ldots\}$

We define an infinite set of random variables $\{Z_i\}$, $i = 1, \ldots$ on

$$\prod_{i=1}^{\infty} \otimes (X \otimes \Omega)_i$$

by $Z_i(x_\infty, \omega_\infty) = -\log p(x_i/\omega_i)$, that is, Z_i is a function only of the ith coordinate of $(x_\infty, \omega_\infty)$. Clearly the Z_i are independent and identically distributed; we shall put $E(Z_i)$ for their mean value. From section 4.4 we know that the Z_i have moments of the first order. (One can similarly show, using the fact that

$$\infty > \sum_{i=0}^{\infty} \frac{(i+1)^n}{2^i} \quad \text{for any } n > 0,$$

that they have moments of all positive orders.)

Let $S_n = \sum_{i=1}^{n} Z_i$. Then the weak law of large numbers says that for any ϵ_1, δ_1, there is an $n(\epsilon_1, \delta_1)$ such that for $n \geq n(\epsilon_1, \delta_1)$ the set of points $(x_\infty, \omega_\infty)$ on which $\left|\frac{S_n}{n} - E(Z_1)\right| \geq \epsilon_1$ has $p_\infty(\ ,\ )$ measure less than δ_1. Now, in the notation of section I, $S_n(X_\infty, \omega_\infty) = -\log p(u/v)$ where $u = \{x_1, \ldots, x_n\}$ and $v = \{\omega_1, \ldots, \omega_n\}$, while $H(X/Y) = \sum_X \int_\Omega -\log p(x/\omega)\, p(x, d\omega) = E(Z_1)$. What we have stated, then, is simply lemma 1.

Now, we are interested in obtaining an upper bound for

$$\text{Prob}\left\{\frac{S_n}{n} - E(Z_1) \geq \epsilon_1\right\}$$

More precisely we shall find sequences $\epsilon_1(n)$ and $\delta_1^+(n)$ such that, as $n \to \infty$, $\epsilon_1(n) \to 0$, $\delta_1^+(n) \to 0$ faster than $1/n$, and $n(\epsilon_1(n), \delta_1^+(n)) = n$.

Let $Z_i^{(r)} = Z_i$ whenever $Z_i < r$, and $Z_i^{(r)} = 0$ otherwise. By section 4.3, $Z_i^{(r)}$ and Z differ on a set of probability $\leq 1/2^r$. Let $S_n^{(r)} = \sum_{i=1}^{n} Z_i^{(r)}$; then S_n and $S_n^{(r)}$ differ on a set of probability $\leq 1 - (1 - 2^{-r})^n < n/2^r$. Furthermore

$$E(Z_1) - E\left(Z_1^{(r)}\right) \leq \sum_{i=0}^{\infty} \frac{r + 1 + i}{2^{r+i}}$$

by the same argument which led to section 4.4. We thus have:

$$\text{Prob}\left\{\frac{S_n}{n} - E(Z_1) \geq \epsilon_1(n)\right\} \leq \text{Prob}\left\{\frac{S_n^{(r)}}{n} - E(Z_1) \geq \epsilon_1(n)\right\} + \frac{n}{2^r}$$

$$\leq \text{Prob}\left\{\frac{S_n^{(r)}}{n} - E\left(Z_1^{(r)}\right) \geq \epsilon_1(n)\right\} + \frac{n}{2^r},$$

since $E(Z_1) \geq E\left(Z_1^{(r)}\right)$. In order to estimate $\text{Prob}\left\{\frac{S_n^{(r)}}{n} - E\left(Z_1^{(r)}\right) \geq \epsilon_1(n)\right\}$ we use a theorem of Feller (11) which, for our purposes, may be stated as follows:

THEOREM: Let $\{X_i\}$, $i = 1, \ldots, n$ be a set of independent, identically distributed, bounded random variables. Let $S = \sum_{i=1}^{n} X_i$ and let

$$F(x) = \text{Prob}\{S - n\,E(X_1) \leq x\}$$

Put $\sigma^2 = E([X_1 - E(X_1)]^2)$ and take $\lambda > \frac{\sup|X_1 - E(X_1)|}{\sigma n^{1/2}}$. Then if $0 < \lambda x < 1/12$ we have

$$1 - F(x\sigma n^{1/2}) = \exp[-1/2\,x^2\,Q(x)]\,[\{1 - \Phi(x)\} + \theta\lambda\exp(-1/2\,x^2)]$$

where

$$|\theta| < 9, \quad |Q(x)| \leq \frac{1}{7}\left(\frac{12\lambda x}{1 - 12\lambda x}\right) \text{ and } \Phi(x) = \frac{1}{(2\pi)^{1/2}}\int_{-\infty}^{x} \exp[-y^2/2]\,dy$$

In order to apply this theorem, we take $r = r(n)$. Now

$$\sigma\left(Z_1^{(r)}\right) = E\left(\left[Z_1^{(r)} - E\left(Z_1^{(r)}\right)\right]^2\right)^{1/2} \to \sigma(Z_1) \quad \text{as } r \to \infty$$

Hence for suitably large n_0, $\frac{3}{2}\sigma(Z_1) > \sigma\left(Z_1^{(r)}\right) > \frac{1}{2}\sigma(Z_1)$ for $n \geq n_0$. We can now take $\lambda \equiv \lambda(n)\ \ \frac{2n^{-1/2}}{\sigma(Z_1)}\, r(n)$.

We henceforth consider $n \geq n_0$. We now have:

$$\text{Prob}\left\{\frac{S_n^{(r)}}{n} - E\left(Z_1^{(r)}\right) \geq \epsilon_1(n)\right\} = \text{Prob}\left\{S_n^{(r)} - n\,E\left(Z_1^{(r)}\right) \geq n\,\epsilon_1(n)\right\}$$

$$= \text{Prob}\left\{S_n^{(r)} - n\,E\left(Z_1^{(r)}\right) \geq \sigma\left(Z_1^{(r)}\right) n^{1/2}\left[n^{1/2}\frac{\epsilon_1(n)}{\sigma\left(Z_1^{(r)}\right)}\right]\right\}$$

$$\leq \text{Prob}\left\{S_n^{(r)} - n\,E\left(Z_1^{(r)}\right) \geq \sigma\left(Z_1^{(r)}\right) n^{1/2}\left[n^{1/2}\frac{2\epsilon_1(n)}{3(Z_1)}\right]\right\}$$

$$\leq \exp\left[\frac{1}{14}x^2\left(\frac{12\lambda x}{1 - 12\lambda x}\right)\right]\left[\{1 - \Phi(x)\} + 9\lambda\exp\left(-\frac{x^2}{2}\right)\right].$$

Using

$$1 - \Phi(x) \sim \frac{1}{(2\pi)^{1/2}\,x}\exp\left(-\frac{x^2}{2}\right)$$

or

$$1 - \Phi(x) \lesssim \frac{2}{(2\pi)^{1/2}\, x} \exp\left(-\frac{x^2}{2}\right)$$

we may rewrite the above as

$$\exp\left(x^2\left[\frac{6\lambda x}{7(1-12\lambda x)} - \frac{1}{2}\right]\right) \cdot \left\{9\lambda + \frac{2}{(2\pi)^{1/2}\, x}\right\}$$

Now $\lambda \equiv \lambda(n) = \dfrac{2n^{1/2}}{\sigma(Z_1)}\, r(n)$ and $x = n^{1/2}\, \dfrac{2\epsilon_1(n)}{3\sigma(Z_1)}$, while

$$\delta_1^+(n) \leq \exp\left(x^2\left[\frac{6\lambda x}{7(1-12\lambda x)} - \frac{1}{2}\right]\right) \cdot \left\{9\lambda(n) + \frac{2}{(2\pi)^{1/2}\, x}\right\} + \frac{n}{2^{r(n)}}$$

It is now clear that we can pick $\epsilon_1(n)$ and $r(n)$ so that $\lambda(n) \to 0$, $x = x(n) \to \infty$, $\lambda x \to 0$ and $\delta_1^+(n) \to 0$ faster than $1/n$.

Let us point out that by using the approximation theorem of section III and thus having to deal with $-\log p(x/\Lambda_j)$, which is bounded, we can eliminate the term $n/2^{r(n)}$. This makes it likely that Feller's theorem can be proven, in our case, without the restriction that the random variables be bounded. There is in fact a remark by Feller that the boundedness condition can be replaced by the condition that Prob $\{|X_i| > n\}$ is a sufficiently rapidly decreasing function of n. But any further discussion would take us too far afield.

VI. We have, up to this point, insisted that the set X of messages be finite. We wish to relax this condition now so that the preceding work can be applied to the continuous channels considered by Shannon (1) and others. However, any attempt to simply replace finite sums by denumerable sums or integrals at once leads to serious difficulties. One can readily find simple examples for which H(X), H(X/Y) and H(X) - H(X/Y) are all infinite.

On the other hand, we may well ask what point there is in trying to work with infinite message ensembles. In any communication system there are always only a finite number of message symbols to be sent, that is, the transmitter intends to send only a finite variety of message symbols. It is quite true that, for example, an atrociously bad telegrapher, despite his intention of sending a dot, dash, or pause, will actually transmit any one of an infinite variety of waveforms only a small number of which resemble intelligible signals. But we can account for this by saying that the "channel" between the telegrapher's mind and hand is "noisy," and, what is more to the point, it is a simple matter to determine all the statistical properties that are relevant to the capacity of this "channel." The channel whose message ensemble consists of the finite number of "intentions" of the telegrapher and whose received signal ensemble is an infinite set of waveforms resulting from the telegrapher's incompetence and noise in the wire is thus of the type considered in section IV.

The case in which one is led to the consideration of so-called continuous channels is typified by the following example. In transmitting printed English via some teletype system one could represent each letter by a waveform, or each pair by a waveform, or every letter and certain pairs by a waveform, and so on. We have here an arbitrariness both in the number of message symbols and in the waveforms by which they are to be represented. It is now clear that we should extend the definition of a channel and its capacity in order to include the case given above.

DEFINITION. Let X be a set of points x and Ω a set of points ω. Let F be a Borel field of subsets Λ of Ω, and let $p(\ /x)$ be, for each $x \in X$, a probability measure on F. For each finite subset R of X the corresponding channel and its capacity C_R is well defined by section IV. The quantity $C = \text{l.u.b. } C_R$ over all finite subsets R of X will be called the capacity of the channel $\{X, p(\ /x), \Omega\}$.

Now for any $H < C$ there is a C_R with $H < C_R \leq C$, so that all our previous results are immediately applicable.

We shall now show that the channel capacity defined above is, under suitable restrictions, identical with that defined by Shannon (1).

Let X be the whole real line, and Ω, ω, F, and Λ as usual. Let $p(x)$ be a continuous probability density over X and for each $\Lambda \in F$, let $p(\Lambda/x)$ satisfy a suitable continuity condition. (See the Appendix for this and subsequent mathematical details.) Then $p(\Lambda) \equiv \int_{-\infty}^{\infty} p(x)p(\Lambda/x)\,dx$ is a probability measure. Since $p(x, \Lambda) \equiv p(x)p(\Lambda/x)$ is, for each x, absolutely continuous with respect to $p(\Lambda)$ we can define the Radon-Nikodym derivative $p(x/\omega)$ by $p(x, \Lambda) = \int_{\Lambda} p(x/\omega)p(d\omega)$. Then, with the x-integral taken as improper, we can define

$$C_p \equiv \int_{-\infty}^{\infty} dx \int_{\Omega} p(x, d\omega) \log \frac{p(x/\omega)}{p(x)} \geq 0$$

If we put $C_s = \text{l.u.b. } C_p$ over all continuous probability densities $p(x)$, then C_s is Shannon's definition of the channel capacity. The demonstration of the equivalence of C, as defined above, and C_s is now essentially a matter of approximating an integral by a finite sum, as follows:

If C_s is finite, then we can find a C_p arbitrarily close to C_s; if $C_s = +\infty$ we can find C_p arbitrarily large. We can further require that $p(x)$ shall vanish outside a suitably large interval, say $[-A, A]$. We can now find step-functions $g(x)$ defined over $[-A, A]$ that approximate $p(x)$ uniformly to any desired degree of accuracy, and whose integral is 1. For such a step-function, C_g is well defined and approximates C_p as closely as desired by suitable choice of $g(x)$.

Let $g(x)$ have n steps, with area p_i, and of course $\sum_1^n p_i = 1$. By suitably choosing positive numbers a_{ij}, integers N_i, and points x_{ij}, with x_{ij} lying in the ith step of $g(x)$ and $\sum_{j=1}^{N_i} a_{ij} = p_i$, we can approximate

$$p(\Lambda) \equiv \int_{-A}^{A} g(x)\,p(\Lambda/x)\,dx \quad \text{by} \quad \sum_{i=1}^{n} \sum_{j=1}^{N_i} a_{ij}\,p(\Lambda/x_{ij})$$

and hence C_g by C_R, where $R = \{x_{ij}\}$. Thus $C \geq C_s$. On the other hand, let $R = \{x_i\}$, not as taken above. Let $p(x_i)$ be such that $H(X) - H(X/Y) = C_R$. Then the singular function $\sum_i p(x_i)\,\delta(x - x_i)$, where $\delta(\ )$ is the Dirac delta-function, can be approximated by continuous probability densities $p(x)$ such that C_p approximates C_R. Hence $C_s \geq C$, or $C = C_s$.

This can clearly be generalized to the case in which X is n-dimensional Euclidean space.

VII. We now wish to relax the condition of independence between successive transmitted symbols. Our definitions will be those of Shannon, as generalized by McMillan, whose paper (1) we now follow.

By an alphabet we mean a finite abstract set. Let A be an alphabet and I the set of all integers, positive, zero, and negative. Denote by A^I the set of all sequences $x = (\ldots, x_{-1}, x_o, x_1, \ldots)$ with $x_t \in A$, $t \in I$.

A cylinder set in A^I is a subset of A^I defined by specifying an integer $n \geq 1$, a finite sequence $\alpha_o, \ldots, \alpha_{n-1}$, of letters of A, and an integer t. The cylinder set corresponding to these specifications is $\{x \in A^I / x_{t+k} = \alpha_k, \ k = 0, \ldots, n-1\}$. We denote by F_A the Borel field generated by the cylinder sets.

An information source $[A, \mu]$ consists of an alphabet A and a probability measure μ defined on F_A. Let T be defined by $T(\ldots, x_{-1}, x_o, x_1, x_2, \ldots) = (\ldots, x'_{-1}, x'_o, x'_1 \ldots)$ where $x'_t = x_{t+1}$. Then $[A, \mu]$ will be called stationary if, for $S \in F_A$, $\mu(S) = \mu(TS)$ (clearly T preserves measurability) and will be called ergodic if it is stationary and $S = TS$ implies that $\mu(S) = 1$ or 0.

By a channel we mean the system consisting of:

1. a finite alphabet A and an abstract space B.
2. a Borel field of subsets of B, designated by β, with $B \in \beta$
3. the Borel field of subsets of $B^I \equiv \prod_{-\infty}^{\infty} \otimes B_i$ (where $B_i = B$) which we define in the usual way, $\prod_{-\infty}^{\infty} \otimes \beta$, and designate F_β.
4. a function ν_x which is, for each $x \in A^I$, a probability measure on F_β, and which has the property that if $x_t^1 = x_t^2$ for $t \leq n$, then $\nu_{x^1}(S) = \nu_{x^2}(S)$ for any $S \in F_\beta$ of the form $S = S_1 \otimes S_2$, where $S_1 \in \prod_{-\infty}^{n} \otimes \beta$ and $S_2 = \prod_{n+1}^{\infty} \otimes B$.

Consider a stationary channel whose input A is a stationary source $[A, \mu]$. Let $C^I = A^I \otimes B^I$ and $F_C = F_A \otimes F_\beta$. We can define a probability measure on F_C by $p(R, S) \equiv p(R \otimes S) = \int_R \nu_x(S)\, d\mu(x)$ for $R \in F_A$, $S \in F_\beta$, assuming certain measurability conditions for $\nu_x(S)$. It is then possible to define the information rate of the channel source, the equivocation of the channel, and the channel capacity in a manner analogous to that of section I. Assuming that $\mu(\ )$ and $p(\ ,\ )$ are ergodic, McMillan proves lemma 1 of section I in this more general framework. Hence the proof of section III remains completely valid, except for the demonstration that the theorem cannot hold for $H > C$.

The difficulty that we wish to discuss arises in the interpretation of $p(\ /u)$. A glance at McMillan's definitions shows that $p(B/u)$ no longer can be interpreted as "the probability of receiving a sequence lying in B, given that the sequence u was sent." This direct causal interpretation is valid only for $\nu_x(\ )$. But the result of the theorem of section II is the existance of a set u_i and disjoint sets B_i such that $p(B_i/u_i) > 1 - e$. Under what conditions can we derive from this an analogous statement for $\nu_{u_i}(B_i)$?

Suppose that for a given integer N we are given, for each sequence $x_1, \ldots, x_{N+1}$ of message symbols, a probability measure $\nu(\ /x_1, \ldots, x_{N+1})$

on the Borel field β of received signals (not sequences of signals). We envisage here the situation in which the received signal depends not only upon the transmitted symbol x_{N+1} but also upon the preceding N symbols which were transmitted.

If $u = \{x_1, \ldots, x_n\}$ then

$$p(\ /u) \equiv \sum_{[x_{-N+1}, \ldots, x_o]} \frac{p(x_{-N+1}, \ldots, x_n)}{p(x_1, \ldots, x_n)}$$

$$\times [\nu(\ /x_{n-N}, \ldots, x_n) \otimes \ldots \otimes \nu(\ /x_{-N+1}, \ldots, x_1)]$$

Let us write the bracket term, which is a probability measure on received sequences of length n, as $\nu_n(\ /x_{-N+1}, \ldots, x_n)$. Now if $p(B_i/u_i) > 1 - e$, then, since

$$\sum_{[x_{-N+1}, \ldots, x_o]} \frac{p(x_{-N+1}, \ldots, x_n)}{p(x_1, \ldots, x_n)} = 1$$

there must be at least one sequence $\{x_{-N+1}, \ldots, x_n\}$ for which

$$\nu_n(B_i/x_{-N+1}, \ldots, x_n) > 1 - e$$

A minor point still remains: we had 2^{nH} sequences u_i and we now have the same number of sequences, but of length n + N. In other words, we are transmitting at a rate $H' = (n/n+N) H$. But since N is fixed we can make H' as near as we choose to H by taking n sufficiently large; hence we can still transmit at a rate as close as desired to the channel capacity.

It is evident that by imposing suitable restrictions on $\nu_x(\ )$ we can do the same sort of thing in a more general context. These restrictions would amount to saying that the channel characteristics are sufficiently insensitive to the remote past history of the channel.

In this connection some interesting mathematical questions arise. If we define the capacity following McMillan for the $\nu(\ /x_1, \ldots, x_{N+1})$ as above, is the capacity actually achieved? It seems reasonable that it is, and that the channel source that attains the capacity will automatically be of the mixing type (see ref. 12, p. 36, Def. 11.1; also p. 57) and hence ergodic. Because of the special form of $\nu_x(\ )$ it easily follows that the joint probability measure would likewise be of mixing type and hence ergodic.

The question of whether or not the equivocation vanishes in this more general setup is also unsettled. Presumably one might be able to extend Feller's theorem to the case of nonindependent random variables that approach independence, or perhaps actually attain independence when far enough apart. To my knowledge nothing of this sort appears in the literature.

Finally there is the question of whether or not, in the more general cases, the assertion that for H > C the main theorem cannot hold is still true. While this seems likely, at least in the case of a channel with finite memory, it is to my knowledge unproven.

APPENDIX

It is our purpose here to supply various proofs that were omitted in the body of the work.

1. $H(X) - H(X/Y)$ is a continuous function of the $p(x_i)$, $i = 1, \ldots, a$.

PROOF. $H(X)$ is clearly continuous. To show the same for $H(X/Y)$ we need only show that for each i, $-p(x_i) \int_\Omega \log p(x_i/\omega)\, p(d\omega/x_i)$ is a continuous function of $p(x_1), \ldots, p(x_a)$. Now

$$p(x_i/\omega) = \frac{p(x_i, d\omega)}{p(d\omega)} = p(x_i) \frac{p(d\omega/x_i)}{p(d\omega)}$$

But since $\sum_i p(\Lambda/x_i) \geq p(\Lambda)$, we have (see ref. 13, p. 133)

$$\frac{p(d\omega/x_i)}{\sum_i p(d\omega/x_i)} = \frac{p(d\omega/x_i)}{p(d\omega)} \cdot \frac{p(d\omega)}{\sum_i p(d\omega/x_i)}$$

$$= \frac{p(d\omega/x_i)}{p(d\omega)} \cdot \frac{\sum_i p(x_i)\, p(d\omega/x_i)}{\sum_i p(d\omega/x_i)}$$

almost everywhere with respect to $\sum_i p(\ /x_i)$ and hence, certainly, almost everywhere with respect to each $p(\ /x_i)$. Thus

$$\frac{p(d\omega/x_i)}{p(d\omega)} = \frac{p(d\omega/x_i)}{\sum_i p(d\omega/x_i)} \Bigg/ \frac{\sum_i p(x_i)\, p(d\omega/x_i)}{\sum_i p(d\omega/x_i)}$$

almost everywhere with respect to $p(\ )$. The dependence on the $p(x_i)$ is now explicitly continuous, so that each $p(x_i/\omega)$ is a continuous function of $p(x_1), \ldots, p(x_a)$ almost everywhere with respect to each $p(\ /x_i)$. We now wish to show that $-p(x_i) \int_\Omega \log p(x_i/\omega)\, p/d\omega(x_i)$ is a continuous function of the $p(x_i)$.

To this end let $\{p_j(x_1), \ldots, p_j(x_a)\}$, $j = 1, \ldots$ be a convergent sequence of points in a-dimensional Euclidean space R_a, with limit $\{p_o(x_1), \ldots, p_o(x_a)\}$. Then we have $\lim_{j \to \infty} p_j(x_i/\omega) = p_o(x_i/\omega)$ almost everywhere with respect to each $p(\ /x)$. We must now show that

$$-p_j(x_i) \int_\Omega \log p_j(x_i/\omega)\, p(d\omega/x_i) \to p_o(x_i) \int_\Omega \log p_o(x_i/\omega)\, p(d\omega/x_i).$$

Suppose, first, that $p_o(x_i) \neq 0$. Now from section IV we have

$$\int_\Omega -\log \left\{ \frac{p(d\omega/x_i)}{\sum_i p(d\omega/x_i)} \Bigg/ \frac{\sum_i p(x_i)\, p(d\omega/x_i)}{\sum_i p(d\omega/x_i)} \right\} p(d\omega/x_i) < \infty$$

whenever $p(x_i) \neq 0$. Take $p(x_i) = p(x_2) = \ldots p(x_a) = 1/a$. Then

$$\int_\Omega -\log a \frac{p(d\omega/x_i)}{\sum_i p(d\omega/x_i)}\, p(d\omega/x_i) < \infty \quad \text{or clearly}$$

$$\int_\Omega -\log \frac{p(d\omega/x_i)}{\sum_i p(d\omega/x_i)}\, p(d\omega/x_i) < \infty. \quad \text{But}$$

$$-\log \frac{p(d\omega/x_i)}{\sum_i p(d\omega/x_i)} \geq -\log \left\{ \frac{p(d\omega/x_i)}{\sum_i p(d\omega/x_i)} \Big/ \frac{\sum_i p_j(x_i)\, p(d\omega/x_i)}{\sum_i p(d\omega/x_i)} \right\}, \quad j = 1, 2, \ldots$$

Since the last term is also bounded below by $\log p_j(x_i)$, then by reference 14, p. 110, we have

$$\lim_{j\to\infty} \int_\Omega -\log \left\{ \frac{p(d\omega/x_i)}{\sum_i p(d\omega/x_i)} \Big/ \frac{\sum_i p_j(x_i)\, p(d\omega/x_i)}{\sum_i p(d\omega/x_i)} \right\} p(d\omega/x_i)$$

$$= \int_\Omega -\log \left\{ \frac{p(d\omega/x_i)}{\sum_i p(d\omega/x_i)} \Big/ \frac{\sum_i p_j(x_i)\, p(d\omega/x_i)}{\sum_i p(d\omega/x_i)} \right\} p(d\omega/x_i)$$

Since $p_o(x_i) \neq 0$, $-p_j(x_i) \int_\Omega \log p(x_i/\omega)\, p(d\omega/x_i) = p_j(x_i)$

$$\int_\Omega -\log \left[p_j(x_i) \cdot \left\{ \frac{p(d\omega/x_i)}{\sum_i p(d\omega/x_i)} \Big/ \frac{\sum_i p_j(x_i)\, p(d\omega/x_i)}{\sum_i p(d\omega/x_i)} \right\} \right] p(d\omega/x_i) \to p_o(x_i)$$

$$\int_\Omega -\log \left[p_o(x_i) \left\{ \frac{p(d\omega/x_i)}{\sum_i p(d\omega/x_i)} \Big/ \frac{\sum_i p_o(x_i)\, p(d\omega/x_i)}{\sum_i p(d\omega/x_i)} \right\} \right] p(d\omega/x_i)$$

$$= -p_o(x_i) \int_\Omega \log p(x_i/\omega)\, p(d\omega/x_i)$$

If $p_o(x_i) = 0$, we can clearly assume $p_j(x_i) \neq 0$, since we have to show that $-p_j(x_i) \int_\Omega \log p_j(x_i/\omega)\, p(d\omega/x_i) \to 0$. As before we have

$$-\log \frac{p_j(x_i/\omega)}{p_j(x_i)} \leq -\log \frac{p(d\omega/x_i)}{\sum_i p(d\omega/x_i)}, \text{ therefore}$$

$$-p_j(x_i) \int_\Omega \log p_j(x_i/\omega)\, p(d\omega/x_i) \leq p_j(x_i) \int_\Omega -\log \frac{p(d\omega/x_i)}{\sum_i p(d\omega/x_i)}\, p(d\omega/x_i)$$

$$+ p_j(x_i) \log \frac{1}{p_j(x_i)} \to 0 \quad \text{as } p_j(x_i) \to 0 \text{ (i.e., as } j \to \infty).$$

2. We wish here to rigorize the discussion of section VI.

We assume that $p(\ /x)$ satisfies the following continuity condition: For any finite closed interval I and any ϵ there is a $\delta(I, \epsilon)$ such that

$$\left| \frac{p(\Lambda/x_2)}{p(\Lambda/x_2)} - 1 \right| < \epsilon \quad \text{for } |x_1 - x_2| \leq \delta \text{ and } x_1, x_2 \in I,$$

whenever $p(\Lambda/x_2) \neq 0$. It follows that if, for $x_1 \in I$, $p(\Lambda/x_1) = 0$, then for $x_2 \in I$ and $|x_1 - x_2| < \delta$, $p(\Lambda/x_2) = 0$. (Indeed, since $\{x/p(\Lambda/x) = 0\}$ is evidently both open and closed, for any Λ, $p(\Lambda/x)$ either vanishes everywhere or nowhere.) That $p(\Lambda) \equiv \int_{-\infty}^{\infty} p(x)\, p(\Lambda/x)\, dx$ is a probability measure is a simple consequence of reference 14, p. 112, Theorem B. Since $p(x)\, p(\Lambda/x)$ is continuous, $p(\Lambda)$ can vanish only if $p(x)\, p(\Lambda/x)$ is zero for all x. Hence, for all x, $p(x)\,p(\Lambda/x)$ is absolutely continuous with respect to $p(\Lambda)$.

We can sharpen this result as follows: Let I be a closed interval over which $p(x) \neq 0$. Then for a given ϵ we can find a δ such that $p(x_1) > p(x_2)/2$ and $p(\Lambda/x_1) \geq (1-\epsilon)\,p(\Lambda/x_2)$, for $x_1, x_2 \in I$ and $|x_1 - x_2| < \delta$. We thus have

$$\int_{-\infty}^{\infty} p(x)\,p(\Lambda/x)\,dx \geq 2\delta \frac{p(x_2)}{2} p(\Lambda/x_2)(1-\epsilon) = \delta\, p(x_2)(1-\epsilon)\,p(\Lambda/x_2)$$

Thus for any $x_2 \in I$,

$$p(x_2)\cdot p(\Lambda/x_2) \leq \frac{1}{(1-\epsilon)\delta}\, p(\Lambda) \equiv k(x_2)\,p(\Lambda)$$

which defines $k(x_2) < \infty$. As in section IV, we can easily show that

$$-\infty < -\int_{\Omega} \log p(x/\omega)\,p(x, d\omega) < \infty \quad \text{for all } x. \text{ Now}$$

$$\int_{\Omega} p(x, d\omega) \log \frac{p(x)}{p(x/\omega)} \leq \int_{\Omega} p(x, d\omega) \left\{ \frac{p(x)}{p(x/\omega)} - 1 \right\} \log e$$

$$= \int_{\Omega} p(x/\omega)\,p(d\omega) \left\{ \frac{p(x)}{p(x/\omega)} - 1 \right\} \log e = 0,$$

the next to last equality being justified by reference 14, p. 133. Therefore, if $\int_{\Omega} p(x, d\omega) \log \frac{p(x/\omega)}{p(x)}$ is, say, continuous in x, then

$$\lim_{a\to\infty} \int_{-a}^{a} \int_{\Omega} p(x, d\omega) \log \frac{p(x/\omega)}{p(x)}\, dx$$

is meaningful and is either positive or equal to $+\infty$.

We shall now show that $\int_{\Omega} p(x, d\omega) \log \frac{p(x/\omega)}{p(x)}$ is indeed continuous. To this end let x_i be a convergent sequence of real numbers with limit x_o. We shall show that $p(x_i/\omega) \to p(x_o/\omega)$ almost everywhere with respect to $p(x_o, \ )$. (Since for $p(x_o) = 0$ this assertion is trivially true, we assume that $p(x_o \neq 0)$.) Let $A_{in}^{+} = \{p(x_i/\omega) - p(x_o/\omega) > 1/n\}$ and $A_{in}^{-} = \{p(x_i/\omega) - p(x_o/\omega) < -1/n\}$. Now

$$p(x_i)\,p(A_{in}^{+}/x_i) - p(x_o)\,p(A_{in}^{+}/x_o) = \int_{A_{in}^{+}} (p(x_i/\omega) - p(x_o/\omega))\,p(d\omega) > 1/n\; p(A_{in}^{+}).$$

There is clearly no loss in generality in assuming $p(x_i) \neq 0$. Then

$$p\left(A_{in}^{+}/x_i\right) - p\left(A_{in}^{+}/x_o\right) > \frac{p(x_o)}{k(x_o)\,n\,p(x_i)}\, p\left(A_{in}^{+}/x_o\right) + \frac{p(x_o) - p(x_i)}{p(x_i)}\, p\left(A_{in}^{+}/x_o\right).$$

Now $\frac{p(k_o)}{k(x_o)\,n\,p(x_i)} + \frac{p(x_o) - p(x_i)}{p(x_i)}$ is positive and bounded away from zero for all i sufficiently large. By the continuity condition on $p(\ /x)$ we therefore have $p\left(A_{in}^{+}/x_o\right) = 0$ for $i > i(n)$ suitably chosen. We get a similar result for $p\left(A_{in}^{-}/x_o\right)$. Let A_n^{+} be the set of points ω which lie in infinitely many A_{in}^{+}, and similarly for A_n^{-}. Then $p\left(A_n^{\pm}/x_o\right) = 0$, and so,

$$p\left(\sum_n A_n^+ + \sum_n A_n^-/x_0\right) = p\left(x_0, \sum_n (A_n^+ + A_n^-)\right) = 0$$

But for any $\omega \in \Omega - \sum_n A_n^+ - \sum_n A_n^-$, $p(x_i/\omega) \to p(x_0/\omega)$, which was to be shown.

As before, let x_i be a convergent sequence with limit x_0.

a. Let us assume first that $p(x_0) \neq 0$. Now

$$\left|\int_\Omega -\log p(x_0/\omega)\, p(d\omega/x_0) - \int_\Omega -\log p(x_i/\omega)\, p(d\omega/x_i)\right|$$

$$= \left|\int_\Omega [-\log p(x_0/\omega) + \log p(x_i/\omega)]\, p(d\omega/x_0) - \int_\Omega -\log p(x_i/\omega)\, p(d\omega/x_i)\right.$$

$$\left. + \int_\Omega -\log p(x_i/\omega)\, p(d\omega/x_0)\right| \leq \left|\int_\Omega [-\log p(x_0/\omega) + \log p(x_i/\omega)]\, p(d\omega/x_0)\right|$$

$$+ \left|\int_\Omega -\log p(x_i/\omega)\, p(d\omega/x_0) - \int_\Omega -\log p(x_i/\omega)\, p(d\omega/x_i)\right|$$

To show that the first of the last two terms goes to zero, we remark, first, that since $p(x_0) \neq 0$ and $p(\Lambda/x_0) \leq (1+a)\, p(\Lambda/x_i)$ for any Λ, for i suitably large, it follows, as in section IV, that

$$\int_{\{-\log p(x_i/\omega) \geq R\}} -\log p(x_i/\omega)\, p(d\omega/x_0)$$

is uniformly bounded for all i, where we use the previously shown result that $p(x)\, p(\Lambda/x) \leq k(x)\, p(\Lambda) \leq M\, p(\Lambda)$ for M suitably chosen and x in a closed interval containing x_0. It is now a simple exercise, by using reference 14, p. 110, to justify the interchange of limit and integration, so that the term in question vanishes as $i \to \infty$. The relation $p(\ /x_0) \leq (1+a)\, p(\ /x_i) \leq (1+a)^2\, p(\ /x_0)$, with $a \to 0$ as $i \to \infty$, at once shows that the second term likewise vanishes as $i \to \infty$.

b. Now suppose that $p(x_0) = 0$. Then by definition we take

$$\int_\Omega -\log p(x_0/\omega)\, p(x_0, d\omega) = 0$$

If $p(x)$ is identically zero in some neighborhood of x_0 there is nothing to be proven. We can then assume that $p(x_i) \neq 0$. For a closed interval containing x_0 and the x_i, we have, for $|x_i - x_j|$ sufficiently small (or equivalently, i and j sufficiently large) that $p(\ /x_i) \geq (1-\epsilon)\, p(\ /x_j)$. Thus

$$p(x_i/\omega) \geq p(x_i) \frac{p(d\omega/x_j)}{p(d\omega)} (1-\epsilon). \quad \text{Hence}$$

$$-\log p(x_i/\omega) \leq -\log [p(x_i)(1-\epsilon)] - \log \frac{p(d\omega/x_j)}{p(d\omega)}$$

for fixed j and any i, both sufficiently large. Further, since $p(x_j) \neq 0$, $p(x_j)\, p(\Lambda/x_j) = M\, p(\Lambda)$ for suitable M. Hence

$$p(x_i)\, p(\Lambda/x_i) \leq \frac{1}{1-\epsilon} \frac{p(x_i)}{p(x_j)} M\, p(\Lambda)$$

Since $p(x_i) \to 0$, we have, for sufficiently large i, $p(x_i)\, p(\Lambda/x_i) \leq p(\Lambda)$, so that $p(x_i/\omega) \leq 1$ or $-\log p(x_i/\omega) \geq 0$. Therefore

$$\int_\Omega -\log p(x_i/\omega)\, p(x_i, d\omega) \leq -p(x_i) \log \left[p(x_i)(1-\epsilon)\right]$$

$$+ p(x_i) \int_\Omega -\log \frac{p(d\omega/x_j)}{p(d\omega)}\, p(d\omega/x_i)$$

As i approaches ∞, the last integral approaches

$$\int_\Omega -\log \frac{p(d\omega/x_j)}{p(d\omega)}\, p(d\omega/x_o), \quad \text{which is} < \infty,$$

using arguments as in section IV. Since $p(x_i) \to 0$, we have, finally,

$$\int_\Omega -\log p(x_i/\omega)\, p(x_i, d\omega) \to 0 \quad \text{as } i \to \infty.$$

References and Footnotes

1. C. E. Shannon, A mathematical theory of communication, Bell System Tech. J. 27, 379-423, 623-656; also B. McMillan, The basic theorems of information theory, Ann. Math. Stat. 24, 196-219.
2. That it is indeed sufficient will be shown in section III.
3. R. M. Fano, Lecture notes on statistical theory of information, Massachusetts Institute of Technology, spring, 1952. This statement asserts that if the channel is considered as transmitting sequence by sequence its capacity per symbol is still bounded by C. Using the fact that the reception rate per symbol may be written as

   $$\frac{H(V) - H(V/U)}{n}$$

   the statement follows upon noticing that H(V/U) depends only upon single-received-symbol probabilities and that H(V) is a maximum when those probabilities are independent. The expression H(V) - H(V/U) then reduces to a sum of single-symbol channel rates, from which the assertion follows at once.
4. It is not difficult to see that H(X) - H(X/Y) is a continuous function of the "variables" $r_i = p(x_i)$, $i = 1, \ldots, a$. This is true also in the context of section IV (c.f. Appendix). Since the set of points in a-dimensional cartesian space R_a defined by $r_i \geq 0$ and $\sum_{i=1}^{a} r_i = 1$ is a closed set, H(X) - H(X/Y) attains a maximum value. This point is, however, not critical, for, given $H < C$ we can certainly find p( ) such that $H < H(X) - H(X/Y) < C$ and then use H(X) - H(X/Y) in place of C.
5. This condition appears to be superfluous. It is, however, strongly indicated by the immediately preceding result and is, in fact, essential for the proof.
6. E. M. Gilbert, A comparison of signalling alphabets, Bell System Tech. J. 31, in particular p. 506.

7. Up to here, the possibility that certain quantities are not integers can be seen not to invalidate any of the various inequalities. In what follows, the modifications needed to account for this possibility are obvious and insignificant and are therefore omitted.

8. Word-wise, this string of inequalities states simply: (a) that in order to minimize the probability of misidentifying the transmitted s we should guess the s with greatest conditional probability as the one actually transmitted; (b) if instead of the above recipe, we assume that s was sent, whenever $r \epsilon A_s$ is received, for all s except s_o, and that in all other circumstances we shall assume s_o to have been sent, then the probability of error is less than e; (c) hence, since P_e is the error obtained by the best method of guessing, $P_e \leq e$.

9. See reference 13, pp. 144-5. This was pointed out by Professor R. M. Fano.

10. J. L. Doob, Stochastic Processes (John Wiley and Sons, Inc., New York, 1953).

11. W. Feller, Generalization of a probability limit theorem of Cramer, Trans. Amer. Math. Soc. 54, 361-372 (1943).

12. E. Hopf, Ergodentheorie (Julius Springer, Berlin, 1937).

13. W. Feller, An Introduction to Probability Theory (John Wiley and Sons, Inc., New York, 1950).

14. P. R. Halmos, Measure Theory (D. Van Nostrand, New York, 1950).

CODING FOR NOISY CHANNELS*

Peter Elias
Department of Electrical Engineering and Research Laboratory of Electronics
Massachusetts Institute of Technology
Cambridge, Massachusetts

Summary: Shannon's and Feinstein's versions of the channel capacity theorem, specialized to the binary symmetric channel, are presented. A much stronger version is proved for this channel. It is shown that the error probability as a function of delay is bounded above and below by exponentials, whose exponents agree for a considerable range of values of the channel and the code parameters. In this range the average behavior of all codes is essentially optimum, but for small transmission rates this is not true. The results of this analysis are shown to apply to check-symbol codes of four kinds which have progressively simpler coding procedures. The last of these is error-free, and makes it possible to transmit information at a rate equal to the channel capacity with a probability one that no decoded symbol will be in error.

Introduction

Since Shannon[1, 2] showed that information could be transmitted over a noisy channel at a positive rate with an arbitrarily low probability of error at the receiver, there has been considerable interest in constructing specific transmission schemes that exhibit such behavior.

For a signal transmitted over a channel perturbed by Gaussian noise, Golay[3] and Fano[4] found schemes which in the limit had the desired behavior, but it was a limit of infinite bandwidth or vanishing transmission rate. Rice[5] investigated the characteristics of transmission using randomly selected noise waveforms, and got an indication of exponential decrease in error probability with increasing time delay. Feinstein[6] showed that the same sort of behavior, at least as an upper bound, held true for more general channels.

For the binary channel, Hamming[7], Gilbert[8], Plotkin[9], and Golay[10] investigated a variety of codes, and found some basic properties of the binary symmetric channel. Laemmel[11], Muller[12], and Reed[13] also constructed specific codes and classes of codes. The first constructive code for transmission at a nonzero rate over a noisy binary channel was discovered recently by the author[14]. The investigation reported in the present paper started as a continuation of that work, and an investigation of the rate at which the error probability decreased with delay originally developed from a comparison of check-symbol codes with codes of less restricted types. It seems more sensible to present the results in reverse order. After a definition of the channel and general coding procedures, Shannon's and Feinstein's channel capacity theorems are stated, and a stronger theorem is given for the binary symmetric channel, which shows in considerable detail the behavior of error probability at the receiver as a function of the parameters of the channel and the code, and the delay time. It is then shown that most of these results carry over to a variety of kinds of check-symbol codes. One of these, of primarily academic interest, is error-free[14], and permits the transmission of an infinite sequence of message symbols at an average rate equal to the channel capacity with a probablity one that no decoded digit is in error.

The Channel

The coding problem we will discuss is illustrated in Fig. 1. The problem is to match the output of an ideal binary message source to a binary symmetric noisy channel.

The message source generates a sequence of binary symbols, say the binary digits zero and one. Zeros and ones are selected with equal probability, and successive selections are statistically independent.

The channel accepts binary symbols as an input and produces binary symbols as an output. Each input symbol has a probability $p_o < 1/2$ of being received in error, and a probability $q_o = 1 - p_o$ of being received as transmitted. The transmission error probability p_o is a constant, independent of the value of the symbol being transmitted: the channel is as likely to turn a one into a zero as to turn a zero into a one. The channel, in effect, adds a noise sequence to the input sequence to produce the output sequence; the noise is a random sequence of zeros and ones, synchronous with the signal sequence, in which the ones have probability p_o, and the addition is addition modulo two of each signal digit to the corresponding noise digit ($1+1 = 0+0 = 0$, $0+1 = 1+0 = 1$).

If the message source were connected directly to the channel, a fraction p_o of the received symbols would be in error. A coding procedure for reducing the effect of the errors is shown in Figs. 1 and 2. The output of the message source is segmented into consecutive blocks of length M. There are 2^M such blocks, and they are selected by the source with equal probability. To each input block of M binary symbols is assigned an output block of N binary symbols, $N > M$.

The input sequences of length M are the messages to be sent; the output sequences of length N are the transmitted signals, and the correspondence between input and output blocks is the code used. The use of the word "code" is justified

*This work was supported in part by the Signal Corps; the Office of Scientific Research, Air Research and Development Command; and the Office of Naval Research.

Reprinted from *IRE Conv. Rec.*, pt. 4, pp. 37–46, Mar. 1955.

by Fig. 2, where the correspondence between input and output blocks is given in the form of a codebook. On the left is a column of the 2^M possible messages, listed as M-digit binary numbers in numerical order. Following each message is the N-digit binary number which is the corresponding signal, so that the codebook has 2^M entries, each of which lists a message and the corresponding signal.

The system in operation is shown in Fig. 1. The source selects a message that is coded into a transmitted signal and sent over the noisy channel. The received block of N -- the received, or noisy, signal -- differs from the transmitted signal in about $p_o N$ of its N symbol values. The decoder receives this noisy signal and reproduces one of the 2^M possible messages, with an average probability P_e of making an incorrect choice.

The most general type of decoder is shown in Fig. 3. It is a codebook with 2^N entries, one for each of the possible received signals. The left column is the received sequence, arranged as an N-digit binary number in numerical order. This is followed by the M-symbol message block that will be reproduced when that sequence is received as a noisy signal.

In order to minimize P_e, the codebook must be so constructed that the message that is selected when a given noisy signal is received is the one corresponding to the signal most likely to have been transmitted. For the binary symmetric channel, the signal most likely to have been transmitted is the one that differs from the received signal in the smallest number of symbol positions. This follows from the fact that a particular group of k errors has probability $p_o^k q_o^{N-k}$ of being introduced by the channel; this probability decreases as k increases, for $p_o < 1/2$.

For this channel, the codebook may be simplified. In fact, the transmitter codebook may be used in reverse order. The noisy signal is compared with each of the possible transmitted signals, and the number of positions in which they differ is counted. The signal with the lowest count is assumed to have been transmitted, and the corresponding message block is reproduced as the best guess at the transmitted message. This guess may, of course, be incorrect, and will be if the noise has altered more than half of the positions in which the transmitted signal differs from some other listed signal.

This decoding procedure may be described in a geometrical language introduced by Hamming[7]. Each signal is taken as a point or a vector in an N-dimensional space, with coordinates equal to the values (zero or one) of its N binary symbols. The distance between two points is defined as the number of coordinates in which they differ. In this language, the noisy signal is decoded as the nearest of the signal points, and the corresponding message is chosen.

For given M and N, the error probability P_e depends on the set of points that are used as signals. If these are clustered in a small part of the space, P_e will be large; if they are far from one another, P_e will be small. As specialized to this channel, Shannon's second coding theorem states an asymptotic relationship between M, N, and P_e for a suitable selection of signal points.

Channel Capacity and Error Probability

First, some definitions are required. Given a binary symmetric channel with transmission error probability p_o and $q_o = 1 - p_o$, the equivocation $E_o = E(p_o)$ and the capacity $C_o = C(p_o)$ of the channel, both measured in bits per symbol, are given by

$$E_o = -p_o \log p_o - q_o \log q_o \qquad C_o = 1 - E_o \tag{1}$$

(Here and later, all logarithms are to the base two.)

Given a coding procedure like that illustrated by Figs. 1, 2, and 3, the redundancy E_1 and the transmission rate C_1, also in bits per symbols, are given by

$$E_1 = \frac{N - M}{N} \qquad C_1 = 1 - E_1 = \frac{M}{N} \tag{2}$$

It is convenient to introduce the probability p_1 which is the upper bound of the transmission error probabilities for which this particular code can be expected to work, and $q_1 = 1 - p_1$. These are uniquely defined by

$$p_1 < \frac{1}{2} \qquad E_1 = E(p_1) = -p_1 \log p_1 - q_1 \log q_1 \tag{3}$$

since a plot of E(p) or C(p) is monotonic for $0 \leq p \leq 1/2$.

Finally, the average probability of an error in decoding, which was written as P_e above, will in general be a function of the block length N, the channel capacity C_o or error probability p_o, and the transmission rate C_1 or the probability p_1. It will be written as $P_e(N, p_o, p_1)$.

Shannon's second coding theorem[1], as applied to this channel, follows.

Theorem 1. Given any fixed $C_1 < C_o$ and any fixed $\epsilon > 0$, for all sufficiently large N there are codes which will transmit information at the rate C_1 bits per symbol and will decode it with an error probability per block of N, $P_e(N, p_o, p_1) < \epsilon$. This cannot be done for $C_1 > C_o$.

Shannon's proof of the theorem proves more than the theorem states. A code is a selection of

2^{NC_1} signal sequences from among 2^N possibilities. Including those codes that select the same signal two or more times to represent several different messages, there are $2^{N \cdot 2^{NC_1}}$ different codes. Each of these will have an average decoding error probability (averaged over the different messages, with equal weights). Shannon shows that the average of all of these (averaged over the different codes, with equal weights) is less than ϵ. Since the error probability for each code is positive, it follows that at least one code has an average error probability less than ϵ; and it also follows, as Shannon remarks, that, at most, a fraction f of the codes can have an average error probability as great as ϵ/f, so that almost all of the codes have arbitrarily small error probability; that is, almost all codes are "good" codes, although some "bad" codes do exist. By the same argument, in any one good code the error probability for most of the individual messages is less than ϵ/f, so that by discarding a few of the signal sequences and transmitting at a very slightly slower rate, any good code can be made into a uniformly good code. This result has considerable practical importance, since a uniformly good code will transmit with the specified small error probability, regardless of the probabilities with which message sequences are selected, and there are many information sources whose statistics are not known in detail.

The major question left open by this theorem is how large N must be for given p_o, p_1, and ϵ. Feinstein[6] has proved a stronger version which provides an upper bound for $P_e(N, p_o, p_1)$. As specialized to the binary symmetric channel it may be written as:

Theorem 2. Given any $C_1 < C_o$, an $\epsilon(p_o, p_1) > 0$ can be found. For any sufficiently large N, a code may be constructed which will transmit information at the rate C_1 bits per symbol which can be decoded with $P_e(N, p_o, p_1) < 2^{-\epsilon N}$.

Feinstein's proof consists of the construction of a code that satisfies the requirements of the theorem and is uniform in the sense that all signals are good signals. Some indication of the relation of ϵ to the channel and code parameters is also given.

The next theorem is much stronger than this, but unlike Shannon's and Feinstein's it does not apply to the general discrete noisy channel without memory, but only to the binary symmetric case. Some more definitions are needed. It turns out that the error probability P_e is bounded not only above but below by exponentials in N, and that for a considerable range of channel and code parameters the exponents of the two bounds agree. The error exponent for the best possible code is defined as

$$a_{opt}(N, p_o, p_1) = \frac{-\log P_e(N, p_o, p_1)}{N} \quad (4)$$

and $a_{avg}(N, p_o, p_1)$ is defined as the same function of the average of the error probabilities of all codes.

An additional probability value is also needed, along with the values of a, C, and E which go with it:

$$p_{crit} = \frac{p^{1/2}}{p^{1/2} + q^{1/2}}, \quad q_{crit} = 1 - p_{crit}$$
$$E_{crit} = E(p_{crit}), \quad C_{crit} = 1 - E_{crit} \quad (5)$$
$$a_{crit} = \lim_{N\to\infty} a_{opt}(N, p_o, p_{crit})$$

Finally, the margin in error probability and the margin in channel capacity need labeling:

$$\delta = p_1 - p_o$$
$$\Delta = C_o - C_1 \quad (6)$$

For a binary symmetric channel with capacity C_o and transmission rate C_1, the following statements hold.

Theorem 3. (a) For $p_o < p_1 < p_{crit}$, $C_o > C_1 > C_{crit}$, the average code is essentially as good as the best code:

$$a(p_o, p_1) = \lim_{N\to\infty} a_{opt}(N, p_o, p_1) = \lim_{N\to\infty} a_{avg}(N, p_o, p_1)$$
$$= -\Delta - \delta \log \frac{p_o}{q_o} \quad (7)$$

(b) For $p_{crit} < p_1 < 1/2$, the average code is not necessarily optimum; for p_1 near 1/2 it is certainly not. Specifically,

$$a_{avg}(p_o, p_1) = \lim_{N\to\infty} a_{avg}(N, p_o, p_1)$$
$$= a_{crit} + C_{crit} - C_1 \quad (8)$$

where a_{crit} is the $a(p_o, p_1)$ of Eq. (5) with $p_1 = p_{crit}$, while for a_{opt} there are two upper and two lower bounds:

$$\liminf a_{opt}(N, p_o, p_1) \geq \begin{cases} a_{crit} + C_{crit} - C_1 \\ \frac{p_1}{2} \log \frac{1}{4pq} - C_1 \end{cases} \quad (9)$$

$$\limsup a_{opt}(N, p_o, p_1) \leq \begin{cases} -\Delta - \delta \log \frac{p_o}{q_o} \\ \frac{E_1}{4} \log \frac{1}{4pq} \end{cases} \quad (10)$$

As $p_1 \to 1/2$, the second bound in (9) approaches the second bound in (10);

$$\lim_{N\to\infty} \lim_{p_1\to 1/2} \alpha_{opt}(N, p_o, p_1) = \frac{1}{4}\log\frac{1}{4pq} \tag{11}$$

which is always greater than

$$\alpha_{avg}\left(p_o, \frac{1}{2}\right) = \alpha_{crit} + C_{crit} \tag{12}$$

The content of this theorem is illustrated by Fig. 4. This is a plot of the channel capacity $C(p)$ vs. transmission error probability p for a binary symmetric channel. A dashed line is drawn tangent to the curve at the point given by the channel parameters p_o, C_o. This tangent line has the slope $\log(p_o/q_o)$. The critical point p_{crit}, C_{crit} is the point at which the slope of the curve is $(1/2)\log(p_o/q_o)$. For $p_o < p_1 < p_{crit}$, the $\alpha(p_o, p_1)$ of (7), which is both the average and the optimum error exponent, is the length of a vertical dropped from the channel capacity curve to the tangent line at the ordinate p_1.

At $p_1 = p_{crit}$, the dotted line that determines $\alpha_{avg}(p_o, p_1)$ diverges from the tangent line. For $p_{crit} < p_1 < 1/2$ the exact value of $\alpha_{opt}(N, p_o, p_1)$ is not known, but is given by the length of a vertical at ordinate p_1, dropped from the channel capacity curve and terminating in the shaded region. The upper and lower bounds of this region provide lower and upper bounds, respectively, on the value of α_{opt}. These bounds converge to $(1/4)\log(1/4pq)$ at $p_1 = 1/2$, and near this point α_{opt} is definitely $> \alpha_{avg}$.

The value of α given by the tangent line at $p_1 = 1/2$, although not approached for the transmission of information at any nonzero rate, is the correct value of α_{opt} for transmission of one bit per block of N symbols.

An outline of the proof of Theorem 3 appears in the Appendix. A more detailed presentation, giving bounds on $P_e(N, p_o, p_1)$, as well as on α, will appear elsewhere.

Check-Symbol Codes

The preceding three theorems are interesting in theory but discouraging in practice. They imply that a good code will require a transmitting codebook containing $N \cdot 2^{NC_1}$ binary digits in all, and either a receiver codebook containing $N \cdot 2^N$ binary digits or another copy of the transmitter codebook and 2^{NC_1} comparisons of the received signal with the possible transmitted signals. Since in interesting cases NC_1 may be of the order of 100, the requirements in time and space are unmanageable. Furthermore, it would be quite consistent with these theorems if no code with any simplicity or symmetry properties were a good code.

The theorems that follow show that this is fortunately not the case. Four kinds of codes of increasing simplicity and convenience from the point of view of realization are demonstrated to have essentially the same behavior, from both a channel capacity and an error probability point of view, as the optimum code. The last of the four is of theoretical interest as well, since it permits the receiver to set the decoding error probability arbitrarily low without consulting the transmitter.

A check-symbol code of block length N is a code in which the 2^{NC_1} signal sequences have in their initial NC_1 positions all 2^{NC_1} possible combinations of symbol values. The first NC_1 positions will be called information positions and the last NE_1 will be called check positions. The signal corresponding to a message sequence is that one of the signal sequences whose initial symbols are the message.

A parity check-symbol (pcs) code is a check-symbol code in which the check positions are filled in with digits each of which completes a parity check of some of the information positions. Such codes were discussed in detail first by Hamming[1], who calls them systematic codes. A pcs code is specified by an $NC_1 \times NE_1$ matrix of zeros and ones, the ones in a row giving the locations of the information symbols whose sum modulo two is the check digit corresponding to that row. The process is illustrated in Fig. 5. Such a code requires $NC_1 \times NE_1 = N^2 C_1 E_1 \leq \frac{1}{4} N^2$ binary digits in its codebook, these being the digits in the check-sum matrix.

A sliding pcs code is defined as a pcs code in which the check-sum matrix is constructed from a sequence of N binary symbols by using the first NC_1 of them for the first row, the second to $(NC_1 + 1)$st for the second row ..., the NE_1th to the Nth for the NE_1th row. This code requires only an N-binary-digit codebook.

Finally a convolutional pcs code is defined as one in which check symbols are interspersed with information symbols, and the check symbols check a fixed pattern of the preceding NC_1 information positions if $C_1 \geq 1/2$; if $E_1 > 1/2$, the information symbols add a fixed pattern of zeros and ones to the succeeding NE_1 check positions. Such a code requires $\max(NC_1, NE_1) \leq N$ binary digits in its codebook. It is illustrated by Fig. 6.

Theorem 4. All the results of Theorem 3 apply to check-symbol codes and to pcs codes. The results of part (a) of that theorem apply to sliding pcs codes.

In reading Theorem 3 into Theorem 4, the average involved in α_{avg} is the average of all codes of the appropriate type; that is, all combinations of check symbols for the check-symbol codes, all check-sum matrices for the pcs codes, all sequences of N binary digits for the sliding pcs code.

Theorem 5. The results of part (a) of

Theorem 3 apply to convolutional pcs codes, if $P_e(N, p_o, p_1)$ is interpreted as the error probability per decoded symbol. For infinite memory (each check symbol checking a set of prior information symbols extending back to the start of transmission over the channel) the N in $P_e(N, p_o, p_1)$ for a particular decoded information symbol is the number of symbols which have been received since it was received.

This theorem shows that error-free coding can be attained at no loss either in channel capacity or in error probability, a question raised by the author when the first error-free code was introduced[14]. By waiting long enough, the receiver can obtain as low a probability of error per digit as is desired, without a change of code being necessary. By gradually reducing the ratio of check to information symbols toward E_o/C_o, using the law of the iterated logarithm for binary sequences, it can be shown that in an infinite sequence of message digits transmission is obtained at average rate C_o with probability one of no errors in the decoded message.

Conclusion

An appreciable gain in simplicity has been achieved in going from an arbitrary average code to a convolutional or sliding pcs code. It is possible to encode and decode either of these codes with a codebook of only N or fewer binary digits. However, the decoding operation will require 2^{NC_1} or 2^{NE_1} (whichever is smaller) comparisons, which will take a great deal of time in interesting cases. No decoding procedure that replaces this operation by a small amount of computing has yet been discovered, although the iterated Hamming code, which is error-free[14], gives hope that it may be possible to manage this while still keeping optimum behavior in terms of channel capacity and error probability -- a feature which the iterated Hamming code lacks.

Acknowledgments

After the analytical work reported in this paper was done, but before it had been organized for presentation, I discovered that Dr. Shannon was also working on the problem of error probability, and was to present his results at the same meeting. In discussing the results with Dr. Shannon, he mentioned the geometric relationship between the tangent line and the capacity curve, illustrated in Fig. 4, in the region $p_1 < p_{crit}$. I do not know whether this would have occurred to me in organizing my results, but I do know that it is vital; the information to the right of p_{crit} is my own, but is impossible to present in any other fashion without getting lost in numbers of families of curves.

It is a pleasure to acknowledge my indebtedness to the atmosphere at the Research Laboratory of Electronics, without which this work would not have gotten started; and to my colleagues, Professors Fano, Huffman, and Yngve, who provided that part of the atmosphere most relevent to this specific project.

Appendix

1. Outline Proof of Theorem 3

Using the symbols and definitions of Eqs. (1), (3), (5), and (6), let $k_1 = Np_1$ be an integer. Define $V_N(k)$, the volume of an N-dimensional sphere of radius k, by

$$V_N(k) = \sum_{j=0}^{k} \binom{N}{j} = \sum_{j=0}^{k} \frac{N!}{j!\,(N-j)!} \tag{A.1}$$

Select $2^N/V_N(k_1)$ sequences as signaling sequences. Then the signaling rate is

$$\frac{1}{N} \log \left\{2^N/V_N(k_1)\right\} = 1 - \frac{1}{N} \log V_N(k_1) \tag{A.2}$$

If the selection of signal sequences can be made so that every possible received sequence differs from one (and only one) signal sequence in k_1 or fewer positions, then the probability of a detection error will be just the probability $P_I(N, p_o, p_1)$ that more than k_1 out of N errors are made in transmission. This is the tail of the binomial distribution:

$$P_I(N, p_o, p_1) = \sum_{j=k_1+1}^{N} p_o^j q_o^{N-j} \binom{N}{j} \tag{A.3}$$

Such a selection is not, in general, possible. However, $P_I(N, p_o, p_1)$ of (A.3) is a lower bound to the average decoding error probability $P_e(N, p_o, p_1)$ for any actual selection of signal points: this follows directly from the fact that $p_o^j q_o^{N-j}$ is a monotonically decreasing function of j.

The average of all possible codes is used to provide an upper bound to the decoding error probability of the best code. The average probability of a detection error, $P_{III}(N, p_o, p_1)$, is the probability $P_{II}(N, j, k_1)$ of a decoding error when just j transmission errors have occurred, averaged over the binomial distribution of j. With Eq. (A.3) this gives

$$P_{III}(N,p_o,p_1) = \sum_{j=0}^{N} P_{II}(N,j,k_1)\, p_o^j q_o^{N-j} \binom{N}{j}$$

$$\leq \sum_{j=0}^{k_1} P_{II}(N, j, k_1)\, p_o^j q_o^{N-j} \binom{N}{j} + P_I(N,p_o,p_1) \tag{A.4}$$

The probability $P_{II}(N, j, k_1)$ of a decoding error when just j transmission errors have occurred is the probability that one of the $\{2^N/V_N(k_1)\} - 1$ incorrect signal sequences differs in j or fewer places from the received sequence. There are a total of $V_N(j)$ sequences which differ from the received sequence in as few as j positions, and the probability of missing all of them in $\{2^N/V_N(k_1)\} - 1$ tries is, for $j < k_1$, bounded by

$$\left(1 - \frac{V_N(j)}{2^N}\right)^{\{2^N/V_N(k_1)\}-1} \geq 1 - \frac{V_N(j)}{V_N(k_1)} \qquad (A.5)$$

Equation (A.5) gives the probability of no decoding error: P_{II} is the probability of a decoding error, so

$$P_{II}(N, j, k_1) \leq \frac{V_N(j)}{V_N(k_1)} \leq \frac{\binom{N}{j}}{\binom{N}{k_1}} \qquad (A.6)$$

Equations (A.4) and (A.6) give

$$P_{III}(N, p_o, p_1) \leq \sum_{j=0}^{k_1} p_o^j q_o^{N-j} \frac{\binom{N}{j}^2}{\binom{N}{k_1}} + P_I(N, p_o, p_1) \qquad (A.7)$$

Now the sums in Eqs. (A.1) and (A.3) are bounded below by the value of their largest term and above by a geometric series multiplied by that term -- the last term in Eq. (A.1), the first in Eq. (A.3). (See Feller[15], p. 126 for the bounds for Eq. (A.3).) The sum in Eq. (A.7) is similarly bounded above and below, if p_1 is less than p_{crit}, which is the condition guaranteeing that the last term in the sum is the largest. Using these results, taking logarithms, and using Stirling's approximation for the binomial coefficients gives, from (A.2),

$$\lim_{N\to\infty} \left\{1 - \frac{1}{N} \log V_N(k_1)\right\} = 1 - E_1 = C_1 \qquad (A.8)$$

from (A.3), for $p_o < p_1 < \frac{1}{2}$,

$$\limsup_N \alpha_{opt}(N, p_o, p_1) \leq \lim_{N\to\infty} \frac{-\log P_{II}(N, p_o, p_1)}{N}$$

$$= -\Delta - \delta \log \frac{p_o}{q_o} \qquad (A.9)$$

and from (A.7), for $p_o < p_1 < p_{crit}$,

$$\liminf_N \alpha_{opt}(N, p_o, p_1) \geq \lim_{N\to\infty} \frac{-\log P_{III}(N, p_o, p_1)}{N}$$

$$= -\Delta - \delta \log \frac{p_o}{q_o} \qquad (A.10)$$

Together, Eqs. (A.9) and (A.10) prove the first part of the theorem, and cover the region in which the dashed-line and the dotted curves of Fig. 4 coincide.

Since the length represented by α_{opt} in this region is the difference between the curve and its tangent, it is second-order in δ or Δ. In fact, for small δ and Δ,

$$\alpha_{opt}(p_o, p_1) \approx \frac{\delta^2}{2pq} \log e \approx \frac{\Delta^2}{2pq\left(\log \frac{p}{q}\right)^2} \log e$$

For $p_1 > p_{crit}$, the largest term in the sum in Eq. (A.7) is not the last, but is that term for which $j^2/(N-j)^2$ is most nearly equal to p_o/q_o. This term is larger than $P_I(N, p_o, p_1)$ for large N, and the sum is bounded above by k_1, the number of terms, times the largest term. Taking the limit of $(1/N)$ multiplied by the logarithm and using Stirling's approximation gives upper and lower bounds for $\alpha_{avg}(N, p_o, p_1)$ which coincide, giving for $p_{crit} < p_1 < \frac{1}{2}$,

$$\liminf_N \alpha_{opt}(N, p_o, p_1) \geq \lim_{N\to\infty} \alpha_{avg}(N, p_o, q_o)$$

$$= \lim_{N\to\infty} \frac{-\log P_{III}(N, p_o, p_1)}{N}$$

$$= C_{crit} + \alpha_{crit} - C \qquad (A.11)$$

This gives the remainder of the dotted curve in Fig. 4.

For p_1 less than p_{crit}, the probability of a detection error as computed above is essentially the probability of escaping from a sphere of radius $k_1 = Np_1$. For p_1 near 1/2, a different point of view is possible and leads to the two solid curves in Fig. 4.

The probability that transmission errors will cause one transmitted signal to be decoded as another is the probability that the noise will alter half or more of the positions in which they differ. If they differ in $Np_1 = k_1$ positions, this probability is just the upper half of the binomial, $P_I\left(Np_1, p_o, \frac{1}{2}\right)$ as given in Eq. (A.3). This is the probability of a particular transition; the total error probability is certainly less than this multiplied by the number of signal sequences. Gilbert[8] has shown that it is always possible to find $2^N/V_N(k_1 - 1)$ signal sequences each of which differs in at least k_1

positions from every other. For large N, by Eq. (A.8), this corresponds to a signaling rate of C_1 bits per symbol, or 2^{NC_1} signal points. Thus

$$P_e(N, p_o, p_1) \leq \frac{2^N}{V_N(k_1 - 1)} P_I\left(Np_1, p_o, \frac{1}{2}\right) \quad (A.12)$$

and asymptotically, from Eqs. (A.8) and (A.3),

$$\lim_{N\to\infty} \frac{-\log P_e(N,p_o,p_1)}{N} = -p_1 \left\{C - 0 + \left(\frac{1}{2} - p_o\right) \log \frac{p_o}{q_o}\right\}$$

$$= p_1 \cdot \frac{1}{2} \left\{\log \frac{1}{2p_o} + \log \frac{1}{2q_o}\right\}$$

$$= \frac{p_1}{2} \log \frac{1}{4pq} \quad (A.13)$$

and

$$\lim \inf_N \alpha_{opt}(N,p_o,p_1) \geq -C_1 + \frac{p_1}{2} \log \frac{1}{4pq} \quad (A.14)$$

This is the upper solid curve in Fig. 4. For an upper bound to α_{opt} we use a result of Plotkin[9] which shows that there are at most 2N signal points whose mutual minimum distance is as great as N/2. This means that the transmission rate for signal points at this distance is $(1 + \log N)/N$ and approaches zero for large N. This result sets a limit to the number of signal points at smaller distances as well. As Plotkin pointed out, if B(N,k) is the number of signal points at mutual distance $\geq k$, then at least half of these agree in their first coordinate. Eliminating the n first coordinates gives

$$B(N-n, k) \geq 2^{-n} B(N, k) \quad (A.15)$$

Using Eqs. (A.14) and (A.15), let n = N - 2k. Then

$$B(N, k) \leq 2^{N-2k} B(2k, k) = 4k \cdot 2^{N-2k} \quad (A.16)$$

For a transmission rate C_1 this determines k:

$$C_1 = 1 - E_1 = \lim_{N\to\infty} \frac{\log B(N,k)}{N} \leq 1 - 2\frac{k}{N},$$

$$\text{or} \quad k \leq \frac{N}{2} E_1 \quad (A.17)$$

Now the error probability for such a set of signal points is certainly greater than the probability of a single transition, which is, in turn, at least as great as the upper half of the binomial

$$P_I\left(\frac{NE_1}{2}, p_o, \frac{1}{2}\right)$$

Thus

$$\lim \sup_N \alpha_{opt}(N,p_o,p_1) \leq \lim_{N\to\infty} \frac{-\log P_I\left(\frac{NE_1}{2}, p_o, \frac{1}{2}\right)}{N}$$

$$= \frac{E_1}{4} \log \frac{1}{4pq} \quad (A.18)$$

which gives the lower solid curve in Fig. 4. At $p_1 = 1/2$, Eqs. (A.18) and (A.14) give the same value, so that

$$\lim_{N\to\infty} \lim_{p_1\to 1/2} \alpha_{opt}(N, p_o, p_1) = \frac{1}{4} \log \frac{1}{4pq} \quad (A.19)$$

These results prove the remainder of the theorem. It should be noted that Eq. (A.19) does not imply that it is impossible to transmit any information with an error probability less than approximately $2^{-\frac{N}{4}\log\frac{1}{4pq}}$ for finite N. It is only impossible to do so while transmitting at a positive rate in the limit of large N. The transmission of one bit per block of N symbols can be accomplished by picking two signal sequences that differ in every position, with an error probability equal to $P_I(N, p_o, 1/2)$ for which, from Eq. (A.3),

$$\lim_{N\to\infty} \frac{-\log P_I\left(N, p_o, \frac{1}{2}\right)}{N} = \frac{1}{2} \log \frac{1}{4pq} \quad (A.20)$$

This error exponent is twice as great as that for the limit (as $p_1 \to 1/2$) of α_{opt} for positive transmission rates. Other points for the transmission of 2, 3, ... log N bits per block of N symbols fall between the value of Eq. (A.20) and that of Eq. (A.19).

2. Outline Proof of Theorems 4 and 5

A large part of the proof of Theorem 3 carries over directly for Theorems 4 and 5. Any upper bound on α_{opt} for the best possible code is automatically an upper bound for the more restricted class of check-symbol codes. Thus Eqs. (A.9) and (A.18), the tangent line and the lower solid curve in Fig. 4, still apply. To get the upper solid curve, Eq. (A.14), it is necessary to show that Gilbert's result, and thus Eq. (A.12), holds for the kind of code considered. This is obvious for pcs codes; Gilbert's proof requires only trivial modifications in this case. Since the pcs codes are a special case of check-symbol codes, the result follows for these as well. For sliding and convolutional pcs codes Gilbert's result is not obvious, although probably still true; that is why only the first part of Theorem 3 is extended to these cases.

The difficult point in Theorems 4 and 5 is the demonstration that the average of all possible codes, of each of the four types considered, is still given by Eqs. (A.10) and (A.11) and the dotted curve in Fig. 4. This requires a demonstration that the inequality of Eq. (A.6) still applies; that is, that the probability of a decoding error when j transmission errors have occurred is essentially the same, on the average, for the different types of check-symbol codes as for the average of all codes. The remainder of the derivation then follows as before.

This will be done for the pcs codes. When a noisy signal is received, the parity-check sums are recomputed at the receiver and added modulo two per position to the received check symbols, as in the Hamming code[7]. The resulting check-symbol pattern is the pattern caused by the transmission errors alone. The probability that this check-symbol pattern will be misinterpreted when j transmission errors have occurred is the probability that some other collection of j or fewer errors has the same check-symbol pattern. There are $V_N(j) - 1$ other patterns of j or fewer errors, and the probability that one of these has the same check-sum pattern is the probability that one of the $V_N(j) - 1$ differences has a check-sum pattern of all zeros. Now, if the check-sum matrix is filled in at random, any error pattern may produce any check-symbol pattern with equal probability. Therefore the probability of any one error pattern having a check-symbol pattern that vanishes is the reciprocal of the total number of possible check-symbol patterns. This number is $V_N(k_1)$, since $2^N/V_N(k_1)$ messages are being sent, and the total probability of a decoding error when j transmission errors have been made is less than this multiplied by the number of difference patterns:

$$P_{II}(N, j, k_1) \leqslant \frac{V_N(j) - 1}{V_N(k_1)} \leqslant \frac{V_N(j)}{V_N(k_1)} \leqslant \frac{\binom{N}{j}}{\binom{N}{k_1}} \qquad (A.21)$$

which is the inequality of (A.6) obtained by a different route.

The essential point in this argument is that every transmission error pattern, in the ensemble of possible codes, may cause every check-symbol pattern, with equal probability. Given this, the rest of the argument presented above follows. This is easy, but tedious, to show for sliding and convolutional pcs coding; the proofs will be omitted here.

References

1. C. E. Shannon, "A mathematical theory of communication," Bell System Tech. J. 27, 379-423, 623-656 (1948).
2. C. E. Shannon, "Communication in the presence of noise," Proc. I.R.E. 37, 10-21 (1949).
3. M. J. E. Golay, "Note on the theoretical efficiency of information reception with PPM," Proc. I.R.E. 37, 1031 (1949).
4. R. M. Fano, "Communication in the presence of additive Gaussian noise," "Communication Theory," Willis Jackson Ed. (Butterworths, London, 1953) 169-182.
5. S. O. Rice, "Communication in the presence of noise -- probability of error of two encoding schemes," Bell System Tech. J. 29, 60-93 (1950).
6. A. Feinstein, "A new basic theorem of information theory," Trans. I.R.E. (PGIT) 4, 2-22 (1954).
7. R. W. Hamming, "Error detecting and error correcting codes," Bell System Tech. J. 29, 147-160 (1950).
8. E. N. Gilbert, "A comparison of signalling alphabets," Bell System Tech. J. 31, 504-522 (1952).
9. M. Plotkin, "Binary codes with specified minimum distance," Univ. of Penna., Moore School Research Division Report 51-20 (1951).
10. M. J. E. Golay, "Binary coding," Trans. I.R.E. (PGIT) 4, 23-28 (1954).
11. A. E. Laemmel, "Efficiency of noise-reducing codes," pp. 111-118 in "Communication Theory," reference 4 above.
12. D. E. Muller, "Metric properties of Boolean algebra and their application to switching circuits," University of Illinois, Digital Computer Laboratory Report No. 46.
13. I. S. Reed, "A class of multiple error-correcting codes and the decoding scheme," Trans. I.R.E. (PGIT) 4, 38-49 (1954).
14. P. Elias, "Error-free coding," Trans. I.R.E. (PGIT) 4, 30-37 (1954).
15. W. Feller, "An Introduction to Probability Theory and Its Applications" (John Wiley and Sons, Inc., New York, 1950).

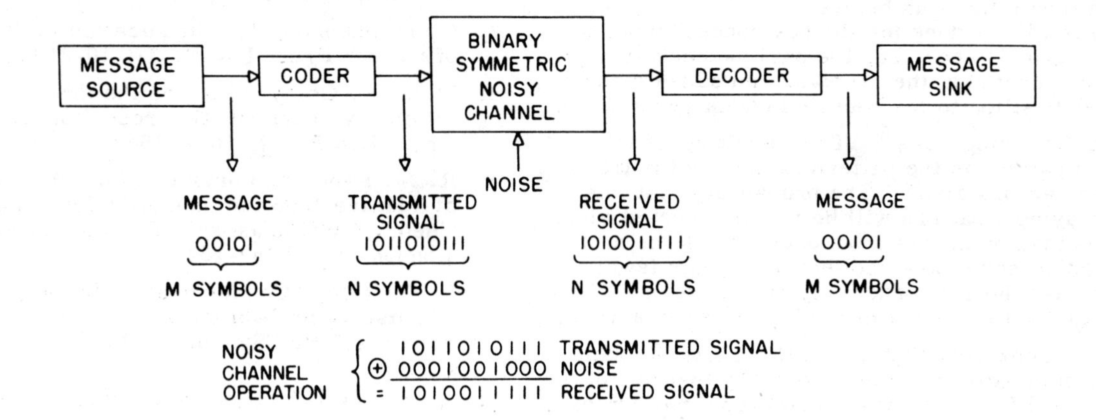

Fig. 1
Noisy communication system

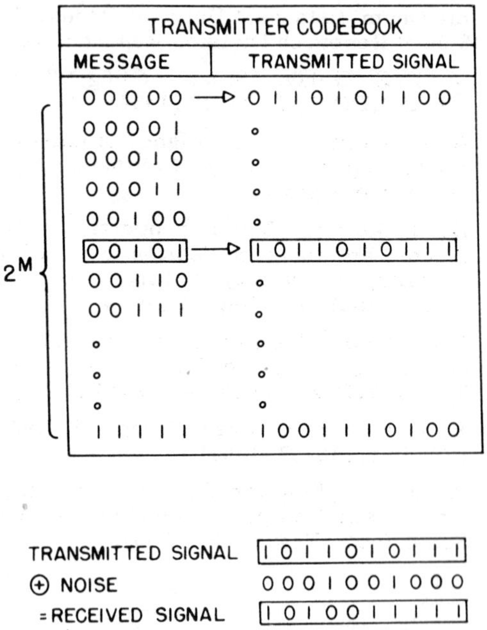

TRANSMITTER CODEBOOK	
MESSAGE	TRANSMITTED SIGNAL
0 0 0 0 0 →	0 1 1 0 1 0 1 1 0 0
0 0 0 0 1	∘
0 0 0 1 0	∘
0 0 0 1 1	∘
0 0 1 0 0	∘
0 0 1 0 1 →	1 0 1 1 0 1 0 1 1 1
0 0 1 1 0	∘
0 0 1 1 1	∘
∘	∘
∘	∘
∘	∘
1 1 1 1 1	1 0 0 1 1 1 0 1 0 0

2^M

TRANSMITTED SIGNAL	1 0 1 1 0 1 0 1 1 1
⊕ NOISE	0 0 0 1 0 0 1 0 0 0
= RECEIVED SIGNAL	1 0 1 0 0 1 1 1 1 1

Fig. 2
Codebook coding

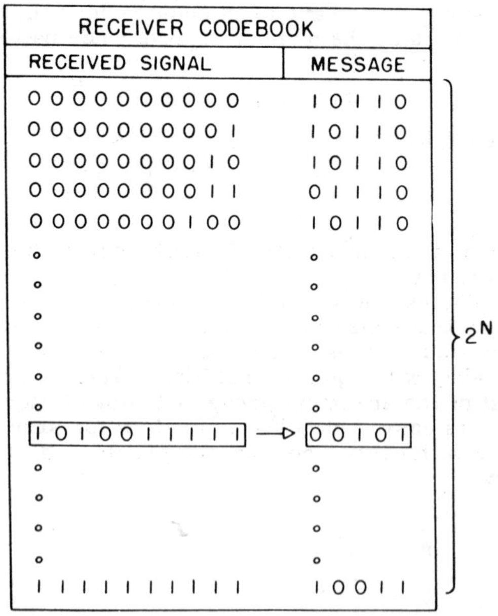

RECEIVER CODEBOOK	
RECEIVED SIGNAL	MESSAGE
0 0 0 0 0 0 0 0 0 0	1 0 1 1 0
0 0 0 0 0 0 0 0 0 1	1 0 1 1 0
0 0 0 0 0 0 0 0 1 0	1 0 1 1 0
0 0 0 0 0 0 0 0 1 1	0 1 1 1 0
0 0 0 0 0 0 0 1 0 0	1 0 1 1 0
∘	∘
∘	∘
∘	∘
∘	∘
∘	∘
∘	∘
1 0 1 0 0 1 1 1 1 1 →	0 0 1 0 1
∘	∘
∘	∘
∘	∘
∘	∘
1 1 1 1 1 1 1 1 1 1	1 0 0 1 1

2^N

Fig. 3
Codebook decoding

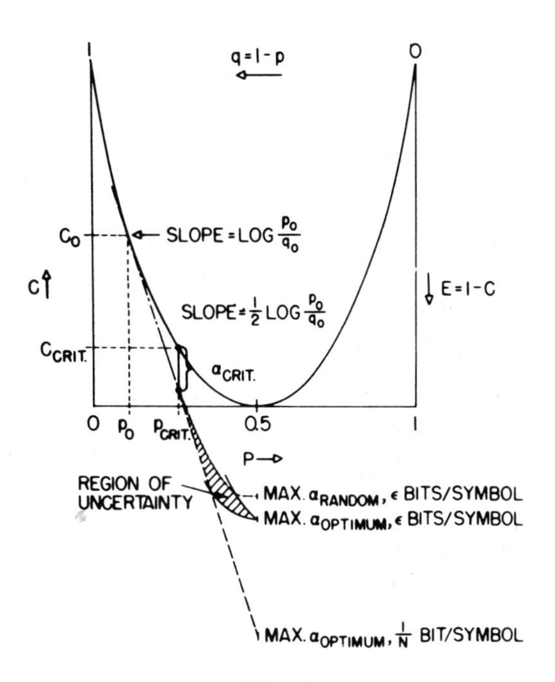

Fig. 4
The error exponent α_{optimum}

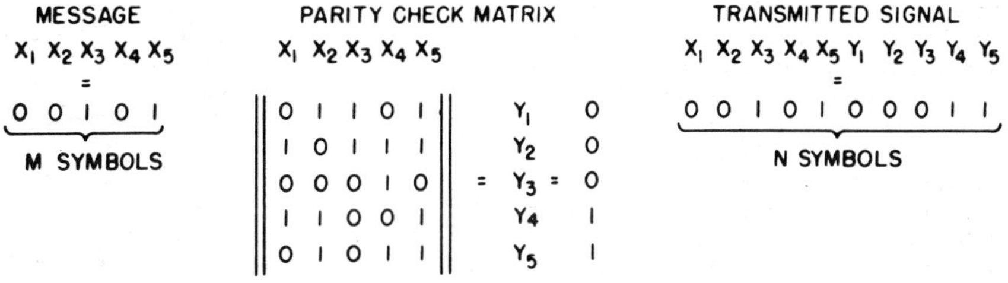

MESSAGE

$X_1\ X_2\ X_3\ X_4\ X_5$
=
0 0 1 0 1
M SYMBOLS

PARITY CHECK MATRIX

$$
\begin{array}{c}
X_1\ X_2\ X_3\ X_4\ X_5 \\
\left\| \begin{array}{ccccc} 0 & 1 & 1 & 0 & 1 \\ 1 & 0 & 1 & 1 & 1 \\ 0 & 0 & 0 & 1 & 0 \\ 1 & 1 & 0 & 0 & 1 \\ 0 & 1 & 0 & 1 & 1 \end{array} \right\|
\end{array}
= \begin{array}{c} Y_1 \\ Y_2 \\ Y_3 \\ Y_4 \\ Y_5 \end{array} = \begin{array}{c} 0 \\ 0 \\ 0 \\ 1 \\ 1 \end{array}
$$

TRANSMITTED SIGNAL

$X_1\ X_2\ X_3\ X_4\ X_5\ Y_1\ Y_2\ Y_3\ Y_4\ Y_5$
=
0 0 1 0 1 0 0 0 1 1
N SYMBOLS

Fig. 5
Parity check coding

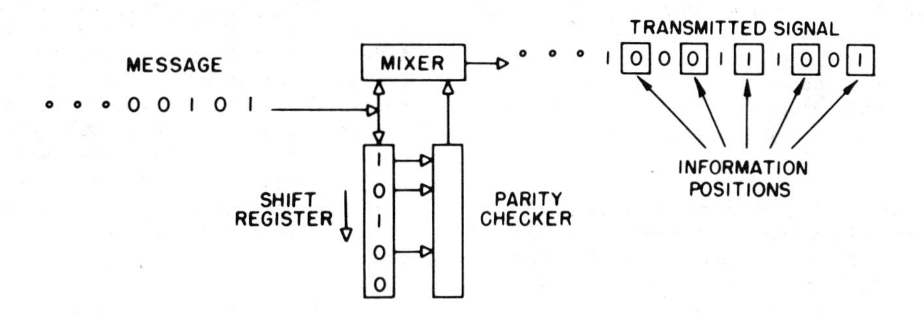

Fig. 6
Convolutional parity check coding

THE ZERO ERROR CAPACITY OF A NOISY CHANNEL

Claude E. Shannon
Bell Telephone Laboratories, Murray Hill, New Jersey
Massachusetts Institute of Technology, Cambridge, Mass.

Abstract

The zero error capacity C_0 of a noisy channel is defined as the least upper bound of rates at which it is possible to transmit information with zero probability of error. Various properties of C_0 are studied; upper and lower bounds and methods of evaluation of C_0 are given. Inequalities are obtained for the C_0 relating to the "sum" and "product" of two given channels. The analogous problem of zero error capacity C_{0F} for a channel with a feedback link is considered. It is shown that while the ordinary capacity of a memoryless channel with feedback is equal to that of the same channel without feedback, the zero error capacity may be greater. A solution is given to the problem of evaluating C_{0F}.

Introduction

The ordinary capacity C of a noisy channel may be thought of as follows. There exists a sequence of codes for the channel of increasing block length such that the input rate of transmission approaches C and the probability of error in decoding at the receiving point approaches zero. Furthermore, this is not true for any value higher than C. In some situations it may be of interest to consider, rather than codes with probability of error approaching zero, codes for which the probability is zero and to investigate the highest possible rate of transmission (or the least upper bound of these rates for such codes. This rate, C_0, is the main object of investigation of the present paper. It is interesting that while C_0 would appear to be a simpler property of a channel than C, it is in fact more difficult to calculate and leads to a number of as yet unsolved problems.

We shall consider only finite discrete memoryless channels. Such a channel is specified by a finite transition matrix $\|p_i(j)\|$ where $p_i(j)$ is the probability of input letter i being received as output letter j ($i = 1,2,\ldots,a$; $j = 1,2,\ldots,b$) and $\sum_j p_i(j) = 1$. Equivalently, such a channel may be represented by a line diagram such as Fig. 1.

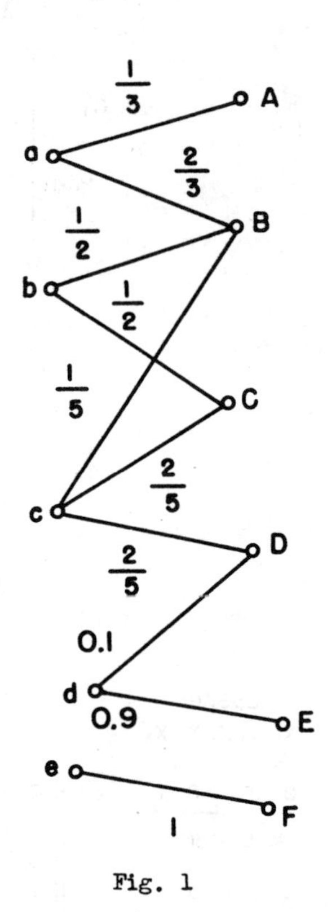

Fig. 1

The channel being memoryless means that successive operations are independent. If the input letters i and j are used, the probability of output letters k and l will be $p_i(k)\,p_j(l)$. A sequence of input letters will be called an **input word**, a sequence of output letters an output word. A mapping of M messages (which we may take to be the integers $1,2,\ldots,M$) into a subset of input words of length n will be called a block code of length n. $R = \frac{1}{n} \log M$ will be called the input rate for this code. Unless otherwise specified, a code will mean such a block code. We will, throughout, use natural logarithms and natural (rather than binary) units of information, since this simplifies the analytical processes that will be employed.

A decoding system for a block code of length n is a method of associating a unique input message (integer from 1 to M) with each possible output word of length n, that is, a function from output words of length n to the integers 1 to M. The probability of error for a code is the probability when the M input messages are used each with probability 1/M that the noise and the decoding system will lead to an input message different from the one that actually occurred.

If we have two given channels, it is possible to form a single channel from them in two natural ways which we call the sum and product of the two channels. The sum of two

Reprinted from *IRE Trans. Inform. Theory*, vol. IT-2, pp. 8–19, Sept. 1956.

channels is the channel formed by using inputs from either of the two given channels with the same transition probabilities to the set of output letters consisting of the logical sum of the two output alphabets. Thus the sum channel is defined by a transition matrix formed by placing the matrix of one channel below and to the right of that for the other channel and filling the remaining two rectangles with zeros. If $p_i(j)$ and $\|p_i'(j)\|$ are the individual matrices, the sum has the following matrix:

$$\begin{array}{cccccccc}
p_1(1) & \cdots & p_1(r) & 0 & \cdots & 0 \\
\vdots & & \vdots & \vdots & & \vdots \\
p_t(1) & \cdots & p_t(r) & 0 & \cdots & 0 \\
0 & \cdots & 0 & p_1'(1) & \cdots & p_1'(r') \\
\vdots & & \vdots & \vdots & & \vdots \\
0 & \cdots & 0 & p_{t'}'(1) & \cdots & p_{t'}'(r')
\end{array}$$

The product of two channels is the channel whose input alphabet consists of all ordered pairs (i, i') where i is a letter from the first channel alphabet and i' from the second, whose output alphabet is the similar set of ordered pairs of letters from the two individual output alphabets and whose transition probability from (i, i') to (j, j') is $p_i(j)\, p_{i'}'(j')$.

The sum of channels corresponds physically to a situation where either of two channels may be used (but not both), a new choice being made for each transmitted letter. The product channel corresponds to a situation where both channels are used each unit of time. It is interesting to note that multiplication and addition of channels are both associative and commutative, and that the product distributes over a sum. Thus one can develop a kind of algebra for channels in which it is possible to write, for example, a polynomial $\sum a_n K^n$, where the a_n are non-negative integers a and K is a channel. We shall not, however, investigate here the algebraic properties of this system.

The Zero Error Capacity

In a discrete channel we will say that two input letters are adjacent if there is an output letter which can be caused by either of these two. Thus, i and j are adjacent if there exists a t such that both $p_i(t)$ and $p_j(t)$ do not vanish. In Fig. 1, a and c are adjacent, while a and d are not.

If all input letters are adjacent to each other, any code with more than one word has a probability of error at the receiving point greater than zero. In fact, the probability of error in decoding words satisfies

$$P_e \geqslant \frac{M-1}{M}\; p_{min}^n$$

where p_{min} is the smallest (non-vanishing) among the $p_i(j)$, n is the length of the code and M is the number of words in the code. To prove this, note that any two words have a possible output word in common, namely the word consisting of the sequence of common output letters when the two input words are compared letter by letter. Each of the two input words has a probability at least p_{min}^n of producing this common output word. In using the code, the two particular input words will each occur $\frac{1}{M}$ of the time and will cause the common output $\frac{1}{M}\; p_{min}^n$ of the time. This output can be decoded in only one way. Hence at least one of these situations leads to an error. This error, $\frac{1}{M}\, p_{min}^n$, is assigned to this code word, and from the remaining $M - 1$ code words another pair is chosen. A source of error to the amount $\frac{1}{M}\, p_{min}^n$ is assigned in similar fashion to one of these, and this is a disjoint event. Continuing in this manner, we obtain a total of at least $\frac{M-1}{M}\; p_{min}^n$ as probability of error.

If it is not true that the input letters are all adjacent to each other, it is possible to transmit at a positive rate with zero probability of error. The least upper bound of all rates which can be achieved with zero probability of error will be called the zero error capacity of the channel and denoted by C_0. If we let $M_0(n)$ be the largest number of words in a code of length n, no two of which are adjacent, then C_0 is the least upper bound of the numbers $\frac{1}{n}\log M_0(n)$ when n varies through all positive integers.

One might expect that C_0 would be equal to $\log M_0(1)$, that is, that if we choose the largest possible set of non-adjacent letters and form all sequences of these of length n, then this would be the best error free code of length n. This is not, in general, true, although it holds in many cases, particularly when the number of input letters is small. The first failure occurs with five input letters with the channel in Fig. 2. In this channel, it is possible to choose at most two non-adjacent letters, for example 0 and 2. Using sequences of these, 00, 02, 20, and 22 we obtain four words in a code of length two. However, it is possible to construct a code of length two with five members no two of which are adjacent as follows: 00, 12, 24, 31, 43. It is readily verified that no two of these are adjacent. Thus, C_0 for this channel is at least $\frac{1}{2}\log 5$.

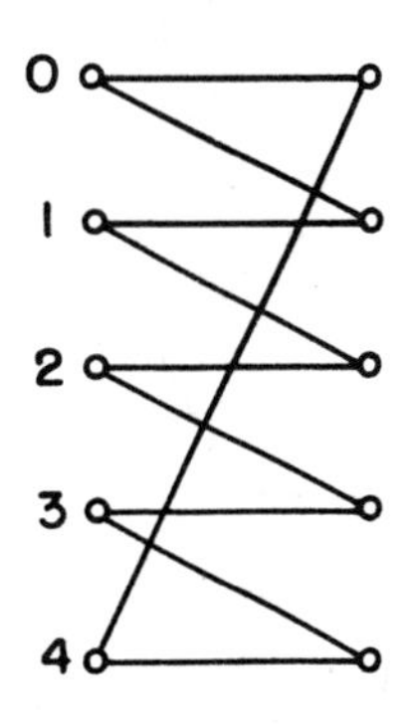

Fig. 2

No method has been found for determining C_0 for the general discrete channel, and this we propose as an interesting unsolved problem in coding theory. We shall develop a number of results which enable one to determine C_0 in many special cases, for example, in all channels with five or less input letters with the single exception of the channel of Fig. 2 (or channels equivalent in adjacency structure to it). We will also develop some general inequalities enabling one to estimate C_0 quite closely in most cases.

It may be seen, in the first place, that the value of C_0 depends only on which input letters are adjacent to each other. Let us define the <u>adjacency matrix</u> for a channel, A_{ij}, as follows.

$$A_{ij} = \begin{cases} 1 & \text{if input letter } i \text{ is adjacent to } j \text{ or if } i = j \\ 0 & \text{otherwise} \end{cases}$$

Suppose two channels have the same adjacency matrix (possibly after renumbering the input letters of one of them). Then it is obvious that a zero error code for one will be a zero error code for the other and, hence, that the zero error capacity C_0 for one will also apply to the other.

The adjacency structure contained in the adjacency matrix can also be represented as a linear graph. Construct a graph with as many vertices as there are input letters, and connect two distinct vertices with a line or branch of the graph if the corresponding input letters are adjacent. Two examples are shown in Fig. 3, corresponding to the channels of Figs. 1 and 2.

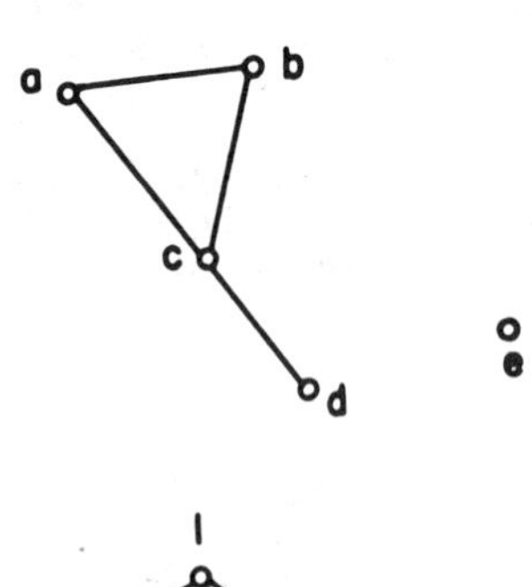

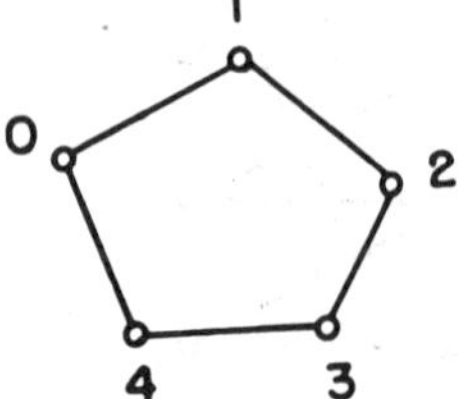

Fig. 3

<u>Theorem 1</u>: The zero error capacity C_0 of a discrete memoryless channel is bounded by the inequalities

$$- \log \min_{P_i} \sum_{ij} A_{ij} P_i P_j \leq C_0 \leq \min_{p_i(j)} C$$

$$\sum_i P_i = 1, \quad P_i \geq 0$$

$$\sum_j p_i(j) = 1, \quad p_i(j) \geq 0$$

where C is the capacity of any channel with transition probabilities $p_i(j)$ and having the adjacency matrix A_{ij}.

The upper bound is fairly obvious. The zero error capacity is certainly less than or equal to the ordinary capacity for any channel with the same adjacency matrix since the former requires codes with zero probability of error while the latter requires only codes approaching zero probability of error. By minimizing the capacity through variation of the $p_i(j)$ we find the lowest upper bound available through this argument. Since the capacity is a continuous function of the $p_i(j)$ in the closed region defined by $p_i(j) \geq 0$, $\sum_j p_i(j) = 1$, we may write min instead of greatest lower bound.

It is worth noting that it is only necessary to consider a particular channel in performing this minimization, although there are an infinite number with the same adjacency matrix. This one particular channel is obtained as follows from the adjacency matrix. If $A_{ik} = 1$ for a pair i k, define an output letter j with $p_i(j)$ and $p_k(j)$ both differing from zero. Now if there are any three input letters, say i k l, all adjacent to each other, define an output letter, say m, with $p_i(m)$ $p_k(m)$ $p_l(m)$ all different from zero. In the adjacency graph this corresponds to a complete sub-graph with three vertices. Next, subsets of four letters or complete subgraphs of four vertices, say i k l m, are given an output letter, each being connected to it, and so on. It is evident that any channel with the same adjacency matrix differs from that just described only by variation in the number of output symbols for some of the pairs, triplets, etc., of adjacent input letters. If a channel has more than one output symbol for an adjacent subset of input letters, then its capacity is reduced by identifying these. If a channel contains no element, say for a triplet i k l of adjacent input letters, this will occur as a special case of our canonical channel which has output letter m for this triplet when $p_i(m)$, $p_k(m)$ and $p_l(m)$ all vanish.

The lower bound of the theorem will now be proved. We use the procedure of random codes based on probabilities for the letters P_i, these being chosen to minimize the quadratic form $\sum_{ij} A_{ij} P_i P_j$. Construct an ensemble of codes each containing M words, each word n letters long. The words in a code are chosen by the following stochastic method. Each letter of each word is chosen independently of all others and is the letter i with probability P_i. We now compute the probability in the ensemble that any particular word is not adjacent to any other word in its code. The probability that the first letter of one word is adjacent to the first letter of a second word is $\sum_{ij} A_{ij}P_iP_j$, since this sums the cases of adjacency with coefficient 1 and those of non-adjacency with coefficient 0. The probability that two words are adjacent in all letters, and therefore adjacent as words, is $(\sum_{ij} A_{ij}P_iP_j)^n$. The probability of non-adjacency is therefore $1 - (\sum_{ij} A_{ij}P_iP_j)^n$. The probability that all M - 1 other words in a code are not adjacent to a given word is, since they are chosen independently,

$$\left[1 - (\sum_{ij} A_{ij}P_iP_j)^n\right]^{M-1}$$

which is, by a well known inequality, greater than $1 - (M - 1)(\sum_{ij} A_{ij}P_iP_j)^n$, which in turn is greater than $1 - M(\sum_{ij} A_{ij}P_iP_j)^n$. If we set $M = (1 - \epsilon)^n(\sum_{ij} A_{ij}P_iP_j)^{-n}$, we then have, by taking ϵ small, a rate as close as desired to $-\log \sum_{ij} A_{ij}P_iP_j$. Furthermore, once ϵ is chosen, by taking n sufficiently large, we can insure that $M(\sum_{ij} A_{ij}P_iP_j)^n = (1 - \epsilon)^n$ is as small as desired, say, less than δ. The probability in the ensemble of codes of a particular word being adjacent to any other in its own code is now less than δ. This implies that there are codes in the ensemble for which the ratio of the number of such undesired words to the total number in the code is less than or equal to δ. For, if not, the ensemble average would be worse than δ. Select such a code and delete from it the words having this property. We have reduced our rate only by at most $\log (1 - \delta)^{-1}$. Since ϵ and δ were both arbitrarily small, we obtain error-free codes arbitrarily close to the rate-log $\min_{P_i} \sum_{ij} A_{ij}P_iP_j$ as stated in the theorem.

In connection with the upper bound of Theorem 1, the following result is useful in evaluating the minimum C. It is also interesting in its own right and will prove useful later in connection with channels having a feedback link.

Theorem 2: In a discrete memoryless channel with transition probabilities $p_i(j)$ and input letter probabilities P_i the following three statements are equivalent.

1) The rate of transmission

$$R = \sum_{i,j} P_ip_i(j) \log (p_i(j)/\sum_k P_kp_k(j))$$

is stationary under variation of all non-vanishing P_i subject to $\sum_i P_i = 1$ and under variation of $p_i(j)$ for those $p_i(j)$ such that $P_ip_i(j) > 0$ and subject to $\sum_j p_i(j) = 1$.

2) The mutual information between input-output pairs $I_{ij} = \log (p_i(j)/\sum_k P_kp_k(j))$ is constant, $I_{ij} = I$, for all ij pairs of non-vanishing probability (i.e. pairs for which $P_ip_i(j) > 0$).

3) We have $p_i(j) = r_j$ a function of j only whenever $P_ip_i(j) > 0$; and also $\sum_{i \epsilon S_j} P_i = h$, a constant independent of j where S_j is the set of input letters that can produce output letter j with probability greater than zero. We also have $I = \log h^{-1}$.

The $p_i(j)$ and P_i corresponding to the maximum and minimum capacity when the $p_i(j)$ are varied (keeping, however, any $p_i(j)$ that are zero fixed at zero) satisfy 1), 2) and 3).

Proof: We will show first that 1) and 2) are equivalent and then that 2) and 3) are equivalent.

R is a bounded continouus function of its arguments P_i and $p_i(j)$ in the (bounded) region of allowed values defined by $\sum P_i = 1$, $P_i \geq 0$, $\sum_j p_i(j) = 1$, $p_i(j) \geq 0$. R has a finite partial derivative with respect to any $p_i(j) > 0$. In fact, we readily calculate

$$\frac{\partial R}{\partial p_i(j)} = P_i \log (p_i(j)/\sum_k P_kp_k(j))$$

A necessary and sufficient condition that R be stationary for small variation of the non-vanishing $p_i(j)$ subject to the conditions given is that

$$\frac{\partial R}{\partial p_i(j)} = \frac{\partial R}{\partial p_i(k)}$$

for all i, j, k such that P_i, $p_i(j)$, $p_i(k)$ do not vanish. This requires that

$$P_i \log p_i(j) / \sum_m P_mp_m(j) =$$

$$P_i \log p_i(k) / \sum_m P_mp_m(k)$$

If we let $Q_j = \sum_m P_mp_m(j)$, the probability of output letter j, then this is equivalent to

$$\frac{p_i(j)}{Q_j} = \frac{p_i(k)}{Q_k}$$

In other words, $p_i(j)/Q_j$ is independent of j, a function of i only whenever $P_i > 0$ and $p_i(j) > 0$. This function of i we call α_i. Thus

$$p_i(j) = \alpha_i Q_j$$

unless $P_i p_i(j) = 0$.

Now, taking the partial derivative of R with respect to P_i we obtain:

$$\frac{\partial R}{\partial P_i} = \sum_j p_i(j) \log \frac{p_i(j)}{Q_j} - 1$$

For R to be stationary subject to $\sum_i P_i = 1$ we must have $\partial R/\partial P_i = \partial R/\partial P_k$. Thus

$$\sum_j p_i(j) \log \frac{p_i(j)}{Q_j} = \sum_j p_k(j) \log \frac{p_k(j)}{Q_j}$$

Since for $P_i p_i(j) > 0$ we have $p_i(j)/Q_j = \alpha_i$, this becomes

$$\sum_j p_i(j) \log \alpha_i = \sum_j p_k(j) \log \alpha_k$$

$$\log \alpha_i = \log \alpha_k$$

Thus α_i is independent of i and may be written α. Consequently

$$\frac{p_i(j)}{Q_j} = \alpha$$

$$\log \frac{p_i(j)}{Q_j} = \log \alpha = I$$

whenever $P_i p_i(j) > 0$.

The converse result is an easy reversal of the above argument. If

$$\log \frac{p_i(j)}{Q_j} = I, \text{ then}$$

$\partial R/\partial P_i = I - 1$, by a simple substitution in the $\partial R/\partial P_i$ formula. Hence R is stationary under variation of P_i constrained by $\sum P_i = 1$. Further, $\partial R/\partial p_i(j) = P_i I = \partial R/\partial p_i(k)$, and hence the variation of R also vanishes subject to $\sum_j p_i(j) = 1$.

We now prove that 2) implies 3). Suppose $\log \frac{p_i(j)}{Q_j} = I$ whenever $P_i p_i(j) > 0$. Then $p_i(j) = e^I Q_j$, a function of j only under this same condition. Also, if $q_j(i)$ is the conditional probability of i given j, then

$$\frac{Q_j\, q_j(i)}{P_i\, Q_j} = e^I$$

$$q_j(i) = e^I P_i$$

$$1 = \sum_{i \in S_j} q_j(i) = e^I \sum_{i \in S_j} P_i$$

To prove that 3) implies 2) we assume

$$p_i(j) = r_j$$

when $P_i p_i(j) > 0$. Then

$$\frac{P_i p_i(j)}{P_i Q_j} = \frac{r_j}{Q_j} = \lambda_j \text{ (say)} = \frac{Q_j q_j(i)}{P_i\, Q_j} = \frac{q_j(i)}{P_i}$$

Now, summing the equation $P_i \lambda_j = q_j(i)$ over $i \in S_j$ and using the assumption from 3) that $\sum_{S_j} P_i = h$ we obtain

$$h\, \lambda_j = 1$$

so λ_j is h^{-1} and independent of j. Hence $I_{ij} = I = \log h^{-1}$.

The last statement of the theorem concerning minimum and maximum capacity under variation of $p_i(j)$ follows from the fact that R at these points must be stationary under variation of all non-vanishing P_i and $p_i(j)$, and hence the corresponding P_i and $p_i(j)$ satisfy condition 1) of the theorem.

For simple channels it is usually more convenient to apply particular tricks in trying to evaluate C_0 instead of the bounds given in Theorem 1, which involve maximizing and minimizing processes. The simplest lower bound, as mentioned before, is obtained by merely finding the logarithm of the maximum number of non-adjacent input letters.

A very useful device for determining C_0 which works in many cases may be described using the notion of an adjacency-reducing mapping.

By this we mean a mapping of letters into other letters, $i \rightarrow \alpha(i)$, with the property that if i and j are not adjacent in the channel (or graph) then $\alpha(i)$ and $\alpha(j)$ are not adjacent. If we have a zero-error code, then we may apply such a mapping letter by letter to the code and obtain a new code which will also be of the zero-error type, since no adjacencies can be produced by the mapping.

Theorem 3: If all the input letters i can be mapped by an adjacency-reducing mapping $i \rightarrow \alpha(i)$ into a subset of the letters no two of which are adjacent, then the zero-error capacity C_0 of the channel is equal to the logarithm of the number of letters in this subset.

For, in the first place, by forming all sequences of these letters we obtain a zero-error code at this rate. Secondly, any zero error code for the channel can be mapped into a code using only these letters and containing, therefore, at most $e^{C_0 n}$ non-adjacent words.

The zero-error capacities, or, more exactly, the equivalent numbers of input letters for all adjacency graphs up to five vertices are shown in Fig. 4. These can all be found readily by the method of Theorem 3, except for the channel of Fig. 2 mentioned previously, for which we know only that the zero-error capacity lies in the range $\frac{1}{2} \log 5 \leq C_0 \leq \log \frac{5}{2}$.

All graphs with six vertices have been examined and the capacities of all of these can also be found by this theorem, with the exception of four. These four can be given in terms of the capacity of Fig. 2, so that this case is essentially the only unsolved problem up to seven vertices. Graphs with seven vertices have not been completely examined but at least one new situation arises, the analog of Fig. 2 with seven input letters.

As examples of how the N_0 values were computed by the method of adjacency-reducing mappings, several of the graphs in Fig. 4 have been labelled to show a suitable mapping. The scheme is as follows. All nodes labelled a are mapped into node α as well as α itself. All nodes labelled b and also β are mapped into node β. All nodes labelled c and γ are mapped into node γ. It is readily verified that no new adjacencies are produced by the mappings indicated and that the α, β, γ nodes are non-adjacent.

C_0 for Sum and Product Channels

Theorem 4. If two memoryless channels have zero-error capacities $C_0' = \log A$ and $C_0'' = \log B$, their sum has a zero-error capacity greater than or equal to $\log (A + B)$ and their product a zero error capacity greater than or equal to $C_0' + C_0''$. If the graph of either of the two channels can be reduced to non-adjacent points by the mapping method (Theorem 3), then these inequalities can be replaced by equalities.

Proof: It is clear that in the case of the product, the zero error capacity is at least $C_0' + C_0''$, since we may form a product code from two codes with rates close to C_0' and C_0''. If these codes are not of the same length, we use for the new code the least common multiple of the individual lengths and form all sequences of the code words of each of the codes up to this length. To prove equality in case one of the graphs, say that for the first channel, can be mapped into A non-adjacent points, suppose we have a code for the product channel. The letters for the product code, of course, are ordered pairs of letters corresponding to the original channels. Replace the first letter in each pair in all code words by the letter corresponding to reduction by the mapping method. This reduces or preserves adjacency between words in the code. Now sort the code words into A^n subsets according to the sequences of first letters in the ordered pairs. Each of these subsets can contain at most B^n members, since this is the largest possible number of codes for the second channel of this length. Thus, in total, there are at most $A^n B^n$ words in the code, giving the desired result.

In the case of the sum of the two channels, we first show how, from two given codes for the two channels, to construct a code for the sum channel with equivalent number of letters equal to $A^{1-\delta} + B^{1-\delta}$, where δ is arbitrarily small and A and B are the equivalent number of letters for the two codes. Let the two codes have lengths n_1 and n_2. The new code will have length n where n is the smallest integer greater than both $\frac{n_1}{\delta}$ and $\frac{n_2}{\delta}$. Now form codes for the first channel and for the second channel for all lengths k from zero to n as follows. Let k equal $an_1 + b$, where a and b are integers and $b < n_1$. We form all sequences of a words from the given code for the first channel and fill in the remaining b letters arbitrarily, say all with the first letter in the code alphabet. We achieve at least $A^{k - \delta n}$ different words of length k none of which is adjacent to any other. In the same way we form codes for the second channel and achieve $B^{k - \delta n}$ words in this code of length k. We now intermingle the k code for the first channel with the n - k code for the second channel in all $\binom{n}{k}$ possible ways and do this for each value of k. This produces a code n letters long with at least $\sum_{k=0}^{n} \binom{n}{k} A^{k - n\delta} B^{n-k-n\delta}$

$= (AB)^{-\delta n}(A + B)^n$ different words. It is readily seen that no two of these different words are adjacent. The rate is at least $\log (A + B) - \delta \log AB$, and since δ was arbitrarily small, we can achieve a rate arbitrarily close to $\log (A + B)$.

To show that it is not possible, when one of the graphs reduces by mapping to non-adjacent points, to exceed the rate corresponding to the number of letters A + B, consider any given code of length n for the sum channel. The words in this consist of sequences of letters each letter corresponding to one or the other of the two

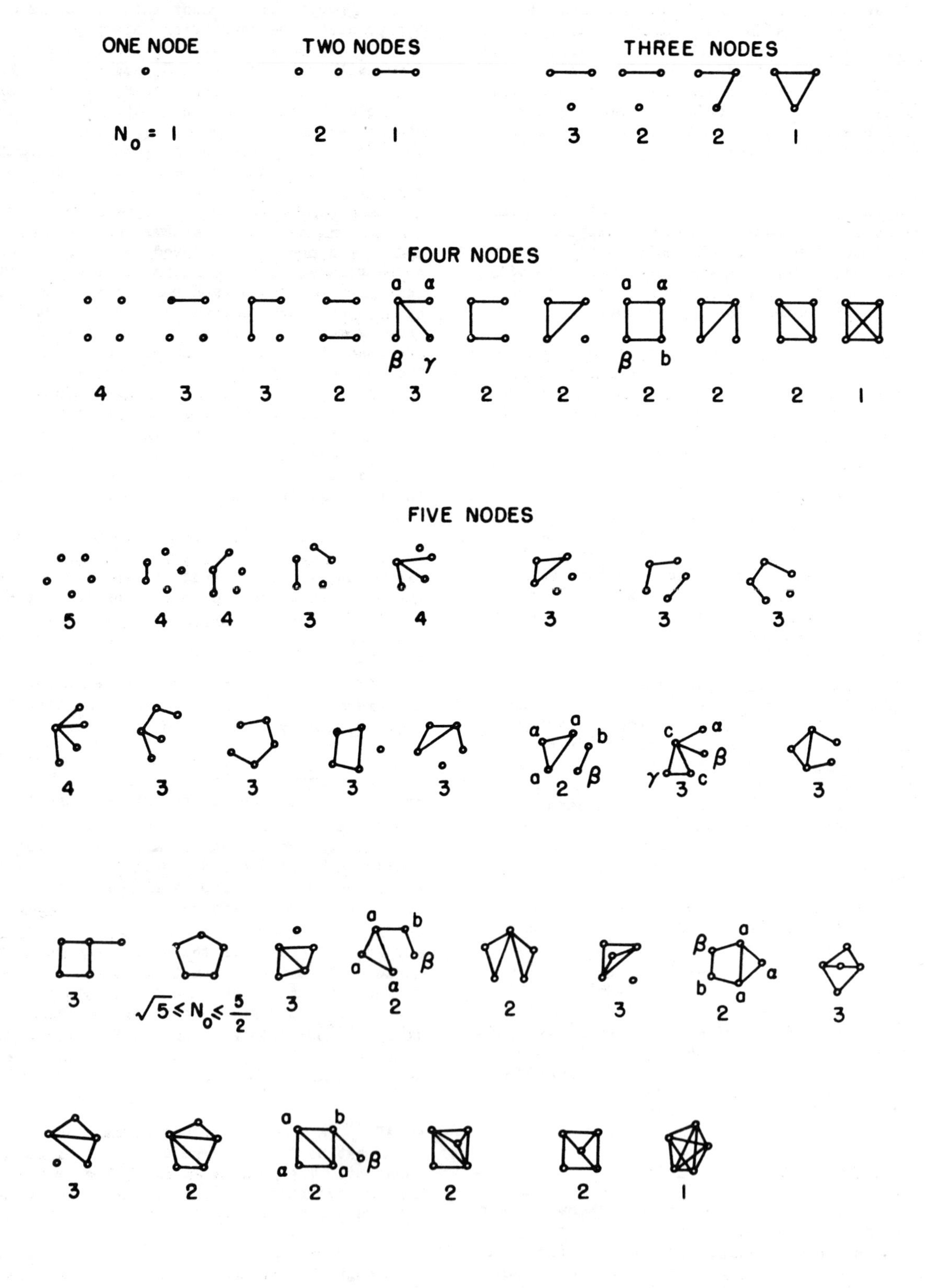

Fig. 4 - All graphs with 1, 2, 3, 4, 5 nodes and the corresponding N_0 for channels with these as adjacency graphs (note $C_0 = \log N_0$)

channels. The words may be subdivided into classes corresponding to the pattern of the choices of letters between the two channels. There are 2^n such classes with $\binom{n}{k}$ classes in which exactly k of the letters are from the first channel and n - k from the second. Consider now a particular class of words of this type. Replace the letters from the first channel alphabet by the corresponding non-adjacent letters. This does not harm the adjacency relations between words in the code. Now, as in the product case, partition the code words according to the sequence of letters involved from the first channel. This produces at most A^k subsets. Each of these subsets contains at most B^{n-k} members, since this is the greatest possible number of non-adjacent words for the second channel of length n - k. In total, then, summing over all values of k and taking account of the $\binom{n}{k}$ classes for each k, there are at most $\sum_k \binom{n}{k} A^k B^{n-k}$ $=(A + B)^n$ words in the code for the sum channel. This proves the desired result.

Theorem 4, of course, is analogous to known results for ordinary capacity C, where the product channel has the sum of the ordinary capacities and the sum channel has an equivalent number of letters equal to the sum of the equivalent numbers of letters for the individual channels. We conjecture but have not been able to prove that the equalities in Theorem 4 hold in general, not just under the conditions given. We now prove a lower bound for the probability of error when transmitting at a rate greater than C_o.

Theorem 5: In any code of length n and rate $R > C_o$, $C_o > 0$, the probability of error P_e will satisfy $P_e \geq (1 - e^{-n(C_o - R)})\, p_{min}^n$, where p_{min} is the minimum non-vanishing $p_i(j)$.

Proof: By definition of C_o there are not more than e^{nC_o} non-adjacent words of length n. With $R > C_o$, among e^{nR} words there must, therefore, be an adjacent pair. The adjacent pair has a common output word which either can cause with a probability at least p_{min}^n. This output word cannot be decoded into both inputs. At least one, therefore, must cause an error when it leads to this output word. This gives a contribution at least $e^{-nR}\, p_{min}^n$ to the probability of error P_e. Now omit this word from consideration and apply the same argument to the remaining $e^{nR} - 1$ words of the code. This will give another adjacent pair and another contribution of error of at least $e^{-nR}\, p_{min}^n$. The process may be continued until the number of code points remaining is just e^{nC_o}. At this time, the computed probability of error must be at least $(e^{nR} - e^{nC_o})e^{-nR}\, p_{min}^n$

$$= (1 - e^{n(C_o - R)})\, p_{min}^n.$$

Channels with a Feedback Link

We now consider the corresponding problem for channels with complete feedback. By this we mean that there exists a return channel sending back from the receiving point to the transmitting point, without error, the letters actually received. It is assumed that this information is received at the transmitting point before the next letter is transmitted, and can be used, therefore, if desired, in choosing the next transmitted letter.

It is interesting that for a memoryless channel the ordinary forward capacity is the same with or without feedback. This will be shown in Theorem 6. On the other hand, the zero error capacity may, in some cases, be greater with feedback than without. In the channel shown in Fig. 5, for example, $C_o = \log 2$. However, we will see as a result of Theorem 7 that with feedback the zero error capacity $C_{oF} = \log 2.5$.

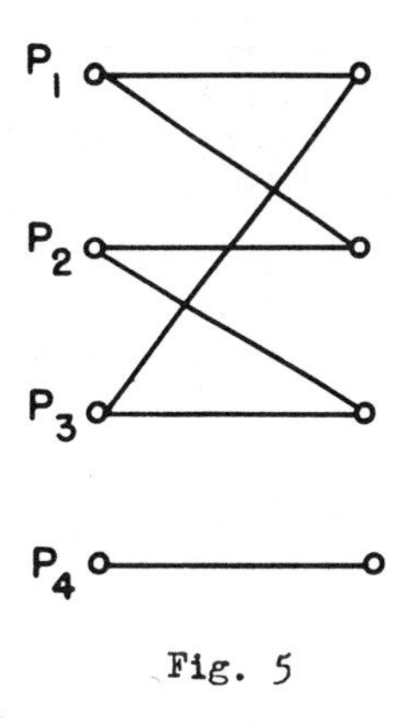

Fig. 5

We first define a block code of length n for a feedback system. This means that at the transmitting point there is a device with two inputs, or, mathematically, a function with two arguments. One argument is the message to be transmitted, the other, the past received letters (which have come in over the feedback link). The value of the function is the next letter to be transmitted. Thus, the function may be thought of as $x_{j+1} = f(k, v_j)$ where x_{j+1} is the j + 1 transmitted letter in a block, k is an index ranging from 1 to M, and represents the specific message, and v_j is a received word of length j. Thus j ranges from 0 to n - 1 and v_j over all received words of these lengths.

In operation, if message m_k is to be sent f is evaluated for f(k -) where the - means "no word" and this is sent as the first transmitted letter. If the feedback link sends back α, say, as the first received letter, the next transmitted letter will be $f(k, \alpha)$. If this is received as β. the next transmitted letter will be $f(k, \alpha \beta)$, etc.

Theorem 6: In a memoryless discrete channel with feedback, the forward capacity is equal to the ordinary capacity C (without feedback). The average change in mutual information I_{vm} between received sequence v and message m for a letter of text is not greater than C.

Proof: Let v be the received sequence to date of a block, m the message, x the next transmitted letter and y the next received letter. These are all random variables and, also, x is a function of m and v. This function, namely, is the one which defines the encoding procedure with feedback whereby the next transmitted letter x is determined by the message m and the feedback information v from the previous received signals. The channel being memoryless implies that the next operation is independent of the past, in particular, $\Pr[y/x] = \Pr[y/x,v]$.

The average change in mutual information, when a particular v has been received, due to the x,y pair is given by (we are averaging over messages m and next received letters y, for a given v):

$$\overline{\Delta I} = \overline{I_{m,vy}} - \overline{I_{m,v}} = \sum_{y,m} \Pr[y,m/v] \cdot \log \frac{\Pr[v,y,m]}{\Pr[v,y]\Pr[m]} - \sum_{m} \Pr[m/v] \cdot \log \frac{\Pr[v,m]}{\Pr[v]\Pr[m]}$$

Since $\Pr[m/v] = \sum_{y} \Pr[y,m/v]$, the second sum may be rewritten as $\sum_{y,m} \Pr[y,m/v] \log \frac{\Pr[v,m]}{\Pr[v]\Pr[m]}$

The two sums then combine to give

$$\overline{\Delta I} = \sum_{y,m} \Pr[y,m/v] \log \frac{\Pr[v,y,m]\Pr[v]}{\Pr[v,m]\Pr[v,y]}$$

$$= \sum_{y,m} \Pr[y,m/v] \log \frac{\Pr[y/v,m]\Pr[v]}{\Pr[v,y]}$$

The sum on m may be thought of as summed first on the m's which result in the same x (for the given v), recalling that x is a function of m and v, and then summing on the different x's. In the first summation, the term $\Pr[y/v,m]$ is constant at $\Pr[y/x]$ and the coefficient of the logarithm sums to $\Pr[x,y/v]$. Thus we can write

$$\Delta I = \sum_{x,y} \Pr[x,y/v] \log \frac{\Pr[y/x]}{\Pr[y/v]}$$

Now consider the rate for the channel (in the ordinary sense without feedback) if we should assign to the x's the probabilities $q(x) = \Pr[x/v]$. The probabilities for pairs, $r(x,y)$, and for the y's alone, $w(y)$, in this situation would then be

$$\begin{aligned} r(x,y) &= q(x)\Pr[y/x] \\ &= \Pr[x/v]\Pr[y/x] \\ &= \Pr[x,y/v] \\ w(y) &= \sum_{x} r(x,y) \\ &= \sum_{x} \Pr[x,y/v] \\ &= \Pr[y/v] \end{aligned}$$

Hence the rate would be

$$\begin{aligned} R &= \sum_{x,y} r(x,y) \log \frac{\Pr[y/x]}{w(y)} \\ &= \sum_{x,y} \Pr[x,y/v] \log \frac{\Pr[y/x]}{\Pr[y/v]} \\ &= \Delta I \end{aligned}$$

Since $R \leq C$, the channel capacity (C being the maximum possible R for all q(x) assignments), we conclude that

$$\Delta I \leq C.$$

Since the average change in I per letter is not greater than C, the average change in n letters is not greater than nC. Hence, in a block code of length n with input rate R, if $R > C$ then the equivocation at the end of a block will be at least $R - C$, just as in the non-feedback case. In other words, it is not possible to approach zero equivocation (or, as easily follows, zero probability of error) at a rate exceeding the channel capacity. It is, of course, possible to do this at rates less than C, since certainly anything that can be done without feedback can be done with feedback.

It is interesting that the first sentence of Theorem 6 can be generalized readily to channels with memory provided they are of such a nature that the internal state of the channel can be calculated at the transmitting point from the initial state and the sequence of letters that have been transmitted. If this is not the case, the conclusion of the theorem will not always be true, that is, there exist channels of a more complex sort for which the forward capacity with feedback exceeds that without feedback. We shall not, however, give the details of these generalizations here.

Returning now to the zero-error problem, we define a zero error capacity C_{oF} for a channel with feedback in the obvious way--the least upper bound of rates for block codes with no errors. The next theorem solves the problem of evaluating C_{oF} for memoryless channels with feedback, and indicates how rapidly C_{oF} may be approached as the block length n increases.

Theorem 7: In a memoryless discrete channel with complete feedback of received letters to the transmitting point, the zero error capacity C_{oF} is zero if all pairs of input letters are adjacent. Otherwise $C_{oF} = \log P_o^{-1}$ where

$$P_o = \min_{P_i} \max_{j} \sum_{i \in S_j} P_i$$

P_i being a probability assigned to input letter i ($\sum_i P_i = 1$) and S_j the set of input letters which can cause output letter j with probability greater than zero. A zero error block code of length n can be found for such a feedback channel which transmits at a rate $R \geq C_{oF} (1 - \frac{2}{n} \log_2 2t)$ where t is the number of input letters.

The P_o occuring in this theorem has the following meaning. For any given assignment of probabilities P_i to the input letters one may calculate, for each output letter j, the total probability of all input letters that can (with positive probability) cause j. This is $\sum_{i \in S_j} P_i$. Output letters for which this is large may be thought of as "bad" in that when received there is a large uncertainty as to the cause. To obtain P_o one adjusts the P_i so that worst output letter in this sense is as good as possible.

We first show that if all letters are adjacent to each other $C_{oF} = 0$. In fact, in any coding system, any two messages, say m_1 and m_2 can lead to the same received sequence with positive probability. Namely, the first transmitted letters corresponding to m_1 and m_2 have a possible received letter in common. Assuming this occurs, calculate the next transmitted letters in the coding system for m_1 and m_2. These also have a possible received letter in common. Continuing in this manner we establish a received word which could be produced by either m_1 or m_2 and therefore they cannot be distinguished with certainty.

Now consider the case where not all pairs are adjacent. We will first prove, by induction on the block length n, that the rate $\log P_o^{-1}$ cannot be exceeded with a zero error code. For n = o the result is certainly true. The inductive hypothesis will be that no block code of length n - 1 transmits at a rate greater than $\log P_o^{-1}$, or, in other words, can resolve with certainty more than

$$e^{(n-1) \log P_o^{-1}} = P_o^{-(n-1)}$$

different messages. Now suppose (in contradiction to the desired result) we have a block code of length n resolving M messages with $M > P_o^{-n}$. The first transmitted letter for the code partitions these M messages among the input letters for the channel. Let F_i be the fraction of the messages assigned to letter i (that is, for which i is the first transmitted letter). Now these F_i are like probability assignments to the different letters and therefore by definition of P_o, there is some output letter, say letter k, such that $\sum_{i \in S_k} F_i \geq P_o$. Consider the set of messages for which the first transmitted letter belongs to S_k. The number of messages in this set is at least $P_o M$. Any of these can cause output letter k as first received letter. When this happens there are n - 1 letters yet to be transmitted and since $M > P_o^{-n}$ we have $P_o M > P_o^{-(n-1)}$. Thus we have a zero error code of block length n - 1 transmitting at a rate greater than $\log P_i^{-1}$, contradicting the inductive assumption. Note that the coding function for this code of length n - 1 is formally defined from the original coding function by fixing the first received letter at k.

We must now show that the rate $\log P_o^{-1}$ can actually be approached as closely as desired with zero error codes. Let P_i be the set of probabilities which, when assigned to the input letters, give P_o for $\min_{P_i} \max_{j} \sum_{i \in S_j} P_i$. The general scheme of the code will be to divide the M original messages into t different groups corresponding to the first transmitted letter. The number of messages in these groups will be approximately proportional to $P_1, P_2, \ldots P_t$. The first transmitted letter, then, will correspond to the group containing the message to be transmitted. Whatever letter is received, the number of possible messages compatible with this received letter will be approximately P_oM. This subset of possible messages is known both at the receiver and (after the received letter is sent back to the transmitter) at the transmitting point.

The code system next subdivides this subset of messages into t groups, again approximately in proportion to the probabilities P_i. The second letter transmitted is that corresponding to the group containing the actual message. Whatever letter is received, the number of messages compatible with the two received letters is now, roughly, P_o^2M.

This process is continued until only a few messages (less than t^2) are compatible with all the received letters. The ambiguity among these is then resolved by using a pair of non-adjacent letters in a simple binary code. The code thus constructed will be a zero error code for the channel.

Our first concern is to estimate carefully the approximation involved in subdividing the messages into the t groups. We will show that for any M and any set of P_i $\sum P_i = 1$, it is possible to subdivide the M messages into groups of $m_1, m_2, \ldots m_t$ such that $m_i = o$ whenever $P_i = o$ and

$$\left| \frac{m_i}{M} - P_i \right| \leqslant \frac{1}{M} \qquad i = 1, \ldots, t$$

We assume without loss of generality that $P_1, P_2, \ldots P_s$ are the non-vanishing P_i. Choose m_1 to be the largest integer such that $\frac{m_1}{M} \leqslant P_1$. Let $P_1 - \frac{m_1}{M} = \delta_1$. Clearly $|\delta_1| \leqslant \frac{1}{M}$. Next choose m_2 to be the smallest integer such that $\frac{m_2}{M} \geqslant P_2$ and let $P_2 - \frac{m_2}{M} = \delta_2$. We have $|\delta_2| \leqslant \frac{1}{M}$. Also $|\delta_1 + \delta_2| \leqslant \frac{1}{M}$, since δ_1 and δ_2 are opposite in sign and each less than $\frac{1}{M}$ in absolute value. Next, m_3 is chosen so that $\frac{m_3}{M}$ approximates, to within $\frac{1}{M}$, to P_3. If $\delta_1 + \delta_2 \geqslant 0$, then $\frac{m_3}{M}$ is chosen less than or equal to P_3. If $\delta_1 = \delta_2 < 0$, then $\frac{m_3}{M}$ is chosen greater than or equal to P_3. Thus again $P_3 - \frac{m_3}{M} = \delta_3 \leqslant \frac{1}{M}$ and $|\delta_1 + \delta_2 + \delta_3| \leqslant \frac{1}{M}$. Continuing in this manner through P_{s-1} we obtain approximationsfor $P_1, P_2, \ldots, P_{s-1}$ with the property that $|\delta_1 + \delta_2 + \ldots + \delta_{s-1}| \leqslant \frac{1}{M}$, or $|M(P_1 + P_2 + \ldots + P_{s-1}) - (m_1 + m_2 + \ldots + m_{s-1})| \leqslant 1$. If we now define m_s as $M - \sum_1^{s-1} m_i$ then this inequality can be written $|M(1 - P_s) - (M - m_s)| \leqslant 1$. Hence $\left| \frac{m_s}{M} - P_s \right| \leqslant \frac{1}{M}$. Thus we have achieved the objective of keeping all approximation $\frac{m_i}{M}$ to within $\frac{1}{M}$ of P_i and having $\sum m_i = M$.

Returning now to our main problem note first that if $P_o = 1$ then $C_{oF} = 0$ and the theorem is trivially true. We assume, then, that $P_o < 1$. We wish to show that $P_o \leqslant (1 - \frac{1}{t})$. Consider the set of input letters which have the maximum value of P_i. This maximum is certainly greater than or equal to the average $\frac{1}{t}$. Furthermore, we can arrange to have at least one of these input letters not connected to some output letter. For suppose this is not the case. Then either there are no other input letters beside this set and we contradict the assumption that $P_o < 1$, or there are other input letters with smaller values of P_i. In this case, by reducing the P_i for one input letter in the maximum set and increasing correspondingly that for some input letter which does not connect to all output letters, we do not increase the value of P_o (for any S_j) and create an input letter of the desired type. By consideration of an output letter to which this input letter does not connect we see that $P_o \leqslant 1 - \frac{1}{t}$.

Now suppose we start with M messages and subdivide into groups approximating proportionality to the P_i as described above. Then when a letter has been received, the set of possible messages (compatible with this received letter) will be reduced to those in the groups corresponding to letters which connect to the actual received letter. Each output letter connects to not more than t - 1 input letters (otherwise we would have $P_o = 1$). For each of the connecting groups, the error in approximating P_i has been less than or equal to $\frac{1}{M}$. Hence the total relative number in all connecting groups for any output letter is less than or equal to $P_o + \frac{t-1}{M}$. The total number of possible messages after receiving the first letter consequently drops from M to a number less than or equal to $P_oM + t - 1$.

In the coding system to be used, this remaining possible subset of messages is subdivided again among the input letters to approximate in the same fashion the probabilities P_i. This subdivision can be carried out both at receiving point and transmitting point using the same standard procedure (say, exactly the one described above) since with the feedback both terminals have available the required data, namely the first received letter.

The second transmitted letter obtained by this procedure will again reduce at the receiving point the number of possible messages to a value not greater than $P_o (P_o M + t - 1) + t - 1$. This same process continues with each transmitted letter. If the upper bound on the number of possible remaining messages after k letters is M_k, then $M_{k+1} = P_o M_k + t - 1$. The solution of this difference equation is

$$M_k = A P_o^k + \frac{t-1}{1-P_o}$$

This may be readily verified by substitution in the difference equation. To satisfy the initial conditions $M_o = M$ requires $A = M - \frac{t-1}{1-P_o}$. Thus the solution becomes

$$M_k = \left(M - \frac{t-1}{1-P_o}\right) P_o^k + \frac{t-1}{1-P_o}$$

$$= M P_o^k + \frac{t-1}{1-P_o}(1 - P_o^k)$$

$$\leq M P_o^k + t(t-1)$$

since we have seen above that $1 - P_o \geq \frac{1}{t}$.

If the process described is carried out for n_1 steps, where n_1 is the smallest integer $\geq d$ where d is the solution of $MP_o^d = 1$, then the number of possible messages left consistent with the received sequence will be not greater than $1 + t(t-1) \leq t^2$ (since $t > 1$, otherwise we should have $C_{oF} = 0$). Now the pair of non-adjacent letters assumed in the theorem may be used to resolve the ambiguity among these t^2 or less messages. This will require not more than $1 + \log_2 t^2 = \log_2 2t^2$ additional letters. Thus, in total, we have used not more than $d + 1 + \log_2 2t^2 = d + \log_2 4t^2 = n$ say as block length. We have transmitted in this block length a choice from $M = P_o^{-d}$ messages. Thus the zero error rate we have achieved is

$$R = \frac{1}{n} \log M \geq \frac{d \log P_o^{-1}}{d + \log_2 4t^2}$$

$$= \left(1 - \frac{1}{n} \log 4t^2\right) \log P_o^{-1}$$

$$= \left(1 - \frac{1}{n} \log 4t^2\right) C_{oF}$$

Thus we can approximate to C_{oF} as closely as desired with zero error codes.

As an example of Theorem 7 consider the channel in Fig. 5. We wish to evaluate P_o. It is easily seen that we may take $P_1 = P_2 = P_3$ in forming the min max of Theorem 7, for if they are unequal the maximum $\sum_{i \in S_j} P_i$ for the corresponding three output letters would be reduced by equalizing. Also it is evident, then, that $P_4 = P_1 + P_2$, since otherwise a shift of probability one way or the other would reduce the maximum. We conclude, then, that $P_1 = P_2 = P_3 = 1/5$ and $P_4 = 2/5$. Finally, the zero error capacity with feedback is $\log P_o^{-1} = \log 5/2$.

There is a close connection between the min max process of Theorem 7 and the process of finding the minimum capacity for the channel under variation of the non-vanishing transition probabilities $p_i(j)$ as in Theorem 2. It was noted there that at the minimum capacity each output letter can be caused by the same total probability of input letters. Indeed, it seems very likely that the probabilities of input letters to attain the minimum capacity are exactly those which solve the min max problem of Theorem 7, and, if this is so, the $C_{min} = \log P_o^{-1}$.

Acknowledgement

I am indebted to Peter Elias for first pointing out that a feedback link could increase the zero-error capacity, as well as for several suggestions that were helpful in the proof of Theorem 7.

Certain Results in Coding Theory for Noisy Channels*

CLAUDE E. SHANNON

Massachusetts Institute of Technology, Cambridge, Massachusetts

In this paper we will develop certain extensions and refinements of coding theory for noisy communication channels. First, a refinement of the argument based on "random" coding will be used to obtain an upper bound on the probability of error for an optimal code in the memoryless finite discrete channel. Next, an equation is obtained for the capacity of a finite state channel when the state can be calculated at both transmitting and receiving terminals. An analysis is also made of the more complex case where the state is calculable at the transmitting point but not necessarily at the receiving point.

PROBABILITY OF ERROR BOUND FOR THE DISCRETE FINITE MEMORYLESS CHANNEL

A discrete finite memoryless channel with finite input and output alphabets is defined by a set of transition probabilities $p_i(j)$,

$$i = 1, 2, \cdots, a; \qquad j = 1, 2, \cdots, b,$$

with $\sum_j p_i(j) = 1$ $(i = 1, 2, \cdots, a)$ and all $p_i(j) \geqq 0$. Here $p_i(j)$ is the probability, if input letter i is used, that output letter j will be received. A *code word* of length n is a sequence of n input letters (that is, n integers each chosen from $1, 2, \cdots, a$). A *block code of length n* with M words is a mapping of the integers from 1 to M (messages) into a set of code words each of length n. A *decoding system* for such a code is a mapping of all sequences of output words of length n into the integers from 1 to M (that is, a procedure for deciding on an original integer or message when any particular output word is received). We will be considering situations in which all integers from 1 to M are used with the same probability $1/M$. The probability of error P_e for a code and decoding system is the probability of an integer being transmitted and received as a word which is mapped into a different integer (that is, decoded as another message).

Thus:

$$P_e = \sum_u \sum_{v \epsilon S_u} \frac{1}{M} Pr(v \mid u)$$

where u ranges over all input integers $1, 2, \cdots, M$; v ranges over the received words of length n; and S_u is the set of received words that are not decoded as u. $Pr(v \mid u)$ is of course the probability of receiving v if the message is u. Thus if u is mapped into input word $(i_1, i_2, \cdots, i_n)$ and v is word $(j_1, j_2, \cdots, j_n)$, then

$$Pr(v \mid u) = p_{i_1}(j_1)\, p_{i_2}(j_2) \cdots p_{i_n}(j_n).$$

While we assume all messages in a code to be used with equal probabilities $1/M$, it is useful, in studying a channel, to consider the assignment of different probabilities to input words. Suppose, in fact, that in a given channel we assign arbitrary probabilities to the different input words u of length n, probability $P(u)$ for word u. We then have probabilities for all input-output word pairs of length n,

$$Pr(u, v) = P(u)\, Pr(v \mid u),$$

where u and v are input and output words of length n and $Pr(v \mid u)$ is the probability of output word v if input word u is used. (This is the product of the transition probabilities for corresponding letters of u and v). Given $P(u)$ then, any numerical function of u and v becomes a random variable. In particular, the mutual information (per letter), $I(u, v)$ is a random variable

$$I(u, v) = \frac{1}{n} \log \frac{Pr(u, v)}{P(u)Pr(v)} = \frac{1}{n} \log \frac{Pr(v \mid u)}{\sum_u P(u)Pr(v \mid u)}$$

The distribution function for this random variable will be denoted by $\rho(x)$. Thus

$$\rho(x) = Pr[I(u, v) \leqq x]$$

The function $\rho(x)$ of course depends on the arbitrary assignment of probabilities $P(u)$. We will now prove a theorem bounding the probability of error for a possible code in terms of the function $\rho(x)$.

THEOREM 1: *Suppose some $P(u)$ for input words u of length n gives rise to a distribution of information per letter $\rho(I)$. Then given any integer M and any $\theta > 0$ there exists a block code with M messages and a decoding system such that if these messages are used with equal probability, the probability of error P_e is bounded by*

$$P_e \leqq \rho(R + \theta) + e^{-n\theta}$$

where $R = (1/n)\log M$.

PROOF: For a given M and θ consider the pairs (u, v) of input and output words and define the set T to consist of those pairs for which $\log Pr(u, v)/P(u)Pr(v) > n(R + \theta)$. When the u's are chosen with probabilities $P(u)$, then the probability that the (u, v) pair will belong to the set T is, by definition of ρ, equal to $1 - \rho(R + \theta)$.

Now consider the ensemble of codes obtained in the following manner. The integers $1, 2, 3, \cdots, M = e^{nR}$ are associated independently with the different possible input words $u_1, u_2, \cdots, u_B$ with probabilities $P(u_1), P(u_2), \cdots P(u_B)$. This produces an ensemble of codes each using M (or less) input words. If there are B different input words u_i, there will be exactly B^M different codes in this ensemble corresponding to the B^M different ways we can associate M integers with B input words. These codes have different probabilities. Thus the (highly degenerate) code in which all integers are mapped into input word u_1 has probability $P(u_1)^M$. A code in which d_k of the integers are mapped into u_k has probability $\prod_k P(u_k)^{d_k}$. We will be concerned with the average probability of error for this ensemble of codes. By this we mean the average probability of error when these codes are weighted according to the probabilities we have just defined. We imagine that in using any one of these codes, each integer is used with probability $1/M$. Note that, for some particular selections, several integers may fall on the same input word. This input word is then used with higher probability than the others.

In any particular code of the ensemble, our decoding procedure will be defined as follows. Any received v is decoded as the integer with greatest probability conditional on the received v. If several integers have the same conditional probability we decode (conventionally) as the smallest such integer. Since all integers have unconditional probability $1/M$, this decoding procedure chooses one of those having the greatest probability of causing the received v.

We now wish to compute the average probability of error or "ambiguity" P_a in this ensemble of codes where we pessimistically include with the errors all cases where there are several equally probable causes of the received v.

In any particular code of the ensemble an input word u or a pair (u, v) will not, in general, occur with the probabilities $P(u)$ or $Pr(u, v)$. In the ensemble average, however, each word u has probability $P(u)$ and each (u, v) pair probability $Pr(u, v)$, since integers are mapped into u with just this probability. Indeed, a particular message, say the integer 1, will be mapped into u with probability $P(u)$. A particular case of integer 1, say, mapped into u and resulting in received v will result in an error or ambiguity if there are, in the code in question, one or more integers mapped into the set $S_v(u)$ of input of words which have a probability of causing v higher than are qual to that of u. Because of the independence in placing the other integers, it is easy to calculate the fraction of codes in which this occurs. In fact, let

$$Q_v(u) = \sum_{u' \epsilon S_v(u)} P(u')$$

* This work was carried out at the Research Laboratory of Electronics, Massachusetts Institute of Technology, and was supported in part by the United States Army (Signal Corps), the United States Air Force (Office of Scientific Research, Air Research and Development Command), and the United States Navy (Office of Naval Research); and in part by Bell Telephone Laboratories, Inc.

Reprinted with permission from *Inform. Contr.*, vol. 1, pp. 6–25, Sept. 1957.

Thus $Q_v(u)$ is the probability associated with all words more probable or as probable conditioned on the received word v as u is. The fraction of codes in which integer 2 is not in $S_v(u)$ is (because of the independence of placing of the integers) equal to $1 - Q_v(u)$. The fraction of codes in which $S_v(u)$ is free of *all* other integers is $(1 - Q_v(u))^{M-1}$. A similar argument applies to any other integer as well as 1. Thus, in the ensemble, the probability of error or ambiguity due to cases where the message is mapped into input word u and received as v is given exactly by

$$Pr(u, v)[1 - (1 - Q_v(u))^{M-1}].$$

The average probability of error or ambiguity, then, is given by

$$P_a = \sum_{u,v} Pr(u, v)[1 - (1 - Q_v(u))^{M-1}]. \tag{1}$$

We now wish to place a bound on this in terms of the information distribution ρ. First, break the sum into two parts, a sum over the (u, v) set T defined above where $\log Pr(u, v)/P(u)Pr(v) > n(R + \theta)$ and over the complementary set $\bar{T}$.

$$P_a = \sum_{\bar{T}} Pr(u, v)[1 - (1 - Qv(u))^{M-1}] + \sum_T Pr(u, v)[1 - (1 - Q_v(u))^{M-1}].$$

Since $[1 - (1 - Q_v(u))]^{M-1}$ is a probability, we may replace it by 1 in the first sum, increasing the quantity. This term becomes, then, $\sum_{\bar{T}} Pr(u, v)$ which by definition is $\rho(R + \theta)$. In the second sum, note first that $(1 - Q_v(u))^{M-1} \geqq 1 - (M - 1)Q_v(u)$ by a well-known inequality. Hence, the second sum is increased by replacing

$$[1 - (1 - Q_v(u))^{M-1}]$$

by $(M - 1)Q_v(u)$ and even more so by $MQ_v(u)$.

$$P_e \leqq P_a \leqq \rho(R + \theta) + M \sum_T Pr(u, v)Q_v(u).$$

We now show that for u, v in T, $Q_v(u) \leqq e^{-n(R+\theta)}$. In fact, with u, v in T

$$\log \frac{Pr(v|u)}{Pr(v)} > n(R + \theta),$$

$$Pr(v \mid u) > Pr(v)e^{n(R+\theta)}.$$

If $u' \epsilon S_v(u)$,

$$Pr(v \mid u') \geqq Pr(v \mid u) > Pr(v)e^{n(R+\theta)}$$

$$Pr(u', v) > Pr(u')Pr(v)e^{n(R+\theta)}$$

$$Pr(u' \mid v) > Pr(u')e^{n(R+\theta)}$$

Summing each side over $u' \epsilon S_v(u)$ gives

$$1 \geq \sum_{u' \epsilon S_v(u)} P_r(u' \mid v) > e^{n(R+\theta)} Q_v(u)$$

The left inequality holds because the sum of a set of disjoint probabilities cannot exceed 1. We obtain

$$Q_v(u) < e^{-n(R+\theta)} \qquad (u, v) \epsilon T$$

Using this in our estimate of P_e we have

$$P_e < \rho(R + \theta) + e^{nR}e^{-n(R+\theta)}\sum_T Pr(u, v)$$

$$\leqq \rho(R + \theta) + e^{-n\theta}$$

using again the fact that the sum of a set of disjoint probabilities cannot exceed one. Since the average P_e over the ensemble of codes satisfies $P_e \leqq \rho(R + \theta) + e^{-n\theta}$, there must exist a particular code satisfying the same inequality. This concludes the proof.

Theorem 1 is one of a number of results which show a close relation between the probability of error in codes for noisy channels and the distribution of mutal information $\rho(x)$. Theorem 1 shows that if, by associating probabilities $P(u)$ with input words, a certain $\rho(x)$ can be obtained, then codes can be constructed with a probability of error bounded in terms of this $\rho(x)$. We now develop a kind of converse relation: given a code, there will be a related $\rho(x)$. It will be shown that the probability of error for the code (with optimal decoding) is closely related to this $\rho(x)$.

THEOREM 2: *Suppose a particular code has* $M = e^{nR}$ *messages and the distribution function for the mutual information* I *(per letter) between messages and received words is* $\rho(x)$ *(the messages being used with equal probability). Then the optimal detection system for this code gives a probability of error* P_e *satisfying the inequalities*

$$\tfrac{1}{2}\rho\left(R - \frac{1}{n}\log 2\right) \leqq P_e \leqq \rho\left(R - \frac{1}{n}\log 2\right)$$

It should be noted that ρ has a slightly different meaning here than in Theorem 1. Here it relates to mutual information between messages and received words—in Theorem 1, between *input words* and received words. If, as would usually be the case, all messages of a code are mapped into distinct input words, these reduce to the same quantity.

PROOF: We first prove the lower bound. By definition of the function ρ, the probability is equal to $\rho(R - (1/n) \log 2)$, that

$$\frac{1}{n} \log \frac{Pr(u, v)}{Pr(u)Pr(v)} \leqq R - \frac{1}{n} \log 2,$$

where u is a message and v a received word. Equivalently,

$$Pr(u \mid v) \leqq Pr(u)e^{Rn}\tfrac{1}{2}$$

or (using the fact that $Pr(u) = e^{-nR}$)

$$Pr(u \mid v) \leqq \tfrac{1}{2}$$

Now fix attention on these pairs (u, v) for which this inequality

$$Pr(u \mid v) \leqq \tfrac{1}{2}$$

is true, and imagine the corresponding (u, v) lines to be marked in black and all other (u, v) connecting lines marked in red. We divide the v points into two classes: C_1 consists of those v's which are decoded into u's connected by a red line (and also any v's which are decoded into u's not connected to the v's): C_2 consists of v's which are decoded into u's connected by a black line. We have established that with probability $\rho(R - (1/n) \log 2)$ the (u, v) pair will be connected by a black line. The v's involved will fall into the two classes C_1 and C_2 with probability ρ_1, say and $\rho_2 = \rho(R - (1/n) \log 2) - \rho_1$. Whenever the v is in C_1 an error is produced since the actual u was one connected by a black line and the decoding is to a u connected by a red line (or to a disconnected u). Thus these cases give rise to a probability ρ_1 of error. When the v in question is in class C_2, we have $Pr(u \mid v) \leqq \frac{1}{2}$. This means that with at least an equal probability these v's can be obtained through other u's than the one in question. If we sum for these v's the probabilities of all pairs $Pr(u, v)$ except that corresponding to the decoding system, then we will have a probability at least $\rho_2/2$ and all of these cases correspond to incorrect decoding. In total, then, we have a probability of error given by

$$\rho_e \geqq \rho_1 + \rho_2/2 \geqq \tfrac{1}{2}\rho(R - (1/n) \log 2)$$

We now prove the upper bound. Consider the decoding system defined as follows. If for any received v there exists a u such that $Pr(u \mid v) > \frac{1}{2}$, then the v is decoded into that u. Obviously there cannot be more than one such u for a given v, since, if there were, the sum of these would imply a probability greater than one. If there is no such u for a given v, the decoding is irrelevant to our argument. We may, for example, let such u's all be decoded into the first message in the input code. The probability of error, with this decoding, is then less than or equal to the probability of all (u, v) pairs for which $Pr(u \mid v) \leqq \frac{1}{2}$. That is,

$$P_e \leqq \sum_S Pr(u, v) \quad \text{(where } S \text{ is the set of pairs } (u, v) \text{ with } Pr(u \mid v) \leqq \tfrac{1}{2}).$$

The condition $Pr(u \mid v) \leqq \frac{1}{2}$ *is equivalent to* $Pr(u, v)/Pr(v) \leqq \frac{1}{2}$, or, again, to $Pr(u, v)/Pr(u)\, Pr(v) \leqq \frac{1}{2} P_r(u)^{-1} = \frac{1}{2} e^{nR}$. This is equivalent to the condition

$$(1/n) \log Pr(u, v)/Pr(u)Pr(v) \leqq R - (1/n) \log 2.$$

The sum $\sum_S Pr(u, v)$ where this is true is, by definition, the distribution function of $(1/n) \log Pr(u, v)/Pr(u)Pr(v)$ evaluated at $R - (1/n) \log 2$, that is,

$$P_e \leqq \sum_S Pr(u, v) = \rho(R - (1/n) \log 2).$$

PROBABILITY OF ERROR BOUND IN TERMS OF MOMENT GENERATING FUNCTION

We will now develop from the bound of Theorem 1 another expression that can be more easily evaluated in terms of the channel parameters. Suppose first that the probabilities $P(u)$ assigned to words in Theorem 1 are equal to the product of probabilities for letters making up the words.

Thus, suppose u consists of the sequence of letters $i_1, i_2, \cdots, i_n$ and $P(u)$ is then $P_{i_1} \cdot P_{i_2} \cdot P_{i_3} \cdots P_{i_n}$. If v consists of letters $j_1, j_2, \cdots, j_n$ then $Pr(v) = Pr(j_1). Pr(j_2) \cdots Pr(j_n)$ and $Pr(u, v) = Pr(i_1, j_1) \cdot Pr(i_2, j_2) \cdots Pr(i_n, j_n)$. Also

$$I(u, v) = \frac{1}{n}\left[\log \frac{Pr(i_1 j_2)}{Pr(i_1)Pr(j_2)} + \log \frac{Pr(i_2 j_2)}{Pr(i_2)Pr(j_2)} + \cdots\right]$$
$$= \frac{1}{n}[I_1 + I_2 + \cdots + I_n]$$

where I_k is the mutual information between the kth letters of u and v.

The different I's are here independent random variables all with the same distribution. We therefore have a central limit theorem type of situation; $nI(u, v)$ is the sum of n independent random variables with identical distributions. $\rho(x)$ can be bounded by any of the inequalities which are known for the distribution of such a sum. In particular, we may use an inequality due to Chernov on the "tail" of such a distribution (Chernov, 1952). He has shown, by a simple argument using the generalized Chebycheff inequality, that the distribution of such sums can be bounded in terms of the moment generating function for a single one of the random variables, say $\varphi(s)$. Thus let

$$\varphi(s) = E[e^{sI}]$$
$$= \sum_{ij} P_i p_i(j) \exp\left[s \log \frac{p_i(j)}{\sum_k P_k p_k(j)}\right]$$
$$= \sum_{ij} P_i p_i(j) \left[\frac{p_i(j)}{\sum_k P_k p_k(j)}\right]^s$$

It is convenient for our purposes to use the log of the moment generating function $\mu(s) = \log \varphi(s)$, (sometimes called the semi-invariant generating function). Chernov's result translated into our notation states that

$$\rho(\mu'(s)) \leqq e^{[\mu(s) - s\mu'(s)]n} \qquad s \leqq 0$$

Thus by choosing the parameter s at any negative value we obtain a bound on the information distribution ρ of exponential form in n. It is easily shown, also, that if the variance of the original distribution is positive then $\mu'(s)$ is a strictly monotone increasing function of s and so also is the coefficient of n in the exponent, $\mu(s) - s\mu'(s)$ (for negative s). Indeed the derivatives of these quantities exist and are $\mu''(s)$ and $-s\mu''(s)$, respectively. $\mu''(s)$ is readily shown to be positive by a Schwartz inequality.

THEOREM 3: *In a memoryless channel with finite input and output alphabets, let $\mu(s)$ be the semi-invariant generating function for mutual information with some assignment of input letter probabilities, P_i for letter i, and with channel transition probabilities $p_i(j)$, that is:*

$$\mu(s) = \log \sum_{i,j} P_i p_i(j) \left[\frac{p_i(j)}{\sum_i P_i p_i(j)}\right]^s$$

Then there exists a code and decoding system of length n, rate R and probability of error P_e satisfying the inequalities

$$R \geqq \mu(s) - (s - 1)\mu'(s)$$
$$P_e \leqq 2e^{(\mu(s) - s\mu'(s))n} \qquad s \leqq 0$$

If as $s \to -\infty$, $\mu(s) - (s - 1)\mu'(s) \to R^* > 0$ then for $R \leqq R^*$

$$P_e \leqq e^{(E^* + R^* - R)n}$$

where $E^* = \lim (\mu(s) - s\mu'(s))$ as $s \to -\infty$.

PROOF: We have, from *Theorem* 1, that

$$P_e \leqq \rho(R + \theta) + e^{-n\theta}$$
$$\leqq e^{[\mu(s) - s\mu'(s)]n} + e^{-n\theta} \qquad s \leqq 0$$

where s is chosen so that $\mu'(s) = R + \theta$. This will hold when θ is such that the resulting s is negative. We choose θ (which is otherwise arbitrary) to make the coefficients of n in the exponents equal. (Since the first term is monotone increasing in θ and the second monotone decreasing, it is easily seen that this choice of θ is quite good to minimize the bound. In fact, the bound can never be less than half its value for this particular θ.) This relation requires that

$$\mu(s) - s\mu'(s) = -\theta$$
$$= R - \mu'(s)$$
$$R = \mu(s) + (1 - s)\mu'(s)$$

Since the exponents are now equal, the probability of error is bounded by twice the first term:

$$P_e \leqq 2e^{[\mu(s) - s\mu'(s)]n}$$

These relations are true for all negative s and give the first results of the theorem.

However, in ome cases, as $s \to -\infty$ the rate R approaches a positive limiting value. In fact, $R \to I_{\min} + \log Pr[I_{\min}]$ and the exponent in the P_e bound approaches $\log Pr[I_{\min}]$. For rates R lower than this limiting value the exponents cannot be made equal by any choice of s. We may, however, now choose θ in such a way that $R + \theta$ is just smaller than $I_{\min}$, say $I_{\min} - \epsilon$. Since $\rho(I_{\min} - \epsilon) = 0$ the probability of error is now bounded by $P_e \leqq e^{-n\theta} = e^{-n(I_{\min} - R - \epsilon)}$. This being true for any ϵ we can construct codes for which it is true with $\epsilon = 0$. That is

$$P_e \leqq e^{-n(I_{\min} - R)}$$

for $R < I_{\min}$. Notice that as R approaches its limiting value in the first bound, $I_{\min} + \log Pr[I_{\min}]$, the exponents in both bounds approach the same value, namely $\log Pr[I_{\min}]$. The coefficient, however, improves from 2 to 1.

These bounds can be written in another form that is perhaps more revealing. Define a set of "tilted" probabilities $Q_s(I)$ for different values of information I by the following:

$$Q_s(I) = \frac{Pr(I)e^{sI}}{\sum_I Pr(I)e^{sI}}$$

In other words the original probability of a value I is increased or decreased by a factor e^{sI} and the resulting values normalized to sum to unity. For large positive values of s, this tilted set of probabilities $Q_s(I)$ tend to emphasize the probabilities $Pr(I)$ for positive I and reduce those for negative I. At $s = 0$ $Q_0(I) = Pr(I)$. At negative s the negative I values have enhanced probabilities at the expense of positive I values. As $s \to \infty$, $Q_s(I) \to 0$ except for $I = I_{\max}$ the largest value of I with positive probability (since the set of u, v pairs is finite, $I_{\max}$ exists), and $Q_s(I_{\max}) \to 1$. These tilted probabilities are convenient in evaluating the "tails" of distribution that are sums of other distributions. In terms of $Q_s(I)$ we may write

$$\mu(s) = \log \sum Pr(I)e^{sI}$$
$$= \sum_{I'} Q_s(I') \log \sum_I Pr(I)e^{sI}$$
$$\mu'(s) = \sum_I Pr(I)e^{sI}I / \sum_I Pr(I)e^{sI}$$
$$= \sum_I Q_s(I)I$$
$$\mu(s) - s\mu'(s) = \sum_I Q_s(I) \log (Pr(I)/Q_s(I))$$
$$\mu - (s - 1)\mu'(s) = \sum_I Q_s(I)[I + \log Pr(I)/Q_s(I)]$$

The coefficients of n in these exponents are of some interest. They relate to the rapidity of approach of P_e to zero as n increases. Plotted as a function of R, the behavior is typically as shown in Fig. 1. Here we have assumed the P_i for the letters to be the P_i which give channel capacity. The coefficient E of n for the first bound in the theorem is a curve tangent to the axis at C (here $s = 0$), convex downward and ending ($s = -\infty$) at $R = I_{\min} + \log Pr[I_{\min}]$ and $E = \log Pr[I_{\min}]$. The second bound in the theorem gives an E curve which is a straight line of slope -1 passing through this point and intersecting the axes at $I_{\min}$, 0 and 0, $I_{\min}$. In the neighborhood of $R = C$ the curve behaves as

$$E = \frac{(C - R)^2}{2\mu''(0)}$$

Here $\mu''(0)$ is the variance of I. These properties all follow directly from the formulas for the curves.

The limiting exponent (as $n \to \infty$) satisfies $E = \mu(s) - (s - 1)\mu'(s)$. We have

$$\frac{dE}{dR} = \frac{dE}{ds} \Big/ \frac{dR}{ds}$$
$$= \frac{s}{1 - s}$$

so the slope of the ER curve is monotone decreasing as s ranges from 0 to $-\infty$, the slope going from 0 to -1. Since the second bound corresponds to a straight line of slope -1 in the ER plot, the two bounds not only join in value but have the same slope as shown in Fig. 1.

The curve would be as indicated if the P_i are those which maximize the rate at the channel capacity, for then

$$R(0) = \mu(0) - (0 - 1)\mu'(0) = \mu'(0) = C.$$

The bound, however, of the theorem applies for any set of P_i when the corresponding $\mu(s)$ is used. To obtain the strongest result the bound should be optimized for each value of R under variation of P_i. The same applies to the straight line portion where we maximize $I_{\min}$. If this were done a curve would be obtained which is the envelope of all possible curves of this type with different values of P_i. Since each individual curve is convex downward the envelope is also convex downward. The equations for this envelope may be found by the Lagrange method maximizing $R + \lambda E + \eta \sum_i P_i$. It must be remembered, of course, that the P_i must be non-negative. The problem is similar to that involved in calculating the channel capacity. The equations for the envelope will be

$$E = \mu(s) - s\mu'(s)$$

$$R = \mu(s) - (s - 1)\mu'(s)$$

$$(1 + \lambda)\frac{\partial \mu}{\partial P_i} - (1 + \lambda)s\frac{\partial \mu'}{\partial P_i} + \frac{\partial \mu''}{\partial P_i} + \eta = 0 \quad \text{for all } i \text{ except a set for which } P_i = 0$$

and subject to:

$$\sum P_i = 1$$

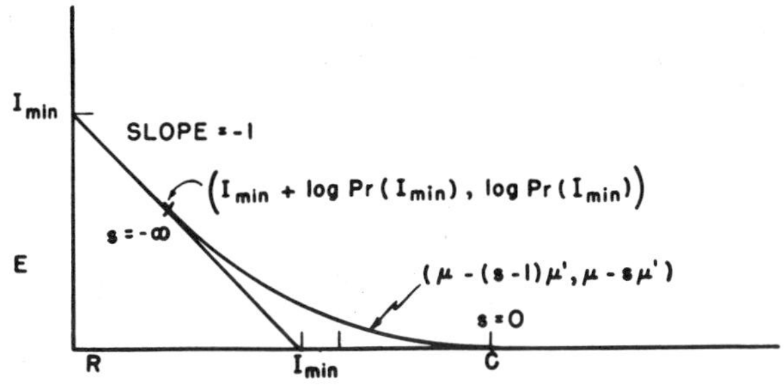

Fig. 1

The bound here should be maximized by choosing different subsets of the P_i for the nonvanishing set.

The upper bound obtained in Theorem 3 is by no means the strongest that can be found. As $n \to \infty$ even the coefficients of n in the exponent can be, in general, improved by more refined arguments. We hope in another paper to develop these further results, and also to give corresponding *lower* bounds on the probability of error of the same exponential type. The upper bound in Theorem 3 is, however, both simple and useful. It has a universality lacking in some of the stronger results (which only assume simple form when n is large).

CAPACITY OF THE FINITE STATE CHANNEL WITH STATE CALCULABLE AT BOTH TERMINALS

In certain channels with memory, the internal state of the channel can be calculated from the initial state (assumed known) at the beginning of transmission and the sequence of transmitted letters. It may also be possible to determine the state at any time at the receiving terminal from the initial state and the sequence of received letters. For such channels we shall say the state is *calculable at both terminals.*

To satisfy the first requirement it is clearly necessary that for any (attainable) internal state s, the next state t must be a function of s and x, $t = f(s, x)$, where x is the transmitted letter.

For the state to be calculable at the receiving point it is necessary that, for all attainable states s, the next stage t must be a function of s and the received letter y, $t = g(s, y)$.

For each possible s, t pair we may find the subset $A(s, t)$ of x's leading from s to t and the subset $B(s, t)$ of y's which correspond to a state transition from s to t. For each input letter x in the set $A(s, t)$ the output letter y will necessarily be in the set $B(s, t)$ and there will be a transition probability, the probability (in state s), if x is transmitted, that y will be received. For a particular s, t pair, the sets of letters $A(s, t)$ and $B(s, t)$ and the corresponding transition probabilities can be thought of as defining a memoryless discrete channel corresponding to the s, t pair. Namely, we consider the memoryless channel with input alphabet the letters from $A(s, t)$, output letters from $B(s, t)$ and the corresponding transition probabilities.

This channel would be physically realized from the given channel as follows. The given channel is first placed in state s, one letter is transmitted from set $A(s, t)$ (resulting in state t), the channel is then returned to state s and a second letter from set $A(s, t)$ transmitted, etc. The capacity of such a discrete memoryless channel can be found by the standard methods. Let the capacity from state s to state t be C_{st} (in natural units) and let $N_{st} = e^{C_{st}}$. Thus N_{st} is the number of equivalent noiseless letters for the s, t sub-channel. If the set $A(s, t)$ is empty, we set $N_{st} = 0$.

The states of such a channel can be grouped into equivalence classes as follows. States s and s' are in the same class if there is a sequence of input letters which, starting with state s, ends in s', and conversely a sequence leading from s' to s. The equivalence classes can be partially ordered as follows. If there is a sequence leading from a member of one class to a member of a second class, the first class is higher in the ordering than the second class.

Within an equivalence class one may consider various possible closed sequences of states; various possible ways, starting with a state, to choose a sequence of input letters which return to this state. The number of states around such a cycle will be called the cycle length. The greatest common divisor of all cycle lengths in a particular equivalence class will be called the basic period of that class. These structural properties are analogous to those of finite state markoff processes, in which "transition with positive probability" takes the place of a "possible transition for some input letter."

We shall consider only channels in which there is just one equivalence class. That is, it is possible to go from any state s to any state t by some sequence of input letters (i.e., any state is accessible from any other). The more general case of several equivalence classes is more complex without being significantly more difficult.

THEOREM 4: *Let K be a finite state channel with finite alphabets, with state calculable at both terminals, and any state accessible from any other state. Let N_{st} be the number of equivalent letters for the sub-channel relating to transitions from state s to state t. Let N be the (unique) positive real eigenvalue of the matrix N_{st}, that is, the positive real root of*

$$| N_{st} - N\delta_s | = 0.$$

Then N is the equivalent number of letters for the given channel K; its capacity is $C = \log N$.

PROOF: We will first show that there exist block codes which transmit at any rate $R < C$ and with probability of error arbitrarily small. Consider the matrix N_{st}. If this is raised to the nth power we obtain a matrix with elements, say, $N_{st}^{(n)}$. The element $N_{st}^{(n)}$ can be thought of as a sum of products, each product corresponding to some path n steps long from state s to state t, the product being the product of the original matrix elements along this path, and the sum being the sum of such products for all such possible paths. This follows immediately by mathematical induction and the definition of matrix multiplication.

Furthermore, $N_{st}^{(n)}$ can be interpreted as the equivalent number of letters for the memoryless channel defined as follows. Imagine starting the original channel in state s and using as input "letters" sequences of length n of the original letters allowing just those sequences which will end in state t after the sequence of n. The output "letters" are sequences of received letters of length n that could be produced under these conditions. This channel can be thought of as a "sum" of channels (corresponding to the different state sequences from s to t in n steps) each of which is a "product" of channels (corresponding to simple transitions from one state to another). (The sum of two channels is a channel in which a letter from either of the two channels may be used; the product is the channel in which a letter from both given channels is used, this ordered pair being an input letter of the product channel). The equivalent number of noise free letters for the sum of channels is additive, and for the product, multiplicative. Consequently the channel we have just described, corresponding to sequences from state s to state t in n steps, has an equivalent number of letters equal to the matrix element $N_{st}^{(n)}$

The original matrix N_{st} is a matrix with non-negative elements. Consequently it has a positive real eigenvalue which is greater than or equal to all other eigenvalues in absolute value. Furthermore, under our assumption that it be possible to pass from any state to any other state by some sequence of letters, there is only one positive real eigenvalue. If d is the greatest common divisor of closed path lengths (through sequences of states), then there will be d eigenvalues equal to the positive real root multiplied by the different dth roots of unity. When the matrix N_{st} is raised to the nth power, a term $N_{st}^{(n)}$ is either zero (if it is impossible to go from s to t in exactly n steps) or is asymptotic to a constant times $N^{(n)}$.

In particular, for n congruent to zero, mod d, the diagonal terms $N_{tt}^{(n)}$ are asymptotic to a constant times N^n, while if this congruence is not satisfied the terms are zero. These statements are all well known results in the Frobenius theory of matrices with non-negative elements, and will not be justified here (Frobenius, 1912).

If we take n a sufficiently large multiple of d we will have, then, $N_{11}^{(n)} > k\, N^n$ with k positive. By taking n sufficiently large, then, the capacity of the channel whose input "letters" are from state 1 to state 1 in n steps can be made greater than $(1/n)\log kN^n = \log N + (1/n)\log k$. Since the latter term can be made arbitrarily small we obtain a capacity as close as we wish to $\log N$. Since we may certainly use the original channel in this restricted way (going from state 1 to state 1 in blocks of n) the original channel has a capacity at least equal to $\log N$.

To show that this capacity cannot be exceeded, consider the channel K_n defined as follows for sequences of length n. At the beginning of a block of length n the channel K_n can be put into an arbitrary state chosen from a set of states corresponding to the states of K. This is done by choice of a "state letter" at the transmitting point and this "state letter" is transmitted noiselessly to the receiving point. For the next n symbols the channel behaves as the given channel K with the same constraints and probabilities. At the end of this block a new state can be freely chosen at the transmitter for the next block. Considering a block of length n (including its initial state information) as a single letter and the corresponding y block including the received "state letter," as a received letter we have a memoryless channel K_n.

For any particular initial-final state pair s, t, the corresponding capacity is equal to $\log N_{st}^{(n)}$. Since we have the "sum" of these channels available, the capacity of K_n is equal to $\log \sum_{s,t} N_{st}^{(n)}$. Each term in this sum is bounded by a constant times N^n, and since there are only a finite number of terms (because there are only a finite number of states) we may assume one constant for all the terms, that is $N_{st}^{(n)} < kN^n$ (all n, s, t). By taking n sufficiently large we clearly have the capacity of K_n per letter, bounded by $\log N + \epsilon$ for any positive ϵ. But now any code that can be used in the original channel can also be used in the K_n channel for any n since the latter has identical constraints except at the ends of n blocks at which point all constraints are eliminated. Consequently the capacity of the original channel is less than or equal to that of K_n for all n and therefore is less than or equal to $\log N$. This completes the proof of the theorem.

This result can be generalized in a number of directions. In the first place, the finiteness of the alphabets is not essential to the argument. In effect, the channel from state s to t can be a general memoryless channel rather than a discrete finite alphabet channel.

A second slight generalization is that it is not necessary that the state be calculable at the receiver after each received letter, provided it is eventually possible at the receiver to determine all previous states. Thus, in place of requiring that the next state be a function of the preceding state and the received letter, we need only require that there should not be two different sequences of states from any state s to any state t compatible with the same sequence of received letters.

THE CAPACITY OF A FINITE STATE CHANNEL WITH STATE CALCULABLE AT TRANSMITTER BUT NOT NECESSARILY AT RECEIVER

Consider now a channel with a finite input alphabet, a finite output alphabet, and a finite number of internal states with the further property that the state is known at the beginning and can be calculated at the transmitter for each possible sequence of input letters. That is, the next state is a function of the current state and the current input letter. Such a channel is defined by this state transition function $s_{n+1} = f(s_n, x_n)$, (the $n+1$ state as a function of state s_n and nth input symbol), and the conditional probabilities in state s, if letter x is transmitted, that the output letter will be y, $p_{sx}(y)$. We do not assume that the state is calculable at the receiving point.

As before, the states of such a channel can be grouped into a partially ordered set of equivalence classes. We shall consider again only channels in which there is just one equivalence class. That is, it is possible to go from any state s to any state t by some sequence of input letters.

We first define a capacity for a particular state s. Let the channel be in state s and let $X_1 = (x_1, x_2, \cdots, x_n)$ be a sequence of n input letters which cause the channel to end in the same state s. If the channel is in state s and the sequence X_1 is used, we can calculate the conditional probabilities of the various possible output sequences Y of length n. Thus, if the sequence X_1 leads through states $s, s_2, s_3, \cdots, s_n, s$ the conditional probability of $Y_1 = (y_1, y_2, \cdots, y_n)$ will be $Pr(Y_1/X_1) = Ps_{x_1}(y_1)Ps_2x_2(y_2)\cdots Ps_nx_n(y_n)$. Consider the X's (leading from s to s in n steps) as individual input letters in a memoryless channel with the y sequences Y as output letters and the conditional probabilities as the transition probabilities. Let $C(n, s)$ be the capacity of this channel. Let $C(s)$ be the least upper bound of $(1/n)C(n, s)$ when n varies over the positive integers. We note the following properties:

1. $C(kn, s) \geqq kC(n, s)$. This follows since in choosing probabilities to assign the X letters of length kn to achieve channel capacity one may at least do as well as the product probabilities for a sequence of kX's each of length n. It follows that if we approximate to $C(s)$ within ϵ at some particular n (i.e. $|C(s) - C(n, s)| < \epsilon$) we will approximate equally well along the infinite sequence $2n, 3n, 4n, \cdots$.
2. $C(s) = C$ is independent of the state s. This is proved as follows. Select a sequence of input letters U leading from state s' to state s and a second sequence V leading from s to s'. Neither of these need contain more than m letters where m is the (finite) number of states in the channel. Select an n_1 for which $C(n_1, s) > C(s) - \epsilon/2$ and with n_1 large enough so that:

$$(C(s) - \epsilon/2)\frac{n_1}{n_1 + 2m} \geq C(s) - \epsilon$$

This is possible since by the remark 1 above $C(s)$ is approximated as closely as desired with arbitrarily large n_1. A set of X sequences for the s' state is constructed by using the sequences for the s state and annexing the U sequence at the beginning and the V sequence at the end. If each of these is given a probability equal to that used for the X sequences in the s state to achieve $C(n, s)$, then this gives a rate for the s' sequences of exactly $C(n, s)$ but with sequences of length at most $n_1 + 2m$ rather than n_1. It follows that $C(s') \geqq (C(s) - \epsilon/2)(n_1/n_1 + 2m) \geqq C(s) - \epsilon$. Of course, interchanging s and s' gives the reverse result $C(s) \geqq C(s') - \epsilon$ and consequently $C(s) = C(s')$. (Note that, if there were several equivalence classes, we would have a C for each class, not necessarily equal).

3. Let $C(n, s, s')$ be the capacity calculated for sequences starting at s and ending at s' after n steps. Let $C(s, s') = \lim_{n\to\infty}(1/n)C(n, s, s')$. Then $C(s, s') = C(s) = C$. This is true since we can change sequences from s to s' into sequences from s to s by a sequence of length at most m added at the end. By taking n sufficiently large in the lim the effect of an added m can be made arbitrarily small, (as in the above remark 2) so that $C(s, s') \geqq C(s) - \epsilon$. Likewise, the s to s sequences which approximate $C(s)$ and can be made arbitrarily long can be translated into s to s' sequences with at most m added letters. This implies $C(s) \geqq C(s, s') - \epsilon$. Hence $C(s) = C(s, s')$.

We wish to show first that starting in state s_1 it is possible to signal with arbitrarily small probability of error at any rate $R < C$ where C is the quantity above in remark 3. More strongly, we will prove the following.

THEOREM 5: *Given any $R < C$ there exists $E(R) > 0$ such that for any $n = k\,d$ (an integer multiple of d, the basic cycle length) there are block codes of length n having M words with $(1/n)\log M \geqq R$ and with probability of error $P_e \leqq e^{-E(R)n}$. There does not exist a sequence of codes of increasing block length with probability of error approaching zero and rate greater than C.*

PROOF: The affirmative part of the result is proved as follows. Let $R_1 = (R + C)/2$. Let s_1 be the initial state of the channel and consider sequences of letters which take the state from s_1 to s_1 in n_1 steps. Choose n_1 so that $C(n_1, s_1) > (3C + R)/4$. Use these sequences as input letters and construct codes for the rate R_1. By Theorem 2 the probability of error will go down exponentially in the length of the code. The codes here are of length $n_1, 2n_1, 3n_1, \cdots$ in terms of the original letters, but this merely changes the coefficient of n by a factor $1/n_1$. Thus, for multiples of n_1 the affirmative part of the theorem is proved. To prove it for all multiples of d, first note that it is true for all sufficiently large multiples of d, since by going out to a sufficiently large multiple of n_1 the effect of a suffix on the code words bringing the state back to s_1 after multiples of d, can be made small (so that the rate is not substantially altered). But now for smaller multiples of d one may use any desired code with a probability of error less than 1 (e.g., interpret any received word as message 1, with $P_e = 1 - 1/M < 1$). We have then a finite set of codes up to some multiple of d at which a uniform exponential bound takes over. Thus, one may choose a coefficient $E(R)$ such that $P_e < e^{-E(R)n}$ for n *any* integer multiple of d.

The negative part of our result, that the capacity C cannot be exceeded, is proved by an argument similar to that used for the case where the state was calculable at the receiver. Namely, consider the channel K_n defined as follows. The given channel K may be put at the beginning into any state and the name of this state transmitted noiselessly to the receiving point. Then n letters are transmitted with the constraints and probabilities of the given channel K. The final state is then also transmitted to the receiver point. This process is then repeated in blocks of n. We have here a memoryless channel which for any n "includes" the given channel. Any code for the given channel K could be used if desired in K_n with equally good probability of error. Hence the capacity of the given channel K must be less than or equal to that of K_n for every n. On the other hand K_n is actually the "sum" of a set of channels corresponding to sequences from state s to state t in n steps; channels with capacities previously denoted by $C(n, s, t)$. For all sufficiently large n, and for all s, t, we have $(1/n)C(n, s, t) < C + \epsilon$ as we have seen above. Hence for all $n > n_0$, say, the capacity of K_n is bounded by $C + \epsilon + (1/n) \log m^2$ where m is the number of states. It follows that the capacity of K is not greater than C.

It is interesting to compare the results of this section where the state is calculable at the transmitter only with those of the preceding section where the state is calculable at both terminals. In the latter case, a fairly explicit formula is given for the capacity, involving only the calculation of capacities of memoryless channels and the solution of an algebraic equation. In the former case, the solution is far less explicit, involving as it does the evaluation of certain limits of a rather complex type.

RECEIVED: April 22, 1957.

REFERENCES

CHERNOV, H., (1952). A Measure of Asymptotic Efficiency for Tests of a Hypothesis Based on the Sum of Observations. *Ann. Math. Stat.* **23,** 493–507.

ELIAS, P. (1956). *In* "Information Theory" (C. Cherry, ed.). Academic Press, New York.

FEINSTEIN, A. (1955). Error Bounds in Noisy Channels Without Memory. *IRE Trans. on Inform. Theory* **IT-1,** 13–14 (Sept.).

FROBENIUS, G. (1912). Über Matrizen aus nichtnegativen Elementen. *Akad. Wiss. Sitzber. Berlin*, pp. 456–477.

SHANNON, C. E. (1948). Mathematical Theory of Communication. *Bell System Tech. J.* **27,** 379–423.

SHANNON, C. E. (1956). The Zero Error Capacity of a Noisy Channel. *IRE Trans. on Inform. Theory* **IT-2,** 8–19 (Sept.).

THE CODING OF MESSAGES SUBJECT TO CHANCE ERRORS[1]

BY J. WOLFOWITZ

1. The transmission of messages

Throughout this paper we assume that all "alphabets" involved contain exactly two symbols, say 0 and 1. What this means will be apparent in a moment. This assumption is made only in the interest of simplicity of exposition, and the changes needed when this assumption is not fulfilled will be obvious.

Suppose that a person has a vocabulary of S words (or messages), any or all of which he may want to transmit, in any frequency and in any order, over a "noisy channel". For example, S could be the number of words in the dictionary of a language, provided that it is forbidden to coin words not in the dictionary. What a "noisy channel" is will be described in a moment. Here we want to emphasize that we do not assume anything about the frequency with which particular words are transmitted, nor do we assume that the words to be transmitted are selected by any random process (let alone that the distribution function of the random process is known). Let the words be numbered in some fixed manner. Thus transmitting a word is equivalent to transmitting one of the integers $1, 2, \cdots, S$.

We shall now explain what is meant by a "noisy channel" of memory m. A sequence of $(m + 1)$ elements, each zero or one, will be called an α-sequence. A function p, defined on the set of all α-sequences, and such that always $0 \leqq p \leqq 1$, is associated with the channel and called the channel probability function. A sequence of n elements, each of which is zero or one, will be call an x-sequence. To describe the channel, it will be sufficient to describe how it transmits any given x-sequence, say x_1. Let α_1 be the α-sequence of the first $(m + 1)$ elements of x_1. The channel "performs" a chance experiment with possible outcomes 1 and 0 and respective probabilities $p(\alpha_1)$ and $(1 - p(\alpha_1))$, and transmits the outcome of this chance experiment. It then performs another chance experiment, independently of the first, with possible outcomes 1 and 0 and respective probabilities $p(\alpha_2)$ and $(1 - p(\alpha_2))$, where α_2 is the α-sequence of the $2^{\text{nd}}, 3^{\text{rd}}, \cdots, (m + 2)^{\text{nd}}$ elements of the sequence x_1. This is repeated until $(n - m)$ independent experiments have been performed. The probability of the outcome one in the ith experiment is $p(\alpha_i)$, where α_i is the α-sequence of the $i^{\text{th}}, (i + 1)^{\text{st}}, \cdots, (i + m)^{\text{th}}$ elements of x_1. The x-sequence x_1 is called the transmitted sequence. The chance sequence $Y(x_1)$ of outcomes of the experiments in consecutive order is called the received sequence. Any sequence of $(n - m)$ elements, each zero or one, will be called a y-sequence. Let y_1 be any y-sequence. If $P\{Y(x_1) = y_1\} > 0$ (the symbol

---

Received January 26, 1957; received in revised form May 15, 1957.

[1] Research under contract with the Office of Naval Research.

Reprinted with permission from *Illinois J. Math.*, vol. 1, pp. 591–606, Dec. 1957.

$P\{\ \}$ denotes the probability of the relation in braces), we shall say that y_1 is a possible received sequence when x_1 is the transmitted sequence.

Let λ be a positive number which it will usually be desired to have small. A "code" of length t is a set $\{(x_i, A_i)\}$, $i = 1, \cdots, t$, where (a) each x_i is an x-sequence, (b) each A_i is a set of y-sequences, (c) for each i

$$P\{Y(x_i) \in A_i\} \geqq 1 - \lambda,$$

(d) $A_1, \cdots, A_t$ are disjoint sets. The coding problem which is a central concern of the theory of transmission of messages may be described as follows: For given S, to find an n and then a code of length S. The practical applications of this will be as follows: When one wishes to transmit the ith word, one transmits the x-sequence x_i. Whenever the receiver receives a y-sequence which is in A_j, he always concludes that the jth word has been sent. When the receiver receives a y-sequence not in $A_1 \cup A_2 \cup \cdots \cup A_S$, he may draw any conclusion he wishes about the word that has been sent. The probability that any word transmitted will be correctly received is $\geqq 1 - \lambda$.

When such a code is used, s/n is called the "rate of transmission," where $s = \log S$. (All logarithms which occur in the present paper are to the base 2.) Except for certain special[2] functions p, one can find a code for any s, provided that one is willing to transmit at a sufficiently small rate; for the law of large numbers obviously applies, and by sufficient repetition of the word to be transmitted, one can insure that the probability of its correct reception exceeds $1 - \lambda$. The practical advantages of a high rate of transmission are obvious. If there were no "noise" (error in transmission) and signals were received exactly as sent, then s symbols zero or one would suffice to transmit any word in the vocabulary, and one could transmit at the rate one. The existence of an error of transmission means that the sequences to be sent must not be too similar in some reasonable sense, lest they be confused as a result of transmission errors. When n is sufficiently large, we can find $S = 2^s$ sufficiently dissimilar sequences. The highest possible rate of transmission obviously depends on the channel probability function.

2. The contents of this paper

The fundamental ideas of the present subject and paper are due to the fundamental and already classical paper [1] of Shannon. Theorem 1 below was stated and proved by Shannon. However, the latter permits the use of what are called "random codes," and indeed proves Theorem 1 by demonstrating the existence of a random code with the desired property. It seems to the present writer questionable whether random codes are properly codes at all. The definition of a code given in Section 1 of the present paper does not admit random codes as codes; what we have called a code is called in the literature of communication theory an "error correcting" code. In any case,

[2] For example, if $p(\alpha_1) = p(\alpha_2)$, then α_1 and α_2 are indistinguishable in transmission.

the desirability of proving the existence of an error correcting code which would satisfy the conclusion of Shannon's Theorem 1 has always been recognized and well understood (see, for example, [8], Section 3).

The achievement of such a proof is due to Feinstein [2] and Khintchine [4]. The latter utilized an idea from the earlier, not entirely rigorous and without gaps, work of Feinstein, to prove, in full rigor, the general Theorem 3 below. In the present paper, starting from first principles in Section 3, we give already in Section 5 a short and simple proof of Theorem 1. We then return to the subject in Section 8 to prove Theorem 3. Even after allowance is made for the fact that Lemmas 8.2 and 8.3 are not proved here, it seems that our proofs have something to offer in simplicity and brevity.

Theorem 2 for general memory m was stated by Shannon in [1]. Khintchine in [4] pointed out that neither the argument of [1] nor any of the arguments to be found in the literature constitute a proof or even the outline of a proof; he also pointed out the desirability of proving the result and mentioned some of the difficulties. In the present paper we give what seems to be the first proof of Theorem 2. We have reason to believe that it is possible to treat the case of general finite memory along the same lines.[3]

The notion of extending the result for stationary Markov chains (Theorem 1) to stationary, not necessarily Markovian processes (Theorem 3) is due to McMillan [5]. The difficult achievement of carrying out this program correctly and without gaps is due to Khintchine [4]. The theorem we cite below as Lemma 8.3 is due to McMillan. Lemma 8.2 is due to Khintchine.

In [4] Khintchine acknowledges his debt to the paper [2] of Feinstein, although he states that its argument is not exact and that it deals largely with the case of zero memory (and only with Theorem 1 of course). The main idea of [2] seems, to the present writer, to be the ingenious one of proving an inequality like (5.4) below. This pretty idea is employed in the present paper; we find it possible to dispense with many of the details which occur in this connection in [2] and [4].

Shannon and all other writers cited above employ the law of large numbers. The simple notion of δx-sequences and the sequences they generate, which so simplifies our proof below and makes the proof of Theorem 2 possible, also enables us to use Chebyshev's inequality instead of the law of large numbers in Theorems 1 and 2. This has the incidental effect of slightly improving Theorems 1 and 2 over Shannon's original formulation by replacing $o(n)$ terms by $O(n^{1/2})$ terms.

This entire paper is self-contained except for the following incidental remark which we make here in passing: The quantity called $e(n)$ in [2] (the maximum

[3] *Added in proof.* A sequel to the present paper, which has been accepted for publication by this Journal, gives an upper bound on the length of a code for any memory m. When $m = 0$ this bound is the same as that given by Theorem 2. The proof of this result is different from that of Theorem 2.

probability of incorrectly receiving any word) is shown there, for $m = 0$, to approach zero "faster than $1/n$". Using the arguments of the present paper and the inequality (96) of page 288 of [9], one can prove easily for any m that

$$e(n) < c_1 n^{-1/2} e^{-c_2 n},$$

where c_1 and c_2 are positive constants.

3. Combinatorial preliminaries

Let x be any x-sequence and α be any α-sequence. Let $N(\alpha \mid x)$ be the number of elements in x such that each, together with the m elements of x which follow it, constitute the sequence α. Let δ and δ_2 be fixed positive numbers. Let π be any nonnegative function defined on the set of all α-sequences such that

$$\sum_\alpha \pi(\alpha) = 1.$$

We shall say that an x-sequence x is a $\delta\pi x$-sequence if

$$(3.1) \qquad | N(\alpha \mid x) - n\pi(\alpha) | \leqq \delta n^{1/2}$$

for every α-sequence.

A y-sequence y will be said to be generated by the x-sequence x if (1) y is a possible received sequence when x is the transmitted sequence, (2) for any α-sequence α_1 the following is satisfied: Let $j(1), \cdots, j(N(\alpha_1 \mid x))$ be the serial numbers of the elements of x which begin the sequence α_1 (e.g., the elements in the places with serial numbers $j(1), j(1) + 1, \cdots, j(1) + m$, constitute the sequence α_1). Then the number $N(\alpha_1, y \mid x)$ of elements one among the elements of y with serial numbers $j(1), \cdots, j(N(\alpha_1 \mid x))$ satisfies

$$(3.2) \qquad | N(\alpha_1, y \mid x) - N(\alpha_1 \mid x)p(\alpha_1) | \leqq \delta_2[N(\alpha_1 \mid x)(p(\alpha_1))(1 - p(\alpha_1))]^{1/2}.$$

Let $M(x)$ denote the number of y-sequences generated by x.

Whenever in this paper the expression 0 log 0 occurs, it is always to be understood as equal to zero. We remind the reader that all logarithms occurring in this paper are to the base 2. For any x-sequence x we define $H_x(Y)$, the conditional entropy of $Y(x)$, by

$$(3.3) \qquad \begin{aligned} H_x(Y) = & -(1/n)\sum_\alpha N(\alpha \mid x)p(\alpha) \log p(\alpha) \\ & -(1/n)\sum_\alpha N(\alpha \mid x)(1 - p(\alpha)) \log (1 - p(\alpha)). \end{aligned}$$

LEMMA 3.1. *For any δ_2 there exists a $K_1 > 0$ such that, for any n and any x-sequence x,*

$$(3.4) \qquad M(x) < 2^{nH_x(Y) + K_1 n^{1/2}}.$$

Proof. Let θ_2 be a generic real number $\leqq \delta_2$ in absolute value. Let y be any y-sequence generated by x. Then

$$
\begin{aligned}
\log P\{Y(x) = y\} &= \textstyle\sum_\alpha N(\alpha \mid x)p(\alpha) \log p(\alpha) \\
&+ \textstyle\sum_\alpha N(\alpha \mid x)(1 - p(\alpha)) \log (1 - p(\alpha)) \\
&+ \textstyle\sum_\alpha \theta_2 (\alpha)[N(\alpha \mid x)p(\alpha)(1 - p(\alpha))]^{1/2} \log p(\alpha) \\
&- \textstyle\sum_\alpha \theta_2 (\alpha) [N(\alpha \mid x)p(\alpha)(1 - p(\alpha))]^{1/2} \log(1 - p(\alpha)) \\
&> -nH_x(Y) + \delta_2 n^{1/2} \textstyle\sum_\alpha \log p(\alpha) \\
&+ \delta_2 n^{1/2} \textstyle\sum_\alpha \log (1 - p(\alpha)) = -nH_x(Y) - K_1 n^{1/2},
\end{aligned} \tag{3.5}
$$

with

$$K_1 = -\delta_2 \textstyle\sum_\alpha \log p(\alpha) - \delta_2 \textstyle\sum_\alpha \log (1 - p(\alpha)).$$

(Here the first summation is over all α such that $p(\alpha) > 0$, and the second summation is over all α such that $p(\alpha) < 1$.) The lemma follows at once from (3.5)

LEMMA 3.2. *Let $\lambda > 0$ be any number. Then, for δ_2 larger than a bound which depends only upon λ, we have, for any n and any x-sequence x,*

$$P\{Y(x) \text{ is a sequence generated by } x\} > 1 - \tfrac{1}{2}\lambda. \tag{3.6}$$

There then exists a $K_2 > 0$ which depends only on δ_2 such that, for any n and any x-sequence x,

$$M(x) > 2^{nH_x(Y) - K_2 n^{1/2}}. \tag{3.7}$$

Proof. (3.6) follows at once from Chebyshev's inequality. As in (3.5) we have, for any y-sequence y generated by x,

$$\log P\{Y(x) = y\} < -nH_x(Y) + K_1 n^{1/2}. \tag{3.8}$$

From (3.6) and (3.8) we have at once that

$$M(x) > (1 - \tfrac{1}{2}\lambda)2^{nH_x(Y) - K_1 n^{1/2}}. \tag{3.9}$$

Then (3.7) follows at once from (3.9).

4. Preliminaries on Markov chains[4]

Let $X_1, X_2, \cdots$ be a stationary, metrically transitive Markov chain with two possible states, 0 and 1; we shall call this the X process, for short. Suppose

[4] Since we do not assume that the words to be transmitted are chosen by any random process or sent with any particular frequency, the introduction of the X process is not a necessity. The lemmas which involve the X process are of purely combinatorial character (e.g., Lemmas 4.1 and 6.1). The X process serves merely as a device for stating or proving certain combinatorial facts. The reader is invited to verify that this entire paper could be written without the introduction of the X process. In that case the $Y(x)$ process would take the place of the Y process. Only the entropies $H(Y)$ and $H_X(Y)$ need be introduced, and this can be done by means of the $Y(x)$ process and δQx-sequences.

(4.1) $$Q_i = P\{X_1 = i\}, \qquad i = 0, 1,$$
and
(4.2) $$q_{ij} = P\{X_{k+1} = j \mid X_k = i\}$$
is the probability of a transition from state i to state j; $i, j = 0, 1$. For any α-sequence α define

(4.3) $$Q(\alpha) = P\{(X_1, \cdots, X_{m+1}) = \alpha\}.$$

The function Q is a function which satisfies the requirements on the function π of Section 3. Let $\gamma < 1$ be any number. It follows at once from Chebyshev's inequality that, for any n and any δ greater than a lower bound which is a function only of γ and the q_{ij},

(4.4) $$P\{(X_1, \cdots, X_n) \text{ is a } \delta Qx\text{-sequence}\} > \gamma.$$

By the Y process we shall mean the sequence $Y_1, Y_2, \cdots$, where Y_i is a chance variable which assumes only the values zero and one, and the conditional probability that $Y_i = 1$, given the values of $X_1, X_2, \cdots, Y_1, \cdots, Y_{i-1}$, is $p(X_i, \cdots, X_{i+m})$. Henceforth we write for short $X = (X_1, \cdots, X_n)$. Then the conditional distribution of $Y = (Y_1, \cdots, Y_{n-m})$, given X, is the same as that of the sequence received when X is the sequence transmitted. The Y process is obviously stationary, and, by Lemma 8.2 below (proof in [4], page 53), metrically transitive. The conditional entropy $H_X(Y)$ of the Y process relative to the X process is defined by

$$H_X(Y) = -\sum_\alpha Q(\alpha)p(\alpha) \log p(\alpha) - \sum_\alpha Q(\alpha)(1 - p(\alpha)) \log (1 - p(\alpha)).$$

One verifies easily that there exists a $K_3 > 0$ such that, for any δQx-sequence x,
(4.5) $$| H_x(Y) - H_X(Y) | < K_3 \delta/n^{1/2}.$$
We at once obtain

LEMMA 4.1. *For any n and any δQx-sequence x, the inequalities (3.4) and (3.7) hold with $H_x(Y)$ replaced by $H_X(Y)$ and K_1 and K_2 replaced by K_1' and K_2', where K_1' and K_2' are positive numbers which depend only upon δ and δ_2.*

We define the chance variable (function of X) $P\{X\}$ as follows: when $X = x$, $P\{X\} = P\{X = x\}$. Similarly we define the chance variable (function of Y) $P\{Y\}$ as follows: when $Y = y$, $P\{Y\} = P\{Y = y\}$. We define the entropy $H(X)$ of the X process by

$$H(X) = \lim_{n\to\infty} -\frac{1}{n} E [\log P\{X\}].$$

This limit obviously exists.

Let the symbol $\sigma^2(\ )$ denote the variance of the chance variable in parentheses. We now prove

LEMMA 4.2. *We have*

$$E[\log P\{Y\}] = -Dn + D_0, \tag{4.6}$$

where D is a nonnegative constant and D_0 is a bounded function of n. Also,

$$\sigma^2(\log P\{Y\}) = O(n). \tag{4.7}$$

The quantity

$$D = \lim_{n\to\infty} -\frac{1}{n} E[\log P\{Y\}]$$

is called the entropy $H(Y)$ of the Y process.

Proof. We have

$$\log P\{Y\} = \sum_{i=1}^{n} \log P\{Y_i \mid Y_1, \cdots, Y_{i-1}\}. \tag{4.8}$$

Let α^* be some fixed α-sequence such that $Q(\alpha^*) > 0$. In the sequence $X_1, X_2, \cdots$, let $j(1), j(2), \cdots$ be the indices such that

$$(X_{j(i)}, X_{j(i)+1}, \cdots, X_{j(i)+m}) = \alpha^*, \qquad i = 1, 2, \cdots.$$

(These exist with probability one.) Let l^* be the smallest integer such that $j(l^*) \geqq n + 1$. (Again l^* is defined with probability one.) Define symbols such as

$$C_1 = \log P\{Y_1, \cdots, Y_{j(1)-1}\}$$

in the obvious manner analogous to that in which $\log P\{Y\}$ was defined. Since C_1 is a sum of quantities which enter into (4.8) and which are all zero or negative, it follows that $B_1 = EC_1$ could fail to exist only if it were $-\infty$. It will be seen that the latter cannot be. Define

$$C_2 = \sum_{i=n-m+1}^{j(l^*)-1} \log P\{Y_i \mid Y_1, \cdots, Y_{i-1}\}.$$

As before, $B_2 = EC_2$ either exists or is $-\infty$. It will be seen that $B_2 \neq -\infty$.

It is easy to see that

$$Ej(1) \leqq \text{a constant, independent of } n \tag{4.9}$$

$$El^* = 1 + nQ(\alpha^*) \tag{4.10}$$

$$E(j(i) - j(i-1)) = \text{a constant, independent of } n \text{ and } i. \tag{4.11}$$

$$E(j(l^*) - n) \leqq \text{a constant, independent of } n. \tag{4.12}$$

From the construction of $j(1), j(2), \cdots$ it follows that the chance variables

$$W_i = \log P\{Y_{j(i)}, \cdots, Y_{j(i+1)-1} \mid Y_1, \cdots, Y_{j(i)-1}\}, \qquad i = 1, 2, \cdots \tag{4.13}$$

are independently and identically distributed. Actually

$$W_i = \log P\{Y_{j(i)}, \cdots, Y_{j(i+1)-1}\}.$$

From Wald's equation ([10], Theorems 7.1 and 7.4), (4.9), (4.11), (4.12), and the fact that the chance variables W_i, C_1, and C_2 are sums of always non-positive and bounded chance variables which appear in the right member of (4.8), it follows that B_1, B_2, and $EW_i = w$ are all finite, and that B_1 and B_2 are bounded uniformly in n. Applying Wald's equation again we obtain, using (4.10), that

$$E \log P\{Y_1, \cdots, Y_{j(l^*)-1}\} = B_1 + nwQ(\alpha^*). \tag{4.14}$$

Hence

$$E \log P\{Y\} = B_1 + nwQ(\alpha^*) - B_2, \tag{4.15}$$

which proves (4.6).

Now

$$\begin{aligned} \log P\{Y\} - E \log P\{Y\} &= (C_1 - B_1) \\ &\quad + \left(\sum_{i=1}^{l^*-1} W_i - nwQ(\alpha^*)\right) - (C_2 - B_2) \\ &= (C_1 - B_1) + \left(\sum_{i=1}^{l^*-1} W_i - (l^* - 1)w\right) \\ &\quad + ((l^* - 1)w - nwQ(\alpha^*)) - (C_2 - B_2). \end{aligned} \tag{4.16}$$

Now we note that

$$\sigma^2(j(i) - j(i - 1)) = \text{a constant, independent of } n. \tag{4.17}$$

Applying an argument like that which leads to Theorem 7.2 of [10], together with (4.17) and Schwarz's inequality, we obtain first that

$$\sigma^2(W_i) = \text{(a finite) constant}, \tag{4.18}$$

and then that

$$E\left(\sum_{i=1}^{l^*-1} W_i - (l^* - 1)w\right)^2 = \sigma^2(W_1)nQ(\alpha^*) \tag{4.19}$$

by (4.10) and

$$E([l^* - 1]w - E[l^* - 1]w)^2 = O(n) \tag{4.20}$$

(see [6], p. 263, equation (8.10)). Obviously

$$\sigma^2(C_1) \leqq \text{a constant independent of } n, \tag{4.21}$$

$$\sigma^2(C_2) \leqq \text{a constant independent of } n. \tag{4.22}$$

Now take the expected value of the squares of the first and third members of (4.16). Using (4.19), (4.20), (4.21), and (4.22) we obtain that the sum of the expected values of the squares which occur after squaring the third member of (4.16) is $O(n)$. The cross products have expected value $O(n)$ by the Schwarz inequality. This proves (4.7) and completes the proof of the lemma.

Another proof of the fact that the variance of $\log P\{Y\}$ is $O(n)$ can be based on the following: It is known from the theory of Markov chains ([7], page 173, equation (2.2)) that there exists a number h, $0 < h < 1$, such that the absolute value of the correlation coefficient between X_i and X_j is less than $h^{|i-j|}$. From the distribution of the Y_i it follows that a similar statement is true of the correlation coefficient between Y_i and Y_j and also of the correlation coefficient between

$$\log P\{Y_i \mid Y_1, \cdots, Y_{i-1}\}$$

and

$$\log P\{Y_j \mid Y_1, \cdots, Y_{j-1}\}.$$

Since $\log P\{Y\}$ can be written in the form (4.8), the desired conclusion can be deduced from the above.

An immediate consequence of Lemma 4.2 and Chebyshev's inequality is that, for any $\varepsilon' > 0$, there exists a $K_4 > 0$ such that, for any n,

$$P\{-nH(Y)-K_4 n^{1/2} < \log P\{Y\} < -nH(Y) + K_4 n^{1/2}\} > 1 - \varepsilon'. \tag{4.23}$$

The following lemma is now an immediate consequence of (4.23):

LEMMA 4.3. *Let $\varepsilon' > 0$ be any number, and $K_4 > 0$ be a number which, for any n, satisfies (4.23). For any n let B be any set of y-sequences such that*

$$P\{Y \epsilon B\} > \gamma_1 > \varepsilon'.$$

Then the set B must contain at least

$$(\gamma_1 - \varepsilon')2^{nH(Y)-K_4 n^{1/2}}$$

y-sequences.

Proof. From (4.23) it follows that the y-sequences in B which satisfy the relationship in braces in (4.23) have probability greater than $\gamma_1 - \varepsilon'$. Since the probability of each such sequence is bounded above by $2^{-nH(Y)+K_4 n^{1/2}}$, the desired result follows.

5. The coding theorem

THEOREM 1. *Let $X_1, X_2, \cdots$ be a stationary, metrically transitive Markov chain with states 0 and 1 and notation as in Section 4. Let the Y process be as defined in Section 4. Let λ be an arbitrary positive number. There exists a $K > 0$ such that, for any n, there is a code of length at least*[5]

$$2^{n(H(Y)-H_X(Y))-Kn^{1/2}}. \tag{5.1}$$

The probability that any word transmitted according to this code will be incorrectly received is less than λ.

[5] An alternate and perhaps more graphic way to state Theorem 1 is to replace $(H(Y)-H_X(Y))$ in (5.1) by $C_1 = \max (H(Y)-H_X(Y))$, where the maximum is over all Markov processes X and their associated Y processes as defined in the statement of Theorem 1. It is obvious that this is an equivalent way of stating Theorem 1.

Proof. We may take $\lambda < \frac{1}{2}$. Let $\gamma < 1$ be any positive number. Let δ be sufficiently large so that (4.4) holds, and choose δ_2 sufficiently large so that (3.6) holds.

Let x_1 be any δQx-sequence, and A_1 any set of y-sequences generated by x_1 such that the following is satisfied for $i = 1$:

$$P\{Y(x_i) \text{ is a sequence generated by } x_i \text{ and not in } A_i\} < \tfrac{1}{2}\lambda. \tag{5.2}$$

Let x_2 be any other δQx-sequence for which we can find a set A_2 of y-sequences generated by x_2 such that A_1 and A_2 are disjoint and (5.2) is satisfied for $i = 2$. Continue in this manner as long as possible, i.e., as long as there exists another δQx-sequence, say x_i, and a set A_i of y-sequences generated by x_i such that $A_1, A_2, \cdots, A_i$ are all disjoint and A_i satisfies (5.2). Let

$$(x_1, A_1), \cdots\cdots, (x_N, A_N)$$

be the resulting code. We have to show that N is large enough.

Let x^* be any δQx-sequence (if one exists) not in the set $x_1, \cdots, x_N$. Then

$$P\{Y(x^*) \text{ is a sequence generated by } x^* \text{ and belongs to } (A_1 \cup A_2 \cup \cdots \cup A_N)\} \geqq \tfrac{1}{2}\lambda. \tag{5.3}$$

If this were not so, we could prolong the code by adding (x^*, A^*), where A^* is the totality of y-sequences generated by x^* and not in $A_1 \cup A_2 \cup \cdots \cup A_N$; this would violate the definition of N. From (4.4), (3.6), (5.2), and (5.3) it follows that

$$P\{Y \in (A_1 \cup A_2 \cup \cdots \cup A_N)\} > \tfrac{1}{2}\gamma\lambda. \tag{5.4}$$

Let the ε' of (4.23) and Lemma 4.3 be equal to $\frac{1}{4}\gamma\lambda$ and let $K_4 > 0$ be any number for which (4.23) is satisfied. It follows from Lemma 4.3 that the set $A_1 \cup A_2 \cup \cdots \cup A_N$ contains at least

$$\tfrac{1}{4}\gamma\lambda \cdot 2^{nH(Y) - K_4 n^{1/2}} \tag{5.5}$$

y-sequences. By Lemma 4.1 the number of y-sequences in $A_1 \cup A_2 \cup \cdots \cup A_N$ is at most

$$N \cdot 2^{nH_X(Y) + K_1' n^{1/2}}. \tag{5.6}$$

The desired result follows at once from (5.5) and (5.6), with

$$K = K_1' + K_4 - \log\left(\tfrac{1}{4}\gamma\lambda\right).$$

6. Further preliminaries[6]

The essential part of the present section is the second part of the inequality (6.13) below, which is basic in the proof of Theorem 2 of Section 7. Neither

[6] All the lemmas of this section are of purely combinatorial character. Lemma 6.3 could be easily proved by a purely combinatorial argument without any use of the Y process. This entire section is a concession to the conventional treatment of the subject. All that is needed for the statement and proof of Theorem 2 is the second part of (6.13) and a formal analytic definition of capacity. See also footnote 4.

Lemma 6.1 nor Lemma 6.2 is used in the sequel, and both are given only for completeness. The proof of Lemma 6.1 is omitted because it is very simple, and the proof of Lemma 6.2 is omitted because it involves some computation.

Let the X and Y processes be as defined in Section 4. Obviously

$$H(X) = -\sum_{i,j} Q_i \, q_{ij} \log q_{ij} . \tag{6.1}$$

We define the chance variables (functions of X and Y) $P\{Y \mid X\}$ and $P\{X \mid Y\}$ as follows: when $X = x$ and $Y = y$, $P\{Y \mid X\} = P\{Y = y \mid X = x\}$, and $P\{X \mid Y\} = P\{X = x \mid Y = y\}$. We verify easily that

$$H_X(Y) = \lim_{n\to\infty} -\frac{1}{n} E\,[\log P\{Y \mid X\}]. \tag{6.2}$$

We define $H_Y(X)$, the conditional entropy of the X process relative to the Y process, by

$$H_Y(X) = \lim_{n\to\infty} -\frac{1}{n} E\,[\log P\{X \mid Y\}]. \tag{6.3}$$

(We shall see in a moment ((6.5)) that this limit exists.) From the obvious relation

$$\log P\{X\} + \log P\{Y \mid X\} = \log P\{Y\} + \log P\{X \mid Y\}, \tag{6.4}$$

we obtain

$$H(X) + H_X(Y) = H(Y) + H_Y(X). \tag{6.5}$$

Throughout the rest of this section we assume that the memory $m = 0$, and that the X_i are independent, identically distributed chance variables. Hence

$$Q_i = q_{ji} , \qquad i, j = 0, 1. \tag{6.6}$$

Since $m = 0$, there are only two α-sequences, namely, (0) and (1), and $Q(i) = Q_i$, $i = 0, 1$. Write for short $Q(1) = q$. We assume that $0 < q < 1$. It seems reasonable in this case to denote what was called in Section 4 a "δQx-sequence" by the term "δqx-sequence", and we shall employ this usage (when $m = 0$ and the chance variables X_i are independent). We now give the values of the various entropies, inserting a zero in the symbol for entropy to indicate that $m = 0$ and the X_i are independently (and identically) distributed.

$$H(X_0) = -q \log q - (1 - q) \log (1 - q). \tag{6.7}$$

$$\begin{aligned} H(Y_0) = &- [qp(1) + (1 - q)p(0)] \log [qp(1) + (1 - q)p(0)] \\ &- [(1 - q)\,(1 - p(0)) + q\,(1 - p(1))] \cdot \\ &\cdot \log[(1 - q)\,(1 - p(0)) + q\,(1 - p(1))]. \end{aligned} \tag{6.8}$$

$$
\begin{aligned}
H_X(Y_0) = & - qp(1) \log p(1) - (1 - q)p(0) \log p(0) \\
& - q(1 - p(1)) \log (1 - p(1)) \\
& - (1 - q)(1 - p(0)) \log (1 - p(0)).
\end{aligned} \tag{6.9}
$$

From (6.5) we obtain

$$
\begin{aligned}
H_Y(X_0) = & -(1 - q)p(0) \log \frac{(1 - q)p(0)}{[qp(1) + (1 - q)p(0)]} \\
& - qp(1) \log \frac{qp(1)}{[qp(1) + (1 - q)p(0)]} \\
& - q(1 - p(1)) \log \frac{q(1 - p(1))}{[q(1 - p(1)) + (1 - q)(1 - p(0))]} \\
& - (1 - q)(1 - p(0)) \log \frac{(1 - q)(1 - p(0))}{[q(1 - p(1)) + (1 - q)(1 - p(0))]} .
\end{aligned} \tag{6.10}
$$

The maximum, with respect to q, of

$$
H(X_0) - H_Y(X_0) = H(Y_0) - H_X(Y_0)
$$

is called the capacity (when $m = 0$) C_0 of the channel.

LEMMA 6.1. *There exists a $K_5 > 0$ such that, for any n, the number $M(\delta q)$ of δqx-sequences satisfies*

$$
2^{nH(X_0) - K_5 n^{1/2}} < M(\delta q) < 2^{nH(X_0) + K_5 n^{1/2}}. \tag{6.11}
$$

Let δ' be some fixed positive number such that any y-sequence which is generated by a δqx-sequence, cannot be generated by an x-sequence which is not a $\delta' qx$-sequence. Such a δ' exists; we have only to take δ' larger than a lower bound which is a function of q, δ, δ_2, $p(0)$, and $p(1)$. We have

LEMMA 6.2. *There exists a $K_6 > 0$ with the following property: Let y be any y-sequence which is generated by some δqx-sequence. Then the number $M'(y)$ of $\delta' qx$-sequences which generate y satisfies*

$$
2^{nH_Y(X_0) - K_6 n^{1/2}} < M'(y) < 2^{nH_Y(X_0) + K_6 n^{1/2}}. \tag{6.12}
$$

We now prove

LEMMA 6.3. *There exists a $K_7 > 0$ such that, for any n, the number $M''(\delta q)$ of different y-sequences generated by all δqx-sequences satisfies*

$$
2^{nH(Y_0) - K_7 n^{1/2}} < M''(\delta q) < 2^{nH(Y_0) + K_7 n^{1/2}}. \tag{6.13}
$$

Proof. Let θ, with any subscript, denote a number not greater than one in absolute value. The chance variables $Y_1, Y_2, \cdots$ are independently and identically distributed. We have

$$
P\{Y_1 = 1\} = qp(1) + (1 - q)p(0) = u, \text{ say}.
$$

If y is generated by a δqx-sequence, then the number V_1 of elements one in y is given by

$$V_1 = n(qp(1) + (1 - q)p(0)) + n^{1/2}(\theta_1\delta + 2\theta_2\delta_2) + 0(1).$$

Since

$$P\{Y = y\} = u^{V_1}(1 - u)^{n-V_1} > 2^{-nH(Y_0)-K_7 n^{1/2}}$$

for a suitable $K_7 > 0$, the second part of (6.13) follows at once.

The first part of (6.13) follows from Lemma 4.3. It may be necessary to increase the above K_7.

When $p(1) = p(0)$, $C_0 = 0$. One can verify that otherwise $C_0 > 0$. Incidentally, it follows from Lemmas 6.1 and 6.2 that $H(X_0) \geqq H_Y(X_0)$.

7. Impossibility of a rate of transmission greater than the capacity when $m = 0$

In this section we prove the following

THEOREM 2. *Let $m = 0$, and let λ, $1 > \lambda > 0$, be any given number. There exists a $K' > 0$ such that, for any n, any code with the property that the probability of transmitting any word incorrectly is $< \lambda$, cannot have a length greater than*

$$2^{nC_0+K'n^{1/2}} \tag{7.1}$$

If $p(1) = p(0)$ and therefore $C_0 = 0$, the theorem is trivial. For then it makes no difference whether one transmits a zero or a one, and it is impossible to infer from the sequence received what sequence has been transmitted. We therefore assume henceforth that $C_0 > 0$.

It will be convenient to divide the proof into several steps. Let q_0 be the value of q which maximizes $H(Y_0) - H_X(Y_0)$. We shall have occasion to consider the various entropies as functions of q, which, in this section, we shall always exhibit explicitly, e.g., $H(Y; q)$.[7] Let δ and δ_2 be positive constants. Throughout this section it is to be understood that by the word "code" we always mean a code with the property that the probability of transmitting any word incorrectly is $< \lambda$.

LEMMA 7.1. *There exists a $K_8 > 0$ with the following property: Let n be any integer. Let $(x_1, A_1), \cdots, (x_N, A_N)$ be any code such that $x_1, \cdots, x_N$ are $\delta q_0 x$-sequences, and A_i, $i = 1, \cdots, N$, contains only y-sequences generated by x_i. Then*

$$N < 2^{nC_0+K_8 n^{1/2}}. \tag{7.2}$$

Proof. It follows from (3.8) and (4.5) that there exists a $K_8' > 0$ such that the set $(A_1 \cup A_2 \cup \cdots \cup A_N)$ contains at least

[7] Naturally, this is the entropy of Y when the X's are independently distributed and $m = 0$.

(7.3)
$$N \cdot 2^{nH_X(Y;q_0)-K_8' n^{1/2}}$$

sequences. By Lemma 6.3 it cannot contain more than

(7.4)
$$2^{nH(Y;q_0)+K_7 n^{1/2}}$$

sequences. The lemma follows at once with $K_8 = K_7 + K_8'$.

Lemma 7.2. *There exists a $K_9 > 0$ with the following property: Let n be any integer. Let $(x_1, A_1), \cdots, (x_N, A_N)$ be any code such that $x_1, \cdots, x_N$ are $\delta q_0 x$-sequences. Then*

(7.5)
$$N < 2^{nC_0+K_9 n^{1/2}}.$$

(In other words, the conclusion of Lemma 7.1 holds even if the A_i, $i = 1, \cdots, N$, are not required to consist only of sequences generated by x_i.)

Proof. Let δ_2 be so large that (3.6) holds. From A_i, $i = 1, \cdots, N$, delete the y-sequences not generated by x_i; call the resulting set A_i'. The A_i', $i = 1, \cdots, N$, are of course disjoint. The set $(x_1, A_i'), \cdots, (x_N, A_N')$ fulfills all the requirements of a code except perhaps the one that the probability of correctly transmitting any word is $> 1 - \lambda$. However, from (3.6) it follows that the probability of correctly transmitting any word when this latter set is used is $> 1 - 3\lambda/2$. But now the result of Lemma 7.1 applies,[8] and the present lemma follows. (Of course the constant K_8 of Lemma 7.1 depends on λ, but this does not affect our conclusion.)

Lemma 7.3. *There exists a constant $K_{10} > 0$ with the following property: Let q be any point in the closed interval* [0, 1], let *n be any integer, and let $(x_1, A_1), \cdots, (x_N, A_N)$ be any code such that $x_1, \cdots, x_N$ are $\delta q x$-sequences. Then*

(7.6)
$$N < 2^{nC_0+K_{10} n^{1/2}}.$$

Proof. Let q', $0 < q' < \frac{1}{2}$, be such that $H(X; q) < \frac{1}{2} C_0$ if $q < q'$ or $q > 1 - q'$. If $q < q'$ or $q > 1 - q'$, the total number of all $\delta q x$-sequences is less than the right member of (7.6) for suitable $K_{10} > 0$. Then (7.6) holds a fortiori.

It remains to consider the case $q' \leqq q \leqq 1 - q'$. If now one applies the argument of Lemma 7.2 and considers how K_9 depends upon q, one obtains that there exists a positive continuous function $K_9(q)$ of q, $q' \leqq q \leqq 1 - q'$, such that, for any n,

$$\begin{aligned} N &< 2^{n(H(Y;q)-H_X(Y;q))+K_9(q) n^{1/2}} \\ &\leqq 2^{nC_0+K_9(q) n^{1/2}}. \end{aligned}$$

We now increase, if necessary, the constant K_{10} of the previous paragraph so that it is not less than the maximum of $K_9(q)$ in the closed interval $[q', 1 - q']$, and obtain the desired result (7.6).

[8] Except when $3\lambda/2 \geqq 1$. In that case we choose δ_2 so large that the right member of (3.6) is $1 - \lambda/a$, where $a > 0$ is such that $\lambda + \lambda/a < 1$.

Proof of Theorem 2. Divide the interval [0, 1] into $J = n^{1/2}/2\delta$ intervals of length $2\delta/n^{1/2}$ and let $t_1, \cdots, t_J$ be the midpoints of these intervals. Let $(x_1, A_1), \cdots, (x_N, A_N)$ be any code. Then this code is the union of J codes $W_1, \cdots, W_J$ as follows: For $i = 1, \cdots, J$, W_i is that subset of the original code all of whose x-sequences are δt_i x-sequences. By Lemma 7.3 the length of W_i, $i = 1, \cdots, J$, is less than $2^{nC_0+K_{10}n^{1/2}}$. Hence the length N of the original code is less than

$$J \cdot 2^{nC_0+K_{10}n^{1/2}}.$$

The theorem follows at once if K' is sufficiently large.

8. Extension to stationary processes

Throughout this section let $X_1, X_2, \cdots$ be a stationary, metrically transitive stochastic process such that X_i, $i = 1, 2, \cdots$, takes only the values one and zero. Define the Y process, $Q(\alpha)$, and $H_X(Y)$ exactly as in Section 4. Let $\varepsilon^* > 0$ be any number, no matter how small, and write $\delta^* = \varepsilon^* n^{1/2}$. Let $\gamma < 1$ be any positive number. From the ergodic theorem we obtain at once the following analogue of (4.4): For n sufficiently large,

$$P\{(X_1, \cdots, X_n) \text{ is a } \delta^*Qx\text{-sequence}\} > \gamma. \tag{8.1}$$

For δ_2 sufficiently large the inequalities (3.6), (3.4), and (3.7) hold exactly as before, and we obtain the following analogue of Lemma 4.1:

LEMMA 8.1. *For any $\varepsilon > 0$, ε^* sufficiently small, and δ_2 sufficiently large, we have, for n sufficiently large and any δ^*Qx-sequence x,*

$$2^{n(H_X(Y)-\varepsilon)} < M(x) < 2^{n(H_X(Y)+\varepsilon)}. \tag{8.2}$$

The following lemmas are proved in [4]:

LEMMA 8.2. *The process $Y_1, Y_2, \cdots$ is metrically transitive.*

LEMMA 8.3. *Let $Z_1, Z_2, \cdots$ be any stationary, metrically transitive stochastic process such that Z_i can take only finitely many values. Let $Z = (Z_1, \cdots, Z_n)$. Define the chance variable (function of Z) $P\{Z\}$ as follows: When Z is the sequence z, $P\{Z\} = P\{Z = z\}$. Then $-(1/n) \log P\{Z\}$ converges stochastically to a constant.*

For our Y process the constant limit of Lemma 8.3 is called the entropy $H(Y)$ of the Y process. This definition of $H(Y)$ is easily verified to be consistent with that of Section 4.

Lemma 8.3 implies the following analogue of (4.23) for our Y process: Let $\varepsilon' > 0$ be any number. Then, for n sufficiently large,

$$P\{-n(H(Y)+\varepsilon') < \log P\{Y\} < -n(H(Y)-\varepsilon')\} > 1 - \varepsilon'. \tag{8.3}$$

Exactly as (4.23) easily implies Lemma 4.3, so (8.3) implies

LEMMA 8.4. *Let $\varepsilon' > 0$ be any number and let n be sufficiently large for* (8.3) *to hold. Let B be any set of y-sequences such that*

$$P\{Y \epsilon B\} > \gamma_1 > \varepsilon'.$$

Then the set B must contain at least

$$(\gamma_1 - \varepsilon')2^{n(H(Y)-\varepsilon')}$$

y-sequences.

Now the analogues of all the preliminaries needed to prove Theorem 1 have been established, and we have, by exactly the same proof,

THEOREM 3. *Let $X_1, X_2, \cdots$ be a stationary, metrically transitive stochastic process with states 0 and 1. Let the Y process be as defined in Section* 4. *Let λ and ε be arbitrary positive numbers. For any n sufficiently large there exists a code of length at least*

$$2^{n(H(Y)-H_X(Y)-\varepsilon)}. \tag{8.4}$$

The probability that any word transmitted according to this code will be incorrectly received is less than λ.[9]

The author is grateful to Professor K. L. Chung and Professor J. Kiefer for their kindness in reading the manuscript and for interesting comments.

REFERENCES

1. C. E. SHANNON, *A mathematical theory of communication*, Bell System Tech. J., vol. 27 (1948), pp. 379–423, 623–656.
2. A. FEINSTEIN, *A new basic theorem of information theory*, Transactions of the Institute of Radio Engineers, Professional Group on Information Theory, 1954 Symposium on Information Theory, pp. 2–22.
3. A. KHINTCHINE, *The concept of entropy in the theory of probability*, Uspehi Matem. Nauk (N.S.), vol. 8 no. 3 (55), (1953), pp. 3–20.
4. ———, *On the fundamental theorems of the theory of information*, Uspehi Matem. Nauk (N.S.), vol. 11 no. 1 (67), (1956), pp. 17–75.
5. B. McMILLAN, *The basic theorems of information theory*, Ann. Math. Statistics, vol. 24 (1953), pp. 196–219.
6. W. FELLER, *An introduction to probability theory and its applications*, New York, John Wiley and Sons, 1950.
7. J. L. DOOB, *Stochastic processes*, New York, John Wiley and Sons, 1953.
8. E. N. GILBERT, *A Comparison of signaling alphabets*, Bell System Tech. J., vol. 31 (1952), pp. 504–522.
9. PAUL LÉVY, *Théorie de l'addition des variables aléatoires*, Paris, Gauthier-Villars, 1937.
10. J. WOLFOWITZ, *The efficiency of sequential estimates and Wald's equation for sequential processes*, Ann. Math. Statistics, vol. 18 (1947), pp. 215–230.

CORNELL UNIVERSITY
ITHACA, NEW YORK

[9] An equivalent way of stating Theorem 3 is to replace $(H(Y) - H_X(Y))$ in (8.4) by $C_2 = \sup\,(H(Y) - H_X(Y))$, where the supremum operation is over all X processes as described in the theorem, each X process with its associated Y process. Obviously $C_0 \leqq C_1 \leqq C_2$ (C_1 is defined in footnote 5). When $m = 0$ it follows from Theorem 2 that $C_0 = C_1 = C_2$. Hence, when $m = 0$, Theorem 3 is actually weaker than Theorem 1.

A Note on a Partial Ordering for Communication Channels

Claude E. Shannon*

Center for Advanced Study in the Behavioral Sciences, Stanford, California

A partial ordering is defined for discrete memoryless channels. It is transitive and is preserved under channel operations of addition and multiplication. The main result proved is that if K_1 and K_2 are such channels, and $K_1 \supseteq K_2$, then if a code exists for K_2, there exists at least as good a code for K_1, in the sense of probability of error.

Consider the three discrete memoryless channels shown in Fig. 1. The first may be said to include the second, since by the use at the input of only the letters A, B, and C, the channel reduces to the second channel. Anything that could be done in the way of signaling with the second channel could be done with the first channel by this artificial restriction (and of course, in general, more, by using the full alphabet). The second channel in a sense includes the third, since if at the receiving point we ignore the difference between received letters A' and B' the third channel results. We could imagine a device added to the output which produces letter A' if either A' or B' goes in, and lets C' go through without change.

These are examples of a concept of channel inclusion we wish to define and study. Another example is the pair of binary symmetric channels in Fig. 2. Here we can reduce the first channel to the second one not by identification of letters in the input or output alphabets but by addition of a statistical device at either input or output; namely, if we place before (or after) the first channel a binary symmetric channel, as shown in Fig. 3, with value p_2 such that $p_1 = p\,p_2 + q\,q_2$, then this over-all arrangement acts like the second channel of Fig. 2. Physically this could be done by a suitable device involving a random element. We might be inclined, therefore, to define a channel K_1 with transition probability matrix $\| p_i(j) \|$ to include channel K_2 with matrix $\| q_i(j) \|$ if there exist stochastic matrices A and B such that

$$A \| p_i(j) \| B = \| q_i(j) \|.$$

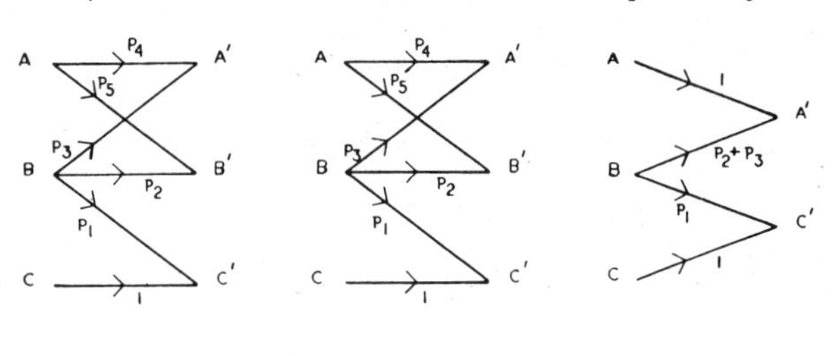

Fig. 1. Examples of channels illustrating inclusion relation.

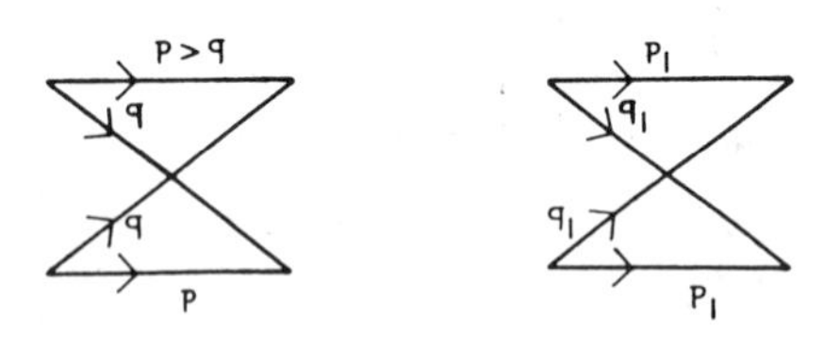

Fig. 2. A further example of inclusion.

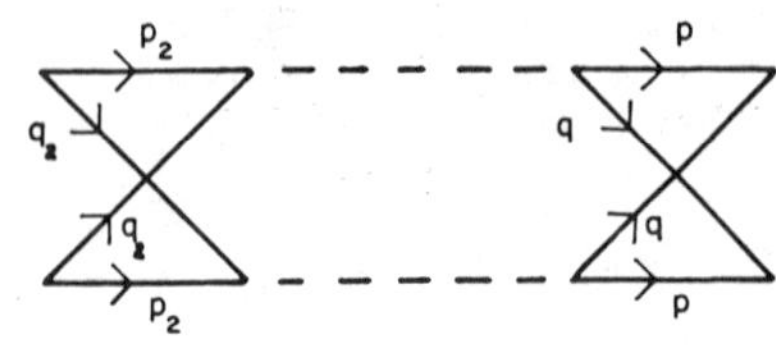

Fig. 3. Reduction of the left channel of Fig. 2 to the right, by a preceding channel.

* On leave of absence from Massachusetts Institute of Technology.

This is a possible definition, but actually we can generalize this somewhat and still obtain the properties we would like for channel inclusion. Namely, we may consider fore and aft stochastic operations which are correlated. Physically one can imagine devices placed at the transmitting end and at the receiving end which involve random but not necessarily independent elements. For example, they may obtain their random choices from tapes which were prepared together with certain correlations. Physically this would be a perfectly feasible process. Mathematically this corresponds, in the simplest case, to the following.

Definition. Let $p_i(j)$ $(i = 1, \cdots, a;\ j = 1, \cdots, b)$ be the transition probabilities for a discrete memoryless channel K_1 and $q_k(l)$ $(k = 1, \cdots c;\ l = 1, \cdots, d)$ be those for K_2. We shall say that K_1 includes K_2, $K_1 \supseteq K_2$, if and only if there exist two sets of transition probabilities $r_{\alpha k}(i)$ and $t_{\alpha j}(l)$, with

$$r_{\alpha k}(i) \geqq 0, \qquad \textstyle\sum_i r_{\alpha k}(i) = 1,$$

and

$$t_{\alpha j}(l) \geqq 0, \qquad \textstyle\sum_l t_{\alpha j}(l) = 1,$$

and there exists

$$g_\alpha \geqq 0, \qquad \textstyle\sum_\alpha g_\alpha = 1$$

with

$$\textstyle\sum_{\alpha,\, i,\, j} g_\alpha r_{\alpha k}(i)\, p_i(j)\, t_{\alpha j}(l) = q_k(l). \tag{1}$$

Roughly speaking, this requires a set of pre- and post-channels R_α and T_α, say, which are used in pairs, g_α being the probability for the pair with subscript α. When this sort of operation is applied the channel K_1 looks like K_2.

Let us define a *pure channel* as one in which all the transitions have probability either 0 or 1; thus each input letter is carried with certainty into some output letter. Any particular pre- and post-channel R_α and T_α in (1) can be thought of as a weighted sum of pure pre- and post-channels operating on K_1. Namely, consider all ways of mapping the input letters of R_α into its output letters and associate probabilities with these to give the equilavent of R_α. The mapping where letter k is mapped into m_k is given probability $\prod_k r_{\alpha k}(m_k)$. A similar reduction can be carried out for the post-channel T_α and combinations of the pre- and post-pure channels are given the corresponding product probabilities.

This reduction to pure components can be carried out for each α and the probabilities added for the same components with different α. In this way, the entire operation in (1) can be expressed as the same sort of operation, where the R_α and T_α are now pure channels, In other words, channel inclusion can be defined equivalently to the above definition but with the added condition that the $r_{\alpha k}(i)$ and $t_{\alpha j}(l)$ correspond to pure channels.

The relation of channel inclusion is transitive. If $K_1 \supseteq K_2$ and $K_2 \supseteq K_3$, then $K_1 \supseteq K_3$. Indeed, if g_α, R_α, T_α are the probabilities and pre- and post-channels for the first inclusion relation, and g'_β, R'_β, T'_β those for the second, then the probabilities $g_\alpha g'_\beta$ with channels $R'_\beta \cup R_\alpha$ for premultiplier and $T_\alpha \cup T'_\beta$ for postmultiplier (the $\cup$ means tandem connection or matrix product) will produce K_3 from K_1. If $K_1 \supseteq K_2$ and $K_2 \supseteq K_1$, we will say these are equivalent channels and write $K_1 \equiv K_2$. Note that always $K_1 \equiv K_1$. Grouping channels into these equivalence classes we have, then, a partial ordering of discrete memoryless channels. There is a universal lower bound of all channels, namely, the channel with one input letter and one output letter with probability 1 for the transition. There is no (finite) universal upper bound of all channels. However, if we restrict ourselves to channels with at most n input and n output letters (or channels equivalent to these) we can give an upper bound to this subset, namely, the pure channel with n inputs and n outputs, the inputs mapped one-to-one into the outputs.

Reprinted with permission from *Inform. Contr.*, vol. 1, pp. 390–398, Dec. 1958.

The ordering relation is preserved under channel operations of addition and multiplication. If K_1, K_1', K_2, and K_2' are channels and $K_1 \supseteq K_1'$ and $K_2 \supseteq K_2'$, then

$$K_1 + K_2 \supseteq K_1' + K_2'$$

$$K_1K_2 \supseteq K_1'K_2'.$$

The sum and product of channels as defined in an earlier paper (Shannon, 1956) correspond to a channel in which either K_1 or K_2 may be used (for the sum) or to a channel where both K_1 and K_2 are used (for the product). To prove the product relationship suppose $(g_\alpha, R_\alpha, T_\alpha)$ produce K_1' from K_1 and $(g_\beta', R_\beta', T_\beta')$ produce K_2' from K_2. Then $(g_\alpha g_\beta', R_\alpha R_\beta', T_\alpha T)_\beta'$ produces $K_1'K_2'$ from K_1K_2, where the product $R_\alpha R_\beta'$ means the product of the channels. The sum case works similarly. The sum $K_1' + K_2'$ can be produced from $K_1 + K_2$ by $(g_\alpha g_\beta', R_\alpha + R_\beta', T_\alpha + T_\beta')$ where the plus means sum of channels and α and β range over all pairs.

If in a memoryless discrete channel K we consider blocks of n letters, then we have another memoryless discrete channel, one in which the input "letters" are input words of length n for the original channel and the output "letters" are output words of length n for the original channel. This channel is clearly equivalent to K^n. Consequently, if $K_1 \supseteq K_2$ the channel K_1^n for words of length n from K_1 includes K_2^n, the channel for words of length n from K_2.

Suppose $K_1 \supseteq K_2$ and $K_1 \supseteq K_3$ and that K_1 and K_3 have matrices $\| p_i(j) \|$ and $\| q_i(j) \|$, respectively. Then K_1 also includes the channel whose matrix is

$$\lambda \| p_i(j) \| + (1 - \lambda) \| q_i(j) \| \qquad (0 \leqq \lambda \leqq 1).$$

Thus, in the transition probability space the set of channels included in K_1 form a convex body. The $\lambda \| p_i(j) \| + (1 - \lambda) \| q_i(j) \|$ channel can in fact be obtained from K_1 by the union of $(\lambda g_\alpha, R_\alpha, T_\alpha)$ and

$$((1 - \lambda)g_{\beta}', R_{\beta}', T_{\beta}').$$

Our most important result and the chief reason for considering the relation of channel inclusion connects this concept with coding theory. We shall show, in fact, that if a code exists for K_2 and $K_1 \supseteq K_2$, at least as good a code exists for K_1 in the sense of low probability of error.

THEOREM. Suppose $K_1 \supseteq K_2$ and there is a set of code words of length n for K_2, $W_1, W_2, \cdots, W_m$, and a decoding system such that if the W_i are used with probabilities P_i then the average probability of error in decoding is P_e. Then there exists for channel K_1 a set of m code words of length n and a decoding system which if used with the same probabilities P_i given an average probability of error $P_e' \leqq P_e$. Consequently, the capacity of K_1 is greater than or equal to that of K_2.

PROOF. If $K_1 \supseteq K_2$ a set $(g_\alpha, R_\alpha, T_\alpha)$ makes K_1 like K_2, where R_α and T_α are *pure* channels. For any particular α, R_α defines a mapping of input words from the K_2^n code into input words from the K_1^n dictionary (namely, the words into which the R_α transforms the code). Furthermore, T_α definies a mapping of K_1 output words into K_2 output words. From a code and decoding system for K_2 we can obtain, for any particular α, a code and decoding system for K_1. Take as the code the set of words obtained by the mapping R_α from the code words for K_2. For the decoding system, decode a K_1 word as the given system decodes the word into which a K_1 word is transformed by T_α. Such a code will have a probability of error for K_1 of, say, $P_{e\alpha}$. Now it is clear that

$$P_e = \sum\nolimits_\alpha g_\alpha P_{e\alpha} \tag{2}$$

since the channel K_2 acts like these different codes with probability g_α. Since this equation says that a (weighted) average of the $P_{e\alpha}$ is equal to P_e, there must be at least one particular $P_{e\alpha}$ that is equal to or greater than P_e. (If all the $P_{e\alpha}$ were less than P_e the right-hand side would necessarily be less than P_e.) The code and decoding system defined above for this particular α then give the main result for the theorem. It follows, then, that $P_{e\,\mathrm{opt}}(M, n)$, the minimum probability of error for M equally probable words of length n, will be at least as good for K_1 as for K_2. Similarly, the channel capacity, the greatest lower bound of rates such that P_e can be made to approach zero, will be at least as high for K_1 as for K_2.

It is interesting to examine geometrically the relation of channel inclusion in the simple case of channels with two inputs and two outputs (the general binary channel). Such a channel is defined by two probabilities p_1 and p_2 (Fig. 4) and can be represented by a point in the unit square. In this connection, see Silverman (1955) where channel capacity and other parameters are plotted as contour lines in such a square. In Fig. 5 the channel with $p_1 = 1/4$, $p_2 = 1/2$ is plotted together with the three other equivalent channels with probabilities p_2, p_1; $1 - p_2, 1 - p_1$; and $1 - p_1, 1 - p_2$. Adding the two points (0,0) and (1,1) gives a total of six points. The hexagon defined by these—i.e., their convex hull—includes all the points which correspond to channels included in the given channel. This is clear since all pairs of pure pre- and post-channels produce from the given channel one of these six. This is readily verified by examination of cases. Hence any mixture with probabilities g_α will correspond to a point within the convex hull.

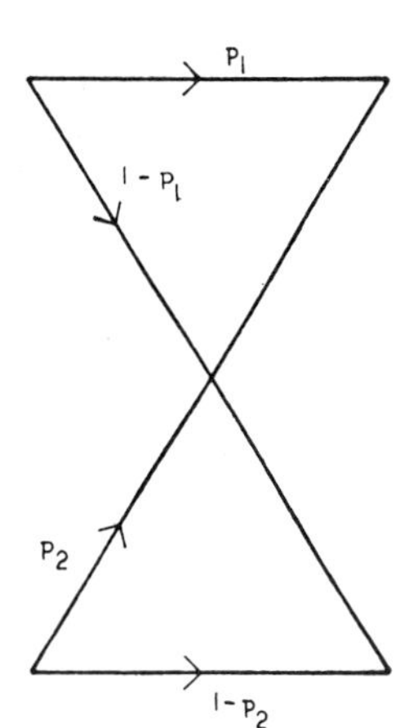

FIG. 4. The general binary channel.

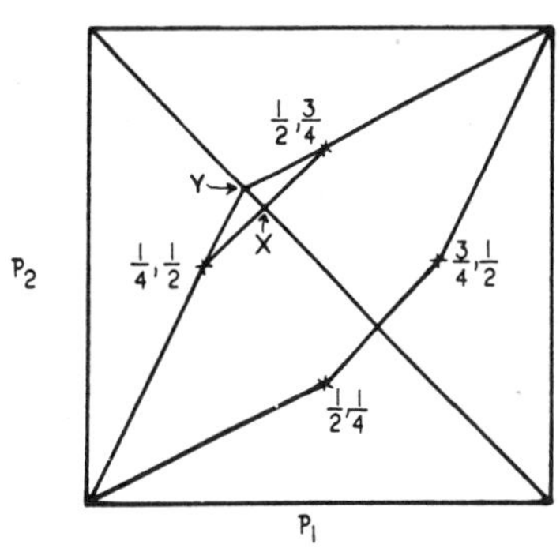

FIG. 5. The hexagon of binary channels included in a typical binary channel.

In Fig. 5, binary symmetric channels lie on the square diagonal from (1,0) to (0,1). Thus the given channel includes in particular the binary

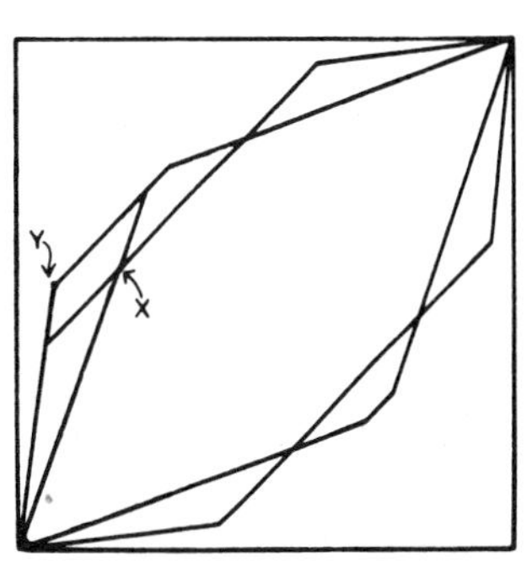

FIG. 6. The greatest lower bound X and the least upper bound Y of two comparable binary channels.

symmetric channel X whose point is $\frac{1}{2}[p_1 + (1 - p_2)], \frac{1}{2}[p_2 + (1 - p_1)]$, in our example $(\frac{3}{8}, \frac{5}{8})$. The channel is *included* in the binary symmetric channel Y with coordinates $(p_1/p_1 + p_2$ and $p_2/p_1 + p_2)$, in our particular case $(\frac{1}{3}, \frac{2}{3})$. These inclusions give simple upper and lower bounds on the capacity of a general binary channel in terms of the more easily calculated binary symmetric channel.

If we have two channels neither of which includes the other, the situation will be that of Fig. 6, with two hexagons. In this case there is a greatest lower bound and a least upper bound of the two channels, namely, the channels represented by X and Y in Fig. 6. Thus, in the binary channel case we have more than a partial ordering; we have a *lattice*.

FURTHER GENERALIZATIONS AND CONJECTURES

We have not been able to determine whether or not the partial ordering defines a lattice in the case of channels with n letters. The set of points included in a given channel can be found by a construction quite similar to Fig. 5, namely, the convex hull of points obtained from the channel by pure pre- and post-channels; but it is not clear, for example, that the intersection of two such convex bodies corresponds to a channel.

Another question relates to a converse of the coding theorem above. Can one show that in some sense the ordering we have defined is the most general for which such a coding theorem will hold?

The notion of channel inclusion can be generalized in various ways to channels with memory and indeed in another paper (Shannon, 1957) we used this sort of notion at a very simple level to obtain some results in coding theory. It is not clear, however, what the most natural generalization will be in all cases.

RECEIVED: March 24, 1958.

REFERENCES

SHANNON, C. E. (1956). Zero error capacity of a noisy channel. *IRE Trans. on Inform. Theory* **IT-2**, 8–19.

SILVERMAN, R. A. (1955). On binary channels and their cascades. *IRE Trans. on Inform. Theory*, **IT-1**, 19–27.

SHANNON, C. E. (1957). Certain results in coding theory for noisy channels. *Inform. and Control* **1**, 6–25.

THE CAPACITY OF A CLASS OF CHANNELS[1]

By David Blackwell, Leo Breiman, and A. J. Thomasian

University of California, Berkeley

1. Summary. Shannon's basic theorem on the capacity of a channel is generalized to the case of a class of memoryless channels. A generalized capacity is defined and is shown to be the supremum of attainable transmission rates when the coding and decoding procedure must be satisfactory for every channel in the class.

2. Definitions and Introduction. For any positive integer n and any set $\mathcal{A}$ we denote by $\mathcal{A}^{(n)}$ the set of all n-tuples $(x_1, \cdots, x_n)$ with each $x_i \varepsilon \mathcal{A}$.

A channel, denoted by $(\mathcal{A}, \mathcal{B}, P(y \mid x))$ or by $P(y \mid x)$, consists of two finite sets $\mathcal{A}$, $\mathcal{B}$ having $a \geqq 2$, $b \geqq 2$ elements, respectively, and a set of probability distributions $P(\cdot \mid x)$ on $\mathcal{B}$, one for each $x \varepsilon \mathcal{A}$. $P(y \mid x)$ is interpreted as the probability of receiving $y \varepsilon \mathcal{B}$ given that $x \varepsilon \mathcal{A}$ was transmitted.

The n-extension of a channel $(\mathcal{A}, \mathcal{B}, P(y \mid x))$ is the channel $(\mathcal{A}^{(n)}, \mathcal{B}^{(n)}, P(v \mid u))$ where $v = (y_1, \cdots, y_n) \varepsilon \mathcal{B}^{(n)}$, $u = (x_1, \cdots, x_n) \varepsilon \mathcal{A}^{(n)}$ and $P(v \mid u) = \prod_{i=1}^{n} P(y_i \mid x_i)$.

When considering a class of channels, $(\mathcal{A}, \mathcal{B}, P_\gamma(y \mid x))$ for $\gamma \varepsilon \mathcal{C}$, where $\mathcal{C}$ is an index set, we shall always assume that the $\mathcal{A}$, $\mathcal{B}$ sets are the same for each channel in the class. We shall sometimes denote such a class of channels by $\mathcal{C}$, the index set.

A (G, ϵ_n, n) code for a class $\mathcal{C}$ of channels for $G \geqq 1$, $\epsilon_n \geqq 0$, n a positive integer, is a sequence of $[G]$ distinct elements of $\mathcal{A}^{(n)}$; $u_1, \cdots, u_{[G]}$; where $[G]$ is the largest integer $\leqq G$, and a sequence of $[G]$ disjoint subsets of $\mathcal{B}^{(n)}$; $B_1, \cdots, B_{[G]}$; such that

$$P_\gamma(B_i^c \mid u_i) \leqq \epsilon_n \quad \text{for} \quad i = 1, \cdots, [G] \quad \text{and all} \quad \gamma \varepsilon \mathcal{C}.$$

The set $\{u_1, \cdots, u_{[G]}\}$ is called the set of input messages of the code and B_i is called the decoding set for u_i. We think of an input letter u_i of the code as being selected arbitrarily and transmitted over an unknown one of the channels P_γ, $\gamma \varepsilon \mathcal{C}$. The letter v is received with probability $P_\gamma(v \mid u)$ and if $v \varepsilon B_j$ it is decoded as u_j. Thus, the probability is $\leqq \epsilon_n$ that any input message u_i will be transmitted so as to be not decoded as u_i; regardless of which channel in the class $\mathcal{C}$ is used.

An $R \geqq 0$ is an attainable transmission rate for a class $\mathcal{C}$ of channels if there exists a sequence of (e^{Rn}, ϵ_n, n) codes for $\mathcal{C}$ with $\epsilon_n \to 0$. Since $\mathcal{A}^{(n)}$ has only a^n points we know that any attainable rate $R \leqq \log a$. Clearly 0 is an attainable rate for any class of channels. For any class of channels $\mathcal{C}$ we define $T = T(\mathcal{C})$ to be the supremum of the set of attainable rates for $\mathcal{C}$.

Received February 16, 1959.

[1] This research was supported by the Office of Naval Research under Contract Nonr-222(53).

Reprinted with permission from *Ann. Math. Stat.*, vol. 30, pp. 1229–1241, Dec. 1959.

If $(\mathcal{A}, \mathcal{B}, P_\gamma(y \mid x))$ for $\gamma \varepsilon \mathcal{C}$ is a class of channels and $Q(x)$ is a given probability distribution on $\mathcal{A}$ then for each $\gamma \varepsilon \mathcal{C}$ we let $P_\gamma(x, y) = P_\gamma(y \mid x)Q(x)$ and we define on $\mathcal{A} \times \mathcal{B}$ the random variable J_γ by

$$J_\gamma(x, y) = \log \frac{P_\gamma(x, y)}{P_\gamma(x)P_\gamma(y)} \quad \text{if} \quad P_\gamma(x, y) > 0$$

$$= 0 \quad \text{if} \quad P_\gamma(x, y) = 0.$$

The dependence of P_γ and J_γ on Q will usually not be exhibited. Since we will often be interested in expressions of the form $x \log x$ it is natural to define $\log 0 = 0$. We will denote the expectation of a random variable X with respect to the P_γ distribution by $E_\gamma X$. If $\mathcal{C}$ has only one element we may drop the subscript γ. Finally for any class $\mathcal{C}$ of channels we define the capacity of the class $\mathcal{C}$ by

$$C(\mathcal{C}) = C = \sup_{Q(x)} \inf_{\gamma \varepsilon \mathcal{C}} E_\gamma J_\gamma$$

where the sup is over all distributions Q on $\mathcal{A}$.

In the case considered by Shannon, $\mathcal{C}$ has only one element and our formula reduces to $C = \sup_Q EJ$, which is the usual formula for the capacity of a memoryless channel. Shannon's theorem then states that $T = C$. $T \geqq C$, $T \leqq C$ are called the direct and converse halves, respectively. This theorem for a single channel has been proved in various ways and under various conditions by Shannon [12], [13], McMillan [11], Feinstein [6], Khinchin [9], Wolfowitz [14], Blackwell, Breiman, and Thomasian [1]. We will show that within the framework that has been set up

$$T(\mathcal{C}) = C(\mathcal{C})$$

always holds true. This result follows immediately from Theorem 1 which also gives an exponential error bound for any rate $R < C$.

THEOREM 1: *Let $(\mathcal{A}, \mathcal{B}, P_\gamma(y \mid x))$ for $\gamma \varepsilon \mathcal{C}$ be any class of channels.*

(a) *For any integer n and any $R > 0$ such that $0 \leqq C - R \leqq 1/2$ there is an (e^{Rn}, ϵ_n, n) code for $\mathcal{C}$ with*

$$\epsilon_n = Ae^{-\frac{(C-R)^2}{B}n}$$

where

$$A = \left[\frac{2^{10}ab^3}{(C - R)^2}\right]^{2ab} \quad \text{and} \quad B = 2^7ab.$$

(b) *For any integer n and $R > C$ if $e^{Rn} \geqq 2$ then any (e^{Rn}, ϵ_n, n) code for $\mathcal{C}$ must satisfy*

$$\epsilon_n \geqq 1 - \frac{C + \dfrac{\log 2}{n}}{R - \dfrac{\log 2}{n}}.$$

The sequence of steps used in proving Theorem 1 will be outlined. Theorem 2 presents a basic inequality, for a single channel, which is contained implicitly in Feinstein [8]. This inequality is of independent interest since it gives the same bound for the maximum probability of error that Shannon [13] gives for the average probability of error. Theorem 2 permits a simple proof of $T \geqq C$ for a single channel. Lemma 2 shows that $\sup_Q$ in the definition of $C(\mathcal{C})$ can be replaced by $\max_Q$. Theorem 3 gives an exponential bound on the error of a code for one channel, which depends only on a, b, $(C - R)^2$. This is convenient in that the particular probabilities $P(y \mid x)$ may not be known and, in any case, need not be computed with. Results related to Theorem 3 have been given by Elias [3] and [4], Feinstein [7], Shannon [13], and Wolfowitz [14].

Lemma 3 generalizes the inequality of Theorem 2 to the case when $\mathcal{C}$ has a finite number of elements, and Theorem 4 generalizes the exponential error bound of Theorem 3 to this case.

Lemma 4 shows that for a given $\mathcal{A}$, $\mathcal{B}$ there is a large finite number of channels on $\mathcal{A}$, $\mathcal{B}$ such that any channel on $\mathcal{A}$, $\mathcal{B}$ is close, in several senses, to one of them. Lemma 5 shows that if a channel has a sequence of codes (e^{Rn}, ϵ_n, n) with $\epsilon_n = e^{-Bn}$ for large n, with $B > 0$, then this same sequence of codes can be used for all channels in a certain neighborhood of the channel. This result justifies some of our attention to exponential error bounds. The technique of Lemma 5 can also be used to get some similar results when the channel probabilities vary from letter to letter.

At this point the direct half of Theorem 1 is demonstrated by approximating the class $\mathcal{C}$ of channels by a certain finite set of channels $\mathcal{C}'$ from Lemma 4; obtaining an exponential error bound code for $\mathcal{C}'$ from Theorem 4; and using Lemma 5 to show that such a code must be satisfactory for $\mathcal{C}$.

The converse half of Theorem 1 is then proved.

Before proceeding to the proofs we pause to clear up one point. It is obvious that

$$C(\mathcal{C}) \leqq \inf_{\gamma \varepsilon \mathcal{C}} \sup_{Q(x)} E_\gamma J_\gamma ,$$

i.e., $C(\mathcal{C}) \leqq$ the capacity of every channel in $\mathcal{C}$. We now exhibit an example where $C(\mathcal{C}) \neq$ inf of the capacities of channels in $\mathcal{C}$. Let $\mathcal{A} = \mathcal{B} = \{1, 2, 3, 4\}$, $\mathcal{C} = \{1, 2\}$, and let $P_1(y \mid x)$ and $P_2(y \mid x)$ be defined by the left and right following matrices, respectively.

$$\begin{pmatrix} \frac{1}{2} & \frac{1}{2} & 0 & 0 \\ 0 & 0 & \frac{1}{2} & \frac{1}{2} \\ \frac{1}{4} & \frac{1}{4} & \frac{1}{4} & \frac{1}{4} \\ \frac{1}{4} & \frac{1}{4} & \frac{1}{4} & \frac{1}{4} \end{pmatrix} \begin{pmatrix} \frac{1}{4} & \frac{1}{4} & \frac{1}{4} & \frac{1}{4} \\ \frac{1}{4} & \frac{1}{4} & \frac{1}{4} & \frac{1}{4} \\ \frac{1}{2} & \frac{1}{2} & 0 & 0 \\ 0 & 0 & \frac{1}{2} & \frac{1}{2} \end{pmatrix}$$

Let $Q(x)$ be any distribution on $\mathcal{A}$ and let $H_i(Y) = -\sum_y P_i(y) \log P_i(y)$, $H_i(Y \mid X) = -\sum_x Q(x) \sum_y P_i(y \mid x) \log P_i(y \mid x)$. Using the fact that $\log x = (\log 2) \log_2 x$ we see that $(\log 2)^{-1} H_1(Y \mid X) = Q(1) + Q(2) + 2Q(3) + 2Q(4) = 1 + Q(3) + Q(4)$. Also from Feinstein [8], p. 15 we have $(\log 2)^{-1} H_1(Y) \leqq 2$ so that $E_1 J_1 = H_1(Y) - H_1(Y \mid X) \leqq (\log 2)(Q(1) + Q(2))$. Similarly $E_2 J_2 \leqq (\log 2)(Q(3) + Q(4))$ so that $C(\mathcal{C}) \leqq (1/2) \log 2$. The case $Q(i) = 1/4$ for $i = 1, \cdots, 4$ shows that $C(\mathcal{C}) = (1/2) \log 2$; the case

$Q(1) = Q(2) = 1/2$ shows the capacity of channel one to be log 2; the case $Q(3) = Q(4) = 1/2$ shows the capacity of channel two to be log 2. Thus for this example

$$\tfrac{1}{2}\log 2 = C(\mathcal{C}) < \inf_{\gamma \epsilon \mathcal{C}} \sup_{Q(x)} E_\gamma J_\gamma = \log 2.$$

3. A basic inequality.

THEOREM 2: *For any channel* $(\mathcal{A}, \mathcal{B}, P(y \mid x))$, *any distribution* $Q(x)$ *on* $\mathcal{A}$, $\alpha > 0$, $G \geqq 1$ *there is a* $(G, \epsilon, 1)$ *code for the channel with* $\epsilon = Ge^{-\alpha} + P(J \leqq \alpha)$.

PROOF: It is clearly sufficient to construct an $(M, \epsilon, 1)$ code with the same ϵ as in the theorem and with $M \geqq G$. Let $A = [J > \alpha]$ and for any $x_0 \epsilon \mathcal{A}$ let $A_{x_0} = \{(x, y) \mid (x_0, y) \epsilon A\}$. $P(J \leqq \alpha) \leqq \epsilon$ so that $P(A) \geqq 1 - \epsilon$, hence there is an x_1 such that $P(A \mid x_1) \geqq 1 - \epsilon$. Let $B_1 = A_{x_1}$. (Each B_k will be a cylinder set with base in $\mathcal{B}$. The base of B_k will be the decoding set for x_k.) At the kth step select x_k such that $P(B_k \mid x_k) \geqq 1 - \epsilon$ where

$$B_k = \bigcup_1^k A_{x_i} - \bigcup_1^{k-1} A_{x_i}.$$

This process will terminate at some $M \geqq 1$. For every x

$$P\left(A - A \cap \left(\bigcup_1^M A_{x_i}\right) \middle| x\right) < 1 - \epsilon$$

otherwise we could add this x to $x_1, \cdots, x_M$ contradicting the definition of M. Thus

$$P(A) = P\left(A \cap \left(\bigcup_1^M A_{x_i}\right)\right) + P\left(A - A \cap \left(\bigcup_1^M A_{x_i}\right)\right)$$
$$\leqq \sum_1^M P(A_{x_i}) + 1 - \epsilon.$$

Now if $(x, y) \epsilon A$ then $J(x, y) > \alpha$ so that $P(y \mid x) > P(y)e^\alpha$. For fixed x sum both sides of this inequality over all y such that $(x, y) \varepsilon A$. Then

$$1 \geqq P(A \mid x) \geqq P(A_x)e^\alpha.$$

Thus $P(A_x) \leqq e^{-\alpha}$ for any $x \varepsilon \mathcal{A}$ so that $P(A) \leqq Me^{-\alpha} + 1 - \epsilon$. Since $P(A) = Ge^{-\alpha} + 1 - \epsilon$, we have $M \geqq G$. Clearly the $B_1, \cdots, B_M$ are disjoint and

$$P(B_k \mid x_k) \geqq 1 - \epsilon$$

for $k = 1, \cdots, M$ so the proof is completed.

Consider a single channel $(\mathcal{A}, \mathcal{B}, P(y \mid x))$ and let $Q(x)$ be specified and determine $P(x, y)$, $J(x, y)$. Applying Theorem 2 to $(\mathcal{A}^{(n)}, \mathcal{B}^{(n)}, P(v \mid u))$ and $Q(u) = Q(x_1) \cdots Q(x_n)$ with $\alpha = n(R + EJ)/2$, $G = e^{Rn}$ we see that for any R such that $0 < R < EJ$ there is an (e^{Rn}, ϵ_n, n) code for $(\mathcal{A}, \mathcal{B}, P(y \mid x))$ with

$$\epsilon_n = e^{-(EJ-R)n/2} + P\left(\frac{1}{n} J' \leqq \frac{R + EJ}{2}\right).$$

Now

$$J'(u, v) = \log \frac{P(u, v)}{P(u)P(v)} \qquad \text{if} \quad P(u, v) > 0$$

$$= 0 \qquad \text{otherwise.}$$

Let $J''(u, v) = \sum_1^n J_i(x_i, y_i)$ where

$$J_i(x_i, y_i) = \log \frac{P(x_i, y_i)}{P(x_1)P(y_1)} \qquad \text{if } P(x_i, y_i) > 0$$

$$= 0 \qquad \text{otherwise.}$$

Clearly $P(J' = J'') = 1$ and J'' is the sum of n independent random variables each having the distribution of $J(x, y)$. Since $EJ > (R + EJ)/2$ we see that $\epsilon_n \to 0$. Now it is easily seen (and we will shortly prove even more) that for a fixed channel EJ is a continuous function of $(Q(x_1), Q(x_2), \cdots, Q(x_n))$ and since the domain of the function is a closed bounded subset of Euclidean space the supremum is actually achieved. Thus for any channel $(\mathfrak{A}, \mathfrak{B}, P(y \mid x))$ there is a distribution $Q(x)$ on $\mathfrak{A}$ such that $C = EJ$. Using this $Q(x)$ in the earlier portions of this paragraph we obtain the direct half of Shannon's theorem for a memoryless channel: $T \geqq C$.

By introducing a brief epsilon argument in the proof of the direct half of Shannon's theorem we could clearly have ignored the question of whether or not there is a maximizing $Q(x)$. Although the fact that there is a maximizing $Q(x)$ in the general case of a class of channels is not vital in the following work, we will pause to prove this fact now. The proof is based on Lemma 1 which will be needed later.

LEMMA 1: *Let $Q(x)$, $Q'(x)$ be any two distributions on $\mathfrak{A}$ such that*

$$| Q(x) - Q'(x) | \leqq \epsilon \leqq 1/e \text{ for all } x \in \mathfrak{A}.$$

Then

$$| H(X) - H'(X) | \leqq a\epsilon^{1/2}$$

where $H(X) = -\sum_x Q(x) \log Q(x)$ *and* $H'(X) = -\sum_x Q'(x) \log Q'(x)$.

PROOF: Let

$$f(y) = [-(y + \epsilon) \log (y + \epsilon)] - [-y \log y]$$

where $0 < \epsilon \leqq 1/e$ and $0 \leqq y \leqq 1 - \epsilon$. Then $f(0) = -\epsilon \log \epsilon > 0$ and $f(1 - \epsilon) = (1 - \epsilon) \log (1 = \epsilon) < 0$ also

$$f'(y) = -\log (y + \epsilon) - 1 + \log y + 1 = \log \frac{y}{y + \epsilon} < 0$$

so that $| f(y) | \leqq \max \{-\epsilon \log \epsilon, -(1 - \epsilon) \log(1 - \epsilon)\}$. Now

$$(1 - \epsilon) \log \frac{1}{1 - \epsilon} \leqq (1 - \epsilon) \left(\frac{1}{1 - \epsilon} - 1\right) = \epsilon \leqq -\epsilon \log \epsilon$$

since $\epsilon \leqq 1/e$. Thus

$$|f(y)| \leqq -\epsilon \log \epsilon = \frac{\epsilon^{\frac{1}{2}\log\frac{1}{\epsilon}}}{\left(\frac{1}{\epsilon}\right)^{\frac{1}{2}}} \leqq \epsilon^{\frac{1}{2}}$$

since $x^{1/2} - \log x \geqq 2 - \log 4 > 0$ for $x > 0$. Applying the result $|f(y)| \leqq \epsilon^{1/2}$ to $y = p$, $\epsilon = q - p$ where $0 \leqq p \leqq q \leqq 1$ and $|q - p| \leqq 1/e$ we see that

$$|[-p \log p] - [-q \log q]| \leqq (|p - q|)^{1/2}$$

which easily gives us the bound on $|H(X) - H'(X)|$ completing the proof.

LEMMA 2: *For any class of channels* $(\mathfrak{A}, \mathfrak{B}, P_\gamma(y \mid x))$ *for* $\gamma \varepsilon \mathfrak{C}$,

$$C = \max_{Q(x)} \inf_{\gamma \varepsilon \mathfrak{C}} E_\gamma J_\gamma .$$

PROOF: Let $(\mathfrak{A}, \mathfrak{B}, P(y \mid x))$ be a channel and $Q(x)$ a distribution on $\mathfrak{A}$ determining $P(x, y) = P(y \mid x)Q(x)$ and $J(x, y)$. Clearly $EJ = H(X) + H(Y) - H(X, Y)$ where $H(X) = -\sum_x P(x) \log P(x)$, $H(Y) = -\sum_y P(y) \log P(y)$, $H(X, Y) = -\sum_{x,y} P(x, y) \log P(x, y)$. Let $Q'(x)$ be another distribution on $\mathfrak{A}$ determining $P'(x, y) = P(y \mid x)Q'(x)$ and $J'(x, y)$, and note that $E'J' = H'(X) + H'(Y) - H'(X, Y)$ where the primed quantities have analogous definitions. Assume that $|Q(x) - Q'(x)| \leqq \epsilon \leqq 1/e$ for all $x \varepsilon \mathfrak{A}$. Clearly $|P(x, y) - P'(x, y)| \leqq P(y \mid x) |Q(x) - Q'(x)| \leqq \epsilon$ and $|P(y) - P'(y)| \leqq \sum_x |P(x, y) - P'(x, y)| \leqq a\epsilon$. Applying Lemma 1 we get

$$\begin{aligned} |EJ - E'J'| &\leqq |H(X) - H'(X)| + |H(Y) - H'(Y)| \\ &\qquad + |H(X, Y) - H'(X, Y)| \\ &\leqq a\epsilon^{1/2} + b(a\epsilon)^{1/2} + ab\epsilon^{1/2} \leqq (a + 2ab)\epsilon^{1/2}. \end{aligned}$$

Thus not only is EJ continuous in $Q(x)$ but it is continuous in $Q(x)$ uniformly in $Q(x)$ and $P(y \mid x)$. We easily take $\inf_{\gamma \varepsilon \mathfrak{C}}$ on the inequalities

$$E'_\gamma J'_\gamma - (a + 2\,ab)\epsilon^{1/2} \leqq E_\gamma J_\gamma \leqq E'_\gamma J'_\gamma + (a + 2ab)\epsilon^{1/2}$$

and see that $\inf_{\gamma \varepsilon \mathfrak{C}} E_\gamma J_\gamma$ is continuous in $Q(x)$ so that once again there is a maximizing $Q(x)$ and Lemma 2 is proved.

4. The error bound for one channel.

THEOREM 3: *Let* $(\mathfrak{A}, \mathfrak{B}, P(y \mid x))$ *be any channel. For any integer* n *and any* $R > 0$ *such that* $0 \leqq C - R \leqq 1/2$, *there is an* (e^{Rn}, ϵ_n, n) *code for the channel with*

$$\epsilon_n = 2e^{-\frac{(C-R)^2}{16ab} n}.$$

PROOF: Applying Theorem 2 to $(\mathfrak{A}^{(n)}, \mathfrak{B}^{(n)}, P(v \mid u))$ with $Q(u) = Q(x_1) \cdots Q(x_n)$, where $Q(x)$ is any distribution on $\mathfrak{A}$, $G = e^{Rn}$, $\alpha = (R + \theta)n$ we see that for any $R > 0$, $\theta > 0$ there is an (e^{Rn}, ϵ_n, n) code for $(\mathfrak{A}, \mathfrak{B}, P(y \mid x))$ with

$$\epsilon_n = e^{-n\theta} + P(J'' \leqq n(R + \theta))$$

where, as shown in Section 3, J'' is the sum of n independent random variables, each having the distribution of $J(x, y)$. Select $R > 0$, $0 \leqq EJ - R \leqq 1/2$ and let $\theta = (EJ - R)^2$. Then $R + \theta \leqq R + (EJ - R)/2 = (EJ + R)/2$.

Thus it remains only to show that

$$P(J'' \leqq n(EJ + R)\tfrac{1}{2}) \leqq e^{-\frac{(EJ-R)^2}{16ab}n}$$

(we will need this result later) for we can then choose Q so that $C = EJ$.

A method due to Chernoff [2] will be used to bound the probability in question. Let $0 \leqq t \leqq 1$, then

$$P\left(0 \leqq \frac{n(EJ + R)}{2} - J''\right) \leqq Ee^{t\left[\frac{n(EJ+R)}{2} - J''\right]} = e^{\frac{tn(EJ+R)}{2}} Ee^{-J''}$$

$$= [e^{\frac{t(EJ+R)}{2}} Ee^{-tJ}]^n$$

so that we need show only that for a proper selection of t,

$$e^{\frac{t(EJ+R)}{2}} Ee^{-tJ} \leqq e^{-\frac{(EJ-R)^2}{16ab}}.$$

Now

$$Ee^{-tJ} = 1 - tEJ + \frac{t^2}{2} EJ^2 e^{-\theta tJ}, \qquad 0 < \theta < 1.$$

We need consider only (x, y) with $P(x, y) > 0$. Terms in $EJ^2 e^{-\theta tJ}$ are of the form

$$P(x, y)\left(\frac{P(x)P(y)}{P(x, y)}\right)^{\theta t} \log^2 \frac{P(x, y)}{P(x)P(y)} \leqq P(x, y)\left(\frac{1}{P(x, y)}\right)^{\theta t} \log^2 \frac{P(x, y)}{P(x)P(y)}$$

$$\leqq (P(x, y))^{1-t} \log^2 \frac{P(x, y)}{P(x)P(y)} \leqq (P(x, y))^{1-t} \log^2 P(x, y)$$

where the last inequality followed from $P(x, y) \leqq P(x)P(y)/P(x, y) \leqq 1/P(x, y)$. Also

$$[(P(x, y))^{\frac{1-t}{2}} \log P(x, y)]^2 = \left(\frac{2}{1 - t}\right)^2 [(P(x, y))^{\frac{1-t}{2}} \log P(x, y))^{\frac{1-t}{2}}]^2$$

$$\leqq \left(\frac{2}{1 - t}\right)^2 \frac{1}{e^2} \leqq \frac{1}{(1 - t)^2}.$$

Thus

$$Ee^{-tJ} \leqq 1 - tEJ + \frac{t^2}{2}\frac{ab}{(1 - t)^2} \leqq e^{-tEJ + \frac{t^2}{2}\frac{ab}{(1-t)^2}}$$

so that

$$e^{\frac{t(EJ+R)}{2}} Ee^{-tJ} \leqq e^{-\frac{1}{2}f(t)}$$

where

$$f(t) = (EJ - R)t - t^2 \frac{ab}{(1-t)^2}.$$

Let $t = (EJ - R)/4ab \leqq 1/8$ so that $1/(1-t)^2 \leqq (8/7)^2$, then

$$f\left(\frac{EJ - R}{4ab}\right) \geqq \frac{(EJ - R)^2}{4ab}\left[1 - \left(\frac{8}{7}\right)^2 \frac{1}{4}\right] \geqq \frac{(EJ - R)^2}{8ab}$$

completing the proof.

5. The error bound for a finite set of channels. Lemma 3 is needed in the proof of Theorem 4.

LEMMA 3: *Let* $(\mathfrak{A}, \mathfrak{B}, P_\gamma(y \mid x))$ *for* $\gamma \in \mathfrak{C} = \{1, 2, \cdots, L\}$ *be a finite class of channels and let* $Q(x)$ *be a distribution on* $\mathfrak{A}$, *determining* $P_\gamma(x, y)$, $J_\gamma(x, y)$.

(a) *Define a channel* $(\mathfrak{A}, \mathfrak{B}, P(y \mid x))$ *by* $P(y \mid x) = (1/L)\sum_{\gamma=1}^{L} P_\gamma(y \mid x)$ *and let* $Q(x)$ *determine* $P(x, y)$, $J(x, y)$. *Then for all* α, δ

$$P(J \leqq \alpha) \leqq \frac{1}{L}\sum_{\gamma=1}^{L} P_\gamma(J_\gamma \leqq \alpha + \delta) + Le^{-\delta}.$$

(b) *For any* $\alpha > 0$, $G \geqq 1$, $\delta > 0$ *there is a* $(G, \epsilon, 1)$ *code for* $\mathfrak{C}$ *with*

$$\epsilon = LGe^{-\alpha} + L^2e^{-\delta} + \sum_{1}^{L} P_\gamma(J_\gamma \leqq \alpha + \delta).$$

PROOF: We first prove part (a).

$$P(J \leqq \alpha) = \frac{1}{L}\sum P_\gamma(J \leqq \alpha) \leqq \frac{1}{L}\sum [P_\gamma(J_\gamma \leqq \alpha + \delta)$$
$$+ P_\gamma(J_\gamma > \alpha + \delta; J \leqq \alpha)]$$

so that we need only prove that $P_\gamma(A_\gamma) \leqq Le^{-\delta}$ where $A_\gamma = (J_\gamma \alpha + \delta; J \leqq \alpha)$. For any $(x, y) \in A_\gamma$ with $P_\gamma(x, y) > 0$ we have

$$e^\alpha P(y) \geqq P(y \mid x) \geqq \frac{1}{L} P_\gamma(y \mid x) \geqq \frac{1}{L} e^{\alpha+\delta} P_\gamma(y)$$

so that $P_\gamma(y) \leqq Le^{-\delta}P(y)$. Summing this last inequality over all y such that there is an x with $(x, y) \in A_\gamma$ we get $P_\gamma(A_\gamma) \leqq \sum P_\gamma(y) \leqq Le^{-\delta}$ which completes the proof of part (a).

Applying Theorem 2 to the channel $P(y \mid x)$ defined in part (a) and then using part (a) to bound $P(J \leqq \alpha)$ we find that there is a $(G, \epsilon_0, 1)$ code for $P(y \mid x)$ with

$$\epsilon_0 = Ge^{-\alpha} + P(J \leqq \alpha) \leqq Ge^{-\alpha} + \frac{1}{L}\sum_\gamma P_\gamma(J_\gamma \leqq \alpha + \delta) + Le^{-\delta}.$$

Now $P_\gamma(y \mid x) \leqq LP(y \mid x)$ so that if x_i is an input letter for the $(G, \epsilon_0, 1)$ code and B_i is its decoding set, then $P_\gamma(B_i^c \mid x_i) \leqq L\ P_\gamma(B_i^c \mid x_i) \leqq L\epsilon_0$. Thus the $(G, \epsilon_0, 1)$ code for $P(y \mid x)$ is a $(G, L\epsilon_0, 1)$ code for $\mathfrak{C}$ and the lemma is proved.

THEOREM 4: *Let* $(\mathfrak{A}, \mathfrak{B}, P_\gamma(y \mid x))$ *for* $\gamma \varepsilon \mathfrak{C} = \{1, 2, \cdots, L\}$ *be a finite class of channels. For any* $R > 0$ *such that* $0 \leqq C - R \leqq 1/2$ *there is an* (e^{Rn}, ϵ_n, n) *code with*

$$\epsilon = 2L^2 e^{-\frac{(C-R)^2}{16ab}n}.$$

PROOF. Applying part (b) of Lemma 3 to the class of channels $(\mathfrak{A}^{(n)}, \mathfrak{B}^{(n)}, P_\gamma(v \mid u))$ with $Q(u) = Q(x_1) \cdots Q(x_n)$ and $Q(x)$ a distribution for which $C = \inf_{\gamma \varepsilon \mathfrak{C}} E_\gamma J_\gamma$ and $G = e^{Rn}$, $\alpha = (R + \theta/2)n$, $\delta = \theta n/2$ we see that there is an (e^{Rn}, ϵ_n, n) code for $\mathfrak{C}$ with

$$\epsilon_n = (L + L^2)e^{-\frac{\theta}{2}n} + \sum_1^L P_\gamma\left(\frac{1}{n} J_\gamma \leqq R + \theta\right).$$

Let $\theta = (C - R)^2$ and note that $R + (C - R)^2 \leqq R + (C - R)/2 \leqq R + (E_\gamma J_\gamma - R)/2 = (E_\gamma J_\gamma + R)/2$. Thus,

$$\epsilon_n \leqq (L + L^2)e^{-\frac{(C-R)^2}{16ab}n} + \sum_1^L P_\gamma\left(\frac{1}{n} J_\gamma \leqq \tfrac{1}{2}(R + E_\gamma J_\gamma)\right).$$

Now

$$P_\gamma\left(\frac{1}{n} J_\gamma \leqq \tfrac{1}{2}(R + E_\gamma J_\gamma)\right) \leqq P_\gamma\left(\frac{1}{n} J_\gamma \leqq \tfrac{1}{2}(R' + E_\gamma J_\gamma)\right)$$

where $R' = E_\gamma J_\gamma - (C - R) \geqq R$ and $0 \leqq E_\gamma J_\gamma - R' \leqq 1/2$. Therefore, we can apply the result obtained in the proof of Theorem 3 and get

$$P_\gamma\left(\frac{1}{n} J_\gamma \leqq \tfrac{1}{2}(R' + E_\gamma J_\gamma)\right) \leqq e^{-\frac{(E_\gamma J_\gamma - R')^2}{16ab}n} = e^{-\frac{(C-R)^2}{16ab}n}.$$

Now $L \geqq 2$ so that $2L + L^2 = L(L + 2) \leqq 2L^2$ and since Theorem 4 reduces to Theorem 3 for $L = 1$, the proof is completed.

6. The direct half of Theorem 1. Lemmas 4 and 5 are needed for the proof of part (a) of Theorem 1.

LEMMA 4: *Let* $\mathfrak{A}, \mathfrak{B}$ *be given. For every integer* $M \geqq 2b^2$ *there is a class of channels* $(\mathfrak{A}, \mathfrak{B}, P_j(y \mid x))$ *with* $\varepsilon\, \mathfrak{D}_M$, *where* $\mathfrak{D}_M$ *has at most* $(M + 1)^{ab}$ *elements, such that for any channel* $(\mathfrak{A}, \mathfrak{B}, P(y \mid x))$ *there is a channel* $(\mathfrak{A}, \mathfrak{B}, P'(y \mid x))$ *in* $\mathfrak{D}_M$ *such that:*

(a) $\mid P(y \mid x) - P'(y \mid x) \mid \leqq b/M$ *for all* x, y.

(b) $P(y \mid x) \leqq e^{2b^2/M} P'(y \mid x)$ *for all* x, y.

(c) *For any distribution* $Q(x)$ *on* $\mathfrak{A}$ *let* $P(x, y) = P(y \mid x)Q(x)$, $P'(x, y) = P'(y \mid x)Q(x)$, *then*

$$\mid EJ - E'J' \mid \leqq 2b\left(\frac{b}{M}\right)^{1/2}.$$

PROOF. Let $\mathfrak{D}_M$ be the class of channels $(\mathfrak{A}, \mathfrak{B}, P(y \mid x))$ such that for all x, y we have $MP(y \mid x) =$ an integer. Clearly $\mathfrak{D}_M$ has at most $(M + 1)^{ab}$ elements. Given the distributions $P(y \mid x)$ we will first construct $P'(y \mid x)$ and prove (a), (b). For this purpose it is enough to carry out the construction for one x_0. Arrange the "b" numbers $P(y \mid x_0)$ in ascending order and designate them by $p_1 \leqq p_2 \leqq \cdots \leqq p_b$. For $i = 1, \cdots, (b-1)$ select p_i' uniquely by $p_i \leqq p_i' < p_i + 1/M$, $Mp_i' =$ an integer. p_i' will be $P'(y \mid x_0)$ with the y being the one corresponding to p_i. Clearly

$$p_i \leqq e^{\frac{2b^2}{M}} p_i' \quad \text{and} \quad |p_i - p_i'| \leqq \frac{b}{M}$$

for $i = 1, \cdots, (b-1)$. It remains to show that if $p_b' = 1 - \sum_1^{b-1} p_i'$ then $p_b' \geqq 0$ and p_b', p_b satisfy the same relations. Now

$$p_b' \geqq 1 - \sum_1^{b-1} \left(p_i + \frac{1}{M}\right) \geqq p_b - \frac{b}{M} \geqq \frac{1}{b} - \frac{b}{M} \geqq \frac{1}{b} - \frac{1}{2b} = \frac{1}{2b}.$$

Thus $p_1', \cdots, p_b'$ form a distribution and $p_b \geqq p_b' \geqq p_b - b/M$ so that

$$|p_b - p_b'| \leqq b/M.$$

Also

$$p_b \leqq p_b' + \frac{b}{M} \leqq p_b' + \frac{2b^2}{M}\frac{1}{2b} \leqq p_b'\left(1 + \frac{2b^2}{M}\right) \leqq e^{\frac{2b^2}{M}} p_b'$$

completing the proof of parts (a) and (b).

In the proof of part (c) we will use part (a) and Lemma 1. In order to use Lemma 1 we observe that $b/M \leqq 1/2b \leqq 1/4 < 1/e$. We also note that

$$|P(y) - P'(y)| \leqq \sum_x |P(y \mid x) - P'(y \mid x)| Q(x) \leqq b/M.$$

Now

$$\begin{aligned} |EJ - E'J'| &\leqq \Big|\Big[-\sum_y P(y) \log P(y)\Big] - \Big[-\sum_y P'(y) \log P'(y)\Big]\Big| \\ &+ \Big|\Big[-\sum_{x,y} P(x,y) \log P(x,y)\Big] - \Big[-\sum_{x,y} P'(x,y) \log P'(x,y)\Big]\Big| \leqq b\left(\frac{b}{M}\right)^{1/2} \\ &+ \sum_x Q(x) \Big|\Big[-\sum_y P(y \mid x) \log P(y \mid x)\Big] - \Big[-\sum_y P'(y \mid x) \log P'(y \mid x)\Big]\Big| \\ &\leqq b\left(\frac{b}{M}\right)^{1/2} + b\left(\frac{b}{M}\right)^{1/2} \end{aligned}$$

and the lemma is proved.

LEMMA 5: *Let $(\mathfrak{A}, \mathfrak{B}, P'(y \mid x))$, $(\mathfrak{A}, \mathfrak{B}, P(y \mid x))$ be two channels and A a non-negative number such that $P(y \mid x) \leqq e^A P'(y \mid x)$ for all x, y. Any (e^{Rn}, ϵ_n, n) code for $(\mathfrak{A}, \mathfrak{B}, P'(y \mid x))$ is an $(e^{Rn}, \epsilon_n e^{An}, n)$ code for $(\mathfrak{A}, \mathfrak{B}, P(y \mid x))$.*

PROOF: Let $u = (x_1, \cdots, x_n) \in \mathfrak{A}^{(n)}$, $v = (y_1, \cdots, y_n) \in \mathfrak{B}^{(n)}$. Then

$$P(v \mid u) = \prod_1^n P(y_i \mid x_i) \leqq e^{An} \prod_1^n P'(y_i \mid x_i) = e^{An} P'(v \mid u).$$

Thus for any subset D of $\mathfrak{B}^{(n)}$ and any $u \in \mathfrak{A}^{(n)}$ we have

$$P(D \mid u) \leqq e^{An} P'(D \mid u).$$

Let $u_i \varepsilon \mathfrak{A}^{(n)}$ be an input message and B_i the corresponding decoding set of an (e^{Rn}, ϵ_n, n) code for $(\mathfrak{A}, \mathfrak{B}, P'(y \mid x))$. Then

$$P(B_i^c \mid u_i) \leqq e^{An} P'(B_i^c \mid u_i) \leqq e^{An} \epsilon_n$$

and the proof is completed.

We turn now to the proof of part (a) of Theorem 1. For each $P(y \mid x) \varepsilon \mathfrak{C}$ select a $P'(y \mid x) \varepsilon \mathfrak{D}_M$ according to Lemma 4 and let $\mathfrak{C}'$ denote this set of channels. Let $C' = C(\mathfrak{C}')$. Since $\mathfrak{C}'$ has at most $(M + 1)^{ab}$ elements we know from Theorem 4 that if $R' > 0$, $0 \leqq C' - R' \leqq 1/2$ then there is an $(e^{R'n}, \epsilon_n', n)$ code for $\mathfrak{C}'$ with

$$\epsilon_n' = 2(M + 1)^{2ab} e^{-\frac{(C'-R')^2}{16ab}n}.$$

For each $P(y \mid x) \varepsilon \mathfrak{C}$ there is a $P'(y \mid x) \varepsilon \mathfrak{C}'$ such that

$$P(y \mid x) \leqq e^{\frac{2b^2}{M}} P'(y \mid x)$$

so that from Lemma 5 the code which we have for $\mathfrak{C}'$ is an $(e^{R'n}, \epsilon_n, n)$ code for $\mathfrak{C}$ with

$$\epsilon_n = 2(M + 1)^{2ab} \exp -\left\{\frac{(C' - R')^2}{16ab} - \frac{2b^2}{M}\right\} n.$$

Let $C = C(\mathfrak{C})$ and let $Q(x)$ be a maximizing distribution for $\mathfrak{C}$. We wish to show that C' cannot be very much smaller than C. For every $P'(y \mid x) \varepsilon \mathfrak{C}'$ there is a $P(y \mid x) \varepsilon \mathfrak{C}$ such that $EJ \leqq E'J' + 2b(b/M)^{1/2}$ where we use $Q(x)$ in both cases. Thus for every $P'(y \mid x) \epsilon \mathfrak{C}$

$$C = \inf_{\mathfrak{C}} EJ \leqq E'J' + 2b \left(\frac{b}{M}\right)^{1/2}$$

so that

$$C \leqq \inf_{\mathfrak{C}'} E'J' + 2b \left(\frac{b}{M}\right)^{1/2} \leqq C' + 2b \left(\frac{b}{M}\right)^{1/2}.$$

Let $R > 0$ be given such that $0 < C - R \leqq 1/2$. We must show how to select R' and M to get our result into the final form.

We select an integer M such that

$$\frac{2^8 ab^3}{(C - R)^2} \leqq M \quad \text{and} \quad (M + 1) \leqq \frac{2^9 ab^3}{(C - R)^2}$$

so that

$$2b \left(\frac{b}{M}\right)^{1/2} \leqq \frac{C - R}{2} \quad \text{and} \quad \frac{2b^2}{M} \leqq \frac{(C - R)^2}{2^7 ab}.$$

We define R' by

$$C' - R' = C - R - 2b \left(\frac{b}{M}\right)^{1/2} \geqq \frac{C - R}{2} > 0.$$

Clearly $C' - R' \leqq 1/2$ so that we have an $(e^{R'n}, \epsilon_n, n)$ code for $\mathcal{C}$ with

$$\epsilon_n \leqq 2(M+1)^{2ab} \exp -\left\{\frac{(C-R)^2}{4(16ab)} - \frac{(C-R)^2}{2^7ab}\right\}$$

$$\leqq 2\left[\frac{2^9ab^3}{(C-R)^2}\right]^{2ab} \exp -\left\{\frac{(C-R)^2}{2^7ab}\right\}.$$

The inequality $C \leqq C' + 2b(b/M)^{1/2}$ shows that $R' \geqq R$ and an $(e^{R'n}, \epsilon_n, n)$ code for $\mathcal{C}$ can easily be reduced to an (e^{Rn}, ϵ_n, n) code for $\mathcal{C}$ so that part (a) of Theorem 1 is proved.

7. Converse half of Theorem 1. The proof is based on Lemma 6.

LEMMA 6: *Let G be an integer, $\mathcal{A}$ a finite set and let $u_1, \cdots, u_G$ be distinct elements of $\mathcal{A}^{(n)}$. Define $Q(x)$ on $\mathcal{A}$ by*

$$Q(x) = \frac{1}{nG}\sum_{j=1}^{B} (\text{the number of times that } x \text{ appears in } u_j).$$

Then any (G, ϵ, n) code, for a channel $(\mathcal{A}, \mathcal{B}, P(y \mid x))$ which uses these $u_1, \cdots, u_G$ for inputs must satisfy

$$(1-\epsilon)\log G - \log 2 \leqq nEJ$$

where $Q(x)$ is used to define $P(x, y)$ and $J(x, y)$.

PROOF: Define a distribution $\nu(u)$ on $\mathcal{A}^{(n)}$ by $\nu(u) = 1/G$ if u is one of $u_1, \cdots, u_G$ and $\nu(u) = 0$ otherwise. Define a distribution $P(u, v)$ on $\mathcal{A}^{(n)} \times \mathcal{B}^{(n)}$ by $P(u, v) = P(v \mid u)\nu(u)$ where $P(v \mid u)$ is obtained from the n-extension of $(\mathcal{A}, \mathcal{B}, P(y \mid x))$. Now define n distributions on $\mathcal{A} \times \mathcal{B}$ by

$$P^{(i)}(x, y) = P(y \mid x)\nu^{(i)}(x)$$

for $i = 1, \cdots, n$ where

$$\nu^{(i)}(x) = \sum_{x_1,\cdots,x_{i-1},x_{i+1},\cdots,x_n} \nu(x_1, \cdots, x_{i-1}, x, x_{i+1}, \cdots, x_n)$$

and observe that $Q(x) = (1/n)\sum_{i=1}^{n} \nu^{(i)}(x)$. Thus, the lemma will be proved if the following chain of inequalities is proved.

$$(1-\epsilon)\log G - \log 2 \leqq \sum_{u,v} P(u, v) \log \frac{P(u, v)}{P(u)P(v)}$$

$$\leqq \sum_{i=1}^{n}\sum_{x,y} P^{(i)}(x, y) \log \frac{P^{(i)}(x, y)}{P^{(i)}(x)P^{(i)}(y)} \leqq n\sum_{x,y} P(x, y) \log \frac{P(x, y)}{P(x)P(y)}.$$

Using $\log x = (\log 2)\log_2 x$ to convert a result from Feinstein [8], pp. 29, 39, 44; which is due to Fano [5]; we obtain the first inequality. The second inequality follows from page 30 of Feinstein [8]. We proceed to prove the third inequality. Now

$$\frac{1}{n}\sum_{i=1}^{n}\sum_{x,y} P^{(i)}(x, y)\,[\log P(y \mid x) - \log P^{(i)}(y)]$$

$$= \sum_{x,y} P(x, y) \log P(y \mid x) - \frac{1}{n}\sum_{i=1}^{n}\sum_{y} P^{(i)}(y) \log P^{(i)}(y)$$

but

$$-\frac{1}{n}\sum_{i=1}^{n}\sum_{y} P^{(i)}(y)\log P^{(i)}(y) \leqq -\sum_{y}\left(\frac{1}{n}\sum_{i=1}^{n} P^{(i)}(y)\right)\log\left(\frac{1}{n}\sum_{i=1}^{n} P^{(i)}(y)\right)$$

$$= -\sum_{y} P(y)\log P(y) = -\sum_{x,y} P(x,y)\log P(y)$$

where this last inequality follows from Lemma 4 on page 16 of Feinstein [8]. Combining the above, we complete the proof of the third inequality and hence of the lemma.

From Lemma 6 we immediately obtain that if G is an integer then for any (G, ϵ, n) code for a class $\mathcal{C}$ of channels there is a $Q(x)$ on $\mathcal{A}$ such that

$$(1-\epsilon)\log G - \log 2 \leqq n \inf_{\gamma\epsilon\mathcal{C}} E_\gamma J_\gamma \leqq nC.$$

Now e^{Rn} may not be an integer but

$$\log\,[e^{Rn}] \geqq \log\,(e^{Rn}-1) \geqq nR + \log(1-e^{-Rn}) \geqq nR - \log 2$$

so that

$$(1-\epsilon)\,(nR-\log 2) \leqq nC + \log 2$$

which completes the proof of part (b) of Theorem 1.

REFERENCES

[1] David Blackwell, Leo Breiman, A. J. Thomasian, "Proof of Shannon's transmission theorem for finite-state indecomposable channels," *Ann. Math. Stat.*, Vol. 29 (1958), pp. 1209–1220.

[2] Herman Chernoff, "A measure of asymptotic efficiency for tests of a hypothesis based on the sum of observations," *Ann. Math. Stat.*, Vol. 23 (1952), pp. 493–507.

[3] Peter Elias, "Coding for noisy channels," *I.R.E. Convention Record* (1955), part 4, pp. 37–44.

[4] Peter Elias, "Coding for two noisy channels," *Proceedings of the London Symposium on Information Theory*, Butterworth Scientific Publications, London, 1955.

[5] R. M. Fano, "Statistical theory of communication," notes on a course given at the Massachusetts Institute of Technology, 1952, 1954.

[6] Amiel Feinstein, "A new basic theorem of information theory," *I.R.E. Trans. P.G.I.T.*, September, 1954, pp. 2–22.

[7] Amiel Feinstein, "Error bounds in noisy channels without memory," *I.R.E. Trans. P.G.I.T.*, September, 1955, pp. 13–14.

[8] Amiel Feinstein, *Foundations of Information Theory*, McGraw-Hill, New York, 1958.

[9] A. I. Khinchin, "On the fundamental theorems of information theory," *Uspekhi Mathematicheskikh Nauk.*, Vol. 21 (1956), pp. 17–75.

[10] A. I. Khinchin, *Mathematical Foundations of Information Theory*, Dover Publications, Inc., 1957.

[11] Brockway McMillan, "The basic theorems of information theory," *Ann. Math. Stat.*, Vol. 24(1953), pp. 196–219.

[12] C. E. Shannon, "A mathematical theory of communication," *Bell System Technical Journal*, Vol. 27 (1948), pp. 379–423, and 623–656.

[13] Claude E. Shannon, "Certain results in coding theory for noisy channels," *Information and Control*, Vol. 1 (1957), pp. 6–25.

[14] J. Wolfowitz, "The coding of messages subject to chance errors," *Illinois J. Math.*, Vol. 1 (1957), pp. 591–606.

A Note on the Strong Converse of the Coding Theorem for the General Discrete Finite-Memory Channel*

J. Wolfowitz

Department of Mathematics, Cornell University, Ithaca, New York

The strong converse for the discrete memoryless channel was proved by the author (Wolfowitz, 1957). (The result actually proved is stronger than the strong converse because of the $O(\sqrt{n})$ term in the exponent). Subsequently the author (1958) and Feinstein (1959) independently gave the capacity C of a discrete finite-memory channel, and proved the strong converse of the coding theorem for the special discrete finite-memory channel studied (Wolfowitz, 1957, 1958). In the present note we prove the strong converse for the general discrete finite-memory channel. Thus our result includes that of Wolfowitz (1958) and Feinstein (1959) as a special case. The proof is a slight modification of the proof of Wolfowitz (1958), whose notation and definitions are hereby assumed. For a definition of the capacity C see (Wolfowitz, 1958) or (Feinstein, 1959); for a definition of the general discrete finite-memory channel see (Feinstein, 1959) or (Feinstein, 1958, p. 90).

We shall assume without essential loss of generality that both the input and output alphabets consist of two letters, say 0 and 1; extension to the case where each alphabet contains any finite number of symbols is trivial. Any sequence of n zeros or ones will be called an n-sequence.

A code (N, λ) is a set

$$\{(u_1, A_1), \cdots, (u_N, A_N)\}$$

where the u_i are n-sequences, the A_i are *disjoint* sets of n-sequences, and

$$P\{v(u_i) \in A_i\} \geq 1 - \lambda, \qquad i = 1, \cdots, N$$

where $v(u_i)$ is the chance received sequence when u_i is the sequence sent, and $P\{\ \}$ denotes the probability of the relation in braces.

Theorem 1. *Strong converse of the coding theorem for the general discrete finite-memory channel.* Let C be the capacity of a general discrete finite-memory channel. Let $\epsilon > 0$ and λ, $0 \leq \lambda < 1$, be arbitrary. For n sufficiently large any code (N, λ) must satisfy

$$N < 2^{n(C+\epsilon)}$$

Proof. In the proof of the theorem we assume that $\lambda > 0$; the theorem is then *a fortiori* true for $\lambda = 0$. Let l be an integer to be chosen later. We shall first prove the theorem for n of the form $k(l + m)$ with k sufficiently large, and then for all n sufficiently large.

Let $v = (y_1, \cdots, y_n)$ be any n-sequence in A_i, $i = 1, \cdots, N$. Adjoin to A_i all n-sequences which have the same elements as v in the places whose indices are congruent to $(m + 1), (m + 2), \cdots, (m + l)$, mod $(l + m)$. These sequences form a cylinder set $D(\bar{v})$ over the sequence $\bar{v}$ (say) which has kl elements; these kl elements are of course the same as the corresponding elements of v. Denote by A_i^* the set A_i after this adjunction has taken place. The A_i^* are no longer necessarily disjoint. However, *a fortiori*

$$P\{v(u_i) \in A_i^*\} \geq 1 - \lambda, \qquad i = 1, \cdots, N$$

It follows from conditions m.1 and m.2 of (Feinstein, 1959, p. 29) or (Feinstein, 1958, p. 90), that the sequences $u_1, \cdots, u_N$ and the received sequences of the form of $\bar{v}$ are connected through a discrete memoryless channel $K(l + m)$ (say) of capacity $C(l + m)$. Let λ' be any number such that $\lambda < \lambda' < 1$. Exactly as in Lemma 3.2 of Wolfowitz (1957) we can choose δ_2 so large that, after deleting from A_i^*, $i = 1, \cdots, N$, all $D(\bar{v})$ whose $\bar{v}$ are not generated by u_i with respect to the channel probability function of $K(l + m)$, we still have

$$P\{\bar{v}(u_i) \in \bar{A}_i^*\} \geq 1 - \lambda', \qquad i = 1, \cdots, N$$

* This research was supported by the U. S. Air Force under contract No. AF 18(600)-685 monitored by the Office of Scientific Research.

where $\bar{A}_i^*$ is $\bar{A}_i^*$ after the deletion, and $\bar{v}(u_i)$ is the chance received sequence when u_i is sent over the channel $K(l + m)$. The $\bar{A}_i^*$ are still not necessarily disjoint.

The letters of the input alphabet of the channel $K(l + m)$ are the 2^{l+m} sequences of $(l + m)$ elements zero or one. Divide up the given code (N, λ) into not necessarily disjoint subcodes, as follows:

In each subcode, all the u-sequences (n-sequences used for transmission) have the same proportions $\{\pi\}$ of the various $(l + m)$ − sequences. Since these proportions must be integral multiples of $1/k$, we have that the number of subcodes is at most

$$[k + 1]^{(2^{l+m})}$$

Obviously

$$N \leq [k + 1]^{(2^{l+m})} \cdot T < n^{(2^{l+m})} \cdot T \tag{1}$$

where T is the maximum length of a subcode.

Let $\{\pi\}$ be any set of proportions of $(l + m)$-sequences in the u-sequences of a subcode; we shall then say that the subcode belongs to $\{\pi\}$. Consider $\{\pi\}$ as a stochastic input for the channel $K(l + m)$, and for this channel denote by $H(Y \mid \pi)$ and $H(Y \mid X \mid \pi)$, respectively, the entropy of the output and the conditional entropy of the output, given the input. From Lemma 4.1 of Wolfowitz (1957) we have that, for any $\{\pi\}$ and k sufficiently large, each $\bar{A}_i^*$ of the subcode which belongs to $\{\pi\}$ contains at least

$$2^{k[H(Y|X|\pi)-(\epsilon/3)]} \tag{2}$$

sequences. The total number of all sequences generated with respect to the channel probability function of $K(l + m)$ by some n-sequence with proportions $\{\pi\}$ is, by Lemma 6.3 of Wolfowitz (1957), for k sufficiently large, less than

$$2^{k[H(Y|\pi)+(\epsilon/3)]} \tag{3}$$

Then (3) is an upper bound on the number of distinct sequences in the union of the $\bar{A}_i^*$ of the subcode which belongs to $\{\pi\}$. We now proceed to give a lower bound.

The $\bar{A}_i^*$ are not necessarily disjoint. However, each cylinder set $D(\bar{v})$ contains 2^{km} sequences. Hence no $\bar{v}$-sequence can belong to more than 2^{km} of the $\bar{A}_i^*$. Consequently, if $T(\pi)$ is the length of the subcode which belongs to $\{\pi\}$, a lower bound on the number of distinct sequences in the union of the $\bar{A}_i^*$ of the subcode which belongs to $\{\pi\}$ is

$$2^{-km} \cdot T(\pi) \cdot 2^{k[H(Y|X|\pi)-(\epsilon/3)]} \tag{4}$$

for all k sufficiently large (uniformly in $\{\pi\}$). From (3) and (4) we have that, for any $\{\pi\}$ and all sufficiently large k,

$$T(\pi) < 2^{km} \cdot 2^{k[H(Y|\pi)-H(Y|X|\pi)+(2\epsilon/3)]} \tag{5}$$

Hence

$$T < 2^{km} \cdot 2^{k[C(l+m)+(2\epsilon/3)]} \leq 2^{n(m/m+l)} \cdot 2^{n[C+(2\epsilon/3)]} \tag{6}$$

Finally from (1) we have that, for all n sufficiently large,

$$N < n^{(2^{l+m})} \cdot 2^{n(m/m+l)} \cdot 2^{n[C+(2\epsilon/3)]} \tag{7}$$

Now choose l so large that $m/(m + l) < \epsilon/12$. Then, for n sufficiently large,

$$N < 2^{n[C+(5\epsilon/6)]} \tag{8}$$

This proves the theorem for n of the form $k(l + m)$.

Suppose $n = k(l + m) + s$, with $1 \leq s < (l + m)$. Then, writing $n' = (k + 1)(l + m)$, we have, for large n,

$$N < 2^{n'[C+(5\epsilon/6)]} < 2^{n[1+(l+m/n)]\ [C+(5\epsilon/6)]} < 2^{n(C+\epsilon)} \tag{9}$$

which proves the theorem.

Instead of proceeding from (6) to (7) as we did let us examine (6) a little more carefully. Instead of the right member of (6) we can write the finer inequality

$$T < 2^{n(m/m+l)} \cdot 2^{n[C(l+m)/(l+m)+2\epsilon/3(m+l)]} \tag{10}$$

Reprinted with permission from *Inform. Contr.*, vol. 3, pp. 89–93, Mar. 1960.

Hence

$$N < n^{(2^{l+m})} \cdot 2^{n(m/m+l)} \cdot 2^{n[C(l+m)/(l+m)+(2\epsilon/3)]} \tag{11}$$

Now (11) holds for *any* l, all n sufficiently large, and any code (N, λ). The quantity ϵ was arbitrary and only the lower bound on n depends upon it. For any fixed l and sufficiently large n,

$$n^{(2^{l+m})} < 2^{n\epsilon/3}.$$

Thus, for any fixed l, any ϵ, and all n sufficiently large,

$$N < 2^{n(m/m+l)} \cdot 2^{n[C(l+m)/(l+m)+\epsilon]}. \tag{12}$$

On the other hand, it follows from either the coding theorem for the general discrete finite-memory channel (Feinstein, 1959) or the proof of the coding theorem of Wolfowitz (1957) and conditions m.1 and m.2 (loc. cit.) that, for n sufficiently large, there exists a code (N, λ) for the general discrete finite-memory channel such that

$$N > 2^{n(C-\epsilon)}. \tag{13}$$

Since ϵ is arbitrary (12) and (13) would contradict each other unless

$$C \leq \frac{m}{l+m} + \frac{C(l+m)}{l+m}$$

This extends the result, obtained by the author (1958) for the special channel of {2}, to the general discrete finite-memory channel. We recapitulate it as

THEOREM 2. The capacity C of the general discrete finite-memory channel satisfies, for every positive integer l,

$$\frac{C(l+m)}{(l+m)} \leq C \leq \frac{C(l+m)}{(l+m)} + \frac{m}{l+m}$$

where $C(l+m)$ is defined in (Wolfowitz, 1958) and (Feinstein, 1959) as the capacity of a discrete memoryless channel.

Finally we remark that, by the argument of the present note, the results, which are to appear later (Wolfowitz, 1960), on simultaneous channels and on channels where only the sender or only the receiver knows the channel probability function being employed, can all be extended to the corresponding general discrete finite-memory channels. The extension is quite routine and straightforward.

RECEIVED: August 24, 1959.

REFERENCES

FEINSTEIN, A. (1959). On the coding theorem and its converse for finite-memory channels. *Inform. and Control* **2**, 25–44.

FEINSTEIN, A. (1958). "Foundation; of Information Theory." McGraw-Hill, New York.

WOLFOWITZ, J. (1957). The coding of messages subject to chance errors. *Illinois J. Math.* **1**, 591–606.

WOLFOWITZ, J. (1958). "The maximum achievable length of an error correcting code," *Illinois J. Math.* **2**, 454–458.

WOLFOWITZ, J. (1960). Simultaneous channels. *Arch. Rational Mech. and Analysis*, to be published.

EXPONENTIAL ERROR BOUNDS FOR FINITE STATE CHANNELS

DAVID BLACKWELL
UNIVERSITY OF CALIFORNIA, BERKELEY

1. Introduction and summary

A *finite state channel* is defined by (1) a finite nonempty set A, the set of *inputs*, (2) a finite nonempty set B, the set of *outputs*, (3) a finite nonempty set T, the set of (channel) states, (4) a *transition law* $p = p(t'|t, a)$, specifying the probability that, if the channel is in state t and is given input a, the resulting state is t', and (5) a function ψ from T to B, specifying the output $b = \psi(t)$ of the channel when it is in state t.

For any sequence $\{a_n, n = 1, 2, \cdots\}$ of random variables with values in A, we may consider the process $\{a_n\}$ as supplying the inputs for the channel, as follows: an initial channel state t_0 is selected with a uniform distribution over T. The input a_1 is then given the channel. The channel then selects a state t_1, with

$$P\{t_1 = t|t_0, a_1\} = p(t|t_0, a_1) \tag{1}$$

and produces output $b_1 = \psi(t_1)$. The channel is then given input a_2 and selects state t_2, with

$$P\{t_2 = t|t_0, t_1, a_1, a_2, b_1\} = p(t|t_1, a_2), \tag{2}$$

and so on. In general, for $n \geqq 0$,

$$\begin{aligned} P\{a_{n+1} = a, t_{n+1} = t, b_{n+1} = b|a_i, 1 \leqq i \leqq n, t_i, 0 \leqq i \leqq n, b_i, 1 \leqq i \leqq n\} \\ = P\{a_{n+1} = a|a_i, i \leqq n\} p(t|t_n, a)\chi(t, b), \end{aligned} \tag{3}$$

where $\chi(t, b) = 1$ if $\psi(t) = b$ and 0 otherwise.

For any random variable x with a finite set of values and any random variable y, the (nonnegative) random variable whose value when $x = x_0$ and $y = y_0$ is

$$-\log P\{x = x_0|y = y_0\} \tag{4}$$

(all logs are base 2) is called the (conditional) entropy of x given y and will be denoted by $i(x|y)$. Its expected value, which cannot exceed the log of the number of values of x, will be denoted by $I(x|y)$. For y a constant, $i(x|y)$ and $I(x|y)$ will be denoted by $i(x)$, $I(x)$ respectively. If each of x, y has only finitely many values, the random variable

$$j(x, y) = i(x) + i(y) - i(x, y) = i(x) - i(x|y) = i(y) - i(y|x) \tag{5}$$

This paper was prepared with the partial support of the Office of Naval Research (Nonr-222-53).

Reprinted from *Proc. 4th Berkeley Symp. Math. Stat. and Prob.*, vol. 1, pp. 57-63, 1961, with permission of the Regents of the University of California. Originally published by the University of California Press.

is called the mutual information between x and y. Its expected value will be denoted by $J(x, y)$. For any stationary process $\{x_n, -\infty < n < \infty\}$ whose variables have only finitely many values, we write $I^*(x)$ for $I(x_0|x_{-1}, x_{-2}, \cdots)$.

An inequality of Shannon [8] and Feinstein [3] relates the existence of codes to the distribution of $j\{(a_1, \cdots, a_N), (b_1, \cdots, b_N)\} = j_N$, as follows.

Shannon-Feinstein inequality. For any integer D, and any number γ there are two functions f, g, where f maps $(1, \cdots, D)$ into the set U of sequences of length N of elements of A, g maps the set V of sequences of length N of elements of B into $(1, \cdots, D)$, for which

$$P\{g(b_1, \cdots, b_N) \neq d|(a_1, \cdots, a_N) = f(d)\} \leqq P\{j_N < \gamma\} + \frac{D}{2\gamma} \tag{6}$$

for $d = 1, \cdots, D$.

Note that the left side of (6) is independent of the distribution of $\{a_n\}$; it is simply the probability that, when an initial state for the channel is selected with a uniform distribution and the channel is then given the input sequence $f(d)$, the resulting output sequence $b_1, \cdots, b_N$ will be one for which $g(b_1, \cdots, b_N) \neq d$. The pair (f, g) can be considered as a code, with which we can transmit any of D messages over our channel in N transmission periods; when message d is presented to the sender, he gives the channel input sequence $f(d)$; the receiver then observes some output sequence v and decides that message $g(v)$ is intended.

For a given number $c \geqq 0$, let $D = [2^{Nc}]$ and write

$$\theta_N(c) = \min_{f,g} \max_{1 \leqq d \leqq D} P\{g(b_1, \cdots, b_N) \neq d|(a_1, \cdots, a_N) = f(d)\}. \tag{7}$$

Thus $\theta_N(c)$ is small if and only if we can transmit any binary sequence of length Nc, by using the channel for N periods, with small error probability.

Shannon [7] associated with each channel a number C, called the capacity of the channel, and proved that, for certain channels, $\theta_N(c) \rightarrow 0$ as $N \rightarrow \infty$ for every $c < C$ but not for any $c > C$. His original work has been considerably simplified and extended by several writers, including Shannon himself [8], McMillan [6], Feinstein [2], [3], Khinchin [5], and Wolfowitz [9], [10]. In particular, for certain channels, Wolfowitz has shown that $\theta_N(c) \rightarrow 1$ as $N \rightarrow \infty$ for every $c > C$.

For a certain class of finite state channels, the indecomposable channels defined below, the fact that $\theta_N(c) \rightarrow 0$ as $N \rightarrow \infty$ for $c < C$ was first proved by Breiman, Thomasian, and the writer [1]. We present in this paper a simpler proof of the slightly stronger fact that for these channels $\theta_N(c) \rightarrow 0$ exponentially: for any $c < C$ there are constants $\alpha > 0$, $\beta < 1$ for which, for all N,

$$\theta_N(c) < \alpha\beta^N. \tag{8}$$

The Shannon-Feinstein inequality reduces (8) at once to the study of the distribution of j_N for large N, as follows: if for a given c we can find an input sequence $\{a_n\}$ for which, for some $\alpha_1 > 0$, $\beta_1 < 1$,

$$P\{j_N \leqq Nc\} \leqq \alpha_1\beta_1^N \tag{9}$$

for all N, the Shannon-Feinstein inequality yields, for every $\epsilon > 0$ and all N,

$$\theta_N(c - \epsilon) \leqq \alpha_1\beta_1^N + 2^{-N\epsilon} \leqq \alpha_2\beta_2^N, \tag{10}$$

where $\alpha_2 = \alpha_1 + 1$, $\beta_2 = \max(\beta_1, 2^{-\epsilon})$. Thus our result (8) is implied by: for every $c < C$, there is an input sequence $\{a_N\}$ for which, for some $\alpha_1 > 0$, $\beta_1 < 1$, (9) holds.

We now define indecomposable channels and the number C. Let

$$\{x_n, n = 1, 2, \cdots\}$$

be any Markov process with a finite number R of states $r = 1, 2, \cdots, R$ and indecomposable transition matrix $\pi = \pi(r'|r) = P\{x_{n+1} = r'|x_n = r\}$. Let ϕ be any function from $(1, \cdots, R)$ to A, and let $a_n = \phi(x_n)$. We consider the source process $\{a_n\}$ as driving the channel, as described above. The process $\{z_n = (x_n, t_n)\}$ is then a Markov process, with transition matrix

$$m = m(r', t'|r, t) = \pi(r'|r)p[t'|t, \phi(r')].$$

If for every indecomposable π and every ϕ, the matrix m is also indecomposable, the finite state channel (A, B, T, p, ψ) is called *indecomposable*. There is then, for each m, a unique stationary Markov process $\{z_n^* = (x_n^*, t_n^*), -\infty < n < \infty\}$ with transition matrix m. Define $a_n^* = \phi(x_n^*)$, $b_n^* = \psi(t_n^*)$, $-\infty < n < \infty$, and let $J^*(\pi, \phi) = I^*(a) + I^*(b) - I^*(a, b)$. The number

$$C = \sup_{\pi,\phi} J^*(\pi, \phi), \tag{11}$$

where the sup is over all indecomposable π and all ϕ, is called the capacity of the channel. The main result of this paper is

THEOREM 1. *Let (A, B, T, p, ψ) be an indecomposable channel of capacity C. For every $c < C$ there is an input sequence $\{a_n, n = 1, 2, \cdots\}$ and there are numbers $\alpha > 0$, $\beta < 1$ for which, for all N,*

$$P\{j[(a_1, \cdots, a_N), (b_1, \cdots, b_N)] \leqq Nc\} \leqq \alpha\beta^N. \tag{12}$$

2. Preliminary reduction of theorem 1

To prove theorem 1, we choose π, ϕ for which $J^* = J^*(\pi, \phi) > c$. Let $z_n = (x_n, t_n)$, with $n = 0, 1, 2, \cdots$ be a Markov process with the transition matrix m and some initial distribution for which the initial distribution of t_0 is uniform. Let $a_n = \phi(x_n)$, $b_n = \psi(t_n)$, with $n = 1, 2, \cdots$. We shall show that the input sequence $\{a_n\}$ has the property specified by theorem 1. Let us write $u_N = (a_1, \cdots, a_N)$, $v_N = (b_1, \cdots, b_N)$. Since $j(u_N, v_N) = i(u_N) + i(v_N) - i(u_N, v_N)$ and $J^* = I^*(a) + I^*(b) - I^*(a, b)$, theorem 1 would be proved if we could bound the probability of each of the events

$$\begin{aligned} &\{i(u_N) \leqq N(I^*(a) - \delta)\}, \\ &\{i(v_N) \leqq N[I^*(b) - \delta]\}, \\ &\{i(u_N, v_N) \geqq N[I^*(a, b) + \delta]\} \end{aligned} \tag{13}$$

above by $\alpha\beta^N$ for some $\alpha > 0$, $\beta < 1$, where $J^* - c = 3\delta$. That we can do this is the assertion of

THEOREM 2. *There are functions* $\alpha = \alpha(R, w, \epsilon)$, $\beta = \beta(R, w, \epsilon)$, *defined for* $R = 2, 3, \cdots, w > 0, \epsilon > 0$, *continuous in* w, ϵ, *increasing in* R *and decreasing in* w, ϵ *with* $a > 0$ *and* $0 < \beta < 1$ *such that, for any Markov process*

$$\{z_n, n = 1, 2, \cdots\}$$

with R *states* $r = 1, 2, \cdots, R$, *indecomposable transition matrix* $\pi = \pi(r'|r) = P\{z_{n+1} = r'|z_n = r\}$ *with smallest positive element* $\geqq w$ (π *may have some elements* 0) *and any function* ϕ *from* $1, \cdots, R$ *into a finite set* A,

$$P\{|i(a_1, \cdots, a_N) - NI^*(a)| \geqq N\epsilon\} \leqq \alpha(R, w, \epsilon)\beta^N(R, w, \epsilon) \tag{14}$$

for all N, *where* $a_n = \phi(z_n)$, *and* $I^*(a)$ *is as defined in section* 1, *namely if* $\{z_n^*, -\infty < n < \infty\}$ *is a stationary Markov process with transition matrix* π *and* $a_n^* = \phi(z_n^*)$, *then* $I^*(a) = I(a_0^*|a_{-1}^*, a_{-2}^*, \cdots)$.

Theorem 2 is a form of the equipartition theorem (Shannon [7], McMillan [6]) for "finitary" processes, with an exponential bound on the probability of exceptional sequences.

3. Proof of theorem 2 for ϕ the identity

For ϕ the identity function, so that $a_n(= z_n)$ is itself a Markov process, we have

$$i_N = i(z_1, \cdots, z_N) = -\log \lambda(z_1) - \sum_{n=1}^{N-1} \log \pi(z_{n+1}|z_n). \tag{15}$$

We use the following inequality of Katz and Thomasian [4].

Katz-Thomasian inequality. For $\{z_n\}$, w *as in theorem* 2 *and* ϕ *real-valued,* $P|\{|\phi(z_1) + \cdots + \phi(z_N) - N\mu| \geqq N\epsilon\} \leqq \alpha_1\beta_1^N$ *where*

$$\begin{aligned} \beta_1 &= \beta_1(R, w, \epsilon, M) = \exp - \left(\frac{w^{3R}\epsilon^2}{2^8M^2r^2}\right), \\ \alpha_1 &= \alpha_1(R, w, \epsilon, M) = \frac{8R}{w^R}\frac{1}{1 - \beta_1}, \\ M &= \max_{r'} \phi(r') - \min_{r} \phi(r), \end{aligned} \tag{16}$$

and $\mu = \sum \lambda(r)\phi(r)$, *where* λ *is the stationary distribution for* π.

We apply the Katz-Thomasian inequality to $z_n' = (z_n, z_{n+1})$, with $\phi' = -\log \pi(r'|r)$, so that $\mu = -\sum_{r,r'} \lambda(r)\pi(r'|r) \log \pi(r'|r) = I^*(z)$, and $M \leqq -\log w$ [we may exclude from z_n' the pairs r, r' with $\pi(r'|r) = 0$], obtaining

$$\begin{aligned} P\left\{\left|\sum_{n=1}^{N-1} \log \pi(z_{n+1}|z_n) + (N-1)I^*(z)\right| \geqq (N-1)\epsilon\right\} \\ \leqq \alpha_1(R^2, w, \epsilon, -\log w)\beta_1^{N-1}(R^2, w, \epsilon, -\log w) \\ = \alpha_2(R, w, \epsilon)\beta_2^N(R, w, \epsilon), \end{aligned} \tag{17}$$

say. Thus

$$P\{|i_N - NI^*(z) + \log \lambda(z_1) + I^*(z)| \geqq N\epsilon\} \leqq \alpha_2\beta_2^N. \tag{18}$$

Since $0 \leqq I^*(z) \leqq \log R$ and, for every $\delta > 0$,

$$P\{|\log \lambda(z_1)| \geqq N\delta\} = \sum_{r:\lambda(r) \leqq 2^{-N\delta}} \lambda(r) \leqq R2^{-N\delta} \tag{19}$$

we easily obtain $\alpha_3(R, w, \epsilon)$, $\beta_3(R, w, \epsilon)$ for which

$$P\{|i_N - NI^*(z)| \geqq N\epsilon\} \leqq \alpha_3\beta_3^N. \tag{20}$$

4. Proof of theorem 2, general case

We prove the general case by approximating the process $\{a_n\}$, in blocks, by a suitable Markov process, and using the fact that we have already proved the theorem for Markov processes. The idea is this: if, in addition to observing the $a_n = \phi(z_n)$ process we observe periodically, say every k trials, the current state of the underlying z_n process, the process now observed, with observations grouped in blocks of k, is a Markov process, so that all long sequences, except a set of exponentially small probability, have about the correct probability. We can choose k so large that (1) this correct probability is nearly the correct probability for the corresponding a sequence and (2) except with exponentially small probability, the probability of the actual a sequence will be near the probability of the actual observed sequence.

Thus choose a positive integer k, and let $x_1 = (a_1, \cdots, a_{k-1}, z_k)$, $x_2 = (a_{k+1}, \cdots, a_{2k-1}, z_{2k})$, $\cdots$, $x_n = (a_{(n-1)k+1}, \cdots, a_{nk-1}, z_{nk})$, $\cdots$. The $\{x_n\}$ process is Markov and, for k relatively prime to the period of $\{z_n\}$, is indecomposable. It has at most R^k states, and the smallest positive element in its transition matrix is at least w^k.

Thus, from the preceding section,

$$P\{|i(x_1, \cdots, x_N) - NI_k| \geqq \epsilon N\} \leqq \alpha_4\beta_4^N, \tag{21}$$

where

$$\begin{aligned} \alpha_4 &= \alpha_4(R, w, \epsilon, k) = \alpha_3(R^k, w^k, \epsilon), \\ \beta_4 &= \beta_4(R, w, \epsilon, k) = \beta_3(R^k, w^k, \epsilon), \end{aligned} \tag{22}$$

and $I_k = I^*(x)$. Now $kI^*(a) \leqq I_k \leqq kI^*(a) + \log R$ and $i(x_1, \cdots, x_N) = i(a_1, \cdots, a_{Nk}) + i(z_k, \cdots, z_{Nk}|a_1, \cdots, a_{Nk})$, which we write $i_N(x) = i_{Nk}(a) + i_N(z|a)$. Then

$$\begin{aligned} P\{i_{Nk}(a) \geqq Nk[I^*(a) + \epsilon]\} &\leqq P\left\{i_N(x) \geqq Nk\left(\frac{I_k - \log R}{k} + \epsilon\right)\right\} \\ &= P\{i_N(x) \geqq N[I_k + (k\epsilon - \log R)]\} \leqq \alpha_4\beta_4^N, \end{aligned} \tag{23}$$

provided $k\epsilon - \log R \geqq \epsilon$, that is, $k \geqq 1 + (1/\epsilon) \log R$. Similarly,

$$(24)\qquad P\{i_{Nk}(a) \leqq Nk[I^*(a) - 4\epsilon]\} \leqq P\left\{i_N(x) - i_N(z|a) \leqq Nk\left(\frac{I_k}{k} - 4\epsilon\right)\right\}$$
$$\leqq P\{i_N(x) \leqq N(I_k - k\epsilon)\} + P\{i_N(z|a) \geqq 3Nk\epsilon\} = P_1 + P_2.$$

As above, $P_1 \leqq \alpha_4\beta_4^N$. To bound P_2, write $i_N(z) = i(z_k, z_{2k}, \cdots, z_{Nk})$. Then

$$(25)\qquad P_2 \leqq P\{i_N(z) \geqq Nk\epsilon\} + P\{i_N(z|a) \geqq i_N(z) + Nk\epsilon\} = P_3 + P_4.$$

The process $\{z_{nk}, n = 1, 2, \cdots, k \text{ fixed}\}$ is a Markov process with R states and (k is relatively prime to the period of $\{z_n\}$) indecomposable transition matrix. For k so large that $k\epsilon \geqq \log R + \epsilon$, that is, $k \geqq 1 + (1/\epsilon) \log R$, we have $P_3 \leqq \alpha_3\beta_3^N$.

For P_4 we use

LEMMA 1. *For any two random variables a, z each with a finite set of values, and any $\delta \geqq 0$,*

$$(26)\qquad P\{i(z|a) \geqq i(z) + \delta\} \leqq 2^{-\delta}.$$

PROOF. A pair (z_0, a_0) of values of z, a for which $i(z_0|a_0) \geqq (z_0) + \delta$ is one for which

$$(27)\qquad \frac{P\{z = z_0|a = a_0\}}{P\{z = z_0\}} \leqq 2^{-\delta},$$

that is, $P\{z = z_0, a = a_0\} \leqq 2^{-\delta}P\{z = z_0\}P\{a = a_0\}$. Summing over all pairs (z_0, a_0) for which the inequality is satisfied yields the lemma.

From the lemma, we obtain $P_4 \leqq 2^{-Nk\epsilon}$. Thus

$$(28)\qquad P\{i_{Nk}(a) \leqq NkI^*(a) - 4\epsilon\} \leqq \alpha_5\beta_5^N,$$

where $\alpha_5 = \alpha_5(R, w, \epsilon, k) = \alpha_4 + \alpha_3 + 1$ and $\beta_5 = \max(\beta_4, \beta_3, 2^{-k\epsilon})$.

Combining (23) and (28) we obtain $\alpha_6(R, w, \epsilon, k)$, $\beta_6(R, w, \epsilon, k)$ for which

$$(29)\qquad P\{|i_{Nk} - NkI^*(a)| \geqq Nk\epsilon\} \leqq \alpha_6\beta_6^N.$$

The block size k is still at our disposal, subject to $k \geqq 1 + (1/\epsilon) \log R$ and relatively prime to the period of $\{z_n\}$. We can find such a

$$k \leqq k^* = [R + 1 + (1/\epsilon) \log R]$$

and obtain, for this k,

$$(30)\qquad P\{|i_{Nk} - NkI^*(a)| \geqq Nk\epsilon\} \leqq \alpha_7\beta_7^N,$$

where

$$(31)\qquad \begin{aligned} \alpha_7 &= \alpha_7(R, w, \epsilon) = \alpha_6(R, w, \epsilon, k^*), \\ \beta_7 &= \beta_7(R, w, \epsilon) = \beta_6(R, w, \epsilon, k^*). \end{aligned}$$

Finally, for any n, say, $n = Nk + d$, with $0 \leqq d \leqq k - 1$, we have

$$(32)\qquad i_n - n[I^*(a) + \epsilon] \leqq i_{(N+1)k} - (N + 1)k\left\{I^*(a) + \epsilon - \frac{I^*(a) + \epsilon}{N + 1}\right\}$$
$$\leqq i_{(N+1)k} - (N + 1)k\left\{I^*(a) + \frac{\epsilon}{2}\right\}$$

for

$$\frac{I^*(a) + \epsilon}{N + 1} \leqq \frac{\epsilon}{2}, \tag{33}$$

which, since $I^*(a) \leqq \log R$, will certainly hold for

$$N \geqq 2\left(\frac{\log R}{\epsilon} + 1\right) = N_0, \tag{34}$$

say, and similarly $i_n - n\{I^*(a) - \epsilon\} \geqq i_{Nk} - Nk\{I^*(a) - \epsilon/2\}$ for

$$N \geqq 2\left(\frac{\log R}{\epsilon} - 1\right). \tag{35}$$

Thus

$$P\{|i_n - nI^*(a)| \geqq n\epsilon\} \leqq \alpha_8\beta_8^{N-N_0}, \tag{36}$$

where

$$\begin{aligned} \alpha_8 &= \alpha_8(R, w, \epsilon) = 2\alpha_7\left(R, w, \frac{\epsilon}{2}\right) \\ \beta_8 &= \beta_8(R, w, \epsilon) = \beta_7\left(R, w, \frac{\epsilon}{2}\right). \end{aligned} \tag{37}$$

Finally, with

$$\alpha_9 = \alpha_8\beta_8^{-N_0}, \qquad \beta_9 = \beta_8^{1/k^*}, \tag{38}$$

we obtain

$$P\{|i_n - nI^*(a)| \geqq n\epsilon\} \leqq \alpha_9\beta_9^n$$

for all n, completing the proof.

REFERENCES

[1] D. BLACKWELL, L. BREIMAN, and A. J. THOMASIAN, "Proof of Shannon's transmission theorem for finite state indecomposable channels," *Ann. Math. Statist.*, Vol. 29 (1958), pp. 1209–1220.

[2] A. FEINSTEIN, "A new basic theorem of information theory," *IRE Transactions P.G.I.T.*, 1954, pp. 2–22.

[3] ———, *Foundations of Information Theory*, New York, McGraw-Hill, 1958.

[4] M. KATZ, JR., and A. J. THOMASIAN, "An exponential bound for functions of a Markov chain," *Ann. Math. Statist.*, Vol. 31 (1960), pp. 470–474.

[5] A. I. KHINCHIN, *Mathematical Foundations of Information Theory*, New York, Dover, 1957.

[6] B. MCMILLAN, "The basic theorems of information theory," *Ann. Math. Statist.*, Vol. 24 (1953), pp. 196–219.

[7] C. E. SHANNON, "Mathematical theory of communication," *Bell System Tech. J.*, Vol. 27 (1948), pp. 379–423.

[8] ———, "Certain results in coding theory for noisy channels," *Information and Control*, Vol. 1 (1957), pp. 6–25.

[9] J. WOLFOWITZ, "The coding of messages subject to chance errors," *Illinois J. Math.*, Vol. 1 (1957), pp. 591–606.

[10] ———, "Strong converse of the coding theorem for semicontinuous channels," *Illinois J. Math.*, Vol. 3 (1959), pp. 477–489.

A Simple Proof of an Inequality of McMillan*

Let l_i, $i = 1, \cdots, b$, be the length of the ith word of a list of b words, each word being a string of letters from an alphabet of a letters. Assume that distinct strings of words from the list, when written out without additional space marks to separate the words, determine distinct strings of letters, so that the total number of strings of words of letter-length k is $\leqq a^k$. Let $l = \max_i l_i$. Then for all integers $n > 0$,

$$\left(\sum_{i=1}^{b} \frac{1}{a^{l_i}}\right)^n = \sum_{i_1, i_2, \cdots, i_n = 1}^{b} \frac{1}{a^{l_{i_1}} a^{l_{i_2}} \cdots a^{l_{i_n}}} = \sum_{j=1}^{b^n} \frac{1}{a^{L_j}},$$

where j runs over the b^n strings of n words and L_j is the number of letters in the jth string. Since $\max_j L_j = nl$ and $\min_j L_j \geqq n$, we have, grouping j's with $L_j = k$,

$$\left(\sum_{i=1}^{b} \frac{1}{a^{l_i}}\right)^n = \sum_{k=n}^{nl} \frac{\text{number of strings of } n \text{ words having } k \text{ letters}}{a^k}$$

$$\leqq \sum_{k=n}^{nl} \frac{a^k}{a^k} \leqq nl.$$

Since $x > 1$ implies $x^n > nl$ for sufficiently large n, it follows that $\sum_{i=1}^{b} a^{-l_i} \leqq 1$, which is McMillan's[1] result.

Jack Karush
Dept. of Statistics
University of California
Berkeley, Calif.

* Received by the PGIT, October 17, 1960. This note was prepared with the partial support of the Office of Ordnance Research, U. S. Army, under Contract DA-04-200-ORD-171, Task Order 3.

[1] B. McMillan, "Two inequalities implied by unique decipherability," IRE Trans. on Information Theory, vol. IT-2, pp. 115–116; December, 1956.

Reprinted from *IEEE Trans. Inform. Theory*, vol. IT-7, p. 118, Apr. 1961.

A Note on "Very Noisy" Channels

BARNEY REIFFEN

Lincoln Laboratory, Massachusetts Institute of Technology*
Lexington, Massachusetts

A "very noisy" channel is defined. This definition corresponds to many physical channels operating at low signal-to-noise ratio. For "very noisy" discrete input memoryless channels, the computation cutoff rate for sequential decoding, R_{comp}, is shown to be one-half the capacity, C. Furthermore, that choice of input probabilities which achieves C also maximizes R_{comp}, and vice versa.

I. INTRODUCTION

A memoryless channel is defined by:

(a) an input alphabet X,
(b) an output alphabet Y,
(c) a transition probability density $p(y/x)$.

For any probability density $p(x)$ defined on the input space, the density on the product space XY is well defined, and the random variable mutual information $I(x; y)$ may be defined:

$$I(x;y) = \ln \frac{p(y \mid x)}{p(y)} \tag{1}$$

where

$$p(y) = \int p(x)p(y \mid x)\, dx \tag{2}$$

The average of the mutual information, denoted by $I(X; Y)$, can be written

$$I(X;Y) = H(Y) - H(Y|X) \tag{3}$$

where

$$H(Y) = -\int p(y) \ln p(y)\, dy \tag{4}$$

$$H(Y \mid X) = -\iint p(x)p(y \mid x) \ln p(y \mid x)\, dx\, dy \tag{5}$$

In this formulation, the Y space is assumed to be continuous. The X space may be continuous or discrete. In the latter case $p(x)$ contains delta functions. The results of this note are trivially extended to discrete input-discrete output channels.

In order to obtain the desired results, we define a channel as "very noisy" if for all $x \in X$

$$\frac{p(y) - p(y \mid x)}{p(y)} = \epsilon_x(y) \ll 1. \tag{6}$$

This description corresponds to many physical channels operating at low signal-to-noise ratio.

Observe that

$$\int p(x)\epsilon_x(y)\, dx = 0 \tag{7}$$

and

$$\int p(y)\epsilon_x(y)\, dy = 0 \tag{8}$$

II. EVALUATION OF C FOR "VERY NOISY" CHANNELS

We first evaluate $H(Y|X)$.

$$H(Y \mid X) = -\int p(x) \left[\int p(y \mid x) \ln p(y \mid x)\, dy\right] dx$$
$$= -\int p(x) \left[\int dy p(y)[1 - \epsilon_x(y)] \times \ln \{p(y)[1 - \epsilon_x(y)]\}\right] dx \tag{9}$$

Using the inequality

$$\ln(1 + \alpha) \geqq \alpha - \tfrac{1}{2}\alpha^2, \quad \alpha^2 < 1 \tag{10}$$

the bracketed expression of Eq. (9) reduces to

$$[\ \ ] \geqq \int p(y) \ln p(y)\, dy - \int p(y)\epsilon_x(y) \ln p(y)\, dy - \int p(y)\epsilon_x(y)\, dy + \frac{1}{2}\int p(y)\epsilon_x^2(y)\, dy + \frac{1}{2}\int p(y)\epsilon_x^3(y)\, dy \tag{11}$$

In Eq. (11), the first integral is $-H(Y)$. The third integral is zero (Eq (8)). The fifth integral will be neglected, since it is a third order term in $\epsilon_x(y)$ which is assumed very much less than unity. Combining these observations, we have, to within terms of second order in $\epsilon_x(y)$,

$$H(Y \mid X) = H(Y) + \iint p(x)p(y)\epsilon_x(y) \ln p(y)\, dx\, dy - \frac{1}{2}\iint p(x)p(y)\epsilon_x^2(y)\, dx\, dy \tag{12}$$

The first integral in Eq. (12) is zero (Eq. (7)). Thus, we have from Eqs. (3) and (12)

$$I(X;Y) = H(Y) - H(Y \mid X) = \frac{1}{2}\iint p(x)p(y)\epsilon_x^2(y)\, dx\, dy \tag{13}$$

Equation (13) is correct to within terms of second order of $\epsilon_x|(y)$.

The channel capacity C is defined as

$$C = \max_{p(x)} I(X;Y) \tag{14}$$

For that $p(x)$ which yields capacity,

$$C = \frac{1}{2}\iint p(x)p(y)\epsilon_x^2(y)\, dx\, dx \tag{15}$$

where Eq. (15) is correct to within terms of second order of $\epsilon_x(y)$.

III. EVALUATION OF R_{comp} FOR "VERY NOISY" CHANNELS

The quantity

$$E_0 = -\ln \int \left[\int p(x)[p(y \mid x)]^{1/2}\, dx\right]^2 dy \tag{16}$$

has considerable physical significance. E_0 is the exponent in the upper bound to error probability corresponding to zero rate. This has been shown by Fano (1961) for the case of discrete channels, but the generalization to discrete input-continuous output channels is straightforward.

For the sequential decoding procedure (Wozencraft and Reiffen, 1961; Reiffen, 1962), a computation cutoff rate, R_{comp}, is defined. For rates less than R_{comp}, the average number of decoding computations does not grow exponentially with constraint length, but algebraically. Reiffen (1962) has shown that for discrete input channels,

$$R_{comp} \leqq E_0 \tag{17}$$

It is conjectured in the reference that Eq. (17) is an equality rather than an inequality. This conjecture has been verified by Fano in unpublished work. Thus, restricting our attention to discrete input channels, we may write

$$R_{comp} = -\ln \int \left[\int p(x)[p(y \mid x)]^{1/2}\, dx\right]^2 dy \tag{18}$$

where it is understood that the density $p(x)$ contains delta functions.

Starting with Eq. (18) we obtain

$$\int \left[\int p(x)[p(y \mid x)]^{1/2}\, dx\right]^2 dy$$

* Operated with support from the United States Army, Navy, and Air Force.

Reprinted with permission from *Inform. Contr.*, vol. 6, pp. 126–130, June 1963.

$$= \int \left[\int p(x)\{p(y)[1 - \epsilon_x(y)]\}^{1/2} \, dx \right]^2 dy \qquad (19)$$

$$\geqq \int p(y) \left[\int p(x)[1 - \tfrac{1}{2}\epsilon_x(y) - \tfrac{1}{8}\epsilon_x^2(y)] \, dx \right]^2 dy$$

where we have used the inequality

$$(1 - \alpha)^{1/2} \geqq 1 - \tfrac{1}{2}\alpha - \tfrac{1}{8}\alpha^2, \qquad \alpha^2 < 1 \qquad (20)$$

The bracketed expression on the right-hand side of Eq. (19) then becomes

$$[\ ] = \int p(x)\, dx - \frac{1}{2} \int p(x)\epsilon_x(y)\, dx - \frac{1}{8} \int p(x)\epsilon_x^2(y)\, dx \qquad (21)$$

The second integral in Eq. (21) is zero (Eq. (7)). Thus,

$$[\ ]^2 = \left[1 - \frac{1}{8} \int p(x)\epsilon_x^2(y)\, dx \right]^2 \geqq 1 - \frac{1}{4} \int p(x)\epsilon_x^2(y)\, dx \qquad (22)$$

Combining Eqs. (18), (19), and (22), we obtain

$$e^{-R_{comp}} \geqq 1 - \frac{1}{4} \iint p(x)p(y)\epsilon_x^2(y)\, dx\, dy \qquad (23)$$

From Eq. (23) we conclude that, within second order terms of $\epsilon_x(y)$,

$$R_{comp} = \frac{1}{4} \iint p(x)p(y)\epsilon_x^2(y)\, dx\, dy \qquad (24)$$

Comparing Eqs. (13) and (24), we see that for any $p(x)$,

$$R_{comp} = \tfrac{1}{2} I(X;Y) \qquad (25)$$

Thus, that choice of $p(x)$ which maximizes $I(X;Y)$ and thereby achieves capacity also maximizes R_{comp}, and vice versa.

IV. CONCLUSIONS

For "very noisy" discrete input memoryless channels

(a) $R_{comp} = \tfrac{1}{2}$ C.

(b) That choice of input probabilities which achieves C also maximizes R_{comp}, and vice versa.

(c) The capacity may be evaluated indirectly by calculating R_{comp} and doubling it. This might be easier than a direct evaluation of C since $\exp(-R_{comp})$ is an algebraic function of $p(x)$, while $I(X; Y)$ is a transcendental function of $p(x)$.

RECEIVED: December 26, 1962

REFERENCES

FANO, R. M. (1961), "Transmission of Information," p. 333. M.I.T. Press and Wiley, New York.

WOZENCRAFT, J. M., AND REIFFEN, B. (1961), "Sequential Decoding." M.I.T. Press and Wiley, New York.

REIFFEN, B. (1962), Sequential decoding for discrete input memoryless channels. *Trans. IRE, PGIT* **IT-8**, 208–220.

A Simple Derivation of the Coding Theorem and Some Applications

ROBERT G. GALLAGER, MEMBER, IEEE

Abstract—**Upper bounds are derived on the probability of error that can be achieved by using block codes on general time-discrete memoryless channels. Both amplitude-discrete and amplitude-continuous channels are treated, both with and without input constraints. The major advantages of the present approach are the simplicity of the derivations and the relative simplicity of the results; on the other hand, the exponential behavior of the bounds with block length is the best known for all transmission rates between 0 and capacity. The results are applied to a number of special channels, including the binary symmetric channel and the additive Gaussian noise channel.**

I. Introduction

THE CODING THEOREM, discovered by Shannon [1] in 1948, states that for a broad class of communication channel models there is a maximum rate, capacity, at which information can be transmitted over the channel, and that for rates below capacity, information can be transmitted with arbitrarily low probability of error.

For discrete memoryless channels, the strongest known form of the theorem was stated by Fano [2] in 1961. In this result, the minimum probability of error P_e for codes of block length N is bounded for any rate below capacity[1] between the limits

$$e^{-N[E_L(R)+0(N)]} \leq P_e \leq 2e^{-NE(R)} \tag{1}$$

In this expression, $E_L(R)$ and $E(R)$ are positive functions of the channel transition probabilities and of the rate R; $0(N)$ is a function going to 0 with increasing N. For a range of rates immediately beneath channel capacity, $E_L(R) = E(R)$.

The function $E(R)$, especially in the range in which $E(R) = E_L(R)$, appears to yield a fundamental characterization of a channel for coding purposes. It brings out clearly and simply the relationships between error probability, data rate, constraint length, and channel behavior. Recent advances in coding theory have yielded a number of effective and economically feasible coding techniques, and (1) provides a theoretical framework within which to discuss intelligently the relative merits of these techniques. Even more important, the function $E(R)$ provides a more meaningful comparison between different channels than can be made on the basis of capacity or SNR. For example, if one is to use coding on a physical communication link, one of the first questions to be answered involves the type of digital modulation systems to use. Considering the modulation system as part of the channel, one can compare modulation systems for coding applications on the basis of their $E(R)$ curves. For an example of such a comparison, see Wozencraft and Kennedy [4].

In Section II of this paper, a simple proof is given that $P_e < e^{-NE(R)}$. In Section III, we establish a number of properties of $E(R)$ and show explicitly how the function $E(R)$ can be calculated. This calculation is just slightly more complicated than the calculation of channel capacity. In Section IV, we give a number of applications of the theory developed in Sections II and III. First, as an example, we derive $E(R)$ for a binary symmetric channel; then we derive a universal $E(R)$ curve for very noisy channels; and finally, we relate $E(R)$ for a set of parallel channels to the $E(R)$ curves of the individual channels.

In Section V, we derive an improved upper bound to P_e for low rates; this yields a larger value of $E(R)$ than was derived in Section II. There is some reason to suspect that the combination of these two bounds produces the true exponential behavior with block length of the best codes. In Section VI, these results are extended to channels with constraints on the input and to channels with continuous input and output alphabets. Finally, the results are applied to the additive Gaussian noise channel as an example.

II. Derivation of the Coding Theorem

Let X_N be the set of all sequences of length N that can be transmitted on a given channel, and let Y_N be the set of all sequences of length N that can be received. We assume that both X_N and Y_N are finite sets. Let $Pr\ (\mathbf{y} \mid \mathbf{x})$, for $\mathbf{y} \in Y_N$ and $\mathbf{x} \in X_N$, be the conditional probability of receiving sequence $\mathbf{y}$, given that $\mathbf{x}$ was transmitted. We assume that we have a code consisting of M code words; that is, a mapping of the integers from 1 to M into a set of code words $\mathbf{x}_1, \cdots, \mathbf{x}_M$, where $\mathbf{x}_m \in X_N$; $1 \leq m \leq M$. We assume that maximum likelihood decoding is performed at the receiver; that is, the decoder decodes the output sequence $\mathbf{y}$ into the integer m if

$$Pr(\mathbf{y}|\mathbf{x}_m) > Pr(\mathbf{y}_m|\mathbf{x}) \text{ for all } m' \neq m,\ 1 \leq m' \leq M \tag{2}$$

Manuscript received March 11, 1964. This work was supported in part by the U. S. Army, Navy, and Air Force under Contract DA36-039-AMC-03200(E); and in part by the National Science Foundation (Grant GP-2495), the National Institutes of Health (Grant MH-04737-04), and the National Aeronautics and Space Administration (Grants NsG-334 and NsG-496).

The author is with the Dept. of Electrical Engineering and the Research Lab. of Electronics, Massachusetts Institute of Technology, Cambridge, Mass.

[1] This paper deals only with error probabilities at rates below capacity. For the strongest known results at rates above capacity, see Gallager [3], Section 6.

Reprinted from *IEEE Trans. Inform. Theory*, vol. IT-11, pp. 3-18, Jan. 1965.

For purposes of overbounding the probability of decoding error, we regard any situation in which no m satisfies (2) as a decoding error. Also, of course, a decoding error is made if the decoded integer is different from the input integer. Now let P_{em} be the probability of decoding error when $\mathbf{x}_m$ is transmitted. A decoding error will be made if a $\mathbf{y}$ is received such that (2) is not satisfied. Thus we can express P_{em} as

$$P_{em} = \sum_{\mathbf{y} \epsilon Y_N} Pr(\mathbf{y}|\mathbf{x}_m)\phi_m(\mathbf{y}) \tag{3}$$

where we define the function $\phi_m(\mathbf{y})$ as

$$\phi_m(\mathbf{y}) = 1 \text{ if } Pr(\mathbf{y}|\mathbf{x}_m) \leq Pr(\mathbf{y}|\mathbf{x}_{m'}) \text{ for some } m' \neq m \tag{4}$$

$$\phi_m(\mathbf{y}) = 0 \text{ otherwise} \tag{5}$$

We shall now upperbound P_{em} by upperbounding the function $\phi_m(\mathbf{y})$:

$$\phi_m(\mathbf{y}) \leq \left[\frac{\sum_{m' \neq m} Pr(\mathbf{y}|\mathbf{x}_{m'})^{1/(1+\rho)}}{Pr(\mathbf{y}|\mathbf{x}_m)^{1/(1+\rho)}}\right]^\rho \quad \rho > 0 \tag{6}$$

The reason for using (6) is not at all obvious intuitively, but we can at least establish its validity by noting that the right-hand side of (6) is always non-negative, thereby satisfying the inequality when $\phi_m(\mathbf{y}) = 0$. When $\phi_m(\mathbf{y}) = 1$, some term in the numerator is greater than or equal to the denominator, thus the numerator is greater than or equal to the denominator; raising the fraction to the ρ power keeps it greater than or equal to 1. Substituting (6) in (3), we have

$$P_{em} \leq \sum_{\mathbf{y} \epsilon Y_N} Pr(\mathbf{y}|\mathbf{x}_m)^{1/(1+\rho)} \left[\sum_{m' \neq m} Pr(\mathbf{y}|\mathbf{x}_{m'})^{1/(1+\rho)}\right]^\rho \quad \text{for any } \rho > 0 \tag{7}$$

Equation (7) yields a bound to P_{em} for a particular set of code words. Aside from certain special cases, this bound is too complicated to be useful if the number of code words is large. We will simplify (7) by averaging over an appropriately chosen ensemble of codes. Let us suppose that we define a probability measure $P(\mathbf{x})$ on the set X_N of possible input sequences to the channel. We can now generate an ensemble of codes by picking each code word, independently, according to the probability measure $P(\mathbf{x})$. Thus the probability associated with a code consisting of the code words $\mathbf{x}_1, \cdots, \mathbf{x}_M$ is $\prod_{m=1}^M P(\mathbf{x}_m)$. Clearly, at least one code in the ensemble will have a probability of error that is as small as the ensemble-average probability of error. Using a bar to represent code ensemble averages, we now have

$$\bar{P}_{em} \leq \overline{\sum_{\mathbf{y} \epsilon Y_N} Pr(\mathbf{y}|\mathbf{x}_m)^{1/(1+\rho)} \left[\sum_{m' \neq m} Pr(\mathbf{y}|\mathbf{x}_{m'})^{1+\rho}\right]^\rho} \tag{8}$$

(the averaging bar in (8) is marked with numbered portions 1, 2, 4, 3)

We now impose the additional restriction that $\rho \leq 1$, and proceed to remove the numbered portions of the averaging bar in (8). First, observe that all of the terms of (8) under the bar are random variables; that is, they are real valued functions of the set of randomly chosen words. Thus we can remove part 1 of the bar in (8), since the average of a sum of random variables is equal to the sum of the averages. Likewise, we can remove part 2, because the average of the product of *independent* random variables is equal to the product of the averages. The independence comes from the fact that the code words are chosen independently.

To remove part 3, let ξ be the random variable in brackets; we wish to show that $\overline{\xi^\rho} \leq \bar{\xi}^\rho$. Figure 1 shows ξ^ρ, and it is clear that for $0 < \rho \leq 1$, ξ^ρ is a convex upward function of ξ; i.e., a function whose chords all lie on or beneath the function.[2] Figure 1 illustrates that $\overline{\xi^\rho} \leq \bar{\xi}^\rho$ for the special case in which ξ takes on only two values, and the general case is a well-known result.[3] Part 4 of the averaging bar can be removed by the interchange of sum and average. Thus[4]

$$\bar{P}_{em} \leq \sum_{\mathbf{y} \epsilon Y_N} \overline{Pr(\mathbf{y}|\mathbf{x}_m)^{1/(1+\rho)}} \left[\sum_{m' \neq m} \overline{Pr(\mathbf{y}|\mathbf{x}_{m'})^{1/(1+\rho)}}\right]^\rho \tag{9}$$

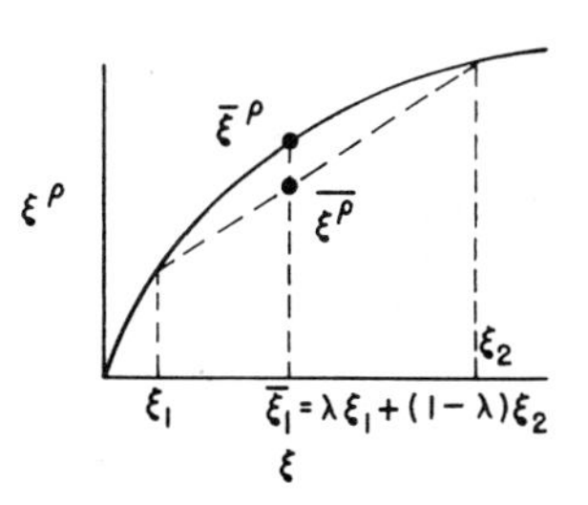

Fig. 1. Convexity of ξ^ρ.

Since the code words are chosen with the probability $P(\mathbf{x})$,

$$\overline{Pr(\mathbf{y}|\mathbf{x}_m)^{1/(1+\rho)}} = \sum_{\mathbf{x} \epsilon X_N} P(\mathbf{x}) Pr(\mathbf{y}|\mathbf{x})^{1/(1+\rho)} \tag{10}$$

Observing that the right-hand side of (10) is independent of m, we can substitute (10) in both the m and m' term in (9). Since the summation in (9) is over $M - 1$ terms, this yields

$$\bar{P}_{em} \leq (M-1)^\rho \sum_{\mathbf{y} \epsilon Y_N} \left[\sum_{\mathbf{x} \epsilon X_N} P(\mathbf{x}) Pr(\mathbf{y}|\mathbf{x})^{1/(1+\rho)}\right]^{1+\rho} \quad \text{for any } \rho,\ 0 < \rho \leq 1 \tag{11}$$

The bound in (11) applies to any discrete channel,

[2] Let $f(\mathbf{x})$ be a real valued function of a vector $\mathbf{x}$ over a region R. We call the region convex if for any $\mathbf{x}_1 \epsilon R$, $\mathbf{x}_2 \epsilon R$; and λ, $0 < \lambda < 1$, we have $\lambda \mathbf{x}_1 + (1-\lambda)\mathbf{x}_2 \epsilon R$. The function $f(\mathbf{x})$ is convex upward over the convex region if for any $\mathbf{x}_1 \epsilon R$, $\mathbf{x}_2 \epsilon R$, and $0 < \lambda < 1$ we have $\lambda f(\mathbf{x}_1) + (1-\lambda) f(\mathbf{x}_2) \leq f[\lambda \mathbf{x}_1 + (1-\lambda)\mathbf{x}_2]$. The function is strictly convex if the inequality, $\leq$, can be replaced with strict inequality, $<$.

[3] See, for example, Blackwell and Girshick [5], page 38.

[4] By a minor modification of the argument used here, only pair-wise independence in the code-word selection is necessary to get from (8) to (9). This makes it possible to apply the bounds developed here to special ensembles of codes such as parity check code ensembles.

memoryless or not, for which $Pr(\mathbf{y} \mid \mathbf{x})$ can be defined. It is valid for all choices of $P(\mathbf{x})$ and all ρ, $0 < \rho \leq 1$.

We shall now assume that the channel is memoryless so as to simplify the bound in (11). Let $x_1, \cdots, x_n, \cdots, x_N$ be the individual letters in an input sequence $\mathbf{x}$, and let $y_1, \cdots, y_n, \cdots, y_N$ be the letters in a sequence $\mathbf{y}$. By a memoryless channel, we mean a channel that satisfies

$$Pr(\mathbf{y}|\mathbf{x}) = \prod_{n=1}^{N} Pr(y_n|x_n) \tag{12}$$

for all $\mathbf{x} \in X_N$ and $\mathbf{y} \in Y_N$ and all N. Now we restrict the class of ensembles of codes under consideration to those in which each letter of each code word is chosen independently of all other letters with a probability measure $p(x)$; $x \in X_1$.

$$P(x) = \prod_{n=1}^{N} p(x_n); \qquad x = (x_1, \cdots, x_n, \cdots, x_N) \tag{13}$$

Substituting (12) and (13) in (11), we get

$$\bar{P}_{em} \leq (M-1)^\rho \sum_{\mathbf{y} \in Y_N} \left[\sum_{\mathbf{x} \in X_N} \prod_{n=1}^{N} p(x_n)\, Pr(y_n \mid x_n)^{1/(1+\rho)} \right]^{1+\rho} \tag{14}$$

We can rewrite the bracketed term in (14) to get

$$\bar{P}_{em} \leq (M-1)^\rho \sum_{\mathbf{y} \in Y_N} \left[\prod_{n=1}^{N} \sum_{x_n \in X_1} p(x_n)\, Pr(y_n \mid x_n)^{1/(1+\rho)} \right]^{1+\rho} \tag{15}$$

Note that the bracketed term in (15) is a product of sums and is equal to the bracketed term in (14) by the usual arithmetic rule for multiplying products of sums. Finally, taking the product outside the brackets in (15), we can apply the same rule again to get

$$\bar{P}_{em} \leq (M-1)^\rho \prod_{n=1}^{N} \sum_{y_n \in Y_1} \left[\sum_{x_n \in X_1} p(x_n)\, Pr(y_n \mid x_n)^{1/(1+\rho)} \right]^{1+\rho} \qquad 0 < \rho \leq 1 \tag{16}$$

We can simplify the notation in (16) somewhat by observing that X_1 is the set of input letters, which is denoted $a_1, \cdots a_k, \cdots a_K$, where K is the size of the channel input alphabet. Also, Y_1 is the set of output letters, denoted $b_1, \cdots, b_j, \cdots b_J$, where J is the size of the output alphabet. Now let P_{jk} denote the channel transition probability $Pr(b_j \mid a_k)$ and let $p(a_k) = p_k$ denote the probability with which letter a_k is chosen in the code ensemble. Substituting this notation in (16), noting that all terms in the product are identical, and including the trivial case $\rho = 0$, we get

$$\bar{P}_{em} \leq (M-1)^\rho \left[\sum_{j=1}^{J} \left(\sum_{k=1}^{K} p_k P_{jk}^{1/(1+\rho)} \right)^{1+\rho} \right]^N \qquad 0 \leq \rho \leq 1 \tag{17}$$

If we now upperbound $M - 1$ by $M = e^{NR}$, where R is the code rate in nats per channel symbol, (17) can be rewritten as

$$\bar{P}_{em} \leq \exp - N \left[-\rho R - \ln \sum_{j=1}^{J} \left(\sum_{k=1}^{K} p_k P_{jk}^{1/(1+\rho)} \right)^{1+\rho} \right] \tag{18}$$

Since the right-hand side of (18) is independent of m, it is a bound on the ensemble probability of decoding error and is independent of the probabilities with which the code words are used. Since at least one code in the ensemble must have an error probability as small as the average,[5] we have proved the following fundamental theorem:

Theorem 1

Consider a discrete memoryless channel with an input alphabet of K symbols, $a_1, \cdots a_K$; an output alphabet of J symbols, $b_1, \cdots b_J$; and transition probabilities $P_{jk} = Pr(b_j \mid a_k)$. For any block length N, any number of code words $M = e^{NR}$, and any probability distribution on the use of the code words, there exists a code for which the probability of decoding error is bounded by

$$P_e \leq \exp - N[-\rho R + E_0(\rho, \mathbf{p})] \tag{19}$$

$$E_0(\rho, \mathbf{p}) = -\ln \sum_{j=1}^{J} \left(\sum_{k=1}^{K} p_k P_{jk}^{1/(1+\rho)} \right)^{1+\rho} \tag{20}$$

where ρ is an arbitrary number, $0 \leq \rho \leq 1$, and $\mathbf{p} = (p_1, p_2, \cdots, p_K)$ is an arbitrary probability vector.[6]

Theorem 1 is valid for all ρ, $0 \leq \rho \leq 1$, and all probability vectors $\mathbf{p} = (p_1, \cdots, p_K)$; thus we get the tightest bound on P_e by minimizing over ρ and $\mathbf{p}$. This gives us the trivial corollary:

Corollary 1: Under the same conditions as Theorem 1, there exists a code for which

$$P_e \leq \exp - NE(R) \tag{21}$$

$$E(R) = \max_{\rho, \mathbf{p}} [-\rho R + E_0(\rho, \mathbf{p})] \tag{22}$$

where the maximization is over all ρ, $0 \leq \rho \leq 1$, and all probability vectors, $\mathbf{p}$.

The function $E(R)$ is the reliability curve discussed in the last section. Except for small values of R (see Section V), Corollary 1 provides the tightest known general bound on error probability for the discrete memoryless channel. We discuss the properties of $E(R)$ in Section III and, in particular, show that $E(R) > 0$ for $0 \leq R < C$, where C is the channel capacity.

It is sometimes convenient to have a bound on error probability that applies to each code word separately rather than to the average.

Corollary 2: Under the same conditions as Theorem 1, there exists a code such that, for all m, $1 \leq m \leq M$, the probability of error when the mth code word is transmitted is bounded by

[5] The same code might not satisfy (18) for all choices of probabilities with which to use the code words; see Corollary 2.

[6] A probability vector is a vector whose components are all non-negative and sum to one.

$$P_{em} \leq 4e^{-NE(R)} \tag{23}$$

where $E(R)$ is given by (22).

Proof: Pick a code with $M' = 2M$ code words which satisfies Corollary 1 when the source uses the $2M$ code words with equal probability. [The rate, R' in (21) and (22) is now $(\ln 2M)/N$.] Remove the M words in the code for which P_{em} is largest. It is impossible for over half the words in the code to have an error probability greater than twice the average; therefore the remaining code words must satisfy

$$P_{em} \leq 2e^{-NE(R')} \tag{24}$$

Since $R' = (\ln 2M)/N = R + (\ln 2)/N$, and since $0 \leq \rho \leq 1$, (22) gives us

$$E(R') \geq E(R) - \frac{\ln 2}{N} \tag{25}$$

Substituting (25) in (24) gives us (23), thereby completing the proof.

III. Properties of the Reliability Curve, $E(R)$

The maximization of (22) over ρ and $\mathbf{p}$ depends on the behavior of the function $E_0(\rho, \mathbf{p})$. Theorem 2 describes $E_0(\rho, \mathbf{p})$ as a function of ρ, and Theorem 3 describes $E_0(\rho, \mathbf{p})$ as a function of $\mathbf{p}$. Both theorems are proved in the Appendix.

Theorem 2

Consider a channel with K inputs, J outputs, and transition probabilities

$$P_{jk},\ 1 \leq k \leq K$$

Let $\mathbf{p} = (p_1, \cdots, p_K)$ be a probability vector on the channel inputs, and assume that the average mutual information

$$I(\mathbf{p}) = \sum_{k=1}^{K} \sum_{j=1}^{J} p_k P_{jk} \ln \frac{P_{jk}}{\sum_{i=1}^{K} p_i P_{ji}}$$

is nonzero. Then, for $\rho \geq 0$,

$$E_0(\rho, \mathbf{p}) = 0 \quad \text{for} \quad \rho = 0 \tag{26}$$

$$E_0(\rho, \mathbf{p}) > 0 \quad \text{for} \quad \rho > 0 \tag{27}$$

$$\frac{\partial E_0(\rho, \mathbf{p})}{\partial \rho} > 0 \quad \text{for} \quad \rho > 0 \tag{28}$$

$$\left.\frac{\partial E_0(\rho, \mathbf{p})}{\partial \rho}\right|_{\rho=0} = I(\mathbf{p}) \tag{29}$$

$$\frac{\partial^2 E_0(\rho, \mathbf{p})}{\partial \rho^2} \leq 0 \tag{30}$$

with equality in (30) if and only if both of the following conditions are satisfied:

1) P_{jk} is independent of k for j, k such that $p_k P_{jk} \neq 0$
2) $\sum_{k; P_{jk} \neq 0} p_k$ is independent of j.

Using this theorem, we can easily perform the maximization of (22) over ρ for a given $\mathbf{p}$. Define

$$E(R, \mathbf{p}) = \max_{0 \leq \rho \leq 1} [-\rho R + E_0(\rho, \mathbf{p})] \tag{31}$$

Setting the partial derivative of the bracketed part of (31) equal to 0, we get

$$R = \frac{\partial E_0(\rho, \mathbf{p})}{\partial \rho} \tag{32}$$

From (30), if some ρ in the range $0 \leq \rho \leq 1$ satisfies (32), then that ρ must maximize (31). Furthermore, from (30) $\partial E_0(\rho, \mathbf{p})/\partial \rho$ is nonincreasing with ρ, so that a solution to (32) exists if R lies in the range

$$\left.\frac{\partial E_0(\rho, \mathbf{p})}{\partial \rho}\right|_{\rho=1} \leq R \leq I(\mathbf{p}) \tag{33}$$

In this range it is most convenient to use (32) to relate $E(R, \mathbf{p})$ and R parametrically as functions of ρ. This gives us

$$E(R, \mathbf{p}) = E_0(\rho, \mathbf{p}) - \rho \frac{\partial E_0(\rho, \mathbf{p})}{\partial \rho} \tag{34}$$

$$R = \frac{\partial E_0(\rho, \mathbf{p})}{\partial \rho} \qquad 0 \leq \rho \leq 1 \tag{35}$$

Figure 2 gives a graphical construction for the solution of these parametric equations.

For $R < \partial E_0(\rho, \mathbf{p})/\partial \rho\, |_{\rho=1}$, the parametric equations (34) and (35) are not valid. In this case, the function $-\rho R + E_0(\rho, \mathbf{p})$ increases with ρ in the range $0 \leq \rho \leq 1$, and therefore the maximum occurs at $\rho = 1$. Thus

$$E(R, \mathbf{p}) = E_0(1, \mathbf{p}) - R \quad \text{for} \quad R < \left.\frac{\partial E_0(\rho, \mathbf{p})}{\partial \rho}\right|_{\rho=1} \tag{36}$$

The behavior of $E(R, \mathbf{p})$ as a function of R given by (34)–(36) is shown in Fig. 3; this behavior depends upon whether $\partial^2 E_0(\rho, \mathbf{p})/\partial \rho^2$ is negative or 0. If it is negative, then R as given by (35) is strictly decreasing with ρ. Differentiating (34) with respect to ρ, we get $-\rho\, \partial^2 E_0(\rho, \mathbf{p})/\partial \rho^2$; thus $E(R, \mathbf{p})$ is strictly increasing with ρ for $\rho \geq 0$, and is equal to 0 for $\rho = 0$. Thus if $R < I(\mathbf{p})$, then $E(R, \mathbf{p}) > 0$. If $\mathbf{p}$ is chosen to achieve capacity, C, then for $R < C$, $E(R, \mathbf{p}) > 0$, and the error probability can be made to vanish exponentially with the block length.

Taking the ratio of the derivatives of (34) and (35), we see that

$$\frac{\partial E(R, \mathbf{p})}{\partial R} = -\rho \tag{37}$$

Thus the parameter ρ in (34) and (35) has significance as the negative slope of the E, R curve.

From the conditions following (30), it is clear that if $\partial^2 E_0(\rho, \mathbf{p})/\partial \rho^2 = 0$ for one value of $\rho > 0$, it is 0 for all $\rho > 0$. Under these circumstances, R and $E(R, \mathbf{p})$ as given by (34) and (35) simply specify the point at which $R = I(\mathbf{p})$, $E(R, \mathbf{p}) = 0$. The rest of the curve, as shown in Fig. 4, comes from (36).

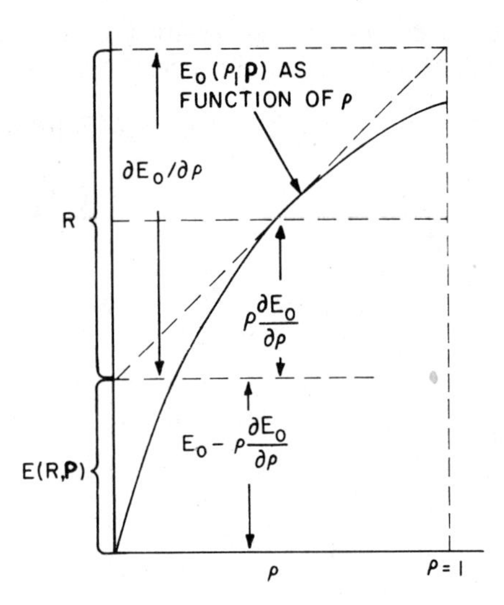

Fig. 2. Geometric construction of $E(R, \mathbf{p})$.

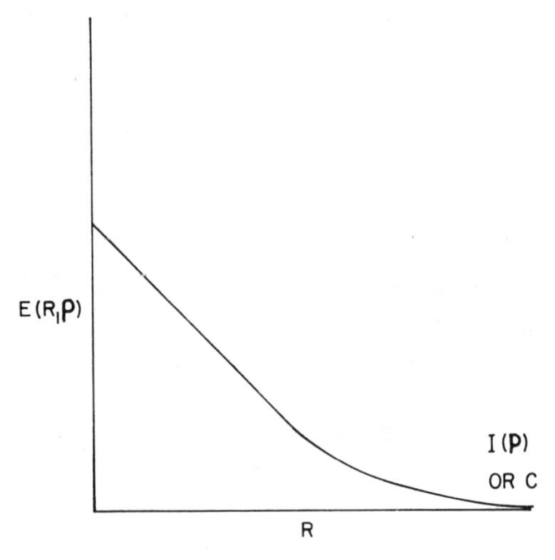

Fig. 3. Exponent, rate curve.

The class of channels for which $\partial^2 E_0(\rho, \mathbf{p})/\partial \rho^2 = 0$ is somewhat pathological. It includes noiseless channels, for which one can clearly achieve zero error probability at rates below capacity. The exponential bounds here simply reflect the probability of assigning more than one message to the same code word. The bound in Section V yields zero error probability in these cases. As an example of a noisy channel with $\partial^2 E_0(\rho, \mathbf{p})/\partial \rho^2 = 0$, see Fig. 4.

An alternative approach to the maximization of (31) over ρ is to regard the function $-\rho R + E_0(\rho, \mathbf{p})$ as a linear function of R with slope $-\rho$ and intercept $E_0(\rho, \mathbf{p})$ for fixed ρ. Thus $E(R, \mathbf{p})$ as a function of R is simply the upper envelope of this set of straight lines (see Fig. 5). (In this paper, the upper envelope of a set of lines will be taken to mean the lowest upper bound to that set of lines.) This picture also interprets $E_0(\rho, \mathbf{p})$ as the E-axis intercept of the tangent of slope $-\rho$ to the E, R curve.

Since $E(R, \mathbf{p})$ is the upper envelope of a set of straight lines, it must be a convex downward function of R; i.e., a function whose chords never lie below the function. This fact, of course, also follows from $\partial E(R, \mathbf{p})/\partial R$ decreasing with ρ and thus increasing with R.

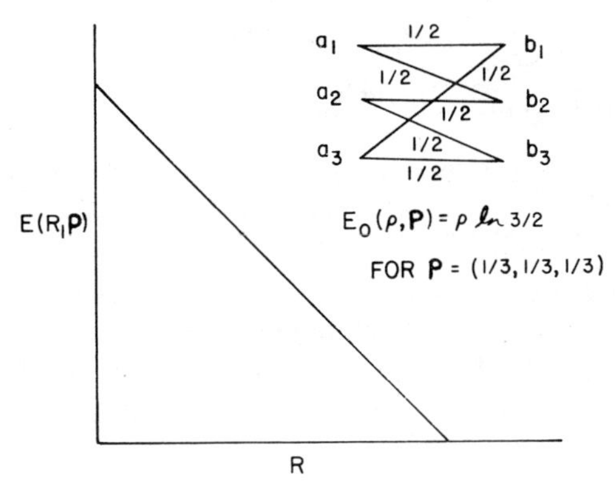

Fig. 4. Exponent, rate curve for channel with $\partial^2 E_0(\rho, \mathbf{p})/\partial \rho^2 = 0$.

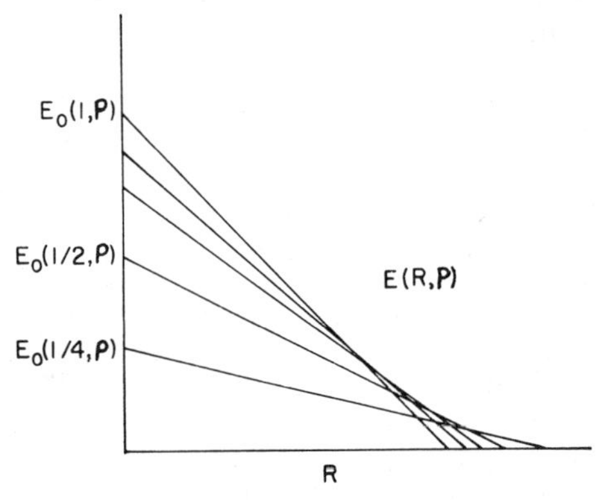

Fig. 5. Exponent, rate curve as envelope of straight lines.

All of the results in this section thus far have dealt with the function $E(R, \mathbf{p})$ defined in (31). The function, $E(R)$, in (22) can be expressed as

$$E(R) = \max_{\mathbf{p}} E(R, \mathbf{p}) \tag{38}$$

where the maximization is over all K-dimensional probability vectors, $\mathbf{p}$. Thus $E(R)$ is the upper envelope of all of the $E(R, \mathbf{p})$ curves, and we have the following theorem.

Theorem 3

For every discrete memoryless channel, the function $E(R)$ is positive, continuous, and convex downward for all R in the range $0 \leq R < C$. Thus the error probability bound $P_e \leq \exp - NE(R)$ is an exponentially decreasing function of the block length for $0 \leq R < C$.

Proof: If $C = 0$, the theorem is trivially true. Otherwise, for the $\mathbf{p}$ that achieves capacity, we have shown that $E(R, \mathbf{p})$ is positive for $0 \leq R < C$, and thus $E(R)$ is positive in the same range. Also, for every probability vector, $\mathbf{p}$, we have shown that $E(R, \mathbf{p})$ is continuous and convex downward with a slope between 0 and -1 and therefore the upper envelope is continuous and convex downward.

One might further conjecture that $E(R)$ has a continuous slope, but this is not true, as we shall show later. $E(R, \mathbf{p})$ has a continuous slope for any $\mathbf{p}$, but the $\mathbf{p}$ that maximizes $E(R, \mathbf{p})$ can change with R and this can lead to discontinuities in the slope of $E(R)$.

We next turn our attention to the actual maximization of $E(R, \mathbf{p})$ over $\mathbf{p}$. We may rewrite (22) as

$$E(R) = \max_{0 \leq \rho \leq 1} [-\rho R + \max_{\mathbf{p}} E_0(\rho, \mathbf{p})] \tag{39}$$

Now define

$$F(\rho, \mathbf{p}) = \sum_j \left(\sum_k p_k P_{jk}^{1/(1+\rho)} \right)^{1+\rho} \tag{40}$$

From (20), $E_0(\rho, \mathbf{p}) = -\ln F(\rho, \mathbf{p})$, so that minimizing $F(\rho, \mathbf{p})$ over $\mathbf{p}$ will maximize $E_0(\rho, \mathbf{p})$.

Theorem 4

$F(\rho, \mathbf{p})$, as given by (40), is a convex downward function of $\mathbf{p}$ over the region in which $\mathbf{p} = (p_1, \cdots, p_K)$ is a probability vector. Necessary and sufficient conditions on the vector (or vectors) $\mathbf{p}$ that minimize $F(\rho, \mathbf{p})$ [and maximize $E_0(\rho, \mathbf{p})$] are

$$\sum_j P_{jk}^{1/(1+\rho)} \alpha_j^{\rho} \geq \sum_j \alpha_j^{1+\rho} \qquad \text{for all } k \tag{41}$$

with equality if $p_k \neq 0$, where $\alpha_j = \sum_k p_k P_{jk}^{1/(1+\rho)}$

This theorem is proved in the Appendix. If all the p_k are positive, (41) is simply the result of applying a Lagrange multiplier to minimize $F(\rho, \mathbf{p})$ subject to $\sum p_k = 1$. As shown in the Appendix, the same technique can be used to get the necessary and sufficient conditions on $\mathbf{p}$ to maximize $I(\mathbf{p})$. The result, which has also been derived independently by Eisenberg, is

$$\sum_j P_{jk} \ln \frac{P_{jk}}{\sum_i p_i P_{ji}} \leq I(\mathbf{p}) \qquad \text{for all } k \tag{42}$$

with equality if $p_k \neq 0$.

Neither (41) nor (42) is very useful in finding the maximum of $E_0(\rho, \mathbf{p})$ or $I(\mathbf{p})$, but both are useful theoretical tools and are useful in verifying that a hypothesized solution is indeed a solution. For channels of any complexity, $E_0(\rho, \mathbf{p})$ and $I(\mathbf{p})$ can usually be maximized most easily by numerical techniques.

Given the maximization of $E_0(\rho, \mathbf{p})$ over $\mathbf{p}$, we can find the function $E(R)$ in any of three ways. First, from (39), $E(R)$ is the upper envelope of the family of straight lines given by

$$-\rho R + \max_{\mathbf{p}} E_0(\rho, \mathbf{p}) \qquad 0 \leq \rho \leq 1 \tag{43}$$

Also, we can plot $\max_{\mathbf{p}} E_0(\rho, \mathbf{p})$ as a function of ρ, and use the graphical technique of Fig. 2 to find $E(R)$.

Finally, we can use the parametric equations, (34) and (35), and for each ρ use the $\mathbf{p}$ that maximizes $E_0(\rho, \mathbf{p})$. To see that this generates the curved portion of $E(R)$, let ρ_0 be a fixed value of ρ, $0 < \rho_0 < 1$, and let $\mathbf{p}_0$ maximize $E_0(\rho_0, \mathbf{p})$. We have already seen that the only point on the straight line $-\rho_0 R + E_0(\rho_0, \mathbf{p}_0)$ that lies on the curve $E(R, \mathbf{p}_0)$, and thus that can lie on $E(R)$, is that given by (34) and (35). Since the straight lines $-\rho R + \max_{\mathbf{p}} E_0(\rho, \mathbf{p})$ generate all points on the $E(R)$ curve, we see that (34)–(36), with $E_0(\rho, \mathbf{p})$ maximized over $\mathbf{p}$, generate all points on the $E(R)$ curve. These parametric equations can, under pathological conditions, also lead to some points not on the $E(R)$ curve. To see this, consider the channel of Fig. 6. From (41), we can verify that for $\rho \leq 0.51$, $E_0(\rho, \mathbf{p})$ is maximized by $p_1 = p_2 = p_3 = p_4 = \frac{1}{4}$. For $\rho > 0.51$, $E_0(\rho, \mathbf{p})$ is maximized by $p_5 = p_6 = \frac{1}{2}$. The parametric equations are discontinuous at $\rho = 0.51$ where the input distribution suddenly changes. Figure 6 shows the $E(R)$ curve for this channel and the spurious points generated by (34) and (35).

	b_1	b_2	b_3	b_4
a_1	0.82	0.06	0.06	0.06
a_2	0.06	0.82	0.06	0.06
a_3	0.06	0.06	0.82	0.06
a_4	0.06	0.06	0.06	0.82
a_5	0.49	0.49	0.01	0.01
a_6	0.01	0.01	0.49	0.49

TRANSITION PROBABILITY MATRIX

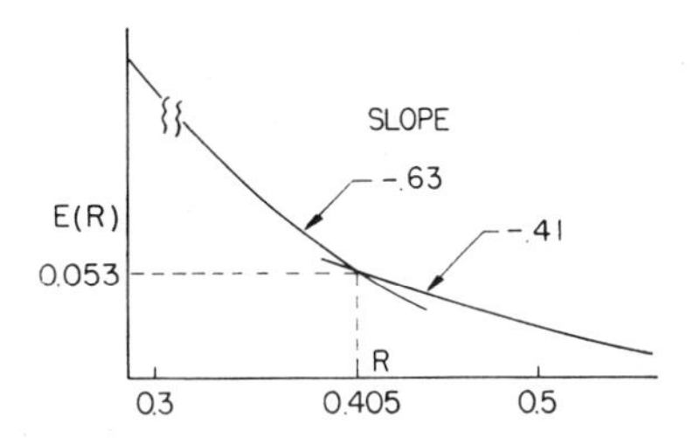

Fig. 6. A pathological channel.

The preceding discussion has described in detail the exponent $E(R)$ controlling the upper bound to error probability described in Section I. It can be shown that the exponent $E_L(R)$ controlling the lower bound to error probability in (1) is given by

$$E_L(R) = \underset{0<\rho<\infty}{\text{g.l.b.}} [-\rho R + \max_{\mathbf{p}} E_0(\rho, \mathbf{p})] \tag{44}$$

Comparing (39) and (44), we see that the only difference is in the range in which ρ is to be maximized. Interpreting $E(R)$ and $E_L(R)$ as the upper envelopes of a family of straight lines of slope $-\rho$, we see that $E(R) = E_L(R)$ for $R_{crit} \leq R < C$, where the critical rate R_{crit} is defined as the g.l.b. of R values for which the slope of $E_L(R)$ is not less than -1. This is a nonzero range of R unless, for the $\mathbf{p}$ that maximizes $E_0(\rho, \mathbf{p})$, we have $\partial^2 E_0(\rho, \mathbf{p})/\partial \rho^2 = 0$ for $0 < \rho \leq 1$; the channel in Fig. 5 is such a channel.

IV. Examples and Applications

Binary Symmetric Channel

A binary symmetric channel has 2 inputs, 2 outputs, and transition probabilities $P_{12} = P_{21} = q$, and $P_{11} =$

$P_{22} = 1 - q$. Thus q is the probability of error on the channel with no coding. Clearly, the input probability vector that maximizes $E_0(\rho, \mathbf{p})$ is $p_1 = p_2 = \frac{1}{2}$. [Formally this can be shown by substitution in (41).] For this choice of $\mathbf{p}$,

$$E_0(\rho, \mathbf{p}) = -\ln \sum_{j=1}^{2} \left(\sum_{k=1}^{2} p_k P_{jk}^{1/(1+\rho)} \right)^{1+\rho}$$
$$= \rho \ln 2 - (1 + \rho) \ln [q^{1/(1+\rho)} + (1 - q)^{1/(1+\rho)}] \tag{45}$$

We now differentiate (45) and evaluate the parametric expressions for exponent and rate, (34) and (35). After some manipulation, we obtain

$$R = \ln 2 - H(q_\rho) \tag{46}$$

$$H(q_\rho) = -q_\rho \ln q_\rho - (1 - q_\rho) \ln (1 - q_\rho) \tag{47}$$

$$E(R, \mathbf{p}) = q_\rho \ln \frac{q_\rho}{q} + (1 - q_\rho) \ln \frac{1 - q_\rho}{1 - q} \tag{48}$$

where

$$q_\rho = \frac{q^{1/(1+\rho)}}{q^{1/(1+\rho)} + (1 - q)^{1/(1+\rho)}} \tag{49}$$

These equations are valid for $0 \leq \rho \leq 1$, or for $\ln 2 - [H\sqrt{q}/(\sqrt{q} + \sqrt{1 - q})] \leq R \leq C$. For rates below this, we have (36), which becomes

$$E(R, \mathbf{p}) = \ln 2 - 2 \ln (\sqrt{q} + \sqrt{1 - q}) - R \tag{50}$$

Except for the lack of a coefficient, $P_e \leq e^{-NE(R,\mathbf{p})}$ [where $E(R, \mathbf{p})$ is given by (46), (48), and (50)] is the random coding bound to error probability on the binary symmetric channel, first derived by Elias [6].

Very Noisy Channels

In this section, we shall consider channels that are very noisy in the sense that the probability of receiving a given output is almost independent of the input. We shall see that a universal exponent, rate curve exists for these channels in the limit. It will be assumed that the channel is discrete and memoryless, although the result can be easily extended to continuous channels. Let $q_1, \cdots, q_J$ be a set of probabilities defined on the channel outputs, and define ϵ_{jk} by

$$P_{jk} = q_j(1 + \epsilon_{jk}) \tag{51}$$

We assume that $|\epsilon_{jk}| \ll 1$ for all j, k. Note that if (51) is multiplied by p_k and summed over k, we get

$$\sum_j q_j \epsilon_{jk} = 0 \qquad \text{for all } k \tag{52}$$

We now compute $E_0(\rho, \mathbf{p})$ for this channel:

$$E_0(\rho, \mathbf{p}) = -\ln \sum_j \left[\sum_k p_k q_j^{1/(1+\rho)} (1 + \epsilon_{jk})^{1/(1+\rho)} \right]^{1+\rho} \tag{53}$$

Expanding $(1 + \epsilon_{jk})^{1/(1+\rho)}$ in a power series in ϵ_{jk}, and dropping terms of higher order than the second, we get

$$E_0(\rho, \mathbf{p}) \approx -\ln \sum_j q_j \left[\sum_k p_k \left(1 + \frac{\epsilon_{jk}}{1 + \rho} - \frac{\rho \epsilon_{jk}^2}{2(1 + \rho)^2} \right) \right]^{1+\rho} \tag{54}$$

The bracketed term to the $(1 + \rho)$ power in (54) may be again expanded as a power series in the ϵ_{jk} to give

$$E_0(\rho, \mathbf{p}) \approx -\ln \sum_j q_j \left[1 + \sum_k p_k \epsilon_{jk} - \frac{\rho}{2(1 + \rho)} \sum_k p_k \epsilon_{jk}^2 + \frac{\rho}{2(1 + \rho)} \left(\sum_k p_k \epsilon_{jk} \right)^2 \right] \tag{55}$$

Using (52), this becomes

$$E_0(\rho, \mathbf{p}) \approx -\ln \left\{ 1 - \frac{\rho}{2(1 + \rho)} \cdot \sum_j q_j \left[\sum_k p_k \epsilon_{jk}^2 - \left(\sum_k p_k \epsilon_{jk} \right)^2 \right] \right\} \tag{56}$$

Finally, expanding (56) and dropping terms of higher than second order in ϵ_{jk}, we have

$$E_0(\rho, \mathbf{p}) \approx \frac{\rho}{1 + \rho} f(\mathbf{p}) \tag{57}$$

where the constant $f(\mathbf{p})$ is given by

$$f(\mathbf{p}) = \frac{1}{2} \sum_j q_j \left[\sum_k p_k \epsilon_{jk}^2 - \left(\sum_k p_k \epsilon_{jk} \right)^2 \right] \tag{58}$$

If we take the mutual information, $I(\mathbf{p})$, use (51) for the transition probabilities, and expand $I(\mathbf{p})$ as a power series in the ϵ_{jk}, dropping terms of higher than second order, we get $f(\mathbf{p})$. Thus channel capacity C is given approximately by

$$C \approx \max_{\mathbf{p}} f(\mathbf{p}) \tag{59}$$

$$\max_{\mathbf{p}} E_0(\rho, \mathbf{p}) \approx \frac{\rho}{1 + \rho} C \tag{60}$$

We can now solve explicitly for ρ to find $E(R) = \max_{0<\rho\leq 1} [-\rho R + \max_{\mathbf{p}} E_0(\rho, \mathbf{p})]$. The solution is

$$P_e \leq e^{-NE(R)}$$

$$E(R) \approx (\sqrt{C} - \sqrt{R})^2 \qquad R \geq \frac{C}{4} \tag{61}$$

$$E(R) \approx \frac{C}{2} - R \qquad R < \frac{C}{4} \tag{62}$$

It is to be observed that the exponent rate curve given by (61) and (62) is identical to that for orthogonal signals in white Gaussian noise [2].

Noisy channels, as defined in this way, were first considered by Reiffen [7], who showed that the exponent corresponding to zero rate was $C/2$.

Parallel Channels

Consider two discrete memoryless channels, the first with K inputs, J outputs, and transition probabilities P_{jk}, and the second with I inputs, L outputs, and transition probabilities Q_{li}. Let

$$E_0^*(\rho, \mathbf{p}) = -\ln \sum_{j=1}^{J} \left(\sum_{k=1}^{K} p_k P_{jk}^{1/(1+\rho)} \right)^{1+\rho}$$

$$E_0^{**}(\rho, \mathbf{q}) = -\ln \sum_{l=1}^{L} \left(\sum_{i=1}^{I} q_i Q_{li}^{1/(1+\rho)} \right)^{1+\rho}$$

where $\mathbf{p} = (p_1, \cdots, p_K)$ and $\mathbf{q} = (q_1, \cdots, q_I)$ represent arbitrary probability assignments on the inputs to the first and second channel, respectively.

Let us consider using these two channels in parallel; that is, in each unit of time, the transmitter sends one symbol over the first channel and one symbol over the second channel. If we consider this pair of channels as a single channel with KI inputs, JL outputs, and transition probabilities $P_{jk}Q_{li}$, then we can find an upper bound to the probability of error achievable through coding on the two channels together. The following theorem, the first half of which is due to R. M. Fano [8], relates the $E(R)$ curve for the parallel combination to the $E(R)$ curves for the individual channels.

Theorem 5

The minimum error probability achievable through coding can be upperbounded by

$$P_e \leq \exp - N[-\rho R + E_0(\rho, \mathbf{pq})] \quad \text{for any } \rho, 0 \leq \rho \leq 1 \tag{63}$$

where

$$E_0(\rho, \mathbf{pq}) = E_0^*(\rho, \mathbf{p}) + E_0^{**}(\rho, \mathbf{q}) \tag{64}$$

Furthermore, if we choose $\mathbf{p}$ and $\mathbf{q}$ to maximize $E_0^*(\rho, \mathbf{p})$ and $E_0^{**}(\rho, \mathbf{q})$, respectively, for a given ρ, then

$$E_0(\rho, \mathbf{pq}) = \max_{\mathbf{r}} E_0(\rho, \mathbf{r}) \tag{65}$$

where $\mathbf{r} = (r_{11}, r_{12}, \cdots, r_{1I}, r_{21}, \cdots, r_{2I}, \cdots, r_{KI})$ represents an arbitrary probability assignment to an input pair and $E_0(\rho, \mathbf{r})$ is the usual E_0 function [see (20)] as applied to the parallel channel combination.

Proof: Regarding the parallel channels as a single channel with input probability vector $\mathbf{r}$, we get

$$E_0(\rho, \mathbf{r}) = -\ln \sum_{j,l} \left[\sum_{k,i} r_{ki} (P_{jk}Q_{li})^{1/(1+\rho)} \right]^{1+\rho} \tag{66}$$

Now assume that the input probability assignment uses letters from the two channels independently; i.e., $r_{ki} = p_k q_i$, where $p_1, \cdots, p_K$ and $q_1, \cdots, q_I$ are probability assignments on the individual channels. Substituting $r_{ki} = p_k q_i$ in (66) and separating the sum on k and i, we get

$$E_0(\rho, \mathbf{r}) = -\ln \sum_{j,l} \left[\sum_k p_k P_{jk}^{1/(1+\rho)} \right]^{1+\rho} \cdot \left[\sum_i q_i q_{li}^{1/(1+\rho)} \right]^{1+\rho}$$

Separating the sum on j and l, we obtain

$$E_0(\rho, \mathbf{r}) = E_0^*(\rho, \mathbf{p}) + E_0^{**}(\rho, \mathbf{q}) \tag{67}$$

Next, we must show that $r_{ki} = p_k q_i$ maximizes $E_0(\rho, \mathbf{r})$ when $\mathbf{p}$ and $\mathbf{q}$ maximize E_0^* and E_0^{**}. We know from (41) that the $\mathbf{p}$ and $\mathbf{q}$ that maximize E_0^* and E_0^{**} must satisfy

$$\sum_j P_{jk}^{1/(1+\rho)} \alpha_j^\rho \geq \sum_j \alpha_j^{1+\rho} \ ; \text{all } k \tag{68}$$

with equality if $p_k \neq 0$, where $\alpha_j = \sum_k p_k P_{jk}^{1/(1+\rho)}$, and

$$\sum_l Q_{li}^{1/(1+\rho)} \beta_l^\rho \geq \sum_l \beta_l^{1+\rho} \ ; \text{all } i \tag{69}$$

with equality if $q_i \neq 0$, where $\beta_l = \sum_i q_i Q_{li}^{1/(1+\rho)}$.

Multiplying (68) and (69) together, we get

$$\sum_{j,l} (P_{jk}Q_{li})^{1/(1+\rho)} (\alpha_j \beta_l)^\rho \geq \sum_{j,l} (\alpha_j \beta_j)^{1+\rho} \tag{70}$$

with equality if $r_{ki} = p_k q_i \neq 0$.

We observe that (70) is the same as (41) applied to the parallel channel combination. Thus this choice of $\mathbf{r}$ maximizes $E_0(\rho, \mathbf{r})$, and the theorem is proven.

Theorem 5 has an interesting geometrical interpretation. Let $E(\rho)$ and $R(\rho)$ be the exponent and rate for the parallel combination, as parametrically related by (34) and (35) with the optimum choice of $\mathbf{r}$ for each ρ. Let $E^*(\rho)$, $R^*(\rho)$, $E^{**}(\rho)$, $R^{**}(\rho)$ be the equivalent quantities for the individual channels. From (64)

$$E(\rho) = E^*(\rho) + E^{**}(\rho) \tag{71}$$

$$R(\rho) = R^*(\rho) + R^{**}(\rho) \tag{72}$$

Thus the parallel combination is formed by vector addition of points of the same slope from the individual $E(\rho)$, $R(\rho)$ curves.

Theorem 5 clearly applies to any number of channels in parallel. If we consider a block code of length N as a single use of N identical parallel channels, then Theorem 5 justifies our choice of independent identically distributed symbols in the ensemble of codes.

V. Improvement of Bound for Low Rates

At low rates, the exponent $E(R)$ derived in Section III does not yield a tight bound on error probability. The exponent is so large at low rates that previously negligible effects such as assigning the same code word to two messages suddenly become important. In this section, we avoid this problem by expurgating those code words for which the error probability is high. Equation (7) gives a bound on error probability for a particular code when the mth word is transmitted. With $\rho = 1$, this is

$$P_{em} \leq \sum_{\mathbf{y} \varepsilon Y_N} \sqrt{Pr(\mathbf{y}|\mathbf{x}_m)} \sum_{m' \neq m} \sqrt{Pr(\mathbf{y}|\mathbf{x}_{m'})} \tag{73}$$

This can be rewritten in the form

$$P_{em} \le \sum_{m' \ne m} q(\mathbf{x}_m, \mathbf{x}_{m'}) \tag{74}$$

$$q(\mathbf{x}_m, \mathbf{x}_{m'}) = \sum_{\mathbf{y}} \sqrt{Pr(\mathbf{y}|\mathbf{x}_m)\, Pr(\mathbf{y}|\mathbf{x}_{m'})} \tag{75}$$

$$= \prod_{n=1}^{N} \sum_{j=1}^{J} \sqrt{Pr(b_j|x_{mn})\, Pr(b_j|x_{m'n})} \tag{76}$$

Equations (75) and (76) are equivalent through the usual arithmetic rule for the product of a sum, where $(b_1, \cdots, b_J)$ is the channel output alphabet. We define $-\ln q(\mathbf{x}_m, \mathbf{x}_{m'})$ as the discrepancy between $\mathbf{x}_m$ and $\mathbf{x}_{m'}$; this forms a useful generalization of Hamming distance on binary symmetric channels to general memoryless channels.

Since P_{em} in (72) is a function of a particular code, it is a random variable over an ensemble of codes. In this section we upperbound $Pr\ (P_{em} \ge B)$, where B is a number to be chosen later, and then expurgate code words for which $P_{em} \ge B$. Using a bar to represent an average over the ensemble of codes, we obtain

$$Pr(P_{em} \ge B) = \overline{\phi_m(\text{code})} \tag{77}$$

$$\phi_m(\text{code}) = \begin{cases} 1 \text{ if } P_{em} \ge B \\ 0 \text{ otherwise} \end{cases} \tag{78}$$

We upperbound ϕ_m by

$$\phi_m(\text{code}) \le \sum_{m' \ne m} \frac{q(\mathbf{x}_m, \mathbf{x}_{m'})^s}{B^s} \qquad 0 < s \le 1 \tag{79}$$

Equation (79) is obvious for $\phi_m = 0$. If $\phi_m = 1$ and $s = 1$, (79) follows from (78) and (74). Decreasing s increases all the terms in (79) that are less than 1, and if any term is greater than 1, (79) is valid anyway. Substituting (79) in (77), we have

$$Pr(P_{em} \ge B) \le B^{-s} \sum_{m' \ne m} \overline{q(\mathbf{x}_m, \mathbf{x}_{m'})^s} \tag{80}$$

Let the letters of the code words in the ensemble of codes be chosen independently by using the probabilities $p_1, \cdots, p_K$ so that $Pr\ (\mathbf{x}_m) = \prod_{n=1}^{N} Pr(x_{mn})$, where $Pr\ (x_{mn}) = p_k$ for $x_{mn} = p_k$. Then using (76), we have

$$\overline{q(\mathbf{x}_m, \mathbf{x}_{m'})^s} = \sum_{\mathbf{x}_m, \mathbf{x}_{m'}} Pr(\mathbf{x}_m) Pr(\mathbf{x}_{m'}) \cdot \prod_{n=1}^{N} \left[\sum_{j=1}^{J} \sqrt{Pr(b_j|x_{mn})\, Pr(b_j|x_{m'n})} \right]^s \tag{81}$$

$$= \prod_{n=1}^{N} \sum_{k=1}^{K} \sum_{i=1}^{K} p_k p_i \left[\sum_{j=1}^{J} \sqrt{Pr(b_j|a_k) Pr(b_j|a_i)} \right]^s \tag{82}$$

Since (82) is independent of m and m', we can substitute it in (80) to get

$$Pr(P_{em} \ge B) \le (M-1)B^{-s} \left[\sum_{k=1}^{K} \sum_{i=1}^{K} p_k p_i \left(\sum_{j=1}^{J} \sqrt{P_{jk} P_{ji}} \right)^s \right]^N \quad \text{for any } s,\ 0 < s \le 1 \tag{83}$$

Now choose B so that the right-hand side of (83) is equal to $\frac{1}{2}$. Then

$$Pr(P_{em} \ge B) \le 1/2$$

$$B = [2(M-1)]^{1/s} \left[\sum_{k,i} p_k p_i \left(\sum_j \sqrt{P_{jk} P_{ji}} \right)^s \right]^{N/s} \tag{84}$$

If we expurgate all code words in the ensemble for which $P_{em} \ge B$, where B is given by (84), the average number of code words remaining in a code is at least $M/2$, since the probability of expurgation is at most $\frac{1}{2}$. Thus there exists a code with $M' \ge M/2$ code words with the error probability for each code word bounded by

$$P_{em} < B < (4M')^{1/s} \left[\sum_{k,i} p_k p_i \left(\sum_j \sqrt{P_{jk} P_{ji}} \right)^s \right]^{N/s} \tag{85}$$

Note that removing a code word from a code cannot increase the error probability associated with any other code word. If we let $M' = e^{NR}$ and define $\rho = 1/s$, (85) can be written

$$P_{em} < \exp - N \left[-\rho R + E_x(\rho, \mathbf{p}) - \rho \frac{\ln 4}{N} \right] \quad \text{for any } \rho \ge 1 \tag{86}$$

$$E_x(\rho, \mathbf{p}) = -\rho \ln \sum_{k,i} p_k p_i \left(\sum_j \sqrt{P_{jk} P_{ji}} \right)^{1/\rho} \tag{87}$$

We can summarize the preceding results in the following theorem.

Theorem 6

Consider a discrete memoryless channel with input alphabet $a_1, \cdots, a_K$, output alphabet $b_1, \cdots, b_J$, and transition probabilities $P_{jk} = Pr\ (b_j \mid a_k)$. Then for any block length N and any number of code words $M' = e^{NR}$, there exists a code such that, for all m, $1 \le m \le M'$, the probability of decoding error when the mth code word is transmitted is bounded by (86) and (87), where $\mathbf{p} = (p_1, \cdots, p_K)$ in (87) is an arbitrary probability vector.

The expurgation technique leading to Theorem 6 is somewhat similar to an earlier expurgation technique applied by Elias [6] to the binary symmetric channel and by Shannon [9] to the additive Gaussian noise channel. The final bound here is somewhat tighter than those bounds and, in fact, the difference between the exponent derived here and the earlier exponents is equal to the rate, R.

The interpretation of Theorem 6 is almost identical to that of Theorem 1. The exponent rate curve given by (86) is the upper envelope of a set of straight lines; the line corresponding to each value of $\rho \ge 1$ has slope $-\rho$ and intercept $E_x(\rho, \mathbf{p})$ on the E axis. The following theorem, which is proved in the Appendix, gives the properties of $E_x(\rho, \mathbf{p})$.

Theorem 7

Let P_{jk} be the transition probabilities of a discrete memoryless channel and let $\mathbf{p} = (p_1, \cdots, p_K)$ be a probability vector on the channel inputs. Assume that

$$I(\mathbf{p}) = \sum_{k,j} p_k P_{jk} \ln \frac{P_{jk}}{\sum_i p_i P_{ji}} \neq 0$$

Then for $\rho > 0$, $E_x(\rho, \mathbf{p})$ as given by (87) is strictly increasing with ρ. Also, $E_x(1, \mathbf{p}) = E_0(1, \mathbf{p})$, where E_0 is given by (20). Finally, $E_x(\rho, \mathbf{p})$ is strictly convex upward with ρ unless the channel is noiseless in the sense that for each pair of inputs, a_k and a_i, for which $p_k \neq 0$ and $p_i \neq 0$, we have either $P_{jk}P_{ji} = 0$ for all j or $P_{jk} = P_{ji}$ for all j.

This theorem can be used in exactly the same way as Theorem 2 to obtain a parametric form for the exponent, rate curve at low rates. Let

$$E(R, \mathbf{p}) = \max_{\rho} \left[-\rho R + E_x(\rho, \mathbf{p}) - \rho \frac{\ln 4}{N} \right] \tag{88}$$

Then, for R in the range

$$\lim_{\rho \to \infty} \frac{\partial E_x(\rho, \mathbf{p})}{\partial \rho} \leq R + \frac{\ln 4}{N} \leq \left. \frac{\partial E_x(\rho, \mathbf{p})}{\partial \rho} \right|_{\rho = 1} \tag{89}$$

we have the parametric equations in ρ

$$\left. \begin{aligned} R + \frac{\ln 4}{N} &= \frac{\partial E_x(\rho, \mathbf{p})}{\partial \rho} \\ E(R, \mathbf{p}) &= -\rho \frac{\partial E_x(\rho, \mathbf{p})}{\partial \rho} + E_x(\rho, \mathbf{p}) \end{aligned} \right\} \tag{90}$$

If $E_x(\rho, \mathbf{p})$ is a strictly convex function of ρ, then (90) represents a convex downward curve with a continuous slope given by $-\rho$.

The smallest rate for which (90) is applicable is

$$\lim_{\rho \to \infty} \frac{\partial E_x(\rho, \mathbf{p})}{\partial \rho} = \lim_{\rho \to \infty} -\ln \sum_{k,i} p_k p_i \left(\sum_j \sqrt{P_{jk} P_{ji}} \right)^{1/\rho}$$

$$= -\ln \sum_{k,i} p_k p_i \phi_{ki} \tag{91}$$

where

$$\phi_{ki} = \begin{cases} 1 & \text{if } \sum_j P_{jk} P_{ji} \neq 0 \\ 0 & \text{if } \sum_j P_{jk} P_{ji} = 0 \end{cases}$$

If there are two inputs in use, k and i, for which there are no common outputs (i.e., for which $\sum_j P_{jk}P_{ji} = 0$), then the right-hand side of (91) is strictly positive. If $R + \ln 4/N$ is less than this quantity, then $E(R, \mathbf{p})$ is infinite. This can be seen most easily by regarding the $E(R, \mathbf{p})$, R curve as the upper envelope of straight lines of slope $-\rho$; the right-hand side of (91) is the limit of the R intercepts of these lines as the slope approaches $-\infty$. Shannon [10] has defined the zero error capacity of a channel as the greatest rate at which transmission is possible with no errors; the right-hand side of (91) thus gives a lower bound to zero error capacity. Fig. 7 shows the exponent, rate curves given by Theorem 6 for some typical channels.

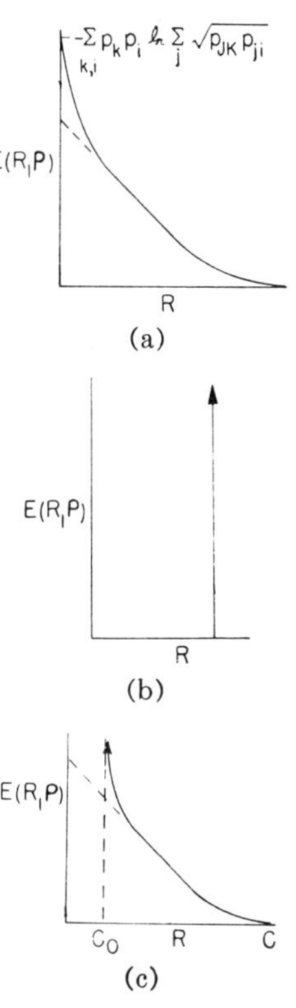

Fig. 7. Typical exponent, rate curves obtained by using low-rate improvement. (a) Ordinary channel. (b) Noiseless channel. (c) Channel with zero error capacity.

If the channel is noiseless in the sense of Theorem 7, then it is not hard to see that $E(R, \mathbf{p})$, as given by (88), is infinite for $R + (\ln 4)/N < I(\mathbf{p})$. It is no great achievement to show that zero error probability is possible on noiseless channels below capacity, but it is satisfying to see that this result comes naturally out of the general formulation.

Very little can be said about the maximization of $E_x(\rho, \mathbf{p})$ over the input probability vector $\mathbf{p}$. It is possible for a number of local maxima to exist, and no general maximization techniques are known.

These low-rate results can be applied to parallel channels by the same procedure as used in Section IV. If the input probability vector $\mathbf{p}$ for the parallel channels chooses letters from the two channels independently, then $E_x(\rho, \mathbf{p})$ for the parallel combination is the sum of the $E_x(\rho, \mathbf{p})$ functions for the individual channels. Unfortunately, $E_x(\rho, \mathbf{p})$ is not always maximized by using the channels independently. An example of this, which is due to Shannon [10], is found by putting the channel in Fig. 8 in parallel with itself. The zero error capacity bound, $\lim_{\rho \to \infty} E_x(\rho, \mathbf{p})/\rho$, for the single channel is ln 2 achieved by using inputs 1 and 4 with equal probability in (91). For the parallel channels, (91) yields ln 5, achieved

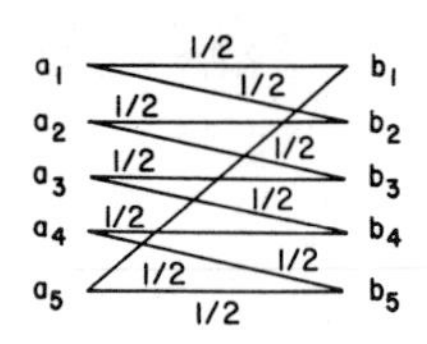

Fig. 8. Transition probabilities for a channel with zero error capacity.

by using the five pairs of inputs, (1, 1), (2, 3), (3, 5), (4, 2), (5, 4), with equal probability.

Theorem 6 yields a rather interesting result when applied to the binary symmetric channel. Letting q be the channel crossover probability and letting $\mathbf{p} = (\frac{1}{2}, \frac{1}{2})$, we rewrite (87)

$$E_x(\rho, \mathbf{p}) = -\rho \ln \{\tfrac{1}{2} + \tfrac{1}{2} [4q(1-q)]^{1/2\rho}\} \tag{92}$$

Using (92) in the parametric equations, (90), and going through some algebraic manipulation, we get

$$R + \frac{\ln 4}{N} = \ln 2 - H(\delta) \tag{93}$$

$$E(R, \mathbf{p}) = \frac{\delta}{2} \ln \frac{1}{4q(1-q)} \tag{94}$$

where the parameter δ is related to ρ by $\delta/(1-\delta) = [4q(1-q)]^{1/2\rho}$, and $H(\delta) = -\delta \ln \delta - (1-\delta) \ln (1-\delta)$. Equations (93) and (94) are valid for $\delta \geq \sqrt{4q(1-q)} / (1 + \sqrt{4q(1-q)})$.

For large N, δ in (93) approaches $D_{\min}/N$, where $D_{\min}$ is the Gilbert bound [11] on minimum distance for a binary code of rate R. The exponent given by (94) turns out to be the same as the exponent for probability of confusion between two code words at the Gilbert distance from each other. This result has also been established for parity-check codes [12].

VI. Continuous Channels and Input Constraints

A time-discrete amplitude-continuous channel is a channel whose input and output alphabets are the set of real numbers. It is usually necessary or convenient to impose a constraint on the code words of such channels to reflect the physical power limitations of the transmitter. Thus, before discussing continuous channels, we discuss the effects of constraints on discrete channels and then generalize the results to continuous channels.

It is possible to include constraints in Theorem 1 by choosing the code ensemble in such a way that the average code word will satisfy the constraint. There are two difficulties with such a procedure. One is the mathematical technicality that not all of the words satisfy the constraint. The other, more important, difficulty is that those code words that satisfy the constraint with a considerable amount to spare sometimes have such a high error probability that the upper bound given by Theorem 1 is not exponentially the tightest bound that can be derived.

In this section, we modify Theorem 1 to get the best exponential bound for discrete channels with input constraints. Then we extend the bound to the continuous channel, and, finally, we use the additive Gaussian noise channel as an example.

Let $f_1 = f(a_1), \cdots, f_K = f(a_K)$ be a real-valued (positive and negative) function of the input letters, $a_1, \cdots, a_K$. We wish to consider codes for which each code word $\mathbf{x} = (x_1, \cdots, x_N)$, is constrained to satisfy

$$\sum_{n=1}^{N} f(x_n) \leq 0 \tag{95}$$

If the input letters are voltage levels and if $f(a_k) = a_k^2 - S_0$, then (95) is a power constraint, constraining each code word to an average power of S_0 per letter. Let $\mathbf{p} = (p_1, \cdots, p_K)$ be a probability vector whose components satisfy

$$\sum_{k=1}^{K} p_k f_k \leq 0 \tag{96}$$

We now define an ensemble of codes in which the probability of a code word $P(\mathbf{x})$ is the conditional probability of picking the letters according to $\mathbf{p}$, given that the constraint $-\delta \leq \sum_{n=1}^{N} f(x_n) \leq 0$, where δ is a number to be chosen later, is satisfied. Mathematically,

$$P(\mathbf{x}) = q^{-1}\phi(\mathbf{x}) \prod_{n=1}^{N} p(x_n) \tag{97}$$

$$\phi(\mathbf{x}) = \begin{cases} 1 & \text{if } -\delta \leq \sum_n f(x_n) \leq 0 \\ 0 & \text{otherwise} \end{cases} \tag{98}$$

$$q = \sum_{\mathbf{x}} \phi(\mathbf{x}) \prod_{n=1}^{N} p(x_n) \tag{99}$$

where $p(x_n) = p_k$ for $x_n = a_k$. We can think of q as a normalizing factor that makes $P(\mathbf{x})$ sum to 1.

We now substitute (99) in (11), remembering that (11) is valid for any ensemble of codes.

$$P_{em} \leq (M-1)^\rho \sum_{\mathbf{y}} \Big[\sum_{\mathbf{x}} q^{-1}\phi(\mathbf{x}) \cdot \prod_{n=1}^{N} p(x_n) \, Pr(\mathbf{y}|\mathbf{x})^{1/(1+\rho)} \Big]^{1+\rho} \qquad 0 \leq \rho \leq 1 \tag{100}$$

Before simplifying (100), we upperbound $\phi(\mathbf{x})$.

$$\phi(\mathbf{x}) \leq \exp r \Big[\sum_{n=1}^{N} f(x_n) + \delta \Big] \quad \text{for } r \geq 0 \tag{101}$$

Equation (101) is obviously valid for $\phi(\mathbf{x}) = 0$; for $\phi(\mathbf{x}) = 1$, we have $\sum_{n=1}^{N} f(x_n) + \delta \geq 0$, and (101) is still valid. The right-hand side of (101) is mathematically more tractable than the left, but still avoids large contributions to P_{em} from sequences for which $\sum_n f(x_n)$ is too small.

Substituting (101) in (100) and going through the same set of steps that were used from (11) to (20), we have proved the following theorem.

Theorem 8

Under the same conditions as Theorem 1, there exists a code in which each code word satisfies the constraint $\sum_{n=1}^{N} f(x_n) \le 0$ and the probability of decoding error is bounded by

$$P_e \le B \exp - N[E_0(\rho, \mathbf{p}, r) - \rho R] \tag{102}$$

$$E_0(\rho, \mathbf{p}, r) = -\ln \sum_{j=1}^{J} \left[\sum_{k=1}^{K} p_k P_{jk}^{1/(1+\rho)} e^{r f_k} \right]^{1+\rho} \tag{103}$$

$$B = \left(\frac{e^{r\delta}}{q} \right)^{1+\rho} \tag{104}$$

where q satisfies (99). Equations (102)–(104) are valid for any ρ, $0 \le \rho \le 1$, any $r \ge 0$, any $\delta > 0$, and any $\mathbf{p}$ satisfying $\sum p_k f_k \le 0$.

We note that if $r = 0$, (103) is the same as (20) except for the coefficient B. If the $\mathbf{p}$ that maximizes $E_0(\rho, \mathbf{p})$ in (20) satisfies the constraint $\sum_{k=1}^{K} p_k f_k \le 0$, we can set $r = 0$ and get the same exponential error behavior as in the unconstrained case. Under these circumstances, if we choose δ large enough, then q will approach $\frac{1}{2}$ with increasing N if $\sum_k p_k f_k = 0$ and will approach 1 for $\sum p_k f_k < 0$.

The more interesting application of Theorem 8 is to cases in which the $\mathbf{p}$ that maximizes $E_0(\rho, \mathbf{p})$ in (20) does not satisfy the constraint $\sum p_k f_k \le 0$; it turns out in this case that $E_0(\rho, \mathbf{p}, r)$ is maximized by choosing $r > 0$.

The engineering approach to the maximization of $E_0(\rho, \mathbf{p}, r)$ over $\mathbf{p}$, r is to conclude that, since the unconstrained maximum is by hypothesis outside the constraint region, the constrained maximum is at the constraint boundary, $\sum_k p_k f_k = 0$. We can then find a stationary point to the quantity inside the logarithm in (104) by using Lagrange multipliers for the constraints $\sum_k p_k = 1$, $\sum_k p_k f_k = 0$. This procedure gives us

$$(1 + \rho) \sum_{j=1}^{J} \alpha_j^{\rho} P_{jk}^{1/(1+\rho)} e^{r f_k} + \lambda + \gamma f_k \ge 0 \quad \text{for all } k \tag{105}$$

with equality if $p_k \ne 0$.

$$\alpha_j = \sum_k p_k P_{jk}^{1/(1+\rho)} e^{r f_k} \tag{106}$$

The inequality in (105) is to account for maxima where some of the $p_k = 0$, as in Theorem 4. We also require a stationary point with respect to r, which gives us

$$(1 + \rho) \sum_j \alpha_j^{\rho} \sum_k p_k f_k P_{jk}^{1/(1+\rho)} e^{r f_k} = 0 \tag{107}$$

If we multiply (105) by p_k and sum over k, we find that $\lambda = -(1 + \rho) \sum_j \alpha_j^{1+\rho}$. If we multiply (105) by $p_k f_k$, sum over k, and compare with (107), we find that $\gamma = 0$. Combining these results, we obtain

$$\sum_j \alpha_j^{\rho} P_{jk}^{1/(1+\rho)} e^{r f_k} \ge \sum_j \alpha_j^{1+\rho} \text{ ; for all } k \tag{108}$$

with equality if $p_k \ne 0$.

It can be shown, although the proof is more involved than that of Theorem 4, that (108) and the constraints $\sum_k p_k = 1$ and $\sum_k p_k f_k = 0$ are necessary and sufficient conditions on the r and $\mathbf{p}$ that maximize $E_0(\rho, \mathbf{p}, r)$ when the unconstrained maximum does not satisfy $\sum p_k f_k \le 0$. When $\mathbf{p}$ and r are maximized in this way, and (102) is maximized over ρ, it can then be shown that for $R \ge R_{\text{crit}}$ (102) has the true exponential behavior with N of the best code of block length N satisfying the given constraint.

The quantity B in (102) and (103) is difficult to bound, but it can be estimated quite easily for large N from the central-limit theorem. Let $S = \sum_{n=1}^{N} \xi_n$, where the ξ_n are independent and $\xi_n = f_k$ with probability p_k. Then $q = Pr\,[-\delta \le S \le 0]$, and it follows from the central-limit theorem[7] that, for fixed δ,

$$\lim_{N \to \infty} \sqrt{N}\, q = \frac{\delta}{\sqrt{2\pi}\, \sigma_f} \tag{109}$$

$$\sigma_f^2 = \sum_k p_k f_k^2 \tag{110}$$

Using (109) in (103), we see that $e^{r\delta}/q$ is approximately minimized by choosing $\delta = 1/r$, with the result

$$\frac{e^{r\delta}}{q} \approx \sqrt{2\pi N}\, \sigma_f e r \tag{111}$$

Here, $\approx$ means that the ratio of the two sides approaches 1 as $N \to \infty$. If the ξ_n are lattice distributions,[7] then δ must be a multiple of the span, and (111) is not valid, although B is still proportional to $N^{(1+\rho)/2}$.

Input Constraints at Low Rates

At low rates, the bound given by (102) and (103) can be improved upon in the same way as Theorem 6 improved upon Theorem 1. In order to do this, we simply choose $Pr\,(\mathbf{x}_m)$ and $Pr\,(\mathbf{x}_{m'})$ in (81) to be given by (97). Applying the bound in (101) to (97), and substituting in (81), we can simplify the expression to get

$$\overline{q(\mathbf{x}_m, \mathbf{x}_{m'})^s} = \frac{e^{2r\delta}}{q^2} \left[\sum_{k=1}^{K} \sum_{i=1}^{K} p_k p_i e^{r(f_k + f_i)} \left(\sum_{j=1}^{J} \sqrt{P_{jk} P_{ji}} \right)^s \right]^N \tag{112}$$

Using (112) in place of (82) and carrying through the same argument used in going from (82) to (87), we get the following theorem.

Theorem 9

Under the same conditions as in Theorem 6, there exists a code for which each code word satisfies both $\sum_n f(x_n) \le 0$ and

[7] If the ξ_n have a nonlattice distribution, (109) follows after a little algebra from Theorem 2, page 210, of Gnedenko and Kolmogorov [13]. If the ξ_n have a lattice distribution, (109) follows from the theorem on page 233, Gnedenko and Kolmogorov [13]. (A lattice distribution is a distribution in which the allowable values of ξ_n can be written in the form $d_k = h \cdot j(k) + a$, where a and h are independent of k and $j(k)$ is an integer for each k. The largest h satisfying this equation is the span of the distribution.) For nonlattice distributions, ξ_n must have a third absolute moment; this is trivial for finite input alphabets and sufficiently general for the continuous inputs that we wish to consider.

$$P_{em} < \exp - N\left\{E_x(\rho, \mathbf{p}, r,) - \rho\left[R + \frac{2}{N}\ln\frac{2e^{r\delta}}{q}\right]\right\} \quad (113)$$

$$E_x(\rho, \mathbf{p}, r) = -\rho \ln \sum_{k,i} p_k p_i e^{r(f_k+f_i)}\left(\sum_j \sqrt{P_{jk}P_{ji}}\right)^{1/\rho} \quad (114)$$

for any $\rho \geq 1$, $r \geq 0$, $\delta > 0$, and $\mathbf{p}$ satisfying $\sum_k p_k f_k \leq 0$. For $\sum_k p_k f_k = 0$, $e^{r\delta}/q$ is given by (109)–(111).

Continuous Channels

Consider a channel in which the input and output alphabets are the set of real numbers. Let $P(y \mid x)$ be the probability density of receiving y when x is transmitted. Let $p(x)$ be an arbitrary probability density on the channel inputs, and let $f(x)$ be an arbitrary real function of the channel inputs; assume that each code word is constrained to satisfy $\sum_{n=1}^{N} f(x_n) \leq 0$, and assume that $\int_{-\infty}^{\infty} p(x)f(x)\,dx = 0$.

Let the input space be divided into K intervals and the output space be divided in J intervals. For each k, let a_k be a point in the kth input interval, and let p_k be the integral of $p(x)$ over that interval. Let P_{jk} be the integral of $P(y \mid a_k)$ over the jth output interval. Then Theorems 8 and 9 can be applied to this quantized channel. By letting K and J approach infinity in such a way that the interval around each point approaches 0, the sums over k and j in Theorems 8 and 9 become Riemann integrals, and the bounds are still valid if the integrals exist.[8] Thus we have proved the following theorem.

Theorem 10

Let $P(y \mid x)$ be the transition probability density of an amplitude-continuous channel and assume that each code word is constrained to satisfy $\sum_{n=1}^{N} f(x_n) \leq 0$. Then for any block length N, any number of code words, $M = e^{NR}$, and any probability distribution on the use of the code words, there exists a code for which

$$P_e \leq B \exp\left[-N\{E_0(\rho, \mathbf{p}, r) - \rho R\}\right] \quad (115)$$

$$E_0(\rho, \mathbf{p}, r) = -\ln \int_{-\infty}^{\infty}\left[\int_{-\infty}^{\infty} p(x)P(y|x)^{1/(1+\rho)}e^{rf(x)}dx\right]^{1+\rho} dy \quad (116)$$

$$B = \left(\frac{e^{r\delta}}{q}\right)^{1+\rho} \quad \text{for any } \rho, \quad 0 \leq \rho \leq 1 \quad (117)$$

Also, for any $\rho \geq 1$, we have for each code word

$$P_{em} < \exp - N\left\{E_x(\rho, \mathbf{p}, r) - \rho\left[R + \frac{2}{N}\ln\frac{2e^{r\delta}}{q}\right]\right\} \quad (118)$$

$$E_x(\rho, \mathbf{p}, r) = -\rho \ln \int_{-\infty}^{\infty}\int_{-\infty}^{\infty} p(x)p(x')e^{rf(x)+rf(x')} \left(\int_{-\infty}^{\infty} \sqrt{P(y|x)P(y|x')}\,dy\right)^{1/\rho} dx\,dx' \quad (119)$$

[8] For the details of this limiting argument, see Gallager [3], Section 8.

Equations (115)–(119) are valid if the Riemann integrals exist for any $r \geq 0$, $\delta > 0$, and $p(x)$ satisfying $\int_{-\infty}^{\infty} p(x)f(x)\,dx \leq 0$. If $\int p(x)f(x)\,dx = 0$ and $\int p(x)\,|f(x)|^3\,dx < \infty$, then $e^{r\delta}/q \approx \sqrt{2\pi N}\sigma_f re$ [see (111)].

In the absence of any input constraint, (115)–(119) still can be applied by setting $r = 0$, and $q = 1$.

Additive Gaussian Noise

As an example of the use of (115)–(119), we consider a time-discrete, amplitude-continuous, additive Gaussian noise channel. For such a channel, when $\mathbf{x}_m = (x_{m1}, \cdots, x_{mN})$ is transmitted, the received sequence, $\mathbf{y}$, can be represented as $(x_{m1} + z_1, \cdots, x_{mN} + z_N)$ when the z_n are Gaussian random variables, statistically independent of each other and of the input. We can assume without loss of generality that the scales of x and y are chosen so that each z_n has mean 0 and unit variance. Thus

$$P(y|x) = \frac{1}{\sqrt{2\pi}} e^{-(y-x)^2/2} \quad (120)$$

We assume that each code word is power-constrained to satisfy

$$\sum_{n=1}^{N} x_{mn}^2 \leq NA \quad (121)$$

or

$$\sum_{n=1}^{N} f(x_{mn}) \leq 0; \qquad f(x) = x^2 - A \quad (122)$$

The quantity A in (121) and (122) is the *power* SNR per degree of freedom. One's intuition at this point would suggest choosing $p(x)$ to be Gaussian with variance A, and it turns out that this choice of $p(x)$ with an appropriate r yields a stationary point of $E_0(\rho, \mathbf{p}, r)$. Thus

$$p(x) = \frac{1}{\sqrt{2\pi A}} e^{-x^2/2A} \quad (123)$$

If we substitute (120), (122), and (123) in (116), the integrations are straightforward, and we get

$$E_0(\rho, \mathbf{p}, r) = rA(1 + \rho) + 1/2 \ln(1 - 2rA) + \frac{\rho}{2}\ln\left(1 - 2rA + \frac{A}{1+\rho}\right) \quad (124)$$

Making the substitution $\beta = 1 - 2rA$ in (124) for simplicity and maximizing (124) over β, we get

$$\beta = 1/2\left\{1 - \frac{A}{1+\rho} + \sqrt{\left(1 - \frac{A}{1+\rho}\right)^2 + \frac{4A}{(1+\rho)^2}}\right\} \quad (125)$$

Using (124) to maximize (116) over ρ, we get

$$P_e \leq B \exp - NE(R) \tag{126}$$

where $E(R)$ is given by the parametric equations in ρ, $0 \leq \rho \leq 1$,

$$E(R) = \tfrac{1}{2} \ln \beta + \frac{(1+\rho)(1-\beta)}{2} \tag{127}$$

$$R = \tfrac{1}{2} \ln \left(\beta + \frac{A}{1+\rho}\right) \tag{128}$$

where, for each ρ, β is given by (125).

The constant B in (126) is given by (104) and (111), where δ_f, from (122), is $\sqrt{2A}$. Thus, as $N \to \infty$

$$B \approx \left[4\pi N A^2 \left(\frac{1-\beta}{2A}\right)^2 e^2\right]^{(1+\rho)/2} = [\pi N e^2 (1-\beta)^2]^{(1+\rho)/2} \tag{129}$$

Equations (127) and (128) are applicable for $0 \leq \rho \leq 1$, or by substituting (125) in (128), for

$$\frac{1}{2} \ln \frac{1}{2}\left\{1 + \frac{A}{2} + \sqrt{1 + \frac{A^2}{4}}\right\} \leq R \leq \tfrac{1}{2} \ln (1 + A) \tag{130}$$

The left-hand side of (130) is R_{crit} and the right-hand side is channel capacity. In this region, $E(R)$ is the same exponent, rate curve derived by Shannon,[9] and this is the region in which Shannon's upper and lower bound exponent agree. Shannon's coefficient, however, is considerably tighter than the one given here.

In order to get the low-rate expurgated bound on error probability, we substitute (120), (122), and (123) in (119). After a straightforward integration, we get

$$E_x(\rho, \mathbf{p}, r) = 2r\rho A + \frac{\rho}{2} \ln (1 - 2rA) + \frac{\rho}{2} \ln \left(1 - 2rA + \frac{A}{2\rho}\right) \tag{131}$$

Letting $\beta_x = 1 - 2rA$, we find that $E_x(\rho, \mathbf{p}, r)$ is maximized by

$$\beta_x = \frac{1}{2}\left[1 - \frac{A}{2\rho} + \sqrt{1 + \frac{A^2}{4\rho^2}}\right] \tag{132}$$

Finally, optimizing (118) over ρ, we find

$$P_{em} < \exp - NE(R)$$

where $E(R)$ is given by the parametric equations for $\rho \geq 1$,

$$E(R) = \rho(1 - \beta_x)$$

$$R = \tfrac{1}{2} \ln \left(\beta_x + \frac{A}{4\rho}\right) - \frac{2}{N} \ln \frac{2e^{r\delta}}{q} \tag{133}$$

[9] The equivalence of (128) and (129) to Shannon's [9] equations (5) and (11) is seen only after a certain amount of algebra. The correspondence between the various parameters is as follows: we put Shannon's quantities on the right and use A_s for Shannon's A:

$$A = A_s^2 \quad \rho = \frac{A_s G(\theta_1) \sin^2 \theta_1}{\cos \theta_1} - 1 \; ; \beta = \frac{1}{[G(\theta_1)]^2 \sin^2 \theta_1}$$

Here, as before, for N large

$$\frac{e^{r\delta}}{q} \approx \sqrt{\pi N}\, e(1 - \beta_x) \tag{134}$$

If we let

$$R' = R + \frac{2}{N} \ln \frac{2e^{r\delta}}{q} \tag{135}$$

we can solve (132)–(134) explicitly to get

$$E(R) = \frac{A}{4}\left(1 - \sqrt{1 - e^{-2R'}}\right) \tag{136}$$

for

$$R' \leq \frac{1}{2} \ln \left(\frac{1}{2} + \frac{1}{2}\sqrt{1 + \frac{A^2}{4}}\right) \tag{137}$$

The exponent given by (136) is larger than the low-rate exponent given by Shannon [9], the difference being equal to R'.

For rates between those specified by (130) and (137), we can use either (124) or (131) with $\rho = 1$. Either way, we get

$$P_e < B \exp - N\left[(1 - \beta) + \frac{1}{2} \ln \left(\beta + \frac{A}{4} - R\right)\right] \tag{138}$$

$$\beta = \frac{1}{2}\left[1 - \frac{A}{2} + \sqrt{1 + \frac{A^2}{4}}\right] \tag{139}$$

Figure 9 shows the $E(R)$ curve given by these equations for various SNR's.

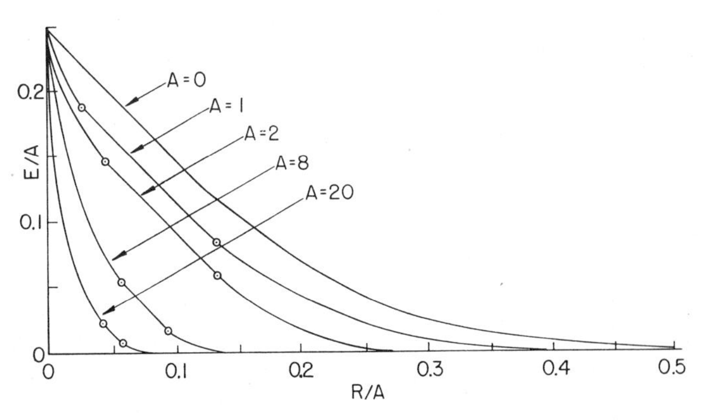

Fig. 9. Exponent, rate curve for additive Gaussian noise. A, power SNR.

Appendix

Both Theorems 2 and 7 require the following lemma in their proofs:

Lemma

Let $a_1, \cdots, a_L$ be a set of non-negative numbers and let $q_1, \cdots, q_L$ be a set of probabilities. Then

$$f(x) = \ln \left(\sum_l q_l a_l^{1/x}\right)^x \tag{140}$$

is nonincreasing with $x > 0$ and is strictly decreasing unless the a_l for which $q_l \neq 0$ are all equal. Also, $f(x)$

is convex downward for $x > 0$ and is strictly convex downward unless all the nonzero a_l for which $q_l \neq 0$ are equal.

Proof: It is a well-known property of weighted means (see Hardy, et al. [14]) that $(\sum_l q_l a_l^r)^{1/r}$ is a nondecreasing function of r for $r > 0$ and is strictly increasing unless the a_l for which $q_l \neq 0$ are all equal. Let $x = 1/r$; this implies that $f(x)$ is nonincreasing or strictly decreasing with $x > 0$. Another property of weighted means[10] stemming from Holder's inequality is that if r and t are unequal positive numbers, θ satisfies $0 < \theta < 1$, and $s = \theta r + (1 - \theta)t$, then

$$\sum_l q_l a_l^s \leq \left(\sum_l q_l a_l^r\right)^\theta \left(\sum q_l a_l^t\right)^{1-\theta} \tag{141}$$

with equality only if all of the nonzero a_l for which $q_l \neq 0$ are equal. Let λ be defined by

$$\lambda = \frac{r\theta}{r\theta + t(1 - \theta)}$$
$$\frac{1}{s} = \frac{\lambda}{r} + \frac{1 - \lambda}{t} \tag{142}$$

Substituting (142) in (141) and taking the $1/s$ power of each side, we get

$$\left(\sum_l q_l a_l^s\right)^{1/s} \leq \left(\sum q_l a_l^r\right)^{\lambda/r} \left(\sum q_l a_l^t\right)^{(1-\lambda)/t} \tag{143}$$

Taking the logarithm of both sides of (143) and interpreting $1/r$ and $1/t$ as two different values of x, we find that $f(x)$ is convex downward with strict convexity under the stated conditions.

Proof of Theorem 2:[11]

$$E_0(\rho, \mathbf{p}) = -\ln \sum_{j=1}^{J} \left(\sum_{k=1}^{K} p_k P_{jk}^{1/(1+\rho)}\right)^{1+\rho}$$

From the lemma, $(\sum_k p_k P_{jk}^{1/(1+\rho)})^{1+\rho}$ is nonincreasing with ρ. Since $I(\mathbf{p}) \neq 0$ by assumption, there is at least one j for which P_{jk} changes with k for $p_k \neq 0$; for that j, $(\sum_k p_k P_{jk}^{1/(1+\rho)})^{1+\rho}$ is strictly decreasing, and thus $E_0(\rho, \mathbf{p})$ is strictly increasing with ρ. From direct calculation we see that $E_0(0, \mathbf{p}) = 0$ and consequently it follows that for $\rho > 0$ $E_0(\rho, \mathbf{p}) > 0$ and $\partial E_0(\rho, \mathbf{p})/\partial\rho > 0$. By direct differentiation, it is also seen that $\partial E_0/\partial\rho\,|_{\rho=0} = I(\mathbf{p})$. Next, let ρ_1 and ρ_2 be unequal positive numbers, let λ satisfy $0 < \lambda < 1$, and let $\rho_3 = \lambda\rho_1 + (1 - \lambda)\rho_2$. From the lemma,

$$\left(\sum_k p_k P_{jk}^{1/(1+\rho_3)}\right)^{1+\rho_3} \leq \left(\sum_k p_k P_{jk}^{1/(1+\rho_1)}\right)^{\lambda(1+\rho_1)} \left(\sum_k p_k P_{jk}^{1/(1+\rho_2)}\right)^{(1-\lambda)(1+\rho_2)} \tag{144}$$

We now apply Holder's inequality,[12] which states that if a_j and b_j are sets of non-negative numbers, then

$$\sum_j a_j b_j \leq \left(\sum_j a_j^{1/\lambda}\right)^\lambda \left(\sum_j b_j^{1/(1-\lambda)}\right)^{1-\lambda} \tag{145}$$

with equality only if the a_j and b_j are proportional. Summing (144) over j, letting a_j and b_j be the two terms on the right, and using (145), we obtain

$$\sum_j \left(\sum_k p_k P_{jk}^{1/(1+\rho_3)}\right)^{1+\rho_3} \leq \left[\sum_j \left(\sum_k p_k P_{jk}^{1/(1+\rho_1)}\right)^{1+\rho_1}\right]^\lambda \cdot \left[\sum_j \left(\sum_k p_k P_{jk}^{1/(1+\rho_2)}\right)^{1+\rho_2}\right]^{1-\lambda} \tag{146}$$

Taking the logarithm of (146) establishes that $E_0(\rho, \mathbf{p})$ is convex upward and thus that $\partial^2 E_0/\partial\rho^2 \leq 0$. The convexity is strict unless both (144) and (145) are satisfied with equality. But condition 1) of Theorem 2 is the condition for (144) to be satisfied with equality and condition 2) is the condition for a_j and b_j to be proportional when condition 1) is satisfied.

Proof of Theorem 4: We begin by showing that $F(\rho, \mathbf{p})$ is a convex downward function of $\mathbf{p}$ for $\rho \geq 0$. From (40) we can rewrite $f(\rho, \mathbf{p})$ as

$$F(\rho, \mathbf{p}) = \sum_j \alpha_j^{1+\rho}; \qquad \alpha_j = \sum_k p_k P_{jk}^{1/(1+\rho)} \tag{147}$$

Let $\mathbf{p} = (p_1, \cdots, p_K)$ and $\mathbf{q} = (q_1, \cdots, q_K)$ be arbitrary probability vectors, and let

$$\alpha_j = \sum_k p_k P_{jk}^{1/(1+\rho)}, \quad \text{and} \quad \beta_j = \sum_k q_k P_{jk}^{1/(1+\rho)}$$

For any λ, $0 < \lambda < 1$, we have

$$f(\rho, \lambda\mathbf{p} + (1 - \lambda)\mathbf{q}) = \sum_j \left[\sum_k (\lambda p_k + (1 - \lambda)q_k) P_{jk}^{1/(1+\rho)}\right]^{1+\rho} = \sum_j [\lambda\alpha_j + (1 - \lambda)\beta_j]^{1+\rho} \tag{148}$$

Since α_j and β_j must be non-negative, and since $x^{1+\rho}$ is a convex downward function of x for $\rho \geq 0$, $x \geq 0$, we can upperbound the right-hand side of (148)

$$F(\rho, \lambda\mathbf{p} + (1 - \lambda)\mathbf{q}) \leq \sum_j \lambda\alpha_j^{1+\rho} + (1 - \lambda)\beta_j^{1+\rho}$$

$$F(\rho, \lambda\mathbf{p} + (1 - \lambda)\mathbf{q}) \leq \lambda F(\rho, \mathbf{p}) + (1 - \lambda)F(\rho, \mathbf{q}) \tag{149}$$

Thus $F(\rho, \mathbf{p})$ is convex downward with $\mathbf{p}$ for $\rho \geq 0$.

The general problem of finding necessary and sufficient conditions for the vector that minimizes a differentiable convex downward function over a convex region of vector space defined by a set of inequalities has been solved by Kuhn and Tucker [17]. For the special case in which the

[10] Hardy, et al. [14], Theorem 17.

[11] The proof of convexity given here is due primarily to H. L. Yudkin.

[12] Hardy, et al., *op. cit.* [14], Theorem 17.

region is constrained by $p_k \geq 0$ for $1 \leq k \leq K$ and $\sum_k p_k = 1$, their solution reduces to

$$\frac{\partial F(\rho, \mathbf{p})}{\partial p_k} \geq u \quad \text{for all } k \text{ with equality if } p_k \neq 0 \tag{150}$$

Differentiating $F(\rho, \mathbf{p})$ and solving for the constant u, we immediately get (41). Similarly, if we substitute the convex downward function, $-I(\mathbf{p})$, in (150), then (42) follows.

Finally, we observe that $F(\rho, \mathbf{p})$ is a continuous function of $\mathbf{p}$ in the closed bounded region in which $\mathbf{p}$ is a probability vector. Therefore, $F(\rho, \mathbf{p})$ has a minimum, and thus (41) has a solution.

Proof of Theorem 7:

$$E_x(\rho, \mathbf{p}) = -\rho \ln \sum_{k,i} p_k p_i \left(\sum_j \sqrt{P_{jk} P_{ji}} \right)^{1/\rho}$$

If we make the associations $p_k p_i = q_l$, $\sum_j \sqrt{P_{jk} P_{ji}} = a_l$, and $\rho = x$, we see that the lemma applies immediately to $-E_x(\rho, \mathbf{p})$. Since $I(\mathbf{p}) \neq 0$ by assumption, $\sum_j \sqrt{P_{jk} P_{ji}}$ cannot be independent of k and i, and $E_x(\rho, \mathbf{p})$ is strictly increasing with ρ. Also, $E_x(\rho, \mathbf{p})$ is convex upward with ρ, and the convexity is strict unless $\sum_j \sqrt{P_{jk} P_{ji}}$ is always 1 or 0 for $p_k p_i \neq 0$. But $\sum_j \sqrt{P_{jk} P_{ji}} = 1$ only if $P_{jk} = P_{ji}$ for all j, and $\sum_j \sqrt{P_{jk} P_{ji}} = 0$ only if $P_{jk} P_{ji} = 0$ for all j.

References

[1] Shannon, C. E., A mathematical theory of communication, *Bell Sys. Tech. J.*, vol. 27, pp. 379, 623, 1948. See also book by same title, University of Illinois Press, Urbana, 1949.

[2] Fano, R. M., *Transmission of information*, The M.I.T. Press, Cambridge, Mass., and John Wiley & Sons, Inc., New York, N. Y., 1961.

[3] Gallager, R. G., Information theory, in *The mathematics of physics and chemistry*, H. Margenau and G. M. Murphy; Eds., D. Van Nostrand Co., Princeton, N. J., vol. 2, 1964.

[4] Wozencraft, J. M., and R. S. Kennedy, Coding and communication, presented at the URSI Conf., Tokyo, Japan, Sep 1963.

[5] Blackwell, D., and M. A. Girshick, *Theory of games and statistical decision*, John Wiley & Sons, Inc., New York, N. Y., 1954.

[6] Elias, P., Coding for two noisy channels, in *Third London Sumposium on Information Theory*, C. Cherry, Ed., Butterworth's Scientific Publications, London, England, 1955.

[7] Reiffen, B., A note on very noisy channels, *Inform. Control*, vol. 6, p. 126, 1963.

[8] Fano, R. M., private communication, 1963.

[9] Shannon, C. E., Probability of error for optimal codes in a Gaussian channel, *Bell System Tech. J.*, vol. 38, p. 611, 1959.

[10] ——, The zero-error capacity of a noisy channel, *IRE Trans. on Information Theory*, vol. IT-2, pp. 8–19, Sep 1956.

[11] Peterson, W. W., *Error correcting codes*, The M.I.T. Press, Cambridge, Mass., and John Wiley & Sons, Inc., New York, N. Y., 1961.

[12] Gallager, R. G., *Low density parity check codes*, The M.I.T. Press, Cambridge, Mass., 1963.

[13] Gnedenko, B. V., and A. N. Kolmogorov, *Limit distributions for sums of independent random variables*, Addison-Wesley Publishing Co., Cambridge, Mass., 1954.

[14] Hardy, G. H., J. E. Littlewood, and G. Polya, *Inequalities*, Cambridge University Press, Cambridge, England, Theorem 16, 1959.

[15] Kuhn, H. W., and A. W. Tucker, Nonlinear programming, in *Second Berkeley Symposium on Mathematical Statistics and Probability*, J. Neyman, Ed., University of California Press, Berkeley, p. 481, 1951.

The Capacity of the Band-Limited Gaussian Channel

By A. D. WYNER

(Manuscript received December 27, 1965)

Shannon's celebrated formula $W \ln(1 + P_o/N_oW)$ for the capacity of a time-continuous communication channel with bandwidth W cps, average signal power P_o, and additive Gaussian noise with flat spectral density N_o has never been justified by a coding theorem (and "converse"). Such a theorem is necessary to establish $W \ln(1 + P_o/N_oW)$ as the supremum of those transmission rates at which one may communicate over this channel with arbitrarily high reliability as the coding and decoding delay becomes large.

In this paper, a number of physically consistent models for this time-continuous channel are proposed. For each model the capacity is established as $W \ln(1 + P_o/N_oW)$ by means of a coding theorem and converse.

I. INTRODUCTION

As an idealized model for the time-continuous Gaussian channel (with bandwidth W cycles per second, two-sided noise spectral density $N_o/2$, and average power P_o), Shannon[1,2] employed the mathematical time-discrete channel which passes $2W$ real numbers x per second, with the average of x^2 restricted to be P_o. Each input x is perturbed by an independent "noise" random variable which is Gaussian with mean zero and variance N_oW. If by "channel capacity" we mean the maximum rate at which a channel is capable of transmitting information with arbitrarily small error probability as the coding and decoding delay becomes large, then the capacity of this time-discrete channel is given by the celebrated formula $W \log_2 (1 + P_o/N_oW)$ bits per second (or $W \ln (1 + P_o/N_oW)$ nats per second).

In order to show that the capacity is given by this formula, it is necessary to prove a coding theorem (showing the possibility of achieving "error-free" communication at any rate less than $W \log_2 (1 + P_o/N_oW))$, and a "converse" (showing the impossibility of achieving "error-free" coding at a rate exceeding this quantity). For this — purely mathematical — channel these theorems have been proved, and there is no question as to the meaning and validity of the capacity formula.

The way in which Shannon arrived at this time-discrete model for a "physical" time-continuous channel is described in detail in Section II. It will suffice to remark here that there remain questions as to the relation of this time-discrete model (and the resulting capacity formula) to a physically meaningful time-continuous channel. These difficulties center on the fact that the inputs and outputs of the time-continuous channel are band-limited signals which are not physically realizable. As we shall see in Section II, such assumptions lead to a number of anomalies and absurdities.

Our purpose in this paper is to find physically consistant mathematical models for the time-continuous band-limited Gaussian channel, and to establish their capacity by means of a coding theorem and converse. Schematically our results are of the following form:

Let $a(T,W,P_o)$ be a class of functions which are "approximately band-limited to W cycles per second and approximately time-limited to T seconds", and which have "average power" P_o. The channel inputs must be members of a. The noise is additive, stationary, and Gaussian with flat two-sided spectral density $N_o/2$ in the band 0 — W cycles per second (or "approximately" given as above). Then the channel capacity, defined as the maximum rate for which arbitrarily high reliability is possible (using signals from a) as T becomes large, is given "approximately" by $W \log_2 (1 + P_o/N_oW)$. The term "approximately" used here will, of course, be given a precise meaning below.

In Section II, Shannon's model and results are discussed, and in Section III our models and results are stated completely and discussed. Our proofs follow in Sections IV and V. A glossary is included at the end of the paper.

II. THE SHANNON MODEL

2.1 *The Time-Discrete Channel*

In order to fix ideas as well as to review some results which will be required subsequently, let us consider the following class of (time-discrete) channels: Every T seconds the input to the channel is a sequence of $n = [\alpha T]$ real numbers $\mathbf{x} = (x_1, x_2, \cdots, x_n)$, where $\alpha (0 < \alpha \leqq \infty)$ is a fixed parameter. Further, the input sequence must satisfy the "energy" constraint

$$E(\mathbf{x}) = \sum_{k=1}^{n} x_k^2 \leqq PT, \tag{1}$$

where $P > 0$ is another fixed parameter, and where $E(\mathbf{x})$ is, as indicated, the sum of the squares of the components of $\mathbf{x}$.

The channel output is also a real n-sequence $\mathbf{y} = (y_1, y_2, \cdots, y_n)$, where

$$y_k = x_k + z_k, \qquad k = 1,2, \cdots, n, \tag{2}$$

and the noise digits z_k $(k = 1,2, \cdots, n)$ are independent, normally distributed random variables with mean zero and variance N.

Let us assume that this channel is to be used in the communication system of Fig. 1. The output of the message source is a sequence of independent and equally likely binary digits which appear at the input of the coder at the rate of R_b digits (bits) per second. Every T seconds the coder input is one of $M = 2^{R_bT}$ binary sequences, each sequence being equally likely. Let us number the possible messages as $1, 2, \cdots, M$. The coder contains a mapping of the message set $\{1, 2, \cdots, M\}$ to a set (called a *code*) of M real n-sequences $\{\mathbf{x}_1, \mathbf{x}_2, \cdots, \mathbf{x}_M\}$ (called *code words*) satisfying (1). If message $i (i = 1, 2, \cdots, M)$ is the coder input, then the coder output (and hence channel input) is the code word $\mathbf{x}_i$. Since it takes T seconds to transmit a code word, the system can process information continuously without a "backup" at the coder input. The transmission rate is R_b bits per second or $R = (\ln 2)R_b$ nats per second.

It is the task of the receiver (or decoder) to examine the received sequence $\mathbf{y}$, and determine which of the M code words was actually transmitted. Thus, we may think of the decoder as a rule which assigns to each possible received sequence $\mathbf{y}$, a code word $\mathbf{x}_i$. Let us denote by P_{ei} the probability that the decoder chooses the wrong code word given that $\mathbf{x}_i$ was transmitted. The over-all error probability is then

$$P_e = \frac{1}{M} \sum_{i=1}^{M} P_{ei}. \tag{3}$$

A transmission rate R (nats per second) is said to be ***permissible*** if for every $\lambda > 0$ one can find a T sufficiently large and a code with parameter T with $M = [e^{RT}]$ code words and $P_e \leqq \lambda$. With such a code, the system could process $R_b = R/\ln 2$ bits per second. We define the *channel capacity* C as the supremum of permissible rates. For the channel under discussion the channel capacity is given by the celebrated formula

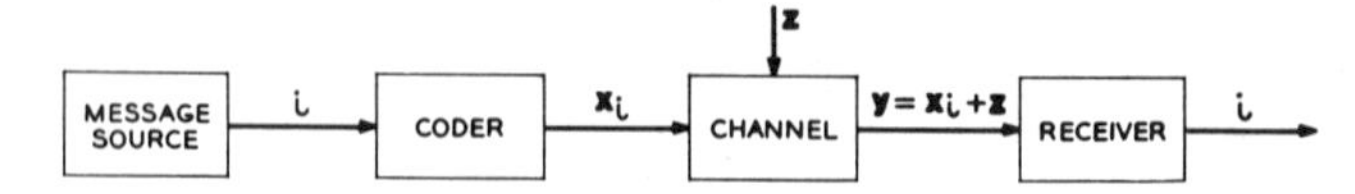

Fig. 1. — Time-discrete channel.

$$C = C_\alpha = \frac{\alpha}{2} \ln \left(1 + \frac{P}{\alpha N}\right). \tag{4}$$

In order to establish C as the capacity, one must prove two theorems. The first ("direct half") states that any $R < C$ is a permissible rate; that is, there exist codes with vanishingly small P_e as $T \to \infty$. The second theorem ("weak converse") states that no $R > C$ is a permissible rate; that is, for any sequence of codes with rate $R > C$, P_e is bounded away from zero. This has been done for the present channel for the case of a finite α by Shannon.[1,2,3] Let us observe that if we let $\alpha \to$

Reprinted with permission from *Bell Syst. Tech. J.*, vol. 45, pp. 359-371, Mar. 1965. Copyright © 1965, The American Telephone and Telegraph Company.

∞ in (4), we have $C_\alpha \xrightarrow{\alpha} P/2N$. The fact that $C_\infty = P/2N$ has been established by Ash.[4] The reader is referred to Ash [Ref. 5, Chapter 8] for a complete discussion of the above. The significance of the channel capacity then, is that it is the maximum rate for which arbitrarily high reliability is possible using signals in a certain class (i.e., those which satisfy (1)) with sufficiently long delay T.

2.2 *Application to the Band-Limited Gaussian Channel*

Shannon[1,2] has applied the above results to the communication system of Fig. 2. As above, the message source emits binary digits at the rate of R_b per second, and after T seconds, one of $M = 2^{R_b T}$ possible messages appears at the coder input. Corresponding to the ith message ($i = 1,2, \cdots, M$) the coder output is the function

$$x_i(t) = \sum_{k=1}^{n} x_{ik}\delta(t - k/2W), \tag{5a}$$

where $\delta(t)$ is the unit impulse, $n = [2WT]$, and the $\{x_{ik}\}_{k=1}^{n}$ satisfy

$$\sum_{k=1}^{n} x_{ik}^2 \leq 2WP_oT, \qquad i = 1,2,\cdots, M. \tag{5b}$$

As for the time-discrete channel, the coder must contain a set of M real n-sequences. The channel input $s_i(t)$ is the result of passing $x_i(t)$ through an ideal low-pass filter with transfer function

$$H(\omega) = \begin{cases} \dfrac{1}{2W} & |\omega| \leq 2\pi W, \\ 0 & |\omega| > 2\pi W, \end{cases} \tag{6}$$

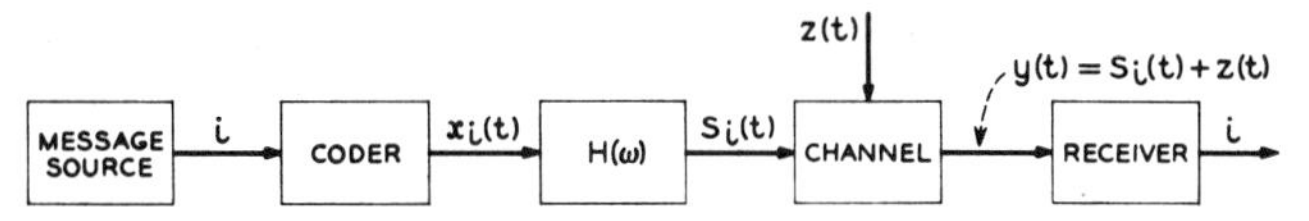

Fig. 2 — Shannon's time-continuous band-limited channel.

so that

$$s_i(t) = \sum_{k=1}^{n} x_{ik}\left[\frac{\sin 2\pi W(t - k/2W)}{2\pi W(t - k/2W)}\right]. \tag{7}$$

Thus, it takes T seconds to generate the filter input, and the system can process information at a rate of $R = (\ln 2)R_b$ nats per second without a "backup" at the coder input. Let us also remark that although the signal $s_i(t)$ is generated in T seconds, due to the physical unrealizability of $H(\omega)$, $s_i(t)$ is nonzero almost everywhere on $(-\infty,\infty)$. This leads to a fundamental difficulty which we shall discuss later.

Let $s(t)$ be the input to the channel due to a repeated application of the coding process (every T seconds). Then $s(t)$ is bandlimited to W cycles per second, and

$$\operatorname*{limit}_{T_o \to \infty} \frac{1}{T_o}\int_{-T_o/2}^{T_o/2} s^2(t)dt \leq P_o. \tag{8}$$

Inequality (8) follows from (5b) and the orthogonality of

$$\frac{\sin 2\pi W(t - k/2W)}{2\pi W(t - k/2W)} \quad \text{and} \quad \frac{\sin 2\pi W(t - k'/2W)}{2\pi W(t - k'/2W)}$$

$(-\infty < k < k' < \infty)$ on the infinite interval $(-\infty,\infty)$. Thus, the channel input is a bandlimited signal with "average power" not exceeding P_o.

Again turning our attention to Fig. 2, the channel output is a function $y(t) = s(t) + z(t)$, where $z(t)$ is a sample from a Gaussian random process with spectral density

$$N(\omega) = \begin{cases} N_o/2 & |\omega| \leq 2\pi W, \\ 0 & |\omega| > 2\pi W. \end{cases} \tag{9a}$$

The corresponding autocorrelation function of the noise is

$$R(\tau) = \mathcal{E}[z(t)z(t + \tau)] = N_oW\frac{\sin 2\pi W\tau}{2\pi W\tau}, \tag{9b}$$

where $\mathcal{E}$ denotes expectation.

Again it is the function of the receiver (or decoder) to examine $y(t)$ and determine what the input information was. Let us consider the signal $s_i(t)$ (7), which was generated during the interval $[0,T]$. The coefficients $\{x_{ik}\}_{k=1}^{n}$ are the values of $s_i(t)$ at the "sampling instants" $t = k/2W$, $k = 1,2, \cdots, n$. Since the noise is also bandlimited, the received signal $y(t)$ is bandlimited and may be completely characterized by its values at the sampling instants $y_k = y(k/2W)$, $k = 0, \pm1, \pm2, \cdots$. Clearly

$$y_k = x_{ik} + z_k, \qquad k = 1,2, \cdots, n, \tag{10}$$

where $z_k = z(k/2W)$ is the value of the noise $z(t)$ at the sampling instant $t = k/2W$. Since $s_i(k/2W) = 0$, for $k < 1$ and $k > n$, the only useful samples of y are $\{y_k\}_{k=1}^{n}$. Further it follows directly from (9b) that the z_k are independent, normally distributed random variables with mean zero and variance N_oW. Thus, it suffices to consider the input and output as n-sequences $\mathbf{x}_i = (x_{i1}, x_{i2}, \cdots, x_{in})$ and $\mathbf{y} = (y_1, \cdots, y_n)$ $(n = 2WT)$ related by (10). Let us remark here, that the code words corresponding to previous and successive intervals will not cause any interference with the code word corresponding to the interval $[0,T]$, since these other code words are zero at the sampling instants.

Inequality (5b) and (10) permit us to apply the results for the time-discrete Gaussian channel discussed above with parameters $\alpha = 2W$, $P = 2WP_o$, and $N = N_oW$. We conclude that this communication system (in Fig. 2) is capable of processing information at any rate R less than

$$C = W \ln\left(1 + \frac{P_o}{N_oW}\right), \tag{11}$$

with vanishingly small error probability as T becomes large. Since the channel inputs are bandlimited to W cycles per second, and by (8) have average power not exceeding P_o, it is generally believed that the *capacity* (taken as the maximum "error-free rate") of a channel which admits only bandlimited signals with average power P_o is given by (11). In fact, it has only been shown that it is possible to do at least as well as C (using the system of Fig. 2), and no converse has been proven. This is the first difficulty with the Shannon model which we shall attempt to remedy.

Further, there are other difficulties inherent in the use of this model. We are taking "capacity" to be a (maximum) transmission *rate*, but what is the rate for the system of Fig. 2? We have said merely that the *coder* can process information at a rate of R nats per second. However, because of the physical unrealizability of $H(\omega)$, we must discard all temporal notions about the channel input $s_i(t)$ as well as the output $y(t)$. The notion of *rate*, therefore, has only a limited meaning. In fact, since the received signal $y(t)$ is an entire function, it is perfectly predictable for all time from observations over a finite interval. Thus the receiver, by observing $y(t)$ in a tiny interval, could extrapolate $y(t)$ for all time and obtain sample values at an arbitrarily high rate. This anomaly is the second difficulty with the Shannon model.

It is the purpose of this paper to present a model for the time-continuous band-limited Gaussian channel for which the capacity (defined as the maximum "error-free rate") is given by (11). This will necessitate proving a "direct half" and "converse" to a coding theorem. Further, the model should avoid the second difficulty mentioned above. We shall obtain results of the following form:

Let $a(T,W,P_o)$ be a class of functions which are "approximately bandlimited to W cycles per second and approximately time-limited to T seconds", and which have total "energy" not exceeding P_oT. The noise is taken to be stationary and Gaussian with spectral density given (or "approximately" given) by (9a). Then the channel capacity, defined as the maximum rate for which arbitrarily high reliability is possible (using signals from a) as T becomes large, is given "approximately" by $W \ln (1 + P_o/N_oW)$. The term "approximately" used here will, of course, be given a precise meaning below.

III. SUMMARY OF RESULTS

We shall propose four models for the channel and find the capacity of each. Each model is of the following form:

(*i*) Definition of a suitable class of allowable signal functions, $a(T,W,P_o)$, which are "approximately bandlimited to W cycles per second, approximately time-limited to T seconds", and with total energy not exceeding P_oT.

(*ii*) Definition of the noise — taken to be stationary additive Gaussian noise with spectral density $N(\omega)$, which is "approximately" given by (9a).

We shall take W and P_o to be fixed parameters. A *code* with parameter T is a set of M functions (called *code words*) in $a(T,W,P_o)$. The transmission rate R is defined by $R = (1/T) \ln M$, so that $M = e^{RT}$. A decoding scheme is a mapping of the space of possible received signals (code word plus a noise sample) onto the code. If code word i ($i = 1, 2, \cdots, M$) is transmitted, we take P_{ei} to be the conditional probability that the decoder chooses a code word other than i, and hence makes an error. Since all code words are equally likely to be transmitted, the over-all error probability P_e is given by (3), i.e.,

$$P_e = \frac{1}{M} \sum_{i=1}^{M} P_{ei}.$$

A transmission rate R is said to be *permissible*, if for every $\lambda > 0$ one can find a T sufficiently large and a code with $M = [e^{RT}]$ code words for which $P_e \leqq \lambda$. The *channel capacity* C is defined as the supremum of permissible rates. We shall find the capacity corresponding to a number of different $a(T,W,P_o)$ and $N(\omega)$. This will, as for the time-discrete channel, necessitate proving two coding theorems — a "direct half" and a "weak converse".

Before beginning the summary we shall need the following definitions. Let $s(t)$, $-\infty < t < \infty$, be a real-valued square-integrable function and $S(\omega)$ be its Fourier transform. Let the norm of $s(t)$ be

$$\| s \| = \left[\int_{-\infty}^{\infty} s^2(t) dt \right]^{\frac{1}{2}}. \tag{12}$$

The frequency and time "concentration" of s are

$$K_B(s, 2\pi W) = \frac{1}{2\pi} \int_{-2\pi W}^{2\pi W} | S(\omega) |^2 d\omega / \| s \|^2, \tag{13a}$$

and

$$K_D(s, T) = \int_{-T/2}^{T/2} s^2(t) dt / \| s \|^2, \tag{13b}$$

respectively. Further, let D_T be the "time-truncation" operator defined by

$$D_T s = \begin{cases} s(t) & |t| \leqq T/2, \\ 0 & |t| > T/2. \end{cases} \tag{14}$$

With these definitions in hand, we are able to state our results. In each case we shall define the channel model and then give the channel capacity. Although there are some difficulties inherent in these models, each model leads to a *mathematical theorem* which justifies Shannon's capacity formula.

Model 1: To begin with, let us take for the set a of "allowable" inputs. $a_1(T,W,P_o)$, the set of functions $s(t)$ satisfying

$$s(t) = 0, \quad | t | > T/2, \tag{15a}$$

$$\| s \|^2 \leqq P_o T, \tag{15b}$$

$$K_B(s, 2\pi W) \geqq 1 - \eta \quad (0 < \eta < 1). \tag{15c}$$

Hence, our allowable signals are functions which are strictly time-limited and approximately band-limited. As $\eta \to 0$, the allowable signals become more perfectly bandlimited. The noise spectrum is taken to be

$$N(\omega) = \begin{cases} N_o/2 & | \omega | \leqq 2\pi W, \\ \nu N_o/2 & | \omega | > 2\pi W, \end{cases} \tag{16}$$

where $0 < \nu \leqq 1$. As $\nu \to 0$, (16) is in some sense "approximately" the same as (9a). The average noise power outside the band ($| \omega | > 2\pi W$), however, is infinite. In this case, Theorem 3 establishes

$$C = C_\eta = W \ln \left(1 + (1 - \eta) \frac{P_o}{N_o W} \right) + \eta \frac{P_o}{\nu N_o} \tag{17}$$

as the channel capacity. As $\eta \to 0$, the capacity approaches the classical formula $W \ln (1 + P_o/N_o W)$.

The principal difficulty with this model is the assumption of infinite average noise power, which is hardly a physically acceptable notion. Further, there are mathematical difficulties inherent in a spectral density given by (16) which implies a covariance containing an impulse function. Often the assumption of a spectrum in (16) can be justified by the fact that it can be approximated as closely as desired in the frequency range of interest by a spectrum with finite power. However, the following theorem, the proof of which is Appendix B, renders this justification meaningless in this case.

Theorem 5: Let $a(T,W,P_o)$ be as in (15) and let the noise be additive and Gaussian with spectral density $N(\omega)$, where

$$\int_{-\infty}^{\infty} N(\omega) d\omega < \infty.$$

Then the capacity $C_\eta = \infty$ regardless of how small η may be.

Intuitively, we may see that this is true by observing that, since the above integral exists, $N(\omega)$ must be arbitrarily small in some frequency range. Hence, by placing some signal energy into this frequency range, we can make the "signal-to-noise" ratio arbitrarily large, and therefore, the permissible rate of transmission arbitrarily high.

Accordingly, we shall assume for the remaining models that the noise is additive, Gaussian, with spectral density

$$N(\omega) = \begin{cases} N_o/2 & | \omega | \leqq 2\pi W, \\ 0 & | \omega | > 2\pi W. \end{cases} \tag{18}$$

This corresponds more closely with the usual formulation of a band-limited channel. It remains to find a suitable class of input signals, $a(T,W,P_o)$. We consider some possibilities.

Model 2: This model defines $a = a_2(T,W,P_o)$ as the set of functions $s(t)$ satisfying

$$S(\omega) = 0, \quad | \omega | > 2\pi W, \tag{19a}$$

$$\| s \|^2 \leqq P_o T, \tag{19b}$$

$$K_D(s, T) \geqq 1 - \eta \quad (0 < \eta < 1). \tag{19c}$$

Thus, a_2 is a set of strictly band-limited, approximately time-limited functions. As $\eta \to 0$, the allowable signals become more perfectly time-limited. With the noise as defined in (18), Theorem 2 establishes

$$C = C_\eta = W \ln \left(1 + (1 - \eta) \frac{P_o}{N_o} \right) + \eta \frac{P_o}{N_o} \tag{20}$$

as the channel capacity. Again, as $\eta \to 0$, C_η approaches the classical formula $W \ln [1 + (P_o/N_o W)]$.

Model 2 is an intuitively plausible model for the band-limited channel, and Theorem 2 which establishes its capacity is a mathematically rigorous result which, in the limit, yields the desired capacity formula. There are, however, two difficulties inherent in this formulation. The first is that since the allowable signals $s(t)$ are band-limited, it is not possible to generate them in finite time. Thus the central idea of a transmission *rate* has, at best, a limited meaning. The Shannon model (Fig. 2) also suffers from this difficulty (see Section II). The other problem with this formulation is that if code words are transmitted sequentially, we will have an interference problem (i.e., the tails of successive signals will overlap), the resolution of which is not known at present. The following two models contain neither of these difficulties.

Model 3: This model avoids the difficulties of Model 2 by letting the code words be strictly time-limited and approximately band-limited. However, as we have seen in Theorem 5, the definition of approximately band-limited functions employed above (15) yields an infinite capacity. Thus we seek an alternate way of characterizing "approximately" band-limited or "slowly changing" functions. We proceed as follows. Let $x(t)$ be a function satisfying $x(t) = 0$, $| t | > T/2$, and $\| x \|^2 < \infty$. If $x = D_T \hat{x}$, where $\hat{x}$ is a strictly bandlimited function and D_T is defined by (14), we may define a "frequency concentration" of x by

$$K_B'(x, 2\pi W) = \frac{\| x \|^2}{\| \hat{x} \|^2}. \tag{21}$$

If we cannot express x as $D_T \hat{x}$, we take $K_B' = 0$. For example, if $x(t)$

or any of its derivatives has even a small discontinuity then we cannot write $x = D_T\hat{x}$, so that $K_B'(x,2\pi W) = 0$ and x is not approximately bandlimited in this sense. This is so no matter how large $K_B(x,2\pi W)$ may be. Conversely, it is shown in Appendix C that for any function x

$$K_B(x,2\pi W) \geqq 1 - 2\sqrt{\frac{1 - K_B'(x,2\pi W)}{K_B'(x,2\pi W)}}, \tag{22}$$

so that a K_B' close to unity implies a K_B close to unity. Thus, saying that a function x has a K_B' close to unity implies that x is "slowly changing" and that K_B is also close to unity.

We now choose that set $a = a_3(T,W,P_o)$ of allowable inputs as the set of functions $s(t)$ for which

$$s(t) = 0, \quad |t| > T/2, \tag{23a}$$

$$\| s \|^2 \leqq P_oT, \tag{23b}$$

$$K_B'(s,2\pi W) \geqq 1 - \eta \quad (0 < \eta < 1). \tag{23c}$$

Thus a_3 is a set of strictly time-limited, and approximately band-limited functions. In this case, Theorem 4 establishes

$$C = C_\eta = W \ln\left(1 + \frac{P_o}{N_oW}\right) + \frac{\eta}{1-\eta}\frac{P_o}{N_o} \tag{24}$$

as the channel capacity. Again $C_\eta \to W \ln [1 + (P_o/N_oW)]$ as $\eta \to 0$.

The significance of constraint (23c) is that it makes it impossible for the communicator to make any *use* of the high-frequency components which must of necessity be included in the signal (since it is time-limited). Model 3, therefore, provides a mathematically rigorous theorem which does not involve any complications concerning physical realizability, and yields the desired capacity.

Our final formulation is as follows:

Model 4: Let $a = a_4(T,W,P_o)$ be the set of strictly time-limited, approximately band-limited functions $s(t)$ which satisfy

$$s(t) = 0, \quad |t| \geqq T/2, \tag{25a}$$

$$\| s \|^2 \leqq P_oT, \tag{25b}$$

$$K_B(s,2\pi W) \geqq 1 - \eta. \tag{25c}$$

Now Theorem 5 (stated above) tells us that if the noise were as in (18), then the capacity is infinite. In actuality one could not be sure that the noise was absolutely band-limited. In fact, whether or not the noise is strictly band-limited is not verifiable in the laboratory. It is reasonable, therefore, to assume that the noise is given by $z(t) = z_1(t) + z_2(t)$, where $z_1(t)$ is a sample from a Gaussian random process with spectral density (18). For $z_2(t)$ we require only that

$$\int_{-T/2}^{T/2} z_2^2(t)dt \leqq \nu N_oWT, \tag{26}$$

where $\nu > 0$ is small. We place no other restrictions on the spectrum of z_2 or on its probability structure. Since the expected value of the energy of $z_1(t)$ in $[-T/2,T/2]$ is N_oWT, (26) implies that the energy of $z(t)$ is nearly all in $z_1(t)$ ($\nu \ll 1$). We shall assume that $z_2(t)$ may depend on the code and decoding rule used, on the code word transmitted, and the sample $z_1(t)$. We require our communication system to perform well no matter what $z_2(t)$ may be.

Let us say that a code (satisfying (25)) and a decoding rule have been chosen. Let us also assume that the rule for selecting $z_2(t)$ has been chosen. Let $P_e(z_2)$ be the resulting error probability. Then define

$$P_e = \max_{z_2} P_e(z_2), \tag{27}$$

where the maximization in (27) is over all rules for choosing $z_2(t)$ — with the code and decoding rule fixed. The channel capacity is the supremum of those rates for which P_e may be made to vanish as $T \to \infty$.

It can be shown (see Appendix D) that the capacity C is given by

$$C = C_{\eta,\nu} = W \ln\left(1 + \frac{P_o}{N_oW}\right) + \varepsilon(\eta,\nu), \tag{28}$$

where $\varepsilon(\eta,\nu) \to 0$ as $\eta,\nu \to 0$ provided $\nu/\eta > P_o/N_oW$, the signal-to-noise ratio. Since we may consider η and ν to be limits on the accuracy of our measuring equipment, the former on measuring the signal* and the latter on measuring the noise, it is reasonable to assume, as we did in (28), that η and ν go to zero at the same rate.

An alternate and mathematically equivalent formulation of Model 4 is as follows: Let the signals $s(t)$ be as in (25) and the noise $z(t)$ be as in (18). Now in reality one could not expect the decoder to be capable of infinitesimally accurate measurements. It is reasonable, therefore, to assume that there is an inherent uncertainty in all measurements made by the decoder, and to require that the communication system perform well despite this uncertainty. Specifically, we require that the decoding regions satisfy the following condition: If $y_1(t)$ is decoded as s_i, and $y_2(t)$ is decoded as $s_j (i \neq j)$, then

$$\int_{-T/2}^{T/2} (y_1(t) - y_2(t))^2\, dt \geqq 2\nu N_oWT. \tag{29}$$

In other words, if a received signal $y(t)$ is close to the "border" between decoding regions, we cannot, because of the uncertainty in the accuracy of our measurements, be sure to which region $y(t)$ belongs. Condition (29) forces the decoder to give up on such a $y(t)$ and to announce an error. The capacity for this alternate model is also given by (28). Let us remark that here ν is again a measure of the accuracy of our measuring instruments, this time at the decoder, so that again it is reasonable to expect η and ν to tend to zero at the same rate.

* *I.e.*, η represents a limit on the measurement of the frequency component of the signal outside the band.

Due to its unusual length, only the first three sections of this paper are reprinted here. The interested reader should consult the original for proofs of the stated results.

Lower Bounds to Error Probability for Coding on Discrete Memoryless Channels. I

C. E. SHANNON* AND R. G. GALLAGER*

Departments of Electrical Engineering and Mathematics, Research Laboratory of Electronics, Massachusetts Institute of Technology, Cambridge, Massachusetts

AND

E. R. BERLEKAMP†

Department of Electrical Engineering, University of California, Berkeley, California

New lower bounds are presented for the minimum error probability that can be achieved through the use of block coding on noisy discrete memoryless channels. Like previous upper bounds, these lower bounds decrease exponentially with the block length N. The coefficient of N in the exponent is a convex function of the rate. From a certain rate of transmission up to channel capacity, the exponents of the upper and lower bounds coincide. Below this particular rate, the exponents of the upper and lower bounds differ, although they approach the same limit as the rate approaches zero. Examples are given and various incidental results and techniques relating to coding theory are developed. The paper is presented in two parts: the first, appearing here, summarizes the major results and treats the case of high transmission rates in detail; the second, to appear in the subsequent issue, treats the case of low transmission rates.

I. INTRODUCTION AND SUMMARY OF RESULTS

The noisy channel coding theorem (Shannon, 1948) states that for a broad class of communication channels, data can be transmitted over the channel in appropriately coded form at any rate less than channel capacity with arbitrarily small error probability. Naturally there is a rub in such a delightful sounding theorem, and the rub here is that the error probability can, in general, be made small only by making the coding constraint length large; this, in turn, introduces complexity into the encoder and decoder. Thus, if one wishes to employ coding on a particular channel, it is of interest to know not only the capacity but also how quickly the error probability can be made to approach zero with increasing constraint length. Feinstein (1955), Shannon (1958), Fano (1961), and Gallager (1965) have shown that for discrete memoryless channels, block coding and decoding schemes exist for which the error probability approaches zero exponentially with increasing block length for any given data rate less than channel capacity.

This paper is concerned primarily with the magnitude of this exponential dependence. We derive some lower bounds on achievable error probability, summarized in Theorems 1 to 4 below, and compare these bounds with the tightest known general upper bounds on error probability.

A *discrete channel* is a channel for which the input and output are sequences of letters from finite alphabets. Without loss of generality, we can take the input alphabet to be the set of integers $(1, \cdots, K)$ and the output alphabet to be the set of integers $(1, \cdots, J)$. A *discrete memoryless channel* is a discrete channel in which each letter of the output sequence is statistically dependent only on the corresponding letter of the input sequence. A discrete memoryless channel is specified by its set of transition probabilities $P(j \mid k)$, $1 \leq j \leq J$, $1 \leq k \leq K$, where $P(j \mid k)$ is the probability of receiving digit j given that digit k was transmitted. If $\mathbf{x} = (k_1, k_2, \cdots, k_N)$ is a sequence of N input letters and $\mathbf{y} = (j_1, \cdots, j_N)$ is a corresponding sequence of N output letters, then for a memoryless channel

$$\Pr(\mathbf{y} \mid \mathbf{x}) = \prod_{n=1}^{N} P(j_n \mid k_n) \tag{1.1}$$

* The work of these authors was supported by the National Aeronautics and Space Administration (Grants NsG-334 and NsG-496), the Joint Services Electronics Program (contract DA-36-039-AMC-03200 (EE)), and the National Science Foundation (Grant GP-2495).

† The work of this author is supported by the Air Force Office of Scientific Research (Grant, AF-AFOSR-639-65).

A *block code* with M code words of length N is a mapping from a set of M *source messages*, denoted by the integers 1 to M, onto a set of M *code words*, $\mathbf{x}_1, \cdots, \mathbf{x}_M$, where each code word is a sequence of N letters from the channel input alphabet. A *decoding scheme* for such a code is a mapping from the set of output sequences of length N into the integers 1 to M. If the source attempts to transmit message m over the channel via this coding and decoding scheme, message m is encoded into sequence $\mathbf{x}_m$; after transmitting $\mathbf{x}_m$, some sequence $\mathbf{y}$ is received which is mapped into an integer m'. If $m' \neq m$, we say that a decoding error has occurred.

It is convenient here to consider a somewhat more general problem, *list decoding*, where the decoder, rather than mapping the received sequence into a single integer, maps it into a list of integers each between 1 and M. If the transmitted source message is not on the list of decoded integers, we say that a *list decoding error* has occurred.

List decoding was first considered by Elias (1955) for the Binary Symmetric Channel. Most of the known bounds on error probability extend readily with simple alterations to list decoding and the concept has been very useful both in providing additional insight about ordinary decoding and as a tool in proving theorems (see, for example, Jacobs and Berlekamp (1967)).

For a given code and list decoding scheme, let Y_m be the set of received sequences for which message m is on the list of decoded integers and let Y_m^c be the complement of the set Y_m. Then the probability of a list decoding error, given that the source message is m, is the conditional probability that $\mathbf{y}$ is in Y_m^c, or

$$P_{e,m} = \sum_{\mathbf{y} \in Y_m^c} \Pr(\mathbf{y} \mid \mathbf{x}_m) \tag{1.2}$$

The error probability for a given code and list decoding scheme is then defined as the average $P_{e,m}$ over m assuming that the messages are equally likely,

$$P_e = \frac{1}{M} \sum_{m=1}^{M} P_{e,m} \tag{1.3}$$

We define $P_e(N, M, L)$ as the minimum error probability for the given channel minimized over all codes with M code words of length N and all list decoding schemes where the size of the list is limited to L. $P_e(N, M, 1)$ is thus the minimum error probability using ordinary decoding. Finally the *rate* R of a code with list decoding is defined as

$$R = \frac{\ln M/L}{N} = \frac{\ln M}{N} - \frac{\ln L}{N} \tag{1.4}$$

For ordinary decoding where $L = 1$, this is the usual definition of rate and is the source entropy per channel digit for equally likely messages. For larger L, we may think of $(\ln L)/N$ as a correction term to account for the fact that the receiver is only asserting the message to be one of a list of L. For example, if $M = L$, (1.4) asserts that $R = 0$, and indeed no channel is required.

With these definitions, we can proceed to summarize the major results of the paper. The major result of Section II is Theorem 1 below, which lower bounds the error probability of a code in terms of the minimum achievable error probability at 2 shorter blocklengths.

THEOREM 1. *Let N_1, N_2 be arbitrary blocklengths and let M, L_1, and L_2 be arbitrary positive integers. Then the minimum error probability achievable for a code of M code words of length $N_1 + N_2$ is bounded by*

$$P_e(N_1 + N_2, M, L_2) \geq P_e(N_1, M, L_1) P_e(N_2, L_1 + 1, L_2) \tag{1.5}$$

In Section VI this theorem leads directly to an exponential type lower bound on error probability which for low transmission rates is considerably tighter than any previously known bound.

In Section III, codes containing only two code words are analyzed in detail. We find the trade-offs between the error probability when the first word is sent and the error probability when the second word is sent. The results, which are used in Sections IV and V, are summarized in Section III by Theorem 5 and Fig. 3.1.

The major result of Section IV is the "sphere packing" bound on error probability, given below as Theorem 2. This theorem, in slightly different

Reprinted with permission from *Inform. Contr.*, vol. 10, pp. 65–103, Jan. 1967.

form, was discovered by Fano (1961) but has not been rigorously proven before.

THEOREM 2. *Given a discrete memoryless channel with transition probabilities* $P(j \mid k)$; $1 \leqq k \leqq K$, $1 \leqq j \leqq J$; $P_e(N, M, L)$ *is lower bounded by*

$$P_e(N, M, L) \geqq \exp - N\{E_{sp}[R - o_1(N)] + o_2(N)\} \quad (1.6)$$

where the function E_{sp} *is defined by*

$$E_{sp}(R) = \underset{\rho \geqq 0}{\text{L.U.B.}} [E_0(\rho) - \rho R] \quad (1.7)$$

$$E_0(\rho) = \max_{\mathbf{q}} E_0(\rho, \mathbf{q}) \quad (1.8)$$

$$E_0(\rho, \mathbf{q}) = -\ln \sum_{j=1}^{J} \left[\sum_{k=1}^{K} q_k P(j \mid k)^{1/(1+\rho)} \right]^{1+\rho} \quad (1.9)$$

The maximum in (1.8) *is over all probability vectors* $\mathbf{q} = (q_1, \cdots, q_K)$; *that is, over all* $\mathbf{q}$ *with nonnegative components summing to* 1. *The quantities* $o(N)$ *go to* 0 *with increasing* N *and can be taken as*

$$o_1(N) = \frac{\ln 8}{N} + \frac{K \ln N}{N} \quad \text{and} \quad o_2(N) = \sqrt{\frac{8}{N}} \ln \frac{e}{\sqrt{P_{\min}}} + \frac{\ln 8}{N} \quad (1.10)$$

where $P_{\min}$ *is the smallest nonzero* $P(j \mid k)$ *for the channel and* K *and* J *are the sizes of the input and output alphabets respectively.*

The quantity in braces in (1.6) can be found graphically from $E_{sp}(R)$ by taking each point on the $E_{sp}(R)$ curve, moving to the right $o_1(N)$ and moving upward $o_2(N)$. Thus the major problem in understanding the implication of the theorem is understanding the behavior of $E_{sp}(R)$. Figure 1 sketches $E_{sp}(R)$ for a number of channels. Figure 1(a) is the typical behavior; the other sketches are examples of the rather peculiar curves that can occur if some of the $P(j \mid k)$ are zero.

For a given ρ, $E_0(\rho) - \rho R$ is a linear function of R with slope $-\rho$. Thus, as shown in Fig. 2, $E_{sp}(R)$ is the least upper bound of this family of straight lines. It is obvious geometrically, and easy to prove analytically, that $E_{sp}(R)$ is nonincreasing in R and is convex $\cup$[1] (see Fig. 2). It is shown in the appendix that $E_{sp}(R) = 0$ for $R \geqq C$ where C is channel capacity and that $E_{sp}(R) > 0$ for $0 \leqq R < C$. It sometimes happens that $E_{sp}(R) = \infty$ for sufficiently small values of R (see Fig. 1(b), (c), (d), (e)). To investigate this, we observe that for fixed ρ, $E_0(\rho) - \rho R$ intercepts the R axis at $E_0(\rho)/\rho$. As $\rho \to \infty$ this line will approach a vertical line at $R = \lim_{\rho\to\infty} E_0(\rho)/\rho$ (see Fig. 2(b)). This limiting rate is called R_∞ and $E_{sp}(R)$ is finite for $R \geqq R_\infty$ and infinite for $R < R_\infty$.

$$R_\infty = \lim_{\rho\to\infty} \max_{\mathbf{q}} \frac{-\ln \sum_j \left[\sum_k q_k P(j \mid k)^{1/(1+\rho)}\right]^{1+\rho}}{\rho}$$

Finding the limit either by expanding in a Taylor series in $1/(1 + \rho)$ or by using L'Hospitals rule,

$$R_\infty = \max_{\mathbf{q}} - \ln \max_{1 \leqq j \leqq J} \sum_k q_k \varphi(j \mid k) \quad (1.11)$$

$$\varphi(j \mid k) = \begin{cases} 1; & P(j \mid k) \neq 0 \\ 0; & P(j \mid k) = 0 \end{cases}$$

That is, for each output, we sum the input probabilities q_k that lead to that output. We then adjust the q_k to minimize the largest of these sums; R_∞ is minus the logarithm of that min-max sum. It can be seen from this that $R_\infty > 0$ iff each output is unreachable from at least one input.

[1] We will use convex $\cup$ (read convex cup) and concave $\cap$ (concave cap) as mnemonic aids to the reader for convex and concave functions. It seems as difficult for the nonspecialist to remember which is which as to remember the difference between stalagmites and stalactites.

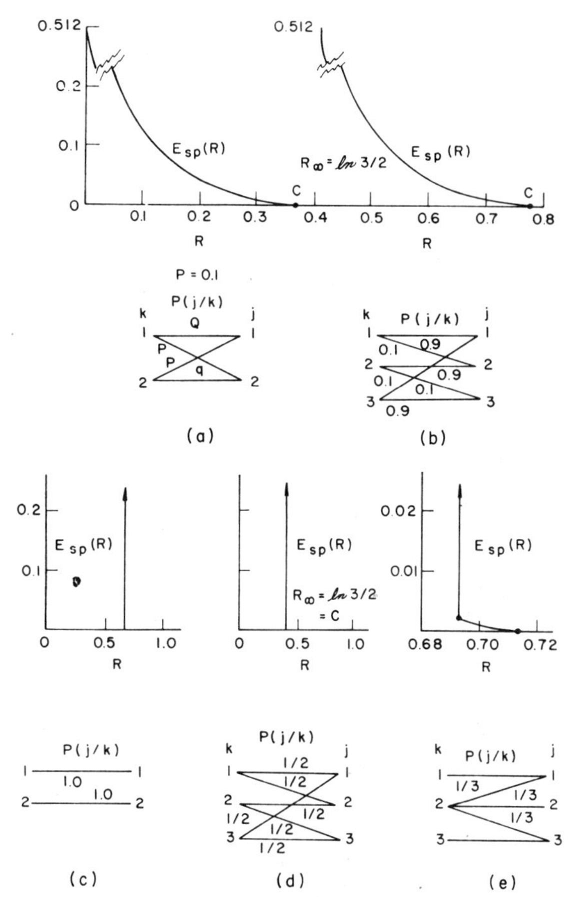

FIG. 1. The sphere packing exponent for several channels.

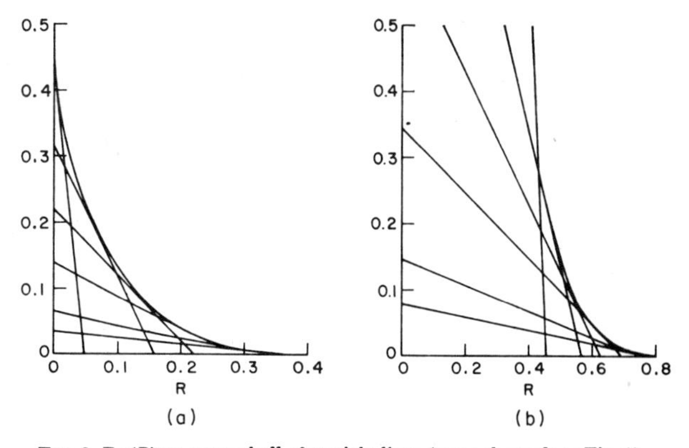

FIG. 2. $E_{sp}(R)$ as convex hull of straight lines (same channels as Fig. 1).

R_∞ is an upper bound to the zero error capacity of the channel, C_0. Shannon (1956) has defined C_0 as the least upper bound of rates at which information can be transmitted with no possibility of errors. C_0 is greater than 0 iff there are two or more inputs from which no common output can be reached and thus it is possible to have $R_\infty > 0$ and $C_0 = 0$ (see Fig. 1(b) for such a channel). Shannon (1956) has shown that if $C_0 > 0$, then the expression in (1.11) for R_∞ is equal to the zero error capacity of the channel with noiseless feedback.

If it happens that R_∞ equals channel capacity C, then the sphere packing bound merely states the true but uninteresting result that $P_e \geqq 0$ for $R < C$. It is shown in the appendix that this occurs iff the following relations are satisfied for the input probability assignment $\mathbf{q} = (q_1, \cdots, q_K)$ that yields capacity.

(a) All transition probabilities that lead to a given output with nonzero probability are the same (i.e., $P(j \mid k)$ is independent of k for those j, k such that $q_k P(j \mid k) \neq 0$).

(b) The sum of the q_k over inputs leading to a given output j is independent of the output j.

These conditions are satisfied by all noiseless channels and also a few noisy channels such as that in Fig. 1(c). For all other channels, $R_\infty < C$. It is shown in the appendix that $E_{sp}(R)$ is strictly convex ∪ and strictly decreasing in this region. $E_{sp}(R)$ need not have a continuous derivative however (see Gallager (1965), Fig. 6).

The sphere packing bound above bears a striking resemblance to the "random coding" *upper* bound on error probability of Fano (1961) and Gallager (1965). That bound, as stated by Gallager, is

$$P_e(N, M, 1) \leqq \exp - NE_r(R) \tag{1.12}$$

where

$$E_r(R) = \max_{0 \leqq \rho \leqq 1} [E_0(\rho) - \rho R] \tag{1.13}$$

Comparing $E_r(R)$ and $E_{sp}(R)$, we see that $E_{sp}(R) \geqq E_r(R)$. Equality holds iff the value of $\rho \geqq 0$ that maximizes $E_0(\rho) - \rho R$ is between 0 and 1. It can be seen from Fig. 2 that the value of $\rho \geqq 0$ that maximizes $E_0(\rho) - \rho R$ is nonincreasing with R. Consequently there exists a number called the *critical rate*, R_{crit}, such that $E_{sp}(R) = E_r(R)$ iff $R \geqq R_{crit}$. R_{crit} lies between R_∞ and C and it is shown in the appendix that $R_{crit} = C$ iff $R_\infty = C$ (i.e., if conditions (a) and (b) above are satisfied). For all other channels there is a nonzero range of rates, $R_{crit} \leqq R \leqq C$, where the upper and lower bounds on error probability agree except for the $o(N)$ terms (see Fig. 3).

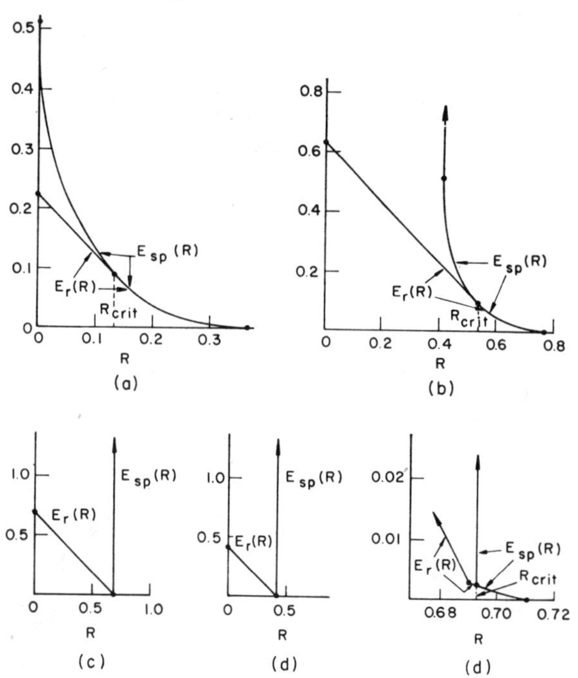

FIG. 3. Comparison of sphere packing exponent with random coding exponent (same channels as Fig. 1).

This completes our discussion of Theorem 2. For a more complete discussion of how to calculate $E_{sp}(R)$ and $E_r(R)$ see Gallager (1965). One additional result needed here, however, is the following (Gallager (1965), Theorem 4): any local maximum of (1.8) over the probability vector $\mathbf{q}$ is a global maximum, and necessary and sufficient conditions on $\mathbf{q}$ to maximize (1.8) for a given ρ are

$$\sum_j [P(j \mid k)]^{1/(1+\rho)} \alpha_j^\rho \geqq \sum_j \alpha_j^{1+\rho} \quad \text{for all} \quad k, 1 \leqq k \leqq K \tag{1.14}$$

where

$$\alpha_j = \sum_k q_k [P(j \mid k)]^{1/(1+\rho)} \tag{1.15}$$

Equation (1.14) must be satisfied with equality except for those k for which $q_k = 0$; this can be seen by multiplying both sides of (1.14) by q_k and summing over k.

In Section V (contained in Part II) we find bounds on error probability for codes with a fixed number of code words in the limit as the block length becomes large. The exponent E_M for a code with M code words is defined as

$$E_M = \limsup_{N\to\infty} \frac{-\ln P_e(N, M, 1)}{N} \tag{1.16}$$

The major result of the section is the following theorem concerning the exponents, E_M.

THEOREM 3. *Given a discrete memoryless channel with transition probabilities* $P(j \mid k)$; $1 \leqq k \leqq K$, $1 \leqq j \leqq J$, *and given that the zero error capacity is zero,* $P_e(N, M, 1)$ *is lower bounded by*

$$P_e(N, M, 1) \geqq \exp - N[E_M + o_3(N)] \tag{1.17}$$

The exponents approach a limit, $E_\infty = \lim_{M\to\infty} E_M$, *given by*

$$E_\infty = \max_{\mathbf{q}} - \sum_{i=1}^{K} \sum_{k=1}^{K} q_i q_k \ln \sum_{j=1}^{J} \sqrt{P(j \mid i) P(j \mid k)} \tag{1.18}$$

The maximum in (1.18) *is over all probability vectors* $\mathbf{q} = (q_1, \cdots, q_K)$. *The exponents* E_M *are bounded by*

$$E_\infty \leqq E_M \leqq E_\infty + 2\sqrt{K}A/\sqrt{[\log_2(\log_2 M)]^-} \tag{1.19}$$

where

$$A = \max_{i,k} - 2 \ln \sum_{j=1}^{J} \sqrt{P(j \mid i) P(j \mid k)} \tag{1.20}$$

and $[x]^-$ *denotes the largest integer less than or equal to* x. *The quantity* $o_3(N)$ *in* (1.17) *can be taken as*

$$o_3(N) = \frac{\ln 4M}{N} - \sqrt{\frac{2}{N}} \ln P_{\min} \tag{1.21}$$

where $P_{\min}$ *is the smallest nonzero* $P(j \mid k)$.

Theorem 3 again requires some interpretation. Since $C_0 = 0$ by assumption, every pair of inputs has at least one output in common so that $\sum_{j=1}^{J} \sqrt{P(j \mid i) P(j \mid k)} > 0$ for all i, k; thus E_∞ and A in (1.18) and (1.20) must be finite.

Each of the exponents E_M can be interpreted as an exponent corresponding to zero rate since for fixed M, the rate of a code $R = (\ln M)/N$ approaches zero as N approaches infinity. On the other hand, if we choose M as a function of N in such a way that $\lim_{N\to\infty} M(N) = \infty$; $\lim_{N\to\infty} (\ln M(N))/N = 0$, then (1.17) becomes

$$P_e(N, M(N), 1) \geqq \exp - N[E_\infty + o_4(N)] \tag{1.22}$$

where $o_4(N)$ approaches zero as N approaches infinity.

For channels with a symmetry condition called pairwise reversibility, the exponents E_M can be uniquely determined. A channel is defined to be pairwise reversible iff for each pair of inputs, i and k,

$$\sum_{j=1}^{J} \sqrt{P(j \mid k) P(j \mid i)} \ln P(j \mid k) = \sum_{j=1}^{J} \sqrt{P(j \mid k) P(j \mid i)} \ln P(j \mid i) \tag{1.23}$$

This condition will be discussed more fully in Section V, but it is satisfied by such common channels as the binary symmetric channel and the binary symmetric erasure channel. For any channel satisfying (1.23), it is shown that

$$E_M = \frac{M}{M-1} \max_{M_1, \cdots, M_K} - \sum_i \sum_k \frac{M_i}{M} \frac{M_k}{M} \ln \sum_j \sqrt{P(j \mid i) P(j \mid k)} \tag{1.24}$$

where the $M_k \geqq 0$ are integers summing to M.

In Section VI, Theorems 1, 2, and 3 are combined to yield a new lower bound on error probability. The sphere packing bound is applied to $P_e(N_1, M, L)$ in (1.5) and the zero rate bound is applied to $P_e(N_2, L+1, 1)$. The result is given by the following theorem.

THEOREM 4. *Let* $E_{sp}(R)$ *and* E_∞ *be given by* (1.7) *and* (1.18) *for an arbitrary discrete memoryless channel for which* $C_0 = 0$. *Let* $E_{sl}(R)$ *be the smallest linear function of* R *which touches the curve* $E_{sp}(R)$ *and which satisfies* $E_{sl}(0) = E_\infty$. *Let* R_1 *be the point where* $E_{sl}(R)$ *touches* $E_{sp}(R)$. *Then for any code with a rate* $R < R_1$,

$$P_e(N, M, 1) \geq \exp - N[E_{sl}(R - o_5(N)) + o_6(N)] \quad (1.25)$$

where $o_5(N)$ and $o_6(N)$ are given by (6.6) and (6.7) and approach zero as N approaches infinity.

The function $E_{sl}(R)$ is sketched for a number of channels in Fig. 4. E_∞ is always strictly less than $E_{sp}(0^+)$ unless channel capacity C is zero. Thus the straight line bound of Theorem 4 is always tighter than the sphere packing bound at low rates for sufficiently large block lengths whenever $C > 0$, $C_0 = 0$.

Theorem 4 can be compared with an upper bound to error probability derived by Gallager (1965, Theorem 6) using expurgated randomly chosen codes. That result states that for any N, M,

$$P_e(N, M, 1) \leq \exp - N\left[E_{ex}\left(R + \frac{\ln 4}{N}\right)\right] \quad (1.26)$$

where the function E_{ex} is given by

$$E_{ex}(R) = \underset{\rho \geq 1}{\text{L.U.B.}} [E_x(\rho) - \rho R] \quad (1.27)$$

$$E_x(\rho) = \max_{\mathbf{q}} - \rho \ln \sum_{k=1}^{K} \sum_{i=1}^{K} q_k q_i \left[\sum_{j=1}^{J} \sqrt{P(j \mid k) P(j \mid i)}\right]^{1/\rho} \quad (1.28)$$

The maximization in (1.28) is again over probability vectors $\mathbf{q}$.

The function $E_{ex}(R)$ is sketched for several channels in Fig. 4. It can

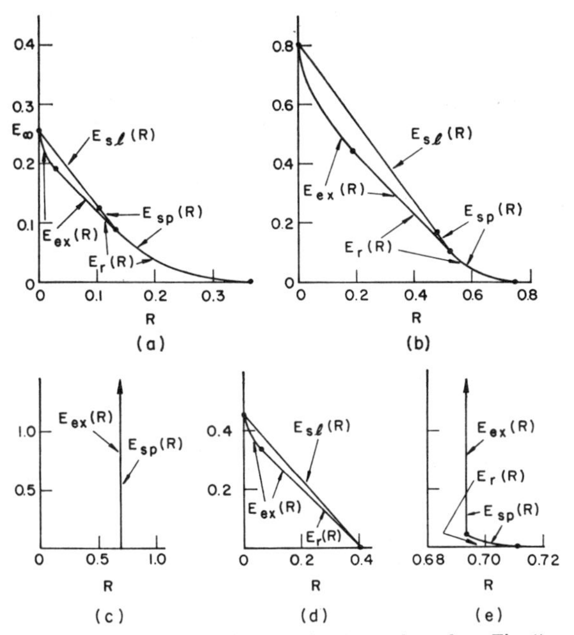

FIG. 4. Bounds on reliability function (same channels as Fig. 1)

be interpreted as the least upper bound of a set of straight lines where the lines have slope $-\rho$ and zero intercept $E_x(\rho)$. The function $E_x(\rho)$ is increasing with ρ and if $C_0 = 0$, we can calculate $\lim_{\rho\to\infty} E_x(\rho)$ as

$$\lim_{\rho\to\infty} E_x(\rho) = \max_{\mathbf{q}} \sum_{k=1}^{K} \sum_{i=1}^{K} q_k q_i \ln \sum_{j=1}^{J} \sqrt{P(j \mid k) P(j \mid i)} \quad (1.29)$$

Also it can be seen from (1.27) that

$$\lim_{R\to 0} E_{ex}(R) = \lim_{\rho\to\infty} E_x(\rho) \quad (1.30)$$

Combining (1.18), (1.29), and (1.30), we see that

$$\lim_{R\to 0} E_{ex}(R) = E_\infty \quad (1.31)$$

Thus, in the limit as $R \to 0$ our upper and lower bounds on P_e have the same exponential dependence on the block length.

It is to be observed that all the upper and lower bounds to error probability discussed so far have an exponential dependence on block length for fixed rate. The correct value of this exponential dependence, as a function of rate, is of fundamental importance in coding theory and is defined as the *reliability function*, $E(R)$, of the channel. More precisely,

$$E(R) = \limsup_{N\to\infty} \frac{-\ln P_e(N, [e^{NR}]^+, 1)}{N} \quad (1.32)$$

where $[x]^+$ is the smallest integer greater than or equal to x. We see that $E_{sp}(R)$ and $E_{sl}(R)$ are upper bounds to $E(R)$, and $E_r(R)$ and $E_{ex}(R)$ are lower bounds. The bounds are identical for the rather uninteresting case of noiseless channels and for some rather peculiar channels such as Fig. 1(e), but for typical channels there is a region of uncertainty for rates between 0 and R_{crit}. Although the bounds are close enough to give considerable insight into the behavior of a channel with coding, it is still interesting to speculate on the value of $E(R)$ in this region of uncertainty, $0 < R \leq R_{crit}$. For the binary symmetric channel, we improve on $E_{sl}(R)$ in Section VI by using a bound on minimum distance derived by Elias, but the technique does not generalize to arbitrary discrete memoryless channels. The authors would all tend to conjecture that $E(R)$ is equal to $E_{ex}(R)$ for $R \leq R_{crit}$ if the maximization in (1.29) is performed on a block basis rather than a letter basis (i.e., using Pr $(\mathbf{y} \mid \mathbf{x})$ in place of $P(j \mid k)$ and $q(\mathbf{x})$ in place of q) (see Gallager (1965)). As yet there is little concrete evidence for this conjecture.

II. PROOF OF THEOREM 1

Theorem 1 establishes a lower bound on error probability for a code in terms of the error probabilities for two codes of shorter block lengths. Let N_1 and N_2 be arbitrary block lengths and consider a code with M code words of block length $N_1 + N_2$. We shall be interested in considering each code word as consisting of two subsequences, the first of length N_1 and the second of length N_2. Let $\mathbf{x}_m$ be the mth code word and let the *prefix* $\mathbf{x}_{m,1}$ be the first N_1 letters of $\mathbf{x}_m$ and let the *suffix* $\mathbf{x}_{m,2}$ be the final N_2 letters of $\mathbf{x}_m$. Likewise, we separate the received sequence $\mathbf{y}$ into the prefix $\mathbf{y}_1$ and the suffix $\mathbf{y}_2$, consisting of N_1 and N_2 letters respectively.

We can visualize a list decoder of size L_2 as first observing $\mathbf{y}_1$, then $\mathbf{y}_2$, and decoding on the basis of these observations. Suppose that on the basis of $\mathbf{y}_1$ alone, there is a given number, say L_1, of messages that are more likely at the decoder than the actual transmitted message. If L_2 of these L_1 messages are also more likely than the transmitted message on the basis of $\mathbf{y}_2$ above, then a list decoding error should surely be made. Reasoning heuristically, it appears that the probability of the first event above is the probability of a list decoding error for a code of M code words of length N_1 with a list size of L_1. Similarly, given the first event, the probability of the second event should be lower bounded by the probability of a list decoding error for a code of block length N_2 consisting of the L_1 most likely messages plus the actual transmitted message. We thus conclude heuristically that

$$P_e(N_1 + N_2, M, L_2) \geq P_e(N_1, M, L_1) P_e(N_2, L_1 + 1, L_2) \quad (2.1)$$

This is the result of Theorem 1, and we now turn to a rigorous proof.

For a given code with M code words of length $N_1 + N_2$, and a list decoding scheme of size L_2, let Y_m be the set of received sequences $\mathbf{y}$ for which message m is on the decoding list. Also, for any given received prefix, $\mathbf{y}_1$, let $Y_{m,2}(\mathbf{y}_1)$ be the set of suffixes $\mathbf{y}_2$ for which m is on the list when $\mathbf{y}_1\mathbf{y}_2$ is received. Using (1.2) and (1.3) the error probability for the code is given by

$$P_e = \frac{1}{M} \sum_{m=1}^{M} \sum_{\mathbf{y} \in Y_m^c} \Pr(\mathbf{y} \mid \mathbf{x}_m) \quad (2.2)$$

For a discrete memoryless channel, Pr $(\mathbf{y} \mid \mathbf{x}_m) = $ Pr $(\mathbf{y}_1 \mid \mathbf{x}_{m,1})$ Pr $(\mathbf{y}_2 \mid \mathbf{x}_{m,2})$ and we can rewrite (2.2) as

$$P_e = \frac{1}{M} \sum_{m=1}^{M} \sum_{\mathbf{y}_1} \Pr(\mathbf{y}_1 \mid \mathbf{x}_{m,1}) \sum_{\mathbf{y}_2 \in Y_{m,2}^c(\mathbf{y}_1)} \Pr(\mathbf{y}_2 \mid \mathbf{x}_{m,2}) \quad (2.3)$$

Now consider the set of code word suffixes, $\mathbf{x}_{1,2}, \cdots, \mathbf{x}_{m,2}, \cdots, \mathbf{x}_{M,2}$. Pick any subset of $L_1 + 1$ of the messages and consider the associated $L_1 + 1$ suffixes as a set of $L_1 + 1$ code words of block length N_2. For any given $\mathbf{y}_1$, the associated $L_1 + 1$ decoding regions $Y_{m,2}(\mathbf{y}_1)$ form a list decoding rule of size L_2. Presumably some suffixes $\mathbf{y}_2$ are mapped into fewer than L_2 messages from the given subset, so that this is not the best set of decoding regions, but it is certainly a valid set. Now $P_e(N_2, L_1 + 1, L_2)$ is a lower bound to the error probability for any set of $L_1 + 1$ code words of length N_2 with any list decoding scheme of size L_2, and at least one code word in any such code must have an error probability that large. Thus, for at least one value of m in any given subset of $L_1 + 1$ suffixes, we have

$$\sum_{y_2 \in Y^c_{m,2}(y_1)} \Pr(y_2 \mid x_{m,2}) \geqq P_e(N_2, L_1 + 1, L_2) \tag{2.4}$$

For any given y_1, consider the entire set of M messages again. Let $m_1(y_1), m_2(y_1), \cdots, m_l(y_1)$ be the set of messages for which (2.4) is *not* satisfied. This set must contain at most L_1 messages since otherwise we would have a subset of $L_1 + 1$ messages for which no member satisfied (2.4). We can then lower bound the left hand side of (2.4) for any m by

$$\sum_{y_2 \in Y^c_{m,2}(y_1)} \Pr(y_2 \mid x_{m,2}) \geqq \begin{cases} 0; & m = m_1(y_1), \cdots, m_l(y_1) \\ P_e(N_2, L_1 + 1, L_2); & m \neq m_1(y_1), \cdots, m_l(y_1) \end{cases} \tag{2.5}$$

where l depends on y_1 but always satisfies $l \leqq L_1$.

Interchanging the order of summation between m and y_1 in (2.3) and substituting (2.5) into (2.3), we obtain

$$P_e \geqq \frac{1}{M} \sum_{y_1} \sum_{\substack{m \neq m_i(y_1) \\ i=1,\cdots,l}} \Pr(y_1 \mid x_{m,1}) P_e(N_2, L_1 + 1, L_2) \tag{2.6}$$

$$P_e \geqq P_e(N_2, L_1 + 1, L_2) \left[\frac{1}{M} \sum_{y_1} \sum_{\substack{m \neq m_i(y_1) \\ i=1,\cdots,l}} \Pr(y_1 \mid x_{m,1}) \right] \tag{2.7}$$

Finally, to complete the proof, we can consider the set of prefixes $x_{1,1}, \cdots, x_{M,1}$ as a set of M code words of length N_1, and the sets $m_1(y_1), \cdots, m_l(y_1)$ as a list decoding rule of size L_1 (recall that $l \leqq L_1$ for all y_1). Let $Y_{m,1}$ be the set of y_1 for which m is on the list $m_1(y_1), \cdots, m_l(y_1)$. Interchanging the sum over m and y_1 in (2.7), we obtain

$$P_e \geqq P_e(N_2, L_1 + 1, L_2) \left[\frac{1}{M} \sum_{m=1}^{M} \sum_{y_1 \in Y^c_{m,1}} \Pr(y_1 \mid x_{m,1}) \right] \tag{2.8}$$

The quantity in brackets is the probability of list decoding error for this code of length N_1 and is lower bounded by $P_e(N_1, M, L_1)$

$$P_e \geqq P_e(N_2, L_1 + 1, L_2) P_e(N_1, M, L_1) \tag{2.9}$$

Thus any code with M code words of length $N_1 + N_2$ and any list decoding scheme of size L_2 has an error probability satisfying (2.9) and this establishes (2.1).

The above theorem can be generalized considerably. First we note that the assumption of a discrete channel was used only in writing sums over the output sequences. For continuous channels, these sums are replaced by integrals. The theorem can also be modified to apply to a broad class of channels with memory. Also, if there is feedback from the receiver to transmitter, the theorem is still valid. The encoder can then change the code word suffixes depending on which y_1 is received, but (2.5) is valid independent of the choice of the set $\{x_{m,2}\}$. Finally the theorem can be extended to the case where two independent channels are available and $x_{m,1}$ is sent over one channel and $x_{m,2}$ is sent over the other channel.

III. ERROR PROBABILITY FOR TWO CODE WORDS

In this section we shall derive both upper and lower bounds to the probability of decoding error for a block code with two code words of length N. Surprisingly enough, the results are fundamental to both Sections IV and V.

Let $P_m(y)$, $m = 1, 2$, be the probability of receiving sequence y when message m is transmitted. If Y_m is the set of sequences decoded into message m, then from (1.2), the probability of decoding error when message m is transmitted is

$$P_{e,m} = \sum_{y \in Y_m^c} P_m(y); \qquad m = 1, 2 \tag{3.1}$$

For initial motivation, suppose that the decoder adopts a maximum likelihood decision rule: decode y into message 1 if $P_1(y) > P_2(y)$ and decode into message 2 otherwise. Under these circumstances $P_m(y)$ in (3.1) is equal to $\min_{m'=1,2} P_{m'}(y)$. Summing (3.1) over m, we then get

$$P_{e,1} + P_{e,2} = \sum_y \min_{m=1,2} P_m(y) \tag{3.2}$$

For any s in the interval $0 < s < 1$, a simple bound on $\min P_m(y)$ is given by

$$\min_{m=1,2} P_m(y) \leqq P_1(y)^{1-s} P_2(y)^s \leqq \max_{m=1,2} P_m(y) \tag{3.3}$$

Thus,

$$P_{e,1} + P_{e,2} \leqq \sum_y P_1(y)^{1-s} P_2(y)^s; \qquad 0 < s < 1 \tag{3.4}$$

We shall see later that when the right hand side of (3.4) is minimized over s, the bound is quite tight despite its apparent simplicity.

The logarithm of the right side of (3.4) is a fundamental quantity in most of the remainder of this paper; we denote it by

$$\mu(s) \triangleq \ln \sum_y P_1(y)^{1-s} P_2(y)^s; \qquad 0 < s < 1 \tag{3.5}$$

It is convenient to extend this definition to cover $s = 0$ and $s = 1$.

$$\mu(0) \triangleq \lim_{s \to 0^+} \mu(s); \qquad \mu(1) \triangleq \lim_{s \to 1^-} \mu(s) \tag{3.6}$$

Then we can rewrite (3.4), minimized over s, as

$$P_{e,1} + P_{e,2} \leqq \min_{0 \leq s \leq 1} \exp \mu(s) \tag{3.7}$$

Some typical modes of behavior of $\mu(s)$ are shown in Fig. 5. The block length in these figures is one and the first code word is the input letter 1 and the second is the input letter 2. It is shown later that $\mu(s)$ is always nonpositive and convex ∪.

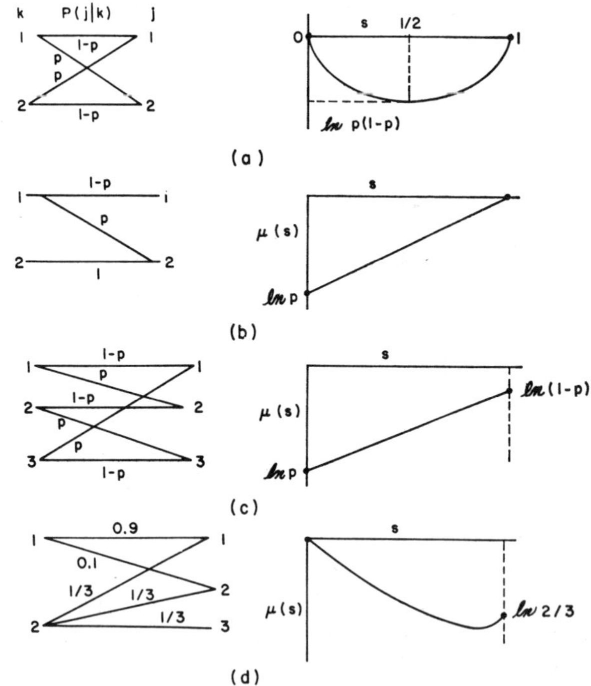

FIG. 5. The functions $\mu(s) = \ln \sum_j P(j \mid 1)^{1-s} P(j \mid 2)^s$ for several channels.

We next show that when the block length is greater than one, $\mu(s)$ can be written as a sum over the individual letters in the block. Let the code words be denoted by $x_m = (k_{m,1}, \cdots, k_{m,N})$, $m = 1, 2$, and let the received sequence be $y = (j_1, \cdots, j_N)$. Then, using (1.1), we have $P_m(y) = \prod_n P(j_n \mid k_{m,n})$, and $\mu(s)$ becomes

$$\mu(s) = \ln \sum_{j_1=1}^{J} \cdots \sum_{j_N=1}^{J} \prod_{n=1}^{N} P(j_n \mid k_{1,n})^{1-s} P(j_n \mid k_{2,n})^s \tag{3.8}$$

$$\mu(s) = \ln \sum_{j_1=1}^{J} P(j_1 \mid k_{1,1})^{1-s} P(j_1 \mid k_{2,1})^s \sum_{j_2=1}^{J} P(j_2 \mid k_{1,2})^{1-s} \cdot P(j_2 \mid k_{2,2})^s \cdots \sum_{j_N=1}^{J} P(j_N \mid k_{1,N})^{1-s} P(j_N \mid k_{2,N})^s \tag{3.9}$$

$$\mu(s) = \sum_{n=1}^{N} \mu_n(s); \qquad \mu_n(s) = \ln \sum_{j=1}^{J} P(j \mid k_{1,n})^{1-s} P(j \mid k_{2,n})^s \tag{3.10}$$

We now generalize the bound in (3.7) in two directions. First we want both upper and lower bounds on $P_{e,1}$ and $P_{e,2}$. Second, for reasons that will be clear in Section IV, we want to allow ourselves the flexibility of making $P_{e,1}$ very much larger than $P_{e,2}$ or vice versa. The following theorem achieves both of these objectives.

Theorem 5. *Let $P_1(\mathbf{y})$ and $P_2(\mathbf{y})$ be two probability assignments on a discrete set of sequences, let Y_1 and Y_2 be disjoint decision regions for these sequences, let $P_{e,1}$ and $P_{e,2}$ be given by (3.1) and assume that $P_1(\mathbf{y})P_2(\mathbf{y}) \neq 0$ for at least one sequence* **y**. *Then, for any s, $0 < s < 1$, either*

$$P_{e,1} > \tfrac{1}{4} \exp\left[\mu(s) - s\mu'(s) - s\sqrt{2\mu''(s)}\right] \tag{3.11}$$

or

$$P_{e,2} > \tfrac{1}{4} \exp\left[\mu(s) + (1-s)\mu'(s) - (1-s)\sqrt{2\mu''(s)}\right] \tag{3.12}$$

Furthermore, for an appropriate choice of Y_1, Y_2,

$$P_{e,1} \leqq \exp\left[\mu(s) - s\mu'(s)\right] \qquad \textit{and} \tag{3.13}$$

$$P_{e,2} \leqq \exp\left[\mu(s) + (1-s)\mu'(s)\right] \tag{3.14}$$

Finally $\mu(s)$ is nonpositive and convex ∪ *for $0 < s < 1$. The convexity is strict unless $P_1(\mathbf{y})/P_2(\mathbf{y})$ is constant over all* **y** *for which $P_1(\mathbf{y})P_2(\mathbf{y}) \neq 0$. Also $\mu(s)$ is strictly negative for $0 < s < 1$ unless $P_1(\mathbf{y}) = P_2(\mathbf{y})$ for all* **y**.

Remarks: The probabilities $P_1(\mathbf{y})$ and $P_2(\mathbf{y})$ do not have to correspond to two code words and in Section IV the theorem will be used where $P_2(\mathbf{y})$ does not correspond to a code word. For interpretation, however, we shall consider only the problem of two code words on a memoryless channel, in which case (3.10) is valid. Taking the derivatives of (3.10), we have

$$\mu'(s) = \sum_{n=1}^{N} \mu_n'(s); \qquad \mu''(s) = \sum_{n=1}^{N} \mu_n''(s) \tag{3.15}$$

Therefore, for any s, $0 < s < 1$, the first part of the theorem states that either

$$P_{e,1} > \tfrac{1}{4} \exp\left\{\sum_{n=1}^{N} [\mu_n(s) - s\mu_n'(s)] - s\sqrt{\sum_n 2\mu_n''(s)}\right\} \tag{3.16}$$

or

$$P_{e,2} > \tfrac{1}{4} \exp\left\{\sum_{n=1}^{N} [\mu_n(s) + (1-s)\mu_n'(s)] - (1-s) \cdot \sqrt{\sum_n \mu_n''(s)}\right\} \tag{3.17}$$

We see from this that in some sense the terms involving $\mu(s)$ and $\mu'(s)$ are proportional to the block length N and that the term involving $\sqrt{\mu''(s)}$ is proportional to $\sqrt{N}$. It follows that for large N we should focus our attention primarily on $\mu(s)$ and $\mu'(s)$.

Figure 6 gives a graphical interpretation of the terms $\mu(s) - \mu'(s)$ and $\mu(s) + (1-s)\mu'(s)$. It is seen that they are the endpoints, at 0 and 1, of the tangent at s to the curve $\mu(s)$. As s increases, the tangent see-saws around, decreasing $\mu(s) - s\mu'(s)$ and increasing $\mu(s) + (1-s)\mu'(s)$. In the special case where $\mu(s)$ is a straight line, of course, this see-sawing does not occur and $\mu(s) - s\mu'(s)$ and $\mu(s) + (1-s)\mu'(s)$ do not vary with s.

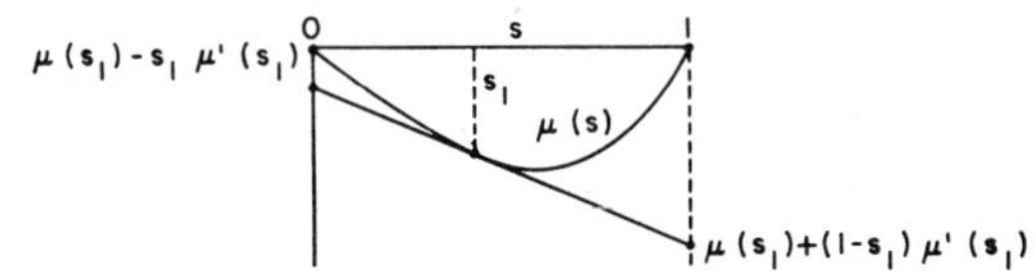

Fig. 6. Geometric interpretation of the exponents $\mu(s) - s\mu'(s)$ and $\mu(s) + (1-s)\mu'(s)$.

Since $\mu(s)$ is convex ∪ over $0 < s < 1$, any tangent to μ in this range will lie on or beneath μ. Furthermore, since $\mu(s) \leqq 0$ in this range, we have in general, for $0 < s < 1$,

$$\mu(s) - s\mu'(s) \leqq 0 \tag{3.18}$$

$$\mu(s) + (1-s)\mu'(s) \leqq 0 \tag{3.19}$$

A particularly important special case of the theorem is that in which s is chosen to minimize $\mu(s)$. In that case we get the following corollary.

Corollary. *Let s^* minimize $\mu(s)$ over $0 \leqq s \leqq 1$. Then either*

$$P_{e,1} \geqq \tfrac{1}{4} \exp\left[\mu(s^*) - s^*\sqrt{2\mu''(s^*)}\right] \tag{3.20}$$

or

$$P_{e,2} \geqq \tfrac{1}{4} \exp\left[\mu(s^*) - (1-s^*)\sqrt{2\mu''(s^*)}\right] \tag{3.21}$$

where if $s^* = 0$ or 1, $\mu''(s^*)$ is the limit of $\mu''(s)$ from the interior of the interval.

Proof of Corollary: If s^* is within the interval $0 < s^* < 1$, then $\mu'(s^*) = 0$ and (3.20) and (3.21) follow immediately from (3.11) and (3.12). If $s^* = 0$, then $\mu'(0^+) \geqq 0$, and

$$\begin{aligned}\lim_{s\to 0^+} \mu(s) - s\mu'(s) &= \mu(0^+) = \mu(s^*)\\ \lim_{s\to 0^+} \mu(s) + (1-s)\mu'(s) &\geqq \mu(0^+) = \mu(s^*)\end{aligned} \tag{3.22}$$

Likewise if $s^* = 1$, *then* $\mu'(1^-) \leqq 0$, and

$$\begin{aligned}\lim_{s\to 1^-} \mu(s) - s\mu'(s) &\geqq \mu(1^-) = \mu(s^*)\\ \lim_{s\to 1^-} \mu(s) + (1-s)\mu'(s) &= \mu(1^-) = \mu(s^*)\end{aligned} \tag{3.23}$$

Substituting these relations into (3.11) and (3.12) completes the proof.

Notice that the exponent $\mu(s^*)$ appearing in (3.20) and (3.21) is the same as the exponent in the upper bound to $P_{e,1} + P_{e,2}$ of (3.7).

Proof of Theorem 5: The sum over **y** in $\mu(s)$ as given by (3.5) can either be considered to be over all output sequences **y** or over only those sequences in the overlap region where $P_1(\mathbf{y})$ and $P_2(\mathbf{y})$ are both nonzero. For the remainder of the proof, we consider all sums over **y** to be over only the overlap region.

Taking the derivatives of $\mu(s)$, we get

$$\mu'(s) = \left\{\sum_{\mathbf{y}} \frac{P_1(\mathbf{y})^{1-s}P_2(\mathbf{y})^s}{\sum_{\mathbf{y}'} P_1(\mathbf{y}')^{1-s}P_2(\mathbf{y}')^s} \ln \frac{P_2(\mathbf{y})}{P_1(\mathbf{y})}\right. \tag{3.24}$$

$$\mu''(s) = \left\{\sum_{\mathbf{y}} \frac{P_1(\mathbf{y})^{1-s}P_2(\mathbf{y})^s}{\sum_{\mathbf{y}'} P_1(\mathbf{y}')^{1-s}P_2(\mathbf{y}')^s} \left[\ln \frac{P_2(\mathbf{y})}{P_1(\mathbf{y})}\right]^2\right\} - [\mu'(s)]^2 \tag{3.25}$$

Let $D(\mathbf{y})$ be the log likelihood ratio,

$$D(\mathbf{y}) = \ln \frac{P_2(\mathbf{y})}{P_1(\mathbf{y})} \tag{3.26}$$

and for $0 < s < 1$, define

$$Q_s(\mathbf{y}) = \frac{P_1(\mathbf{y})^{1-s}P_2(\mathbf{y})^s}{\sum_{\mathbf{y}'} P_1(\mathbf{y}')^{1-s}P_2(\mathbf{y}')^s} \tag{3.27}$$

It will be seen later that $Q_s(\mathbf{y})$ is large for those **y** that are likely to cause errors; thus this probability assignment allows us to focus our attention on the region of interest.

If we consider $D(\mathbf{y})$ to be a random variable with probability assignment $Q_s(\mathbf{y})$, then we see from (3.24) and (3.25) that $\mu'(s)$ and $\mu''(s)$ are the mean and variance of $D(\mathbf{y})$ respectively. Since $\mu''(s)$ is a variance, it is nonnegative and therefore $\mu(s)$ is convex ∪. It can also be seen from this that $\mu(s)$ will be strictly convex ∪ unless $P_2(\mathbf{y})/P_1(\mathbf{y})$ is a constant for all **y** in the overlap region. Since $\mu(0)$ and $\mu(1)$ are nonpositive (see (3.5) and (3.6)), it follows from convexity that μ is nonpositive for all s, $0 \leqq s \leqq 1$. Furthermore, for $\mu(s)$ to be 0 at any point within the interval (0, 1) it is necessary for $\mu(0)$, $\mu(1)$, and $\mu''(s)$ all to be zero. It is easy to see that this can happen only if $P_1(\mathbf{y}) = P_2(\mathbf{y})$ for all **y**.

It can be verified easily by substituting (3.26) and (3.5) into (3.27) that

$$P_1(\mathbf{y}) = \{\exp[\mu(s) - sD(\mathbf{y})]\}Q_s(\mathbf{y}) \tag{3.28}$$

$$P_2(\mathbf{y}) = \{\exp[\mu(s) + (1-s)D(\mathbf{y})]\}Q_s(\mathbf{y}) \tag{3.29}$$

We shall now establish the second part of the theorem, (3.13) and (3.14). For a given s, define the decision region Y_1 to be

$$Y_1 = \{\mathbf{y}: D(\mathbf{y}) < \mu'(s)\} \tag{3.30}$$

Then for $\mathbf{y} \in Y_1^c$, we have $-sD(\mathbf{y}) \leq -s\mu'(s)$, and (3.28) is bounded by

$$P_1(\mathbf{y}) \leq \{\exp[\mu(s) - s\mu'(s)]\}Q_s(\mathbf{y}); \qquad \mathbf{y} \in Y_1^c \tag{3.31}$$

Substituting (3.31) into (3.1), we have

$$P_{e,1} \leq \{\exp[\mu(s) - s\mu'(s)]\} \sum_{\mathbf{y} \in Y_1^c} Q_s(\mathbf{y}) \tag{3.32}$$

Equation (3.13) follows upon upper bounding the sum of probabilities in (3.32) by 1. Equation (3.14) follows in the same way upon recognizing that $(1 - s)D(\mathbf{y}) \leq (1 - s)\mu'(s)$ for $\mathbf{y} \in Y_2^c$.

We now turn to the proof of the first part of the theorem. Define Y_s as the set of sequences for which $D(\mathbf{y})$ is within $\sqrt{2}$ standard deviations of its mean according to $Q_s(\mathbf{y})$.

$$Y_s = \{\mathbf{y}: | D(\mathbf{y}) - \mu'(s) | \leq \sqrt{2\mu''(s)}\} \tag{3.33}$$

From the Chebychev inequality,

$$\sum_{\mathbf{y} \in Y_s} Q_s(\mathbf{y}) > \tfrac{1}{2} \tag{3.34}$$

We now lower bound $P_{e,1}$ and $P_{e,2}$ by considering only those $\mathbf{y}$ in the set Y_s. This is motivated by the fact that for the decision rule (3.30), most of the errors presumably occur when $| D(\mathbf{y}) - \mu'(s)|$ is small.

$$P_{e,1} = \sum_{\mathbf{y} \in Y_1^c} P_1(\mathbf{y}) \geq \sum_{\mathbf{y} \in Y_1^c \cap Y_s} P_1(\mathbf{y}) \tag{3.35}$$

$$P_{e,2} = \sum_{\mathbf{y} \in Y_2^c} P_2(\mathbf{y}) \geq \sum_{\mathbf{y} \in Y_2^c \cap Y_s} P_2(\mathbf{y}) \tag{3.36}$$

For $\mathbf{y} \in Y_s$, (3.33) gives us

$$\mu'(s) - \sqrt{2\mu''(s)} \leq D(\mathbf{y}) \leq \mu'(s) + \sqrt{2\mu''(s)} \tag{3.37}$$

Thus, for $\mathbf{y} \in Y_s$, (3.28) and (3.29) are bounded by

$$P_1(\mathbf{y}) \geq \{\exp[\mu(s) - s\mu'(s) - s\sqrt{2\mu''(s)}]\}Q_s(\mathbf{y}) \tag{3.38}$$

$$P_2(\mathbf{y}) \geq \{\exp[\mu(s) - (1 - s)\mu'(s) - (1 - s)\sqrt{2\mu''(s)}]\} \cdot Q_s(\mathbf{y}) \tag{3.39}$$

Substituting (3.38) into (3.35) and (3.39) into (3.36) leaves only $Q_s(\mathbf{y})$ under the summation signs.

$$P_{e,1} \geq \{\exp[\mu(s) - s\mu'(s) - s\sqrt{2\mu''(s)}]\} \sum_{\mathbf{y} \in Y_1^c \cap Y_s} Q_s(\mathbf{y}) \tag{3.40}$$

$$P_{e,2} \geq \{\exp[\mu(s) + (1 - s)\mu'(s) - (1 - s)\sqrt{2\mu''(s)}]\} \cdot \sum_{\mathbf{y} \in Y_2^c \cap Y_s} Q_s(\mathbf{y}) \tag{3.41}$$

Since Y_1 and Y_2 are disjoint, (3.34) yields

$$\sum_{\mathbf{y} \in Y_1^c \cap Y_s} Q_s(\mathbf{y}) + \sum_{\mathbf{y} \in Y_2^c \cap Y_s} Q_s(\mathbf{y}) > \tfrac{1}{2} \tag{3.42}$$

Thus, either

$$\sum_{\mathbf{y} \in Y_1^c \cap Y_s} Q_s(\mathbf{y}) > \tfrac{1}{4} \tag{3.43}$$

or

$$\sum_{\mathbf{y} \in Y_2^c \cap Y_s} Q_s(\mathbf{y}) > \tfrac{1}{4} \tag{3.44}$$

Substituting these inequalities into (3.40) and (3.41) completes the proof of the theorem.

There are a number of other approaches that could have been taken to prove theorems essentially equivalent to Theorem 3. The theorem treats a simple statistical decision theory problem with 2 hypotheses. According to the Neyman-Pearson (1928) theorem, we can minimize $P_{e,1}$ for a given value of $P_{e,2}$ by letting Y_1 be the set of $\mathbf{y}$ for which $D(\mathbf{y})$ is less than a constant chosen to give $P_{e,2}$ its given value. Then $P_{e,1}$ is the probability according to $P_1(\mathbf{y})$ that $D(\mathbf{y})$, which is the sum of N independent random variables, is greater than or equal to that constant. Likewise, $P_{e,2}$ is the probability according to $P_2(\mathbf{y})$ that $D(\mathbf{y})$ is less than the constant. A number of estimates and bounds on the probability that a sum of independent random variables will be far from the mean are given by Feller (1943), Chernoff (1952), Chapter 8 of Fano (1961), and Gallager (1965b). The particular theorem chosen here was selected primarily for the simplicity of the result and for its generality. Observe that Theorem 5 is applicable whenever $\mu(s)$ and its first two derivatives exist. For example $\mathbf{y}$ may be a sequence of real numbers and $P_1(\mathbf{y})$ and $P_2(\mathbf{y})$ may be replaced with probability densities.

IV. THE SPHERE PACKING BOUND

Let $\mathbf{x}_1, \mathbf{x}_2, \cdots, \mathbf{x}_M$ be a set of M code words each of length N for use on a discrete memoryless channel with transition probabilities $P(j \mid k)$. Assume a list decoding scheme in which for each received sequence $\mathbf{y}$, the decoder produces a list of at most L integers from 1 to M. If Y_m is the set of output sequences $\mathbf{y}$ for which message m is on the decoding list, then, as in (1.2), the probability of list decoding error when message m is sent is

$$P_{e,m} = \sum_{\mathbf{y} \in Y_m^c} \Pr(\mathbf{y} \mid \mathbf{x}_m) \tag{4.1}$$

Let $P_{e,\max}$ be the maximum over m of $P_{e,m}$ for the code and list decoding scheme under consideration. In this section we first find a lower bound on $P_{e,\max}$ for a special class of codes called fixed composition codes. We then generalize the results to arbitrary codes, and prove Theorem 2 of the introduction.

For any given m, $P_{e,m}$ can generally be reduced by enlarging the size of the decoding set Y_m; this will decrease the size of $Y_{m'}$, for some $m' \neq m$, however, and thus generally increase $P_{e,m'}$. In order to keep some control over the size of Y_m without specifically considering the other code words, we define an arbitrary product probability measure on the output sequences $\mathbf{y} = (j_1, \cdots, j_N)$,

$$f_N(\mathbf{y}) = \prod_{n=1}^{N} f(j_n) \tag{4.2}$$

where $\mathbf{f} = \{f(1), \cdots, f(J)\}$ is an arbitrary probability assignment on the output letters 1 to J. The size of Y_m is now defined as

$$F(Y_m) = \sum_{\mathbf{y} \in Y_m} f_N(\mathbf{y}) \tag{4.3}$$

Theorem 5 can be used to relate $P_{e,m}$ and $F(Y_m)$ if we let $\Pr(\mathbf{y} \mid \mathbf{x}_m)$ correspond to $P_1(\mathbf{y})$ in the theorem and let $f_N(\mathbf{y})$ correspond to $P_2(\mathbf{y})$. The function $\mu(s)$ of Theorem 5 corresponding to $\Pr(\mathbf{y} \mid \mathbf{x}_m)$ and $f_N(\mathbf{y})$ is given by

$$\mu(s) = \ln \sum_{\mathbf{y}} \Pr(\mathbf{y} \mid \mathbf{x}_m)^{1-s} f_N(\mathbf{y})^s \tag{4.4}$$

Assume that $\Pr(\mathbf{y}/\mathbf{x}_m)f_N(\mathbf{y}) \neq 0$ for at least one $\mathbf{y}$. Theorem 5 then states that for each s, $0 < s < 1$, either

$$P_{e,m} > \tfrac{1}{4} \exp[\mu(s) - s\mu'(s) - s\sqrt{2\mu''(s)}] \tag{4.5}$$

or

$$F(Y_m) > \tfrac{1}{4} \exp[\mu(s) + (1 - s)\mu'(s) - (1 - s)\sqrt{2\mu''(s)}] \tag{4.6}$$

Since $f_N(\mathbf{y}) = \prod_n f(j_n)$, $\mu(s)$ can be broken up into a sum of terms as in (3.10). If $\mathbf{x}_m = (k_{m,1}, \cdots, k_{m,N})$, we have

$$\mu(s) = \sum_{n=1}^{N} \mu_{k_{m,n}}(s, \mathbf{f}) \tag{4.7}$$

$$\mu_k(s, \mathbf{f}) = \ln \sum_j P(j/k)^{1-s} f(j)^s \tag{4.8}$$

The function $\mu(s)$ depends on $\mathbf{x}_m$ only through the number of appearances of each alphabet letter in $\mathbf{x}_m$. Let

$$q_k(m) = \frac{\text{Number of times input letter } k \text{ appears in } \mathbf{x}_m}{N} \tag{4.9}$$

The vector $\mathbf{q}(m) = (q_1(m), \cdots, q_K(m))$ is called the composition of the mth code word. In terms of $\mathbf{q}(m)$, $\mu(s)$ becomes

$$\mu(s) = N \sum_{k=1}^{K} q_k(m)\mu_k(s, \mathbf{f}) \tag{4.10}$$

Let us restrict our attention temporarily to codes in which all code words have the same composition. Then the m can be dropped from $q_k(m)$ in (4.10), and (4.5) and (4.6) become: for each s, $0 < s < 1$, either

$$P_{e,m} > \tfrac{1}{4} \exp N\left\{\sum_k q_k[\mu_k(s, \mathbf{f}) - s\mu_k'(s, \mathbf{f})] - \frac{s}{\sqrt{N}}\sqrt{2\sum_k q_k\mu_k''(s, \mathbf{f})}\right\} \tag{4.11}$$

or

$$F(Y_m) > \tfrac{1}{4} \exp N\left\{\sum_k q_k[\mu_k(s, \mathbf{f}) + (1-s)\mu_k'(s, \mathbf{f})] - \frac{(1-s)}{\sqrt{N}} \sqrt{2\sum_k q_k\mu_k''(s, \mathbf{f})}\right\} \quad (4.12)$$

The square root terms in (4.11) and (4.12) turn out to be unimportant for large N. Thus we simplify the expressions by the following loose but general bound on μ_k'' (see appendix).

$$s\sqrt{\mu_k''(s, \mathbf{f})} \leq \ln \frac{e}{\sqrt{P_{\min}}} \quad (4.13)$$

$P_{\min}$ is the smallest nonzero transition probability on the channel.

We can now relate $F(Y_m)$ to the number of code words M and the list size L by observing that

$$\sum_{m=1}^{M} F(Y_m) = \sum_{m=1}^{M} \sum_{\mathbf{y} \in Y_m} f_N(\mathbf{y}) \leq L \quad (4.14)$$

Equation (4.14) follows from the facts that each $\mathbf{y}$ appears in at most L decoding subsets and that $\sum_{\mathbf{y}} f_N(\mathbf{y}) = 1$. As a consequence of (4.14), there must be some m for which

$$F(Y_m) \leq L/M \quad (4.15)$$

For this m, we can substitute (4.13) and (4.15) into (4.11) and (4.12). Bringing the factors of $\frac{1}{4}$ inside the exponents and upper bounding $P_{e,m}$ by $P_{e,\max}$, (4.11) and (4.12) become: either

$$P_{e,\max} > \exp N\left\{\sum_{k=1}^{K} q_k[\mu_k(s, \mathbf{f}) - s\mu_k'(s, \mathbf{f})] - \sqrt{\frac{2}{N}} \ln \frac{e}{\sqrt{P_{\min}}} - \frac{\ln 4}{N}\right\} \quad (4.16)$$

or

$$\frac{L}{M} > \exp N\left\{\sum_{k=1}^{K} q_k[\mu_k(s, \mathbf{f}) + (1-s)\mu_k'(s, \mathbf{f})] - \frac{1-s}{s}\sqrt{\frac{2}{N}} \ln \frac{e}{\sqrt{P_{\min}}} - \frac{\ln 4}{N}\right\} \quad (4.17)$$

Equations (4.16) and (4.17) provide a parametric lower bound on $P_{e,\max}$ for a given L/M in terms of the parameter s in the same way that Theorem 5 provided a parametric lower bound on $P_{e,1}$ for a given $P_{e,2}$. The bound is valid for any fixed composition code of composition $\mathbf{q}$ with M code words of length N and for any list decoding scheme with lists of size L.

The reason for calling this a sphere packing bound is somewhat historical, but also adds some insight into what we have done. From the discussion following Theorem 5, we see that $P_{e,m}$ can be minimized for a decoding subset of given size by picking the set Y_m to be those $\mathbf{y}$ for which $\ln [f_N(\mathbf{y})/\Pr(\mathbf{y} \mid \mathbf{x}_m)]$ is less than a constant. If we think of $\ln [f_N(\mathbf{y})/\Pr(\mathbf{y} \mid \mathbf{x}_m)]$ as a generalized type of distance from $\mathbf{x}_m$ to $\mathbf{y}$, then we can think of the Y_m that minimizes $P_{e,m}$ as being a sphere around $\mathbf{x}_m$. Thus our bound on $P_{e,\max}$ in terms of M would be a very tight bound if we could pick the Y_m as a set of spheres, each sphere around one code word, with spheres packed into the space of output sequences.

The bound of (4.16) and (4.17) is a function of the arbitrary probability assignment $\mathbf{f}$. The straightforward approach now would be to find that $\mathbf{f}$ which yields the tightest bound on $P_{e,\max}$, i.e., that *maximizes* the lower bound for a given composition. We could then look for the best composition, i.e., the $\mathbf{q}$ that *minimizes* the lower bound on $P_{e,\max}$. Such a procedure turns out to be both tedious and unenlightening. We shall instead simply state the resulting $\mathbf{f}$ and $\mathbf{q}$ as functions of the parameter s and then show that this choice gives us the bound of Theorem 2.

For a given s, $0 < s < 1$, let $\mathbf{q}_s = (q_{1,s}, \cdots, q_{K,s})$ satisfy the equations

$$\sum_j P(j \mid k)^{1-s} \alpha_{j,s}^{s/(1-s)} \geq \sum_j \alpha_{j,s}^{1/(1-s)}; \qquad \text{all } k \quad (4.18)$$

where

$$\alpha_{j,s} = \sum_{k=1}^{K} q_{k,s} P(j \mid k)^{1-s} \quad (4.19)$$

Let $\mathbf{f}_s = (f_s(1), \cdots, f_s(J))$ be given by

$$f_s(j) = \frac{\alpha_{j,s}^{1/(1-s)}}{\sum_{j'=1}^{J} \alpha_{j',s}^{1/(1-s)}} \quad (4.20)$$

This is a rather formidable looking set of equations, but the solutions have some remarkable properties. If we set $\rho = s/(1-s)$, (4.18) and (4.19) are identical to the necessary and sufficient conditions (1.14) and (1.15) on $\mathbf{q}$ to maximize the function $E_0(\rho, \mathbf{q})$ discussed in Section I. Thus (4.18) is satisfied with equality for those k with $q_{k,s} > 0$. Since $E_0(\rho, \mathbf{q})$ must have a maximum over the probability vectors $\mathbf{q}$, (4.18) and (4.19) must have a solution (though it need not be unique).

The fact that $\mathbf{f}$ is chosen here as a function of s in no way changes the validity of the lower bound to $P_{e,\max}$ given by (4.16) and (4.17). We must remember, however, that $\mu_k'(s, \mathbf{f}_s)$ is the partial derivative of μ_k with respect to s holding $\mathbf{f}_s$ fixed. The condition that for each $k, f(j)P(j \mid k) \neq 0$ for some j is clearly met by $\mathbf{f}_s$, since the left side of (4.18) must be strictly positive.

Next we show that $\mathbf{f}_s$ has the property that $\mu_k(s, \mathbf{f}_s)$ is independent of k for those inputs with $q_{k,s} \neq 0$. Substituting (4.20) into the expression (4.8) for μ_k, we have

$$\mu_k(s, \mathbf{f}_s) = \ln \sum_{j=1}^{J} P(j \mid k)^{1-s} \alpha_{j,s}^{s/(1-s)} - s \ln \sum_{j=1}^{J} \alpha_{j,s}^{1/(1-s)} \quad (4.21)$$

Using (4.18) in (4.21),

$$\mu_k(s, \mathbf{f}_s) \geq (1-s) \ln \sum_{j=1}^{J} \alpha_{j,s}^{1/(1-s)}$$

with equality if $q_{k,s} \neq 0$. Finally, using (4.19) for $\alpha_{j,s}$, we have the expression for $E_0(\rho)$ in (1.8) and (1.9). Thus

$$\mu_k(s, \mathbf{f}_s) \geq -(1-s)E_0\left(\frac{s}{1-s}\right); \qquad \text{equality if } q_{k,s} \neq 0 \quad (4.22)$$

One final property of $\mathbf{q}_s$ and $\mathbf{f}_s$, which we shall not need but which gives some insight into why $\mathbf{f}_s$ yields the tightest bound on $P_{e,\max}$ for the "best" composition $\mathbf{q}_s$, is that $\mathbf{q}_s$, $\mathbf{f}_s$ yields a min-max point for the function $\sum_k q_k\mu_k(s, \mathbf{f})$. That is, for all $\mathbf{q}$, $\mathbf{f}$,

$$\sum_k q_{k,s}\mu_k(s, \mathbf{f}) \leq \sum_k q_{k,s}\mu_k(s, \mathbf{f}_s) \leq \sum_k q_k\mu_k(s, \mathbf{f}_s) \quad (4.23)$$

This relation is established in the appendix.

We can now state a theorem that is equivalent to Theorem 2 in the introduction, with the exception that the theorem here applies only to fixed composition codes.

THEOREM 6. *Let $P(j \mid k)$ be the transition probabilities for a discrete memoryless channel and let a fixed composition code for the channel have M code words of length N with a list decoding scheme of list size L. Then at least one code word will have a probability of list decoding error bounded by*

$$P_{e,\max} \geq \exp -N\left\{E_{sp}\left(R - \frac{\ln 4}{N} - \epsilon\right) + \sqrt{\frac{8}{N}} \ln \frac{e}{\sqrt{P_{\min}}} + \frac{\ln 4}{N}\right\} \quad (4.24)$$

where $R = (1/N)\ln(M/L)$, the function E_{sp} is given by (1.7), and ϵ is an arbitrarily small positive number.

Proof: We shall first express the parametric lower bound on $P_{e,\max}$ of (4.16) and (4.17) in a more convenient way. Define $R(s, \mathbf{q})$ as minus the quantity in braces in (4.17), using $\mathbf{f}_s$ for $\mathbf{f}$.

$$R(s, \mathbf{q}) = \sum_k - q_k[\mu_k(s, \mathbf{f}_s) + (1-s)\mu_k'(s, \mathbf{f}_s)] + \frac{1-s}{s}\sqrt{\frac{2}{N}} \ln \frac{e}{\sqrt{P_{\min}}} + \frac{\ln 4}{N} \quad (4.25)$$

Then (4.17) can be rewritten

$$R = \frac{\ln M/L}{N} < R(s, \mathbf{q}) \tag{4.26}$$

Also we can use (4.25) to eliminate the μ_k' term in (4.16), getting

$$P_{e,\max} > \exp N\left\{\sum_k q_k\left(1 + \frac{s}{1-s}\right)\mu_k(s, \mathbf{f}_s) + \frac{s}{1-s}R(s, \mathbf{q}) - \sqrt{\frac{8}{N}}\ln\frac{e}{\sqrt{P_{\min}}} - \left(1 + \frac{s}{1-s}\right)\frac{\ln 4}{N}\right\} \tag{4.27}$$

Thus, for every s, $0 < s < 1$, either (4.26) or (4.27) is satisfied.

We now consider two separate cases

(a) $\qquad R = R(s, \mathbf{q})$ for some s, $0 < s < 1 \qquad$ (4.28)

(b) $\qquad R < R(s, \mathbf{q})$ for all s, $0 < s < 1 \qquad$ (4.29)

It is shown in the appendix that $R(s, \mathbf{q})$ is a continuous function of s for $0 < s < 1$, and it can be seen from the term containing $(1 - s)/s$ in (4.25) that $\lim_{s\to 0} R(s, \mathbf{q}) = \infty$. Thus either (a) or (b) above must be satisfied. If (a) is satisfied for some s, then (4.26) is unsatisfied and (4.27) must be satisfied for that s; substituting (4.22) and (4.28) into (4.27), we have

$$P_{e,\max} > \exp N\left\{-E_0\left(\frac{s}{1-s}\right) + \frac{s}{1-s}\left(R - \frac{\ln 4}{N}\right) - \sqrt{\frac{8}{N}}\ln\frac{e}{\sqrt{P_{\min}}} - \frac{\ln 4}{N}\right\} \tag{4.30}$$

Using ρ for $s/(1 - s)$ and further lower bounding by taking the lowest upper bound of the negative exponent over ρ, we have

$$P_{e,\max} > \exp - N\left\{\underset{\rho\geq 0}{\text{L.U.B.}}\left[E_0(\rho) - \rho\left(R - \frac{\ln 4}{N}\right)\right] + \sqrt{\frac{8}{N}}\ln\frac{e}{\sqrt{P_{\min}}} + \frac{\ln 4}{N}\right\} \tag{4.31}$$

$$> \exp - N\left\{\underset{\rho\geq 0}{\text{L.U.B.}}\left[E_0(\rho) - \rho\left(R - \frac{\ln 4}{N} - \epsilon\right)\right] + \sqrt{\frac{8}{N}}\ln\frac{e}{\sqrt{P_{\min}}} + \frac{\ln 4}{N}\right\} \tag{4.32}$$

Using the definition of E_{sp} in (1.7), this is equivalent to (4.24) and proves the theorem for case (a).

Next we show that for case (b), (4.24) reduces to $P_{e,\max} \geq 0$ which is trivially true. From (3.18),

$$\mu_k(s, \mathbf{f}_s) - s\mu_k'(s, \mathbf{f}_s) \leq 0; \quad -\mu_k'(s, \mathbf{f}_s) \leq \frac{-\mu_k(s, \mathbf{f}_s)}{s} \tag{4.33}$$

Substituting (4.33) into (4.25), we obtain for all s, $0 < s < 1$,

$$R < R(s, \mathbf{q}) \leq -\sum_{k=1}^{K} q_k\left(1 + \frac{1-s}{s}\right)\mu_k(s, \mathbf{f}_s) + \frac{1-s}{s}\sqrt{\frac{2}{N}}\ln\frac{e}{\sqrt{P_{\min}}} + \frac{\ln 4}{N}$$

Using (4.22) again and letting $\rho = s/(1 - s)$, this becomes

$$R < \frac{E_0(\rho)}{\rho} + \frac{1}{\rho}\sqrt{\frac{2}{N}}\ln\frac{e}{\sqrt{P_{\min}}} + \frac{\ln 4}{N}; \qquad \text{all } \rho > 0 \tag{4.34}$$

Using (1.7) and (4.34), we have

$$E_{sp}\left(R - \frac{\ln 4}{N} - \epsilon\right) = \underset{\rho\geq 0}{\text{L.U.B.}}\left[E_0(\rho) - \rho\left(R - \frac{\ln 4}{N} - \epsilon\right)\right] \geq \underset{\rho\geq 0}{\text{L.U.B.}} - \sqrt{\frac{2}{N}}\ln\frac{e}{\sqrt{P_{\min}}} + \rho\epsilon \tag{4.35}$$

Thus E_{sp} is infinite here and (4.24) reduces to $P_{e,\max} \geq 0$, completing the proof.

The theorem will now be generalized to lower bound the error probability for an arbitrary set of code words rather than a fixed composition set. The number of different ways to choose the composition of a code word is the number of ways of picking K nonnegative integers, $N_1, N_2, \cdots, N_K$ such that $\sum_k N_k = N$, where K is the input alphabet size and N is the block length. Thus there are $\binom{N+K-1}{K-1}$ different compositions, and it follows that in any code of M code words, there must be some composition containing a number of code words M' bounded by

$$M' \geq M \Big/ \binom{N+K-1}{K-1} \tag{4.36}$$

Consider the messages corresponding to this set of M' words as a fixed composition code and assume that the same list decoding scheme is used as for the original code. Thus for each m in the fixed composition set, Y_m is the same as for the original code and $P_{e,m}$ is the same. This is presumably a rather foolish decoding scheme for the fixed composition code since the decoding lists might contain fewer than L integers from the fixed composition set. None the less, Theorem 6 applies here, and using $\ln(M'/L)/N$ for R, there is some m in the fixed composition set for which $P_{e,m}$ satisfies

$$P_{e,m} > \exp - N\left\{E_{sp}\left[\frac{\ln(M'/L)}{N} - \frac{\ln 4}{N} - \epsilon\right] + \sqrt{\frac{8}{N}}\ln\frac{e}{\sqrt{P_{\min}}} + \frac{\ln 4}{N}\right\} \tag{4.37}$$

Since E_{sp} is a decreasing function of its argument, we can substitute (4.36) into (4.37). Also $P_{e,m} \leq P_{e,\max}$ for the original code, so that

$$P_{e,\max} > \exp - N\left\{E_{sp}\left[\frac{\ln(M/L) - \ln\binom{N+K-1}{K-1}}{N} - \frac{\ln 4}{N} - \epsilon\right] + \sqrt{\frac{8}{N}}\ln\frac{e}{\sqrt{P_{\min}}} + \frac{\ln 4}{N}\right\} \tag{4.38}$$

For the given channel, define $P_{e,\max}(N, M, L)$ as the minimum $P_{e,\max}$ over all codes of M code words of length N and all list decoding schemes of list size L. Equation (4.38) clearly applies to the code and decoding scheme that achieves $P_{e,\max}(N, M, L)$. Finally, since

$$\binom{N+K-1}{K-1} < N^K \tag{4.39}$$

We can rewrite (4.38) as

$$P_{e,\max}(N, M, L) > \exp - N\left\{E_{sp}\left[\frac{\ln(M/L)}{N} - \frac{K\ln N}{N} - \frac{\ln 4}{N}\right] + \sqrt{\frac{8}{N}}\ln\frac{e}{\sqrt{P_{\min}}} + \frac{\ln 4}{N}\right\} \tag{4.40}$$

We have chosen $\epsilon > 0$ to absorb the inequality in (4.39).

One more step will now complete the proof of Theorem 2. We show that, in general,

$$P_e(N, M, L) \geq \tfrac{1}{2}P_{e,\max}(N, [M/2]^+, L) \tag{4.41}$$

To see this, consider the code that achieves the minimum average error probability $P_e(N, M, L)$. At least $M/2$ of these words must have $P_{e,m} \leq 2P_e(N, M, L)$. This set of $[M/2]^+$ code words with the original decoding scheme then has $P_{e,\max} \leq 2P_e(N, M, L)$. By definition, however, this $P_{e,\max}$ is greater than or equal to $P_{e,\max}(N, [M/2]^+, L)$, thus establishing (4.41).

Substituting (4.40) into (4.41), we obtain (1.6), thus completing the proof of Theorem 2.

In the proof of Theorem 2, it was not made quite clear why the artifice of fixed composition codes had to be introduced. We started the derivation of the bound by relating the error probability for a given message, m, to the size of the decoding subset $F(Y_m)$, and then observing that at least one $F(Y_m)$ must be at most L/M. This last observation, however, required that all Y_m be measured with the same probability assignment $\mathbf{f}$. Unfortunately, a good choice of $\mathbf{f}$ for one code word composition is often a very poor choice for some other composition, and in general, no choice of $\mathbf{f}$ is uniformly good. We eventually chose $\mathbf{f}$ as a function of the parameter s, but the appropriate value of s (i.e., that which satisfies (4.28) with equality) is a function of the code word composition $\mathbf{q}$, making $\mathbf{f}_s$ also implicitly dependent upon $\mathbf{q}$.

The reliance of the bound on fixed composition codes is particularly unfortunate in that it prevents us from extending the bound to continuous channels, channels with memory, and channels with feedback. In the first case the size of the input alphabet K becomes infinite, and in the other cases $\mu(s)$ in (4.4) depends on more than just the composition of a code word. One way to avoid these difficulties is to classify code words by the value of s for which (4.28) is satisfied with equality but, so far, no *general* theorem has been proved using this approach. These extensions to more general channels are possible, however, if the channel has sufficient symmetry and we conjecture that the exponential bound $E_{sp}(R)$ is valid under much broader conditions than we have assumed here.

APPENDIX

PROPERTIES OF $E_{sp}(R)$

Using (1.7) and (1.8) we can rewrite $E_{sp}(R)$ as

$$E_{sp}(R) = \max_{\mathbf{q}} E(R, \mathbf{q}) \tag{A.1}$$

$$E(R, \mathbf{q}) = \underset{\rho \geq 0}{\text{L.U.B.}} [E_0(\rho, \mathbf{q}) - \rho R] \tag{A.2}$$

Define $I(\mathbf{q})$ as the average mutual information on the channel using the input probabilities $(q_1, \cdots, q_K)$,

$$I(\mathbf{q}) = \sum_{k=1}^{K} \sum_{j=1}^{J} q_k P(j \mid k) \ln \frac{P(j \mid k)}{\sum_{i=1}^{K} q_i P(j \mid i)} \tag{A.3}$$

It has been shown by Gallager (1965, Theorem 2), that if $I(\mathbf{q}) \neq 0$, then

$$E_0(\rho, \mathbf{q}) \geqq 0 \tag{A.4}$$

$$0 < \frac{\partial E_0(\rho, \mathbf{q})}{\partial \rho} \leqq I(\mathbf{q}) \tag{A.5}$$

$$\frac{\partial^2 E_0(\rho, \mathbf{q})}{\partial \rho^2} \leqq 0 \tag{A.6}$$

with equality in (A.4) iff $\rho = 0$; in (A.5) if $\rho = 0$; and in (A.6) iff the following conditions are satisfied:

(a) $P(j \mid k)$ is independent of k for those j, k such that $q_k P(j \mid k) \neq 0$.

(b) The sum of the q_k over inputs leading to output j with nonzero probability is independent of j. It follows trivially from the same proof that $E_0(\rho, \mathbf{q}) = 0$ for all $\rho \geqq 0$ if $I(\mathbf{q}) = 0$.

Using these results, we can give $E(R, \mathbf{q})$ parametrically as

$$E(R, \mathbf{q}) = E_0(\rho, \mathbf{q}) - \rho \frac{\partial E_0(\rho, \mathbf{q})}{\partial \rho} \tag{A.7}$$

$$R = \frac{\partial E_0(\rho, \mathbf{q})}{\partial \rho} \tag{A.8}$$

Equations (A.7) and (A.8) are valid for

$$\lim_{\rho \to \infty} \frac{\partial E_0(\rho, \mathbf{q})}{\partial \rho} < R < \left. \frac{\partial E_0(\rho, \mathbf{q})}{\partial \rho} \right|_{\rho=0} = I(\mathbf{q}) \tag{A.9}$$

also,

$$E(R, \mathbf{q}) = 0 \quad \text{if} \quad R \geqq I(\mathbf{q}) \tag{A.10}$$

$$E(R, \mathbf{q}) = \infty \quad \text{if} \quad R < \lim_{\rho \to \infty} \frac{\partial E_0(\rho, \mathbf{q})}{\partial \rho} \tag{A.11}$$

From (A.7) and (A.8), we have

$$\frac{\partial E(R, \mathbf{q})}{\partial R} = -\rho; \qquad R = \frac{\partial E_0(\rho, \mathbf{q})}{\partial \rho} \tag{A.12}$$

If (A.6) is satisfied with strict inequality, then R in (A.8) is strictly decreasing with ρ and from (A.12), $E(R, \mathbf{q})$ is strictly decreasing with R and is strictly convex $\cup$ over the range of R given by (A.9).

We now observe from (A.10) that if $R \geqq C = \max_{\mathbf{q}} I(\mathbf{q})$, then $E(R, \mathbf{q}) = 0$ for all $\mathbf{q}$ and $E_{sp}(R) = 0$. Also if $R < C$, then for the $\mathbf{q}$ that yields capacity, $E(R, \mathbf{q}) > 0$ and thus $E_{sp}(R) > 0$. Finally, for a given R in the range $R_\infty < R < C$ the $\mathbf{q}$ that maximizes $E(R, \mathbf{q})$ satisfies (A.9), and thus $E_{sp}(R)$ is strictly decreasing and strictly convex $\cup$ in this range.

Next suppose that $R_{crit} = C$. Then for some $\rho^* \geqq 1$, $E_0(\rho^*)/\rho^* = C$, and thus for some $\mathbf{q}$, $E_0(\rho^*, \mathbf{q})/\rho^* = C$. But since $\partial E_0(\rho, \mathbf{q})/\partial \rho \leqq C$, this implies that $\partial E_0(\rho, \mathbf{q})/\partial \rho = C$ for $0 \leqq \rho \leqq \rho^*$ and $\partial^2 E_0(\rho, \mathbf{q})/\partial \rho^2 = 0$ for $0 \leqq \rho \leqq \rho^*$. From (A.6) this implies that conditions (a) and (b) above are satisfied for $\mathbf{q}$ yielding capacity. This in turn implies that $\partial E_0(\rho, \mathbf{q})/\partial \rho = C$ for all ρ and thus $R_\infty = C$. The same argument shows that if $R_\infty = C$, conditions (a) and (b) above must be satisfied.

A BOUND ON μ_k''

From (3.25), $\mu_k''(s)$ is the variance of the random variable $D_k(j) = \ln [f(j)/P(j/k)]$ with the probability assignment

$$Q_{sk}(j) = \frac{P(j/k)^{1-s} f(j)^s}{\sum_i P(i/k)^{1-s} f(i)^s} \tag{A.13}$$

If follows that $s^2 \mu_k''(s)$ is the variance of $sD_k(j)$ with the same probability assignment. From (A.13), however, we see that

$$sD_k(j) = \ln \frac{Q_{sk}(j)}{P(j/k)} + \mu_k(s) \tag{A.14}$$

Thus $s^2 \mu_k''(s)$ is also the variance of the random variable $\ln [Q_{sk}(j)/P(j/k)]$ with the probability assignment $Q_{sk}(j)$. Since a variance can be upper bounded by a second moment around any point, we have

$$s^2 \mu_k''(s) \leqq \sum_j Q_{sk}(j) \left[\ln \frac{Q_{sk}(j)}{P(j/k)} - \ln \frac{e}{\sqrt{P_{\min}}} \right]^2 \tag{A.15}$$

where $P_{\min}$ is the smallest nonzero transition probability on the channel and the sum is over those j for which $P(j/k) > 0$.

We next upper bound the right hand side of (A.15) by maximizing over all choices of the probability vector $Q_{sk}(j)$. There must be a maximum since the function is continuous and the region is closed and bounded. The function cannot be maximized when any of the $Q_{sk}(j) = 0$, for the derivative with respect to such a $Q_{sk}(j)$ is infinite. Thus the maximum must be at a stationary point within the region, and any stationary point can be found by the LaGrange multiplier technique. This gives us, for each j,

$$\left[\ln \frac{Q_{sk}(j)\sqrt{P_{\min}}}{P(j/k)e} \right]^2 + 2 \ln \frac{Q_{sk}(j)\sqrt{P_{\min}}}{P(j/k)e} + \lambda = 0 \tag{A.16}$$

Solving for the logarithmic term, we obtain

$$\ln \frac{Q_{sk}(j)\sqrt{P_{\min}}}{P(j/k)e} = -1 \pm \sqrt{1 - \lambda}; \quad \text{each } j \tag{A.17}$$

There are two cases to consider: first where the same sign is used for the square root for each j; and second when the positive square root is used for some j and the negative for others. In the first case, all terms on the left are equal, and $Q_{sk}(j) = P(j/k)$ to satisfy the constraint that $Q_{sk}(j)$ is a probability vector. Then (A.15) reduces to

$$s^2 \mu_k''(s) \leqq \left[\ln \frac{e}{\sqrt{P_{\min}}} \right]^2 \tag{A.18}$$

In the second case, the left hand side of (A.17) is upper bounded by $Q_{sk}(j) = 1$, $P(j/k) = P_{\min}$, yielding $-\ln e\sqrt{P_{\min}}$. From the right hand side of (A.17), the terms using the negative square root can have a magnitude at most 2 larger than the positive term. Thus

$$\left| \ln \frac{Q_{sk}(j)\sqrt{P_{\min}}}{P(j/k)e} \right| \leqq 2 - \ln e\sqrt{P_{\min}} = \ln \frac{e}{\sqrt{P_{\min}}} \tag{A.19}$$

Substituting (A.19) into (A.15) again yields (A.18) completing the proof.

PROOF THAT $\mathbf{q}_s$, $\mathbf{f}_s$ YIELDS A SADDLE POINT FOR $q_k \mu_k(s, \mathbf{f})$ (SEE (4.23))

From (4.22), we see that the right side of (4.23) is valid and also that

$$\sum_k q_{ks} \mu_k(s, \mathbf{f}_s) = (1 - s) \ln \sum_j \alpha_{js}^{1/(1-s)}. \tag{A.20}$$

In order to establish the left side of (4.23) we must show that

$$\sum_k q_{ks} \ln [\sum_j P(j \mid k)^{1-s} f(j)^s] - (1 - s) \ln \sum_j \alpha_{js}^{1/(1-s)} \leqq 0 \quad (A.21)$$

Combining the logarithm terms, and using the inequality $\ln z \leqq z - 1$ for $z \geqq 0$ (taking $\ln 0$ as $-\infty$), the left side of (A.21) becomes

$$\sum_k q_{ks} \ln \frac{\sum_j P(j \mid k)^{1-s} f(j)^s}{(\sum_j \alpha_{js}^{1/(1-s)})^{1-s}} \leqq \frac{\sum_{k,j} q_{ks} P(j \mid k)^{1-s} f(j)^s}{(\sum_j \alpha_{js}^{1/(1-s)})^{1-s}} - 1 \quad (A.22)$$

$$\leqq \sum_j f_s(j)^{1-s} f(j)^s - 1 \quad (A.23)$$

$$\leqq 0 \quad (A.24)$$

when we have used (4.19) and then (4.20) to go from (A.22) to (A.23), and used Holder's inequality to go from (A.23) to (A.24). This completes the proof.

PROOF THAT $R(s, \mathbf{q})$ (SEE (4.25)) IS CONTINUOUS IN s, $0 < s < 1$

The problem here is to show that $\mathbf{f}_s$ is a continuous vector function of s. It will then follow immediately that $\mu_k(s, \mathbf{f}_s)$ and $\mu_k'(s, \mathbf{f}_s)$ are continuous functions of s, and then from (4.25) that $R(s, \mathbf{q})$ is a continuous function of s for fixed $\mathbf{q}$.

$E_0(\rho, \mathbf{q})$ as given by (1.9) can be rewritten as

$$E_0\left(\frac{s}{1-s}, \mathbf{q}\right) = -\ln \sum_j \alpha_j(s, \mathbf{q})^{1/(1-s)} \quad (A.25)$$

$$\alpha_j(s, \mathbf{q}) = \sum_k q_k P(j \mid k)^{1-s} \quad (A.26)$$

Let $\mathbf{q}_s$ be a choice of probability vector $\mathbf{q}$ that maximizes $E_0(s/(1 - s), \mathbf{q})$. We show that $\alpha_j(s, \mathbf{q}_s)$, which is α_{js} as defined in (4.19), is a continuous function of s, and it then follows from (4.20) that $\mathbf{f}_s$ is a continuous function of s. Since $\mathbf{q}_s$ maximizes $E_0(s/(1 - s), \mathbf{q})$, we have

$$E_0\left(\frac{s}{1-s}, \mathbf{q}_s\right) = -\ln \min_{\alpha(s,\mathbf{q})} \sum_j \alpha_j(s, \mathbf{q})^{1/(1-s)} \quad (A.27)$$

where the minimization is over the set of vectors $\boldsymbol{\alpha}$ whose components satisfy (A.26) for some choice of probability vector $\mathbf{q}$. Since this is a convex set of vectors and since $\sum_j \alpha_j^{1/(1-s)}$ is a *strictly* convex ∪ function of $\boldsymbol{\alpha}$ for $0 < s < 1$, the minimizing $\boldsymbol{\alpha}$ in (A.27) is unique and the strict convexity tells us that for any s, $0 < s < 1$ and for any $\epsilon > 0$ there exists a $\delta > 0$ such that if

$$\mid \alpha_j(s, \mathbf{q}) - \alpha_j(s, \mathbf{q}_s) \mid \geqq \epsilon/2; \quad \text{any } j \quad (A.28)$$

then

$$\sum_j \alpha_j(s, \mathbf{q})^{1/(1-s)} \geqq \sum_j \alpha_j(s, \mathbf{q}_s)^{1/(1-s)} + \delta \quad (A.29)$$

Next we observe that $E_0(s/(1 - s), \mathbf{q})$ is a continuous function of s with the continuity being uniform in $\mathbf{q}$. It follows from this that $E_0(s/(1 - s), \mathbf{q}_s)$ is also continuous in s. Also $\alpha_j(s, \mathbf{q})$ is continuous in s, uniformly in $\mathbf{q}$. It follows from these three statements that for a given s, $0 < s < 1$, and for the given ϵ, δ above, there exists a $\delta_1 > 0$ such that for $\mid s_1 - s \mid < \delta_1$,

$$\left| \sum_j \alpha_j(s_1, \mathbf{q}_{s_1})^{1/(1-s_1)} - \sum_j \alpha_j(s, \mathbf{q}_{s_1})^{1/(1-s)} \right| < \delta/2 \quad (A.30)$$

$$\left| \sum_j \alpha_j(s_1, \mathbf{q}_{s_1})^{1/(1-s_1)} - \sum_j \alpha_j(s, \mathbf{q}_s)^{1/(1-s)} \right| < \delta/2 \quad (A.31)$$

$$\mid \alpha_j(s_1, \mathbf{q}_{s_1}) - \alpha_j(s, \mathbf{q}_{s_1}) \mid < \epsilon/2; \quad \text{all } j \quad (A.32)$$

Combining (A.30) and (A.31), we see that (A.29) is unsatisfied for $\mathbf{q} = \mathbf{q}_{s_1}$; thus (A.28) must be unsatisfied for all j and

$$\mid \alpha_j(s, \mathbf{q}_{s_1}) - \alpha_j(s, \mathbf{q}_s) \mid < \epsilon/2; \quad \text{all } j, \mid s - s_1 \mid < \delta_1 \quad (A.33)$$

Combining (A.32) and (A.33), we then have for all j

$$\mid \alpha_j(s_1, \mathbf{q}_{s_1}) - \alpha_j(s, \mathbf{q}_s) \mid < \epsilon; \quad \mid s - s_1 \mid < \delta_1 \quad (A.34)$$

Thus $\alpha_j(s, \mathbf{q}_s)$ is continuous in s, completing the proof. Using other methods, it can be shown that $\alpha_j(s, \mathbf{q}_s)$ is a piecewise analytic function of s.

RECEIVED: January 18, 1966

REFERENCES

ASH, R. B. (1965), "Information Theory." Interscience, New York.

BERLEKAMP, E. R. (1964), "Block Coding with Noiseless Feedback." Ph.D. Thesis, Department of Electrical Engineering, M.I.T.

BHATTACHARYYA, A. (1943), On a measure of divergence between two statistical populations defined by their probability distributions. *Bull. Calcutta Math. Soc.* **35,** No. 3, 99–110.

CHERNOFF, H. (1952), A measure of asymptotic efficiency for tests of an hypothesis based on the sum of observations. *Ann. Math. Statist.* **23,** 493.

ELIAS, P. (1955), "List Decoding for Noisy Channels." Tech. Rept. 335, Research Laboratory of Electronics, M.I.T.

FANO, R. M. (1961), "Transmission of Information." M.I.T. Press, and Wiley, New York.

FEINSTEIN, A. (1955), Error bounds in noisy channels without memory. *IEEE Trans. Inform. Theory* **IT-1,** 13–14.

FELLER, W. (1943), Generalizations of a probability limit theorem of Cramer. *Trans. Am. Math. Soc.* **54,** 361.

GALLAGER, R. (1963), "Low Density Parity Check Codes." M.I.T. Press.

GALLAGER, R. (1965a), A simple derivation of the coding theorem and some applications. *IEEE Trans. Inform. Theory* **IT-11,** 3–18.

GALLAGER, R. (1965), "Lower Bounds on the Tails of Probability Distributions." M.I.T. Research Laboratory of Electronics. OPR **77,** pp. 277–291.

GILBERT, E. N. (1952), A comparison of signalling alphabets. *Bell System Tech. J.* **3,** 504–522.

HAMMING, R. W. (1950), Error detecting and error correcting codes. *Bell System Tech. J.* **29,** 47–160.

HELLIGER, E. (1909), Neue Begrundung der Theorie quadratischer Formen von unendlichvielen Veranderlichen. *J. reine angew. Math.* **136,** 210–271.

JACOBS, I. M., AND BERLEKAMP, E. R. (1967), A lower bound to the distribution of computation for sequential decoding. *IEEE Trans. Inform. Theory* **IT-13,** in press.

NEYMAN, J. AND PEARSON, E. S. (1928), On the use and interpretation of certain test criterion for purposes of statistical inference, *Biometrica* **20A,** 175, 263.

PETERSON, W. W. (1961), "Error-Correcting Codes." M.I.T. Press, and Wiley, New York.

PLOTKIN, M. (1960), Research Division Report 51–20, University of Pennsylvania. Published in 1960 as: Binary codes with specified minimum distance. *IEEE Trans. Inform. Theory* **IT-6,** 445–450.

REIFFEN, B. (1963), A note on "very noisy" channels. *Inform. and Control* **6,** 126–130.

SHANNON, C. E. (1948), A mathematical theory of communication. *Bell System Tech. J.* **27,** 379, 623. Also in book form with postscript by W. Weaver, Univ. of Illinois Press, Urbana, Illinois.

SHANNON, C. E. (1956), Zero error capacity of noisy channels. *IEEE Trans. Inform. Theory* **IT-2,** 8.

SHANNON, C. E. (1958), Certain results in coding theory for noisy channels. *Inform. Control* **1,** 6.

SUN, M. (1965), Asymptotic bounds on the probability of error for the optimal transmission of information in the channel without memory which is symmetric in pairs of input symbols for small rates of transmission. *Theory Probab. Appl.* (Russian) **10,** no. 1, 167–175.

Lower Bounds to Error Probability for Coding on Discrete Memoryless Channels. II

C. E. Shannon* and R. G. Gallager*

Departments of Electrical Engineering and Mathematics, Research Laboratory of Electronics, Massachusetts Institute of Technology, Cambridge, Massachusetts 02139

AND

E. R. Berlekamp†

Department of Electrical Engineering, University of California, Berkeley, California

New lower bounds are presented for the minimum error probability that can be achieved through the use of block coding on noisy discrete memoryless channels. Like previous upper bounds, these lower bounds decrease exponentially with the block length N. The coefficient of N in the exponent is a convex function of the rate. From a certain rate of transmission up to channel capacity, the exponents of the upper and lower bounds coincide. Below this particular rate, the exponents of the upper and lower bounds differ, although they approach the same limit as the rate approaches zero. Examples are given and various incidental results and techniques relating to coding theory are developed. The paper is presented in two parts: the first, appearing in the January issue, summarizes the major results and treats the case of high transmission rates in detail; the second, appearing here, treats the case of low transmission rates.

1. ZERO RATE EXPONENTS

In this section we shall investigate the error probability for codes whose block length is much larger than the number of codewords, $N \gg M$. We assume throughout this section that the zero error capacity of the channel, C_0, is zero. We also assume that ordinary decoding is to be used rather than list decoding, i.e., that the list size L is one.

Our basic technique will be to bound the error probability for a given set of code words in terms of the error probability between any pair of the words, say $\underline{x}_m$ and $\underline{x}_{m'}$. We can apply the corollary to Theorem I-5, given by (I-3.20) and (I-3.21), as follows.[1] Let $P_1(\underline{y})$ and $P_2(\underline{y})$ in Theorem I-5 correspond to $\Pr(\underline{y} \mid \underline{x}_m)$ and $\Pr(\underline{y} \mid \underline{x}_{m'})$ here, and let Y_1 and Y_2 in Theorem I-5 correspond to the decoding regions Y_m and $Y_{m'}$ for the given decoding scheme here. The fact that some output sequences are decoded into messages other than m or m' in no way effects the validity of Theorem 5 or its corollary. From (I-3.20) and (I-3.21), the error probabilities $P_{e,m}$ and $P_{e,m'}$ for the given decoding scheme are bounded by either

$$P_{e,m} \geqq \tfrac{1}{4} \exp\left[\mu(s^*) - s^* \sqrt{2\mu''(s^*)}\right] \qquad (1.01)$$

or

$$P_{e,m'} \geqq \tfrac{1}{4} \exp\left[\mu(s^*) \pm (1 - s^*) \sqrt{2\mu''(s^*)}\right], \qquad (1.02)$$

where

$$\mu(s) = \ln \sum_{\underline{y}} \Pr(\underline{y} \mid \underline{x}_m)^{1-s} \Pr(\underline{y} \mid \underline{x}_{m'})^{s} \qquad (1.03)$$

and s^* minimizes $\mu(s)$ over $0 \leqq s \leqq 1$.

This result can be put into a more convenient form with the aid of the following definitions.

[1] References to equations, sections and theorems of the first part of this paper will be prefixed by I.

* The work of these authors is supported by the National Aeronautics and Space Administration (Grants NsG-334 and NsG-496), the Joint Services Electronics Program (contract DA-36-039-AMC-03200 (EE)), and the National Science Foundation (Grant GP-2495).

† The work of this author is supported by the Air Force Office of Scientific Research (Grant AF-AFOSR-639-65).

The joint composition of $\underline{x}_m$ and $\underline{x}_{m'}$, $q_{i,k}(m, m')$ is the fraction of the positions in the block in which the ith channel input occurs in codeword $\underline{x}_m$ and the kth channel input occurs in $\underline{x}_{m'}$.

The function $\mu_{i,k}(s)$ is defined for $0 < s < 1$ by

$$\mu_{i,k}(s) \triangleq \ln \sum_j P(j \mid i)^{1-s} P(j \mid k)^{s}. \qquad (1.04)$$

As before,

$$\mu_{i,k}(0) = \lim_{s \to 0^+} \mu_{i,k}(s)$$

and

$$\mu_{i,k}(1) = \lim_{s \to 1^-} \mu_{i,k}(s).$$

Using (I-3.10), $\mu(s)$ in (1.03) can be expressed in terms of these definitions by

$$\mu(s) = N \sum_i \sum_k q_{i,k}(m, m') \mu_{i,k}(s). \qquad (1.05)$$

The *discrepancy* between $\underline{x}_m$ and $\underline{x}_{m'}$, $D(m, m')$, is defined by

$$D(m, m') \triangleq - \min_{0 \leqq s \leqq 1} \sum_i \sum_k q_{i,k}(m, m') \mu_{i,k}(s). \qquad (1.06)$$

It can be seen that the quantity $\mu(s^*)$ appearing in (1.01) and (1.02) is given by $-ND(m, m')$. The discrepancy plays a role similar to that of the conventional Hamming distance for binary symmetric channels.

The *minimum discrepancy* for a code $D_{\min}$ is the minimum value of $D(m, m')$ over all pairs of code words of a particular code.

The *maximum minimum discrepancy*, $D_{\min}(N, M)$ is the maximum value of $D_{\min}$ over all codes containing M code words of block-length N.

Theorem 1. *If $\underline{x}_m$ and $\underline{x}_{m'}$ are a pair of code words in a code of block-length N, then either*

$$P_{e,m} \geqq \frac{1}{4} \exp -N\left[D(m, m') + \sqrt{\frac{2}{N}} \ln (1/P_{\min})\right] \qquad (1.07)$$

or

$$P_{e,m'} \geqq \frac{1}{4} \exp -N\left[D(m, m') + \sqrt{\frac{2}{N}} \ln (1/P_{\min})\right], \qquad (1.08)$$

where $P_{\min}$ is the smallest nonzero transition probability for the channel.

Proof. We shall show that $\mu''(s)$ is bounded by

$$\mu''(s) \leqq N\left[\ln \frac{1}{P_{\min}}\right]^2. \qquad (1.09)$$

Then the theorem will follow from (1.01) and (1.02) by upper bounding s^* and $(1 - s^*)$ by 1. To establish (1.09), we use (I-3.25), obtaining

$$\mu''_{i,k}(s) = \sum_j Q_s(j) \left[\ln \frac{P(j \mid k)}{P(j \mid i)}\right]^2 - [\mu'_{i,k}(s)]^2, \qquad (1.10)$$

where $Q_s(j)$ is a probability assignment over the outputs for which $P(j \mid k)$ and $P(j \mid i)$ are nonzero. Observing that

$$|\ln P(j \mid k)/P(j \mid i)| \leqq \ln (1/P_{\min}),$$

we can ignore the last term in (1.10), getting

$$\mu''_{i,k}(s) \leqq \sum_j Q_s(j)[\ln (1/P_{\min})]^2 = [\ln (1/P_{\min})]^2. \qquad (1.11)$$

Combining (1.11) with (1.05), we have (1.09), completing the proof.

Since the probability of error for the entire code of M code words is lower bounded by $P_e \geqq P_{e,m}/M$ for any m, it follows from the theorem that

$$P_e \geqq \frac{1}{4M} \exp - N\left[D_{\min} + \sqrt{\frac{2}{N}} \ln \frac{1}{P_{\min}}\right]. \qquad (1.12)$$

Conversely, we now show that there exist decoding regions such that

$$P_{e,m} \leqq (M - 1) \exp -ND_{\min} \quad \text{for all } m. \qquad (1.13)$$

Reprinted with permission from *Inform. Contr.*, vol. 10, pp. 522–552, May 1967.

These regions may be chosen as follows: From Theorem I-5, there exist decoding regions $Y_m(m, m')$ and $Y_{m'}(m, m')$ for the code containing only the codewords m and m' such that both $P_{e,m}$ and $P_{e,m'}$ are no greater than $\exp -ND_{\min}$. To decode the larger code, set $Y_m = \bigcap_{m'} Y_m(m, m')$. Since the sets Y_m are not overlapping, they are legitimate decoding sets. Also, $Y_m^c = \bigcup_{m'} Y_m^c(m, m')$, and since the probability of a union of events cannot exceed the sum of their probabilities, we have

$$P_{e,m} \leqq \sum_{\mathbf{y} \in Y_m^c} \Pr(\mathbf{y} \mid \mathbf{x}_m) \leqq \sum_{m' \neq m} \sum_{\mathbf{y} \in Y_m^c(m,m')} \Pr(\mathbf{y} \mid \mathbf{x}_m) \quad (1.14)$$

$$\leqq (M - 1) \exp - ND_{\min}. \quad (1.15)$$

Combining (1.12) and (1.15) yields the first part of the following theorem:

THEOREM 2. *Let E_M be defined by*

$$\limsup_{N \to \infty} -\frac{1}{N} \ln P_e(N, M, 1).$$

Then

$$E_M = \limsup_{N \to \infty} D_{\min}(N, M) = \underset{N}{\text{l.u.b.}}\, D_{\min}(N, M) = \lim_{N \to \infty} D_{\min}(N, M). \quad (1.16)$$

The second part of the theorem follows from the observation that we can construct a code of block length AN from a code of blocklength N by repeating every word of the original code A times. The two codes have equal $q_{i,k}(m, m')$ for all i, k, m, m', and hence they have equal $D_{\min}$.

Thus

$$D_{\min}(AN, M) \geqq D_{\min}(N, M). \quad (1.17)$$

This implies the second part of the theorem. The third part follows from (1.17) and the fact that $P_e(N, M, 1)$ is nonincreasing with N.

Theorem 2 reduces the problem of computing E_M to the problem of computing $D_{\min}(N, M)$. This computation is always easy for $M = 2$, so we treat that case first. Recall from (1.06) that $-D(m, m')$ is the minimum over s of a weighted sum of the $\mu_{i,k}(s)$. This can be lower bounded by the weighted sum of the minimums, yielding

$$-D(m, m') \geqq \sum_i \sum_k q_{i,k}(m, m') \min_{0 \leq s \leq 1} \mu_{i,k}(s). \quad (1.18)$$

with equality iff the same value of s simultaneously minimizes all $\mu_{i,k}(s)$ for which $q_{i,k}(m, m') > 0$. If we, set $q_{i,k}(m, m') = 1$ for the i, k pair that minimizes $\min_{0 \leq s \leq 1} \mu_{i,k}(s)$, then (1.18) is satisfied with equality and at the same time the right-hand side is minimized. We thus have

$$E_2 = D_{\min}(N, 2) = \max_{i,k} [- \min_{0 \leq s \leq 1} \mu_{i,k}(s)]. \quad (1.19)$$

It is interesting to compare this expression with the sphere packing exponent $E_{sp}(R)$ in the limit as $R \to 0$. If $R_\infty = 0$, some manipulation on (I-1.7), (I-1.8), and (I-1.9) yields

$$E_{sp}(0^+) = \lim_{\rho \to \infty} E_0(\rho) = \max_{\mathbf{q}} - \ln \sum_j \prod_k P(j \mid k)^{q_k} \quad (1.20)$$

Comparing (1.20) with the definition of $\mu_{i,k}(s)$ in (1.04), we see that $E_2 \leqq E_{sp}(0^+)$ with equality iff the probability vector $\mathbf{q}$ that maximizes (1.20) has only 2 nonzero components.

Having found the pair of input letters i, k that yield E_2, it clearly does not matter whether we set $q_{i,k}(1, 2) = 1$ or $q_{k,i}(1, 2) = 1$. However, we must *not* attempt to form some linear combination of these two optimum solutions, for by making both $q_{i,k}(1, 2)$ and $q_{k,i}(1, 2)$ nonzero we may violate the condition for equality in (1.18). For example, suppose we compare the following two codes of block length N for the completely asymmetric binary channel of Fig. I-56. The disastrous result is depicted below:

Code 1: $\quad \mathbf{x}_1 = 1\,1\,1\,1\,1 \quad 1\,1\,1\,1\,1$

$\qquad\qquad \mathbf{x}_2 = 2\,2\,2\,2\,2 \quad 2\,2\,2\,2\,2$

$\qquad\qquad\quad \leftarrow N/2 \rightarrow \quad \leftarrow N/2 \rightarrow$

Code 2: $\quad \mathbf{x}_1 = 1\,1\,1\,1\,1 \quad 2\,2\,2\,2\,2$

$\qquad\qquad \mathbf{x}_2 = 2\,2\,2\,2\,2 \quad 1\,1\,1\,1\,1.$

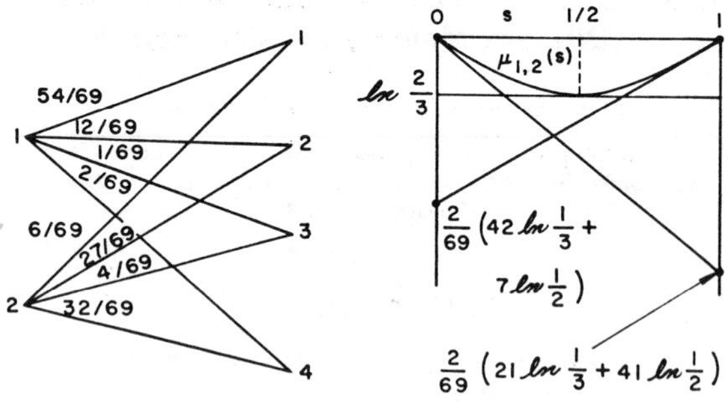

FIG. 1. A pairwise reversible binary input channel.

Using either code, an error will occur only if the received sequence consists entirely of output letter 2. For Code 1, $P_e = \frac{1}{2}p^N$; for Code 2, $P_e = \frac{1}{2}p^{N/2}$.

For a class of channels to be defined as pairwise reversible channels, this sensitivity to interchanging letters does not occur, and for these channels we shall soon see that the calculation of E_M is relatively straightforward. A channel is pairwise reversible iff, for each i, k, $\mu'_{i,k}(\frac{1}{2}) = 0$. Differentiating (1.04), this is equivalent to

$$\sum_j \sqrt{P(j \mid i)P(j \mid k)} \ln P(j \mid i) = \sum_j \sqrt{P(j \mid i)P(j \mid k)} \ln P(j \mid k); \quad \text{all } i, k. \quad (1.21)$$

Equation (1.21) is equivalent to $\mu_{i,k}(s)$ being minimized at $s = \frac{1}{2}$ for all i, k. This guarantees that (1.18) is satisfied with equality and that a pair of inputs in the same position in a pair of code words, $\mathbf{x}_m$ and $\mathbf{x}_{m'}$, can be reversed without changing $D(m, m')$.

The class of pairwise reversible channels includes all of the symmetric binary input channels considered by Sun (1965) and Dobrushin (1962) (which are defined in a manner that guarantees that $\mu_{i,k}(s) = \mu_{k,i}(s)$

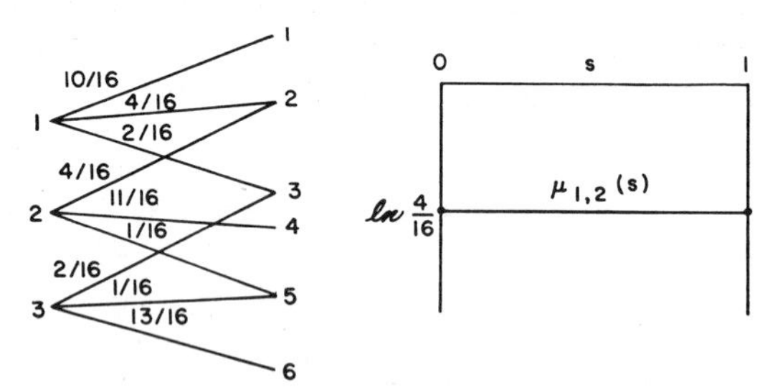

FIG. 2. A pairwise erasing ternary input channel (nonuniform but pairwise reversible).

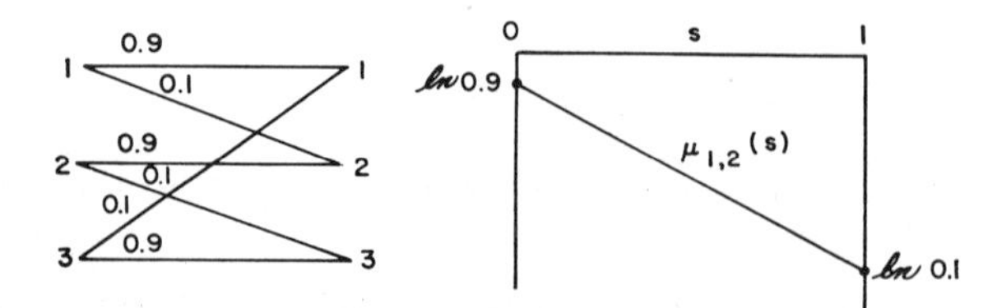

FIG. 3. A ternary unilateral channel (TUC) (uniform but not pairwise reversible).

for all s), and many other binary input channels, such as the one in Fig. 1 (as the reader is invited to verify). For multi-input channels, there is no relationship between the class of pairwise reversible channels and the uniform channels discussed by Fano (1961, p. 126). The channel of Fig. 2 is pairwise reversible but nonuniform; from any pair of inputs it looks like a binary erasure channel. The channel of Fig. 3 is not pairwise reversible even though it is uniform; from any pair of inputs it looks like an asymmetric binary erasure channel.

For pairwise reversible channels, we may compute an exact expression for E_M. To do this, we obtain a lower bound on $D_{\min}(N, M)$ which can be attained for certain values of N. The bound is derived by a method first introduced by Plotkin (1951). For any pair of code words for a pairwise reversible channel, we have[2]

$$D(m, m') = -\sum_i \sum_k q_{i,k}(m, m')\mu_{i,k}(\tfrac{1}{2}). \quad (1.22)$$

Since the minimum discrepancy cannot exceed the average discrepancy,

$$D_{\min}(N, M) \leqq \frac{1}{M(M-1)} \sum_{n \neq m'} \sum D(m, m'). \quad (1.23)$$

The total discrepancy can be computed on a column by column basis.

$$\sum_{m=1}^{M} \sum_{m'=1}^{M} D(m, m') = -\sum_{n=1}^{N} \sum_{i=1}^{K} \sum_{k=1}^{K} M_i(n) M_k(n) \mu_{i,k}(\tfrac{1}{2}), \quad (1.24)$$

where $M_k(n)$ is the number of times the kth channel input occurs in the nth column. Let M_k^* denote the number of times the kth channel input occurs in the best possible column,

$$\max_{\Sigma M_k = M} [-\sum_i \sum_k M_i M_k \mu_{i,k}(\tfrac{1}{2})] = -\sum_i \sum_k M_i^* M_k^* \mu_{i,k}(\tfrac{1}{2}) \quad (1.25)$$

Combining (1.23) through (1.25) results in a bound for pairwise reversible channels.

$$D_{\min}(N, M) \leqq -1/(M(M-1)) \sum_i \sum_k M_i^* M_k^* \mu_{i,k}(\tfrac{1}{2}) \quad (1.26)$$

We now show that this bound can be achieved when $N = M!/\prod_k M_k^*!$ To do this, we select the first column of the code so that it has the prescribed composition, the kth channel input occurring M_k^* times. Then we choose as subsequent columns of the code all possible permutations of the first column. In the constructed code, every column contributes the same maximum amount to the total discrepancy, assuring equality between (1.24) and (1.25). Every pair of codewords is the same distance apart, assuring equality in (1.23). Because of these two facts, (1.26) holds with equality when $N = M!/(\prod_k M_k^*!)$.

This construction can likewise be used for channels that are not pairwise reversible. The constructed code has the property that $q_{i,k}(m, m') = q_{k,i}(m, m') = q_{i,k}$ independent of m and m'. This guarantees that, for this code, (1.06) is optimized by setting $s = \frac{1}{2}$, for $\mu_{i,k}(s) + \mu_{k,i}(s)$ always attains its minimum at $s = \frac{1}{2}$, even when $\mu_{i,k}(s)$ does not.

However, it may be possible to improve upon this construction for channels which are not pairwise reversible. We summarize these results in a theorem, whose proof follows directly from Theorem 2, (1.26), and the construction discussed in the preceding two paragraphs.

This general inequality also follows directly from the definitions of E_∞ and $E_{ex}(0^+)$ without invoking Corollary 3.2.

We now proceed to show that Corollary 3.3 holds with equality even for channels which are not pairwise reversible.

THEOREM 4. *For any discrete memoryless channel* $E_\infty = E_{ex}(0^+)$.

Remarks. The natural approach in attempting to prove Theorem 4 would be to attempt to calculate the average discrepancy on a column by column basis as in (1.24). This direct approach does not work for channels that are not pairwise reversible, however, the difficulty being that the value of s that determines $D(m, m')$ in (1.06) is not the same as the value of s that minimizes $\mu_{i,k}(s)$ for the pairs of letters in the two code words.

We shall circumvent this difficulty by going through some manipulations on a particular subset of the code words in a code. The argument is rather lengthy and will be carried out as a sequence of 5 Lemmas. For motivation, the reader is advised to keep the ternary unilateral channel (TUC) of Figure 3 in mind throughout the proof. We begin by defining a relation of dominance between code words.

DEFINITION. $\underline{x}_m$ dominates $\underline{x}_{m'}$ iff

$$-\sum_i \sum_k q_{i,k}(m, m')\mu'_{i,k}(\tfrac{1}{2}) \geqq 0. \quad (1.28)$$

Notice that either $\underline{x}_m$ dominates $\underline{x}_{m'}$, or $\underline{x}_{m'}$ dominates $\underline{x}_m$, or both. This follows because

$$\mu'_{i,k}(\tfrac{1}{2}) = -\mu'_{k,i}(\tfrac{1}{2}); \qquad q_{i,k}(m, m') = q_{k,i}(m', m) \quad (1.29)$$

$$\sum_i \sum_k q_{i,k}(m', m)\mu'_{i,k}(\tfrac{1}{2}) = -\sum_i \sum_k q_{i,k}(m, m')\mu'_{i,k}(\tfrac{1}{2}). \quad (1.30)$$

For the TUC the codeword consisting of all 1's dominates any other codeword which contains at least as many 2's as 3's, but it is dominated by any other codeword which contains at least as many 3's as 2's.

Notice that dominance is *not* necessarily transitive except when the input alphabet is binary. In general, we may have $\underline{x}$ dominate $\underline{x}'$ and $\underline{x}'$ dominate $\underline{x}''$ without having $\underline{x}$ dominate $\underline{x}''$.

LEMMA 4.1. *If $\underline{x}_m$ dominates $\underline{x}_{m'}$, then*

$$D(m, m') \leqq \sum_i \sum_k q_{i,k}(m, m')[-\mu_{i,k}(\tfrac{1}{2}) - \tfrac{1}{2}\mu'_{i,k}(\tfrac{1}{2})].$$

Proof. Recall from (1.06) that

$$D(m, m') = -\min_{0 \leq s \leq 1} \sum_i \sum_k q_{i,k}(m, m')\mu_{i,k}(s). \quad (1.06)$$

THEOREM 3.

$$E_M \geqq 1/(M(M-1)) \max_{\Sigma M_k = M} \sum_i \sum_k M_i M_k (-\ln \sum_j \sqrt{P(j/i)\ P(j/k)})$$

with equality for channels which are pairwise reversible.

We next compare this result with $E_{ex}(0^+)$, Gallager's (1965) lower bound to $E(0^+)$, the error exponent at infinitesimal rates. $E_{ex}(0^+)$ is given by (I-1.29) and (I-1.30) as

$$E_{ex}(0^+) = \max_q \sum_i \sum_k q_i q_k (-\ln \sum_j \sqrt{P(j/i)P(j/k)}), \quad (1.27)$$

where q is the probability vector specifying the composition of the code. The vector q is unrestricted by the Diophantine constraints placed on the vector $\underline{M}^*/M$. (Here M_k^* is the kth component of $\underline{M}^*$). This additional freedom can only increase $E_{ex}(0^+)$. This proves the first of the three corollaries.

COROLLARY 3.1. *For pairwise reversible channels,*

$$E_M \leqq (M/(M-1))E_{ex}(0^+)$$

The evaluation of the expression on the right of Theorem 3 is complicated by the Diophantine constraints on the components of the vector M. To first order in M, however, these constraints may be ignored, as indicated by the following corollary.

COROLLARY 3.2. *For any channel,*

$$E_M \geqq M/(M-1)E_{ex}(0^+) - 0(1/M^2)$$

where

$$0(1/M^2) \leqq \frac{-K\mu_{\max} - \sum\sum_{i \neq k} (\mu_{i,k}(\tfrac{1}{2}) - \mu_{\max})}{4M(M-1)}.$$

Here K is the number of channel inputs and $\mu_{\max} = \max_{i \neq k} \mu_{i,k}(\tfrac{1}{2})$.

Since this corollary is not essential to the proof of Theorem 4, we omit its proof. The details of the straightforward but tedious calculation are given by Berlekamp (1964).

For the remainder of this section, we shall be primarily concerned with the behavior of E_M for very large M. We are especially interested in the limit of E_M as M goes to infinity, which we denote by the symbol E_∞.

Since E_M is a monotonic nonincreasing function of M, it is clear that the limit exists. As a consequence of Corollaries 3.1 and 3.2, we have

COROLLARY 3.3. $E_\infty \geqq E_{ex}(0^+)$ *with equality for channels which are pairwise reversible.*

[2] Readers who are familiar with the statistical literature will recognize the expression for $\mu_{i,k}(\frac{1}{2})$ as the measure of the difference between the distributions $P(j/i)$ and $P(j/k)$ which was first suggested by Helliger (1909) and later developed by Bhattacharyya (1943).

The tangent line to a convex U function is a lower bound to the function. Taking this tangent to $\mu_{i,k}(s)$ at $s = \frac{1}{2}$ yields

$$\min_{0\leq s\leq 1} \sum_i \sum_k q_{i,k}(m, m')\mu_{i,k}(s) \geq \min_{0\leq s\leq 1} \sum_i \sum_k q_{i,k}(m, m')[\mu_{i,k}(\tfrac{1}{2}) + (s - \tfrac{1}{2})\mu'_{i,k}(\tfrac{1}{2})]. \tag{1.31}$$

From the definition of dominance, (1.28), this linear function of s is minimized at $s^* = 1$.

q.e.d.

LEMMA 4.2. *From an original code containing M codewords, we may extract a subset of at least $\log_2 M$ codewords which form an "ordered" code, in which each word dominates every subsequent word.*

Proof. We first select the word in the original code which dominates the most others. According to the remarks following (1.28), this word must dominate at least half of the other words in the original code. We select this word as $\underline{x}_1$ in the ordered code. All words in the original code which are not dominated by $\underline{x}_1$ are then discarded. From the remaining words in the original code, we select the word which dominates the most others and choose it as $\underline{x}_2$ in the ordered code. The words which are not dominated by $\underline{x}_2$ are then discarded from the original code. This process is continued until all words of the original code are either placed in the ordered code or discarded. Since no more than half of the remaining words in the original code are discarded as each new word is placed in the ordered code, the ordered code contains at least $\log_2 M$ codewords.

q.e.d.

Within an ordered code, every word dominates each succeeding word. In particular, every word in the top half of the code dominates every word in the bottom half of the code. This fact enables us to bound the average discrepancy between words in the top half of the code and words in the bottom half of the code on a column by column basis. Using this technique, Lemma 4.3 gives us a bound to the minimum discrepancy of any ordered code in terms of $E_{ex}(0^+)$ and another term which must be investigated further in subsequent lemmas.

LEMMA 4.3. *Consider any ordered code having $2M$ words of block length N. The minimum discrepancy of this code is bounded by*

$$D_{\min} \leq \sum_{m=1}^{M} \sum_{m'=M+1}^{2M} D(m, m')/M^2$$
$$\leq E_{ex}(0^+) + 2d_{\max}\sqrt{K}\sqrt{\frac{1}{4N}\sum_{n=1}^{N}\sum_{k=1}^{K}(q_k^t(n) - q_k^b(n))^2},$$

where

$$d_{\max} \triangleq \max_{i,k} |\mu_{i,k}(\tfrac{1}{2}) + \tfrac{1}{2}\mu'_{i,k}(\tfrac{1}{2})| \tag{1.32}$$

and $q^t(n) = [q_1^t(n), \cdots, q_K^t(n)]$ is the composition of the nth column of the top half of the code (i.e., the kth channel input letter occurs $Mq_k^t(n)$ times in the nth column of the first M codewords). Similarly, $q^b(n) = [q_1^b(n), \cdots, q_K^b(n)]$ is the composition of the nth column of the bottom half of the code.

Proof.

$$D_{\min} \leq \sum_{m=1}^{M}\sum_{m'=M+1}^{2M} \frac{D(m, m')}{M^2} \tag{1.33}$$

$$\leq \sum_{m=1}^{M}\sum_{m'=M+1}^{2M}\sum_{i=1}^{K}\sum_{k=1}^{K} \frac{q_{i,k}(m, m')}{M^2}\left[-u_{i,k}(1/2) - \frac{1}{2}u'_{i,k}(1/2)\right]. \tag{1.34}$$

Now for any values of i and k,

$$\sum_{m=1}^{M}\sum_{m'=M+1}^{2M} \frac{q_{i,k}(m, m')}{M^2} = \sum_{n=1}^{N} \frac{q_i^t(n)q_k^b(n)}{N} \tag{1.35}$$

because both sides represent the average number of occurrences of the ith letter in the top half of the code opposite the kth letter in the same column of the bottom half of the code. Using this fact gives

$$D_{\min} \leq \frac{1}{N}\sum_{n=1}^{N}\sum_{i=1}^{K}\sum_{k=1}^{K} q_i^t(n)q_k^b(n)\left[-u_{i,k}\left(\frac{1}{2}\right) - \frac{1}{2}u'_{i,k}\left(\frac{1}{2}\right)\right]. \tag{1.36}$$

This bounds $D_{\min}$ in terms of the vectors $q^t(n)$ and $q^b(n)$. We now introduce the vectors $\underline{q}(n)$ and $\underline{r}(n)$ defined by

$$\underline{q}(n) \triangleq \tfrac{1}{2}[q^t(n) + q^b(n)]$$
$$\underline{r}(n) \triangleq \tfrac{1}{2}[q^t(n) - q^b(n)]. \tag{1.37}$$

$$q^t(n) = \underline{q}(n) + \underline{r}(n)$$
$$q^b(n) = \underline{q}(n) - \underline{r}(n) \tag{1.38}$$

$$q_i^t(n)q_k^b(n) = [q_i(n) + r_i(n)][q_k(n) - r_k(n)]$$
$$= q_i(n)q_k(n) + r_i(n)q_k(n) - q_i^t(n)r_k(n). \tag{1.39}$$

Since $\underline{q}(n)$ is an average of the probability vectors $q^t(n)$ and $q^b(n)$, $\underline{q}(n)$ is itself a probability vector. In fact, $\underline{q}(n)$ is just the composition vector for the nth column of the whole code. Since $\underline{q}(n)$ is a probability vector.

$$-\sum_i\sum_k q_i(n)q_k(n)\mu_{i,k}(\tfrac{1}{2}) \leq \max_{\underline{q}} -\sum_i\sum_k q_iq_k\mu_{i,k}(\tfrac{1}{2})$$
$$= E_{ex}(0^+). \tag{1.40}$$

Equation (1.40) follows from (1.27) and the definition of $\mu_{i,k}$ in (1.06). Furthermore, since $\mu'_{i,k}(\frac{1}{2}) = -\mu'_{k,i}(\frac{1}{2})$, we have

$$\sum_i\sum_k q_i(n)q_k(n)\mu'_{i,k}(\tfrac{1}{2}) = 0. \tag{1.41}$$

Substituting (1.39), (1.40), and (1.41) into (1.36) gives

$$D_{\min} \leq E_{ex}(0^+) + \frac{1}{N}\sum_{n=1}^{N}\sum_i\sum_k \left| r_i(n)q_k(n) - q_i^t(n)r_k(n)\right| \cdot \left|\mu_{i,k}\left(\frac{1}{2}\right) + \frac{1}{2}\mu'_{i,k}\left(\frac{1}{2}\right)\right| \tag{1.42}$$

$$\leq E_{ex}(0^+) + \frac{d_{\max}}{N}\sum_{n=1}^{N}\sum_i\sum_k |r_i(n)q_k(n) - q_i^t(n)r_k(n)|, \tag{1.43}$$

where we have used the definition of $d_{\max}$ in (1.32). The remainder term is bounded as follows:

$$\sum_i\sum_k |r_i(n)q_k(n) - q_i^t(n)r_k(n)|$$
$$\leq \sum_i\sum_k |r_i(n)q_k(n)| + |q_i^t(n)r_k(n)|$$
$$= \sum_k |r_k(n)| \sum_i |q_i(n)| + |q_i^t(n)|$$
$$= 2\sum_k |r_k(n)| \tag{1.44}$$
$$\leq 2\sqrt{K}\sqrt{\sum_k r_k^2(n)}. \tag{1.45}$$

Equation (1.45) follows from Cauchy's inequality which states that

$$\sum a_kb_k \leq \sqrt{\sum_k a_k^2 \sum_k b_k^2}.$$

We have used $a_k = 1$, $b_k = |r_k(n)|$. Averaging (1.45) over all N columns gives

$$1/N\sum_{n=1}^{N}\sum_i\sum_k |r_i(n)q_k(n) - q_i^t(n)r_k(n)|$$
$$\leq \frac{2\sqrt{K}}{N}\sum_{n=1}^{N}\sqrt{\sum_{k=1}^{K} r_k^2(n)} \tag{1.46}$$
$$\leq 2\sqrt{K}\sqrt{\frac{1}{N}\sum_{n=1}^{N}\sum_{k=1}^{K} r_k^2(n)}$$

by Cauchy. Substituting (1.37) into (1.46) completes the proof of Lemma 4.3.

Lemma 4.3 bounds the minimum discrepancy in terms of the quantity

$$\frac{1}{4N}\sum_{n=1}^{N}\sum_{k=1}^{K}(q_k^t(n) - q_k^b(n))^2 = 1/N\sum_{n,k} r_k(n)^2 = 1/N\sum_{n=1}^{N} \underline{r}(n)^2,$$

where we let $\underline{r}(n)^2$ denote the dot product of the K-dimensional vector $\underline{r}(n)$ with itself.

To complete the proof of Theorem 4, we would like to show that $1/N \sum_{n=1}^{N} r(n)^2$ can be made arbitrarily small. Unfortunately, however, the direct approach fails, because many columns may have substantially different compositions in their top halves and their bottom halves. Nor can this difficulty be resolved by merely tightening the bound in the latter half of Lemma 4.3, for columns which are very inhomogeneous may actually make undeservedly large contributions to the total discrepancy between the two halves of the code. For example, consider a code for the *TUC* of Fig. 3. A column whose top fourth contains ones, whose middle half contains twos, and whose bottom fourth contains threes contributes $-\frac{1}{2} \ln \frac{1}{10} - \frac{1}{2} \ln \frac{9}{10}$ to the average discrepancy. We wish to show that the minimum discrepancy for this channel is actually not much better than $-\frac{1}{3} \ln \frac{1}{10} - \frac{1}{3} \ln \frac{9}{10}$. This cannot be done directly because of columns of the type just mentioned. We note, however, that this column which contributes so heavily to the average discrepancy between the top and bottom halves of the code contributes nothing to discrepancies between words in the same quarter of the block. It happens that all abnormally good columns have some fatal weakness of this sort, which we exploit by the following construction.

LEMMA 4.4. *Given an ordered code with 2M words of block length N, we can form a new code with M words of block length 2N by annexing the*

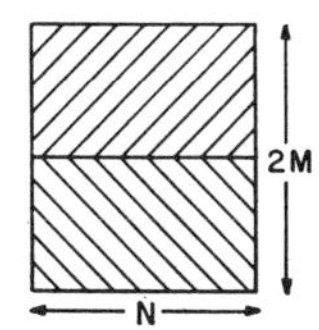

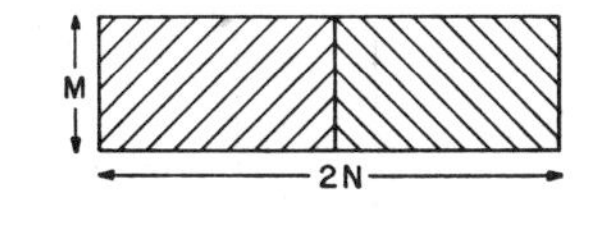

FIG. 4. Halving an ordered code.

$(M + i)$th word to the ith word for all $i = 1, \cdots, M$ as shown in Fig. 4. The new code has the following properties.

(1) The new code is ordered.

(2) The minimum discrepancy of the new code is no smaller than the minimum discrepancy of the original code.

$$(3) \quad \text{Var}(q') - \text{Var}(q) = \left(\frac{1}{4N}\right) \sum_{n=1}^{N} (q'(n) - q'(n+N))^2$$

$$(4) \qquad \text{Var}(q) \leqq \text{Var}(q') < 1$$

where:

$q(n)$ is the composition of the nth column of the original code, $n = 1, 2, \cdots, N$.

$q'(n)$ is the composition of the nth column of the new code, $n = 1, 2, \cdots, 2N$.

$$\bar{q} = 1/N \sum_{n=1}^{N} q(n)$$

$$\bar{q}' = 1/2N \sum_{n=1}^{2N} q'(n)$$

$$\text{Var}(q) = 1/N \sum_{n=1}^{N} (q(n) - \bar{q})^2 = \left[\frac{1}{N} \sum_{n=1}^{N} q(n)^2\right] - \bar{q}^2$$

$$\text{Var}(q') = 1/2N \sum_{n=1}^{2N} (q'(n) - \bar{q}')^2 = \left[\frac{1}{2N} \sum_{n=1}^{2N} q'(n)^2\right] - \bar{q}'^2.$$

Proof of Property 1. Let $q'_{i,k}(m, m')$ be the joint composition of the mth and m'th words in the new code, i.e., the fraction of times that the ith channel input letter occurs in the mth word of the new code opposite the kth channel input letter in the m'th word. By the halving construction which generated the new code (Fig. 4),

$$q'_{i,k}(m, m') = \tfrac{1}{2}[q_{i,k}(m, m') + q_{i,k}(m + M, m' + M)]. \quad (1.47)$$

If $m < m'$, then, in the original code

$$-\sum_i \sum_k q_{i,k}(m, m')\mu'_{i,k}(\tfrac{1}{2}) \geqq 0$$

$$-\sum_i \sum_k q_{i,k}(m + M, m' + M)\mu'_{i,k}(\tfrac{1}{2}) \geqq 0$$

Proof of Property 2. In the new code,

$$D'(m, m') = \tfrac{1}{2}[D(m, m') + D(m + M, m' + M)].$$

Thus $D'(m, m')$ can not be smaller than both $D(m, m')$ and $D(m + M, m' + M)$.

Proof of Property 3. $q(n) = \frac{1}{2}[q'(n) + q'(n + N)]$

$$\bar{q} = \frac{1}{2N} \sum_{n=1}^{N} [q'(n) + q'(n + N)] = \bar{q}' \quad (1.49)$$

$$\begin{aligned}\text{Var}(q') - \text{Var}(q) &= \left(\frac{1}{2N} \sum_{n=1}^{2N} q'(n)^2\right) - \left(\frac{1}{N} \sum_{n=1}^{N} q(n)^2\right) \\ &= \frac{1}{4N} \sum_{n=1}^{N} \{2[q'(n)^2 + q'(n + N)^2] - (q'(n) + q'(n + N))^2\} \quad (1.50) \\ &= \frac{1}{4N} \sum_{n=1}^{N} (q'(n) - q'(n + N))^2.\end{aligned}$$

Proof of Property 4. From Property 3, $\text{Var}(q) \leqq \text{Var}(q')$. Also, for every n,

$$[q'(n)]^2 = \sum_k [q_k'(n)]^2 \leqq 1 \quad (1.51)$$

$$\text{Var}(q') \leqq \frac{1}{2N} \sum_{n=1}^{2N} [q'(n)]^2 \leqq 1. \quad (1.52)$$

We may now complete the proof of the theorem by iterating the halving construction to prove Lemma 4.5.

LEMMA 4.5.

$$D_{\min}(N, M) < E_{ex}(0^+) + \frac{2d_{\max}\sqrt{K}}{\sqrt{[\log(\log M)]^-}} \quad (1.53)$$

Proof. Starting from any original code containing M codewords of block length N, we may extract a subset of $2^{[\log(\log M)]^-}$ code words which form an ordered code. This follows from Lemma 4.2 and the observation that $2^{[\log(\log M)]^-} \leqq \log M$. (Here $[\log(\log M)]^-$ is the largest integer less than or equal to $\log(\log M)$.)

We next halve the ordered code $[\log(\log M)]^-$ times. This gives us a sequence of $[\log(\log M)]^- + 1$ codes, starting with the original ordered code and terminating with a degenerate code containing only one codeword of block length $N2^{[\log(\log M)]^-}$. Since the properties of Lemma 4.4 are hereditary, every code in the sequence is ordered and each code has a minimum discrepancy no smaller than any of its ancestors (except the final degenerate code, for which the minimum discrepancy is undefined). The average variance of the column compositions of each of these codes is at least as great as the average variance of the column compositions of the preceding codes; yet the average variance of each code in the sequence must be between zero and one. Consequently, this sequence of $[\log(\log M)]^- + 1$ codes must contain two consecutive codes for which the difference in the variance of column compositions is less than $1/[\log(\log M)]^-$. The former of these two consecutive codes is non-degenerate, and Lemma 4.3 applies, with

$$\begin{aligned}\frac{1}{4N} \sum_{n=1}^{N} \sum_{k=1}^{K} (q_k^t(n) - q_k^b(n))^2 &= \frac{1}{4N} \sum_{n=1}^{N} (q'(n) - q'(n+N))^2 \\ &= \text{Var}(q') - \text{Var}(q) < 1/[\log(\log M)]^-\end{aligned} \quad (1.54)$$

q.e.d.

Theorem 4 follows directly from Lemma 4.5 and Theorem 2.

q.e.d.

Combining (1.53) and (1.12), we obtain an explicit bound on $P_e(N, M, 1)$.

$$P_e(N, M, 1) \geqq \exp - N\left[E_{ex}(0^+) + \frac{2d_{\max}\sqrt{K}}{\sqrt{[\log(\log M)]^-}} + \sqrt{\frac{2}{N}} \ln \frac{1}{P_{\min}} + \frac{\ln 4M}{N}\right] \quad (1.55)$$

If we upper bound $d_{\max}$, as given by (1.32) by

$$d_{\max} \leqq 2 \max_{i,k} |\mu_{i,k}(\tfrac{1}{2})|,$$

then (1.55) becomes equivalent to (I-1.17) and we have completed the proof of Theorem I-3.

Equation (1.55) has a rather peculiar behavior with M. On the other hand, $P_e(N, M, 1)$ must be a monotone nondecreasing function of M, and thus for any M greater than some given value, we can use (1.55) evaluated at that given M. It is convenient to choose this given M as $2^{\sqrt{N}^-}$, yielding

$$P_e(N, M, 1) \geqq \exp - N[E_{ex}(0^+) + o_4(N)]; \qquad M \geqq 2^{\sqrt{N}} \tag{1.56}$$

where

$$o_4(N) = \frac{2d_{\max}\sqrt{K}}{\sqrt{[\log N]^-}} + \sqrt{\frac{2}{N}} \ln \frac{1}{P_{\min}} + \frac{\ln 2}{\sqrt{N}} + \frac{2 \ln 2}{N}. \tag{1.57}$$

These equations can now be restated in a form similar to our other bounds on $P_e(N, M, 1)$.

THEOREM 5.

$$P_e(N, M, 1) \geqq \exp -N[E_{lr}(R - o_3(N)) + o_4(N)], \tag{1.58}$$

where

$$E_{lr}(R) = \begin{cases} E_{ex}(0^+); & R \geqq 0 \\ \infty; & R < 0 \end{cases} \tag{1.59}$$

$$o_3(N) = \frac{\ln 2}{\sqrt{N}}. \tag{1.60}$$

Proof. Observe that when $M \geqq 2^{\sqrt{N}}$ we have $R = (\ln M)/N \geqq (\ln 2)/\sqrt{N}$ and (1.58) reduces to (1.56). For $M < 2^{\sqrt{N}}$, (1.58) simply states that $P_e(N, M, 1) \geqq 0$.

2. THE STRAIGHT LINE BOUND

We have seen that the sphere packing bound (Theorem I-2) specifies the reliability of a channel at rates above R_{crit} and that the zero rate bound (Theorem I-3 or Theorem 5) specifies the reliability in the limit as the rate approaches zero. In this section, we shall couple these results with Theorem I-1 to establish the straight line bound on reliability given in Theorem I-4. Actually we shall prove a somewhat stronger theorem here which allows us to upper bound the reliability of a channel by a straight line between the sphere packing exponent and any low rate, exponential bound on error probability.

THEOREM 6. *Let $E_{lr}(R)$ be a nonincreasing function of R (not necessarily that given by (1.59)), let $o_3(N)$ and $o_4(N)$ be nonincreasing with N and let $No_3(N)$ and $No_4(N)$ be nondecreasing with N. Let $R_2 < R_1$ be nonnegative numbers and define the linear function*

$$E_{sl}(R_0) = \lambda E_{sp}(R_1) + (1 - \lambda)E_{lr}(R_2), \tag{2.01}$$

where E_{sp} is given by (I-1.07) and λ is given by

$$R_0 = \lambda R_1 + (1 - \lambda)R_2. \tag{2.02}$$

If

$$P_e(N, M, 1) \geqq \exp - N[E_{lr}(R - o_3(N)) + o_4(N)] \tag{2.03}$$

is valid for arbitrary positive M, N, then

$$P_e(N, M, 1) \geqq \exp - N\{E_{sl}[R - o_5(N)] + o_6(N)\} \tag{2.04}$$

is valid for

$$R_2 \leqq R - o_5(N) \leqq R_1, \tag{2.05}$$

where

$$o_5(N) = o_1(N) + o_3(N) + R_2/N \tag{2.06}$$

$$o_6(N) = o_2(N) + o_4(N) + \frac{1}{N} E_{lr}(R_2) \tag{2.07}$$

and $o_1(N)$ and $o_2(N)$ are given by (I-1.10) and $R = (\ln M)/N$.

Remarks. As shown in Figs. 5–8, $E_{sl}(R)$ is a straight line joining $E_{lr}(R_2)$ at R_2 to $E_{sp}(R_1)$ at R_1. It is clearly desirable, in achieving the best bound, to choose R_1 and R_2 so as to minimize $E_{sl}(R)$. If $E_{lr}(R)$ is not convex ∪, it may happen, as in Fig. 8 that the best choice of R_1, R_2 depends on R.

Theorem I-4 of the introduction is an immediate consequence of Theorem 6, obtained by choosing $E_{lr}(R)$ as in Theorem 5 and choosing

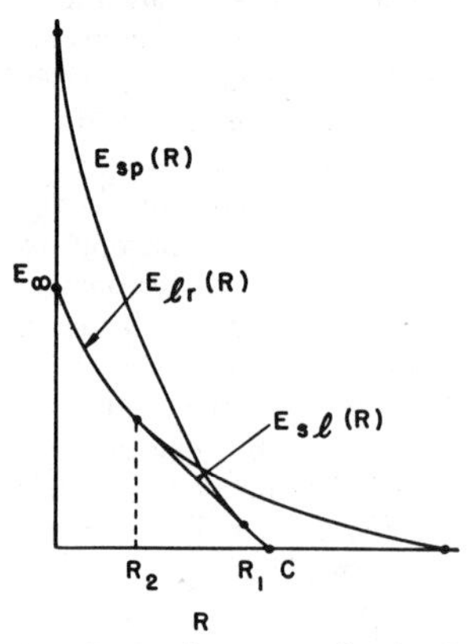

FIGS. 5–8. Geometric construction for $E_{sl}(R)$.

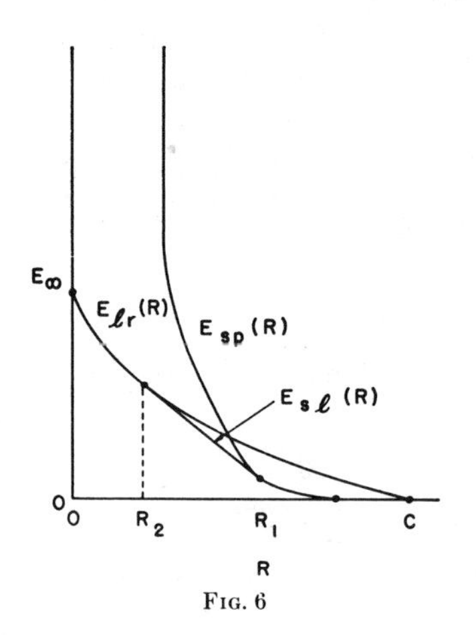

FIG. 6

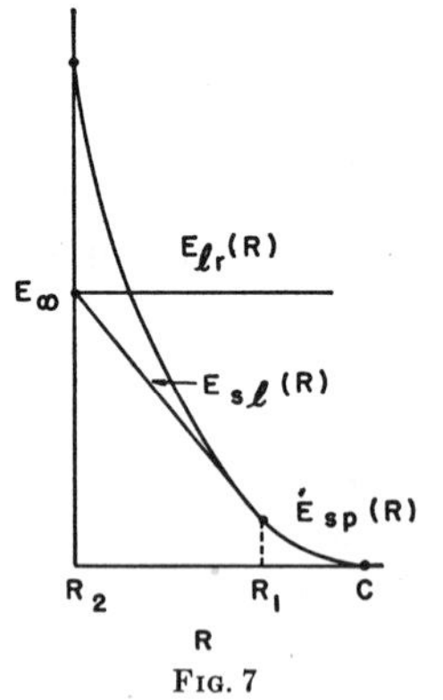

FIG. 7

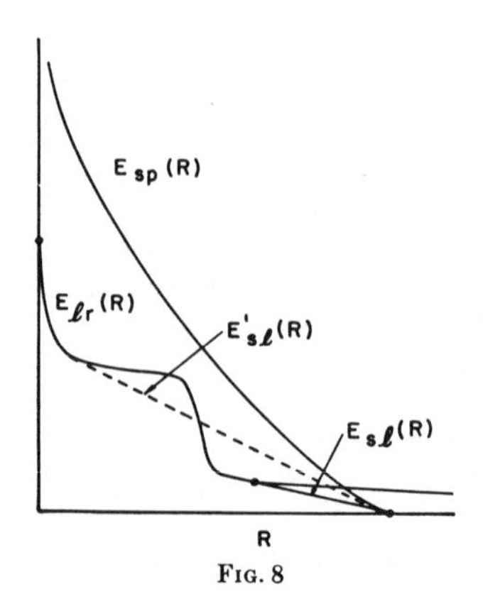

FIG. 8

$R_2 = 0$. The increased generality of Theorem 6 over Theorem I-4 is nonempty, however. In Theorem 8 we shall give an example of a low rate bound for the binary symmetric channel in which $E_{lr}(R)$ behaves as in Fig. 5.

The restriction in the theorem that $E_{lr}(R)$ be nonincreasing with R is no real restriction. Since $P_e(N, M, 1)$ is nonincreasing with M, any bound in which $E_{lr}(R)$ is increasing with R can be tightened to a bound in which $E_{lr}(R)$ is not increasing. Likewise the restriction that $No_3(N)$ and $No_4(N)$ be increasing with N is not serious since any bound can be weakened slightly to satisfy this restriction.

Proof. By Theorem I-1, we have

$$P_e(N, M, 1) \geqq P_e(N_1, M, L)P_e(N_2, L+1, 1), \quad (2.08)$$

where $N_1 + N_2 = N$ and L is an arbitrary positive integer. Applying the sphere packing bound, Theorem I-2, to $P_e(N_1, M, L)$ and applying (2.03) to $P_e(N_2, L+1, 1)$, we have

$$P_e(N, M, 1) \geqq \exp\left\{-N_1\left[E_{sp}\left(\frac{\ln M/L}{N_1} - o_1(N_1)\right) + o_2(N_1)\right] - N_2\left[E_{lr}\left(\frac{\ln (L+1)}{N_2} - o_3(N_2)\right) + o_4(N_2)\right]\right\}. \quad (2.09)$$

Using the expressions for $o_1(N)$ and $o_2(N)$ in (I-1.10), we see that $No_i(N)$ is increasing with N for $i = 1, 2, 3, 4$. Thus we can lower bound (2.09) by

$$P_e(N, M, 1) \geqq \exp\left\{-N_1E_{sp}\left(\frac{\ln M/L}{N_1} - \frac{No_1(N)}{N_1}\right) - No_2(N) - N_2E_{lr}\left(\frac{\ln (L+1)}{N_2} - \frac{No_3(N)}{N_2}\right) - No_4(N)\right\} \quad (2.10)$$

This is valid for any positive integers N_1 and N_2 summing to N, and we observe that it is trivially valid if either N_1 or N_2 is 0.

We next get rid of the restrictions that L, N_1, and N_2 be integers. Let $\tilde{L}$ be an arbitrary real number between L and $L + 1$. We can lower bound the right-hand side of (2.10) by replacing $\ln M/L$ with $\ln M/\tilde{L}$ and $\ln (L+1)$ with $\ln \tilde{L}$. Similarly, let $\tilde{N}_1$ be an arbitrary real number between N_1 and $N_1 + 1$. The right-hand side of (2.10) can be lower bounded by replacing N_1 with $\tilde{N}_1$. Finally, since $N_2 \leqq N - \tilde{N}_1 + 1$, we can lower bound (2.10) by replacing N_2 with $N - \tilde{N}_1 + 1$. Making these changes, we have

$$P_e(N, M, 1) \geqq \exp\left\{-\tilde{N}_1E_{sp}\left(\frac{\ln (M/\tilde{L}) - No_1(N)}{\tilde{N}_1}\right) - N[o_2(N) + o_4(N)] - (N - \tilde{N}_1 + 1)E_{lr}\left(\frac{\ln \tilde{L} - No_3(N)}{N - \tilde{N}_1 + 1}\right)\right\} \quad (2.11)$$

Define λ to satisfy

$$R - o_5(N) = \lambda R_1 + (1 - \lambda)R_2 \quad (2.12)$$

From the restriction (2.05), λ satisfies $0 \leqq \lambda \leqq 1$. Now choose $\tilde{N}_1$ and $\tilde{L}$ by

$$\tilde{N}_1 = \lambda N \quad (2.13)$$

$$\ln \tilde{L} = R_2(N - \tilde{N}_1 + 1) + No_3(N). \quad (2.14)$$

By rearranging (2.14), we see that the argument of E_{lr} in (2.11) satisfies

$$\frac{\ln \tilde{L} - No_3(N)}{N - \tilde{N}_1 + 1} = R_2 \quad (2.15)$$

Likewise, using (2.12), (2.13), (2.14), and (2.06), the argument of E_{sp} in (2.11) is given by

$$\frac{\ln(M/\tilde{L}) - No_1(N)}{\tilde{N}_1} = \frac{1}{\lambda}\left[\frac{\ln M}{N} - \frac{\ln \tilde{L}}{N} - o_1(N)\right] = \frac{1}{\lambda}\left[R - R_2\left(1 - \lambda + \frac{1}{N}\right) - o_1(N) - o_2(N)\right] = \frac{1}{\lambda}[R - R_2(1-\lambda) - o_5(N)] = R_1. \quad (2.16)$$

Substituting (2.15) and (2.16) into (2.11), we have

$$P_e(N, M, 1) \geqq \exp - N\left\{\lambda E_{sp}(R_1) + \left(1 - \lambda + \frac{1}{N}\right)E_{lr}(R_2) + o_2(N) + o_4(N)\right\} \quad (2.17)$$

Combining (2.12), (2.02), and (2.01), we have

$$E_{sl}(R - o_5(N)) = \lambda E_{sp}(R_1) + (1 - \lambda)E_{lr}(R_2) \quad (2.18)$$

Finally, substituting (2.18) and (2.07) into (2.17), we have (2.04), completing the proof.

The straight line bound $E_{sl}(R)$ depends critically on the low rate bound $E_{lr}(R)$ to which it is joined. If the low rate bound is chosen as E_∞, then the resulting straight line bound $E_{sl}(R)$ is given by Theorem I-4. Plots of this bound for several channels are shown in Figure I-4.

From the discussion following (1.20), we see that if $C \neq 0$ and $C_0 = 0$, then E_∞ is strictly less than $E_{sp}(0^+)$, and the straight line bound $E_{sl}(R)$ of Theorem 4 exists over a nonzero range of rates. Also it follows from Theorem 7 of Gallager (1965) that $E_{ex}(R)$ is strictly convex ∪ and therefore is strictly less than $E_{sl}(R)$ in the interior of this range of rates.

There is an interesting limiting situation, however, in which $E_{sl}(R)$ and $E_{ex}(R)$ virtually coincide. These are the very noisy channels, first introduced by Reiffen (1963) and extended by Gallager (1965). A very noisy channel is a channel whose transition probabilities may be expressed by

$$P(j \mid k) = r_j(1 + \epsilon_{j,k}), \quad (2.19)$$

where r_j is an appropriate probability distribution defined on the channel outputs and $|\epsilon_{j,k}| \ll 1$ for all j and k. The function $E_0(\rho)$ for such a channel can be expanded as a power series in $\epsilon_{j,k}$. By neglecting all terms of higher than second order, Gallager (1965) obtained

$$E_0(\rho) = \frac{\rho}{1+\rho}C, \quad (2.20)$$

where the capacity C is given by

$$C = \max_q \tfrac{1}{2} \sum_j r_j\left[\sum_k q_k\epsilon_{j,k}^2 - \left(\sum_k q_k\epsilon_{j,k}\right)^2\right] \quad (2.21)$$

$$= \max_q \tfrac{1}{4} \sum_i \sum_k q_iq_k \sum_j r_j(\epsilon_{j,i}^2 + \epsilon_{j,k}^2 - 2\epsilon_{j,i}\epsilon_{j,k}). \quad (2.22)$$

The resulting random coding exponent is given by

$$E_r(R) = (\sqrt{C} - \sqrt{R})^2 \quad \text{for } C/4 \leqq R \leqq C \quad (2.23)$$

$$= C/2 - R \quad \text{for } R < C/4. \quad (2.24)$$

We can calculate E_∞ in the same way

$$E_\infty = \max_q - \sum_i \sum_k q_iq_k \ln \sum_j \sqrt{P(j \mid i)P(j \mid k)}. \quad \text{(I-1.18)}$$

Using (2.19) and expanding to second order in ϵ, gives

$$\sum_j \sqrt{P(j \mid i)P(j \mid k)} = \sum_j r_j(1 + \epsilon_{j,i}/2 - \epsilon_{j,i}^2/8) \cdot (1 + \epsilon_{j,k}/2 - \epsilon_{j,k}^2/8). \quad (2.25)$$

From (2.19) we observe that

$$\sum_j r_j\epsilon_{j,k} = 0 \quad \text{for all } k \quad (2.26)$$

$$\sum_j \sqrt{P(j \mid i)P(j \mid k)} = 1 - \tfrac{1}{8}\sum_j r_j(\epsilon_{j,i}^2 + \epsilon_{j,k}^2 - 2\epsilon_{j,i}\epsilon_{j,k}). \quad (2.27)$$

From (2.27), (I-1.18), and (2.22), we conclude that

$$E_\infty = C/2 = E_r(0). \quad (2.28)$$

Thus in the limit as the $\epsilon_{j,k}$ approach 0, the upper and lower bounds to the reliability $E(R)$ come together at all rates and (2.23) and (2.24) give the reliability function of a very noisy channel.

For channels which are not very noisy, the actual reliability may lie well below the straight line bound from E_∞ to the sphere packing bound. As a specific case in which these bounds may be improved, we consider the binary symmetric channel.

This channel has received a great deal of attention in the literature, primarily because it provides the simplest context within which most coding problems can be considered. The minimum distance of a code, d_{min}, is defined as the least number of positions in which any two code words differ. We further define $d(N, M)$ as the maximum value of d_{min} over all codes with M code words of length N. Here we are interested primarily in the asymptotic behavior of $d(N, M)$ for large N and M and fixed $R = (\ln M)/N$. The *asymptotic distance ratio* is defined as

$$\delta(R) \triangleq \limsup_{N\to\infty} \frac{1}{N} d(N, [e^{RN}]^+). \tag{2.29}$$

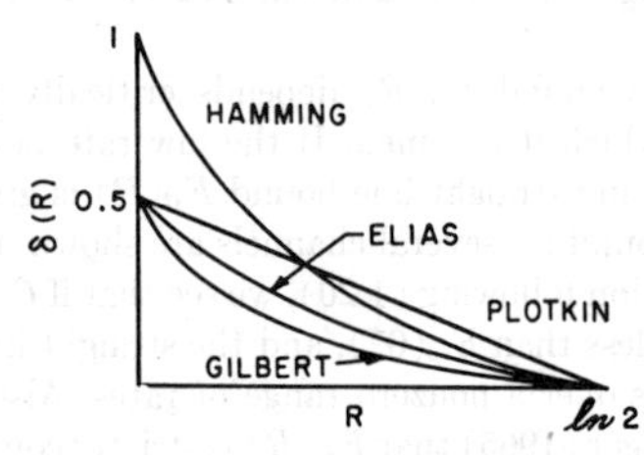

FIG. 9. Comparison of bounds on minimum distance for a binary symmetric channel.

There are two well known upper bounds to $\delta(R)$, due to Hamming (1950) and Plotkin (1951), and one well known lower bound due to Gilbert (1952). These are given implicitly by

$$\ln 2 - H(\delta(R)/2) \geqq R \quad \text{(Hamming)} \tag{2.30}$$

$$\ln 2 - 2\delta(R) \ln 2 \geqq R \quad \text{(Plotkin)} \tag{2.31}$$

$$\ln 2 - H(\delta(R)) \leqq R \quad \text{(Gilbert)}, \tag{2.32}$$

where

$$H(\delta) = -\delta \ln \delta - (1 - \delta) \ln (1 - \delta). \tag{2.33}$$

See Peterson (1961) for an excellent discussion of these bounds.

Here we shall derive a third upper bound to $\delta(R)$, derived by Elias in 1960 but as yet unpublished. As shown in Fig. 9 the Elias bound is stronger than either the Hamming or Plotkin bounds for $0 < R < \ln 2$. It should be observed, however, that this superiority applies only to the asymptotic quantity, $\delta(R)$. For sufficiently small values of N, M there are a number of bounds on $d(N, M)$ which are stronger than the Elias bound.

THEOREM 7 (Elias).

$$\delta(R) \leqq 2\lambda_R(1 - \lambda_R), \tag{2.34}$$

where λ_R *is given by*

$$\ln 2 - H(\lambda_R) = R; \quad 0 \leqq \lambda_R \leqq \tfrac{1}{2}. \tag{2.35}$$

Before proving this theorem, we shall discuss the relationship between $\delta(R)$ and the reliability function $E(R)$. Suppose that a code contains two code words at a distance d apart. From I-3.10, $\mu(s)$ for these two words is given by $d \ln [p^s q^{1-s} + q^s p^{1-s}]$, where p is the cross-over probability of the channel (see Fig. I-5a) and $q = 1 - p$. This is minimized at $s = \frac{1}{2}$, and from (I-3.20) and (I-3.21), one of the code words has an error probability bounded by

$$P_{e,m} \geqq \frac{1}{4} \exp\left[d \ln 2\sqrt{pq} - \sqrt{\frac{d}{2}} \ln \frac{1}{p}\right], \tag{2.36}$$

where we have used (1.11) in bounding $\mu''(\frac{1}{2})$.

Next, for a code with $2M$ code words of block length N, we see by expurgating M of the worst words that at least M code words have a distance at most $d(N, M)$ from some other code word. For such a code

$$P_e \geqq \frac{1}{8} \exp\left[-d(N, M) \ln 2\sqrt{pq} - \sqrt{d(N, M)/2} \ln \frac{1}{p}\right]. \tag{2.37}$$

Combining (2.37) with (2.29), we obtain

$$P_e(N, M, 1) \geqq \exp -N[\delta(R) \ln 2\sqrt{pq} + o(N)] \tag{2.38}$$

$$E(R) \leqq \frac{\delta(R)}{2} \ln 4pq. \tag{2.39}$$

Conversely, if a code of block length N has minimum distance $\delta(R)N$, then it is always possible to decode correctly when fewer than $\frac{1}{2}\delta(R)N$ errors occur. By using the Chernov (1952) bound, if $p < \frac{1}{2}\delta(R)$, the probability of $\frac{1}{2}\delta(R)N$ or more errors is bounded by

$$P_e \leqq \exp - N\left[-\frac{\delta(R)}{2} \ln p - \left(1 - \frac{\delta(R)}{2}\right) \ln q - H\left(\frac{\delta(R)}{2}\right)\right] \tag{2.40}$$

$$E(R) \geqq -\frac{\delta(R)}{2} \ln p - \left(1 - \frac{\delta(R)}{2}\right) \ln q - H\left(\frac{\delta(R)}{2}\right). \tag{2.41}$$

For more complete discussions of techniques for bounding the error probability on a binary symmetric channel, see Fano (1961), Chap. 7 or Gallager (1963), Chap. 3. The bounds on reliability given by (2.39) and (2.41) are quite different, primarily because it is usually possible to decode correctly when many more than $\frac{1}{2}\delta(R)N$ errors occur. As p becomes very small, however, the minimum distance of the code becomes increasingly important, and dividing (2.39) and (2.41) by $-\ln p$, we see that

$$\frac{\delta(R)}{2} = \lim_{p\to 0} \frac{E(R)}{-\ln p}. \tag{2.42}$$

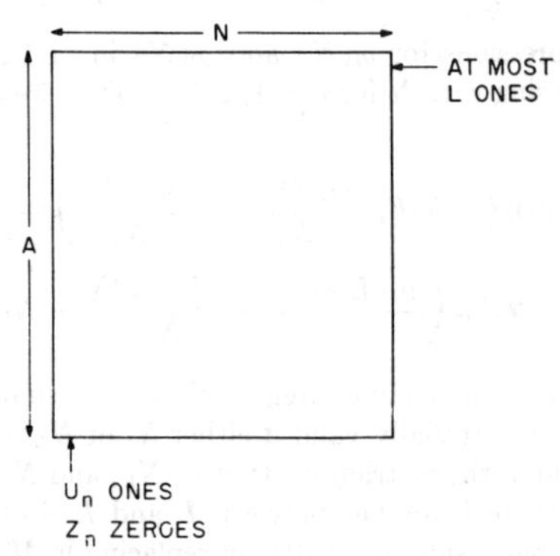

FIG. 10. Construction for Elias bound.

Along with (2.42), there are several other interesting connections between $E(R)$ and $\delta(R)$. For example, if one could show that $\delta(R)$ was given by the Gilbert bound (2.32) with equality, then upon substituting (2.32) into (2.39) one would find an *upper* bound for reliability which is equal to the lower bound $E_{ex}(R)$ over the range of rates for which $E_{ex}(R) > E_r(R)$. By combining this with Theorem 6, $E(R)$ would be determined for all rates and would be equal to the known lower bound to $E(R)$. Thus the question of determining $E(R)$ for the *BSC* hinges around the problem of determining $\delta(R)$.

Proof of Theorem 7. The proof of the Elias bound combines the arguments of Plotkin and Hamming in an ingenious way. For any integer L, $0 \leqq L \leqq N/2$, there are $\sum_{i=0}^{L} \binom{N}{i}$ binary N-tuples within a sphere of radius L around any given code word (i.e., N-tuples that have a distance L or less from the code word). For M code words, these spheres contain $M \sum_{i=0}^{L} \binom{N}{i}$ N-tuples, counting an N-tuple once for each appearance in a sphere. Since there are only 2^N different binary N-tuples, some critical N-tuple must appear in at least A spheres where

$$A \triangleq \left[2^{-N} M \sum_{i=0}^{L} \binom{N}{i}\right]^+. \tag{2.43}$$

Thus this critical N-tuple contains at least A code words within a sphere of radius L around itself.

For the remainder of the proof, we consider only these A code words and we assume that L is chosen so that $A \geqq 2$. For convenience we translate these code words by subtracting the critical word from each of them. Each of the A translated code words then has at most L ones.

We next list the A translated code words as in Fig. 10. Let U_n denote the number of ones in the nth column Z_n, the number of zeroes. The total number of ones in the $A \times N$ matrix of Fig. 10 may be computed either by summing along the columns or along the rows. This gives

$$\sum_n U_n \leqq AL. \tag{2.44}$$

We now compute the total distance among the $\binom{A}{2}$ pairs of translated code words. The contribution to the total distance from the nth column is $U_n Z_n$. Consequently,

$$d_{\text{tot}} = \sum_n U_n Z_n. \tag{2.45}$$

Since the minimum distance cannot exceed the average distance, we have

$$d_{\min} \leqq d_{\text{tot}} \Big/ \binom{A}{2} = \sum_{n=1}^{N} U_n(A - U_n) \Big/ \binom{A}{2}. \tag{2.46}$$

The function $\sum_{n=1}^{N} U_n(A - U_n)$ is a concave function of the U_n, and is therefore maximized, subject to the constraint (2.44), by making the partial derivation with respect to U_n a constant. Thus the maximum occurs with $U_n = AL/N$ for all n:

$$d_{\min} \leqq \frac{2NA^2 \left(\frac{L}{N}\right)\left(1 - \frac{L}{N}\right)}{A(A-1)} = 2N(L/N)(1 - L/N) \left(1 + \frac{1}{A-1}\right) \tag{2.47}$$

$$\frac{d_{\min}}{N} \leqq 2(L/N)(1 - L/N) + \frac{1}{2(A-1)}. \tag{2.48}$$

Since (2.48) is valid for any L such that $A \geqq 2$, L can be chosen so as to optimize the bound. In the theorem, however, we are interested in asymptotic results for fixed R, large N, and $M = [e^{NR}]^+$. First we lower bound A.

Shannon[3] has shown that

$$\binom{N}{L} \geqq [8L(N - L)/N]^{-1/2} \exp NH(L/N). \tag{2.49}$$

The first term is lower bounded by taking $L = N/2$, yielding

$$\sum_{i=0}^{L} \binom{N}{i} > \binom{N}{L} \geqq \frac{1}{\sqrt{2N}} \exp NH(L/N). \tag{2.50}$$

Next, choose L to satisfy

$$H\left(\frac{L-1}{N}\right) < \ln 2 - \frac{\ln M}{N} + \frac{3}{2}\frac{\ln N}{N} \leqq H\left(\frac{L}{N}\right). \tag{2.51}$$

Observe that for any fixed $R > 0$, this will have a solution for large enough N. Combining (2.43), (2.50), and (2.51) we obtain

$$A > \sqrt{\frac{1}{2N}} \exp\left[\frac{3}{2} \ln N\right] = \frac{N}{\sqrt{2}} \tag{2.52}$$

Next recalling the definition of λ_R in (2.35), the left-hand side of (2.51) becomes

$$H\left(\frac{L-1}{N}\right) < H(\lambda_R) + \frac{3}{2}\frac{\ln N}{N}. \tag{2.53}$$

Since H is a concave $\cap$ function, we can combine (2.53) with the result that $H(\frac{1}{2}) = \ln 2$ to obtain

$$\frac{L-1}{N} < \lambda_R + \left(\frac{3}{2}\frac{\ln N}{N}\right)\left[\frac{\ln 2 - H(\lambda_R)}{\frac{1}{2} - \lambda_R}\right] \tag{2.54}$$

[3] C. E. Shannon, unpublished seminar notes, M. I. T., 1956. For a published derivation, see Ash (1965), p. 113.

Substituting (2.52) and (2.54) into (2.48), we have

$$\frac{d(N, M)}{N} \leqq 2\lambda_R(1 - \lambda_R) + o(N), \tag{2.55}$$

where $o(N)$ can be taken as

$$o(N) = 3\frac{\ln N}{N}\left(\frac{\ln 2 - H(\lambda_R)}{\frac{1}{2} - \lambda_R}\right) + \frac{2}{N} + \frac{1}{\sqrt{2N} - 2}. \tag{2.56}$$

If we now substitute the Elias bound (2.34) into (2.39), we get a new upper bound on reliability given by:

THEOREM 8. *For a binary symmetric channel, an upper bound on reliability is given by*

$$E(R) \leqq E_{lr}(R) = -\lambda_R(1 - \lambda_R) \ln 4pq, \tag{2.57}$$

where λ_R *is given by* (2.35).

RECEIVED: January 18, 1966

REFERENCES

ASH, R. B. (1965), "Information Theory." Wiley (Interscience), New York.

BERLEKAMP, E. R. (1964), Block coding with noiseless feedback. Ph.D. Thesis. Department of Electrical Engineering. M.I.T.

BHATTACHARYYA, A. (1943), On a measure of divergence between two statistical populations defined by their probability distributions. *Bull. Calcutta Math. Soc.* **35**(3), 99–110.

CHERNOFF, H. (1952), A measure of asymptotic efficiency for tests of an hypothesis based on the sum of observations. *Ann. Math. Stat.* **23**, 493.

DOBRUSHIN, (1962), "Optimal binary codes for small rates of transmission of information," *Theory of Probability and its Applications*, Vol. 7, p. 199–204.

ELIAS, P. (1955), List decoding for noisy channels. *Tech. Report 335*. Research Laboratory of Electronics, M.I.T., Cambridge.

FANO, R. M. (1961), "Transmission of Information." M.I.T. Press, Cambridge, and Wiley, New York.

FEINSTEIN, A. (1955), Error bounds in noisy channels without memory. *IEEE Trans.* **IT-1**, 13–14.

FELLER, W. (1943), Generalizations of a probability limit theorem of Cramer. *Trans. Am. Math. Soc.* **54**, 361.

GALLAGER, R. (1963), "Low Density Parity Check Codes." M.I.T. Press, Cambridge.

GALLAGER, R. (1965), A simple derivation of the coding theorem and some applications. *IEEE Trans.* **IT-11**, 3–18.

GALLAGER, R. (1965), Lower bounds on the tails of probability distributions. M.I.T. Research Laboratory of Electronics, OPR 77, 277–291.

GILBERT, E. N. (1952), A comparison of signalling alphabets. *BSTJ* **3**, 504–522.

HAMMING, R. W. (1950), Error detecting and error correcting codes. *BSTJ* **29**, 47–160.

HELLIGER, E. (1909), Neue Begrundung der Theorie quadratischer Formen von unendlichvielen Veranderlichen. *J. Reine Angew. Math.* **136**, 210–271.

JACOBS, I. M., AND BERLEKAMP, E. R. (1967), A lower bound to the distribution of computation for sequential decoding. *IEEE Trans.* **IT-13** (to appear).

NEYMAN, J., AND PEARSON, E. S. (1928), On the use and interpretation of certain test criterion for purposes of statistical inference. *Biometrica* **20A**, 175, 263.

PETERSON, W. W. (1961), "Error-Correcting Codes." M.I.T. Press, Cambridge, and Wiley, New York.

PLOTKIN, M. (1960), Research Division Report 51-20, University of Pennsylvania. Eventually published in 1960 as: Binary codes with specified minimum distance. *PGIT* **IT6**, 445–450.

REIFFEN, B. (1963), A note on 'very noisy' channels. *Inform. Control* **6**, 126–130.

SHANNON, C. E. (1948), A mathematical theory of communication. *BSTJ* **27**, 379, 623. Also in book form with postscript by W. Weaver, University of Illinois Press, Urbane, Illinois.

SHANNON, C. E. (1956), Zero error capacity of noisy channels. *IEEE Trans.* **IT-2**, 8.

SHANNON, C. E. (1958), Certain results in coding theory for noisy channels. *Inform. Control* **1**, 6.

SHANNON, C. E., GALLAGER, R. G., AND BERLEKAMP, E. R. (1967), Lower bounds to error probability for coding on discrete memoryless channels. I. *Inform. Control.* **10**, 65–103.

SUN, M. (1965), Asymptotic bounds on the probability of error for the optimal transmission of information in the channel without memory which is symmetric in pairs of input symbols for small rates of transmission. *Theory Prob. Appl.* (Russian) **10** (1), 167–175.

Exponential Error Bounds for Erasure, List, and Decision Feedback Schemes

G. DAVID FORNEY, JR., MEMBER, IEEE

Abstract—By an extension of Gallager's bounding methods, exponential error bounds applicable to coding schemes involving erasures, variable-size lists, and decision feedback are obtained. The bounds are everywhere the tightest known.

Introduction

ONE of the central problems of coding theory is the question of the error probability obtainable by coding on memoryless channels. Coding theory's first result, and still its most important, was Shannon's[1] proof that every memoryless channel has a capacity C, such that arbitrarily small error probabilities can be obtained if and only if the code rate R is less than C. Many years of continuing attempts to make more precise statements about error probabilities[2] culminated in Gallager's elegant derivation[3] of an exponential upper bound on attainable error probabilities, and in a nearly identical lower bound.[4]

The communication situation to which this work has been addressed is one in which the decoder makes a single "hard" decision about which code word was sent, the event of error being the event in which this decision, or estimate, is wrong. In the abstract, this is indeed the archetypical communications problem. However, experience with codes as part of larger systems indicates that it is often useful for the decoder to produce more than just an estimate alone. This paper contains results bearing on the two following situations.

1) The decoder has the option of not deciding at all, of rejecting all estimates. The resulting output is called an erasure. Only if the decoder does make an estimate, and it is wrong, do we have an undetected error.

2) The decoder has the option of putting out more than one estimate. The resulting output is called a list. Only if the correct code word is not on the list do we have a list error.

In the first case, it is clear that by allowing the erasure probability to increase, the undetected error probability can be reduced. In the second, by allowing the average list size to increase, the probability of list error can be reduced. Our results consist of exponential bounds *à la* Gallager[3] on the obtainable tradeoffs between these quantities; these bounds we believe to be quite tight.

In what follows, we shall first discuss at greater length the situations in which the erasure and list options may be useful. We shall then present the results themselves with geometrical and analytical interpretations of their character. Some of the less instructive proofs are relegated to the Appendix. Finally, we shall return to a consideration of the implications of these results in the communications situations now to be described.

Manuscript received April 3, 1967. This paper was presented at the 1967 International Symposium on Information Theory, San Remo, Italy. This work was supported by the Rome Air Development Center, Griffiss AFB, N. Y., under Contract AF30(602)-4071.

The author is with Codex Corporation, Watertown, Mass. 02172

Use of the Erasure and List Options

We now discuss circumstances in which the erasure and list options may be useful: when the transmitted data contains some redundancy, when a feedback channel is available, or when further stages of coding (concatenation) are contemplated.

A. Redundant Data

In the usual theoretical communications situation the incoming data are imagined to be completely random as, for example, a stream of independent bits each equally likely to be zero or one. In theory, if there were redundancy in the data, it could be removed by suitable source encoding. In practice, source encoding is frequently not considered worth the effort, and the communications system is given redundant data.

The erasure option is then useful whenever an erased code word can be reconstituted from knowledge of other code words. For example, in telemetry, each code word may contain samples from one or a number of sensors; if one code word is erased, the missing sensor readings can be interpolated from neighboring readings whenever the process being measured is not changing rapidly. Such redundancy may be called external redundancy.

On the other hand, the list option is useful whenever incorrect code words can be recognized by internal evidence. For example, in the transmission of English text, any incorrect code words on the list will generally look like garble and thus can be discarded. Such redundancy may be called internal redundancy.

With either internal or external redundancy, the list or erasure option therefore permits a considerable reduction in the probability of actual error, without reduction of the effective rate of information transfer, or equivalently an increase in rate, with no increase in error probability.

B. Feedback Channel

If the receiver has some way of signaling to the transmitter, the erasure option can be used in a simple strategy called decision feedback. Here an erasure results in a request for a repeat transmission, whereupon the transmitter

Reprinted from *IEEE Trans. Inform. Theory*, vol. IT-14, pp. 206–220, Mar. 1968.

sends the same code word again. Although it has been shown[5] that feedback cannot increase the capacity of a memoryless channel, our results indicate that the error probability can be dramatically decreased for rates near capacity, without any sacrifice in the effective rate of information transfer. Furthermore, decision feedback is rather simple to implement, simpler in many cases than one-way transmission with ordinary decoding.

One could also use the occurrence of two or more code words on a list as a signal for repeat request but, in general, of the erasure criteria one could use, this will not be the optimum one. Another way of using lists would be in a kind of information feedback in which the transmitter would be made aware of the entries on the decoder's list, either by their direct transmission, or by feeding back what was actually received and letting the transmitter deduce by decoding what the decoder's list must be; the transmitter would then transmit only enough information to resolve the uncertainty. We shall see, however, that this additional complication fails to improve on the decision feedback error probability.

C. *Concatenation*

In concatenation[6] the code words of one code are used as symbols in a higher-order second-stage code. For example, on a binary-input channel, the basic code might be the (15, 5) Bose–Chaudhuri code, say, which has 32 code words. The second-stage code would then have symbols from the finite field with 32 elements, $GF(32)$, say, a (24, 46) Reed–Solomon code, which has minimum distance 9. Up to 4 decoding errors in 32 decodings of the (15, 5) code could then be corrected by the second-stage code. (This amounts to introduction of systematic external redundancy, rather than fortuitous, as mentioned earlier.) We have shown[6] that such a two-stage approach may yield the same performance as a single code, with less complexity.

If we allow the basic decoder an erasure option, then we generally reduce the number of decoding errors at the cost of introducing some decoding erasures. Algebraic erasure and error-correcting algorithms[7] can then be used to pick up the resulting erasures and errors. The resulting overall error probability is lower than that attainable with error correction alone, though in at least one interesting case[6] the improvement is small.

Additional improvements can be obtained, at the cost of increased complexity, by the use of a method called generalized minimum distance decoding.[8] Here the basic decoder puts out not an erasure or estimate alone, but an estimate with a reliability indicator, for which the optimum choice is the threshold for which the word would have been erased, had an erasure option been used. Therefore, our bounds give us information on this scheme as well.

Finally, if the basic decoder uses the list option, the second-stage decoder may in a number of trials attempt error correction with all the possibilities on all the lists. Obviously, if the average list size becomes large, the number of trials required becomes staggering; however, our results show that the average list size remains manageable for an interesting range of parameters.

Decision Regions and Optimum Criteria

We are concerned only with block codes on discrete memoryless channels; generalizations may be expected. A discrete memoryless channel has K inputs x_k, $1 \leq k \leq K$, and J outputs y_j, $1 \leq j \leq J$, and is characterized by its transition probability matrix $p_{jk} = \Pr(y_j \mid x_k)$, which gives the probability that y_j will be the output when x_k is the input. A block code of rate R (nats) and length N for such a channel consists of $M = \exp NR$ sequences $\mathbf{x}_m$ of N inputs x_{mi}, $1 \leq i \leq N$, called code words. A sequence of N outputs y_i is called a received word $\mathbf{y}$. The probability of receiving $\mathbf{y}$ given the transmission of $\mathbf{x}_m$ is then given by

$$\Pr(\mathbf{y} \mid \mathbf{x}_m) = \prod_{i=1}^{N} \Pr(y_i \mid x_{mi}), \tag{1}$$

where the product expression follows from the fact that successive probabilities are independent, which in turn follows by definition from the memorylessness of the channel.

Any of the decoding options we have described may be characterized in terms of decision regions R_m defined over the space of received words, which have the significance that the decoder puts out the code word $\mathbf{x}_m$ as an estimate if the received word $\mathbf{y}$ is in the region R_m. Fig. 1 is a schematic illustration of typical decision region configurations for ordinary decoding, decoding with an erasure option, and decoding with a list option. In the former case, the decision regions are disjoint and exhaust the space; each received word $\mathbf{y}$ is in one and only one region, so that the output is always a single estimate. With the erasure option, the regions remain disjoint, but not exhaustive; some received words therefore lie in no decision region and result in erasures. With the list option, the regions are no longer disjoint; a received word may lie in zero, one, or two or more decision regions, and thus may be decoded into a list of any size.

With ordinary decoding, an error occurs if $\mathbf{x}_m$ is sent and the received word lies in some decision region $R_{m'}$, where $m' \neq m$. The probability of error is therefore

$$\Pr(\mathcal{E}) = \sum_{m} \sum_{m' \neq m} \sum_{\mathbf{y} \varepsilon R_{m'}} \Pr(\mathbf{y}, \mathbf{x}_m), \tag{2}$$

where the sum over $(\mathbf{y} \varepsilon R_{m'})$ means over all $\mathbf{y}$ in $R_{m'}$. Since $\Pr(\mathbf{y}, \mathbf{x}_m) = \Pr(\mathbf{x}_m \mid \mathbf{y}) \Pr(\mathbf{y})$, the standard problem of choosing the decision regions R_m to minimize the probability of error $\Pr(\mathcal{E})$ is obviously solved by maximum a posteriori probability decoding, in which

$$\mathbf{y} \varepsilon R_m \text{ iff } \Pr(\mathbf{x}_m \mid \mathbf{y}) > \Pr(\mathbf{x}_{m'} \mid \mathbf{y}), \quad \text{all} \quad m' \neq m. \tag{3}$$

Any $\mathbf{y}$ for which two or more $\Pr(\mathbf{x}_m \mid \mathbf{y})$ tie for largest is a boundary point.

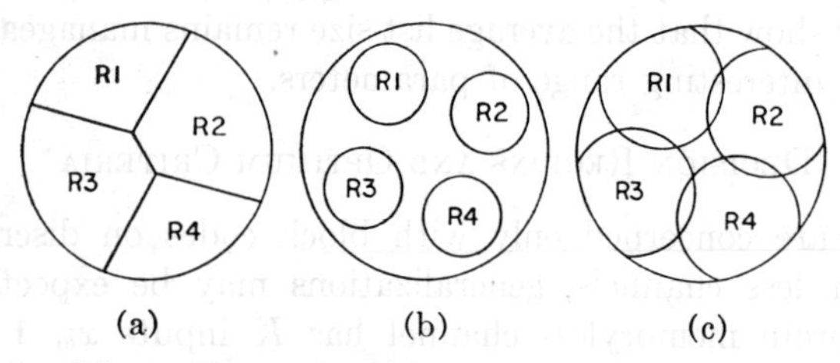

Fig. 1. Schematic representation of typical decision regions: (a) ordinary decoding, (b) erasure option, and (c) list option.

Equivalently, one can say that an error occurs if $\mathbf{y}$ lies in R_m and some code word $\mathbf{x}_{m'}$, $m' \neq m$, was actually transmitted. For our later convenience, we define this event as the event E_2; then

$$\Pr(E_2) = \sum_m \sum_{\mathbf{y} \epsilon R_m} \sum_{m' \neq m} \Pr(\mathbf{y}, \mathbf{x}_{m'}). \tag{4}$$

Of course $\Pr(\mathcal{E}) = \Pr(E_2)$, the transformation between (2) and (4) being effected by a rearrangement of the order of summation and an interchange of the dummy variables m and m'.

When the erasure option is introduced, E_2 is the event of undetected error, so that (2) and (4) continue to give the undetected error probability. However, we must now contend with a nonzero erasure probability. Let us define the event E_1 as the event in which the received word $\mathbf{y}$ does not fall in the decision region R_m corresponding to the transmitted code word $\mathbf{x}_m$; the probability of E_1 is

$$\Pr(E_1) = \sum_m \sum_{\mathbf{y} \not\epsilon R_m} \Pr(\mathbf{y}, \mathbf{x}_m), \tag{5}$$

where the sum over $(\mathbf{y} \not\epsilon R_m)$ means over all $\mathbf{y}$ not in R_m. If E_1 occurs, either an undetected error or an erasure must ensue; therefore, the probability of an erasure X is simply

$$\Pr(X) = \Pr(E_1) - \Pr(\mathcal{E}). \tag{6}$$

As the probability of undetected error will normally be much less than the probability of erasure, $\Pr(E_1)$ is not only an upper bound but also an excellent approximation to $\Pr(X)$.

The question of how to choose the decision regions R_m so as to minimize $\Pr(E_1)$ for a given $\Pr(E_2)$, or vice versa, may be seen by comparison of (4) and (5) to be the standard decision theoretic one of minimizing one quantity within a region and another outside it. The solution would be given by the Neyman–Pearson theorem, except that one of the quantities is not a probability. We therefore in the Appendix trivially generalize the Neyman–Pearson theorem as follows.

Theorem 1

Let $f_0(y)$ and $f_1(y)$ be two non-negative functions defined on the discrete space Y, and define

$$P_0(R) = \sum_{y \epsilon R} f_0(y),$$
$$P_1(Y - R) = \sum_{y \not\epsilon R} f_1(y), \tag{7}$$

where R is any region of Y. Let R_1 then be the region containing all y such that

$$\frac{f_1(y)}{f_0(y)} \geq \eta, \tag{8}$$

and let R_2 be any other region such that

$$P_0(R_2) \leq P_0(R_1). \tag{9}$$

Then

$$P_1(Y - R_2) \geq P_1(Y - R_1). \tag{10}$$

Corollary: Let the decision regions R_m be defined by

$$\mathbf{y} \epsilon R_m \text{ iff } \frac{\Pr(\mathbf{y}, \mathbf{x}_m)}{\sum_{m' \neq m} \Pr(\mathbf{y}, \mathbf{x}_{m'})} \geq \exp NT, \tag{11}$$

where N is the code length and T is an arbitrary parameter. Then there is no other set of decision regions which gives both a lower $\Pr(E_1)$ and a lower $\Pr(E_2)$ than this set does.

The proof of the corollary simply involves substituting $f_1(y) = \Pr(\mathbf{y}, \mathbf{x}_m)$ and $f_0(y) = \sum_{m' \neq m} \Pr(\mathbf{y}, \mathbf{x}_{m'})$ in (5) and (4), and using Theorem 1 for each m.

The corollary proves that the criterion (11) gives the optimum tradeoff between $\Pr(E_1)$ and $\Pr(E_2)$, or equivalently between $\Pr(\mathcal{E})$ and $\Pr(X)$ [since with $\Pr(\mathcal{E}) = \Pr(E_2)$ fixed, minimizing $\Pr(E_1)$ minimizes $\Pr(X)$]. The arbitrary parameter T governs the relative magnitudes of $\Pr(E_1)$ and $\Pr(E_2)$; clearly as T increases, $\Pr(E_1)$ increases while $\Pr(E_2)$ decreases, since the decision regions R_m shrink. We observe that in order for the decision regions to be necessarily disjoint T must be positive, for then (11) cannot be satisfied by more than one received word $\mathbf{y}$.

A suboptimum criterion of some practical utility comes from the observation that the probability $\Pr(\mathbf{y}, \mathbf{x}_{m2})$ of the second most likely code word $\mathbf{x}_{m2}$ is usually much larger than that of all the rest, excluding the first, so that the criterion

$$\mathbf{y} \epsilon R_m \text{ iff } \frac{\Pr(\mathbf{y}, \mathbf{x}_m)}{\Pr(\mathbf{y}, \mathbf{x}_{m2})} \geq \exp NT \tag{11a}$$

is a fairly good approximation to (11) for erasure schemes.

Now let us consider the list option, where the decision regions overlap. The event of a list error is the event in which the transmitted code word is not on the list, or thus in which the received word $\mathbf{y}$ is not in the decision region R_m corresponding to the transmitted code word $\mathbf{x}_m$. But this is precisely the event E_1. On the other hand, the probability that some code word $\mathbf{x}_m$ will be on the list although some other code word $\mathbf{x}_{m'}$, $m' \neq m$, was sent, is

$$\Pr(\mathbf{x}_m \text{ on list and incorrect}) = \sum_{\mathbf{y} \epsilon R_m} \sum_{m' \neq m} \Pr(\mathbf{y}, \mathbf{x}_{m'}). \tag{12}$$

The average number of incorrect words on the list $\bar{N}_I$ is then

$$\bar{N}_I = \sum_m \Pr(\mathbf{x}_m \text{ on list and incorrect})$$
$$= \sum_m \sum_{\mathbf{y}\epsilon R_m} \sum_{m' \neq m} \Pr(\mathbf{y}, \mathbf{x}_{m'}). \quad (13)$$

We observe that if we extend the definition (4) of $\Pr(E_2)$ to include overlapping decision regions, where $\Pr(E_2)$ is no longer a probability, then $\bar{N}_I = \Pr(E_2)$. Therefore, the problem of choosing decision regions for the optimum tradeoff between list-error probability and average number of incorrect words on the list is identical to that already considered; again (11) is the optimum criterion. The distinction is that in order to obtain lists, T must be allowed to become negative. The list option and the erasure option are two sides of the same coin, the only difference being in the sign of T.

The likelihood ratio criterion of (11) is therefore the one to use both with erasures and lists. Since

$$\frac{\Pr(\mathbf{y}, \mathbf{x}_m)}{\sum_{m' \neq m} \Pr(\mathbf{y}, \mathbf{x}_{m'})} = \frac{\Pr(\mathbf{y}, \mathbf{x}_m)}{\Pr(\mathbf{y}) - \Pr(\mathbf{y}, \mathbf{x}_m)} = \frac{\Pr(\mathbf{x}_m \mid \mathbf{y})}{1 - \Pr(\mathbf{x}_m \mid \mathbf{y})}, \quad (14)$$

an equivalent criterion is

$$\mathbf{y} \ \epsilon \ R_m \text{ iff } \Pr(\mathbf{x}_m \mid \mathbf{y}) \geq \eta. \quad (15)$$

where

$$\eta = \frac{\exp NT}{1 + \exp NT}. \quad (16)$$

The a posteriori probability is, therefore, an optimum erasure or list criterion. To recapitulate, in ordinary decoding, guess that code word $\mathbf{x}_m$ for which $\Pr(\mathbf{x}_m \mid \mathbf{y})$ is greatest. With the erasure option, guess that code word $\mathbf{x}_m$ for which $\Pr(\mathbf{x}_m \mid \mathbf{y})$ is greatest, as long as $\Pr(\mathbf{x}_m \mid \mathbf{y}) \geq \eta$, $\eta \geq \frac{1}{2}$; otherwise erase. With list decoding, to minimize the average list size (the average number of incorrect words on the list plus one) for a given list-error probability, put on the list all code words $\mathbf{x}_m$ for which $\Pr(\mathbf{x}_m \mid \mathbf{y}) \geq \eta$, $\eta < \frac{1}{2}$.

Review of Gallager's Results

For ordinary decoding, Gallager[3] has proved that there exists a block code of length N and rate R nats such that with maximum a posteriori probability decoding the error probability is bounded by

$$\Pr(\mathcal{E}) < \exp[-NE(R)], \quad (17)$$

where $E(R)$, the error exponent, is given at high rates by

$$E(R) = \max_{0 \leq \rho \leq 1, \mathbf{p}} E_0(\rho, \mathbf{p}) - \rho R,$$
$$E_0(\rho, \mathbf{p}) = -\ln \sum_i \left[\sum_k p_k p_{ik}^{1/1+\rho}\right]^{1+\rho}, \quad (18)$$

where $\mathbf{p}$ is any vector of input probabilities p_k. At low rates $E(R)$ is given by the "expurgated" exponent

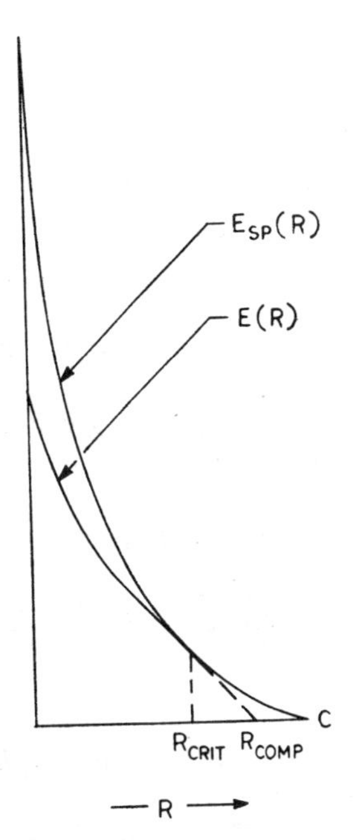

Fig. 2. Ordinary error and sphere-packing exponents.

$$E(R) = \max_{\rho \geq 1, \mathbf{p}} E_x(\rho, \mathbf{p}) - \rho R,$$
$$E_x(\rho, \mathbf{p}) = -\rho \ln \sum_k \sum_{k'} p_k p_{k'} \left[\sum_i p_{ik}^{1/2} p_{ik'}^{1/2}\right]^{1/\rho}. \quad (19)$$

We note that

$$E_0(1, \mathbf{p}) = E_x(1, \mathbf{p}); \quad (20)$$

at intermediate rates, therefore, $E(R)$ is given by the straight line

$$E(R) = R_{\text{comp}} - R,$$
$$R_{\text{comp}} = \max_{\mathbf{p}} E_0(1, \mathbf{p}), \quad (21)$$

where R_{comp} is the limiting computational cutoff rate of sequential decoding.[9] These bounds are known[4] to be exponentially tight at zero rate, and at all rates for which $E(R)$ equals the sphere-packing exponent

$$E_{sp}(R) = \max_{\rho \geq 0, \mathbf{p}} E_0(\rho, \mathbf{p}) - \rho R, \quad (22)$$

which, in general, will be at all rates between some critical rate R_{crit} and capacity C.

Fig. 2 depicts $E(R)$ and $E_{sp}(R)$ for a typical case, that of a binary symmetric channel with crossover probability $p = 0.01$.

The Bounds

In the Appendix, we show that similar, two-parameter exponential error bounds govern the quantities $\Pr(E_1)$ and $\Pr(E_2)$, which we recall are the erasure and undetected-error probabilities for erasure schemes, or the list-error probability and average number of incorrect words on the list for list schemes. These bounds are given by Theorem 2.

Theorem 2

There is a block code of length N and rate R nats such that when the likelihood ratio criterion of (11) is used with a threshold T, one can simultaneously obtain

$$\Pr(E_1) < \exp[-NE_1(R, T)],$$
$$\Pr(E_2) < \exp[-NE_2(R, T)], \tag{23}$$

where $\Pr(E_1)$ and $\Pr(E_2)$ are given by (5) and (4), $E_1(R, T)$ is given at high rates by

$$E_1(R, T) = \max_{0\le s\le\rho\le 1, \mathbf{p}} E_0(s, \rho, \mathbf{p}) - \rho R - sT,$$
$$E_0(s, \rho, \mathbf{p}) = -\ln \sum_i \Big[\sum_k p_k p_{ik}^{1-s}\Big] \Big[\sum_{k'} p_{k'} p_{ik'}^{s/\rho}\Big]^\rho, \tag{24}$$

and at low rates by

$$E_1(R, T) = \max_{0\le s\le 1, \rho\ge 1, \mathbf{p}} E_x(s, \rho, \mathbf{p}) - \rho R - sT,$$
$$E_x(s, \rho, \mathbf{p}) = -\rho \ln \sum_k \sum_{k'} p_k p_{k'} \Big[\sum_i p_{ik}^{1-s} p_{ik'}^{s}\Big]^{1/\rho}, \tag{25}$$

and $E_2(R, T)$ is given by

$$E_2(R, T) = E_1(R, T) + T. \tag{26}$$

As with the ordinary error exponent, we note that

$$E_0(s, 1, \mathbf{p}) = E_x(s, 1, \mathbf{p}), \tag{27}$$

so that for fixed T the two bounds are joined at intermediate rates by a straight line of slope -1.

Properties of the Bounds

We can easily show that when $T = 0$ the optimum value of s is $\rho/(1+\rho)$ in $E_0(s, \rho, \mathbf{p})$ and $\frac{1}{2}$ in $E_x(s, \rho, \mathbf{p})$. With these substitutions our exponents reduce to Gallager's ordinary error exponent $E(R)$, as one would hope. Actually, our result is a bit stronger, in that the error exponent $E(R)$ has been shown to be attainable not only with maximum a posteriori probability decoding (or, since all code words are assumed equally likely in the proof, maximum likelihood decoding, i.e., choose $\mathbf{x}_m$ if $\Pr(\mathbf{y} \mid \mathbf{x}_m) > \Pr(\mathbf{y} \mid \mathbf{x}_{m'})$, all $m' \neq m$), but also with a slightly weaker erasure-type decoding (choose $\mathbf{x}_m$ if $\Pr(\mathbf{y} \mid \mathbf{x}_m) \geq \sum_{m'\neq m} \Pr(\mathbf{y} \mid \mathbf{x}_{m'})$; otherwise erase). However, the stronger result is also obtained easily with a small change in Gallager's derivation.

A lower bound to these exponents, which is a good approximation for T near zero, comes from leaving s at the value which is optimum for $T = 0$; at high rates

$$E_1(R, T) = \max_{0\le s\le\rho\le 1, \mathbf{p}} E_0(s, \rho, \mathbf{p}) - \rho R - sT$$
$$\geq \max_{0\le\rho\le 1, \mathbf{p}} E_0\Big(\frac{\rho}{1+\rho}, \rho, \mathbf{p}\Big) - \rho R - \frac{\rho}{1+\rho} T$$
$$= E(R) - \frac{\rho}{1+\rho} T;$$

$$E_2(R, T) \geq E(R) + \frac{1}{1+\rho} T. \tag{28}$$

Similarly, at low rates,

$$E_1(R, T) \geq E(R) - \tfrac{1}{2}T;$$
$$E_2(R, T) \geq E(R) + \tfrac{1}{2}T. \tag{29}$$

These results had been proved earlier[6] by a variation in Gallager's derivation. They show that as T increases, $E_2(R, T)$ (the undetected-error exponent) increases at least as fast as $E_1(R, T)$ (the erasure exponent) decreases, which implies that if we permit the erasure probability to increase by a factor of 10, say, the undetected-error probability will decrease by more than a factor of 10. Since ρ goes to zero as R approaches the capacity C, we see that the tradeoff is expecially favorable for rates near capacity. Conversely, at least for small T, the list error exponent increases more slowly than the average incorrect list-size exponent decreases as T decreases below zero, with the tradeoff being especially unfavorable near capacity.

In the Appendix, we derive sharper bounds on the achievable tradeoff between the exponents given by (24) and (26) in the high-rate region. These bounds are expressed in terms of the fixed-$\mathbf{p}$ sphere-packing exponent $E_{sp}(R, \mathbf{p})$, defined as

$$E_{sp}(R, \mathbf{p}) = \max_{\rho\ge 0} E_0(\rho, \mathbf{p}) - \rho R. \tag{30}$$

For many channel models commonly used, there will be a single optimum $\mathbf{p}$ for all R; for this $\mathbf{p}$, $E_{sp}(R, \mathbf{p})$ equals the sphere-packing exponent $E_{sp}(R)$ of (22), and the latter may be used in the bound. Unfortunately, this is not the case for all channels. Also, if $R(\rho, \mathbf{p})$ is the rate at which $E_{sp}(R, \mathbf{p})$ has slope $-\rho$ [the rate for which ρ is optimum in (30)], then we define the conjugate rate $\bar{R}$ of $R(\rho, \mathbf{p})$ to be $R(1/\rho, \mathbf{p})$. We then have the following theorem.

Theorem 3

For any $\mathbf{p}$ and any R' greater than the conjugate rate $\bar{R}$ of R and less than $R(0, \mathbf{p})$, there is a threshold T such that

$$E_1(R, T) = E_{sp}(R', \mathbf{p}),$$
$$E_2(R, T) \geq E_{sp}(R, \mathbf{p}) - R + R'. \tag{31}$$

The proof of Theorem 3 shows that equality holds when the quantity

$$q_j(x) = \sum_k p_k p_{ik}^x \tag{32}$$

is independent of j. In particular this is true for channels symmetric from input and output, such as the binary symmetric channel, where the $\mathbf{p}$ is the optimum equiprobable input distribution. Of course, for this $\mathbf{p}$ $E_{sp}(R, \mathbf{p}) = E_{sp}(R)$. For such totally symmetric channels, we therefore have Theorem 3(a).

Theorem 3(a)

For totally symmetric channels, there is a threshold T such that for any rate R' greater than the conjugate rate $\bar{R}$ of R and less than C,

$$E_1(R, T) = E_{sp}(R'),$$
$$E_2(R, T) = E_{sp}(R) - R + R'. \qquad (33)$$

Theorem 3 suggests the following construction. Draw the sphere-packing exponent $E_{sp}(R', \mathbf{p})$, and through it draw a straight line of slope $+1$ intersecting $E_{sp}(R', \mathbf{p})$ at the rate R. Then as we vary R' in the range $\bar{R} \leq R' \leq R(0, \mathbf{p})$, the two curves bound the obtainable tradeoff between $E_1(R, T)$ and $E_2(R, T)$. Fig. 3 illustrates this construction with the sphere-packing exponent of the binary symmetric channel of Fig. 2. The construction is performed for two rates conjugate to one another.

The following heuristic picture may help to understand Theorem 3. Suppose that we had a code of rate R' that was effectively sphere-packed, with error exponent $E_{sp}(R')$. Suppose further that it were possible to use a subset of words in the code as an effectively sphere-packed code of rate $R < R'$. Then if the decision regions belonging to the original code continued to be used for the new code, the erasure probability would continue to be exp $[-NE_{sp}(R')]$. An error could not be made, however, unless the received word were outside the new code's proper sphere, an event of probability exp $[-NE_{sp}(R)]$. Even in the latter event, no error would occur unless the received word fell in some other decision region. But the percentage of the total space covered by decision regions proper to code words in the new code is the ratio of the number of words in the new code to those in the original, or exp $[-N(R' - R)]$. Assuming that this is indeed approximately the probability of error, given that the received word falls outside the correct exclusive sphere, we would arrive at an undetected-error probability of exp $\{-N[E_{sp}(R) - R + R']\}$.

We obtain similar results for the very noisy channel, defined as any channel with transition probability matrix

$$p_{ik} = (1 + \epsilon_{jk})q_i, \qquad (34)$$

where $\epsilon_{jk} \ll 1$ for all j and k. To second order in the ϵ_{jk}, the capacity of the very noisy channel is[10]

$$C = \max_{\mathbf{p}} \tfrac{1}{2} \sum_j q_j[\sum_k p_k\epsilon_{jk}^2 - (\sum_k p_k\epsilon_{jk})^2]. \qquad (35)$$

By substitution of (34) into (24) and (25), we find that for the optimum $\mathbf{p}$ and to second order in the ϵ_{jk},

$$E_0(s, \rho) = \left(2s - s^2 - \frac{s^2}{\rho}\right) C,$$
$$E_x(s, \rho) = (2s - 2s^2)C. \qquad (36)$$

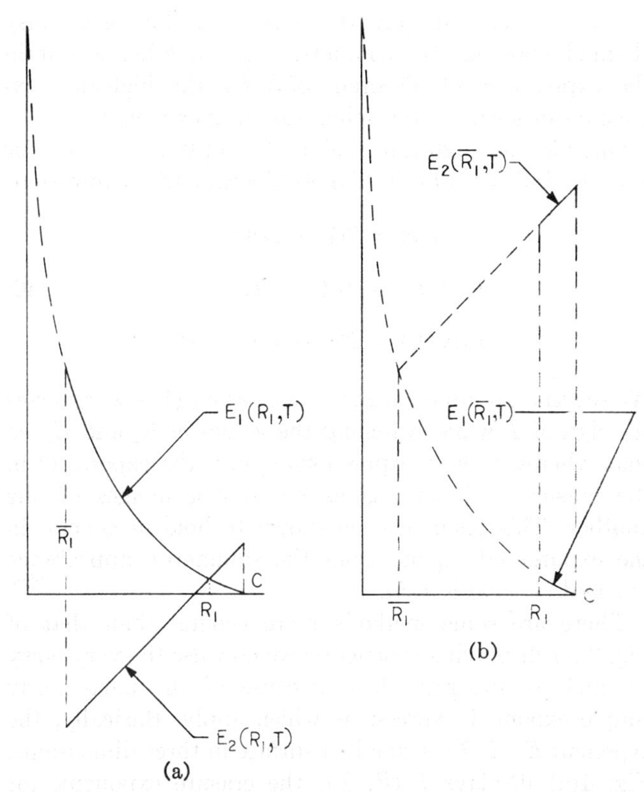

Fig. 3. Construction of exponents E_1 and E_2 from sphere-packing exponents in high-rate region for two conjugate rates R_1 and $\bar{R}_1$.

We see immediately that the expurgated exponent is of no use, since $E_x(s, \rho) = E_x(s, 1) = E_0(s, 1)$. In the high-rate region, we obtain the parametric equations

$$R(s, \rho) = \frac{\partial E_0}{\partial \rho} = \frac{s^2}{\rho^2} C,$$
$$T(s, \rho) = \frac{\partial E_0}{\partial s} = \left[2 - 2s - \frac{2s}{\rho}\right] C,$$
$$E_1(s, \rho) = s^2C,$$
$$E_2(s, \rho) = \left[s^2 - 2s + 2 - \frac{2s}{\rho}\right] C. \qquad (37)$$

The sphere-packing bound is obtained in parametric form from the substitution $s = \rho/(1 + \rho)$

$$E_{sp}(\rho) = \left(\frac{\rho}{1 + \rho}\right)^2 C,$$
$$R(\rho) = \left(\frac{1}{1 + \rho}\right)^2 C. \qquad (38)$$

We can therefore write

$$E_1(s, \rho) = E_{sp}\left(\frac{s}{1 - s}\right),$$
$$E_2(s, \rho) = E_{sp}\left(\frac{\rho - s}{s}\right) - R\left(\frac{\rho - s}{s}\right) + R\left(\frac{s}{1 - s}\right), \qquad (39)$$
$$R(s, \rho) = R\left(\frac{\rho - s}{s}\right).$$

It follows that, despite the fact that the very noisy channel may not be symmetric, we nonetheless obtain the expressions of Theorem 3(a) for the high-rate exponents in terms of the sphere-packing exponents.

Outside the high-rate region, the optimum value for ρ is 1, and we obtain the semiparametric expressions

$$\begin{aligned} T(s) &= 2(1 - 2s)C, \\ E_1(s, R) &= 2s^2C - R, \\ E_2(s, R) &= 2(1 - s)^2C - R. \end{aligned} \tag{40}$$

We see that an interchange between s and $(1 - s)$ reverses the sign of T while switching the values of E_1 and E_2, so that wherever these expressions apply, the exponents in the erasure and list regions are mirror images of one another. This result can be shown to hold in general in the expurgated region, from the symmetric appearance of s and $(1 - s)$ in (25).

There are some methods more general than that of Fig. 3 for displaying exponents; we now use the very noisy channel to illustrate them because of the particularly simple exponent expressions which apply. Basically, the exponent $E_1(R, T)$ is simply a surface in three dimensions. Fig. 4(a) displays $E_1(R, T)$, the erasure exponent, for $T \geq 0$, and $E_2(R, T)$, the average incorrect list-size exponent, for $T \leq 0$; the complementary exponents must be imagined by adding the magnitude of T to the illustrated exponent. The exponents are shown in the list region only for $E_2(R, T) \geq 0$, though of course the average incorrect list size exponent may validly go negative.

Since T is not a basic coding parameter, we can alternatively solve for T and obtain a surface in the three dimensions R, E_1, and E_2, thus giving the attainable tradeoff between the two exponents E_1 and E_2 directly for every R. Fig. 4(b) illustrates this approach.

We might mention one last way of expressing these relationships which seems to add nothing but a certain symmetry. It could be called a thermodynamic approach, since it involves defining a state function

$$E_0'(\lambda, s', \rho', \mathbf{p}) = \lambda E_0\left(\frac{s'}{\lambda}, \frac{\rho'}{\lambda}, \mathbf{p}\right) \tag{41}$$

from which all the significant variables can be determined by partial differentiation:

$$R = \frac{\partial E_0'}{\partial \rho'}, T = \frac{\partial E_0'}{\partial s'}, E = \frac{\partial E_0'}{\partial \lambda}. \tag{42}$$

We have yet to find any use for this curiosity.

Finally, we wish to mention a class of channels for which the exponents assume very different form. This class consists of all channels with transition probability matrix satisfying

$$p_{jk} = \beta_j \, \delta_{jk}, \tag{43}$$

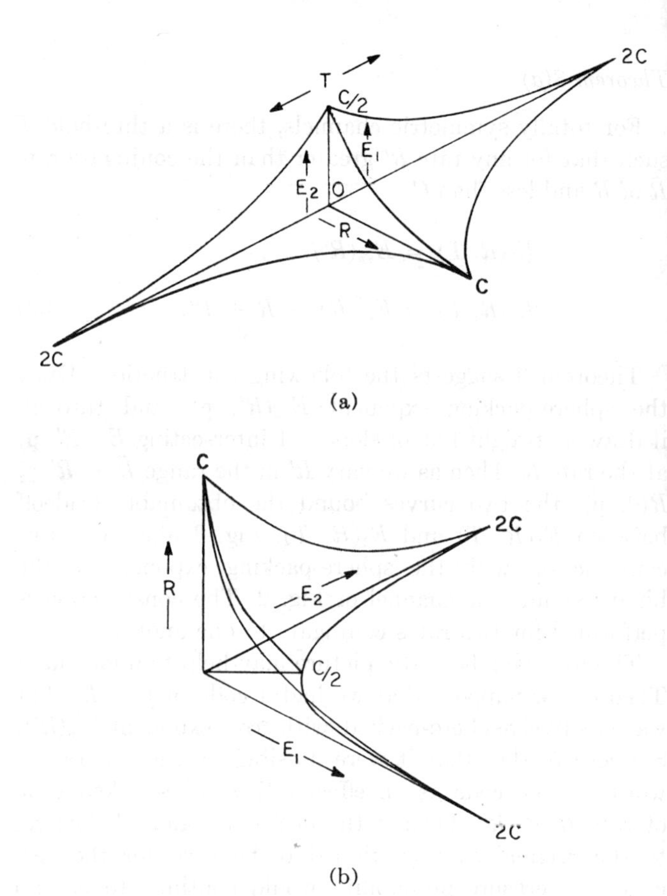

Fig. 4. Very noisy channel exponents.

where δ_{jk} equals either zero or one. We call these channels erasure-type channels, since their archetype is the binary erasure channel, for which the erasure output has $\beta_j = p$, the erasure probability, with δ_{jk} equal to one for both inputs x_k; while for the other two outputs only one δ_{jk} equals one and $\beta_j = 1 - p$. In the erasure-type channel, the occurrence of an output y_j rules out all code words which have in that position any input x_k such that $\delta_{jk} = 0$, but in no way discriminates between the remaining code words. Decoding with such a channel is then a sieve-like process, in which each output allows the discard of a certain percentage of the code words previously under consideration. After N outputs, either only one code word will be left, in which case that will be the correct one, or there will be a list of several, with no basis for choosing between them.

When we substitute (43) into (24) and (25) to obtain exponents for erasure-type channels, we find that

$$\begin{aligned} E_0(s, \rho, \mathbf{p}) &= -\ln \sum_j \left[\sum_k p_k \delta_{jk}\right]^{1+\rho}, \\ E_x(s, \rho, \mathbf{p}) &= -\rho \ln \sum_k \sum_{k'} p_k p_{k'} \left[\sum_j \beta_j \delta_{jk} \delta_{jk'}\right]^{1/\rho}. \end{aligned} \tag{44}$$

The salient observation to be made is that neither of these expressions depends on s. Therefore, for $T > 0$, the optimum s is the smallest one, $s = 0$, while for $T < 0$, the optimum s is the largest one, $s = \rho$ or $s = 1$ as the case may be, and thus

$$E_1(R, T) = \max_{0 \le s \le \rho \le 1, \mathbf{p}} E_0(s, \rho, \mathbf{p}) - \rho R - sT$$
$$= \max_{0 \le \rho \le 1, \mathbf{p}} E_0(\rho, \mathbf{p}) - \rho R \tag{45}$$
$$= E(R), \quad T \ge 0 \quad \text{and} \quad R \ge R_{\text{crit}}.$$

Similarly,

$$E_1(R, T) = E(R), \quad T \ge 0 \quad \text{and} \quad R \le R_{\text{crit}}. \tag{46}$$

Thus by increasing T, the undetected-error exponent $E_2(R, T) = E_1(R, T) + T$ can be made as small as desired, without affecting the erasure exponent $E_1(R, T)$, which is simply the ordinary error exponent $E(R)$.

In the list region,

$$E_2(R, T)$$
$$= \max_{0 \le \rho \le 1, \mathbf{p}} E_0(\rho, \mathbf{p}) - \rho(R + T) + T$$
$$= \begin{cases} E(R + T) + T, & T \le 0 \quad \text{and} \quad R + T \ge R_{\text{crit}}, \\ R_{\text{crit}} - R, & T \le 0 \quad \text{and} \quad R_{\text{crit}} \le R \le R_{\text{crit}} - T, \\ E(R), & T \le 0 \quad \text{and} \quad R \le R_{\text{crit}}. \end{cases} \tag{47}$$

The typical appearance of these curves is illustrated in Fig. 5.

A plausible interpretation of this result is readily found. In the erasure region, by letting T become infinite, we can get an undetected-error probability of zero, while the erasure probability remains the same as the ordinary error probability. The explanation is that all ordinary decoding errors are detectable as erasures, since they correspond to decoded lists of size greater than one. On the list side, the average incorrect list size is equal to the erasure probability for rates below R_{crit}, which says that when decoded lists do occur they are small, with size increasing less than exponentially with N. At higher rates, given that a decoded list does occur, the average list size increases exponentially with N; in general, the average list size, given a list, seems to go as exp $N[R - R_{\text{comp}} + E(R)]$.

Implications for Decision Feedback

One of the most interesting applications of our bounds is the determination of the performance attainable with decision feedback. We recall that when a feedback channel is available, each erasure triggers a request for repeat of a code word. If the erasure probability is Pr (X), then the probability that a code word will be repeated i or more times is $[\Pr(X)]^i$, and the average number of times a code word is repeated is

$$1 + \Pr(X) + [\Pr(X)]^2 + \cdots = \frac{1}{1 - \Pr(X)}. \tag{48}$$

Therefore, if the code rate is R, the effective rate of information transfer is reduced to

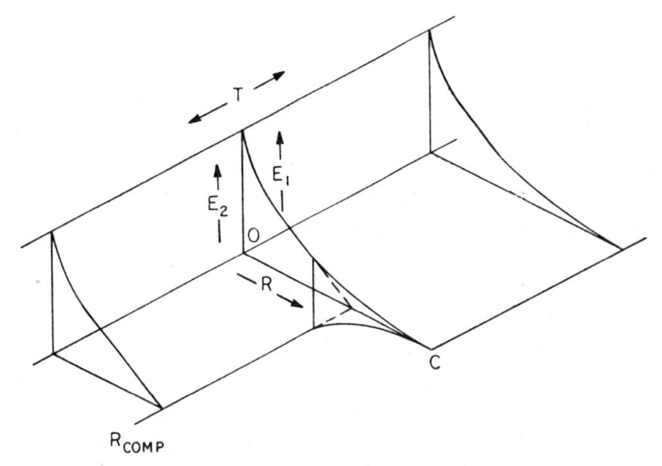

Fig. 5. Exponents for erasure-type channels.

$$R[1 - \Pr(X)]. \tag{49}$$

Clearly, for any T such that $E_1(R, T) > 0$, the effective rate of information transfer can be made as close to R as desired by increasing N, since

$$\Pr(X) < \exp[-NE_1(R, T)]. \tag{50}$$

Meanwhile, the undetected-error probability per code word is bounded by

$$\Pr(\mathcal{E}) < \exp[-NE_2(R, T)]. \tag{51}$$

These considerations lead to the definition of the feedback exponent $E_f(R)$ as the limiting value of $E_2(R, T)$ as $E_1(R, T)$ approaches zero. This value is determined by solving for T in (24) and (26), obtaining

$$E_2 = \max_{0 \le s \le \rho \le 1, \mathbf{p}} \frac{1}{s} E_0(s, \rho, \mathbf{p}) - \frac{\rho}{s} R - \frac{1 - s}{s} E_1. \tag{52}$$

As E_1 approaches zero, this maximum is attained for fixed R by letting s and ρ both go to zero at the constant ratio $\rho/s = \nu \ge 1$. Since $E_0(s, \rho, \mathbf{p})$ also goes to zero as this happens, we must use L' Hôpital's rule

$$\lim_{s \to 0} \frac{1}{s} E_0(s, \rho, \mathbf{p})$$
$$= \frac{\partial}{\partial s} E_0(s, \nu s, \mathbf{p}) \bigg|_{s=0}$$
$$= \sum_i \sum_k p_k p_{jk} [\ln p_{jk} - \ln (\sum_{k'} p_{k'} p_{jk'}^{1/\nu})^\nu] \tag{53}$$
$$= E_{0f}(\nu, \mathbf{p}),$$

where the last step defines $E_{0f}(\nu, \mathbf{p})$. Then by substitution in (52)

$$E_f(R) = \max_{\nu \ge 1, \mathbf{p}} E_{0f}(\nu, \mathbf{p}) - \nu R. \tag{54}$$

A careful examination shows that all limits are approached continuously and with continuous derivatives, so that if $E_2(R, T) = E_f(R) - \delta$, then $E_1(R, T) > 0$ with strict inequality. One can then make a statement of the following type. For any ϵ and δ greater than zero there exists an N such that there is a decision-feedback scheme involving a block code of length N and rate R nats in which the effective rate of information transfer exceeds $R(1 - \epsilon)$, while the probability of error is bounded by

$$\Pr(\mathcal{E}) < \exp\{-N[E_f(R) - \delta]\}. \quad (55)$$

We observe that while the feedback exponent $E_f(R)$ approaches zero as R approaches capacity, as it must, it approaches 0 with a slope of -1, in contrast to the ordinary error exponent $E(R)$, whose slope is generally zero at capacity. This implies that the ratio of the feedback exponent $E_f(R)$ to the ordinary error exponent $E(R)$ becomes infinite as the rate approaches capacity, or, in other words, that near capacity the achievement of low error probabilities is dramatically simplified by decision feedback.

For the totally symmetric and very noisy channels discussed earlier, the feedback exponent may be found at all rates by allowing the R' of Theorem 3(a) to approach C; then $E_{sp}(R')$ goes to zero, and

$$E_f(R) = E_{sp}(R) - R + C, \; C_{0,sp} \leq R \leq C, \quad (56)$$

where $C_{0,sp} = R(\infty)$ is the rate at which the sphere-packing bound becomes infinite. This is a large exponent, even in comparison to the sphere-packing bound, which is the true ordinary error exponent only at high rates. Moreover, by Theorem 3 we see that for any channel, if we choose $\mathbf{p}$ as the distribution which gives $R(0, \mathbf{p}) = C$, then

$$E_f(R) \geq E_{sp}(R, \mathbf{p}) - R + C, \qquad R(\infty, \mathbf{p}) \leq R \leq C. \quad (57)$$

Viterbi and Gallager had shown earlier in unpublished work that $E_f(R) \geq C - R$.

As an example of a case in which the bound of (57) is exceeded, we might cite the erasure-type channels; from Fig. 5 it is clear that $E_f(R)$ is infinite at all rates less than C. In other words, if repeats are demanded whenever ambiguity in decoding remains, any rate less than capacity can be maintained with no errors. In fact, on the binary erasure channel, where $C = 1 - p$ (p being the erasure probability), we can achieve this performance with codes of block length one, simply by asking for a repeat whenever a bit is erased.

For any channel with at least one transition probability equal to zero, there may be a rate greater than zero for which $E_f(R)$ becomes infinite. We call such a rate the feedback zero-error capacity $C_{0,f}$. From these results, $C_{0,f} \leq C$, with equality for erasure-type channels. $C_{0,f}$ must be the limit of $\nu^{-1}E_{0f}(\nu, \mathbf{p})$ as ν becomes large, for then $E_f(R)$ becomes $\nu(C_{0,f} - R)$ for large ν, and in the limit infinite if $R < C_{0,f}$. Again L'Hôpital's rule is required:

$$\begin{aligned} C_{0,f} &= \max_{\mathbf{p}} \frac{\partial}{\partial \nu} E_{0f}(\nu, \mathbf{p}) \Big|_{\nu=\infty} \\ &= \max_{\mathbf{p}} \left[- \sum_i q_i \ln q_i(0)\right], \end{aligned} \quad (58)$$

where

$$\begin{aligned} q_i &= q_i(1) = \sum_k p_k p_{ik}, \\ q_i(0) &= \sum_k p_k \theta_{ik}, \\ \theta_{ik} &= \begin{cases} 0, & \text{if } p_{ik} = 0, \\ 1, & \text{otherwise.} \end{cases} \end{aligned} \quad (59)$$

This result may be obtained directly by the following argument. Let us consider the following method of communicating a single one of $\exp NR$ messages. First a code word of length N is chosen by picking each letter independently and with probability p_k of being x_k. This represents the correct message, and is transmitted, resulting in the received word $\mathbf{y} = [y_{i(1)}, y_{i(2)}, \cdots, y_{i(N)}]$. The probability of the output y_i occurring in any position in the received word is the q_i of (59), independently of what appears in all other positions. Now the other $\exp NR - 1$ code words are chosen, exactly as the correct one was.

To ensure zero-error probability, a sieve-like decoding procedure is necessary, as with erasure-type channels. Each occurrence of the output y_i justifies the discard of all code words which in that position have any letter x_k such that $p_{ik} = 0$; all others must be retained. Given a received output y_i, therefore, the probability that any code word other than the one transmitted will be consistent with y_i is the $q_i(0)$ defined by (59). Given a whole received word $\mathbf{y}$, the probability of retaining a particular incorrect word is then

$$p_r(\mathbf{y}) = \prod_{i=1}^{N} q_{i(i)}(0). \quad (60)$$

If $p_r(\mathbf{y})$ is less than $\exp(-NR)$, then chances are that all the incorrect code words will be discarded, with probability approaching one as N becomes large, by the law of large numbers. Conversely, if $p_r(\mathbf{y})$ is greater than $\exp(-NR)$, then chances are that at least one of the incorrect code words will not be discarded, again with probability approaching one as N becomes large. The probability that $p_r(\mathbf{y})$ is less than $\exp(-NR)$ is

$$\begin{aligned} \Pr\left[\prod_{i=1}^{N} q_{i(i)}(0) < \exp(-NR)\right] \\ = \Pr\left(\sum_{i=1}^{N} \ln q_{i(i)}(0) < -NR\right), \end{aligned} \quad (61)$$

where the probability is over the ensemble of all $\mathbf{y}$. But the probability that a sum of N independent random variables exceeds N times a constant goes to one with large N, if the constant is less than the mean of the variables, and to zero otherwise. But the mean of $\ln q_i(0)$ is

$$\overline{\ln q_i(0)} = \sum_i q_i \ln q_i(0) = C_{0,f}, \tag{62}$$

where the latter equality holds if we choose $\mathbf{p}$ to maximize the mean, as is obviously desirable. Thus if $R < C_{0,f}$, the probability of (61) goes to one with large N. But

$$\begin{aligned}\Pr(\text{repeat}) &= \Pr[\text{repeat} \mid p_r(\mathbf{y}) < \exp(-NR)] \cdot \Pr[p_r(\mathbf{y}) < \exp(-NR)] \\ &\quad + \Pr[\text{repeat} \mid p_r(\mathbf{y}) \geq \exp(-NR)] \cdot \Pr[p_r(\mathbf{y}) \geq \exp(-NR)] \\ &= \begin{cases} 0\cdot 1 + 1\cdot 0 = 0, & R < C_{0,f}, \quad N \text{ large}, \\ 0\cdot 0 + 1\cdot 1 = 1, & R > C_{0,f}, \quad N \text{ large}. \end{cases}\end{aligned} \tag{63}$$

This argument, therefore, shows not only that arbitrarily small repeat probabilities (actually decreasing exponentially with N) can be obtained with zero-error probability when $R < C_{0,f}$, but also that when $R > C_{0,f}$ the repeat probability must approach one if zero-error probability is demanded. (A rigorous converse has been developed by Gallager in unpublished work.)

Implications for List Decoding

Previous work on list decoding has assumed a decoding list of fixed size. It has been shown that for a list size which is large but not exponential in N, a list-error exponent equal everywhere to the sphere-packing exponent could be obtained,[11] and that for list sizes increasing as $\exp NL$, a list-error exponent of $E_{sp}(R - L)$ was achievable.[4] Such large list sizes are of course unattractive for implementation, however. The principal significance of the present results is that by allowing a list of variable size, the average number of words on the list can be kept very small. In fact, we see from Figs. 4 and 5 that over a large range of $R < C$ and $T < 0$, the average number of incorrect words on the list is bounded by a quantity which decreases exponentially with N, and, therefore, that for large N the average list size is effectively one. Of course, the list for any one transmission may be as large as $\exp N(-T)$, so considerable buffering may be required. If provision of such buffering is feasible, however, and the list output useful, then one may obtain a list-error exponent well above the ordinary error exponent attainable at the same rate.

The question of how large a list-error exponent can be attained at a given rate R for a positive average incorrect list-size exponent is of some interest. Let us define $E_l(R)$ as the limiting value of $E_2(R, T)$, as $E_1(R, T)$ approaches zero. Rather than writing out parametric expressions, we note only a few properties which give the general character of $E_l(R)$. First, at zero rate, the parameter ρ may be allowed to become infinite, so the list exponent equals the feedback exponent

$$\begin{aligned}E_l(0) &= \lim_{\substack{s\to 1\\ \rho\to\infty}} E_x(s, \rho, \mathbf{p}) \\ &= -\sum_k \sum_{k'} p_k p_{k'} \sum_i p_{ik} \ln \frac{p_{ik'}}{p_{ik}} \\ &= E_f(0),\end{aligned} \tag{64}$$

as follows from the mirror symmetry of exponents in the expurgated region. Second, to get an idea of behavior at high rates, let us first examine the list exponent for erasure-type channels, where $E_0(s, \rho, \mathbf{p})$ is independent of s. We recall from (47) that for such channels

$$\begin{aligned}E_1(R, T) &= E(R + T), \\ E_2(R, T) &= E(R + T) + T, \\ T \leq 0 \quad &\text{and} \quad R + T \geq R_{\text{crit}}.\end{aligned} \tag{65}$$

Clearly, the threshold T for which $E_2(R, T)$ equals zero can be determined from the ordinary error exponent as that $T(R)$ which solves

$$-T(R) = E[R + T(R)], \tag{66}$$

the list exponent is then simply

$$E_l(R) = E[R + T(R)] = -T(R). \tag{67}$$

Graphically, $E_l(R)$ is therefore the intercept of a line of slope -1 drawn through R with the $E(R)$ curve; Fig. 6 illustrates the construction of $E_l(R)$ in the high-rate region. We note, as was clear from Fig. 5, that $E_l(R)$ is infinite at rates below R_{comp}. However, at rates above R_{comp}, $E_l(R)$ rapidly approaches the $E(R)$ curve, so that little improvement is obtained in this region.

For the general channel, the construction of Fig. 6 represents an upper bound to the attainable list exponent, as we now show. Since from (24) and (26) or (52), in the high-rate region

$$E_1 = \max_{0\leq s\leq\rho\leq 1,\mathbf{p}} \frac{1}{1-s} E_0(s, \rho, \mathbf{p}) - \frac{\rho}{1-s} R - \frac{s}{1-s} E_2, \tag{68}$$

we have

$$E_l(R) = \max_{0\leq s\leq\rho\leq 1,\mathbf{p}} \frac{E_0(s, \rho, \mathbf{p}) - \rho R}{1 - s}. \tag{69}$$

Let

$$E_{l0}(R) = \max_{0\leq\rho\leq 1,\mathbf{p}} \frac{E_0(\rho, \mathbf{p}) - \rho R}{1 - \rho}, \tag{70}$$

which is the curve given by the construction of Fig. 6;

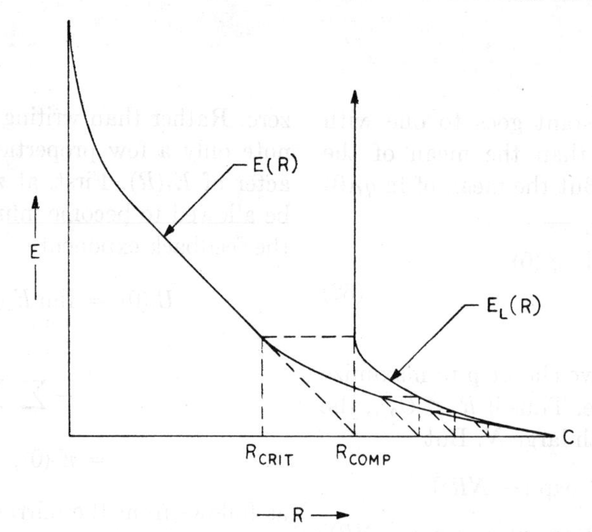

Fig. 6. Construction of list exponent from ordinary error exponent for erasure-type channels.

then we can bound (69) by loosening the restrictions on s:

$$\begin{aligned} E_l(R) &\leq \max_{0\leq s_1\leq \rho\leq 1, 0\leq s_2\leq \rho, \mathbf{p}} \frac{E_0(s_1, \rho, \mathbf{p}) - \rho R}{1 - s_2} \\ &= \max_{0\leq\rho\leq 1, \mathbf{p}} \frac{E_0(\rho, \mathbf{p}) - \rho R}{1 - \rho} \\ &= E_{l0}(R) \end{aligned} \tag{71}$$

with equality for erasure-type channels.

The moral we draw is that at rates above R_{comp}, not much improvement is obtained by going to lists over ordinary decoding. At rates below R_{comp}, however, significant improvement can be achieved, approaching that attainable with feedback at very low rates. Thus, as with sequential decoding,[9] we find that R_{comp} represents a kind of complexity limit for high-performance schemes without feedback.

Finally, we may inquire about the existence of a list zero-error capacity $C_{0,l}$, namely a rate for which with a sieve-like decoding procedure we obtain an average number of incorrect words on the decoding list decreasing exponentially with N, and of course no errors. We find that the $E_l(R)$ of (69) goes to infinity as $s \to \rho \to 1$, where the rate becomes

$$C_{0,l} = \max_{\mathbf{p}} \left[- \ln \sum_i q_i q_i(0)\right], \tag{72}$$

where q_i and $q_i(0)$ are defined as in (59). From the fact that $E_{l0}(R)$ bounds $E_l(R)$ above, we have that for any channel

$$C_{0,l} \leq R_{\text{comp}}. \tag{73}$$

Like the feedback zero-error capacity, the list zero-error capacity can be obtained directly. We have already shown that with the random selection procedure of that earlier argument, the probability of an incorrect code word being retained, given $\mathbf{y}$, is the $p_r(\mathbf{y})$ of (60); the average number of incorrect words retained is therefore

$$\begin{aligned} &(\exp NR - 1) \overline{\prod_{i=1}^{N} q_{i(i)}(0)} \\ &= (\exp NR - 1) \prod_{i=1}^{N} \overline{q_{i(i)}(0)} \\ &= (\exp NR - 1) \left[\sum_i q_i q_i(0)\right]^N \\ &< \exp N[R - C_{0,l}], \end{aligned} \tag{74}$$

where the second step depends on the $q_i(0)$ being independent, and the last on the use of the optimum $\mathbf{p}$. Thus the average number of incorrect words on the decoder's list decreases exponentially with N whenever $R < C_{0,l}$.

An interesting inequality links the sphere-packing, list, and feedback zero-error capacities. By letting ρ approach infinity in (22), we find that

$$C_{0,sp} = \max_{\mathbf{p}} \{- \ln [\max q_i(0)]\}. \tag{75}$$

We can then express all three capacities in terms of the quantity $M(r)$ defined by

$$M(r) = \left[\sum_i q_i q_i(0)^r\right]^{1/r}. \tag{76}$$

It is well known[12] that if $r > r'$, then $M(r) \geq M(r')$, with equality only when $q_i(0)$ is independent of j, as with totally symmetric channels. We observe that

$$\begin{aligned} C_{0,sp} &= \max_{\mathbf{p}} \lim_{r\to\infty} [- \ln M(r)], \\ C_{0,l} &= \max_{\mathbf{p}} [- \ln M(1)], \\ C_{0,f} &= \max_{\mathbf{p}} \lim_{r\to 0} [- \ln M(r)], \end{aligned} \tag{77}$$

from which there follows immediately the inequality

$$C_{0,sp} \leq C_{0,l} \leq C_{0,f}, \tag{78}$$

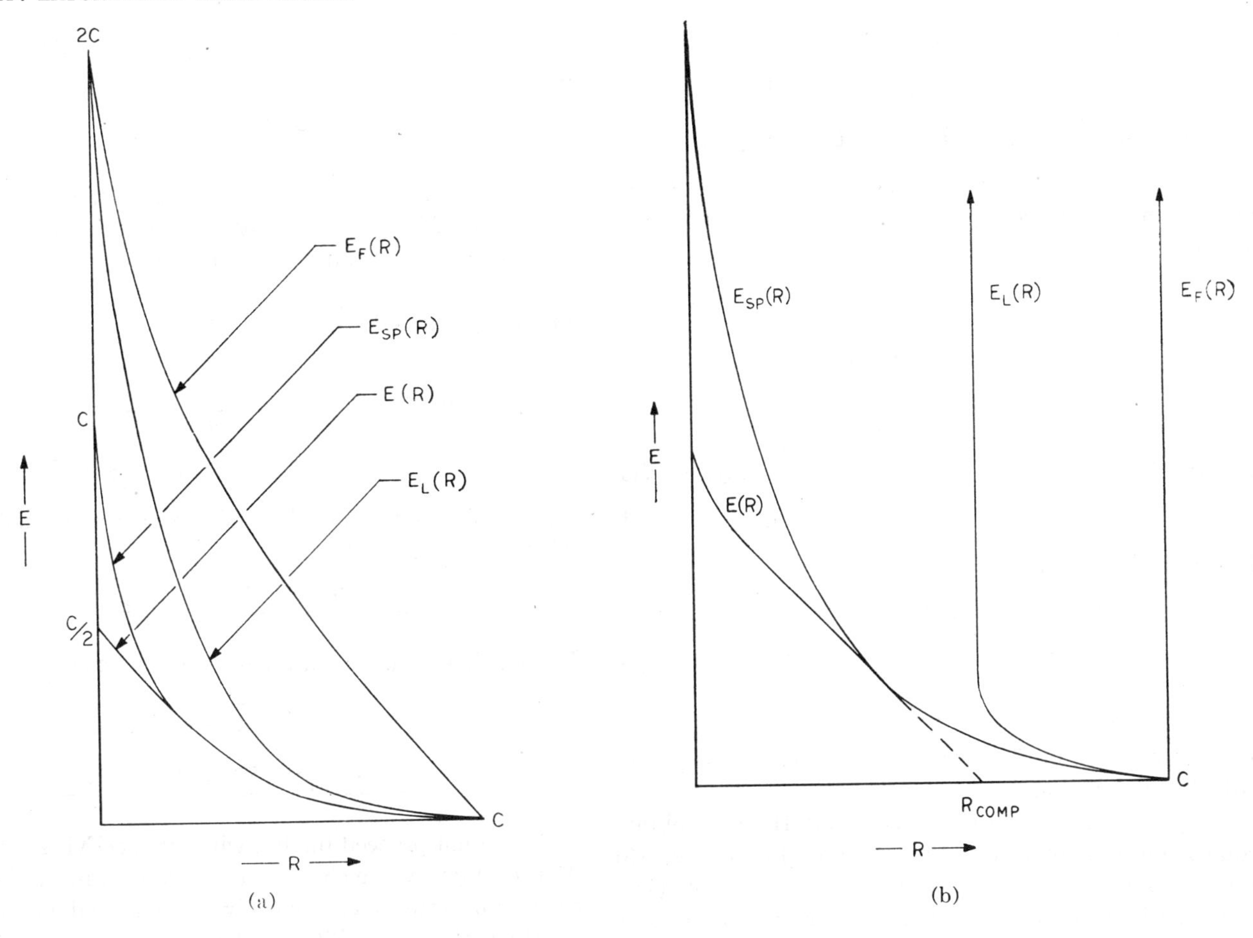

Fig. 7. (a) Very noisy channel exponents:

$$E_f(R) = 2C - 2\sqrt{RC};$$

$$E_l(R) = \begin{cases} 2C - 2\sqrt{2RC}, & R \leq (6 - 4\sqrt{2})C \\ 2\sqrt{RC} - 2C\sqrt{2\sqrt{R/C} - 1}, & R \geq (6 - 4\sqrt{2})C; \end{cases}$$

$$E_{sp}(R) = (\sqrt{C} - \sqrt{R})^2;$$

$$E(R) = \begin{cases} E_{sp}(R), & R \geq C/4 \\ C/2 - R, & R \leq C/4. \end{cases}$$

(b) Erasure-type channel exponents.

where the equalities hold only when $q_i(0)$ is independent of j.

Finally, in Fig. 7 we have plotted the four exponents $E(R)$, $E_{sp}(R)$, $E_l(R)$, and $E_f(R)$ for the very noisy channel and for an erasure-type channel to give some graphic feeling for their relative magnitudes.

Open Questions

One would hope that, at least in the high-rate range, the upper bounds developed here were exponentially tight. One's hopes are encouraged by the similarity of this derivation to that which yields the ordinary error exponent, known to be tight in the high-rate region, and also by the fact that these bounds are equal to, or better than the best previously known in all cases, with no counterexamples yet sighted. Finding a complementary lower bound on attainable probabilities is therefore a theoretical task of considerable interest.

Secondly, we hope that these results, which demonstrate the performance improvements obtainable when the decoder is given flexible decoding options, will stimulate the development of applications. We are pursuing applications in concatenation schemes; many others have been interested in various uses of feedback. Undoubtedly these efforts do not exhaust the potential uses of the ideas treated here.

Appendix

In this Appendix we collect proofs which we felt would have unduly cluttered the main text.

Proof of Theorem 1 (Neyman–Pearson): Our proof follows Davenport and Root.[13] Let U be the region common to the regions R_1 and R_2 defined in the theorem. Then

$$P_1(Y - R_1) = P_1(Y) - P_1(R_1) = P_1(Y) - P_1(R_1 - U) - P_1(U), \tag{79}$$
$$P_1(Y - R_2) = P_1(Y) - P_1(R_2 - U) - P_1(U),$$

where Y is the whole space and $P_1(\cdot)$ is the measure defined in (7). Since $R_1 - U$ is contained in R_1, (8) applies, so

$$P_1(R_1 - U) = \sum_{y \varepsilon R_1 - U} f_1(y) \geq \sum_{y \varepsilon R_1 - U} \eta f_0(y) = \eta P_0(R_1 - U). \tag{80}$$

On the other hand, $R_2 - U$ is wholly outside R_1, so that

$$P_1(R_2 - U) \leq \eta P_0(R_2 - U). \tag{81}$$

But with the use of (9) we have that

$$P_0(R_2 - U) = P_0(R_2) - P_0(U) \leq P_0(R_1) - P_0(U) = P_0(R_1 - U). \tag{82}$$

The combination of (79) through (82) gives the inequality (10) of the theorem.

Proof of Theorem 2 (General Bounds): Here we obtain bounds on the quantities Pr (E_1) and Pr (E_2) of (5) and (4) under the decision criterion expressed by (11). For notational economy, we define for this proof only the shorthand notation

$$P_m = \Pr(\mathbf{y} \mid \mathbf{x}_m), \quad Q_m = \sum_{m' \neq m} \Pr(\mathbf{y} \mid \mathbf{x}_{m'}), \tag{83}$$

and from the beginning let all code words be equiprobable ($\Pr(\mathbf{x}_m) = 1/M$), though this can be delayed for a considerable time in the proof. Then the decision criterion becomes

$$\mathbf{y} \, \varepsilon \, R_m \text{ iff } P_m/Q_m \geq \exp NT, \tag{84}$$

and

$$\Pr(E_1) = \frac{1}{M} \sum_m \sum_{\mathbf{y} \notin R_m} P_m, \quad \Pr(E_2) = \frac{1}{M} \sum_m \sum_{\mathbf{y} \varepsilon R_m} Q_m. \tag{85}$$

Our first step is to use (84) in (85)

$$\Pr(E_1) = \frac{1}{M} \sum_m \sum_{\mathbf{y} \notin R_m} P_m^{1-s} Q_m^s (P_m/Q_m)^s \leq \frac{1}{M} \sum_m \sum_{\mathbf{y} \notin R_m} P_m^{1-s} Q_m^s \exp NTs, \quad s \geq 0,$$
$$\Pr(E_2) = \frac{1}{M} \sum_m \sum_{\mathbf{y} \varepsilon R_m} P_m^{1-s} Q_m^s (Q_m/P_m)^{1-s} \leq \frac{1}{M} \sum_m \sum_{\mathbf{y} \varepsilon R_m} P_m^{1-s} Q_m^s \exp[-NT(1-s)], \quad s \leq 1, \tag{86}$$

where we have introduced s as a parameter. Defining the moment generating function $g_m(-s)$ by

$$g_m(-s) = \sum_{\mathbf{y}} P_m (P_m/Q_m)^{-s}, \tag{87}$$

and observing that $g_m(-s)$ is non-negative, we can further bound Pr (E_1) and Pr (E_2) by allowing the sum to extend over the entire output space, whence

$$\Pr(E_1) \leq \exp NTs \frac{1}{M} \sum_m g_m(-s), \quad \Pr(E_2) \leq \exp[-NT(1-s)] \frac{1}{M} \sum_m g_m(-s), \quad 0 \leq s \leq 1. \tag{88}$$

The key to the proof is now Jensen's inequality,[12] which states that

$$[\sum_j a_j^\nu]^{1/\nu} \leq \sum_j a_j, \quad \nu \geq 1. \tag{89}$$

Introducing a second parameter ρ, we have that

$$g_m(-s) = \sum_{\mathbf{y}} P_m^{1-s} [(\sum_{m' \neq m} P_{m'})^{s/\rho}]^\rho \leq \sum_{\mathbf{y}} P_m^{1-s} (\sum_{m' \neq m} P_{m'}^{s/\rho})^\rho, \quad \rho \geq s. \tag{90}$$

We cannot proceed further without specifying the code. As is customary, we choose a code at random by choosing each input letter of each code word by a random selection in which the probability of choosing input x_k is p_k. Denoting an average over the ensemble of all such codes by an overbar, we can then bound the average moment generating function $\overline{g_m(-s)}$ by Gallager's methods:[3]

$$\overline{g_m(-s)} \leq \overline{\sum_{\mathbf{y}} P_m^{1-s} (\sum_{m' \neq m} P_{m'}^{s/\rho})^\rho} \tag{91a}$$

from (90);

$$= \sum_{\mathbf{y}} \overline{P_m^{1-s} (\sum_{m' \neq m} P_{m'}^{s/\rho})^\rho} \tag{91b}$$

average of a sum equals sum of the averages;

$$= \sum_{\mathbf{y}} \overline{P_m^{1-s}} \; \overline{(\sum_{m' \neq m} P_{m'}^{s/\rho})^\rho} \tag{91c}$$

average of a product of independent variables equals product of the averages; independence ensured by our choice of ensemble;

$$\leq \sum_{\mathbf{y}} \overline{P_m^{1-s}} \; (\overline{\sum_{m' \neq m} P_{m'}^{s/\rho}})^\rho, \quad \rho \leq 1 \tag{91d}$$

$[\overline{f^r}]^{1/r} \leq \bar{f}$, $r \leq 1$; see under (76) in the text;

$$= \sum_{\mathbf{y}} \overline{P_m^{1-s}} \; (\sum_{m'} \overline{P_{m'}^{s/\rho}})^\rho \tag{91e}$$

average of sum equals sum of averages;

$$< \exp \rho NR \sum_{\mathbf{y}} \overline{P_m^{1-s}} \; (\overline{P_{m'}^{s/\rho}})^\rho \tag{91f}$$

exp $NR - 1$ identical terms.

If we now recall the useful rule for interchanging sum aud product signs,

$$\sum_{\mathbf{y}} \prod_{i=1}^{N} f(y_i) = \prod_{i=1}^{N} \sum_{y_i} f(y_i), \tag{92}$$

we see that (91f) breaks down into an average of a product of N identically distributed random variables, which is the product of the averages, and which can be written out as

$$\overline{g_m(-s)} < \exp \rho NR[\sum_j (\sum_k p_k p_{jk}^{1-s})(\sum_{k'} p_{k'} p_{jk'}^{s/\rho})^\rho]^N, \tag{91g}$$

or

$$\overline{g_m(-s)} < \exp \{-N[E_0(s, \rho, \mathbf{p}) - \rho R]\},$$
$$0 \le s \le \rho \le 1, \tag{91h}$$

where $E_0(s, \rho, \mathbf{p})$ is defined as in (24). Since this bound is independent of m, it also applies to the quantity $M^{-1} \sum_m \overline{g_m(-s)}$, and since it applies to the average, there must be at least one code in the ensemble for which it applies to the quantity $M^{-1} \sum_m g_m(-s)$. For this code, by substitution in (88), we obtain simultaneously

$$\Pr(E_1) < \exp \{-N[E_0(s, \rho, \mathbf{p}) - \rho R - sT]\},$$
$$\Pr(E_2) < \exp \{-N[E_0(s, \rho, \mathbf{p}) - \rho R + (1-s)T]\},$$
$$0 \le s \le \rho \le 1. \tag{93}$$

For fixed R and T, the bounds differ by T for any s, ρ, and $\mathbf{p}$, so that they are both maximized by the same values of these parameters, as stated in (24) and (26).

Gallager's expurgation technique improves this bound at low rates. Define

$$q(\mathbf{x}_m, \mathbf{x}_{m'}, s) = \sum_{\mathbf{y}} P_m^{1-s} P_{m'}^s, \tag{94}$$

then from (87)

$$g_m(-s) = \sum_{m' \ne m} q(\mathbf{x}_m, \mathbf{x}_{m'}, s). \tag{95}$$

If we choose a code from a random ensemble as before, the probability that $g_m(-s)$ will equal or exceed some number B is

$$\Pr[g_m(-s) \ge B] = \overline{\phi_{\text{code}}(s)}, \tag{96}$$

where

$$\phi_{\text{code}}(s) = \begin{cases} 1, & g_m(-s) \ge B, \\ 0, & g_m(-s) < B. \end{cases} \tag{97}$$

We upper bound $\phi_{\text{code}}(s)$ by

$$\phi_{\text{code}}(s) \le \sum_{m' \ne m} \frac{q(\mathbf{x}_m, \mathbf{x}_{m'}, s)^{1/\rho}}{B^{1/\rho}}, \qquad \rho \ge 1. \tag{98}$$

Equation (98) is obvious for $\phi_{\text{code}}(s) = 0$; for $\rho = 1$ and $\phi_{\text{code}}(s) = 1$, (98) follows from (95) and (97); increasing ρ increases all terms in the sum less than 1, and if any term is greater than 1, (98) is true anyway. Thus

$$\Pr[g_m(-s) \ge B] \le B^{-1/\rho} \sum_{m' \ne m} \overline{(\sum_{\mathbf{y}} P_m^{1-s} P_{m'}^s)^{1/\rho}},$$
$$\rho \ge 1 \tag{99a}$$

substituting (94) and (98) into (96);

$$< B^{-1/\rho} \exp NR \overline{(\sum_{\mathbf{y}} P_m^{1-s} P_{m'}^s)^{1/\rho}} \tag{99b}$$

$\exp NR - 1$ equal terms;

$$= B^{-1/\rho} \exp NR[\sum_k \sum_{k'} p_k p_{k'} (\sum_j p_{jk}^{1-s} p_{jk'}^s)^{1/\rho}]^N \tag{99c}$$

using (92), and writing out the indicated average, a product of N identical averages. With $E_x(s, \rho, \mathbf{p})$ defined as in (25), we have

$$\Pr[g_m(-s) \ge B] < B^{-1/\rho} \exp\left\{-\frac{N}{\rho}[E_x(s, \rho, \mathbf{p}) - \rho R]\right\},$$
$$0 \le s \le 1, \qquad \rho \ge 1. \tag{100}$$

Let us then choose B to be

$$B_0 = 2^\rho \exp \{-N[E_x(s, \rho, \mathbf{p}) - \rho R]\}, \tag{101}$$

then

$$\Pr[g_m(-s) \ge B_0] < \tfrac{1}{2}. \tag{102}$$

If we then expurgate all code words $\mathbf{x}_m$ for which $g_m(-s) \ge B_0$, the average number of code words remaining in a code will be greater than $M' = M/2$, from (102), so there must be at least one code with M' code words and thus rate R', where

$$R' > \frac{1}{N} \ln M' = R - \frac{\ln 2}{N}. \tag{103}$$

For every word in this code

$$g_m(-s) < B_0$$
$$= \exp\left\{-N\left[E_x(s, \rho, \mathbf{p}) - \rho R - \frac{\rho \ln 2}{N}\right]\right\}, \tag{104}$$

using (101). Substituting back in (88), we find that we can achieve simultaneously

$$\Pr(E_1) < \exp \{-N[E_x(s, \rho, \mathbf{p}) - \rho R - sT - o(1)]\}$$

$$\Pr(E_2) < \exp \{-N[E_x(s, \rho, \mathbf{p}) - \rho R$$
$$+ (1-s)T - o(1)]\}, \qquad 0 \le s \le 1, \quad \rho \ge 1, \tag{105}$$

where $o(1)$ is a function which goes to zero as N goes to infinity. Again the bounds are jointly optimized by a single set of s, ρ, and $\mathbf{p}$, giving the bounds of (25) and (26). The $o(1)$ is ignored in the text as inconsequential.

Proof of Theorem 3 (Construction from Sphere-Packing Exponent): An alternative statement of Theorem 2 comes from solving for T in (24) and (26); then we can say that there exists a code of length N and rate R nats and a decoding scheme such that

$$\Pr(E_1) < \exp(-NE_1),$$

$$\Pr(E_2) < \exp[-NE_2(R, E_1)], \tag{106}$$

$$E_2(R, E_1) = \max_{0 \le s \le \rho \le 1, \mathbf{p}} \frac{1}{s} E_0(s, \rho, \mathbf{p}) - \frac{\rho}{s} R - \frac{1-s}{s} E_1,$$

where $E_0(s, \rho, \mathbf{p})$ is as in (24). It is now convenient to perform the change of variables

$$\rho_1 = \frac{\rho - s}{s}$$
$$\rho_2 = \frac{s}{1-s} \tag{107}$$

and to restrict ourselves to some particular value of $\mathbf{p}$; then

$$E_2(R, E_1) \ge \max_{\rho_1 \ge 0, 0 \le \rho_2 \le 1/\rho_1} \frac{1+\rho_2}{\rho_2} E_0(\rho_1, \rho_2, \mathbf{p}) - (1+\rho_1)R - \frac{1}{\rho_2} E_1, \tag{108}$$

$$E_0(\rho_1, \rho_2, \mathbf{p}) = -\ln \sum_i \left(\sum_k p_k p_{ik}^{1/(1+\rho_2)}\right) \left(\sum_{k'} p_{k'} p_{ik'}^{1/(1+\rho_1)}\right)^{\rho_2(1+\rho_1)/(1+\rho_2)},$$

where equality holds when we use the optimum $\mathbf{p}$. Now take some rate R' which satisfies $R(1/\rho_1, \mathbf{p}) \le R' \le R(0, \mathbf{p})$, and let

$$E_1 = E_{sp}(R', \mathbf{p}) = \max_{0 \le \rho \le 1/\rho_1} E_0(\rho, \mathbf{p}) - \rho R', \tag{109}$$

where the definition of the fixed-rate sphere-packing exponent is that of (30). From the convexity of $E_{sp}(R', \mathbf{p})$,[3] we have the inverse relationship

$$R' = \max_{0 \le \rho \le 1/\rho_1} \frac{E_0(\rho, \mathbf{p}) - E_1}{\rho}. \tag{110}$$

Now we bound $E_0(\rho_1, \rho_2, \mathbf{p})$ by Hölder's inequality:[12]

$$E_0(\rho_1, \rho_2, \mathbf{p}) \ge -\ln \left[\sum_i \left(\sum_k p_k p_{ik}^{1/(1+\rho_2)}\right)^{1+\rho_2}\right]^{1/(1+\rho_2)} \cdot \left[\sum_i \left(\sum_k p_k p_{ik}^{1/(1+\rho_1)}\right)^{1+\rho_1}\right]^{\rho_2/(1+\rho_2)}$$
$$= \frac{1}{1+\rho_2} E_0(\rho_2, \mathbf{p}) + \frac{\rho_2}{1+\rho_2} E_0(\rho_1, \mathbf{p}), \tag{111}$$

where $E_0(\rho, \mathbf{p})$ is defined by (18); equality holds if

$$q_i(x) = \sum_k p_k p_{ik}^x \tag{112}$$

is independent of j. Now substitution of (111) into (108) yields

$$E_2(R, E_1) \ge \max_{\rho_1 \ge 0, 0 \le \rho_2 \le 1/\rho_1} E_0(\rho_1, \mathbf{p}) - (1+\rho_1)R + \frac{E_0(\rho_2, \mathbf{p}) - E_1}{\rho_2}. \tag{113}$$

The maximizations can be carried out separately, and with the use of (109) and (110) become

$$E_2(R, E_1) \ge E_{sp}(R, \mathbf{p}) - R + R'. \tag{114}$$

This and (109) combine to give the statement of Theorem 3. Theorem 3(a), for totally symmetric channels, follows from tracing the cases of equality.

Acknowledgment

The author is indebted to R. G. Gallager for his intellectual leadership and invaluable criticism.

References

[1] C. E. Shannon and W. Weaver, *The Mathematical Theory of Communication.* Urbana, Ill.: University of Illinois Press, 1949.
[2] R. M. Fano, *Transmission of Information.* Cambridge, Mass.: M.I.T. Press, 1961.
[3] R. G. Gallager, "A simple derivation of the coding theorem and some applications," *IEEE Trans. Information Theory,* vol. IT-11, pp. 3–18, January 1965.
[4] C. E. Shannon, R. G. Gallager, and E. Berlekamp, "Lower bounds to error probability for coding on discrete memoryless channels," *Information Control,* vol. 10, pp. 65–103, 1967.
[5] C. E. Shannon, "The zero-error capacity of a noisy channel," *IRE Trans. Information Theory,* vol. IT-2, pp. 8–19, September 1956.
[6] G. D. Forney, *Concatenated Codes.* Cambridge, Mass.: M.I.T. Press, 1966.
[7] G. D. Forney, "On decoding BCH codes," *IEEE Trans. Information Theory,* vol. IT-11, pp. 549–558, October 1965.
[8] G. D. Forney, "Generalized minimum distance decoding," *IEEE Trans. Information Theory,* vol. IT-12, pp. 125–132, April 1966.
[9] J. E. Savage, "Sequential decoding—the computation problem," *Bell Sys. Tech. J.,* vol. 45, pp. 149–177, 1966.
[10] B. Reiffen, "A note on very noisy channels," *Information Control,* vol. 6, p. 126, 1963.
[11] P. M. Ebert, "Error bounds for parallel communication channels," M.I.T. Research Laboratory of Electronics, Cambridge, Mass., Tech. Rept. 448, 1966, Appendix C.
[12] G. H. Hardy, J. E. Littlewood, and G. Polya, *Inequalities.* Cambridge, England: Cambridge University Press, 1959.
[13] W. B. Davenport, Jr., and W. L. Root, *Random Signals and Noise.* New York: McGraw-Hill, 1958, pp. 322–324.

Error Bounds for the White Gaussian and Other Very Noisy Memoryless Channels With Generalized Decision Regions

ANDREW J. VITERBI, SENIOR MEMBER, IEEE

Abstract—For a class of generalized decision strategies, which afford the possibility of erasure or variable-size list decoding, asymptotically tight upper and lower error bounds are obtained for orthogonal signals in additive white Gaussian noise channels. Under the hypothesis that a unique signal set is asymptotically optimal for the entire class of strategies, these bounds are shown to hold for the optimal set in both the white Gaussian channel and the class of input-discrete very noisy memoryless channels.

I. Introduction

IN CLASSICAL decision theory with M hypotheses, the space of observations[1] is divided into M disjoint regions chosen so as to maximize some criterion of optimality. The term generalized decision regions will be used to describe regions which are not disjoint, or whose union does not cover the entire space of observations, or both. We shall concern ourselves with the two classes of strategies as described by the decision regions of Fig. 1:

A) The decision regions are disjoint, and their union does not cover the entire space of observations.

B) The decision regions are not disjoint. Their union may or may not cover the space.

Forney [1] first suggested such strategies for decoding the output of a discrete memoryless channel transmitting one of M equiprobable messages, and he attached the following interpretations to the two classes of strategies, assuming a finite-dimensional observation space. In Class A, if the observed (received) vector falls outside all of the M decision regions, no decision is made, and as a result the message is erased. If a feedback channel is available, transmission of the lack of decision to the transmitter provides the possibility of repeating the messages. Thus, Class A is referred to as an erasure or single-decision feedback strategy.

In Class B, if the observed vector is such that it lies in the intersection of two or more of the overlapping decision regions, a list of decisions is made. That is, rather than making a decision, a list of alternatives is generated corresponding to the regions in whose intersection the observed vector lies. Of course, if the union of the regions does not cover the space, occasionally the list may be empty. This variable-size list decision or decoding is to be distinguished from conventional fixed-size list decoding.

The foregoing description was given quantitative substance by Forney [1], who obtained general expressions for the probabilities of error and of erasure in Class A strategies and for the probability of error and the average list

Manuscript received April 22, 1968; revised August 29, 1968. This research was sponsored in part by the U. S. Army Electronic Command, Grant DAA B07-67-C0540 and by NASA Grant NsG-237-62.
The author is with the Department of Engineering, University of California, Los Angeles, Calif. 90024.

[1] In information and coding theory we refer to the observations as components of the received signals or as the symbols of the received sequences.

Reprinted from *IEEE Trans. Inform. Theory*, vol. IT-15, pp. 279-287, Mar. 1969.

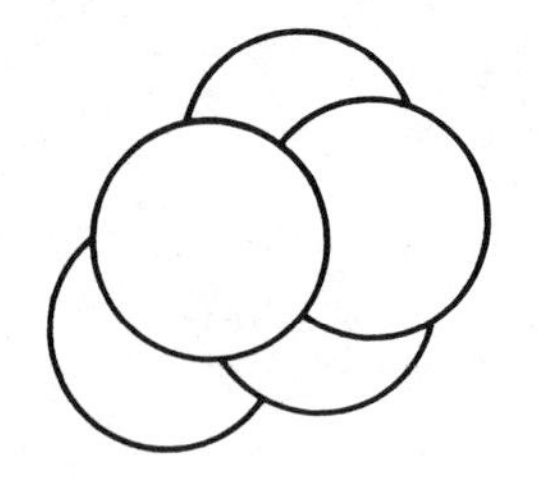

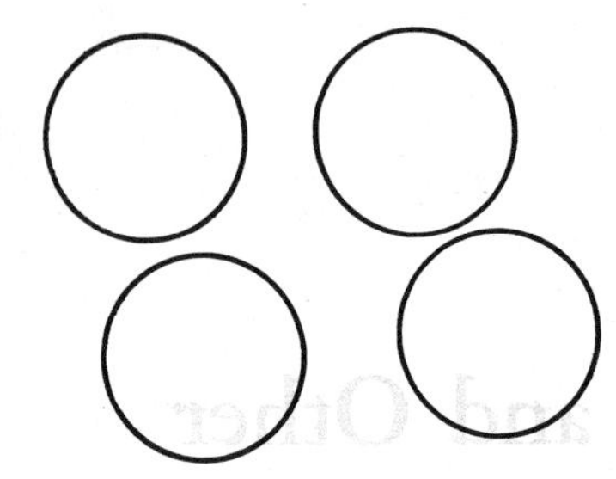

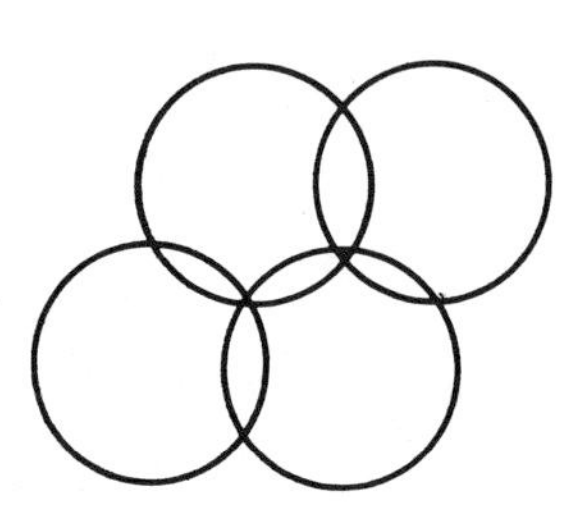

Fig. 1. Generalized decision regions.

size in Class B strategies. Assuming equiprobable messages and an N-dimensional continuous[2] observation space, letting $\mathbf{y}$ denote the observation vector, H_m denote the hypothesis that message m was sent, R_m denote the decision region for H_m, with $\overline{H_m}$ and $\overline{R_m}$ their respective complements, and $p(\mathbf{y} \mid H_m)$ denote the probability density function of $\mathbf{y}$ given H_m, the probabilities of error of the first and second kind are defined to be

$$P_a = \frac{1}{M} \sum_{m=1}^{M} P_a(H_m); \qquad P_b = \frac{1}{M} \sum_{m=1}^{M} P_b(H_m) \tag{1}$$

where

$$P_a(H_m) = \int_{\overline{R_m}} p(\mathbf{y} \mid H_m) \, dy$$

$$P_b(H_m) = \int_{R_m} \sum_{m' \neq m} p(\mathbf{y} \mid H_{m'}) \, dy.$$

The interpretation of these probabilities is different for Class A and Class B strategies. It is easily seen to be as follows:

$$\text{A)} \quad P_a = P_E + P_x \tag{2}$$

$$P_b = P_E \tag{3}$$

where P_E is the probability of error (i.e., of making an incorrect decision) and P_x is the probability of erasure.

$$\text{B)} \quad P_a = P_E + P_x \tag{4}$$

$$P_b = \mathbf{E}(L) \tag{5}$$

where P_E is the probability that the correct decision is not on the list and the list is not empty, P_x is the probability of erasure (the list is empty), and $\mathbf{E}(L)$ is the expected number of incorrect decisions on the list. Note that the expected total list size is $\mathbf{E}(L) + 1 - P_E$.

From this point Forney [1] proceeded to obtain the optimum decision rule for any strategy in either class according to a Neyman–Pearson criterion which minimizes P_b for a given fixed P_a, or conversely. He then determined upper bounds on P_a and P_b for memoryless channels using optimum rule in each case.

We shall use a more direct approach that leads not only to the same upper bounds, but also to expressions for lower bounds. In the special case of the additive Gaussian noise channel we shall obtain simple explicit upper and lower bounds when an orthogonal signal set is used. These upper and lower bounds, which decrease exponentially with signal transmission time, have asymptotically equal exponents; consequently, the exact asymptotic performance is obtained. Arguments will then be given for the asymptotic optimality of orthogonal signals, thus implying that the exact asymptotic performance is the best achievable for the white Gaussian channel.

Finally, we shall extend these methods to treat the class of input-discrete very noisy memoryless channels and obtain upper and lower bounds which coincide everywhere with the upper bounds obtained by Forney [1].

II. General Upper and Lower Bounds for Memoryless Channels

Both upper and lower bounds follow from a theorem of Shannon *et al.* [2], which reworded in our terminology is as follows.

Theorem

Let $p(\mathbf{y})$ and $q(\mathbf{y})$ be arbitrary density functions[3] defined on the observation space Y and let Y_a and Y_b be disjoint regions in Y. Then provided for at least one $\mathbf{y} \in Y$, $p(\mathbf{y})\, q(\mathbf{y}) \neq 0$, at least one of the following inequalities must hold for each s, $0 \leq s \leq 1$:

$$P_a \triangleq \int_{\overline{Y_a}} p(\mathbf{y}) \, dy > \tfrac{1}{4} \exp\left[\mu(s) - s\mu'(s) - s\sqrt{2\mu''(s)}\right]$$

$$P_b \triangleq \int_{\overline{Y_b}} q(\mathbf{y}) \, dy > \tfrac{1}{4} \exp\left[\mu(s) + (1 - s)\mu (s) - (1 - s)\sqrt{2\mu''(s)}\right] \tag{6}$$

[2] The extension to discrete observation spaces is obvious; integrals are replaced by summations and density functions by distributions.

[3] In the original theorem these are taken as probability density functions. The proof is the same in general, the only difference being that restriction to probability density leads to $\mu(s)$ being nonpositive on the unit interval, whereas here $\mu(s)$ may be positive.

where

$$\mu(s) \triangleq \ln \int_Y p(\mathbf{y})^{1-s} q(\mathbf{y})^s \, d\mathbf{y} \tag{7}$$

is a convex function in the interval $0 \leq s \leq 1$. Furthermore, if

$$Y_a = \{\mathbf{y} : \ln [p(\mathbf{y})/q(\mathbf{y})] > -\mu'(s)\} \tag{8}$$

and $Y_b = \bar{Y}_a$, then we also have

$$P_a \leq \exp [\mu(s) - s\mu'(s)] \tag{9}$$

and

$$P_b \leq \exp [\mu(s) + (1 - s)\mu'(s)].$$

In the present application we take Y as Euclidean N space, $Y_a = R_m$, $Y_b = \bar{R}_m$, $p(\mathbf{y}) = p(\mathbf{y} \mid H_m)$ and $q(\mathbf{y}) = \sum_{m' \neq m} p(\mathbf{y} \mid H_{m'})$. With these substitutions P_a and P_b of (6) and (9) become the same as the probabilities $P_a(H_m)$ and $P_b(H_m)$ of (1). Then (7) becomes

$$\mu_m(s) = \ln \int_Y p(\mathbf{y} \mid H_m)^{1-s} [\sum_{m' \neq m} p(\mathbf{y} \mid H_{m'})]^s \, d\mathbf{y} \tag{10}$$

and consequently, using (1) we have

$$\frac{1}{4M} \sum_{m=1}^{M} \exp [\mu_m(s) - s\mu_m'(s) - s\sqrt{2\mu_m''(s)}] < P_a$$

$$\leq \frac{1}{M} \sum_{m=1}^{M} \exp [\mu_m(s) - s\mu_m'(s)] \tag{11}$$

$$\frac{1}{4M} \sum_{m=1}^{M} \exp [\mu_m(s) + (1 - s)\mu_m'(s) - (1 - s)\sqrt{2\mu_m''(s)}] < P_b$$

$$\leq \frac{1}{M} \sum_{m=1}^{M} \exp [\mu_m(s) + (1 - s)\mu_m'(s)].$$

The lower bounds are to be taken as alternatives; that is, at least one of the two must hold. The upper bounds correspond to those obtained in [1]. [Inequalities (A.10) correspond to the upper bounds of (11) with $\exp \mu_m(s) \triangleq g_m(-s)$ and $-\mu_m'(s)$ taken as the threshold in the definition of R_m.] The upper and lower bound exponents are asymptotically equal for large transmission times T since, as will be shown, $\mu_m(s)$ and its first two derivatives are proportional to T.

Thus, the main problem is the determination of $\mu_m(s)$ for each m and for the particular signal set in question. In [1], upper bounds on $\mu_m(s)$ are obtained for general memoryless channels by random coding arguments. In the next two sections we shall limit our attention to orthogonal signal sets transmitted over the additive white Gaussian channel and obtain upper and lower bounds on $\mu_m(s)$ that are asymptotically equal.

III. Upper Bound on $\mu(s)$ for Orthogonal Signals in White Gaussian Noise

It is well known [3], [4] that the normalized observables required for optimum decision of M orthogonal signals $s^{(m)}(t)$ each of energy $\mathcal{E}$ in white Gaussian noise of one-sided density N_0 are

$$y_m = \frac{1}{\sqrt{\mathcal{E}N_0}} \int_0^T y(t) s^{(m)}(t) \, dt \qquad m = 1, 2, \cdots, M$$

and the likelihood functions are

$$p(\mathbf{y} \mid H_m) = (2\pi)^{-M/2} \exp [(y_m - \sqrt{\mathcal{E}/N_0})^2/2] \prod_{m' \neq m}^{M} \exp (-y_{m'}^2/2)$$

$$= (2\pi)^{-M/2} \exp \left(-\sum_{k=1}^{M} y_k^2/2\right) \exp (-CT + \sqrt{2CT}\, y_{m'}) \tag{12}$$

where $C \triangleq \mathcal{E}/TN_0$ is the capacity of the white Gaussian channel. It follows from (8) that the decision regions are

$$R_m = \{\mathbf{y} : \ln [\exp (\sqrt{2CT}\, y_m)/ \sum_{m' \neq m} \exp (\sqrt{2CT}\, y_{m'})] > -\mu_m'(s)\}. \tag{13}$$

It also follows from (10) and (12) that

$$\mu_m(s) = \ln (2\pi)^{-M/2} \int_{-\infty}^{\infty} \cdots \int_{-\infty}^{\infty} \exp \left(-\sum_{k=1}^{M} y_k^2/2\right) \cdot \exp (-CT) \exp (\sqrt{2CT}\,(1 - s)y_m) \cdot [\sum_{m' \neq m} \exp (\sqrt{2CT}\, y_{m'})]^s \, d\mathbf{y}$$

$$= \ln \{\exp (-CT)\mathbf{E}[\exp (\sqrt{2CT}\,(1 - s)y_m)] \cdot \mathbf{E}[\sum_{m' \neq m} \exp (\sqrt{2CT}\, y_{m'})]^s\}$$

where the expectations are with respect to the M independent, zero-mean unit variance Gaussian random variables. The first expectation is immediately obtained to be $\exp [CT(1 - s)^2]$ and consequently

$$\mu_m(s) = -CT(2s - s^2) + \ln \mathbf{E}[\sum_{m' \neq m} \exp (\sqrt{2CT}\, y_{m'})]^s. \tag{14}$$

It is the second expression that is difficult to evaluate, but we shall obtain upper and lower bounds that are asymptotically equal with large T.

For the upper bound we use essentially the same approach as for the general case [1]. First of all, applying Jensen's inequality

$$(\sum_i a_i)^\nu \leq \sum_i a_i^\nu; \qquad a_i > 0 \quad \nu \leq 1$$

we have

$$\mathbf{E}\{[\sum_{m' \neq m} \exp (\sqrt{2CT}\, y_{m'})]^{s/\rho}\}^\rho \leq \mathbf{E}\{\sum_{m' \neq m} \exp [\sqrt{2CT}\,(s/\rho)y_{m'}]\}^\rho, \qquad s \leq \rho. \tag{15}$$

Next, letting $\rho \leq 1$ we have, using the convexity of the function $f(z) = z^{\rho}$,

$$\mathbf{E}\{\sum_{m' \neq m} \exp [\sqrt{2CT}\,(s/\rho)y_{m'}]\}^{\rho}$$

$$\leq \{\mathbf{E} \sum_{m' \neq m} \exp [\sqrt{2CT}\,(s/\rho)y_{m'}]\}^{\rho}$$

$$= (M-1)^{\rho} \exp (CTs^2/\rho)$$

$$< \exp [CT(s^2/\rho) + \rho RT], \quad s \leq \rho \leq 1 \tag{16}$$

where $R \triangleq \ln M/T$ nats per second is the data rate.[4] Thus, combining (14)–(16), we have

$$\mu_m(s) < -CT[2s - s^2(1 + \rho/\rho)] + \rho RT, \; s \leq \rho \leq 1. \tag{17}$$

Minimizing the right side of (17) to obtain the tightest bound, we have

$$\rho = \begin{cases} s\sqrt{C/R} & 0 \leq s \leq \sqrt{R/C} \\ 1 & \sqrt{R/C} \leq s \leq 1. \end{cases}$$

Thus,

$$\mu_m(s) = \mu(s)$$

$$< \begin{cases} -T[-Cs^2 + 2s(C - \sqrt{RC})] & 0 \leq s \leq \sqrt{R/C} \\ -T[2C(s - s^2) - R] & \sqrt{R/C} \leq s \leq 1. \end{cases} \tag{18}$$

This bound on $\mu(s)/CT$ is plotted in Fig. 2 for $R/C = \frac{1}{8}, \frac{1}{4}, \frac{1}{2}$, and $\frac{3}{4}$.

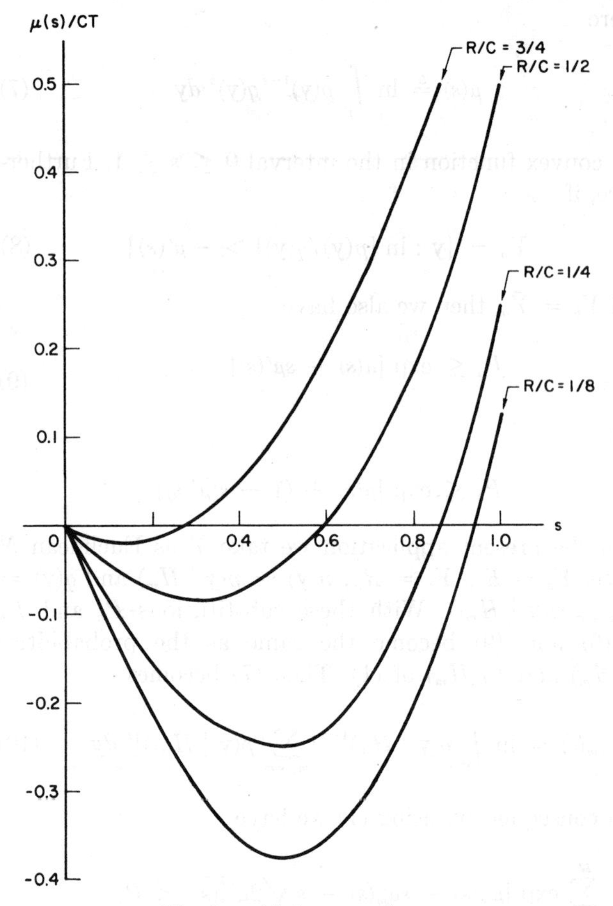

Fig. 2. $\mu(s)$ for several rates.

IV. Lower Bound on $\mu(s)$ for Othogonal Signals in White Gaussian Noise

We begin by observing that

$$\mathbf{E}[\sum_{m' \neq m} \exp (\sqrt{2CT}\, y_{m'})]^s > \mathbf{E}[\max_{m' \neq m} \exp (\sqrt{2CT}\, y_{m'})]^s$$

$$= \mathbf{E}[\exp (\sqrt{2CT}\, s \max_{m' \neq m} y_{m'})]. \tag{19}$$

Since the variables are independent zero-mean unit variance Gaussian random variables, letting $\lambda \triangleq CTs^2$, we obtain for the right side of (19),

$$\mathbf{E} \exp \{\sqrt{2\lambda} \max_{m' \neq m} y_{m'}\}$$

$$= \int_{-\infty}^{\infty} \exp (\sqrt{2\lambda}\, x) \frac{d}{dx}\left[\int_{-\infty}^{x} e^{-y^2/2}/\sqrt{2\pi}\right]^{M-1} dx$$

$$= (M-1)e^{\lambda}(2\pi)^{(M-1)/2} \int_{-\infty}^{\infty} \exp [-(x - \sqrt{2\lambda})^2/2] \cdot \left[\int_{-\infty}^{x} e^{-y^2/2}\right]^{M-2} dx$$

$$= (M-1)e^{\lambda}P_C(M-1, \lambda) \tag{20}$$

where we have recognized [5] that the integral is just the probability of correct decision among $M - 1$ orthogonal signals in white Gaussian noise with a maximum-likelihood decision rule and an energy-to-noise density ratio equal to λ. It is also well known [6] that this probability is lower bounded by

$$P_C = 1 - P_E > 1 - \exp [-T\epsilon(\tilde{R})] \tag{21}$$

where $\epsilon(\tilde{R}) > 0$ for $\tilde{R} \triangleq [\ln (M-1)]/T < \lambda/T = Cs^2$. Thus, combining (19)–(21) and recognizing that $\tilde{R} = R - O(T^{-1})$ and $\lambda/T = Cs^2$, we obtain

$$\mathbf{E}[\sum_{m' \neq m} \exp (\sqrt{2CT}\, y_{m'})]^s$$

$$> \exp \{T[\tilde{R} + Cs^2 + \ln (1 - \exp [-T\epsilon(\tilde{R})])/T]\}$$

$$= \exp \{T[R + Cs^2 - O(T^{-1})]\} \quad \sqrt{R/C} \leq s \leq 1. \tag{22}$$

To extend the bound to lower values of s, we proceed by using a technique of Cramér [7]. We begin by defining the transformation

$$1 - \frac{\xi}{M-1} = F(x) \triangleq \int_{-\infty}^{x} e^{-y^2/2}\, dy/\sqrt{2\pi} \tag{23}$$

which has the inverse [7]

$$x = F^{-1}[1 - \xi/(M-1)]$$

$$= \sqrt{2 \ln (M-1)} - \frac{\ln 4\pi \ln (M-1)}{2\sqrt{2 \ln (M-1)}} - \frac{\ln \xi}{\sqrt{2 \ln (M-1)}} + O\left(\frac{1}{\ln (M-1)}\right). \tag{24}$$

[4] Where the nat is the natural-logarithmic equivalent of the bit.

Substituting (23) and (24) in (20) and letting $\tilde{R} = \ln (M-1)/T$ and $\lambda = CTs^2$ we obtain

$$\mathbf{E} \exp (\sqrt{2\lambda} \max_{m' \neq m} y_{m'})$$
$$= (M-1) \int_{-\infty}^{\infty} \exp (\sqrt{2\lambda}\, x)[F(x)]^{M-2}\, dF(x)$$
$$= \int_0^{M-1} \exp \{\sqrt{2\lambda}\, F^{-1}[1 - \xi/(M-1)]\} \cdot [1 - \xi/(M-1)]^{M-2}\, d\xi$$
$$= \exp \left\{ sT\left[2\sqrt{\tilde{R}C} - \frac{\ln 4\pi \tilde{R}T}{2\sqrt{2(\tilde{R}/C)T}} + O(T^{-3/2}) \right]\right\} \cdot \int_0^{M-1} \xi^{-s\sqrt{C/\tilde{R}}}[1 - \xi/(M-1)]^{M-2}\, d\xi. \quad (25)$$

This last integral is bounded from below by $[e(1 - s\sqrt{c/\tilde{R}})]^{-1}$ for $s < \sqrt{\tilde{R}/C}$. Therefore, combining (19) and (25) we have

$$\mathbf{E}[\sum_{m' \neq m} \exp \sqrt{2CT}\, y_{m'}]^s > \exp \{T[2s\sqrt{RC} - O(\ln T/T)]\}, \quad 0 \leq s < \sqrt{R/C}. \quad (26)$$

Thus, combining (14), (22), and (26) we obtain

$$\mu_m(s) = \mu(s) > \begin{cases} -T[-Cs^2 + 2s(C - \sqrt{RC}) + O(\ln T/T)], & 0 \leq s \leq \sqrt{R/C} \\ -T[2C(s - s^2) - R + O(T^{-1})], & \sqrt{R/C} < s \leq 1. \end{cases} \quad (27)$$

Except for the asymptotically negligible terms the lower bound (27) is identical to the upper bound (18). Using these results in the general error bound expressions (11) will lead to asymptotically exact expressions.

V. Probability Bounds for the Additive White Gaussian Noise Channel

As was shown in the two preceding sections

$$\mu(s) \lesssim \begin{cases} -T[-Cs^2 + 2s(C - \sqrt{RC})] & 0 \leq s \leq \sqrt{R/C} \\ -T[2C(s - s^2) - R] & \sqrt{R/C} < s \leq 1 \end{cases} \quad (28)$$

where $\lesssim$ denotes "less than but asymptotically equal to." Thus,

$$\mu'(s) \sim \begin{cases} -T[-2Cs + 2(C - \sqrt{RC})] & s \leq \sqrt{R/C} \\ -T[2C(1 - 2s)] & s > \sqrt{R/C} \end{cases} \quad (29)$$

and

$$\mu''(s) \sim \begin{cases} 2CT & s \leq \sqrt{R/C} \\ 4CT & s > \sqrt{R/C}. \end{cases} \quad (30)$$

From (29) and (30) it is clear that the minimum of $\mu_m(s)$ occurs at $s_0 = 1 - \sqrt{R/C}$ if $s_0 \leq \sqrt{R/C}$ or $R/C \geq \frac{1}{4}$ and at $s_0 = \frac{1}{2}$ if $s_0 > \sqrt{R/C}$ or $R/C < \frac{1}{4}$. Thus,

$$\min_s \mu(s) \lesssim \begin{cases} -T(\sqrt{C} - \sqrt{R})^2 & \text{for } \frac{1}{4} \leq R/C \leq 1 \\ -T(C/2 - R) & \text{for } 0 \leq R/C < \frac{1}{4} \end{cases} \quad (31)$$

Note also that since $\mu'(s_0) = 0$, the decision regions (13) are in this case

$$R_m = \left\{ \mathbf{y} : \ln \left[\frac{\exp (\sqrt{2CT}\, y_m)}{\sum_{m' \neq m} \exp (\sqrt{2CT}\, y_{m'})} \right] > 0 \right\}. \quad (32)$$

This must correspond to disjoint decision regions since this threshold can be exceeded by only one y_m. Furthermore, for $s < s_0$, $-\mu'(s) > 0$, so the threshold of (13) is positive, and by the same argument the regions must be disjoint. On the other hand, for $s > s_0$, $-\mu'(s) < 0$ and as a result the threshold may be exceeded by two or more y_m, resulting in overlapping decision regions. Consequently, the region $0 \leq s \leq s_0$ corresponds to all optimal strategies in Class A, while $s_0 < s \leq 1$ corresponds to all strategies in Class B.

We note from (2), (3), and (11) that if we choose s_0 to minimize $\mu(s)$ then

$$P_E \leq P_E + P_x \leq \exp \min_s [\mu(s)]. \quad (33)$$

Thus we have from (33) and (31)

$$P_E + P_x < \exp [-TE(R)]$$

where

$$E(R) = \begin{cases} C/2 - R & 0 \leq R/C < \frac{1}{4} \\ (\sqrt{C} - \sqrt{R})^2 & \frac{1}{4} \leq R/C < 1 \end{cases} \quad (34)$$

Furthermore, we have from the lower bound part of the theorem if $P_E = \exp [-TE(R)]$, then

$$P_E + P_x > \exp \{-T[E(R) + O(1/\sqrt{T})]\}. \quad (35)$$

Thus, if we were to treat an erasure as an error (for example, by arbitrarily adjoining the uncovered observation space to any one decision region), (34) and (35) could be regarded as bounds on the error probability for conventional maximum-likelihood decision. These in fact coincide with the well-known result for this case [6] and represent an alternative derivation thereof.

We have already noted that all strategies with $s < s_0$ are in Class A, corresponding to error-erasure situations. As is shown in Fig. 3, once the value of s^* is chosen, the two exponents of $P_a = P_E + P_x$ and of $P_b = P_E$ are obtained from the intercepts of the line tangent to $\mu(s)$ at s^* with the vertical lines $s = 0$ and $s = 1$. Thus, by making $P_E + P_x$ somewhat greater than (34) we can ensure a P_E which is somewhat smaller. In general we may state that if

$$P_E + P_x \leq \exp (-T\epsilon),$$

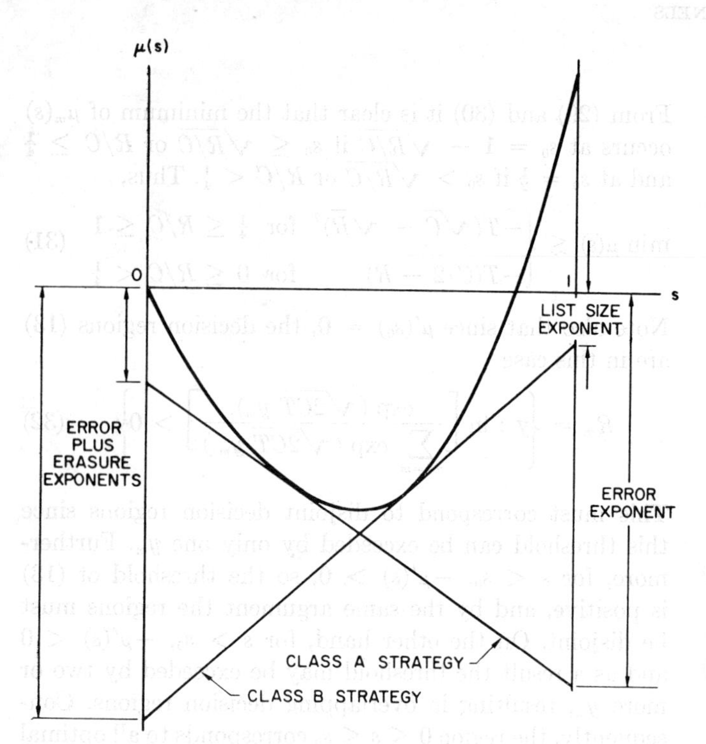

Fig. 3. Determination of erasure and error exponents or expected list size and error exponents from $\mu(s)$.

then

$$\exp \{-T[\mu(s^*) + (1 - s^*)\mu'(s^*) + O(T^{-1/2})]\}$$

$$< P_E < \exp \{-T[\mu(s^*) + (1 - s^*)\mu'(s^*)]\} \quad (36)$$

where s^* is such that

$$\epsilon = \mu(s^*) - s^*\mu'(s^*). \quad (37)$$

In particular, as $\epsilon \to 0$, $s^* \to 0$. Thus, since $\mu(0) = 0$, and $\mu'(0) \sim 2CT(1 - \sqrt{R/C})$,

$$\exp -T[E_f(R) + O(1/\sqrt{T})] < \lim_{\epsilon \to 0} P_E < \exp -TE_f(R)$$

where

$$E_f(R) = \mu'(0)/T = 2(C - \sqrt{RC})$$
$$= C - R + (\sqrt{C} - \sqrt{R})^2. \quad (38)$$

$E_f(R)$ has been termed the feedback exponent [1] because it corresponds to the limiting exponent for single-decision feedback. Of even greater practical interest is the facility with which one can obtain both the erasure exponent and the error exponent from the graphical construction of Fig. 3.

On the other hand, for $s > s_0$ we have overlapping decision regions and consequently variable-size list decoding. For these strategies, we have from (4) and (5), $P_a = P_E + P_x$ and $P_b = \mathbf{E}(L)$, the expected number of incorrect decisions on the list. Now proceeding as for the error-erasure case (see Fig. 3), we have that if

$$\mathbf{E}(L) \leq \exp(-T\epsilon)$$

then

$$\exp \{-T[\mu(s^*) - s^*\mu'(s^*) + O(1/\sqrt{T})]\}$$

$$< P_E + P_x < \exp \{-T[\mu(s^*) - s^*\mu'(s^*)]\} \quad (39)$$

where s^* is such that

$$\epsilon = \mu(s^*) + (1 - s^*)\mu'(s^*). \quad (40)$$

As we let $\epsilon \to 0$ in this case, the solution of (40) becomes

$$s^* = 1 - \sqrt{R/2C}$$

when

$$s^* > \sqrt{R/C} \quad \text{or} \quad R/C < 6 - 4\sqrt{2} \quad (41)$$

and

$$s^* = 1 - \sqrt{2RC - 1}$$

when

$$s^* < \sqrt{R/C} \quad \text{or} \quad R/C > 6 - 4\sqrt{2}. \quad (42)$$

Thus, inserting (41) and (42) in (39) we have

$$\exp \{-T[E_l(R) + O(1/\sqrt{T})]\}$$

$$< \lim_{\epsilon \to 0} (P_E + P_x) < \exp[-TE_l(R)]$$

where

$$E_l(R) = \begin{cases} 2(C - \sqrt{2RC}) & R \leq (6 - 4\sqrt{2})C \\ 2\sqrt{RC} - C\sqrt{2\sqrt{R/C} - 1} & R > (6 - 4\sqrt{2})C. \end{cases} \quad (43)$$

$E_l(R)$ has been termed the list exponent [1]. Again, as can be seen in Fig. 3, the improvement in error (and erasure) probability and the corresponding cost in expected list size can be readily determined graphically.

The results (38) and (43) are identical to the upper-bound exponents for single-decision feedback and list decoding obtained by Forney [1] for the class of very noisy channels. That this coincides with the upper bound for orthogonal signals in white Gaussian noise is hardly surprising. On the other hand, we have shown that for this channel and signal set these are also exponentially tight lower bounds.

We still have not removed the limitation to orthogonal signals. We can, however, make the following statement regarding the optimality of such signals: If for all optimal generalized decision strategies there exists a unique asymptotically optimum signal set, it must be orthogonal. For otherwise, if there were a uniformly better signal set, this would imply that at $s = s_0$, corresponding to the minimum of $\mu(s)$ for orthogonal signals, some other signal set would be better that would lead to a $P_E + P_x$ lower than the P_E using maximum-likelihood decision regions. Thus, treating erasures as errors we should have found a signal set asymptotically better than the orthogonal one using maximum-likelihood decision and thus have a contradiction to the well-known asymptotic optimality of orthogonal signals in the conventional case [8]. Of course, this does not exclude the possibility that there is no unique asymptotically optimum signal set for all generalized decision strategies, but intuition supported by geometric reasoning seems to deny this possibility.

VI. Upper and Lower Bounds for Input-Discrete Very Noisy Memoryless Channels

We now proceed to generalize the results of the preceding section to the wider class of very noisy memoryless channels. We must return to the general bounds (11) and obtain upper and lower bounds on $\mu_m(s)$ of (10). The upper bound can be obtained for any memoryless channel by random coding arguments. These are essentially contained in the Appendix to [1] and then specialized to very noisy channels, resulting in expressions identical to (28) (with N, the dimensionality of the observation vector, replacing T). We shall therefore consider only the lower bound.

We begin by generalizing the approach in (19) to obtain from (10)

$$\mu_m(s) = \ln \int_{Y^N} p(\mathbf{y} \mid H_m)^{1-s} \sum_{m' \neq m} p(\mathbf{y} \mid H'_m)^s \, d\mathbf{y}$$

$$> \ln \int_{Y^N} p(\mathbf{y} \mid H'_m)^{1-s} \max_{m' \neq m} p(\mathbf{y} \mid H'_m)^s \, d\mathbf{y}, \qquad 0 \leq s \leq 1. \tag{44}$$

Now for the class of very noisy memoryless channels we have [9]

$$p(\mathbf{y} \mid H_m) = \prod_{n=1}^{N} p(y_n \mid x_n^{(m)})$$

where $x_n^{(m)}$ is the nth component of the mth codeword,

$$p(y_n \mid x_n^{(m)}) = p(y_n)[1 + \epsilon(x_n^{(m)}, y_n)] \tag{45}$$

$p(y_n)$ is the a priori channel output distribution,[5] and

$$|\epsilon(x_n^{(m)}, y_n)| \ll 1 \qquad \text{for all } n \text{ and } m \tag{46}$$

and where we assume that we may neglect terms higher than quadratic in ϵ in any series expansions. Thus,

$$\mu_m(s) > \ln \int_{Y^N} \prod_{n=1}^{N} p(y_n) \cdot \exp\left\{(1-s) \sum_{n=1}^{N} \ln [1 + \epsilon(x_n^{(m)}, y_n)]\right\} \cdot \max_{m' \neq m} \exp\left\{s \sum_{n=1}^{N} \ln [1 + \epsilon(x_n^{(m')}, y_n)]\right\} d\mathbf{y}.$$

Now making use of condition (46) to discard terms above quadratic in the series expansion of $\ln(1 + \epsilon)$ we have

$$\mu_m(s) > \ln \mathbf{E}_y \exp\left\{\left[(1-s) \sum_{n=1}^{N} \epsilon(x_n^{(m)}, y_n) - \epsilon^2(x_n^{(m)}, y_n)/2\right] \cdot \max_{m' \neq m} \exp\left[s \sum_{n=1}^{N} \epsilon(x_n^{(m')}, y_n) - \epsilon^2(x_n^{(m')}, y_n)/2\right]\right\} \tag{47}$$

where $\mathbf{y}$ is a set of independent random variables with measure $\prod_n p(y_n)$. To proceed, we specialize first to fixed-composition codes; that is, to codes in which each word has the same number of each symbol of a finite-input-alphabet. We shall subsequently remove the fixed-composition constraint, but not the finite-input-alphabet conditions; thus, all our results will be for input-discrete channels. For fixed-composition codes we define the relative frequencies of symbol x as

$$q(x) \triangleq (\text{number of occurrences of } x \text{ in any codeword})/N. \tag{48}$$

Now letting

$$\alpha_m \triangleq \sum_{n=1}^{N} \epsilon(x_n^{(m)}, y_n)$$
$$\beta_m \triangleq \sum_{n=1}^{N} \epsilon^2(x_n^{(m)}, y_n) \tag{49}$$

we have, using the conditions of fixed composition codes and memoryless channels,

$$\mathbf{E}_y(\alpha_m) = \sum_{n=1}^{N} \int_Y p(y_n)\epsilon(x_n^{(m)}, y_n) \, dy_n = N \sum_X q(x) \int_Y p(y)\epsilon(x, y) \, dy = 0 \tag{50}$$

since

$$0 = \int_Y p(y \mid x) \, dy - 1 = \int_Y p(y)[1 + \epsilon(x, y)] \, dy - 1 = \int_Y p(y)\epsilon(x, y) \, dy.$$

Also

$$\text{var}_y(\alpha_m) = \mathbf{E}_y(\beta_m) = \sum_{n=1}^{N} \int_Y p(y_n)\epsilon^2(x_n^{(m)}, y_n) \, dy_n = N \sum_X q(x) \int_Y p(y)\epsilon^2(x, y) \, dy = 2NI(X, Y) \tag{51}$$

where we have used the independence of the random variables $\epsilon(x_n^{(m)}, y_n)$, $n = 1, 2, \cdots, N$ under the metric of $\mathbf{y}$, as well as the fact that for very noisy channels with input distribution (or relative frequency) $q(x)$,

$$I(X, Y) \triangleq \sum_X q(x) \int_Y p(y \mid x) \ln\left[\frac{p(y \mid x)}{p(y)}\right] dy = \sum_X q(x) \int_Y p(y)[1 + \epsilon(x, y)] \ln [1 + \epsilon(x, y)] \, dy \approx \sum_X q(x) \int_Y p(y)\epsilon^2(x, y)/2$$

where we have neglected terms higher than quadratic in the expansion of $\ln(1 + \epsilon)$. Furthermore

$$\text{var}_y(\beta_m) = \sum_{n=1}^{N} \int_Y p(y_n)\epsilon^4(x_n^{(m)}, y_n) \, dy_n \approx 0 \tag{52}$$

since we neglect powers of ϵ greater than quadratic. Thus we may treat β_m as a constant, and assuming the best composition $q(x)$ which maximizes $I(X, Y)$, we have

$$\beta_m = 2N \max_{q(x)} I(X, Y) = 2NC \tag{53}$$

so that for a code with this composition

$$\mu_m(s) > -NC + \ln \mathbf{E}_y\{\exp [(1-s)\sqrt{2NC}\, \zeta_m] \cdot \max_{m' \neq m} \exp (s\sqrt{2NC}\, \zeta'_m)\} \tag{54}$$

[5] Actually, in the original definition of very noisy channels, $p(y_n)$ need not be the a priori output distribution. With this added condition, however, the results are obtained more directly.

where

$$\zeta_m = \alpha_m/\sqrt{2NC} \tag{55}$$

and we have from (50) and (51) that $\mathbf{E}(\zeta_m) = 0$, var $(\zeta_m) = 1$. Furthermore, from the central limit theorem it follows that ζ_m is asymptotically a zero-mean unit-variance Gaussian variable, since the third moment of $\epsilon(x_n^{(m)}, y_n)$ is bounded for all n.

We now assume that there exists a code for which the variables α_m (or ζ_m) are independent. As we shall show, if there exists a unique code that is asymptotically optimum for all decision strategies under consideration, its asymptotic performance will be no better than that code which results in independent α_m.

Proceeding on this basis we have that for the best fixed composition code

$$\mu_m(s) > -NC + \ln \mathbf{E}_y\{\exp[(1-s)\alpha_m]\} \qquad 0 \le s \le 1$$
$$+ \ln \mathbf{E}_y\{\max_{m' \ne m} \exp(s\,\alpha_{m'})\} \tag{56}$$

where $\{\alpha_m\}$ are independent Gaussian variables. The first expectation is readily obtained for very noisy channels using (50) to (53) for

$$\mathbf{E}_y \exp[(1-s)\alpha_m]$$
$$= \mathbf{E}_y\left[1 + (1-s)\alpha_m + \frac{(1-s)^2}{2}\alpha_m^2 + \cdots\right]$$
$$= 1 + (1-s)^2 NC + \frac{(1-s)^4}{2!}(NC)^2 + \frac{(1-s)^6}{3!}(NC)^3 + \cdots$$
$$= \exp[(1-s)^2 NC]. \tag{57}$$

For the second expectation we use a precise form of the central limit theorem for ζ_m as defined in (55). From Loève [10], for

$$\zeta_m = \frac{1}{\sqrt{2NC}} \sum_{n=1}^{N} \epsilon(x_n^{(m)}, y_n)$$

the distribution function $F_\zeta(x)$ differs from the normalized Gaussian distribution function

$$G(x) \triangleq \int_{-\infty}^{x} e^{-u^2/2}\, du/\sqrt{2\pi}$$

by at most

$$|F_\zeta(x) - G(x)| \le \frac{\theta}{(2NC)^{3/2}} \sum_{n=1}^{N} \mathbf{E}\, |\epsilon(x_n^{(m)}, y_n)|^3 \approx 0 \tag{58}$$

since we neglect terms above quadratic in ϵ.

Thus, applying the same arguments for the independent asymptotically Gaussian variables $\alpha_{m'} = \sqrt{2CN}\,\zeta_{m'}$ as were used in Section IV for the variables $\sqrt{2CT}\,y_{m'}$, and defining rate in this case $R = (\ln M)/N$ nats/dimension, we obtain

$$\ln \mathbf{E}_y[\max_{m' \ne m} \exp(s\alpha_{m'})]$$
$$> \begin{cases} N[2s\sqrt{RC} - O(\ln N/N)] & 0 \le s \le \sqrt{R/C} \\ N[Cs^2 + R - O(N^{-1})] & \sqrt{R/C} < s \le 1. \end{cases} \tag{59}$$

Thus, combining (56), (57), and (59) we have

$$\mu_m(s) = \mu(s)$$
$$> \begin{cases} -N[-Cs^2 + 2s(C - \sqrt{RC}) + O(\ln N/N)] \\ \qquad\qquad 0 \le s \le \sqrt{R/C} \\ -N[2C(s - s^2) - R + O(N^{-1})] \\ \qquad\qquad \sqrt{R/C} < s \le 1. \end{cases} \tag{60}$$

To remove the limitation to fixed-composition codes we can proceed exactly as in the proof of the sphere packing [2] lower bound that introduces the additional term $O(\ln N/N)$.

To justify the independence condition on the α_m preceding (56) we note that if a unique code is asymptotically optimum for all decision strategies under consideration then it will be optimum also for $s = s_0$, where $\mu'(s_0) = 0$, which corresponds to the error probability for conventional maximum-likelihood decision. But for this case, the code that results in independent α_m yields

$$\mu(s_0) = \min_s \mu(s) = \begin{cases} -N(C/2 - R) & 0 \le R/C \le \frac{1}{4} \\ -N(\sqrt{C} - \sqrt{R})^2 & \frac{1}{4} \le R/C \le 1 \end{cases}$$

which corresponds to the best code for a very noisy channel [2], [9]. Thus, no code can be uniformly better than that for which the α_m are asymptotically independent.

VIII. Conclusions

We have obtained exact asymptotic results for error and erasure probabilities or average list sizes with a class of generalized decision strategies for orthogonal signals in white Gaussian noise. These results hold for the optimum signals in white Gaussian noise, and even in the wider class of input-discrete very noisy memoryless channels, provided there exists a unique signal set which is asymptotically optimum for the entire class of decision strategies under consideration.

References

[1] G. D. Forney, Jr., "Exponential error bounds for erasure, list, and decision feedback schemes," *IEEE Trans. Information Theory*, vol. IT-14, pp. 206–220, March 1968.

[2] C. E. Shannon, R. G. Gallager, and E. R. Berlekamp, "Lower bounds to error probability for coding on discrete memoryless channels, I," *Inform. and Control*, vol. 10, pp. 81–83, January 1967.

[3] J. M. Wozencraft and I. M. Jacobs, *Principles of Communication Engineering.* New York: Wiley, 1965, ch. 4.

[4] A. J. Viterbi, *Principles of Coherent Communication.* New York: McGraw-Hill, 1966, ch. 8.

[5] *Ibid.*, p. 221.

[6] J. M. Wozencraft and I. M. Jacobs, *supra*, p. 345.

[7] H. Cramér, *Mathematical Methods of Statistics.* Princeton, N. J.: Princeton University Press, 1946, pp. 374–376.

[8] A. V. Balakrishnan, "A contribution to the sphere-packing problem of communication theory," *J. Math. Anal. Appl.*, vol. 3, pp. 485–506, December 1961.

[9] R. G. Gallager, "A simple derivation of the coding theorem and some applications," *IEEE Trans. Information Theory*, vol. IT-11, pp. 3–18, January 1965.

[10] M. Loève, *Probability Theory.* Princeton, N. J.: Van Nostrand, 1955, p. 288.

Part II
Rate Distortion Theory

On the Shannon Theory of Information Transmission in the Case of Continuous Signals*

ANDREI N. KOLMOGOROV†

I. Introduction

THE ROLE of the entropy of a random object ξ, capable of taking the values $x_1, x_2, \cdots, x_n$ with the probabilities $p_1, p_2, \cdots, p_n$,

$$H(\xi) = -\sum_k p_k \log p_k$$

in information theory and in the theory of information transmission using discrete signals, can be considered to have been explained sufficiently. Furthermore, I insist that the fundamental concept, which admits of generalization to perfectly arbitrary continuous information and signals, is not directly the entropy concept but the concept of the quantity of information $I(\xi, \eta)$ in the random object ξ relative to the object η. In the discrete case this quantity is evaluated correctly according to the well-known Shannon formula:[1]

$$I(\xi, \eta) = H(\eta) - MH(\eta/\xi).$$

For a finite-dimensional distribution, possessing density, the quantity $I(\xi, \eta)$ is determined, according to Shannon, by the analogous formula

$$I(\xi, \eta) = h(\eta) - Mh(\eta/\xi),$$

where $h(\eta)$ is the "differential entropy"

$$h(\eta) = -\int p(y) \log p(y)\, dy,$$

and $h(\eta/\xi)$ is the conditional differential entropy defined in an analogous manner. It is well known that the quantity $h(\xi)$ has no direct real interpretation and is not even invariant with respect to coordinate transformation in the space of the x's. For an infinitely-dimensional distribution, the analog of $h(\xi)$ is nonexistent, in general.

According to the proper meaning of the word, the entropy of the object ξ with a continuous distribution is always infinite. If the continuous signals can, nevertheless, serve to transmit finitely great information, then it is only because they are always observed with bounded accuracy. Consequently, it is natural to define the appropriate "ϵ-entropy" $H_\epsilon(\xi)$ of the object ξ by giving the accuracy of observation ϵ. Shannon did thus under the designation "rate of creating information with respect to a fidelity criteron." Although choosing a new name for this quantity does not alter the situation, I decided to call your attention to that proposition by underlining the more widespread interest in the concept and its deep analogy to the ordinary exact entropy. I imply, beforehand, that, as remarked in Section IV, the theorem on the extremal role of the normal distribution (both in the finite-dimensional and the infinite-dimensional cases) is retained for the ϵ entropy. Furthermore, I give in Sections II and III, without pretending to its unconditional newness, an abstract formulation of the definition and fundamental properties of $I(\xi, \eta)$ and a survey of the fundamental problems of the Shannon theory of information transmission. Certain specific results obtained recently by Soviet investigators are explained in Sections IV to VI. I wish to emphasize, especially, the very significant interest, as it appears to me, in the investigations of the asymptotic behavior of the ϵ entropy as $\epsilon \to 0$. The cases investigated earlier

$$H_\epsilon(\xi) \sim n \log \frac{1}{\epsilon}; \qquad \overline{H}_\epsilon(\xi) = 2w$$

where n is the number of measurements and w is the bandwidth of the spectrum, are only very particular cases of the rules which can be encountered here. In order to understand the perspectives disclosed here, my note,[2] explained in another terminology, might be of interest; hence, I am placing a certain number of reprints at the disposal of the participants of the symposium.

To a considerable degree, my report reproduces the contents of a report presented jointly with Iaglom and Gel'fand at the Third All-Union Mathematics Conference.

* Presented at 1956 Symposium on Information Theory at Mass. Inst. Tech., Cambridge, Mass., September 10-12, 1956. Translated by Morris D. Friedman.

† Academician, Academy of Science, USSR.

[1] It seems expedient to me that the notation $H(\eta/x)$ is the conditional entropy of η for $\xi = x$ and $MH(\eta/\xi)$ is the mathematical expectation of this conditional entropy for the variable ξ.

[2] A. N. Kolmogorov, *Doklady, AN USSR*, vol. 108, no. 3, pp. 385–388; 1956.

Reprinted from *IEEE Trans. Inform. Theory*, vol. IT-2, pp. 102–108, Sept. 1956.

However, since the present symposium is of a more engineering character, I omitted a number of mathematical details. The work of Khinchin on the logical foundations of the theory remains beyond the limits of my survey.

As regards the work of Soviet radio engineers, you will hear about some of them from the other speakers. In the note itself, I will have occasion to note only the interest, in principle, of certain early work of Kotel'nikov, circa 1933 (see Section VI, further).

II. Quantity of Information in One Random Object Relative to Another

Let ξ and η be random objects with regions of possible values X and Y,

$$P_\xi(A) = P(\xi \varepsilon A); \qquad P_\eta(B) = P(\eta \varepsilon B)$$

the appropriate probability distributions, and

$$P_{\xi\eta}(C) = P((\xi, \eta) \varepsilon C)$$

the joint probability distribution of the objects ξ and η. By definition, the quantity of information in the random object ξ relative to the random object η is given by the formula

$$I(\xi, \eta) = \int_X \int_Y P_{\xi\eta}(dx\, dy) \log \frac{P_{\xi\eta}(dx\, dy)}{P_\xi(dx) P_\eta(dy)}. \tag{1}$$

The exact meaning of this formula requires certain elucidation and the general properties of $I(\xi, \eta)$, given later, are correct only for certain limitations of a set-theoretical character on the distributions P_ξ, P_η and $P_{\xi\eta}$, but I will not dwell on this here. In every case, the general theory can be explained, without great difficulty, in such a way that it will be applicable to random objects ξ and η of very general nature (vectors, functions, generalized functions, etc.).

Eq. (1) can be considered to be due to Shannon although he was limited to the case

$$P_\xi(A) = \int_A p_\xi(x)\, dx; \qquad P_\eta(B) = \int_B p_\eta(y)\, dy$$

$$P_{\xi\eta}(C) = \iint_C p_{\xi\eta}(x, y)\, dx\, dy$$

when (1) transforms into

$$I(\xi, \eta) = \int_X \int_Y P_{\xi\eta}(x, y) \log \frac{P_{\xi\eta}(x, y)}{P_\xi(x) P_\eta(y)}\, dx\, dy.$$

Sometimes, it is useful to represent the distribution P as

$$P_{\xi\eta}(C) = \iint_C a(x, y) P_\xi(dx) P_\eta(dy) + S(C) \tag{2}$$

where the function $S(C)$ is singular relative to the product

$$P_\xi \times P_\eta.$$

If the singular component of S is lacking, then the formula

$$\alpha_{\xi\eta} = a(\xi, \eta) \tag{3}$$

determines the random quantity $\alpha_{\xi\eta}$ uniquely to the accuracy of probability zero. Sometimes, the following theorem formulated by Gel'fand and Iaglom[3] is useful.

Theorem: If $S(X \times Y) > 0$, then $I(\xi, \eta) = \infty$. If $S(X \times Y) = 0$, then

$$\begin{aligned} I(\xi, \eta) &= \int_X \int_Y a(x, y) \log a(x, y) P_\xi(dx) P_\eta(dy) \\ &= \int_X \int_Y \log a(x, y) P_{\xi\eta}(dx\, dy) \\ &= M \log \alpha_{\xi\eta}. \end{aligned} \tag{4}$$

Let us enumerate certain fundamental properties of $I(\xi, \eta)$.

1) $I(\xi, \eta) = I(\eta, \xi)$.
2) $I(\xi, \eta) \geq 0$; $I(\xi, \eta) = 0$ only if ξ and η are independent.
3) If the pair (ξ_1, η_1) and (ξ_2, η_2) are independent, then
$$I[(\xi_1, \xi_2), (\eta_1, \eta_2)] = I(\xi_1, \eta_1) + I(\xi_2, \eta_2).$$
4) $I[(\xi, \eta), \zeta] \geq I(\xi, \zeta)$.
5) $I[(\xi, \eta), \zeta] = I(\eta, \zeta)$, if and only if ξ, η, ζ is a Markov sequence, *i.e.*, if the conditional distribution of ζ depends only on η for fixed ξ and η.

Apropos property 4, it is useful to note the following. In the case of the entropy

$$H(\xi) = I(\xi, \xi),$$

there is the bound on the entropy of the (ξ, η) pair from above:

$$H(\xi, \eta) \leq H(\xi) + H(\eta)$$

as well as the bound from below which results from 1 and 4.

$$H(\xi, \eta) \geq H(\xi); \qquad H(\xi, \eta) \geq H(\eta).$$

A similar estimate for the quantity of information in ζ relative to the (ξ, η) pair does not exist. From

$$I(\xi, \zeta) = 0; \quad I(\eta, \zeta) = 0$$

there still does not result the equality

$$I[(\xi, \eta), \zeta] = 0,$$

as can be shown by elementary examples.

For later use, let us note the special case when ξ and η are the random vectors:

$$\xi = (\xi_1, \cdots, \xi_m)$$

$$\eta = (\eta_1, \cdots, \eta_n) = (\xi_{m+1}, \cdots, \xi_{m+n}),$$

and the quantities

$$\xi_1, \xi_2, \cdots, \xi_{m+n}$$

are distributed normally with the second central moments

$$s_{ij} = M[(\xi_i - M\xi_i)(\xi_j - M\xi_j)].$$

[3] A. N. Kolmogorov, A. M. Iaglom, and I. M. Gel'fand, "Quantity of Information and Entropy for Continuous Distributions," Report at Third All-Union Math. Conf., 1956.

If the determinant

$$C = | s_{ij} |_{1 \le i, j \le m+n}$$

is not zero, then, as was calculated by Gel'fand and Iaglom

$$I(\xi, \eta) = \frac{1}{2} \log \frac{AB}{C} \quad (5)$$

where

$$A = | s_{ij} |_{1 \le i, j \le m}; \qquad B = | s_{ij} |_{m < i, j \le m+n}.$$

It is often more expedient, however, to use another approach without the $C > 0$ limitation. As is known,[4] all the second moments s_{ij} except those for which $i = j$ or $j = m + i$ go to zero after a suitable linear coordinate transformation in the X and Y spaces. For such a choice of coordinates

$$I(\xi, \eta) = -\tfrac{1}{2} \sum [1 - r^2(\xi_k, \eta_k)] \quad (6)$$

where the summation is taken over those

$$k \le \min (m, n)$$

for which the denominator in the expression of the correlation coefficient

$$r(\xi_k, \eta_k) = \frac{s_{k,m+k}}{\sqrt{s_{kk} \cdot s_{m+k,m+k}}}$$

is not zero.

III. Abstract Explanation of the Principles of the Shannon Theory

Shannon considers the transmission of information according to the scheme

$$\xi \rightarrow \eta \rightarrow \eta' \rightarrow \xi'$$

where the "transmitting apparatus"

$$\eta \rightarrow \eta'$$

is characterized by the conditional distribution

$$P_{\eta'/\eta}(B'/y) = P(\eta' \varepsilon B'/\eta = y)$$

of the "output signal" η' for a given "input signal" η and a certain limitation

$$P_\eta \varepsilon V$$

of the input signal distribution P_η. The "coding"

$$\xi \rightarrow \eta$$

and "decoding" operations

$$\eta' \rightarrow \xi'$$

are characterized by the conditional distributions

$$P_{\eta/\xi}(B/x) = P(\eta \varepsilon B/\xi = x)$$

$$P_{\xi'/\eta'}(A'/y') = P(\xi' \varepsilon A'/\eta' = y').$$

[4] A. M. Obukhov, *Izv. AN USSR, Phys.-Math. Series*, pp. 339–370; 1938.

The fundamental Shannon problem is the following. Given the spaces X, X', Y, Y' of possible values of the "input message" ξ, the "output message" ξ', the input signal η, and the output signal η'; given the characteristics of the transmitter, *i.e.*, the conditional distribution $P_{\eta'/\eta}$ and the class V of admissible input signal distributions P_η; finally, given the distribution

$$P_\xi(A) = P(\xi \varepsilon A)$$

of the input message and the "fidelity criterion"

$$P_{\xi\xi'} \varepsilon W$$

where W is a certain class of joint distributions

$$P_{\xi\xi'}(C) = P[(\xi, \xi') \varepsilon C]$$

of the input and output communications. To find: Is it possible, and if it is, by what means, to give a coding and decoding rule (*i.e.*, the conditional distributions $P_{\eta/\xi}$ and $P_{\xi'/\eta'}$) in such a manner that by calculating the distribution $P_{\xi\xi'}$ in terms of the distributions $P_\xi, P_{\eta/\xi}, P_{\eta'/\eta}, P_{\xi'/\xi}$ under the assumption that the sequence

$$\xi, \eta, \xi', \eta'$$

is Markovian, we will obtain

$$P_{\xi\xi'} \varepsilon W?$$

As does Shannon, so let us define the "capacity" of the transmitter thus

$$C = \sup_{P_\eta \varepsilon V} I(\eta, \eta')$$

and let us introduce the quantity

$$H_W(\xi) = \inf_{P_{\xi\xi'} \varepsilon W} I(\xi, \xi')$$

which Shannon calls the "rate of creating information relative to a fidelity criterion" when computed per unit time. Then, the *necessary condition of the possibility of transmission*

$$H_W(\xi) \le C \quad (7)$$

results at once from property 5 of Section II.

The incomparably deep idea of Shannon is that (7), when applied to the continuous operation of a "communication channel," is "almost sufficient" in a certain sense and under certain very broad conditions. From the mathematical point of view, it is a matter here of proving a limit theorem of the following type. It is assumed that the space X, X', Y, Y' of the distributions P_ξ and $P_{\eta'/\eta}$ of the classes V and W, and therefore, of the quantities C and $H_W(\xi)$, depend on the parameter T (which plays the role, in applications, of the duration of transmitter operation). It is required to establish that the condition

$$\liminf_{T \to \infty} \frac{C^T}{H_W^T(\xi)} > 1 \quad (8)$$

is sufficient, under a certain sufficiently general character of the assumptions, for the possibility of transmission, satisfying the conditions formulated above, for sufficiently large T. Naturally, in such a formulation the problem is somewhat indistinct (for example, similar to the general problem of studying possible limit distributions for a sum of large numbers of "small" components). However, I intended to avoid any return to the terminology of the theory of stationary, random processes here, since it was shown in the note of the young Romanian mathematician Rozenblat-Rot Milu[5] that interesting results can be obtained in the designated direction without the assumption of stationariness.

Many remarkable works have been devoted to the derivation of a limit theorem of the kind indicated. The work of Khinchin[6] is the contribution of a USSR mathematician in this research direction. It appears to me that much remains to be done here. Namely, results of this kind are intended to give a foundation to the widespread conviction that the expression $I(\xi, \eta)$ is not just one of the possible methods of measuring the "quantity of information" but it is a measure of the quantity of information having an advantage, in principle, over the others, actually. Since the "information," by its original nature, is not a scalar, then axiomatic investigations, permitting $I(\xi, \eta)$ [or the entropy $H(\xi)$] to be characterized uniquely by using simple formal properties, in this respect have lesser value, in my opinion. The situation here seems to me to be similar to our being ready to assign, at once, the greatest value to that method, out of all those proposed by Gauss to give a foundation to the normal law of error distribution, which starts from the limit theorem for the sum of a large number of small components. Other methods (for example, based on the principle of the arithmetic mean) demonstrate only why *any other* error distribution law could not be as acceptable and suitable as the normal law but they do not answer the question of why the normal law is actually encountered often in real problems. Similarly, the beautiful formal properties of the expressions $H(\xi)$ and $I(\xi, \eta)$ cannot demonstrate why they are sufficient for the complete (albeit from the asymptotic point of view) solution of many problems in many cases.

IV. Calculation and Estimation of the ϵ Entropy in Certain Particular Cases

If the condition

$$P_{\xi\xi'} \in W$$

is chosen as the certainty of exact coincidence of ξ and ξ'

$$P(\xi = \xi') = 1$$

then

$$H_W(\xi) = H(\xi).$$

In conformance with this, it seems to be natural to designate $H_W(\xi)$ in the general case as the "entropy of the random object ξ for the accuracy of reproduction W."

[5] Rozenblat-Rot Milu, *Trudy, Third All-Union Math. Conf.*, vol. 2, pp. 132–133; 1956.

[6] A. Ia. Khinchin, *Usp. Mate. Nauk*, vol. 11, no. 1 (67), pp. 17–75; 1956.

Now, let us assume that X is a metric space and that the space X' coincides with X, *i.e.*, methods are investigated of the approximate transmission of information from the point $\xi \in X$ by using the indication of the point ξ' of the same space X. It seems natural to require that

$$P\{\rho(\xi, \xi') \leq \epsilon\} = 1 \qquad (W_\epsilon^0)$$

or that

$$M\rho^2(\xi, \xi') \leq \epsilon^2 \qquad (W_\epsilon).$$

We will denote these two forms of the "ϵ entropy" of the distribution P_ξ by

$$H_{W_\epsilon^0}(\xi) = H_\epsilon^0(\xi)$$

$$H_{W_\epsilon}(\xi) = H_\epsilon(\xi).$$

As regards the ϵ entropy H_ϵ^0, I shall only note here a certain estimate for

$$H_\epsilon^0(X) = \sup_{P_\xi} H_\epsilon^0(\xi)$$

where the upper bound is taken over all the probability distributions P_ξ in the space X. As is known, for $\epsilon = 0$,

$$H_0^0(X) = \sup_{P_\xi} H(\xi) = \log N_x$$

where N_x *is the number of elements of the manifold* X. For $\epsilon > 0$,

$$\log N_x^c(2\epsilon) \leq H_\epsilon^0(x) \leq \log N_x^a(\epsilon)$$

where $N_x^a(\epsilon)$ and $N_x^c(\epsilon)$ are characteristics of the space X which are introduced in my note.[2] The asymptotic properties of the function $N_x(\epsilon)$ as $\epsilon \to 0$, studied in my work[2] for a number of specific spaces X, are interesting analogs of the properties, explained later, of the asymptotic behavior of the function $H_\epsilon(\xi)$.

Let us now turn to the ϵ entropy $H_\epsilon(\xi)$. If X is an n-dimensional Euclidean space and if

$$P_\xi(A) = \int_A P_\xi(x)\, dx_1\, dx_2 \cdots dx_n$$

then, at least in the case of the sufficiently smooth function $p_\xi(x)$, the following well-known formula holds:

$$H_\epsilon(\xi) = n \log \frac{1}{\epsilon} + [h(\xi) - n \log \sqrt{2\pi e}] + o(1) \qquad (9)$$

where

$$h(\xi) = -\int_X p_\xi(x) \log p_\xi(x)\, dx_1 \cdots dx_n$$

is the "differential entropy," already introduced in the first Shannon works. Hence, the asymptotic behavior of $H_\epsilon(\xi)$ in the case of sufficiently smooth continuous distributions in n-dimensional space is determined, *to a first approximation*, by the dimensionality of the space and the differential entropy $h(\xi)$ only enters as the second term in the expression for $H_\epsilon(\xi)$.

It is natural to expect that the growth of $H_\epsilon(\xi)$ as $\epsilon \to 0$ will be substantially more rapid for typical distributions in infinite-dimensional spaces. As the simplest example, let us consider the Wiener random function $\xi(t)$, defined for $0 \leq t \leq 1$, with the normally-distributed independent increments:

$$\Delta\xi = \xi(t + \Delta t) - \xi(t)$$

for which

$$\xi(0) = 0; \qquad M\Delta\xi = 0; \qquad M(\Delta\xi)^2 = \Delta t.$$

Iaglom found that in this case, in the L^2 metric

$$H_\epsilon(\xi) = \frac{4}{\pi}\frac{1}{\epsilon^2} + o\left(\frac{1}{\epsilon^2}\right) \tag{10}$$

Under certain natural assumptions, the formula

$$H_\epsilon(\xi) = \frac{4}{\pi}\chi\left(\frac{1}{\epsilon^2}\right) + o\left(\frac{1}{\epsilon^2}\right) \tag{11}$$

where

$$\chi(\xi) = \int_{t_0}^{t_1} Mb \mid t, \xi(t) \mid dt$$

can be obtained in a more general way for the diffuse kind of Markov process on the $t_0 \leq t \leq t_1$ time segment with

$$M_{\Delta\xi} = A[t, \xi(t)]\,\Delta t + o(\Delta t);$$

$$M(\Delta\xi)^2 = B[t, \xi(t)]\,\Delta t + o(\Delta t).$$

The ϵ-entropy H_ϵ can be calculated exactly for the case of the normal distribution in an n-dimensional space or in Hilbert space. After a suitable orthogonal coordinate transformation, the n-dimensional vector ξ assumes the form

$$\xi = (\xi_1, \xi_2, \cdots, \xi_n)$$

where the coordinates ξ_k are mutually independent and distributed normally. The parameter θ, for given ϵ, is determined from the equation

$$\epsilon^2 = \sum \min(\theta^2, D^2\xi_k)$$

and in the case of ξ distributed normally

$$H_\epsilon(\xi) = \frac{1}{2}\sum_{D^2\xi_k \to \theta^2} \log\frac{D^2\xi_k}{\theta^2}. \tag{12}$$

The approximating vector

$$\xi' = (\xi_1', \xi_2', \cdots, \xi_n')$$

should be chosen such that

$$\xi_k' = 0$$

for $D^2\xi_k \leq \theta^2$ and

$$\xi_k = \xi_k' + \Delta_k; \qquad D^2\Delta_k = \theta^2; \qquad D^2\xi_k' = D^2\xi_k - \theta^2$$

for $D^2\xi_k > \theta$ and the vectors ξ_k and Δ_k are mutually independent. The infinite-dimensional case is in no way different from the finite dimensional.

Finally, it is very essential that the *maximum value of* $H_\epsilon(\xi)$ *for the vector* ξ (n dimensional or infinite dimensional) *be attained in the normal distribution case for given second central moments*. This result can be obtained directly or from the following proposition of Pinsker.[7]

Theorem: Let the positive-definite symmetric matrix of the s_{ij}, $0 \leq i, j \leq m + n$ quantities and the distribution P_ξ of the vector be given

$$\xi = (\xi_1, \xi_2, \cdots, \xi_m)$$

for which the central second moments equal $s_{i,j}$ (for $0 \leq i, j \leq m$). Let the condition W on the joint distribution $P_{\xi\xi'}$ of the vector ξ and the vector

$$\xi' = (\xi_{m+1}, \xi_{m+2}, \cdots, \xi_{m+n})$$

be that the central second moments of the quantities

$$\xi_1, \xi_2, \cdots, \xi_{m+n}$$

equal s_{ij} (for $0 \leq i, j \leq m + n$). Then

$$H_W(\xi) \leq \frac{1}{2}\log\frac{AB}{C}. \tag{13}$$

The notation in (13) corresponds to the explanation of Section II. It is seen from a comparison with the results of Section II, that inequality (13) becomes the equality in the case of the normal distribution P_ξ.

The principles of solving the variational problems arising in the calculation of the "rate of creating information" were indicated sufficiently long ago by Shannon. Shannon and Weaver[8] write: "Unfortunately these formal solutions are difficult to evaluate in particular cases and seem to be of little value."[9] In substance, however, many problems of this kind are simple enough, as is seen from the above. It is possible that the slow development of investigations in this direction is related to insufficient understanding of the fact that the solution of the variation problem often appears to be degenerate in typical cases: For example, the evaluation of $H_\epsilon(\xi)$ in the problem selected above, for the normally distributed vector ξ in the n-dimensional case, the vector ξ' often appears to be not n dimensional but only k dimensional with $k < n$; in the infinite-dimensional case, the vector ξ' always appears to be finite dimensional.

V. Quantity of Information and Rate of Creating Information in the Stationary Process Case

Let us consider two stationary and stationarily-related processes,

$$\xi(t), \eta(t) \qquad -\infty < t < +\infty.$$

Let us denote by ξ_T and η_T the segments of the ξ and η processes in the time $0 < t \leq T$ and by ξ_- and η_- the flow of the ξ and η processes on the negative semiaxis $-\infty < t \leq 0$. To give the pair (ξ, η) of the stationarily-related ξ and η processes means to give the probability distribution

[7] M. S. Pinsker, *Trudy, Third All-Union Math. Conf.*, vol. 1, p. 125; 1956.

[8] C. E. Shannon and W. Weaver, "The Mathematical Theory of Communication," University of Illinois Press; 1949.

[9] *Ibid.*, sec. 28, p. 79 of the Russian translation.

$P_{\xi\eta}$, invariant to shift along the t axis, in the space of the function pair $\{x(t), y(t)\}$. If ξ_- is fixed, then the following conditional probability

$$P_{\xi_T\eta/\xi_-}(C/x_-) = P\{(\xi_T, \eta) \in C/\xi_- = x_-\}$$

arises from the distribution $P_{\xi\eta}$. Using this distribution, the conditional quantity of information

$$I(\xi_T, \eta/x)$$

is calculated in conformance with Section II. If the mathematical expectation

$$MI(\xi_T, \eta/\xi_-)$$

is finite for any $T > 0$, then it is finite for all other $T > 0$ and

$$MI(\xi_T, \eta/\xi_-) = T\vec{I}(\xi, \eta).$$

It is natural to call the quantity $\vec{I}(\xi, \eta)$ the "rate of creating information of the process η for compliance with the process ξ." If the process ξ can be extrapolated with complete accuracy to future occurrences, then

$$\vec{I}(\xi, \eta) = 0.$$

In particular, this will be so if the process ξ has a bounded spectrum. Generally speaking, the following equality

$$\vec{I}(\xi, \eta) = \vec{I}(\eta, \xi) \tag{14}$$

does not hold. However, under sufficiently broad conditions on the "regularity" of the process ξ,[10] the equality

$$\vec{I}(\xi, \eta) = \bar{I}(\xi, \eta)$$

holds, where

$$\bar{I}(\xi, \eta) = \lim_{T\to\infty} \frac{1}{T} I(\xi_T, \eta_T).$$

Since $I(\xi_T, \eta_T) = I(\eta_T, \xi_T)$, then always

$$\bar{I}(\xi, \eta) = \bar{I}(\eta, \xi)$$

and, therefore, when both equalities $\vec{I}(\xi, \eta) = \bar{I}(\xi, \eta)$ and $\vec{I}(\eta, \xi) = \bar{I}(\eta, \xi)$ are correct, equality (14) holds. Now, let W be a certain class of joint distributions $P_{\xi\xi'}$ of two stationary and stationarily-related processes ξ and ξ'. It is natural to call the equantity

$$\vec{H}_W(\xi) = \inf_{P_{\xi\xi'} \in W} \vec{I}(\xi', \xi)$$

the "rate of creating information in the process ξ under the accuracy of reproduction W." It can be shown that

$$\vec{H}_W(\xi) = \bar{H}_W(\xi)$$

where

$$\bar{H}_W = \inf_{P_{\xi\xi'} \in W} \bar{I}(\xi, \xi')$$

[10] Here and later, the regularity of the process means, roughly speaking, that the segments of the process, corresponding to two segments of the t axis sufficiently removed from each other, are almost independent. In the case of Gaussian processes, the well-known definition of regularity introduced in my work[14] is applicable here.

under certain assumptions on the regularity of the process ξ and for certain natural types of conditions W.

VI. Calculation and Estimation of the Amount of Information and the Rate of Creating Information in Terms of the Spectrum

Pinsker[11] established the formula:

$$\bar{I}(\xi, \eta) = -\frac{1}{4\pi} \int_{-\infty}^{\infty} \log [1 - r^2(\lambda)]\, d\lambda \tag{15}$$

where

$$r^2(\lambda) = \frac{|f_{\xi\eta}(\lambda)|^2}{f_{\xi\xi}(\lambda) f_{\eta\eta}(\lambda)}$$

and $f_{\xi\xi}$, $f_{\xi\eta}$, $f_{\eta\eta}$ are spectral densities; for the case when the distribution $P_{\xi\eta}$ is normal and at least one of the processes ξ or η is regular. In connection with the review by Doob,[12] we would like to note that the novelty, in principle, of the Pinsker result is somewhat greater than can be expected on the basis of this review. The expression

$$\bar{h}(\xi) = \log (2\pi\sqrt{e}) + \frac{1}{4\pi} \int_{-\pi}^{\pi} \log f_{\xi\xi}(\lambda)\, d\lambda \tag{16}$$

is known in the case of processes with discrete time t for the differential entropy of a normal process per unit time:

$$\bar{h}(\xi) = \lim_{T\to\infty} \frac{1}{T} h(\xi_1, \xi_2, \cdots, \xi_T).$$

However, no analog of the expression $\bar{h}(\xi)$ exists in the continuous time and unbounded spectrum case and the Pinsker formula requires independent derivation.

It is natural to characterize the accuracy of reproducing the stationary process ξ, using the stationary process ξ' stationarily related to ξ, by the quantity

$$\sigma^2 = M[\xi(\alpha) - \xi'(\alpha)]^2$$

and in the case of a W condition of the form

$$\sigma^2 \leq \epsilon^2$$

it is natural to call the quantity

$$\bar{H}_\epsilon(\xi) = \bar{H}_W(\xi)$$

the ϵ entropy per unit time of the process ξ and under the assumption that

$$\vec{H}_W(\xi) = \bar{H}_W(\xi)$$

the rate of creating information in the process ξ for average accuracy of transmission ϵ. It can be concluded from the appropriate statement for finite-dimensional distributions (see Section IV) that the quantity $\bar{H}_\epsilon(\xi)$ attains a maximum in the case of the normal process ξ for a given spectral density $f_{\xi\xi}(\lambda)$. In the normal case, $\bar{H}_\epsilon(\xi)$ can be calculated easily in terms of the spectral density exactly as was explained in Section IV applied to $H_\epsilon(\xi)$ for the n-di-

[11] M. S. Pinsker, *Doklady, AN USSR*, vol. 98, 213–216; 1954.
[12] J. L. Doob, *Math. Revs.*, vol. 16, p. 495; 1955.

mensional distribution. The parameter θ is determined from the equation

$$\epsilon^2 = \int_{-\infty}^{\infty} \min\,[\theta^2, f_{\xi\xi}(\lambda)]\, d\lambda. \qquad (17)$$

Using this parameter, the quantity $\overline{H}_\epsilon(\xi)$ is found from the formula

$$\overline{H}_\epsilon(\xi) = \frac{1}{2} \int_{f_{\xi\xi}(\lambda) \theta^2} \log \frac{f(\lambda)}{\theta^2}\, d\lambda. \qquad (18)$$

Spectral densities of the kind shown in Fig. 1 which are approximated well by the function:

$$\varphi(\lambda) \doteq \begin{cases} a^2 & \text{for } A \leq |\lambda| \leq A + W \\ 0 & \text{in the remaining cases} \end{cases}$$

are of practical interest. It is easy to calculate that

$$\theta^2 \sim \frac{\epsilon^2}{2W}$$

$$\overline{H}_\epsilon(\xi) \sim W \log \frac{2Wa^2}{\epsilon^2} \qquad (19)$$

approximately, in this case for *not too small* ϵ for a normal process.

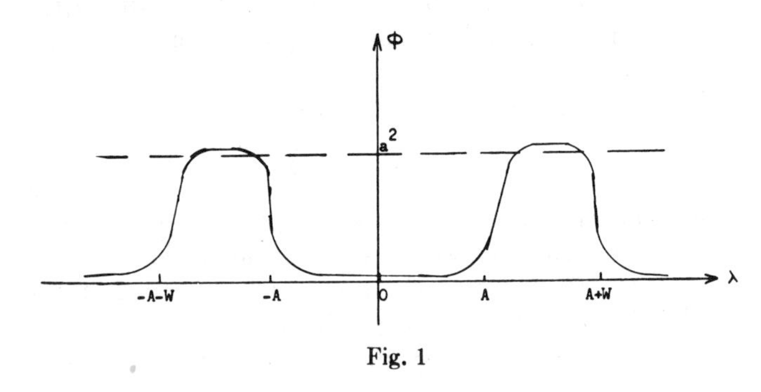

Fig. 1

Certainly, (19) is none other than the well-known Shannon formula

$$R = W \log \frac{Q}{N}. \qquad (20)$$

Here, however, the novelty, in principle, is that now we see why and within what limits (for not too small ϵ) this formula can be used for a process with an unbounded spectrum and such are all the processes in the theory of information transmission which really interest us.

Writing (19) thus

$$\overline{H}_\epsilon(\xi) \sim 2W \log (a\sqrt{2W}) \log \frac{1}{\epsilon} \qquad (21)$$

and comparing with (9), we see that the double width $2W$ of the useful frequency band plays the role of the number of measurements. This idea of the equivalence of twice the frequency bandwidth to the number of measurements, occurring in a certain sense of the word, per unit time, was apparently first expressed by Kotel'nikov.[13] On the basis of this idea, Kotel'nikov indicated the fact that a function, for which the spectrum is limited to bandwidth $2W$, is determined uniquely by the values of the function at the points

$$\cdots, -\frac{2}{2W}, -\frac{1}{2W}, 0, \frac{1}{2W}, \frac{2}{2W}, \cdots, \frac{k}{2W}, \cdots.$$

Shannon retained this argumentation, using the representation obtained in this manner to derive (20). Since a function with a bounded spectrum is always singular in the sense of my work[14] and the observation of such a function is not related, generally, to the stationary flow of new information, then the sense of this kind of argumentation does not remain completely clear so that the new derivation of the *approximate* formula (21), cited here, seems to me to be not devoid of interest.

The growth of $\overline{H}_\epsilon(\xi)$ as ϵ decreases occurs, for small ϵ for any normally distributed regular random function, substantially more rapidly than would be obtained according to (21). In particular, if $f_{\xi\xi}(\lambda)$ has order $1/\lambda\beta$ as $\lambda \to \infty$, then $\bar{h}_\epsilon(\xi)$ has order $1/(2/_\epsilon(\beta - 1))$.

[13] V. A. Kotel'nikov, Material for the First All-Union Conf. on questions of communications; 1933.

[14] A. N. Kolmogorov, Bulletin, Moscow Univ. I, no. 6; 1941. English translation available.

CODING THEOREMS FOR A DISCRETE SOURCE WITH A FIDELITY CRITERION*

Claude E. Shannon

Departments of Mathematics and Electrical Engineering
and
Research Laboratory of Electronics
Massachusetts Institute of Technology
Cambridge, Massachusetts

Summary

Consider a discrete source producing a sequence of message letters from a finite alphabet. A single letter distortion measure is given by a non-negative matrix (d_{ij}). The entry d_{ij} measures the "cost" or "distortion" if letter i is reproduced at the receiver as letter j. The average distortion of a communications system (source-coder-noisy channel-decoder) is taken to be $d = \sum_{i,j} P_{ij} d_{ij}$ where P_{ij} is the probability of i being reproduced as j. It is shown that there is a function R(d) that measures the "equivalent rate" of the source for a given level of distortion. For coding purposes where a level d of distortion can be tolerated, the source acts like one with information rate R(d). Methods are given for calculating R(d), and various properties discussed. Finally, generalizations to ergodic sources, to continuous sources, and to distortion measures involving blocks of letters are developed.

In this paper a study is made of the problem of coding a discrete source of information, given a <u>fidelity criterion</u> or a <u>measure of the distortion</u> of the final recovered message at the receiving point relative to the actual transmitted message. In a particular case there might be a certain tolerable level of distortion as determined by this measure. It is desired to so encode the information that the maximum possible signaling rate is obtained without exceeding the tolerable distortion level. This work is an expansion and detailed elaboration of ideas presented earlier[1], with particular reference to the discrete case.

We shall show that for a wide class of distortion measures and discrete sources of information there exists a function R(d) (depending on the particular distortion measure and source) which measures, in a sense, the equivalent rate R of the source (in bits per letter produced) when d is the allowed distortion level. Methods will be given for evaluating R(d) explicitly in certain simple cases and for evaluating R(d) by a limiting process in more complex cases. The basic results are roughly that it is impossible to signal at a rate faster than C/R(d) (source letters per second) over a memoryless channel of capacity C (bits per second) with a distortion measure less than or equal to d. On the other hand, by sufficiently long block codes it is possible to approach as closely as desired the rate C/R(d) with distortion level d.

Finally, some particular examples, using error probability per letter of message and other simple distortion measures, are worked out in detail.

The Single-Letter Distortion Measure

Suppose that we have a discrete information source producing a sequence of letters or "word" $m = m_1, m_2, m_3, \ldots, m_t$, each chosen from a finite alphabet. These are to be transmitted over a channel and reproduced, at least approximately, at a receiving point. Let the reproduced word be $Z = z_1, z_2, \ldots, z_t$. The z_i letters may be from the same alphabet as the m_i letters or from an enlarged alphabet including, perhaps, special symbols for unknown or semi-unknown letters. In a noisy telegraph situation m and Z might be as

*This work was supported in part by the U. S. Army (Signal Corps), the U. S. Air Force (Office of Scientific Research, Air Research and Development Command), and the U. S. Navy (Office of Naval Research).

Reprinted from *IRE Nat. Conv. Rec.*, pt. 4, pp. 142-163, Mar. 1959.

follows:

m = I HAVE HEARD THE MERMAIDS SINGING...

Z = I H?VT HEA?D TSE B?RMAIDZ ??NGING...

(with "?" marks printed above the letters in "H?VT" and "??NGING")

In this case, the Z alphabet consists of the ordinary letters and space of the m alphabet, together with additional symbols "?", "$\overset{?}{A}$", "$\overset{?}{B}$", etc., indicating less certain identification. Even more generally, the Z alphabet might be entirely different from the m alphabet.

Consider a situation in which there is a measure of the fidelity of transmission or the "distortion" between the original and final words. We shall assume first that this distortion measure is of a very simple and special type and later we shall generalize considerably on the basis of the special case.

A single-letter distortion measure is defined as follows. There is given a matrix d_{ij} with $d_{ij} \geq 0$. Here i ranges over the letters of the m alphabet of, say, a letters (assumed given a numerical ordering), while j ranges over the Z alphabet. The quantity d_{ij} may be thought of as a "cost" if letter i is reproduced as letter j.

If the Z alphabet includes the m alphabet, we will assume the distortion between an m letter and its correct reproduction to be zero and all incorrect reproductions to have positive distortion. It is convenient in this case to assume that the alphabets are arranged in the same indexing order so that $d_{ii} = 0$, $d_{ij} > 0$ $(i \neq j)$.

The distortion d, if word m is reproduced as word Z, is to be measured by

$$d(m, Z) = \frac{1}{t} \sum_{k=1}^{t} d_{m_k Z_k}$$

If in a communication system, word m occurs with probability P(m) and the conditional probability, if m is transmitted, that word Z will be reproduced, is $P(Z|m)$, then we assume that the over-all distortion of the system is given by

$$d = \sum_{m, Z} P(m)\, P(Z|m)\, d(m, Z)$$

Here we are supposing that all messages and reproduced words are of the same length t. In variable-length coding systems the analogous measure is merely the over-all probability that letter i is reproduced as j, multiplied by d_{ij} and summed on i and j. Note that d = 0 if and only if each word is correctly reproduced with probability 1, otherwise d > 0 (in cases where the Z alphabet includes the m alphabet).

Some Simple Examples

A distortion measure may be represented by giving the matrix of its elements, all terms of which are non-negative. An alternative representation is in terms of a line diagram similar to those used for representing a memoryless noisy channel. The lines are now labeled, however, with the values d_{ij} rather than probabilities.

A simple example of a distortion measure, with identical m and Z alphabets, is the error probability per letter. In this case, if the alphabets are ordered similarly, $d_{ij} = 1 - \delta_{ij}$. If there were three letters in the m and Z alphabets, the line diagram would be that shown in Fig. 1(a). Such a distortion measure might be appropriate in measuring the fidelity of a teletype or a remote typesetting system.

Another example is that of transmitting the quantized position of a wheel or shaft. Suppose that the circumference is divided into five equal arcs. It might be only half as costly to have an error of plus or minus one segment as larger errors. Thus the distortion measure might be

$$d_{ij} = \begin{cases} 0 & i = j \\ 1/2 & |i-j| = 1 \mod 5 \\ 1 & |i-j| > 1 \mod 5 \end{cases}$$

A third example might be a binary system sending information each second, either "all's well" or "emergency," for some situation. Generally, it would be considerably more important that the "emergency" signal be correctly received than that the "all's well" signal be correctly received. Thus if these were weighted 10 to 1, the diagram would be as shown in Fig. 1(b).

A fourth example with entirely distinct m and Z alphabets is a case in which the m alphabet consists of three possible readings, -1, 0, and +1. Perhaps, for some reasons of economy, it is desired to work with a reproduced alphabet of two letters, $-\frac{1}{2}$ and $+\frac{1}{2}$. One might then have the matrix that is shown in Fig. 2.

The Rate-Distortion Function R(d)

Now suppose that successive letters of the message are statistically independent but chosen with the same probabilities, P_i the probability of letter i from the alphabet. This type of source we call an independent letter source.

Given such a set of probabilities P_i and a distortion measure d_{ij}, we define a rate-distortion curve as follows. Assign an arbitrary set of transition probabilities $q_i(j)$ for transitions from i to j. (Of course, $q_i(j) \geq 0$ and $\sum_j q_i(j) = 1$.) One could calculate for this assignment two things: first, the distortion measure $d(q_i(j)) = \sum_{ij} P_i\, q_i(j)\, d_{ij}$ if letter i were reproduced as j with conditional probability $q_i(j)$, and, second, the average mutual information between i and j if this were the case, namely

$$R(q_i(j)) = E \log \frac{q_i(j)}{\sum_k P_k\, q_k(j)}$$

$$= \sum_{i,j} P_i\, q_i(j) \log \frac{q_i(j)}{\sum_k P_k\, q_k(j)}$$

The rate-distortion function $R(d^*)$ is defined as the greatest lower bound of $R(q_i(j))$ when the $q_i(j)$ are varied subject to their probability limitations and subject to the average distortion d being less than or equal to d^*.

Note that $R(q_i(j))$ is a continuous function of the $q_i(j)$ in the allowed region of variation of $q_i(j)$ which is closed. Consequently, the greatest lower bound of R is actually attained as a minimum for each value of R that can occur at all. Further, from its definition it is clear that R(d) is a monotonically decreasing function of d.

Convexity of the R(d) Curve

Suppose that two points on the R(d) curve are (R, d) obtained with assignment $q_i(j)$ and (R', d') attained with assignment $q_i'(j)$. Consider a mixture of these assignments $\lambda q_i(j) + (1-\lambda)\, q_i'(j)$. This produces a d" (because of the linearity of d) not greater than $\lambda d + (1-\lambda)\, d'$. On the other hand, $R(q_i(j))$ is known to be a convex downward function (the rate for a channel as a function of its transition probabilities). Hence $R'' \leq \lambda R + (1-\lambda)\, R'$. The minimizing $q_i''(j)$ for d" must give at least this low a value of R". Hence the curve R as a function of d (or conversely) is convex downward.

The minimum possible d value clearly occurs if, for each i, $q_i(j)$ is assigned the value 1 for the j having the minimum d_{ij}. Thus, the lowest possible d is given by

$$d_{min} = \sum_i P_i \min_j d_{ij}$$

If the m alphabet is imaged in the Z alphabet, then $d_{min} = 0$, and the corresponding R value is the ordinary entropy or rate for the source. In the more general situation, $R(d_{min})$ may be readily evaluated if there is a unique $\min_j d_{ij}$ by evaluating R for the assignment mentioned. Otherwise, the evaluation of $R(d_{min})$ is a bit more complex.

On the other hand, R = 0 is obtained if and only if $q_i(j) = Q_j$, a function of j only. This is because an average mutual information is positive unless the events are independent. For a given Q_j giving R = 0, the d is then $\sum_{ij} P_i\, Q_j\, d_{ij} = \sum_j Q_j \sum_i P_i\, d_{ij}$. The inner sum is non-negative. If we wish the minimum d for R = 0, this would result by finding a j that gives a minimum $\sum_i P_i\, d_{ij}$ (say j^*) and making $Q_{j}^* = 1$. This can be done by assigning $q_i(j^*) = 1$ (all other $q_i(j)$ are made 0).

Summarizing, then, R(d) is a convex downward function as shown in Fig. 3 running from $R(d_{min})$ at $d_{min} = \sum_i P_i \min_j d_{ij}$ to zero at $d_{max} = \min_j \sum_i P_i\, d_{ij}$. It is continuous both ways (R as a function of d or d as a function of R) interior to this interval because of its convexity. For $d \geq d_{max}$, we have R = 0. The curve is strictly monotonically decreasing from d_{min} to d_{max}. Also, it is easily seen that in this interval the assignment of $q_i(j)$ to obtain any point $R(d^*)$ must give a d satisfying the equality $d = d^*$ (not the inequality $d < d^*$). For $d^* > d_{max}$ the inequality will occur for the minimizing $q_i(j)$. Thus the minimizing problem can be limited to a consideration of minima in the subspace where $d = d^*$, except in the range $d^* > d_{max}$ (where $R(d^*) = 0$).

The convex downward nature of R as a function of the assigned $q_i(j)$ is helpful in evaluating the R(d) in specific cases. It implies that any local minimum (in the subspace for a fixed d) is the absolute minimum in this subspace. For otherwise we could connect the local and absolute minima by a straight line and find a continuous series of points lower than the local minimum along this line. This would contradict its being a local minimum.

Furthermore, the functions $R(q_i(j))$ and $d(q_i(j))$ have continuous derivatives interior to the allowed $q_i(j)$ set. Hence ordinary calculus methods (e. g., Lagrangian multipliers) may be used to locate the minimum. In general, however, this still involves the solution of a set of simultaneous equations.

Solution for R(d) in Certain Simple Cases

One special type of situation leads to a simple explicit solution for the R(d) curve. Suppose that all a input letters are equiprobable: $P_i = 1/a$. Suppose further that the d_{ij} matrix is square and is such that each row has the same set of entries and each column also has the same set of entries, although, of course, in different order.

An example of this type is the positioning of a wheel mentioned earlier if all positions are equally likely. Another example is the simple error probability distortion measure if all letters are equally likely.

In general, let the entries in any row or column be $d_1, d_2, d_3, \ldots, d_a$. Then we shall show that the minimizing R for a given d occurs when all lines with distortion assignment d_k are given the probability assignment

$$q_k = \frac{e^{-\lambda d_k}}{\sum_i e^{-\lambda d_i}}$$

Here λ is a parameter ranging from 0 to ∞ which determines the value of d. With this minimizing assignment, d and R are given parametrically in terms of λ:

$$d = \frac{\sum_i d_i e^{-\lambda d_i}}{\sum_i e^{-\lambda d_i}}$$

$$R = \log \frac{a}{\sum_i e^{-\lambda d_i}} - \lambda d$$

When $\lambda = 0$ it can be seen that $d = \frac{1}{a} \sum_i d_i$ and $R = 0$. When $\lambda \to \infty$, $d \to d_{min}$ and $R \to \log \frac{a}{k}$ where k is the number of d_i with value d_{min}.

This solution is proved as follows. Suppose that we have an assignment $q_i(j)$ giving a certain d^* and a certain R^*. Consider now a new assignment where each line with d_{ij} value d_1 is assigned the average of the assignments for these lines in the original assignment. Similarly, each line labeled d_2 is given the average of all the d_2 original assignments, and so on. Because of the linearity of d, this new assignment has the same d value, namely d^*. The new R is the same as or smaller than R^*. This is shown as follows. R may be written $H(m) - H(m|Z)$. $H(m)$ is not changed, and $H(m|Z)$ can only be increased by this averaging. The latter fact can be seen by observing that because of the convexity of $-\sum_i x_i \log x_i$ we have

$$-\sum_j \alpha_j \sum_t x_j^{(t)} \log x_j^{(t)}$$

$$\geq -\sum_t \left(\sum_j \alpha_j x_j^{(t)} \right) \log \sum_j \alpha_j x_j^{(t)}$$

where for a given t, $x_j^{(t)}$ is a set of probabilities, and α_j is a set of weighting factors. In particular

$$-\sum_j \frac{\sum_s q_j^{(s)}}{\sum_{s,i} q_i^{(s)}} \sum_t \frac{q_j^{(t)}}{\sum_s q_j^{(s)}} \log \frac{q_j^{(t)}}{\sum_s q_j^{(s)}}$$

$$\geq -\sum_t \frac{\sum_j q_j^{(t)}}{\sum_{s,i} q_i^{(s)}} \log \frac{\sum_j q_j^{(t)}}{\sum_{s,i} q_i^{(s)}}$$

where $q_j^{(s)}$ is the original assignment to the line of value d_j from letter s. But this inequality can be interpreted on the left as $H(m|Z)$ after the averaging process, while the right-hand side is $H(m|Z)$ before the averaging. The desired result then follows.

Hence, for the minimizing assignment, all lines with the same d value will have equal probability assignments. We denote these by q_i corresponding to a line labeled d_i. The rate R and distortion d can now be written

$$d = \sum_i q_i d_i$$

$$R = \log a + \sum_i q_i \log q_i$$

since all z's are now equiprobable, and $H(m) = \log a$, $H(m|Z) = -\sum_i q_i \log q_i$. We wish, by proper choice of the q_i, to minimize R for a given d and subject to $\sum_i q_i = 1$. Consider then, using Lagrange multipliers,

$$U = \log a + \sum_i q_i \log q_i + \lambda \sum_i q_i d_i + \mu \sum_i q_i$$

$$\frac{\partial U}{\partial q_i} = 1 + \log q_i + \lambda d_i + \mu = 0$$

$$q_i = A\, e^{-\lambda d_i}$$

If we choose $A = \frac{1}{\sum_i e^{-\lambda d_i}}$ we satisfy $\sum_i q_i = 1$.

This then gives a stationary point and by the convexity properties mentioned above it must be the absolute minimum for the corresponding value of d. By substituting this probability assignment in the formulas for d and R we obtain the results stated above.

Rate for a Product Source with a Sum Distortion Measure

Suppose that we have two independent sources each with its own distortion measure d_{ij} and $d'_{i'j'}$ and resulting in rate distortion functions $R_1(d_1)$ and $R_2(d_2)$. Suppose that each source produces one letter each second. Considering ordered pairs of letters as single letters the combined system may be called the product source. If the total distortion is to be measured by the sum of the individual distortions, $d = d_1 + d_2$, then there is a simple method of determining the function R(d) for the product source. In fact, we shall show that R(d) is obtained by adding both coordinates of the curves $R_1(d_1)$ and $R_2(d_2)$ at points on the two curves having the same slope. The set of points obtained in this manner is the curve R(d). Furthermore, a probability assignment to obtain any point of R(d) is the product of the assignments for the component points.

We shall first show that given any assignment $q_{i,i'}(j,j')$ for the product source, we can do at least as well in the minimizing process using an assignment of the form $q_i(j)\, q'_{i'}(j')$ where the q and q' are derived from the given $q_{i,i'}(j,j')$. Namely, let

$$q_i(j) = \sum_{i',j'} P'_{i'}\, q_{i,i'}(j,j')$$

$$q'_{i'}(j') = \sum_{i,j} P_i\, q_{i,i'}(j,j')$$

We see that these are non-negative and, summed on j and j', respectively, give 1, so they are satisfactory transition probabilities. Also the assignment $q_i(j)\, q'_{i'}(j')$ gives the same total distortion as the assignment $q_{i,i'}(j,j')$. The former is

$$\sum_{i,i',j,j'} P_i P'_{i'} q_i(j)\, q'_{i'}(j')[d_{ij} + d'_{i'j'}]$$

$$= \sum_{i,j} P_i q_i(j)\, d_{ij} + \sum_{i',j'} P'_i\, q'_i(j')\, d'_{i'j'}$$

$$= \sum_{\substack{i,i'\\ j,j'}} P_i P'_{i'}\, q_{i,i'}(j,j')\, [d_{ij} + d'_{i',j'}]$$

This last may be recognized as the distortion with $q_{i,i'}(j, j')$.

On the other hand, the mutual information R is decreased or left constant if we use $q_i(j)\ q'_{i'}(j)$ instead of $q_{i,i'}(j, j')$. In fact, this average mutual information can be written in terms of entropies as follows (using asterisks for entropies with the assignment $q_i(j)\ q'_{i'}(j')$ and none for the assignment $q_{i,i'}(j, j')$). We have

$$R = H(i, i') - H(i, i' | j, j')$$
$$\geq H(i, i') - H(i | j) - H(i' | j')$$
$$= H(i, i') - H^*(i | j) - H^*(i' | j')$$

Here we use the fact that with our definition of $q_i(j)$ and $q'_{i'}(j')$ we have $Pr^*(i|j) = Pr(i|j)$ and $Pr^*(i'|j') = Pr(i'|j')$. (This follows immediately on writing out these probabilities.) Now, using the fact that the sources are independent, $H(i, i') = H(i) + H(i') = H^*(i) + H^*(i')$. Hence our last reduction above is equal to R^*. This is the desired conclusion.

It follows that any point on the R(d) curve for the product source is obtained by an independent or product assignment $q_i(j)\ q'_{i'}(j')$ and consequently is the sum in both coordinates of a pair of points on the two curves. The best choice for a given distortion d is clearly given by

$$R(d) = \min_t [R_1(t) + R_2(d-t)]$$

and this minimum will occur when

$$\frac{d}{dt} R_1(t) = \frac{d}{dt} R_2(d-t)$$

Thus the component points to be added are points where the component curves have the same slope. The convexity of these curves insures the uniqueness of this pair for any particular d.

The Lower Bound on Distortion for a Given Channel Capacity

The importance of the R(d) function is that it determines the channel capacity required to send at a certain rate and with a certain minimum distortion. Consider the following situation. We have given an independent letter source with probabilities P_i for the different possible letters. We have given a single-letter distortion measure d_{ij} which leads to the rate distortion function R(d). Finally, there is a memoryless discrete channel K of capacity C bits per second (we assume that this channel may be used once each second). We wish to transmit words of length t from the source over the channel with a block code. The length of the code words in the channel is n. What is the lowest distortion d that might be obtained with a code and a decoding system of this sort?

Theorem 1. Under the assumptions given above it is not possible to have a code with distortion d smaller than the (minimum) d* satisfying

$$R(d^*) = \frac{n}{t} C$$

or, equivalently, in any code, $d \geq \phi\left(\frac{n}{t} C\right)$, where ϕ is the function inverse to R(d).

This theorem, and a converse positive result to be given later, show that R(d) may be thought of as the equivalent rate of the source for a given distortion d. Theorem 1 asserts that for the distortion d and t letters of text, one must supply in the channel at least t R(d) total bits of capacity spread over the n uses of the channel in the code. The converse theorem will show that by taking n and t sufficiently large and with suitable codes it is possible to approach this limiting curve.

To prove Theorem 1, suppose that we have given a block code which encodes all message words of length t into channel words of length n and a decoding procedure for interpreting channel output words of length n into Z words of length t. Let a message word be represented by $m = m_1, m_2, \ldots, m_t$. A channel input word is $X = x_1, x_2, \ldots, x_n$. A channel output word is $Y = y_1, y_2, \ldots, y_n$ and a reproduced, or Z, word is $Z = z_1, z_2, \ldots, z_t$. By the given code and decoding system, X is a function of m and Z is a function of Y. The m_i are chosen independently according to the letter probabilities, and the channel transition probabilities give a set of conditional probabilities $P(y|x)$ applying to each x_i, y_i pair. Finally, the source and channel are independent in the sense that $P(Y|m,X) = P(Y|X)$.

We wish first to show that $H(m|Z) \geq H(m) - nC$. We have that $H(m|Z) \geq H(m|Y)$ (since Z is a function of Y) amd also that $H(m|Y) \geq H(X|Y) - H(X) + H(m)$. This last is because, from the independence condition above, $H(Y|m,X) = H(Y|X)$, so $H(Y,m,X) - H(m,X) = H(X,Y) - H(X)$. But $H(m,X) = H(m)$, since X is a function of m, and for the same reason $H(m,X,Y) = H(m,Y)$. Hence, rearranging, we have

$$H(X,Y) = H(m,Y) + H(X) - H(m,X)$$

$$= H(m,Y) + H(X) - H(m)$$

$$H(X|Y) \leq H(m|Y) + H(X) - H(m)$$

Here we used $H(m,x) = H(m)$ and then subtracted $H(Y)$ from each side. Hence $H(m|Z) \geq H(X|Y) - H(X) + H(m)$.

Now we show that $H(X|Y) \geq nC$. This follows from a method we have used in other similar situations, by considering the change in $H(X|Y)$ with each received letter. Thus (using Y_k for the first k y letters, etc.),

$$\Delta H(X|Y)$$

$$= H(X|y_1, y_2, \ldots, y_k) - H(X|y_1, y_2, \ldots, y_{k+1})$$

$$= H(X, Y_k) - H(Y_k) - H(X, Y_k, y_{k+1}) + H(Y_k, y_{k+1})$$

$$= H(y_{k+1}|Y_k) - H(y_{k+1}|X, Y_k)$$

$$= H(y_{k+1}|Y_k) - H(y_{k+1}|x_{k+1})$$

$$\leq H(y_{k+1}) - H(y_{k+1}|x_{k+1})$$

$$\leq C$$

Here we used the fact that the channel is memoryless, so $P(y_{k+1}|X, Y_k) = P(y_{k+1}|x_{k+1})$ and therefore $H(y_{k+1}|X, Y_k) = H(y_{k+1}|x_{k+1})$. Finally, C is the maximum possible $H(y) - H(y|X)$ giving the last inequality.

Since the incremental change in $H(X|Y_k)$ is bounded by C, the total change after n steps is bounded by nC. Consequently, the final $H(X|Y)$ is at least the initial value $H(X)$ less nC.

$$H(m|Z) \geq H(X|Y) - H(X) + H(m)$$

$$\geq H(X) - nC - H(X) + H(m)$$

$$H(m|Z) \geq H(m) - nC \qquad (1)$$

We now wish to overbound $H(m|Z)$ in terms of the distortion d. We have

$$H(m|Z) = H(m_1 m_2 \ldots m_t | z_1 z_2 \ldots z_t)$$

$$\leq \sum_i H(m_i|z_i)$$

$$= \sum_i H(m_i) - \sum_i (H(m_i) - H(m_i|z_i))$$

The quantity $H(m_i) - H(m_i|z_i)$ is the average mutual information between original message letter m_i and the reproduced letter z_i. If we let d_i be the distortion between these letters, then $R(d_i)$ (the rate-distortion function evaluated for this d_i) satisfies

$$R(d_i) \leq H(m_i) - H(m_i|z_i)$$

since $R(d_i)$ is the minimum mutual information for the distortion d_i. Hence our inequality may be written

$$H(m|Z) \leq \sum_{i=1}^{t} H(m_i) - \sum_{i=1}^{t} R(d_i)$$

Using now the fact that $R(d)$ is a convex downward function, we have

$$H(m|Z) \leq \sum_i H(m_i) - t\, R\left(\sum_i \frac{d_i}{t}\right)$$

But $\sum_i \frac{d_i}{t} = d$, the over-all distortion of the system, so

$$H(m|Z) \leq \sum_i H(m_i) - t\, R(d)$$

Combining this with our previous inequality (1) and using the independent letter assumption, so $H(m) = \sum_i H(m_i)$, we have

$$H(m) - nC \leq H(m) - t\, R(d)$$

$$nC \geq t\, R(d)$$

This is essentially the result stated in Theorem 1.

It should be noted that the result in the theorem is an assertion about the minimum distortion after any finite number n of uses of the channel. It is not an asymptotic result for large n. Also, it applies as seen by the method of proof, for any code, block or variable length, provided only that after n uses of the channel, t (or more) letters are reproduced at the receiving point, whatever the received sequence may be.

The Coding Theorem for a Single-Letter Distortion Measure

We now prove a positive coding theorem corresponding to the negative statements of Theorem 1; namely, that it is possible to approach the lower bound of distortion for a given ratio of number n of channel letters to t message letters. We consider then a source of message letters and single-letter distortion measure d_{ij}. More generally than Theorem 1, however, this source may be ergodic; it is not necessarily an independent letter source. This more general situation will be helpful in a later generalization of the theorem. For an ergodic source there will still, of course, be letter probabilities P_i, and we could determine the rate distortion function R(d) based on these probabilities as though it were an independent letter source.

We first establish the following result.

Lemma 1. Suppose that we have an ergodic source with letter probabilities P_i, a single-letter distortion measure d_{ij}, and a set of assigned transition probabilities $q_i(j)$ such that

$$\sum_{i,j} P_i\, q_i(j)\, d_{ij} = d^*$$

$$\sum_{i,j} P_i\, q_i(j) \log \frac{q_i(j)}{\sum_k P_k\, q_k(j)} = R$$

Let Q(Z) be the probability measure of a sequence Z in the space of reproduced sequences if successive source letters had independent transition probabilities $q_i(j)$ into the Z alphabet. Then, given $\epsilon > 0$, for all sufficiently large block length t, there exists a set α of messages of length t from the source with total source probability $P(\alpha) \geq 1 - \epsilon$, and for each m belonging to α a set of Z blocks of length t, say β_m, such that

1) $d(m, Z) \leq d^* + \epsilon$ for $m \in \alpha$ and $Z \in \beta_m$

2) $Q(\beta_m) \geq e^{-t(R+\epsilon)}$ for any $m \in \alpha$

In other words, and somewhat roughly, long messages will, with high probability, fall in a certain subset α. Each member m of this subset has an associated set of Z sequences β_m. The members of β_m have only (at most) slightly more than d* distortion with m and the logarithm of the total probability of β_m in the Q measure is underbounded by $e^{-t(R+\epsilon)}$.

To prove the lemma, consider source blocks of length t and the Z blocks of length t. Consider the two random variables, the distortion d between an m block and a Z block and the (unaveraged) mutual information type of expression below:

$$d = \frac{1}{t} \sum_i d_{m_i z_i}$$

$$I(m;Z) = \frac{1}{t} \log \frac{\Pr(Z|m)}{Q(Z)} = \frac{1}{t} \sum_i \log \frac{\Pr(z_i|m_i)}{Q(z_i)}$$

Here m_i is the ith letter of a source block m, and z_i is the ith letter of a Z block. Both R and d are random variables, taking on different values corresponding to different choices of m and Z. They are both the sum of t random variables which are identical functions of the joint m Z process except for shifting along over t positions.

Since the joint process is ergodic, we may apply the ergodic theorem and assert that when t is large, d and R will, with probability nearly 1, be close to their expected values. In particular, for any given ϵ_1 and δ, if t is sufficiently large, we will have with probability $\geq 1 - \frac{\delta^2}{2}$ that

$$d \leq \sum_{i,j} P_i\, q_i(j)\, d_{ij} + \epsilon_1 = d^* + \epsilon_1 .$$

Also, with probability at least $1 - \frac{\delta^2}{2}$ we will have

$$I \leq \sum_{i,j} P_i\, q_i(j) \log \frac{q_i(j)}{Q_j} + \epsilon_1 = R(d^*) + \epsilon_1$$

Let γ be the set of (m, Z) pairs for which both inequalities hold. Then $Pr(\gamma) \geq 1 - \delta^2$ because each of the conditions can exclude, at most, a set of probability $\delta^2/2$. Now for any m_1 define β_{m_1} as the set of Z such that (m_1, Z) belongs to γ.

We have

$$Pr(\beta_m | m) \geq 1 - \delta$$

on a set α of m whose total probability satisfies $Pr(\alpha) \geq 1 - \delta$. This is true, since if it were not we would have a total probability in the set complementary to γ of at least $\delta \cdot \delta = \delta^2$, a contradiction. The first δ would be the probability of m not being in α, and the second δ the conditional probability for such m's of Z not being in β_m. The product gives a lower bound on the probability of the complementary set to γ.

If $Z \in \beta_{m_1}$, then

$$\frac{1}{t} \log \frac{Pr(Z|m_1)}{Q(Z)} \leq R(d^*) + \epsilon_1$$

$$Pr(Z|m_1) \leq Q(Z)\, e^{t(R(d^*)+\epsilon_1)}$$

$$Q(Z) \geq Pr(Z|m_1)\, e^{-t(R(d^*)+\epsilon_1)}$$

Sum this inequality over all $Z \in \beta_{m_1}$.

$$Q(\beta_m) = \sum_{Z \in \beta_{m_1}} Q(Z)$$

$$\geq e^{-t(R+\epsilon_1)} \sum_{Z \in \beta_{m_1}} Pr(Z|m_1)$$

If $m_1 \in \alpha$ then $\sum_{Z \in \beta_{m_1}} Pr(Z|m_1) \geq 1 - \delta$ as seen above. Hence the inequality can be continued to give

$$Q(\beta_{m_1}) \geq (1-\delta)\, e^{-t(R+\epsilon_1)}, \quad m_1 \in \alpha$$

We have now established that for any $\epsilon_1 > 0$ and $\delta > 0$ there exists a set α of m's and sets β_m of Z's defined for each m with the three properties

1) $Pr(\alpha) \geq 1 - \delta$

2) $d(Z, m) \leq d^* + \epsilon_1$ if $Z \in \beta_m$

3) $Q(\beta_m) \geq (1-\delta)\, e^{-t(R+\epsilon_1)}$ if $m \in \alpha$

provided that the block length t is sufficiently large. Clearly, this implies that for any $\epsilon > 0$ and sufficiently large t we will have

1) $Pr(\alpha) \geq 1 - \epsilon$

2) $d(Z, m) \leq d^* + \epsilon$ if $Z \in \beta_m$

3) $Q(\beta_m) \geq e^{-t(R+\epsilon)}$

since we may take the ϵ_1 and δ sufficiently small to satisfy these simplified conditions in which we use the same ϵ. This concludes the proof of the lemma.

Before attacking the general coding problem, we consider the problem indicated schematically in Fig. 4. We have an ergodic source and a single-letter distortion measure that gives the rate distortion function R(d). It is desired to encode this by a coder into sequences u in such a way that the original messages can be reproduced by the reproducer with an average distortion that does not exceed d* (d* being some fixed tolerable distortion level). We are considering here block coding devices for both boxes. Thus the coder takes as input successive blocks of length t produced by the source and has, as output, corresponding to each possible m block, a block from a u alphabet.

The aim here is to do the coding in such a way as to keep the entropy of the u sequences as low as possible, subject to this requirement of reproducibility with distortion d* or less. Here the entropy to which we are referring is the entropy per letter of the original source. Alternatively, we might think of the source as producing one letter per second and we are then interested in the u entropy per second.

We shall show that for any d* and any $\epsilon > 0$ coders and reproducers can be found that are such that $H(u) \leq R(d^*) + \epsilon$. As $\epsilon \to 0$ the block length involved in the code in general increases.

This result, of course, is closely related to our interpretation of $R(d*)$ as the equivalent rate of the source for distortion $d*$. It will follow readily from the following theorem.

Theorem 2. Given an ergodic source, a distortion measure d_{ij}, and rate distortion function $R(d)$ (based on the single-letter frequencies of the source), given $d* \geq d_{min}$ and $\delta > 0$, for any sufficiently large t there exists a set Λ containing M words of length t in the Z alphabet with the following properties:

1) $\frac{1}{t} \log M \leq R(d*) + \delta$

2) The average distortion between an m word of length t and its nearest (i.e., least distortion) word in the set Λ is less than or equal to $d* + \delta$.

This theorem implies (except for the δ in property (2) which will later be eliminated) the results mentioned above. Namely, for the coder, one merely uses a device that maps any m word into its nearest member of Λ. The reproducer is then merely an identity transformation. The entropy per source letter of the coded sequence cannot exceed $R(d*) + \delta$, since this would be maximized at $\frac{1}{t} \log M$ if all of the M members of Λ were equally probable and $\frac{1}{t} \log M$ is by the theorem to be less than or equal to $R(d*) + \delta$.

This theorem will be proved by a random coding argument. We shall consider an ensemble of ways of selecting the members of Λ and estimate the average distortion for this ensemble. From the bounds on the average it will follow that at least one code exists in the ensemble with the desired properties.

The ensemble of codes is defined as follows. For the given $d*$ there will be a set of transition probabilities $q_i(j)$ that result in the minimum R, that is, $R(d*)$. The set of letter probabilities, together with these transition probabilities, induce a measure $Q(Z)$ in the space of reproduced words. The Q measure for a single Z letter, say letter j, is $\sum_i P_i q_i(j)$. The Q measure for a Z word consisting of letters $j_1, j_2, \ldots, j_t$ is $Q(Z) = \prod_{k=1}^{t} \left(\sum_i P_i q_i(j_k) \right)$.

In the ensemble of codes of length t, the integers from 1 to M are mapped into Z words of length t in all possible ways. An integer is mapped into a particular word Z_1, say, with probability $Q(Z_1)$, and the probabilities for different integers are statistically independent. This is exactly the same process as that of constructing a random code ensemble for a memoryless channel, except that here the integers are mapped into the Z space by using the $Q(Z)$ measure. Thus we arrive at a set of codes (if there are f letters in the Z alphabet there will be f^{tM} different codes in the ensemble) and each code will have an associated probability. The code in which integer i is mapped into Z_i has probability $\prod_{i=1}^{M} Q(Z_i)$.

We now use Lemma 1 to bound the average distortion for this ensemble of codes (using the probabilities associated with the codes in calculating the average). Note, first, that in the ensemble of codes if $Q(\beta)$ is the Q measure of a set β of Z words, then the probability that this set contains no code words is $[1 - Q(\beta)]^M$, that is, the product of the probability that code word 1 is not in β, that for code word 2, etc. Hence the probability that β contains at least one code word is $1 - [1 - Q(\beta)]^M$. Now, referring to Lemma 1, the average distortion may be bounded by

$$\overline{d} \leq \epsilon d_{max} + [1 - Q(\beta_m)]^M d_{max} + (d* + \epsilon)$$

Here d_{max} is the largest possible distortion between an m letter and a Z letter. The first term, ϵd_{max}, arises from message words m which are not in the set α. These have total probability less than or equal to ϵ and, when they occur, average distortion less than or equal to d_{max}. The second term overbounds the contribution that is due to cases in which the set β_m for the message m does not contain at least one code word. The probability in the ensemble of this is certainly bounded by $[1 - Q(\beta_m)]^M$, and the distortion is necessarily bounded by d_{max}. Finally, if the message is in α and there is at least one code word in β_m, the distortion is bounded by $d* + \epsilon$, according to Lemma 1. Now,

$Q(\beta_m) \geq e^{-t[R(d*)+\epsilon]}$. Also, for $0 < x \leq 1$,

$$(1-x)^{\frac{1}{x}} = e^{\frac{1}{x}\log(1-x)} \leq e^{\frac{1}{x}\left(-x+\frac{x^2}{2}\right)}$$

$$= e^{-1+\frac{x}{2}} \leq e^{-\frac{1}{2}}$$

(using the alternating and monotonically decreasing nature of the terms of the logarithmic expansion). Hence

$$[1 - Q(\beta_m)]^M$$

$$\leq (1 - e^{-t[R(d*)+\epsilon]})^M$$

$$= \left[1 - e^{-t[R(d*)+\epsilon]}\right]^{e^{t[R(d*)+\epsilon]}\, e^{-t[R(d*)+\epsilon]}M}$$

$$\leq e^{-\frac{1}{2}e^{-t[R(d*)+\epsilon]}M}$$

If we choose for M, the number of points, the value $e^{t[R(d*)+2\epsilon]}$ (or, if this is not an integer, the smallest integer exceeding this quantity), then the expression given above is bounded by $e^{-\frac{1}{2}e^{t\epsilon}}$. Thus the average distortion is bounded with this choice of M by

$$\bar{d} \leq \epsilon d_{max} + e^{-\frac{1}{2}e^{t\epsilon}} d_{max} + d* + \epsilon$$

$$\leq d* + \delta$$

provided that ϵ in Lemma 1 is chosen small enough to make $(\epsilon d_{max} + 1) \leq \frac{\delta}{2}$ and then t is chosen large enough to make $e^{-\frac{1}{2}e^{t\epsilon}} d_{max} \leq \frac{\delta}{2}$. We also require that ϵ be small enough and t large enough to make M, the integer just greater than or equal to $e^{t[R(d*)+2\epsilon]}$, less than or equal to $e^{t[R(d*)+\delta]}$. Since Lemma 1 holds for all sufficiently large t and any positive ϵ, these can all be simultaneously satisfied.

We have shown, then, that the conditions of the theorem are satisifed by the average distortion of the ensemble of codes. It follows that there exists at least one specific code in the ensemble whose average distortion is bounded by $d* + \epsilon$. This concludes the proof.

Corollary: Theorem 2 remains true if δ is replaced by 0 in property (1). It also remains true if the δ in property (1) is retained and the δ in property (2) is replaced by 0, provided in this case that $d* > d_{min}$, the smallest d for which $R(d)$ is defined.

This corollary asserts that we can attain (or do better than) one coordinate of the $R(d)$ curve and approximate, as closely as desired, the other, except possibly for the d_{min} point. To prove the first statement of the corollary, note first that it is true for $d* \geq d_1$, the value for which $R(d_1) = 0$. Indeed, we may achieve the point $\bar{d} = d_1$ with $M = 1$ and a code of length 1, using only the Z word consisting of the single Z letter which gives this point of the curve. For $d_{min} \leq d* < d_1$, apply Theorem 2 to approximate $d** = d* + \frac{\delta}{2}$. Since the curve is strictly decreasing, this approximation will lead to codes with $\bar{d} \leq d* + \delta$ and $\frac{1}{t}\log M \leq R(d*)$, if the δ in Theorem 2 is made sufficiently small.

The second simplification in the corollary is carried out in a similar fashion, by choosing a $d**$ slightly smaller than the desired $d*$ that is such that $R(d**) = R(d*) + \frac{\delta}{2}$, and by using Theorem 2 to approximate this point of the curve.

Now suppose we have a memoryless channel of capacity C. By the coding theorem for such channels it is possible to construct codes and decoding systems with rate approximating C (per use of the channel) and error probability $\leq \epsilon_1$ for any $\epsilon_1 > 0$. We may combine such a code for a channel with a code of the type mentioned above for a source at a given distortion level $d*$ and obtain the following result.

Theorem 3. Given a source characterized by $R(d)$ and a memoryless channel with capacity $C > 0$, given $\epsilon > 0$ and $d* > d_{min}$, there exists, for sufficiently large t and n, a block code that maps source words of length t into channel words of length n and a decoding system that maps channel output words of length n into reproduced words of length t which satisfy

1) $\bar{d} \leq d^*$

2) $\frac{nC}{t} \leq R(d^*) + \epsilon$

Thus we may attain a desired distortion level d^* (greater than d_{min}) and at the same time approximate using the channel at a rate corresponding to $R(d^*)$. This is done, as in the corollary stated above, by approximating the $R(d)$ curve slightly to the left of d^*, say, at $R(d^*) - \delta$. Such a code will have $M = e^{t[R(d^*-\delta)+\delta_1]}$ words, where δ_1 can be made small by taking t large. A code for the channel is constructed with M words and of length n, the largest integer satisfying $\frac{nC}{t} \leq R(d^* - \delta) + \delta_1$. By choosing t sufficiently large, this will approach zero error probability, since it corresponds to a rate less than channel capacity. If these two codes are combined, it produces an over-all code with average distortion $\leq d^*$.

Duality of a Source and a Channel

There is a curious and provocative duality between the properties of a source with a distortion measure and those of a channel. This duality is enhanced if we consider channels in which there is a "cost" associated with the different input letters, and it is desired to find the capacity subject to the constraint that the expected cost not exceed a certain quantity. Thus input letter i might have cost a_i and we wish to find the capacity with the side condition $\sum_i P_i a_i \leq a$, say, where P_i is the probability of using input letter i. This problem amounts, mathematically, to maximizing a mutual information under variation of the P_i with a linear inequality as constraint. The solution of this problem leads to a capacity cost function $C(a)$ for the channel. It can be shown readily that this function is concave downward. Solving this problem corresponds, in a sense, to finding a source that is just right for the channel and the desired cost.

In a somewhat dual way, evaluating the rate distortion function $R(d)$ for a source amounts, mathematically, to minimizing a mutual information under variation of the $q_i(j)$, again with a linear inequality as constraint. The solution leads to a function $R(d)$ which is convex downward. Solving this problem corresponds to finding a channel that is just right for the source and allowed distortion level. This duality can be pursued further and is related to a duality between past and future and the notions of control and knowledge. Thus, we may have knowledge of the past but cannot control it; we may control the future but not have knowledge of it.

Numerical Results for Some Simple Channels

In this section some numerical results will be given for certain simple channels and sources. Consider, first, the binary independent letter source with equiprobable letters and suppose that the distortion measure is the error probability (per digit). This falls into the class for which a simple explicit solution can be given. The $R(d)$ curve, in fact, is

$$R(d) = 1 + d \log_2 d + (1-d) \log_2 (1-d)$$

This, of course, is the capacity of a symmetric binary channel with probabilities d and $(1-d)$, the reason being that these are the probability assignment $q_i(j)$ which solves the minimizing problem.

This $R(d)$ curve is shown in Fig. 5. Also plotted are a number of points corresponding to specific simple codes, with the assumption of a noiseless binary channel. These will give some idea of how well the lower bound may be approximated by simple means. One point, $d = 0$, is obtained at rate $R = 1$ simply by sending the binary digits through the channel. Other simple codes which encode 2, 3, 4, and 5 message letters into one channel letter are the following. For the ratio 3 or 5, encode message sequences of three or five digits into 0 or 1 accordingly as the sequence contains more than half zeros or more than half ones. For the ratios 2 and 4, the same procedure is followed, while sequences with half zeros and half ones are encoded into 0.

At the receiving point, a 0 is decoded into a

sequence of zeros of the appropriate length and a 1 into a sequence of ones. These rather degenerate codes are plotted in Fig. 5 with crosses. Simple though they are, with block length of the channel sequences only one, they still approximate to some extent the lower bound.

Plotted on the same curve are square points corresponding to the well-known single-error correcting codes with block lengths 3, 7, 15, and 31. These codes are used backwards here – any message in the 15-dimensional cube, for example, is transmitted over the channel as the eleven message digits of its nearest code point. At the receiving point, the corresponding fifteen-digit message is reconstructed. This can differ at most in one place from the original message. Thus for this case the ratio of channel to message letters is $\frac{11}{15}$, and the error probability is easily found to be $\frac{1}{16}$. This series of points gives a closer approximation to the lower bound.

It is possible to fill in densely between points of these discrete series by a technique of mixing codes. For example, one may alternate in using two codes. More generally, one may mix them in proportions λ and $1-\lambda$, where λ is any rational number. Such a mixture gives a code with a new ratio R of message to channel letters, given by $\frac{1}{R} = \frac{\lambda}{R_1} + \frac{(1-\lambda)}{R_2}$, where R_1 and R_2 are the ratios for the given codes, and with new error probability

$$P_e = \frac{\lambda R_1 P_{e1} + (1-\lambda) R_2 P_{e2}}{\lambda R_1 + (1-\lambda) R_2}$$

This interpolation gives a convex upward curve between any two code points. When applied to the series of simple codes and single-error correcting codes in Fig. 5, it produces the dotted-line interpolations indicated.

Another channel was also considered in this connection, namely, the binary symmetric channel of capacity $C = \frac{1}{2}$. This has probabilities 0.89 that a digit is received correctly and 0.11 incorrectly. Here the series of points (Fig. 6) for simple codes actually touches the lower bound at the point $R = \frac{1}{2}$. This is because the channel itself, without coding, produces just this error probability. Any symmetric binary channel will have one point that can be attained exactly by means of straight transmission.

Figure 7 shows the $R(d)$ curve for another simple situation, a binary independent letter source but with the reproduced Z alphabet consisting of three letters, 0, 1, and ?. The distortion measure is zero for a correct digit, one for an incorrect digit, and 0.25 for ?. In the same figure is shown, for comparison, the $R(d)$ curve without the ? option.

Figure 8 shows the $R(d)$ curves for independent letter sources with various numbers of equiprobable letters in the alphabet (2, 3, 4, 5, 10, 100). Here again the distortion measure is taken to be error probability (per digit). With b letters in the alphabet the $R(d, b)$ curve is given by

$$R(d, b) = \log_2 b + d \log_2 d + (1-d) \log_2 \frac{1-d}{b-1}$$

Generalization to Continuous Cases

We will now sketch briefly a generalization of the single-letter distortion measure to cases where the input and output alphabets are not restricted to finite sets but vary over arbitrary spaces.

Assume a message alphabet $A = \{m\}$ and a reproduced letter alphabet $B = \{z\}$. For each pair (m, z) in these alphabets let $d(m, z)$ be a non-negative number, the distortion if m is reproduced as z. Further, we assume a probability measure P defined over a Borel field of subsets of the A space. Finally, we require that, for each z belonging to B, $d(m, z)$ is a measurable function with finite expectation.

Consider a finite selection of points $z_i (i = 1, 2, \ldots, \ell)$ from the B space, and a measurable assignment of transition probabilities $q(z_i | m)$. (That is, for each i, $q(z_i | m)$ is a measurable function in the A space.) For such a choice of z_i and assignment $q(z_i | m)$, a mutual information and an average distortion are determined.

$$R = \sum_i \int q(z_i|m) \log \frac{q(z_i|m)}{\int q(z_i|m)\, dP(m)} dP(m)$$

$$d = \sum_i \int d(m, z_i)\, q(z_i|m)\, dP(m)$$

We define the rate distortion function $R(d^*)$ for such a case as the greatest lower bound of R when the set of points z_i is varied (both in choice and number) and the $q(z_i|m)$ is varied over measurable transition probabilities, subject to keeping the distortion at the level d^* or less.

Most of the results we have found for the finite alphabet case carry through easily under this generalization. In particular, the convexity property of the R(d) curve still holds. In fact, if R(d) can be approximated to within ϵ by a choice z_i and $q(z_i|m)$ and $R(d')$ by a choice z_i' and $q'(z_i'|m)$, then one considers the choice z_i'' consisting of the union of the points z_i and z_i', together with $q''(z_i''|m) = \frac{1}{2}[q(z_i''|m) + q'(z_i''|m)]$, (using zero if $q(z''|m)$ or $q'(z''|m)$ is undefined). This leads, by the convexity of R and by the linearity of d, to an assignment for $d'' = \frac{1}{2}d + \frac{1}{2}d'$, giving an R'' within ϵ of the midpoint of the line joining dR(d) and d'R(d'). It follows, since ϵ can be made arbitrarily small, that the greatest lower bound of $R(d'')$ is on or below this midpoint.

In the general case it is, however, not necessarily true that the R(d) curve approaches a finite end-point when d decreases toward its minimum possible value. The behavior may be as indicated in Fig. 9 with R(d) going to infinity as d goes to d_{min}. On the other hand, under the conditions we have stated, there is a finite d_{max} for which $R(d_{max}) = 0$. This value of d is given by

$$d_{max} = \underset{z}{\text{g.}\ell\text{.b.}}\ E[d(m, z)]$$

The negative part of the coding theorem goes through in a manner essentially the same as the finite alphabet case, it being assumed that the only allowed coding functions from the source sequences to channel inputs correspond to measurable subsets of the source space. (If this assumption were not made, the average distortion would not, in general, even be defined.) The various inequalities may be followed through, changing the appropriate sums in the A space to integrals and resulting in the corresponding negative theorem.

For the positive coding theorem, also, substantially the same argument may be used with an additional ϵ involved to account for the approximation to the greatest lower bound of R(d) with a finite selection of z_i points. Thus, one chooses a set of z_i to approximate, within ϵ, the R(d) curve and then proceeds with the random coding method. The only point to be noted is that the d_{max} term must now be handled in a slightly different fashion. To each code in the ensemble one may add a particular point, say z_o, and replace d_{max} by $E(d(m, z_o))$, a finite quantity. The results of the theorem then follow.

Difference Distortion Measure

A special class of distortion measures for certain continuous cases of some importance and for which more explicit results can be obtained will now be considered. For these the m and z spaces are both the sets of all real numbers. The distortion measure d(m, z) will be called a <u>difference distortion measure</u> if it is a function only of the difference m − z, thus d(m, z) = d(m−z). A common example is the squared error measure, $d(m, z) = (m-z)^2$ or, again, the absolute error criterion $d(m, z) = |m-z|$.

We will develop a lower bound on R(d) for a difference distortion measure. First we define a function $\phi(d)$ for a given difference measure d(u) as follows. Consider an arbitrary distribution function G(u) and let H be its entropy and d the average distortion between a random variable with a given distribution and zero. Thus

$$H = -\int_{-\infty}^{\infty} \log dG(u)\, dG(u)$$

$$d = \int_{-\infty}^{\infty} d(u)\, dG(u)$$

We wish to vary the distribution $G(u)$, keeping $d \leq d^*$ and seek the maximum H. The least upper bound, if finite, is clearly actually attained as a maximum for some distribution. This maximum H for a given d^* we call $\phi(d^*)$, and a corresponding distribution function is called a maximizing distribution for this d^*.

Now suppose we have a distribution function for the m space (generalized letter probabilities) $P(m)$, with entropy $H(m)$. We wish to show that

$$R(d) \geq H(m) - \phi(d)$$

Let z_i be a set of z points and $q(z_i|m)$ an assignment of transition probabilities. Then the mutual information between m and z may be written

$$R = H(m) - \sum_i Q_i \, H(m|z_i)$$

where Q_i is the resulting probability of z_i. If we let d_i be the average distortion between m and z_i, then

$$H(m|z_i) \leq \phi(d_i)$$

This is because $\phi(d)$ was the maximum H for a given average distortion and also because the distortion is a function only of the difference between m and z, so that this maximizing value applies for any z_i. Thus

$$R \geq H(m) - \sum_i Q_i \, \phi(d_i)$$

Now $\phi(d)$ is a concave function. This is a consequence of the concavity of entropy considered as a function of a distribution function and the linearity of d in the same space of distribution functions, by an argument identical with that used previously. Hence, $\sum_i Q_i \, \phi(d_i) \leq \phi(\sum_i Q_i \, d_i) = \phi(d)$, where d is the average distortion with the choice z_i and the assigned transition probabilities. It follows that

$$R \geq H(m) - \phi(d)$$

This being true for any assignment z_i and $q(z_i|m)$ proves the desired result.

If, for a particular $P(m)$ and $d(u)$, assignments can be made which approach this lower bound, then, of course, this is the $R(d)$ function. Such is the case, for example, if $P(m)$ is gaussian and $d(u) = u^2$ (mean square error measure of distortion). Suppose that the message has variance σ^2, and consider a gaussian distribution of mean zero and variance $\sigma^2 - d$ in the z space. (If this is zero or negative, clearly $R(d) = 0$ by using only the z point zero.) Let the conditional probabilities $q(m|z)$ be gaussian with variance d. This is consistent with the gaussian character of $P(m)$, since normal distributions convolve to give normal distributions with the sum of the individual variances. These assignments determine the conditional probability measure $q(z|m)$, also then normal.

A simple calculation shows that this assignment attains the lower bound given above. The resulting $R(d)$ curve is

$$R(d) = \begin{cases} \log \dfrac{\sigma}{\sqrt{d}} & d \leq \sigma^2 \\ 0 & d > \sigma^2 \end{cases}$$

This is shown for $\sigma^2 = 1$ in Fig. 9.

Definition of a Local Distortion Measure

Thus far we have considered only a distortion measure d_{ij} (or $d(m, z)$) which depends upon comparison of a message letter with the corresponding reproduced letter, this letter-to-letter distortion to be averaged over the length of message and over the set of possible messages and possible reproduced messages. In many practical cases, however, this type of measure is not sufficiently general. The seriousness of a particular type of error often depends on the context.

Thus in transmitting a stock market quotation, say: "A.T.&T. 5900 shares, closing 194," an error in the 9 of 5900 shares would normally be much less serious than an error in the 9 of the closing price.

We shall now consider a distortion measure that depends upon local context and, in fact, compares blocks of g message letters with the corresponding blocks of g letters of the reproduced message.

A local distortion measure of span g is a function $d(m_1, m_2, \ldots, m_g;\ z_1, z_2, \ldots, z_g)$ of message sequences of length g and reproduced message sequences of length g (from a possibly different or larger alphabet) with the property that $d \geq 0$. The distortion between $m = m_1, m_2, \ldots, m_t$ and $z = z_1, z_2, \ldots, z_t$ $(t \geq g)$ is defined by

$$d(m, Z) = \frac{1}{t-g} \sum_{k=1}^{t-g+1} d(m_k, m_{k+1}, \ldots, m_{k+g-1};\ z_k, z_{k+1}, \ldots, z_{k+g-1})$$

The distortion of a block code in which message m and reproduced version Z occur with probability $P(m, Z)$ is defined by

$$d = \sum_{m, Z} P(m, Z)\, d(m, Z)$$

In other words, we assume, with a local distortion measure, that the evaluation of an entire system is obtained by averaging the distortions for all block comparisons of length g each with its probability of occurrence a weighting factor.

The Functions $R_n(d)$ and $R(d)$ for a Local Distortion Measure and Ergodic Source

Assume that we have given an ergodic message source and a local distortion measure. Consider blocks of n message letters with their associated probabilities (as determined by the source) together with possible blocks Z of reproduced message of length n. Let an arbitrary assignment of transition probabilities from the m blocks to the Z blocks, $q(Z|m)$, be made. For this assignment we can calculate two quantities: 1) the average mutual information per letter $R = \frac{1}{n} E\left(\log \frac{q(Z|m)}{Q(Z)}\right)$ and 2) the average distortion if the m's were reproduced as Z's with the probabilities $q(Z|m)$. This is $d = \sum_{m, Z} P(m, Z)\, d(m, Z)$. By variation of $q(Z|m)$, while holding $d \leq d^*$, we can, in principle, find the minimum R for each d^*. This we call $R_n(d^*)$.

The minimizing problem here is identical with that discussed previously if we think of m and Z as individual letters in a (large) alphabet, and various results relating to this minimum can be applied. In particular, $R_n(d)$ is a convex downward function.

We now define the rate distortion function for the given source relative to the distortion measure as

$$R(d) = \liminf_{n \to \infty} R_n(d)$$

It can be shown, by a direct but tedious argument that we shall omit, that the "inf" may be deleted from this definition. In other words, $R_n(d)$ approaches a limit as $n \to \infty$.

We are now in a position to prove coding theorems for a general ergodic source with a local distortion measure.

The Positive Coding Theorem for a Local Distortion Measure

Theorem 4. Suppose that we are given an ergodic source and a local distortion measure with rate distortion function $R(d)$. Let K be a memoryless discrete channel with capacity C, let d^* be a value of distortion, and let ϵ be a positive number. Then there exists a block code with distortion less than or equal to $d^* + \epsilon$, and a signaling rate at least $\left(\frac{C}{R} - \epsilon\right)$ message letters per channel letter.

Proof. Choose an n_1 so that $R_{n_1}(d^*) - R(d^*) < \frac{\epsilon}{3}$ and, also, so large that $\frac{g}{n_1} d_{max} < \frac{\epsilon}{3}$. Now consider blocks of length n_1 as "letters" of an enlarged alphabet. Using Theorem 3 we can construct a block code using sufficiently long sequences of these "letters" signaling at a rate close to (say within $\epsilon/3$ of) $R_{n_1}(d^*)/C$ (in terms of original message letters) and with distortion less than $d^* + \frac{\epsilon}{3}$. It must be remembered that this distortion is based on a single "letter" comparison. However, the distortion by the given local distortion measure will differ from this only because of overlap comparisons (g for each n_1 letters of message) and hence the discrepancy is, at most, $\frac{g}{n_1} d_{max} < \frac{\epsilon}{3}$. It follows that this code signals at a rate within ϵ of $R(d^*)$ and at a distortion within ϵ of d^*.

The Converse Coding Theorem

Theorem 5. Suppose that we are given an ergodic source and a local distortion measure with rate distortion function $R(d)$. Let K be a memoryless discrete channel with capacity C, let d^* be a value of distortion, and let ϵ be a positive number. Then there exists t_0 which is such that any code transmitting $t \geq t_0$ message letters with n uses of the channel at distortion d^*, or less, satisfies

$$\frac{n}{t} C \geq R(d^*) - \epsilon$$

That is, the channel capacity bits used per message letter must be nearly $R(d^*)$ for long transmissions.

Proof. Choose t_0 so that for $t \geq t_0$ we have $R_t(d) \geq R(d) - \epsilon$. Since $R(d)$ was defined as $\liminf_{t \to \infty} R_t(d)$, this is possible. Suppose that we have a code for such a $t \geq t_0$ which maps sequences m consisting of t message letters into sequences X of n channel letters and decodes sequences Y of n channel output letters into sequences Z of reproduced messages. The channel will have, from its transition probabilities, some $P(Y|X)$. Furthermore, from the encoding and decoding functions, we shall have $X = f(m)$ and $Z = g(Y)$. Finally, there will be, from the source, probabilities for the message sequences $P(m)$. By the encoding function $f(m)$ this will induce a set of probabilities $P(X)$ for input sequences. If the channel capacity is C, the average mutual information $R(X, Y)$ between input and output sequences must satisfy

$$R(X, Y) = E \log \frac{P(X|Y)}{P(X)} \leq nC$$

since nC is the maximum possible value of this quantity when $P(X)$ is varied. Also, since X is a function of m and Z is a function of Y, we have

$$R(m, Z) = E \log \frac{P(m|Z)}{P(m)} \leq R(X, Y) \leq nC$$

The coding system in question amounts, over-all, to a set of conditional probabilities from m sequences to Z sequences as determined by the two coding functions and the transition probabilities. If the distortion of the over-all system is less than or equal to d^*, then $tR_t(d^*) = \min_{P(Z|m)} R(m, Z)$ is certainly less than or equal to the particular $R(m, Z)$ obtained with the probabilities given by the channel and coding system. (The t factor is because $R_t(d)$ is measured on on a per message letter basis, while the $R(m, Z)$ quantities are for sequences of length t.) Thus

$$tR_t(d^*) \leq R(m, Z) \leq nC$$

$$t(R(d^*) - \epsilon) \leq nC$$

$$\frac{n}{t} C \geq R(d^*) - \epsilon$$

This is the conclusion of the theorem.

Notice from the method of proof that the code used again need not be a block code, provided only that after n uses of the channel t recovered letters are written down. If one has some kind of variable-length code and, starting at time zero, uses this code continually, the inequality of the theorem will hold for any finite time after t_0 message letters have been recovered; and of course as longer and longer blocks are compared, $\epsilon \to 0$. It is even possible to generalize this to variable-length codes in which, after n uses of the channel, the number of recovered message letters is a random variable depending, perhaps, on the particular message and the particular chance operation of the channel. If, as is usually the case in such codes, there exists an average signaling rate with the properties that after n uses of the channel then, with probability nearly one, t letters will be written down, with t lying between $t_1(1-\delta)$ and $t_1(1+\delta)$ (the $\delta \to 0$ as $n \to \infty$), then essentially the same theorem applies, using the mean t_1 for t.

Channels with Memory

Finally, we mention that while we have, in the above discussion, assumed the channel to be memoryless, very similar results, both of positive and negative type, can be obtained for channels with memory.

For a channel with memory one may define a capacity C_n for the first n uses of the channel starting at state s_o. This C_n is $\frac{1}{n}$ times the maximum average mutual information between input sequences of length n and resulting output sequences when the probabilities assigned the input sequences of length n are varied. The lower bound on P_e after n uses of the channel is now of the form $P_e \geq \phi\left(\frac{R}{C_n}\right)$.

We can also define the capacity C for such a channel as $C = \limsup_{n \to \infty} C_n$. The positive parts of the theorem then state that one can find arbitrarily long block codes with rate $\geq R$, and digit error probability $\leq \phi\left(\frac{R}{C}\right) + \epsilon$. In most channels of interest, of course, historical influences die out in such a way as to make $C_n \to C$ as $n \to \infty$. For memoryless channels, $C_n = C$ for all n.

References

1. C. E. Shannon and Warren Weaver, The Mathematical Theory of Communication, University of Illinois Press, 1949.

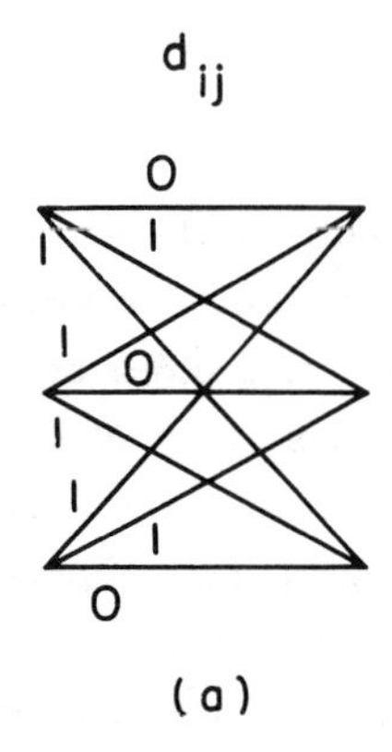

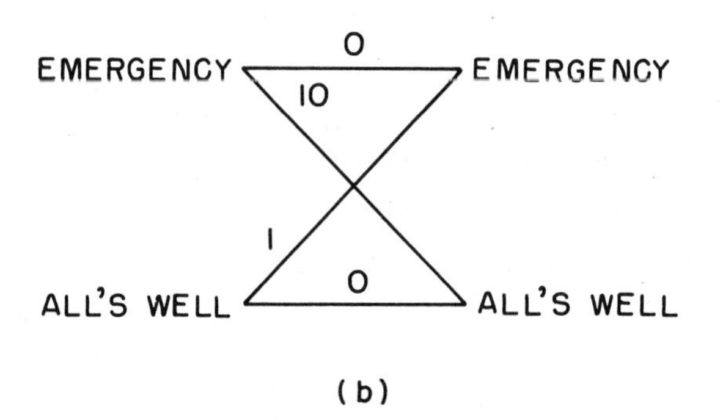

Fig. 1.

	$-\frac{1}{2}$	$+\frac{1}{2}$
-1	1	2
0	1	1
1	2	1

Fig. 2.

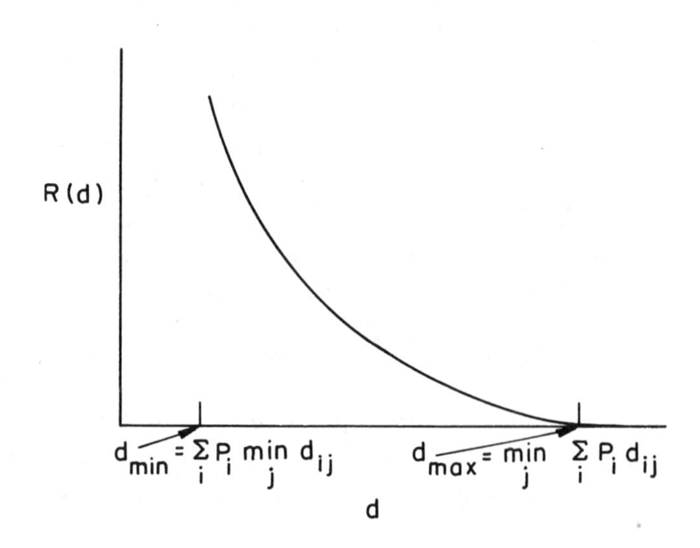

Fig. 3.

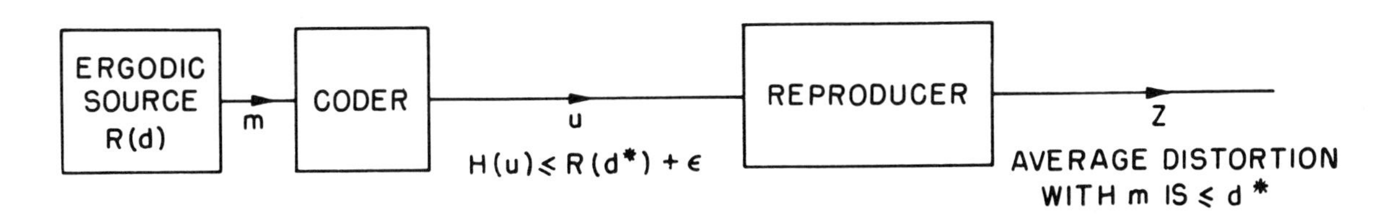

Fig. 4.

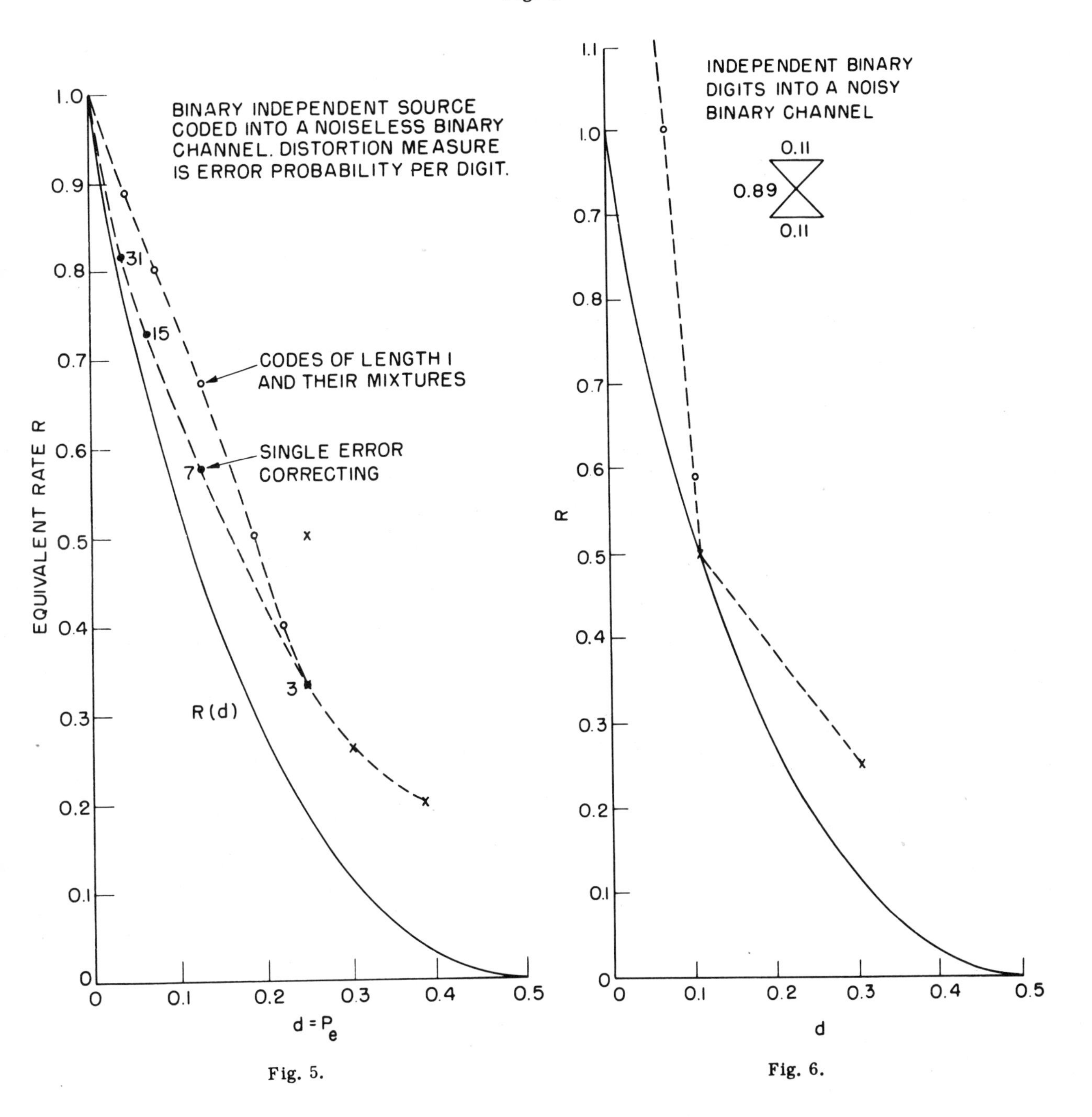

Fig. 5.

Fig. 6.

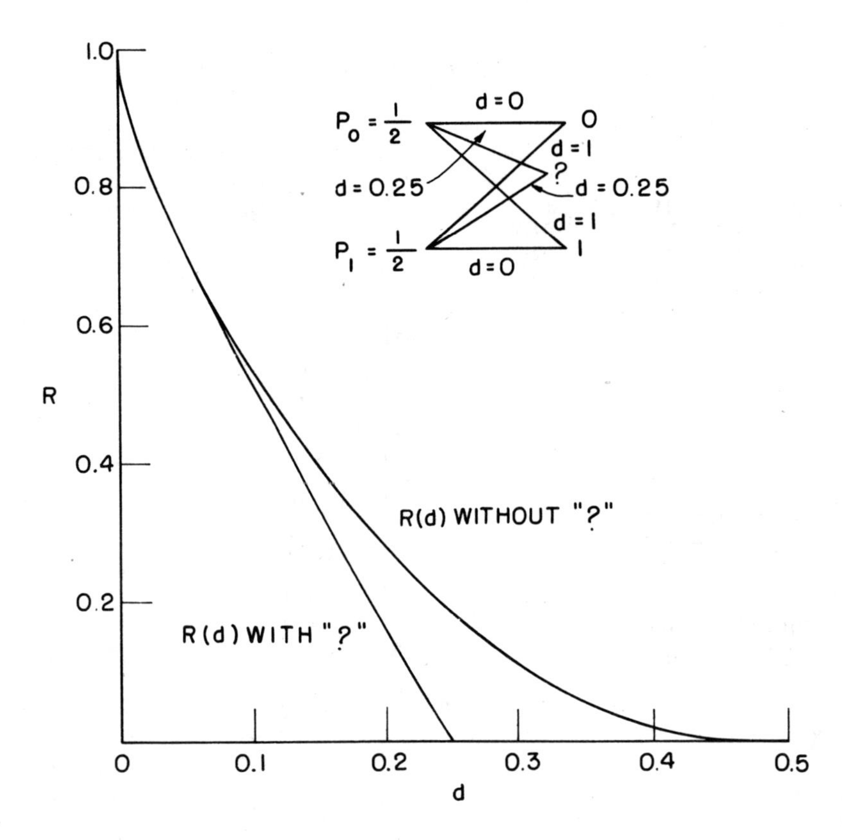
1.0
0.8
0.6
0.4
0.2
R
0
0.1
0.2
0.3
0.4
0.5
d
$P_0 = \frac{1}{2}$
$P_1 = \frac{1}{2}$
d = 0
d = 0
d = 1
d = 1
d = 0.25
d = 0.25
0
?
1
R(d) WITHOUT "?"
R(d) WITH "?"

Fig. 7.

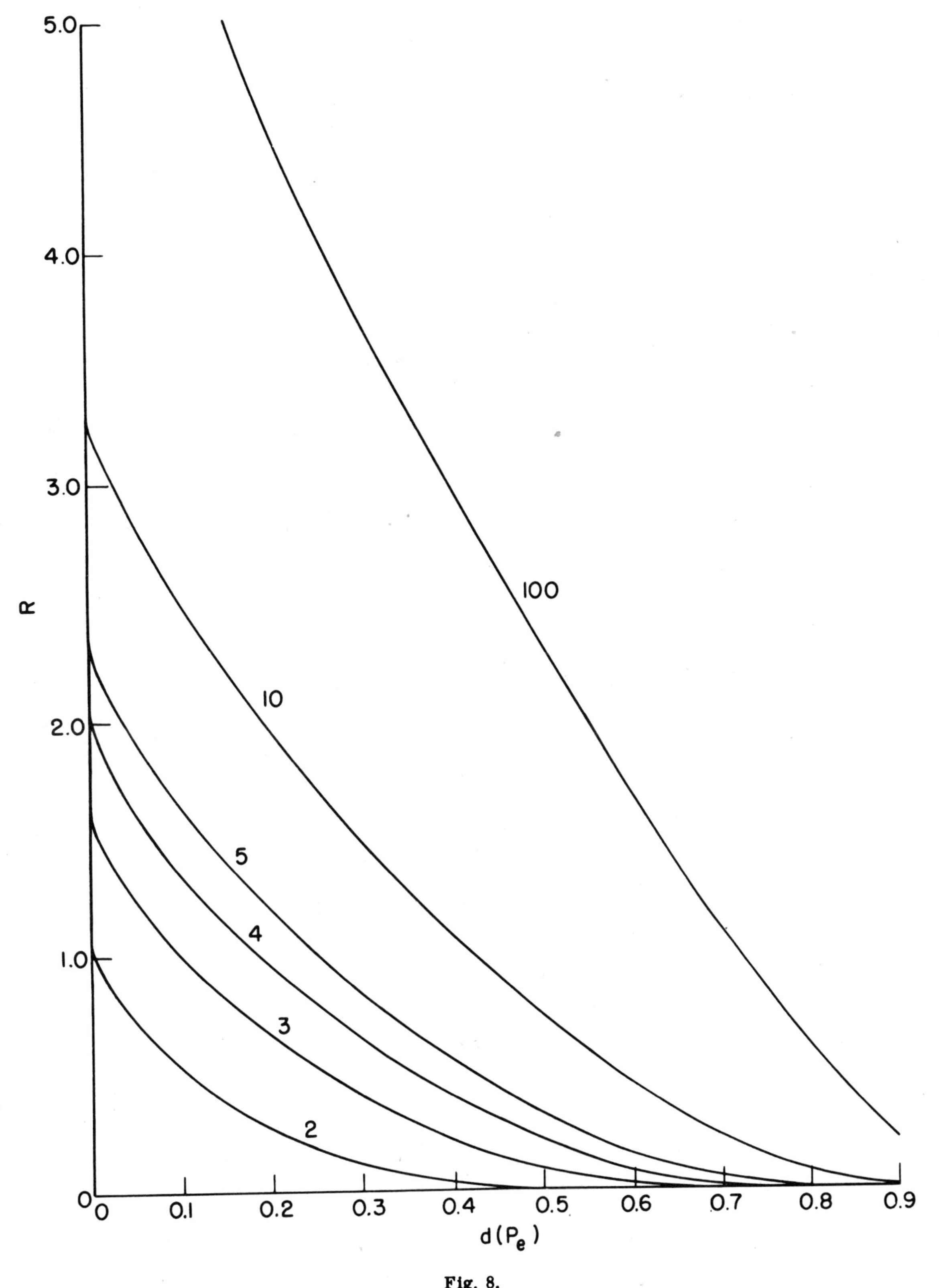

Fig. 8.

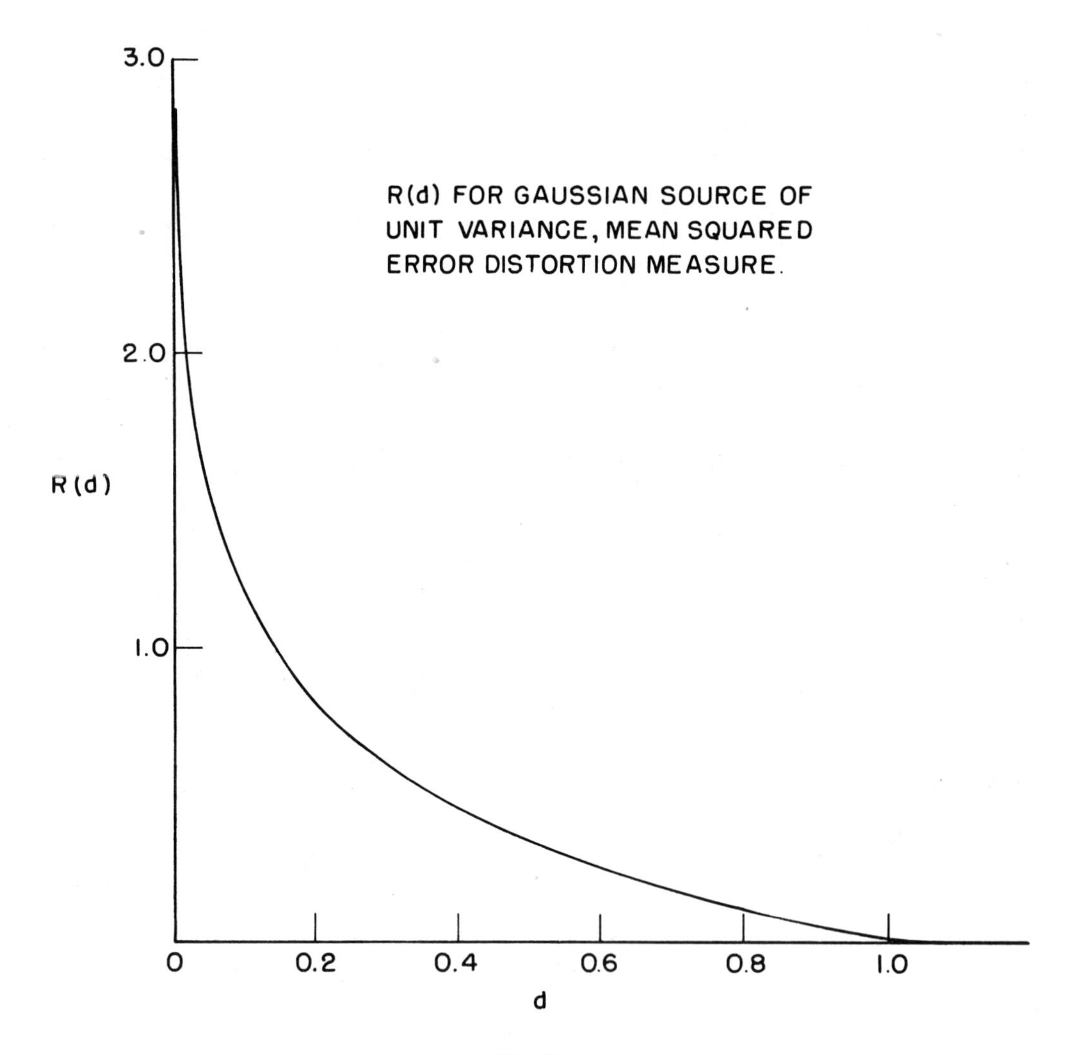
R(d) FOR GAUSSIAN SOURCE OF
UNIT VARIANCE, MEAN SQUARED
ERROR DISTORTION MEASURE.
3.0
2.0
1.0
R(d)
0
0.2
0.4
0.6
0.8
1.0
d

Fig. 9.

Quantizing for Minimum Distortion*

JOEL MAX†

Summary—**This paper discusses the problem of the minimization of the distortion of a signal by a quantizer when the number of output levels of the quantizer is fixed. The distortion is defined as the expected value of some function of the error between the input and the output of the quantizer. Equations are derived for the parameters of a quantizer with minimum distortion. The equations are not soluble without recourse to numerical methods, so an algorithm is developed to simplify their numerical solution. The case of an input signal with normally distributed amplitude and an expected squared error distortion measure is explicitly computed and values of the optimum quantizer parameters are tabulated. The optimization of a quantizer subject to the restriction that both input and output levels be equally spaced is also treated, and appropriate parameters are tabulated for the same case as above.**

* Manuscript received by the PGIT, September 25, 1959. This work was performed by the Lincoln Lab., Mass. Inst. Tech., Lexington, Mass., with the joint support of the U. S. Army, Navy, and Air Force.

† Lincoln Lab., Mass. Inst. Tech., Lexington, Mass.

IN MANY data-transmission systems, analog input signals are first converted to digital form at the transmitter, transmitted in digital form, and finally reconstituted at the receiver as analog signals. The resulting output normally resembles the input signal but is not precisely the same since the quantizer at the transmitter produces the same digits for all input amplitudes which lie in each of a finite number of amplitude ranges. The receiver must assign to each combination of digits a single value which will be the amplitude of the reconstituted signal for an original input anywhere within the quantized range. The difference between input and output signals, assuming errorless transmission of the digits, is the quantization error. Since the digital transmission rate of any system is finite, one has to use a quantizer which sorts the input into a finite number of ranges, N. For a given N, the system is described by specifying the end

Reprinted from *IRE Trans. Inform. Theory*, vol. IT-6, pp. 7–12, Mar. 1960.

points, x_k, of the N input ranges, and an output level, y_k, corresponding to each input range. If the amplitude probability density of the signal which is the quantizer input is given, then the quantizer output is a quantity whose amplitude probability density may easily be determined as a function of the x_k's and y_k's. Often it is appropriate to define a distortion measure for the quantization process, which will be some statistic of the quantization error. Then one would like to choose the N y_k's and the associated x_k's so as to minimize the distortion. If we define the distortion, D, as the expected value of $f(\epsilon)$, where f is some function (differentiable), and ϵ is the quantization error, and call the input amplitude probability density $p(x)$, then

$$D = E[f(s_{\text{in}} - s_{\text{out}})]$$
$$= \sum_{i=1}^{N} \int_{x_i}^{x_{i+1}} f(x - y_i)p(x)\, dx$$

where $x_{N+1} = \infty$, $x_1 = -\infty$, and the convention is that an input between x_i and x_{i+1} has a corresponding output y_i.

If we wish to minimize D for fixed N, we get necessary conditions by differentiating D with respect to the x_i's and y_i's and setting derivatives equal to zero:

$$\frac{\partial D}{\partial x_j} = f(x_j - y_{j-1})p(x_j) - f(x_j - y_j)p(x_j) = 0 \qquad j = 2, \cdots, N \qquad (1)$$

$$\frac{\partial D}{\partial y_j} = -\int_{x_j}^{x_{j+1}} f'(x - y_j)p(x)\, dx = 0 \qquad j = 1, \cdots, N \qquad (2)$$

(1) becomes (for $p(x_j) \neq 0$)

$$f(x_j - y_{j-1}) = f(x_j - y_j) \qquad j = 2, \cdots, N \qquad (3)$$

(2) becomes

$$\int_{x_j}^{x_{j+1}} f'(x - y_j)p(x)\, dx = 0 \qquad j = 1, \cdots, N. \qquad (4)$$

We may ask when these are sufficient conditions. The best answer one can manage in a general case is that if all the second partial derivatives of D with respect to the x_i's and y_i's exist, then the critical point determined by conditions (3) and (4) is a minimum if the matrix whose ith row and jth column element is

$$\left.\frac{\partial^2 D}{\partial p_i\, \partial p_j}\right|_{\text{critical point}},$$

where the p's are the x's and y's, is positive definite. In a specific case, one may determine whether or not the matrix is positive definite or one may simply find all the critical points (*i.e.*, those satisfying necessary conditions) and evaluate D at each. The absolute minimum must be at one of the critical points since "end points" can be easily ruled out.

The sort of f one would want to use would be a good metric function, *i.e.*, $f(x)$ is monotonically nondecreasing

$$f(0) = 0$$
$$f(x) = f(-x).$$

If we require that $f(x)$ be *monotonically increasing* (with x) then (1) implies

$$|x_j - y_{j-1}| = |x_j - y_j| \qquad j = 2, \cdots, N$$

which implies (since y_{j-1} and y_j should not coincide) that

$$x_j = (y_j + y_{j-1})/2 \qquad j = 2, \cdots, N$$

(x_j is halfway between y_j and y_{j-1}).

We now take a specific example of $f(x)$ to further illuminate the situation.

Let $f(x) = x^2$

(3) implies

$$x_j = (y_j + y_{j-1})/2 \quad \text{or} \quad y_j = 2x_j - y_{j-1} \qquad j = 2, \cdots, N, \qquad (5)$$

(4) implies

$$\int_{x_j}^{x_{j+1}} (x - y_j)p(x)\, dx = 0 \qquad j = 1, \cdots, N. \qquad (6)$$

That is, y_j is the centroid of the area of $p(x)$ between x_j and x_{j+1}.

Because of the complicated functional relationships which are likely to be induced by $p(x)$ in (6), this is not a set of simultaneous equations we can hope to solve with any ease. Note, however, that if we choose y_1 correctly we can generate the succeeding x_i's and y_i's by (5) and (6), the latter being an implicit equation for x_{j+1} in terms of x_j and y_j.

A method of solving (5) and (6) is to pick y_1, calculate the succeeding x_i's and y_i's by (5) and (6) and then if y_N is the centroid of the area of $p(x)$ between x_N and ∞, y_1 was chosen correctly. (Of course, a different choice is appropriate to each value of N.) If y_N is not the appropriate centroid, then of course y_1 must be chosen again. This search may be systematized so that it can be performed on a computer in quite a short time.[1]

This procedure has been carried out numerically on the IBM 709 for the distribution $p(x) = 1/\sqrt{2\pi}\, e^{-x^2/2}$, under the restriction that $x_{N/2+1} = 0$ for N even, and $y_{(N+1)/2} = 0$ for N odd. This procedure gives symmetric results, *i.e.*,

[1] Obtaining *explicit* solutions to the quantizer problem for a nontrivial $p(x)$ is easily the most difficult part of the problem. The problem may be solved analytically where $p(x) = 1/\sqrt{2\pi}\, e^{-x^2/2}$ only for $N = 1$, $N = 2$. For $N = 1$, $x_1 = -\infty$, $y_1 = 0$, $x_2 = +\infty$. For $N = 2$, $x_1 = -\infty$, $y_1 = -\sqrt{2/\pi}$, $x_2 = 0$, $y_2 = \sqrt{2/\pi}$, $x_3 = +\infty$, ($\sqrt{2/\pi}$ is the centroid of the portion of $1/\sqrt{2\pi}\, e^{-x^2/2}$ between the origin and $+\infty$.) For $N \geq 3$, some sort of numerical estimation is required. A somewhat different approach, which yields results somewhat short of the optimum, is to be found in V. A. Gamash, "Quantization of signals with non-uniform steps," *Electrosvyaz*, vol. 10, pp. 11–13; October, 1957.

if a signal amplitude x is quantized as y_k, then $-x$ is quantized as $-y_k$. The answers appear in Table I on page 11.

An attempt has been made to determine the functional dependence of the distortion on the number of output levels. A log-log plot of the distortion vs the number of output levels is in Fig. 1. The curve is not a straight line. The tangent to the curve at $N = 4$ has the equation $D = 1.32\ N^{-1.74}$ and the tangent at $N = 36$ has the equation $D = 2.21\ x^{-1.96}$. One would expect this sort of behavior for large N. When N is large, the amplitude probability density does not vary appreciably from one end of a single input range to another, except for very large amplitudes, which are sufficiently improbable so that their influence is slight. Hence, most of the output levels are very near to being the means of the end points of the corresponding input ranges. Now, the best way of quantizing a uniformly distributed input signal is to space the output levels uniformly and to put the end points of the input ranges halfway between the output levels, as in Fig. 2, shown for $N = 1$. The best way of producing a quantizer with $2N$ output levels for this distribution is to divide each input range in half and put the new output levels at the midpoints of these ranges, as in Fig. 3. It is easy to see that the distortion in the second case is $\frac{1}{4}$ that in the first. Hence, $D = kN^{-2}$ where k is some constant. In fact, k is the variance of the distribution.

If this sort of equal division process is performed on each input range of the optimum quantizer for a normally distributed signal with N output levels where N is large then again a reduction in distortion by a factor of 4 is expected. Asymptotically then, the equation for the tangent to the curve of distortion vs the number of output levels should be $D = kN^{-2}$ where k is some constant.

Commercial high-speed analog-to-digital conversion equipment is at present limited to transforming equal input ranges to outputs midway between the ends of the input ranges. In many applications one would like to know the best interval length to use, *i.e.*, the one yielding minimum distortion for a given number of output levels, N. This is an easier problem than the first, since it is only two-dimensional (for $N \geqq 2$), *i.e.*, D is a function of the common length r of the intervals and of any particular output level, y_k. If the input has a symmetric distribution and a symmetric answer is desired, the problem becomes one dimensional. If $p(x)$ is the input amplitude probability density and $f(x)$ is the function such that the distortion D is $E[f(s_{out} - s_{in})]$, then, for an even number $2N$ of outputs,

$$D = 2 \sum_{i=1}^{N-1} \int_{(i-1)r}^{ir} f\left(x - \left[\frac{2i-1}{2}\right]r\right) p(x)\,dx + 2 \int_{(N-1)r}^{\infty} f\left(x - \left[\frac{2N-1}{2}\right]r\right) p(x)\,dx. \quad (7)$$

For a minimum we require

$$\frac{dD}{dr} = -\sum_{i=1}^{N-1} (2i-1) \int_{(i-1)r}^{ir} f'\left(x - \left[\frac{2i-1}{2}\right]r\right) p(x)\,dx - (2N-1) \int_{(N-1)r}^{\infty} f'\left(x - \left[\frac{2N-1}{2}\right]r\right) p(x)\,dx = 0. \quad (8)$$

A similar expression exists for the case of an odd number of output levels. In either case the problem is quite susceptible to machine computation when $f(x)$, $p(x)$ and N are specified. Results have been obtained for $f(x) = x^2$, $p(x) = 1/\sqrt{2\pi}\, e^{-x^2/2}$, $N = 2$ to 36. They are indicated in Table II on page 12.

A log-log plot of distortion vs number of output levels appears in Fig. 1. This curve is not a straight line. The tangent to the curve at $N = 36$ has the equation $D = 1.47\ N^{-1.74}$. A log-log plot of output level spacing vs number of outputs for the equal spacing which yields lowest distortion is shown in Fig. 4. This curve is also not a straight line. Lastly, a plot of the ratio of the distortion for the optimum quantizer to that for the optimum equally spaced level quantizer can be seen in Fig. 5.

Key to the Tables

The numbering system for the table of output levels, y_i, and input interval end points, x_i, for the minimum mean-squared error quantization scheme for inputs with a normal amplitude probability density with standard deviation unity and mean zero is as follows:

> For the number of output levels, N, even, x_1 is the first end point of an input range to the right of the origin. An input between x_j and x_{j+1} produces an output y_j.
>
> For the number of output levels, N, odd, y_1 is the smallest non-negative output. An input between x_{j-1} and x_j produces an output y_j.

This description, illustrated in Fig. 6, is sufficient because of the symmetry of the quantizer. The expected squared error of the quantization process and informational entropy of the output of the quantizer are also tabulated for the optimal quantizers calculated.[2] (If p_k is the probability of the kth output, then the informational entropy is defined as $-\sum_{k=1}^{N} p_k \log_2 p_k$.)

Table II also pertains to a normally distributed input with standard deviation equal to unity. The meaning of the entries is self-explanatory.

[2] The values of informational entropy given show the minimum average number of binary digits required to code the quantizer output. It can be seen from the tables that this number is always a rather large fraction of $\log_2 N$, and in most cases quite near 0.9 $\log_2 N$. In the cases where $N = 2^n$, n an integer, a simple n binary digit code for the outputs of the quantizer makes near optimum use of the digital transmission capacity of the system.

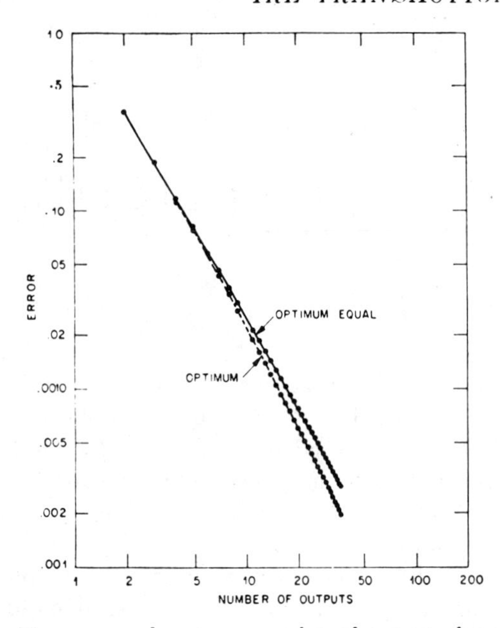

Fig. 1—Mean squared error vs number of outputs for optimum quantizer and optimum equally spaced level quantizer. (Minimum mean squared error for normally distributed input with $\sigma = 1$.)

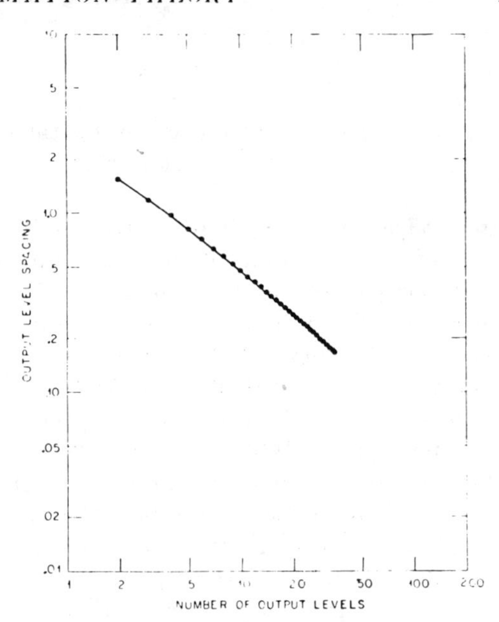

Fig. 4—Output level spacing vs number of output levels for equal optimum case. (Minimum mean squared error for normally distributed input with $\sigma = 1$.)

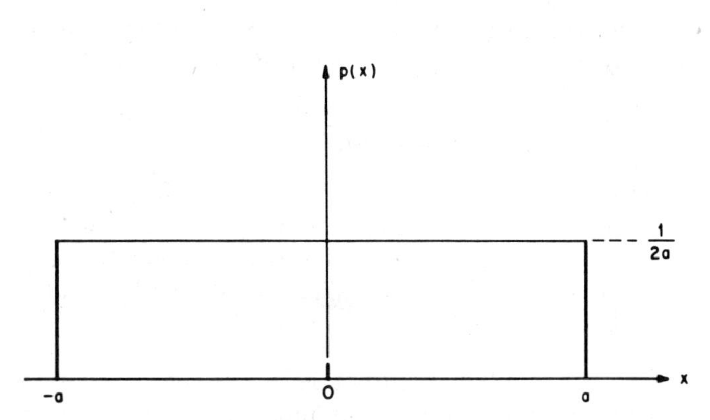

Fig. 2—Optimum quantization for the uniformly distributed case, $N = 1$. (Short strokes mark output levels and long strokes mark end points of corresponding input ranges.)

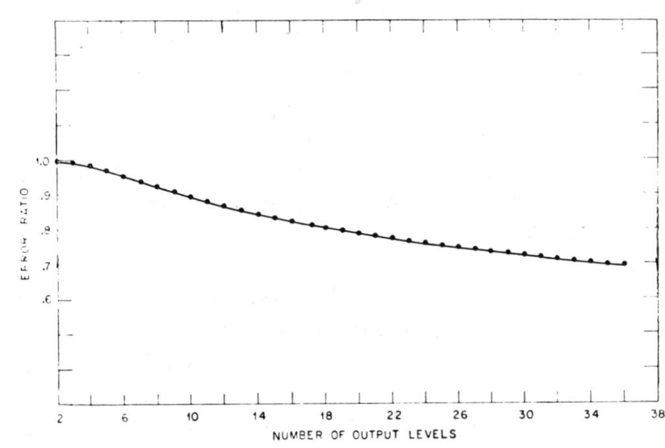

Fig. 5—Ratio of error for optimum quantizer to error for optimum equally spaced level quantizer vs number of outputs. (Minimum mean squared error for normally distribured input with $\sigma = 1$).

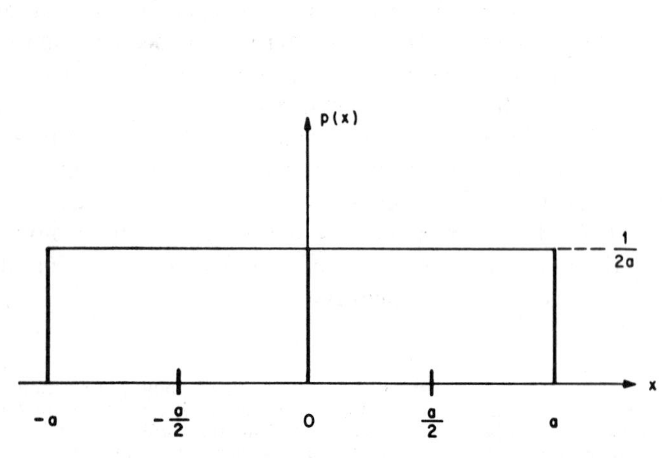

Fig. 3—Optimum quantization for the uniformly distributed case, $N = 2$. (Short strokes mark output levels and long strokes mark end points of corresponding input ranges.)

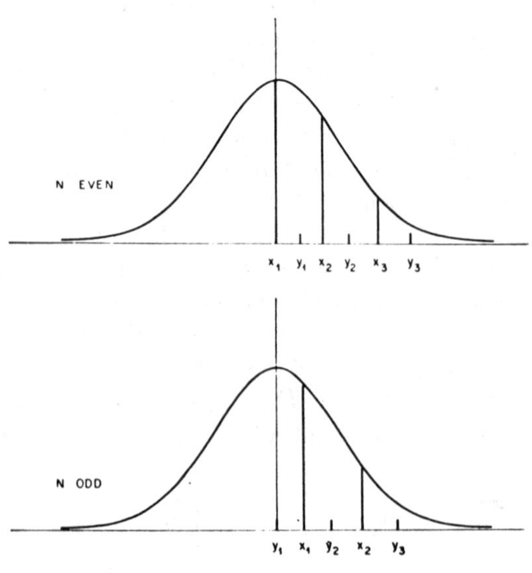

Fig. 6—Labeling of input range end points and output levels for the optimum quantizer. (Short strokes mark output levels and long strokes mark input range end points.)

TABLE I

Parameters for the Optimum Quantizer

	$N = 1$		$N = 2$		$N = 3$	
	x_j	y_j	x_j	y_j	x_j	y_j
$j = 1$	—	0.0	0.0	0.7980	0.0	1.224
2					0.6120	
Error	1.000		0.3634		0.1902	
Entropy	0.0		1.000		1.536	

	$N = 4$		$N = 5$		$N = 6$	
	x_j	y_j	x_j	y_j	x_j	y_j
$j = 1$	0.0	0.4528	0.3823	0.0	0.0	0.3177
2	0.9816	1.510	1.244	0.7646	0.6589	1.000
3				1.724	1.447	1.894
Error	0.1175		0.07994		0.05798	
Entropy	1.911		2.203		2.443	

	$N = 7$		$N = 8$		$N = 9$	
	x_j	y_j	x_j	y_j	x_j	y_j
$j = 1$	0.2803	0.0	0.0	0.2451	0.2218	0.0
2	0.8744	0.5606	0.5006	0.7560	0.6812	0.4436
3	1.611	1.188	1.050	1.344	1.198	0.9188
4		2.033	1.748	2.152	1.866	1.476
						2.255
Error	0.04400		0.03454		0.02785	
Entropy	2.647		2.825		2.983	

	$N = 10$		$N = 11$		$N = 12$	
	x_j	y_j	x_j	y_j	x_j	y_j
$j = 1$	0.0	0.1996	0.1837	0.0	0.0	0.1684
2	0.4047	0.6099	0.5599	0.3675	0.3401	0.5119
3	0.8339	1.058	0.9656	0.7524	0.6943	0.8768
4	1.325	1.591	1.436	1.179	1.081	1.286
5	1.968	2.345	2.059	1.693	1.534	1.783
6				2.426	2.141	2.499
Error	0.02293		0.01922		0.01634	
Entropy	3.125		3.253		3.372	

	$N = 13$		$N = 14$		$N = 15$	
	x_j	y_j	x_j	y_j	x_j	y_j
$j = 1$	0.1569	0.0	0.0	0.1457	0.1369	0.0
2	0.4760	0.3138	0.2935	0.4413	0.4143	0.2739
3	0.8126	0.6383	0.5959	0.7505	0.7030	0.5548
4	1.184	0.9870	0.9181	1.086	1.013	0.8512
5	1.623	1.381	1.277	1.468	1.361	1.175
6	2.215	1.865	1.703	1.939	1.776	1.546
7		2.565	2.282	2.625	2.344	2.007
8						2.681
Error	0.01406		0.01223		0.01073	
Entropy	3.481		3.582		3.677	

	$N = 16$		$N = 17$		$N = 18$	
	x_j	y_j	x_j	y_j	x_j	y_j
$j = 1$	0.0	0.1284	0.1215	0.0	0.0	0.1148
2	0.2582	0.3881	0.3670	0.2430	0.2306	0.3464
3	0.5224	0.6568	0.6201	0.4909	0.4653	0.5843
4	0.7996	0.9424	0.8875	0.7493	0.7091	0.8339
5	1.099	1.256	1.178	1.026	0.9680	1.102
6	1.437	1.618	1.508	1.331	1.251	1.400
7	1.844	2.069	1.906	1.685	1.573	1.746
8	2.401	2.733	2.454	2.127	1.964	2.181
				2.781	2.504	2.826
Error	0.009497		0.008463		0.007589	
Entropy	3.765		3.849		3.928	

	$N = 19$		$N = 20$		$N = 21$	
	x_j	y_j	x_j	y_j	x_j	y_j
$j = 1$	0.1092	0.0	0.0	0.1038	0.09918	0.0
2	0.3294	0.2184	0.2083	0.3128	0.2989	0.1984
3	0.5551	0.4404	0.4197	0.5265	0.5027	0.3994
4	0.7908	0.6698	0.6375	0.7486	0.7137	0.6059
5	1.042	0.9117	0.8661	0.9837	0.9361	0.8215
6	1.318	1.173	1.111	1.239	1.175	1.051
7	1.634	1.464	1.381	1.524	1.440	1.300
8	2.018	1.803	1.690	1.857	1.743	1.579
9	2.550	2.232	2.068	2.279	2.116	1.908
10		2.869	2.594	2.908	2.635	2.324
11						2.946
Error	0.006844		0.006203		0.005648	
Entropy	4.002		4.074		4.141	

	$N = 22$		$N = 23$		$N = 24$	
	x_j	y_j	x_j	y_j	x_j	y_j
$j = 1$	0.0	0.09469	0.09085	0.0	0.0	0.08708
2	0.1900	0.2852	0.2736	0.1817	0.1746	0.2621
3	0.3822	0.4793	0.4594	0.3654	0.3510	0.4399
4	0.5794	0.6795	0.6507	0.5534	0.5312	0.6224
5	0.7844	0.8893	0.8504	0.7481	0.7173	0.8122
6	1.001	1.113	1.062	0.9527	0.9122	1.012
7	1.235	1.357	1.291	1.172	1.119	1.227
8	1.495	1.632	1.546	1.411	1.344	1.462
9	1.793	1.955	1.841	1.681	1.595	1.728
10	2.160	2.366	2.203	2.000	1.885	2.042
11	2.674	2.982	2.711	2.406	2.243	2.444
12				3.016	2.746	3.048
Error	0.005165		0.004741		0.004367	
Entropy	4.206		4.268		4.327	

	$N = 25$		$N = 26$		$N = 27$	
	x_j	y_j	x_j	y_j	x_j	y_j
$j = 1$	0.08381	0.0	0.0	0.08060	0.07779	0.0
2	0.2522	0.1676	0.1616	0.2425	0.2340	0.1556
3	0.4231	0.3368	0.3245	0.4066	0.3921	0.3124
4	0.5982	0.5093	0.4905	0.5743	0.5537	0.4719
5	0.7797	0.6870	0.6610	0.7477	0.7202	0.6354
6	0.9702	0.8723	0.8383	0.9289	0.8936	0.8049
7	1.173	1.068	1.025	1.121	1.077	0.9824
8	1.394	1.279	1.224	1.328	1.273	1.171
9	1.641	1.510	1.442	1.556	1.487	1.374
10	1.927	1.772	1.685	1.814	1.727	1.599
11	2.281	2.083	1.968	2.121	2.006	1.854
12	2.779	2.480	2.318	2.514	2.352	2.158
13		3.079	2.811	3.109	2.842	2.547
14						3.137
Error	0.004036		0.003741		0.003477	
Entropy	4.384		4.439		4.491	

	$N = 28$		$N = 29$		$N = 30$	
	x_j	y_j	x_j	y_j	x_j	y_j
$j = 1$	0.0	0.07502	0.07257	0.0	0.0	0.07016
2	0.1503	0.2256	0.2182	0.1451	0.1406	0.2110
3	0.3018	0.3780	0.3655	0.2913	0.2821	0.3532
4	0.4556	0.5333	0.5154	0.4396	0.4255	0.4978
5	0.6132	0.6930	0.6693	0.5912	0.5719	0.6460
6	0.7760	0.8589	0.8287	0.7475	0.7225	0.7990
7	0.9460	1.033	0.9956	0.9100	0.8788	0.9586
8	1.126	1.218	1.172	1.081	1.043	1.127
9	1.319	1.419	1.362	1.263	1.217	1.306
10	1.529	1.640	1.570	1.461	1.404	1.501
11	1.766	1.892	1.804	1.680	1.609	1.717
12	2.042	2.193	2.077	1.929	1.840	1.964
13	2.385	2.578	2.417	2.226	2.111	2.258
14	2.871	3.164	2.899	2.609	2.448	2.638
15				3.190	2.926	3.215
Error	0.003240		0.003027		0.002834	
Entropy	4.542		4.591		4.639	

Cont'd next page

TABLE I, *Cont'd*

	$N = 31$		$N = 32$		$N = 33$	
	x_i	y_i	x_i	y_i	x_i	y_i
$j = 1$	0.06802	0.0	0.0	0.06590	0.06400	0.0
2	0.2045	0.1360	0.1320	0.1981	0.1924	0.1280
3	0.3422	0.2729	0.2648	0.3314	0.3218	0.2567
4	0.4822	0.4115	0.3991	0.4668	0.4530	0.3868
5	0.6254	0.5528	0.5359	0.6050	0.5869	0.5192
6	0.7730	0.6979	0.6761	0.7473	0.7245	0.6547
7	0.9265	0.8481	0.8210	0.8947	0.8667	0.7943
8	1.088	1.005	0.9718	1.049	1.015	0.9392
9	1.259	1.170	1.130	1.212	1.171	1.091
10	1.444	1.347	1.299	1.387	1.338	1.252
11	1.646	1.540	1.482	1.577	1.518	1.424
12	1.875	1.753	1.682	1.788	1.716	1.612
13	2.143	1.997	1.908	2.029	1.940	1.821
14	2.477	2.289	2.174	2.319	2.204	2.060
15	2.952	2.665	2.505	2.692	2.533	2.347
16		3.239	2.977	3.263	3.002	2.718
17						3.285
Error	0.002658		0.002499		0.002354	
Entropy	4.685		4.730		4.773	

	$N = 34$		$N = 35$		$N = 36$	
	x_i	y_i	x_i	y_i	x_i	y_i
$j = 1$	0.0	0.06212	0.06043	0.0	0.0	0.05876
2	0.1244	0.1867	0.1816	0.1209	0.1177	0.1765
3	0.2495	0.3122	0.3036	0.2423	0.2359	0.2952
4	0.3758	0.4394	0.4272	0.3650	0.3552	0.4152
5	0.5043	0.5691	0.5530	0.4895	0.4762	0.5372
6	0.6355	0.7020	0.6819	0.6166	0.5996	0.6620
7	0.7705	0.8391	0.8146	0.7471	0.7261	0.7903
8	0.9104	0.9818	0.9523	0.8820	0.8567	0.9231
9	1.057	1.131	1.096	1.023	0.9923	1.062
10	1.211	1.290	1.248	1.170	1.134	1.207
11	1.375	1.460	1.411	1.327	1.285	1.362
12	1.553	1.646	1.587	1.495	1.445	1.528
13	1.749	1.853	1.781	1.679	1.619	1.710
14	1.971	2.090	2.001	1.883	1.812	1.913
15	2.232	2.375	2.260	2.119	2.030	2.146
16	2.559	2.743	2.584	2.401	2.287	2.427
17	3.025	3.307	3.048	2.767	2.609	2.791
18				3.328	3.070	3.349
Error	0.002220		0.002097		0.001985	
Entropy	4.815		4.856		4.895	

TABLE II

PARAMETERS FOR THE OPTIMUM EQUALLY SPACED LEVEL QUANTIZER

Number Output Levels	Output Level Spacing	Mean Squared Error	Informational Entropy
1	—	1.000	0.0
2	1.596	0.3634	1.000
3	1.224	0.1902	1.536
4	0.9957	0.1188	1.904
5	0.8430	0.08218	2.183
6	0.7334	0.06065	2.409
7	0.6508	0.04686	2.598
8	0.5860	0.03744	2.761
9	0.5338	0.03069	2.904
10	0.4908	0.02568	3.032
11	0.4546	0.02185	3.148
12	0.4238	0.01885	3.253
13	0.3972	0.01645	3.350
14	0.3739	0.01450	3.440
15	0.3534	0.01289	3.524
16	0.3352	0.01154	3.602
17	0.3189	0.01040	3.676
18	0.3042	0.009430	3.746
19	0.2909	0.008594	3.811
20	0.2788	0.007869	3.874
21	0.2678	0.007235	3.933
22	0.2576	0.006678	3.990
23	0.2482	0.006185	4.045
24	0.2396	0.005747	4.097
25	0.2315	0.005355	4.146
26	0.2240	0.005004	4.194
27	0.2171	0.004687	4.241
28	0.2105	0.004401	4.285
29	0.2044	0.004141	4.328
30	0.1987	0.003905	4.370
31	0.1932	0.003688	4.410
32	0.1881	0.003490	4.449
33	0.1833	0.003308	4.487
34	0.1787	0.003141	4.524
35	0.1744	0.002986	4.560
36	0.1703	0.002843	4.594

INFORMATION TRANSMISSION WITH ADDITIONAL NOISE

R.L.Dobrushin
The Moscow State University,
Moscow, USSR

B.S.Tsybakov
The Institute of Radiotechnics
and Electronics of the Academy
of Sciences, Moscow, USSR

1. The usual handling of the main problem

The usual handling of the main problem of Shannon theory of optimal information coding (cf. [1-3]) consists in the following: there is a message to be transmitted which is a random variable ξ whose values are set over a measurable sample space X. (In most applications X is assumed to be the space of functions $x(t)$ where the argument t takes values within a specific interval or is identified with a discrete series of values so that the variable ξ may be treated as a random process with continuous or discrete time respectively. Further there is the space Y of signal values on the communication channel input (in most applications it is also a space of functions).

The coding operation is given simply by a function $f(x)$, $x \in X$ having values within the Y. (We shall abstain, for the purposes of this article, from making reservations concerning the measurability of the space, the functions and the sets considered. The orthodoxically mathematical reader may well reconstruct these for himself). Thus the message ξ is mapped onto the random signal $\eta = f(\xi)$ on the channel input */. It is sometimes necessary to impose certain limits on the variable η (for example the physically natural requirement of the limitation of the average input signal power). In general case these limitations may be expressed by the requirement that the distribution of the input signal η should belong to a certain selected class of distributions $\mathcal{P}_\eta$.

*/ The definition of coding may be generalized by admitting the operation of randomisation, i.e. admitting that for a fixed ξ the signal η could take random values. Such a generalisation would not change the results of this study.

Besides, there is given the space of output signal values $\tilde{Y}$ and the channel transition function $Q(y, B)$, $y \in Y$, $B \subset \tilde{Y}$ which is for any fixed y the distribution of probabilities in the measurable space $\tilde{Y}$. A communication over the channel brings about on its output a random value $\tilde{\eta}$ which is called the channel output signal being related to the message ξ and the input signal η by the following conditions: a) that with any $y \in Y$ and the measurable $B \subset \tilde{Y}$ the conditional probability

$$\mathcal{P}\{\tilde{\eta} \in B / \eta = y\} = Q(y, B) \qquad (1)$$

and b) that with the given value $\eta = y$ the conditional distribution $\tilde{\eta}$ does not depend on the value assumed by the variable ξ, i.e. the three variables $\xi, \eta, \tilde{\eta}$ form the Markov chain. This last condition b) is often shadowed in the explanatory expositions of the theory. Yet, it is very important since it implies that the probability characteristics of the output signal are fully defined by the value of the input signal transmitted over the channel and are in no other indirect way related to the value of the message to be transmitted. The conditions a) and b) may be collocated in the following relation: for any $x \in X, y \in Y, B \subset \tilde{Y}$

$$\mathcal{P}\{\tilde{\eta} \in B / \eta = y, \xi = x\} = Q(y, B). \qquad (2)$$

Finally there is the space $\tilde{X}$ of the message values on the output. The decoding is defined as the function $g(\tilde{y})$, $\tilde{y} \in \tilde{Y}$ having values in the space $\tilde{X}$. The random variable $\tilde{\xi}$ which is called the output message is determined by the relation $\tilde{\xi} = g(\tilde{\eta})$. The selection of the coding $f(x)$; the decoding $g(\tilde{y})$; and the condition (2) defines uniquely the mutual probabilities distribution of all four variables ξ, $\eta = f(\xi)$, $\tilde{\eta}$, $\tilde{\xi} = g(\tilde{\eta})$
The transmission quality for a given method of communication occuring with fixed

Reprinted from *IRE Trans. Inform. Theory*, vol. IT-8, pp. 293-304, July 1962.

coding $f(x)$ and decoding $g(\tilde{y})$ is defined by the mutual distribution $P_{\xi\tilde{\xi}}$ of input and output messages. We shall extract from the set of possible mutual distributions for a pair of variables taking values over spaces X and $\tilde{X}$, respectively, a certain subset W which we shall call a condition of the transmission precision. We shall assume that our main object is to select the coding and decoding in such a way that the mutual distribution $P_{\xi\tilde{\xi}} \in W$ or, the condition of the transmission precision to be fulfilled. For example, in many cases it would be natural to suppose a cost function $\rho(x, \tilde{x})$, $x \in X$, $\tilde{x} \in \tilde{X}$ to be given, showing the extent of the loss of the input message x being interpreted on the output as $\tilde{x}$. Thus if X and $\tilde{X}$ are spaces of functions $x(t)$ and $\tilde{x}(t)$, $a \leq t \leq b$, with real values, then we can write

$$\rho(x(t), \tilde{x}(t)) = \int_a^b [x(t) - \tilde{x}(t)]^2 dt$$

or

$$\rho(x(t), \tilde{x}(t)) = \sup_{a \leq t \leq b} |x(t) - \tilde{x}(t)|.$$

The precision condition W is then defined as the set of mutual distributions of variables $\xi, \tilde{\xi}$ such that the ensemble average

$$M\rho(\xi, \tilde{\xi}) \leq \rho_0$$

where ρ_0 is a fixed constant. Sometimes the precision condition is defined by several functions $\rho_i(x, \tilde{x})$ and a set of limitations of the last-mentioned inequality type.

The main Shannon problem consists in the investigation of possibilities resulting from the best optimum choice of coding and decoding. Specifically, it consists in the following: there are given fixed spaces $X, Y, \tilde{Y}, \tilde{X}$ the input message distribution ξ, the channel transition function $Q(y, B)$ the channel input condition $\mathcal{P}_\eta$ and the condition of transmission precision W. The question to be answered is this: can there be such choices of coding $f(x)$ and decoding $g(\tilde{y})$ that: $\eta = f(\xi)$, $\tilde{\xi} = g(\tilde{\eta})$ distribution η belongs to $\mathcal{P}_\eta$, the condition (2) is met and, for the variable pair $\xi, \tilde{\xi}$ the precision condition W is fulfilled, In this case we shall say that the transmission is possible.

The solution of this problem was suggested in the main even by Shannon himself [1]. Later it was elaborated by his successors. This solution consists in the following. For any two random variables α and $\tilde{\alpha}$ taking values over spaces A and $\tilde{A}$ the expression $J(\alpha, \tilde{\alpha})$ is found which is called the amount of information in α relative to $\tilde{\alpha}$. If the variables α and $\tilde{\alpha}$ assume each a finite number of values and P_{ij}, $P_i = \sum P_{ij}$, $\tilde{P}_j = \sum P_{ij}$ are their mutual and individual probability distributions, then

$$J(\alpha, \tilde{\alpha}) = \sum_{ij} P_{ij} \log \frac{P_{ij}}{P_i \tilde{P}_j}. \tag{3}$$

In the general case should be considered all possible measurable functions $\varphi(a)$, $a \in A$, $\psi(\tilde{a})$, $\tilde{a} \in \tilde{A}$ assuming each a finite number of values and it should be accepted that

$$J(\alpha, \tilde{\alpha}) = \sup_{\varphi, \psi} J(\varphi(\alpha), \psi(\tilde{\alpha})), \tag{4}$$

where the upper bound is taken over all possible pairs φ, ψ while the information $J(\varphi(\alpha), \psi(\tilde{\alpha}))$ is derived from the equation (3). The channel capacity C is determined as

$$C = \sup J(\eta, \tilde{\eta}) \tag{5}$$

where the upper bound is taken over all pairs of random variables $\eta, \tilde{\eta}$ assuming values over spaces $Y, \tilde{Y}$ and such that condition (1) is true and the distribution η belongs to $\mathcal{P}_\eta$. The entropy of the message ξ under given condition W is defined as

$$H(W) = \inf J(\xi, \tilde{\xi}) \tag{6}$$

where ξ has the given distribution and the lower bound is taken over all variables $\tilde{\xi}$ such that the mutual distribution ξ and $\tilde{\xi}$ belongs to W.

The first statement of Shannon theorem is that given

$$H(W) > C \tag{7}$$

the transmission is impossible. This true without any stipulations. The reversed statement runs that provided

$$H(W) < C \tag{8}$$

the transmission is possible. This one is true only in an asymptotic sence and with substantial limitations. We shall not enter here upon that complicated problem but direct the reader to special literature (which is reviewed in [5]). For the present purposes we will just note that the main constraint is in $H(W) \to \infty$ (implying $C \to \infty$) and therefore the theorem is applicable only when a large amount of information is transmitted. The rest of the constraints which are many all require as the only premise that the random processes under consideration have some properties of the ergodic type or just guarantee the nondegeneration in some sence of the objects considered. These constraints may easily be supposed to be met in the generality of physically interesting cases.

2. Situation with additional moise. Handling of the problem. General results

The present study treats of a handling of the problem which generalizes the original handling by Shannon. It seems that in numerous actual cases not the original message ξ is coded which is of interest to the recepient at the channel output, but merely some preliminarily distorbed (noised) variant, i.e. some random variable ζ which stands in some relation to ξ.

For instance, imagine our aim to be to transmit over the communication channel the value of some physical variable ξ (say, pressure in a certain point). Admit further that to measure ξ an instrument (a gauge) is used which has a mean square operating error σ^2.

In this case rather than ξ (the message to be transmitted) it is $\zeta = \xi + \gamma$ which enters on the channel input, where γ is the random measuring error with σ^2 variance.

Here is an example taken from image transmission. It is known that the image to be transmitted is first fed onto the mozaic of the transmitting tube and then scanned from it. As a result of various fluctuations to which transmitting tubes subject the mozaic elements potential occasionally deviates from the value dictated to it by the brightness of a particular point of the object handled. Due to these deviations, the message entering the transmiter and then being coded is not an exact copy of the object but rather some noise-bearing variant of it.

Finally an example where we have a communication exchange centre from which long-distance information transmission is undertaken and in this exchange centre information coding equipment is maintained! However the particular messages are generated at points other than the centre which are connected to it by local transmission lines. It would be not reasonable to install powerful coding equipment on the transmitting ends of these lines, so there is no problem of optimal coding choice just there. Suppose then that the mode of treatment of the message on the local input end be a g given one, then the message ξ will differ from the value of the message ζ on the recieving end of that line and we might assume as given their common distribution (for instance we could considere noise on this line to be additive, then $\zeta = \xi + \gamma$ where γ is a discrete variable independent of ξ).

Following out these considerations we shall assume that besides space X of the input message values we have given space Z, of noised input message values and also given mutual distribution of variables ξ and ζ which take values over spaces X and Z, respectively. This mutual distribution can be conveniently defined through the input message distribution ξ and the transient function for any $x \in X$, $C \subset Z$

$$Q_I(x; C) = \mathcal{P}\{\zeta \in C / \xi = x\} \qquad (9)$$

which is the conditional distribution of the noised message for a fixed value of the original message. Further the coding $f(z)$ is a function of which the argument is $z \in Z$ and the values $f(z) \in Y$.

Similarly to the original handling of Shannon problem the channel output signal $\tilde{\eta}$ must satisfy condition (1). As to the condition b) it must be replaced by a stronger condition: b') for fixed variable $\eta = y$ the conditional distribution for $\tilde{\eta}$ is independent of values taken by the pair of random variables (ξ, ζ). The explicit meaning of this again is that the output signal is defined by the input signal and is in no other way related either to the message itself or to its noised variant. Conditions (1) and b') can be correlated as follows: for any $x \in X$, $z \in Z$, $y \in Y$, $B \subset \tilde{Y}$

$$\mathcal{P}\{\tilde{\eta} \in B / \eta = y, \zeta = z, \xi = x\} = Q(y, B) \qquad (10)$$

In addition, as previously, we shall assume that distribution η belongs to $\mathcal{P}_\eta$.

A second generalisation we shall introduce into the Shannon scheme is like this. Suppose that decoding $g(\tilde{y})$, $\tilde{y} \in \tilde{Y}$ reduces the output signal not to the mesage ξ itself, which is supplied to the user, but merely to a "decoded signal" $\tilde{\zeta} = g(\tilde{\eta})$ taking values over a space $\tilde{Z}$. The message $\tilde{\xi}$ itself is a random variable stochastically related to $\tilde{\zeta}$. Thus elaborating on the above example we can assume that decoding equipment is operated at a communication exchange centre on the recieving end of the long distance communication line and that the decoded signal is transmitted from this centre over local lines to information users while subject to additional distortion in this process. On the other hand, at the receiving end of the local communication line there is no processing of the recieved signal at all or such processing is only elementary so as to claim no optimum improvment. Then the message ξ on the local line output will be stochastically related to the decoded signal $\tilde{\zeta}$ at the exchange centre. (For example we can assume additive noise and $\tilde{\xi}=\tilde{\zeta}+\gamma$ where γ is independent of $\tilde{\zeta}$).

In following out the above said we shall assume that there is an additionally given measurable space $\tilde{Z}$ of decoded signal values and also a given transient function $Q_{II}(\tilde{z},A)$ which with fixed $\tilde{z}\in\tilde{Z}$ is a probability distribution over $\tilde{X}$. It is supposed that there is given decoding $g(\tilde{y})$, $\tilde{y}\in\tilde{Y}$ which is a function with values within $\tilde{Z}$, and the random variable $\tilde{\zeta}=g(\tilde{\eta})$. It is further assumed that the random a output message $\tilde{\xi}$ is such that a) with any $\tilde{z}\in\tilde{Z}$, $\tilde{A}\subset\tilde{X}$

$$\mathcal{P}\{\tilde{\xi}\in B/\tilde{\zeta}=\tilde{z}\}=Q_{II}(\tilde{z},\tilde{A}) \qquad (11)$$

and that b) with a fixed value $\tilde{\zeta}=\tilde{z}$ the conditional distribution for the output message $\tilde{\xi}$ is independent of values taken by random variables $\xi,\zeta,\eta,\tilde{\eta}$. The apparent meaning of the above assumption is that all data in the $\tilde{\xi}$ concerning what occurred on the input end and concerning the value of the output signal $\tilde{\eta}$ are derived from the value of the decoded signal $\tilde{\zeta}$. Both conditions can be put together like this: with any $x\in X$, $z\in Z, y\in Y, \tilde{y}\in\tilde{Y}, \tilde{z}\in\tilde{Z}, \tilde{B}\subset\tilde{X}$

$$\mathcal{P}\{\tilde{\xi}\in\tilde{B}/\tilde{\zeta}=\tilde{z},\tilde{\eta}=\tilde{y},\eta=y,\zeta=z,\xi=x\}=Q_{II}(\tilde{z},\tilde{B}) \qquad (12)$$

The choice of coding f and decoding g, together with conditions (9),(10) and (12) defines uniquely the common distribution of the variable system ξ,ζ, $\eta=f(\zeta)$, $\tilde{\eta}$, $\tilde{\zeta}=g(\eta)$, $\tilde{\xi}$.

Furthermore we shall assume that as previously there is the condition of transmission precision W superimposed on the common distribution of ξ and $\tilde{\xi}$. We can now formulate the main problem which is of interest to us: given spaces $X, Z, Y, \tilde{Y}, \tilde{Z}, \tilde{X}$ distribution of ξ, transient functions $Q(y,B)$, $Q_I(x,C)$ $Q_{II}(\tilde{z},\tilde{A})$ it is inquired whether coding $f(z)$ and decoding $g(\tilde{y})$ can be so chosen that having the conditions $\eta=f(\zeta)$, $\tilde{\zeta}=g(\eta)$, distribution of η within $\mathcal{P}_\eta$ and relations (9), (10) and (12) satisfied, the mutual distribution of variables ξ and $\tilde{\xi}$ should be consistent with the precision condition W. In this case we will say that the transmission is possible. The system under discussion is illustrated on Fig. 1.

In formulation the solution we will again need the channel capacity as defined by relation (5). As to the definition (6) of enthropy $H(W)$, it must be considerably.complicated. We shall say that the set of four random variables $\xi,\zeta,\tilde{\zeta},\tilde{\xi}$ adopting values over $X, Z, \tilde{Z}, \tilde{X}$ respectively, is acceptable provided, 1) ξ has a given message input distribution; 2) the conditional probability

$$\mathcal{P}\{\zeta\in C/\xi=x\}=Q_I(x,C) \qquad (13)$$

with any $x\in X$ and $C\subset Z$, 3) the conditional probability

$$\mathcal{P}\{\tilde{\zeta}\in\tilde{C}/\xi=x,\zeta=z\}=\mathcal{P}\{\tilde{\zeta}\in\tilde{C}/\zeta=z\} \qquad (14)$$

with any $x\in X$, $z\in Z$ and $\tilde{C}\subset\tilde{Z}$, 4) the conditional probability

$$\mathcal{P}\{\tilde{\xi}\in\tilde{A}/\xi=x,\zeta=z,\tilde{\zeta}=\tilde{z}\}=Q_{II}(\tilde{z},\tilde{A}) \qquad (15)$$

with any $x\in X, z\in Z, \tilde{z}\in\tilde{Z}, \tilde{A}\subset\tilde{X}$ and finally, 5) the mutual distribution of variables ξ and $\tilde{\xi}$ belongs to W. Now we will determine

$$H(W,Q_I,Q_{II})=\inf J(\zeta,\tilde{\zeta}) \qquad (16)$$

where the lower bound is taken over all pairs of random variables ζ and $\tilde{\zeta}$ such that there are additional variables ξ and $\tilde{\xi}$ such that the set of variables $\xi,\zeta,\tilde{\zeta},\tilde{\xi}$ is acceptable.

Though the definition (16) seems to be rather bulky the entropy calculation $H(W,Q_I,Q_{II})$ for important specific cases is only a little more complicated than Shannon's usual enthropy $H(W)$ calculation, this we shall see below (section 5). The apparent meaning of definition (16) is also rather clear and almost coincides with the apparent meaning of the usual definition (6). We treat under the lower bound sign in (16) the variables ζ and $\tilde{\zeta}$ and not ξ and $\tilde{\xi}$ as in (6) since in our new situation it is exactly variables ζ and $\tilde{\zeta}$ respectively which are directly coded or present the the result of decoding. However, the set of pairs in which the upper bound is taken appears as previously with the requirement that the pair $(\xi,\tilde{\xi})$ have its distribution within W. As to conditions (13) and (15) they mean that ζ is derived from ξ, and $\tilde{\xi}$ from $\tilde{\zeta}$ with a noise contribution of given statistical characteristics. In the particular case when there is no additional noise contribution (i.e.under our new terms; the spaces X and Z coincide and with any $x\in X$ the distribution $Q_I(x,C)$ is concentrated in point x and, similarly, $\tilde{Z}$ and $\tilde{X}$ coincide, the distribution $Q_{II}(z,C)$ with any $\tilde{z}\in\tilde{Z}$ being concentrated in point $\tilde{z}$), the definition (16) is reduced to the Shannon definition (6).

Now the main problem solution is quite analogous to the solution of usual Shannon problem. We will show that if

$$H(W, Q_I, Q_{II}) > C \qquad (17)$$

then the transmission is impossible. If, on the other hand

$$H(W, Q_I, Q_{II}) < C \qquad (18)$$

then the transmission is possible. Naturally, in the conventional Shannon theorem this latter statement can only be true under additional limitations and in asymptotic sense. We shall go into this further in sect. 4. For the present we will just note that these conditions are quite identical to those of the Shannon theorem.

3. Statement on the impossibility of the transmission proved

Now for a proof of the impossibility of transmission if relation (17) is true. We shall need two known general properties of information quantity.

A) Let α, β be arbitrary random variables and $h(\alpha)$ a function defined within the space of variable of α. Then

$$J(\alpha, \beta) \geq J(h(\alpha), \beta). \qquad (19)$$

If there are two random variables β and γ adopting values over spaces B and C we can, if we wish, consider the pair (β, γ) as one random variable of which the values are pairs (b, c), $b \in B$, $c \in C$. Hence the information $J(\alpha, (\beta, \gamma))$ is defined. Since in giving value to the variable pair (β, γ) the value of β is also given uniquely we can consider β a function of the pair and by reference to (19) write that

$$J(\alpha, (\beta, \gamma)) \geq J(\alpha, \beta). \qquad (20)$$

B) Let α, β and γ be random variables taking values within A, B and C respectively, and such that with any $a \in A$, $b \in B$ and $\tilde{C} \subset C$

$$\mathcal{P}\{\gamma \in \tilde{C} / \beta = b, \alpha = a\} = \mathcal{P}\{\gamma \in \tilde{C} / \beta = b\} \qquad (21)$$

i.e. in mathematical language α, β, γ form the Markov chain. Then

$$J(\alpha, (\beta, \gamma)) = J(\alpha, \beta). \qquad (22)$$

In point of proofs of the A) and B) properties let it be noticed that for discrete variables they are elementary and are given for example in [4]. In the general case it is similarly performed. For rigorous mathematical demonstration it is given in [3].

As we proceed to prove the statement in question, let us assume that the transmission is possible, i.e. a system of variables $\xi, \zeta, \eta, \tilde{\eta}, \tilde{\zeta}, \tilde{\xi}$ exists which satisfies all requirements mentioned in sect. 2. Then, for the four variables $\xi, \zeta, \tilde{\zeta}, \tilde{\xi}$ the requirements 1)-5) are satisfied, which are part of the enthropy (16) determination.(Indeed, 1) is valid since by definition ξ has a given distribution, 2) flows from relation (9), 3) flows from relation (10) and the fact that $\tilde{\zeta}$ is not a random function of $\tilde{\eta}$, 4) is derived from (12) and, finally, 5) is also true by definition). Thus,

$$J(\zeta, \tilde{\zeta}) \geq H(W, Q_I, Q_{II}). \qquad (23)$$

It follows further from condition (10) that (1) is valid and since it was assumed that the distribution η belongs $\mathcal{P}_\eta$, it is apparent from definition (5) that

$$J(\eta, \tilde{\eta}) \leq C. \qquad (24)$$

It also derives from condition (10) that to the three variables $\zeta, \eta, \tilde{\eta}$ the property B) is applicable, hence

$$J((\zeta, \eta), \tilde{\eta}) = J(\eta, \tilde{\eta}). \qquad (25)$$

Using inequality (20) we abtain

$$J(\zeta, \tilde{\eta}) \leq J(\eta, \tilde{\eta}). \qquad (26)$$

At last, due to $\tilde{\zeta} = f(\tilde{\eta})$ we have, using property A), that

$$J(\zeta, \tilde{\eta}) \geq J(\zeta, \tilde{\zeta}). \qquad (27)$$

Collecting inequalities (23), (24), (26),(27) together we arrive at the conclusion that if the transmission is possible then

$$H(W, Q_I, Q_{II}) \leq C \qquad (28)$$

which proves the point in question.

4. Setting up the equivalent Shannon scheme

Our vindicatation of the impossibility of transmission if condition (18) is true will be based on a demonstration that the problem put forward in Sect.2 can be reduced to the conventional Shannon problem if a proper choice of the condition of transmission precision is made. By the same procedure the result of the previous section could also have been obtained but there we preferred to advance the complete proving method without relying on alien results*/.

We shall start by comparing precision condition W - which by definition is a set of mutual probability distributions

---

*/ Further exposition of this section may possibly seem too complicated and can thus be omitted without however hampering the understanding of the rest of the sections.

over spaces X and $\tilde{X}$ - another set $\tilde{W}$ of mutual probabilities distributions over spaces Z and $\tilde{Z}$. Specifically, we introduced in sect. 3 the concept of an acceptable system of random variables $\xi, \zeta, \tilde{\zeta}, \tilde{\xi}$ as a system of random variables satisfying conditions 1)-5). We now will suppose that a distribution $P_{\zeta\tilde{\zeta}}$ being the mutual distribution of two random variables ζ and $\tilde{\zeta}$ belongs within $\tilde{W}$ if $\xi, \tilde{\xi}$ can be so given that the variables system $\xi, \zeta, \tilde{\zeta}, \tilde{\xi}$ is acceptable. This definition is of course not very effective but from the mathematical standpoint it specifies uniquely the distribution set *7. Furthermore, since we have given mutual distribution of the input message ξ and the noised message ζ we can also consider as a given one-dimensional distribution P_ζ of the variable ζ which adopts values over space Z.

If we now define $H(\tilde{W})$ as

$$H(\tilde{W}) = \inf J(\zeta, \tilde{\zeta}) \qquad (29)$$

where the lower bound over all pairs $\zeta, \tilde{\zeta}$ is taken such that ζ has a distribution P_ζ and the mutual distribution $(\zeta, \tilde{\zeta})$ belongs to $\tilde{W}$, then $H(\tilde{W})$ will be the entropy of the message ζ under the provision of $\tilde{W}$ in the sense of Sect.1 (cf. equation 6). To go on, it follows from the definition of $\tilde{W}$ that the set of pairs $(\zeta, \tilde{\zeta})$ over which the lower bound is taken in (29) coincides with the set of pairs over which the lower bound is taken in (16) (it follows from (13) that for acceptable variables the distribution ζ coincides with P_ζ) and therefore

$$H(\tilde{W}) = H(W, Q_I, Q_{II}) \qquad (30)$$

and this in turn, on the basis of (28) brings us to

$$H(\tilde{W}) < C. \qquad (31)$$

Assuming that to the channel under consideration and the message ζ with the precision condition W the second statement of the Shannon theory (cf.1) is applicable (the reasonableness of this assumption shall be discussed later) there must exist coding $f(z)$, $z \in Z$ - a function with values in Y, decoding $g(\tilde{y})$, $\tilde{y} \in \tilde{Y}$ - a function with values in $\tilde{Z}$ such that for $\eta = f(\zeta)$, $\tilde{\zeta} = g(\tilde{\eta})$ the distribution η belongs to $\mathcal{P}_\eta$ condition (2) is satisfied and the common distribution of pair $(\zeta, \tilde{\zeta})$ belongs to $\tilde{W}$

---

*/ In Sect. 5 below the conditions W and $\tilde{W}$ will be illustrated by a case discussion.

We will now show that the coding $f(x)$ and decoding $g(\tilde{y})$ as set up by us provide the solution of the problem under consideration as they permit to build a variables system $\xi, \zeta, \eta = f(\zeta), \tilde{\eta}, \tilde{\zeta} = g(\tilde{\eta}), \tilde{\xi}$ appearing in the definition of the possibility of transmission with additional noise. As was noted in Sect.2 the conditions (9), (10),and (12) define uniquely the mutual distribution of this variables system. From (9) it follows that ζ will have a distribution P_ζ. Hence $\eta = f(\zeta)$ will have the same distribution as in the previously devised transmissionmethodfor which the precision condition W is satisfied and hence the requirement is satisfied that distribution η belongs to $\mathcal{P}_\eta$. Accordingly, to obtain the ultimate result it only remains to be ascertained that the mutual distribution of pair $(\xi, \tilde{\xi})$ belongs to W. Note that condition (9) coincides with condition (13). Further there follows from (10) condition (14). Indeed, because $\eta = f(\zeta)$ condition $\{\eta = y, \zeta = x\}$ is equivalent to condition $\{\zeta = x\}$ and from $\tilde{\zeta} = g(\tilde{\eta})$ it is true that $\{\tilde{\zeta} \in \tilde{C}\}$ is equivalent to $\{\tilde{\eta} \in g^{-1}(\tilde{C})\}$ where g^{-1} is the inversed function g. Hence,averaging over all possible $\tilde{y} \in \tilde{Y}$ and $y \in Y$ we can obtain from (12) the formula (15). Thus for variables $\xi, \zeta, \tilde{\zeta}, \tilde{\xi}$ conditions 1)-4) defining the acceptable system are fulfilled. Having a given mutual distribution of ζ and $\tilde{\zeta}$ the conditional probability (14) is uniquely defined and hence likewise the mutual distribution of random variables' system satisfying conditions 1)-4). However we have shown that the pair $\zeta, \tilde{\zeta}$ satisfies condition $\tilde{W}$. Therefore a system of variables must exist for which all requirements 1)-5) defining acceptability are fulfilled. The unique definition mentioned above determines that such system is the one we have set build up and thus for for the pair $(\xi, \tilde{\xi})$ the common distribution belongs to W.

In this manner the problem of the validity of the second statement of the Shannon theorem for message transmission with additional noise has been reduced to the problem of the validity of this statement in the usual handling of the Shannon theorem for the same channel and a precision condition $\tilde{W}$. As a result of this we can use here the well-elaborated general theory (cf. for example [3]) which proves that this statement is true for a sufficiently broad range of conditions. The only difficulty is in the possibility of a situation where the condition W being a simple and natural one, the condition $\tilde{W}$ will be described in a rather complicated way and the provision of proving for it the assumption under which the main Shannon theorem is true might give some difficulties. Accordingly, more research on the mathematical plane is needed here. We shall venture but one additional remark in this context.

As we stated in Sect.1 the most frequent is the case when condition W is given by the function $\rho(x,\tilde{x})$ and a limitation on its ensemble average. Then the condition $\tilde{W}$ will also be given in this manner. Specifically, let function $\tilde{\rho}(z,\tilde{z})$, $z\in Z$, $\tilde{z}\in\tilde{Z}$ be introduced in the following way. Suppose

$$\hat{\rho}(x,\tilde{z}) = \int_Z \rho(x,\tilde{x})\, Q_{II}(z, d\tilde{x}), \quad (32)$$

and further

$$\tilde{\rho}(z,\tilde{z}) = M\{\hat{\rho}(\xi,\tilde{z}) / \zeta = z\}. \quad (33)$$

(Here we use the advantage of having a given mutual distribution of variables ξ and ζ and so we can always calculate the conditonal distribution $\hat{\rho}(\xi,\tilde{z})$ if $\zeta = z$). Then for any acceptable system of random variables $\xi, \zeta, \tilde{\zeta}, \tilde{\xi}$ the equations will hold

$$M\tilde{\rho}(\zeta,\tilde{\zeta}) = M\{M\{\hat{\rho}(\xi,\tilde{\zeta})/\zeta = z\}\} = \quad (34)$$
$$= M\{\int_Z \rho(\xi,\tilde{x})\, Q_{II}(\zeta, d\tilde{x})\} = M\rho(\xi,\tilde{\xi}).$$

In the lather transition we used condition (15) and therefore condition $\tilde{W}$ will result, if W looks like $M\rho(\xi,\tilde{\xi}) \leq \rho_0$, in condition

$$M\tilde{\rho}(\zeta,\tilde{\zeta}) \leq \rho_0. \quad (35)$$

5. Entropy in Gaussian case

In this section we shall find an explicit expression for entropy $H(W, Q_I, Q_{II})$ in Gaussian case.

Before giving the conditions of the problem examined below, we shall specially single out the following well-known property of the pair of Gaussian random processes which will be much used below.

If (α,β) is a random stationary Gaussian pair of processes, then - as it will be shown immediately-there holds

$$\alpha = K\beta + \chi, \quad (36)$$

where K is linear operator and χ - a random stationary Gaussian process independent of β.

An important particular case of linear operator K is the one when it is given as

$$K\beta = \int_{-\infty}^{\infty} K(\tau)\beta(t-\tau)\, d\tau \quad (37)$$

for continuous time and as

$$K\beta = \sum_{i=-\infty}^{\infty} K(i)\beta(n-i) \quad (38)$$

for discrete time where $K(\tau)$ and $K(i)$ are transient characteristics of the filters.

Other forms of operator K (for example, differentiation $K\beta = \frac{d}{dt}\beta(t)$) are also possible, which in the most general form are given as sequence limits of the operators of type (37)-(38).

Let representation (36) obtained. Then spectral characteristic $k(\lambda)$ of operator K and spectral density $f_{\chi\chi}(\lambda)$ of process χ which give completely K and χ respectively, are determined from equalities

$$f_{\alpha\alpha}(\lambda) = |k(\lambda)|^2 f_{\beta\beta}(\lambda) + f_{\chi\chi}(\lambda) \quad (39)$$

and

$$f_{\alpha\beta}(\lambda) = k(\lambda) f_{\beta\beta}(\lambda) \quad (40)$$

which are the effect of relation (36). In equalities (39) and (40) stands for individual and mutual spectral densities of processes designated by subscript indexes. Equalities (39) and (40) lead to formulas

$$k(\lambda) = \frac{f_{\alpha\beta}(\lambda)}{f_{\beta\beta}(\lambda)} \quad (41)$$

and

$$f_{\chi\chi}(\lambda) = f_{\alpha\alpha}(\lambda) - \frac{|f_{\alpha\beta}(\lambda)|^2}{f_{\beta\beta}(\lambda)}. \quad (42)$$

Let us show now that if we choose $k(\lambda)$ and $f_{\chi\chi}(\lambda)$ according to formulas (41) and (42), then with linear operator K and Gaussian process χ independent of β, which are determined by these $k(\lambda)$ and $f_{\chi\chi}(\lambda)$ relation (36) will obtain.

Indeed, as (α,β) is a pair of Gaussian processes

$$\begin{vmatrix} f_{\alpha\alpha}(\lambda) & f_{\alpha\beta}(\lambda) \\ f_{\beta\alpha}(\lambda) & f_{\beta\beta}(\lambda) \end{vmatrix} = f_{\alpha\alpha}(\lambda) f_{\beta\beta}(\lambda) - |f_{\alpha\beta}(\lambda)|^2 \geq 0 \quad (43)$$

From inequality (43) it follows that

$$f_{\alpha\alpha}(\lambda) \geq \frac{|f_{\alpha\beta}(\lambda)|^2}{f_{\beta\beta}(\lambda)}. \quad (44)$$

Let it be noted now that Gaussian process χ with spectral density defined by formula (42) does exist because

$$\int f_{\chi\chi}(\lambda)\, d\lambda = \int f_{\alpha\alpha}(\lambda)\, d\lambda - \int \frac{|f_{\alpha\beta}(\lambda)|^2}{f_{\beta\beta}(\lambda)}\, d\lambda \leq \infty \quad (45)$$

where integration is run over au interval $(-\pi,\pi)$ in a discrete case and $(-\infty,\infty)$ in a continuous case. The inequality in (45) arises from the facts that the integral of $f_{\alpha\alpha}(\lambda)$ is limited and that because of inequality (44) it represents the upper bound for the integral of function $\frac{|f_{\alpha\beta}(\lambda)|^2}{f_{\beta\beta}(\lambda)}$.

Operator K is applicable to process β because for the same reasons

$$\int |K(\lambda)|^2 f_{\beta\beta}(\lambda)\, d\lambda = \int \frac{|f_{\alpha\beta}(\lambda)|^2}{f_{\beta\beta}(\lambda)}\, d\lambda \leq \infty \qquad (46)$$

If now with the help of operator K and process χ we set up the process

$$\alpha' = K\beta + \chi \qquad (47)$$

and using formulas (41) and (42) calculate spectral densities $f_{\alpha'\alpha'}(\lambda)$, $f_{\alpha'\beta}(\lambda)$ and $f_{\beta\alpha'}(\lambda)$ we can find that matrices

$$\begin{Vmatrix} f_{\alpha'\alpha'}(\lambda) & f_{\alpha'\beta}(\lambda) \\ f_{\beta\alpha'}(\lambda) & f_{\beta\beta}(\lambda) \end{Vmatrix} \quad \text{and} \quad \begin{Vmatrix} f_{\alpha\alpha}(\lambda) & f_{\alpha\beta}(\lambda) \\ f_{\beta\alpha}(\lambda) & f_{\beta\beta}(\lambda) \end{Vmatrix}$$

coincide. From this it follows that Gaussian pair (α', β) coincides with pair (α, β) and thus relation (36) is entirely valid.

Now for the formulation of the problem under consideration.

Let the input message $\xi = \xi(t)$ be a Gaussian random stationary process with discrete or continuous time t.

Suppose that the noised message $\zeta = \zeta(t)$ together with message ξ forms a pair of Gaussian stationary and stationary-connected processes. This means that the conditional distribution Q_I (see (11)) is in this case a conditional Gaussian distribution

Let us also take the decoded signal $\tilde{\zeta} = \tilde{\zeta}(t)$ and output message $\tilde{\xi} = \tilde{\xi}(t)$ to be random processes related to each other by

$$\tilde{\xi} = B\tilde{\zeta} + \nu \qquad (48)$$

Here ν is a Gaussian random process independent of ζ and B - linear operator.

Equality (48) means that conditional probability Q_{II} (see (12)) is a conditional Gaussian one.

As a criterion of transmission quality let us accept the frequency-weighted mean square criterion. We will consider that condition of transmission precision W consists in the pair $(\xi, \tilde{\xi})$ meeting inequality

$$M[\Psi(\tilde{\xi} - \xi)]^2 \leq \varepsilon^2 \qquad (49)$$

where Ψ is given linear operator.

Like in sect.2, the sequence of random processes $\xi, \zeta, \tilde{\zeta}, \tilde{\xi}$ will be regarded as the Markov chain.

The task consists in that having given Gaussian message and given Gaussian conditional probabilities Q_I and Q_{II} and condition of transmission precision (40) we find entropy $H(W, Q_I, Q_{II})$ determined by equality (16).

Since according to (16) in finding $H(W, Q_I, Q_{II})$ variation occurs over distributions of pair $(\zeta, \tilde{\zeta})$, we must ascertain what limitations condition (49) puts on pair $(\zeta, \tilde{\zeta})$, in other words, we must find condition $\tilde{W}$.

For that purpose we will point out first of all that from the fact that the pair of random processes (ξ, ζ) is a Gaussian one there follows that process ξ can be presented by

$$\xi = \mathcal{D}\zeta + \gamma \qquad (50)$$

where $\mathcal{D}$ is linear operator and γ is a random Gaussian process independent of ζ.

Let us further show that as $\xi, \zeta, \tilde{\zeta}, \tilde{\xi}$ form the Markov chain, process γ is independent of processes $\tilde{\zeta}$, $\tilde{\xi}$ and ν and process ν is independent of processes ξ and ζ.

Indeed, let Γ_1 be some subset of a set of sampling functions Γ of process γ and let $\Gamma_1 + \mathcal{D}z$ mean that to each function belonging to Γ_1 there is added function $\mathcal{D}z$ where z is a sampling function of process ζ. Set $\Gamma_1 + \mathcal{D}z$ according to equality (50) is a subset of set X of sampling functions of process ξ. Let us consider $\mathcal{P}\{\xi \in \Gamma_1 + \mathcal{D}z / \zeta = z, \tilde{\zeta} = \tilde{z}, \tilde{\xi} = \tilde{x}\}$ the conditional probability of ξ belonging to subset $\Gamma_1 + \mathcal{D}z$ with fixed $\zeta, \tilde{\zeta}$ and $\tilde{\xi}$. On the grounds of equality (50) we shall have

$$\begin{aligned} &\mathcal{P}\{\xi \in (\Gamma_1 + \mathcal{D}z) / \zeta = z, \tilde{\zeta} = \tilde{z}, \tilde{\xi} = \tilde{x}\} = \\ &= \mathcal{P}\{(\mathcal{D}z + \gamma) \in (\Gamma_1 + \mathcal{D}z) / \zeta = z, \tilde{\zeta} = \tilde{z}, \tilde{\xi} = \tilde{x}\} = \\ &= \mathcal{P}\{\gamma \in \Gamma_1 / \zeta = z, \tilde{\zeta} = \tilde{z}, \tilde{\xi} = \tilde{x}\}. \end{aligned} \qquad (51)$$

From equality (50) and the independence of processes γ and ζ it also follows that

$$\mathcal{P}\{\xi \in (\Gamma_1 + \mathcal{D}z) / \zeta = z\} = \mathcal{P}\{\gamma \in \Gamma_1 / \zeta = z\} = \mathcal{P}\{\gamma \in \Gamma_1\} \qquad (52)$$

Using the property of the Markov chain we can write

$$\mathcal{P}\{\xi \in \Gamma_1 + \mathcal{D}z / \zeta = z, \tilde{\zeta} = \tilde{z}, \tilde{\xi} = \tilde{x}\} = \mathcal{P}\{\xi \in (\Gamma_1 + \mathcal{D}z) / \zeta = z\} \qquad (53)$$

Now substituting $\mathcal{P}\{\gamma \in \Gamma_1 / \zeta = z, \tilde{\zeta} = \tilde{z}, \tilde{\xi} = \tilde{x}\}$ for the left part of equality (53) by means of (51) and $\mathcal{P}\{\gamma \in \Gamma_1\}$ for the right part of (53) by means of (52) we shall have:

$$\mathcal{P}\{\gamma \in \Gamma_1 / \zeta = z, \tilde{\zeta} = \tilde{z}, \tilde{\xi} = \tilde{x}\} = \mathcal{P}\{\gamma \in \Gamma_1\}. \qquad (54)$$

The above equality stands for the independence of γ-process both from process ζ (which was assumed earlier), and from $\tilde{\zeta}$ and $\tilde{\xi}$ processes also. Because

ν-process, is a linear combination of $\tilde{\zeta}$ and $\tilde{\xi}$, the γ-process is independent of ν also.

Such resoning may also be done to prove the independence of ν-process from ξ and ζ.

Now, taking into account equalities (48) and (50), we transform the condition of precision (49) and shall have:

$$M[\Psi(B\tilde{\zeta}+\nu-\mathcal{D}\zeta-\gamma)]^2 \le \varepsilon^2. \qquad (55)$$

Condition (55) is the unknown precision condition $\tilde{W}$. As may be seen in the situation considered conditions W and $\tilde{W}$ are both frequency-weighted mean square conditions.

Condition (55) may be expressed through spectral densities of random processes involved in it. For this purpose let as insert spectral characteristics $\psi(\lambda)$, $b(\lambda)$ and $d(\lambda)$ of operators Ψ, B and $\mathcal{D}$, respectively. Using these spectral characteristics we may represent precision condition (55) in this form:

$$\int|\psi(\lambda)|^2\big[f_{\nu\nu}(\lambda)+f_{\gamma\gamma}(\lambda)+|b(\lambda)|^2 f_{\tilde{\zeta}\tilde{\zeta}}(\lambda)+|d(\lambda)|^2 f_{\zeta\zeta}(\lambda)- -2\,\mathrm{Re}\, b(\lambda)d(\lambda)f_{\zeta\tilde{\zeta}}(\lambda)\big]d\lambda \le \varepsilon^2 \qquad (56)$$

Inequality (56) shows, that the condition imposed on pair $(\zeta,\tilde{\zeta})$, when finding entropy $H(W,Q_I,Q_{II})$ is only a limitation on spectral densities of ζ and $\tilde{\zeta}$ processes This circumstance gives an opportunity to use the results of [6], when determining the lower bound $J(\zeta,\tilde{\zeta})$ appearing in (16). In this paper it is shown that when a random process ζ is a Gaussian one, and when there are given spectral density of $f_{\tilde{\zeta}\tilde{\zeta}}(\lambda)$ process $\tilde{\zeta}$ and

a mutual spectral density $f_{\zeta\tilde{\zeta}}(\lambda)$ of ζ and $\tilde{\zeta}$ processes, the lower bound $J_m(\zeta,\tilde{\zeta})$ of information amount $J(\zeta,\tilde{\zeta})$ is obtained on Gaussian pair $(\zeta,\tilde{\zeta})$. In this case

$$J_m(\zeta,\tilde{\zeta}) = -\frac{1}{4\pi}\int \log\frac{f_{\zeta\zeta}(\lambda)f_{\tilde{\zeta}\tilde{\zeta}}(\lambda)-|f_{\zeta\tilde{\zeta}}(\lambda)|^2}{f_{\zeta\zeta}(\lambda)\,f_{\tilde{\zeta}\tilde{\zeta}}(\lambda)}\,d\lambda. \qquad (57)$$

From the results of [6] mentioned above and also from the fact that processes $\xi, \zeta, \tilde{\zeta}, \tilde{\xi}$ form Markov chain it follows:

1) random processes $\xi, \zeta, \tilde{\zeta}, \tilde{\xi}$ are Gaussian in totality;

2) for entropy $H(W, Q_I, Q_{II})$ the formula

$$H(W,Q_I,Q_{II}) = \min J_m(\zeta,\tilde{\zeta}), \qquad (58)$$

is true, where the minimum is found over functions $f_{\tilde{\zeta}\tilde{\zeta}}(\lambda)$ and $f_{\zeta\tilde{\zeta}}(\lambda)$ which satisfy condition (56).

The variation over function $f_{\zeta\tilde{\zeta}}(\lambda)$ in integral $J_m(\zeta,\tilde{\zeta})$ can not be convinently accomplished directly, because function $f_{\zeta\tilde{\zeta}}(\lambda)$ (besides term (56)) must be the spectral density of the given process ζ and the varied process $\tilde{\zeta}$. To acide these difficulties let us make some changes in condition (56) and integral (57).

Firstly, it is to be noted that because of the Gaussian type of the pair of random processes $(\zeta,\tilde{\zeta})$, we may represent $\tilde{\zeta}$ by

$$\tilde{\zeta} = A\zeta + \mu \qquad (59)$$

where A is linear operator; μ – Gaussian random process, independent of ζ.

In accordance with this equality, having a fixed ζ we may perform variation in three ways: 1) varying of linear operator A and process μ; 2) varying of $\tilde{\zeta}$ and operator A, and 3) varying of $\tilde{\zeta}$ and μ-process. It is natural that all limitations imposed on the functions being varied must be met with any one method. We shall use below the second method.

Because processes $\xi, \zeta, \tilde{\zeta}, \tilde{\xi}$ form Markov chain, it may be shown that μ-process is independent of ξ.

Equality (59) induces the following spectral relation:

$$f_{\zeta\tilde{\zeta}}(\lambda) = a(\lambda) f_{\zeta\zeta}(\lambda). \qquad (60)$$

For further transformations we shall use (besides (60)) the following relations

$$f_{\xi\zeta}(\lambda) = d(\lambda) f_{\zeta\zeta}(\lambda) \qquad (61)$$

and

$$f_{\xi\xi}(\lambda) = |d(\lambda)|^2 f_{\zeta\zeta}(\lambda) + f_{\gamma\gamma}(\lambda) \qquad (62)$$

which are the results of equality (50).

Using the above relations (60),(61),(62 (62) between spectral densities we may reduce condition (56) to the form:

$$\int|\psi(\lambda)|^2\big[f_{\nu\nu}(\lambda)+f_{\xi\xi}(\lambda)+|b(\lambda)|^2 f_{\tilde{\zeta}\tilde{\zeta}}(\lambda)- -2\,\mathrm{Re}\, a(\lambda)b(\lambda)f_{\xi\zeta}(\lambda)\big]d\lambda \le \varepsilon^2 \qquad (63)$$

and equation (57) for $J_m(\zeta,\tilde{\zeta})$ to the form

$$J_m(\zeta,\tilde{\zeta}) = -\frac{1}{4\pi}\int \log\frac{f_{\zeta\zeta}(\lambda)f_{\tilde{\zeta}\tilde{\zeta}}(\lambda)-|a(\lambda)|^2 f^2_{\zeta\zeta}(\lambda)}{f_{\zeta\zeta}(\lambda)\,f_{\tilde{\zeta}\tilde{\zeta}}(\lambda)}\,d\lambda \qquad (64)$$

Thus, the unknown entropy $H(W,Q_I,Q_{II})$ may be found as a minimum $J_m(\varsigma,\tilde{\varsigma})$ by varying functions $f_{\tilde{\xi}\tilde{\xi}}(\lambda)$ and $a(\lambda)$ which satisfy condition (63). To find the minimum $J_m(\varsigma,\tilde{\varsigma})$ we will use the Lagrange method. For this purpose let $a(\lambda)$ be reformed as

$$a(\lambda) = a_1(\lambda) + i a_2(\lambda) \quad (65)$$

and set up a function

$$G(\lambda) = \log \frac{f_{\varsigma\varsigma}(\lambda) f_{\tilde{\xi}\tilde{\xi}}(\lambda) - [a_1^2(\lambda)+a_2^2(\lambda)] f^2_{\varsigma\varsigma}(\lambda)}{f_{\varsigma\varsigma}(\lambda) f_{\tilde{\xi}\tilde{\xi}}(\lambda)} + \quad (66)$$

$$+ h_1 \Big[f_{\nu\nu}(\lambda) + f_{\xi\xi}(\lambda) + |B(\lambda)|^2 f_{\tilde{\xi}\tilde{\xi}}(\lambda) - a_1(\lambda) \operatorname{Re} f_{\xi\varsigma}(\lambda) + a_2(\lambda) \operatorname{Im} f_{\xi\varsigma}(\lambda) \Big]$$

where h_1 is Lagrange Constant. For the definition of extremes $a_1(\lambda)$, $a_2(\lambda)$ and $f_{\tilde{\xi}\tilde{\xi}}(\lambda)$ Euler equations should be set up

$$\frac{\partial G(\lambda)}{\partial a_1(\lambda)} = 0; \quad \frac{\partial G(\lambda)}{\partial a_2(\lambda)} = 0; \quad \frac{\partial G(\lambda)}{\partial f_{\varsigma\varsigma}(\lambda)} = 0. \quad (67)$$

For resolving this system of equations relative to functions $f_{\tilde{\xi}\tilde{\xi}}(\lambda)$, $a_1(\lambda)$ and $a_2(\lambda)$ it should be taken into account that the varied functions should satisfy additional conditions $f_{\tilde{\xi}\tilde{\xi}}(\lambda) \geq 0$ and

$$f_{\mu\mu}(\lambda) = f_{\tilde{\xi}\tilde{\xi}}(\lambda) - |a(\lambda)|^2 f_{\varsigma\varsigma}(\lambda) \geq 0. \quad (68)$$

The above inequality is deduced from relation (59).

The solution of (67) which we omit, leads to the following functions $f_{\tilde{\xi}\tilde{\xi}}(\lambda)$, $a_1(\lambda)$ and $a_2(\lambda)$ which minimize the functional (64):

$$f_{\tilde{\xi}\tilde{\xi}}(\lambda) = \begin{cases} \dfrac{|f_{\xi\varsigma}(\lambda)|^2}{|B(\lambda)|^2 f_{\varsigma\varsigma}(\lambda)} - \dfrac{h}{|B(\lambda)\psi(\lambda)|^2}, & \text{if } \dfrac{|\psi(\lambda) f_{\xi\varsigma}(\lambda)|^2}{f_{\varsigma\varsigma}(\lambda)} \geq h \\ 0, & \text{if } \dfrac{|\psi(\lambda) f_{\xi\varsigma}(\lambda)|^2}{f_{\varsigma\varsigma}(\lambda)} < h, \end{cases} \quad (69)$$

$$a_1(\lambda) = \frac{B(\lambda) f_{\tilde{\xi}\tilde{\xi}}(\lambda)}{f_{\xi\varsigma}(\lambda)} \quad (70)$$

and

$$a_2(\lambda) = 0 \quad (71)$$

where h is a positive constant.

If we put solutions (69),(70) and (71) into (63), which should be changed for equality, then to find the constant h we have the equation:

$$\varepsilon^2 = \int |\psi(\lambda)|^2 \frac{f_{\xi\xi}(\lambda) f_{\varsigma\varsigma}(\lambda) - |f_{\xi\varsigma}(\lambda)|^2}{f_{\varsigma\varsigma}(\lambda)} d\lambda + \quad (72)$$

$$+ \int |\psi(\lambda)|^2 f_{\nu\nu}(\lambda) d\lambda +$$

$$+ \int \min\left\{ h, \frac{|\psi(\lambda) f_{\xi\varsigma}(\lambda)|^2}{f_{\varsigma\varsigma}(\lambda)} \right\} d\lambda.$$

The first integral in the right part of (72) is the least frequency-weighted mean-square error ocurring in optimum filtration of noised information ς in order to extract message ξ. Let it be denoted by σ_1^2.

The second integral in the right part of (72) stands for the mean-square (frequency-weighted) power of additive noise ν existing on the output of a communication system. We denote it by σ_2^2. Taking into accound definitions thus introduced, rewrite (72) in the form:

$$\varepsilon^2 = \sigma_1^2 + \sigma_2^2 + \int \min\left\{ h, \frac{|\psi(\lambda) f_{\xi\varsigma}(\lambda)|^2}{f_{\varsigma\varsigma}(\lambda)} \right\} d\lambda. \quad (73)$$

Now, substituting (69),(70) and (71) into functional (64) we shall have for entropy

$$H(W,Q_I,Q_{II}) = \frac{1}{4\pi} \int \log \max\left\{ \frac{|\psi(\lambda) f_{\xi\varsigma}(\lambda)|^2}{h f_{\varsigma\varsigma}(\lambda)}, 1 \right\} d\lambda. \quad (74)$$

Formula (74) with equation (73) gives the final expression for entropy $H(W,Q_I,Q_{II})$ in the Gaussian case under consideration.

Formulas (73) and (74) are a generalisation of expressions, obtained previously by Pinsker [6] for the usual Shannon scheme without additional noise. Indeed, supposing $\xi = \varsigma$ and $\tilde{\varsigma} = \tilde{\xi}$ we shall obtain

$$f_{\xi\varsigma}(\lambda) = f_{\xi\xi}(\lambda); \quad f_{\varsigma\varsigma}(\lambda) = f_{\xi\xi}(\lambda); \quad \sigma_1^2 = \sigma_2^2 = 0 \quad (75)$$

and

$$\varepsilon^2 = \int \min\{ h, |\psi(\lambda)|^2 f_{\xi\xi}(\lambda) \} d\lambda, \quad (76)$$

but

$$H(W) = \frac{1}{4\pi} \int \log \max\left\{ \frac{|\psi(\lambda)|^2 f_{\xi\xi}(\lambda)}{h}, 1 \right\} d\lambda. \quad (77)$$

Formula (77) with equality (76) represents entropy $H(W)$ for Gaussian source in the scheme without additional noise.

Comparing expressions (76) and (77) with (73) and (74) one can see that the role of spectral density of message $f_{\xi\xi}(\lambda)$ in formulas (73),(74) is plagd by $|f_{\xi\varsigma}(\lambda)|^2 / f_{\varsigma\varsigma}(\lambda)$ signifying spectral density of a noised message after filtration. This filtration is optimal in the sense of frequency-weighted mean square criterion.

Further, the error ε^2 in (75) corresponds to the error $\varepsilon_1^2 = \varepsilon^2 - \sigma_1^2 - \sigma_2^2$ in (73). The number ε_1^2 is maximum error which may be allowed in transmission from ς to $\tilde{\varsigma}$ in the scheme on Fig. 1.

Let us how consider an example: assume that

$$\varsigma = \xi + \alpha \quad (78)$$

where α is independent of ξ and

$$\tilde{\xi} = \varsigma + \nu \quad (79)$$

where $\tilde{\varsigma}$ and ν are both independent. Let all processes depende on continuous time and assume that:

$$f_{\xi\xi}(\lambda) = \frac{Q}{\Lambda_1}; \quad f_{\alpha\alpha}(\lambda) = \frac{N_1}{\Lambda_1}; \quad f_{\nu\nu}(\lambda) = \frac{N_2}{\Lambda_1} \quad (80)$$

for $|\lambda| \leq \Lambda_1$ and $f_{\xi\xi}(\lambda) = f_{\alpha\alpha}(\lambda) = f_{\nu\nu}(\lambda) = 0$ when $|\lambda| > \Lambda_1$. Quantities Q, N_1 and N_2 stand for mean powers of input message ξ noise α and noise ν respectively, and Λ_1 is the bandwidth of this message.

Besides, assume that

$$|\psi(\lambda)|^2 = 1. \quad (81)$$

In this case formulas (73) and (74) result in

$$H(w, Q_I, Q_{II}) = \begin{cases} \Lambda \log \frac{Q_1}{\varepsilon_1^2}, & \text{if } Q_1 > \varepsilon_1^2 \\ 0, & \text{if } Q_1 \leq \varepsilon_1^2 \end{cases} \quad (82)$$

where

$$\Lambda = \frac{\Lambda_1}{2\pi}; \quad Q_1 = \frac{Q^2}{Q+N_1}; \quad \varepsilon_1^2 = \varepsilon^2 - \frac{QN_1}{Q+N_1} - N_2. \quad (83)$$

Formula (82) is analogous to the known Shannon formula [1], the only difference being that instead of Q_1 it has Q, and instead of ε_1^2 it has ε^2.

6. Entropy evaluation

To make some comments conserning the possibility of using the previous results in other situations.

In sect.5 we have assumed ν to be a random gaussian process (see (48)). This assumption is not important in the sense that if ν is a nongaussian process, then all reasoning, as well as formulas (73) and (74) is true.

Now assume that input message ξ (not necessarily gaussian) is related with noised message ς by

$$\varsigma = E\xi + \delta, \quad (84)$$

where δ is also a random, not necessarily Gaussian process, not correlated with ξ, and E is a linear operator.

In this case condition $\tilde{w}$, expressed through equality (56) remains valid and expression (74) is a bound for of entropy from above.

7. A transmission method

In conclusion we submit for consideration a certain natural point of view concerning transmission of noised message. For illustration we shall analize a case of mean square criterion without output noising $\tilde{\varsigma} = \tilde{\xi}$.

Let us set up the transmission as follows. First we extract message ξ from noised message ς by optimal method from the point of view of mean square criterion. The extracted message, denoted as ξ' will evidently look like

$$\xi' = M(\xi/\varsigma) \quad (85)$$

and will be a certain function (or an operator) of the noised message ς.

The mean square deviation of ξ from ξ' will be

$$\sigma^2 = M(\xi - \xi')^2 \leq M(\xi - \varsigma)^2. \quad (86)$$

After finishing the operation, of optimal extracting we shall consider ξ' as the message to be transmitted. Condition of transmission precision will be

$$M(\xi' - \tilde{\xi})^2 \leq \delta^2. \quad (87)$$

If we assume that

$$\delta^2 = \varepsilon^2 - \sigma^2, \quad (88)$$

where ε^2 - is the given maximum permissible mean square deviation of ξ from $\tilde{\xi}$ then considering the evident inequality

$$M(\xi - \xi')^2 + M(\xi' - \tilde{\xi})^2 \geq M(\xi - \tilde{\xi})^2 \quad (89)$$

it can be seen that condition (87) gives

$$M(\xi - \tilde{\xi})^2 \leq \varepsilon^2. \quad (90)$$

This shows that with (88) condition (87) meets the requirement of precision condition for transmission of original message ξ. Therefore we shall consider that equality (88) obtains. The condition of transmission precision of ξ' will be

$$M(\xi' - \tilde{\xi})^2 \leq \varepsilon^2 - \sigma^2. \quad (91)$$

For the future let it be noted that from inequality (91) follows inequality (90) and it means that, in general, inequality (90) determines wider group of pairs $(\xi', \tilde{\xi})$, than inequality (91) does.

Entropy $\overline{H}(w)$, corresponding to the method of transmission described will be

$$\overline{H}(w) = \inf J(\xi', \tilde{\xi}) \quad (92)$$

where the lower bound can be defined if ξ' has given preselected distribution and condition (91) is true.

The method of transmission described above consists of two optimal operations: optimal extraction of message and optimal transmission of the extracted information. This method is only one of those which are possible and the question arises whether is it the optimal one, that is whether the combination of these two optimal operations also presents the optimal procedure.

The non-conformity of expressions (16) and (92) for entropies $H(W, Q_I, Q_{II})$ and $\bar{H}(W)$ respectively suggests that we have no reason to consider the method described above to be optimal. A more detailed study reveals that there obtains

$$H(W, Q_I, Q_{II}) \leq \bar{H}(W) \qquad (93)$$

which is evident from the general point of view. Indeed, suppose, for example, that inequality (85) gives a unique relation between ζ and ξ'. Then using the general properties of information we have

$$J(\xi', \tilde{\xi}) = J(M(\xi/\zeta), \tilde{\xi}) = J(\zeta, \tilde{\xi}). \qquad (94)$$

Using this equality we present now

$$H(W, Q_I, Q_{II}) = \inf J(\xi', \tilde{\xi}) \qquad (95)$$

where the lower bound can be found over all pairs $(\xi', \tilde{\xi})$ if ξ' has given distribution and condition (90) is true.

Expression (95) differs from (92) only by limitations on the varied pairs' class $(\xi', \tilde{\xi})$. Applying the above note that in general condition (90) defines a wider class of pairs $(\xi', \tilde{\xi})$ we obtain the inequality (93) in question.

We should note that the result obtained does not exclude the possibility that in some particular cases the method of transmission described is the optimal one. Moreover, calculation made in Sect. 5 can be considered as a proof that in a Gaussian case, with frequency-weighted mean-square criterion of quality, the method of transmission described is the optimal one among all others possible. It is easy to see from a comparison of formulas (73) and (74) with the corresponding results obtained for transmission of a noiseless message that in this particular case inequality (93) changes into an equality.

It would be nice to have some interesting examples where rigorous inequality exists in (93).

Literature

1. Shannon C.E. Weaver W., The mathermatical theory of communication. University of Illinois. Press, 1949.

2. Kolmogorov A.N., On the Shannon theory of information transmission in the case of continuous signals, IRE Trans., on Inform. Theory. I -2, 1956, N 4, p.p. 102-108.

3. Dobrushin R.L., A general formulation of the basic Shannon Theorem in information theory, Uspekhi matematicheskih nauk, V.XIV, N 6, p.p. 3-104, 1959, (Russian).

4. Dobrushin R.L., Information transmission through a channel with feedback. Theory of probability and its application. (Russian), vol.3. 1958, p.p.395-412.

5. Dobrushin R.L., Mathematical problems in the Shannon theory of optimal coding of information. Proc. of the fourth. Berkeley sympos. on mathematical stat. and prob. Ed. by Neiman J.v.I, Univ. California Press, 1961, p.p. 211-252.

6. Pinsker M.S. Calculation of the speed of information creation by means of stationary random process and of the stationary channel capacity. Proceeding of the Academy of Sciences of the USSR, 1956, vol. 111, N 4, p.p. 753-756, (Russian).

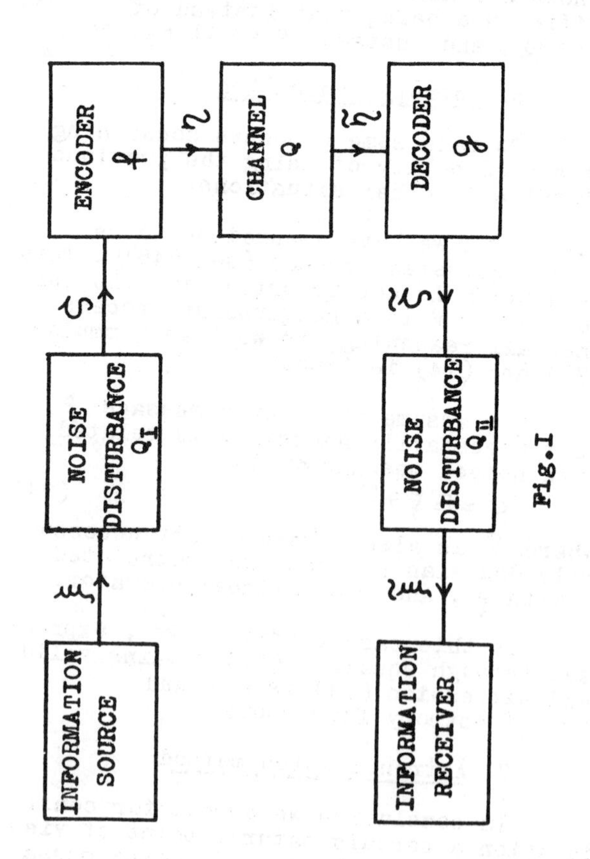

Fig. I

Information Rates of Gaussian Signals Under Criteria Constraining the Error Spectrum

Shannon[1] introduced the concept of information rate under a specified fidelity criterion and obtained explicit results for the case of Gaussian white noise under a mean square error criterion. The information rate of a general Gaussian signal under a mean square error criterion was computed by Kolmogorov.[2] By taking an approach somewhat different from that chosen by Kolmogorov, it is a simple matter to obtain results for the case where not only the total mean square error but also some of its spectral properties are prescribed.

Consider a stationary Gaussian signal $x(t)$ with power spectral density $S_x(f)$. Suppose $x(t)$ is applied to an idealized set of band-pass filters F_i, with adjacent but non-overlapping passbands, each of width Δf and free of frequency and phase distortion. In other words, the postulated filters divide $x(t)$ into a series of components $g_i(t)$ which are statistically independent and, if Δf is taken small enough, have essentially flat power spectra. Thus Shannon's results for information rate under a mean square error criterion become applicable to each component. If the mean square error allowed in the ith band is N_i, the information rate of $g_i(t)$ is

$$r_i \approx \begin{cases} \Delta f \log \dfrac{\Delta f\, S_x(f_i)}{N_i}, & N_i \le \Delta f\, S_x(f_i) \\ 0, & N_i > \Delta f\, S_x(f_i) \end{cases} \quad (1)$$

Eq. (1) becomes exact when $\Delta f \to 0$. If, instead of fixing the maximum mean square errors N_i, one prescribes the maximum value of the error spectrum $N(f)$, then (for sufficiently small Δf) $N_i \approx N(f_i)\Delta f$. Because of the independence of the various $g_i(t)$, the total information rate under the specified restraint on the error spectrum is simply the sum of the various r_i. Thus (1) leads to

$$R = \lim_{\Delta f \to 0} \sum_{i=0}^{\infty} r_i = \int_0^\infty df \log \frac{S_x(f)}{N_1(f)} \quad (2)$$

where

$$N_1(f) = \begin{cases} N(f) \text{ where } N(f) \le S_x(f) \\ S_x(f) \text{ where } N(f) > S_x(f) \end{cases} \quad (3)$$

Note that the integrand in (2) is non-negative and that it vanishes in frequency ranges where allowed error spectrum exceeds the signal spectrum.

Formal justification of the limiting operations implied in the above argument is somewhat tedious, but involves no conceptual difficulties.[3]

If average properties of the error spectrum rather than its complete frequency structure are specified, one can proceed simply by minimizing (2) over the range of the allowed error spectra. Thus the rate under a mean square error criterion is simply

$$R_1 = \min_{N(f)} \{R\} = \min_{N(f)} \left\{ \int_0^\infty df \log \frac{S_x(f)}{N_1(f)} \right\} \quad (4)$$

under the constraint

$$\int_0^\infty df\, N(f) \le E. \quad (5)$$

E is the allowed mean square error.

Since the results of this particular computation are not new, the formal minimization is not presented here. It is, in any case, quite straightforward.[3] However, a graphical interpretation of the conclusion may be of some interest. The positive frequency axis may be divided into two sets, α and β, such that

$$f \in \alpha \text{ if } S_x(f) < A$$

and

$$f \in \beta \text{ if } S_x(f) \ge A \quad (6)$$

where A is a number so chosen that

$$\int_0^\infty df N(f) = E = A\mu(\beta) + \int_{f\in\alpha} df S_x(f) \quad (7)$$

and $\mu(\beta)$ is the Lebesque measure of the set β.[4]

In other words, the total area E of the noise spectrum "fills up" part of the area under the signal spectrum much as a liquid fills an irregular container. This is illustrated in Fig. 1. Frequency ranges which are completely "filled up" make no contribution to the information rate. The desired rate R_1 is given by

$$R_1 = \int_{f\in\beta} df \log \frac{S_x(f)}{A} \quad (8)$$

which is, of course, equivalent to the expression previously given by Kolmogorov.[2]

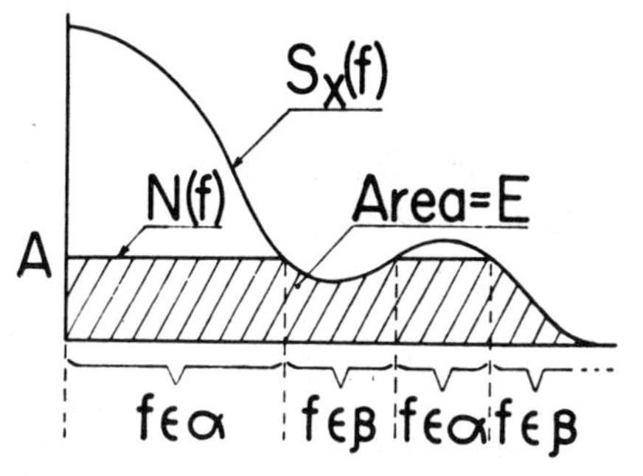

Fig. 1—The relationship of noise and signal spectra.

A case of some practical interest is the calculation of information rates under a frequency weighted mean square error criterion. If $L(f)$ is a non-negative weighting function on the error spectrum and the allowable weighted mean square error is T, the desired rate R_2 is given by

$$R_2 = \min_{N(f)} \left\{ \int_0^\infty df \log \frac{S_x(f)}{N_1(f)} \right\} = \min_{L(f)N(f)} \left\{ \int_0^\infty df \log \frac{L(f)\, S_x(f)}{L(f)\, N_1(f)} \right\} \quad (9)$$

under the constraint

$$\int_0^\infty df\, L(f) N(f) \le T. \quad (10)$$

It is clear that the minimization problem defined by (9) and (10) is identical to the one defined by (4) and (5), if in the latter one replaces signal and noise spectra with their weighted versions, $L(f)S_x(f)$ and $L(f)N(f)$, respectively. The results given by (6)–(8) therefore remain applicable with the same modifications. It is interesting to note that the minimizing error spectrum is, in this case

$$N(f) = \begin{cases} \dfrac{A}{L(f)} & f \in \beta \\ S_x(f) & f \in \alpha. \end{cases} \quad (11)$$

In other words, the error is distributed over the frequency scale in such a manner that relatively large contributions occur in frequency ranges where the weight $L(f)$ is small.

R. A. McDonald
P. M. Schultheiss
Dept. of Engrg. and Appl. Science
Yale University
New Haven, Conn.

Manuscript received October 30, 1963. The work reported here was performed at Yale University, New Haven, Conn., and was supported by a contract between Yale University and Bell Telephone Laboratories, Murray Hill, N. J.

[1] C. E. Shannon, "The Mathematical Theory of Communication," University of Illinois Press, Urbana, Ill.; 1949.

[2] A. N. Kolmogorov, "On the Shannon theory of information transmission in the case of continuous signals," IRE Trans. on Information Theory, vol. 2, pp. 102–108; December, 1956.

[3] R. A. McDonald, "Information Rates . . . ," D. Engr. thesis, School of Engineering, Yale University, New Haven, Conn.; 1961.

[4] If $\int_0^\infty S_x(f)\, df \ge E$. Otherwise R_1 is clearly zero.

Reprinted from *Proc. IEEE*, vol. 52, pp. 415–416, Apr. 1964.

Analog Source Digitization: A Comparison of Theory and Practice

I. Introduction

We consider the transmission of an analog waveform $u(t)$ to a sink (final destination of the source output) by first converting this waveform to a digital representation, which is in turn transmitted over the channel by a digital communication system. Since it is impossible to represent the set of all possible waveforms $u(t)$ in some interval of time with a finite number of binary digits, the receiver then cannot infer the exact source output from the digits it receives, and hence distortions result between $u(t)$ and the final receiver output even if the digits are transmitted over the channel without error.

Throughout this correspondence we will assume that the digital representation of $u(t)$ is transmitted over the channel *without error*. This assumption is consistent with situations in which a coded digital communication system is used to transmit reliably over a noisy channel. This work also applies to the storage of analog waveforms in digital memories with no errors in read-in and read-out. We model an analog source as a stationary Gaussian process with zero mean and known power density spectrum $S(f)$. We use as a fidelity criterion the mean-squared error between source output and the reconstructed waveform at the final output of the receiver. This fidelity criterion is mathematically tractable and is meaningful in many applications. The fundamental limitations on minimizing the number of binary digits in the digital representation of $u(t)$ for a fixed mean-squared error are presented. Some simple digitization schemes are also compared to the theoretical limits for several different source spectra.

II. Quantizers

Perhaps the most common method of converting analog data samples to digital form is a simple amplitude quantizer. Consider a time-discrete Gaussian source producing $1/T_0$ independent samples per second and consider converting (encoding) these samples into digital form with a quantizer having M uniformly spaced output levels $\{v_i\}$, $i = 1, \cdots M$, (a uniform, M level quantizer in our terminology). Suppose further that a sample u is mapped by the quantizer into the nearest output level (in a euclidian distance sense) (see Fig. 1), and that the set of output levels is centered about the mean of the samples (which is always taken to be zero in this work). The mean-squared error ϵ^2 between quantizer inputs and outputs is now a function only of M, the number of output levels, and $v_{\max}$, the maximum output level. The number of binary digits needed to represent the set of quantizer output levels is r, where r is the smallest integer satisfying

$$M \leq 2^r.$$

This is one method of converting analog samples to binary digits.

A second method of digitizing an analog quantity would be to represent quantizer outputs by symbols from an M-symbol alphabet by simply numbering the output levels from 1 to M. A trivial sort of block coding could be used to convert sequences of such M-ary symbols to sequences of binary digits, requiring a minimum of $\log_2 M$ bits per M-ary symbol, and hence $\log_2 M$ bits per sample.

A third and more complex digitization scheme is as follows. The quantizer outputs will not be equiprobable in the case of Gaussian samples, and so the entropy $H_Q(V)$ of these outputs satisfies

$$H_Q(V) = -\sum_{i=1}^{M} p_i \log_2 p_i < \log_2 M \text{ bits per sample}$$

where p_i, the probability of the output level v_i, is easy to calculate for a uniform quantizer. The quantizer outputs, regarded as outputs of a discrete memoryless source, could theoretically be coded to require a minimum of only $H_Q(V)$ bits per sample to be transmitted to the sink to achieve vanishing probability of erroneous reconstruction of the sequence of quantizer outputs from the coded version.[1,2] We will refer henceforth to such coding as "entropy coding" of the quantizer outputs.

The point to be made is that the data rate required to send the quantizer outputs to the receiver depends on the amount of processing that the quantizer outputs undergo. If each quantizer output is expressed directly in binary digits, the required data rate is r bits per sample where $M \leq 2^r$. If sequences of the M-ary quantizer outputs are mapped into sequences of binary digits,[3] the data rate can be reduced to $\log_2 M$ bits per sample. If the quantizer outputs are entropy coded, a data rate of $H_Q(V)$ bits per sample is required. It remains to optimize performance for each of these methods of transmitting the quantizer outputs to the receiver.

The information rate of a source producing independent Gaussian samples has been derived by Shannon for the fidelity criterion of ϵ^2 mean-squared error between the source output and its facsimile at the sink as[1,4]

$$R(\epsilon^2) = \tfrac{1}{2} \log_2 \frac{A}{\epsilon^2} \text{ bits per sample}$$

where A is the variance of the source samples. Part of the significance of the information rate of a source with a fidelity criterion is as

[1] C. E. Shannon, "A mathematical theory of communication," *Bell Sys. Tech. J.*, vol. 37, pp. 379–623, 1948.
[2] R. M. Fano, *Transmission of Information*. Cambridge, Mass.: M.I.T. Press, and New York: Wiley, 1961.
[3] This is clearly not necessary when M is a power of 2.
[4] C. E. Shannon, "Coding theorems for a discrete source with a fidelity criterion," *IRE Nat'l Conv. Rec.*, pt. 4, pp. 142–163, 1959.

Reprinted from *IEEE Trans. Inform. Theory*, vol. IT-13, pp. 323-326, Apr. 1967.

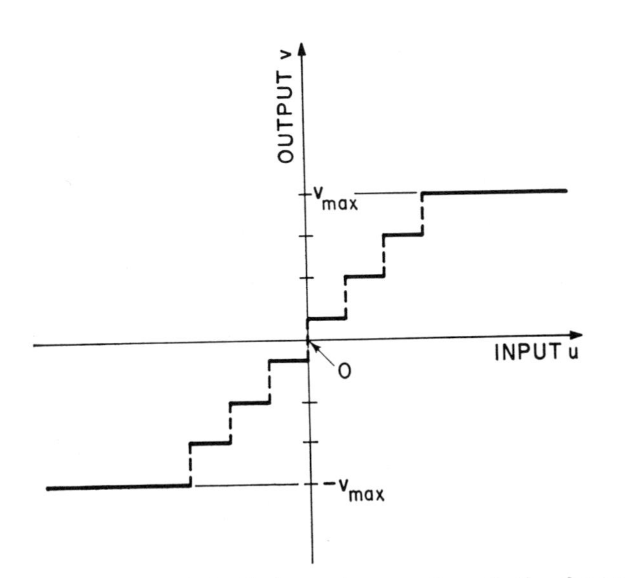

Fig. 1. Input-output relationship for a uniform quantizer with $M = 8$ output levels.

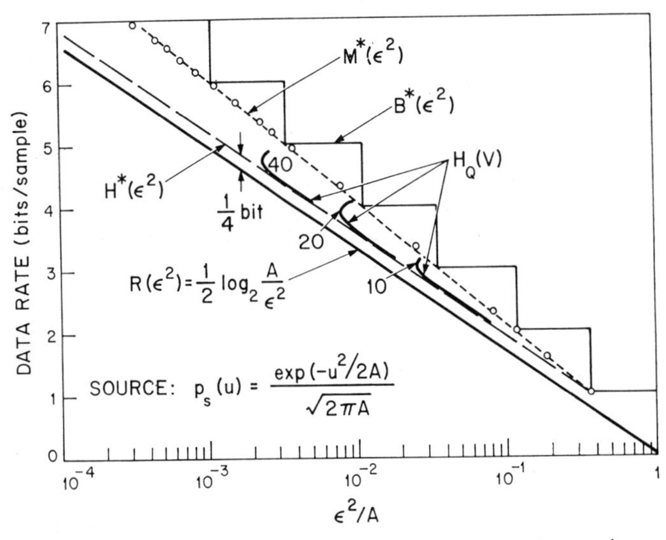

Fig. 2. Data rates required to send uniform quantizer outputs to a receiver as a function of ϵ^2/A. Also shown are curves of quantizer output entropy $H_Q(V)$ vs. ϵ^2/A as v_{max} is varied for $M = 10, 20, 40$.

follows. It is impossible to satisfy the fidelity criterion (ϵ^2 mean-squared error, in our case) by transmitting over a channel of capacity $C < R(\epsilon^2)$ between source and sink. If we consider using a noiseless binary channel between source and sink, the capacity of this channel must be at least $R(\epsilon^2)$ bits per source sample to achieve ϵ^2 mean-squared error. That is, no fewer than $R(\epsilon^2)$ bits transmitted per source sample could achieve ϵ^2 mean-squared error. This statement holds for any source encoder which translates the source samples to binary digits, whether it is a uniform quantizer or a device which maps blocks of source samples into sequences of binary digits. $R(\epsilon^2)$ thus gives the limiting performance that one could expect for any scheme which transmits the source output to the sink in the form of binary digits.

In Fig. 2 we show several curves of $H_Q(V)$ vs. ϵ^2/A (where $A = E(u^2)$) as v_{max} is varied for quantizers having 10, 20, and 40 output levels. For values of ϵ^2/A less than 0.5, the lower envelope of such curves for M ranging from 2 to 140 is fitted quite accurately by the function

$$H^*(\epsilon^2) = \tfrac{1}{4} + \tfrac{1}{2}\log_2\frac{A}{\epsilon^2}\text{ bits per sample.}$$

The curve of $H^*(\epsilon^2)$ vs. ϵ^2/A represents the best obtainable performance for uniform quantizers used with Gaussian samples, and this performance would be achievable only by entropy coding of the quantizer levels.

In Fig. 2 we also plot $R(\epsilon^2)$ for independent Gaussian samples. It is quite striking that the uniform quantizer with entropy coding requires only about $\frac{1}{4}$ bits per sample more than the minimum data rate achievable by *any* digitization scheme. It is also worth noting that the adjustment of quantizer parameters is not sensitive in that the curves of $H_Q(V)$ vs. ϵ^2/A for a given M are very close to the $H^*(\epsilon^2)$ curve over a wide range of ϵ^2/A.

When sequences of M-ary quantizer outputs are expressed in binary digits, the data rate ($\log_2 M$ bits per sample) is independent of the quantizer level spacing, and performance is optimized by simply minimizing the mean-squared error. The circles in Fig. 2 represent values of $\log_2 M$ plotted against the minimum mean-squared error obtainable by a uniform quantizer with M output levels. These points are fitted closely by the function

$$M^*(\epsilon^2) = 0.125 + 0.6\log\frac{A}{\epsilon^2}\text{ bits per sample.}$$

Thus $M^*(\epsilon^2)$ represents the achievable data rate (bits per sample) if each of the quantizer outputs is converted directly to an M-ary digit. It may not be surprising that $M^*(\epsilon^2)$ diverges from $R(\epsilon^2)$ as ϵ^2/A decreases, but even this simple digitization scheme performs respectably relative to theoretical limits over a useful range of ϵ^2/A. The staircase curve in Fig. 2 is denoted $B^*(\epsilon^2)$ and represents the required data rate when each of the quantizer outputs is converted directly to binary digits with no further coding.

The circled points in Fig. 2 were computed by J. Max[5] for uniform quantizers for Gaussian samples. He also minimized the mean-squared error by optimizing the choice of the M output levels and the range of input amplitudes mapped into each level. He gave the entropy of the M output levels for the quantizers so optimized, as well as for quantizers with uniform output level spacing. In contrast to this, we have minimized the mean-squared error for a *fixed output entropy*, or from another viewpoint, minimized the output entropy for a fixed mean-squared error for uniform quantizers. The difference between our work on uniform quantizers and Max's is mainly a philosophical one, but an important one. We have obtained a curve $H^*(\epsilon^2)$ which gives the trade-off between output entropy and mean-squared error for uniform quantizers. Our results indicate that this trade-off for uniform quantizers is very close to the ideal relationship for Gaussian samples.

III. Gaussian Processes

A more general class of analog sources for which the information rate is known (under a mean-squared error fidelity criterion) is the class of stationary Gaussian processes.[6] Therefore, we are in a position to compare theory with practice in digitization of this class of sources. The information rate $R(\epsilon^2)$ for a Gaussian process with power spectrum $S(f)$ is given by the parametric relations

$$R(\varphi) = \int_{\mathfrak{F}} \log_2\frac{S(f)}{\varphi}\,df\text{ bits per second}^7$$

$$\epsilon^2(\varphi) = 2\varphi\int_{\mathfrak{F}} df + 2\int_{\bar{\mathfrak{F}}} S(f)\,df$$

[5] J. Max, "Quantizing for minimum distortion," *IRE Trans. on Information Theory*, vol. IT-6, pp. 7–12, March 1960.

[6] T. J. Goblick, Jr., "Theoretical limitations on the transmission of data from analog sources," *IEEE Trans. on Information Theory*, vol. IT-11, pp. 558–567, October 1965. References on this development are given; note that ϵ is used to denote the mean-squared error, while ϵ^2 is used here.

[7] Henceforth, we will abbreviate bits per second as bps.

and

$$\mathfrak{F} = \{f \geq 0 \text{ such that } S(f) \geq \varphi\}$$

$$\bar{\mathfrak{F}} = \{f \geq 0 \text{ such that } S(f) < \varphi\}, \qquad \varphi > 0.$$

In the case of a bandlimited white Gaussian source of bandwidth W_s, the output can be represented by (independent) samples spaced $1/2W_s$ seconds apart, and from $R(\epsilon^2)$ for the time-discrete source we have in this case

$$R(\epsilon^2) = W_s \log_2 \frac{A}{\epsilon^2} \text{ bps}$$

where

$$A = \int_{-\infty}^{\infty} S(f)\, df.$$

Suppose the output waveform $u(t)$ of a white bandlimited Gaussian source is digitized by sampling and quantizing with a uniform quantizer, and that entropy coding is done on the quantizer outputs to reduce the average number of bits necessary to specify these levels to the receiver. The minimum data rate necessary to achieve ϵ^2 mean-squared error with this digitization system is gotten from the results of the previous section as

$$H(\epsilon^2) = 2W_s H^*(\epsilon^2) \text{ bps}.$$

If the quantizer outputs are not coded, but directly expressed in M-ary digits, the data rate becomes

$$M(\epsilon^2) = 2W_s M^*(\epsilon^2) \text{ bps}.$$

Likewise, if the quantizer outpucs are each mapped into binary digits, the data rate is

$$B(\epsilon^2) = 2W_s B^*(\epsilon^2) \text{ bps}.$$

The digitization systems described above can clearly be used with the same results with any Gaussian process strictly bandlimited to W_s Hz and with average power A. The samples for such sources will be correlated in general, but the above systems achieve their stated performance while ignoring these correlations between samples. Of course, this is not the best obtainable performance for an arbitrary band limited Gaussian process because we have ignored these correlations, and this costs something in performance.

A stationary Gaussian process with an arbitrary power density spectrum $S(f)$ could be digitized by the above system provided that it was first bandlimited by an ideal lowpass filter with cutoff frequency W_F Hz. Since the waveform $v(t)$ reconstructed from the quantized samples of $u(t)$ will be bandlimited to W_F Hz, the spectrum of the error $e(t) = u(t) - v(t)$ will be equal to $S(f)$ for $|f| > W_F$. Therefore, the bandlimiting filter introduces a mean-squared error of

$$\epsilon_F^2 = 2\int_{W_F}^{\infty} S(f)\, df.$$

If the samples of $u(t)$ are quantized to yield a mean-squared error of ϵ_Q^2, the total mean-squared error is

$$\epsilon^2 = \epsilon_F^2 + \epsilon_Q^2$$

and the resulting data rate is

$$H(\epsilon^2) = 2W_F H^*(\epsilon_Q^2) \text{ bps},$$

$$M(\epsilon^2) = 2W_F M^*(\epsilon_Q^2) \text{ bps},$$

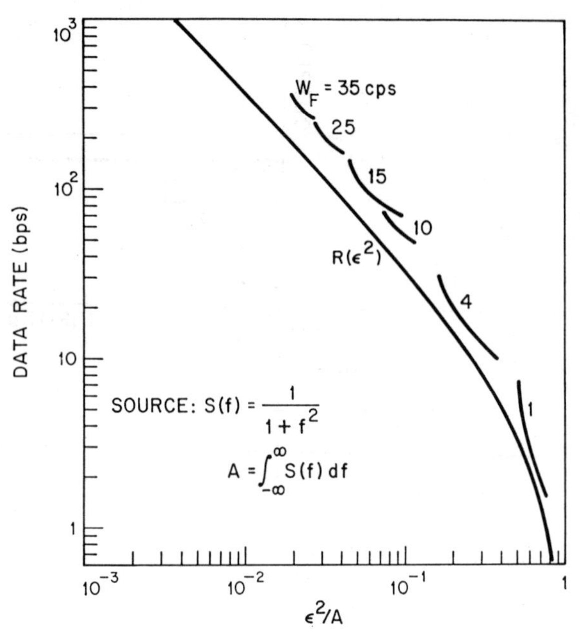

Fig. 3. Data rate $H(\epsilon^2)$ vs. ϵ^2/A for the BSQC encoder corresponding to severa values of filter bandwidth W_F for a Gaussian process.

or

$$B(\epsilon^2) = 2W_F B^*(\epsilon_Q^2) \text{ bps},$$

depending upon whether the quantizer outputs are entropy coded, expressed in M-ary symbols, or binary digits, respectively.

The above data rates are clearly functions of both W_F and ϵ_Q^2. If W_F is first fixed and ϵ_Q^2 varied, a curve of $H(\epsilon^2)$ vs. ϵ^2 will result. An example of such curves is shown for various values of W_F in Fig. 3 which were computed for the nonbandlimited source power spectrum

$$S(f) = \frac{1}{1 + f^2}.$$

The lower envelope of these curves, denoted $H_L(\epsilon^2)$, is clearly significant in characterizing the digitization system consisting of the 1) bandlimiting filter, 2) sampler, 3) uniform quantizer, and 4) entropy encoder (abbreviated as the BSQC encoder or digitizer). $H_L(\epsilon^2)$ represents the lowest data rate (in bps) necessary to achieve ϵ^2 mean-squared error with the BSQC encoder.

In Fig. 4 we show curves of $R(\epsilon^2)/W_s$ (solid lines) and $H_L(\epsilon^2)/W_s$ (dashed lines) vs. ϵ^2/A for sources with power spectra given by

$$S(f) = \frac{1}{1 + \left(\frac{f}{W_s}\right)^{2n}}$$

where

$$A = \int_{-\infty}^{\infty} S(f)\, df.$$

For the range of ϵ^2/A included in this figure, $H_L(\epsilon^2)$ is no more than 2 times $R(\epsilon^2)$ for the source corresponding to $n = 1$, and this margin between $H_L(\epsilon^2)$ and $R(\epsilon^2)$ decreases significantly as n increases. The circles in the figure represent values of $B(\epsilon^2)$ resulting from the variation of W_F and ϵ_Q^2 that were closest to the $R(\epsilon^2)$ curve. These points represent performance achievable by mapping each quantizer output directly into binary digits (termed the BSQB encoder in Fig. 4). A significant saving in data rate is seen to be attributable

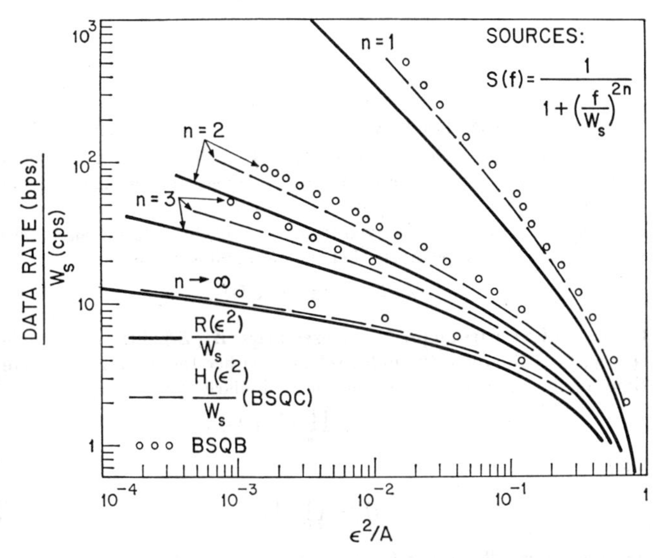

Fig. 4. Normalized data rates vs. ϵ^2/A for BSQC and BSQB encoders for several different Gaussian source spectra.

to the coding of quantizer outputs, as compared to merely converting them directly to binary digits. The values of $M(\epsilon^2)$ are not significantly different from the values of $B(\epsilon^2)$ and are not shown in the figure.

It can be shown that if the output of the ideal lowpass filter is "whitened" prior to sampling and quantizing, with "recoloring" done by a filter following the interpolation of the quantized samples, the performance of both encoders is unchanged. This is, of course, due to the fact mentioned earlier that the results of sampling and quantizing a bandlimited source are independent of the source spectrum.

V. Summary

1) The information rate of a time-discrete, amplitude continuous source relative to a fidelity criterion provides a bound on the performance of any kind of quantizer.
2) The information rate of an analog source relative to a fidelity criterion provides a lower bound to the digital data rate achievable by *any* digitization scheme that satisfies the fidelity criterion.
3) In the case of the Gaussian processes studied here, simple filtering, sampling, and quantizing schemes are compared to the theoretically ideal performance to give a feeling for the gap between theory and practice.

Thus if we know the information rate of a source, we are in a position to compare the data rates of existing digitization systems with the theoretical minimum corresponding to a certain level of fidelity. Such a comparison not only provides a measure of the worth of existing systems, but large gaps between theoretical and practical data rates would provide legitimate motivation for further research into new digitization techniques.

Acknowledgment

The authors are deeply indebted to Mrs. M. Simon for writing many computer programs to obtain the numerical data used in the figures.

T. J. Goblick, Jr.
M.I.T. Lincoln Laboratory[8]
Cambridge, Mass.
J. L. Holsinger
Defense Research Corp.
Santa Barbara, Calif.

[8] Operated with support from the U. S. Air Force.

Rate Distortion Theory for Sources with Abstract Alphabets and Memory

TOBY BERGER†

Advanced Development Laboratory, Raytheon Company
Wayland, Massachusetts 01778

This paper is devoted to the formulation and proof of an abstract alphabet version of the fundamental theorem of Shannon's rate distortion theory. The validity of the theorem is established for both discrete and continuous parameter information sources satisfying a certain regularity condition intermediate in restrictiveness between ergodicity and weak mixing. For ease of presentation, only single letter fidelity criteria are considered during the main development after which various generalizations are indicated.

I. INTRODUCTION

The rate distortion function was introduced by Shannon (1948, 1959) to specify the rate at which a source produces information relative to a fidelity criterion. Rate distortion theory is an essential tool in the study of analog data sources and, more generally, is applicable to any communication problem in which the entropy rate of the source exceeds the capacity of the channel over which it must be transmitted. Since perfect transmission is impossible in such situations, it becomes fruitful to study the problem of approximating the given source with one of lower entropy in such a way that the least possible distortion results relative to some prescribed fidelity criterion.

For the case of a bounded distortion measure and a discrete parameter source producing statistically independent samples, Shannon (1959) defined the rate distortion function, $R(\cdot)$, and then proved that $R(D)$ represents the minimum capacity a channel must have in order for it to be possible to reproduce the source at the channel output with an average distortion no greater than D. The extension to discrete parameter sources with memory was briefly sketched also, though no consideration was given to the statistical dependence between successive blocks of source letters used in the random coding argument.

In the present paper we provide a detailed proof of an abstract alphabet formulation of the fundamental theorem of rate distortion theory for discrete or continuous parameter sources and unbounded distortion measures. The statistical dependence between successive blocks of letters produced by the source necessitates our imposing a regularity condition on the source called block ergodicity (see Definition 1), which is intermediate in restrictiveness between ordinary ergodicity and weak mixing. Our main result (Theorem 2) may be interpreted as formally establishing that the source-fidelity criterion combinations in question possess the property of information stability required for application of the basic theorems presented by Dobrushin (1959) in his generalized formulation of information theory.

While this paper was undergoing review Goblick (1967) reported similar results for time discrete strongly mixing sources. Still more recently Gallager (1968) has successfully treated time discrete finite alphabet ergodic sources by means of a decomposition technique devised by Nedoma (1963).

The terminology and notation employed are presented in Section II together with certain preliminary results. The rate distortion function of an information source with respect to a (single letter) fidelity criterion is defined in Section III. The fundamental theorem of rate distortion theory, which imbues the rate distortion function with its operational significance, is stated and proved in Section IV. Some straightforward generalizations of the results are indicated in the final section.

† Present address: School of Electrical Engineering, Cornell University, Ithaca, New York 14850.

II. TERMINOLOGY, NOTATION, AND PRELIMINARIES

Let the space Z_0, called the joint alphabet, be the Cartesian product of two non-empty abstract spaces X_0 and Y_0 called the message alphabet and the approximating alphabet, respectively. Let $\mathfrak{Z}_0$ be the product σ-algebra of subsets of Z_0 derived from the σ-algebras $\mathfrak{X}_0$ and $\mathfrak{Y}_0$ of subsets of X_0 and Y_0, respectively. Let the measurable space

$$(Z, \mathfrak{Z}) = \prod_{t \in M} (Z_t, \mathfrak{Z}_t)$$

be the infinite Cartesian product of exemplars $(Z_t, \mathfrak{Z}_t)$ of the measurable space $(Z_0, \mathfrak{Z}_0)$, where the index set M is either the integers (discrete time) or the real line (continuous time). Similarly, let

$$(X, \mathfrak{X}) = \prod_{t \in M} (X_t, \mathfrak{X}_t)$$

and

$$(Y, \mathfrak{Y}) = \prod_{t \in M} (Y_t, \mathfrak{Y}_t),$$

from which it follows that $(Z, \mathfrak{Z}) = (X \times Y, \mathfrak{X} \times \mathfrak{Y})$.

Let $\mu(\cdot)$ be a probability on the σ-algebra $\mathfrak{X}$, that is, a nonnegative, countably additive set function such that $\mu(X) = 1$. We call the triple $[X, \mathfrak{X}, \mu]$ the message source, or simply the source whenever no ambiguity results. Following McMillan (1953), we abbreviate the symbol for the source to $[X, \mu]$. Elements $x \in X$ are called realizations of the source. We use the symbol "$\pi\ [\mu]$" to indicate that the proposition $\pi(x)$ is true for almost every x with respect to μ.

The value assumed by the projection of a realization $x \in X$ into the tth coordinate space is denoted by $x_t \in X_t$. Similarly, the tth coordinates of $y \in Y$ and $z = (x, y) \in Z$ are denoted by y_t and z_t, respectively.

For any $r, t \in M$, with $r < t$, we define the probability space $(X_{r,t}, \mathfrak{X}_{r,t}, \mu_{r,t})$ by

$$(X_{r,t}, \mathfrak{X}_{r,t}) = \prod_{s \in [r,t)} (X_s, \mathfrak{X}_s) \tag{1}$$

with $\mu_{r,t}$ being the restriction of μ to $\mathfrak{X}_{r,t}$. The measurable spaces $(Y_{r,t}, \mathfrak{Y}_{r,t})$ and $(Z_{r,t}, \mathfrak{Z}_{r,t})$ are defined by expressions analogous to (1). For completeness, we set $(X_{t,t}, \mathfrak{X}_{t,t}) = (X_t, \mathfrak{X}_t)$, with analogous notation for the approximating and joint spaces, too. For the particular case of $r = 0$ a single, superscript t is used, e.g.,

$$(X_{0,t}, \mathfrak{X}_{0,t}, \mu_{0,t}) = (X^t, \mathfrak{X}^t, \mu^t).$$

Typical elements of $X^t[Y^t, Z^t]$ are called message [approximating, joint] t-blocks and are denoted by $x^t[y^t, z^t = (x^t, y^t)]$.

We indulge in the following abuses of notation. The symbol x_s used to denote the sth component of a realization $x \in X$ also is used to denote the sth component of a typical t-block, x^t, $0 \leq s < t$. Moreover, the superscript t of x^t, y^t and z^t often is suppressed, since one or two judiciousy placed t's generally suffice to permit relatively lengthy expressions to be interpreted unambiguously.

Let $\{T^t, t \in M\}$ be the group of shift transformations from X to X defined by $(T^t x)_s = x_{s+t}$.[1] A source $[X, \mu]$ is called t-stationary if $\mu(T^t E) = \mu(E)$ for all $E \in \mathfrak{X}$, where $T^t E = \{T^t x : x \in E\}$. We say $[X, \mu]$ is stationary if it is t-stationary for all $t \in M$. A $\mathfrak{X}$-measurable function g is called t-invariant if $g(T^t x) = g(x)$ for all $x \in X$, and invariant if it is t-invariant for every $t \in M$. A set $E \in \mathfrak{X}$ is t-invariant (invariant) if its characteristic function is t-invariant (invariant). A source $[X, \mu]$ is t-ergodic (ergodic) if $\mu(E) = \mu^2(E)$ for every t-invariant (invariant) set E.

[1] The symbol T^t also will be used to denote the shift by t coordinates on the spaces Y and Z.

Reprinted with permission from *Inform. Contr.*, vol. 13, pp. 254–273, Sept. 1968.

We shall subsequently be concerned with encoding the source $[X, \mu]$ by means of operations on successive message blocks. In this regard consider segmenting each realization $x \in X$ into an infinite sequence of τ-blocks. The space of all such infinite sequences of τ-blocks, together with the σ-algebra and probability naturally induced by $\mathfrak{X}$ and μ, will be denoted by $[X, \mu]_\tau$ and called the τ-block source derived from $[X, \mu]$. The source $[X, \mu]_\tau$ is time discrete, producing one message τ-block every τ seconds.

DEFINITION 1. A source $[X, \mu]$ is *block ergodic* if it is τ-ergodic for every positive $\tau \in M$.

We show in the appendix that block ergodicity lies between ergodicity and weak mixing in restrictiveness. Note that τ-ergodicity of $[X, \mu]$ and ergodicity of $[X, \mu]_\tau$ are equivalent.

Let the mapping $q^t: X^t \times \mathcal{Y}^t \to [0, 1]$ be a transition probability, by which we mean that q^t satisfies the following two conditions:

(a) For every $x \in X^t$, the set function $q^t(x, \cdot)$ is a probability on $\mathcal{Y}^t$

(b) For every $F \in \mathcal{Y}^t$, the function $q^t(\cdot, F)$ is $\mathfrak{X}^t$-measurable.

Let ω^t denote the joint probability induced on $\mathcal{Z}^t$ by μ^t and q^t. That is, for any set $G \in \mathcal{Z}^t$, we have[2]

$$\omega^t(G) = \int q^t(x, G_x)\, d\mu^t, \tag{2}$$

where $G_x = \{y: (x, y) \in G\}$. We denote the marginal probability that ω^t induces on $\mathcal{Y}^t$ by

$$\nu^t(F) = \omega^t(X^t \times F) = \int q^t(x, F)\, d\mu^t,$$

and the product probability on $\mathcal{Z}^t$ derived from the marginals of ω^t by

$$w^t = \mu^t \times \nu^t.$$

If ω^t is absolutely continuous with respect to w^t (henceforth written $\omega^t \ll w^t$), then we denote the associated Radon–Nikodym derivative by

$$f_t = d\omega^t/dw^t.$$

If $\omega^t \ll w^t$, then for any $G \in \mathcal{Z}^t$ we have

$$\omega^t(G) = \int_G f_t\, dw^t = \int_G f_t\, d(\mu^t \times \nu^t) = \int \left(\int_{G_x} f_t(x, y)\, d\nu^t \right) d\mu^t. \tag{3}$$

Comparing (2) and (3), we note for future reference that

$$q^t(x, G_x) = \int_{G_x} f_t(x, y)\, d\nu^t\ [\mu^t]. \tag{4}$$

DEFINITION 2. The *average information*, I_t, of the joint probability space $(Z^t, \mathcal{Z}^t, \omega^t)$ is defined by[3]

$$I_t = \sup \sum_{i=1}^{\infty} \omega^t(G_i) \log [\omega^t(G_i)/w^t(G_i)], \tag{5}$$

where the supremum is taken with respect to all partitions $\{G_i\}$ of the space Z^t by countably many rectangles $G_i = E_i \times F_i$ with $E_i \in \mathfrak{X}^t$ and $F_i \in \mathcal{Y}^t$. Dobrushin (1959) shows that the value of I_t always is nonnegative and remains unchanged if the restriction to rectangular partitions is relaxed. When ω^t and w^t are induced by a transition probability q^t, then I_t becomes a functional of q^t which we shall denote by $I_t(q)$.

LEMMA 1. *If I_t is finite, then $\omega^t \ll w^t$ and*

$$I_t = \int \log f_t\, d\omega^t. \tag{6}$$

Proof. See the translator's remarks by Feinstein in the book by Pinsker (1960).

We shall need the so-called generalized Shannon–McMillan limit theorem and its corollary, the generalized asymptotic equipartition property (AEP). The following version of the theorem, due to Perez (1964), is essentially the most general to be established to date.

THEOREM 1. (*Perez's Theorem*). *Let the joint source $[Z, \mathcal{Z}, \omega]$ be stationary, and let $(Z^t, \mathcal{Z}^t, \omega^t)$ be its restriction to joint t-blocks. Then the finiteness of the information rate,*

$$R \triangleq \lim_{t\to\infty} (1/t) I_t, \tag{7}$$

is a necessary and sufficient condition for the existence of an invariant, ω-integrable function $h(z)$ such that[4]

$$\lim_{t\to\infty} (1/t) \log f_t(z) = h(z)\ [\omega]. \tag{8}$$

If $[Z, \mathcal{Z}, \omega]$ is ergodic as well, then $h(z)$ is a constant, namely the information rate R of (7).

COROLLARY 1. (*Generalized* AEP). *If the joint source $[Z, \mathcal{Z}, \omega]$ of Theorem 1 also is ergodic and R of (7) is finite, then for all $\epsilon > 0$*

$$\lim_{t\to\infty} \omega^t\{z: |(1/t) \log f_t(z) - R| > \epsilon\} = 0. \tag{9}$$

III. THE RATE DISTORTION FUNCTION

In this section we define the rate distortion function of an information source with respect to a single letter fidelity criterion.

Any $\mathcal{Z}_0$-measurable function ρ mapping Z_0 into $[0, \infty)$ is called a single letter distortion measure. Its physical significance is that $\rho(\alpha, \beta)$ specifies the distortion that results when $\alpha \in X_0$ is approximated by $\beta \in Y_0$. We associate with ρ a family of distortion measures,

$$F_\rho = \{\rho_t, 0 \leqq t < \infty\}, \tag{10}$$

each member of which maps Z into $[0, \infty)$. Specifically, we set $\rho_0(z) = \rho(z_0)$, and for $t > 0$ we define

$$\rho_t(z) = \frac{1}{t} \sum_{s=0}^{t-1} \rho_0(T^s z) \quad \text{or} \quad \frac{1}{t} \int_0^t \rho_0(T^s z)\, ds \tag{11}$$

for discrete or continuous time, respectively. Since $\rho_t(z)$ depends only on those components of z that belong to z^t, it is also possible to interpret it as a $\mathcal{Z}^t$-measurable mapping from Z^t into $[0, \infty)$ and we often shall.

DEFINITION 3. The family of distortion measures F_ρ of (10) is called a *single letter fidelity criterion*.

We proceed to define the rate distortion function of a message source with respect to a single letter fidelity criterion. Toward this end let us introduce the family of functions, $\{R_t(D), 0 \leqq t < \infty\}$, defined as follows:

For each $D \in [0, \infty)$, let $\mathcal{Q}_t(D)$ be the class of transition probabilities $q^t: X^t \times \mathcal{Y}^t \to [0, 1]$ for which

$$\mathfrak{D}(q) \triangleq \int \rho_t(z)\, d\omega^t \leqq D. \tag{12}$$

That is, q^t belongs to $\mathcal{Q}_t(D)$ only if it induces a joint probability ω^t on $\mathcal{Z}_t$ for which the average distortion $\mathfrak{D}(q)$ between message t-blocks and approximating t-blocks does not exceed D. We also introduce the average information rate $\mathcal{R}_t(q)$ defined, with reference to (5), to be

$$\mathcal{R}_t(q) = (1/t)[I_t(q)], \tag{13}$$

where the value $+\infty$ is not excluded. The function $R_t(D)$ then is defined by

$$R_t(D) = \inf_{\mathcal{Q}_t(D)} \mathcal{R}_t(q). \tag{14}$$

If $\mathcal{Q}_t(D)$ is empty, $R_t(D)$ is taken to be $+\infty$.

DEFINITION 4. The *rate distortion function* $R(D)$ of the message source $[X, \mu]$ with respect to the single letter fidelity F_ρ is given by the prescription

$$R(D) = \liminf_{t\to\infty} R_t(D). \tag{15}$$

For completeness we record here the following facts concerning $R(D)$ (Shannon, 1959). First, the "inf" may be deleted in (15) provided $[X, \mu]$ is stationary (see also Reiffen, 1966). Second, $R(D)$ is a nonnegative, monotonic nonincreasing, convex downward function of D over the interval in which it is finite. Third, if there is an element $y \in Y$ for which

$$\lim_{t\to\infty} \int \rho_t(x, y)\, d\mu < \infty,$$

then there is a distortion value $D_{\max}$ such that $R(D)$ vanishes identically for $D \geqq D_{\max}$.

It is heuristically clear from Definition 4 that $R(D)$, if finite, in some sense represents the minimum average information per unit time that must be supplied about a realization $x \in X$ in order to permit specifica-

[2] Unless subscripted by a particular set, integrals with respect to a probability extend over the entire space governed by that probability.

[3] All logarithms in this paper are to the base e.

[4] The symbol "f_t" in (8) denotes a function defined on Z, not Z^t. By rights this function should be given a different symbol, say $\tilde{f}_t$, and then defined by $\tilde{f}_t(z) = f_t(z^t)$, but it is customary to use the same symbol for both functions.

tion of a $y \in Y$ that approximates this x with an average distortion with respect to F_ρ that does not exceed D. A precise mathematical statement of the sense in which this heuristic interpretation of $R(D)$ is correct is provided by the fundamental theorem of rate distortion theory.

IV. THE FUNDAMENTAL THEOREM OF RATE DISTORTION THEORY

In this section we prove our main result (Theorem 2) and then discuss some of its implications.

Theorem 2. *If there exists a $\beta \in Y_0$ such that*

$$\int \rho_0(x, \beta)\, d\mu = \hat{\rho} < \infty, \tag{16}$$

then the following two statements are valid for any $\epsilon > 0$ and any $D \geq 0$:

Positive Statement. If $[X, \mu]$ *is both stationary and block ergodic and $R(D)$ is finite, then there exists a value of t and a subset S of Y^t containing N approximating t-blocks such that* $\log N \leq t[R(D) + \epsilon]$ *and*

$$D(S) \triangleq \int \rho_t(x \mid S)\, d\mu^t \leq D + \epsilon, \tag{17}$$

where

$$\rho_t(x \mid S) = \min_{y \in S} \rho_t(x, y).^5 \tag{18}$$

Negative Statement. For all t any set $\tilde{S} \subset Y^t$ that contains only $\tilde{N}$ elements, where $\log \tilde{N} \leq t[R(D) - \epsilon]$, *must satisfy the inequality $D(\tilde{S}) > D$.*

Proof. We shall prove the positive statement by a random coding argument. Consider an ensemble of subsets S of Y^t each of which contains N elements, where N is the largest integer in $\exp\{t[R(D) + \epsilon]\}$. If we can show that the average over this ensemble of $D(S)$ as defined in (17) does not exceed $D + \epsilon$, then at least one S in the ensemble must satisfy inequality (17) and the proof will be complete.

We define our ensemble of subsets of Y^t as follows. The choice of a particular set of N elements of Y^t may be envisioned as the choice of a single element $S \in Y^{tN}$. We choose each such S independently of all the others according to a common probability measure λ^{tN} defined on $\mathcal{Y}^{tN}$. Furthermore we select the N elements of each S independently, i.e.,

$$\lambda^{tN} = \nu_1^t \times \nu_2^t \times \cdots \times \nu_N^t, \tag{19}$$

where each ν_i^t is a probability on $\mathcal{Y}^t$. Moreover, for $i = 1, 2, \cdots, N - 1$, we set each ν_i^t equal to a common probability, ν^t. The probability ν_N^t that governs the selection of the last element of each set is a degenerate one concentrated at a special element $b \in Y^t$, where b may be taken as any approximating t-block for which

$$\int \rho_t(x, b)\, d\mu^t < \infty. \tag{20}$$

In particular, one candidate for b is that element every component of which is the β that appears in (16), in which case the integral in (20) equals $\hat{\rho}$. Since this choice of b simplifies subsequent arguments, we set $b_s = \beta$, $0 \leq s < t$. Recalling (17) and (18), we see that the average of $D(S)$ over the ensemble of sets S may be written

$$\bar{D} \triangleq \int D(S)\, d\lambda^{tN} = \int\left(\int \rho_t(x \mid S)\, d\mu^t\right) d\lambda^{tN}.$$

We now define the set

$$A = \{x \in X^t : \rho_t(x, b) \leq \hat{\rho} + \delta\}, \tag{21}$$

where δ is an arbitrary positive constant. The fact that b belongs to every S in the ensemble guarantees that

$$\rho_t(x \mid S) \leq \rho_t(x, b). \tag{22}$$

Accordingly, where $\bar{A}$ denotes the complement of A, we have

$$\bar{D} \leq \int_{\bar{A}} \rho_t(x, b)\, d\mu^t + \int\left(\int_A \rho_t(x \mid S)\, d\mu^t\right) d\lambda^{tN}. \tag{23}$$

Now (20) and (22) imply that the iterated integral appearing on the far right of (23) is finite, so Fubini's theorem yields

$$\int\left(\int_A \rho_t(x \mid S)\, d\mu^t\right) d\lambda^{tN} = \int_A\left(\int \rho_t(x \mid S)\, d\lambda^{tN}\right) d\mu \tag{24}$$

[5] Either, but not both, of the ϵ's in the positive statement may be set to zero by a straightforward extension of the proof given below.

The inner integral on the right of (24) is the ensemble average of $\rho_t(x \mid S)$ for fixed x. Hence, letting

$$P_t = P_t(u \mid x) = \lambda^{tN}\{S : \rho_t(x \mid S) \leq u\}$$

denote the cumulative probability distribution function of $\rho_t(x \mid S)$ for fixed x, we may write

$$\int\left(\int_A \rho_t(x \mid S)\, d\mu^t\right) d\lambda^{tN} = \int_A\left(\int_0^{\hat{\rho}+\delta} u\, dP_t\right) d\mu^t, \tag{25}$$

where the finite upper limit of $\hat{\rho} + \delta$ follows from (22) and the definitions of A and P_t. Upon introducing the complementary distribution function

$$\bar{P}_t = 1 - P_t,$$

we note that

$$\int_0^{\hat{\rho}+\delta} u\, dP_t = -\int_0^{\hat{\rho}+\delta} u\, d\bar{P}_t = -(\hat{\rho} + \delta)\bar{P}_t(\hat{\rho} + \delta \mid x) + \int_0^{\hat{\rho}+\delta} \bar{P}_t\, du,$$

the last integral being with respect to ordinary Lebesgue measure. Thus,

$$\begin{aligned}\int_0^{\hat{\rho}+\delta} u\, dP_t &\leq \int_0^{\hat{\rho}+\delta} \bar{P}_t\, du \leq (D + \delta) \\ &\quad + \int_{D+\delta}^{\hat{\rho}+\delta} \bar{P}_t\, du \leq D + \delta + (\hat{\rho} - D)\bar{P}_t(D + \delta \mid x),\end{aligned} \tag{26}$$

where we have used the fact that $\bar{P}_t(u \mid x)$ is a monotonic nonincreasing function of u that never exceeds unity. (Of course, we are also assuming that $\hat{\rho} > D$, since the theorem is trivial for $\hat{\rho} \leq D$.) Substituting (26) into (25) and upper bounding the integration over A by integration over X^t allows us to replace (23) by

$$\bar{D} \leq D + \delta + \int_{\bar{A}} \rho_t(x, b)\, d\mu^t + (\hat{\rho} - D)\int \bar{P}_t(D + \delta \mid x)\, d\mu^t. \tag{27}$$

We complete the proof by showing that both integrals in (27) vanish in the limit of infinite t. The integral over $\bar{A}$ may be dispensed with by means of the ergodic theorem. Toward this end we write

$$\begin{aligned}\int_{\bar{A}} \rho_t(x, b)\, d\mu^t &= \int_{\bar{A}} [\rho_t(x, b) - \hat{\rho}]\, d\mu^t + \hat{\rho}\mu^t(\bar{A}) \\ &= \int [\rho_t(x, b) - \hat{\rho}]\, d\mu^t \\ &\qquad - \int_A [\rho_t(x, b) - \hat{\rho}]\, d\mu^t + \hat{\rho}\mu^t(\bar{A}) \\ &= \int_A [\hat{\rho} - \rho_t(x, b)]\, d\mu^t + \hat{\rho}\mu^t(\bar{A}),\end{aligned} \tag{28}$$

where we have used the fact that the expected value of $\rho_t(x, b)$ is $\hat{\rho}$ for our choice of b. Let

$$B = \{x \in X^t : \rho_t(x, b) \geq \hat{\rho} - \delta\},$$

and observe that

$$\begin{aligned}\int_A [\hat{\rho} - \rho_t(x, b)]\, d\mu^t &= \int_{A \cap B} [\hat{\rho} - \rho_t(x, b)]\, d\mu^t \\ &\quad + \int_{A \cap \bar{B}} [\hat{\rho} - \rho_t(x, b)]\, d\mu^t \\ &\leq \delta\mu^t(A \cap B) + \hat{\rho}\mu^t(A \cap \bar{B}) \leq \delta + \hat{\rho}\mu^t(\bar{B}).\end{aligned} \tag{29}$$

Combining (28) and (29) and noting that A and $\bar{B}$ are disjoint yields

$$\int_{\bar{A}} \rho_t(x, b)\, d\mu^t \leq \delta + \hat{\rho}\mu^t(\bar{A} \cup \bar{B}). \tag{30}$$

Since $\bar{A} \cup \bar{B} = \{x : |\rho_t(x, b) - \hat{\rho}| > \delta\}$, it follows in the case of, say, continuous time that

$$\mu^t(\bar{A} \cup \bar{B}) = \mu\left\{x \in X : \left|\frac{1}{t}\int_0^t \rho_0(T^s x, \beta)\, ds - \hat{\rho}\right| > \delta\right\}. \tag{31}$$

Because $[X, \mu]$ is ergodic, the first term within the absolute value brackets in (31) converges to the expected value of $\rho_0(x, \beta)$, namely $\hat{\rho}$ of (16). Thus, $\mu^t(\bar{A} \cup \bar{B}) \to 0$, and the integral in (30) vanishes for large t since δ is arbitrary.

We now show that the other integral in (27), namely

$$J_t \triangleq \int \bar{P}_t(D+\delta \mid x)\, d\mu^t, \tag{32}$$

also vanishes in the limit of large t. We begin by noting that excluding the special element b from consideration when determining the minimum in (18) yields the inequality

$$\begin{aligned}\bar{P}_t(D+\delta \mid x) &= \lambda^{tN}\{S:\rho_t(x \mid S) > D+\delta\} \\ &\leqq [1 - \nu^t\{y:\rho_t(x, y) \leqq D+\delta\}]^{N-1},\end{aligned} \tag{33}$$

where ν^t is the common probability governing the independent selection of the other $N-1$ elements of S.

At this point it becomes necessary to discuss in greater detail the manner in which we select ν^t. First, we choose a block length τ large enough to ensure that

$$R_\tau(D) < R(D) + \delta/2. \tag{34}$$

Next, recalling (13) and (14), we choose a transition probability $q^\tau \in \mathcal{Q}_\tau(D)$ for which

$$\mathcal{R}_\tau(q) < R_\tau(D) + \delta/2. \tag{35}$$

In the ensuing discussion q^τ is referred to as the "τ-block channel." As usual, ω^τ and ν^τ denote the probabilities that q^τ induces on $\mathcal{Z}^\tau$ and $\mathcal{Y}^\tau$, respectively.

Since our concern is with the behavior of J_t as $t \to \infty$, no loss in generality results from confining attention solely to values of t of the form $t = n\tau$, $n = 1, 2, \cdots$. For such t consider the transition probability

$$q^t = q_1^\tau \times q_2^\tau \times \cdots \times q_n^\tau,$$

where each q_k^τ equals q^τ. That is, q^t is the n-fold product of the τ-block channel with itself, and corresponds physically to segmenting each source t-block into n successive τ-blocks and then transforming these independently with the τ-block channel. The probability ν^t induced by q^t on $\mathcal{Y}^t$ is the one we use to randomly generate the approximating t-blocks of our ensemble. Although q^t operates on successive message τ-blocks independently of one another, the source $[X, \mu]$ in general produces message τ-blocks that are mutually dependent. As a result, ν^t is *not* the n-fold product of ν^τ with itself. Since this dependence can only decrease the average information $I_t(q)$ below the value $nI_\tau(q)$ it would assume if successive message τ-blocks were indeed statistically independent and identically distributed in accordance with μ^τ, we have

$$(1/t)I_t(q) \leqq (1/\tau)I_\tau(q) = \mathcal{R}_\tau(q) < R(D) + \delta. \tag{36}$$

We have assumed in the theorem statement that $R(D)$ is finite, so (36) implies that I_t is finite for every t. This permits us to conclude from Lemma 1 that $\omega^t \ll w^t$ and, therefore, that the Radon–Nikodym derivative $f_t = d\omega^t/dw^t$ exists and is unique $[w^t]$.

We now introduce the sets

$$\Delta = \{z \in Z^t:\rho_t(z) \leqq D+\delta\}$$

and

$$\Gamma = \{z \in Z^t:f_t(z) \leqq e^{t[R(D)+\delta]}\},$$

and their respective $\mathcal{Y}^t$-measurable sections $\Delta_x = \{y:(x, y) \in \Delta\}$ and $\Gamma_x = \{y:(x, y) \in \Gamma\}$. Returning to (33) we deduce with the aid of (4) that

$$\begin{aligned}\nu^t\{y:\rho_t(x, y) \leqq D+\delta\} &= \nu^t(\Delta_x) = \int_{\Delta_x} d\nu^t \\ &\geqq e^{-t[R(D)+\delta]} \int_{\Delta_x \cap \Gamma_x} f_t(x, y)\, d\nu^t = e^{-t[R(D)+\delta]} q^t(x, (\Delta \cap \Gamma)_x).\end{aligned} \tag{37}$$

where we have used that fact that $\Delta_x \cap \Gamma_x = (\Delta \cap \Gamma)_x$. Inequality (37) may be used to cast (33) in the form

$$\bar{P}(D+\delta \mid x) \leqq [1 - e^{-t[R(D)+\delta]} q^t(x, (\Delta \cap \Gamma)_x)]^{N-1}. \tag{38}$$

At this point we digress momentarily to establish an inequality which we formulate as

LEMMA 2. *If* $0 \leqq \xi, \gamma \leqq 1$ *and* $K \geqq 1$, *then*

$$p(\xi, \gamma) = (1 - \xi\gamma)^K \leqq 1 - \gamma + (1 - \xi)^K.$$

Proof. $p(\xi, \gamma)$ is a convex downward function of $\gamma \in [0, 1]$ for fixed $\xi \in [0, 1]$ because $\partial^2 p/\partial\gamma^2 = (K)(K-1)\xi^2(1-\xi\gamma)^{K-2} \geqq 0$. Accordingly, $p(\xi, \gamma) \leqq (1-\gamma)p(\xi, 0) + \gamma p(\xi, 1) \leqq 1 - \gamma + (1-\xi)^K$.

Application of Lemma 2 and the well-known inequality $\log x \leqq x - 1$ to (38) yields

$$\begin{aligned}&\bar{P}(D+\delta \mid x) \\ &\quad \leqq 1 - q^t(x, (\Delta \cap \Gamma)_x) + \exp\{-(N-1)e^{-t[R(D)+\delta]}\}.\end{aligned} \tag{39}$$

Since N has been chosen as the largest integer in $\exp\{-t[R(D)+\epsilon]\}$, choosing $\delta < \epsilon$ drives the last term in (39) to zero at a double exponential rate as $t \to \infty$. Hence, if we can show that

$$\lim_{t\to\infty} \int [1 - q^t(x, (\Delta \cap \Gamma)_x)]\, d\mu^t = 0,$$

then it will follow from (38) and (47) that $J_t \to 0$. Toward this end we observe from (2) that

$$\int [1 - q^t(x, (\Delta \cap \Gamma)_x)]\, d\mu^t = 1 - \omega^t(\Delta \cap \Gamma) \leqq \omega^t(\bar{\Delta}) + \omega^t(\bar{\Gamma}).$$

Thus, it suffices to show that the joint probabilities of the sets $\bar{\Delta}$ and $\bar{\Gamma}$ vanish in the limit of large t. In this regard, consider the joint τ-block source $[Z, \omega]_\tau$ that results from repeated use of the τ-block channel to transform the successive τ-blocks produced by $[X, \mu]_\tau$. Our assumption that $[X, \mu]$ is stationary and block ergodic assures that $[X, \mu]_\tau$ is, too. Therefore, $[Z, \omega]_\tau$ is both stationary and ergodic because it is the result of a memoryless operation on the time discrete stationary ergodic source, $[X, \mu]_\tau$ (Alder, 1961). Since $t = n\tau$, we have

$$\begin{aligned}\omega^t(\bar{\Delta}) &= \omega\{z \in Z:\rho_{n\tau}(z) > D+\delta\} \\ &= \omega\left\{z \in Z:\frac{1}{n}\sum_{k=0}^{n-1} \rho_\tau(T^{k\tau}z) > D+\delta\right\}.\end{aligned} \tag{40}$$

The ergodic theorem applied to $[Z, \omega]_\tau$ implies that in the limit of large n (large t) the normalized sum in (40) converges in probability to the expected value of $\rho_\tau(z)$, i.e., to the average distortion of a single τ-block. This, of course, is simply the average distortion $\mathfrak{D}_\tau(q)$ associated with the τ-block channel, q^τ. But $\mathfrak{D}_\tau(q) \leqq D$ because $q^\tau \in \mathcal{Q}_\tau(D)$, so $\omega^t(\bar{\Delta}) \to 0$ as desired.

As regards $\bar{\Gamma}$ we may write

$$\omega^t(\bar{\Gamma}) = \omega^{n\tau}\left\{z:\frac{1}{n\tau}\log f_{n\tau}(z) > R(D) + \delta\right\}. \tag{41}$$

The next step clearly is to apply the AEP. Note, however, that the generalized AEP as given by Corollary 1 is not directly applicable to the joint source $[Z, \omega]$ induced by $[X, \mu]$ and repeated use of the τ-block channel because $[Z, \omega]$ is not stationary. On the other hand, Corollary 1 does apply to $[Z, \omega]_\tau$ which, as we have previously noted, is both stationary and ergodic under our assumptions. Hence, with $[Z, \omega]_\tau$ playing the role of the joint source in Corollary 1, we see that $(n\tau)^{-1}\log f_{n\tau}(z)$ converges in probability to the constant

$$R \triangleq \lim_{n\to\infty} \frac{I_{n\tau}(q)}{n\tau} = \lim_{t\to\infty} \frac{I_t(q)}{t} < R(D) + \delta,$$

where we have used inequality (36). It follows from (41) that $\omega^t(\bar{\Gamma}) \to 0$, which completes the proof of the positive statement.

The negative statement is easily substantiated via proof by contradiction. Indeed, suppose there exists a set $\tilde{S} \subset Y^t$ for which $D(\tilde{S}) \leqq D$ but which contains only $\tilde{N}$ elements, where $\log \tilde{N} \leqq t[R(D) - \epsilon]$. Let $\tilde{q}^t$ denote the degenerate transition probability that deterministically maps each $x \in X^t$ into whichever $y \in \tilde{S}$ minimizes $\rho_t(x, y)$. Then $\mathfrak{D}_t(\tilde{q}) = D(\tilde{S}) \leqq D$, so $\tilde{q}^t \in \mathcal{Q}_t(D)$. Since $\tilde{S}$ contains only $\tilde{N}$ elements, we obtain $R_t(D) \leqq \mathcal{R}_t(\tilde{q}) \leqq t^{-1}\log \tilde{N} \leqq R(D) - \epsilon$. Moreover, this inequality chain must hold for all integral multiples of t, too, because repeated use of $\tilde{q}^t$ for successive t-blocks leaves the average distortion unchanged and can only decrease the average information rate for a stationary source. The validity of $R_{nt}(D) \leqq R(D) - \epsilon$ for arbitrarily large n, however, contradicts the definition of $R(D)$.

The practical significance of Theorem 2 resides in the following considerations. If we segment each realization of $[X, \mu]$ into successive t-blocks and then map each t-block, x, so obtained into whichever $y \in S$ minimizes $\rho_t(x, y)$, the distortion that results is $\rho_t(x \mid S)$ of (18). For stationary $[X, \mu]$ the average value of this distortion will be the same for each successive t-block, namely $D(S)$ of (17). This segmenting and mapping process deterministically associates with each realization $x \in X$ a particular element of Y, call it $y_S(x)$. If we let ν_S denote the probability induced on $\mathcal{Y}$ by $[X, \mu]$ and the mapping $y_S(\cdot)$, then the associated t-block source $[Y, \nu_S]_t$ approximates $[X, \mu]$ with an average distortion of $D(S) \leqq D + \epsilon$. Since each of the approximating t-blocks that comprise $y_S(x)$ is an element of the finite set S, $[Y, \nu_S]_t$ is a time discrete stationary source with a finite alaphabet. It is well known that the entropy per symbol of such a source never exceeds the logarithm of the number of elements in the source alphabet (Fano, 1961). Therefore, Theorem 2 guarantees the existence of a source that approximates $[X, \mu]$ with an average distortion of $D + \epsilon$ and also possesses an entropy rate that does not exceed $t^{-1} \log N \leqq R(D) + \epsilon$. Now, the channel coding theorem (Shannon, 1948) states that any discrete source with entropy rate H can, with proper encoding and decoding, be transmitted over any channel of capacity $C > H$ with an arbitrarily small frequency of errors. (See also Dobrushin, 1959, and Wolfowitz, 1964.) The increase in average distortion that results from transmission of $[Y, \nu_S]_t$ over any channel of capacity $C > R(D) + \epsilon$ therefore can be made arbitrarily small provided ρ is bounded. (For unbounded ρ even a single channel error might be disastrous.) Hence, we have

COROLLARY 2. *Let $\epsilon > 0$ and $D \geqq 0$ be given. If $[X, \mu]$ is stationary and block ergodic and ρ is bounded, then it is possible to reproduce the source output at the receiving end of any channel of capacity $C > R(D) + \epsilon$ with an average distortion with respect to F_ρ that does not exceed $D + \epsilon$.*

Either, but not both, of the ϵ's in Corollary 2 may be set to zero. The following converse of Corollary 2 also is valid.

COROLLARY 3. *If $[X, \mu]$ has rate distortion function $R(D)$ with respect to F_ρ, then its output cannot be reproduced with an average distortion of D or less at the receiving end of any memoryless channel of capacity $C < R(D)$.*

Proof. Both the proof given by Shannon (1959, Theorem 5) and his subsequent comments regarding the extension to channels with memory can be extended in the usual way to account for abstract alphabets and continuous time. We omit the details.

V. CONCLUSIONS AND EXTENSIONS

In the preceding sections we have formulated and proved the fundamental theorem of rate distortion theory for block ergodic sources with abstract alphabets and single letter fidelity criteria. This is tantamount to establishing the information stability of the source-fidelity criterion combinations in question, an essential requirement for the principal theorems of the generalized theory of information transmission as given by Dobrushin (1959).

It is possible to extend the above results in several directions. First the extension from single letter distortion measures to distortion measures of span g (Shannon, 1959) is reasonably straightforward. A distortion measure of span $g > 0$ is any nonnegative measurable function $\eta(z)$ that depends on z only through z^g. The associated fidelity criterion $F_\eta = \{\eta_t, g \leqq t < \infty\}$ is specified, say for continuous time, by

$$\eta_t(z) = \frac{1}{t-g}\int_0^{t-g} \eta_g(T^s z)\, ds, \quad t > g.$$

That is, $\eta_t(z)$ is the sliding average of η over all successive joint g-blocks in the joint t-block, z^t. If η is bounded, this extension is trivial. If not, then it suffices to replace (16) with the assumption that there exists a time index $t \geqq g$ and an element $b \in Y^t$ such that

$$\int \eta_{2t}(x, 2b)\, d\mu < \infty, \tag{42}$$

where $2b \in Y^{2t}$ denotes the element obtained by cascading two copies of b.[6]

Another extension concerns the simultaneous imposition of several fidelity criteria, F_{η_k}, $k = 1, 2, \cdots, K$. Let η_k be of span g_k and suppose that its sliding average is required not to exceed D_k. If we replace η_k by $\eta_k' = (D/D_k)\eta_k$, then the average of η_k' must not exceed D for each k.

[6] It does not suffice for $\eta_t(x, b)$ to be integrable. The objective is to construct a realization $y \in Y$ for which $\eta_\tau(x, y)$ is integrable for all τ. If y is to consist of infinite repetitions of some $b \in Y^t$, then it is also necessary not to incur infinite average distortion when the span of length g overlaps the end of one b and the beginning of the next one.

This condition specifies the set $\mathcal{Q}_t(D)$ of permissible transition probabilities q^t, and we continue to define $R_t(D)$ and $R(D)$ as in Section III. Condition (16) is replaced by the existence of an element $b \in Y^t$, $t \geqq g_k$, which simultaneously satisfies (42) for $k = 1, 2, \cdots, K$.

The block ergodicity assumption clearly may be replaced by the requirement that there exists a divergent sequence $\{\tau_i\}$ of positive time indices for which $[X, \mu]$ is τ_i-ergodic. Furthermore, there is good reason to suppose that the results can be extended to arbitrary almost periodic ergodic message sources via still more general versions of Theorem 1 and Corollary 1 suggested by Perez (1964) on the basis of work by Jacobs (1959); this matter is presently under investigation.

Another point worthy of further study is whether or not condition (16) (or (42)) can be relaxed. For example, if the source and approximating alphabets both are the real line, the source produces independent Cauchy variates, and the distortion measure is mean squared error, then (16) is not satisfied for any choice of β. $R(D)$ is still defined, however, although there no longer is a distortion value $D_{\max}$ above which it vanishes identically. At present there is no guarantee in such cases that the function $R(D)$ can be meaningfully interpreted as specifying the rate at which the source produces information relative to the fidelity criterion.

APPENDIX. ERGODICITY, BLOCK ERGODICITY, AND WEAK MIXING

A block ergodic source is ergodic by definition, but an ergodic source need not be block ergodic. For example, if we concentrate μ with equal weights of $\frac{1}{2}$ on the two alternating sequences $x = \cdots 0101010 \cdots$ and Tx, then $[X, \mu]$ is τ-ergodic if and only if τ is odd.

A time continuous source $[X, \mathfrak{X}, \mu]$ is said to be weakly mixing if for any two sets $E_1, E_2 \in \mathcal{Y}$,

$$\lim_{t\to\infty} \frac{1}{t}\int_0^t |\mu(E_1 \cap T^{-s}E_2) - \mu(E_1)\mu(E_2)|\, ds = 0. \tag{A.1}$$

(For time discrete sources the integral is replaced by a sum from 0 to $t - 1$.) Following Pinsker (1960, p. 70), we restrict attention to those time continuous sources $[X, \mu]$ for which

$$\lim_{\tau\to 0} \mu(E \Delta T^{-\tau}E) = 0 \tag{A.2}$$

for every $E \in \mathfrak{X}$, where $A\Delta B = (A \cap \bar{B}) \cup (\bar{A} \cap B)$, and call such sources continuous random processes.

LEMMA. *All weakly mixing continuous random processes are block ergodic.*

Proof. We must show that any weakly mixing source that satisfies (A.2) is τ-ergodic for all $\tau > 0$. Choose $\tau > 0$, let $E \in \mathfrak{X}$ be any τ-invariant set, and let t in (A.1) tend to infinity in increments of τ. Then it follows from (A.1) that

$$0 = \frac{1}{\tau}\lim_{n\to\infty}\frac{1}{n}\sum_{k=0}^{n-1}\int_{k\tau}^{(k+1)\tau} |\mu(E \cap T^{-s}E) - \mu^2(E)|\, ds.$$

Changing integration variable to $r = s - k\tau$ and noting from τ-invariance of E that

$$E \cap T^{-s}E = E \cap T^{-r}(T^{-k\tau}E) = E \cap T^{-r}E,$$

we conclude that

$$\begin{aligned} 0 &= \lim_{n\to\infty}\frac{1}{n}\sum_{k=0}^{n-1}\int_0^\tau |\mu(E \cap T^{-r}E) - \mu^2(E)|\, dr \\ &= \int_0^\tau |\mu(E \cap T^{-r}E) - \mu^2(E)|\, dr. \end{aligned} \tag{A.3}$$

Thus, $\mu(E \cap T^{-r}E,) = \mu^2(E)$ for almost all $r \in [0, \tau]$. Since $E\Delta T^{-r}E$ contains $E \cap \overline{T^{-r}E}$ we have

$$\mu(E\Delta T^{-r}E) \geqq \mu(E) - \mu(E \cap T^{-r}E) = \mu(E) - \mu^2(E)$$

for almost all $r \in [0, \tau]$. It follows from (A.2) that $\mu(E) = \mu^2(E)$ and, hence, that $[X, \mu]$ is τ-ergodic as was to be shown.

If a time discrete source $[X, \mu]$ is weakly mixing, then retaining only every τth term in the sum corresponding to (A.1) shows that $[X, \mu]_\tau$ also is weakly mixing and thus a fortiori ergodic. Hence, weak mixing always implies block ergodicity.

However, block ergodicity does not imply weak mixing. To show this we first exhibit a measure space $[\Omega, \mathfrak{F}, P]$ on which a P-preserving transformation can be defined that fails to be weakly mixing even though all its powers are ergodic. For example, let Ω be the unit circle in the complex plane, $\mathfrak{F}$ be the σ-field generated by the open arcs, and P be radian Lebesgue measure. If $c \in \Omega$ and $V\omega = c\omega$ defines the transformation V, then it is well known that V is ergodic if and only if c is not a root of unity. But if c is not a root of unity, then neither is c^n for any positive integer n, so V^n is ergodic for all n. However, V is not weakly mixing. To see this, let $f(\omega) = \omega$ and observe that $U^n f = c^n f$, where U is the isometry induced on $L_2(P)$ by V. Employing the notation and weak mixing criterion of Halmos (1956, p. 38), we have $(U^n f, f) = c^n$ but $(f, 1) = (1, f) = 0$, so

$$\lim_{n\to\infty} \frac{1}{n} \sum_{j=0}^{n-1} |(U^j f, f) - (f, 1)(1, f)| = 1 \neq 0$$

and therefore V is not weakly mixing.

Next, we establish an isomorphism between the rotation V on Ω and the shift T on the space X of binary sequences as follows. Let $\phi : \Omega \to X$ take the point ω into the binary sequence $x = (\cdots, x_{-1}, x_0, x_1, \cdots)$ specified by the condition $x_n = 0$ if and only if $V^n\omega$ belongs to the upper half of the unit circle, $n = 0, \pm 1, \pm 2, \cdots$. By using the fact that the orbit of each $\omega \in \Omega$ is dense when c is not a root of unity, it is easy to show that ϕ is one-to-one. However, ϕ is not onto, the sequence comprised entirely of zeros being an example of an x that is not the image under ϕ of any $\omega \in \Omega$. Nevertheless, $(\Omega, \mathfrak{F}, P, V)$ is isomorphic to $(X, \mathfrak{X}, \mu, T)$ in the sense of Billingsley (1965, p. 53) if μ is defined by $\mu(E) = P(\phi^{-1}E)$ and $\mathfrak{X}$ is the σ-field generated by the image of $\mathfrak{F}$ under ϕ. Since (weak) mixing and (block) ergodicity are invariants under such an isomorphism, $[X, \mu]$ is a stationary block ergodic source that is not weakly mixing.

We conclude from the above that block ergodicity is more restrictive than ergodicity but less stringent than weak mixing.

ACKNOWLEDGMENT

I am indebted to Dr. I. G. Stiglitz for pointing out the applicability of Lemma 2 which greatly simplifies the random coding argument. I also wish to thank the referee for his highly constructive comments regarding both content and style.

RECEIVED: August 14, 1967; revised May 31, 1968.

REFERENCES

ADLER, R. L. (1961), Ergodic and mixing properties of infinite memory channels. *Proc. Amer. Math. Soc.* **12**, 924–930.

BILLINGSLEY, P. (1965), "Ergodic Theory and Information" p. 53. Wiley, New York.

DOBRUSHIN, R. L. (1959), A general formulation of the fundamental Shannon theorem in information theory. *Uspehi Mat. Akad. Nauk.* **14**, 3–104. [English translation, *Transl. Am. Math. Soc., Series* **2 33**, 323–438].

FANO, R. M. (1961), "Transmission of Information." M.I.T. Press and Wiley, New York.

GALLAGER, R. G. (1968), "Information Theory and Reliable Communication." Wiley, New York. [to be published].

GOBLICK, T. J. (1967), A coding theorem for time-discrete analog data sources. Presented at IEEE International Information Theory Conference, San Remo, Italy.

HALMOS, P. R. (1956), "*Ergodic Theory.*" Chelsea, New York.

JACOBS, K. (1959), The transmission of discrete information through periodic and almost periodic channels. *Math. Annalen.* **137**, 125–135 (In German).

MCMILLAN, B., (1953), The basic theorems of information theory. *Ann. Math. Stat.*, **24**, 196–219.

PEREZ, A. (1964), Extensions of Shannon-McMillan's limit theorem to more general stochastic processes. "Trans. Third Prague Conf. on Inform. Theory, Statist. Decision Functions and Random Processes," 545–574, Publ. House Czech. Akad. Sci., Prague. and Academic Press, New York.

PINSKER, M. S. (1960), "Information and Information Stability of Random Variables and Processes." Izd. Akad. Nauk SSSR, Moscow. [English translation, Holden-Day, San Francisco, 1964].

REIFFEN, B. (1966), A per letter converse to the channel coding theorem. *IEEE Trans. on Information Theory* **IT-12**, 475–480.

SHANNON, C. E. (1948), A mathematical theory of communication. *Bell System Tech. J.* **27**, 379–423, 623–656. [Also in book form with postscript by W. Weaver, Univ. of Illinois Press, Urbana, Illinois, 1949].

SHANNON, C. E. (1959), Coding theorems for a discrete source with a fidelity criterion. *IRE Nat'l. Conv. Rec.*, Part 4 142–163. (Also in "Information and Decision Processes," edited by R. E. Machol. McGraw Hill, New York, 1960, 93–126).

WOLFOWITZ, J. (1964), "Coding Theorems of Information Theory," Second Edition. Springer-Verlag, New York.

An Application of Rate-Distortion Theory to a Converse to the Coding Theorem

JOHN T. PINKSTON, MEMBER, IEEE

Abstract—**A lower bound to the information rate $R(D)$ for a discrete memoryless source with a fidelity criterion is presented for the case in which the distortion matrix contains the same set of entries, perhaps permuted, in each column. A necessary and sufficient condition for $R(D)$ to equal this bound is given. In particular, if the smallest column element is zero and occurs once in each row, then there is a range of D, $0 \leq D \leq D_1$, in which equality holds. These results are then applied to the special case of $d_{ij} = 1 - \delta_{ij}$, for which the average distortion is just the probability of incorrectly reproducing the source output. We show how to construct $R(D)$ for this case, from which one can solve for the minimum achievable probability of error when transmitting over a channel of known capacity.**

Manuscript received June 1, 1966; revised July 15, 1968. This work was performed at the M.I.T. Research Laboratory of Electronics, Department of Electrical Engineering, Massachusetts Institute of Technology, supported by an NSF graduate fellowship. It formed part of the author's Ph.D. dissertation.

The author is with the Department of Defense, Fort George G. Meade, Md. 20755

I. Introduction

IN HIS well-known coding theorem, Shannon [1] proved that a source could be reliably transmitted over a channel if $H(X)$, the source entropy, were less than C, the channel capacity. Conversely, such behavior is not possible if $H(X) > C$; in this case Fano [2] has shown that the probability of decoding error for a block code is bounded away from zero for all values of block length, and Wolfowitz [3] proved that this error

Reprinted from *IEEE Trans. Inform. Theory*, vol. IT-15, pp. 66–71, Jan. 1969.

probability must indeed approach 1 as the block length increases. Furthermore, for memoryless sources, Gallager [4] has demonstrated a lower bound on the per letter probability of error which is positive if $H(X) > C$, and Reiffen [6] has extended this bound to more general sources. These per letter results are not implied by the others, which apply only to block errors.

In this paper, the minimum achievable error rate per letter is calculated exactly in terms of C and $H(X)$ by using the rate-distortion function $R(D)$ for the source to be transmitted and the distortion matrix $d_{ij} = 1 - \delta_{ij}$. We show that the lower bound of Gallager is achievable as long as the error probability is below $(M - 1)p_{\min}$, where M is the number of source letters, and $p_{\min}$ is the smallest source letter probability.

The methods used to find $R(D)$ involve derivation of a lower bound to this function which is applicable for a more general class of distortion matrices, and which is shown in general to be equal to $R(D)$ over some nonzero range $0 < D < D_1$.

II. Background and Notation

The results on which this work is based were first presented by Shannon [5], but are less well known than his work on channel coding, so we shall summarize them. We shall consider a discrete memoryless source X, with M letters, the ith letter occurring with probability p_i. We map this source into some representation alphabet Y (possibly, although not necessarily, identical to the source alphabet), and suppose that a matrix $\{d_{ij}\}$ is specified which tells the amount of distortion when source letter i is reproduced as output letter j. The rate-distortion function corresponding to this matrix and source X is defined by

$$R(D) = \min_{p(j\mid i)} I(X; Y)$$

subject to the constraint that

$$\sum_{ij} p_i p(j \mid i)\, d_{ij} \leq D,$$

where $p(j \mid i)$ is defined as the conditional probability of the representation j, given that the ith source letter occurred. We also define the probabilities $\{q_j\}$ and $\{q(i \mid j)\}$ as the unconditional probability of the output letters, and the conditional probability of the input i, given that output j was observed. That is,

$$q_j = \sum_i p_i p(j \mid i)$$

and

$$p_i = \sum q_j q(i \mid j).$$

Also, all information measure notation is identical with that of Fano.

Thus, to get $R(D)$, $I(X; Y)$, the average mutual information between the source and reproduction alphabets, is minimized over all sets of transition probabilities (or test channels) between these alphabets for which the average distortion is, at most, D.

Shannon proved that it is not possible with any coding scheme to transmit the source under consideration through any channel of capacity less than $R(D)$ with average distortion less than D. Conversely, given any channel of capacity $C > R(D)$, coding schemes exist which result in average distortion arbitrarily close to D when used over this channel. $R(D)$ is thus the minimum channel capacity needed to reproduce the given source at a receiver with average distortion at most D, and for any channel of capacity C, we can calculate the minimum possible distortion that can result from its use. We simply solve

$$R(D) = C \tag{1}$$

for D. Shannon also showed that the $R(D)$ is a continuous, strictly decreasing, convex downward function defined for $D_{\min} < D < D_{\max}$, where $D_{\min}$ is the smallest possible distortion, given by

$$D_{\min} = \sum_i p_i (\min d_{ij}),$$

and $D_{\max}$ is the distortion at zero rate,

$$D_{\max} = \min_j \left(\sum_i p_i\, d_{ij}\right).$$

For $D > D_{\max}$, $R(D)$ can be considered to be 0. Also, for $D_{\min} \leq D \leq D_{\max}$, the strict monotonicity of $R(D)$ implies that the test channel achieving $R(D)$ yields an average distortion exactly equal to D.

We shall derive a lower bound to $R(D)$ for the class of square distortion matrices that have the same set of entries, perhaps permuted, in each column. If we assume that $d_{ii} = 0$ and $d_{ij} > 0$ for $i \neq j$, then we can prove that there exists a $D_1 > 0$ such that $R(D)$ is equal to the lower bound for $0 \leq D \leq D_1$. Using this lower bound, we can find $R(D)$ exactly for the case

$$d_{ij} = \begin{cases} 0 & \text{if } i = j \\ 1 & i \neq j. \end{cases}$$

Since for this matrix, the average distortion is just the probability that a source letter will be reproduced as some other letter, we can solve (1) for the minimum achievable probability of error in transmitting a discrete source over a channel of capacity C.

III. Lower Bound to $R(D)$

Let $\{p_i\}$ be the letter probabilities of a discrete memoryless M-letter source, and suppose that the distortion matrix has the same set of entries $\{d_i\}$ in each column, although perhaps permuted. We also assume that $d_{ii} = 0$ and $d_{ij} > 0$ if $i \neq j$, so that $D_{\min} = 0$. Lemma 1 in Section VII will show that there is very little loss of generality involved in this assumption. Define

$$\Phi(D) = \max_{z} H(Z)$$

over all M-dimensional probability vectors z for which

$$\sum z_i\, d_i = D,$$

where $\{d_i\}$ is the set of column entries. Now suppose that we are operating with the test channel between source and reproduction that minimizes $I(X; Y)$ for distortion level D, which we shall assume to be less than $D_{\max}$. Then

$$R(D) = I(X; Y) \\ = H(X) - \sum_j q_j H(X \mid y_j),$$

where q_j = Pr [output letter j]. But by definition,

$$H(X \mid y_j) \leq \Phi(D_j),$$

where

$$D_j = E\{d_{ij} \mid y = j\} \\ = \sum_i q(i \mid j)\, d_{ij}.$$

Thus

$$R(D) \geq H(X) - \sum_j q_j \Phi(D_j).$$

Finally, we use the fact that Φ is concave $\cap$, which is proved by Shannon and follows from the fact that $H(X)$ is concave $\cap$ and D is linear as a function of the probability vector $\underline{z}$. Thus

$$R(D) \geq H(X) - \Phi(\sum q_j D_j)$$

or

$$R(D) \geq H(X) - \Phi(D)$$

since $\sum q_j D_j = D \leq D_{\max}$.

Fig. 1 shows a typical $\Phi(D)$ curve.

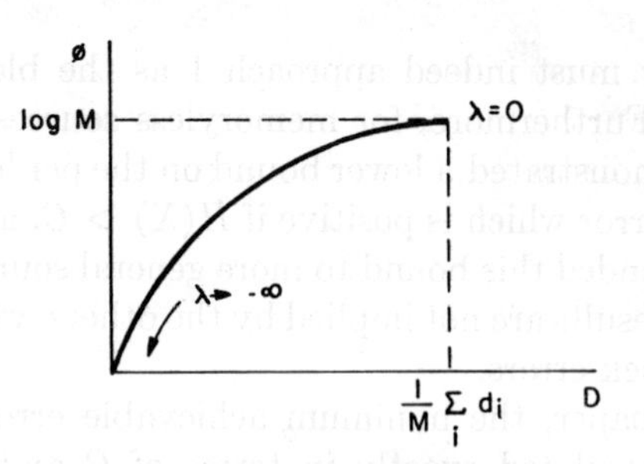

Fig. 1. A typical $\Phi(D)$ curve.

IV. Calculation of $\Phi(D)$

To find $\Phi(D)$, we want to maximize $\sum_i (-z_i \log z_i + \lambda z_i d_i + \mu z_i)$ where the undetermined multipliers λ and μ have been inserted to take care of the constraints $\sum_i z_i d_i = D$, and $\sum_i z_i = 1$, respectively. Setting the derivative with respect to z_i equal to zero, and solving for z_i, we obtain

$$z_i = \frac{e^{\lambda d_i}}{A(\lambda)}$$

where μ has been absorbed into the normalizing constant

$$A(\lambda) = \sum_i e^{\lambda d_i}$$

and λ is still to be determined from the condition that $\sum_i z_i d_i = D$ or

$$D = \frac{A'(\lambda)}{A(\lambda)}. \tag{2}$$

Our result for $\Phi(D)$ then comes out to be

$$\Phi = -\sum z_i \log z_i \\ = \sum z_i(\log A(\lambda) - \lambda\, d_i) \\ = \log A(\lambda) - \lambda \frac{A'(\lambda)}{A(\lambda)}. \tag{3}$$

Thus we have Φ and D parametrically in λ. Note that the convexity of Φ can also be shown by straightforward calculation of its second derivative. It is easy to see from (2) and (3) that as $\lambda \to -\infty$, $D \to 0$, and $\Phi \to 0$, and also that for $\lambda < 0$, Φ is monotonically increasing. Since $D \to 1/M \sum d_i = D_{\max}$ as $\lambda \to 0$, and since D is monotonic in λ, we are never interested in positive λ.

V. Conditions Under Which $R(D)$ is Equal to the Lower Bound

Suppose there exists a channel for which the conditional probability of input i, given output j, is

$$p_{x|y}(i \mid j) = \frac{e^{\lambda d_{ij}}}{A(\lambda)}$$

when $\{p_i\}$ are the input probabilities. Let $\{q_i\}$ be the output probabilities. As in the previous two sections, we assume that $\{d_{ij}\}$ has the same set of entries in each column. The average mutual information for the channel is

$$I(X; Y) = H(X) - \sum_j q_j H(X \mid y_j)$$

and

$$H(X \mid y_j) = -\sum p_{x|y}(i \mid j) \log p_{x|y}(i \mid j) \\ = \log A(\lambda) - \lambda \frac{A'(\lambda)}{A(\lambda)}$$

by virtue of the assumption on $\{d_{ij}\}$. Thus $H(X \mid y_j)$ is independent of j, and so

$$I(X; Y) = H(X) - \log A(\lambda) + \lambda \frac{A'(\lambda)}{A(\lambda)}.$$

The average distortion is then

$$D = \sum_{ij} q_j p_{x|y}(i \mid j)\, d_{ij} \\ = \frac{A'(\lambda)}{A(\lambda)}.$$

Thus $I(X; Y) = H(X) - \Phi(D)$, the lower bound of the previous section, so $R(D)$ is equal to this lower bound as long as such a channel exists. This will occur if a set of probabilities $\{q_j\}$ exists such that

$$p_i = \sum_j q_j p_{x|y}(i \mid j).$$

Therefore a sufficient condition for equality of $R(D)$ with the lower bound is that

$$\Lambda q = p$$

must have a solution $q = \Lambda^{-1}p$ with all non-negative entries, where Λ is an $M \times M$ matrix with

$$\Lambda_{ij} = \frac{e^{\lambda d_{ij}}}{A(\lambda)}.$$

Clearly, if the distortion matrix has $d_{ii} = 0$ and $d_{ij} > 0$, if $i \neq j$, then $\Lambda \to Id$ as $\lambda \to -\infty$, and clearly $q = p$ is a non-negative solution for this case. This corresponds to zero distortion, and thus we have $R(D)$ equal to the lower bound at $D = 0$. Furthermore, since the elements of Λ are continuous functions of λ, the elements of the inverse will be likewise, and so q will vary continuously with λ. Since D is bounded away from zero if any $q_j = 0$, there must be a range of λ, $-\infty \leq \lambda \leq \lambda_1$ for which (5) has a valid solution. Thus there is a $D_1 = A'(\lambda_1)/A(\lambda_1) > 0$ such that $R(D) = H(X) - \Phi(D)$ for all $D \leq D_1$.

It is interesting to note that if $\{d_{ij}\}$ satisfies the condition that all rows as well as all columns have the same entries, and if the source letters are equiprobable, then $q = p = (1/M \cdots 1/M)^T$ is always a solution, and $R(D)$ is therefore given by

$$R(D) = H(X) - \Phi(D)$$

for all $D \leq D_{\max}$.

VI. Application to the Converse of the Coding Theorem

Suppose that our distortion matrix is $d_{ij} = i - \delta_{ij}$, where δ_{ij} is the Kronecker delta. The average distortion is then just the probability that a source letter is not reproduced correctly. Knowing $R(D)$ for this matrix and some source, we can find the minimum achievable probability of error when transmitting this source over a channel of capacity $C < H(X)$ simply by solving $R(D) = C$ for D, the desired minimum probability. It will be convenient to consider the source probabilities ordered, so that $p_1 \leq p_2 \leq \cdots \leq p_m$ which involves no loss of generality.

We shall now find the lower bound to $R(D)$ for this case and see where equality holds. First

$$A(\lambda) = \sum e^{\lambda d_i} = 1 + (M - 1)e^{\lambda};$$

therefore,

$$D = \frac{A'(\lambda)}{A(\lambda)} = \frac{(M-1)e^{\lambda}}{1 + (M-1)e^{\lambda}}.$$

Solving for λ, we get

$$\lambda = \log D - \log (M - 1) - \log (1 - D)$$

so that

$$A(\lambda) = \frac{1}{1 - D}.$$

Now

$$\Phi = \log A(\lambda) - \lambda D$$
$$= H_2(D) + D \log (M - 1)$$

where $H_2(D) = -D \log D - (1 - D) \log (1 - D)$; this result is identical with that obtained by Gallager. So the bound is

$$R(D) \geq H(X) - H_2(D) - D \log (M - 1).$$

We now want to find the region of equality. Writing

$$\Lambda = \frac{1}{A(\lambda)} \begin{bmatrix} 1 & \alpha & & \cdots & \alpha \\ \alpha & 1 & \alpha & \cdots & \vdots \\ \vdots & & & & \alpha \\ \alpha & & \cdots & \alpha & 1 \end{bmatrix}$$

where $\alpha = e^{\lambda}$ and $\alpha < 1$ since $\lambda < 0$, we see that the equations to be solved are

$$\frac{1}{A(\lambda)} [q_i + \alpha(1 - q_i)] = p_i; \qquad \text{for all } j.$$

The solution is

$$q_i = \frac{[1 + (M-1)\alpha]p_i - \alpha}{1 - \alpha}. \tag{6}$$

Now α is monotonic as a function of λ, and therefore also of D, and $\alpha \to 0$ as $\lambda \to -\infty$ and $D \to 0$. It is clear that for small enough α, all the q_i are positive, and as α increases, the first one to go negative will be the one corresponding to p_1, the smallest p. We therefore solve

$$q_1(\alpha) = 0$$

for the α below which all the q's are positive. This becomes

$$p_1 + \alpha(M - 1)p_1 - \alpha = 0$$

or

$$\alpha = \frac{p_1}{1 - (M-1)p_1}$$

from which

$$D_1 = \frac{(M-1)\alpha}{1 + (M-1)\alpha} = (M - 1)p_1$$

is the value of D below which $R(D) = H(X) - H_2(D) - D \log (M - 1)$. Clearly, $D_1 \leq D_{\max} = 1 - p_{\max} = \sum_{i=1}^{M-1} p_i$, with equality if and only if the $M - 1$ smallest p_i are equal. As we have seen, equality holds if all p_i are equal, and now we can relax this condition to only the $M - 1$ smallest. For example, for $p = (\frac{1}{4}, \frac{1}{4}, \frac{1}{2})^T$,

$$R(D) = H(X) - H_2(D) - D$$
$$= 1.5 - H_2(D) - D \text{ bits}; \qquad 0 \leq D \leq \tfrac{1}{2}.$$

In general, we now have $R(D)$ for $D \leq D_1 = (M - 1) p_{\min}$. Before we proceed to the task of finding $R(D)$ for $D > D_1$, let us remark that for a binary source, our development indicates that $D_1 = D_{\max}$. Therefore

$$R(D) = H(X) - H_2(D)$$

for all binary sources and the distortion $d_{ij} = 1 - \delta_{ij}$.

VII. Construction of $R(D)$ for $d_{ij} = 1 - \delta_{ij}$, for all D

For this task we shall need two lemmas, and it is convenient to present them at this time.

Lemma 1

Suppose $R(D)$ for some source p and distortion matrix $\{d_{ij}\}$ is known, and suppose we form the new distortions $\hat{d}_{ij} = d_{ij} + w_i$ (i.e., each row has a constant added to it). Then

$$\hat{R}(D) = R(D - \bar{W})$$

where $\bar{W} = \sum w_i p_i$ and $\hat{R}$ corresponds to the source p and distortion $\hat{d}_{ij}$.

Proof:

$$\hat{R}(D) = \min I(X; Y)$$

subject to

$$\sum_{ij} p_i p(j \mid i)\, \hat{d}_{ij} \le D$$

or

$$\sum_{ij} p_i p(j \mid i)(d_{ij} + w_i) \le D$$

or

$$\sum_{ij} p_i p(j \mid i)\, d_{ij} \le D - \bar{W}$$

which by definition is $R(D - \bar{W})$. Q.E.D.

Lemma 2

Suppose a distortion matrix has a row that is all zeros, and p_1 is the probability of the source letter corresponding to that row. Then

$$R(D) = (1 - p_1)\hat{R}\left(\frac{D}{1 - p_1}\right)$$

where $\hat{R}(\cdot)$ corresponds to the distortion matrix with the row of zeros deleted, and the source with letter 1 deleted and probability vector

$$\underline{p}' = \left(\frac{p_2}{1 - p_1}, \cdots, \frac{p_M}{1 - p_1}\right).$$

Proof: The average distortion is not affected by the way in which source letter 1 is reproduced. Thus to minimize $I(X; Y)$, we must choose $p_{y|x}(j \mid 1)$ so that $I(x_1; Y) = 0$. With this choice,

$$\begin{aligned} R(D) &= \min I(X; Y) \\ &= \min \left[p_1 I(x_1; Y) + \sum_{i=2}^{M} p_i I(x_i; Y) \right] \\ &= \min \left[(1 - p_1) \sum_{i=2}^{M} \frac{p_i}{1 - p_1} I(x_i; Y) \right]. \end{aligned}$$

The constraint is

$$\sum_i p_i E_y\{d_{ij} \mid x = i\} \le D$$

but $E_y\{d_{ij} \mid x = 1\} = 0$, so the constraint becomes

$$\sum_{i=2}^{M} \frac{p_i}{1 - p_1} E_y\{d_{ij} \mid x = i\} \le \frac{D}{1 - p_1}.$$

Now, by definition of $R(D)$, the assertion follows.

This lemma roughly states that if a row is all zeros, we need not expend any information on the transmission of the corresponding source letter.

We are finally prepared to find $R(D)$ for the distortion $d_{ij} = 1 - \delta_{ij}$ for all values of D. We simply note that the solutions for q_j of (6) are monotonic in α, and therefore in D. Furthermore, it can be shown (Gallager [7]) that once a q_j goes to 0, it never becomes positive for any larger D. Since p_1 is the smallest p_i, q_1 is the first output probability to become 0, which occurs at $D = D_1 = (M - 1)p_1$. Then we know that for $D > D_1$, output 1 will never be used, and we can therefore remove it from the output alphabet and delete the corresponding column from $\{d_{ij}\}$ without affecting $R(D)$. Thus for $D > D_1$, we might as well have the distortion matrix

$$d_{ij} = \begin{bmatrix} 1 & 1 & & \cdots & 1 \\ 0 & 1 & & \cdots & 1 \\ 1 & 0 & 1 & \cdots & 1 \\ \vdots & & & & \vdots \\ 1 & \cdots & & 1 & 0 \end{bmatrix}$$

with M rows and $M - 1$ columns, and the first row is all 1's. Now, by Lemma 1, we can subtract 1 from the top row, and write

$$R(D) = R^{(1)}(D - p_1),$$

where $R^{(1)}$ corresponds to the matrix

$$d_{ij}^{(1)} = \begin{bmatrix} 0 & 0 & & \cdots & 0 \\ 0 & 1 & & \cdots & 1 \\ 1 & 0 & 1 & \cdots & 1 \\ \vdots & & & & \vdots \\ 1 & \cdots & & 1 & 0 \end{bmatrix}.$$

Now we have a matrix with an all zero row, so by Lemma 2

$$R^{(1)}(D) = (1 - p_1)R^{(2)}\left(\frac{D}{1 - p_1}\right).$$

Combining these results, we have

$$R(D) = (1 - p_1)R^{(2)}\left(\frac{D - p_1}{1 - p_1}\right)$$

for $D > D_1$, and where $R^{(2)}$ corresponds to the $M - 1$ by $M - 1$ matrix with $d_{ij} = 1 - \delta_{ij}$, and the source

$$\underline{p}^{(2)} = \left(\frac{p_2}{1 - p_1}, \cdots, \frac{p_M}{1 - p_1}\right).$$

$R^{(2)}(D)$ can now be lower bounded exactly as before, and this lower bound is valid for values of the argument up to

$(M-2)p_2/(1-p_1)$, where p_2 is the second smallest source probability. Thus the second breakpoint comes at D_2, where

$$\frac{D_2 - p_1}{1 - p_1} = \frac{(M-2)p_2}{1-p_1}$$

or

$$D_2 - p_1 + (M-2)p_2 \geq D_1.$$

So we have

$$R(D) = H(X) - H(D) - D \log (M-1) \qquad \text{for} \quad 0 \leq D \leq (M-1)p_1$$

and

$$R(D) = (1-p_1)\left[H_{M-1}(X) - H_2\left(\frac{D-p_1}{1-p_1}\right) - \left(\frac{D-p_1}{1-p_1}\right)\log(M-2)\right] \qquad \text{for} \quad (M-1)p_1 < D \leq p_1 + (M-2)p_2,$$

where

$$H_{M-1}(X) = -\sum_{i=2}^{M} \frac{p_i}{1-p_1} \log \frac{p_i}{1-p_1}.$$

It is clear that this process can be continued at the expense of some algebraic complexity. The result is

$$R(D) = (1-S_k)\left[H_{M-k}(X) - H_2\left(\frac{D-S_k}{1-S_k}\right) - \left(\frac{D-S_k}{1-S_k}\right)\log(M-k-1)\right] \quad \text{for} \quad D_{k-1} \leq D \leq D_k,$$

where $S_k = \sum_{i=1}^{k} p_i$, $D_k = S_{k-1} + (M-k)p_k$, and $H_{M-k}(X)$ is the entropy of the source

$$\left(\frac{p_{k+1}}{1-S_k}, \cdots, \frac{p_M}{1-S_k}\right).$$

By inverting this function to get $D(R)$, one can find the minimum achievable error probability at any signaling rate. Fig. 2 shows $R(D)$ for the source $p = (\frac{1}{8}, \frac{3}{8}, \frac{1}{2})$, for which $D_1 = \frac{1}{4}$ and $D_{max} = \frac{1}{2}$.

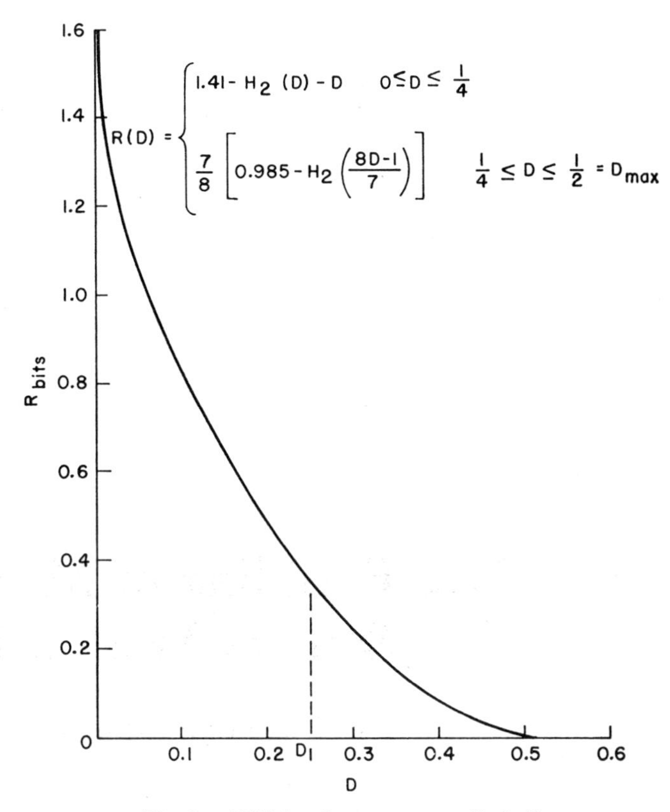

Fig. 2. $R(D)$ for the source $\underline{p} = (\frac{1}{8}, \frac{3}{8}, \frac{1}{2})$.

VIII. Conclusion

We have demonstrated a lower bound to $R(D)$ for discrete sources and distortion matrices satisfying the column condition. A necessary and sufficient condition for $R(D)$ to equal this lower bound is that the equation $\Lambda \underline{q} = \underline{p}$ have a solution $\underline{q}$ with all non-negative entries, where $\Lambda_{ij} = e^{\lambda d_{ij}}/A(\lambda)$. In particular, if the smallest column element is zero, and it occurs once in each row, then there is a range of D, $0 \leq D \leq D_1$ in which $R(D)$ is equal to the lower bound.

This bound was applied to the special case of $d_{ij} = 1 - \delta_{ij}$, for which the distortion is just probability of error in reproducing the source. We have shown how to construct $R(D)$ for this case, from which one can solve for the minimum achievable probability of error when transmitting over a channel of known capacity.

References

[1] C. E. Shannon, "A mathematical theory of communication," *Bell Sys. Tech. J.*, vol. 27, pp. 379–423, 623–656, 1946.
[2] R. M. Fano, *Transmission of Information.* Cambridge, Mass.: M.I.T. Press, 1961, pp. 186–187.
[3] J. Wolfowitz, *Coding Theorems of Information Theory*, 2nd ed. Berlin: Springer, 1964.
[4] R. G. Gallager, "Information Theory," in *Mathematics of Physics and Chemistry*, H. Margenau and G. M. Murphy, Eds. Princeton, N. J.: Van Nostrand, 1964, vol. 2, pp. 190–248.
[5] C. E. Shannon, "Coding theorems for a discrete source with a fidelity criterion," *1959 IRE Nat'l Conv. Rec.*, pt. 4, pp. 142–163.
[6] B. Reiffen, "A per letter converse to the coding theorem," *IEEE Trans. Information Theory*, vol. IT-12, pp. 475–480, October 1966.
[7] R. G. Gallager, *Information Theory and Reliable Communication.* New York: Wiley, 1968, ch. 9.

Tree Encoding of Memoryless Time-Discrete Sources With a Fidelity Criterion

FREDERICK JELINEK, SENIOR MEMBER, IEEE

Abstract—**In this paper we show that for memoryless time-discrete sources with a bounded fidelity criterion, the limiting average distortion achievable by tree codes of rate R is D, the solution of the equation $R = R(D)$, where $R(\ )$ denotes the usual rate distortion function. Thus the performance of tree codes is as good as that of block codes. Some theoretical and experimental results are also discussed indicating that tree codes and corresponding encoding algorithms exist having for given values of R and D an implementation complexity that is far smaller than the one obtainable from block codes.**

I. Introduction

THE PROBLEM of implementation complexity of codes appropriate for time-discrete memoryless sources with a fidelity criterion [1] has remained almost unexamined, and present theoretical analysis deals only with block codes that have no structure that would facilitate noncodebook encoding. The success of the method of sequential decoding of transmitted information [3, ch. 10] has demonstrated the advantages of a tree structure of codes. Goblick [2] suggested that tree codes might also be desirable for source coding, and indeed his simulation for memoryless binary symmetrical sources indicated that a payoff lies in that direction. This paper contains the first theoretical analysis of the properties of tree codes. We prove that for sources with a bounded distortion function the limiting minimal average distortion achievable by tree codes of rate R is D, the solution of the equation $R = R(D)$, where $R(\ )$ denotes the usual rate distortion function. Thus in this respect tree codes are as good as block codes.

The above result does not, of course, settle the question of whether tree codes can really be simply implemented. Some theorems discussed in the last section of this paper indicate that encoding strategies that take advantage of the tree structure exist which for the same distortion performance lend themselves to much simpler instrumentation than do block codes. The proofs due to the author of these theorems can be found in a recent report [7]. They have been left out of this paper because the most recent experimental investigation reveals that actual performance of even very simple algorithms is far superior to the performance bounds derived (see also Section III).

We shall now introduce some notation. In the pursuit of brevity, we will not provide a rationale for our interest in coding with a fidelity criterion. Instead, we direct the reader to the original work of Shannon [1] and/or Jelinek [3, ch. 11]. Those interested in more practical applications of these ideas might profitably consult [4] and [5].

The outputs of a time-discrete memoryless source are the values $z_1, z_2, z_3, \cdots$ assumed by the terms of a sequence $Z_1, Z_2, Z_3, \cdots$ of independent, identically distrib-

Manuscript received June 24, 1968; revised February 7, 1969. This work was supported in part by the National Aeronautics and Space Administration under Contract NSR 33-010-026.

The author is with the School of Electrical Engineering and Center for Radiophysics and Space Research, Cornell University, Ithaca, N. Y. 14850.

Reprinted from *IEEE Trans. Inform. Theory*, vol. IT-15, pp. 584–590, Sept. 1969.

uted random variables (throughout this paper, unless stated otherwise, upper case letters will denote sample values, and lower case letters the corresponding random variables). We will assume that the range of Z_i is a subset $\mathcal{A}$ (called the *source alphabet*) of the real line, and that Z_i is a random variable of the *mixed type*, i.e., one having a generalized density function

$$q(z) = q^*(z) + \sum_{i=1}^{\infty} q_i \,\delta(z - \omega_i) \qquad z \in \mathcal{A} \quad (1)$$

where $q^*(\ )$ is an integrable, nonnegative function, the real numbers ω_i are distinct and have no limit points, $q_i \geq 0$ for all i, and $\delta(\ )$ is Dirac's delta function. Thus, as a special case, Z_i can be discrete or continuous (the restriction to mixed type random variables is due to our desire to avoid entanglement in measure-theoretic arguments, and is not essential). The set $\mathcal{A}$ will be considered to be such that

$$q(z) > 0 \qquad \text{for all} \quad z \in \mathcal{A}. \quad (2)$$

Let $\mathcal{B}$ be another subset of the real line, called the *reproducer alphabet.* The *distortion* between a source sequence

$$\boldsymbol{z}^l \triangleq (z_1, \cdots, z_l), \qquad z_i \in \mathcal{A} \quad (3a)$$

and a reproducer sequence

$$\boldsymbol{y}^l \triangleq (y_1, \cdots, y_l), \qquad y_i \in \mathcal{B} \quad (3b)$$

is defined by

$$d(\boldsymbol{z}^l, \boldsymbol{y}^l) \triangleq \sum_{i=1}^{l} d(z_i, y_i) \quad (4)$$

where $d(,)$ can without loss of generality [6, Theorem 1, p. 8] be assumed to satisfy

$$0 \leq d(z, y) < \infty \qquad \text{for all} \quad z \in \mathcal{A}, \quad y \in \mathcal{B} \quad (5a)$$

$$d(z, y(z)) = 0 \qquad \text{for some} \quad y(z) \in \mathcal{B}. \quad (5b)$$

In this paper we will restrict our attention to distortion measures satisfying the additional condition

$$d(z, y) < d_M < \infty \qquad \text{for all} \quad z \in \mathcal{A}, \quad y \in \mathcal{B}. \quad (6)$$

Restriction (6), although theoretically a significant one, is from a practical point of view entirely unimportant. In every real situation there will obviously be a maximal penalty on a mismatch between a source output and its representation. The much abused distance criterion $(z - y)^2$ is popular because of its mathematical convenience and *in spite* of the excess weight it assigns to large deviations.

Shannon [1] (discrete alphabets $\mathcal{A}$ and $\mathcal{B}$) and Goblick [2] have shown that it is possible to find a set $\mathcal{V}_l$ of $M = e^{lR}$ codewords $\{\boldsymbol{v}_1, \boldsymbol{v}_2, \cdots, \boldsymbol{v}_M\}$ of length l and a mapping Ψ from source sequences $\boldsymbol{z}^l$ into $\mathcal{V}_l$ such that the distortion averaged over the source output sequences satisfies

$$\boldsymbol{E}[d(\boldsymbol{Z}^l, \Psi(\boldsymbol{Z}^l))] \leq l[D + f(l, R, D)]. \quad (7)$$

In (7), $f(l, R, D) \to 0$ as $l \to \infty$ (R and D fixed), provided $R \geq R(D)$, where

$$R(D) \triangleq \inf_{w \in \mathcal{L}(D)} \int_{\mathcal{A}} \int_{\mathcal{B}} \left[q(z)w(y/z) \ln \frac{w(y/z)}{w(y)} \right] dy\, dz. \quad (8)$$

$$\mathcal{L}(D) \triangleq \Big\{ \boldsymbol{w} : w(\ /z) \text{ a generalized density over } \mathcal{B} \text{ for all } z \in \mathcal{A}, \text{ and } \int_{\mathcal{A}} \int_{\mathcal{B}} [q(z)w(y/z)\, d(z, y)]\, dy\, dz \leq D \Big\} \quad (9)$$

and

$$w(y) \triangleq \int_{\mathcal{A}} w(y/z) q(z)\, dz. \quad (10)$$

Shannon and Goblick also proved a converse theorem to the effect that if $R < R(D)$ then for all block codes $\mathcal{V}_l$ of rate R and word length l, and all mapping functions Ψ from $\boldsymbol{z}^l$ into $\mathcal{V}_l$,

$$\boldsymbol{E}[d(\boldsymbol{Z}^l, \Psi(\boldsymbol{Z}^l))] > lD. \quad (11)$$

It follows from (5) and (6) that only average distortions $D \in (0, D_M)$ are of interest, where

$$D_M \triangleq \inf_{y \in \mathcal{B}} \boldsymbol{E}[d(Z, y)] \leq d_M, \quad (12)$$

and it follows from (8) that $R(D_M) = 0$. It can be shown that $R(D)$ is a decreasing convex function of $D \in (0, D_M)$. As the notation of (12) indicates, the expectation operation involves averaging with respect to the source-defined random variable Z while the sample value y of the reproducer alphabet is held fixed.

II. Average Distortion of Optimal Tree Codes

The trouble with the result (7) is that it was shown to hold for codes $\mathcal{V}_l$ and mapping functions Ψ whose structure is in no way restricted, so that in general it would take an exponentially large search (of the order of e^{lR}) to determine the codeword $\Psi(\boldsymbol{z}^l)$ pertaining to a particular source output sequence $\boldsymbol{z}^l$. It was Goblick [2] who first suggested that the search problem could be alleviated if the codeword sets $\{\boldsymbol{v}_1, \boldsymbol{v}_2, \cdots, \boldsymbol{v}_M\}$ had a tree structure. He ran simulations which indicated that at least for binary symmetrical sources such codes would perform in accordance with the result (7). However, this section contains the first proof of the corresponding theorem.

Consider the tree of Fig. 1. It can be regarded as a representation of a tree code with $M = 16$ words of length $l = 12$ selected from the alphabet $\mathcal{B} = \{0, 1\}$. Thus the digits lying on the path marked by the thick line constitute the codeword $\boldsymbol{v}_4 = (0\,0\,0\,1\,0\,1\,0\,1\,1\,0\,1\,0)$. Similarly, the twelfth codeword is $\boldsymbol{v}_{12} = (1\,1\,1\,0\,1\,0\,0\,0\,0\,0\,1\,0)$.

In general, a tree code of word length $l = kn_0$ would be based on a k-level tree with g branches leaving each node, each branch being associated with a sequence of n_0 digits belonging to the alphabet $\mathcal{B}$ (in Fig. 1, $k = 4$, $n_0 = 3$, $g = 2$). The rate of such a code (in natural units)

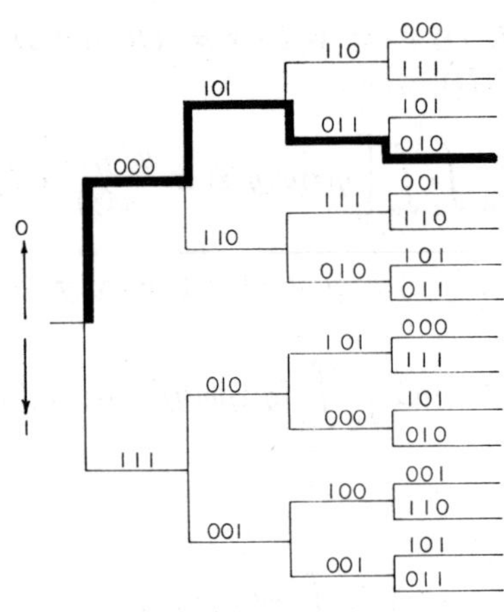

Fig. 1.

would be

$$R \triangleq \frac{\ln g}{n_0}. \qquad (13)$$

The optimal encoding function Ψ will satisfy the following condition for all source output sequences $\boldsymbol{z}^l$:

$$d(\boldsymbol{z}^l, \Psi(\boldsymbol{z}^l)) = \min_{i \in \{1, \cdots, M\}} d(\boldsymbol{z}^l, \boldsymbol{v}_i). \qquad (14)$$

With the help of (14) we can then define the minimal distance $d(\boldsymbol{z}^l; \mathcal{V}_l)$ of the source output sequence $\boldsymbol{z}^l$ from the code $\mathcal{V}_l \triangleq \{\boldsymbol{v}_1, \boldsymbol{v}_2, \cdots, \boldsymbol{v}_M\}$:

$$d(\boldsymbol{z}^l; \mathcal{V}_l) \triangleq d(\boldsymbol{z}^l, \Psi(\boldsymbol{z}^l)). \qquad (15)$$

As was pointed out at the end of the preceeding section, we will be interested in the average distortion incurred by optimal tree codes of rates R satisfying

$$0 < R < R(0). \qquad (16)$$

Theorem 1

Let a source with a distortion function satisfying assumptions (4) through (6) of Section I be given. If $R \in (0, R(0))$ satisfies

$$R > R(D_0) \qquad (17)$$

(see (13) and (8)) then there exists a sequence of tree codes $\mathcal{V}_{kn_0}$, $k = 1, 2, \cdots$, of rate at most R, such that

$$\lim_{k \to \infty} \boldsymbol{E}\left[\frac{1}{kn_0} d(\boldsymbol{Z}^{kn_0}; \mathcal{V}_{kn_0})\right] \leq D_0, \qquad (18)$$

where the expectation is with respect to the source output process.

The above theorem will be proven in a series of lemmas. Let

$$\gamma \triangleq R - R(D_0) > 0. \qquad (19)$$

Then it follows from (8) through (10) that there exists a (in general nonunique) conditional density function $w^0(y/z)$ such that

$$R(D_0) + (1/3)\gamma \geq \int_{\mathcal{A}} \int_{\mathcal{B}} \left[q(z) w^0(y/z) \log \frac{w^0(y/z)}{w^0(y)}\right] dy\, dz \qquad (20)$$

and

$$D \triangleq \int_{\mathcal{A}} \int_{\mathcal{B}} [q(z) w^0(y/z)\, d(z, y)]\, dy\, dz \leq D_0 \qquad (21)$$

where

$$w^0(y) \triangleq \int_{\mathcal{A}} w^0(y/z) q(z)\, dz, \qquad y \in \mathcal{B}. \qquad (22)$$

Our proof will be based on a random coding argument. The pertinent code ensemble will be generated as follows:

> All the digits $y \in \mathcal{B}$ associated with the various branches of the tree code (a sequence of n_0 digits to each branch) are selected independently of each other on the basis of the density $w^0(y)$ that satisfies (20) through (22).

Lemma 1

For all positive integers l_1 and l_2 such that $l_1 = k_1 n_0$, $l_2 = k_2 n_0$, and all sequences $\boldsymbol{z}^{l_1+l_2} = (\boldsymbol{z}_1^{l_1}, \boldsymbol{z}_2^{l_2})$,

$$P\{\mathcal{V}_{l_1+l_2} : d(\boldsymbol{z}^{l_1+l_2}; \mathcal{V}_{l_1+l_2}) \geq x\}$$
$$\leq P\{\mathcal{V}_{l_1}, \mathcal{V}_{l_2} : d(\boldsymbol{z}_1^{l_1}; \mathcal{V}_{l_1}) + d(\boldsymbol{z}_2^{l_2}; \mathcal{V}_{l_2}) \geq x\} \quad \text{for all } x, \qquad (23)$$

and

$$\boldsymbol{E}[d(\boldsymbol{z}^{l_1+l_2}; \mathcal{V}_{l_1+l_2})] \leq \boldsymbol{E}[d(\boldsymbol{z}_1^{l_1}; \mathcal{V}_{l_1})] \mid \boldsymbol{E}[d(\boldsymbol{z}_2^{l_2}; \mathcal{V}_{l_2})]. \qquad (24)$$

Remarks: The random event referred to by the left-hand side probability in (23) consists of a selection of a tree $\mathcal{V}_{l_1+l_2}$ from the ensemble of trees with $k_1 + k_2$ levels. The probability on the right-hand side of (23) refers to a simultaneous selection of two trees: $\mathcal{V}_{l_1}$ from the ensemble of trees with k_1 levels, and $\mathcal{V}_{l_2}$ from an independent ensemble of trees with k_2 levels ($l_i = k_i n_0$). The corresponding situation obtains with respect to the averaging in (24). In both (23) and (24) the sequence $\boldsymbol{z}^{l_1+l_2}$ is held fixed.

Proof: Let $\boldsymbol{v}_i^{l_1}$, $i = 1, 2, \cdots, g^{k_1}$ denote the reproducer sequence associated with the ith path of k_1 branches belonging to some tree code $\mathcal{V}_{l_1+l_2}$ [so in Fig. 1, $\boldsymbol{v}_2^6 = (0\ 0\ 0\ 1\ 1\ 0)$, $\boldsymbol{v}_3^6 = (1\ 1\ 1\ 0\ 1\ 0)$], and let $\mathcal{V}_{l_2}^i$ denote the subtree of $\mathcal{V}_{l_1+l_2}$ stemming from the terminal node of $\boldsymbol{v}_i^{l_1}$ (thus Fig. 2 is a representation of $\mathcal{V}_6^3$ stemming from the terminal node of $\boldsymbol{v}_3^6$ in Fig. 1). Then it follows from the additive nature of the d function and from definition (15) that for all fixed source sequences $\boldsymbol{z}^{l_1+l_2}$,

$$P\{d(\boldsymbol{z}^{l_1+l_2}; \mathcal{V}_{l_1+l_2}) \geq x\}$$
$$= P\{d(\boldsymbol{z}_1^{l_1}, \boldsymbol{v}_i^{l_1}) + d(\boldsymbol{z}_2^{l_2}, \mathcal{V}_{l_2}^i) \geq x \quad \text{for all } i = 1, 2, \cdots, g^{k_1}\} \qquad (25)$$

where the random event is the selection of a tree $\mathcal{V}_{l_1+l_2}$. For every particular code $\mathcal{V}_{l_1+l_2}$, let j be a path index such that

$$d(\boldsymbol{z}_1^{l_1}, \boldsymbol{v}_j^{l_1}) \leq d(\boldsymbol{z}_1^{l_1}, \boldsymbol{v}_i^{l_1}) \quad \text{for all } i = 1, 2, \cdots, g^{k_1}$$

(j is a random variable over the code ensemble). Then

$$d(\boldsymbol{z}_1^{l_1}, \boldsymbol{v}_j^{l_1}) = d(\boldsymbol{z}_1^{l_1}, \mathcal{V}_{l_1}) \qquad (26)$$

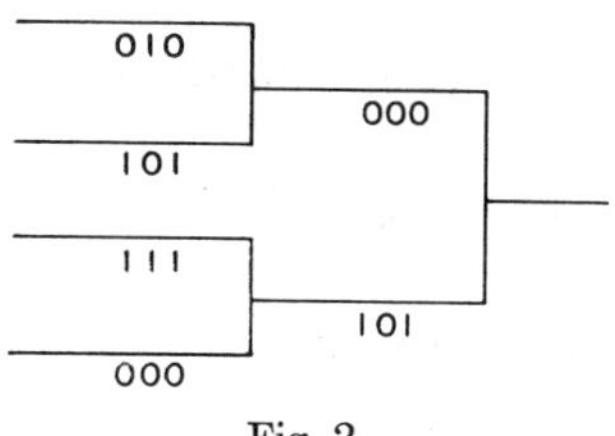

Fig. 2.

where $\mathcal{V}_{l_1}$ denotes the tree code consisting of the words $v_1^{l_1}, v_2^{l_1}, \cdots, v_{h_1}^{l_1} (h_1 = g^{k_1})$. Since

$$d(z^{l_1+l_2}; \mathcal{V}_{l_1+l_2}) \le d(z_1^{l_1}, v_j^{l_1}) + d(z_2^{l_2}; \mathcal{V}_{l_2}^j)$$

it follows from (25) and (26) that

$$P\{d(z^{l_1+l_2}; \mathcal{V}_{l_1+l_2}) \ge x\} \le P\{d(z_1^{l_1}; \mathcal{V}_{l_1}) + d(z_2^{l_2}; \mathcal{V}_{l_2}^j) \ge x\}.$$

But the code ensemble selection rule certainly makes $\mathcal{V}_{l_2}^j$ independent of j. In fact, the subtree $\mathcal{V}_{l_2}^j$ of $\mathcal{V}_{l_1+l_2}$ is identically distributed with the tree $\mathcal{V}_{l_2}$. Hence for every $z^{l_1+l_2}$,

$$P\{d(z^{l_1+l_2}; \mathcal{V}_{l_1+l_2}) \ge x\} \le P\{d(z_1^{l_1}; \mathcal{V}_{l_1}) + d(z_2^{l_2}; \mathcal{V}_{l_2}) \ge x\}. \quad (27)$$

Now, since $d(z, y) \ge 0$,

$$\begin{aligned} E[d(z^{l_1+l_2}; \mathcal{V}_{l_1+l_2})] &= \int_0^\infty P\{d(z^{l_1+l_2}; \mathcal{V}_{l_1+l_2}) \ge x\}\, dx \\ &\le \int_0^\infty P\{d(z_1^{l_1}; \mathcal{V}_{l_1}) + d(z_2^{l_2}; \mathcal{V}_{l_2}) \ge x\}\, dx \\ &= E[d(z_1^{l_1}; \mathcal{V}_{l_1}) + d(z_2^{l_2}; \mathcal{V}_{l_2})] \end{aligned}$$

where the inequality follows from (27). Q.E.D.

Lemma 2

There exists a value D^* such that

$$\lim_{k\to\infty} E\left[\frac{1}{kn_0} d(z^{kn_0}; \mathcal{V}_{kn_0})\right] = D^* \quad (28)$$

and

$$\lim_{k\to\infty} P\left\{Z^{kn_0}, \mathcal{V}_{kn_0} : \left|\frac{1}{kn_0} d(Z^{kn_0}; \mathcal{V}_{kn_0}) - D^*\right| > \epsilon\right\} = 0 \quad \text{for all } \epsilon > 0. \quad (29)$$

Proof: Again, to simplify notation, let $l = kn_0$, and let l vary in steps of size n_0. Define

$$D^* \triangleq \liminf_{l\to\infty} E\left[\frac{1}{l} d(Z^l; \mathcal{V}_l)\right] \quad (30)$$

Such a limit exists since for all l, [see (6)],

$$0 \le E\left[\frac{1}{l} d(Z^l; \mathcal{V}_l)\right] \le d_M < \infty.$$

Let $\delta > 0$ be arbitrary. Then there exists a value l_0 such that

$$E\left[\frac{1}{l_0} d(z^{l_0}; \mathcal{V}_{l_0})\right] \le D^* + \delta.$$

Let $f(l)$ be the number satisfying

$$f(l) = l - tl_0, \qquad 0 \le f(l) < l_0, \qquad t \text{ an integer}.$$

Then, it follows directly by repeated application of (24) that

$$E\left[\frac{1}{l} d(Z^l; \mathcal{V}_l)\right] \le \frac{1}{l} \sum_{i=1}^t E[d(Z_i^{l_0}; \mathcal{V}_{l_0})] + \frac{1}{l} E[d(Z_{t+1}^{f(l)}; \mathcal{V}_{f(l)})] \le D^* + \delta + \frac{l_0}{l} d_M$$

where $Z^l = (Z_1^{l_0}, \cdots, Z_t^{l_0}, Z_{t+1}^{f(l)})$. Hence

$$\limsup_{l\to\infty} E\left[\frac{1}{l} d(Z^l; \mathcal{V}_l)\right] \le D^* + \delta$$

and since δ was arbitrary, (28) follows.

Next, by repeated application of (23) of Lemma 1,

$$\begin{aligned} P\{d(Z^l; \mathcal{V}_l) \ge l(D^* + 3\delta)\} &\le P\left\{\sum_{i=1}^t d(Z_i^{l_0}; \mathcal{V}_{l_0}) + d(Z_{t+1}^{f(l)}; \mathcal{V}_{f(l)}) \ge l(D^* + 3\delta)\right\} \\ &\le P\left\{\sum_{i=1}^t \frac{1}{l_0} d(Z_i^{l_0}; \mathcal{V}_{l_0}) \ge t(D^* + 2\delta)\right\} \end{aligned} \quad (31)$$

where the last inequality holds for t sufficiently large (this follows from the assumed boundedness (6) of the distortion function). But

$$\frac{1}{l_0} d(Z_1^{l_0}; \mathcal{V}_{l_0}), \qquad \frac{1}{l_0} d(Z_2^{l_0}; \mathcal{V}_{l_0}), \cdots$$

are independent, identically distributed random variables whose expectation was just shown to be less than or equal to $D^* + \delta$. Since the value of $\delta > 0$ was arbitrary, it follows from (31) and the weak law of large numbers that

$$\lim_{l\to\infty} P\{d(Z^l; \mathcal{V}_l) > l(D^* + \epsilon)\} = 0 \quad \text{for all } \epsilon > 0. \quad (32)$$

Recalling that $l = kn_0$, it follows directly from (28) and (32) that

$$\lim_{l\to\infty} P\{d(Z^l; \mathcal{V}_l) < l(D^* - \epsilon)\} = 0. \quad (33)$$

Relations (32) and (33) together imply (29). Q.E.D.

Lemma 3

The number D^* of Lemma 2 satisfies

$$D^* \le D_0 \quad (34)$$

whenever $R - R(D_0) = \gamma > 0$.

Proof: Let $Y^l \triangleq (Y_1, Y_2, \cdots, Y_l)$, $l = kn_0$, be a random vector whose components are independent and identically distributed with the density $w^0(y)$ defined in (22). Also, define the sets

$$\mathcal{S}(\delta, z^l) \triangleq \left\{ y^l : \ln \frac{w^0(y^l/z^l)}{w^0(y^l)} \le l[R(D_0) + (2/3)\gamma], \, d(z^l, y^l) \le l[D_0 + \delta] \right\}, \quad (35)$$

where $\delta > 0$ and $\gamma = R - R(D_0)$. Then for all $y^l \in \mathcal{S}(\delta, z^l)$,

$$w^0(y^l) \ge w^0(y^l/z^l) \exp[-l[R(D_0) + (2/3)\gamma]] \quad (36)$$

so that (note that the code selection from the ensemble is independent of z^l)

$$\begin{aligned} P\{Z^l, Y^l : Y^l \in \mathcal{S}(\delta, Z^l)\} &= \int P\{Y^l \in \mathcal{S}(\delta, z^l)\} q(z^l)\, dz^l \\ &= \int q(z^l)\, dz^l \int_{\mathcal{S}(\delta, z^l)} w^0(y^l)\, dy^l \\ &\ge \exp\{-l[R(D_0) + (2/3)\gamma]\} \\ &\quad \int q(z^l)\, dz^l \int_{\mathcal{S}(\delta, z^l)} w^0(y^l/z^l)\, dy^l. \end{aligned} \quad (37)$$

But since

$$\ln \frac{w^0(y^l/z^l)}{w^0(y^l)} = \sum_{i=1}^{l} \ln \frac{w^0(y_i/z_i)}{w^0(y_i)}$$

while (see (20))

$$E\left[\ln \frac{w^0(Y/Z)}{w^0(Y)}\right] \triangleq \int_{\mathcal{A}} \int_{\mathcal{B}} \left[q(z) w^0(y/z) \ln \frac{w^0(y/z)}{w^0(y)}\right] dz\, dy \le R(D_0) + (1/3)\gamma$$

and

$$d(z^l, y^l) = \sum_{i=1}^{l} d(z_i, y_i)$$

while (see (21))

$$E[d(Z, Y)] \triangleq \int_{\mathcal{A}} \int_{\mathcal{B}} [q(z) w^0(y/z)\, d(z, y)]\, dz\, dy \le D_0$$

then by the weak law of large numbers, for every $\eta \in (0, 1)$ there exists an integer l_0 such that

$$\int q(z^l)\, dz^l \int_{\mathcal{S}(\delta, z^l)} w^0(y^l/z^l)\, dy^l \ge \eta \quad \text{for all} \quad l \ge l_0. \quad (38)$$

Hence, substituting (38) into (37),

$$\begin{aligned} P\{Z^l, Y^l : Y^l \in \mathcal{S}(\delta, Z^l)\} &= \int q(z^l)\, dz^l \int_{\mathcal{S}(\delta, z^l)} w^0(y^l)\, dy \\ &\ge \eta \exp\{-l[R(D_0) + (2/3)\gamma]\} \quad \text{if} \quad l \ge l_0. \end{aligned} \quad (39)$$

Let us now return to our tree code and consider the ith node on level $k(l = kn_0)$, corresponding to some codeword v_i. Define the indicator function

$$\phi_k(i, z^l) \triangleq \begin{cases} 1 & \text{if} \quad v_i \in \mathcal{S}(\delta, z^l) \\ 0 & \text{otherwise.} \end{cases} \quad (40)$$

Then, since under the code selection rule, $\phi_k(1, Z^l)$, $\phi_k(2, Z^l)$, $\cdots$ are identically distributed,

$$\begin{aligned} E\left[\sum_{i=1}^{g^k} \phi_k(i, Z^l)\right] &= g^k E[\phi_k(1, Z^l)] \\ &\ge \eta \exp[l[R - R(D_0) - (2/3)\gamma]] \\ &= \eta \exp[(1/3)\gamma l] \qquad l \ge l_0 \end{aligned} \quad (41)$$

where the inequality follows from (39) and (13) and the last equality from the definition (19) of γ. Since $\gamma > 0$, there exists an integer $l_1 = k_1 n_0$ such that

$$E\left[\sum_{i=1}^{g^{k_1}} \phi_{k_1}(i, Z^{l_1})\right] = m \qquad \text{for some} \quad m > 1. \quad (42)$$

Define next the branching process $U_0 \triangleq 1, U_1, U_2, \cdots$ where for $r \ge 1$,

$$U_r \triangleq \begin{cases} \sum_{j=1}^{U_{r-1}} W_j^r & \text{if} \quad U_{r-1} \ge 1 \\ 0 & \text{otherwise} \end{cases} \quad (43)$$

and $\{W_j^r, j = 1, 2, \cdots, r = 1, 2, \cdots\}$ are independent random variables, identically distributed with probability

$$P\{W_j^r = h\} = P\left\{Z^{l_1}, \mathcal{V}_{l_1} : \sum_{i=1}^{g^{k_1}} \phi_{k_1}(i, Z^{l_1}) = h\right\}, \qquad h = 0, 1, 2, \cdots. \quad (44)$$

(Feller [8, p. 272], describes a branching process as follows:

> We consider particles which are able to produce new particles of like kind. A single particle forms the original, or zero, generation. Every particle has probability $p_k(k = 0, 1, 2, \cdots)$of creating exactly k new particles; the direct descendants of the nth generation form the $(n + 1)$st generation. The particles of each generation act independently of each other. We are interested in the size of the successive generations.

Clearly, in our case $p_h = P\{W_j^r = h\}$). Thus $P\{W_j^r = h\}$ is equal to the probability that there are exactly h paths (of length k_1 branches) in the randomly selected tree $\mathcal{V}_l$, that correspond to codewords $y \in \mathcal{S}(\delta, Z^{l_1})$, where Z^{l_1} is the sequence generated by the source. Let $l = tl_1$. If $Z^l = (Z^{l_1}, Z^{l_1}, \cdots, Z_t^{l_1})$, then

$$P\{U_t > 0\} \le P\{d(Z^l; \mathcal{V}_l) \le l[D_0 + \delta]\} \quad (45)$$

since $U_t > 0$ implies that there exists at least one codeword $v = (y_1^{l_1}, y_2^{l_1}, \cdots, y_t^{l_1})$ in the tree such that

$$d(z_r^{l_1}, y_r^{l_1}) < l_1[D_0 + \delta] \quad \text{for all} \quad r = 1, 2, \cdots, t. \quad (46)$$

But by [8, pp. 275–276],

$$P\{U_t > 0\} \le P\{U_{t-1} > 0\},$$

and if $E[W_j^r] > 1$ for all j and r (which holds by (42) and (44)), then

$$\lim_{t \to \infty} P\{U_t > 0\} = \xi > 0.$$

Therefore, for l_1 sufficiently large,

$$\liminf_{t\to\infty} P\{Z^{tl_1}, \mathcal{V}_{tl_1} : d(Z^{tl_1}; \mathcal{V}_{tl_1}) \leq tl_1[D_0 + \delta]\} \geq \xi > 0. \quad (47)$$

But by Lemma 2, for any $\epsilon > 0$,

$$\lim_{t\to\infty} P\{d(Z^{tl_1}, \mathcal{V}_{tl_1}) \leq tl_1 \cdot [D^* - \epsilon]\} = 0. \quad (48)$$

Therefore, for all $\epsilon > 0$, $\delta > 0$,

$$D_0 + \delta > D^* - \epsilon$$

so that

$$D^* \leq D_0. \qquad \text{Q.E.D.}$$

Since (28) of Lemma 2, and Lemma 3 taken together show that over the ensemble of codes,

$$E\left[\frac{1}{kn_0} d(Z^{kn_0}; \mathcal{V}_{kn_0})\right] \leq D_0$$

there must exist a sequence of codes $\mathcal{V}_{n_0}, \mathcal{V}_{2n_0}, \mathcal{V}_{3n_0}, \cdots$ that satisfies (18). This proves Theorem 1.

III. Discussion and Some Additional Results

Theorem 1 does not provide any estimate of the speed of convergence of

$$E\left[\frac{1}{kn_0} d(Z^{kn_0}, \mathcal{V}_{kn_0})\right]$$

to D_0, and does not concern itself with the search effort that the encoder must expend before finding the codeword corresponding to the source output sequence. In fact, in the preceding development no advantage was taken of the tree structure of the codes. To be precise, the solution to the following problem is required:

> 'Let $\mathcal{S}(R, D)$ be the class of codes of rate $R, R > R(D)$, which when used with a given source result in average distortion that is less than or equal to D. Let $f(\mathcal{V})$ be the average encoding effort per source digit associated with a code $\mathcal{V} \in \mathcal{S}(R, D)$. Find $\min[f(\mathcal{V})]$ and the code (together with its implementation) $\mathcal{V}' \in \mathcal{S}(R, D)$ such that $f(\mathcal{V}') = \min[f(\mathcal{V})]$.'

The best known upper bounds for block codes for binary sources with uniform output probabilities and distortion function $d(z, y) = 1 - \delta(z, y)$ give for $\mathcal{V} \in \mathcal{S}(0.5, 0.13)$ $[0.5 \cong R(0.11)]$ the estimate $f(\mathcal{V}) \leq 10^{12}$. On the other hand, the author together with his student Anderson [7] have found an encoding implementation scheme (which is less crude than the one described below) for a tree code $\mathcal{V}' \in \mathcal{S}(0.5, 0.13)$ such that $f(\mathcal{V}') \cong 150$ (this number was obtained by simulation and not by theoretical analysis).

We shall conclude this paper by quoting some theorems that can be found in a report [7] by the author. They have been omitted from this paper because their proof is rather involved and they are weaker than the above experimental results.

It is possible to show that to every source there corresponds a set $\mathcal{G}_t$ of well composed sequences z^t such that

$$P\{Z^t \notin \mathcal{G}_t\} \leq e^{-tA}, \qquad A > 0$$

and having the property that over a suitable ensemble of tree codes of any rate $R > R(D)$, and for any $\rho > 0$ there exist numbers $\beta_0(\delta)$ and $k_0(\delta, \rho)$ such that

$$P\{\mathcal{V}_{kn_0} : d(z^{kn_0}; \mathcal{V}_{kn_0}) - d^*(z^{kn_0}) > kn_0\,\delta\} \leq \exp\left[-k^\rho \beta_0(\delta)\right] \quad (49)$$

for all source output sequences $z^{kn_0} \in \mathcal{G}_{kn_0}$ and all $k > k_0(\delta, \rho)$, where

$$d^*(z^{kn_0}) = \sum_{i=1}^{kn_0} d^*(z_i)$$

and

$$E[d^*(Z)] = D.$$

Therefore the distance between "well composed" source output sequences and the closest codeword in a tree code $\mathcal{V}_{kn_0}$ may be said to converge faster than any power of k to the distortion $kn_0 D$.

The crudest encoding algorithm utilizing the tree structure of a code and the result just quoted is as follows:

1) $i = 1$.

2) If $z^{in_0} \in \mathcal{G}_{in_0}$ go to 3), otherwise go to 5).

3) Search through all g^i codewords corresponding to paths leading to nodes on level i to see if there exists a word $y_*^{in_0}$ such that

$$d(z^{in_0}, y_*^{in_0}) - d^*(z^{in_0}) \leq in_0\epsilon \quad (50)$$

where $\epsilon > 0$ is a suitable constant. If such a word exists, encode z^{in_0} into it and go to 4). If not, go to 5).

4) Encoding of the sequence $z_{in_0+1}, z_{in_0+2}, \cdots$ starts with step 1) (after in_0 has been subtracted from the index of each letter) using the tree whose root node is the last node of the path corresponding to $y_*^{in_0}$ (we have assumed that an unlimited tree is at our disposal).

5) If $i < \nu$, increase i by 1 and go to 2). Otherwise encode $z^{\nu n_0}$ arbitrarily into the codeword $y_*^{\nu n_0}$ corresponding to the first node on the νth level of the tree, and go to 4).

Note from the assumed additive character of the distance measure (4) that in order to extend the comparison of $z^{(i-1)n_0}$ to that of z^{in_0}, the encoder need only compute the sums

$$\sum_{j=(i-1)n_0+1}^{in_0} [d(z_j, y_j) - d^*(z_j) - \epsilon] \quad (51)$$

for the various codeword subsequences

$$(y_{(i-1)n_0+1}, \cdots, y_{in_0}).$$

If as a unit of computation effort one takes the calculation of the sum (51), then it is possible to show that there exists a function $R_T(D)$ such that the above algorithm will lead to average distortion $D + \epsilon$ for coding rates $R > R_T(D)$, and that the first moment of the search effort per source output digit will be bounded by a constant that is independent of $\epsilon > 0$. Now $R_T(D) \geq R(D)$ and the reason for inequality is that z^{kn_0} must belong to

$\mathcal{G}_{kn_0}$ in order that (49) be valid. However, for the following class of completely symmetrical sources, $\mathcal{G}_{kn_0}$ includes all sequences z^{kn_0}, and so $R_T(D) = R(D)$ for all D:

$$P\{Z = z\} = 1/a \qquad \text{for all } z \in \{0, 1, \cdots, a-1\}$$

and the distortion matrix $[d(z, y)]$ $(y = 0, 1, \cdots, a-1)$ is such that the entries of each row and of each column are permutations of the set $\{d(0, 0), d(0, 1), \cdots, d(0, a-1)\}$.

We conjecture that $R_T(D) = R(D)$ for all sources if we employ less crude encoding strategies that do not search out all of the possible paths until an acceptable one is found.

Acknowledgment

The author is deeply indebted to Prof. H. Kesten of Cornell University who suggested the branching process approach of the proof of Lemma 3.

References

[1] C. E. Shannon, "Coding theorems for a discrete source with a fidelity criterion," *IRE Natl. Conv. Rec.* pt., 4, pp. 142–163, 1959.
[2] T. J. Goblick, Jr., "Coding for a discrete information source with a distortion measure," Ph.D. dissertation, Dept. of Elec. Engrg., Massachusetts Institute of Technology, Cambridge, October 1962.
[3] F. Jelinek, *Probabilistic Information Theory*. New York: McGraw-Hill, 1968.
[4] T. J. Goblick, Jr., "Theoretical limitations on the transmission of data from analog sources," *IEEE Trans. Information Theory*, vol. IT-11, pp. 558–567, October 1965.
[5] T. J. Goblick, Jr., and J. L. Holsinger, "Analog source digitization: A comparison of theory and practice," *IEEE Trans. Information Theory* (Correspondence), vol. IT-13, pp. 323–326, April 1967.
[6] J. T. Pinkston, III, "Encoding independent sample information sources," M. I. T. Research Lab. of Electronics, Cambridge, Mass., Tech. Rept. TR 462, October 1967.
[7] F. Jelinek and J. B. Anderson, "Tree encoding of sources with a fidelity criterion and its implemention," Center for Radiophysics and Space Research, Cornell University, Ithaca, N. Y., Tech. Rept., 1969.
[8] W. Feller, *An Introduction to Probability Theory and Its Applications*, vol. 1. New York: Wiley, 1964.

The Rate of a Class of Random Processes

DAVID J. SAKRISON, MEMBER, IEEE

***Abstract*—In certain situations, a single transmission system must be designed to function satisfactorily when used for any source from a class $\mathcal{A}$ of sources. In this situation, the rate-distortion function $R_{\mathcal{A}}(d)$ is the minimum capacity required by any transmission system that can transmit each source from $\mathcal{A}$ with average distortion $\leq d$. One of the most interesting classes of sources is a class of random processes. Here we consider a weighted-square error-distortion measure and the class of all stationary random processes that satisfy a certain strong mixing property, that have zero mean, known power, and a bounded fourth moment, and that satisfy one of the following alternative specifications on the spectrum: 1) the spectrum is known exactly; 2) the amount of power within the band $0 \leq f \leq f_k$ is known for $N - 1$ frequencies $f_1 < f_2 < \cdots < f_{N-1}$; or 3) the fraction of power outside some frequency f_1 is $\leq 1 - \gamma$.**

For the class of sources determined by each of the above three cases and for an arbitrary error-weighting function we evaluate the rate-distortion function.

I. Introduction

For a source α with a known probability distribution, the rate-distortion function $R_\alpha(d)$ specifies the minimum channel capacity required to transmit the source α with average distortion no greater than d. In practice, the encoding system often must be designed when only vague information is available concerning the source distribution, i.e., it is known only that the source is in some class $\mathcal{A}$. In this situation, the appropriate meaning of the rate-distortion function $R_{\mathcal{A}}(d)$ is the minimum capacity required by any single system that is capable of transmitting *each* source from $\mathcal{A}$ with an average distortion no greater than d. In a recent paper [1] we have shown that for what we term a *compact* class of sources, $R_{\mathcal{A}}(d)$ is simply the supremum over the α in $\mathcal{A}$ of $R_\alpha(d)$. The definition of a compact class of sources is given in Section II.

One of the most interesting sources is a random process. It is the purpose of this paper to evaluate the rate-distortion function for several classes of random processes of practical interest under a mean-weighted-square error-distortion measure.

First let us define precisely the distortion measure that we consider. Let A denote a linear operator on $L_2[0, T]$ with impulse response $a(t)$ and transfer function $A(f)$. We assume that $a(t)$ is nonzero on some finite interval $[0, T_a]$. For $U(t)$ a sample of the source output of T seconds' duration and $\tilde{U}(t)$ its approximation, the weighted error is

$$E(t) = \int_0^T a(t - s)[U(s) - \tilde{U}(s)]\, ds \qquad 0 \leq t \leq T + T_a. \tag{1}$$

On the interval $T_a \leq t \leq T$ there are no spurious transient effects, and we take the average distortion to be

$$\begin{aligned} d &= E\left\{\frac{1}{T - T_a} \int_{T_a}^T |E(t)|^2\, dt\right\} \\ &= E\left\{\frac{1}{T - T_a}\, ||A(U - \tilde{U})||^2_{[T_a T]}\right\}. \end{aligned} \tag{2}$$

The purpose of the weighting function $A(f)$ is to tailor the distortion measure to correspond to the user's subjective evaluation of the error in the reproduced signal $\tilde{U}(t)$. For example, if $u(t)$ is a sample of a musical waveform, one would probably want $|A(f)|^2$ roughly inversely proportional to $S_u(f)$, so that high-frequency sounds, which are usually of low power, are not obscured. Further, $|A(f)|^2$ could taper off to zero quickly outside of the audible range of the ear. If $u(t)$ is a sample of a television scan, one would probably want $a(t)$ to contain an approximation to an impulse term plus the derivative of an impulse, so that changes in intensity level are emphasized and the edges of objects accurately preserved. Again, $|A(f)|^2$ could taper off to zero outside the resolution limits of the optical system employed.

Next let us define the classes of processes that we intend to consider In doing so, it will be convenient to refer to the process formed by operating on $U(t)$ with A,

$$V(t) = \int_0^T U(s)a(t - s)\, ds \qquad T_a \leq t \leq T. \tag{3}$$

We make no detailed assumptions concerning the probability distribution of the process $U(t)$; instead we assume only that it is a stationary process satisfying the following partial characterization.

1) $U(t)$ is zero mean,

$$E\{U_t\} = 0. \tag{4}$$

We denote the power spectral density of $U(t)$ by $S_u(f)$. For future convenience we denote the power within f' by

$$G(f') = 2 \int_0^{f'} S_u(f)\, df. \tag{5}$$

Manuscript received October 31, 1968; revised March 21, 1969. This work was sponsored by the National Aeronautics and Space Administration under Grant NGR 05-003-143 and the Joint Services Electronics Program under Grant AFOSR-68-1488.

The author is with the Department of Electrical Engineering and Computer Sciences, Space Science Laboratory, and Electronics Research Laboratory, University of California, Berkeley, Calif. 94720.

Reprinted from *IEEE Trans. Inform. Theory*, vol. IT-16, pp. 10–16, Jan. 1970.

2) The power of $U(t)$ is known,

$$E\{U_t^2\} = G(\infty) = R_u(0) = P. \tag{6}$$

3) The process V_t has a bounded fourth moment,

$$E\{(V_t)^4\} \leq k_1 < \infty. \tag{7}$$

This would be guaranteed if U_t had a bounded fourth moment and $a(t)$ consisted of a finite number of impulses plus an absolutely integrable term.

4) For any T, let $\mathfrak{F}_0^T$ denote the σ field of events that can be defined in terms of the random variables U_t, $0 < t \leq T$. For any N let $E_1, \cdots, E_N$ denote any N events, each of which is in $\mathfrak{F}_0^T$. We denote a time shift of ns seconds by S^n, so that S^nE_n denotes the event E_n defined on the shifted segment $U(t + ns), 0 < t \leq T$. As s becomes large, the events S^nE_n, $n = 1, \cdots, N$, occur on time intervals that are all mutually separated by at least $s - T$ seconds. What we require is the asymptotic independence of these events; specifically we require for all $\alpha \in \mathfrak{A}$ that there exist an $\epsilon(E_1, \cdots, E_N)$ such that

$$\left| P_\alpha\left(\bigcap_{n=1}^{T} S^nE_n\right) - \prod_{n=1}^{N} P_\alpha(E_n) \right| \leq \epsilon(E_1, \cdots, E_N) \tag{8}$$

with

$$\lim_{s\to\infty} \epsilon(E_1, \cdots, E_N) = 0. \tag{9}$$

Note that the above convergence is uniform over the class of sources $\mathfrak{A}$, but need not be uniform over the events in $\mathfrak{F}_0^T$. This mixing property is neither broader than the conventional strong mixing property [2], since it requires N-wise factorization (for arbitrary N) instead of just pair-wise factorization, nor is it narrower, since the events in question need only be in $\mathfrak{F}_0^T$. It is implied by, and broader than, the uniform strong mixing property used by Rosenblatt [3] to prove a central limit theorem.

We then consider three classes of processes defined by 1)–4) above and one of three alternative assumptions concerning how much is known about the spectrum $S_u(f)$:

a) $S_u(f)$ is known exactly;

b) $G(f)$ is known at $N - 1$ frequencies

$$G(f_k) = G_k \qquad k = 1, \cdots, N - 1;$$

$$0 < f_1 < f_2 < \cdots < f_{N-1} < \infty \tag{10}$$

(this could be ascertained by measuring the power at the output of a bank of $N - 1$ low-pass filters);

c) for some frequency f_1 it is known that

$$G(f_1) \geq \gamma G(\infty) \qquad 0 < \gamma \leq 1. \tag{11}$$

(This might be simply based on an intuitive conservative estimate of the bandwidth of $U(t)$.)

In addition, in cases b) and c) we make the assumption that the spectrum $S_u(f)$ is such that

$$|A(f)|^2 \, S_u(f) \leq \frac{k_2}{f^2} \qquad k_2 < \infty. \tag{12}$$

The class of all wide-sense-stationary processes satisfying 1)–4) above and assumption a) on the spectrum we refer to simply as class a; if in addition it satisfies inequality (12) and assumption b) or c) on the spectrum, we refer to it simply as class b or class c, respectively.

Having defined the classes of processes whose rates we wish to evaluate, we now briefly comment on how to obtain the rate-distortion function for these three classes. First, in Section II, we show that each of the above three classes of sources is compact. With this result in hand, it follows from our earlier work [1] that $R_{\mathfrak{A}}(d)$, the rate-distortion function for the class $\mathfrak{A}$, is given by

$$R_{\mathfrak{A}}(d) = \sup_{\alpha\in\mathfrak{A}} R_\alpha(d). \tag{13}$$

Next, we note that the three classes of sources a–c each include the class of Gaussian sources with spectra satisfying assumptions a)–c), respectively. Further, the rate-distortion function of any source with a given spectrum is less than or equal to that of a Gaussian source with the same spectrum; this can be shown [4] to be a direct consequence of the corresponding property derived by Shannon for a random variable with a mean-square-fidelity criterion. Thus for each of the three classes a–c, $R_{\mathfrak{A}}(d)$ is the supremum over the rate-distortion functions of Gaussian sources with spectra falling within a), b), or c). In Section III we give parametric expressions for determining $R_{\mathfrak{A}}(d)$ for each of these three classes of spectra.

Results related to our results for class a were presented by B. S. Tsybakov at the 1969 Information Theory Symposium. Tsybakov considered a discrete-time-vector-valued source with a known covariance matrix for the components of the vector, but assumed no further knowledge concerning the statistics of the process, including the covariance between vectors occurring at different time instants. In the Russian preprint then available, no statement assuming independence of the vectors at distinct times, nor assuming an ergodic or mixing property seemed to be made. The results would be the same, but some such property is required. His results are a finite-dimensional version of our parametric expressions for class a.

II. Compactness of the Classes of Sources

First we must define what we mean by a *compact* class of sources. Let $\mathfrak{U}$ be the space on which samples of the source output take on values. For example, if we consider a sample of a random process of T seconds' duration, $\mathfrak{U}$ might be $L_2[0, T]$. Let $\rho(\cdot, \cdot)$ be a metric on $\mathfrak{U}$. A subset $\mathfrak{U}'$ of $\mathfrak{U}$ is said to be *totally bounded* if for any $\epsilon_2 > 0$ there exists a finite covering of $\mathfrak{U}'$ by neighborhoods of radius at most ϵ_2. We consider the distortion measure formed by taking the distance function $d(\cdot, \cdot)$ to be the metric $\rho(\cdot, \cdot)$

$$d(u, \tilde{u}) = \rho(u, \tilde{u}).$$

Let $E_\alpha\{\cdot\}$ denote expectation with respect to the source distribution α. If given any $\epsilon_1 > 0$, there exists a function $u_p = u_p(u)$ whose range is a totally bounded set $\mathfrak{U}'$ and

$$E_\alpha\{\rho(U, U_p)\} \leq \epsilon_1$$

for *all* $\alpha \in \mathfrak{A}$, we say that the class $\mathfrak{A}$ of sources is compact.

To show that the classes of random processes *a–c* are compact, we first consider a finite-dimensional process that is a projection of $U(t)$ onto an N-dimensional space. This finite-dimensional process then has each of its components truncated to a finite range to yield a process $U_p(t)$. Last, we show that by making N and the truncation range sufficiently large, the distortion between $U(t)$ and $U_p(t)$ can be made arbitrarily small. Since a bounded N-dimensional set is totally bounded, this completes the proof.

In carrying out this proof, the weighted-square error-distortion measure on $U(t)$ is most easily discussed in terms of the process $V(t)$. [See (3).]

$$V(t) = [AU](t) = \int_0^T a(t - s)U(s)\,ds \qquad 0 \leq t \leq T + T_a$$

because the distortion between $U(t)$ and $\tilde{U}(t)$ under the weighted-square error criterion is equal to the integral-square error between the processes $V(t)$ and $\tilde{V}(t) = [A\tilde{U}](t)$ on the interval $[T_a, T]$.

Let us consider class *a* first. The process $V(t)$ can be expanded in a Karhunen–Loève expansion on the interval $[T_a, T]$

$$V(t) = \lim_{N\to\infty} V_N(t) = \lim_{N\to\infty} \sum_{k=1}^{N} V_k\varphi_k(t) \tag{14}$$

in which the $\varphi_k(t)$ are the normalized eigenfunctions of $R_v(t_1, t_2)$, the correlation function of the process $V(t)$. Corresponding to this, there is an expansion of $U(t)$ [4], [5]

$$U_N(s) = \sum_{k=1}^{N} V_k\theta_k(s) \qquad 0 \leq s \leq T \tag{15}$$

in which the functions $\theta_k(s)$ are given by

$$\theta_k = [A^* \circ A]^{-1}A\varphi_k \tag{16}$$

and the operator A^* is defined by

$$[A^*v](s) = \int_{T_a}^{T} a(t - s)v(t)\,dt \qquad 0 \leq s \leq T. \tag{17}$$

The weighted-square error between U and U_N is given by

$$||A(U - U_N)||^2_{[T_a,T]} = ||(V - V_N)||^2_{[T_a,T]} = \sum_{k=N+1}^{\infty} |V_k|^2. \tag{18}$$

Next, consider the fourth moment of one of the expansion coefficients

$$\begin{aligned} E\{(V_k)^4\} &= \int_{T_a}^{T} dt_1\varphi_k^*(t_1) \cdots \int_{T_a}^{T} dt_4\varphi_k^*(t_4)E\{V_{t_1}V_{t_2}V_{t_3}V_{t_4}\} \\ &\leq \int_{T_a}^{T} dt_1\,|\varphi_k^*(t_1)| \cdots \int_{T_a}^{T} dt_4\,|\varphi_k^*(t_4)|\,E\{|V_t|^4\} \\ &\leq k_1\left[\int_{T_a}^{T} dt\,|\varphi_k(t)|\right]^4 \\ &\leq k_1(T - T_a)^2\,||\varphi_k||^4 = k_1(T - T_a)^2. \end{aligned} \tag{19}$$

The interchange of expectation and integration involved in the first equation above is justified by Fubini's theorem, since $E\,\{|V_{t_1} \cdots V_{t_4}|\}$ is bounded by k_1, and φ_k is absolutely integrable. The next-to-last inequality follows simply from the Schwarz inequality.

Let $\tilde{V}_k$ denote the random variable V_k truncated to some finite interval $[-v_0, v_0]$. By a simple argument involving a Chebyshev inequality, it can be shown [1] that if V_k has a bounded fourth moment, the mean-square error $E\{(V_k - \tilde{V}_k)^2\}$ can be made arbitrarily small by making v_0 sufficiently large. If one then approximates $U(t)$ by

$$U_p(t) = \sum_{k=1}^{N} \tilde{V}_k\theta_k(t) \qquad 0 \leq t \leq T \tag{20}$$

the weighted-mean-square error is

$$d = \sum_{k=N+1}^{\infty} E\{V_k^2\} + \sum_{k=1}^{N} E\{(V_k - \tilde{V}_k)^2\}. \tag{21}$$

The first term can be made less than $\epsilon_1/2$ by making N sufficiently large; then, for this value of N, the second term can be made less than $\epsilon_1/2$ by making v_0 sufficiently large. We can thus take $\mathfrak{U}'$ to be the set $|v_k| \leq v_0$, $k = 1, \cdots, N$; $v_k = 0$, $k > N$. By the Heine–Borel theorem, this set is totally bounded and hence calss *a* of sources is compact.

Next, we consider the compactness of the classes *b* and *c*. Here we take the functions φ_k for expanding $V(t)$ to be

$$\varphi_k(t) = \exp\left[\frac{i2\pi kt}{(T - T_a)}\right] \qquad T_a \leq t \leq T; \quad k = 0, \pm1, \pm2, \cdots. \tag{22}$$

The expansion of $U(t)$ and the functions $\theta_k(s)$, $0 \leq s \leq T$, are again defined as in (15) and (16). Our proof of compactness now follows exactly as for class *a*. We need only note two points. First, by the same argument used for class *a*, the V_k again have a bounded fourth moment. Second, we need to show that the truncation error

$$\sum_{k>N} E\{|V_k|^2\}$$

can be made smaller than $\epsilon_1/2$. Note that on the rectangle $T_a \leq t_1, t_2 \leq T$, $R_v(t_1, t_2)$ depends only on $t_1 - t_2$ and is the inverse Fourier transform of $S_v(f) = S_u(f)\,|A(f)|^2$. From theorem 4 of Root and Pitcher [6], it then follows that for $T_0 = T - T_a$

$$\begin{aligned} 2E\{|V_k|^2\} &= \frac{1}{2\pi T_0}\int_{-\infty}^{\infty} S_v\left(\frac{x+k\pi}{2\pi T_0}\right)\left(\frac{\sin x}{x}\right)^2 dx \\ &= \int_{-\infty}^{\infty} S_v(f)\left[\frac{\sin 2\pi T_0(f-k/2T_0)}{2\pi T_0(f-k/2T_0)}\right]^2 df \\ &\le \frac{1}{(2\pi T_0 f_0)^2}\left[\int_{-\infty}^{(k/2T_0)-f_0} S_v(f)\,df + \int_{(k/2T_0)+f_0}^{\infty} S_v(f)\,df\right] \\ &\quad + \int_{(k/2T_0)-f_0}^{(k/2T_0)+f_0} S_v(f)\,df. \end{aligned} \tag{23}$$

The two integrals inside the brackets can be bounded by $d_0 = R_v(0)$, the zero-rate distortion that is finite by inequality (7). By using inequality (12), the third integral can be bounded by

$$\begin{aligned} I_3 &= \int_{(k/2T_0)-f_0}^{(k/2T_0)+f_0} S_v(f)\,df \le k_2\int_{(k/2T_0)-f_0}^{(k/2T_0)+f_0}\frac{df}{f^2} \\ &= k_2\left[\frac{1}{(k/2T_0)-f_0} - \frac{1}{k/2T_0+f_0}\right] \\ &= 2k_2 f_0\left[\left(\frac{k}{2T_0}\right)^2 - f_0^2\right]^{-1}. \end{aligned}$$

Setting $f_0 = +(k/2T_0)^{2/3}$, we have for $k \ge T_0 2^{5/2}$

$$I_3 \le 4k_2\left(\frac{k}{2T_0}\right)^{-4/3}. \tag{24}$$

Inequality (23) then weakens to

$$\begin{aligned} 2E\{\,|V_k|^2\} &\le [\pi^{-2}(2T_0)^{-2/3} \\ &\quad + 4k_2(2T_0)^{4/3}]k^{-4/3} \triangleq C_0 k^{-4/3}. \end{aligned} \tag{25}$$

Thus the weighted-square error in truncating the expansion of $U(t)$ to $2N+1$ terms is bounded for $N \ge T_0 2^{5/2}$ by

$$\sum_{|k|>N} E\{V_k^2\} \le C_0 \sum_{k=N+1} k^{-4/3} \le \frac{3C_0}{N^{1/3}} \tag{26}$$

and the classes b and c are compact.

III. Evaluation of the Rates of the Three Classes of Sources

We now give parametric relations determining the rate-distortion function for the class of processes a–c.

Class a

Since the spectrum is known exactly, $R_\alpha(d)$ is given simply by the parametric relations giving the rate-distortion function for a Gaussian process with a weighted-square error criterion. Kolmogorov [7] first presented these relations for integral-square error, and Pinsker [8] was the first to arrive at the relations for the weighted-error case. Pinsker's results were discussed and extended in a paper in English by Dobrushin and Tsybakov [9]. A derivation of these relations in English appears in [4] and [5]. These parametric relations are

$$d(\mu) = 2\left[\mu\int_{E(\mu)} df + 2\int_{E^c(\mu)} |A(f)|^2\, S_u(f)\, df\right] \tag{27}$$

$$R_\alpha(\mu) = \int_{E(\mu)} \ln\left[|A(f)|^2\,\frac{S_u(f)}{\mu}\right] df \tag{28}$$

in which $E(\mu)$ is the set of nonnegative frequencies

$$E(\mu) = \{f : |A(f)|^2\, S_u(f) > \mu\} \tag{29}$$

and $E^c(\mu)$ is its complement on the set of nonnegative frequencies.

Class b

It is easiest to rephrase the constraints of (10) on the spectrum as constraints on the power within intervals. Let

$$\begin{aligned} I_1 &= \{f : 0 \le f \le f_1\} \\ I_k &= \{f : f_{k-1} < f \le f_k\} \qquad k = 2, \cdots, N-1 \\ I_N &= \{f : f_{N-1} < f < \infty\} \end{aligned} \tag{30}$$

$$\begin{aligned} P_1 &= G_1 \\ P_k &= G_k - G_{k-1} \qquad k = 2, \cdots, N-1 \\ P_N &= P - G_{N-1} \end{aligned} \tag{31}$$

To evaluate $R_\alpha(d)$, we now need to find for a fixed value of d, given by (27), the maximum value that $R_\alpha(d)$ as given by (28) can take on over all spectra $S_u(f)$ satisfying

$$2\int_{I_k} S_u(f)\, df = P_k \qquad k = 1, \cdots, N. \tag{32}$$

The answer is again given parametrically. Let the sets $F_k(\beta_k)$ be defined by

$$F_k(\beta_k) = \{f{:}f \in I_k,\ |A(f)|^2 > \beta_k\} \tag{33}$$

and let $l(F_k)$ denote the "length" (Lebesgue measure) of F_k. For a given value of the parameter μ, let the N β_k be selected such that

$$\frac{P_k\beta_k}{2l[F_k(\beta_k)]} = \mu. \tag{34}$$

Note that β_k is unique since $l[F_k(\beta_k)]$ decreases as β_k increases. Then for the value of distortion given by

$$d = d(\mu) = 2\mu\sum_{k=1}^{N} l[F_k(\beta_k)] = \sum_{k=1}^{N}\beta_k P_k \tag{35}$$

the maximum value assumed by the rate is

$$R_\alpha(\mu) = \sum_{k=1}^{N}\int_{F_k(\beta_k)} \ln\left[\frac{|A(f)|^2}{\beta_k}\right] df. \tag{36}$$

The spectrum yielding the maximum is

$$S_u(f) = \begin{cases} P_k/2l[F_k(\beta_k)] & f \in F_k(\beta_k);\ k = 1, \cdots, N \\ 0 & \text{elsewhere} \end{cases} \tag{37}$$

Equations (35) and (36) thus parametrically determine

$R_a(d)$ for class b. The proof that (35) and (36) determine the maximum of R for a given value of d is given in Appendix I.

Class c

Class c is determined by an inequality constraint on the power outside a given frequency. Here the rate must be determined by a two-step procedure. First using (35)–(37) with $N = 1$, one calculates the rate only under the constraint that the total power on $[0, \infty)$ is P. The spectrum of (37) is then integrated to find the power inside the frequency f_1. If this satisfies the constraint $G(f_1) \geq \gamma G(\infty)$, the rate-distortion values given by (35) and (36) are the correct ones. If the value calculated for $G(f_1)$ is less than $\gamma G(\infty)$, the constrained maximum occurs with equality holding in the constraint (this is shown in Appendix II). In this case we proceed to the second step and calculate the rate according to (35) and (36) with $N = 2$ and

$$P_1 = \gamma P, \qquad P_2 = (1 - \gamma)P, \tag{38}$$

$$I_1 = \{f : 0 \leq f \leq f_1\}, \qquad I_2 = \{f : f > f_1\} \tag{39}$$

Let us now describe the above algorithm more explicitly. First we give the particular expressions, resulting from (35) and (36) with $N = 1$, that give values for the rate and distortion with no constraint on how the power is distributed. Letting

$$F(\beta_0) = \{f : f \geq 0, |A(f)|^2 > \beta_0\} \tag{40}$$

we have, for the distortion level

$$d = \beta_0 P, \tag{41}$$

the maximum rate

$$R_a(d) = \int_{F(\beta_0)} \ln \left[\frac{|A(f)|^2}{\beta_0}\right] df, \tag{42}$$

corresponding to the spectrum

$$S_u(f) = \begin{cases} \dfrac{P}{2l[F(\beta_0)]} & f \in F(\beta_0) \\ 0 & f \notin F(\beta_0). \end{cases} \tag{43}$$

Let $F_1(\beta_0) = I_1 \cap F(\beta_0)$. If the power inside frequency f_1 satisfies the constraint; i.e., if

$$P_1 = \frac{Pl[F_1(\beta_0)]}{l[F(\beta_0)]} \geq \gamma P$$

or

$$\frac{l[F_1(\beta_0)]}{l[F(\beta_0)]} \geq \gamma \tag{44}$$

then the rate for class c for the distortion level $\beta_0 P$ is given by (42). If inequality (44) is not satisfied, $R_a(\mu)$ must be found from (34)–(36) using $N = 2$ and the two intervals and power levels indicated by (38) and (39).

IV. Conclusion

In this paper we have described three classes of random processes defined by only rudimentary statements concerning the probability distribution and by three different assumptions concerning the system designer's knowledge of the spectrum of the process. These classes thus form useful models for sources for which one might have to design a transmission system. For each of the three classes, we have presented in Section III algorithms for calculating $R_a(d)$, the minimum capacity required by a transmission system capable of transmitting each source in the class with a mean-weighted-square error less than or equal to d.

Appendix I

It can be shown that the spectrum of (37) yields the distortion and rate of (35) and (36) simply by substituting from (37) into (27) and (28), respectively. We now show that the rate of (36) is the greatest rate consistent with the value of d given by (35) and the constraints on the power expressed by (32).

Let $E(\mu)$ and the $F_k(\beta_k)$ be defined as before, and let

$$E_k(\mu) = I_k \cap E(\mu) \tag{45}$$

$$\left.\begin{aligned} A_k &= E_k(\mu) \cap F_k(\beta_k) & B_k &= E_k(\mu) \cap F_k^c(\beta_k) \\ C_k &= E_k^c(\mu) \cap F_k(\beta_k) & D_k &= E_k^c(\mu) \cap F_k^c(\beta_k) \end{aligned}\right\} \tag{46}$$

$$P_{Ak} = 2 \int_{A_k} S_u(f)\, df \qquad k = 1, 2, \cdots, N \tag{47}$$

and correspondingly for P_{Bk}, P_{Ck}, P_{Dk};

$$Q_{Ck} = 2 \int_{C_k} \left[\frac{|A(f)|^2}{\beta_k}\right] S_u(f)\, df \tag{48}$$

and correspondingly for Q_{Dk}. Note from the definitions of C_k and D_k that

$$Q_{Ck} > P_{Ck}, \quad Q_{Dk} \leq P_{Dk}. \tag{49}$$

To obtain the value of d given by (35), μ must be picked to satisfy

$$2\mu l[E(\mu)] + \sum_{k=1}^{N} \beta_k (Q_{Ck} + Q_{Dk}) = \sum_{k=1}^{N} \beta_k P_k$$

or

$$\mu = \sum_{k=1}^{N} \frac{\beta_k (P_k - Q_{Ck} - Q_{Dk})}{2l[E(\mu)]}. \tag{50}$$

From (28), the rate is

$$R_a(\mu) = \sum_{k=1}^{N} \int_{E_k(\mu)} \ln \left[\frac{|A(f)|^2}{\mu}\right] + \sum_{k=1}^{N} \int_{E_k(\mu)} \ln [S_u(f)]\, df.$$

Applying the convexity inequality of the logarithm to each of the integrals in the second sum of this equation, adding a $[l(E_k) \ln (1/\beta_k)]$ term in the first sum and subtracting it in the second, and shifting the $\ln (1/\mu)$ term, yields

$$R_a(\mu) \leq \sum_{k=1}^{N} \int_{E_k(\mu)} \ln \left[\frac{|A(f)|^2}{\beta_k} \right] df + l(E) \sum_{k=1}^{N} \frac{l(E_k)}{l(E)} \ln \left[\frac{\beta_k(P_{Ak} + P_{Bk})}{2\mu l(E_k)} \right]. \quad (51)$$

Now applying the convexity inequality of the logarithm to the second sum and substituting for μ from (50), we obtain

$$R_a(\mu) \leq \sum_{k=1}^{N} \int_{E_k(\mu)} \ln \left[\frac{|A(f)|^2}{\beta_k} \right] df + l(E) \ln \left[\frac{\sum_{k=1} \beta_k(P_{Ak} + P_{Bk})}{2\mu l(E)} \right]$$

$$\leq \sum_{k=1}^{N} \int_{A_k} \ln \left[\frac{|A(f)|^2}{\beta_k} \right] df + l(E) \ln \left[\frac{\sum_{k=1}^{N} \beta_k(P_{Ak} + P_{Bk})}{\sum_{k=1}^{N} \beta_k(P_k - Q_{Ck} - Q_{Dk})} \right] \quad (52)$$

in which we have noted that $|A(f)|^2/\beta_k \leq 1$ on B_k. Next, using the inequality

$$\log (1 + x) \leq x$$

and the relation

$$P_k = P_{Ak} + P_{Bk} + P_{Ck} + P_{Dk}$$

on inequality (52), we obtain

$$R_a(\mu) \leq \sum_{k=1}^{N} \int_{A_k} \ln \left[\frac{|A(f)|^2}{\beta_k} \right] df + \frac{l(E) \sum_{k=1}^{N} \beta_k(Q_{Ck} - P_{Ck} + Q_{Dk} - P_{Dk})}{\sum_{k=1}^{N} \beta_k(P_k - Q_{Ck} - Q_{Dk})}$$

$$\leq \sum_{k=1}^{N} \int_{Ak} \ln \left[\frac{|A(f)|^2}{\beta_k} \right] df + \frac{1}{2\mu} \sum_{k=1}^{N} \beta_k(Q_{Ck} - P_{Ck}) \quad (53)$$

use having been made of inequality (49) and (50) in going from the first to the second inequality.

Now we need to find an upper bound on Q_{Ck} for a given value of P_{Ck}. The maximum value that

$$Q_{Ck} = 2 \int_{Ck} \frac{|A(f)|^2}{\beta_k} S_u(f) \, df \quad (54)$$

can take on, subject to the constraints

$$|A(f)|^2 S_u(f) \leq \mu \quad (55)$$

$$2 \int_{Ck} S_u(f) \, df = P_{Ck} \quad (56)$$

is obtained by maximizing $|A(f)|^2$ while using as little power as possible. This is achieved by putting all of P_{Ck} where $|A(f)|^2$ is the largest. Let

$$H_k(\alpha_k) = \{f : f \in C_k, |A(f)|^2 > \alpha_k\} \quad (57)$$

and let $\alpha_k \geq \beta_k$ be picked so that

$$2 \int_{H_k(\alpha_k)} \frac{\mu}{|A(f)|^2} df = P_{Ck}. \quad (58)$$

The quantity Q_{Ck} is then maximized by setting

$$S_u(f) |A(f)|^2 = \begin{cases} \mu & f \in H_k(\alpha_k) \\ 0 & f \in H_k^c(\alpha_k) \cap I_k \end{cases} \quad (59)$$

yielding

$$Q_{Ck} \leq 2 \int_{H_k(\alpha_k)} \left(\frac{\mu}{\beta_k} \right) df. \quad (60)$$

Thus for a fixed value of P_{Ck}

$$Q_{Ck} - P_{Ck} \leq 2 \int_{H_k(\alpha_k)} \left[\frac{\mu}{\beta_k} - \frac{\mu}{|A(f)|^2} \right] df$$

$$= \frac{2\mu}{\beta_k} \int_{H_k(\alpha_k)} \left(1 - \frac{\beta_k}{|A(f)|^2} \right) df$$

$$\leq -\frac{2\mu}{\beta_k} \int_{H_k(\alpha_k)} \ln \left[\frac{\beta_k}{|A(f)|^2} \right] df$$

$$\leq \frac{2\mu}{\beta_k} \int_{C_k} \ln \left[\frac{|A(f)|^2}{\beta_k} \right] df \quad (61)$$

Substituting from inequality (61) into inequality (53) and recalling that $F_k = A_k \cup C_k$, yields finally

$$R_a(\mu) \leq \sum_{k=1}^{N} \int_{F_k(\mu)} \ln \left[\frac{|A(f)|^2}{\beta_k} \right] df. \qquad \text{Q.E.D.} \quad (62)$$

Appendix II

Let the distortion for the unconstrained solution be expressed in the form

$$d = \beta_0 P. \quad (63)$$

Now suppose the power in the I_1 interval is less than the unconstrained value; i.e.,

$$P_1 = 2 \int_0^{f_1} S_u(f) \, df < \gamma P.$$

Then P_1 must be increased. Now for any power assignment P_1, $P_2 = P - P_1$, we know that the maximum rate is obtained with

$$\frac{P_1\beta_1}{l[F_1(\beta_1)]} = \frac{P_2\beta_2}{l[F_2(\beta_2)]} = \frac{(P - P_1)\beta_2}{l[F_2(\beta_2)]}. \quad (64)$$

Note from this equation that as P_1 is increased $\beta_1/l[F_1(\beta)]$ must decrease monotonically and $\beta_2/l[F_2(\beta_2)]$ must increase monotonically.

Since $l[F_k(\beta_k)]$ decreases monotonically with increasing β_k, β_1 decreases monotonically and β_2 increases monotonically as P_1 increases. Note also that for the unconstrained solution we have $S_u(f)$ constant on E, and hence $\beta_1 = \beta_2 = \beta_0$ for this case.

The distortion corresponding to (64) is

$$d = \beta_0 P = \beta_1 P_1 + \beta_2 P_2 = \beta_1 P_1 + (P - P_1)\beta_2. \quad (65)$$

Eliminating P_1 between (64) and (65) yields

$$\frac{l[F_1(\beta_1)]}{\beta_1}(\beta_0 - \beta_1) = \frac{l[F_2(\beta_2)]}{\beta_2}(\beta_2 - \beta_0). \quad (66)$$

Now consider $\beta_1'' < \beta_1' < \beta_0$, $\beta_2'' > \beta_2' > \beta_0$, corresponding to $P_1'' > P_1' > P_1$, with P_1 the unconstrained power in I_1. Writing (67) for β_k'' and for β_k' and subtracting the second equation from the first, yield after some manipulation

$$\Delta\beta_1 \frac{l[F_1(\beta_1'')]}{\beta_1''} + \Delta\beta_2 \frac{l[F_2(\beta_2'')]}{\beta_2''} = (\beta_0 - \beta_1')\left[\frac{l[F_1(\beta_1'')]}{\beta_1''} - \frac{l[F_1(\beta_1')]}{\beta_1'}\right] + (B_2' - \beta_0)\left[\frac{l[F_2(\beta_2')]}{\beta_2'} - \frac{l[F_2(\beta_2'')]}{\beta_2''}\right] > 0 \quad (67)$$

in which we have denoted

$$\Delta\beta_k = \beta_k'' - \beta_k'. \quad (68)$$

Now let $\Delta_1 = F_1(\beta_1'') - F_1(\beta_1')$ and $\Delta_2 = F_2(\beta_2') - F_2(\beta_2'')$ and consider the difference in the corresponding rates

$$\Delta R = R'' - R' = \int_{\Delta_1} \ln |A(f)|^2 \, df - \int_{\Delta_2} \ln |A(f)|^2 \, df + \ln\left(\frac{1}{\beta_1''}\right) l[F_1(\beta_1'')] - \ln\left(\frac{1}{\beta_1'}\right) l[F_1(\beta_1')] + \ln\left(\frac{1}{\beta_2''}\right) l[F_2(\beta_2'')] - \ln\left(\frac{1}{\beta_2'}\right) l[F_2(\beta_2')]. \quad (69)$$

However, on Δ_1, $\beta_1' < |A(f)|^2 \leq \beta_1''$, so that

$$\int_{\Delta_1} \ln |A(f)|^2 \, df \leq \ln \beta_1'' \{l[F_1(\beta_1'')] - l[F_1(\beta_1')]\} \quad (70)$$

and similarly

$$-\int_{\Delta_2} \ln |A(f)|^2 \, df \leq \ln \beta_2' \{l[F_2(\beta_2'')] - l[F_2(\beta_2')]\}. \quad (71)$$

Substituting inequalities (70) and (71) into (69) and cancelling terms yield,

$$\Delta R \leq l[F_1(\beta_1'')] \ln\left[\frac{(\beta_1'' - \Delta\beta_1)}{\beta_1''}\right] + l[F_2(\beta_2'')] \ln\left[\frac{(\beta_2'' - \Delta\beta_2)}{\beta_2''}\right] \quad (72)$$

Using the inequality

$$\ln [1 + x] \leq x$$

and inequality (67) yields finally

$$\Delta R \leq -\beta_1 \frac{l[F_1(\beta_1'')]}{\beta_1''} - \beta_2 \frac{l[F_2(\beta_2'')]}{\beta_2''} < 0. \quad (73)$$

Thus R decreases monotonically as P_1 increases from its unconstrained value. Q.E.D.

Acknowledgment

The author expresses grateful appreciation to one of the reviewers for pointing out the need for some form of ergodic or mixing property in applying the results of [1] to random process sources.

References

[1] D. J. Sakrison, "The rate distortion function for a class of sources," (to be published in *Inform. and Control*).
[2] M. Rosenblatt, *Random Processes.* New York: Oxford, 1962, p. 110.
[3] ——, "A central limit theorem and a strong mixing condition," *Proc. Natl. Acad. Sci.*, vol. 42, pp. 43–47, January 1956.
[4] D. J. Sakrison, "The rate distortion function of a Gaussian process with a weighted square error criterion," *IEEE Trans. Information Theory*, vol. IT-14, pp. 506–508, May 1968.
[5] ——, "Addendum to 'The rate distortion function of a Gaussian process with a weighted square-error criterion' ", *IEEE Trans. Information Theory* (Correspondence), vol. IT-15, pp. 610–611, September 1969.
[6] W. L. Root and T. S. Pitcher, "On the Fourier-series expansion of random functions," *Ann. of Math. Stat.*, vol. 26, pp. 313–318, June 1955.
[7] A. N. Kolmogorov, "On the Shannon theory of information transmission in the case of signals," *IRE Trans. Information Theory*, vol. IT-2, pp. 102–108, December 1956.
[8] M. S. Pinsker, "Calculation of the speed of information creation by means of stationary channel capacity" (in Russian), *Dokl. Akad. Nauk SSR*, vol. 3, pp. 753–756, 1956.
[9] R. L. Dobrushin and B. S. Tsybakov, "Information transmission with additional noise," *IRE Trans. Information Theory*, vol. IT-8, pp. 293–304, September 1962.

Information Rates of Wiener Processes

TOBY BERGER, MEMBER, IEEE

Abstract—Rate distortion functions are calculated for time discrete and time continuous Wiener processes with respect to the mean squared error criterion. In the time discrete case, we find the interesting result that, for $0 \leq D \leq \sigma^2/4$, $R(D)$ for the Wiener process is identical to $R(D)$ for the sequence of zero mean independent normally distributed increments of variance σ^2 whose partial sums form the Wiener process. In the time continuous case, we derive the explicit formula $R(D) = 2\sigma^2/(\pi^2 D)$, where σ^2 is the variance of the increment during a one-second interval. The resulting $R(D)$ curves are compared with the performance of an optimum integrating delta modulation system. Finally, by incorporating a delta modulation scheme in the random coding argument, we prove a source coding theorem that guarantees our $R(D)$ curves are physically significant for information transmission purposes even though Wiener processes are nonstationary.

I. Introduction

In this paper explicit expressions are derived for the mean squared error (MSE) rate distortion functions of time discrete and time continuous Wiener processes. The resulting $R(D)$ curves are compared with the performance of some suboptimum encoding schemes. Also, a source coding theorem is proved that imbues the $R(D)$ curves with physical significance despite the fact that Wiener processes are nonstationary.

Let $X = \{x_k, 0 \leq k < \infty\}$ be a time-discrete Wiener process. That is, $x_0 = 0$ and for $k > 0$ the increments $u_k = x_k - x_{k-1}$ are statistically independent identically distributed normal random variables with mean zero[1] and variance σ^2. Let $\mathbf{x} = (x_1, \cdots, x_n)$, let $\mathbf{y} = (y_1, \cdots, y_n)$ be any element of Euclidean n space, and let

$$\rho(\mathbf{x}, \mathbf{y}) = n^{-1} \sum_{k=1}^{n} (x_k - y_k)^2.$$

With every conditional probability measure, $q(\mathbf{y}|\mathbf{x})$, we associate an MSE value

$$\rho_q = E_q[\rho(\mathbf{x}, \mathbf{y})],$$

where E_q denotes joint expectation over $\mathbf{x}$ and $\mathbf{y}$ when q governs the transition from $\mathbf{x}$ to $\mathbf{y}$.

The MSE rate distortion function of X is defined by the prescription

$$R_X(D) = \lim_{n\to\infty} R_n(D),$$

where

$$R_n(D) = \inf_{q \in Q_D} E_q[n^{-1} I(\mathbf{x}; \mathbf{y})],$$

$$Q_D = \{q : \rho_q \leq D\},$$

and $I(\mathbf{x}; \mathbf{y})$ is the mutual information between $\mathbf{x}$ and $\mathbf{y}$. (For a rigorous definition of $I(\mathbf{x}; \mathbf{y})$ in the case of continuous variates, see Pinsker [1].)

Roughly speaking, $R_X(D)$ is the minimum rate[2] at which one must receive information about a realization of X in order to be able to reproduce it with an MSE per letter of D. Rigorous justification for interpreting $R(D)$ curves in this manner has been provided for independent letter sources by Shannon [2] and for certain classes of stationary sources by Pinsker [3], Goblick [4], Berger [5], and Gallager [6]. It is generally felt that $R(D)$ retains its physical significance for many nonstationary sources, too, although the proof of a coding theorem to this effect has yet to appear in the literature. In Section IV we extend the proof of the source coding theorem to encompass Wiener processes.

II. Calculation of $R(D)$ for Wiener Processes

Kolmogorov [7] reports the following parametric formula for $R_n(D)$ when $\mathbf{x}$ is an n-dimensional zero mean normally distributed random vector:[3]

$$D_\theta = \frac{1}{n} \sum_{k=1}^{n} \min(\theta, \lambda_k), \tag{1a}$$

$$R_n(D_\theta) = \frac{1}{n} \sum_{k=1}^{n} \max(0, \log \sqrt{\lambda_k/\theta}), \tag{1b}$$

where the λ_k are the eigenvalues of the correlation matrix

$$\Phi(i, j) = E(x_i x_j) \qquad 1 \leq i, j \leq n.$$

For the Wiener process

$$\Phi(i, j) = \sigma^2 \min(i, j),$$

and we show in the Appendix that

$$\lambda_k = \sigma^2 \Big/ \left[4 \sin^2\left(\frac{2k-1}{2n+1} \cdot \frac{\pi}{2}\right)\right] \qquad k = 1, \cdots, n. \tag{2}$$

Thus $R_n(D)$ is determined. To compute $R_X(D)$, first observe that

$$\lambda(f) \triangleq \lim_{\substack{k,n\to\infty \\ k/n \to f}} \lambda_k = \frac{\sigma^2}{4 \sin^2(\pi f/2)}. \tag{3}$$

Manuscript received January 24, 1969; revised July 23, 1969. This paper was presented at the International Symposium on Information Theory, Ellenville, N. Y., January 28–31, 1969.

The author is with the School of Electrical Engineering, Cornell University, Ithaca, N. Y. 14850.

[1] The $R(D)$ curves derived in Section II apply equally well to cases in which the mean is any deterministic function. Included in particular are time discrete and time continuous Wiener processes with drift.

[2] We measure $R(D)$ in nats/s. That is, we use natural logarithms in the definition of $I(\mathbf{x}; \mathbf{y})$ and assume that X produces one x_k each second.

[3] Kolmogorov calls $R_n(\epsilon^2)$ the ϵ entropy of $\mathbf{x}$ and denotes it by $H_\epsilon(\mathbf{x})$. For a detailed derivation of (1), see Berger [8], Gallager [6].

Reprinted from *IEEE Trans. Inform. Theory*, vol. IT-16, pp. 134–139, Mar. 1970.

It follows upon taking limits in (1) that $R_X(D)$ is given parametrically by

$$D_\theta = \int_0^1 \min\left[\theta, \frac{\sigma^2}{4\sin^2(\pi f/2)}\right] df, \tag{4a}$$

$$R_X(D_\theta) = \int_0^1 \max\left[0, \log\left(\frac{\sigma}{2\sqrt{\theta}\sin(\pi f/2)}\right)\right] df. \tag{4b}$$

For $\theta \leq \sigma^2/4$, (4) reduces to

$$D_\theta = \theta,$$

$$R_X(D_\theta) = \log\left(\frac{\sigma}{2\sqrt{\theta}}\right) - \int_0^1 \log\left[\sin\left(\frac{\pi f}{2}\right)\right] df$$

$$= \log\left(\frac{\sigma}{2\sqrt{\theta}}\right) + \log 2,$$

and eliminating θ yields the explicit result

$$R_X(D) = \left(\frac{1}{2}\right)\log\left(\frac{\sigma^2}{D}\right) \qquad 0 \leq D \leq \sigma^2/4. \tag{5}$$

For $\theta > \sigma^2/4$, performing the integrations in (4) gives

$$D_\theta = \left(\frac{1}{\pi}\right)\left[2\theta\sin^{-1}\left(\frac{\sigma}{2\sqrt{\theta}}\right) + \left(\frac{\sigma}{2}\right)(4\theta - \sigma^2)^{1/2}\right], \tag{6a}$$

$$R_X(D_\theta) = \left(\frac{1}{\pi}\right)\left[Cl_2\left(2\sin^{-1}\left(\frac{\sigma}{2\sqrt{\theta}}\right)\right.\right.$$

$$\left.\left. + \sin^{-1}\left(\frac{\sigma}{2\sqrt{\theta}}\right)\log\left(\frac{\sigma^2}{\theta}\right)\right], \tag{6b}$$

where

$$Cl_2(x) = -\int_0^x \log\left[2\sin\left(\frac{t}{2}\right)\right] dt$$

is Clausen's (second) integral, which has been tabulated extensively [9]–[11]. The $R_X(D)$ curve, as given by (5) and (6), is shown in Fig. 1. (Differentiation reveals that R_X' and R_X'' are continuous at $D = \sigma^2/4$, but R_X''' is not.) Asymptotic expansion of (6) for large θ (large D) yields

$$R_X(D) \sim 2\sigma^2/(\pi^2 D), \tag{7}$$

which also is depicted in Fig. 1.

It is interesting to note that (5) is identical to Shannon's [2] formula for the rate distortion function $R_U(D)$ of the independent letter source $U = \{u_k\} = \{x_k - x_{k-1}\}$. The only difference is that the formula $R_U(D) = (\frac{1}{2})\log(\sigma^2/D)$ continues to hold until $D = \sigma^2$. Thus, (5) implies that for $D \leq \sigma^2/4$, the information required to reproduce X with fidelity D is no greater than that required to reproduce U with fidelity D. This rather surprising result clearly cannot continue to hold for large D. Indeed $R_U(D) = 0$ for $D \geq \sigma^2$, since guessing that each u_k is zero yields an MSE per letter of σ^2. For X, the best guess is that each x_k is zero, too, but the resulting MSE per letter is

$$\lim_{n\to\infty} n^{-1}\sum_{k=1}^{n}\overline{x_k^2} = \lim_{n\to\infty} n^{-1}\sum_{k=1}^{n} k\sigma^2 = \lim_{n\to\infty}\sigma^2(n+1)/2 = \infty.$$

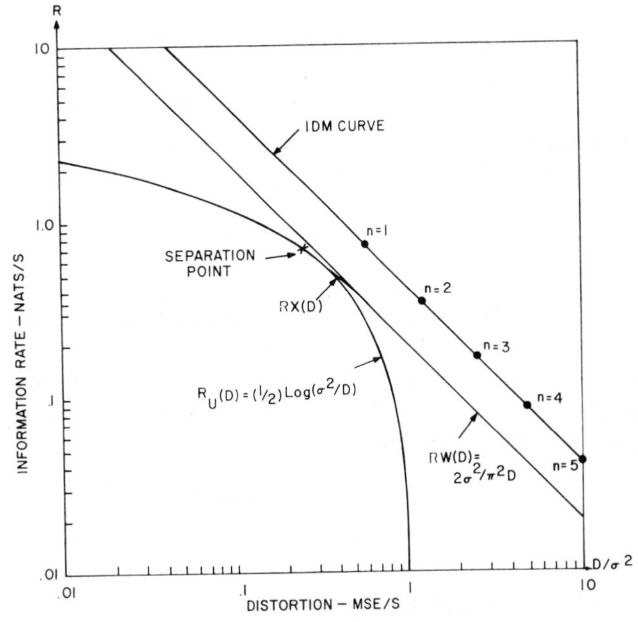

Fig. 1. $R(D)$ curves for X, U, and W.

Thus $R_X(D)$ never reaches zero for any finite D, as the asymptotic behavior given by (7) indicates.

The following discussion provides some insight into why $R_U(D)$ and $R_X(D)$ coincide for $D \leq \sigma^2/4$. The MSE rate-distortion function $R_Z(D)$ of an arbitrary zero-mean *stationary* Gaussian sequence

$$Z = \{z_k, k = 0, \pm 1, \pm 2, \cdots\}$$

is known [12] to possess a lower bound of the form

$$\frac{1}{2}\log\left(\frac{Q_1}{D}\right) \leq R_Z(D).$$

The constant Q_1, called the entropy power of Z, is given by

$$Q_1 = \exp\left(\int_0^1 \log Z(f)\, df\right), \tag{8}$$

where $Z(f)$ is the spectrum of the random sequence Z, namely

$$Z(f) = \sum_{n=-\infty}^{\infty} \zeta_n e^{-j2\pi nf} \qquad \zeta_n = \overline{z_k z_{k+n}}.$$

Moreover, the lower bound is tight for all $D \leq \inf_f Z(f)$,

$$R_Z(D) = \frac{1}{2}\log\left(\frac{Q_1}{D}\right) \qquad D \leq \inf_f Z(f). \tag{9}$$

The next step is to recognize that Q_1 of (8) also is the "one-step prediction error" of Z, i.e., the least MSE attainable in estimating z_n given the infinite past up to and including z_{n-1} (see [13], ch. 10). Now, it is customary to interpret $\lambda(f)$ of (3) as the spectrum of X despite the fact that X is nonstationary. Since the one-step prediction error of X clearly is $Q_1 = \sigma^2$ and

$$\inf_f \lambda(f) = \frac{\sigma^2}{4},$$

(9) provides a mathematical basis for anticipating the result

$$R_X(D) = \frac{1}{2} \log \left(\frac{\sigma^2}{D}\right) \qquad 0 \le D \le \sigma^2/4.$$

For time-continuous stationary normal processes, Kolmogorov [7] shows that $R(D)$ for the criterion of minimum MSE per unit time is given parametrically by

$$D_\theta = \int_0^\infty \min [\theta, \lambda(f)]\, df, \tag{9a}$$

$$R(D_\theta) = \int_0^\infty \max [0, \ln (\sqrt{\lambda(f)/\theta})]\, df, \tag{9b}$$

where $\lambda(f)$ is the power spectrum of the process. It is easy to show [17] that (9) holds for nonstationary procesess, too, provided we define

$$\lambda(f) \triangleq \lim_{\substack{k, b-a \to \infty \\ k/(b-a) \to f}} \lambda_k(a, b)$$

where the $\lambda_k(a, b)$ are the eigenvalues of the Karhunen–Loève (K–L) integral equation

$$\lambda f(t) = \int_a^b \Phi(t, s) f(s)\, ds.$$

In the case of a time continuous Wiener process, $W = \{w_t, 0 \le t < \infty\}$, we have $\Phi(t, s) = E(w_t w_s) = \sigma^2 \min (t, s)$, and we show in the Appendix that

$$\lambda_k(0, T) = \left[\frac{2\sigma T}{(2k-1)\pi}\right]^2 \qquad 1 \le k < \infty. \tag{10}$$

Hence,

$$\lambda(f) = \lim_{\substack{k, T \to \infty \\ k/T \to f}} \lambda_k(0, T) = \left(\frac{\sigma}{\pi f}\right)^2. \tag{11}$$

Substituting (11) into (9) and integrating gives $D_\theta = 2\sigma \sqrt{\theta}/\pi$ and $R_W(D_\theta) = \sigma/(\pi\sqrt{\theta})$. Upon eliminating θ we find that[4]

$$R_W(D) = 2\sigma^2/(\pi^2 D) = 0.203\sigma^2/D \text{ nats/s}. \tag{12}$$

Note that (12) is identical to (7), which gives the asymptotic behavior of $R_X(D)$ for large D (see Fig. 1). Comparing (12) with (5) reveals that as D tends to zero $R_W(D)$ diverges as D^{-1} whereas $R_X(D)$ diverges only as $\log (D^{-1})$. This fact can be used to adjust intelligently the balance between interpolation error and quantization error in digital communication systems designed to transmit realizations of W with small MSE using closely spaced time samples as the data base.

III. Comparison of $R_x(D)$ and $R_W(D)$ With the Performance of Suboptimal Systems

Because the variance of the x_k increases linearly with k, any fixed bit rate PCM system suffers unbounded MSE as the length of the sequence to be reproduced increases indefinitely. However, it is possible to meaningfully compare $R_X(D)$ with the performance of the delta modulation system depicted in Fig. 2. If the receiver of this system is constrained to be a summer, then the optimum linear operator for the transmitter feedback loop also is a summer [14], and the system is called an integrating delta modulator (IDM). Moreover, the quantizer outputs s_k are equally likely to be plus or minus and are nearly independent of one another when the loop gain α is chosen optimally. One therefore needs a channel with a capacity of at least one bit per letter to transmit the sequence $\{s_k\}$ without error. Using the Wiener–Hopf technique, Fine [15] has shown that the asymptotic MSE for the IDM with X as input is

$$D \triangleq \lim_{k\to\infty} \overline{(r_k - x_k)^2} = \frac{\alpha^2}{3} + 2 \sum_{n=1}^{\infty} [-\alpha\sigma\sqrt{n/2\pi} \cdot \exp(-n\alpha^2/2\sigma^2) + (\sigma^2 + n\alpha^2) \operatorname{cerf}(\alpha\sqrt{n}/\sigma)]$$

where

$$\operatorname{cerf}(x) = \sqrt{1/2\pi} \int_x^\infty e^{-\beta^2/2}\, d\beta.$$

Calculation reveals that $\alpha = 1.045\sigma$ yields the minimum D, namely $0.585\sigma^2$. Thus, an optimized IDM realizes the point ($R = 1$ bit $= 0.693$ nat, $D/\sigma^2 = 0.585$) plotted in Fig. 1. Letting the IDM operate only once every n letters yields the sequence of points indexed by n on the IDM curve. Increasing the number of levels at the quantizer output yields smaller values of D/σ^2 at the expense of logarithmically increasing R. Since mathematical techniques for analyzing such multilevel systems are not fully developed at present, the exact (R, D) coordinates of the points associated with them are unknown. However, a preliminary analysis by Gish [16] suggests that these points lie on a curve that essentially parallels $R_X(D)$.

We can compare $R_W(D)$ of (12) with the IDM of Fig. 2 by interpreting the x_k therein as samples of $\{w_t\}$ taken every τ seconds. From our earlier remarks we see that, with the loop gain optimized at $\alpha = 1.045\ \sigma\sqrt{\tau}$, an IDM realizes the point ($R = 0.693/\tau$, $D/\sigma^2 = 0.585\tau$) in the plane of Fig. 1. Eliminating τ yields the IDM performance curve $R = 0.406\ \sigma^2/D$, which is poorer than $R_W(D)$ by roughly a factor of 2. Of course the IDM also must interpolate between the samples in order to reconstruct a realization of W. It can be shown [17] that linear interpolation, which is ideal for W when the endpoints are known exactly, increases the scale factor 0.406 to approximately 0.430.

Recently, Posner *et al.* [18] have reported that if the nth K–L expansion coefficient is quantized into uniform intervals of length ϵ_n such that $\sum_n \epsilon_n^2 \le D$, then the combined output entropy of the quantizers is at least $6.711\ \sigma^2/D$. Note that reconstructing W from the quantized K–L coefficients and the K–L eigenfunctions yields a MSE smaller than $\sum_n \epsilon_n^2$ since one must average over the distribution of the K–L coefficients within each quantization interval. Relaxation of the uniform quantization restriction allegedly reduces the constant 6.711 to

[4] Kolmogorov [7] attributes to Iaglom an equation for the asymptotic form of $R_W(D)$ for small D over a finite interval $[0, T]$. This equation differs from (12) by a factor of 2π that probably stems from the way the L_2 metric is defined.

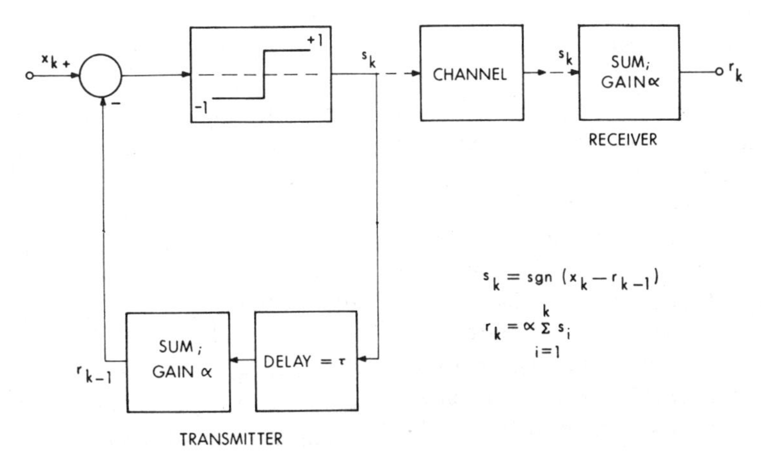

Fig. 2. Block diagram of a delta modulation system.

some as yet undetermined number between 17/32 and 1[18]. These results suggest that conversion to an orthogonal expansion representation is the critical step in efficient encoding of Wiener processes. Subsequent block encoding of the expansion coefficients in order to approach $R(D)$ affords only minor improvement over appropriate quantization techniques at the expense of a much more complex encoder.

IV. Source Coding Theorem for Wiener Processes

Investigations in the theory of source encoding usually proceed as follows. Given a real random process $Z = \{z_t, -\infty < t < \infty\}$, one first defines a rate distortion function $R_Z(\cdot)$ with respect to an appropriate distortion measure (e.g., MSE) and then attempts to prove the following source coding theorem.

> Given $D \geq 0$ and $\epsilon > 0$ there exists a $T > 0$ and a mapping $f(\cdot)$ from the set of possible realizations of the finite-time process $\{z_t, 0 \leq t \leq T\}$ into a set S that contains less than $\exp\{T[R(D) + \epsilon]\}$ elements such that the average distortion between the realizations and their images under $f(\cdot)$ does not exceed $D + \epsilon$.

For stationary sources, the validity of the above theorem implies that arbitrarily long segments of the source output can be reproduced with fidelity $D + \epsilon$ on the basis of information supplied at the rate of $R(D) + \epsilon$ nats/s. One need merely divide the long segment into successive intervals of duration T, and then represent the source output in each such interval by its image in S under $f(\cdot)$.

For *nonstationary* sources, the situation is somewhat more complicated. The set S and the mapping $f(\cdot)$ may have to differ for successive intervals of length T. In general what is needed is a scheme for mapping the source output over the interval from kT to $(k + 1)T$ into one of J_k possible preselected messages, where

$$J_k < \exp\{T[R(D) + \epsilon]\} \qquad k = 0, 1, \cdots,$$

in such a way that one can reconstruct the source output for *all* time with fidelity $D + \epsilon$ given only the infinite sequence $\{n_k\}$ of successive message numbers.

For a source with independent increments, such as the Wiener process W, it is natural to consider encoding the changes in the source output over successive intervals of length T. Since the increment histories $\{w_t - w_{kT}, kT < t \leq (k + 1)T\}$ are statistically independent and identically distributed for different values of k, the canonical time histories represented by the J messages into which they are encoded need not vary with k. The only difficulty is that specification of the increment histories with a MSE per second of $D + \epsilon$ is not in itself sufficient to permit reconstruction of W for all time with fidelity $D + \epsilon$. It is also necessary to specify accurately the absolute source output at the end of each T-second interval. The reason is that when one adds the estimate of the new increment history to the previous endpoint estimate, any error in the estimate of the endpoint becomes an error per unit time over the entire succeeding interval of length T. Such errors are cumulative, so unbounded distortion eventually results.

Let us consider how much information must be supplied to specify the endpoints. Since the kth endpoint w_{kT} is a zero-mean normal random variable with variance $k\sigma^2 T$, at least $(\frac{1}{2}) \log (k\sigma^2 T/\epsilon)$ nats of information are required to specify it with a MSE of ϵ. (This is because $R(D) = (\frac{1}{2}) \log (\sigma^2/D)$ for $\mathfrak{N}(0, \sigma^2)$ random variables, which implies that $(\frac{1}{2}) \log (k\sigma^2 T/\epsilon)$ is the minimum possible mutual information between w_{kT} and any estimate of it that achieves a MSE of ϵ.) Hence, even though an endpoint need be specified only once every T seconds, the information per second required to specify the kth one, namely $(\frac{1}{2}T) \log (k\sigma^2 T/\epsilon)$, diverges with k regardless of how large we choose T. This difficulty can be circumvented by taking advantage of the dependence between successive endpoints. Each endpoint is obtained from the previous one by the addition of an $\mathfrak{N}(0, \sigma^2 T)$ increment, so specifying each of these net increments with an MSE of ϵ requires only $(\frac{1}{2}T) \log (\sigma^2 T/\epsilon)$ nats/s, which becomes negligible as $T \to \infty$. However, with such a scheme successive endpoint errors are cumulative, so the MSE of the estimate of the kth endpoint will be $k\epsilon$. Naturally, this is unacceptable for large k regardless of how small ϵ is chosen. The proper way to track the endpoints is by a scheme based on the IDM of Fig. 2.

Theorem

Given any $\epsilon > 0$ the infinite sequence of endpoints $\{w_{kT}, k = 0, 1, \cdots\}$ can be reproduced with an MSE per symbol of ϵ on the basis of less than ϵ nats/s of information by choosing T large enough.

Proof: Consider an IDM that tracks samples of the Wiener process W taken every τ seconds. As noted in Section III, the IDM achieves an asymptotic MSE of $0.585\ \sigma^2\tau$ per sample when the optimum loop gain of $\alpha = 1.045\sigma\sqrt{\tau}$ is used. Hence, if we choose $0 < \tau < \epsilon/0.585\sigma^2$ and let the IDM run for n samples, its estimate of the nth sample will have a MSE less than ϵ for large enough n. Let $T = n\tau$ for such a τ and n. Since the IDM forms its estimate of each sample by either adding or subtracting α from its estimate of the previous sample, its estimate of the nth sample must be a multiple of α that lies between $-n\alpha$ and $+n\alpha$. Therefore, only $(n\tau)^{-1} \log (2n + 1)$ nats/s are needed to specify the endpoints with an MSE of ϵ. By choosing n (hence T) large enough, we can make $(n\tau)^{-1} \log (2n + 1) < \epsilon$ as was to be shown.

The above theorem leads directly to a proof of the source coding theorem for the Wiener process W. We simply run the IDM in parallel with the K–L expansion coefficient calculator. The proof that the incremental history of the source output over a T-second interval (as represented by the K–L expansion coefficients) can be mapped into a set containing $J < \exp \{T[R(D) + \epsilon]\}$ messages with a resulting MSE per second of less than $D + \epsilon$ is a special case of a more general theorem by the author [5].[5] A code for W then can be constructed with $J' = (2n + 1)J$ codewords, where the factor $2n + 1$ stems from also encoding the IDM estimate of the net increment over the interval of duration $T = n\tau$ seconds. In decoding, or reconstruction, the next increment history estimate is added to the endpoint estimated by the IDM rather than to the value reconstructed at the end of the previous increment history. Although this produces discontinuities in the reconstructed process at intervals of T seconds, it prevents the reconstructed waveform from drifting away from the one actually produced by the source. From the theorem about the endpoints proven above we see that the code in question signals at the rate $T^{-1} \log J' \leq (n\tau)^{-1} \log (2n + 1) + T^{-1} \log J < R(D) + 2\epsilon$ for large enough T. Since the endpoint and increment history errors are additive, the MSE per second achieved by this coding scheme is bounded above by $D + 2\epsilon$ and does not diverge with time. This completes the proof of the source coding theorem for W.

The above results can be extended to a rather broad class of both time discrete and time continuous processes with independent increments and fidelity criteria other than MSE. A separate correspondence devoted to the proof of a source coding theorem to this effect is presently being prepared.

Appendix

Orthogonal Expansion of Wiener Processes

We treat the time continuous Wiener process first. The K–L integral equation on $[0, T]$ with $\Phi(t, s) = \sigma^2 \min (t, s)$ reads

[5] Take the abstract alphabet X_0 of [5] to be the uncountable Cartesian product of real lines swept out by the time parameter during an interval of length T. The sequence of incremental histories over successive T-second intervals then constitutes a time discrete *stationary* source that produces one output from this alphabet every T seconds.

$$\lambda f(t) = \sigma^2 \left(\int_0^t s d(s)\, ds + t \int_t^T f(s)\, ds \right). \tag{13}$$

Differentiating (13) twice yields

$$\lambda f''(t) + \sigma^2 f(t) = 0. \tag{14}$$

Upon applying the boundary conditions

$$f(0) = 0 \quad \text{and} \quad f'(T) = 0 \tag{15}$$

we find that

$$f_k(t) = A \sin [(2k - 1)(\pi t/2T)], \tag{16}$$

where $A = \sqrt{2/T}$ for orthonormality. The eigenvalues given by (10) follow directly from (14) and (16).

For the time-discrete case, the equations analogous to (13)–(16) are, respectively,

$$\lambda f(i) = \sigma^2 \sum_{j=0}^{i-1} jf(j) + i \sum_{j=i}^{n} f(j), \tag{13'}$$

$$\lambda[f(i + 2) - 2f(i + 1) + f(i)] + \sigma^2 f(i + 1) = 0, \tag{14'}$$

$$f(0) = 0 \qquad \lambda[f(n) - f(n - 1)] = \sigma^2 f(n), \tag{15'}$$

$$f_k(i) = A \sin \left[\left(\frac{2k - 1}{2n + 1} \right) \pi i \right]. \tag{16'}$$

Substituting (16′) into (14′), replacing i by $l = i + 1$, and expanding terms of the form $\sin (a \pm b)$ leads to the eigenvalues given by (2).

References

[1] M. S. Pinsker, *Information and Information Stability of Random Variables and Processes.* Moscow: Izd. Akad. Nauk. USSR, 1960. (English transl.). San Francisco: Holden-Day, 1964.
[2] C. E. Shannon, "Coding theorems for a discrete source with a fidelity criterion," *IRE Natl. Conv. Rec.*, pt. 4, pp. 142–163, 1959; (also in *Information and Decision Processes*, R. E. Machol, Ed. New York: McGraw-Hill, 1960, pp. 93–126.)
[3] M. S. Pinsker, "Sources of messages," *Probl. Peredachi Inform.*, vol. 14, pp. 5–20, 1963.
[4] T. J. Goblick, Jr., "A coding theorem for time discrete analog data sources," presented at IEEE Internatl. Symp. on Information Theory, San Remo, Italy, September 1967.
[5] T. Berger, "Rate distortion theory for sources with abstract alphabets and memory," *Inform. and Control*, vol. 13, pp. 254–273, September 1968.
[6] R. G. Gallager, *Information Theory and Reliable Communication.* New York: Wiley, 1968.
[7] A. N. Kolmogorov, "On the Shannon theory of information transmission in the case of continuous signals," *IRE Trans. Information Theory*, vol. IT-2, pp. 102–108, December 1956.
[8] T. Berger, "Nyquist's problem in data transmission," Ph.D. dissertation, Harvard University, Cambridge, Mass., Div. of Engrg. and Appl. Phys., December 1965.
[9] L. Lewin, *Dilogarithms and Associated Functions.* London: Macdonald, 1958.
[10] M. Abramowitz and I. A. Stegun, *Handbook of Mathematical Functions*, NBS Appl. Math. Ser. 55. Washington, D. C.: U. S. Government Printing Office, 1964.
[11] Amsterdam Mathematical Centre, "Polylogarithms," Rept. R24, pt. 1, 1954.
[12] T. Berger, *Rate Distortion Theory: A Mathematical Basis for Data Compression.* Englewood Cliffs, N. J.: Prentice-Hall, (in preparation).
[13] U. Grenander and G. Szegö, *Toeplitz Forms and their Applications.* Berkeley and Los Angeles: University of California Press, 1958.
[14] T. Fine, "Properties of an optimum digital system and applications," *IEEE Trans. Information Theory*, vol. IT-10, pp. 287–296, October 1964.
[15] ——, "The response of a particular nonlinear system with feedback to each of two random processes," *IEEE Trans. Information Theory*, vol. IT-14, pp. 255–264, March 1968.
[16] H. Gish, private communication.
[17] T. Berger, "Rate distortion functions of nonstationary Gaussian processes with applications to diffusion," Raytheon Company, Wayland, Mass., Tech. Memo. TB-63, November 1967.
[18] E. C. Posner, E. R. Rodemich, and H. Rumsey, Jr., "Product entropy of Gaussian distributions," unpublished memo., Jet Propulsion Laboratory, Pasadena, Calif., December 1967.

Transmission of Noisy Information to a Noisy Receiver With Minimum Distortion

JACK K. WOLF, MEMBER, IEEE, AND JACOB ZIV, ASSOCIATE MEMBER, IEEE

Abstract—**This paper is concerned with the transmission of information with a fidelity criterion where the source output may be distorted prior to encoding and, furthermore, where the output of the decoder may be distorted prior to its delivery to the final destination. The criterion for optimality is that the normalized average of the squared norm of the difference between the T-second undistorted source sample and the corresponding T-second sample delivered to the final destination be minimum. The optimal structure of the encoder and decoder is derived for any T.**

I. Introduction

THE classical problem of information transmission with a fidelity criterion [1] assumes that one can directly encode the output of a source for transmission over a noisy channel and, furthermore, that the decoded output of the channel is delivered without further distortion to the final destination (information sink). This paper is concerned with a generalization of this model, first discussed by Dobrushin and Tsybakov [2]. In this generalized model, the source output may be distorted prior to encoding, and furthermore, the output of the decoder may be distorted prior to its delivery to the final destination.

Reasons for considering this generalized model are detailed by Dobrushin and Tsybakov. In brief, some practical situations where this model applies are as follows.

1) The distortions at the source or at the receiver, or both, may be due to uncoded transmission via communication channels over which the communication system designer has no control. For example, a central communication link may service customers who furnish

Manuscript received September 9, 1969; revised January 26, 1970.
J. K. Wolf was with Bell Telephone Laboratories, Inc., Murray Hill, N. J. He is now with the Department of Electrical Engineering, Polytechnic Institute of Brooklyn, Brooklyn, N. Y. 11201.
J. Ziv was with Bell Telephone Laboratories, Inc., Murray Hill, N. J. He is now with Technion, Israel Institute of Technology, Haifa, Israel.

Reprinted from *IEEE Trans. Inform. Theory*, vol. IT-16, pp. 406–411, July 1970.

their own communication links to carry information to and from this central link.

2) The information to be transmitted may be measured data that are affected by measurement errors.

3) Digital processing may be used at the transmitter or receiver, or both, and such processing introduces quantizations and round-off errors.

The problem considered in this paper is the optimal design of an encoder–decoder for transmitting a T-second distorted version of a T-second output of a source to a receiver that also introduces distortion after decoding. The criterion for optimality is that the normalized average of the squared norm of the difference between the T-second undistorted source sample and the corresponding T-second distorted output sample of the decoder be minimum. Dobrushin and Tsybakov gave results for the case where $T \to \infty$ and, in particular, proved a coding theorem and a converse. Our results apply for all T. Furthermore, we show that a structure proposed by Dobrushin and Tsybakov is optimal under very general conditions; Dobrushin and Tsybakov proved the optimality of this structure only in one special case.

Section II of this paper gives a detailed formulation of the problem, a statement of the main result, and interpretation of this result. Section III contains examples of the application of this result. The proof of the main result is given in the Appendix.

II. Statement of Problem and Principal Results

In Section II-A we define a "mapping" (sometimes called a channel) in a general way. We do this since the concept of a mapping occurs in several ways in our discussion of a communication system. We then discuss the concept of a "cascade of mappings."

In Section II-B we describe the communication system that we shall consider in terms of a cascade of mappings. In this section we also formulate the problem of communication with a "fidelity criterion."

In Section II-C we state the main result of this paper concerning the transmission of information with respect to a particular fidelity criterion. The remainder of this section contains interpretations of this result.

A. Mapping

A "w to z mapping" is defined as follows. For every $T > 0$ we have a set W_T of allowable T-second "inputs" and a set Z_T of possible T-second "outputs." Every T seconds, some $\mathbf{w} \in W_T$ is mapped stochastically into an output $\mathbf{z} \in Z_T$. This probabilistic mapping is governed by a probability measure $\mu_{\mathbf{w}}$ on the set Z_T. That is, given an input $\mathbf{w} \in W_T$, the probability that $\mathbf{z} \in B$ (where B is a measurable subset of Z_T) is $\mu_{\mathbf{w}}(B)$. The mapping is said to be "deterministic" if $\mu_{\mathbf{w}}(B)$ is either 0 or 1 for all B and all $\mathbf{w} \in W_T$.

A cascade of mappings is a sequence of mappings where first $\mathbf{w} \in W_T$ is mapped into $\mathbf{z} \in Z_T$, then $\mathbf{z} \in Z_T$ is mapped into $\mathbf{q} \in Q_T$, etc. Furthermore, it is assumed for such a cascade that $\mu_{\mathbf{w},\mathbf{z}}(A) = \mu_{\mathbf{z}}(A)$ where A is a measurable subset of Q_T. That is, the probability that $\mathbf{q} \in A$, given $\mathbf{w}$ and $\mathbf{z}$, depends only on $\mathbf{z}$ and not on $\mathbf{w}$ (see Fig. 1). It is also assumed that the output space Z_T of the first mapping is a subset of the allowable input space for the second mapping.

B. Communication With a Fidelity Criterion

Consider the communication system given in Fig. 2. Blocks 2 through 6 in this diagram are mappings as defined in the previous section. Block 1, the source, is a trivial mapping in that in the time interval $(0, T)$, $T > 0$, it is the generator for $\mathbf{x}$, an element of some measurable sample space X_T. Similarly, block 7, the final destination, is a trivial mapping in that it acts as a sink for $\hat{\mathbf{x}}$, an element of the same measurable sample space X_T.

All blocks, with the important exceptions of blocks 3 and 5, are assumed to be "uncontrollable mappings," mappings that cannot be affected by the designer of the communication system. Blocks 3 and 5 are termed controllable mappings and are deterministic.

The problem of the communication system designer is to specify the deterministic mappings associated with the encoder and decoder so as to optimize some fidelity criterion of transmission to be discussed subsequently. The chain of blocks in Fig. 2 represents a cascade of mappings so that the output space for the ith mapping in the chain must be a subset of the allowable input space for the $(i + 1)$th mapping. For example, the input space to the communication channel often has particular restrictions associated with it, perhaps a power constraint, and this fact must be taken into account in the design of the encoder.

Ideally we should like to choose our controllable mappings so that $\mathbf{x} \equiv \hat{\mathbf{x}}$, but usually this cannot be achieved owing to the distortions introduced by the uncontrollable mappings. As a measure of the goodness of the design of the controllable mappings we adopt a distortion function $d(\mathbf{x}, \hat{\mathbf{x}})$, which tells how well $\hat{\mathbf{x}}$ approximates $\mathbf{x}$. In this paper this distortion function is

$$d_T(\mathbf{x}, \hat{\mathbf{x}}) = \frac{1}{T} (\hat{\mathbf{x}} - \mathbf{x}, \hat{\mathbf{x}} - \mathbf{x}) = \frac{1}{T} ||\hat{\mathbf{x}} - \mathbf{x}||^2 \tag{1}$$

where $(\mathbf{a}, \mathbf{b})$ denotes a suitable inner product and where $|| \; ||$ denotes the corresponding norm. Thus we assume that $\mathbf{x}$ and $\hat{\mathbf{x}}$ are elements of a separable Hilbert space of random vectors. The system distortion measure that we seek to minimize is

$$D_T = E \, d_T(\mathbf{x}, \hat{\mathbf{x}}) \tag{2}$$

where E denotes statistical expectation. For a given T, let D_T^* denote the infimum of D_T with respect to all allowable encoder–decoder pairs.

Explicit formulas are given for this distortion measure for three different types of message sources.

1) If X_T and $\hat{X}_T$ are L_2 spaces defined over the interval $0 \le t \le T$, then $(\mathbf{a}, \mathbf{b}) = \int_0^T a(t)b(t)dt$ and

$$D_T = E \frac{1}{T} \int_0^T (\hat{x}(t) - x(t))^2 \, dt. \tag{3}$$

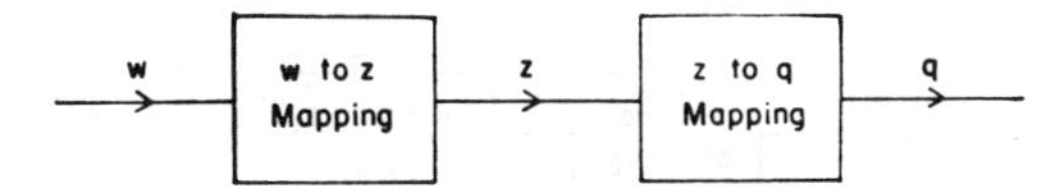

Fig. 1. Cascade of mappings.

2) If X_T and $\hat{X}_T$ are N-dimensional Euclidean spaces of vectors $\mathbf{x} = (x_1, x_2, \cdots, x_N)$, $\hat{\mathbf{x}} = (\hat{x}_1, \hat{x}_2, \cdots, \hat{x}_N)$, then $(\mathbf{a}, \mathbf{b}) = \sum_{i=1}^{N} a_i b_i$ and

$$D_T = E \frac{\rho_s}{N} \sum_{i=1}^{N} (\hat{x}_i - x_i)^2, \tag{4}$$

where $N = \rho_s T$.

3) If X_T and $\hat{X}_T$ are N-dimensional binary spaces of vectors $\mathbf{x} = (x_1, x_2, \cdots, x_N)$ and $\hat{\mathbf{x}} = (\hat{x}_1, \hat{x}_2, \cdots, \hat{x}_N)$ where $x_i, \hat{x}_i \in [0, 1]$, then again $(\mathbf{a}, \mathbf{b}) = \sum_{i=1}^{N} a_i b_i$ and

$$D_T = E \frac{\rho_s}{N} \sum_{i=1}^{N} (\hat{x}_i - x_i)^2 = E \frac{\rho_s}{N} \sum_{i=1}^{N} d_H(x_i, \hat{x}_i) \tag{5}$$

where

$$d_H(x_i, \hat{x}_i) = \begin{cases} 0 & x_i = \hat{x}_i \\ 1 & x_i \neq \hat{x}_i \end{cases}.$$

In this case, $(N/\rho_s)\, d_T(\mathbf{x}, \hat{\mathbf{x}})$ is the Hamming distance [3] between the vectors $\mathbf{x}$ and $\hat{\mathbf{x}}$ and $\rho_s^{-1} E\, d_T(\mathbf{x}, \hat{\mathbf{x}})$ is the average probability of error per bit upon comparing $\mathbf{x}$ and $\hat{\mathbf{x}}$.

It should be noted that D_T^* is related to the minimum probability of error per letter for a source of M letters, where these letters are represented as $(M - 1)$ dimensional vectors that take on values corresponding to the vertices of a simplex (in $M - 1$ dimensions) with unit distance between vertices. The binary source discussed above is a special case of this representation.

C. Statement of Main Result and Some Observations

Let $\mathbf{\mu} = E_s \mathbf{x}$ and let $\mathbf{v} = E_{\hat{y}} \mathbf{\mu}$ (see Fig. 2) where $E_a \mathbf{z}$ denotes the conditional expectation of $\mathbf{z}$ given $\mathbf{a}$. The main result of this paper can then be stated as the following theorem.

Theorem: For any given allowable encoder–decoder pair

$$D_T = E \frac{1}{T} ||\mathbf{\mu} - \mathbf{x}||^2 + E \frac{1}{T} ||\mathbf{v} - \mathbf{\mu}||^2 + E \frac{1}{T} ||\hat{\mathbf{x}} - \mathbf{v}||^2. \tag{6}$$

Hence

$$D_T^* = E \frac{1}{T} ||\mathbf{\mu} - \mathbf{x}||^2 + \inf_{f,g} \left[E \frac{1}{T} ||\mathbf{v} - \mathbf{\mu}||^2 + E \frac{1}{T} ||\hat{\mathbf{x}} - \mathbf{v}||^2 \right]. \tag{7}$$

It should be pointed out that $\mathbf{v}$ depends only on the choice of the encoder f and that $\hat{\mathbf{x}}$ must be an element of the input space X_T.

The proof of this theorem is straightforward and is given in the Appendix. The following observations pertain to this theorem.

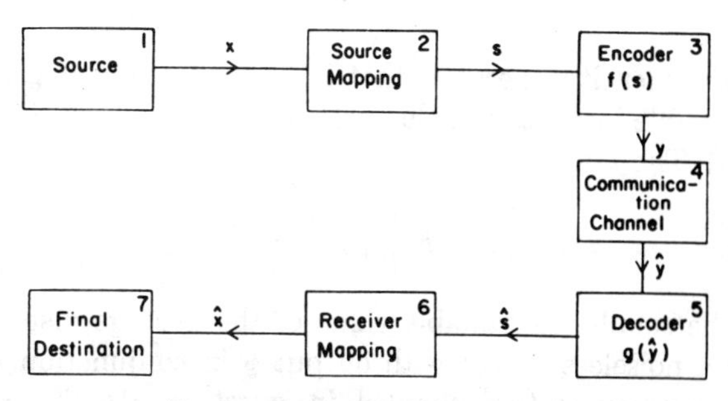

Fig. 2. Communication system.

1) The first term $E(1/T)\, ||\mathbf{\mu} - \mathbf{x}||^2$ is the smallest D_T^* that could possibly be achieved for the communication system irrespective of channel encoder–decoder pair, or receiver mapping [4]. (Note that $\mathbf{\mu}$ is the conditional expectation of $\mathbf{x}$ given $\mathbf{s}$.) This value of D_T^* depends only upon the source mapping.

2) It follows from (49) that

$$D_T^* = E \frac{1}{T} ||\mathbf{\mu} - \mathbf{x}||^2 + \inf_{f,g} E \frac{1}{T} ||\hat{\mathbf{x}} - \mathbf{\mu}||^2. \tag{8}$$

This form is useful in the case where the receiver mapping is the identity operator (i.e., no receiver noise).

3) If the receiver mapping is of the form

$$\hat{\mathbf{x}} = \hat{\mathbf{s}} + \mathbf{n} \tag{9}$$

where $\mathbf{n}$ is zero mean and independent of $\hat{\mathbf{s}}$, then D_T^* reduces to

$$D_T^* = E \frac{1}{T} ||\mathbf{\mu} - \mathbf{x}||^2 + E \frac{1}{T} ||\mathbf{n}||^2 + \inf_{f,g} E \frac{1}{T} ||\hat{\mathbf{s}} - \mathbf{\mu}||^2. \tag{10}$$

This equation results from writing D_T as

$$D_T = E \frac{1}{T} ||\mathbf{\mu} - \mathbf{x}||^2 + E \frac{1}{T} ||\hat{\mathbf{s}} + \mathbf{n} - \mathbf{\mu}||^2$$

$$= E \frac{1}{T} ||\mathbf{\mu} - \mathbf{x}||^2 + E \frac{1}{T} ||\hat{\mathbf{s}} - \mathbf{\mu}||^2 + E \frac{1}{T} ||\mathbf{n}||^2 + 2E \frac{1}{T} (\mathbf{n}, \hat{\mathbf{s}} - \mathbf{\mu})$$

and noting that the inner product is zero since the noise has zero mean and is independent of $\hat{\mathbf{s}}$ and $\mathbf{\mu}$. Note that the infimum over the allowable encoder–decoder pairs does not involve the receiver mapping. Thus the design of the optimum encoder and decoder is independent of the statistics of the receiver noise. The encoder–decoder design does depend, however, upon the statistics of the source mapping even if this mapping is due to additive independent noise. The smallest possible value of D_T^* is then

$$E \frac{1}{T} ||\mathbf{\mu} - \mathbf{x}||^2 + E \frac{1}{T} ||\mathbf{n}||^2$$

and this value is achievable if the communication channel

is the identity operator, that is, if $\mathbf{y} = \hat{\mathbf{y}}$ and if $\boldsymbol{\mu}$ is an allowable input to the channel.

4) The term

$$\inf E \frac{1}{T} ||\mathbf{v} - \boldsymbol{\mu}||^2$$

is the smallest attainable value of the average distortion for a noiseless source with output $\boldsymbol{\mu}$ in conjunction with the communication channel in question. As $T \to \infty$, this average distortion approaches the rate distortion function for the source $\boldsymbol{\mu}$ and the channel, where the distortion measure is the square of the norm of the difference.

5) The term

$$\inf E \frac{1}{T} ||\hat{\mathbf{x}} - \mathbf{v}||^2$$

is bounded from below by the rate distortion function [1] of the channel characterized by the receiver mapping with a message source $\mathbf{v}$. This is a lower bound since in our case we are prohibited from processing the output of this channel.

6) The entire communication system can be reinterpreted in terms of the block diagram given in Fig. 3. The encoder is subdivided into a cascade of two blocks. The first block calculates the conditional mean of $\mathbf{x}$ given $\mathbf{s}$. The second block encodes $\boldsymbol{\mu} = E_s\mathbf{x}$ as if it were the output of a noiseless source. The decoder is also subdivided into the cascade of two blocks. The first block calculates the conditional mean of $\boldsymbol{\mu}$ given $\hat{\mathbf{y}}$. The second block then operates on $\mathbf{v} = E_{\hat{y}}\boldsymbol{\mu}$ in a manner determined only by the receiver mapping. If the receiver mapping is the identity operator or consists of additive independent noise and if $\mathbf{v}$ is an element of X_T, then the second block is omitted; that is, the optimal second block is the identity operator. This structure was considered in the paper by Dobrushin and Tsybakov. They proved the optimality of this structure under the assumptions that

1) the source is Gaussian,
2) the source mapping is additive Gaussian noise,
3) the receiver mapping is independent additive noise,
4) the delay $T \to \infty$.

With the exception of assumption 3) in certain cases, these assumptions were not required in our work.

III. Applications

A. Discrete Time Gaussian Source With Gaussian Source Mapping

Let $\mathbf{x} = (x_1, x_2, \cdots, x_N)$, the T-second output of the source, be described by the pdf

$$p(\mathbf{x}) = \prod_{i=1}^{N} p(x_i) = \prod_{i=1}^{N} (2\pi\sigma_x^2)^{-1/2} \exp(-\tfrac{1}{2}x_i^2/\sigma_x^2). \quad (11)$$

For convenience we assume $N = T$ (i.e., $\rho_s = 1$). The source mapping is assumed to be additive independent

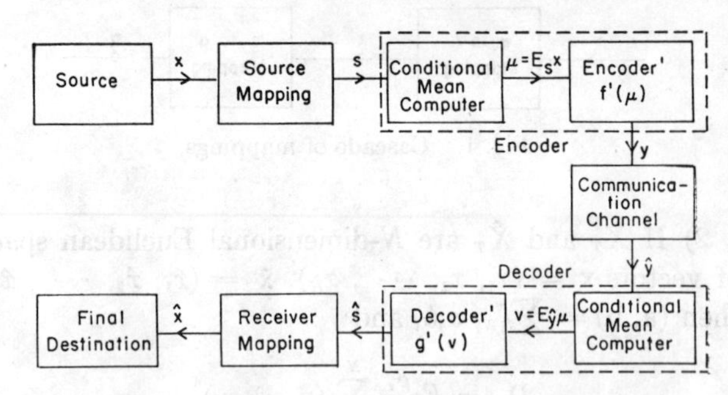

Fig. 3. Equivalent communications system.

Gaussian noise. That is

$$\mathbf{s} = \mathbf{x} + \mathbf{n} = (x_1 + n_1, x_2 + n_2, \cdots, x_N + n_N) \quad (12)$$

where

$$p(\mathbf{x}, \mathbf{n}) = p(\mathbf{x})p(\mathbf{n}) = p(\mathbf{x}) \prod_{i=1}^{N} (2\pi\sigma_n^2)^{-1/2} \exp(-\tfrac{1}{2}n_i^2/\sigma_n^2). \quad (13)$$

The communications channel is characterized by its capacity C defined as

$$C = \sup [I(\mathbf{y}, \hat{\mathbf{y}})/N], \quad (14)$$

where $I(\mathbf{y}, \hat{\mathbf{y}})$ is the mutual information between $\mathbf{y}$ and $\hat{\mathbf{y}}$, and the supremum is taken over N and over all allowable input distributions. The receiver mapping is governed by the equations

$$\hat{\mathbf{x}} = \hat{\mathbf{s}} + \mathbf{z} = (\hat{s}_1 + z_1, \hat{s}_2 + z_2, \cdots, \hat{s}_N + z_N) \quad (15)$$

where

$$p(\hat{\mathbf{s}}, \mathbf{z}) = p(\hat{\mathbf{s}})p(\mathbf{z}) \quad (16)$$

$$E\mathbf{z} = \mathbf{0} \quad (17)$$

$$E \frac{1}{N} ||\mathbf{z}||^2 = E \frac{1}{N} \sum_{i=1}^{N} z_i^2 = \sigma_z^2. \quad (18)$$

Note that the receiver noise is assumed to be additive and independent of the input, but not necessarily Gaussian.

It is easily shown that for this case

$$\boldsymbol{\mu} = E_s\mathbf{x} = \frac{\sigma_x^2}{\sigma_x^2 + \sigma_n^2} \mathbf{s} \quad (19)$$

so that the following are true.

1) The components of $\boldsymbol{\mu}$ are independent Gaussian variates of zero mean and variance

$$\sigma_x^4/(\sigma_x^2 + \sigma_n^2). \quad (20)$$

2) $$E \frac{1}{N} ||\boldsymbol{\mu} - \mathbf{x}||^2 = \sigma_x^2\sigma_n^2/(\sigma_x^2 + \sigma_n^2). \quad (21)$$

If we are interested in the distortion in the limit as $T \to \infty$, we can use the results for the rate-distortion function of a Gaussian source [1] together with (20) to obtain

$$\lim_{N\to\infty} \inf \frac{1}{N} E ||\mathbf{v} - \boldsymbol{\mu}||^2 = \frac{\sigma_x^4}{\sigma_x^2 + \sigma_n^2} e^{-2C}. \quad (22)$$

Finally, the receiver noise is additive and independent of the input; therefore,

$$E \frac{1}{N} \|\hat{\mathbf{x}} - \mathbf{v}\|^2 = \sigma_z^2. \tag{23}$$

Combining (21), (22), and (23) we have

$$\lim_{N\to\infty} D_T^* = \frac{\sigma_x^2\sigma_n^2}{\sigma_x^2 + \sigma_n^2} + \sigma_z^2 + \frac{\sigma_x^4}{\sigma_x^2 + \sigma_n^2} e^{-2C}. \tag{24}$$

This result was obtained previously by Dobrushin and Tsybakov [2]. Typical plots of C versus ($\lim_{N\to\infty} D_T^*$) are given in Fig. 4.

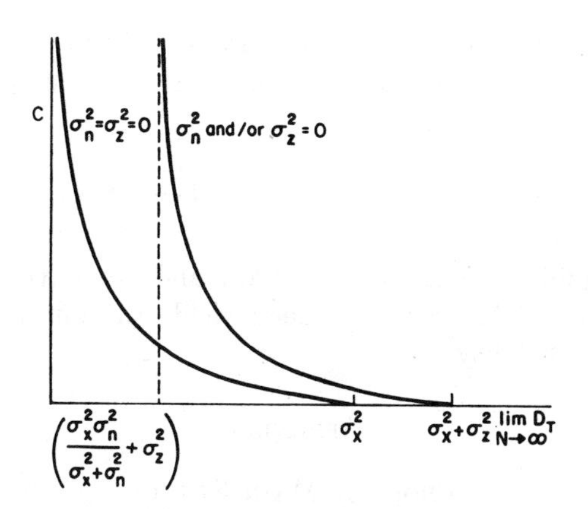

Fig. 4. Rate-distortion function.

B. Binary Source With a Binary Symmetric Channel as the Source Mapping

Let the T-second output of the source be the vector $\mathbf{x}$

$$\mathbf{x} = (x_1, x_2, \cdots, x_N) \tag{25}$$

where $x_i \in (0, 1)$. Furthermore, assume that $P(x_i = 0) = P(x_i = 1) = \frac{1}{2}$ and

$$P(\mathbf{x}) = \prod_{i=1}^{N} P(x_i). \tag{26}$$

For simplicity again assume $\rho_s = 1$ so that $N = T$.

Let the source mapping be a binary symmetric channel with crossover probability p. That is,

$$\mathbf{s} = (s_1, s_2, \cdots, s_N) \tag{27}$$

where

$$P[s_i = 1 \mid x_i = 0] = P[s_i = 0 \mid x_i = 1] = p \qquad i = 1, 2, \cdots, N. \tag{28}$$

Finally, assume there is no noise at the receiver so that

$$\hat{\mathbf{x}} = \hat{\mathbf{s}} = (\hat{x}_i, \hat{x}_2, \cdots, \hat{x}_N) \tag{29}$$

where $\hat{x}_i \in [0, 1]$, $i = 1, 2, \cdots, N$.

An *ad hoc* design procedure is to design the encoder-decoder to minimize the bit error probability between $\hat{\mathbf{x}}$ and $\mathbf{s}$. Let

$$P_e(N) = \frac{1}{N} \sum_{i=1}^{N} P_e^{(i)}$$

where $P_e^{(i)}$ is the probability that $\hat{x}_i \neq s_i$ and let $P_e^*(N)$ be the infimum of $P_e(N)$ taken over all encoder-decoder pairs with coding delay N. Then the resulting average bit error probability is

$$\begin{aligned} P_b &= \frac{1}{N} \sum_{i=1}^{N} P\{\hat{x}_i \neq x_i\} \\ &= p(1 - P_e^*(N)) + (1 - p)P_e^*(N) \\ &= p + (1 - 2p)P_e^*(N). \end{aligned} \tag{30}$$

Using the results of this paper, we now show that this *ad hoc* scheme indeed minimizes the resulting bit error probability between $\mathbf{x}$ and $\hat{\mathbf{x}}$.

For the case of no receiver noise, we have from (8)

$$D_T^* = E \frac{1}{N} \|\boldsymbol{\mu} - \mathbf{x}\|^2 + \inf E \frac{1}{N} \|\hat{\mathbf{x}} - \boldsymbol{\mu}\|^2 \tag{31}$$

where in this case D_T^* is the minimum attainable bit error probability. (See (5) and the discussion following.) To calculate the first term of (31), we note that the components of the conditional mean $\boldsymbol{\mu} = (\mu_1, \mu_2, \cdots, \mu_N)$ are mutually independent and can be written as

$$\mu_i = |s_i - p| \qquad i = 1, 2, \cdots, N \tag{32}$$

where $P[\mu_i = p] = P[\mu_i = 1 - p] = \frac{1}{2}$ for $i = 1, 2, \cdots, N$. Thus

$$\begin{aligned} E \frac{1}{N} \|\boldsymbol{\mu} - \mathbf{x}\|^2 &= \frac{1}{N} \sum_{i=1}^{N} E(\mu_i - x_i)^2 \\ &= \frac{1}{N} \sum_{i=1}^{N} \tfrac{1}{2}((p - 0)^2(1 - p) \\ &\quad + (1 - p - 0)^2 p + (p - 1)^2 p \\ &\quad + (1 - p - 1)^2(1 - p)) \end{aligned} \tag{33}$$

or

$$E \frac{1}{N} \|\boldsymbol{\mu} - \mathbf{x}\|^2 = p(1 - p). \tag{34}$$

For a given encoder-decoder pair define

$$P_1^{(i)} = P(\hat{x}_i = 1 \mid \mu_i = p) \tag{35}$$

$$P_2^{(i)} = P(\hat{x}_i = 0 \mid \mu_i = 1 - p). \tag{36}$$

Then the second term in (31) becomes

$$\begin{aligned} \inf E \frac{1}{N} \|\hat{\mathbf{x}} - \boldsymbol{\mu}\|^2 &= \inf \frac{1}{N} \sum_{i=1}^{N} \tfrac{1}{2}((0 - p)^2(1 - P_1^{(i)}) \\ &\quad + (1 - p)^2 P_1^{(i)} + (0 - (1 - p))^2 P_2^{(i)} \\ &\quad + (1 - (1 - p))^2(1 - P_2^{(i)})) \end{aligned} \tag{37}$$

or

$$\inf E \frac{1}{N} \|\hat{\mathbf{x}} - \boldsymbol{\mu}\|^2 = p^2 + (1 - 2p) \inf \left(\frac{1}{N} \sum_{i=1}^{N} (\tfrac{1}{2}P_1^{(i)} + \tfrac{1}{2}P_2^{(i)})\right). \tag{38}$$

Combining (34) and (38) we have

$$D_T^* = p + (1 - 2p) \inf \left(\frac{1}{N} \sum_{i=1}^{N} (\tfrac{1}{2}P_1^{(i)} + \tfrac{1}{2}P_2^{(i)})\right), \tag{39}$$

which is the desired result. Also it follows from (32) that there is a one to one mapping between $\mathfrak{u}$ and s so that

$$\inf\left(\frac{1}{N}\sum_{i=1}^{N}\left(\tfrac{1}{2}P_1^{(i)}+\tfrac{1}{2}P_2^{(i)}\right)\right)=P_e^*(N). \tag{40}$$

Thus (39) is equivalent to (30), and we have proved that the *ad hoc* scheme indeed yields the minimum bit error probability.

Appendix

Proof of Main Result

From (1) and (2), D_T is defined as

$$D_T = E\frac{1}{T}\,||\hat{\mathbf{x}}-\mathbf{x}||^2 = E\frac{1}{T}\,||\hat{\mathbf{x}}-\mathfrak{u}+\mathfrak{u}-\mathbf{x}||^2 \tag{41}$$

where $\mathfrak{u} = E_s\mathbf{x}$. Expanding (41) we have

$$D_T = E\frac{1}{T}\,||\hat{\mathbf{x}}-\mathfrak{u}||^2 + E\frac{1}{T}\,||\mathfrak{u}-\mathbf{x}||^2 + 2E\frac{1}{T}(\hat{\mathbf{x}}-\mathfrak{u},\mathfrak{u}-\mathbf{x}) \tag{42}$$

where $(\mathbf{a}, \mathbf{b})$ denotes inner product. We will show that this inner product is zero.

Let $\mathbf{a}$ and $\mathbf{b}$ be two random vectors whose values are elements in a separable Hilbert space. Let $\mathbf{h}(\mathbf{a})$ be some single-valued deterministic function of $\mathbf{a}$ in the Hilbert space.

Take any complete orthonormal set of elements ψ_n and let

$$\mathbf{h}(\mathbf{a}) = \sum_n \alpha_n\varphi_n \tag{43}$$

and

$$b = \sum_n \beta_n\varphi_n. \tag{44}$$

Then

$$\begin{aligned} E_{\mathbf{a}}(\mathbf{h}(\mathbf{a}),\mathbf{b}) &= E_{\mathbf{a}}\Big(\sum_n \alpha_n\varphi_n, \sum_n \beta_n\varphi_n\Big) = E_{\mathbf{a}}\Big(\sum_n \alpha_n\beta_n\Big)\\ &= \Big(\sum_n (\alpha_n)(E_{\mathbf{a}}\beta_n)\Big) = \Big(\sum_n \alpha_n\varphi_n, \sum_n (E_{\mathbf{a}}\beta_n)\varphi_n\Big)\\ &= (\mathbf{h}(\mathbf{a}), E_{\mathbf{a}}\mathbf{b}). \end{aligned} \tag{45}$$

Now

$$E(\hat{\mathbf{x}}-\mathfrak{u},\mathfrak{u}-\mathbf{x}) = EE_{\mathbf{s}}E_{\mathbf{s},\mathbf{x}}(\hat{\mathbf{x}},\mathfrak{u}-\mathbf{x}) - EE_{\mathbf{s}}(\mathfrak{u},\mathfrak{u}-\mathbf{x}). \tag{46}$$

Using (45), the first term becomes

$$\begin{aligned} EE_{\mathbf{s}}E_{\mathbf{s},\mathbf{x}}(\hat{\mathbf{x}},\mathfrak{u}-\mathbf{x}) &= EE_{\mathbf{s}}(E_{\mathbf{s},\mathbf{x}}\hat{\mathbf{x}},\mathfrak{u}-\mathbf{x}) = EE_{\mathbf{s}}(E_{\mathbf{s}}\hat{\mathbf{x}},\mathfrak{u}-\mathbf{x})\\ &= E(E_{\mathbf{s}}\hat{\mathbf{x}},\mathfrak{u}-E_{\mathbf{s}}\mathbf{x})\\ &= E(E_{\mathbf{s}}\hat{\mathbf{x}},\mathfrak{u}-\mathfrak{u}) = 0. \end{aligned} \tag{47}$$

The second term becomes

$$EE_{\mathbf{s}}(\mathfrak{u},\mathfrak{u}-\mathbf{x}) = E(\mathfrak{u},\mathfrak{u}-E_{\mathbf{s}}\mathbf{x}) = E(\mathfrak{u},\mathfrak{u}-\mathfrak{u}) = 0. \tag{48}$$

Thus D_T can be written as

$$D_T = E\frac{1}{T}\,||\hat{\mathbf{x}}-\mathfrak{u}||^2 + E\frac{1}{T}\,||\mathfrak{u}-\mathbf{x}||^2. \tag{49}$$

We now write the first term as

$$E\frac{1}{T}\,||\hat{\mathbf{x}}-\mathbf{v}+\mathbf{v}-\mathfrak{u}||^2 = E\frac{1}{T}\,||\hat{\mathbf{x}}-\mathbf{v}||^2 + E\frac{1}{T}\,||\mathbf{v}-\mathfrak{u}||^2 + 2E\frac{1}{T}(\hat{\mathbf{x}}-\mathbf{v},\mathbf{v}-\mathfrak{u}) \tag{50}$$

where $\mathbf{v} = E_{\hat{\mathbf{y}}}\mathfrak{u}$. Again it can be shown that the inner product is zero by writing it as

$$E(\hat{\mathbf{x}}-\mathbf{v},\mathbf{v}-\mathfrak{u}) = EE_{\hat{\mathbf{y}}}E_{\hat{\mathbf{y}},\mathfrak{u}}(\hat{\mathbf{x}},\mathbf{v}-\mathfrak{u}) - EE_{\hat{\mathbf{y}}}(\mathbf{v},\mathbf{v}-\mathfrak{u}) \tag{51}$$

and demonstrating that the two resulting terms are each zero. Combining (41) and (42) we have

$$D_T = E\frac{1}{T}\,||\mathfrak{u}-\mathbf{x}||^2 + E\frac{1}{T}\,||\mathbf{v}-\mathfrak{u}||^2 + E\frac{1}{T}\,||\hat{\mathbf{x}}-\mathbf{v}||^2, \tag{52}$$

which proves the first part of the theorem.

Taking the infimum over all allowable encoder–decoder pairs and noting that the first term is a constant with respect to the choice of the encoder and decoder results in (7), which completes the proof.

Acknowledgment

The authors wish to acknowledge helpful discussions with Dr. T. T. Kadota of Bell Telephone Laboratories, Inc.

References

[1] C. E. Shannon, "Coding theorems for a discrete source with a fidelity criterion," *IRE Natl. Conv. Rec.*, pt. 4, pp. 142–163, 1959.
[2] R. L. Dobrushin and B. S. Tsybakov, "Information transmission with additional noise," *IRE Trans. Information Theory*, vol. IT-8, pp. 293–304, September 1962.
[3] W. W. Peterson, *Error Correcting Coding*. Cambridge, Mass: M.I.T. Press, 1961, p. 7.
[4] A. Papoulis, *Probability, Random Variables and Stochastic Processes*. New York: McGraw-Hill, 1965, pp. 216–217.

Bounds on the Rate-Distortion Function for Stationary Sources With Memory

AARON D. WYNER, MEMBER, IEEE, AND JACOB ZIV, ASSOCIATE MEMBER, IEEE

Abstract—**In this paper, we study discrete-time stationary sources S with memory. The rate $R(\beta)$ of the source relative to a distortion measure is compared with $R^*(\beta)$, the rate of the memoryless source S^* with the same marginal statistics as S. We show that $R^*(\beta) - \Delta \le R(\beta) \le R^*(\beta)$, where Δ is a measure of the memory of the source. A number of interesting applications of these bounds are given.**

I. Introduction

WE DEFINE a (information) *source* S (or perhaps more accurately a *source-user* pair) as consisting of 1) a source output set $\mathscr{W}$, and a user set $\mathscr{Z}$, both subsets of Euclidean m-space, 2) a nonnegative-valued distortion function D, defined on $\mathscr{W} \times \mathscr{Z}$, and 3) a probability law that we will define later.

The source output is assumed to be an infinite sequence $\cdots,W_{k-1},W_k,W_{k+1},\cdots$ $(-\infty < k < \infty)$ of random variables, $W_k \in \mathscr{W}$, which appear at the rate of, say, one each second. We wish to transmit these source outputs through a communication channel that may be either noiseless or noisy and to deliver a sequence $\cdots,Z_{k-1},Z_k,Z_{k+1},\cdots$ $(-\infty < k < \infty)$ to a user, where $Z_k \in \mathscr{Z}$. If $W_k = w \in \mathscr{W}$ and $Z_k = z \in \mathscr{Z}$, we say that the system introduced a distortion of $D(w,z)$. In some sense, to be specified later, we seek to minimize the average of this distortion.

We now define the probability law of the source outputs mentioned in 3) of the definition of the source. We assume that the source output sequence $\{W_k\}_{-\infty}^{\infty}$ is defined probabilistically by a set of consistent probability density functions (pdf)

$$P_S^{(k,n)}(w_k,w_{k+1},\cdots,w_{k+n-1}), \qquad -\infty < k < \infty, 1 \le n < \infty,$$

which is the joint pdf for the n random variables W_k, $W_{k+1},\cdots W_{k+n-1}$. The source is said to be stationary if $P_S^{(k,n)} = P_S^{(n)}$, independent of k. Although we will be dealing exclusively with stationary sources, many of the results can be applied to nonstationary sources. A stationary source is said to be memoryless if

$$P_S^{(n)}(w_1,w_2,\cdots,w_n) = \prod_{k=1}^{n} P_S^{(1)}(w_k). \tag{1}$$

Corresponding to any stationary source S with pdf $P_S^{(n)}$, it is useful to define a memoryless source S^* with pdf

$$P_{S^*}^{(n)}(w_1,w_2,\cdots,w_n) = \prod_{k=1} P_S^{(1)}(w_k). \tag{2}$$

We remark that our definition of a source includes discrete sources, where $\mathscr{W}$ has but a countable number of members, if impulse (delta) functions are allowed to appear in the density functions.

The central problem in coding theory is the transmission of a sequence of N source outputs (starting, say, from W_1) through a communication channel to a user in such a way as to minimize the average distortion

$$\bar{D}^{(N)} = ED^{(N)}(\boldsymbol{W}^{(N)},\boldsymbol{Z}^{(N)}), \tag{3}$$

where

$$\boldsymbol{W}^{(N)} = (W_1,W_2,\cdots,W_N)$$

and

$$\boldsymbol{Z}^{(N)} = (Z_1,Z_2,\cdots,Z_N)$$

are the first N source outputs and user inputs, and

$$D^{(N)}(\boldsymbol{W}^{(N)},\boldsymbol{Z}^{(N)}) = \frac{1}{N}\sum_{k=1}^{N} D(W_k,Z_k) \tag{4}$$

(E denotes expected value). Processing, that is, coding and decoding, is allowed at both ends of the channel. The Shannon theory gives us considerable information about $\bar{D}^{(N)}$. We proceed as follows.

Let S be a source. Consider blocks of N successive source outputs $\boldsymbol{W} = \boldsymbol{W}^{(N)} = (W_1,\cdots,W_N)$. Let $p_t(z \mid w)$ be a conditional pdf on $\mathscr{Z}^N$ given $\boldsymbol{W} = \boldsymbol{w} \in \mathscr{W}^N$. Then the random N-vector $\boldsymbol{Z} = \boldsymbol{Z}^{(N)} = (Z_1,\cdots,Z_N)$ is defined by the joint pdf (for $\boldsymbol{W}$ and $\boldsymbol{Z}$)

$$p_{\boldsymbol{WZ}}(\boldsymbol{w},\boldsymbol{z}) = P_S^{(N)}(\boldsymbol{w})\, p_t(\boldsymbol{z} \mid \boldsymbol{w}).$$

The density $p_t(z \mid w)$ is called a "test channel." Corresponding to each possible test channel, the average distortion is

$$ED^{(N)}(\boldsymbol{W},\boldsymbol{Z}) = \frac{1}{N}\sum_{k=1}^{N} ED(W_k,Z_k).$$

For a given N, and arbitrary $\beta \ge 0$, we define the equivalent rate of the source (rate-distortion function) as

$$R_N(\beta) = \inf_{p_t \in \tau_N(\beta)} \frac{1}{N} I\{\boldsymbol{W},\boldsymbol{Z}\}, \tag{5}$$

where

$$I\{\boldsymbol{W},\boldsymbol{Z}\} = \int_{\mathscr{W}^N}\int_{\mathscr{Z}^N} dw\, dz P_S^{(N)}(\boldsymbol{w})\, p_t(\boldsymbol{z} \mid \boldsymbol{w}) \times \log\left[p_t(\boldsymbol{z} \mid \boldsymbol{w})/p_{\boldsymbol{Z}}(\boldsymbol{z})\right]$$

Manuscript received December 1, 1970; revised February 11, 1971. This research was done while A. D. Wyner was visiting the Weizmann Institute of Science, Rehovot, Israel.
A. D. Wyner is with Bell Telephone Laboratories, Inc., Murray Hill, N.J. 07974.
J. Ziv is with the Faculty of Electrical Engineering, Technion-Israel Institute of Technology, Haifa, Israel.

Reprinted from *IEEE Trans. Inform. Theory*, vol. IT-17, pp. 508-513, Sept. 1971.

is the mutual information between the N-vectors W and Z corresponding to the test channel p_t. The density for Z is $p_Z(z) = \int_{w^N} P_{WZ}(w,z)\, dw$. The set $\tau_N(\beta)$ is the set of test channels p_t (for blocks of length N) such that

$$ED^{(N)}(W,Z) \leq \beta.$$

Now suppose that the channel is stationary and has capacity[1] C. Then the converse to the coding theorem tells us that $\bar{D}^{(N)}$ cannot exceed $\beta_N^* = \beta_N^*(C)$, the smallest solution of

$$R_N(\beta_N^*) \leq C. \tag{6}$$

Further, it is known that $R_N(\beta)$ tends to a limit, $R(\beta) = R_\infty(\beta)$, from above (as $N \to \infty$). Thus we conclude that $\bar{D}^{(N)} \geq \beta^*(C)$, the smallest solution of

$$R(\beta^*) \leq C. \tag{7}$$

Furthermore, the direct half of the coding theorem tells us that it is possible to attain a distortion $\bar{D}^{(N)}$ arbitrarily close to $\beta^*(C)$ provided N is sufficiently large and the source S is ergodic and satisfies some other rather weak conditions [4, p. 500, Theorem 9.8.3], [5].

It is a consequence of Theorem 9.2.1 [4, p. 446] that if a source is memoryless, the rate $R_N(\beta)$ is independent of N. In particular, for the memoryless source S^* (corresponding to the arbitrary source S), we can unambiguously write its rate as $R^*(\beta)$. Furthermore, since the statistics for W_1 are identical for sources S and S^*, we have $R^*(\beta) = R_1(\beta)$, where $R_1(\beta)$ is the rate for source S. The main result of this paper compares for a source S, rates $R_N(\beta)$ and $R^*(\beta)$. This result, together with a number of interesting examples and applications, is given in Section II. The proof follows in Section III. The reader is also referred to Grey [6] and Berger [3] for related results.

II. Results and Applications

We show in Section III that

$$R^*(\beta) - \Delta_N \leq R_N(\beta) \leq R^*(\beta), \tag{8a}$$

where

$$\Delta_N = \frac{1}{N} E \log \frac{P_S^{(N)}(W)}{P_{S^*}^{(N)}(W)} = \frac{1}{N} \int_n dw P_S^{(N)}(w) \log \frac{P_S^{(N)}(w)}{P_{S^*}^{(N)}(w)} \tag{8b}$$

is the entropy of the N successive outputs of source S^* relative to the N successive outputs of S. The upper bound is an expression of the widely held belief that memory decreases the rate of a source. The lower bound, which is our main result, may be interpreted as follows: $R_N(\beta)$ is at least as great as the rate $R^*(\beta)$ of the corresponding memoryless source minus a term Δ_N that is a kind of measure of the memory of the source and is independent of the distortion D and β.

[1] By capacity we mean the maximum error-free rate in the sense of [1, p. 223].

Remarks

Remark 1: $\Delta_N \geq 0$, with equality if and only if $P_S^{(N)} \equiv P_{S^*}^{(N)}$, i.e., the source S is memoryless.

Proof:

$$-\Delta_N = \frac{1}{N} \int P_S^{(N)}(w) \log \frac{P_{S^*}^{(N)}(w)}{P_S^{(N)}(w)}\, dw$$

$$\leq \frac{1}{N} \int P_S^{(N)}(w) \left[\frac{P_{S^*}^{(N)}(w)}{P_S^{(N)}(w)} - 1\right] dw = 0,$$

where the inequality follows from $\log x \leq x - 1$ (with equality if and only if $x = 1$).

Remark 2—Representations of Δ_N: 1) When entropies (either continuous or discrete) exist, we can write

$$\Delta_N = H\{W_k\} - \frac{1}{N} H\{W_1,\cdots,W_N\}, \tag{9}$$

where H is either the continuous or the discrete entropy formula (see [1, p. 18, 236]).

2) We can also write

$$\Delta_N = \frac{1}{N} \sum_{k=2}^{N} I\{W_k; W_1,\cdots,W_{k-1}\}, \tag{10}$$

where $I\{;\}$ is the mutual information defined in the usual way (see [4, p. 29, eq. (2.4.18)]).

Proof: We establish (10) by induction on N. For $N = 2$, that $\Delta_2 = \frac{1}{2} I\{W_2; W_1\}$ follows from the definition of Δ_N in (8b). If (10) is true for $N = N_0$, write

$$\begin{aligned}(N_0 + 1)\Delta_{N_0+1} &= E \log \frac{P_S^{(N_0+1)}(W_1,\cdots,W_{N_0+1})}{\prod_{k=1}^{N_0+1} P_S^{(1)}(W_k)} \\ &= E \log \frac{P_S^{(N_0)}(W_1,\cdots,W_{N_0})}{\prod_{k=1}^{N_0} P_S^{(1)}(W_k)} \\ &\quad \cdot \frac{P_S^{(N_0+1)}(W_1,\cdots,W_{N_0+1})}{P_S^{(1)}(W_{N_0+1}) P_S^{(N_0)}(W_1,\cdots,W_{N_0})} \\ &= N_0 \Delta_{N_0} + I\{W_{N_0+1}; W_1,\cdots,W_{N_0}\} \\ &= \sum_{k=2}^{N_0+1} I\{W_k; W_1,\cdots,W_{k-1}\}.\end{aligned}$$

Remark 3: From representation (10) we conclude 1) Δ_N is increasing in N, and 2)

$$\Delta_\infty = \lim_{N\to\infty} \Delta_N = I\{W_1; W_0, W_{-1},\cdots\}. \tag{11}$$

To prove 1), observe that since $I\{W_k; W_1,\cdots,W_{k-1}\}$ is nondecreasing in k, we have from (10)

$$\Delta_N \leq \frac{(N-1)}{N} I\{W_N; W_1,\cdots,W_{N-1}\} < I\{W_{N+1}; W_1,\cdots,W_N\}.$$

Thus

$$(N+1)\Delta_{N+1} = N\Delta_N + I\{W_{N+1}; W_1,\cdots,W_N\} > (N+1)\Delta_N$$

so that $\Delta_{N+1} > \Delta_N$. Assertion 2) follows immediately from (10) and the stationary of S. Note that $I\{W_1; W_0, W_{-1},\cdots\}$ may be infinite.

Remark 4: We conclude from Remark 3, assertion 1) that

$$R^*(\beta) - \Delta_\infty \leq R(\beta) \leq R^*(\beta), \qquad (12)$$

where Δ_∞ is given by (11). Inequality (12) is probably the most useful form of our result.

Remark 5: When the source has a finite memory of L, that is, when W_k depends on $W_{k-1},W_{k-2},\cdots$ only through $W_{k-1},\cdots,W_{k-L}$, then from (10), for $N \geq L$,

$$\Delta_N = \frac{1}{N}\sum_{k=2}^{L} I\{W_k;W_1,\cdots,W_{k-1}\} + \left(\frac{N-L}{N}\right) I\{W_{L+1};W_1,\cdots,W_L\}. \qquad (13a)$$

Thus

$$\Delta_\infty = I\{W_{L+1};W_1,\cdots,W_L\}. \qquad (13b)$$

Further, in the Markov case ($L = 1$),

$$\Delta_N = \frac{N-1}{N} I\{W_k;W_{k+1}\}, \qquad N = 2,3,\cdots. \qquad (13c)$$

Remark 6: There are cases when the lower bound of (12) holds with equality. One such case is given in example 2) below.

Examples and Applications

1) Gaussian Source: Let the source outputs be a stationary Gaussian time series with $EW_k = 0$ and $EW_kW_{k+n} = \rho_n$. Let the $N \times N$ matrix $\mathcal{R}_N$ have ρ_{i-j} as entry ij. Then the density for $W_1,W_2,\cdots,W_N$ is

$$P_S^{(N)}(w) = (2\pi)^{-n/2}|\mathcal{R}_N|^{-1/2} \exp\{-\tfrac{1}{2}w\mathcal{R}_N^{-1}w^T\}, \qquad (14)$$

where $|\mathcal{R}_N|$ is the determinant of the matrix $\mathcal{R}_N$ and "T" denotes matrix transpose. Thus

$$\frac{1}{N}H\{W_1,W_2,\cdots,W_N\} = -\frac{1}{N}E\log P_S^{(N)}(W) = \tfrac{1}{2}\log[2\pi|\mathcal{R}_N|^{1/N}] + \frac{1}{2N}EW\mathcal{R}_N^{-1}W^T.$$

But if the entry of $\mathcal{R}_N^{-1}$ is ρ_{ij}^{-1}, then

$$EW\mathcal{R}_N^{-1}W^T = E\sum_{i,j=1}^{N} W_i\rho_{ij}^{-1}W_j = \sum_{i,j}\rho_{ij}^{-1}\rho_{ji} = \sum_{i=1}^{N}\sum_{j=1}^{N}\rho_{ij}^{-1}\rho_{ji} = N.$$

Thus

$$\frac{1}{N}H\{W_1,W_2,\cdots,W_N\} = \tfrac{1}{2}\log[2\pi e|\mathcal{R}_N|^{1/N}]. \qquad (15)$$

Further, W_k is a normal variate with variance ρ_0. Thus

$$H\{W_k\} = \tfrac{1}{2}\log 2\pi e\rho_0. \qquad (16)$$

Hence from (9),

$$\Delta_N = \tfrac{1}{2}\log \rho_0/|\mathcal{R}_N|^{1/N}. \qquad (17)$$

Let us define the spectrum of the time series as[2]

$$S(f) = \sum_{n=-\infty}^{\infty}\rho_n \exp(-i2\pi fn), \qquad -\tfrac{1}{2} < f < \tfrac{1}{2}. \qquad (18)$$

It follows from the theory of Fourier series that $S(f)$ is real and symmetric about $f = 0$, and that

$$\rho_n = \int_{-1/2}^{1/2} S(f)\cos 2\pi nf\, df. \qquad (19)$$

Let us remark that the time series $\{W_n\}_{-\infty}^{\infty}$ may be thought of as samples (at the sampling instants $t = n, -\infty < n < \infty$) of a band-limited random process with spectrum $S(f)$ in the band $[-\frac{1}{2},\frac{1}{2}]$.

The Szegö theorem [7] states that, as $N \to \infty$,

$$|\mathcal{R}_N|^{1/N} \to \bar{S} = \exp\left\{\int_{-1/2}^{1/2}\log S(f)\,df\right\}, \qquad (20)$$

the geometric mean of $S(f)$. Since $\rho_0 = \int_{-1/2}^{1/2} S(f)\,df$, we have

$$\Delta_\infty = \tfrac{1}{2}\log\int_{-1/2}^{1/2} S(f)\,df - \tfrac{1}{2}\int_{-1/2}^{1/2}\log S(f)\,df. \qquad (21)$$

2) sth Power Distortion: When $\mathcal{W},\mathcal{Z} \subseteq$ Reals, $P_S^{(n)}(W)$ contains no impulses and $D(w,z) = |w - z|^s$, then we can write the lower bound of (8a) as

$$R_N(\beta) \geq R^*(\beta) - \Delta_N = R^*(\beta) - H\{W_k\} + \frac{1}{N}H\{W_1,\cdots,W_N\}, \qquad (22)$$

where H is the usual continuous entropy. But it is known [8] that

$$R^*(\beta) \geq H\{W_k\} - \frac{1}{s}\log\left[\frac{2^s e\Gamma^s(1/s)\beta}{s^{s-1}}\right]. \qquad (23)$$

The lower bound (22) therefore reduces to

$$R_N(\beta) \geq \frac{1}{N}H\{W_1,W_2,\cdots,W_N\} - \frac{1}{s}\log\left[2^s e\Gamma^s\left(\frac{1}{s}\right)\beta s^{-(s-1)}\right]. \qquad (24)$$

The entropy of the source is defined by

$$H_S = \lim_{N\to\infty}\frac{1}{N}H\{W_1,\cdots,W_N\}, \qquad (25)$$

so that on taking limits as $N \to \infty$ in (24), we have

$$R(\beta) \geq \frac{1}{s}\log\frac{e^{sH_S}s^{s-1}}{2^s e\Gamma^s(1/s)\beta}. \qquad (26)$$

In the special case of the Gaussian source (in example 1) above), we have from (15) and (20) that

$$H_S = \lim_{N\to\infty}\tfrac{1}{2}\log(2\pi e|\mathcal{R}_N|^{1/N}) = \tfrac{1}{2}\log 2\pi e\bar{S}, \qquad (27)$$

[2] We assume that ρ_n is such that the series (18) converges. A sufficient condition that this to be so is $\sum_{n=-\infty}^{\infty}|\rho_n| < \infty$. We also assume that $S(f) \leq M < \infty$, a.e.

so that (26) becomes

$$R(\beta) \geq (1/s) \log \{e^{\frac{1}{2}s-1}(\pi\bar{S}/2)^{s/2}s^{s-1}[\Gamma(1/s)]^{-s}\beta^{-1}\}. \quad (28a)$$

When $s = 2$, (28a) becomes

$$R(\beta) \geq \tfrac{1}{2} \log (\bar{S}/\beta), \quad (28b)$$

which is known to be a tight bound for $\beta < \min S(f)$. Inequality (28b) is stated by Berger [2].

Similar bounds for the case where $D(w,z) = r(w - z)$ can be obtained using the same techniques (see [8]).

3) Stochastic Difference Equation: Say that the time series $\{W_k\}_{-\infty}^{\infty}$ is characterized by the difference equation

$$\Delta W_k = W_{k+1} - W_k = f(W_k) + X_k, \qquad -\infty < k < \infty, \quad (29)$$

where the sequence $\{X_k\}_{-\infty}^{\infty}$ is a sequence of independent random variables with identical density $P_X(x)$. The function f is arbitrary, although only certain choices will result in a stationary time series. $\{W_k\}_{-\infty}^{\infty}$ is Markovian, so that we can apply (13c) to obtain

$$\begin{aligned}\Delta_\infty = I\{W_k; W_{k-1}\} &= H\{W_k\} - H\{W_k \mid W_{k-1}\} \\ &= H\{W_k\} - H\{X_k\}. \quad (30)\end{aligned}$$

If the distortion is again $D(w,z) = |w - z|^s$, we can again make use of (12), (23), and (30) to obtain

$$R(\beta) \geq H\{X_k\} - 1/s \log [2^s e\Gamma^s(1/s)\beta s^{(1-s)}]. \quad (31)$$

In the special case where X_k is Gaussian with variance σ_X^2, and $s = 2$, we obtain

$$R(\beta) \geq \tfrac{1}{2} \log (\sigma_X^2/\beta). \quad (32)$$

Thus, roughly speaking, the rate needed to reproduce W_k to within a mean-squared error of β is at least the rate required to reproduce the X_k to within the same mean-squared error.

4) Discrete Source: Let $\mathscr{W} = \mathscr{Z}$ be a countable set and let

$$D(w,z) = \begin{cases} 1, & w \neq z \\ 0, & w = z \end{cases}$$

be the Hamming distortion. Then it is well known that for $\beta = 0$,

$$R(0) = \lim_{N\to\infty} \frac{1}{N} H\{W_1,W_2,\cdots,W_N\} = H_S$$

and $R^*(0) = H\{W_k\}$. Thus $R^*(0) - R(0) = \Delta_\infty$, so that the lower bound of (12) is satisfied with equality when $\beta = 0$. Inequality (12) is, of course, valid for all β.

5) Asymptotics as $\beta \to 0$: For many sources,

$$\lim_{\beta\to 0} R(\beta) = \infty.$$

From (12) we have $|R^*(\beta) - R(\beta)| < \Delta_\infty$. Thus when $\Delta_\infty < \infty$, we have for these sources as $\beta \to 0$,

$$R(\beta) \sim R^*(\beta). \quad (33a)$$

6) Rate of Approach of $R_N(\beta)$ to $R(\beta)$: Say that we are given a source $S = \{\mathscr{W},\mathscr{Z},D,P_S\}$, as above, with rate $R_N(\beta) \to R(\beta)$ as $N \to \infty$. We shall apply the bounds of (8a) to a supersource S_N, whose outputs are successive blocks of N outputs of source S, to obtain an estimate on the rate of approach of $R_N(\beta)$ to $R(\beta)$. Now the supersource S_N has output set $\mathscr{W}^N$, user set $\mathscr{Z}^N$, distortion between $w = (w_1,\cdots,w_N) \in \mathscr{W}^N$ and $z = (z_1,\cdots,z_N) \in \mathscr{Z}^N$ given by $D^{(N)}(w,z) = 1/N \sum_{k=1}^N D(w_k,z_k)$ and probability law identical to that for the original source S. The rate $R^{(N)}(\beta)$ for the supersource S_N is $NR(\beta)$. Furthermore, the rate $R^{(N)*}(\beta)$ of the memoryless source S_N^* corresponding to S_N is $NR_N(\beta)$. If we apply (8a), we obtain

$$\begin{aligned}|NR_N(\beta) - NR(\beta)| &\leq \Delta_N < \Delta_\infty \\ &= I\{W_1,\cdots,W_N; W_0,W_{-1},\cdots\} \quad (33b)\end{aligned}$$

or

$$|R_N(\beta) - R(\beta)| \leq \frac{1}{N} I\{W_1,\cdots,W_N; W_0,W_{-1},\cdots\}. \quad (33c)$$

Now,

$$\begin{aligned}\frac{1}{N} I\{W_1,\cdots,W_N; W_0,W_{-1},\cdots\} &= \frac{1}{N} H\{W_1,\cdots,W_N\} \\ &\quad - \frac{1}{N} H\{W_1,W_2,\cdots,W_N \mid W_0,W_{-1},\cdots\}, \quad (34)\end{aligned}$$

where H is either continuous or discrete entropy, whichever is appropriate. But from the formula

$$\begin{aligned}H\{X_1,\cdots,X_m \mid Y\} &= H\{X_1 \mid Y\} + H\{X_2 \mid Y,X_1\} \\ &\quad + H\{X_3 \mid Y_1,X_1,X_2\} \\ &\quad + \cdots + H\{X_m \mid Y,X_1,\cdots,X_{m-1}\}\end{aligned}$$

(see, for example, [1, p. 20, Theorem 1.4.4]), the last term in (34) is

$$\frac{1}{N} H\{W_1,\cdots,W_N \mid W_0,W_{-1},\cdots\} = H\{W_1 \mid W_0,W_{-1},\cdots\},$$

so that (34) becomes

$$\begin{aligned}\frac{1}{N} I\{W_1,\cdots,W_N; W_0,W_{-1},\cdots\} &= \frac{1}{N} H\{W_1,\cdots,W_N\} \\ &\quad - H\{W_1 \mid W_0,W_{-1},\cdots\}. \quad (35)\end{aligned}$$

Now, from Theorem 6.4.1 in Ash [1] or Theorem 3.5.1 in Gallager [4], the right member of (35) approaches zero as $N \to \infty$. Thus (33c) is an upper bound on the rate of convergence of $R_N(\beta)$ to $R(\beta)$. Note that the right member of (33c) depends neither on β or the measure D.

Suppose that the source output set $\mathscr{W}$ is finite, and that the source has finite memory L. Then

$$\begin{aligned}\frac{1}{N} I\{W_1,\cdots,W_N; W_0,W_{-1},\cdots\} &= \frac{1}{N} I\{W_1,\cdots,W_N; W_0,\cdots,W_{-(L-1)}\} \\ &\leq \frac{1}{N} H\{W_1,\cdots,W_L\}. \quad (36)\end{aligned}$$

Thus $R_N(\beta)$ approaches $R(\beta)$ at least as fast as $1/N$.

A more interesting case is that of a stationary Gaussian source with $EW_k = 0$ and $EW_kW_{k+n} = \rho_n$. In this case we have (using the formula $H\{X \mid Y\} = H\{X,Y\} - H\{Y\}$ [1, p. 20])

$$\begin{aligned} I\{W_1,\cdots,W_N; W_0,\cdots,W_{-(p-1)}\} &= H\{W_1,\cdots,W_N\} - H\{W_1,\cdots,W_N \mid W_0,\cdots,W_{-(p-1)}\} \\ &= H\{W_1,\cdots,W_N\} + H\{W_0,W_{-1},\cdots,W_{-(p-1)}\} \\ &\quad - H\{W_N,W_{N-1},\cdots,W_{-(p-1)}\} \\ &= \frac{N}{2}\log\,[2\pi e|\mathscr{R}_N|^{1/N}] + \frac{p}{2}\log\,[2\pi e|\mathscr{R}_p|^{1/p}] \\ &\quad - \frac{(N+p)}{2}\log\,[2\pi e|\mathscr{R}_{N+p}|^{1/N+p}] \\ &= \tfrac{1}{2}\log\frac{|\mathscr{R}_N|\,|\mathscr{R}_p|}{|\mathscr{R}_{N+p}|}, \end{aligned} \tag{37}$$

where we have made use of (15). We now employ a refinement of Szegö's theorem, which is proved in Grenander and Szegö [7, p. 76]. Let $S(f)$ be as defined in (18). Then provided $S(f)$ is differentiable and satisfies the Lipschitz condition

$$|S'(f) - S'(f)| < K|f_1 - f_2|^{\alpha}, \qquad K > 0,\ 0 < \alpha < 1, \tag{38}$$

we have

$$\lim_{n\to\infty} |\mathscr{R}_n|/\bar{S}^n = K_0, \tag{39}$$

where K_0 is a constant whose rather complicated formula is given in [7]. Thus with N held fixed, $|\mathscr{R}_p|/|\mathscr{R}_{N+p}| \to \bar{S}^N$ as $p \to \infty$. Thus from (33c) and (37), for this Gaussian source,

$$\begin{aligned} |R_N(\beta) - R(\beta)| &\le \lim_{p\to\infty}\frac{1}{N} \\ &\quad \times I\{W_1,\cdots,W_N \mid W_0,W_{-1},\cdots,W_{-(p-1)}\} \\ &= \frac{1}{2N}\log\,[|\mathscr{R}_N|/\bar{S}^N]. \end{aligned}$$

Again applying (39), we have as $N \to \infty$,

$$|R_N(\beta) - R(\beta)| \le N^{-1}(\log K_0^{1/2})(1 + \varepsilon_N), \tag{40}$$

where $\varepsilon_N \to 0$. Note that the right member of (40) does not depend on the distortion measure D or the value of β.

III. Proof of Main Result

We seek to establish (8a),

$$R^*(\beta) - \Delta_N \le R_N(\beta) \le R^*(\beta).$$

We shall prove each of these inequalities separately.

A. Upper Bound

The inequality $R_N(\beta) \le R^*(\beta)$ will follow from the following lemma.

Lemma 1: Let $p_t^{(1)}(z \mid w) \in \tau_1(\beta)$ be a test channel. Then define the corresponding test channel for blocks of length N by

$$p_t^{(N)}(\boldsymbol{z} \mid \boldsymbol{w}) = \prod_{k=1}^{N} p_t^{(1)}(z_k \mid w_k), \tag{41}$$

where $\boldsymbol{w} = (w_1,\cdots,w_N) \in \mathscr{W}^N$ and $\boldsymbol{z} = (z_1,\cdots,z_N) \in \mathscr{Z}^N$. The density $p_t^{(N)}$ defines a joint density on $\mathscr{W}^N \times \mathscr{Z}^N$, and $I\{\boldsymbol{W};\boldsymbol{Z}\}$ is the corresponding information. The lemma states that 1)

$$N^{-1}I\{\boldsymbol{W};\boldsymbol{Z}\} \ge R_N(\beta)$$

and 2)

$$N^{-1}I\{\boldsymbol{W};\boldsymbol{Z}\} \le \frac{1}{N}\sum_{k=1}^{N} I\{W_k;Z_k\} = I\{W_1;Z_1\}.$$

Proof: 1) Since

$$ED^{(N)}(\boldsymbol{W},\boldsymbol{Z}) = \frac{1}{N}\sum_{k=1}^{N} ED(W_k;Z_k) \le \beta$$

(because $p_t^{(1)} \in \tau_1(\beta)$), then $p_t^{(N)} \in \tau_N(\beta)$. Thus by definition (5), $R_N(\beta) \le N^{-1}I\{\boldsymbol{W};\boldsymbol{Z}\}$.

2) Write

$$I\{\boldsymbol{W};\boldsymbol{Z}\} - \sum_{k=1}^{N} I\{W_k;Z_k\} = E\log\frac{p_t^{(N)}(\boldsymbol{Z} \mid \boldsymbol{W})}{p(\boldsymbol{Z})} - \sum_{k=1}^{N} E\log\frac{p_t^{(1)}(Z_k \mid W_k)}{p(Z_k)},$$

where $p(\boldsymbol{Z})$ and $p(Z_k)$ are the densities for $\boldsymbol{Z}$ and Z_k corresponding to $p_t^{(N)}$. Continuing, we have

$$\begin{aligned} I\{\boldsymbol{W};\boldsymbol{Z}\} - \sum_{k=1}^{N} I\{W_k;Z_k\} &= E\log\frac{\prod_{k=1}^{N} p(Z_k)}{p(\boldsymbol{Z})} \\ &= \int p(\boldsymbol{z})\,d\boldsymbol{z}\left[\log\frac{\prod_{k=1}^{N} p(z_k)}{p(\boldsymbol{z})}\right] \\ &\le \int p(\boldsymbol{z})\,d\boldsymbol{z}\left[\frac{\prod_{k=1}^{N} p(z_k)}{p(\boldsymbol{z})} - 1\right] \\ &= 0, \end{aligned}$$

where the inequality follows from $\log x \le x - 1$. This is the inequality part of 2). That $I\{W_k;Z_k\} \equiv I\{W_1;Z_1\}$, for $1 \le k \le N$, follows from the stationarity of the source output sequence and the definition of $p_t^{(N)}$ (41). This proves the lemma.

The upper bound now follows. From parts 1) and 2) of Lemma 1,

$$R_N(\beta) \le I\{W_1;Z_1\}, \tag{42}$$

where the joint pdf for W_1 and Z_1 corresponds to an arbitrary test channel $p_t^{(1)} \in \tau_1(\beta)$. Taking the infimum of (42) with respect to all such test channels yields

$$R_N(\beta) \le \inf_{p_t^{(1)} \in \tau_1(\beta)} I\{W_1;Z_1\} = R_1(\beta) = R^*(\beta),$$

which is the upper bound.

B. Lower Bound

We begin by establishing the following lemma.

Lemma 2: Let $\boldsymbol{W}^{(N)} = (W_1, W_2, \cdots, W_N)$ be a block of N successive source outputs, and let $\boldsymbol{Z}^{(N)} = (Z_1, Z_2, \cdots, Z_N)$ be the corresponding user sequence for some test channel $p_t(z \mid w)$. Then

$$I\{\boldsymbol{W},\boldsymbol{Z}\} \geq \sum_{k=1}^{N} I\{W_k, Z_k\} - N\,\Delta_N,$$

where Δ_N is given by (8b).

Proof:

$$\sum_{k=1}^{N} I\{W_k;Z_k\} - N\,\Delta_n - I\{\boldsymbol{W},\boldsymbol{Z}\}$$

$$= \sum_{k=1}^{N} E \log \frac{p(W_k \mid Z_k)}{P_S(W_k)} - E \log \frac{P_S^{(N)}(\boldsymbol{W})}{P_{S^*}^{(N)}(\boldsymbol{W})} - E \log \frac{p(\boldsymbol{W} \mid \boldsymbol{Z})}{P_S^{(N)}(\boldsymbol{W})},$$

where $p(W_k \mid Z_k)$ and $p(\boldsymbol{W} \mid \boldsymbol{Z})$ are the conditional pdfs for W_k and $\boldsymbol{W}$ given Z_k and $\boldsymbol{Z}$, respectively. Continuing,

$$\sum_{k=1}^{N} I\{W_k;Z_k\} - N\,\Delta_n - I\{\boldsymbol{W},\boldsymbol{Z}\}$$

$$= E\left[\log \frac{\prod_k p(W_k \mid Z_k)}{P_{S^*}^{(N)}(\boldsymbol{W})} - \log \frac{P_S^{(N)}(\boldsymbol{W})}{P_{S^*}^{(N)}(\boldsymbol{W})} - \log \frac{p(\boldsymbol{W} \mid \boldsymbol{Z})}{P_S^{(N)}(\boldsymbol{W})}\right]$$

$$= E \log \frac{\prod_k p(W_k \mid Z_k)}{p(\boldsymbol{W} \mid \boldsymbol{Z})}$$

$$= \int_{\mathscr{Z}^N} p(z)\, dz \int_{\mathscr{W}^N} p(w \mid z)\, dw \log \left(\frac{\prod_k p(w_k \mid z_k)}{p(w \mid z)}\right)$$

$$\leq \int p(z)\, dz \int p(w \mid z)\, dw \left[\frac{\prod_k p(w_k \mid z_k)}{p(w \mid z)} - 1\right] = 0.$$

Hence the lemma.

The lower bound now follows directly. Let $p_t(z \mid w) \in \tau_N(\beta)$. For this test channel let $\beta_k = ED(W_k, Z_k)$ for $1 \leq k \leq N$. Thus $(1/N) \sum_1^N \beta_k \leq \beta$. Applying the lemma, we get

$$\frac{1}{N} I\{\boldsymbol{W};\boldsymbol{Z}\} \geq \frac{1}{N} \sum_{k=1}^{N} I\{W_k;Z_k\} - \Delta_N.$$

But $I\{W_k;Z_k\} \geq R_1(\beta_k) = R^*(\beta_k)$ for $1 \leq k \leq N$. Thus

$$\frac{1}{N} I\{\boldsymbol{W};\boldsymbol{Z}\} \geq \frac{1}{N} \sum_{k=1}^{N} R^*(\beta_k) - \Delta_N.$$

By the convexity (downward) of $R^*(\beta)$,

$$\frac{1}{N} I\{\boldsymbol{W};\boldsymbol{Z}\} \geq R^*\left(\frac{1}{N} \sum_{k=1}^{N} \beta_k\right) - \Delta_N \geq R^*(\beta) - \Delta_N,$$

where the last inequality follows from the nonincreasing property of $R^*(\beta)$. Thus

$$R_N(\beta) = \inf \frac{1}{N} I\{\boldsymbol{W},\boldsymbol{Z}\} \geq R^*(\beta) - \Delta_N.$$

A necessary and sufficient condition for the lower bound of (8a) to hold with equality is that the test channel $P_t(z \mid w)$, which attains $R_N(\beta)$, be such that the "backward" channel pdf $p(w \mid z) = \prod_{k=1}^{N} p(w_k \mid z_k)$.

References

[1] R. B. Ash, *Information Theory.* New York: Interscience, 1965.
[2] T. Berger, "Information rates of Wiener processes," *IEEE Trans. Inform. Theory*, vol. IT-16, Mar. 1970, pp. 134–139.
[3] ——, *Rate Distortion Theory.* Englewood Cliffs, N.J.: Prentice-Hall, to be published.
[4] R. G. Gallager, *Information Theory and Reliable Communication.* New York: Wiley, 1968.
[5] T. J. Goblick, Jr., "A coding theorem for time-discrete analog data sources," *IEEE Trans. Inform. Theory*, vol. IT-15, May 1969, pp. 401–407.
[6] R. M. Gray, "Information rates of stationary ergodic finite-alphabet sources," this issue, pp. 516–523.
[7] V. Grenander and G. Szegö, *Toeplitz Forms and Their Applications.* Berkeley, Calif.: Univ. Calif. Press, 1958, chs. 3, 5.
[8] A. D. Wyner and J. Ziv, "On communication of analog data from a bounded source space," *Bell Syst. Tech. J.*, vol. 48, Dec. 1969, pp. 3139–3173.

Part III
Many Terminal Channels

C. E. Shannon

Channels with Side Information at the Transmitter

Abstract: In certain communication systems where information is to be transmitted from one point to another, additional side information is available at the transmitting point. This side information relates to the state of the transmission channel and can be used to aid in the coding and transmission of information. In this paper a type of channel with side information is studied and its capacity determined.

Introduction

Channels with feedback[1] from the receiving to the transmitting point are a special case of a situation in which there is additional information available at the transmitter which may be used as an aid in the forward transmission system. In Fig. 1 the channel has an input x and an output y.

Figure 1

There is a second output from the channel, u, available at the transmitting point, which may be used in the coding process. Thus the encoder has as inputs the message to be transmitted, m, and the side information u. The sequence of input letters x to the channel will be a function of the available part (that is, the past up to the current time) of these signals.

The signal u might be the received signal y, it might be a noisy version of this signal, or it might not relate to y but be statistically correlated with the general state of the channel. As a practical example, a transmitting station might have available a receiver for testing the current noise conditions at different frequencies. These results would be used to choose the frequency for transmission.

A simple discrete channel with side information is shown in Fig. 2. In this channel, x, y and u are all binary variables; they can be either zero or one. The channel can be used once each second. Immediately after it is used the random device chooses a zero or one independently of previous choices and with probabilities 1/2, 1/2. This value of u then appears at the transmitting point. The next x that is sent is added in the channel modulo 2 to this value of u to give the received y. If the side information u were *not* available at the transmitter, the channel would be that of Fig. 3, a channel in which input 0 has probabilities 1/2 of being received as 0 and 1/2 as 1 and similarly for input 1.

Figure 2

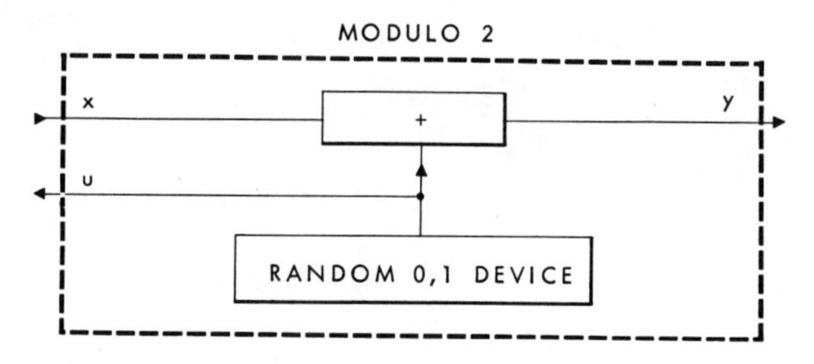

Figure 3

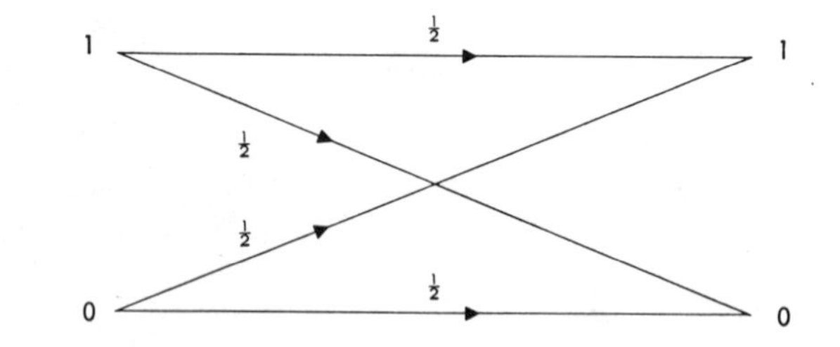

Such a channel has capacity zero. However, with the side information *available*, it is possible to send one bit per second through the channel. The u information is used to compensate for the noise inside by a preliminary reversal of zero and one, as in Fig. 4.

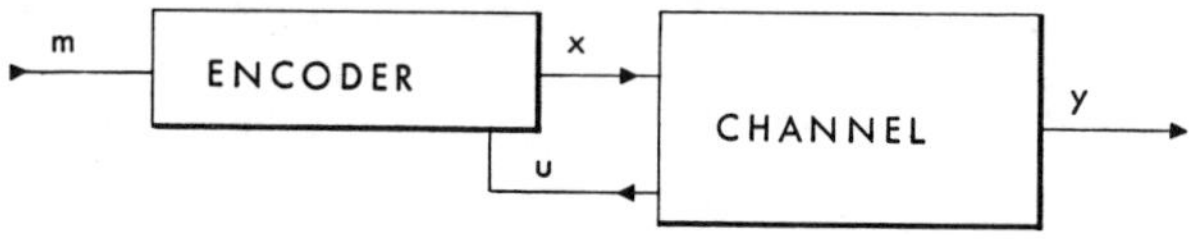

Figure 4

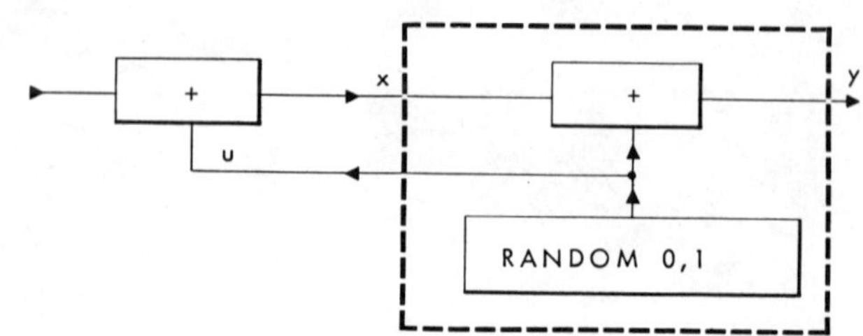

Reprinted with permission from *IBM J. Res. Develop.*, vol. 2, pp. 289–293, Oct. 1958.

Without studying the problem of side information in its fullest generality, which would involve possible historical effects in the channel, possibly infinite input and output alphabets, et cetera, we shall consider a moderately general case for which a simple solution has been found.

The memoryless discrete channel with side state information

Consider a channel which has a finite number of possible states, $s_1, s_2, \ldots, s_h$. At each use of the channel a new state is chosen, probability g_t for state s_t. This choice is statistically independent of previous states and previous input or output letters in the channel. The state is available as side information u at the transmitting point. When in state s_t the channel acts like a particular discrete channel K_t. Thus, its operation is defined by a set of transition probabilities $P_{ti}(j)$, $t=1, 2, \ldots, h$, $i=1, 2, \ldots, a$, $j=1, 2, \ldots, b$, where a is the number of input letters and b the number of output letters. Thus, abstractly, the channel is described by the set of state probabilities g_t and transition probabilities $p_{ti}(j)$, with g_t the probability of state t and $p_{ti}(j)$ the conditional probability, if in state t and i is transmitted, that j will be received.

A *block code* with M messages (the integers $1, 2, \ldots, M$) may be defined as follows for such a channel with side information. This definition, incidentally, is analogous to that for a channel with feedback given previously.[1] If n is the block length of the code, there are n functions $f_1(m;u_1)$, $f_2(m;u_1,u_2)$, $f_3(m;u_1,u_2,u_3), \ldots, f_n(m;u_1,u_2,\ldots,u_n)$. In these functions m ranges over the set of possible messages. Thus $m=1, 2, \ldots, M$. The u_i all range over the possible side information alphabet. In the particular case here each u_i can take values from 1 to g. Each function f_i takes values in the alphabet of input letters x of the channel. The value $f_i(m;u_1,u_2,\ldots,u_i)$ is the input x_i to be used in the code if the message is m and the side information up to the time corresponding to i consisted of $u_1, u_2, \ldots, u_i$. This is the mathematical equivalent of saying that a code consists of a way of determining, for each message m and each history of side information from the beginning of the block up to the present, the next transmitted letter. The important feature here is that only the data available at the time i, namely $m; u_1, u_2, \ldots, u_i$, may be used in deciding the next transmitted letter x_i, not the side information $u_{i+1}, \ldots, u_n$ yet to appear.

A decoding system for such a code consists of a mapping or function $h(y_1, y_2, \ldots, y_n)$ of received blocks of length n into messages m; thus h takes values from 1 to M. It is a way of deciding on a transmitted message given a complete received block $y_1, y_2, \ldots, y_n$.

For a given set of probabilities of the messages, and for a given channel and coding and decoding system, there will exist a calculable probability of error P_e; the probability of a message being encoded and received in such a way that the function h leads to deciding on a different message. We shall be concerned particularly with cases where the messages are equiprobable, each having probability $1/M$. The *rate* for such a code is $(1/n)\log M$. We are interested in the channel capacity C, that is, the largest rate R such that it is possible to construct codes arbitrarily close to rate R and with probability of error P_e arbitrarily small.

It may be noted that if the state information were *not* available at the transmitting point, the channel would act like a memoryless channel with transition probabilities given by

$$p'_i(j) = \sum_t g_t p_{ti}(j).$$

Thus, the capacity C_1 under this condition could be calculated by the ordinary means for memoryless channels. On the other hand, if the state information were available *both* at transmitting and receiving points, it is easily shown that the capacity is then given by $C_2 = \sum_t g_t C_t$, where C_t is the capacity of the memoryless channel with transmission probabilities $p_{ti}(j)$. The situation we are interested in here is intermediate—the state information is available at the transmitting point but not at the receiving point.

Theorem: *The capacity of a memoryless discrete channel K with side state information, defined by g_t and $p_{ti}(j)$, is equal to the capacity of the memoryless channel K' (without side information) with the same output alphabet and an input alphabet with a^h input letters $X=(x_1, x_2, \ldots, x_h)$ where each $x_i=1, 2, \ldots, a$. The transition probabilities $r_X(y)$ for the channel K' are given by*

$$r_X(y) = r_{x_1, x_2, \ldots, x_h}(y) = \sum g_t p_{tx_t}(y).$$

Any code and decoding system for K' can be translated into an equivalent code and decoding system for K with the same probability of error. Any code for K has an equivocation of message (conditional entropy per letter of the message given the received sequence) at least $R-C$, where C is the capacity of K'. Any code with rate $R>C$ has a probability of error bounded away from zero (whatever the block length n)

$$P_e \geq \frac{R-C}{6\left(R + \dfrac{1}{n}\ln\dfrac{R}{R-C}\right)}.$$

It may be noted that this theorem reduces the analysis of the given channel K *with* side information to that for a memoryless channel K' with more input letters but *without* side information. One uses known methods to determine the capacity of this derived channel K' and this gives the capacity of the original channel. Furthermore, codes for the derived channel may be translated into codes for the original channel with identical probability of error. (Indeed, all statistical properties of the codes are identical.)

We first show how codes for K' may be translated into codes for K. A code word for the derived channel K' consists of a sequence of n letters X from the X input alphabet of K'. A particular input *letter* X of this channel may be recognized as a particular *function* from the state alphabet to the input alphabet x of channel K. The full possible alphabet of X consists of the full set of a^h different possible functions from the state alphabet with h values to the input value with a values. Thus, each letter $X=(x_1, x_2, \ldots, x_h)$ of a code word for K' may be interpreted as a function from state u to input alphabet x. The translation of codes consists merely of using the input x given by this function of the state variable. Thus if the state variable u has the value 1, then x_1 is used in channel K; if it were state k, then x_k. In other words, the translation is a simple letter-by-letter translation without memory effects depending on previous states.

The codes for K' are really just another way of describing certain of the codes for K—namely those where the next input letter x is a function only of the message m and the current state u, and does not depend on the previous states.

It might be pointed out also that a simple physical device could be constructed which, placed ahead of the channel K, makes it look like K'. This device would have the X alphabet for one input and the state alphabet for another (this input connected to the u line of Fig. 1). Its output would range over the x alphabet and be connected to the x line of Fig. 1. Its operation would be to give an x output corresponding to the X function of the state u. It is clear that the statistical situations for K and K' with the translated code are identical. The probability of an input word for K' being received as a particular output word is the same as that for the corresponding operation with K. This gives the first part of the theorem.

To prove the second part of the theorem, we will show that in the original channel K, the change in conditional entropy (equivocation) of the message m at the receiving point when a letter is received cannot exceed C (the capacity of the derived channel K'). In Fig. 1, we let m be the message; x, y, u be the next input letter, output letter and state letter. Let U be the past sequence of u states from the beginning of the block code to the present (just before u), and Y the past sequence of output letters up to the current y. We are assuming here a given block code for encoding messages. The messages are chosen from a set with certain probabilities (not necessarily equal). Given the statistics of the message source, the coding system, and the statistics of the channel, these various entities m, x, y, U, Y all belong to a probability space and the various probabilities involved in the following calculation are meaningful. Thus the equivocation of message when Y has been received, $H(m|Y)$, is given by

$$H(m|Y) = -\sum_{m,Y} P(m, Y) \log P(m|Y)$$

$$= -E\left(\log P(m|Y)\right).$$

(The symbol $E(G)$ here and later means the expectation or average of G over the probability space.) The *change* in equivocation when the next letter y is received is

$$H(m|Y)-H(m|Y,y) = -E\left(\log P(m|Y)\right) + E\left(\log P(m|Y,y)\right)$$

$$= E\left(\log \frac{P(m|Y,y)}{P(m|Y)}\right)$$

$$= E\left(\log \frac{P(m,Y,y)P(Y)}{P(Y,y)P(m,Y)}\right)$$

$$= E\left(\log \frac{P(y|m,Y)P(Y)}{P(Y,y)}\right)$$

$$= E\left(\log \frac{P(y|m,Y)}{P(y)}\right) - E\left(\log \frac{P(Y,y)}{P(Y)P(y)}\right)$$

$$H(m|Y)-H(m|Y,y) \le E\left(\log \frac{P(y|m,Y)}{P(y)}\right). \qquad (1)$$

The last reduction is true since the term $E\left(\log \frac{P(Y,y)}{P(Y)P(y)}\right)$ is an average mutual information and therefore non-negative. Now note that by the independence requirements of our original system

$$P(y|x) = P(y|x,m,u,U) = P(y|x,m,u,U,Y).$$

Now since x is a strict function of m, u, and U (by the coding system function) we may omit this in the conditioning variables

$$P(y|m,u,U) = P(y|m,u,U,Y),$$

$$\frac{P(y,m,u,U)}{P(m,u,U)} = \frac{P(y,m,u,U,Y)}{P(m,u,U,Y)}.$$

Since the new state u is independent of the past $P(m,u,U) = P(u)P(m,U)$ and $P(m,u,U,Y) = P(u)P(m,U,Y)$. Substituting and simplifying,

$$P(y,u|m,U) = P(y,u|m,U,Y).$$

Summing on u gives

$$P(y|m,U) = P(y|m,U,Y).$$

Hence:

$$H(y|m,U) = H(y|m,U,Y) \le H(y|m,Y)$$

$$-E\left(\log P(y|m,U)\right) \le -E\left(\log P(y|m,Y)\right).$$

Using this in (1),

$$H(m|Y)-H(m|Y,y) \le E\left(\log \frac{P(y|m,U)}{P(y)}\right). \qquad (2)$$

We now wish to show that $P(y|m,U) = P(y|X)$. Here X is a random variable specifying the *function* from u to x imposed by the encoding operation for the next input x to the channel. Equivalently, X corresponds to an input letter in the derived channel K'. We have $P(y|x,u) = P(y|x,u,m,U)$. Furthermore, the coding system used implies a functional relation for determining the next input letter x, given m, U and u. Thus $x=f(m,U,u)$. If $f(m,U,u) = f(m',U',u)$ for two particular pairs (m,U) and (m',U') but for all u, then it follows that $P(y|m,U,u) = P(y|m',U',u)$ for all u and y; since m, U and u lead to the same x as m', U', and u. From this we obtain $P(y|m, U) = \sum_u P(u)P(y|m, U, u) = \sum_u P(u)P(y|m', U', u) = P(y|m', U')$. In other words, (m, U) pairs which give the same function $f(m, U, u)$ give the same value of $P(y|m, U)$ or, said another way, $P(y|m, U) = P(y|X)$.

Returning now to our inequality (2), we have

$$H(m|Y) - H(m|Y, y) \leq E\left(\log \frac{P(y|X)}{P(y)}\right)$$

$$\leq \max_{P(X)} E\left(\log \frac{P(y|X)}{P(y)}\right)$$

$$H(m|Y) - H(m|Y, y) \leq C.$$

This is the desired inequality on the equivocation. The equivocation cannot be reduced by more than C, the capacity of the derived channel K', for each received letter. In particular in a block code with M equiprobable messages, $R = 1/n \log M$. If $R > C$, then at the end of the block the equivocation must still be at least $nR - nC$, since it starts at nR and can only reduce at most C for each of the n letters.

It is shown in the Appendix that if the equivocation per letter is at least $R - C$ then the probability of error in decoding is bounded by

$$P_e \geq \frac{R - C}{6\left(R + \frac{1}{n} \ln \frac{R}{R - C}\right)}.$$

Thus the probability of error is bounded away from zero regardless of the block length n, if the code attempts to send at a rate $R > C$. This concludes the proof of the theorem.

As an example of this theorem, consider a channel with two output letters, any number a of input letters and any number h of states. Then the derived channel K' has two output letters and a^h input letters. However, in a channel with just two output letters, only two of the input letters need be used to achieve channel capacity, as shown in (2). Namely, we should use in K' only the two letters with maximum and minimum transition probabilities to one of the output letters. These two may be found as follows. The transition probabilities for a particular letter of K' are averages of the corresponding transitions for a set of letters for K, one for each state. To maximize the transition probability to one of the output letters, it is clear that we should choose in each state the letter with the maximum transition to that output letter. Similarly, to minimize, one chooses in each state the letter with the minimum transition probability to that letter. These two resulting letters in K' are the only ones used, and the corresponding channel gives the desired channel capacity. Formally, then, if the given channel has probabilities $p_{ti}(1)$ in state t for input letter i to output letter 1, and $p_{ti}(2) = 1 - p_{ti}(1)$ to the other output letter 2, we calculate:

$$p_1 = \sum_t g_t \max_i p_{ti}(1),$$

$$p_2 = \sum_t g_t \min_i p_{ti}(1).$$

The channel K' with two input letters having transition probabilities p_1 and $1 - p_1$ and p_2, $1 - p_2$ to the two output letters respectively, has the channel capacity of the original channel K.

Another example, with three output letters, two input letters and three states, is the following. With the states assumed to each have probability 1/3, the probability matrices for the three states are:

State 1			*State 2*			*State 3*		
1	0	0	0	1	0	0	0	1
0	1/2	1/2	1/2	0	1/2	1/2	1/2	0

In this case there are $2^3 = 8$ input letters in the derived channel K'. The matrix of these is as follows:

1/2	1/2	0
0	1/2	1/2
1/2	0	1/2
2/3	1/6	1/6
1/6	2/3	1/6
1/6	1/6	2/3
1/3	1/3	1/3
1/3	1/3	1/3

If there are only three output letters, one need use only three input letters to achieve channel capacity, and in this case it is readily shown that the first three can (and in fact must) be used. Because of the symmetry, these three letters must be used with equal probability and the resulting channel capacity is log (3/2).

In the original channel, it is easily seen that, if the state information were *not* available, the channel would act like one with the transition matrix

1/3	1/3	1/3
1/3	1/3	1/3

This channel clearly has zero capacity. On the other hand, if the state information were available at the *receiving* point or at *both* the receiving point and the transmitting point, the two input letters can be perfectly distinguished and the channel capacity is log 2.

Appendix

Lemma: Suppose there are M possible events with probabilities $p_i (i = 1, 2, \ldots, M)$. Given that the entropy H satisfies

$$H = -\sum p_i \ln p_i \geqslant \Delta,$$

then the total probability P_e for all possibilities except the most probable satisfies

$$P_e \geqslant \frac{\Delta}{6 \ln\left(\frac{M \ln M}{\Delta}\right)}.$$

Proof: For a given H, the minimum P_e will occur if all the probabilities except the largest one are equal. This follows from the convexity properties of entropy; equalizing two probabilities increases the entropy. Consequently, we may assume as the worst case a situation where there are $M-1$ possibilities, each with probability q, and one possibility with probability $1-(M-1)q$. Our given condition is then

$$-(M-1)q\ln q-[1-(M-1)q]\ln[1-(M-1)q]\geqslant\Delta.$$

Since $f(x)=-(1-x)\ln(1-x)$ is concave downward with slope 1 at $x=0$, ($f'(x)=1+\ln(1-x)$; $f''(x)=-\frac{1}{1-x}\leqslant 0$ for $0\leqslant x\leqslant 1$), it follows that $f(x)\leqslant x$ and the second term above is dominated by $(M-1)q$. The given condition then implies

$$-(M-1)q\ln q+(M-1)q\geqslant\Delta$$

or

$$(M-1)q\ln\frac{e}{q}\geqslant\Delta.$$

Now assume in contradiction to the conclusion of the lemma that

$$P_e=(M-1)q<\frac{\Delta}{6\left(\ln M+\ln\dfrac{\ln M}{\Delta}\right)}.$$

Since $q\ln\frac{e}{q}$ is monotone increasing in q, this would imply that

$$(M-1)q\ln\frac{e}{q}<\frac{\Delta}{6\left(\ln M+\ln\dfrac{\ln M}{\Delta}\right)}\log\frac{6e(M-1)\left(\ln M+\ln\dfrac{\ln M}{\Delta}\right)}{\Delta}$$

$$=\frac{\Delta}{6}\left[\frac{\ln\dfrac{M-1}{\Delta}}{\ln M+\ln\dfrac{\ln M}{\Delta}}+\frac{\ln 6e}{\ln M+\ln\dfrac{\ln M}{\Delta}}+\frac{\ln\left(\ln M+\ln\dfrac{\ln M}{\Delta}\right)}{\ln M+\ln\dfrac{\ln M}{\Delta}}\right]$$

$$\leqslant\frac{\Delta}{6}\left[1+3+\frac{1}{e}\right]<\Delta \qquad (M>1).$$

The first dominating constant is obtained by writing the corresponding term as $(\ln\ln M-\ln\Delta+\ln(M-1)-\ln\ln M)/(\ln\ln M-\ln\Delta+\ln M)$. Since $\ln M\geqslant\Delta$, this is easily seen to be dominated by 1 for $M\geqslant 2$. (For $M=1$, the lemma is trivially true since then $\Delta=0$.) The term dominated by 3 is obvious. The last term is of the form $\ln Z/Z$. By differentiation we find this takes its maximum at $Z=e$ and the maximum is $1/e$. Since our conclusion contradicts the hypothesis of the lemma, we have proved the desired result.

The chief application of this lemma is in placing a lower bound on probability of error in coding systems. If it is known that in a certain situation the "equivocation," that is, the conditional entropy of the message given a received signal, exceeds Δ, the lemma leads to a lower bound on the probability of error. Actually, the equivocation is an average over a set of received signals. Thus, the $\Delta=\sum P_i\Delta_i$ where P_i is the probability of receiving signal i and Δ_i is the corresponding entropy of message. If $f(\Delta)$ is the lower bound in the lemma, that is,

$$f(\Delta)=\frac{\Delta}{6\ln\left(\dfrac{M\ln M}{\Delta}\right)},$$

then the lower bound on P_e would be $P_e\geqslant\sum P_i f(\Delta_i)$. Now the function $f(\Delta)$ is convex downward (its second derivative is non-negative in the possible range). Consequently $\sum P_i f(\Delta_i)\geqslant f(\sum P_i\Delta_i)=f(\Delta)$ and we conclude that *the bound of the lemma remains valid even in this more general case by merely substituting the averaged value of* Δ.

A common situation for use of this result is in signaling with a code at a rate R greater than channel capacity C. In many types of situation this results in an equivocation of $\Delta=n(R-C)$ after n letters have been sent. In this case we may say that the probability of error for the block sent is bounded by (substituting these values in the lemma)

$$P_e\geqslant\frac{R-C}{6\left(R+\dfrac{1}{n}\ln\dfrac{R}{(R-C)}\right)}=\frac{R-C}{6\left(R-\ln\left(1-\dfrac{C}{R}\right)\right)}$$

This then is a lower bound on probability of error for rates greater than capacity under these conditions.

References

1. C. E. Shannon, "Zero Error Capacity of a Noisy Channel," *IRE Transactions on Information Theory, 1956 Symposium*, **IT-2**, No. 3.
2. C. E. Shannon, "Geometrische Deutung einiger Ergebnisse bei der Berechnung der Kanalkapazität," *Nachrichtentechnische Zeitschrift*, **10**, Heft 1, January 1957.
3. C. E. Shannon, "Certain Results in Coding Theory for Noisy Channels," *Information and Control*, **1**, No. 1, September 1957.

Revised manuscript received September 15, 1958

TWO-WAY COMMUNICATION CHANNELS

CLAUDE E. SHANNON
MASSACHUSETTS INSTITUTE OF TECHNOLOGY
CAMBRIDGE, MASSACHUSETTS

1. Introduction

A two-way communication channel is shown schematically in figure 1. Here x_1 is an input letter to the channel at terminal 1 and y_1 an output while x_2 is an

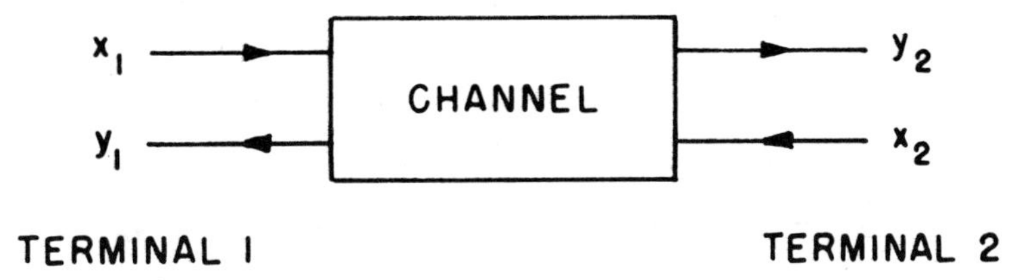

FIGURE 1

input at terminal 2 and y_2 the corresponding output. Once each second, say, new inputs x_1 and x_2 may be chosen from corresponding input alphabets and put into the channel; outputs y_1 and y_2 may then be observed. These outputs will be related statistically to the inputs and perhaps historically to previous inputs and outputs if the channel has memory. The problem is to communicate in both directions through the channel as effectively as possible. Particularly, we wish to determine what pairs of signalling rates R_1 and R_2 for the two directions can be approached with arbitrarily small error probabilities.

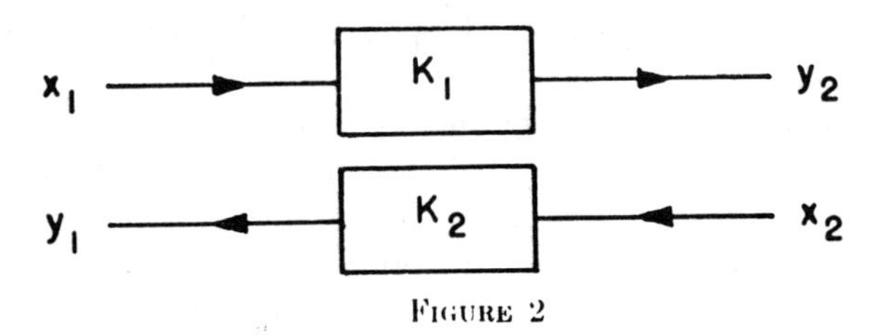

FIGURE 2

Before making these notions precise, we give some simple examples. In figure 2 the two-way channel decomposes into two independent one-way noiseless binary

This work was supported in part by the U.S. Army (Signal Corps), the U.S. Air Force (Office of Scientific Research, Air Research and Development Command), and the U.S. Navy (Office of Naval Research).

Reprinted from *Proc. 4th Berkeley Symp. Math. Stat. and Prob.*, vol. 1, pp. 611–644, 1961, with permission of the Regents of the University of California. Originally published by the University of California Press.

channels K_1 and K_2. Thus x_1, x_2, y_1 and y_2 are all binary variables and the operation of the channel is defined by $y_2 = x_1$ and $y_1 = x_2$. We can here transmit in each direction at rates up to one bit per second. Thus we can find codes whose

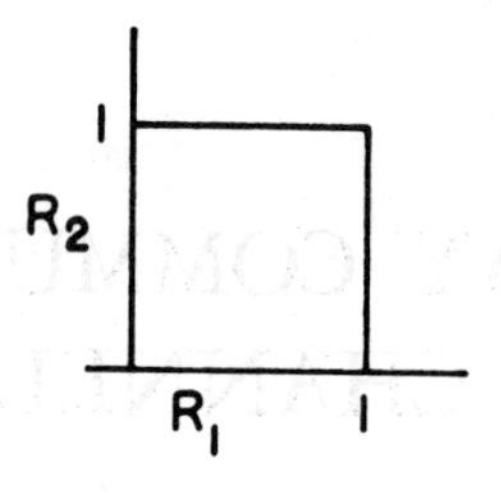

FIGURE 3

rates (R_1, R_2) approximate as closely as desired any point in the square, figure 3, with arbitrarily small (in this case, zero) error probability.

In figure 4 all inputs and outputs are again binary and the operation is defined

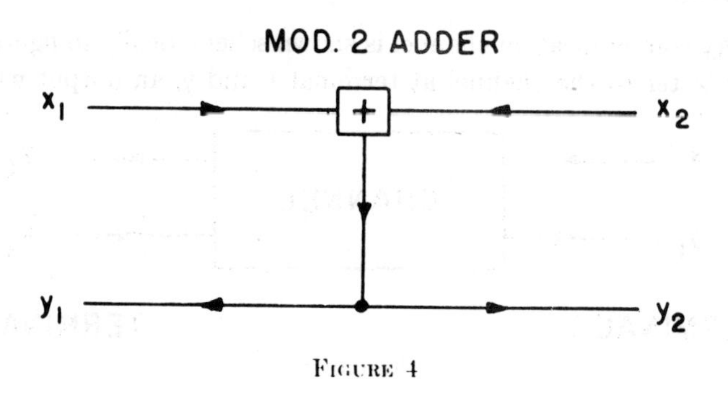

FIGURE 4

by $y_1 = y_2 = x_1 + x_2$ (mod 2). Here again it is possible to transmit one bit per second in each direction simultaneously, but the method is a bit more sophisticated. Arbitrary binary digits may be fed in at x_1 and x_2 but, to decode, the observed y must be corrected to compensate for the influence of the transmitted x. Thus an observed y_1 should be added to the just transmitted x_1 (mod 2) to determine the transmitted x_2. Of course here, too, one may obtain lower rates than the (1, 1) pair and again approximate any point in the square, figure 3.

A third example has inputs x_1 and x_2 each from a *ternary* alphabet and outputs y_1 and y_2 each from a binary alphabet. Suppose that the probabilities of different output pairs (y_1, y_2), conditional on various input pairs (x_1, x_2), are given by table I. It may be seen that by using only $x_1 = 0$ at terminal 1 it is possible to send one bit per second in the $2 - 1$ direction using only the input letters 1 and 2 at terminal 2, which then result with certainty in a and b respectively at terminal 1. Similarly, if x_2 is held at 0, transmission in the $1 - 2$ direction is possible at one bit per second. By dividing the time for use of these two strategies in the ratio λ to $1 - \lambda$ it is possible to transmit in the two directions with

TABLE I

x_1x_2 \ y_1y_2		Output Pair			
		aa	ab	ba	bb
Input Pair	00	1/4	1/4	1/4	1/4
	01	1/2	1/2	0	0
	02	0	0	1/2	1/2
	10	1/2	0	1/2	0
	11	1/4	1/4	1/4	1/4
	12	1/4	1/4	1/4	1/4
	20	0	1/2	0	1/2
	21	1/4	1/4	1/4	1/4
	22	1/4	1/4	1/4	1/4

average rates $R_1 = 1 - \lambda$, $R_2 = \lambda$. Thus we can find codes approaching any point in the triangular region, figure 5. It is not difficult to see, and will follow

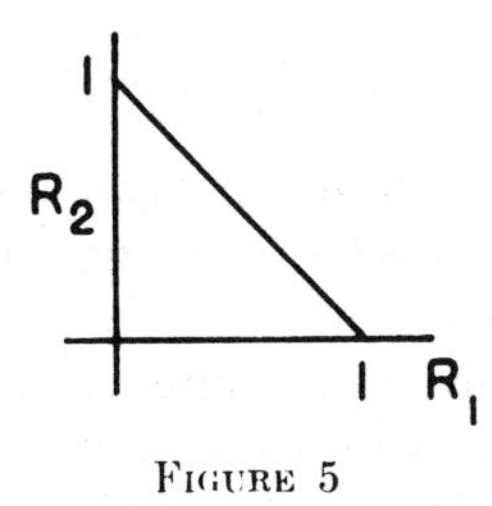

FIGURE 5

from later results, that no point outside this triangle can be approached with codes of arbitrarily low error probability.

In this channel, communication in the two directions might be called incompatible. Forward communication is possible only if x_2 is held at zero. Otherwise, all x_1 letters are completely noisy. Conversely, backward communication is possible only if x_1 is held at zero. The situation is a kind of discrete analogue to a common physical two-way system; a pair of radio telephone stations with "push-to-talk" buttons so arranged that when the button is pushed the local receiver is turned off.

A fourth simple example of a two-way channel, suggested by Blackwell, is the binary multiplying channel. Here all inputs and outputs are binary and the operation is defined $y_1 = y_2 = x_1x_2$. The region of approachable rate pairs for this channel is not known exactly, but we shall later find bounds on it.

In this paper we will study the coding properties of two-way channels. In particular, inner and outer bounds on the region of approachable rate pairs (R_1, R_2) will be found, together with bounds relating to the rate at which zero error probability can be approached. Certain topological properties of these bounds will be discussed and, finally, we will develop an expression describing the region of approachable rates in terms of a limiting process.

2. Summary of results

We will summarize here, briefly and somewhat roughly, the main results of the paper. It will be shown that for a memoryless discrete channel there exists a convex region G of approachable rates. For any point in G, say (R_1, R_2), there exist codes signalling with rates arbitrarily close to the point and with arbitrarily small error probability. This region is of the form shown typically in figure 6,

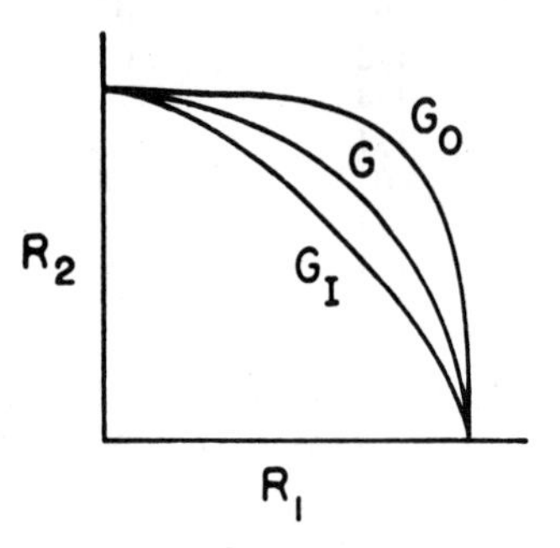

FIGURE 6

bounded by the middle curve G and the two axis segments. This curve can be described by a limiting expression involving mutual informations for long sequences of inputs and outputs.

In addition, we find an inner and outer bound, G_I and G_O, which are more easily evaluated, involving, as they do, only a maximizing process over single letters in the channel. G_O is the set of points (R_{12}, R_{21}) that may be obtained by assigning probabilities $P\{x_1, x_2\}$ to the input letters of the channel (an arbitrary joint distribution) and then evaluating

$$
\begin{aligned}
R_{12} &= E\left(\log \frac{P\{x_1|x_2, y_2\}}{P\{x_1|x_2\}}\right) = \sum_{x_1x_2y_2} P\{x_1x_2y_2\} \log \frac{P\{x_1|x_2, y_2\}}{P\{x_1|x_2\}} \\
R_{21} &= E\left(\log \frac{P\{x_2|x_1, y_1\}}{P\{x_2|x_1\}}\right),
\end{aligned}
\tag{1}
$$

where $E(\mu)$ means expectation of μ. The inner bound G_I is found in a similar way but restricting the distribution to an independent one $P\{x_1, x_2\} = P\{x_1\}P\{x_2\}$. Then G_I is the *convex hull* of (R_{12}, R_{21}) points found under this restriction.

It is shown that in certain important cases these bounds are identical so the capacity region is then completely determined from the bounds. An example is also given (the binary multiplying channel) where there is a discrepancy between the bounds.

The three regions G_I, G and G_O are all convex and have the same intercepts on the axes. These intercepts are the capacities in the two directions when the other input letter is fixed at its best value [for example, x_1 is held at the value which maximizes R_{21} under variation of $P\{x_2\}$]. For any point inside G the error probabilities approach zero exponentially with the block length n. For any point outside G at least one of the error probabilities for the two codes will be bounded away from zero by a bound independent of the block length.

Finally, these results may be partially generalized to channels with certain types of memory. If there exists an internal state of the channel such that it is possible to return to this state in a bounded number of steps (regardless of previous transmission) then there will exist again a capacity region G with similar properties. A limiting expression is given determining this region.

3. Basic definitions

A *discrete memoryless two-way channel* consists of a set of transition probabilities $P\{y_1, y_2|x_1, x_2\}$ where x_1, x_2, y_1, y_2 all range over finite alphabets (not necessarily the same).

A *block code pair* of length n for such a channel with M_1 messages in the forward direction and M_2 in the reverse direction consists of two sets of n functions

$$\begin{gathered} f_0(m_1), f_1(m_1, y_{11}), f_2(m_1, y_{11}, y_{12}), \cdots, f_{n-1}(m_1, y_{11}, \cdots, y_{1,n-1}) \\ g_0(m_2), g_1(m_2, y_{21}), g_2(m_2, y_{21}, y_{22}), \cdots, g_{n-1}(m_2, y_{21}, \cdots, y_{2,n-1}). \end{gathered} \tag{2}$$

Here the f functions all take values in the x_1 alphabet and the g functions in the x_2 alphabet, while m_1 takes values from 1 to M_1 (the forward messages) and m_2 takes values from 1 to M_2 (the backward messages). Finally y_{1i}, for $i = 1, 2, \cdots, n-1$, takes values from the y_1 alphabet and similarly for y_{2i}. The f functions specify how the next input letter at terminal 1 should be chosen as determined by the message m_1 to be transmitted and the observed outputs $y_{11}, y_{12}, \cdots$ at terminal 1 up to the current time. Similarly the g functions determine how message m_2 is encoded as a function of the information available at each time in the process.

A *decoding system* for a block code pair of length n consists of a pair of functions $\phi(m_1, y_{11}, y_{12}, \cdots, y_{1n})$ and $\psi(m_2, y_{21}, y_{22}, \cdots, y_{2n})$. These functions take values from 1 to M_2 and 1 to M_1 respectively.

The decoding function φ represents a way of deciding on the original transmitted message from terminal 2 given the information available at terminal 1 at the end of a block of n received letters, namely, $y_{11}, y_{12}, \cdots, y_{1n}$ together with the transmitted message m_1 at terminal 1. Notice that the transmitted sequence $x_{11}, x_{12}, \cdots, x_{1n}$ although known at terminal 1 need not enter as an argument in the decoding function since it is determined (via the encoding functions) by m_1 and the received sequence.

We will assume, except when the contrary is stated, that all messages m_1 are equiprobable (probability $1/M_1$), that all messages m_2 are equiprobable (probability $1/M_2$), and that these events are statistically independent. We also assume that the successive operations of the channel are independent,

$$P\{y_{11}, y_{12}, \cdots, y_{1n}, y_{21}, y_{22}, \cdots, y_{2n}|x_{11}, x_{12}, \cdots, x_{1n}, x_{21}, x_{22}, \cdots, x_{2n}\} = \prod_{i=1}^{n} P\{y_{1i}, y_{2i}|x_{1i}, x_{2i}\}. \tag{3}$$

This is the meaning of the memoryless condition. This implies that the probability of a set of outputs from the channel, conditional on the corresponding inputs, is the same as this probability conditional on these inputs and any previous inputs.

The *signalling rates* R_1 and R_2 for a block code pair with M_1 and M_2 messages for the two directions are defined by

$$R_1 = \frac{1}{n} \log M_1 \tag{4}$$
$$R_2 = \frac{1}{n} \log M_2.$$

Given a code pair and a decoding system, together with the conditional probabilities defining a channel and our assumptions concerning message probability, it is possible, in principle, to compute error probabilities for a code. Thus one could compute for each message pair the probabilities of the various possible received sequences, if these messages were transmitted by the given coding functions. Applying the decoding functions, the probability of an incorrect decoding could be computed. This could be averaged over all messages for each direction to arrive at final error probabilities P_{e1} and P_{e2} for the two directions.

We will say that a point (R_1, R_2) belongs to the *capacity region* G of a given memoryless channel K if, given any $\epsilon > 0$, there exists a block code and decoding system for the channel with signalling rates R_1^* and R_2^* satisfying $|R_1 - R_1^*| < \epsilon$ and $|R_2 - R_2^*| < \epsilon$ and such that the error probabilities satisfy $P_{e1} < \epsilon$ and $P_{e2} < \epsilon$.

4. Average mutual information rates

The two-way discrete memoryless channel with finite alphabets has been defined by a set of transition probabilities $P\{y_1, y_2|x_1, x_2\}$. Here x_1 and x_2 are the input letters at terminals 1 and 2 and y_1 and y_2 are the output letters. Each of these ranges over its corresponding finite alphabet.

If a set of probabilities $P\{x_1\}$ is assigned (arbitrarily) to the different letters of the input alphabet for x_1 and another set of probabilities $P\{x_2\}$ to the alphabet for x_2 (these two taken statistically independent) then there will be definite corresponding probabilities for y_1 and y_2 and, in fact, for the set of four random variables x_1, x_2, y_1, y_2, namely,

$$P\{x_1, x_2, y_1, y_2\} = P\{x_1\} P\{x_2\} P\{y_1, y_2|x_1, x_2\} \tag{5}$$
$$P\{y_1\} = \sum_{x_1, x_2, y_2} P\{x_1, x_2, y_1, y_2\},$$

and so forth.

Thinking first intuitively, and in analogue to the one-way channel, we might think of the rate of transmission from x_1 to the terminal 2 as given by $H(x_1) - H(x_1|x_2, y_2)$, that is, the uncertainty or entropy of x_1 less its entropy conditional on what is available at terminal 2, namely, y_2 and x_2. Thus, we might write

$$R_{12} = H(x_1) - H(x_1|x_2, y_2) \tag{6}$$
$$= E\left[\log \frac{P\{x_1, x_2, y_2\}}{P\{x_1\}P\{x_2, y_2\}}\right]$$
$$= E\left[\log \frac{P\{x_1|x_2, y_2\}}{P\{x_1\}}\right]$$

$$R_{21} = H(x_2) - H(x_2|x_1, y_1) \tag{7}$$
$$= E\left[\log \frac{P\{x_1, x_2, y_1\}}{P\{x_2\}P\{x_1, y_1\}}\right]$$
$$= E\left[\log \frac{P\{x_2|x_1, y_1\}}{P\{x_2\}}\right].$$

These are the average mutual informations with the assigned input probabilities between the input at one terminal and the input-output pair at the other terminal. We might expect, then, that by suitable coding it should be possible to send in the two directions *simultaneously* with arbitrarily small error probabilities and at rates arbitrarily close to R_{12} and R_{21}. The codes would be based on these probabilities $P\{x_1\}$ and $P\{x_2\}$ in generalization of the one-way channel. We will show that in fact it is possible to find codes based on the probabilities $P\{x_1\}$ and $P\{x_2\}$ which do this.

However the capacity region may be larger than the set of rates available by this means. Roughly speaking, the difference comes about because of the probability of having x_1 and x_2 dependent random variables. In this case the appropriate mutual informations are given by $H(x_2|x_1) - H(x_2|x_1, y_1)$ and $H(x_1|x_2) - H(x_1|x_2, y_2)$. The above expressions for R_{21} and R_{12} of course reduce to these when x_1 and x_2 are independent.

5. The distribution of information

The method we follow is based on random codes using techniques similar to those used in [1] for the one-way channel. Consider a sequence of n uses of the channel or, mathematically, the product probability space. The inputs are $X_1 = (x_{11}, x_{12}, \cdots, x_{1n})$ and $X_2 = (x_{21}, x_{22}, \cdots, x_{2n})$ and the outputs $Y_1 = (y_{11}, y_{12}, \cdots, y_{1n})$ and $Y_2 = (y_{21}, y_{22}, \cdots, y_{2n})$, that is, sequences of n choices from the corresponding alphabets.

The conditional probabilities for these blocks are given by

$$P\{Y_1, Y_2|X_1, X_2\} = \prod_k P\{y_{1k}, y_{2k}|x_{1k}, x_{2k}\}. \tag{8}$$

This uses the assumption that the channel is memoryless, or successive operations independent. We also associate a probability measure with input blocks X_1 and X_2 given by the product measure of that taken for x_1, x_2. Thus

$$(9)\qquad \begin{aligned} P\{X_1\} &= \prod_k P\{x_{1k}\} \\ P\{X_2\} &= \prod_k P\{x_{2k}\}. \end{aligned}$$

It then follows that other probabilities are also the products of those for the individual letters. Thus, for example,

$$(10)\qquad \begin{aligned} P\{X_1, X_2, Y_1, Y_2\} &= \prod_k P\{x_{1k}, x_{2k}, y_{1k}, y_{2k}\} \\ P\{X_2|X_1, Y_1\} &= \prod_k P\{x_{2k}|x_{1k}, y_{1k}\}. \end{aligned}$$

The (unaveraged) mutual information between, say, X_1 and the pair X_2, Y_2 may be written as a sum, as follows:

$$(11)\qquad \begin{aligned} I(X_1; X_2, Y_2) &= \log \frac{P\{X_1, X_2, Y_2\}}{P\{X_1\}P\{X_2, Y_2\}} = \log \frac{\prod_k P\{x_{1k}, x_{2k}, y_{2k}\}}{\prod_k P\{x_{1k}\} \prod_k P\{x_{2k}, y_{2k}\}} \\ &= \sum_k \log \frac{P\{x_{1k}, x_{2k}, y_{2k}\}}{P\{x_{1k}\}P\{x_{2k}, y_{2k}\}} \\ I(X_1; X_2, Y_2) &= \sum_k I(x_{1k}; x_{2k}, y_{2k}). \end{aligned}$$

Thus, the mutual information is, as usual in such independent situations, the sum of the individual mutual informations. Also, as usual, we may think of the mutual information as a random variable. Here $I(X_1; X_2, Y_2)$ takes on different values with probabilities given by $P\{X_1, X_2, Y_2\}$. The *distribution function* for $I(X_1; X_2, Y_2)$ will be denoted by $\rho_{12}(Z)$ and similarly for $I(X_2; X_1, Y_1)$

$$(12)\qquad \begin{aligned} \rho_{12}(Z) &= P\{I(X_1; X_2, Y_2) \leqq Z\} \\ \rho_{21}(Z) &= P\{I(X_2; X_1, Y_1) \leqq Z\}. \end{aligned}$$

Since each of the random variables $I(X_1; X_2, Y_2)$ and $I(X_2; X_1, Y_1)$ is the sum of n independent random variables, each with the same distribution, we have the familiar statistical situation to which one may apply various central limit theorems and laws of large numbers. The mean of the distributions ρ_{12} and ρ_{21} will be nR_{12} and nR_{21} respectively and the variances n times the corresponding variances for one letter. As $n \to \infty$, $\rho_{12}[n(R_{12} - \epsilon)] \to 0$ for any fixed $\epsilon > 0$, and similarly for ρ_{21}. In fact, this approach is exponential in n; $\rho_{12}[n(R_{12} - \epsilon)] \leqq \exp[-A(\epsilon)n]$.

6. Random codes for the two-way channel

After these preliminaries we now wish to prove the existence of codes with certain error probabilities bounded by expressions involving the distribution functions ρ_{12} and ρ_{21}.

We will construct an ensemble of codes or, more precisely, of *code pairs*, one code for the 1 − 2 direction and another for the 2 − 1 direction. Bounds will be established on the error probabilities P_{e1} and P_{e2} *averaged over the ensemble*, and from these will be shown the existence of *particular* codes in the ensemble with related bounds on their error probabilities.

The random ensemble of code pairs for such a two-way channel with M_1 words in the 1 − 2 code and M_2 words in the 2 − 1 code is constructed as follows. The M_1 integers $1, 2, \cdots, M_1$ (the messages of the first code) are mapped in all possible ways into the set of input words X_1 of length n. Similarly the integers $1, 2, \cdots, M_2$ (the messages of the second code) are mapped in all possible ways into the set of input words X_2 of length n.

If there were a_1 possible input *letters* at terminal 1 and a_2 input *letters* at terminal 2, there will be a_1^n and a_2^n input *words* of length n and $a_1^{nM_1}$ *mappings* in the first code and $a_2^{nM_2}$ in the second code. We consider all pairs of these codes, a total of $a_1^{nM_1}a_2^{nM_2}$ pairs.

Each code pair is given a weighting, or probability, equal to the probability of occurrence of that pair if the two mappings were done independently and an integer is mapped into a word with the assigned probability of that word. Thus, a code pair is given a weighting equal to the product of the probabilities associated with all the input words that the integers are mapped into for both codes. This set of code pairs with these associated probabilities we call the *random ensemble of code pairs* based on the assigned probabilities $P\{X_1\}$ and $P\{X_2\}$.

Any particular code pair of the ensemble could be used to transmit information, if we agreed upon a method of decoding. The method of decoding will here consist of two functions $\phi(X_1, Y_1)$ and $\psi(X_2, Y_2)$, a special case of that defined above. Here X_1 varies over the input words of length n at terminal 1, and Y_1 over the possible received blocks of length n. The function ϕ takes values from 1 to M_2 and represents the decoded message for a received Y_1 if X_1 was transmitted. (Of course, X_1 is used in the decoding procedure in general since it may influence Y_1 and is, therefore, pertinent information for best decoding.)

Similarly, $\psi(X_2, Y_2)$ takes values from 1 to M_1 and is a way of deciding on the transmitted message m_1 on the basis of information available at terminal 2. It should be noted here that the decoding functions, ϕ and ψ, need not be the same for all code pairs in the ensemble.

We also point out that the encoding functions for our random ensemble are more specialized than the general case described above. The sequence of input letters X_1 for a given message m_1 do not depend on the received letters at terminal 1. In any particular code of the ensemble there is a strict mapping from messages to input sequences.

Given an ensemble of code pairs as described above and decoding functions, one could compute for each particular code pair two error probabilities for the two codes: P_{e1}, the probability of error in decoding the first code, and P_{e2} that for the second. Here we are assuming that the different messages in the first code occur with equal probability $1/M_1$, and similarly for the second.

By the *average error probabilities for the ensemble* of code pairs we mean the averages $E(P_{e1})$ and $E(P_{e2})$ where each probability of error for a particular code is weighted according to the weighting factor or probability associated with the code pair. We wish to describe a particular method of decoding, that is, a choice of ϕ and ψ, and then place upper bounds on these average error probabilities for the ensemble.

7. Error probability for the ensemble of codes

THEOREM 1. *Suppose probability assignments $P\{X_1\}$ and $P\{X_2\}$ in a discrete memoryless two-way channel produce information distribution functions $\rho_{12}(Z)$ and $\rho_{21}(Z)$. Let $M_1 = \exp(R_1 n)$ and $M_2 = \exp(R_2 n)$ be arbitrary integers and θ_1 and θ_2 be arbitrary positive numbers. Then the random ensemble of code pairs with M_1 and M_2 messages has (with appropriate decoding functions) average error probabilities bounded as follows:*

$$(13)\qquad \begin{aligned} E(P_{e1}) &\leqq \rho_{12}[n(R_1 + \theta_1)] + e^{-n\theta_1} \\ E(P_{e2}) &\leqq \rho_{21}[n(R_2 + \theta_2)] + e^{-n\theta_2}. \end{aligned}$$

There will exist in the ensemble at least one code pair whose individual *error probabilities are bounded by* two times *these expressions, that is, satisfying*

$$(14)\qquad \begin{aligned} P_{e1} &\leqq 2\rho_{12}[n(R_1 + \theta_1)] + 2e^{-n\theta_1} \\ P_{e2} &\leqq 2\rho_{21}[n(R_2 + \theta_2)] + 2e^{-n\theta_2}. \end{aligned}$$

This theorem is a generalization of theorem 1 in [1] which gives a similar bound on P_e for a one-way channel. The proof for the two-way channel is a generalization of that proof.

The statistical situation here is quite complex. There are several statistical events involved: the choice of messages m_1 and m_2, the choice of code pair in the ensemble of code pairs, and finally the statistics of the channel itself which produces the output words Y_1 and Y_2 according to $P\{Y_1, Y_2|X_1, X_2\}$. The ensemble error probabilities we are calculating are averages over *all* these statistical events.

We first define decoding systems for the various codes in the ensemble. For a given θ_2, define for each pair X_1, Y_1 a corresponding set of words in the X_2 space denoted by $S(X_1, Y_1)$ as follows:

$$(15)\qquad S(X_1, Y_1) = \left\{X_2 \middle| \log \frac{P\{X_1, X_2, Y_1\}}{P\{X_2\}P\{X_1, Y_1\}} > n(R_2 + \theta_2)\right\}.$$

That is, $S(X_1, Y_1)$ is the set of X_2 words whose mutual information with the particular pair (X_1, Y_1) exceeds a certain level, $n(R_2 + \theta_2)$. In a similar way, we define a set $S'(X_2, Y_2)$ of X_1 words for each X_2, Y_2 pair as follows:

$$(16)\qquad S'(X_2, Y_2) = \left\{X_1 \middle| \log \frac{P\{X_1, X_2, Y_2\}}{P\{X_1\}P\{X_2, Y_2\}} > n(R_1 + \theta_1)\right\}.$$

We will use these sets S and S' to define the decoding procedure and to aid in overbounding the error probabilities. The decoding process will be as follows. In any particular code pair in the random ensemble, suppose message m_1 is sent and this is mapped into input word X_1. Suppose that Y_1 is received at terminal 1 in the corresponding block of n letters. Consider the subset of X_2 words, $S(X_1, Y_1)$. Several situations may occur. (1) There is no message m_2 mapped into the subset $S(X_1, Y_1)$ for the code pair in question. In this case, X_1, Y_1 is decoded (conventionally) as message number one. (2) There is exactly one message mapped into the subset. In this case, we decode as this particular message. (3) There are more than one such messages. In this case, we decode as the smallest numbered such message.

The error probabilities that we are estimating would normally be thought of as calculated in the following manner. For each code pair one would calculate the error probabilities for all messages m_1 and m_2, and from their averages get the error probabilities for that code pair. Then these error probabilities are averaged over the ensemble of code pairs, using the appropriate weights or probabilities. We may, however, interchange this order of averaging. We may consider the cases where a particular $\overline{m}_1$ and $\overline{m}_2$ are the messages and these are mapped into particular $\overline{X}_1$ and $\overline{X}_2$, and the received words are $\overline{Y}_1$ and $\overline{Y}_2$. There is still, in the statistical picture, the range of possible code pairs, that is, mappings of the other $M_1 - 1$ messages for one code and $M_2 - 1$ for the other. We wish to show that, averaged over this subset of codes, the probabilities of any of these messages being mapped into subsets $S'(\overline{X}_2, \overline{Y}_2)$ and $S(\overline{X}_1, \overline{Y}_1)$ respectively do not exceed $\exp(-n\theta_1)$ and $\exp(-n\theta_2)$.

Note first that if X_1 belongs to the set $S'(\overline{X}_2, \overline{Y}_2)$ then by the definition of this set

$$\log \frac{P\{X_1, \overline{X}_2, \overline{Y}_2\}}{P\{X_1\}P\{\overline{X}_2, \overline{Y}_2\}} > n(R_1 + \theta_1)$$

$$P\{X_1|\overline{X}_2, \overline{Y}_2\} > P\{X_1\}e^{n(R_1+\theta_1)}. \tag{17}$$

Now sum each side over the set of X_1 belonging to $S'(\overline{X}_2, \overline{Y}_2)$ to obtain

$$1 \geqq \sum_{X_1 \in S'(\overline{X}_2, \overline{Y}_2)} P\{X_1|\overline{X}_2, \overline{Y}_2\} > e^{n(R_1+\theta_1)} \sum_{X_1 \in S'(\overline{X}_2, \overline{Y}_2)} P\{X_1\}. \tag{18}$$

The left inequality here holds since a sum of disjoint probabilities cannot exceed one. The sum on the right we may denote by $P\{S'(\overline{X}_2, \overline{Y}_2)\}$. Combining the first and last members of this relation

$$P\{S'(\overline{X}_2, \overline{Y}_2)\} < e^{-n(R_1+\theta_1)}. \tag{19}$$

That is, the total probability associated with any set $S'(\overline{X}_2, \overline{Y}_2)$ is bounded by an expression involving n, R_1 and θ_1 but *independent* of the particular $\overline{X}_2$, $\overline{Y}_2$.

Now recall that the messages were mapped independently into the input words using the probabilities $P\{X_1\}$ and $P\{X_2\}$. The probability of a particular message being mapped into $S'(\overline{X}_2, \overline{Y}_2)$ in the ensemble of code pairs is just $P\{S'(\overline{X}_2, \overline{Y}_2)\}$. The probability of being in the complementary set is $1 -$

$P\{S'(\overline{X}_2, \overline{Y}_2)\}$. The probability that *all* messages other than $\overline{m}_1$ will be mapped into this complementary set is

$$\begin{aligned}(20)\qquad [1 - P\{S'(\overline{X}_2, \overline{Y}_2)\}]^{M_1-1} &\geqq 1 - (M_1 - 1)P\{S'(\overline{X}_2, \overline{Y}_2)\}\\ &\geqq 1 - M_1 P\{S'(\overline{X}_2, \overline{Y}_2)\}\\ &\geqq 1 - M_1 e^{-n(R_1+\theta_1)}\\ &= 1 - e^{-n\theta_1}.\end{aligned}$$

Here we used the inequality $(1 - x)^p \geqq 1 - px$, the relation (19) and finally the fact that $M_1 = \exp(nR_1)$.

We have established, then, that in the subset of cases being considered ($\overline{m}_1$ and $\overline{m}_2$ mapped into $\overline{X}_1$ and $\overline{X}_2$ and received as $\overline{Y}_1$ and $\overline{Y}_2$), with probability at least $1 - \exp(-n\theta_1)$, there will be no other messages mapped into $S'(\overline{X}_2, \overline{Y}_2)$. A similar calculation shows that with probability exceeding $1 - \exp(-n\theta_2)$ there will be no other messages mapped into $S(\overline{X}_1, \overline{Y}_1)$. These bounds, as noted, are independent of the particular $\overline{X}_1$, $\overline{Y}_1$ and $\overline{X}_2$, $\overline{Y}_2$.

We now bound the probability of the actual message $\overline{m}_1$ being within the subset $S'(\overline{X}_2, \overline{Y}_2)$. Recall that from the definition of $\rho_{12}(Z)$

$$(21)\qquad \rho_{12}[n(R_1 + \theta_1)] = P\left\{\log \frac{P\{X_1, X_2, Y_2\}}{P\{X_1\}P\{X_2, Y_2\}} \leq n(R_1 + \theta_1)\right\}.$$

In the ensemble of code pairs a message $\overline{m}_1$, say, is mapped into words X_1 with probabilities just equal to $P\{X_1\}$. Consequently, the probability in the full ensemble of code pairs, message choices and channel statistics, that the actual message is mapped into $S'(\overline{X}_2, \overline{Y}_2)$ is precisely $1 - \rho_{12}[n(R_1 + \theta_1)]$.

The probability that the actual message is mapped *outside* $S'(\overline{X}_2, \overline{Y}_2)$ is therefore given by $\rho_{12}[n(R_1 + \theta_1)]$ and the probability that there are any other messages mapped into $S'(\overline{X}_2, \overline{Y}_2)$ is bounded as shown before by $\exp(-n\theta_1)$. The probability that *either* of these events is true is then certainly bounded by $\rho_{12}[n(R_1 + \theta_1)] + \exp(-n\theta_1)$; but this is then a bound on $E(P_{e1})$, since if neither event occurs the decoding process will correctly decode.

Of course, the same argument with interchanged indices gives the corresponding bound for $E(P_{e2})$. This proves the first part of the theorem.

With regard to the last statement of the theorem, we will first prove a simple combinatorial lemma which is useful not only here but in other situations in coding theory.

LEMMA. *Suppose we have a set of objects $B_1, B_2, \cdots, B_n$ with associated probabilities $P_1, P_2, \cdots, P_n$, and a number of numerically valued properties (functions) of the objects $f_1, f_2, \cdots, f_d$. These are all nonnegative, $f_i(B_j) \geqq 0$, and we know the averages A_i of these properties over the objects,*

$$(22)\qquad \sum_j P_j f_i(B_j) = A_i, \qquad i = 1, 2, \cdots, d.$$

Then there exists an object B_p for which

$$(23)\qquad f_i(B_p) \leqq dA_i, \qquad i = 1, 2, \cdots, d.$$

More generally, given any set of $K_i > 0$ *satisfying* $\sum_{i=1}^{d} (1/K_i) \leqq 1$, *then there exists an object* B_p *with*

$$f_i(B_p) \leqq K_i A_i, \qquad i = 1, 2, \cdots, d. \tag{24}$$

PROOF. The second part implies the first by taking $K_i = d$. To prove the second part let Q_i be the total probability of objects B for which $f_i(B) > K_i A_i$. Now the average $A_i > Q_i K_i A_i$ since $Q_i K_i A_i$ is contributed by the B_i with $f(B) > K_i A_i$ and all the remaining B have f_i values $\geqq 0$. Hence

$$Q_i < \frac{1}{K_i}, \qquad i = 1, 2, \cdots, d. \tag{25}$$

The total probability Q of objects violating *any* of the conditions is less than or equal to the sum of the individual Q_i, so that

$$Q < \sum_{i=1}^{d} \frac{1}{K_i} \leqq 1. \tag{26}$$

Hence there is at least one object not violating any of the conditions, concluding the proof.

For example, suppose we know that a room is occupied by a number of people whose average age is 40 and average height 5 feet. Here $d = 2$, and using the simpler form of the theorem we can assert that there is someone in the room not over 80 years old and not over ten feet tall, even though the room might contain aged midgets and youthful basketball players. Again, using $K_1 = 8/3$, $K_2 = 8/5$, we can assert the existence of an individual not over 8 feet tall and not over 106 2/3 years old.

Returning to the proof of theorem 1, we can now establish the last sentence. We have a set of objects, the code pairs, and two properties of each object, its error probability P_{e1} for the code from 1 to 2 and its error probability P_{e2} for the code from 2 to 1. These are nonnegative and their averages are bounded as in the first part of theorem 1. It follows from the combinatorial result that there exists at least one particular code pair for which simultaneously

$$\begin{aligned} P_{e1} &\leqq 2\{\rho_{12}[n(R_1 + \theta_1)] + e^{-n\theta_1}\} \\ P_{e2} &\leqq 2\{\rho_{21}[n(R_2 + \theta_2)] + e^{-n\theta_2}\}. \end{aligned} \tag{27}$$

This concludes the proof of theorem 1.

It is easily seen that this theorem proves the possibility of code pairs arbitrarily close in rates R_1 and R_2 to the mean mutual information per letter R_{12} and R_{21} for any assigned $P\{x_1\}$ and $P\{x_2\}$ and with arbitrarily small probability of error. In fact, let $R_{12} - R_1 = R_{21} - R_2 = \epsilon > 0$ and in the theorem take $\theta_1 = \theta_2 = \epsilon/2$. Since $\rho_{12}[n(R_{12} - \epsilon/2)] \to 0$ and, in fact, exponentially fast with n (the distribution function $\epsilon n/2$ to the left of the mean, of a sum of n random variables) the bound on P_{e1} approaches zero with increasing n exponentially fast. In a similar way, so does the bound on P_{e2}. By choosing, then, a sequence of the M_1 and M_2 for increasing n which approach the desired rates R_1 and R_2 from below, we obtain the desired result, which may be stated as follows.

THEOREM 2. *Suppose in a two-way memoryless channel K an assignment of probabilities to the input letters $P\{x_1\}$ and $P\{x_2\}$ gives average mutual informations in the two directions*

$$R_{12} = E\left(\log \frac{P\{x_1|x_2, y_2\}}{P\{x_1\}}\right)$$

$$R_{21} = E\left(\log \frac{P\{x_2|x_1, y_1\}}{P\{x_2\}}\right). \tag{28}$$

Then given $\epsilon > 0$ there exists a code pair for all sufficiently large block length n with signalling rates in the two directions greater than $R_{12} - \epsilon$ and $R_{21} - \epsilon$ respectively, and with error probabilities $P_{e1} \leqq \exp[-A(\epsilon)n]$, $P_{e2} \leqq \exp[-A(\epsilon)n]$ where $A(\epsilon)$ is positive and independent of n.

By trying different assignments of letter probabilities and using this result, one obtains various points in the capacity region. Of course, to obtain the best rates available from this theorem we should seek to maximize these rates. This is most naturally done à la Lagrange by maximizing $R_{12} + \lambda R_{21}$ for various positive λ.

8. The convex hull G_1 as an inner bound of the capacity region

In addition to the rates obtained this way we may construct codes which are *mixtures* of codes obtained by this process. Suppose one assignment $P\{x_1\}$, $P\{x_2\}$ gives mean mutual informations R_{12}, R_{21} and a second assignment $P'\{x_1\}$, $P'\{x_2\}$ gives R'_{12}, R'_{21}. Then we may find a code of (sufficiently large) length n for the first assignment with error probabilities $< \delta$ and rate discrepancy less than or equal to ϵ and a second code of length n' based on $P'\{x_1\}$, $P'\{x_2\}$ with the same δ and ϵ. We now consider the code of length $n + n'$ with $M_1M'_1$ words in the forward direction, and $M_2M'_2$ in the reverse, consisting of all words of the first code followed by all words for the same direction in the second code. This has signalling rates R_1^* and R_2^* equal to the weighted average of rates for the original codes $[R_1^* = nR_1/(n + n') + n'R'_1/(n + n'); R_2^* = nR_2/(n + n') + n'R'_2/(n + n')]$ and consequently its rates are within ϵ of the weighted averages, $|R_1^* - nR_{12}/(n + n') - n'R'_{12}/(n + n')| < \epsilon$ and similarly. Furthermore, its error probability is bounded by 2δ, since the probability of either of two events (an error in either of the two parts of the code) is bounded by the sum of the original probabilities. We can construct such a mixed code for *any* sufficiently large n and n'. Hence by taking these large enough we can approach any weighted average of the given rates and simultaneously approach zero error probability exponentially fast. It follows that *we can annex to the set of points found by the assignment of letter probabilities all points in the convex hull of this set.* This actually does add new points in some cases as our example, of a channel (table I) with incompatible transmission in the two directions, shows. By mixing the codes for assignments which give the points (0, 1) and (1, 0) in equal proportions, we obtain the point (1/2, 1/2). There is no single letter assignment giving this pair of rates. We may summarize as follows.

THEOREM 3. *Let G_I be the convex hull of points* (R_{12}, R_{21})

$$R_{12} = E\left(\log \frac{P\{x_1|x_2, y_2\}}{P\{x_1\}}\right) \tag{29}$$
$$R_{21} = E\left(\log \frac{P\{x_2|x_1, y_1\}}{P\{x_2\}}\right)$$

when $P\{x_1\}$ and $P\{x_2\}$ are given various probability assignments. All points of G_I are in the capacity region. For any point (R_1, R_2) in G_I and any $\epsilon > 0$ we can find codes whose signalling rates are within ϵ of R_1 and R_2 and whose error probabilities in both directions are less than $\exp[-A(\epsilon)n]$ for all sufficiently large n, and some positive $A(\epsilon)$.

It may be noted that the convex hull G_I in this theorem is a closed set (contains all its limit points). This follows from the continuity of R_{12} and R_{21} as functions of the probability assignments $P\{x_1\}$ and $P\{x_2\}$. Furthermore if G_I contains a point (R_1, R_2) it contains the projections $(R_1, 0)$ and $(0, R_2)$. This will now be proved.

It will clearly follow if we can show that the projection of any point obtained by a letter probability assignment is also in G_I. To show this, suppose $P\{x_1\}$ and $P\{x_2\}$ give the point (R_{12}, R_{21}). Now R_{12} is the average of the various particular R_{12} when x_2 is given various particular values. Thus

$$R_{12} = \sum_{x_2} P\{x_2\} \sum_{x_1, y_2} P\{x_1, y_2|x_2\} \log \frac{P\{x_1|x_2, y_2\}}{P\{x_1\}}. \tag{30}$$

There must exist, then, a particular x_2, say x_2^*, for which the inner sum is at least as great as the average, that is, for which

$$\sum_{x_1, y_2} P\{x_1, y_2|x_2^*\} \log \frac{P\{x_1|x_2^*, y_2\}}{P\{x_1\}} \geq \sum_{x_2} P\{x_2\} \sum_{x_1, y_2} P\{x_1, y_2|x_2\} \log \frac{P\{x_1|x_2, y_2\}}{P\{x_1\}}. \tag{31}$$

The assignment $P\{x_1|x_2^*\}$ for letter probabilities x_1 and the assignment $P\{x_2\} = 1$ if $x_2 = x_2^*$ and 0 otherwise, now gives a point on the horizontal axis below or to the right of the projection of the given point R_{12}, R_{21}. Similarly, we can find an x_1^* such that the assignment $P\{x_2|x_1^*\}$ for x_2 and $P\{x_1^*\} = 1$ gives a point on the vertical axis equal to or above the projection of R_{12}, R_{21}. Note also that the assignment $P\{x_1^*\} = 1$, $P\{x_2^*\} = 1$ gives the point (0, 0). By suitable mixing of codes obtained for these four assignments one can approach any point of the quadrilateral defined by the corresponding pairs of rates, and in particular any point in the rectangle subtended by R_{12}, R_{21}. It follows from these remarks that the convex hull G_I is a region of the form shown typically in figure 7 bounded by a horizontal segment, a convex curve, a vertical segment, and two segments of the axes. Of course, any of these parts may be of zero length.

The convex hull G_I is, as we have seen, inside the capacity region and we will refer to it as the *inner bound*.

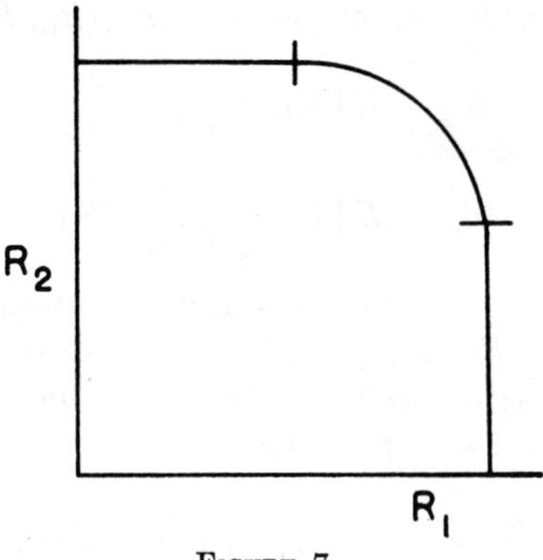

FIGURE 7

It is of some interest to attempt a sharper evaluation of the rate of improvement of error probability with increasing code length n. This is done in the appendix and leads to a generalization of theorem 2 in [1]. The bound we arrive at is based on logarithms of moment generating functions.

9. An outer bound on the capacity region

While in some cases the convex hull G_I, the inner bound defined above, is actually the capacity region this is not always the case. By an involved calculation R. G. Gallager has shown that in the binary multiplying channel the inner bound is strictly interior to the capacity region. However a *partial* converse to theorem 3 and an *outer* bound on the capacity region can be given. Suppose we have a code starting at time zero with messages m_1 and m_2 at the two terminals. After n operations of the channel, let Y_1 and Y_2 be the received blocks at the two terminals (sequences of n letters), and let x_1, x_2, y_1, y_2 be the next transmitted and received letters. Consider the change in "equivocation" of message at the two terminals due to the next received letter. At terminal 2, for example, this change is (making some obvious reductions)

$$
\begin{aligned}
(32)\qquad \Delta &= H(m_1|m_2, Y_2) - H(m_1|m_2, Y_2, y_2)\\
&= E\left[\log \frac{P\{m_2, Y_2\}}{P\{m_1, m_2, Y_2\}}\right] - E\left[\log \frac{P\{m_2, Y_2, y_2\}}{P\{m_1, m_2, Y_2, y_2\}}\right]\\
&= E\left[\log \frac{P\{y_2|m_1, m_2, Y_2\}}{P\{y_2|x_2\}} \frac{P\{y_2|x_2\}}{P\{y_2|Y_2, m_2\}}\right].
\end{aligned}
$$

Now $H(y_2|m_1, m_2, Y_2) \geqq H(y_2|m_1, m_2, Y_1, Y_2) = H(y_2|x_1, x_2)$ since adding a conditioning variable cannot increase an entropy and since $P\{y_2|m_1, m_2, Y_1, Y_2\} = P\{y_2|x_1, x_2\}$.

Also $H(y_2|x_2) \geqq H(y_2|Y_2, m_2)$ since x_2 is a function of Y_2 and m_2 by the coding function. Therefore

$$\Delta \leqq E\left(\log \frac{P\{y_2|x_1, x_2\}}{P\{y_2|x_2\}}\right) + H(y_2|Y_2, m_2) - H(y_2|x_2) \tag{33}$$

$$\Delta \leqq E\left(\log \frac{P\{y_2|x_1, x_2\}}{P\{y_2|x_2\}}\right) = E\left(\log \frac{P\{y_2, x_1, x_2\}P\{x_2\}}{P\{x_2, y_2\}P\{x_1, x_2\}}\right) \tag{34}$$

$$= E\left(\log \frac{P\{x_1|x_2, y_2\}}{P\{x_1|x_2\}}\right).$$

This would actually lead to a converse of theorem 1 if we had independence of the random variables x_1 and x_2. This last expression would then reduce to $E[\log (P\{x_1|x_2, y_2\}/P\{x_1\})]$. Unfortunately in a general code they are not necessarily independent. In fact, the next x_1 and x_2 may be functionally related to received X and Y and hence dependent.

We may, however, at least obtain an outer bound on the capacity surface. Namely, the above inequality together with the similar inequality for the second terminal imply that the vector change in equivocation due to receiving another letter must be a vector with components bounded by

$$E\left(\log \frac{P\{x_1|x_2, y_2\}}{P\{x_1|x_2\}}\right), E\left(\log \frac{P\{x_2|x_1, y_1\}}{P\{x_2|x_1\}}\right) \tag{35}$$

for some $P\{x_1, x_2\}$. Thus the vector change is included in the convex hull of all such vectors G_O (when $P\{x_1, x_2\}$ is varied).

In a code of length n, the *total* change in equivocation from beginning to end of the block cannot exceed the sum of n vectors from this convex hull. Thus this sum will lie in the convex hull nG_O, that is, G_O expanded by a factor n.

Suppose now our given code has signalling rates $R_1 = (1/n) \log M_1$ and $R_2 = (1/n) \log M_2$. Then the initial equivocations of message are nR_1 and nR_2. Suppose the point (nR_1, nR_2) is outside the convex hull nG_O with nearest distance $n\epsilon_1$, figure 8. Construct a line L passing through the nearest point of nG_O and

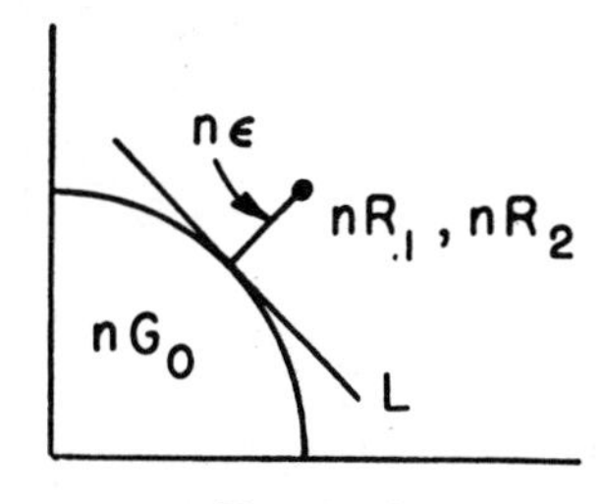

FIGURE 8

perpendicular to the nearest approach segment with nG_O on one side (using the fact that nG_O is a convex region). It is clear that for any point (nR_1^*, nR_2^*) on the nG_O side of L and particularly for any point of nG_O, that we have $|nR_1 - nR_1^*| + |nR_2 - nR_2^*| \geqq n\epsilon$ (since the shortest distance is $n\epsilon$) and furthermore at least one of the $nR_1 - nR_1^*$ and $nR_2 - nR_2^*$ is at least $n\epsilon/\sqrt{2}$. (In a right triangle at least one leg is as great as the hypotenuse divided by $\sqrt{2}$.)

Thus after n uses of the channel, if the signalling rate pair R_1, R_2 is distance ϵ outside the convex hull G_O, at least one of the two final equivocations is at least $\epsilon/\sqrt{2}$, where all equivocations are on a per second basis. Thus for signalling rates ϵ outside of G_O the equivocations per second are bounded from below independent of the code length n. This implies that the error probability is also bounded from below, that is, at least one of the two codes will have error probability $\geqq f(\epsilon) > 0$ independent of n, as shown in [2], appendix.

To summarize, the capacity region G is included in the convex hull G_O of all points R_{12}, R_{21}

$$R_{12} = E\left[\log \frac{P\{x_1|x_2, y_2\}}{P\{x_1|x_2\}}\right] \tag{36}$$

$$R_{21} = E\left[\log \frac{P\{x_2|x_1, y_1\}}{P\{x_2|x_1\}}\right]$$

when arbitrary joint probability assignments $P\{x_1, x_2\}$ are made.

Thus the inner bound G_I and the outer bound G_O are both found by the same process, assigning input probabilities, calculating the resulting average mutual informations R_{12} and R_{21} and then taking the convex hull. The only difference is that for the outer bound a general joint assignment $P\{x_1, x_2\}$ is made, while for the inner bound the assignments are restricted to independent $P\{x_1\}P\{x_2\}$.

We now develop some properties of the outer bound.

10. The concavity of R_{12} and R_{21} as functions of $P(x_1,x_2)$

THEOREM 4. *Given the transition probabilities* $P\{y_1, y_2|x_1, x_2\}$ *for a channel* K, *the rates*

$$R_{12} = E\left[\log \frac{P\{y_2|x_1, x_2\}}{P\{y_2|x_2\}}\right] \tag{37}$$

$$R_{21} = E\left[\log \frac{P\{y_1|x_1, x_2\}}{P\{y_1|x_1\}}\right]$$

are concave downward functions of the assigned input probabilities $P\{x_1, x_2\}$. For example, $R_{12}(P_1\{x_1, x_2\}/2 + P_2\{x_1, x_2\}/2) \geqq R_{12}(P_1\{x_1, x_2\})/2 + R_{12}(P_2\{x_1, x_2\})/2$.

This concave property is a generalization of that given in [3] for a one-way channel. To prove the theorem it suffices, by known results in convex functions, to show that

$$R_{12}\left(\frac{1}{2}P_1\{x_1, x_2\} + \frac{1}{2}P_2\{x_1, x_2\}\right) \geqq \frac{1}{2}R_{12}(P_1\{x_1, x_2\}) + \frac{1}{2}R_{12}(P_2\{x_1, x_2\}). \tag{38}$$

But $R_{12}(P_1\{x_1, x_2\})$ and $R_{12}(P_2\{x_1, x_2\})$ may be written

$$R_{12}(P_1\{x_1, x_2\}) = \sum_{x_2} P_1\{x_2\} \sum_{x_1, y_2} P_1\{x_1, y_2|x_2\} \log \frac{P_2\{y_2|x_1, x_2\}}{P_1\{y_2|x_2\}} \tag{39}$$

$$R_{12}(P_2\{x_1, x_2\}) = \sum_{x_2} P_2\{x_2\} \sum_{x_1, y_2} P_2\{x_1, y_2|x_2\} \log \frac{P_2\{y_2|x_1, x_2\}}{P_2\{y_2|x_2\}}. \tag{40}$$

Here the subscripts 1 on probabilities correspond to those produced with the probability assignment $P_1\{x_1, x_2\}$ to the inputs, and similarly for the subscript 2. The inner sum $\sum_{x_1, y_2} P_1\{x_1, y_2|x_2\} \log (P_1\{y_2|x_1, x_2\}/P_1\{y_2|x_2\})$ may be recognized as the rate for the channel from x_1 to y_2 conditional on x_2 having a particular value and with the x_1 assigned probabilities corresponding to its conditional probability according to $P_1\{x_1, x_2\}$.

The corresponding inner sum with assigned probabilities $P_2\{x_1, x_2\}$ is $\sum_{x_1, y_2} P_2\{x_1, y_2|x_2\} \log (P_2\{y_2|x_1, x_2\}/P_2\{y_2|x_2\})$, which may be viewed as the rate conditional on x_2 for the same one-way channel but with the assignment $P_2\{x_1|x_2\}$ for the input letters.

Viewed this way, we may apply the concavity result of [2]. In particular, the weighted average of these rates with weight assignments $P_1\{x_2\}/(P_1\{x_2\} + P_2\{x_2\})$ and $P_2\{x_2\}/(P_1\{x_2\} + P_2\{x_2\})$ is dominated by the rate for this one-way channel when the probability assignments are the weighted average of the two given assignments. This weighted average of the given assignment is

$$\begin{aligned} P_3\{x_1, x_2\} &= \frac{P_1\{x_2\}}{P_1\{x_2\} + P_2\{x_2\}} P_1\{x_1|x_2\} + \frac{P_2\{x_2\}}{P_1\{x_2\} + P_2\{x_2\}} P_2\{x_1|x_2\} \\ &= \frac{1}{2} \frac{1}{(P_1\{x_2\} + P_2\{x_2\})} 2\,(P_1\{x_1, x_2\} + P_2\{x_1, x_2\}). \end{aligned} \tag{41}$$

Thus the sum of two corresponding terms (the same x_2) from (38) above is dominated by $P_1\{x_2\} + P_2\{x_2\}$ multiplied by the rate for this one-way channel with these averaged probabilities. This latter rate, on substituting the averaged probabilities, is seen to be

$$\sum_{x_1, y_2} P_3\{x_1, y_2|x_2\} \log \frac{P_3\{y_2|x_1, x_2\}}{P_3\{y_2|x_2\}} \tag{42}$$

where the subscript 3 corresponds to probabilities produced by using $P_3\{x_1, x_2\} = (P_1\{x_1, x_2\} + P_2\{x_1, x_2\}/2)/2$. In other words, the sum of (39) and (40) (including the first summation on x_2) is dominated by

$$\begin{aligned} &\sum_{x_2} (P_1\{x_2\} + P_2\{x_2\}) \sum_{x_1, y_2} P_3\{x_1, y_2|x_2\} \log \frac{P_3\{y_2|x_1, x_2\}}{P_3\{y_2|x_2\}} \\ &\qquad = 2 \sum_{x_1, x_2, y_2} P_3\{x_1, y_2, x_2\} \log \frac{P_3\{y_2|x_1, x_2\}}{P_3\{y_2|x_2\}}. \end{aligned} \tag{43}$$

This is the desired result for the theorem.

11. Applications of the concavity property; channels with symmetric structure

Theorem 4 is useful in a number of ways in evaluating the outer bound for particular channels. In the first place, we note that $R_{12} + \lambda R_{21}$ as a function of $P\{x_1, x_2\}$ and for positive λ is also a concave downward function. Consequently any local maximum is the absolute maximum and numerical investigation in locating such maxima by the Lagrange multiplier method is thereby simplified.

In addition, this concavity result is very powerful in helping locate the maxima when "symmetries" exist in a channel. Suppose, for example, that in a given channel the transition probability array $P\{y_1, y_2|x_1, x_2\}$ has the following property. There exists a relabelling of the input letters x_1 and of the output letters y_1 and y_2 which interchanges, say, the first two letters of the x_1 alphabet but leaves the set of probabilities $P\{y_1, y_2|x_1, x_2\}$ the same. Now if some particular assignment $P\{x_1, x_2\}$ gives outer bound rates R_{12} and R_{21}, then if we apply the same permutation to the x alphabet in $P\{x_1, x_2\}$ we obtain a new probability assignment which, however, will give exactly the same outer bound rates R_{12} and R_{21}. By our concavity property, if we average these two probability assignments we obtain a new probability assignment which will give at least as large values of R_{12} and R_{21}. In this averaged assignment for any particular x_2 the first two letters in the x_1 alphabet are assigned equal probability. In other words, in such a case an assignment for maximizing $R_{12} + \lambda R_{21}$, say $P\{x_1, x_2\}$ viewed as a matrix, will have its first two rows identical.

If the channel had sufficiently symmetric structure that *any* pair of x_1 letters might be interchanged by relabelling the x_1 alphabet and the y_1 and y_2 alphabets while preserving $P\{y_1, y_2|x_1, x_2\}$, then a maximizing assignment $P\{x_1, x_2\}$ would exist in which *all* rows are identical. In this case the entries are functions of x_2 only: $P\{x_1, x_2\} = P\{x_2\}/\alpha$ where α is the number of letters in the x_1 alphabet. Thus the maximum for a dependent assignment of $P\{x_1, x_2\}$ is actually obtained with x_1 and x_2 independent. *In other words, in this case of a full set of symmetric interchanges on the x_1 alphabet, the inner and outer bounds are identical.* This gives an important class of channels for which the capacity region can be determined with comparative ease.

An example of this type is the channel with transition probabilities as follows. All inputs and outputs are binary, $y_1 = x_2$ (that is, there is a noiseless binary channel from terminal 2 to terminal 1). If $x_2 = 0$, then $y_2 = x_1$, while if $x_2 = 1$, y_2 has probability .5 of being 0 and .5 of being 1. In other words, if x_2 is 0 the binary channel in the forward direction is noiseless, while if x_2 is 1 it is completely noisy. We note here that if the labels on the x_1 alphabet are interchanged while we simultaneously interchange the y_2 labels, the channel remains unaltered, all conditional probabilities being unaffected. Following the analysis above, then, the inner and outer bounds will be the same and give the capacity region. Furthermore, the surface will be attained with equal rows in the $P\{x_1, x_2\}$ matrix as shown in table II.

TABLE II

		x_2 0	1
x_1	0	$p/2$	$q/2$
	1	$p/2$	$q/2$

For a particular p this assignment gives the rates

$$R_{12} = p, \qquad R_{21} = -(p \log p + q \log q). \tag{44}$$

These come from substituting in the formulas or by noting that in the $1 - 2$ direction the channel is acting like an erasure channel, while in the $2 - 1$ direction it is operating like a binary noiseless channel with unequal probabilities assigned to the letters. This gives the capacity region of figure 9.

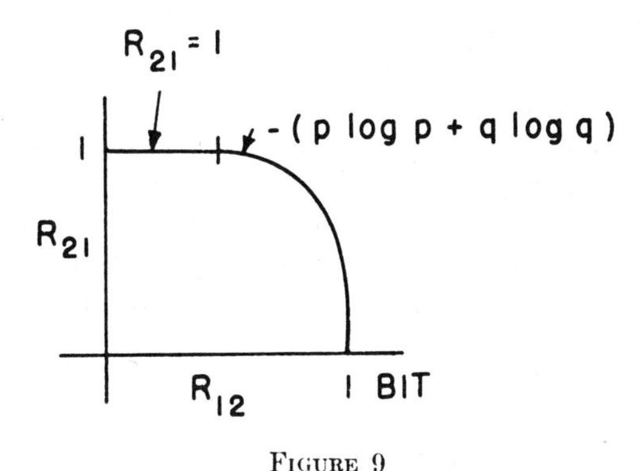

FIGURE 9

There are many variants and applications of these interchange and symmetry tricks for aid in the evaluation of capacity surfaces. For example, if *both* the x_1 and x_2 alphabets have a full set of interchanges leaving the transition probabilities the same, then the maximizing distribution must be identical both in rows and columns and hence all entries are the same, $P\{x_1, x_2\} = 1/\alpha c$ where α and c are the number of letters in the x_1 and x_2 alphabets. In this case, then, all attainable $R_{12}R_{21}$ points are dominated by the particular point obtained from this uniform probability assignment. *In other words, the capacity region is a rectangle in the case of a full set of symmetric interchanges for both x_1 and x_2.*

An example of this type is the channel of figure 2 defined by $y_1 = y_2 = x_1 \oplus x_2$ where $\oplus$ means mod 2 addition.

12. Nature of the region attainable in the outer bound

We now will use the concavity property to establish some results concerning the set Γ of points (R_{12}, R_{21}) that can be obtained by all possible assignments of probabilities $P\{x_1, x_2\}$ in a given channel K, and whose convex hull is G_O. We will show that the set Γ is in fact already convex and therefore identical with G_O and that it consists of all points in or on the boundary of a region of the type shown in figure 10 bounded by a horizontal segment L_1, an outward convex seg-

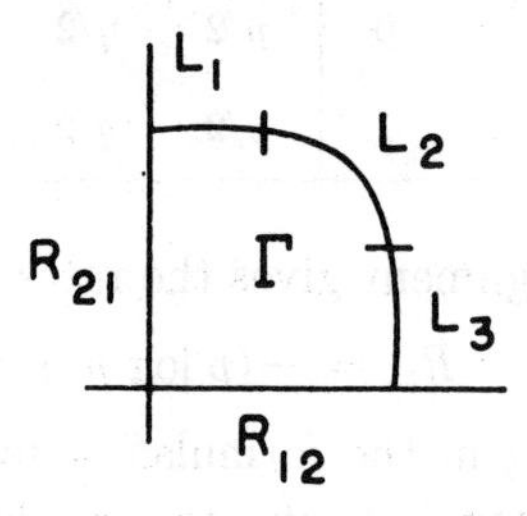

FIGURE 10

ment L_2, a vertical segment L_3 and two segments of the coordinate axes. Thus G_O has a structure similar to G_I.

Suppose some $P\{x_1, x_2\}$ gives a point (R_{12}, R_{21}). Here R_{12} is, as we have observed previously, an average of the different R_{12} which would be obtained by fixing x_2 at different values, that is, using these with probability 1 and applying the conditional probabilities $P\{x_1|x_2\}$ to the x_1 letters. The weighting is according to factors $P\{x_2\}$. It follows that some particular x_2 will do as well at least as this weighted average. If this particular x_2 is x_2^*, the set of probabilities $P\{x_1|x_2^*\}$ gives at least as large a value of R_{12} and simultaneously makes $R_{21} = 0$. In

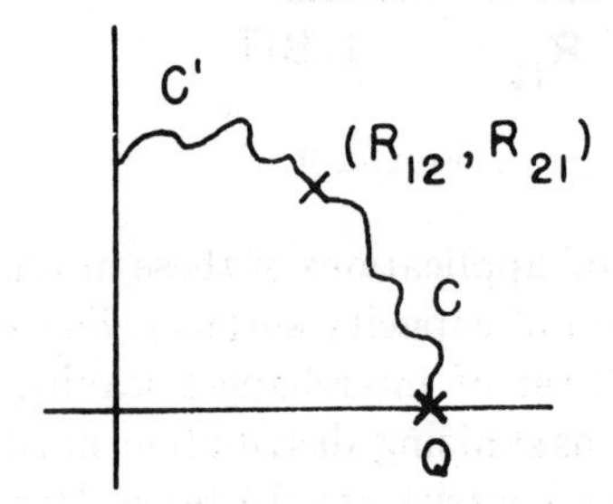

FIGURE 11

figure 11 this means we can find a point in Γ below or to the right of the projection of the given point as indicated (point Q).

Now consider mixtures of these two probability assignments, that is, assignments of the form $\lambda P\{x_1x_2\} + (1 - \lambda)P\{x_1|x_2^*\}$. Here λ is to vary continuously from 0 to 1. Since R_{12} and R_{21} are continuous functions of the assigned probability, this produces a continuous curve C running from the given point to the point Q. Furthermore, this curve lies entirely to the upper right of the connecting line segment. This is because of the concavity property for the R_{12} and R_{21} expressions. In a similar way, we construct a curve C', as indicated, of points belonging to Γ and lying on or above the horizontal straight line through the given point.

Now take all points on the curves C and C' and consider mixing the corresponding probability assignments with the assignment $P\{x_1^*, x_2^*\} = 1$ (all other pairs given zero probability). This last assignment gives the point $(0, 0)$. The fraction of this $(0, 0)$ assignment is gradually increased for 0 up to 1. As this is done the curve of resulting points changes continuously starting at the CC' curve and collapsing into the point $(0, 0)$. The end points stay on the axes during this operation. Consequently by known topological results the curve sweeps through the entire area bounded by C, C' and the axes and in particular *covers the rectangle subtended by the original point* (R_{12}, R_{21}).

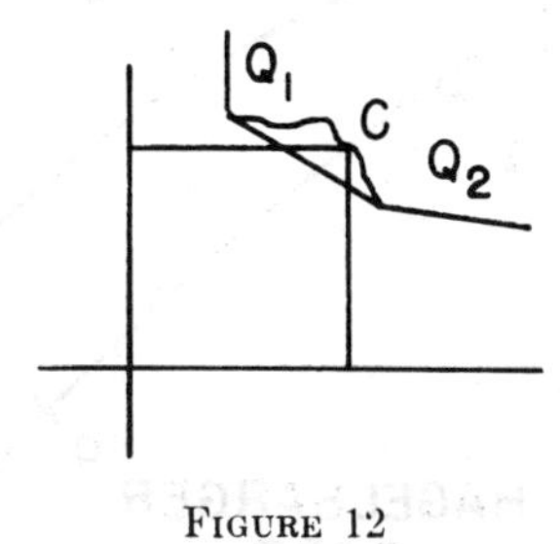

FIGURE 12

We will show that the set of points Γ is a convex set. Suppose Q_1 and Q_2, figure 12, are two points which can be obtained by assignments $P_1\{x_1, x_2\}$ and $P_2\{x_1, x_2\}$.

By taking mixtures of varying proportions one obtains a continuous curve C connecting them, lying, by the concavity property, to the upper right of the connecting line segment. Since these are points of Γ all of their subtended rectangles are, as just shown, points of Γ. It follows that all points of the connecting line segment are points of Γ. Note that if Q_1 and Q_2 are in the first and third quadrants relative to each other the result is trivially true, since then the connecting line segment lies in the rectangle of one of the points.

These results are sufficient to imply the statements at the beginning of this section, namely the set Γ is convex, identical with G_O, and if we take the largest attainable R_{12} and for this R_{12} the largest R_{21}, then points in the subtended rectangle are attainable. Similarly for the largest R_{21}.

It may be recalled here that the set of points attainable by *independent* assignments, $P\{x_1, x_2\} = P\{x_1\}P\{x_2\}$, is not necessarily a convex set. This is shown by the example of table I.

It follows also from the results of this section that *the end points of the outer bound curve* (where it reaches the coordinate axes) *are the same as the end points of the inner bound curve.* This is because, as we have seen, the largest R_{12} can be achieved using only one particular x_2 with probability 1. When this is done, $P\{x_1, x_2\}$ reduces to a product of independent probabilities.

13. An example where the inner and outer bounds differ

The inner and outer bounds on the capacity surface that we have derived above are not always the same. This was shown by David Blackwell for the binary multiplying channel defined by $y_1 = y_2 = x_1x_2$. The inner and outer bounds for this channel have been computed numerically and are plotted in figure 13. It may be seen that they differ considerably, particularly in the middle

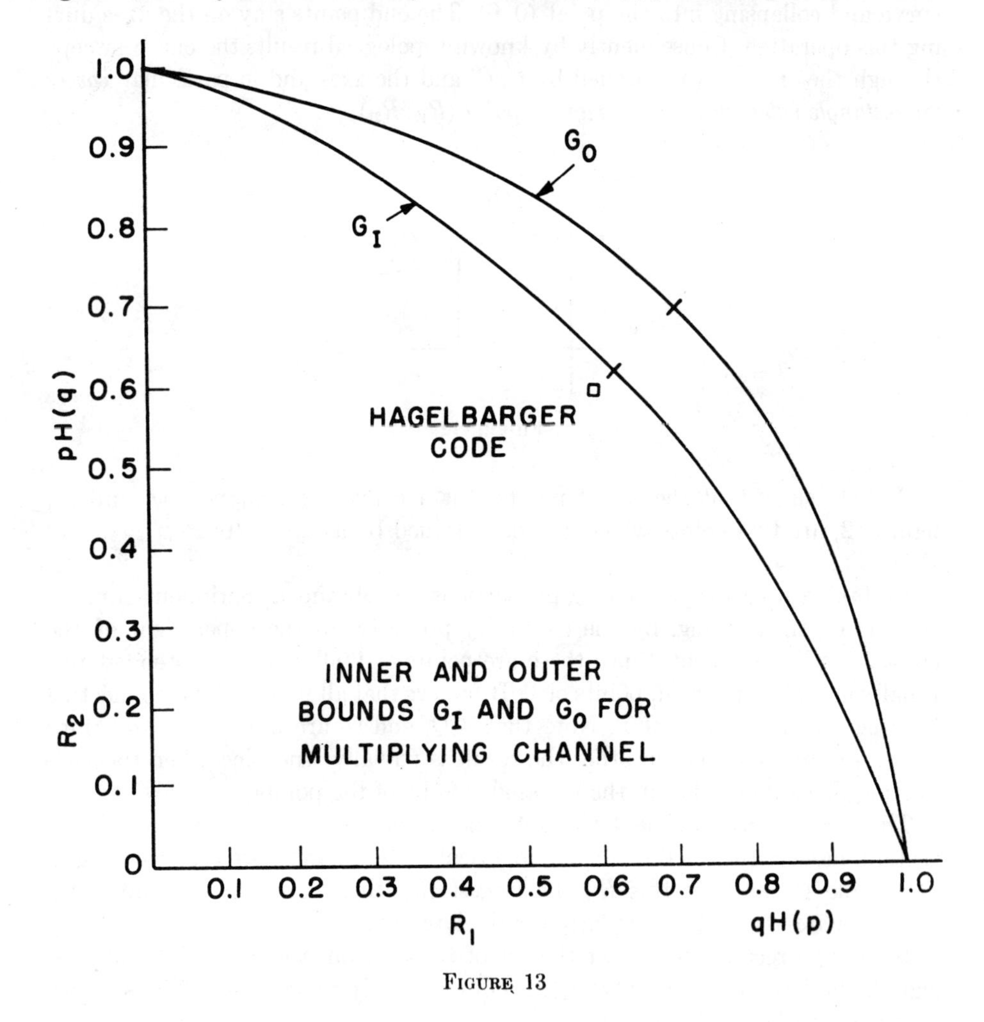

FIGURE 13

of the range. The calculation of the inner bound, in this case, amounts to finding the envelope of points

$$
\begin{aligned}
R_{12} &= -p_2[p_1 \log p_1 + (1 - p_1) \log (1 - p_1)] \\
R_{21} &= -p_1[p_2 \log p_2 + (1 - p_2) \log (1 - p_2)].
\end{aligned}
\tag{45}
$$

These are the rates with independent probability assignments at the two ends:

probability p_1 for using letter 1 at terminal 1 and probability p_2 for using letter 1 at terminal 2. By evaluating these rates for different p_1 and p_2 the envelope shown in the figure was obtained.

For the outer bounds, the envelope of rates for a general dependent assignment of probabilities is required. However it is easily seen that any assignment in which $P\{0, 0\}$ is positive can be improved by transferring this probability to one of the other possible pairs. Hence we again have a two parameter family of points (since the sum of the three other probabilities must be unity). If the probabilities are denoted by $p_1 = P\{1, 0\}$, $p_2 = P\{0, 1\}$, $1 - p_1 - p_2 = P\{1, 1\}$, we find the rates are

$$
\begin{aligned}
R_{12} &= -(1 - p_1)\left[\frac{p_2}{1 - p_1}\log\frac{p_2}{1 - p_1} + \left(1 - \frac{p_2}{1 - p_1}\right)\log\left(1 - \frac{p_2}{1 - p_1}\right)\right] \\
R_{21} &= -(1 - p_2)\left[\frac{p_1}{1 - p_2}\log\frac{p_1}{1 - p_2} + \left(1 - \frac{p_1}{1 - p_2}\right)\log\left(1 - \frac{p_1}{1 - p_2}\right)\right].
\end{aligned}
\tag{46}
$$

Here again a numerical evaluation for various values of p_1 and p_2 led to the envelope shown in the figure.

In connection with this channel, D. W. Hagelbarger has devised an interesting and simple code (not a block code however) which is error free and transmits at average rates $R_{12} = R_{21} = .571$, slightly less than our lower bound. His code operates as follows. A 0 or 1 is sent from each end with independent probabilities 1/2, 1/2. If a 0 is received then the next digit transmitted is the complement of what was just sent. This procedure is followed at both ends. If a 1 is received, both ends progress to the next binary digit of the message. It may be seen that three-fourths of the time on the average the complement procedure is followed and one-fourth of the time a new digit is sent. Thus the average number of channel uses per message digit is $(3/4)(2) + (1/4)(1) = 7/4$. The average rate is $4/7 = .571$ in both directions. Furthermore it is readily seen that the message digits can be calculated without error for each communication direction.

By using message sources at each end with biased probabilities it is possible to improve the Hagelbarger scheme slightly. Thus, if 1's occur as message digits with probability .63 and 0's with probability .37, we obtain rates in both directions

$$
R_{12} = R_{21} = \frac{-.63\log .63 - .37\log .37}{1 - (.63)^2} = .593. \tag{47}
$$

We will, in a later section, develop a result which in principle gives for any channel the exact capacity region. However, the result involves a limiting process over words of increasing length and consequently is difficult to evaluate in most cases. In contrast, the upper and lower bounds involve only maximizing operations relating to a single transmitted letter in each direction. Although sometimes involving considerable calculation, it is possible to actually evaluate them when the channel is not too complex.

14. Attainment of the outer bound with dependent sources

With regard to the outer bound there is an interesting interpretation relating to a somewhat more general communication system. Suppose that the message sources at the two ends of our channel are not independent but statistically dependent. Thus, one might be sending weather information from Boston to New York and from New York to Boston. The weather at these cities is of course not statistically independent. If the dependence were of just the right type for the channel or if the messages could be transformed so that this were the case, then it may be possible to attain transmission at the rates given by the outer bound. For example, in the multiplying channel just discussed, suppose that the messages at the two ends consist of streams of binary digits which occur with the dependent probabilities given by table III. Successive x_1, x_2 pairs

TABLE III

		x_2 0	1
x_1	0	0	.275
	1	.275	.45

are assumed independent. Then by merely sending these streams into the channel (without processing) the outer bound curve is achieved at its midpoint.

It is not known whether this is possible in general. Does there always exist a suitable pair of dependent sources that can be coded to give rates R_1, R_2 within ϵ of any point in the outer bound? This is at least often possible in the noiseless, memoryless case, that is, when y_1 and y_2 are strict functions of x_1 and x_2 (no channel noise). The source pair defined by the assignment $P\{x_1, x_2\}$ that produces the point in question is often suitable in such a case without coding as in the above example.

The inner bound also has an interesting interpretation. If we artificially limit the codes to those where the transmitted sequence at each terminal depends only on the message and not on the received sequence at that terminal, then the inner bound is indeed the capacity region. This results since in this case we have at each stage of the transmission (that is, given the index of the letter being transmitted) independence between the two next transmitted letters. It follows that the total vector change in equivocation is bounded by the sum of n vectors, each corresponding to an independent probability assignment. Details of this proof are left to the reader. The independence required would also occur if the transmission and reception points at each end were at different places with no direct cross communication.

15. General solution for the capacity region in the two-way channel

For a given memoryless two-way channel K we define a series of *derived channels* $K_1, K_2, \cdots$. These will also be memoryless channels and the capacity region for K will be evaluated as a limit in terms of the inner bounds for the series K_n.

The channel K_1 is identical with K. The derived channel K_2 is one whose input letters are actually strategies for working with K for a block of two input letters. Thus the input letters at terminal 1 for K_2 consist of pairs $[x_1^1, f(x_1^1, y_1^1)]$. Here x_1^1 is the first transmitted letter of the pair and ranges therefore over the a possible input letters of K. Now $f(x_1^1, y_1^1)$ represents any function from the first input letter x_1^1 and output letter y_1^1 to the second input letter x_1^2. Thus this function may be thought of as a rule for choosing a second input letter at terminal 1 depending on the first input letter and the observed first output letter. If x_1^1 can assume a values and y_1^1 can assume b values, then the (x_1^1, y_1^1) pair can assume ab values, and since the function f takes values from a possibilities there are a^{ab} possible functions. Hence there are $a \cdot a^{ab}$ possible pairs $[x_1^1, f(x_1^1, y_1^1)]$, or possible input letters to K_2 at terminal 1.

In a similar way, at terminal 2 consider pairs $[x_2^1, g(x_2^1, y_2^1)]$. Here g ranges over functions from the first received and transmitted letters at terminal 2 and takes values from the x_2 alphabet. Thus these pairs have $c \cdot c^{cd}$ values, where c and d are the sizes of the input and output alphabets at terminal 2.

The pairs $[x_1^1, f(x_1^1, y_1^1)]$ and $[x_2^1, g(x_2^1, y_2^1)]$ may be thought of as strategies for using the channel K in two letter sequences, the second letter to be dependent on the first letter sent and the first letter received. The technique here is very similar to that occurring in the theory of games. There one replaces a sequence of moves by a player (whose available information for making a choice is increasing through the series) by a single move in which he chooses a strategy. The strategy describes what the player will do at each stage in each possible contingency. Thus a game with many moves is reduced to a game with a single move chosen from a larger set.

The *output* letters for K_2 are, at terminal 1, pairs (y_1^1, y_1^2) and, at terminal 2, pairs (y_2^1, y_2^2); that is, the pairs of received letters at the two terminals. The transition probabilities for K_2 are the probabilities, if these strategies for introducing a particular pair of letters were used in K, that the output pairs would occur. Thus

$$(48) \qquad P_{K_2}\{(y_1^1, y_1^2), (y_2^1, y_2^2) | [x_1^1, f(x_1^1, y_1^1)], [x_2^1, g(x_2^1, y_2^1)]\}$$
$$= P_K\{y_1^1, y_2^1 | x_1^1, x_2^1\} P_K\{y_1^2, y_2^2 | f(x_1^1, y_1^1), g(x_2^1, y_2^1)\}.$$

In a similar way the channels $K_3, K_4, \cdots$ are defined. Thus K_n may be thought of as a channel corresponding to n uses of K with successive input letters at a terminal functions of previous input and output letters at that terminal. Therefore the input letters at terminal 1 are n-tuples

$$(49)\qquad [x_1^1, f(x_1^1, y_1^1), \cdots, f_{n-1}(x_1^1, x_1^2, \cdots, x_1^{n-1}, y_1^1, y_1^2, \cdots, y_1^{n-1})],$$

a possible alphabet of

$$(50)\qquad aa^{ab}a^{(ab)^2}\cdots a^{(ab)^{n-1}} = a^{[(ab)^n-1]/(ab-1)}$$

possibilities. The output letters at terminal 1 consist of n-tuples

$$(51)\qquad (y_1^1, y_1^2, \cdots, y_1^n)$$

and range therefore over an alphabet of b^n generalized letters. The transition probabilities are defined for K_n in terms of those for K by the generalization of equation (39)

$$(52)\qquad P_{K_n}\{y_1^1, y_1^2, \cdots, y_1^n|(x_1^1, f_1, f_2, \cdots, f_{n-1}), (x_2^1, g_1, g_2, \cdots, g_{n-1})\} = \prod_{i=1}^{n} P_K\{y_1^i|f_{i-1}, g_{i-1}\}.$$

The channel K_n may be thought of, then, as a memoryless channel whose properties are identical with using channel K in blocks of n, allowing transmitted and received letters within a block to influence succeeding choices.

For each of the channels K_n one could, in principle, calculate the lower bound on its capacity region. The lower bound for K_n is to be multiplied by a factor $1/n$ to compare with K, since K_n corresponds to n uses of K.

THEOREM 5. *Let B_n be the lower bound of the capacity region for the derived channel K_n reduced in scale by a factor $1/n$. Then as $n \to \infty$ the regions B_n approach a limit B which includes all the particular regions and is the capacity region of K.*

PROOF. We first show the positive assertion that if (R_{12}, R_{21}) is any point in some B_n and ϵ is any positive number, then we can construct block codes with error probabilities $P_e < \epsilon$ and rates in the two directions at least $R_{12} - \epsilon$ and $R_{21} - \epsilon$. This follows readily from previous results if the derived channel K_n and its associated inner bound B_n are properly understood. K_n is a memoryless channel, and by theorem 3 we can find codes for it transmitting arbitrarily close to the rates R_{12}, R_{21} in B_n with arbitrarily small error probability. These codes are sequences of letters from the K_n alphabet. They correspond, then, to sequences of *strategies* for blocks of n for the original channel K.

Thus these codes can be directly translated into codes for K n times as long, preserving all statistical properties, in particular the error probability. These codes, then, can be interpreted as codes signalling at rates $1/n$ as large for the K channel with the same error probability. In fact, from theorem 3, it follows that for any pair of rates strictly inside B_n we can find codes whose error probability decreases at least exponentially with the code length.

We will now show that the regions B_n approach a limit B as n increases and that B includes all the individual B_n. By a limiting region we mean a set of points B such that for any point P of B, and $\epsilon > 0$, there exists n_0 such that for $n > n_0$ there are points of B_n within ϵ of P, while for any P not in B there exist ϵ and n_0 such that for $n > n_0$ no points of B_n are within ϵ of P. In the first place B_n is included in B_{kn} for any integer k. This is because the strategies for B_{kn} include as special cases strategies where the functional influence only involves subblocks of n. Hence all points obtainable by independent probability assignments with K_n are also obtainable with K_{kn} and the convex hull of the latter set must include the convex hull of the former set.

It follows that the set B_{kn} approaches a limit B, the union of all the B_{kn} plus limit points of this set. Also B includes B_{n_1} for any n_1. For n and n_1 have a common multiple, for example nn_1, and B includes B_{nn_1} while B_{nn_1} includes B_{n_1}.

Furthermore, any point obtainable with K_{kn} can be obtained with $K_{kn+\alpha}$, for $0 \leqq \alpha \leqq n$, reduced in both coordinates by a factor of not more than $k/(k+1)$. This is because we may use the strategies for K_{kn} followed by a series of α of the first letters in the x_1 and x_2 alphabets. (That is, fill out the assignments to the length $kn + \alpha$ with essentially dummy transmitted letters.) The only difference then will be in the normalizing factor, $1/(\text{block length})$. By making k sufficiently large, this discrepancy from a factor of 1, namely $1/(k+1)$, can be made as small as desired. Thus for any $\epsilon > 0$ and any point P of B there is a point of B_{n_1} within ϵ of P for *all* sufficiently large n_1.

With regard to the converse part of the theorem, suppose we have a block code of length n with signalling rates (R_1, R_2) corresponding to a point outside B, closest distance to B equal to ϵ. Then since B includes B_n, the closest distance to B_n is at least ϵ. We may think of this code as a block code of length 1 for the channel K_n. As such, the messages m_1 and m_2 are mapped directly into "input letters" of K_n without functional dependence on the received letters. We have then since m_1 and m_2 are independent the independence of probabilities associated with these input letters sufficient to make the inner bound and outer bound the same. Hence the code in question has error probability bounded away from zero by a quantity dependent on ϵ but not on n.

16. Two-way channels with memory

The general discrete two-way channel with memory is defined by a set of conditional probabilities

$$P\{y_{1n}, y_{2n} | x_{11}, x_{12}, \cdots, x_{1n}; x_{21}, x_{22}, \cdots, x_{2n}; y_{11}, y_{12}, \cdots, y_{1n-1}; y_{21}, y_{22}, \cdots, y_{2n-1}\}. \tag{53}$$

This is the probability of the nth output pair y_{1n}, y_{2n} conditional on the preceding history from time $t = 0$, that is, the input and output sequences from the starting time in using the channel. In such a general case, the probabilities might change in completely arbitrary fashion as n increases. Without further limitation, it is too general to be either useful or interesting. What is needed is some condition of reasonable generality which, however, ensures a certain stability in behavior and allows, thereby, significant coding theorems. For example, one might require finite historical influence so that probabilities of letters depend only on a bounded past history. (Knowing the past d inputs and outputs, earlier inputs and outputs do not influence the conditional probabilities.) We shall, however, use a condition which is, by and large, more general and also more realistic for actual applications.

We will say that a two-way channel has the *recoverable state property* if it satisfies the following condition. There exists an integer d such that for any input and output sequences of length n, X_{1n}, X_{2n}, Y_{1n}, Y_{2n}, there exist two functions $f(X_{1n}, Y_{1n})$, $g(X_{2n}, Y_{2n})$ whose values are sequences of input letters of the same length less than d and such that if these sequences f and g are now sent over the channel it is returned to its original state. Thus, conditional probabilities after this are the same as if the channel were started again at time zero.

The recoverable state property is common in actual physical communication systems where there is often a "zero" input which, if applied for a sufficient period, allows historical influences to die out. Note also that the recoverable state property may hold even in channels with an infinite set of internal states, provided it is possible to return to a "ground" state in a bounded number of steps.

The point of the recoverable state condition is that if we have a block code for such a channel, we may annex to the input words of this code the functions f and g at the two terminals and then repeat the use of the code. Thus, if such a code is of length n and has, for one use of the code, signalling rates R_1 and R_2 and error probabilities P_{e1} and P_{e2}, we may *continuously* signal at rates $R_1' \geqq nR_1/(n + d)$ and $R_2' \geqq nR_2/(n + d)$ with error probabilities $P_{e1}' \leqq P_{e1}$ and $P_{e2}' \leqq P_{e2}$.

For a recoverable state channel we may consider strategies for the first n letters just as we did in the memoryless case, and find the corresponding inner bound B_n on the capacity region (with scale reduced by $1/n$). We define the region B which might be called the limit supremum of the regions B_n. Namely, B consists of all points which belong to an infinite number of B_n together with limit points of this set.

THEOREM 6. *Let (R_1, R_2) be any point in the region B. Let n_0 be any integer and let ϵ_1 and ϵ_2 be any positive numbers. Then there exists a block code of length $n > n_0$ with signalling rates R_1', R_2' satisfying $|R_1 - R_1'| < \epsilon_1$, $|R_2 - R_2'| < \epsilon_1$ and error probabilities satisfying $P_{e1} < \epsilon_2$, $P_{e2} < \epsilon_2$. Conversely, if (R_1, R_2) is not in B then there exist n_0 and $\delta > 0$ such that any block code of length exceeding n_0 has either $P_{e1} > \delta$ or $P_{e2} > \delta$ (or both).*

PROOF. To show the first part of the theorem choose an $n_1 > n_0$ and also large enough to make both $dR_1/(d + n)$ and $dR_2/(d + n)$ less than $\epsilon_1/2$. Since the point (R_1, R_2) is in an infinite sequence of B_n, this is possible. Now construct a block code based on n_1 uses of the channel as individual "letters," within $\epsilon_1/2$ of the rate pair (R_1, R_2) and with error probabilities less than ϵ_2. To each of the "letters" of this code annex the functions which return the channel to its original state. We thus obtain codes with arbitrarily small error probability $< \epsilon_2$ approaching the rates R_1, R_2 and with arbitrarily large block length.

To show the converse statement, suppose (R_1, R_2) is *not* in B. Then for some

n_0 every B_n, where $n > n_0$, is outside a circle of some radius, say ϵ_2, centered on (R_1, R_2). Otherwise (R_1, R_2) would be in a limit point of the B_n. Suppose we have a code of length $n_1 > n_0$. Then its error probability is bounded away from zero since we again have a situation where the independence of "letters" obtains.

The region B may be called the capacity region for such a recoverable state channel. It is readily shown that B has the same convexity properties as had the capacity region G for a memoryless channel. Of course, the actual evaluation of B in specific channels is even more impractical than in the memoryless case.

17. Generalization to T-terminal channels

Many of the tricks and techniques used above may be generalized to channels with three or more terminals. However, some definitely new phenomena appear in these more complex cases. In another paper we will discuss the case of a channel with two or more terminals having inputs only and one terminal with an output only, a case for which a complete and simple solution of the capacity region has been found.

APPENDIX. ERROR PROBABILITY BOUNDS IN TERMS OF MOMENT GENERATING FUNCTIONS

Suppose we assign probabilities $P\{x_1\}$ to input letters at terminal 1 and $P\{x_2\}$ to input letters at terminal 2. (Notice that we are here working with letters, not with words as in theorem 2.) We can then calculate the log of the moment generating functions of the mutual information between input letters at terminal 1 and input letter-output letter pairs at terminal 2. (This is the log of the moment generating function of the distribution ρ_{12} when $n = 1$.) The expressions for this and the similar quantity in the other direction are

$$\mu_1(s) = \log \sum_{x_1,x_2,y_2} P\{x_1, x_2, y_2\} \exp\left(s \log \frac{P\{x_1, x_2, y_2\}}{P\{x_1\}P\{x_2, y_2\}}\right) \tag{54}$$

$$= \log \sum_{x_1,x_2,y_2} \frac{P\{x_1, x_2, y_2\}^{s+1}}{P\{x_1\}^s P\{x_2, y_2\}^s},$$

$$\mu_2(s) = \log \sum_{x_1,x_2,y_1} \frac{P\{x_1, x_2, y_1\}^{s+1}}{P\{x_2\}^s P\{x_1, y_1\}^s}. \tag{55}$$

These functions μ_1 and μ_2 may be used to bound the tails on the distributions ρ_{12} and ρ_{21} obtained by adding n identically distributed samples together. In fact, Chernoff [4] has shown that the tail to the left of a mean may be bounded as follows:

$$\begin{aligned} \rho_{12}[n\mu_1'(s_1)] &\leqq \exp\{n[\mu_1(s_1) - s_1\mu_1'(s_1)]\}, & s_1 &\leqq 0,\\ \rho_{21}[n\mu_2'(s_2)] &\leqq \exp\{n[\mu_2(s_2) - s_2\mu_2'(s_2)]\}, & s_2 &\leqq 0. \end{aligned} \tag{56}$$

Thus, choosing an arbitrary negative s_1, this gives a bound on the distribution function at the value $n\mu_1'(s_1)$. It can be shown that $\mu'(s)$ is a monotone increasing function and that $\mu'(0)$ is the mean of the distribution. The minimum $\mu'(s)$ corresponds to the minimum possible value of the random variable in question, in this case, the minimum $I(x_1; x_2, y_2)$. Thus, an s_1 may be found to place $\mu_1(s_1)$ anywhere between $I_{\min}(x_1; x_2, y_2)$ and $E(I)$. Of course, to the left of $I_{\min}$ the distribution is identically zero and to the right of $E(I)$ the distribution approaches one with increasing n.

We wish to use these results to obtain more explicit bounds on P_{e1} and P_{e2}, using theorem 2. Recalling that in that theorem θ_1 and θ_2 are arbitrary, we attempt to choose them so that the exponentials bounding the two terms are equal. This is a good choice of θ_1 and θ_2 to keep the total bound as small as possible. The first term is bounded by $\exp\{n[\mu_1(s_1) - s_1\mu_1'(s_1)]\}$ where s_1 is such that $\mu_1'(s_1) = R_1 + \theta_1$, and the second term is equal to $\exp(-n\theta_1)$. Setting these equal, we have

$$\mu_1(s_1) - s_1\mu_1'(s_1) = -\theta_1, \qquad R_1 + \theta_1 = \mu_1'(s_1). \tag{57}$$

Eliminating θ_1, we have

$$R_1 = \mu_1(s_1) - (s_1 - 1)\mu_1'(s_1) \tag{58}$$

and

$$E(P_{e1}) \leqq 2\exp\{n[\mu_1(s_1) - s_1\mu_1'(s_1)]\}. \tag{59}$$

This is because the two terms are now equal and each dominated by $\exp\{n[\mu_1(s_1) - s_1\mu_1'(s_1)]\}$. Similarly, for

$$R_2 = \mu_2(s_2) - (s_2 - 1)\mu_2'(s_2) \tag{60}$$

we have

$$E(P_{e2}) \leqq 2\exp\{n[\mu_2(s_2) - s_2\mu_2'(s_2)]\}. \tag{61}$$

These might be called parametric bounds in terms of the parameters s_1 and s_2. One must choose s_1 and s_2 such as to make the rates R_1 and R_2 have the desired values. These s_1 and s_2 values, when substituted in the other formulas, give bounds on the error probabilities.

The derivative of R_1 with respect to s_1 is $-(s_1 - 1)\mu_1''(s_1)$, a quantity always positive when s_1 is negative except for the special case where $\mu''(0) = 0$. Thus, R_1 is a monotone increasing function of s_1 as s_1 goes from $-\infty$ to 0, with R_1 going from $-I_{\min} - \log P\{I_{\min}\}$ to $E(I)$. The bracketed term in the exponent of $E(P_{e1})$, namely $\mu_1(s_1) - s_1\mu_1(s_1)$, meanwhile varies from $\log P\{I_{\min}\}$ up to zero. The rate corresponding to $s_1 = -\infty$, that is, $-I_{\min} - \log P\{I_{\min}\}$, may be positive or negative. If negative (or zero) the entire range of rates is covered from zero up to $E(I)$. However, if it is positive, there is a gap from rate $R_1 = 0$ up to this end point. This means that there is no way to solve the equation for rates in this interval to make the exponents of the two terms equal. The best course here to give a good bound is to choose θ_1 in such a way that $n(R_1 + \theta_1)$ is just smaller than $I_{\min}$, say $I_{\min} - \epsilon$. Then $\rho_{12}[n(R_1 + \theta_1)] = 0$ and only the second term, $\exp(\theta_1 n)$, is left in the bound. Thus $\exp[-n(I_{\min} - R_1 - \epsilon)]$ is a bound on P_e. This is true for any $\epsilon > 0$. Since we can construct such codes for any positive ϵ and since there are only a finite number of codes, this implies that we can construct a code satisfying this inequality with $\epsilon = 0$. Thus, we may say that

$$E(P_{e1}) \leqq \exp[-n(I_{\min} - R_1)], \qquad R_1 \leqq I_{\min}. \tag{62}$$

Of course, exactly similar statements hold for the second code working in the reverse direction. Combining and summarizing these results we have the following.

THEOREM 7. *In a two-way memoryless channel K with finite alphabets, let* $P\{x_1\}$ *and* $P\{x_2\}$ *be assignments of probabilities to the input alphabets, and suppose these lead to the logarithms of moment generating functions for mutual information* $\mu_1(s_1)$ *and* $\mu_2(s_2)$,

$$\begin{aligned} \mu_1(s_1) &= \log \sum_{x_1,x_2,y_2} \frac{P\{x_1, x_2, y_2\}^{s+1}}{P\{x_1\}^s P\{x_2, y_2\}^s} \\ \mu_2(s_2) &= \log \sum_{x_1,x_2,y_2} \frac{P\{x_1, x_2, y_1\}^{s+1}}{P\{x_2\}^s P\{x_1, y_1\}^s}. \end{aligned} \tag{63}$$

Let $M_1 = \exp(R_1 n)$, $M_2 = \exp(R_2 n)$ *be integers, and let* s_1, s_2 *be the solutions (when they exist) of*

$$\begin{aligned} R_1 &= \mu_1(s_1) - (s_1 + 1)\mu_1'(s_1) \\ R_2 &= \mu_2(s_2) - (s_2 + 1)\mu_2'(s_2). \end{aligned} \tag{64}$$

The solution s_1 *will exist if*

$$-I_{\min}(x_1; x_2, y_2) - \log P\{I_{\min}(x_1; x_2, y_2)\} \leqq R_1 \leqq E[I(x_1; x_2, y_2)], \tag{65}$$

and similarly for s_2. *If both* s_1 *and* s_2 *exist, then there is a code pair for the channel K of length n with* M_1 *and* M_2 *messages and error probabilities satisfying*

$$\begin{aligned} P_{e1} &\leqq 4 \exp\{+n[\mu_1(s_1) - s_1\mu_1'(s_1)]\} \\ P_{e2} &\leqq 4 \exp\{+n[\mu_2(s_2) - s_2\mu_2'(s_2)]\}. \end{aligned} \tag{66}$$

If either (or both) of the R is so small that the corresponding s does not exist, a code pair exists with the corresponding error probability bounded by

$$P_{e1} \leqq 2 \exp\{-n[I(x_1; x_2, y_2) - R_1]\} \tag{67}$$

or

$$P_{e2} \leqq 2 \exp\{-n[I(x_2; x_1, y_1) - R_2]\}. \tag{68}$$

Thus, if s_1 *exists and not* s_2, *then inequalities* (66) *would be used. If neither exists,* (67) *and* (68) *hold.*

REFERENCES

[1] C. E. SHANNON, "Certain results in coding theory for noisy channels," *Information and Control*, Vol. 1 (1957), pp. 6–25.

[2] ———, "Channels with side information at the transmitter," *IBM J. Res. Develop.*, Vol. 2 (1958), pp. 289–293.
[3] ———, "Geometrische Deutung einiger Ergebnisse bei der Berechnung der Kanalkapazitat," *Nachrtech. Z.*, Vol. 10 (1957).
[4] H. Chernoff, "A measure of asymptotic efficiency for tests of a hypothesis based on the sum of observations," *Ann. Math. Statist.*, Vol. 23 (1952), pp. 493–507.

INFORMATION TRANSMISSION IN A CHANNEL WITH FEEDBACK

R. L. DOBRUSHIN

(*Translated by R. A. Silverman*)

1. Introduction

In information theory one usually considers the following situation, shown in Figure 1 (see e.g. [5] and [3]). The statistical properties of the channel are assumed to be given, and one studies the possibility of converting a given input message into a given output message using optimum methods of coding and detection.

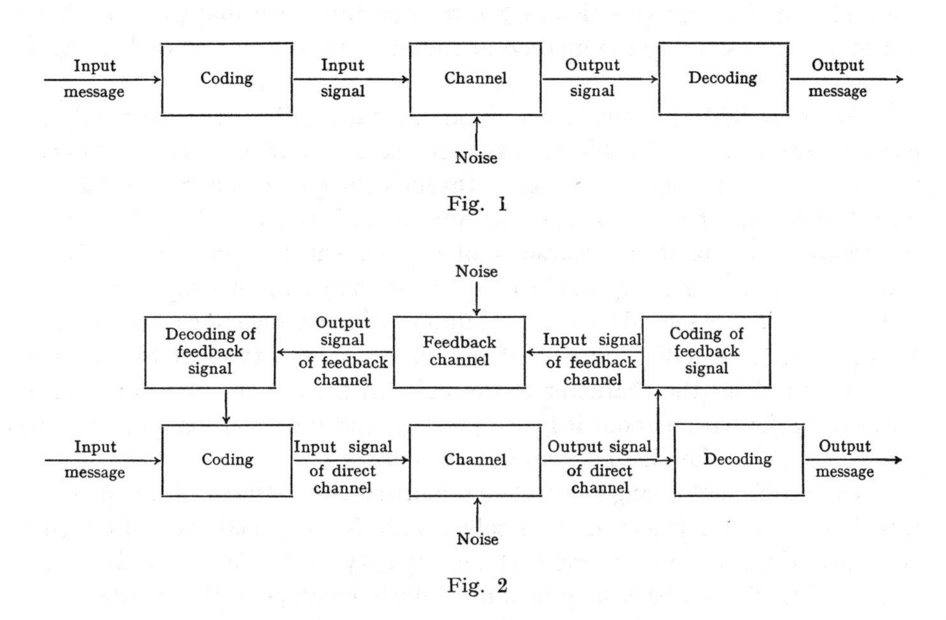

Fig. 1

Fig. 2

In many actual communication systems there is also the possibility of transmitting signals in the reverse direction, from the output to the input. In such cases one speaks of systems with feedback (see Figure 2). The feedback is often used to improve the quality of transmission in the forward direction. Thus, for example, if two parties A and B are talking by telephone, a feedback channel is ordinarily used even in the case where the only purpose of their conversation is to convey information emanating at A and destined for B, and B has no information for A and A does not intend to receive any information. Thus, B interrogates A and asks A to repeat words that he has not heard

Reprinted with permission from *Theory of Prob. and Applications*, vol. 34, pp. 367–383, Dec. 1958.

distinctly. If the information is easily distorted or especially important, then B repeats all that he has heard, so that A can verify that B has received the message correctly. In recent years, telegraph communication systems handling discrete information have been developed which are based on just this principle. The papers [1], [2], [6] and [8] are devoted to a discussion of concrete systems of this kind. They also pose the problem (which arises naturally) of comparing the channel capacities of systems with feedback with systems without feedback for the same noise level, but this problem is not completely solved in these papers.

In the basic part of the present paper (just as in the earlier papers [1], [2], [6], [8]) we consider a communication channel with discrete time, in which the symbols transmitted at different instants of time are perturbed independently of one another. We shall call such a channel a channel without memory. We show that if the information contained in the messages exceeds the channel capacity in the forward direction, then this message cannot be transmitted over a channel with feedback, regardless of the capacity of the feedback channel. Thus, use of feedback does not increase the capacity of a channel without memory. Of course, it should not be forgotten that this result, like all results of information theory, is of an asymptotic character and pertains to optimum methods of coding and detection. The use of feedback does not permit the transmission of a message which cannot be transmitted without using feedback; however, it can simplify the method of coding signals, which is used for transmission.

We show that the situation is fundamentally different if we consider a channel with memory. In this case, the feedback channel can convey information about how the signals have gone through the channel, and therefore also about the character of past noise; this information, which makes more precise our information about the character of the noise in the present and future, can help in selecting an optimum method for coding the messages. However, in this case, it is essential that the communication channel be sufficiently fast-acting, since the ergodic character of real noise leads to the fact that delayed information about the character of the noise in the remote past does not increase our information about it in the present, and the situation may become the same as for the channel without memory.

The problem of giving a complete quantitative treatment of the questions raised here remains unsolved. It is solved only for a special class of channels with memory, where we assume that the capacity of the feedback channel is so large that the feedback loop instantaneously conveys to the channel input complete and errorless information about the input signal and where we assume that the channel is slow-acting; in particular, for a channel with which some random parameter α is associated, the channel must reduce to a channel without memory for any fixed value $\alpha = a_e$ of the parameter. In the case of such a channel, we also find the formula for its asymptotic capacity without using feedback. This formula is evidently new and may also be of independent interest. Calculations of the gain in capacity achieved by using feedback are given for some examples.

Summarizing the above considerations, we can say that using feedback for "interrogation" does not increase the channel capacity, but can simplify the method of transmission. The channel capacity can be increased by using feedback for "tracking the state of the direct channel".

2. Mathematical Formulation of the Problem

For the sake of notational simplicity, we shall assume that every one of the random variables under consideration takes a finite number of values. However, all the constructions which follow can be carried over without change to the case of channels with an arbitrary range of values.

Let $(E_1, \cdots, E_m)$ be the alphabet of symbols at the channel input and let $(\bar{E}_1, \cdots, \bar{E}_{\bar{m}})$ be the alphabet of symbols at the channel output. We fix an integer n, the *length of transmission*. We denote by $\mathscr{E}^{(n)}$ the space of all possible input signals, which consists of m^n sequences of the form $(E_{i_1}, E_{i_2}, \cdots, E_{i_n})$, and we denote by $\bar{\mathscr{E}}^{(n)}$ the space of all possible output signals, which consists of $\bar{m}^n$ sequences of the form $(\bar{E}_{j_1}, \bar{E}_{j_2}, \cdots, \bar{E}_{j_n})$. The communication channel is defined by giving the system of conditional probabilities

$$p(j_1, \cdots, j_n | i_1, \cdots, i_n), \qquad \begin{array}{l} i_k = 1, \cdots, m, \\ j_k = 1, \cdots, \bar{m}, \end{array}$$

for the probability that the input signal $(E_{i_1}, \cdots, E_{i_n})$ is transformed into the output signal $(\bar{E}_{j_1}, \cdots, \bar{E}_{j_n})$. It is obvious that

$$\sum_{j_1, \cdots, j_n} p(j_1, \cdots, j_n | i_1, \cdots, i_n) = 1 \tag{1}$$

for any $i_1, \cdots, i_n$.

Following Khinchin [7], we shall call a communication channel a *channel without anticipation* if

$$\sum_{j_{k+1}, \cdots, j_n} p(j_1, \cdots, j_n | i_1, \cdots, i_n) = q(j_1, \cdots, j_k | i_1, \cdots, i_k), \tag{2}$$

for any fixed integral k, $0 \leq k \leq n$, and indices $j_1, \cdots, j_k, i_1, \cdots, i_k$, i.e., if (2) does not depend on the choice of the indices $i_{k+1}, \cdots, i_n$. The meaning of the condition (2) is that the conditional probability of the first k output symbols taking fixed values $\bar{E}_{j_1}, \cdots, \bar{E}_{j_k}$, given that the input signal is $E_{i_1}, \cdots, E_{i_n}$, should depend only on the first k symbols of this signal. In cases where the parameter has the meaning of physical time, the assumption that the communication channel is a channel without anticipation is natural, since it means that at a given instant of time the output signal does not depend on which values are taken by the symbols of the input signal at future instants of time. In what follows. we shall consider only channels without anticipation.

We shall say that the channel under consideration is a *channel without memory* if we can assign a set of probabilities $p_{ij}^{(k)}$, $i = 1. \cdots, m$, $j = 1, \cdots, \bar{m}$, such that

$$p(j_1, \cdots, j_n | i_1, \cdots, i_n) = p_{i_1 j_1}^{(1)} p_{i_2 j_2}^{(2)} p_{i_3 j_3}^{(3)} \cdots p_{i_n j_n}^{(n)} \tag{3}$$

for all $i_1, \cdots, i_n$ and $j_1, \cdots, j_n$. It is clear that a channel without memory is a channel without anticipation. We shall call a memoryless channel homogeneous in time if $p_{ij}^{(k)}$ does not depend on k.

Now in addition let $\mathscr{F} = (F_1, \cdots, F_e)$ and $\bar{\mathscr{F}} = (\bar{F}_1, \cdots, \bar{F}_{\bar{e}})$ be the spaces of input messages and output messages, respectively. Ordinarily, in the applications, the spaces $\mathscr{F}$ and $\bar{\mathscr{F}}$ have just as complicated a structure as the spaces $\mathscr{E}^{(n)}$, $\bar{\mathscr{E}}^{(n)}$, but this fact is not important for us. We specify a random quantity η which takes values in the space $\mathscr{F}$ of input messages; we shall call η the *input message*. We also consider random quantities ξ, $\bar{\xi}$ and $\bar{\eta}$ which take values in the spaces $\mathscr{E}^{(n)}$, $\bar{\mathscr{E}}^{(n)}$ and $\bar{\mathscr{F}}$, respectively. We call these quantities the input signal, the output signal and the output message, respectively. We set (see (2))

$$r_k(j_k|i_1, \cdots, i_k; j_1, \cdots, j_{k-1}) = \frac{q(j_1, \cdots, j_k|i_1, \cdots, i_k)}{\sum_{j_k=1}^{\bar{m}} q(j_1, \cdots, j_k|i_1, \cdots, i_k)}. \tag{4}$$

The quantity $r_k(j_k|i_1, \cdots, i_k; j_1, \cdots, j_{k-1})$ gives the conditional probability that the k-th output symbol takes the value $\bar{E}_{j_k}$ given the k symbols $E_{i_1}, \cdots, E_{i_k}$ of the input signal and the $k-1$ symbols $\bar{E}_{j_1}, \cdots, \bar{E}_{j_{k-1}}$ of the output signal. In the case of a memoryless channel

$$r_k(j_k|i_1, \cdots, i_k; \; j_1, \cdots, j_{k-1}) = p_{i_k j_k}^{(k)}. \tag{5}$$

We shall say that *the input message η is transformed into the output message $\bar{\eta}$ as a result of transmission through a channel of length n*, if it is possible to choose an input signal $\xi = (\xi_1, \cdots, \xi_n)$ and an output signal $\bar{\xi} = (\bar{\xi}_1, \cdots, \bar{\xi}_n)$ such that the following two conditions are satisfied. First of all, the conditional probability

$$\begin{aligned} &\mathbf{P}\{\bar{\xi}_k = \bar{E}_{j_k}|\xi_1 = E_{i_1}, \cdots, \xi_k = E_{i_k}, \bar{\xi}_1 = \bar{E}_{j_1}, \cdots, \bar{\xi}_{k-1} = \bar{E}_{j_{k-1}}, \eta = F_j\} \\ &\qquad = r(j_k|i_1, \cdots, i_k; \; j_1, \cdots, j_{k-1}) \end{aligned} \tag{6}$$

for any $j = 1, \cdots, e$, $i_r = 1, \cdots, m$, $j_r = 1, \cdots, \bar{m}$. The intuitive meaning of this condition is that the conditional probability of the k-th output symbol taking the value $\bar{E}_{j_k}$ for the k input symbols and the $k-1$ previous output symbols is just what is predetermined by the properties of the given channel, regardless of which particular message is coded into the signal. Secondly, the conditional probability

$$\begin{aligned} &\mathbf{P}\{\bar{\eta} = \bar{F}_\alpha|\bar{\xi} = (\bar{E}_{j_1}, \cdots, \bar{E}_{j_n}), \xi = (E_{i_1}, \cdots, E_{i_n}), \eta = F_j\} \\ &\qquad = \mathbf{P}\{\bar{\eta} = \bar{F}_\alpha|\bar{\xi} = (\bar{E}_{j_1}, \cdots, \bar{E}_{j_n})\} \end{aligned} \tag{7}$$

for any $\alpha = 1, \cdots, \bar{e}$, $j = 1, \cdots, e$, $i_r = 1, \cdots, m$, $j_r = 1, \cdots, \bar{m}$. The intuitive meaning of this second condition is that the particular output message $\bar{F}_\alpha$ which is obtained as a result of decoding the output signal $(\bar{E}_{j_1}, \cdots, \bar{E}_{j_n})$ does not depend on what the input message and input signal were. This condition is natural, since in carrying out the decoding we cannot make use of any additional information about the message and signal at the channel input other than the information contained in the output signal.

We shall say that *the input message η is transformed into the output message*

$\bar{\eta}$ as a result of transmission through a channel of length n without using feedback if the input signal ξ and the output signal $\bar{\xi}$ can be chosen in such a way that besides conditions (6) and (7) the following supplementary condition is also satisfied:

$$\tag{8} \begin{aligned} \mathbf{P}\{\xi_k = E_{i_k} | \xi_1 = E_{i_1}, \cdots, \xi_{k-1} = E_{i_{k-1}}, \bar{\xi}_1 = \bar{E}_{j_1}, \cdots, \bar{\xi}_{k-1} = \bar{E}_{j_{k-1}}, \eta = F_j\} \\ = \mathbf{P}\{\xi_k = E_{i_k} | \xi_1 = E_{i_1}, \cdots, \xi_{k-1} = E_{i_{k-1}}, \eta = F_j\} \end{aligned}$$

for any $k = 1, \cdots, n$, $i_r = 1, \cdots, m$, $j_r = 1, \cdots, \bar{m}$, $j = 1, \cdots, e$. If condition (8) is not satisfied then we shall say that *the message η is transformed into the message $\bar{\eta}$ as a result of transmission through a channel of length n using feedback.* The intuitive meaning of the assumption (8) is that (in the case of transmission without feedback) when it is necessary to specify the k-th symbol of the input signal during the process of coding, we can only use the message F_j and the $k-1$ previous input symbols, whereas the symbols $\bar{E}_{j_1}, \cdots, \bar{E}_{j_{k-1}}$ of the output signal are not available to us. On the other hand, the advantages conferred by using feedback consists in the fact (and only in the fact) that the feedback loop furnishes us information at the input about the symbols $\bar{E}_{j_1}, \cdots, \bar{E}_{j_{k-1}}$ of the output signal at earlier instants of time, information which can be used in coding the k-th symbol. It is just this scheme which comprises the examples of using feedback adduced in Section 1.

3. Formulation of the Basic Theorem

We now recall that if two random quantities ζ and $\bar{\zeta}$ take the values $(G_1, \cdots, G_k)$ and $(\bar{G}_1, \cdots, \bar{G}_k)$, respectively, then by *the information of the pair ζ and $\bar{\zeta}$* is meant the number

$$\tag{9} I(\zeta, \bar{\zeta}) = \sum \mathbf{P}\{\zeta = G_i, \bar{\zeta} = \bar{G}_j\} \log \frac{\mathbf{P}\{\zeta = G_i, \bar{\zeta} = G_j\}}{\mathbf{P}\{\zeta = G_i\}\mathbf{P}\{\bar{\zeta} = G_j\}}.$$

We shall say that *two random quantities $\xi = (\xi_1, \cdots, \xi_n)$ and $\bar{\xi} = (\bar{\xi}_1, \cdots, \bar{\xi}_n)$ are related by a channel of length n* if these quantities take values in the spaces $\mathscr{E}^{(n)}$ and $\bar{\mathscr{E}}^{(n)}$, respectively, and if

$$\tag{10} \mathbf{P}\{\bar{\xi} = (\bar{E}_{j_1}, \cdots, \bar{E}_{j_n}) | \xi = (E_{i_1}, \cdots, E_{i_n})\} = p(j_1, \cdots, j_n | i_1, \cdots, i_n)$$

for any $i_1, \cdots, i_n, j_1, \cdots, j_n$. As usual, by *the capacity of the channel of length n without feedback* we mean the number

$$\tag{10'} C^{(n)} = \sup I(\xi, \bar{\xi}),$$

where the upper bound is taken over all pairs of random quantities ξ, $\bar{\xi}$ related by the channel of length n. If the channel in question is a memoryless channel, then

$$\tag{11} C^{(n)} = D_1 + \cdots + D_n,$$

where

$$\tag{12} D_k = \sup_{\{p_i\}} \sum_{i,j} p_i p_{ij}^{(k)} \log \frac{p_{ij}^{(k)}}{\bar{p}_j}, \quad \bar{p}_j = \sum_{i=1}^{m} p_i p_{ij}^{(k)},$$

and the upper bound is taken over all possible probability distributions $(p_1, \cdots, p_m)$. Equation (11), which is natural from an intuitive point of view, will be proved in Section 6.

We now formulate the first statement of Shannon's well-known theorem about a channel without feedback.

Theorem 1. *If the input message η is transformed into the output message $\tilde{\eta}$ as a result of transmission through a channel of length n without using feedback, then*

$$I(\eta, \tilde{\eta}) \leqq C^{(n)}, \tag{13}$$

where $C^{(n)}$ is the capacity of the channel of length n without feedback.

The familiar proof of this statement will be given in Section 5, since it has not been given heretofore in just the form we require. The second statement of Shannon's theorem says that the condition (13) is also sufficient for the message η to be transformed into the message $\tilde{\eta}$ as a result of transmission through a channel of length n, in a certain asymptotic sense as $n \to \infty$, if certain additional regularity conditions imposed on the ergodic channel are met. We shall give neither a proof nor even a precise formulation of this assertion, since it is not used in what follows and we need it only for general orientation. For details see [3], [5], [7], and [8].

A basic result of this paper is the following theorem about a memoryless channel with feedback.

Theorem 2. *If the input message η is transformed into the output message $\tilde{\eta}$ as a result of transmission through a memoryless channel of length n using feedback, then*

$$I(\eta, \tilde{\eta}) \leqq C^{(n)}, \tag{14}$$

where $C^{(n)}$ is the capacity of the channel of length n without feedback.

Thus, if information can be transmitted through a memoryless channel using feedback, this information cannot exceed the capacity of the channel without feedback. Since, according to the second statement of Shannon's theorem it is true that in a certain asymptotic sense information less than the channel capacity can be transmitted over the channel without using feedback, in an analogous asymptotic sense, we can say that any information which can be transmitted through a memoryless channel using feedback can be transmitted through the same channel without using feedback.

In a subsequent section (Section 8) we consider a special class of channels without memory. We shall call channels of this class random channels without memory. We assume that we are given a set of transition probabilities $p_{ij}(\alpha)$, which specify for any fixed $\alpha = a$ a memoryless channel which is homogeneous in time, and that the parameter α is a random variable. Intuitively, we can imagine that the random memoryless channel is an ordinary memoryless channel, but with a statistical operating regime which is unknown *a priori* and with a given *a priori* probability distribution for this regime. For simplicity, we assume that the random parameter α takes only a finite number of values $a_1, \cdots, a_s$ with probabilities $q(a_1), \cdots, q(a_s)$, respectively. Then we can use the formula

$$(15) \qquad p(j_1, \cdots, j_n | i_1, \cdots, i_n) = \sum_{k=1}^{s} q(a_k) p_{i_1 j_1}(a_k) p_{i_2 j_2}(a_k) \cdots p_{i_n j_n}(a_k)$$

for the conditional probabilities specifying the channel as a formal mathematical definition of the random memoryless channel. It is obvious that we get a channel without anticipation.

In Section 7 we shall establish the following asymptotic expression for the capacity of the random memoryless channel without feedback

$$(16) \qquad C^{(n)} = n\bar{C} + O(1)$$

as $n \to \infty$, where

$$(16') \qquad \bar{C} = \sup \sum_k q(a_k) \sum_{i,j} p_i p_{ij}(a_k) \log \frac{p_{ij}(a_k)}{\bar{p}_j(a_k)}, \quad \bar{p}_j(a_k) = \sum_i p_i p_{ij}(a_k),$$

and the upper bound is taken over all possible probability distributions $p_1, \cdots, p_m$. We emphasize that the p_i unlike the $p_j(\alpha)$ do not depend on the random parameter α. In addition, we shall prove the following theorem about the random memoryless channel with feedback.

Theorem 3. *If the input message η is transformed into the output message $\bar{\eta}$ as a result of transmission through a random memoryless channel of length n using feedback, then the information*

$$(17) \qquad I(\eta, \bar{\eta}) \leqq \bar{K} n,$$

where

$$(18) \qquad \bar{K} = \sum_k q(a_k) D(a_k)$$

and

$$(18') \qquad D(a_k) = \sup_{\{p_i\}} \sum_{i,j} p_i p_{ij}(a_k) \log \frac{p_{ij}(a_k)}{\bar{p}_j}, \quad \bar{p}_j = \sum_i p_i p_{ij}(a_k)$$

is equal to the capacity per unit time without feedback of the memoryless channel specified by the transition probabilities $p_{ij}(a_k)$. (Cf. (11) and (12).)

For a non-random channel, the quantity α takes only one value and $\bar{K}n = C^{(n)}$. Therefore Theorem 3 is a generalization of Theorem 2. It is clear that $\bar{C}$ is always less than or equal to $\bar{K}$; some examples given in Section 8 show that the difference $\bar{K} - \bar{C}$, which characterizes the gain in capacity by using feedback, is generally speaking positive, but usually small. Intuitively, we can explain the difference between $\bar{C}$ and $\bar{K}$ by saying that in the case where feedback is absent the optimum probabilities p_i for the channel input signals should be chosen in common for all the operating regimes of the channel, whereas feedback furnishes information at the input about the operating regime of the channel, so that the input signal probabilities p_i can therefore be chosen to be optimum for the regime in question.

Theorem **3** says nothing about the possibility of transmitting a message with information $I(\eta, \bar{\eta}) < \bar{K}n$, and therefore the question naturally arises as to whether the estimate (17) is optimal. The random communication channel is ergodic only in the case where it reduces to an ordinary channel. Therefore we can get no results for it of the type of the converse of Shannon's theorem. However, by slightly varying the channel considered above, we can make it ergodic. To do so we must assume that at every instant of time there occurs with probability u a random change of operating regime which is independent of the previous operation of the channel. If u is close to 0, then our new channel will "almost coincide" with the random memoryless channel considered above, and the characteristics which we find for the random memoryless channel will be limits of the corresponding characteristics for the new channel. It can be shown that if

$$\frac{I(\eta, \bar{\eta})}{\bar{K}n} < 1,$$

then the message η can be transformed into the message $\bar{\eta}$ as a result of transmission using feedback, if u is small enough and the channel length n is large enough. We shall give here neither a complete proof nor a precise statement of this fact, but we shall indicate the intuitive idea of the proof. If u is small, then for a long interval of time (of order $1/u$) the channel has a constant operating regime. Transmitting various signals for tracking purposes and using the feedback channel, in a small part of this interval we can find in just which regime the channel is operating. Then, we can use the coding method which is optimum for the regime $\alpha = a_k$ which was found, and we can transmit the message at a rate of $D(a_k)$ bits per second. Since for approximately one $q(a_k)$th of the time the channel operates in the regime a_k, the average message transmission rate will be close to

$$\sum_k q(a_k)D(a_k) = \bar{K},$$

as was to be shown.

4. Some Properties of Information

We recall some familiar properties of the information $I(\zeta, \bar{\zeta})$ defined by equation (9). Consider a third random quantity β, taking the values $B_1, \cdots, B_r$. By the conditional information of ζ and $\bar{\zeta}$, under the condition $\beta = B_u$, we mean the number

$$I(\zeta, \bar{\zeta}|\beta = B_u) = \sum_{i,j} \mathbf{P}\{\zeta = G_i, \bar{\zeta} = \bar{G}_j|\beta = B_u\} \log \frac{\mathbf{P}\{\zeta = G_i, \bar{\zeta} = \bar{G}_j|\beta = B_u\}}{\mathbf{P}\{\zeta = G_i|\beta = B_u\}\mathbf{P}\{\bar{\zeta} = \bar{G}_j|\beta = B_u\}}.$$

By the average value of the conditional information of the pair $\zeta, \bar{\zeta}$ for a given β we mean the number

$$\mathbf{M}I(\zeta, \bar{\zeta}|\beta) = \sum_{u=1}^{r} I(\zeta, \bar{\zeta}|\beta = B_u)\mathbf{P}\{\beta = B_u\}. \tag{19}$$

The pair of random quantities ζ, β can be regarded as a single random quantity (ζ, β), the values of which are the kr pairs (G_i, B_u). With this notation, the following important formula for conditional information is valid

$$I((\zeta, \beta), \bar{\zeta}) = I(\beta, \bar{\zeta}) + \mathbf{M}I(\zeta, \bar{\zeta}|\beta). \tag{20}$$

Evidently this formula was first stated explicitly in the work of Kolmogorov [**3**]. In the case which interests us (i.e., discrete quantities) equation (20) is a consequence of the following simple calculation

$$I((\zeta, \beta), \bar{\zeta}) = \sum_{i,j,u} \mathbf{P}\{\zeta = G_i, \bar{\zeta} = \bar{G}_j, \beta = B_u\} \times \log \frac{\mathbf{P}\{\zeta = G_i; \bar{\zeta} = \bar{G}_j, \beta = B_u\}}{\mathbf{P}\{\zeta = G_i, \beta = B_u\}\mathbf{P}\{\bar{\zeta} = \bar{G}_j\}}$$

$$= \sum_{i,j,u} \mathbf{P}\{\zeta = G_i, \bar{\zeta} = \bar{G}_j, \beta = B_u\}$$

$$\times \log \frac{\mathbf{P}\{\zeta = G_i, \bar{\zeta} = \bar{G}_j|\beta = B_u\}\mathbf{P}\{\beta = B_u\}\mathbf{P}\{\bar{\zeta} = \bar{G}_j, \beta = B_u\}}{\mathbf{P}\{\zeta = G_i|\beta = B_u\}\mathbf{P}\{\beta = B_u\}\mathbf{P}\{\bar{\zeta} = \bar{G}_j\}\mathbf{P}\{\bar{\zeta} = \bar{G}_j, \beta = B_u\}}$$

$$= \sum_{i,j,u} \mathbf{P}\{\zeta = G_i, \bar{\zeta} = \bar{G}_j, \beta = B_u\} \log \frac{\mathbf{P}\{\zeta = G_i, \bar{\zeta} = \bar{G}_j|\beta = B_u\}}{\mathbf{P}\{\zeta = G_i|\beta = B_u\}\mathbf{P}\{\bar{\zeta} = \bar{G}_j|\beta = B_u\}}$$

$$+ \sum_{i,j,u} \mathbf{P}\{\zeta = G_i, \bar{\zeta} = \bar{G}_j, \beta = B_u\} \log \frac{\mathbf{P}\{\bar{\zeta} = \bar{G}_j, \beta = B_u\}}{\mathbf{P}\{\bar{\zeta} = \bar{G}_j\}\mathbf{P}\{\beta = B_u\}}$$

$$= I(\beta, \bar{\zeta}) + \mathbf{M}I(\zeta, \bar{\zeta}|\beta).$$

We now note some important consequences of the conditional information formula (20). First of all, we note that since the conditional information $I(\zeta, \bar{\zeta}|\beta) \geqq 0$, then

$$I((\zeta, \beta), \bar{\zeta}) \geqq I(\beta, \bar{\zeta}) \tag{21}$$

for any three random quantities ζ, β, $\bar{\zeta}$. We now observe that

$$I(\gamma, \delta) \leqq I(\gamma, \gamma) \tag{22}$$

for any two random quantities γ and δ. To derive the inequality (22) it is sufficient to note that using the conditional information formula and the inequality (21) we have

$$I(\gamma, \delta) \leqq I(\gamma, (\gamma, \delta)) = I((\gamma, \delta), \gamma) = I(\gamma, \gamma) + \mathbf{M}I(\gamma, \delta|\gamma),$$

and that $\mathbf{M}I(\delta, \gamma|\gamma) = 0$, since the information of any random quantity concerning a constant is zero. We also find useful the widely known fact (see [5], [7]) that if a quantity γ takes a finite number s of values, then its entropy

$$I(\gamma, \gamma) \leqq \log s. \tag{23}$$

We now recall that we say that the three random quantities ζ, β, $\bar{\zeta}$ form a Markov chain if

$$\mathbf{P}\{\bar{\zeta} = \bar{G}_j|\beta = B_u, \zeta = G_i\} = \mathbf{P}\{\bar{\zeta} = \bar{G}_j|\beta = B_u\} \tag{24}$$

for any i, j, u. The condition (24) is equivalent to the following condition:

$$\mathbf{P}\{\bar{\zeta} = \bar{G}_j, \zeta = G_i | \beta = B_u\} = \mathbf{P}\{\bar{\zeta} = \bar{G}_j | \beta = B_u\}\mathbf{P}\{\zeta = G_i | \beta = B_u\} \tag{25}$$

for all i, j, u. The familiar fact that the conditions (25) and (24) are equivalent is not hard to verify by direct calculation. It clearly follows from (25) that if the quantities $\bar{\zeta}$, β, ζ form a Markov chain, then the conditional information $I(\zeta, \bar{\zeta} | \beta = B_u) = 0$, and consequently, the conditional information formula (20) shows that

$$I(\bar{\zeta}, (\zeta, \beta)) = I(\bar{\zeta}, \beta). \tag{26}$$

In what follows this important fact will be used repeatedly.

Finally, we need the following formula for triple information, i.e.,

$$I((\beta, \gamma), \delta) + I(\beta, \gamma) = I(\beta, (\gamma, \delta)) + I(\gamma, \delta) \tag{27}$$

for any three random quantities β, γ, δ. To derive this formula it is sufficient to observe that by direct calculation both the left-hand and right-hand sides of this formula equal

$$\sum_{i,j,k} \mathbf{P}\{\beta = B_i, \gamma = C_j, \delta = D_k\} \log \frac{\mathbf{P}\{\beta = B_i, \gamma = C_j, \delta = D_k\}}{\mathbf{P}\{\beta = B_i\}\mathbf{P}\{\gamma = C_j\}\mathbf{P}\{\delta = D_k\}}.$$

5. Proof of Shannon's Theorem for a Channel Without Feedback

We suppose that the input message η is transformed into the output message $\tilde{\eta}$ as a result of transmission through a channel of length n and we consider the corresponding input signal $\xi = (\xi_1, \cdots, \xi_n)$ and output signal $\tilde{\xi} = (\tilde{\xi}_1, \cdots, \tilde{\xi}_n)$. Applying the formula for total probability to the system of events $\xi = (E_{i_1}, \cdots, E_{i_n})$, we deduce from equation (7) that

$$\mathbf{P}\{\tilde{\eta} = \tilde{F}_\alpha | \tilde{\xi} = (\tilde{E}_{j_1}, \cdots, \tilde{E}_{j_n}), \eta = F_j\} = \mathbf{P}\{\tilde{\eta} = \tilde{F}_\alpha | \tilde{\xi} = (\tilde{E}_{j_1}, \cdots, \tilde{E}_{j_n})\}.$$

Comparing this equality and the definition (24) of a Markov chain, we see that the quantities η, $\tilde{\xi}$, $\tilde{\eta}$ form a Markov chain. Therefore, it follows from equations (21) and (26) that the information

$$I(\eta, \tilde{\eta}) \leqq I((\tilde{\eta}, \tilde{\xi}), \eta) = I(\tilde{\xi}, \eta). \tag{28}$$

We emphasize that we derived the inequality (28) starting with the condition (7), which is true both when feedback is used and when feedback is not used, and therefore (28) is also true for channels with feedback.

We now consider transmission without using feedback. Then equation (8) is true. We wish to show, starting from the general condition (6) and equation (8), that

$$\begin{aligned} \mathbf{P}\{(\tilde{\xi}_1, \cdots, \tilde{\xi}_k) = (\tilde{E}_{j_1}, \cdots, \tilde{E}_{j_k}) | (\xi_1, \cdots, \xi_k) = (E_{i_1}, \cdots, E_{i_k}), \eta = F_j\} \\ = \prod_{\alpha=1}^{k} r(j_\alpha | i_k, \cdots, i_\alpha; j_1, \cdots, j_{\alpha-1}) \end{aligned} \tag{29}$$

for all $j_1, \cdots, j_k, i_1, \cdots, i_k$ and any $1 \leqq k \leqq n$. We shall prove equation (29) by induction on k. For $k = 1$ it reduces at once to equation (6) for $k = 1$.

Suppose it is valid for $k-1$ factors. Now we note that it follows from equation (8) that the events $\{\xi_k = E_{i_k}\}$ and $\{(\bar{\xi}_1, \cdots, \bar{\xi}_{k-1}) = (\bar{E}_{j_1}, \cdots, \bar{E}_{j_{k-1}})\}$ are independent under the condition that $\{(\xi_1, \cdots, \xi_{k-1}) = (E_{i_1}, \cdots, E_{i_{k-1}})\}$ and $\{\eta = F_j\}$. Hence it follows that

$$(30)\quad \begin{aligned} &\mathbf{P}\{(\bar{\xi}_1, \cdots, \bar{\xi}_{k-1}) = (\bar{E}_{j_1}, \cdots, \bar{E}_{j_{k-1}}) | (\xi_1, \cdots, \xi_k) = (E_{i_1}, \cdots, E_{i_k}), \eta = F_j\} \\ &= \mathbf{P}\{(\bar{\xi}_1, \cdots, \bar{\xi}_{k-1}) = (\bar{E}_{j_1}, \cdots, \bar{E}_{j_{k-1}}) | (\xi_1, \cdots, \xi_{k-1}) = (E_{i_1}, \cdots, E_{i_{k-1}}), \eta = F_j\}. \end{aligned}$$

Now applying the conditional probability formula, equation (6), equation (30) and the induction hypothesis, we find that

$$(31)\quad \begin{aligned} &\mathbf{P}\{(\bar{\xi}_1, \cdots, \bar{\xi}_k) = (\bar{E}_{j_1}, \cdots, \bar{E}_{j_k}) | (\xi_1, \cdots, \xi_k) = (E_{i_1}, \cdots, E_{i_k}), \eta = F_j\} \\ &= \mathbf{P}\{\bar{\xi}_k = \bar{E}_{j_k} | (\xi_1, \cdots, \xi_k) = (E_{i_1}, \cdots, E_{i_k}), (\bar{\xi}_1, \cdots, \bar{\xi}_{k-1}) = (\bar{E}_{j_1}, \cdots, \bar{E}_{j_{k-1}}), \eta = F_j\} \\ &\quad \cdot \mathbf{P}\{(\bar{\xi}_1, \cdots, \bar{\xi}_{k-1}) = (\bar{E}_{j_1}, \cdots, \bar{E}_{j_{k-1}}) | (\xi_1, \cdots, \xi_k) = (E_{i_1}, \cdots, E_{i_k}), \eta = F_j\} \\ &= r(j_k | i_1, \cdots, i_k; j_1, \cdots, j_{k-1}) \\ &\quad \cdot \mathbf{P}\{(\bar{\xi}_1, \cdots, \bar{\xi}_{k-1}) = (\bar{E}_{j_1}, \cdots, \bar{E}_{j_{k-1}}) | (\xi_1, \cdots, \xi_{k-1}) = (E_{i_1}, \cdots, E_{i_{k-1}}), \eta = F_j\} \\ &= \prod_{\alpha=1}^{n} r(j_\alpha | i_1, \cdots, i_\alpha; j_1, \cdots, j_{\alpha-1}), \end{aligned}$$

as was to be proved.

Applying equation (29) for $k = n$, we see that the sequence of quantities $\eta, \xi, \bar{\xi}$ forms a Markov chain. Arguing just as in the derivation of equation (28), we discover that

$$(32)\quad I(\eta, \bar{\xi}) \leqq I(\xi, \bar{\xi})$$

in the case of transmission without using feedback. Again applying equation (29) for $k = n$, we find that

$$(33)\quad \begin{aligned} &\mathbf{P}\{(\bar{\xi}_1, \cdots, \bar{\xi}_n) = (\bar{E}_{j_1}, \cdots, \bar{E}_{j_n}) | (\xi_1, \cdots, \xi_n) = (E_{i_1}, \cdots, E_{i_n})\} \\ &= \prod_{k=1}^{n} r(j_k | i_1, \cdots, i_k; j_1, \cdots, j_{k-1}). \end{aligned}$$

It follows from equations (2) and (4) that

$$(34)\quad r(j_k | i_1, \cdots, i_k; j_1, \cdots, j_{k-1}) = \frac{\sum_{j_{k+1}, \cdots, j_n} p(j_1, \cdots, j_n | i_1, \cdots, i_n)}{\sum_{j_k, \cdots, j_n} p(j_1, \cdots, j_n | i_1, \cdots, i_n)}.$$

Setting (34) in (33), we find that

$$(35)\quad \begin{aligned} &\mathbf{P}\{(\bar{\xi}_1, \cdots, \bar{\xi}_n) = (\bar{E}_{j_1}, \cdots, \bar{E}_{j_n}) | (\xi_1, \cdots, \xi_n) = (E_{i_1}, \cdots, E_{i_n})\} \\ &= p(j_1, \cdots, j_n | i_1, \cdots, i_n). \end{aligned}$$

Equation (35) shows that in the case of transmission without using feedback the quantities ξ and $\bar{\xi}$ are related by a channel of length n, and therefore it follows from the definition of capacity (10′) that

$$(36)\quad I(\xi, \bar{\xi}) \leqq C^{(n)}.$$

From the inequalities (36), (32) and (28) we infer that

$$I(\eta, \tilde{\eta}) \leqq C^{(n)},$$

whereby we have proved Shannon's theorem, formulated in Section 3.

6. Derivation of the Formulas for the Capacity of the Memoryless Channel and the Random Memoryless Channel in the Absence of Feedback

We begin by proving the following important fact. If $\xi = (\xi_1, \cdots, \xi_n)$ and $\tilde{\xi} = (\tilde{\xi}_1, \cdots, \tilde{\xi}_n)$ are two random quantities related by a memoryless channel, then

$$I(\xi, \tilde{\xi}) \leqq I(\xi_1, \tilde{\xi}_1) + I(\xi_2, \tilde{\xi}_2) + \cdots + I(\xi_n, \tilde{\xi}_n). \tag{37}$$

Consider first the case $n = 2$. In this case the triple information formula (27) shows that

$$I((\xi_1, \xi_2), (\tilde{\xi}_1, \tilde{\xi}_2)) = I(\tilde{\xi}_1, (\xi_1, \xi_2, \tilde{\xi}_2)) + I((\xi_1, \xi_2), \tilde{\xi}_2) - I(\tilde{\xi}_1, \tilde{\xi}_2). \tag{38}$$

Equation (3) for a memoryless channel shows that

$$\mathbf{P}\{\tilde{\xi}_1 = \tilde{E}_{j_1} | \xi_1 = E_{i_1}, \xi_2 = E_{i_2}, \tilde{\xi}_2 = \tilde{E}_{j_2}\} = p^{(1)}_{i_1 j_1},$$
$$\mathbf{P}\{\tilde{\xi}_2 = \tilde{E}_{j_2} | \xi_1 = E_{i_1}, \xi_2 = E_{i_2}\} = p^{(2)}_{i_2 j_2}.$$

Consequently, the quantities $\tilde{\xi}_1, \xi_1, (\xi_2, \tilde{\xi}_2)$ and the quantities $\tilde{\xi}_2, \xi_2, \xi_1$ form Markov chains. Then according to equation (26)

$$\begin{aligned} I(\tilde{\xi}_1, (\xi_1, \xi_2, \tilde{\xi}_2)) &= I(\xi_1, \tilde{\xi}_1), \\ I((\xi_1, \xi_2), \tilde{\xi}_2) &= I(\xi_2, \tilde{\xi}_2), \end{aligned} \tag{39}$$

and (37) for $n = 2$ follows from (38) and (39). To generalize this formula to arbitrary n it is sufficient to observe that the four quantities $\xi_1' = (\xi_1, \cdots, \xi_{n-1})$, $\tilde{\xi}_1' = (\tilde{\xi}_1, \cdots, \tilde{\xi}_{n-1})$, $\xi_2' = \xi_2$ and $\tilde{\xi}_2' = \tilde{\xi}_2$ form a channel without memory, so that according to what was shown above we have

$$I((\xi_1, \cdots, \xi_n), (\tilde{\xi}_1, \cdots, \tilde{\xi}_n)) \leqq I((\xi_1, \cdots, \xi_{n-1}), (\tilde{\xi}_1, \cdots, \tilde{\xi}_{n-1})) + I(\xi_n, \tilde{\xi}_n). \tag{40}$$

Applying induction, we deduce from (40) the desired inequality (37).

We now use (37) to derive equation (11) for the capacity of a channel without memory. Since if $(\xi_1, \cdots, \xi_n)$ and $(\tilde{\xi}_1, \cdots, \tilde{\xi}_n)$ are related by the channel, we have

$$\mathbf{P}\{\tilde{\xi}_k = E_{j_k} | \xi_k = E_{i_k}\} = p^{(k)}_{ij},$$

then $I(\xi_k, \tilde{\xi}_k) \leqq D_k$ (see (12)), whence it follows according to (37) that

$$I(\xi, \tilde{\xi}) \leqq D_1 + \cdots + D_n = C^{(n)}. \tag{41}$$

To show that the equality sign is achieved in (41), it is sufficient to note that if we denote by $\bar{p}^{(k)}_i$ the set of probabilities p_i for which the upper bound D_k in (12) is achieved and if we take for ξ, $\tilde{\xi}$ quantities with the joint distribution

$$\mathbf{P}\{\xi = (E_{i_1}, \cdots, E_{i_n}), \tilde{\xi} = (E_{j_1}, \cdots, E_{j_n})\} = \prod_{k=1}^{n} \bar{p}^{(k)}_i p^{(k)}_{i_k j_k},$$

then we obtain quantities related by the channel for which

$$I(\xi, \tilde{\xi}) = D_1 + \cdots + D_n = C^{(n)}.$$

Now let $\xi = (\xi_1, \cdots, \xi_n)$ and $\tilde{\xi} = (\tilde{\xi}_1, \cdots, \tilde{\xi}_n)$ be random quantities related by a random memoryless channel of length n, and let α be the random variable used in the definition of the random memoryless channel. Applying the conditional information formula (20), we note that

$$I((\tilde{\xi}, \alpha), \xi) = I(\xi, \tilde{\xi}) + \mathbf{M}I(\alpha, \xi|\tilde{\xi}) = I(\xi, \alpha) + \mathbf{M}I(\xi, \tilde{\xi}|\alpha),$$

whence it follows that

$$I(\xi, \tilde{\xi}) = \mathbf{M}I(\tilde{\xi}, \xi|\xi) + I(\xi, \alpha) - \mathbf{M}I(\alpha, \xi|\tilde{\xi}). \tag{42}$$

Moreover, we note that because of the general inequality (22) and the inequality (23), we have

$$I(\xi, \tilde{\xi}) = \mathbf{M}I(\xi, \tilde{\xi}|\alpha) + O(1), \tag{43}$$

where the term $O(1)$ is uniformly bounded for all n and all pairs ξ and $\tilde{\xi}$. We shall show below that

$$\sup_{(\xi, \tilde{\xi})} \mathbf{M}\, I(\xi, \tilde{\xi}|\alpha) = n\bar{C} \tag{44}$$

(see (16′)), where the upper bound is taken over all pairs of random quantities ξ and $\tilde{\xi}$ related by a random memoryless channel of length n. Equation (16) will follow in an obvious way from (43) and (44). Now we use the fact that for any fixed $\alpha = a_k$ the quantities ξ and $\tilde{\xi}$ are related by a memoryless channel with transition probabilities $p_{ij}^{(l)} = p_{ij}(a_k)$. Applying the inequality (37), we discover that

$$I(\xi, \tilde{\xi}|\alpha = a_k) \leqq I(\xi_1, \tilde{\xi}_1|\alpha = a_k) + I(\xi_2, \tilde{\xi}_2|\alpha = a_k) + \cdots + I(\xi_n, \tilde{\xi}_n|\alpha = a_k). \tag{45}$$

From this it follows that

$$\mathbf{M}I(\xi, \tilde{\xi}|\alpha) \leqq \sum_{e=1}^{n} \sum_{k} I(\xi_e, \tilde{\xi}_e|\alpha = a_k) q(a_k), \tag{46}$$

where the equality sign is achieved in (45) and (46) if the quantities $\xi_1, \cdots, \xi_n$ are independent for any condition of the form $\alpha = a_k$. To derive the desired equality (43) it remains to observe that

$$\sup_{\xi_e} \sum_{k} q(a_k) I(\xi_e, \tilde{\xi}_e|\alpha = a_k) = \bar{C}$$

by definition of the quantity $\bar{C}$ (see (16′)) and that if we take for ξ the set $(\xi_1, \cdots, \xi_n)$ of independent quantities ξ_e which have the distribution $\{p_i\}$ for which the upper bound in (16′) is achieved and if we set

$$\mathbf{P}\{\tilde{\xi}_1 = E_{j_1}, \cdots, \tilde{\xi}_n = E_{j_n}|\xi_1 = E_{i_1}, \cdots, \xi_n = E_{i_n}\} = p(j_1, \cdots, j_n|i_1, \cdots, i_n),$$

(see (15)) then we obtain a pair of quantities $(\xi, \tilde{\xi})$ which are related by a random memoryless channel of length n and which are such that

$$\mathbf{M}I(\xi, \tilde{\xi}|\alpha) = n\bar{C}.$$

7. Information Transmission Using Feedback

First of all we note that the method of proving Shannon's theorem which was used in Section 5 for a channel without feedback is fundamentally inapplicable to the case of a channel with feedback. The point is that in the case of transmission using feedback the input signal ξ and the output signal $\tilde{\xi}$ may not be related by the channel even for a memoryless channel. As an example demonstrating this statement, consider the memoryless channel specified by the transition probabilities

$$p_{ij}^{(k)} = \tfrac{1}{2} \tag{47}$$

for all i, j and k. This means that the output symbol does not depend on the corresponding input symbol. It is clear that the capacity of such a channel is $C^{(n)} = 0$. At the same time, it is easy to verify that if we choose the quantities $\xi = (\xi_1, \cdots, \xi_n)$ and $\tilde{\xi} = (\tilde{\xi}_1, \cdots, \tilde{\xi}_n)$ such that ξ and $\tilde{\xi}$ do not depend on the input message η and the output message $\tilde{\eta}$, such that

$$\mathbf{P}\{\xi_k = E_1\} = \mathbf{P}\{\xi_k = E_2\} = \mathbf{P}\{\tilde{\xi}_k = E_1\} = \mathbf{P}\{\tilde{\xi}_k = E_2\} = \tfrac{1}{2},$$

for all k, and such that $\xi_k = \tilde{\xi}_{k-1}$ and does not depend on the remaining $\xi_i, i \neq k$, and $\tilde{\xi}_j, j \neq k-1$, then the four quantities η, ξ, $\tilde{\xi}$ and $\tilde{\eta}$ will satisfy the conditions (6) and (7). Intuitively, our construction means that at each instant of time we transmit the signal received at the output at the preceding instant of time, independently of the value of the input message. It is clear that in this case

$$I(\xi, \tilde{\xi}) = \sum_{k=2}^{n} I(\xi_k, \tilde{\xi}_{k-1}) = n-1 > C^{(n)} = 0.$$

We now turn to the proof of the basic theorem about the memoryless channel with feedback. We assume that the input message η is transformed into the output message $\tilde{\eta}$ as a result of transmission through a memoryless channel of length n using feedback, and that ξ and $\tilde{\xi}$ are the corresponding input and output signals. As we already noted, the inequality (28) which says that

$$I(\eta, \tilde{\eta}) \leqq I(\eta, \tilde{\xi}). \tag{28'}$$

is also true for the case of transmission using feedback. Because of this we shall evaluate the information

$$I(\eta, \tilde{\xi}) = I(\eta, (\tilde{\xi}_1, \cdots, \tilde{\xi}_n)).$$

Applying n times the conditional information formula (20), we deduce that

$$\begin{aligned} I(\eta, \tilde{\xi}) &= I((\tilde{\xi}_1, \cdots, \tilde{\xi}_n), \eta) = I((\tilde{\xi}_1, \cdots, \tilde{\xi}_{n-1})\eta) + \mathbf{M}I(\tilde{\xi}_n, \eta | (\tilde{\xi}_1, \cdots, \tilde{\xi}_{n-1})) \\ &= I((\tilde{\xi}_1, \cdots, \tilde{\xi}_{n-2}), \eta) + \mathbf{M}I(\tilde{\xi}_{n-1}, \eta | (\tilde{\xi}_1, \cdots, \tilde{\xi}_{n-2})) + \mathbf{M}I(\tilde{\xi}_n, \eta | (\tilde{\xi}_1, \cdots, \tilde{\xi}_{n-1})) \\ &= I(\tilde{\xi}_1, \eta) + \mathbf{M}I((\tilde{\xi}_2, \eta | \tilde{\xi}_1) + \cdots + \mathbf{M}I(\tilde{\xi}_n, \eta | (\tilde{\xi}_1, \cdots, \tilde{\xi}_{n-1})). \end{aligned} \tag{48}$$

Equation (6) shows that under the condition $(\tilde{\xi}_1, = \tilde{E}_{j_1}, \cdots, \tilde{\xi}_{k-1} = \tilde{E}_{j_{k-1}})$ the three random quantities η, $(\xi_1, \cdots, \xi_k)$, $\tilde{\xi}_k$ form a Markov chain. Therefore, it follows from the general formulas (21) and (25) that

$$(49)\quad \begin{aligned} &I(\bar{\xi}_k, \eta|\bar{\xi}_1 = \bar{E}_{j_1}, \cdots, \bar{\xi}_{k-1} = \bar{E}_{j_{k-1}}) \\ &\leq I(\bar{\xi}_k, (\eta, \xi_1, \cdots, \xi_k)|\bar{\xi}_1 = \bar{E}_{j_1}, \cdots, \bar{\xi}_{k-1} = \bar{E}_{j_{k-1}}) \\ &= I(\bar{\xi}_k, (\xi_1, \cdots, \xi_k)|\bar{\xi}_1 = \bar{E}_{j_1}, \cdots, \bar{\xi}_{k-1} = \bar{E}_{j_{k-1}}). \end{aligned}$$

Using equations (6) and (5), we discover that in the case of a memoryless channel

$$(50)\quad \mathbf{P}\{\bar{\xi}_k = \bar{E}_{j_k}|\xi_1 = E_{i_1}, \cdots, \xi_k = E_{i_k}, \bar{\xi}_1 = \bar{E}_{j_1}, \cdots, \bar{\xi}_{k-1} = \bar{E}_{j_{k-1}}\} = p^{(k)}_{i_k j_k}.$$

This equality shows that under the condition $(\bar{\xi}_1 = \bar{E}_{j_1}, \cdots, \bar{\xi}_{k-1} = \bar{E}_{j_{k-1}})$ the three quantities $(\xi_1, \cdots, \xi_{k-1})$, ξ_k, $\bar{\xi}_k$ form a Markov chain, and therefore,

$$(51)\quad I(\bar{\xi}_k(\xi_1,\cdots,\xi_k)|\bar{\xi}_1=\bar{E}_{j_1},\cdots,\bar{\xi}_{k-1}=\bar{E}_{j_{k-1}})=I(\bar{\xi}_k,\xi_k|\bar{\xi}_1=\bar{E}_{j_1},\cdots,\bar{\xi}_{k-1}=\bar{E}_{j_{k-1}}).$$

It follows from equation (50) that

$$(52)\quad \mathbf{P}\{\bar{\xi}_k = \bar{E}_{j_k}|\xi_k = E_{i_k}, \bar{\xi}_1 = \bar{E}_{j_1}, \cdots, \bar{\xi}_{k-1} = \bar{E}_{j_{k-1}}\} = p^{(k)}_{i_k j_k}.$$

Therefore, it is a consequence of the definition (12) that the conditional information

$$(53)\quad I(\xi_k, \bar{\xi}_k|\bar{\xi}_1 = \bar{E}_{j_1}, \cdots, \bar{\xi}_{k-1} = \bar{E}_{j_{k-1}}) \leq D_k.$$

It follows from (53), (51) and (49) that

$$(54)\quad I(\bar{\xi}_k, \eta|\bar{\xi}_1 = \bar{E}_{j_1}, \cdots, \bar{\xi}_{k-1} = \bar{E}_{j_{k-1}}) \leq D_k$$

for all $k = 1, \cdots, n$. Applying (54) and (48), we deduce that

$$(55)\quad I(\eta, \bar{\xi}) \leq D_1 + \cdots + D_n = C^{(n)}.$$

The inequality (28′) shows that the assertion of the basic theorem for a memoryless channel with feedback follows from the inequality (55).

We now turn to a consideration of the random memoryless channel. We wish to prove Theorem 3, formulated in Section 3. Suppose the input message η is transformed into the output message $\bar{\eta}$ as a result of transmission through a random memoryless channel of length n using feedback. Let $\xi = (\xi_1, \cdots, \xi_n)$ and $\bar{\xi} = (\bar{\xi}_1, \cdots, \bar{\xi}_n)$ be the corresponding input and output signals. Here equation (28) is again applicable and shows that

$$(56)\quad I(\eta, \bar{\eta}) \leq I(\eta, \bar{\xi}).$$

Arguing just as in the derivation of equation (42) (we need only replace ξ everwhere by η) we find that

$$(57)\quad I(\eta, \bar{\xi}) = \mathbf{M}I(\eta, \bar{\xi}|\alpha) + I(\eta, \alpha) - \mathbf{M}I(\alpha, \eta|\bar{\xi}).$$

From the intuitive probabilistic interpretation of the quantity α it follows [1]

[1] Equations (6) and (15) serve as a formal definition of transmission over a random memoryless channel with feedback. Therefore, strictly speaking, we must show that for any three quantities η, ξ, $\bar{\xi}$ satisfying the conditions (6) and (15), we can construct a quantity α with distribution $\{q(a_e)\}$ such that η and α are independent and such that equation (58) is true. Because of its simplicity, we do not give this straightforward argument.

in an obvious way that the input message and the parameter α are independent quantities and that the condition (6) can now be replaced by the condition

$$\begin{aligned}\mathbf{P}\{\bar{\xi}_k = \bar{E}_{j_k} \mid \xi_1 = E_{i_1}, \cdots, \xi_k = E_{i_k}, \bar{\xi}_1 = \bar{E}_{j_1}, \cdots, \bar{\xi}_{k-1} = \bar{E}_{j_{k-1}}, \eta = F_j, \alpha = a_e\} \\ = p_{i_k j_k}(a_e).\end{aligned} \tag{58}$$

It follows from the independence of η and α that $I(\eta, \alpha) = 0$ and therefore according to (57)

$$I(\eta, \bar{\xi}) \leqq \mathbf{M}I(\eta, \bar{\xi} \mid \alpha). \tag{59}$$

For any fixed condition $\alpha = a_e$, equation (58) shows that the quantities η, ξ, $\bar{\xi}$ accomplish transmission of the message η through the memoryless channel specified by the transition probabilities $p_{ij}(a_e)$ using feedback. Therefore, repeating the argument given at the beginning of this section, we come to the conclusion that

$$I(\eta, \bar{\xi} \mid \alpha = a_e) \leqq nD(a_e), \tag{60}$$

where $D(a_e)$ is the capacity per unit time of the memoryless channel with transition probabilities $p_{ij}(a_e)$ (see (18)). It follows from (60) that

$$\mathbf{M}I(\eta, \bar{\xi} \mid \alpha) \leqq n \sum_e q(a_e) D(a_e),$$

and this and the inequality (59) yield the assertion of Theorem 3.

8. Some Examples

For some concrete random memoryless channels we compare the channel capacity per unit time $\bar{C}$ without using feedback (see (16′)) and the channel capacity per unit time $\bar{K}$ using feedback (see (18)).

Suppose that the number of states at the input and output equals 2 and that for any value of the parameter a_k the matrix $p_{ij}(a_k)$ is symmetric, i.e., $p_{11}(a_k) \equiv p_{22}(a_k)$, $p_{12}(a_k) \equiv p_{21}(a_k)$. Then for all k the upper bound in the expression (18′) for $D(a_k)$ is attained for $p_1 = p_2 = \frac{1}{2}$. Therefore for $p_1 = p_2 = \frac{1}{2}$, the expression

$$\sum_k q(a_k) \sum_{i,j} p_i p_{ij}(a_k) \log \frac{p_{ij}(a_k)}{\bar{p}_j(a_k)},$$

standing under the upper bound sign in (16′), is equal to $\bar{K}$. Since it is obvious that $\bar{C}$ is always less than $\bar{K}$, we come to the conclusion that for the example in question $\bar{C} = \bar{K}$, i.e., using feedback does not give a gain in channel capacity. An analogous argument holds, of course, in any case where the optimum signal distribution does not depend on the parameter α. If the transition probability matrix is not assumed to be symmetric, then the optimum distribution will depend on α, and the use of feedback will in this case give a gain in capacity.

We now consider an example of a channel in which use of feedback can substantially increase capacity. Let the input states be $E_1, \cdots, E_n$ and the output states $\bar{E}_1, \cdots, \bar{E}_n$. Let the parameter α take the n values $a_1, \cdots, a_n$ each with probability $1/n$. We assume that the transition probabilities $p_{ij}(a_k)$ are given by the formula

$$p_{i1}(a_k) = \begin{cases} 0, & i \neq k, \\ 1, & i = k, \end{cases}$$

$$p_{i2}(a_k) = \begin{cases} 1, & i = k, \\ 0, & i \neq k. \end{cases}$$

Then the capacity $D(a_k)$ will be $1/2$ for any fixed k. Consequently, the capacity per unit time with feedback will be

$$\bar{K} = \log 2.$$

On the other hand (16′) takes its maximum value for $p_1 = \cdots = p_n = 1/n$, and therefore

$$\bar{C} = \frac{1}{n} \log n - \left(1 - \frac{1}{n}\right) \log \left(1 - \frac{1}{n}\right).$$

The gain in capacity $(\bar{K} - \bar{C})/\bar{K}$ is small for small n. As $n \to \infty$, the quantity

$$\bar{C} \sim \frac{\log n + 1}{n}$$

and the gain approaches 1.

Received by the editors
March 6, 1958

REFERENCES

[1] W. B. Bishop and B. L. Buchanan, *Message redundancy vs. feedback for reducing message uncertainty*, IRE National Convention Record, 1957, Part 2, pp. 33–39.

[2] E. S. Gorbunov, *A comparison of some noise-combating codes*, Electrosvyaz, No. 12, 1956, pp. 42–47. (In Russian.)

[3] A. N. Kolmogorov, *Theory of information transmission*, Collection: "Session of the Academy of Sciences of the USSR on scientific problems of automatic production", Plenary sessions, Oct. 15–20, 1956; Izv. Akad. Nauk SSSR, 1957, pp. 66–99. (In Russian.)

[4] A. N. Kolmogorov, *A new metric invariant of transitive dynamical systems and automorphisms of Lebesgue spaces*, Dokl. Akad. Nauk SSSR, 119, 5, 1958, pp. 861–864. (In Russian.)

[5] C. E. Shannon, *The mathematical theory of communication*, Bell System Tech. J., 27, 1948, pp. 379–423, pp. 623–656.

[6] B. Harris, A. Hauptschein, L. S. Schwartz, *Optimum decision feedback systems*, IRE National Convention Record, 1957, Part 2, pp. 3–10.

[7] A. Ya. Khinchin, *On the basic theorems of information theory*, Uspekhi Mat Nauk, 11, 1, 1956, pp. 17–75.

[8] I. P. Tsaregradskii, *A note on the capacity of a stationary channel with finite memory*, Theory Prob. Applications, 3, 1958, pp. 79–91. (English Translation.)

[9] S. S. L. Chang, *Theory of information feedback systems*, I. R. E. Transactions on Information Theory, Vol. IT-2, No. 3, 1956, pp. 29–40.

TRANSMISSION OF INFORMATION IN CHANNELS WITH FEEDBACK

R. L. DOBRUSHIN (*MOSCOW*)

(*Summary*)

In this paper we prove that the use of feedback does not increase the capacity of channels without memory. We also consider some simple channels with memory and compare their capacities when feedback is and is not used.

Signal Design for Sequential Detection Systems with Feedback

GEORGE L. TURIN, SENIOR MEMBER, IEEE

Abstract—**We consider a coherent white Gaussian channel, through which one of two signals is sent to a receiver which operates as a sequential detector. A noiseless delayless feedback link is assumed, which continuously informs the transmitter of the state of the receiver's uncertainty concerning which signal was sent, and which also synchronizes the transmitter when the receiver has reached a decision. The transmitter, in turn, uses the output of the feedback link to modify its transmission so as to hasten the receiver's decision.**

The following problem is posed: Given average- and peak-power constraints on the transmitter and a prescribed error probability for the receiver, what signal waveforms should the transmitter use in order to minimize the average transmission time, and how should it utilize the fedback values of the receiver's uncertainty to modify these waveforms while transmission is in progress? We give partial solutions to these questions. In particular, we have shown that if the peak-to-average power ratio is allowed to be sufficiently large, substantial improvement of performance may be achieved through the use of uncertainty fèedback.

Introduction

CONSIDER A CHANNEL disturbed only by additive white Gaussian noise n, defined on the time interval $(-\infty, \infty)$. Through this channel, one of two signal waveforms s_+ and s_- defined on $[t_0, \infty)$, may be transmitted with a priori probabilities P_+ and P_-, respectively. The output z of the channel may therefore be either of the form $z = s_+ + n$ (hypothesis H_+) or of the form $z = s_- + n$ (hypothesis H_-). Suppose that z is observed over the time interval $[t_0, t]$, where t is a parameter taking on values in $[t_0, \infty)$; denote this observation by z_t.

We shall concern ourselves with the following sequential test for deciding between hypotheses H_+ and H_- at the receiver, given the sequence of observations $\{z_t;\ t_0 \leq t < \infty\}$. We first form

$$y(t) \triangleq \ln \frac{Pr[H_+/z_t]}{Pr[H_-/z_t]} \qquad (t \geq t_0), \tag{1}$$

where Pr $[H_\pm/z_t]$ are the a posteriori probabilities of $H_\pm$, given z_t, and

$$y(t_0) \triangleq \ln \frac{P_+}{P_-} \triangleq y_0. \tag{2}$$

No decision is made for $t_0 \leq t < t_0 + T$, where

$$t_0 + T = \sup \{t : Y_- < y(t) < Y_+\}, \tag{3}$$

$Y_\pm$ being prespecified thresholds. It is decided that H_+ is the true hypothesis if $y(t_0 + T) = Y_+$; it is decided that H_- is true if $y(t_0 + T) = Y_-$. Notice that since $y(t)$ is clearly a random process, T is a random variable.

The thresholds $Y_\pm$ may be related to certain error probabilities as follows. We note that whenever it is decided that H_+ is true, i.e., the test stops with $y = Y_+$, we have from (1):

$$Pr[H_-/z_{t_0+T}] = e^{-Y_+} Pr[H_+/z_{t_0+T}]. \tag{4}$$

Averaging (4) over all observations $z_{t_0} + T$ which lead to decision H_+, we obtain

$$P_- P(e/H_-) = e^{-Y_+} P_+[1 - P(e/H_+)], \tag{5}$$

where $P(e/H_\pm)$ is the probability of an erroneous decision when $s_\pm$ is sent. A similar equation involving Y_- may be obtained. Solving these equations for $Y_\pm$, and using (2), we obtain

$$Y_\pm = y_0 \pm \ln \frac{1 - P(e/H_\pm)}{P(e/H_\mp)}. \tag{6}$$

The average error probability is $P(e) \triangleq P_+ P(e/H_+) + P_- P(e/H_-)$.

For the assumed channel, one has for $y(t)$:[1]

$$y(t) = y_0 + \frac{2}{N_0} \int_{t_0}^{t} \{s_+[y(\tau), \tau] - s_-[y(\tau), \tau]\} z(\tau)\, d\tau - \frac{1}{N_0} \int_{t_0}^{t} \{s_+^2[y(\tau), \tau] - s_-^2[y(\tau), \tau]\}\, d\tau, \tag{7}$$

where N_0 is the (single-ended) noise power density (watts per c/s). In (7), we have written the dependence of $s_\pm$ on t both directly and also through $y(t)$, the latter to allow for what we shall call *uncertainy feedback*. That is, we allow the receiver to feed back to the transmitter the current value of y, thus indicating the state of the receiver's uncertainty concerning what is being sent. The transmitter may then modify its transmission so as to improve the progress of the test; although the receiver does not know which signal is being sent, it knows what modifications the transmitter will make under both hypotheses, and the receiver makes the appropriate modifications of its own structure, according to (7). Note that the feedback link also serves the necessary function of synchronizing the transmitter for the transmission of a new binary digit,

Manuscript received June 24, 1964; revised March 24, 1965. This paper is dedicated to the memory of Dr. Edmund Eisenberg. The work reported herein was supported by the Dept. of the Air Force under Grant AF-AFOSR-230-64.

The author is with the Dept. of Electrical Engineering, University of California, Berkeley, Calif.

[1] Woodward, P. M., *Probability and Information Theory, with Applications to Radar.* New York: McGraw-Hill, 1953, ch 4. Note in (7) that, although $y(t)$ is a sample function of a random process, its values over $[t_0, t]$ are known precisely to the transmitter and receiver alike, so that the integral may be computed as if the signals were deterministic. See[3], however.

Reprinted from *IEEE Trans. Inform. Theory*, vol. IT-11, pp. 401–408, July 1965.

for the transmitter knows that the receiver has made a decision on the current transmission when the fedback value of y reaches one of the thresholds.

A block diagram of the system we have described is given in Fig. 1.

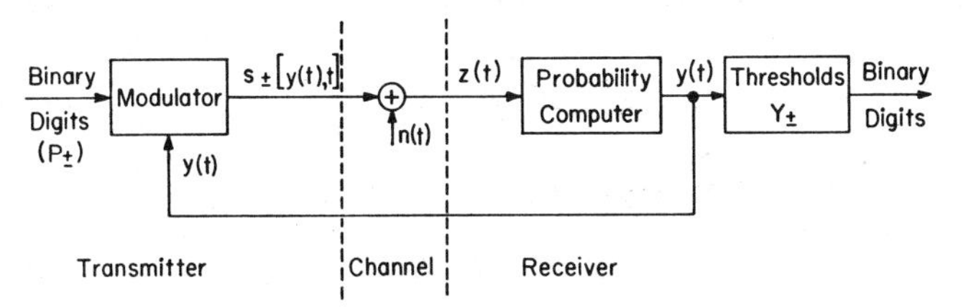

Fig. 1. A sequential detection system with uncertainty feedback.

We shall consider in this paper the problem of choosing the signals $s_\pm$, subject to constraints on their power, so as to minimize $E[T]$, the average time it takes to reach a decision. We shall also evaluate the error probability associated with the use of the optimal signals thus chosen, and compare this probability with that obtained when only the feed-forward link is used, in conjunction with nonsequential detection.

Before proceeding, it is important to note some of the deficiencies of the model we have established. First, the sequential test we use is not necessarily optimum, i.e., does not necessarily minimize $E[T]$ for given error probabilities $P(e/H_\pm)$ and given signals $s_\pm$.[2] Nonetheless, our test is a good one, as we shall see when we evaluate its performance with optimal signals.

Second, we have assumed a noiseless feedback link. This may be justified in such situations as telemetry from a satellite, where the transmitter on the satellite is of low power, but the ground-based receiver may have associated with it a feedback transmitter of extremely high power.

Third, and most serious, we have assumed delayless transmission channels: we have assumed that the signal component of z *instantaneously* reflects the modification of the transmission associated with the *present* value of y. If the actual time delays encountered in the physical channels are small compared with the reciprocals of the signal bandwidths, however, this assumption may be a good approximation.

Finally, we have assumed that the receiver knows the levels of the signals $s_\pm$ *as they appear at its input* [see (7)]. This assumption requires measuring equipment at the receiver.

Formulation of the Problem

If we differentiate (7) with respect to t and substitute $z = s_\pm + n$ in the result, we obtain the two differential equations

$$\frac{dy(t)}{dt} = \pm\frac{1}{N_0}\,\Delta^2(y,t) + \frac{2}{N_0}\,\Delta(y,t)n(t), \tag{8}$$

where $\Delta(y,t) \triangleq s_+(y,t) - s_-(y,t)$. These are generalized Langevin equations,[3] one governing the evolution of the process y when H_+ is true, the other governing the evolution when H_- is true. Since n is a white Gaussian process, y is a Markov process and is described completely by the probability densities $p_\pm(y, t/y_0, t_0)$ of y at time t, given that $y(t_0) = y_0$ and that $H_\pm$ is true. In fact, these densities satisfy, respectively, the Fokker-Planck equations[3]

$$\frac{\partial p_\pm}{\partial t} = \mp\frac{\partial}{\partial y}\left[\frac{\Delta}{N_0}\left(\Delta + \frac{\partial\Delta}{\partial y}\right)p_\pm\right] + \frac{\partial^2}{\partial y^2}\left[\frac{\Delta^2}{N_0}p_\pm\right], \tag{9}$$

subject to the initial conditions $p_\pm(y, t_0/y_0, t_0) = \delta(y - y_0)$ and suitable boundary conditions at $y = \pm\infty$.

We are, however, not interested in the densities $p_\pm$ themselves, but in the densities $q_\pm(y, t/y_0, t_0)$ defined by the statements: $q_\pm(y_1, t_1/y_0, t_0)dy$ = probability that if $H_\pm$ is true and $y(t_0) = y_0$, then $y_1(t)$ lies in $(y_1, y_1 + dy)$ *and* $Y_- < y(t) < Y_+$ for all $t_0 \le t \le t_1$. These are the densities that describe the uncertainty of the receiver at a time t_1 when the receiver is still testing. Notice, however, that they are not proper probability densities, since

$$\begin{aligned}\int_{Y_-}^{Y_+} q_\pm(y, t_1/y_0, t_0)\,dy &= Pr[Y_- < y(t) < Y_+, \text{ all } t_0 \le t \le t_1/y(t_0) = y_0; H_\pm] \\ &\triangleq F_\pm(t_1/y_0, t_0),\end{aligned} \tag{10}$$

which is not generally equal to unity. [$F_\pm$ are the so-called "first-passage" probability distributions of $y(t)$; in the present contex they are the probabilities that the receiver has *not* made a decision by time t_1, given that $H_\pm$ is true, respectively.]

It has been shown[4] that $q_\pm$ satisfy the same differential equations as $p_\pm$, viz.,

$$\frac{\partial q_\pm}{\partial t} = \mp\frac{\partial}{\partial y}\left[\frac{\Delta}{N_0}\left(\Delta + \frac{\partial\Delta}{\partial y}\right)q_\pm\right] + \frac{\partial^2}{\partial y^2}\left[\frac{\Delta^2}{N_0}q_\pm\right], \tag{11}$$

subject to the initial conditions $q_\pm(y, t_0/y_0, t_0) = \delta(y - y_0)$ and the boundary conditions $q_+(Y_\pm, t/y_0, t_0) = q_-(Y_\pm, t/y_0, t_0) = 0$. Thus, if we can solve (11), we can use (10) to compute

$$F(t/y_0, t_0) = P_+F_+(t/y_0, t_0) + P_-F_-(t/y_0, t_0), \tag{12}$$

[2] The test may be shown to be optimum in this sense if the observed process z is stationary, but this is not generally the case here. See Blackwell, D., and M. A. Girshick, *Theory of Games and Statistical Decisions.* New York: Wiley, 1954, chs 9 and 10.

[3] Doob, J. L., *Stochastic Processes.* New York: Wiley, 1953, ch 6, sec 3; also Wong, E., and M. Zakai, On the relation between ordinary and stochastic differential equations, Tech. Rept. 64-26, Electronics Research Lab., University of California, Berkeley, Aug 11, 1964. Equation (9) differs from (3.4′), p 275, of Doob in the term containing $\partial\Delta/\partial y$. The disparity arises from the difference between the stochastic differential equations used by Doob to derive his (3.4′) and our (8), which is a differential equation of the "ordinary" type. In our (8), one should first consider $n(t)$ as a sample function of a well-behaved noise process. If one then allows the noise process to approach a white process, the limiting solution of (8) is a member function of a Markov process with transition densities satisfying (9), rather than Doob's (3.4′); see Wong and Zakai.[3] In the examples considered in the present paper, however, $\partial\Delta/\partial y = 0$, so the disparity vanishes.

[4] Fortet, R., Les functions aléatoires du type de Markoff associées à certaines équations linéares aux dérivées partielles du type parabolique, *J. Math. Pures Appl.*, vol 22, 1943, pp 177–243.

the probability that the test has *not* stopped by time t. Then, since $f(t/y_0, t_0) \triangleq -\partial F(t/y_0, t_0)/\partial t$ is the probability density distribution[5] of the stopping time $t_0 + T$, we have

$$t_0 + E[T/y_0] = \int_{t_0}^{\infty} tf(t/y_0, t_0)\, dt. \tag{13}$$

Our object is to choose $s_\pm$, subject to power constraints, so as to minimize (13).

A Special Class of Signals

Equations (11) are difficult to solve in general, because $\Delta(y, t)$ may depend in a quite complicated way upon its arguments. However, the problem becomes tractable if we restrict ourselves to the class of signals of the form

$$s_\pm(y, t) = \pm U_\pm(y)\sigma(t). \tag{14}$$

We thus restrict the transmitter, on learning the state of the receiver's uncertainty, to an adjustment of the amplitude of its transmission of some waveform $\sigma(t)$. If the transmitter sees the receiver heading toward a wrong decision, it may, for example, increase its instantaneous power drastically, while leaving the power small if the receiver is doing well without the transmitter's help.

If we further assume that $U_+(y) + U_-(y) = 1$, then (11) takes the particularly simple form

$$\frac{N_0}{\sigma^2}\frac{\partial q_\pm}{\partial t} = \mp\frac{\partial q_\pm}{\partial y} + \frac{\partial^2 q_\pm}{\partial y^2}. \tag{15}$$

We may further simplify these equations by making the change of variable

$$t' \triangleq R(t, t_0) \triangleq \frac{1}{N_0}\int_{t_0}^{t} \sigma^2(\tau)\, d\tau, \tag{16}$$

and defining two new functions $\hat{q}_\pm$ by[6]

$$q_\pm(y, t/y_0, t_0) = \hat{q}_\pm[y, R(t, t_0)/y_0] = \hat{q}_\pm(y, t'/y_0). \tag{17}$$

Then (15) becomes

$$\frac{\partial \hat{q}_\pm}{\partial t'} = \mp\frac{\partial \hat{q}_\pm}{\partial y} + \frac{\partial^2 \hat{q}_\pm}{\partial y^2}. \tag{18}$$

These are well-studied equations, whose solutions are given in Appendix I. We emphasize here that these solutions, and the quantities derived from them below, *do not depend on $\sigma(t)$ or on $U_\pm(y)$.*

An Expression for $E[T/y_0]$

Let us now rewrite (13) in terms of the solutions of (18). We first define

$$\hat{F}_\pm(t'/y_0) \triangleq \int_{Y_-}^{Y_+} \hat{q}_\pm(y, t'/y_0)\, dy, \tag{19a}$$

$$\hat{F}(t'/y_0) \triangleq P_+\hat{F}_+(t'/y_0) + P_-\hat{F}_-(t'/y_0), \tag{19b}$$

$$\hat{f}_\pm(t'/y_0) \triangleq -\frac{\partial}{\partial t'}\hat{F}_\pm(t'/y_0), \tag{20a}$$

$$\hat{f}(t'/y_0) \triangleq -\frac{\partial}{\partial t'}\hat{F}(t'/y_0). \tag{20b}$$

Then, from (10), (12), (16), and (17), we have

$$F_\pm(t/y_0, t_0) = \hat{F}_\pm[R(t, t_0)/y_0], \tag{21a}$$

$$F(t/y_0, t_0) = \hat{F}[R(t, t_0)/y_0], \tag{21b}$$

$$f_\pm(t/y_0, t_0) \triangleq -\frac{\partial}{\partial t}F_\pm(t/y_0, t_0) = \hat{f}_\pm[R(t, t_0)/y_0]\frac{\partial}{\partial t}R(t, t_0), \tag{22a}$$

$$f(t/y_0, t_0) = \hat{f}[R(t, t_0)/y_0]\frac{\partial}{\partial t}R(t, t_0). \tag{22b}$$

Equation (13) hence may be written as

$$t_0 + E[T/y_0] = \int_{t_0}^{\infty} t\hat{f}[R(t, t_0)/y_0]\frac{\partial}{\partial t}R(t, t_0)\, dt. \tag{23}$$

Further, if we assume $\sigma^2(t)$ to be almost nowhere zero, then $R(t, t_0)$ is monotone increasing in t, and we may write an inverse of the transformation (16):

$$t = t_0 + r(t'). \tag{24}$$

Then, using (16) and (24), (23) becomes

$$\bar{T}(y_0) \triangleq E[T/y_0] = \int_0^{\infty} r(t')\hat{f}(t'/y_0)\, dt'. \tag{25}$$

An Average-Power Constrait

We imagine the binary transmission we are trying to detect to be a member of an infinite sequence of such transmissions, and impose a constraint which requires that the average power transmitted over this infinite sequence be not greater than some prescribed value P_{av}.[7] That is, we require that

$$\frac{\bar{S}}{\bar{T}} \leq P_{av}, \tag{26}$$

where $\bar{S}$ is the average energy per binary transmission, and $\bar{T}$ [see (25)] is the average duration of a binary transmission.

$\bar{S}$ may be evaluated as follows. We first consider the classes $\mathcal{Y}_\pm^t$ of all member functions of the process y which stop—i.e., first reach one of the thresholds—at time t, given that $H_\pm$ is true. We then compute the average energies expended by members of these classes, viz.,

[5] We assume that $F(\infty) = 0$, so that f is a proper density. This assumption is equivalent to assuming that $R(t, t^0)$ of (16) satisfies $R(\infty, t_0) = \infty$; see Appendix I.

[6] It is easily shown, by application of the usual separation-of-variable techniques to the solution of (15) (see Appendix I), that (17) can indeed be written in the form given. This equation implies that the transition densities, $q_\pm$, of y are stationary—i.e., depend on t and t_0 only through the difference $t - t_0$—if and only if $\sigma(t)$ is a constant.

[7] Other average-power constraints, involving shorter averaging times, may be used at the expense of making the optimization problem considerably more involved. There is reason to believe, however, that the solution of the problem is relatively insensitive to the definition of average power, within reasonable limits. See Turin, G. L., Signal design for sequential detection systems, Tech. Rept. M-69, Electronics Research Lab., University of California, Berkeley, May 6, 1964.

$$S_\pm(t/y_0, t_0) \triangleq E\left[\int_{t_0}^{t} s_\pm^2[y(\tau), \tau]\, d\tau / y \,\varepsilon\, \mathcal{Y}_\pm^t\right]$$

$$= \int_{t_0}^{t} E[U_\pm^2[y(\tau)]/y \,\varepsilon\, \mathcal{Y}_\pm^t]\sigma^2(\tau)\, d\tau$$

$$= \int_{t_0}^{t} d\tau\, \sigma^2(\tau) \int_{Y_-}^{Y_+} dy\, U_\pm^2(y) \frac{f_\pm(t/y, \tau) q_\pm(y, \tau/y_0, t_0)}{f_\pm(t/y_0, t_0)}, \quad (27)$$

where the ratio of densities in the last integrand is recognized as the probability density of y at time $\tau \,\varepsilon\, [t_0, t]$, given that $y \,\varepsilon\, \mathcal{Y}_\pm^t$.

We now average $S_\pm(t/y_0, t_0)$ over the distribution $f_\pm(t/y_0, t_0)$ of stopping times, yielding

$$\bar{S}_\pm(y_0) \triangleq \int_{t_0}^{\infty} dt \int_{t_0}^{t} d\tau\, \sigma^2(\tau) \cdot \int_{Y_-}^{Y_+} dy\, U_\pm^2(y) f_\pm(t/y, \tau) q_\pm(y, \tau/y_0, t_0). \quad (28)$$

Next, transforming the variable t in (28) according to (16) and (24), and also transforming τ according to

$$\tau' = R(\tau, t_0),\ \tau = t_0 + r(\tau', \quad (29)$$

we obtain, using (17) and (22a),

$$\bar{S}_\pm(y_0) = N_0 \int_0^\infty dt' \int_0^{t'} d\tau' \int_{Y_-}^{Y_+} dy\, U_\pm^2(y) \cdot \hat{f}_\pm(t' - \tau'/y) \hat{q}_\pm(y, \tau'/y_0)$$

$$= N_0 \int_0^\infty d\tau' \int_{\tau'}^{\infty} dt' \int_{Y_-}^{Y_+} dy\, U_\pm^2(y) \cdot \hat{f}_\pm(t' - \tau'/y) \hat{q}_\pm(y, \tau'/y_0). \quad (30)$$

The integral on t' is clearly unity [see (19) and (20)], so we have

$$\bar{S}_\pm(y_0) = N_0 \int_0^\infty d\tau' \int_{Y_-}^{Y_+} dy\, U_\pm^2(y) \hat{q}_\pm(y, \tau'/y_0)$$

$$= N_0 \int_{Y_-}^{Y_+} U_\pm^2(y) Q_\pm(y/y_0)\, dy, \quad (31)$$

where

$$Q_\pm(y/y_0) \triangleq \int_0^\infty \hat{q}_\pm(y, \tau'/y_0)\, d\tau'. \quad (32)$$

Finally, we average $\bar{S}_\pm$ over the a priori distribution to obtain

$$\bar{S}(y_0) = P_+\bar{S}_+ + P_-\bar{S}_-. \quad (33)$$

A Peak-Power Constraint

We also impose a peak-power constraint on the transmitted signals, viz.,

$$s_\pm^2(y, t) = \sigma^2(t) U_\pm^2(y) \le P_{\text{peak}}. \quad (34)$$

If we denote by $U_{\max}^2$ the maximum of the values taken on by U_+^2 and U_-^2, i.e.,

$$U_{\max}^2 \triangleq \max\,[\max_y U_+^2(y), \max_y U_-^2(y)], \quad (35)$$

then (34) implies

$$\sigma^2(t) \le \frac{P_{\text{peak}}}{U_{\max}^2}, \quad (36a)$$

or

$$\frac{\partial}{\partial t} R(t, t_0) \le \frac{P_{\text{peak}}}{U_{\max}^2 N_0} \quad (36b)$$

or

$$\frac{d}{dt'} r(t') \ge \frac{N_0 U_{\max}^2}{P_{\text{peak}}}. \quad (36c)$$

Noting from (16) and (24) that $r(0) = 0$, we have, therefore, from (36c)

$$r(t') \ge \frac{N_0 U_{\max}^2 t'}{P_{\text{peak}}}. \quad (36d)$$

Restatement of the Problem

We have already noted that $\hat{f}$ in (25) does not depend on $U_\pm$ or on σ. Further, the other factor r in the integrand of (25) depends only on σ, but not on $U_\pm$ [see (16) and (24)]. Thus, regardless of the choice of $U_\pm$, in the absence of constraints, $\bar{T}(y_0)$ of (25) could be made as small as desired by making $\sigma(t)$ appropriately large [$r(t')$ appropriately small].

However, two lower bounds on $\bar{T}$ are implied by the power constraints given above. First, from (26), we have

$$\bar{T}(y_0) \ge \frac{\bar{S}(y_0)}{P_{\text{av}}} \triangleq T_1, \quad (37)$$

where $\bar{S}$ is given by (33). Second, (25) and (36d) imply

$$\bar{T}(y_0) \ge \frac{N_0 U_{\max}^2}{P_{\text{peak}}} \int_0^\infty t' \hat{f}(t'/y_0)\, dt' \triangleq T_2. \quad (38)$$

Note that both lower bounds depend on U_+, U_-, and y_0, but not on σ. Moreover, since we have assumed that $U_-(y) = 1 - U_+(y)$, we may write the bounds as functions only of U_+ and y_0. Thus, from (37) and (38) we may write

$$\bar{T}(y_0) \ge \max\,[T_1(U_+/y_0), T_2(U_+/y_0)] \triangleq T_0(U_+/y_0). \quad (39)$$

Since, as we have just noted, any non-negative value of $\bar{T}$ can be achieved by appropriate choice of σ, it is clear that T_0 can always be achieved, for any U_+. Thus, the problem becomes one of first finding a U_+ such that T_0 is minimized, and then finding a σ such that $\bar{T} = \min_{U_+} T_0$.

The Symmetric Case

We now further specialize to the symmetric case, in which $P_+ = P_- = \frac{1}{2}$ (hence $y_0 = 0$) and $P(e/H_+) = P(e/H_-) = P(e)$ (hence $Y_+ = -Y_- = Y$). Then, using the results of Appendixes II and III, we have for T_1 and T_2:

$$T_1(U_+/0) = \frac{N_0}{2P_{\text{av}}} \int_{-Y}^{Y} \{U_+^2(y) e^{y/2} + [1 - U_+(y)]^2 e^{-y/2}\} \frac{\sinh \dfrac{Y - |y|}{2}}{\cosh \dfrac{Y}{2}}\, dy, \quad (40)$$

and

$$T_2(U_+/0) = \frac{N_0}{P_{av}} \frac{Y \tanh \frac{Y}{2}}{\alpha} U^2_{max}, \tag{41}$$

where

$$\alpha \triangleq \frac{P_{peak}}{P_{av}}. \tag{42}$$

We note that by definition $P_{peak} \geq P_{av}$, so that $1 \leq \alpha \leq \infty$. It is instructive first to find the optimal signals for the two extreme values of α.

$\alpha = 1$

Through use of (35) and (41) in (40), one may easily show that

$$T_1 \leq \frac{N_0 U^2_{max}}{P_{av}} Y \tanh \frac{Y}{2} = \alpha T_2. \tag{43}$$

Thus, when $\alpha = 1$, $T_1 \leq T_2$ for all U_+; i.e., $T_0 = T_2$. This case therefore corresponds to the purely peak-power-limited case, in which the only constraint on the signals derives from (36).

We must now minimize the functional T_2 of (42) with respect to $U_+ = 1 - U_-$. This involves minimization of the functional U^2_{max} of (35). It is clear that this latter functional is minimized when $U_+(y) = U_-(y) = \frac{1}{2} \triangleq U_+^{(1)}(y)$ for all y. Then (41) becomes

$$T_2(U_+^{(1)}/0) = \min_{U_+} T_2(U_+/0) \Bigg|_{\alpha=1} = \frac{N_0}{4P_{av}} Y \tanh \frac{Y}{2}. \tag{44}$$

It follows from (25), (36), and (38) that $\bar{T}$ achieves the lower bound T_2 of (44) if and only if $\sigma^2(t) = 4P_{peak}$ for all t. That is, the optimum signals for $\alpha = 1$ are [see (14)][8]

$$s_\pm(y, t) = \pm\sqrt{P_{peak}}. \tag{45}$$

This result is not surprising. For, if the peak power is severely limited, being constrained to be equal to the average power, then it is obvious that the transmitter should always operate at peak power. This implies use of the constant signals of (45), with *no* feedback control, for the only possible control would be a reduction of instantaneous power, and this would reduce the average power below the peak power. Note that, although no uncertainty feedback is required in the present case, the feedback link is still needed for synchronization purposes.

If we assume that $P(e) \ll 1$, then [see (6)] $Y \gg 1$, and we have from (44)

$$\bar{T}(0) \cong \frac{N_0 Y}{4P_{av}}, \tag{46a}$$

or, using (6),

$$P(e) \cong e^{-4P_{av}\bar{T}/N_0}. \tag{46b}$$

[8] The signals given here and elsewhere are in low-pass form. The usual bandpass approximations show, however, that there would be little lost by using the equivalent bandpass signals, together with a coherent detector.

For comparison, consider the same channel used in conjunction with a symmetric, nonsequential system, in which one or the other of the signals (45) is transmitted for time T, and then the receiver makes a decision. It is well known that the probability of error for this comparison system is given by[9]

$$P(e) \cong \frac{e^{-P_{av}T/N_0}}{2\sqrt{\pi P_{av}T/N_0}}, \tag{47}$$

where we have defined $P_{av} \triangleq P_{peak}$. [More generally, P_{av} in (47) stands for the energy in $s_\pm$, divided by the transmission time T.] If we compare the two systems by setting $\bar{T}$ in (46b) equal to T in (47) and requiring that the systems operate with the same error probability, we see from (46) and (47) that the sequential system utilizes approximately 6 dB less average power for the same quality of performance. This power advantage has previously been noted by Viterbi.[10]

$\alpha = \infty$

In this case, it is clear from (40) and (41) that $T_2 \leq T_1$ for all U_+; i.e., $T_0 = T_1$. In fact, we are now considering the case in which the only constraint is the average-power constraint deriving from (37).

We must now minimize T_1 of (40) with respect to U_+. This is a straightforward variational problem, the solution of which is

$$U_+^{(\infty)}(y) \triangleq \frac{1}{1 + e^y}. \tag{48a}$$

Correspondingly, $U_-(y)$ is

$$U_-^{(\infty)}(y) \triangleq \frac{1}{1 + e^{-y}}. \tag{48b}$$

Substitution of (48a) into (40) results in

$$T_1(U_+^{(\infty)}/0) = \min_{U_+} T_1(U_+/0) = \frac{N_0}{2P_{av}} \left[Y \tanh \frac{Y}{2} - 2 \ln \cosh \frac{Y}{2} \right]. \tag{49a}$$

For $Y \gg 1$, (49a) becomes

$$T_1(U_+^{(\infty)}/0) \cong \frac{N_0}{P_{av}} \ln 2. \tag{49b}$$

We may easily find a waveform $\sigma(t)$ for which $\bar{T}$ achieves the lower bound of (49b). In fact, if we let

$$\sigma(t) = K, \quad t \geq t_0, \tag{50}$$

where K is a constant, then from (16), (24), (25), and (82) we have

$$\bar{T}(0) = \frac{N_0}{K^2} Y \tanh \frac{Y}{2} \cong \frac{N_0 Y}{K^2}, \tag{51}$$

[9] Helstrom, C. W., The resolution of signals in white, Gaussian noise, *Proc. IRE*, vol 43, Sep 1955, pp 1111–1118, eq. (13).

[10] Viterbi, A. J., Improvement of coherent communication over the Gaussian channel by error-free decision feedback, Jet Propulsion Lab., Space Programs Summary 37–23, vol IV, California Institute of Technology, Pasadena, Oct 31, 1963, pp 179–180.

where the approximation holds for $Y \gg 1$. Thus, if we let

$$K^2 = \frac{P_{av}Y}{\ln 2}, \tag{52}$$

$\bar{T}$ of (51) achieves the lower bound of (49b). Optimal signals for $\alpha = \infty$ may therefore be written in the form

$$s_\pm(y, t) = \pm\frac{\sqrt{P_{av}Y/\ln 2}}{1 + e^{\pm y}}, \tag{53}$$

where Y is evaluated through (6) from the prescribed probability of error.

Notice that the actual peak-to-average power ratio in these signals is just

$$\alpha' = \frac{Y}{\ln 2} \cong \frac{\ln (1/P(e))}{\ln 2}. \tag{54}$$

On the other hand, we are allowed to make α' arbitrarily large, since $\alpha = \infty$. Therefore, we see from (54) that we may prescribe an arbitrarily small probability of error in the present case, and still achieve the average transmission time per bit given by (49); to do this, we must allow the transmitter to transmit at arbitrarily large peak powers.

This limiting result may be checked against the channel capacity theorem of information theory. The channel we are investigating is an infinite bandwidth, additive white Gaussian channel which utilizes average power P_{av}. Such a channel, even with a noiseless feedback link, has channel capacity[11]

$$C = \lim_{W\to\infty} \frac{W \ln\left(1 + \frac{P_{av}}{N_0 W}\right)}{\ln 2} = \frac{P_{av}}{N_0 \ln 2} \text{ bits/sec.} \tag{55}$$

That is, it is possible, according to the fundamental theorem of information theory, to transmit on the average up to one bit every $(N_0/P_{av})\ln 2$ seconds with an arbitrarily small probability of error. In fact, the sequential system we are now considering achieves this limiting behavior, and, further, achieves it without the necessity of coding.

If we do not prescribe an arbitrarily small error probability, then use of the signals of (53) does not entail use of the allowable infinite peak-to-average power ratio, but rather the finite ratio (54). Then, from (6), (51), (52), and (54), we have (for $P(e) \ll 1$)

$$P(e) \cong e^{-\alpha' P_{av}\bar{T}/N_0}. \tag{56}$$

Comparison of (56) with the error probability (47) for the nonsequential system shows that when the two systems operate at the same error probability, and $T = \bar{T}$, the sequential system enjoys a power advantage of α'. Thus, for example, the sequential system can operate at $P(e) = 10^{-4}$ with 11 dB less average power than a nonsequential system which has the same error probability and the same data rate.

[11] This is an extension, due to R. G. Gallager (private communication), of the theorem of Shannon which asserts that, for a memoryless, discrete channel, a noiseless feedback link cannot increase the feed-forward channel capacity; see Shannon, C. E., The zero-error capacity of a noisy channel, *IRE Trans. on Information Theory*, vol IT-2, Sep 1956, pp S8–S19; also see Baghdady, E. J., Ed., *Lectures on Communication System Theory*. New York: McGraw-Hill, 1961, ch 14 (by P. E. Green, Jr., with appendix by P. Elias); and also, Horstein, M., Sequential transmission using noiseless feedback, *IEEE Trans. on Information Theory*, vol IT-9, Jul 1963, pp 137–143, in which the error exponent for a discrete channel with feedback, similar to that studied here, is derived.

The General Case

We now consider the general case $1 \leq \alpha \leq \infty$. One may conjecture from the discussion of the case $\alpha = \infty$ that if the prescribed value of α is merely not less than α' of (54), although not necessarily infinite, then the signals of (53) are optimal. It is in fact easy to prove this conjecture.

We first note from (39) and (49) that

$$T_0(U_+/0) \geq \min_{U_+} T_0(U_+/0)$$
$$\geq \min_{U_+} T_1(U_+/0) = T_1(U_+^{(\infty)}/0). \tag{57}$$

Then, if $T_1(U_+^{(\infty)}/0) \geq T_2(U_+^{(\infty)}/0)$, we will have

$$T_0(U_+^{(\infty)}/0) = T_1(U_+^{(\infty)}/0),$$

i.e., T_0 will actually achieve the lower bound (57) at $U_+ = U_+^{(\infty)}$. But from (41), (48), and (49) we have, for $Y \gg 1$,

$$T_1(U_+^{(\infty)}/0) \cong \frac{N_0}{P_{av}} \ln 2, \tag{58a}$$

$$T_2(U_+^{(\infty)}/0) \cong \frac{N_0 Y}{P_{av}\alpha}. \tag{58b}$$

Thus, if

$$\alpha \geq \frac{Y}{\ln 2} = \frac{\ln\left[\frac{1}{P(e)}\right]}{\ln 2} = \alpha', \tag{59}$$

T_0 indeed is minimized [taken to the bound (57)] by $U_+^{(\infty)}$ of (48); T_0 then has the value given by (58a). Then, by following the argument from (50) onward, we find that this minimum value of T_0 can actually be achieved by use of signals of the form (53), thus proving the conjecture. As previously noted, the actual peak-to-average power ratio of these signals is just $\alpha' \leq \alpha$.

By rewriting (59) as

$$P(e) \geq 2^{-\alpha} \tag{60}$$

we see from the preceding discussion that the average stopping time $\bar{T} = (N_0/P_{av})\ln 2$ can be achieved with any prescribed error probability not less than $2^{-\alpha}$. Again, for comparison purposes we may write $P(e)$ in the form (56), whence it becomes clear [c.f. (47)] that the sequential system has a power advantage of α' over the comparison nonsequential system.

When $\alpha < \alpha'$, the problem of minimizing $T_0(U_+/0)$ appears to be quite difficult. One may approach the problem by limiting the class of allowable feedback functions U_+. Thus, for example, one might consider only those U_+ of the form

$$U_+(y;\lambda) = \frac{1}{1+e^{\lambda y}}, \tag{61}$$

substitute this form into (40), (41), and (39), and then minimize the resulting T_0 with respect to the single parameter λ. The optimization problem would then consist in finding λ_{opt} as a function of α, evaluating the T_0 corresponding to λ_{opt}, and finally finding a waveform $\sigma(t)$ such that $\bar{T}$ achieves this value of T_0.[12] Even such a computation is most involved, and seems, in view of the flexibility of the previous results, not to be worth the effort.

Discussion

The reason that uncertainty feedback can afford such large power advantages may be stated quite simply: the transmitter works in cahoots with the receiver, instead of working in complete ignorance of what the receiver is doing, as it does in the absence of uncertainty feedback. Thus, the transmitter is quite content to let the receiver "do the work" of making a decision, unless the receiver tries to decide wrongly.

More specifically, consider for illustrative purposes the situation in which the signals

$$s_\pm[y(t), t] = \pm\tfrac{1}{2}\sigma(t)[1 \mp \operatorname{sgn} y(t)] \tag{62}$$

are used.[13] Then, if s_+ is being sent, whenever $y > 0$ the transmitter is turned off, and the transmitter is turned on whenever $y < 0$. Conversely, if s_- is being sent, the on and off conditions are reversed. Thus, the transmitter is off whenever the receiver is nearer the correct threshold than the incorrect one.

Now let us consider, through (7), the behavior of $y(t)$ when the signals of (62) are used. To do this, we differentiate (7) with respect to t, keeping separate the terms contributed by the two integrals. We have:

$$\frac{dy}{dt} = \frac{2}{N_0}[s_+(y, t) - s_-(y, t)]z(t) - \frac{1}{N_0}[s_+^2(y, t) - s_-^2(y, t)]. \tag{63}$$

Now suppose that s_+ is being sent and that $y > 0$. Then $z(t) = n(t)$, since the transmitter is turned off. Further, (63) becomes, using (62),

$$\frac{dy}{dt} = \frac{2}{N_0}\sigma(t)n(t) + \frac{1}{N_0}\sigma^2(t). \tag{64}$$

The first term in (64) contributes only a zero-mean noise to dy/dt. The second term, *which is generated totally within the receiver*, contributes a positive mean to dy/dt. Thus, y will tend (in a probabilistic sense) to become more positive, *due to the effort of the receiver alone*. As long as the receiver causes y to progress toward the positive (correct) threshold, the transmitter stands idle.

[12] Notice that $U_+(y; 1) = U_+^{(\infty)}(y)$ and $U_+(y; 0) = U_+^{(1)}(y)$, so the solutions obtained by this procedure would have as special cases those previously obtained for $\alpha = 1$ and $\alpha' \leq \alpha \leq \infty$.

[13] Here $U_+(y) = (1/2)\sigma(t)(1\text{-}sgn\ y)$, which corresponds to $U_+(y; \infty)$ in (61). It may be shown that in this case, with $\alpha = \infty$, one may achieve $\bar{T} = N_0/P_{av}$ by choosing $\sigma(t) = K$; cf. (49b).

On the other hand, suppose that the noise causes y to become negative while s_+ is being transmitted. Then $z(t) = \sigma(t) + n(t)$, and (63) becomes

$$\frac{dy}{dt} = \frac{2}{N_0}\sigma(t)[\sigma(t) + n(t)] - \frac{1}{N_0}\sigma^2(t). \tag{65}$$

Again dy/dt has a zero-mean noise contribution, $(2/N_0)\cdot\sigma(t)n(t)$. Now, however, the contribution to the mean of dy/dt by the receiver is negative: $-(1/N_0)\sigma^2(t)$. But the transmitter is now on, and causes an additional contribution of $(2/N_0)\sigma^2(t)$ to the mean of dy/dt, thus making the mean positive, and again causing y to move toward the positive threshold. That is, the function of the transmitter is to override the receiver when it is progressing incorrectly. Since when $P(e) \ll 1$ the receiver rarely goes wrong, the transmitter is only infrequently called upon to redirect the receiver; that is, the transmitter is only rarely turned on, with the concomitant saving of power.

When signals other than those of (62) are used [e.g., those of (53)], a similar discussion applies, but the transmitter is turned off slowly as the receiver approaches the correct threshold, rather than abruptly.

Appendix I

Solutions of (18)

We assume solutions of (18) of the form $\hat{q}_\pm(y, t'/y_0) = u_\pm(y)v_\pm(t')$, and substitute this form into (18), obtaining

$$\frac{1}{v_\pm}\frac{dv_\pm}{dt'} = \frac{1}{u_\pm}\left[\mp\frac{du_\pm}{dy} + \frac{d^2u_\pm}{dy^2}\right]. \tag{66}$$

Since the left-hand side of (66) is a function only of t', and the right-hand side a function only of y, both sides must equal a constant, say $-\lambda_\pm$. Then $u_\pm$ and $v_\pm$ must satisfy the ordinary differential equations

$$\frac{d^2u_\pm}{dy^2} \mp \frac{du_\pm}{dy} = -\lambda_\pm u_\pm, \tag{67a}$$

$$\frac{dv_\pm}{dt'} = -\lambda_\pm v_\pm. \tag{67b}$$

Equation (67a) must be solved using the boundary conditions $u_\pm(Y_+) = u_\pm(Y_-) = 0$. Discrete sets of solutions thus exist, which are easily shown to have the form

$$u_\pm^{(n)}(y) = a_\pm^{(n)}e^{\pm y/2}\begin{cases}\cos n\omega_0(y - y_1), & n = 1, 3, 5, \cdots \\ \sin n\omega_0(y - y_1), & n = 2, 4, 6, \cdots,\end{cases} \tag{68}$$

where the $a_\pm^{(n)}$ are arbitrary constants,

$$\omega_0 \triangleq \frac{\pi}{Y_+ - Y_-}, \tag{69a}$$

and

$$y_1 \triangleq \frac{Y_+ + Y_-}{2}. \tag{69b}$$

The corresponding eigenvalues of (67a) are

$$\lambda_+^{(n)} = \lambda_-^{(n)} = \tfrac{1}{4} + n^2\omega_0^2 \triangleq \lambda^{(n)} \quad (n = 1, 2, \cdots). \tag{70}$$

Now, for any eigenvalue $\lambda^{(n)}$, (67b) has the solutions

$$v_{\pm}^{(n)}(t') = k_{\pm}^{(n)} e^{-\lambda^{(n)} t'}, \tag{71}$$

where the $k_{\pm}^{(n)}$ are arbitrary constants.

The general solutions of (18) may be written as linear combinations of the solutions $u_{\pm}^{(n)}(y)v_{\pm}^{(n)}(t')$:

$$\hat{q}_{\pm}(y, t'/y_0) = \sum_{n=1}^{\infty} c_{\pm}^{(n)} u_{\pm}^{(n)}(y) e^{-\lambda^{(n)} t'}. \tag{72}$$

Further, the $c_{\pm}^{(n)}$ may be evaluated by invoking the initial conditions $\hat{q}_{\pm}(y, 0/y_0) = \delta(y - y_0)$, i.e.,

$$\sum_{n=1}^{\infty} c_{\pm}^{(n)} u_{\pm}^{(n)}(y) = \delta(y - y_0). \tag{73}$$

For, using (68), we have

$$\begin{aligned} u_{\mp}^{(m)}(y_0) &= \int_{Y_-}^{Y_+} u_{\mp}^{(m)}(y)\,\delta(y - y_0)\,dy \\ &= \sum_{n=1}^{\infty} c_{\pm}^{(n)} \int_{Y_-}^{Y_+} u_{\mp}^{(m)}(y) u_{\pm}^{(n)}(y)\,dy \\ &= \frac{\pi}{2\omega_0} c_{\pm}^{(m)} a_{\pm}^{(m)} a_{\mp}^{(m)}. \end{aligned} \tag{74}$$

Then, solving (74) for $c_{\pm}^{(m)}$ and using (68), (72) becomes

$$\begin{aligned} \hat{q}_{\pm}(y, t'/y_0) = \frac{\omega_0}{\pi} e^{\pm(y-y_0)/2} \sum_{n=1}^{\infty} [&\cos n\omega_0(y - y_0) \\ &- \cos n\omega_0(y + y_0 - 2Y_+)] e^{-\lambda^{(n)} t'}. \end{aligned} \tag{75}$$

Finally, we may easily evaluate $\hat{f}_{\pm}(t'/y_0)$:

$$\begin{aligned} \hat{f}_{\pm}(t'/y_0) &= -\frac{\partial}{\partial t'} \int_{Y_-}^{Y_+} \hat{q}_{\pm}(y, t'/y_0)\,dy \\ &= \frac{2\omega_0^2}{\pi} \sum_{n=1}^{\infty} n e^{-\lambda^{(n)} t'} [e^{\pm(Y_+ - y_0)/2} \sin n\omega_0(Y_+ - y_0] \\ &\quad + e^{\mp(y_0 - Y_-)/2} \sin n\omega_0(y_0 - Y_-)]. \end{aligned} \tag{76}$$

Appendix II

Evaluation of $\bar{S}_{\pm}$

In the symmetric case, $y_0 = 0$, $Y_+ = -Y_- = Y$, and (75) becomes

$$\hat{q}_{\pm}(y, t'/0) = \frac{\omega_0}{\pi} e^{\pm y/2} \sum_{n=1}^{\infty} [1 - \cos n\pi] \cos n\omega_0 y e^{-\lambda^{(n)} t'}. \tag{77}$$

Integrating on t' term by term and using (32) and (70), we have

$$Q_{\pm}(y/0) = \frac{\omega_0}{\pi} e^{\pm y/2} \sum_{n=1}^{\infty} \frac{1 - \cos n\pi}{\frac{1}{4} + n^2\omega_0^2} \cos n\omega_0 y. \tag{78}$$

The sum is a Fourier series which may be evaluated in closed form, yielding

$$Q_{\pm}(y/0) = e^{\pm y/2} \frac{\sinh\left(\dfrac{Y - |y|}{2}\right)}{\cosh \dfrac{Y}{2}}, \tag{79}$$

where we have made use of (69a). Note that $Q_+(y) = Q_-(-y)$.

Appendix III

Evaluation of $\int_0^\infty t'\hat{f}(t'/0)dt'$

In the symmetric case, $y_0 = 0$, $Y_+ = -Y_- = Y$, and, using (69a) and (70), (76) becomes

$$\begin{aligned} \hat{f}_{\pm}(t'/0) = \frac{\pi}{Y^2} \cosh \frac{Y}{2} \sum_{n=1}^{\infty} n \sin \frac{n\pi}{2} \\ \cdot \exp\left[-\frac{1}{4}\left(1 + \frac{n^2\pi^2}{Y^2}\right)t'\right]. \end{aligned} \tag{80}$$

We see that $\hat{f}_+ = \hat{f}_- = \hat{f}$. Integrating term by term, we have

$$\int_0^\infty t'\hat{f}(t'/0)\,dt' = \frac{16\pi}{Y^2} \cosh \frac{Y}{2} \sum_{n=1}^{\infty} \frac{n}{1 + \dfrac{n^2\pi^2}{Y^2}} \sin \frac{n\pi}{2}. \tag{81}$$

The sum is a Fourier series which may be evaluated in closed form, yielding

$$\int_0^\infty t'\hat{f}(t'/0)\,dt' = Y \tanh \frac{Y}{2}. \tag{82}$$

Acknowledgment

The author is grateful to many of his colleagues for tolerating his attempted monologues on this work, and for converting them into fruitful dialogues; he would especially like to acknowledge the comments of Drs. M. Horstein, E. Wong, and M. Zakai.

A Coding Scheme for Additive Noise Channels with Feedback—Part I: No Bandwidth Constraint

J. P. M. SCHALKWIJK, MEMBER, IEEE AND T. KAILATH, MEMBER, IEEE

Abstract—In some communication problems, it is a good assumption that the channel consists of an additive white Gaussian noise forward link and an essentially noiseless feedback link. In this paper, we study channels where no bandwidth constraint is placed on the transmitted signals. Such channels arise in space communications.

It is known that the availability of the feedback link cannot increase the channel capacity of the noisy forward link, but it can considerably reduce the coding effort required to achieve a given level of performance. We present a coding scheme that exploits the feedback to achieve considerable reductions in coding and decoding complexity and delay over what would be needed for comparable performance with the best known (simplex) codes for the one-way channel. Our scheme, which was motivated by the Robbins-Monro stochastic approximation technique, can also be used over channels where the additive noise is not Gaussian but is still independent from instant to instant. An extension of the scheme for channels with limited signal bandwidth is presented in a companion paper (Part II).

I. Introduction

IN CERTAIN COMMUNICATION problems we have the possibility of using a noiseless "feedback" link to improve communication over a noisy forward link. A good example is communication with a space satellite—the power in the ground-to-satellite direction can be so much larger than in the reverse direction that the first link can be taken to be an (essentially) noiseless link. Similar possibilities may also arise in special terrestrial situations.

It seems reasonable that the availability of a noiseless feedback link should substantially improve communication over the noisy forward link. Therefore, Shannon's result [4], Theorem 6, that the channel capacity of a *memoryless* noisy channel is not increased by noiseless feedback, is rather surprising. Still, some advantage should accrue from the presence of a noiseless feedback link and, in fact, the advantage is that noiseless feedback enables a substantial reduction in the complexity of coding and decoding required to achieve a given performance over the noisy link.

In this paper, we shall illustrate the simplifications obtained when the noisy link is an additive white Gaussian noise channel operated under an average power constraint. No restrictions are placed on the usable signal bandwidth. Such channels seem to be typical of those in space communications. In terrestrial communications we are often forced to impose bandwidth limitations on the transmitted signals. Channels with bandwidth constraints are discussed in a companion paper [17]. The communication scheme we shall present can also be used over non-Gaussian white noise channels. However, it is difficult to evaluate the effect of the feedback link in such cases because very few results are available for non-Gaussian (one-way) channels. We shall now briefly review the known results for the Gaussian case.

A. Additive White Gaussian Noise Channels

We shall assume that the noise in the channel is Gaussian and white with a (two-sided) spectral density $N_0/2$. The transmitted signals are required to have an average power P_{av} but no constraints are imposed on their bandwidth and peak power.

For this channel, the channel capacity is given by (e.g., Fano [9], Ch. VI)

$$C = \frac{P_{av}}{N_0 \ln 2} \text{ bits/second} = \frac{P_{av}}{N_0} \text{ nats/second.} \tag{1}$$

If one of M messages is to be transmitted over such a channel, the best code is universally believed to be a "regular-simplex" set of codewords (i.e., a set of M equal-energy signals with mutual cross-correlations of $-1/M - 1$). For large M, an orthogonal signal set (for which the cross-correlations are zero rather than $-1/M - 1$) performs almost as well. The ideal receiver for such signals is a bank of M correlation detectors, whose outputs are scanned to determine the correlator yielding the largest output. The error probability for an orthogonal (or simplex) signal set has been evaluated numerically for values of M from 2 to 10^6. For larger values of M, the following asymptotic expression can be used. If T is the duration of each of the M signals, assumed equally likely a priori, then (cf. Fano [9], Chapter VI, and Zetterberg [12])

$$P_{e,\text{orth}}(M, T) = \frac{\text{constant}}{T^{\beta}} e^{-TE(R)} \tag{2}$$

This work was supported by NASA NSG 337, Tri Service Nonr 225 (83), and AFOSR Contract AF 49(638)-1517.

T. Kailath is with the Department of Electrical Engineering, Stanford University, Stanford, Calif.

J. P. M. Schalkwijk is with the Applied Research Laboratory, Sylvania Electronic Systems, a division of Sylvania Electric Products Inc., Waltham, Mass.

Reprinted from *IEEE Trans. Inform. Theory*, vol. IT-12, pp. 172–182, Apr. 1966.

where

$$E(R) = \begin{cases} C/2 - R, & 0 \le R \le C/4 \\ (\sqrt{C} - \sqrt{R})^2, & \frac{C}{4} \le R \le C \end{cases} \quad (3)$$

$1 \le \beta \le 2$

R = the signalling rate = $\ln M/T$ nats/second
C = the channel capacity = P_{av}/N_0 nats/second.

Equation (2) shows that the error probability for orthogonal codes decreases essentially exponentially with T. As a result, for large T (low P_e), the choice of a suitable pair of values R and T to achieve a given P_e is essentially determined by the quantity $E(R)$. Equations (2) and (3) specify the "tradeoffs" that can be made between the signal duration T and the signaling rate R—for rates near channel capacity, $E(R)$ is small and we need a large T to achieve a given P_e; if we are required to use a small value of T, the rate R must be suitably reduced.

We can now present the results we have obtained by assuming that a noiseless feedback link is available.

B. Summary of Our Results

We have developed a coding scheme that exploits the presence of the noiseless feedback link. The scheme was suggested by the Robbins-Monro stochastic approximation procedure for determining the zero of a function by noisy observations of its values at chosen points [2]. While we do not know if this coding scheme is "optimum," it has the virtue of great simplicity both in encoding and decoding. It also enables us to achieve the channel capacity P_{av}/N_0 of the white Gaussian channel (the capacity is the same, as we mentioned above, whether or not a noiseless feedback link is available). And most important it provides a dramatic reduction in the rate at which the error probability varies with signaling duration. Thus, for our scheme, we have

$$P_{e,fb} = \frac{1}{\sqrt{6\pi \frac{C}{R} e^{2(C-R)T}}} \exp\left[-\frac{3C}{2R} e^{2(C-R)T}\right] \quad (4)$$

showing that the error probability goes down much faster than exponentially with T, which is how the error probability behaves for one-way channels [cf. (2) and (3)]. As a sample comparison between the feedback and nonfeedback cases, consider the values of T required to achieve, for example,

$$P_e = 10^{-7}, \quad R = 0.8C, \quad C = 1 \text{ bit/second.}$$

We have

$$T_{f.b.} = 15 \quad \text{and} \quad T_{orth} = 2030.$$

It is natural to wonder how our results are affected by delay and noise in the feedback link. The effect of the delay is merely a small increase in the error probability. The effects of noise are more serious—we find that, provided the noise is smaller than a certain threshold value, the error probability is essentially unaffected; however, with noisier feedback, our scheme deteriorates rapidly and other coding techniques must be devised.

Our coding scheme does *not* depend upon the Gaussian nature of the additive white noise. It can be applied to any additive white[1] noise channel and will enable us to signal, with arbitrarily low error probability, at rates up to P_{av}/N_0 nats/second. (We recall that $N_0/2$ is defined as the (two-sided) spectral density of the white noise.) This result thus yields a lower bound of P_{av}/N_0 for the channel capacity of additive white noise channels of spectral density $N_0/2$ and transmitter power P_{av}. The actual capacity of such channels may be much larger than this, but the capacity is usually too complicated to evaluate analytically. The capacity is the same with or without feedback. The error probabilities will, however, be considerably different. No results on the error probabilities seem to be available for one-way non-Gaussian white noise channels; when a noiseless feedback channel is available, the expression given above for $P_{e,fb}$ will continue to be valid, for large T, for non-Gaussian channels. Some further remarks on such chanels are made in Section IV.

We should also mention that our coding scheme does not depend upon knowledge of $N_0/2$ (the noise spectral density) for its operation. Of course, this knowledge will be necessary for evaluating the performance of the scheme.

Before proceeding to the derivation of our results, we shall give a brief discussion of related work.

C. Other Studies of Noiseless Feedback Channels

A general discussion of feedback communication systems, with reference to earlier work by Chang and others, is given by Green [10] who distinguishes between post- and predecision feedback systems. In postdecision feedback systems the transmitter is informed only about the receiver's decision; in predecision feedback systems, the state of uncertainty of the receiver as to which message was sent is fed back. Postdecision feedback systems require less capacity in the backward direction; however, the improvement over one-way transmission will also be less than that obtainable with predecision feedback.

Viterbi [17] discusses a postdecision feedback system for the white Gaussian noise channel. The receiver computes the likelihood ratio as a function of time and makes a decision when the value of the likelihood ratio crosses one of a pair of thresholds. The transmitter is informed by means of postdecision feedback that the receiver has made a decision, and it then starts sending the next message. For rates higher than half the channel capacity,

[1] By white noise, we shall mean noise whose values at any two instants of time are statistically independent.

the reliability is increased roughly by a factor of four as compared to one-way communication.

Turin [15] has a predecision feedback scheme applying to the white Gaussian noise channel, and giving an even greater improvement over one-way communication than Viterbi's scheme does. The receiver again computes the likelihood ratio as a function of time, but now the value of the likelihood ratio is fed back continuously to the transmitter. The transmitted signal is a function of the binary digit (that is, 0 or 1) being sent and of the value of the likelihood ratio, and is adjusted so as to make this ratio cross, as quickly as possible, one of a pair of decision thresholds. Average and peak power constraints on the transmitted signals are studied. The average time $\bar{T}$ for deciding on a binary digit turns out to be

$$\bar{T} = (\ln 2)(P_{av}/N_0)^{-1} \text{ seconds,}$$

where P_{av} is the average power and N_0 is the (one-sided) noise power spectral density. The possibility of error P_e vanishes if infinite peak power and infinite bandwidth are allowed. Hence, a rate is achieved that is equal to the channel capacity

$$C = \frac{P_{av}}{N_0} \text{ nats/second.}$$

In this scheme, the actual time required to make a decision is variable (though the mean value is $\bar{T}$). The variance and other parameters of this variable time do not appear to be readily computable. As opposed to this, our feedback scheme is a "block" scheme, with a decision being made after a preassigned interval, the interval being determined by the desired rate and error probability.

II. The Coding Scheme and its Evaluation

Our coding scheme was motivated by the Robbins-Monro [2] stochastic approximation procedure which we shall describe briefly.

A. The Robbins-Monro Procedure

Consider the situation indicated in Fig. 1. One wants to determine θ, a zero of $F(x)$, without knowing the shape of the function $F(x)$. It is possible to measure the values of the function $F(x)$ at any desired point x. The observations are noisy, however, so that instead of $F(x)$ one obtains $Y(x) = F(x) + Z$, where Z is some additive disturbance. The "noise" Z is assumed to be independent and identically distributed from trial to trial. To estimate θ, Robbins and Monro proposed the following recursive scheme. Start with an arbitrary initial guess X_1 and make successive guesses according to

$$X_{n+1} = X_n - a_n Y_n(X_n), \qquad n = 1, 2, \cdots .$$

For the procedure to work, that is, for X_{n+1} to tend to θ, the coefficients a_n must satisfy $a_n \geq 0$, $\sum a_n = \infty$, and $\sum a_n^2 < \infty$. A sequence $\langle a_n \rangle$ fulfilling these requirements is $a_n = 1/an$, $a > 0$.

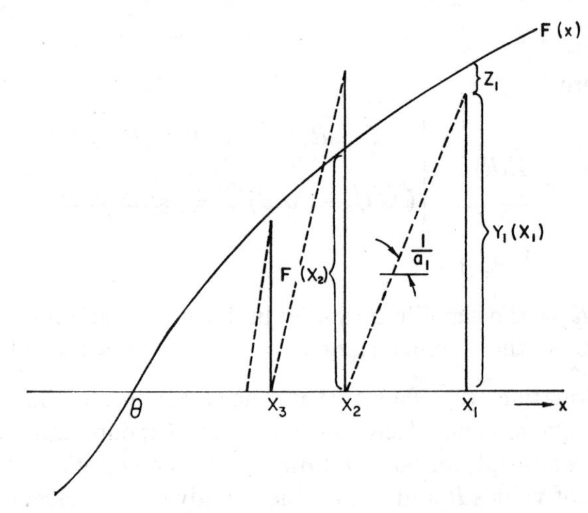

Fig. 1. The Robbins-Monro procedure.

The following additional requirements are needed on the function $F(x)$ and on Z.

1) $F(x) \gtrless 0$ according to $x \gtrless \theta$.

2) $\inf \{|F(x)|; \epsilon < |x - \theta| < 1/\epsilon\} > 0$ for all $\epsilon > 0$.

3) $|F(x)| \leq K_1 |x - \theta| + K_2$, where K_1 and K_2 are constants.

4) If $\sigma^2(x) = E[Y(x) - F(x)]^2$, then $\sup_x \sigma^2(x) = \sigma^2 < \infty$.

With these requirements, the following theorem can be established.

Theorem: When the above conditions on the a_n, the $F(x)$, and the Z_n are met, $Z_n \to \theta$ almost surely; and furthermore, if $E |X_1|^2 < \infty$, then $E |X_n - \theta|^2 \to 0$.

Robbins and Monro proved the convergence in mean square. The "convergence almost surely" was first proved by Wolfowitz [5]. A good proof of the preceding theorem is Dvoretzky's [3], where several types of stochastic-approximation procedures are treated in a unified manner.[2]

The Robbins-Monro procedure is nonparametric, that is, no assumptions concerning the distribution of the additive disturbance, except for zero mean and finite variance, are necessary. However, it was shown by Sacks [6] that $\sqrt{n}(X_{n+1} - \theta)$ is normally distributed for large n. In fact, let the following assumptions, which complement the earlier requirements, be fulfilled.

5) $\sigma^2(x) \to \sigma^2(\theta)$ as $x \to \theta$.

6) $F(x) = \alpha(x - \theta) + \delta(x)$, where $\alpha > 0$ and $\delta(x) = 0(|x - \theta|^{1+\rho})$, $\rho > 0$.

7) There exist $t>0$ and $\delta>0$ such that $\sup \{E |Z(x)|^{2+\delta}; |x - \theta| \leq t\} < \infty$.

8) $a_n = 1/an$, and $2\alpha > a$.

Then we have the theorem [6]: Fulfillment of all the conditions (1–8) yields

$$\sqrt{n}\,(X_{n+1} - \theta) \sim N\left[0, \frac{\sigma^2}{a(2\alpha - a)}\right].$$

This result will be used presently.

[2] A recent general survey of stochastic approximation methods is given by Venter [14].

B. An Equivalent Discrete-Time Channel

To apply the stochastic approximation procedure to the communication channel, we shall need to obtain a discrete-time equivalent of the additive white Gaussian noise channel. This can be done in many ways. We shall present a mathematically convenient method—later we shall comment on its physical viability.

To obtain a discrete-time equivalent, we shall assume that the message information is transmitted by suitably modulating the amplitude of a known basic waveform, $\phi(t)$. The signal in the channel (see Fig. 2) will thus be of the form

$$s(t) = \sum_i x_i \phi(t - i\Delta), \qquad i = 0, 1, \cdots$$

where Δ is a time interval that will be suitably chosen later. We shall require the basic waveform $\phi(t)$ to have unit energy and to be orthogonal for shifts Δ, i.e., $\phi(t)$ should satisfy

$$\int \phi(t - i\Delta)\phi(t - j\Delta)\, dt = \delta_{ij}. \tag{5}$$

The integral extends over all values of t for which the integrand is different from zero.

Reception will be achieved using a filter matched to $\phi(t)$, that is, $h(t) = \phi(-t)$. The output of this matched filter at $t = i\Delta$, $i = 1, 2, \cdots$, will be the sequence $\{Y_i(X_i)\}$ where $Y_i(X_i) = X_i + Z_i$, and

$$Z_i = \int n(t)\phi(t - i\Delta)\, dt.$$

It can be easily be checked that the $\{Z_i\}$ will be uncorrelated zero mean random variables with

$$E[Z_i Z_j] = \frac{N_0}{2}\,\delta_{ij}.$$

When the additive noise is Gaussian, the $\{Z_i\}$ will be Gaussian and, therefore, also independent. In the Gaussian case, it is easy to see that the discrete-time channel thus obtained (where a sequence of numbers $\{X_i\}$ is transmitted and sequence $\{Y_i(X_i) = X_i + Z_i\}$ is received) is completely equivalent to the original continuous-time channel. This follows from the fact that the matched filter for Gaussian white noise channels computes the likelihood ratio, which is a sufficient statistic and therefore preserves all the information in the received waveform that is relevant to the decision making process.

Finally, we note that by virtue of the orthonormality of $\phi(t - i\Delta)$ and $\phi(t - j\Delta)$, $i \neq j$, the transmitted energy in $s(t) = \sum x_i \phi(t - i\Delta)$ is $\sum x_i^2$.

We can now describe our coding scheme.

C. The Coding Scheme

The transmitter has to send one of M possible messages to a receiver. A noiseless feedback channel is available. We shall proceed as follows (see also Fig. 3).

Divide the unit interval into M disjoint, equal-length "message intervals." Pick as the "message point" θ, the midpoint of the message interval corresponding to the particular message being transmitted. Through this message point θ, put a straight line $F(x) = \alpha(x - \theta)$, with slope $\alpha > 0$. Start out with $X_1 = 0.5$ and send to the receiver the 'number" $F(X_1) = \alpha(X_1 - \theta)$, as discussed in Section II-B. At the receiver one obtains the "number" $Y_1(X_1) = \alpha(X_1 - \theta) + Z_1$, where Z_1 is a Gaussian random variable with zero mean and variance $N_0/2 = \sigma^2$, say. The receiver now computes $X_2 = X_1 - (a/1)Y_1(X_1)$, where a is a constant which will be specified soon, and retransmits this value to the transmitter which then sends $F(X_2) = \alpha(X_2 - \theta)$. In general, one receives $Y_n(X_n) = F(X_n) + Z_n$ and computes $X_{n+1} = X_n - (a/n)Y_n(X_n)$. The number X_{n+1} is sent back to the transmitter, which then will send $F(X_{n+1}) = \alpha(X_{n+1} - \theta)$.

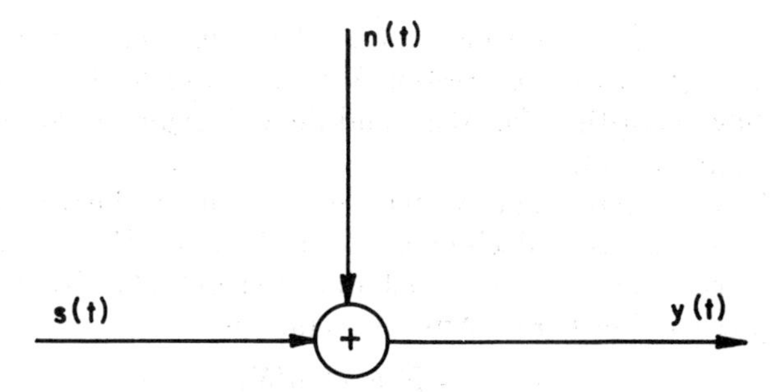

Fig. 2. Model for the additive noise channel.

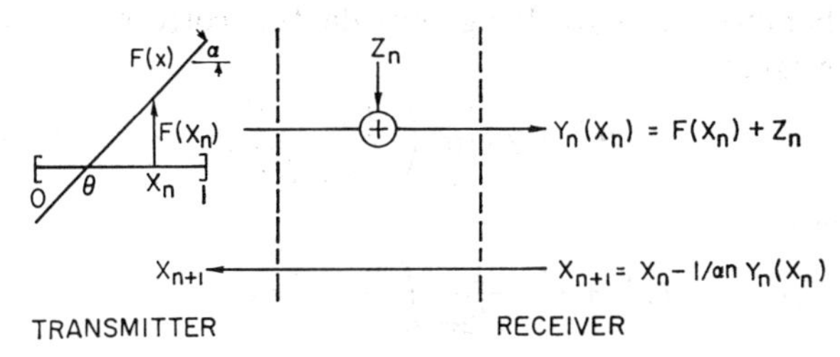

Fig. 3. Proposed coding scheme for wideband signals.

From Sacks' theorem [6], quoted earlier, on asymptotic distributions of stochastic approximation procedures, it follows that the best value for a is $a = 1/\alpha$ and that in this case $\sqrt{n}(X_{n+1} - \theta)$ converges in distribution to a normal random variable with zero mean and variance $(\sigma/\alpha)^2$.

In the Gaussian case, the distribution of $(X_{n+1} - \theta)$ can be computed directly for any n without reference to Sacks' theorem. With $a_n = 1/\alpha n$, the recursion relation

$$X_{n+1} = X_n - \frac{1}{\alpha n} Y_n(X_n), \qquad Y_n(X_n) = \alpha(X_n - \theta) + Z_n$$

is easily solved to yield

$$X_{n+1} = \theta - \frac{1}{\alpha n}\sum_1^n Z_i.$$

Since the Z_i are independent, $N(0, \sigma^2)$, X_{n+1} will be Gaussian with mean θ and variance $\sigma^2/\alpha^2 n$. We may also point out here that X_{n+1} is (in the Gaussian case) the maximum likelihood estimate of θ, given $Y_1(X_1), \cdots Y_n(X_n)$. The estimate X_{n+1} is also unbiased and efficient (i.e., it achieves the Cramér-Rao lower bound on the variance of any estimate of θ). This interpretation of X_{n+1} will be used in Section III-C; it also leads to the coding algorithm for the band-limited signal case (see Schalkwijk [16]).

Now suppose that N iterations are made before the receiver makes its decision as to which of the M messages was sent. What is the probability of error? The situation is presented in Fig. 4. After N iterations,

$$X_{N+1} \sim N(\theta, \sigma^2/\alpha^2 N).$$

The length of the message interval is $1/M$. Hence, the probability of X_{N+1} lying outside the correct message interval is

$$P_e = 2 \operatorname{erfc}\left(\frac{\frac{1}{2}M^{-1}}{\sigma/\alpha\sqrt{N}}\right), \tag{6}$$

where

$$\operatorname{erfc} x = \frac{1}{\sqrt{2\pi}} \int_x^\infty e^{-t^2/2}\, dt.$$

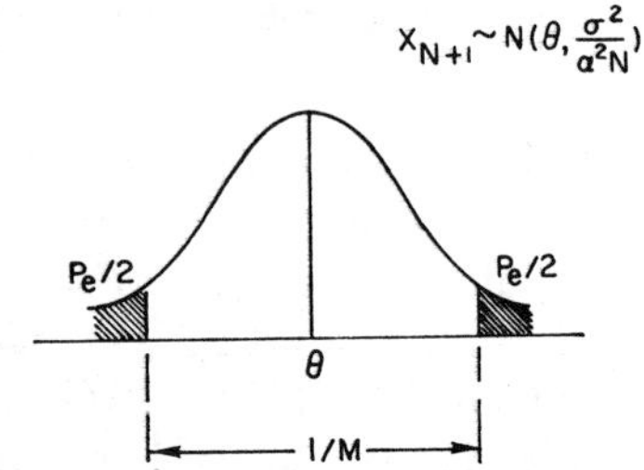

Fig. 4. The error probability is the shaded area.

D. Achieving Channel Capacity

Equation (6) shows, not unexpectedly, that P_e can be driven to zero by increasing the number N of iterations. However, if this is done without increasing M, the signaling rate (which we shall define as $R = \ln M/T$ nats/second) will go to zero. This tradeoff of rate for reliability had seemed quite natural and inevitable, until Shannon pointed out 1) that a constant rate R could be maintained if M was increased along with T (which is monotonically related to N), according to the formula $M = e^{TR}$, 2) that if R were not too high, i.e., M did not increase too rapidly with T (or N), then the degradation in performance introduced by increasing M could be more than compensated for by the good effects of increasing T, and therefore, 3) that for such rates, arbitrarily low error probability could be achieved by taking T (or N) large enough. The largest such rate, at which arbitrarily low error probabilities can be achieved, was called the channel capacity [11].

To apply Shannon's observations to our problem, we inquire how rapidly we can increase M with N while still enabling the probability of error to vanish for increasing N. The distribution in Fig. 4 squeezes in at a rate $1/\sqrt{N}$ (this being the standard deviation). Therefore, if $1/M$ is decreased at a rate slightly less than $1/\sqrt{N}$, one can "trap" the Gaussian distribution within the message interval and thus make the probability of correct detection go to unity. We therefore set

$$M(N) = N^{1/2(1-\epsilon)}. \tag{7}$$

The consequent error probability is

$$P_e = 2 \operatorname{erfc}\left[\frac{\alpha}{2\sigma} N^{\epsilon/2}\right] \tag{8}$$

and as $N \to \infty$,

$$\lim_{N\to\infty} P_e = \begin{cases} 0 & \text{for } \epsilon > 0 \\ 1 & \text{for } \epsilon < 0. \end{cases}$$

The critical rate (determined by $\epsilon = 0$) in nats per second will be

$$R_{crit} = \left[\frac{\ln M(N)}{T}\right]_{\epsilon=0} = \frac{\ln N}{2T} \text{ nats/second.} \tag{9}$$

However, in order to keep R_{crit} finite as $T \to \infty$, N must grow exponentially with T. Thus, setting $N = e^{2AT}$, with A being some constant, gives

$$R_{crit} = \frac{\ln N}{2T} = A \text{ nats/second.} \tag{10}$$

But what prevents us from choosing A arbitrarily large and thereby achieving an arbitrarily high rate of error-free transmission? The answer is that A is limited by the average power constraint P_{av}, which has not as yet been taken into account. The effect of P_{av} on A can be seen by calculating the average transmitted power with the proposed scheme. The transmitted power will depend upon the additive noise. Therefore, using $E[\cdot]$ to denote averaging over the noise process gives

$$P_{av}(N) = \frac{1}{T} E\left[\alpha^2(X_1 - \theta)^2 + \sum_{i=1}^{N-1} \alpha^2(X_{i+1} - \theta)^2\right]. \tag{11}$$

If we assume a uniform prior distribution for the message point θ, $E(X_1 - \theta)^2$ will be $\frac{1}{12}$. Furthermore, we have seen that $E(X_{i+1} - \theta)^2 = \sigma^2/\alpha^2 i$. Substitution in the formula for the average power leads to [also using (10)]

$$P_{av}(N) = \frac{2A}{\ln N}\left(\frac{\alpha^2}{12} + \sigma^2 \sum_{i=1}^{N-1} \frac{1}{i}\right). \tag{12}$$

Therefore,

$$A = \frac{P_{av}}{N_0}\left[\frac{\alpha^2}{6N_0 \ln N} + \frac{\sum_1^N \frac{1}{i}}{\ln N}\right]^{-1} < \frac{P_{av}}{N_0}$$

$$\lim_{N\to\infty} P_{av}(N) = 2\sigma^2 A = N_0 A \quad \text{or} \quad A = \frac{P_{av}}{N_0}. \tag{13}$$

Therefore, A cannot be arbitrarily large but is constrained to be less than or equal to P_{av}/N_0. The critical rate, therefore, is

$$R_{crit} = A = \frac{P_{av}}{N_0} \text{ nats/second}$$

which is just the channel capacity of the (one-way) additive white Gaussian noise channel.

It may be useful to view this result in the following way. The noise variance in our scheme goes down as $1/N$, which is no faster than the rate at which the noise variance goes down with a simple repetitive coding scheme (i.e., sending each message N times). However, with simple repetition the signal power increases with N (and would therefore violate any average power constraint for large enough N), while with our scheme the transmitted power decreases suitably with N so as to meet the average power constraint. Figure 5 is a sketch of the behavior of the expected instantaneous transmitter power as a function of time.

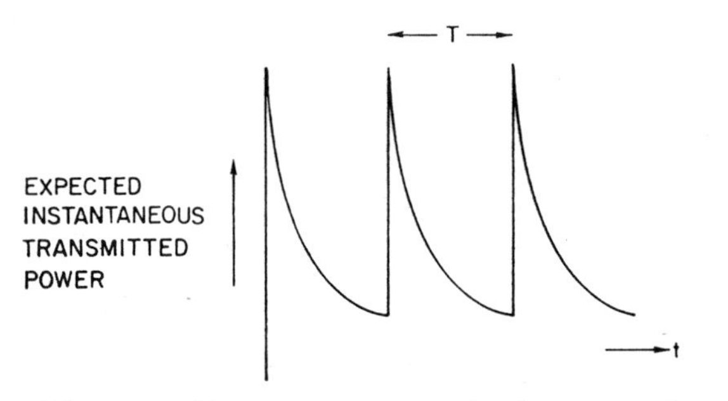

Fig. 5. The expected instantaneous transmitted power as a function of time.

Our feedback scheme cannot signal at a higher rate than is possible for the one-way channel, but it can achieve the same performance with considerably less coding and decoding complexity. As far as the coding goes, our recursive scheme for determining the transmitted waveform is somewhat simpler than the scheme for generating orthogonal waveforms. However, the real simplicity is in the decoding: with M orthogonal signals, ideal decoding requires searching for the largest of M matched filter outputs, a laborious operation for large M; in our scheme, we just have to check in what amplitude range the output of a single matched filter lies.

We can make our complexity comparisons more quantitative by comparing the expressions for the error probability with and without feedback. Before making this comparison we note that we have not as yet specified the slope α of our straight-line coding function. As far as achieving the rate P_{av}/N_0, any value of α will do. However, in evaluating the error probability for a fixed number of iterations N there is an optimum value of α.

E. Optimum P_e for Finite N

The value of the slope α, given R/C and given N, that minimizes the probability of error is easily determined. From (8), minimizing the probability of error is equivalent to maximizing

$$\frac{\alpha^2}{4\sigma^2} N^{\epsilon} = \frac{\alpha^2}{2N_0} N^{\epsilon}.$$

Now, differentiating with respect to α^2, one has for the optimum α,

$$\frac{d}{d(\alpha^2/N_0)}\left(\frac{\alpha^2}{2N_0} N^{\epsilon}\right) = \tfrac{1}{2}N^{\epsilon} + \frac{\alpha^2}{2N_0}\frac{dN^{\epsilon}}{d\epsilon}\frac{d\epsilon}{d(\alpha^2/N_0)} = 0.$$

To compute $d\epsilon/d(\alpha^2/N_0)$, an expression for ϵ is needed. Using (7) to get $R = \ln M(N)/T = (1-\epsilon)A$ and setting $\sigma^2 = N_0/2$, we get from (12),

$$R = (1-\epsilon)A = (1-\epsilon)C\left[\frac{\alpha^2}{6N_0} + \sum_{i=1}^{N-1}\frac{1}{i}\right]^{-1} \ln N$$

from which

$$\epsilon = 1 - \frac{R}{C}\left[\frac{\alpha^2}{6N_0} + \sum_{i=1}^{N-1}\frac{1}{i}\right](\ln N)^{-1}. \tag{14}$$

Hence, $d\epsilon/[d(\alpha^2/N_0)] = -(R/6C)(\ln N)^{-1}$, and using this, we will have

$$\frac{d}{d(\alpha^2/N_0)}\left(\frac{\alpha^2}{2N_0} N^{\epsilon}\right) = \tfrac{1}{2}N^{\epsilon} - \frac{\alpha^2}{2N_0}N^{\epsilon}(\ln N)\frac{R}{C}\frac{1}{6\ln N} = 0.$$

Therefore, the optimum value of α^2, say α_0^2, is

$$\alpha_0^2 = 6N_0\left(\frac{R}{C}\right)^{-1}. \tag{15}$$

Substituting for α_0^2 in the formula for the probability of error we finally have

$$P_e = 2\,\mathrm{erfc}\left[\left(3\frac{C}{R}N^{\epsilon}\right)^{1/2}\right]. \tag{16}$$

Figure 6 gives curves for the probability of error as a function of the number N of iterations. The parameter R/C is the rate relative to channel capacity. The curves start at that value of N beyond which ϵ as given by (14) is positive. Note that for relative rates approaching unity, the number of transmissions per message becomes very high.

Equation (15) gives the optimum value of the slope α as a function of the relative rate R/C and the noise power spectral density $N_0/2$. Figure 7 shows curves of the probability of error vs. the slope squared relative to its optimum value.

We can also write down an asymptotic expression for the probability of error, similar to the expression quoted in Section I for orthogonal codes. From (8), and substituting $\sigma^2 = N_0/2$, the probability of error is

$$P_e = 2\,\mathrm{erfc}\left[\left(\frac{\alpha^2}{2N_0}N^{\epsilon}\right)^{1/2}\right].$$

By using the optimum α_0^2 of α^2 given by (15) and using the well-known asymptotic formula for the *erfc* function, we obtain (asymptotically for large N)

$$P_e \approx \frac{\exp\left[-\frac{3}{2}\left(\frac{R}{C}\right)^{-1}N^{\epsilon}\right]}{\left[6\pi\left(\frac{R}{C}\right)^{-1}N^{\epsilon}\right]^{1/2}}.$$

Furthermore, $N = e^{2AT}$ and $R = (1-\epsilon)A$, where A is asymptotically equal to C. We therefore have

$$P_e \approx \frac{\exp\left[-\frac{3C}{2R}e^{2(C-R)T}\right]}{\left[6\pi\frac{C}{R}e^{2(C-R)T}\right]^{1/2}}. \tag{17}$$

It will be convenient to make the comparison with orthogonal codes on the basis of a "*blocklength L*" in binary digits, which will be defined as follows. Let $2^L = M$.

After N iterations, M will be $N^{\frac{1}{2}(1-\epsilon)}$, and hence, $L = \frac{1}{2}(1 - \epsilon) \log_2 N$. Figure 8 gives curves of the probability of error vs. the blocklength L.

Similar curves can be obtained for orthogonal codes by using the relations

$$\log_{10} P_e \approx -\frac{TE(R)}{\ln 10} \quad \text{and} \quad 2^L = M = e^{RT},$$

which yield

$$\log_{10} P_e \approx L \frac{E(R)}{R} \log_{10} 2. \tag{18}$$

This expression for $\log_{10} P_e$ is plotted in Fig. 9 for several values of R/C.

For example, let the relative rate be $R/C = 0.8$. Suppose a probability of error $P_e = 10^{-7}$ is required. The asymptotic expression for the probability of error for orthogonal codes indicates a blocklength of approximately $L = 1625$ binary digits (see Fig. 9). Figure 8 shows that the WB coding scheme requires a blocklength of only $L = 12$ binary digits. For relative rates closer to unity an even more marked difference is obtained.

If $C = 1$ bit/second, these blocklengths correspond to a coding delay T:

$$T = \begin{cases} \dfrac{1625}{0.8} = 2031 \text{ seconds (orthogonal codes)} \\[2ex] \dfrac{12}{0.8} = 15 \text{ seconds (with our feedback scheme).} \end{cases}$$

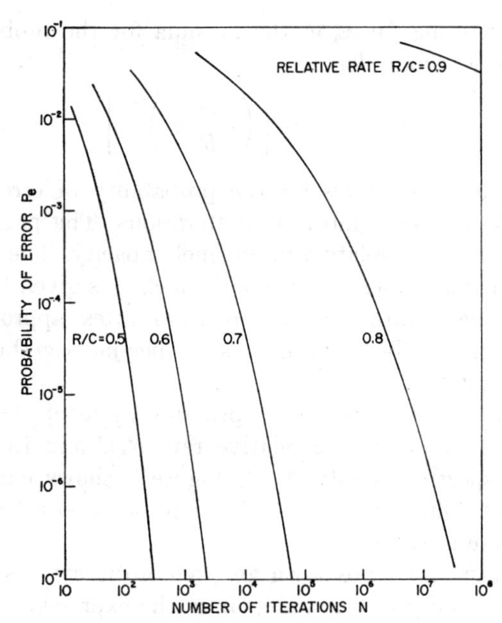

Fig. 6. The probability of error as a function of the number of iterations.

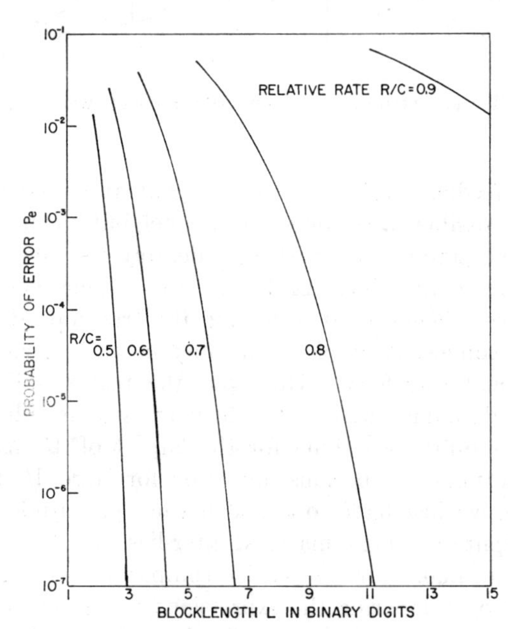

Fig. 8. The probability of error as a function of the blocklength in binary digits.

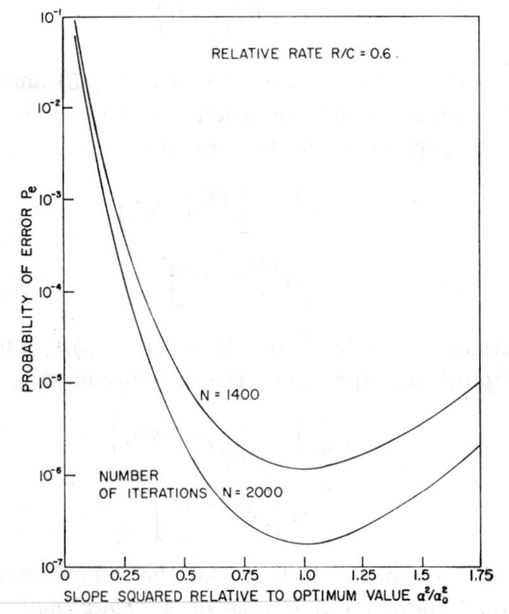

Fig. 7. The probability of error vs. the slope squared relative to its optimum value.

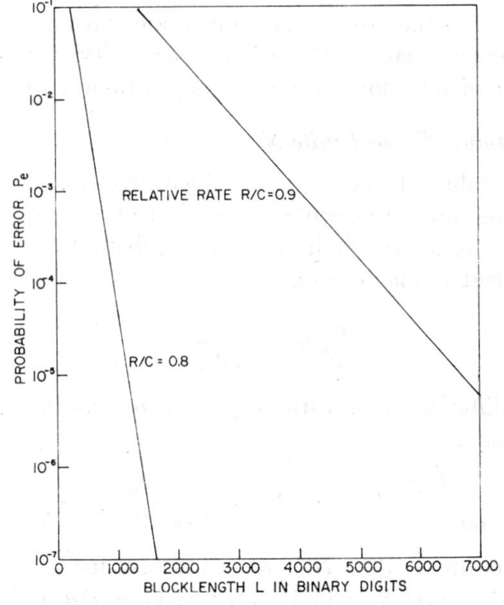

Fig. 9. The asymptotic expression for the probability of error for orthogonal codes as a function of the blocklength in binary digits.

Therefore, the use of feedback provides a considerable reduction in coding delay (and hence, coding and decoding complexity).[3] The savings due to feedback become even more pronounced as we go to lower values of P_e and to rates closer to the channel capacity. However, we should point out two ways in which the comparison above is somewhat unfair. The first, and less important, is that we have obtained the value $T = 15$ seconds for the feedback channel by using the exact formula for $P_{e,\, fb}$, whereas the value $T = 2031$ seconds was obtained from the asymptotic formula for $P_{e,\, orth}$. However, for the error probability (10^{-7}) we are considering, the exact calculation for T (which is difficult to perform) would not give results much different from $T = 2031$. The second, and more serious objection, is that the value for orthogonal codes is based on a strict power limitation of P_{av} for all code words, whereas in the feedback scheme, it is only the *expected* average power that is limited to P_{av}. (The expectation is over the Gaussian noise variables.) The instantaneous average (over the time interval T) power is the sum of a large number of squared Gaussian variates (11); the expected (or mean) value of the average power is P_{av} and using a well-known relation ($Ex^4 = 3Ex^2$ for a Gaussian random variable x), we see that the variance is $3P_{av}$. If we make allowance for this variation by using, say, $P_{av} + 3\sqrt{3P_{av}}$ for the power with the feedback scheme, we will need a larger coding delay T and the reduction will not be in the ratio 2000 to 15. However, without making any recalculations, we feel it is fair to say that the use of feedback definitely produces an "order of magnitude" reduction in the necessary coding delay T.

III. Further Properties and Extensions

In this section we shall examine the bandwidth and peak power requirements of our coding scheme, study the effects of loop delay and feedback noise, and consider extensions to channels where the spectral density $N_0/2$ may not be known and/or where the additive white noise is not Gaussian.

A. Bandwidth of the Transmitted Signals

The feedback communication system described in this section has no constraint on the bandwidth of the transmitted signals. It will be shown presently why it is not possible to cope with a bandwidth constraint.

From Section II-D, $N = e^{2AT}$ iterations are made in T seconds. Suppose the transmitted signals have bandwidth W, then the number of iterations is at most equal to the number of degrees of freedom. The number of degrees of freedom of a waveform of bandwidth W and duration T is approximately equal to $2WT$. Putting $N = 2WT$,

$$W = \frac{1}{2T} e^{2AT}$$

where

$$A = C\left(\frac{\alpha^2}{6N_0} + \sum_{i=1}^{N-1} \frac{1}{i}\right)^{-1} \ln N \tag{19}$$

which follows from substituting $\sigma^2 = N_0/2$ into (12). From (19) we see that A is asymptotically equal to C for large N, and hence, $W \approx 1/(2T)e^{2CT}$. That is, W grows exponentially with T and $\lim_{T\to\infty} W(T) = \infty$.

Substituting $T = 1/(2A) \ln N$ into (19) leads to an expression for W in terms of the number of iterations:

$$W = A \frac{N}{\ln N} \text{ c/s.} \tag{20}$$

B. Peak Power

It is known a priori that θ must lie in the interval [0, 1]. Restricting the Robbins-Monro procedure to this interval will limit the peak power for fixed bandwidth W. This can be done with the aid of the following theorem from Venter [14].

Theorem: Suppose D is a closed convex subset of R^P, P-dimensional Euclidean space, and it is known a priori that $\theta \in D$. Then modify the stochastic approximation procedure in the following way:

$$X_{n+1} = \begin{cases} X_n + a_n Y_n(X_n) & \text{if } X_n + a_n Y_n(X_n) \in D \\ \left[\begin{array}{l}\text{the point on the} \\ \text{boundary of } D \text{ clos-} \\ \text{est to } X_n + a_n Y_n(X_n)\end{array}\right] & \text{if } X_n + a_n Y_n(X_n) \notin D. \end{cases}$$

Whenever the original procedure converges, so does its restriction to D. The asymptotic rate of convergence for both procedures is the same.

A special case of this theorem, in which the closed convex subset is equal to the unit interval [0, 1] and $p = 1$, is applicable to our coding scheme. Hence, the modified procedure is as follows.

$$X_{n+1} = \begin{cases} 0 & \text{if } X_n + a_n Y_n(X_n) \le 0 \\ X_n + a_n Y_n(X_n) & \text{if } 0 < X_n + a_n Y_n(X_n) < 1. \\ 1 & \text{if } 1 \le X_n + a_n Y_n(X_n) \end{cases}$$

In investigating how the peak power P_{peak} depends on the bandwidth W, let us consider a basic signal $\phi(t)$,

$$\phi(t) = \sqrt{2W} \frac{\sin 2\pi W t}{2\pi W t}.$$

This signal has bandwidth W and satisfies the orthonormality condition (5) for $\Delta = 1/2W$. With $N = e^{2AT}$ [A given by (19)],

$$\Delta = \frac{T}{N} = Te^{-2AT}$$

$$P_{peak} = 2W = \frac{1}{\Delta} = \frac{1}{T} e^{2AT}.$$

[3] As we mentioned in the Introduction, with $C = 1$, Turin's scheme [15] would require only an *average* coding delay of 1 second and the *average* signaling rate will be 1 bit/second. However, the actual coding delays may fluctuate considerably from the average value.

Hence, for large T (or N), the P_{peak} goes to infinity while the average power remains finite. A similar phenomenon will be discovered if the basic signal $\phi(t)$ is chosen to have a duration Δ. Of course, this exponentially growing (with T) peak power also occurs with one-way channels if orthogonal signals of finite duration or of the form $\sin 2\pi Wt/\sqrt{2W}\pi t$ are used. Since we are using matched filter reception, we can use pulse compression techniques to alleviate the peak power problem. However, this topic is somewhat apart from the main theme of this paper, and we shall therefore not pursue it any further.

C. Loop Delay

Up to this point, only instantaneous feedback has been considered. In a practical situation there will be feedback delay.

Let $F(x) = \alpha(x - \theta)$, and let the additive random variables Z_n be identically distributed. From the iterative relation,

$$X_{n+1} = X_n - \frac{1}{\alpha n} Y_n(X_n)$$

where $Y_n(X_n) = F(X_n) + Z_n$, it can easily be shown that

$$X_{n+1} - \theta = -\frac{1}{\alpha n} \sum_{i=1}^{n} Z_i. \tag{21}$$

This means that (when the Z_n are Gaussian) X_{n+1} is the maximum likelihood estimate of θ, based on the observations $Y_1(X_1)$ through $Y_n(X_n)$.

Now suppose there are d units of loop delay, so that $Y_n(X_n)$ can first be used to determine X_{n+d+1}. The first time one can use received information is when computing X_{d+2}.

Let us choose as X_{n+d+1} the maximum likelihood estimate of θ, based on observations $Y_1(X_1)$ through $Y_n(X_n)$. The iterative relation now becomes

$$X_{n+d+1} = \frac{(n-1)X_{n+d} + X_n}{n} \quad \frac{1}{\alpha n} Y_n(X_n). \tag{22}$$

It follows easily that

$$X_{n+d+1} - \theta = -\frac{1}{\alpha n} \sum_{i=1}^{n} Z_i. \tag{23}$$

One must complete d more transmissions in order to obtain the same variance as in the case of instantaneous feedback, and thus, the influence of the delay will become negligible for large values of n.

D. Non-Gaussian Noise

If the additive white noise is Gaussian, our coding scheme will permit error-free transmission at any rate less than channel capacity. For the scheme to work it is not necessary to know the noise power spectral density $N_0/2$. However, as shown in Section II, knowledge of $N_0/2$ permits one to choose the slope α in an optimum fashion in the nonasymptotic case.

Stochastic approximation, in general, and the Robbins-Monro procedure, in particular, are nonparametric. Therefore, the coding scheme will also work in the case of non-Gaussian white noise.

What about the probability of error? Sacks' theorem [6] on the asymptotic distribution of X_{n+1} implies that X_{n+1} is asymptotically Gaussian with the required variance. Hence, all the calculations given earlier in this Section are still valid for large N.

Finally, does one achieve channel capacity when the additive noise is non-Gaussian The critical rate of our system is still $R_{\text{crit}} = P_{\text{av}}/N_0$, and this gives a lower bound on the channel capacity for all non-Gaussian white noise channels with noise of spectral density $N_0/2$.

E. Influence on Feedback Noise on Wideband Coding Scheme

In the case of noiseless feedback it is immaterial whether X_{n+1} or $Y_n(X_n)$ is sent back to the transmitter. This is not true in the case of noisy feedback. The following notation is adopted for this case: a single prime refers to the forward direction and a double prime to the feedback link. Thus, $N_0'/2$ is the (two-sided) power spectral density of the additive white Gaussian noise in the forward channel, and we shall write $N_0''/2$ for the spectral density of the white Gaussian noise $n''(t)$ in the feedback link. The noises in the forward and feedback links are assumed to be independent.

The estimates of θ obtained by the receiver and transmitter are denoted by X_n' and X_n'', respectively. $Y_n'(X_n'')$ is the noisy observation made by the receiver. This value is sent back to the transmitter which obtains $Y_n''(X_n'') = Y_n'(X_n'') + Z_n''$, where Z_n'' is the additive noise in the feedback link.

The influence of feedback noise is mainly a reduction in relative rate R/C in the case where the receiver's estimate X_{n+1} is sent back to the transmitter. The probability of error increases only slightly. When the receiver's observation $Y_n'(X_n'')$ is sent back to the transmitter, the feedback noise reduces the rate only slightly and its main effect is an increase in the error probability.

Consider first the case where X_{n+1} is sent back. Equation (12) for the average power changes in that an additional term $\alpha^2(N_0''/2)(N/T)$ due to the feedback noise appears, and also σ^2 changes to $\sigma^2 = \frac{1}{2}(N_0' + \alpha^2 N_0'')$ instead of $\sigma^2 = N_0'/2$. If it is assumed that the feedback noise is small compared to the additive disturbance in the forward channel, then σ^2 will only change slightly. The error probability in (8) will also only change slightly provided that all other quantities in (8) remain the same.

Figure 10 is a plot of the relative rate

$$\frac{R}{C} = (1 - \epsilon)\left\{\frac{1}{\ln N}\left[\frac{\alpha_0^2}{6N_0'} + \sum_{k=1}^{N-1} \frac{1}{k} + \alpha_0^2 \frac{N_0''}{N_0'}\left(N + \sum_{k=1}^{N-1} \frac{1}{k}\right)\right]\right\}^{-1} \tag{24}$$

vs. the number N of iterations for different values of N_0''. The upper curve is for noiseless feedback. The probability of error for noiseless feedback is $P_e' = 10^{-4}$. In the case of noisy feedback it is only slightly higher.

Equation (24) follows from (12), adding the additional term $\alpha^2(N_0''/2)(N/T)$. For α_0^2 the optimum value for noiseless feedback is used, that is, the value given by (15).

It is seen from Fig. 10 that for noiseless feedback the relative rate approaches unity with increasing N; however, in the case of noisy feedback, the curve for noiseless feedback is followed for some time after which the relative rate drops to zero quite suddenly. (Note that no optimization in the presence of feedback noise is attempted. The particular system we use is optimum for $N_0'' = 0$.)

The feedback power P_{fb} is

$$P_{\text{fb}} = \frac{1}{\alpha_0^2 T}\frac{1}{2}(N_0' + \alpha_0^2 N_0'')\sum_{k=1}^{N-1}\frac{1}{k} + \frac{1}{12}\frac{N}{T} \tag{25}$$

and is again hardly affected by the feedback noise.

Now consider the case where $Y_n'(X_n'')$ is sent back. The average transmitted power as given by (12) is only slightly affected in that now $\sigma^2 = \frac{1}{2}(N_0' + N_0'')$ instead of $\sigma^2 = N_0'/2$, and the same is true for the relative rate, assuming N_0'' small compared to N_0'.

What is the influence of the feedback noise on the error probability? X_{n+1}'' as used by the transmitter is equal to

$$X_{n+1}'' = X_n'' - \frac{1}{\alpha n} Y_n''(X_n'')$$

where $Y_N''(X_n'') = Y_n'(X_n'') + Z_n''$ in which $Y_n'(X_n'') = F(X_n'') + Z_n'$ is the noisy observation made by the receiver. A simple derivation shows that

$$X_{n+1}' = X_{n+1}'' + \frac{1}{\alpha}\sum_{i=1}^{n}\frac{1}{i}Z_i'' \tag{26}$$

where X_{n+1}' is the estimate of the message point θ computed by the receiver. Hence,

$$X_{n+1}' \sim N\left[\theta, \frac{\sigma^2}{\alpha^2 n} + \frac{\sigma''^2}{\alpha^2}\sum_{i=1}^{n}\left(\frac{1}{i}\right)^2\right]$$

and the variance, say σ_t^2, of X_{N+1}' is equal to

$$\sigma_t^2 = \frac{\sigma^2}{\alpha^2 N} + \frac{\sigma''^2}{\alpha^2}\sum_{i=1}^{N}\left(\frac{1}{i}\right)^2. \tag{27}$$

The formula for the probability of error is

$$P_e = 2\,\text{erfc}\left[\frac{\frac{1}{2}M(N)^{-1}}{\sigma_t}\right] = 2\,\text{erfc}\left[\frac{N^{-1/2(1-\epsilon)}}{2\sigma_t}\right]. \tag{28}$$

Again, as in the noiseless feedback case, let us find the optimum value of α_0^2 of α^2. (Note that in the earlier case, where the receiver's estimate X_{n+1} is sent back, such an optimization was not attempted for nonzero feedback noise.) As before,

$$\epsilon = 1 - \frac{R}{C}(\ln N)^{-1}\left(\frac{\alpha^2}{6N_0}\sum_{k=1}^{N-1}\frac{1}{k}\right) \tag{14}$$

where now $N_0 = N_0' + N_0''$. It is desired to minimize the probability of error with respect to α^2. From (28), this is equivalent to minimizing $\sigma_t^2 N^{-\epsilon}$. Setting the derivative equal to zero,

$$\frac{d}{d\alpha^2}(\sigma_t^2 N^{-\epsilon}) = -\frac{1}{\alpha^2}\sigma_t^2 N^{-\epsilon} + \sigma_t^2 N^{\epsilon}(\ln N)\left[\frac{R}{C}(\ln N)^{-1}\frac{1}{6N_0}\right] = 0$$

yields

$$\alpha_0^2 = 6N_0\left(\frac{R}{C}\right)^{-1} \tag{29}$$

which has the same form as (15) for noiseless feedback.

Figure 11 shows curves of the probability of error P_e vs. the number of iterations N, with the parameter being the power spectral density N_0''/N_0' of the feedback noise relative to the corresponding quantity for the forward

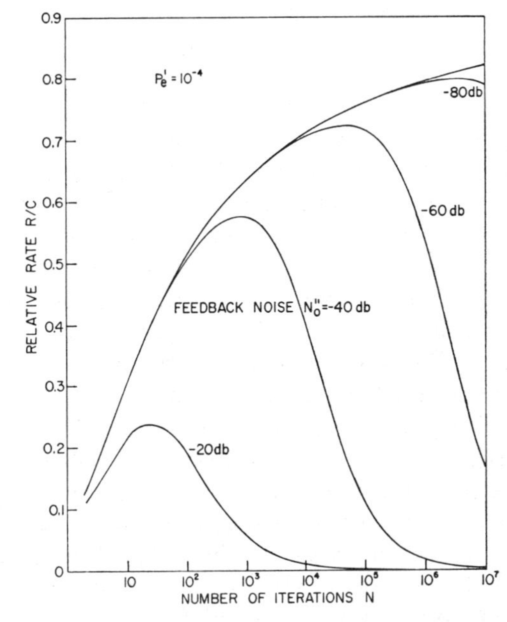

Fig. 10. The relative rate vs. the number of iterations for the case where X_n is sent back.

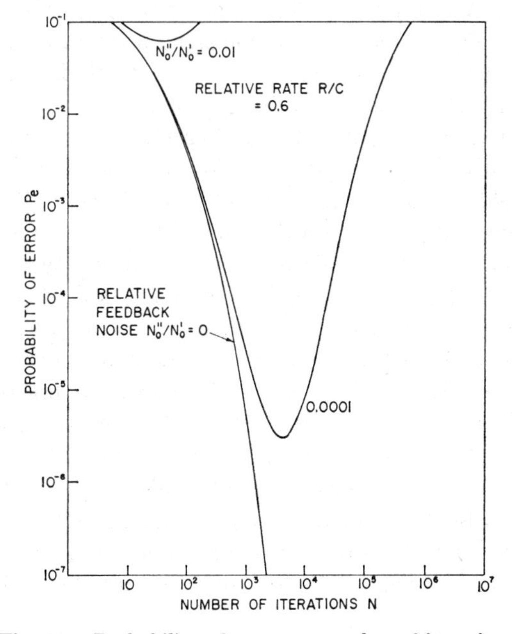

Fig. 11. Probability of error vs. number of iterations.

link. The P_e curves have a minimum for nonzero variance of the feedback noise, and it does not make sense to do more iterations per message than the value indicated by the minimum of the P_e curve.

The average feedback power p_{fb} is

$$P_{fb} = P_{av} - \frac{\alpha_0^2}{12T} + \frac{N_0'}{2}\frac{N}{T}. \quad (30)$$

In conclusion, it should be observed that one can either 1) insist on a vanishing probability of error in which case the rate of signaling will approach zero, or 2) require a nonvanishing rate in which case there is a minimum achievable probability of error different from zero.

We should also point out that by using a differrent scheme, it may be possible to obtain much better results for noisy feedback channels than are yielded by our scheme.

IV. Concluding Remarks

There are several areas for further work that may be investigated. We shall briefly mention some of them. One is the question of whether our method of exploiting the feedback with a linear encoding function is the most efficient. Even if it should turn out to be the most efficient for Gaussian noise, we can ask whether for non-Gaussian additive white noise, we cannot achieve a rate greater than P_{av}/N_0 by using some other form of encoding function. For example, T. Cover of Stanford University has suggested subjecting the straight line to the nonlinear transformation that would convert a non-Gaussian probability density function into a Gaussian density function; if the stochastic approximation technqiue is applicable to the resulting curve, this might yield better results for the non-Gaussian case. (It should be pointed out here that asymptotically the shape of $F(x)$ does not matter, since Sacks' theorem on the asymptotic distribution of the estimate is true under very mild assumptions, given in Section II on $F(x)$. The shape of the regression function is important only in nonasymptotic calculations.)

Another possibility is to use sequential detection in combination with our scheme. Instead of making a decision after a prespecified number of iterations, we could wait until the matched filter output crossed a suitable threshold. Such operation would certainly result in some improvement in the error probability, but we suspect the gain may not be worth the extra complexity.

Our results should also be extendable to other situations where stochastic approximation techniques apply, e.g., channels with unknown gains, slowly changing random delays, etc. It may be mentioned that the coding scheme suggested by the Kiefer-Wolfowitz stochastic approximation technique for determining the minimum of an unknown function does not yield results as good as those obtained for the coding scheme in this paper (which was suggested by the Robbins-Monro procedure). J. Venter (Stanford University, 1965) showed in unpublished work that a coding scheme based on the Kiefer-Wolfowitz procedure cannot achieve channel capacity.

Of course, a major question is the study of communication over noisy feedback links. When more general results, e.g., on the capacity, of such channels are known, it may be easier to look for efficient communication schemes with noisy feedback.

Finally, we mention an extension [16] of the results in this paper to the case of signals with a bandwidth constraint; the scheme to be used under this constraint is more complicated than the one given in this paper, but, of course, it also applies to the non-band-limited-signal case and, in fact, has some advantages—fixed peak power, fewer iterations, etc.

Note added in proof: J. Omura has pointed out a mistake in the coefficient in (17) that arises from an error in arguments based on (10)–(12). The correct formula is

$$P_e \sim \exp\left[-be^{2(C-R)T}\right]/\left[4\pi b e^{2(C-R)T}\right]^{1/2}$$

where $2b = 3e^{-(1+\gamma)}$ and $\gamma = 0.577 \cdots$ is Euler's constant. This causes small changes in the curves, but does not affect the main lines of the argument. The details will appear in Omura's thesis at Stanford University.

References

[1] C. E. Shannon, "A mathematical theory of communication," *Bell Sys. Tech. J.*, vol. 27, pp. 379–424 and 623–657, July-October 1948.

[2] H. Robbins and S. Monro, "A stochastic approximation method," *Ann. Math. Stat.*, vol. 22, pp. 400–407, September 1951.

[3] A. Dvoretzky, "On stochastic approximation," *Proc. Third Berkeley Symposium on Mathematical Statistics and Probability*, pp. 39–55, 1956.

[4] C. E. Shannon, "The zero-error capacity of a noisy channel," *IRE Trans. on Information Theory*, vol. IT-2, pp. 8–19, September 1956.

[5] J. Wolfowitz, "On stochastic approximation methods," *Ann. Math. Stat.*, vol. 27, pp. 1151–1156, December 1956.

[6] J. Sacks, "Asymptotic distributions of stochastic approximation procedures," *Ann. Math. Stat.*, vol. 29, pp. 373–405, June 1958.

[7] C. E. Shannon, "Probability of error for optimal codes in a Gaussian channel," *Bell Sys. Tech. J.*, vol. 38, pp. 611–656, May 1959.

[8] P. Elias, "Channel capacity without coding," in *Lectures on Communication System Theory*, Baghdady, Ed. New York: McGraw-Hill, 1961.

[9] R. M. Fano, *Transmission of Information*. New York: Wiley, 1961.

[10] P. E. Green, "Feedback Communication Systems," in *Lectures on Communication System Theory*, Baghdady, Ed. New York: McGraw-Hill, 1961.

[11] J. Wolfowitz, *Coding Theorems of Information Theory*. Berlin: Springer-Verlag, 1961.

[12] L. H. Zetterberg, "Data transmission over a noisy Gaussian channel," *Trans. Roy. Inst. of Tech.*, no. 184, Stockholm, Sweden, 1961.

[13] M. Horstein, "Sequential transmission using noiseless feedback," *IEEE Trans. on Information Theory*, vol. IT-9, pp. 136–143, July 1963.

[14] J. Venter, "On stochastic approximation methods," Ph.D. dissertation, University of Chicago, Ill., 1963.

[15] G. L. Turin, "Signal design for sequential detection systems with feedback," *IEEE Trans. on Information Theory*, vol. IT-11, pp. 401–408, July 1965.

[16] P. Schalkwijk, "Coding for additive noise channels with feedback—part II: band-limited channels," this issue, page 183.

[17] A. J. Viterbi, "The effect of sequential decision feedback on communication over the Gaussian Channel," *Information and Control*, vol. 8, pp. 80–92, February 1965.

A Coding Scheme for Additive Noise Channels with Feedback

Part II: Band-Limited Signals

J. PIETER M. SCHALKWIJK, MEMBER, IEEE

Abstract—In Part I of this paper, we presented a scheme for effectively exploiting a noiseless feedback link associated with an additive white Gaussian noise channel with *no* signal bandwidth constraints. We now extend the scheme for this channel, which we shall call the wideband (WB) scheme, to a band-limited (BL) channel with signal bandwidth restricted to $(-W, W)$. Our feedback scheme achieves the well-known channel capacity, $C=W \ln(1+P_{av}/N_0W)$, for this system and, in fact, is apparently the first deterministic procedure for doing this. We evaluate the fairly simple exact error probability for our scheme and find that it provides considerable improvements over the best-known results (which are lower bounds on the performance of sphere-packed codes) for the one-way channel. We also study the degradation in performance of our scheme when there is noise in the feedback link.

I. Introduction

IN THIS PAPER a band-limited (BL) channel with feedback is considered. The signal bandwidth is restricted to $(-W, W)$.

A general introduction has been given in Part I, with particular attention to wideband (WB) channels.

The BL coding scheme developed here, is as far as we know, the first deterministic coding procedure to achieve the well-known capacity

$$C = W \ln [1 + (P_{av}/N_0W)].$$

To our knowledge, the only other results pertaining to the band-limited (BL) channel have been published by Elias [3]. He divided the channel into K subchannels of bandwidth $w = W/K$. If noiseless feedback is available and if $K \rightarrow \infty$, information can be sent at a rate equal to $W \ln [1 + (P_{av}/N_0W)]$. However, since the signal bandwidth is w instead of W, the coding and decoding complexity for the feedback scheme becomes an arbitrarily small fraction of that required without feedback.

II. A Feedback Communication System with a Constraint on the Bandwidth

Let T be the time in seconds necessary for the transmission of a particular message. For the WB coding scheme discussed in Part I, as for orthogonal codes in one-way transmission, the bandwidth $W(T)$ of the transmission is an exponential function of the coding delay T. In order to make the probability of error vanish for a fixed relative rate smaller than one, a large bandwidth is required.

Suppose now that one is given a fixed bandwidth W, which the transmission is not supposed to exceed. With this additional transmitter constraint imposed, the channel capacity C is no longer P_{av}/N_0 as before, but it now given by $W \ln[1 + (P_{av}/N_0W)]$, nats/second. For small values of P_{av}/N_0W the latter capacity approaches P_{av}/N_0 as it should, for when $W \rightarrow \infty$, both channels are identical.

Shannon [1] derives the capacity formula, $W \ln [1 + (P_{av}/N_0W)]$, by a random coding argument, and up till now no deterministic way was known for constructing a code achieving the critical rate for a band-limited white Gaussian noise channel with or without feedback. In this part the first such code will be developed for the case where noiseless feedback is available.

As in Part I, an optimization for finite block-length is carried through, the results are compared with bounds on one-way transmission plotted by Slepian [4], and the deterioration of the present scheme due to feedback noise is considered.

A. The BL Coding Scheme

In the WB coding scheme discussed in Part I, the variance of the estimate X_{N+1} for the message point θ was inversely proportional to the number N of iterations. The critical rate was $R_{crit} = (\ln N)/2T$ nats/second, and in order to achieve a constant rate one had to choose $N = e^{2AT}$, that is, the number of transmissions had to increase exponentially with time.

Now suppose one has to meet a bandwidth constraint W in cycles per second. In this case the number of independent transmissions can only increase linearly with time. The highest number of independent transmissions per second is approximately equal to $2W$. Substituting $N = 2WT$ in the equation for the critical rate above gives $R_{crit} = (\ln 2WT)/2T$ nats/second. Hence, $R_{crit} \rightarrow 0$ with increasing T, and so the system discussed in Part I has to be modified in order to achieve a constant rate different from zero in the band-limited case.

Two useful observations can be made at this point. First, while the critical rate approaches zero when we take $2W$ iterations per second the asymptotic relation $R_{crit}(T) \approx P_{av}(T)/N_0$ is still valid. In other words, both the rate and the average power approach zero for

This work was supported by NASA NsG 377.
The author is with the Applied Research Laboratory, Sylvania Electronic Systems, a division of Sylvania Electric Products Inc., Waltham, Mass.

Reprinted from *IEEE Trans. Inform. Theory*, vol. IT-12, pp. 183–189, Apr. 1966.

increasing T. The limit of their ratio, however, is equal to the constant N_0. The second observation is that X_{N+1} can be looked at as the maximum likelihood estimate of θ having observed $Y_1(X_1)$ through $Y_n(X_n)$, and assuming Gaussian noise, as explained in Part I–Section II.

With these two observations in mind, we shall present a coding scheme for the band-limited white Gaussian noise channel.

Suppose that transmissions take place at integer values of time, the time unit being $1/(2W)$ second. Numbers are sent again by amplitude modulating some basic waveform of bandwidth W and unit energy. The disturbance is white Gaussian noise (with spectral density $N_0/2$,) and reception takes place using a matched filter.

The coding scheme starts out the same as in Part I–Section II. At the transmitter:

1) divide the unit interval [0, 1] into M disjoint message intervals of equal length; let θ be the midpoint of the message interval corresponding to the particular message to be transmitted, and

2) at instant one, transmit $\alpha(X_{11} - \theta)$, where $X_{11} = 0.5$ and α is some constant to be determined later.

At the receiver:

1) receive $Y_{11}(X_{11}) = \alpha(X_{11} - \theta) + Z_{11}$, where Z_{11} is as before a Gaussian random variable with mean zero and variance $\sigma^2 = N_0/2$, and

2) compute $X_{12} = X_{11} - \alpha^{-1}Y_{11}(X_{11})$, then set $X_{21} = X_{12}$ and send X_{21} back to the transmitter.

Up to this point everything is the same as for the coding scheme of Part I–Section II-B. In other words, $X_{21} - \theta = -(1/\alpha)Z_{11}$, where X_{21} is the maximum likelihood estimate of θ having observed $Y_{11}(X_{11})$.

Now, in order to prevent the expected power per transmission from decreasing, as it did in the WB coding scheme, we shall take the next transmission as $g\alpha(X_{21} - \theta)$ instead of $\alpha(X_{21} - \theta)$, where the constant g will be determined presently. The receiver obtains the noisy observation

$$Y_{21}(X_{21}) = g\alpha(X_{21} - \theta) + Z_{21}$$

and then computes

$$X_{22} = X_{21} - (g\alpha)^{-1}Y_{21}(X_{21}).$$

We now have two independent estimates of θ:

$$X_{21} = \theta - \frac{1}{\alpha}Z_{11} \quad \text{and} \quad X_{22} = \theta - \frac{1}{g\alpha}Z_{21}.$$

For the value X_{31} to be sent back to the transmitter, we shall take the maximum likelihood estimate of θ having observed $Y_{11}(X_{11})$ and $Y_{21}(X_{21})$, that is, we shall set

$$X_{31} = \frac{\left(\frac{1}{g\alpha}\right)^2 X_{21} + \left(\frac{1}{\alpha}\right)^2 X_{22}}{\left(\frac{1}{\alpha}\right)^2 + \left(\frac{1}{g\alpha}\right)^2} = \frac{X_{21} + g^2X_{22}}{1 + g^2}.$$

What is the variance of our successive maximum likelihood estimates X_{11}, X_{21}, and X_{31}? It is known that

$$X_{21} \sim N\left(\theta, \frac{\sigma^2}{\alpha^2}\right) \quad \text{and} \quad X_{31} \sim N\left(\theta, \frac{\sigma^2}{\alpha^2}\frac{1}{1 + g^2}\right).$$

If, however, $g = (\alpha^2 - 1)^{1/2}$ is chosen, then

$$X_{31} \sim N\left(\theta, \frac{\sigma^2}{(\alpha^2)^2}\right).$$

In general, X_{i1}, $i = 2, 3, \cdots$, is sent back. The next transmission is

$$\alpha^{i-1}(\alpha^2 - 1)^{1/2}(X_{i1} - \theta) \tag{1}$$

but the receiver obtains

$$Y_{i1}(X_{i1}) = \alpha^{i-1}(\alpha^2 - 1)^{1/2}(X_{i1} - \theta) + Z_{i1}$$

and then computes

$$X_{i2} = X_{i1} - [\alpha^{i-1}(\alpha^2 - 1)^{1/2}]^{-1}Y_{i1}(X_{i1})$$

and

$$X_{i+1,1} = \frac{X_{i1} + (\alpha^2 - 1)X_{i2}}{\alpha^2}.$$

The maximum likelihood estimate $X_{i+1,1}$ is normally distributed with mean θ and variance $\sigma^2/[(\alpha^2)^i]$, that is,

$$X_{i+1,1} \sim N\left[\theta, \frac{\sigma^2}{(\alpha^2)^i}\right]. \tag{2}$$

From this point on, the analysis is very similar again to that of Part I. Suppose the transmitter sends one of M possible messages, that is, the interval [0, 1] is divided into M disjoint equal-length message intervals. The message point θ is the midpoint of the message interval corresponding to the particular message being transmitted. The probability of the receiver deciding on the wrong message interval (i.e., the probability of X_{N+1} lying outside the correct interval) is

$$P_e = 2 \operatorname{erfc}\left(\frac{\frac{1}{2}M^{-1}}{\sigma/\alpha^N}\right).$$

Now pick $M = \alpha^{N(1-\epsilon)}$, that is, $R = (\ln M)/N = (1 - \epsilon)\ln \alpha$, nats/dimension (the time unit was $1/(2W)$ seconds). This gives for the probability of error

$$P_e = 2 \operatorname{erfc}\left(\frac{\alpha^{N\epsilon}}{2\sigma}\right) \tag{3}$$

and thus,

$$\lim_{N\to\infty} P_e(N, \epsilon) = \begin{cases} 0 & \text{for } \epsilon > 0 \\ 1 & \text{for } \epsilon < 0. \end{cases}$$

In other words, the critical rate is equal to $R_{crit} = \ln \alpha$, nats/dimension. Putting $\alpha = e^A$ gives $R_{crit} = A$.

Next let us derive an expression for the average power P_{av}.

$$P_{av} = \frac{1}{T} E\Big\{\alpha^2(X_{11} - \theta)^2 + \sum_{i=2}^{N} [\alpha^{i-1}(\alpha^2 - 1)^{1/2}]^2(X_{i,1} - \theta)^2\Big\}$$

$$= \frac{1}{T}\Big\{\alpha^2 E(X_{11} - \theta)^2 + \sum_{i=2}^{N} [\alpha^{i-1}(\alpha^2 - 1)^{1/2}]^2 \frac{\sigma^2}{(\alpha^2)^{i-1}}\Big\}.$$

Substituting $T = N/(2W)$ seconds, $\sigma^2 = N_0/2$, $\alpha = e^A$, and $E(X_{11} - \theta)^2 = 1/12$ (assuming a uniform prior distribution for θ), one gets

$$P_{av} = \frac{We^{2A}}{6N} + \frac{N-1}{N} N_0 W(e^{2A} - 1). \tag{4}$$

Hence, asymptotically,

$$R_{crit} = \begin{cases} A = \frac{1}{2}\ln\left(1 + \frac{P_{av}}{N_0 W}\right) & \text{nats/dimension, or} \\ 2WA = W\ln\left(1 + \frac{P_{av}}{N_0 W}\right) & \text{nats/second} \end{cases} \tag{5}$$

which is the channel capacity as computed by Shannon [1]! This result proves that the BL coding system presented here *achieves capacity for the band-limited white Gaussian noise channel.* It is the first deterministic coding procedure to do so.

B. Optimization for Finite Blocklength

As in Part I, let us now investigate how far one falls short of the ideal when only permitting a finite coding delay N (in time units of $1/(2W)$ second).

In Section I-A, the slope α_i at the ith transmission was taken as

$$\alpha_i = \alpha^{i-1+\delta_{i1}}(\alpha^2 - 1)^{1/2(1-\delta_{i1})}$$

where

$$\delta_{ij} = \begin{cases} 1 & \text{if } i = j \\ 0 & \text{if } i \neq j. \end{cases}$$

In order to make an optimization possible, an additional factor a is introduced, hence,

$$\alpha_i = a\alpha^{i-1+\delta_{i1}}(\alpha^2 - 1)^{1/2(1-\delta_{i1})} \quad i = 1, 2, 3, \cdots. \tag{6}$$

The receiver now has

$$Y_{i1}(X_{i1}) = \alpha_i(X_{i1} - \theta) + Z_{i1}$$

and computes

$$X_{i2} = X_{i1} - \alpha_i^{-1} Y_{i1}(X_{i1}),$$
$$X_{21} = X_{12},$$

and

$$X_{i+1,1} = \frac{X_{i1} + (\alpha^2 - 1)X_{i2}}{\alpha^2} \quad \text{for } i = 2, 3, \cdots.$$

Effectively, the introduction of the factor a reduces N_0 by a factor a^2.

By (3), minimizing the probability of error is equivalent to maximizing the expression

$$a^2 \frac{\alpha^{2N\epsilon}}{2N_0} \tag{7}$$

where ϵ can be obtained from

$$R = (1 - \epsilon)\ln\alpha \text{ nats/dimension} \tag{8}$$

and $\sigma^2 = N_0/2$ was substituted for the variance.

Substituting $\alpha = e^A$ in (4) and allowing for the additional factor a leads to the following expression for the average power:

$$P_{av} = a^2\left[\frac{W\alpha^2}{6N} + \frac{N-1}{N}\frac{N_0}{a^2} W(\alpha^2 - 1)\right]$$

which can be modified as

$$\frac{P_{av}}{N_0 W} = a^2 \frac{\alpha^2}{6NN_0} + \frac{N-1}{N}(\alpha^2 - 1). \tag{9}$$

Now assuming C, R, W, N_0, and N constant, let us maximize (7) with respect to a^2. Note that C and W constant implies P_{av}/N_0W constant, for

$$C = W\ln\left(1 + \frac{P_{av}}{N_0 W}\right).$$

Having gone through these preliminaries, one is now ready to perform the optimization. Set the derivative of $a^2(\alpha^{2N\epsilon}/2N_0)$ equal to zero,

$$\frac{d}{da^2}\left(a^2\frac{\alpha^{2N\epsilon}}{2N_0}\right) = \frac{\alpha^{2N\epsilon}}{2N_0} + a^2\frac{d}{d\alpha^2}\left(\frac{\alpha^{2N\epsilon}}{2N_0}\right)\frac{d\alpha^2}{da^2} + a^2\frac{d}{d\epsilon}\left(\frac{\alpha^{2N\epsilon}}{2N_0}\right)\frac{d\epsilon}{d\alpha^2}\frac{d\alpha^2}{da^2} = 0. \tag{10}$$

From (8) it follows that

$$\epsilon = 1 - \frac{2R}{\ln\alpha^2} \quad \text{and} \quad \frac{d\epsilon}{d\alpha^2} = \frac{2R}{\alpha^2}\frac{1}{(\ln\alpha^2)^2}$$

and from (9) it follows that

$$\frac{d\alpha^2}{da^2} = -\frac{\alpha^2}{a^2 + 6N_0(N-1)}.$$

Making these substitutions in (10) and putting the result equal to zero finally gives, after some algebra, the following simple expression for the optimum value a_0^2 of a^2:

$$a_0^2 = 6N_0. \tag{11}$$

For the probability of error, substituting $\sigma^2 = N_0/2$ and $a_0^2 = 6N_0$ in (3), one has

$$P_e = 2\,\text{erfc}\,[(3\alpha^{2N\epsilon})^{1/2}].$$

Solving for α^2 from (9) and (11) gives

$$\alpha^2 = \frac{N-1}{N} + \frac{P_{av}}{N_0 W}.$$

By (8) one has

$$\frac{R}{2W} = \ln \alpha^{1-\epsilon} \quad \text{or} \quad \alpha^{1-\epsilon} = \exp\left(\frac{R}{2W}\right)$$

where R is now in nats/second. Hence,

$$\alpha^{\epsilon} = \frac{\alpha}{\exp\left(\frac{R}{2W}\right)} = \frac{\left(\frac{N-1}{N} + \frac{P_{av}}{N_0 W}\right)^{1/2}}{\exp\left(\frac{R}{2W}\right)}$$

and finally,

$$P_e = 2 \operatorname{erfc}\left\{\sqrt{3}\left[\frac{\frac{N-1}{N} + \frac{P_{av}}{N_0 W}}{\exp\left(\frac{R}{W}\right)}\right]^{N/2}\right\}. \tag{12}$$

This final result will be compared in the next section with the bounds on one-way communication as obtained by Slepian [4].

C. Comparison with Slepian's Results

In 1963 Slepian [4] plotted lower bounds on communication in the one-way case based on a geometrical approach to the coding problem for band-limited white Gaussian noise channels used by Shannon [2]. That is, there is no one-way communication system whose performance is any better than that plotted by Slepian. Figures 1 through 6 compare Slepian's curves (dashed lines) with the results described by (12) (solid lines). Note that the solid curves are exact, that is, they are not a bound as Slepian's curves are. The graphs presented are described in the following.

1) Figure 1 shows the signal-to-noise ratio $S/N = 10 \log_{10} (P_{av}/N_0 W)$ in decibels vs. the rate R/W in dits/cycle, as given by Shannon's capacity equation, $\ln [1 + (P_{av}/N_0 W)]$.

2) Figures 2(a) to 2(c) indicate the additional signal-to-noise ratio, in decibels above the value indicated in Fig. 1, required for a finite coding delay N, as a function of the rate in dits per cycle. The probability of error for the three figures is, respectively, $P_e = 10^{-2}, 10^{-4}$, and 10^{-6}. It is seen that a large improvement is obtained by going from $N = 5$ to $N = 15$, especially in the feedback scheme. Increasing the coding delay further does not result in much improvement.

3) Figures 3(a), and 3(b) are plots of the additional signal-to-noise ratio in decibels above the ideal value indicated in Fig. 1 vs. the coding delay N, for different values of the probability of error P_e and for a rate of $R/W = 0.2$ dit per cycle. Figure 3(b) represents a plot for the bounds computed by Slepian. Note that the curves for the feedback scheme in Fig. 3(a) indicate a much lower relative (to the ideal, given in Fig. 1) signal-to-noise ratio, except for extremely small values of N.

4) Figures 4(a) and (b) are plots of the probability of error vs. the coding delay N, with the signal-to-noise ratio in decibels above the ideal as the parameter. The rate is $R/W = 0.2$ dit per cycle. Note the difference in shape between the two sets of curves.

5) Figure 5 is a plot of the relative rate R/C vs. the rate R/W in dits per cycle for different values of the coding delay. The probability of error is $P_e = 10^{-4}$.

6) Figures 6(a) and (b) are plots of the relative rate R/C vs. the coding delay N for different values of the signal-to-noise ratio.

D. Influence of Feedback Noise on the BL Coding Scheme

In this section, only the configuration in which $Y_n'(X_n'')$ (the received "number") is sent back will be investigated. The results for the case where X_n (the receiver's estimate) is sent back are similar to those in Part I in that the rate drops off to zero quickly.

Using the same notation as in Part I, it follows easily that

$$X'_{N+1,1} = X''_{N+1,1} + \alpha_1^{-1} Z''_{11} + \frac{\alpha^2 - 1}{\alpha^2} \sum_{i=2}^{N} \alpha_i^{-1} Z_{i1} \tag{13}$$

where $\sum_{i=2}^{1} = 0$, and α_i is given by (6). Hence,

$$X'_{N+1,1} \sim N\left\{\theta, \frac{1}{a^2}\left[\frac{\sigma'^2 + \sigma''^2}{\alpha^{2N}} + \frac{\sigma''^2}{\alpha^2} + \sigma''^2(\alpha^{2(N-1)} - 1)\right]\right\}.$$

The variance σ_t^2 of the estimate X'_{N+1} of θ, as computed by the receiver, is

$$\sigma_t^2 = \frac{1}{a^2}\left[\frac{\sigma'^2 + \sigma''^2}{\alpha^{2N}} + \frac{\sigma''^2}{\alpha^2} + \sigma''^2(\alpha^{2(N-1)} - 1)\right]. \tag{14}$$

For the probability of error one has, from (3),

$$P_e = 2 \operatorname{erfc}\left[\frac{\alpha^{N(1-\epsilon)}}{2\sigma_t}\right] \tag{15}$$

where again $R = (1 - \epsilon) \ln \alpha$, nats/dimension.

The expression for the signal-to-noise ratio in the forward direction is, from (9),

$$\frac{P_{av}}{N_0' W} = a^2 \frac{\alpha^2}{6NN_0'} + \frac{N-1}{N} \frac{N_0' + N_0''}{N_0'} (\alpha^2 - 1). \tag{16}$$

Figure 7 presents curves for the probability of error P_e vs. the coding delay N for $R/W = 0.2$ dit per cycle, and different values of the feedback noise relative to the forward noise, N_0''/N_0'. For a^2 the value $a^2 = 6N_0'$ as given by (11) is used. Hence, the curves present the degradation due to feedback noise of a system that is optimum for the noiseless feedback case.

III. Concluding Remarks

The WB (wideband) coding scheme, discussed in Part I, was suggested by the Robbins–Monro stochastic approximation procedure. In the Gaussian case it turns out that this coding procedure determines the maximum likelihood estimate of the message point θ recursively. Since the maximum likelihood estimate approaches θ

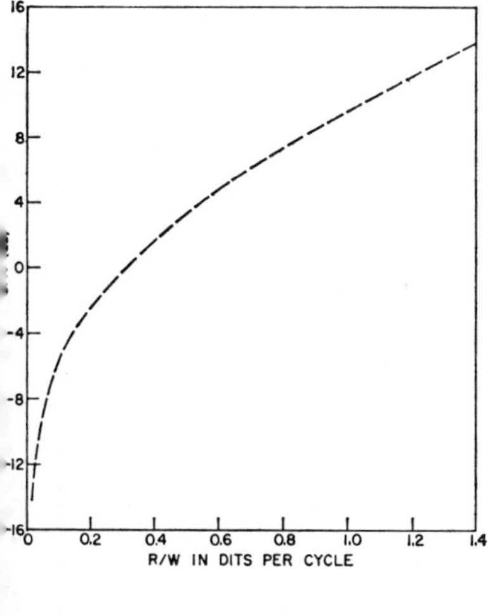

Fig. 1. The signal-to-noise ratio required by Shannon's capacity equation.

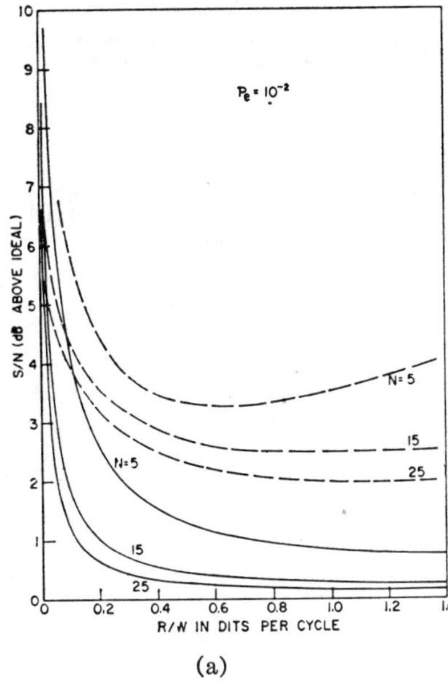

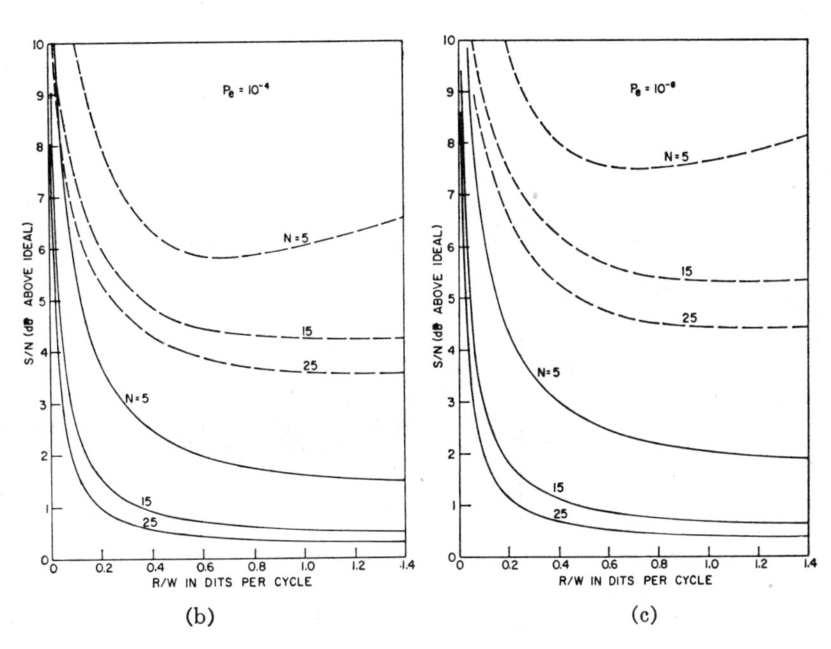

(a) (b) (c)

Fig. 2. The additional signal-to-noise ratio required when using a finite coding delay. (a) $P_e = 10^{-2}$. (b) $P_e = 10^{-4}$. (c) $P_e = 10^{-6}$.

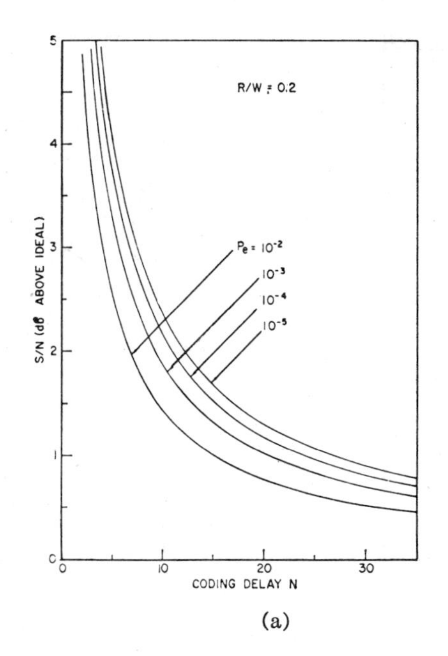

(a)

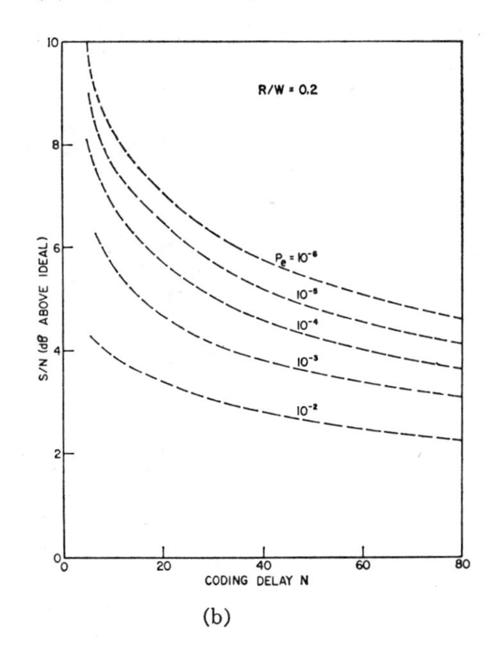

(b)

Fig. 3. The additional signal-to-noise ratio as a function of the coding delay for different values of the probability of error. (a) BL coding scheme. (b) Bounds on one-way communication.

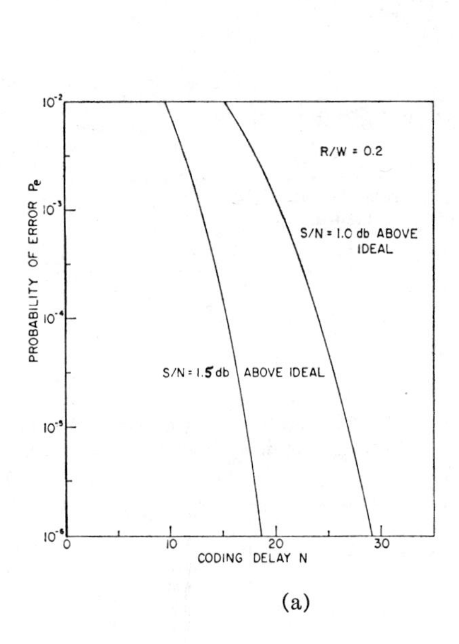

(a)

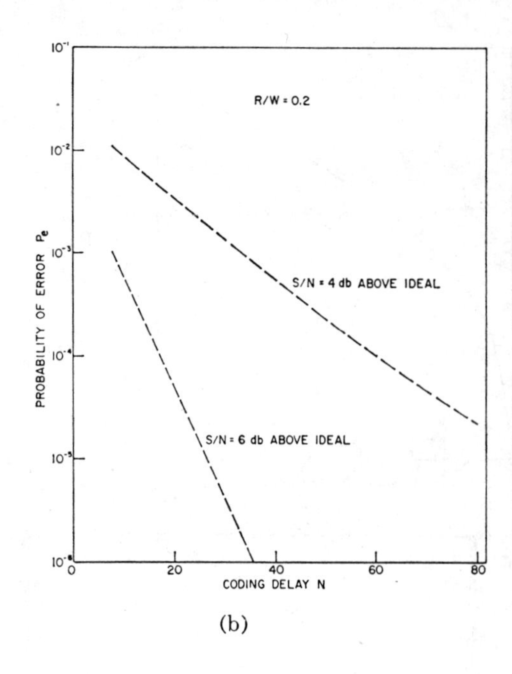

(b)

Fig. 4. The probability of error as a function of the coding delay for different values of the relative signal-to-noise ratio. (a) BL coding scheme. (b) Bounds on one-way communication.

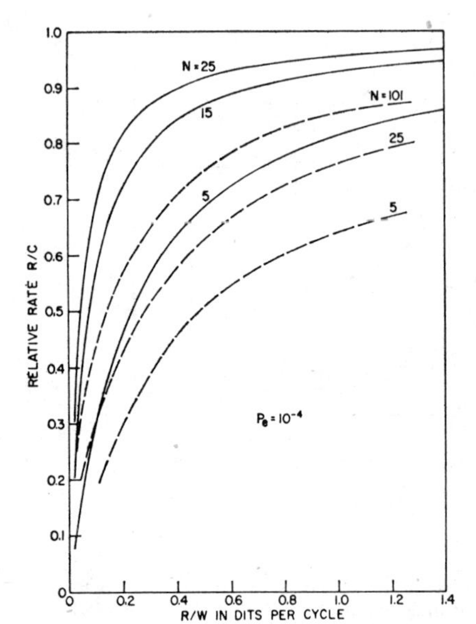

Fig. 5. The relative rate vs. the rate per unit bandwidth for different values of the coding delay.

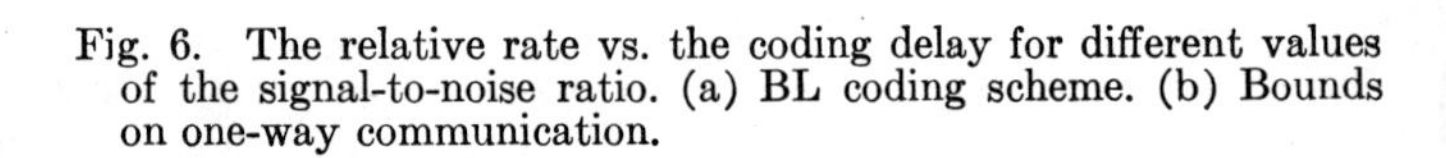
Fig. 6. The relative rate vs. the coding delay for different values of the signal-to-noise ratio. (a) BL coding scheme. (b) Bounds on one-way communication.

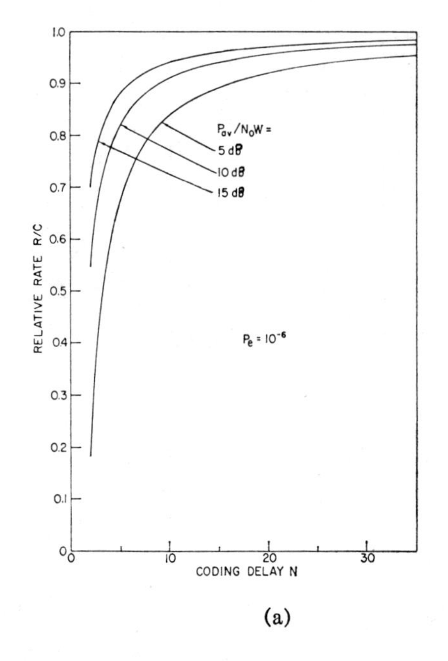

(a)

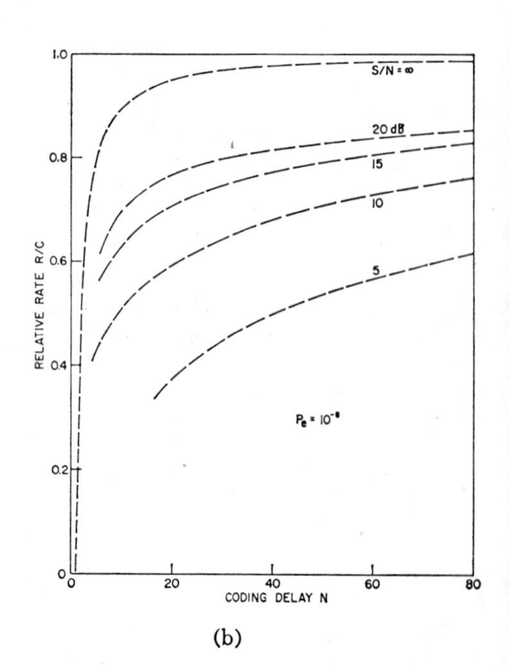

(b)

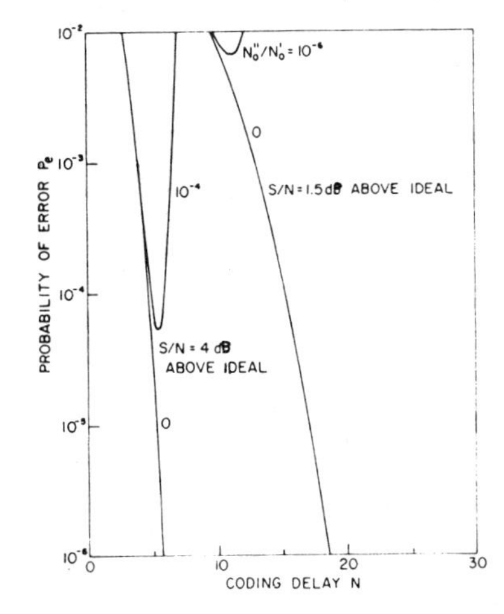

Fig. 7. The probability of error as a function of the coding delay for different values of the relative signal-to-noise ratio in the presence of feedback noise.

and the transmitted power is proportional to the square of the difference, the expected transmitted power per iteration decreases in this scheme. Retaining the maximum likelihood property but making up for the transmitted power in order to make the expected power per transmission a constant, leads to the BL (band-limited) coding schemes. This simple scheme is the first deterministic procedure to achieve the channel capacity, $W \ln [1 + (P_{av}/N_0W)]$, of the band-limited white Gaussian noise channel.

It is believed that this approach of recursive maximum likelihood estimation to the coding problem with feedback has a much wider area of application, for example, channels with unknown parameters, fading channels, dependences between the noises in forward and feedback links, and so on. The method is ideally suited for noiseless feedback and it may well be possible to find an extension that is in some sense optimum for the noisy feedback case.

References

[1] C. E. Shannon, "A mathematical theory of communication," *Bell Sys. Tech. J.*, vol. 27, pp. 379–424 and 623–657, July-October 1948.
[2] ——, "Probability of error for optimal codes in a Gaussian channel," *Bell Sys. Tech. J.*, vol. 38, pp. 611–656, May 1959.
[3] P. Elias, "Channel capacity without coding," in *Lectures on Communication System Theory*, Baghdady, Ed. New York: McGraw-Hill, 1961.
[4] D. Slepian, "Bounds on communication," *Bell Sys. Tech. J.*, vol. 42, pp. 681–707, May 1963.
[5] J. P. M. Schalkwijk and T. Kailath, "A coding scheme for additive noise channels with feedback—Part I: no bandwidth constraint," this issue, page 172.

Networks of Gaussian Channels with Applications to Feedback Systems

PETER ELIAS, FELLOW, IEEE

***Abstract*—This paper discusses networks (directed graphs) having one input node, one output node, and an arbitrary number of intermediate nodes, whose branches are noisy communications channels, in which the input to each channel appears at its output corrupted by additive Gaussian noise. Each branch is labeled by a non-negative real parameter which specified how noisy it is. A branch originating at a node has as input a linear combination of the outputs of the branches terminating at that node.**

The channel capacity of such a network is defined. Its value is bounded in terms of branch parameter values and procedures for computing values for general networks are described. Explicit solutions are given for the class D_0 which includes series-parallel and simple bridge networks and all other networks having r paths, b branches, and v nodes with $r = b - v + 2$, and for the class D_1 of networks which is inductively defined to include D_0 and all networks obtained by replacing a branch of a network in D_1 by a network in D_1.

The general results are applied to the particular networks which arise from the decomposition of a simple feedback system into successive forward and reverse (feedback) channels. When the feedback channels are noiseless, the capacities of the forward channels are shown to add. Some explicit expressions and some bounds are given for the case of noisy feedback channels.

Introduction

THE min-cut max-flow theorem[1]-[3] gives the capacity of a network made up of branches of given capacity. It applies to networks of noisy communications channels if the assumption is made that arbitrarily large delays and arbitrarily complex encoding and decoding operations may take place at each interior node.

This paper presents the theory of networks of another kind of channel—a channel with additive Gaussian noise, for which the only operation which takes place at a node is linear combination of the arriving signal and noise voltages, with no significant delay and no decoding or recoding.

The Problem

Consider the Class D of two-terminal networks like that shown in Fig. 1, in which there are no cycles, each of the b branches B_i is directed, and each branch lies on one of the r paths R_j which go from the input terminal on the left to the output terminal on the right. A signal voltage e_0 of mean-square value P_0 (the *signal power*) is applied to the input terminal, node V_1 at the left. At each interior node, the output (signal plus noise) of each branch B_i arriving at the node is given a (positive or negative) weight, the *branch transmission* t_i, and the resulting linear combination of signal and noise voltages is supplied as input to all the branches leaving that node.

Manuscript received December 14, 1966. This work, which was presented to the International Congress of Mathematicians, Moscow, August, 1966, was supported by Joint Services Electronics Program Contract DA36-039-AMC-03200(E), and National Aeronautics and Space Administration Grant NsG-334.

The author is with the Department of Electrical Engineering and the Research Lab. of Electronics, Massachusetts Institute of Technology, Cambridge, Mass. 02139

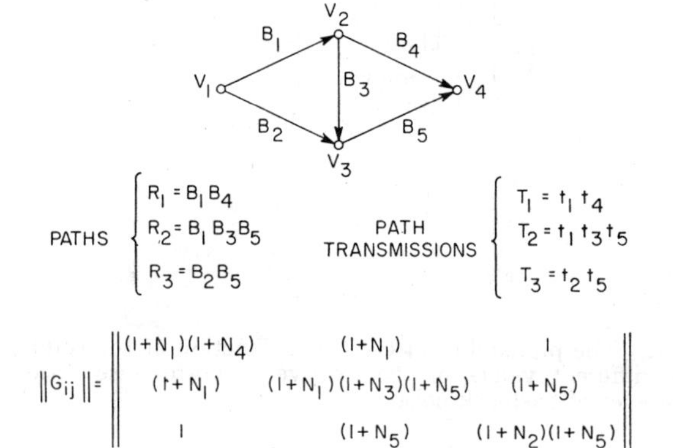

$$\|G_{ij}\| = \begin{Vmatrix} (1+N_1)(1+N_4) & (1+N_1) & 1 \\ (1+N_1) & (1+N_1)(1+N_3)(1+N_5) & (1+N_5) \\ 1 & (1+N_5) & (1+N_2)(1+N_5) \end{Vmatrix}$$

Fig. 1. A network in Class D.

Each branch B_i adds to its input voltage e_i a Gaussian noise voltage n_i whose mean-square value (the *noise power*) is a constant N_i (the *noise-to-signal power ratio,* also called the *parameter* of the branch) times the mean-square value P_i (the *input power*) of its input voltage. The noise voltage in each branch is statistically independent of the noise voltages in other branches and of the signal voltage:

$$\overline{e_i^2} = P_i, \quad 0 \leqq i \leqq b; \quad \overline{n_i^2} = N_iP_i, \quad 1 \leqq i \leqq b$$
$$\overline{n_in_j} = 0, \quad i \neq j; \quad \overline{n_ie_0} = 0. \tag{1}$$

Since the branch input voltage and its noise are uncorrelated, the mean-square value of the branch output voltage (the *output power*) is just

$$\overline{(e_i + n_i)^2} = \overline{e_i^2} + \overline{n_i^2} = P_i + N_iP_i = P_i(1 + N_i). \tag{2}$$

The power output of each branch generator depends on the power level at its input, and thus on the power level of the signal and of all other noise generators which affect its input power, as well as on the values of the branch transmissions. However, once the power levels of the signal and of all noises and the values of the branch transmissions are fixed, the network is linear. The final output at the right-hand output terminal V_v is a linear combination of the b branch noise generator voltages and the signal voltage e_0. We constrain the values of the branch transmissions t_i by requiring that the coefficient of e_0 in this sum be unity.

The network is equivalent to a single branch (noisy channel) of the same kind as the component branches,

Reprinted from *IEEE Trans. Inform. Theory*, vol. IT-13, pp. 493-501, July 1967.

since the linear combination of the b branch noise voltages which appears in the output is a Gaussian noise voltage, and the overall action of the two-terminal network is to receive an input signal and to produce at its output the input signal plus an independent Gaussian noise. The ratio of output noise power to signal power, N_{b+1}, is a function of the branch transmissions as well as the parameters N_i of the network branches. The *optimum noise-to-signal power* of the network, N_{opt}, is defined as the minimum value of N_{b+1} which can be obtained by varying the branch transmissions.

The problem is to find N_{opt} as a function of the given N_i.

Series and Parallel Networks

To express the results most simply in important special cases it is convenient to associate with each branch, not only the parameter N_i, but the *signal-to-noise ratio*,

$$S_i = 1/N_i, \tag{3}$$

and the *capacity* per use of the channel,

$$C_i = \tfrac{1}{2} \log (1 + S_i). \tag{4}$$

Equivalent quantities are defined for the network: S_{opt} is the maximum signal-to-noise ratio attainable by varying the branch transmissions, and C_{opt} is the largest channel capacity so attainable.

We can then state three results.

Series Networks

A network in D in which all b branches are in series has N_{opt} given by

$$1 + N_{\text{opt}} = \prod_{i=1}^{b} (1 + N_i). \tag{5}$$

Parallel Networks

A network in D in which all b branches are in parallel has S_{opt} given by

$$S_{\text{opt}} = \sum_{i=1}^{b} S_i. \tag{6}$$

Duality

Given two channels of capacities C_1 and C_2. Let the optimum capacity of the network consisting of the two channels in series be C_s. Let the optimum capacity of the two channels connected in parallel be C_p. Then

$$C_1 + C_2 = C_s + C_p. \tag{7}$$

The result on series networks expressed by (5) does not seem to have been published. The result for parallel branches expressed by (6) is known as optimum diversity combining, or the ratio squarer[4],[5] and was discovered independently of the general theory. Both follow directly from the general results following. The duality relationship of (7) follows directly from (4), (5), and (6), and also seems not to have been published. We have

$$\begin{aligned}
C_s &= \tfrac{1}{2} \log \left(1 + \frac{1}{N_s}\right) \\
&= \tfrac{1}{2} \log \left(1 + \frac{1}{(1 + N_1)(1 + N_2) - 1}\right) \\
&= \tfrac{1}{2} \log \left(\frac{(1 + N_1)(1 + N_2)}{N_1 + N_2 + N_1 N_2}\right) \\
&= \tfrac{1}{2} \log \left(\frac{(1 + S_1)(1 + S_2)}{1 + S_1 + S_2}\right) \\
&= \tfrac{1}{2} \log (1 + S_1)(1 + S_2) - \tfrac{1}{2} \log (1 + S_1 + S_2) \\
&= C_1 + C_2 - C_p.
\end{aligned}$$

Equation (7), incidentally, also holds for other pairs of channels, such as two binary symmetric channels with different crossover probabilities p_1 and p_2, or a binary symmetric channel in series with a binary erasure channel and in parallel with it. However, the interpretation of parallel channels is different in those cases; it involves having the receiver observe the outputs of both channels when a common input symbol is applied to both. Since the output symbols of the two channels cannot be combined into an input symbol for the same kind of channel without loss of information, there is no tidy network theory for such channels and we discuss them no further.

Feedback Networks

The next results apply to a subset F of networks in D which represent a dissection in space of the time sequence of forward and return signal flows encountered in a feedback system, as shown in Figs. 2 and 3. The transmitter applies a signal voltage to the input node V_1. It proceeds over a noisy branch B_1 to node V_2 at the receiver. The receiver sends it back over B_2 to V_3 at the transmitter. The transmitter forms a linear combination of the original signal and the noisy version of it received at V_3 and transmits it over B_3 to V_4. In Fig. 2 the receiver then takes a linear combination of the outputs at V_2 and V_4 as the output voltage. In Fig. 3 the process continues. In both figures, and for all nets in F, the branches on the left connecting odd-numbered nodes and the branches on the right connecting even-numbered nodes are noiseless. They serve only to provide linear combinations of previously received values for the next transmission and to provide the requisite delay. Odd-numbered branches, from odd to even nodes, are called *forward* channels; even numbered branches, from even to odd nodes, are *feedback* channels.

The Uniform Delay Property

In practice, delays will be introduced by the forward and feedback channels. In order to avoid having signal voltage samples applied at different times getting mixed up at intermediated nodes, we assume that the noiseless branches on the left and the right have delays selected so as to give the network the *uniform delay property* that all paths connecting any two nodes have the same delay. Therefore, at any node only one signal sample and one sample of the output of each earlier noise generator will

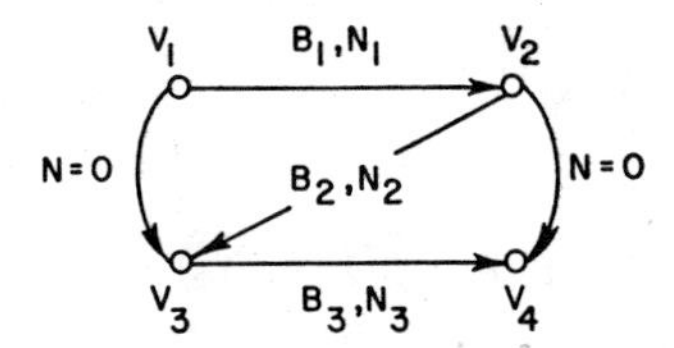

Fig. 2. A network in Class F for $k = 2$.

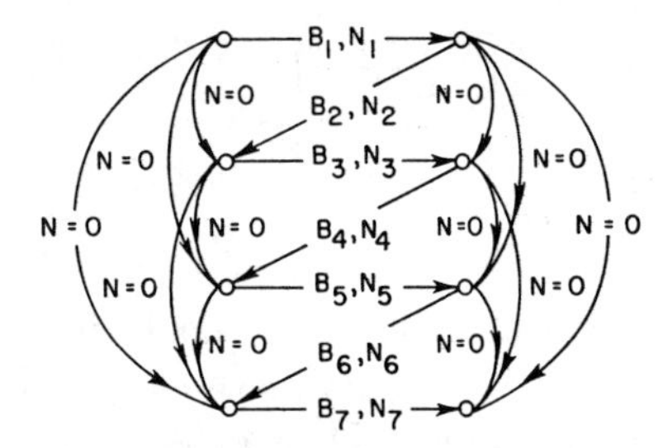

Fig. 3. A network in Class F for $k = 4$.

arrive at a given time over different paths. This can be accomplished for any network in F, or indeed in D, if an initial set of delay values d_i are given for the branches B_i, by increasing some of them in the following fashion. Assign a delay value to each node V_j equal to the maximum delay obtained by adding the delay values of the branches along each path from V_1 to V_j. Then assign to B_i the new delay value d_i' which is the difference between the delay values of its terminal and initial nodes, $d_i' \geqq d_i$.

We will henceforth assume that this process has been carried out for all networks in F or D, and that all have the uniform delay property. It is then not necessary to keep track of the delay values of networks or branches. We now state results for feedback networks.

Noiseless Feedback: For a network in F, if all feedback branches are noiseless, and the k forward branches have capacities C_{2j-1}, $1 \leqq j \leqq k$, then the optimum capacity of the network is given by

$$C_{\text{opt}} = \sum_{j=1}^{k} C_{2j-1} \tag{8}$$

and the optimum signal-to-noise ratio S_{opt} by

$$1 + S_{\text{opt}} = \prod_{j=1}^{k} (1 + S_{2j-1}). \tag{9}$$

In particular, if

$$S = \sum_{j=1}^{k} S_{2j-1}$$

is fixed, but an arbitrarily large k is available, we have

$$1 + S_{\text{opt}} = \lim_{k\to\infty} \prod_{j=1}^{k} \left(1 + \frac{S}{k}\right) = e^S,$$
$$S_{\text{opt}} = e^S - 1. \tag{10}$$

Noisy Feedback, $k = 2$: For a network in F with two noisy forward channels B_1 and B_3, and one noisy feedback channel B_2, the optimum signal-to-noise ratio is

$$S_{\text{opt}} = S_1 + S_3 + \frac{S_1 S_2 S_3}{(1 + S_1)(1 + S_3) + S_2}. \tag{11}$$

Unfortunately, a general formula like (11) for a net in F with $k > 2$ is not available, although the computation of S_{opt} for any particular case is a straightforward numerical analysis problem. However, we do have some inequalities which hold for all nets in F and which provide some insight.

Noisy Feedback, General Case: For a network in F with k noisy forward branches B_{2j-1}, $1 \leqq j \leqq k$, and $k - 1$ noisy feedback branches B_{2j}, $1 \leqq j \leqq k - 1$, the optimum signal to noise ratio S_{opt} is bounded by

$$S_{\text{opt}} \geqq \sum_{j=1}^{k} S_{2j-1} \tag{12}$$

$$1 + S_{\text{opt}} \leqq \prod_{j=1}^{k} (1 + S_{2j-1}) \tag{13}$$

$$S_{\text{opt}} \leqq \sum_{j=1}^{2k-1} S_j. \tag{14}$$

If signal-to-noise ratio costs c_1 per unit for forward channels and c_2 per unit for feedback channels, so that the total cost for a network in F is

$$c = c_1 \sum_{j=1}^{k} S_{2j-1} + c_2 \sum_{j=1}^{k-1} S_{2j},$$

then for sufficiently large S_{opt}, the cost per unit of S_{opt} may be made arbitrarily close to c_2:

$$\frac{c}{S_{\text{opt}}} \leqq c_2(1 + \delta). \tag{15}$$

The results for noiseless feedback and for noisy feedback with $k = 2$ were published by the author some time ago.[6],[7] Schalkwijk and Kailath have recently investigated the noiseless case from the point of view of error probability for the transmission of discrete messages.[8]-[10] Turin[14] has also dealt with a closely related question. The noiseless feedback results of (8) and (9) are remarkable, since they permit the transmission of a continuous signal of fixed bandwidth over a noisy channel at a rate equal to channel capacity, no matter how large the bandwidth of the forward channel. No coding or decoding is needed, provided that a noiseless feedback channel is available. Furthermore, they do so without introducing any of the discontinuities which must occur when a continuous signal is mapped onto a space of higher dimensionality—discontinuities which were pointed out by Shannon[11] and Kotelnikov,[12] and have recently been discussed by Wozencraft and Jacobs.[13] Equation (10) implies that a signal-to-noise ratio of 10 in bandwidth W is equivalent to a signal-to-noise ratio of $e^{10} - 1$, or about 22 000 if the available forward channel is wideband and has white noise, and a noiseless feedback channel is available (see the literature,[6],[7] for further discussion).

The inequality (12) follows from the parallel network result of (6). The result of setting all feedback channel transmissions at zero and using the forward channels in parallel gives the right side of (12). The optimum choice of branch transmissions must do at least as well. The second inequality, (13), says that noise in the feedback channels does not help; the right side is just the noiseless feedback result of (9). It is a consequence of a more general result which will be given, and which shows that increasing N_i in any branch cannot decrease N_{opt}. The third result, (14), says that, given a choice, it is better to use signal-to-noise ratio in the forward rather than the feedback channels. The total S_{opt} attainable by feedback is less than would be attained by taking all of the feedback channels, turning them around, and using them in parallel with the forward channels, which gives the right side of (14) by (6). This will also be derived later. The final result, (15), shows why feedback is interesting even if it does not do as well as the same amount of signal-to-noise ratio in the forward direction, by (14). Signal-to-noise ratio in the feedback direction may be cheaper, as when a satellite is communicating to Earth, and if it is, it is possible by means of feedback to buy forward signal-to-noise ratio at the same cost, if one wants enough of it. Equation (15) is a direct consequence of (11). It is necessary only to choose S_1 equal to S_3, and S_2 so large that it is possible to have $S_1 \ll S_2$ and $S_1^2 \gg S_2$ at the same time. For $k > 2$ the result will be of the same character, but better, i.e., a smaller δ will do. Or a smaller amount of S_{opt} can be bought at the same unit cost—but the absence of a formula makes the demonstration harder.

General Results

To state and prove the theorem from which the above results follow we need some further definitions. For each pair of paths R_i, R_j from V_1 to V_v in a network in D, we define G_{ij} as a product which contains one factor $(1 + N_k)$ for each branch B_k which lies in both paths; if R_i and R_j share no branches, $G_{ij} = 1$. Formally, if we treat the symbol R_i as denoting the set of branches which are contained in the ith path, then $R_i \cap R_j$ is the set of branches which the two paths have in common, and

$$G_{ij} = \prod_{k:B_k \in R_i \cap R_j} (1 + N_k) \qquad (16)$$
$$= 1 \quad \text{for} \quad R_i \cap R_j \text{ empty}.$$

We also define the *path transmission* T_i of path R_i as the product of the branch transmissions t_k for those branches which lie on R_i:

$$T_i = \prod_{k:B_k \in R_i} t_k. \qquad (17)$$

The *network transmission* $T_{0,b+1}$ is the sum of all path transmissions. By the assumption made in the discussion following (2), the branch transmissions t_k are constrained so that the network transmission, which is the coefficient of the signal voltage e_0 in the output, is unity.

$$T_{0,b+1} = \sum_{i=1}^{r} T_i = 1. \qquad (18)$$

Theorem

For any network in D, we have

$$1 + N_{\text{opt}} = \min_{t_k} \left\{ \sum_{i=1}^{r} \sum_{j=1}^{r} G_{ij} T_i T_j \right\} \geqq 1 \Big/ \left\{ \sum_{i=1}^{r} \sum_{j=1}^{r} G_{ij}^{-1} \right\} \qquad (19)$$

and

$$S_{\text{opt}} = 1/\min_{t_k} \left\{ \sum_{i=1}^{r} \sum_{j=1}^{r} (G_{ij} - 1) T_i T_j \right\}$$
$$\leqq \sum_{i=1}^{r} \sum_{j=1}^{r} [G_{ij} - 1]^{-1}, \qquad (20)$$

where the T_i are given in terms of the t_k by (17) and are subject to the constraint (18), and G_{ij}^{-1} and $[G_{ij} - 1]^{-1}$ are elements of the inverses of the matrices $||G_{ij}||$ and $||G_{ij} - 1||$. The inverses of $||G_{ij}||$ and $||G_{ij} - 1||$ always exist unless there is at least one noiseless path from input to output, so that some $G_{ii} = 1$. In this case $N_{\text{opt}} = 0$ and $S_{\text{opt}} = \infty$. These values are attained by setting $T_i = 1$ and all other $T_j = 0$, $j \neq i$.

Equality holds on the right in (19) and (20) for networks in the set D_0, which includes any network in D with r paths, b branches, and v nodes for which

$$r = b - v + 2, \qquad (21)$$

and for networks in the set D_1 which includes the networks in D_0 and, inductively, any network which is constructed from a network in D_1 by replacing any branch by another network in D_1.

Note that D_0 contains simple series networks, for which $r = 1$ and $b = v - 1$, and simple parallel networks, for which $r = b$ and $v = 2$. D_1 therefore contains all series-parallel networks, but it contains others as well—for example, the (topologically equivalent) networks of Figs. 1 and 2, for which $b = 5$, $v = 4$, and $r = 3$, but not the network of Fig. 3, for which $b = 11$, $v = 6$, and $r = 8$, or any network in F with $k > 2$.

Proof: For the proof we need one more definition. T_{ij}, the *transmission from branch i to branch j*, is just the network transmission as defined in (18) for the subnetwork consisting of branch i and all other branches which lie on some directed path which goes through branch i to the initial node of branch j. (Thus, B_i is included in the subnetwork, but B_j is not; and t_i is a common factor of all of the terms in the sum T_{ij}.) If there are no paths through B_i and B_j, or if B_j precedes B_i on such a path, then $T_{ij} = 0$. $T_{0,j}$ is the transmission of a subnetwork with input node V_1 and output node the initial node of B_j, and $T_{i,b+1}$ is the transmission of the subnetwork of paths through branch i to the output node V_v.

We now derive an expression for P_{b+1}, the output power of the network. By the statistical independence of the noise voltage generators from one another and from the signal source, the output power at the right-hand node is the sum of the powers transmitted to that node by these

$b+1$ separate sources. The source in branch i contributes an amount of power equal to its generated power P_iN_i times the square of the transmission from B_i to the output. Thus,

$$P_{b+1} = \sum_{i=0}^{b} P_iN_iT_{i,b+1}^2 = P_0(1 + N_{b+1}), \qquad (22)$$

where the right-most equality follows from the fact that by the constraint of (18), (2) holds for $i = b+1$, and where the signal power contributed to the output is represented in the sum by the term for $i = 0$, with $N_0 = 1$ and $T_{0,b+1} = 1$.

Similarly the input power to any branch B_i may be expressed as the sum of the contributions of the generators which lie to its left:

$$P_i = \sum_{k=0}^{i-1} P_kN_kT_{ki}^2. \qquad (23)$$

Here we have assumed that the branches are numbered in an order such that if B_i precedes B_j on some directed path, $i < j$.

By successive substitution of (23) into (22) and in the resulting expressions, the subscripts on the P's appearing on the right can all be reduced to zero. The result is a sum of terms, all of which have P_0 as a factor. There is one term for each of the 2^b subsets W_m of the b branches which has the property that all of the branches in W_m are included in a path from input to output, i.e., that there is an integer f with $R_f \supseteq W_m$. If W_m is such a subset, say $W_m = (B_i, B_j, B_k)$ with $i < j < k$, then the corresponding term is

$$P_0T_{0i}^2N_iT_{ij}^2N_jT_{jk}^2N_kT_{k,b+1}^2 = P_0(T_{0i}T_{ij}T_{jk}T_{k,b+1})^2N_iN_jN_k. \qquad (24)$$

The product of the transmission terms which appears on the right is just the sum of the transmissions of all paths from input to output which include all three of the branches B_i, B_j, B_k. If there are no such paths, then one or more of the T_{ij} in (24) will vanish. Thus the output power is expressed in terms of the path transmissions T_i and the branch parameters N_i. Dividing through by P_0 gives an expression for $1 + N_{b+1}$

$$1 + N_{b+1} = \sum_{k=0}^{2^b-1} \{ \sum_{i:R_i \supseteq W_k} T_i\}^2 \prod_{j:B_j \varepsilon W_k} N_j, \qquad (25)$$

where W_0 is the null set, for which the product is taken to be 1. The sum is also 1 for $k = 0$, since it is just the square of the network transmission of (18). Thus excluding the term for $k = 0$ gives an expression for N_{b+1} as a sum of products of positive terms, which is monotone nondecreasing in each N_j. We thus have proved Lemma 1.

Lemma 1

For any given set of path transmissions T_i, the network noise-to-signal ratio N_{b+1} is a monotone nondecreasing function of each branch noise-to-signal ratio N_j.

This lemma provides the proof of (13), which was referred to previously.

We have also proved that N_{b+1} can vanish for a nonvanishing set of path transmissions only if there is some path R_i along which every branch is noiseless, so that setting $T_i = 1$ and $T_j = 0$, $j \neq i$ gives a right-hand side in (25) in which only the term for W_0 remains. The matrix $||G_{ij} - 1||$ will be singular if, and only if, there is such a noiseless path since it will then map the transmission vector $\underline{T}$ with $T_i = 1$ and $T_j = 0$, $j \neq i$ into the null vector. The matrix $||G_{ij}||$ can be singular only under the same circumstances, but may not be even when noiseless paths exist.

We next show the equivalence of the right side of (25) to the quadratic form.

$$\sum_{i=1}^{r}\sum_{j=1}^{r} G_{ij}T_iT_j, \qquad (26)$$

where the T_i are still subject to the constraint (18). Substituting into (26) the definition (16) of G_{ij} gives

$$\sum_{i=1}^{r}\sum_{j=1}^{r} T_iT_j\{ \prod_{m:B_m \varepsilon R_i \cap R_j} (1 + N_m)\}. \qquad (27)$$

Expanding the product gives

$$\sum_{i=1}^{r}\sum_{j=1}^{r} T_iT_j \sum_{k:W_k \subseteq R_i \cap R_j} \{ \prod_{m:B_m \varepsilon W_k} N_m\}. \qquad (28)$$

Inverting the order of summation to sum over all W_k,

$$\sum_{k=0}^{2^b-1} \{ \sum_{i,j:W_k \subseteq R_i \cap R_j} T_iT_j\}\{ \prod_{m:B_m \varepsilon W_k} N_m\}. \qquad (29)$$

We then recognize that the parentheses enclose a term which is just the square of the sum of T_i over the i for which W_k is included in R_i:

$$\sum_{k=0}^{2^b-1} \{ \sum_{i:W_k \subseteq R_i} T_i\}^2 \prod_{m:B_m \varepsilon W_k} N_m, \qquad (30)$$

which is just the right side of (25).

We have thus proved that for T_i constrained by (18),

$$1 + N_{b+1} = \sum_{i=1}^{r}\sum_{j=1}^{r} G_{ij}T_iT_j. \qquad (31)$$

Squaring (18) gives

$$1 = 1^2 = \left\{\sum_{i=1}^{r} T_i\right\}^2 = \sum_{i=1}^{r}\sum_{j=1}^{r} T_iT_j, \qquad (32)$$

and subtracting (32) from (31) gives

$$N_{b+1} = \sum_{i=1}^{r}\sum_{j=1}^{r} [G_{ij} - 1]T_iT_j, \qquad (33)$$

or

$$S_{b+1} = 1 \Big/ \sum_{i=1}^{r}\sum_{j=1}^{r} (G_{ij} - 1)T_iT_j. \qquad (34)$$

Now N_{opt}, by definition, is the minimum value of N_{b+1} as the branch transmissions are varied, and S_{opt}

is its reciprocal. We have therefore proved the first part of the theorem: namely, the equalities on the left in (19) and (20).

To obtain the inequalities on the right in (19) and (20), we minimize (31) and (33) by varying the path transmissions T_i independently, subject only to the constraint imposed by (18). The additional constraints imposed by the topology of the network and by (17), which expresses the T_i in terms of the real independent variables t_k, are ignored. The results are lower bounds to the minima which (31) and (33) can actually attain in the network.

Using a Lagrange multiplier $2M$, we set the derivative of

$$\sum_{i=1}^{r} \sum_{j=1}^{r} G_{ij}T_iT_j - 2M \sum_{i=1}^{r} T_i \tag{35}$$

with respect to T_i equal to zero. This gives

$$\sum_{j=1}^{r} G_{ij}T_j = M, \qquad 1 \leqq j \leqq r. \tag{36}$$

Using the minimizing T_j which satisfy (36), we multiply by T_i and sum, using the constraint of (18) and attaining a lower bound to $1 + N_{\text{opt}}$:

$$\sum_{i=1}^{r} \sum_{j=1}^{r} G_{ij}T_iT_j = M \sum_{i=1}^{r} T_i = M \leqq 1 + N_{\text{opt}}. \tag{37}$$

Solving (36) for the minimizing T_i gives

$$T_j = M \sum_{i=1}^{r} G_{ij}^{-1}. \tag{38}$$

Summing on j and using (18),

$$1 = \sum_{j=1}^{r} T_j = M \sum_{i=1}^{r} \sum_{j=1}^{r} G_{ij}^{-1}, \tag{39}$$

or from (37),

$$1 + N_{\text{opt}} \geqq M = 1 \Big/ \left\{ \sum_{i=1}^{r} \sum_{j=1}^{r} G_{ij}^{-1} \right\}. \tag{40}$$

This completes the proof of (19) in the theorem. The derivation of (20) is strictly parallel and will be omitted. It remains only to prove the assertions made for networks in D_0 and in D_1. To prove that equality holds on the right in (19) and (20) for networks in D_0, it is necessary to show that for such networks it is possible to vary path transmissions independently by varying branch transmissions. In fact we prove a stronger result.

Lemma 2

A network in D which has $r = b - v + 2$ has a cutset of r branches each of which is included in just one path. Removal of this cutset divides the network into two parts: a tree connected to V_1 (which may reduce to V_1 alone), and a tree connected to V_v (which may reduce to V_v alone).

Given Lemma 2, we can set the r transmissions of the branches in the cutset as the r desired path transmissions and set the transmissions of all other branches equal to unity.

To prove Lemma 2, assign weights to nodes and branches from the left, assigning weight 1 to node V_1 and then assigning to each branch the weight of its initial node and to each node the sum of the weights of its incoming branches. With this assignment the weight of a node or a branch is clearly the number of routes from the input node V_1 to that node or branch.

Choose from each of the r paths the right-most branch of weight 1. This set of branches, c in number, is a cutset, since it interrupts each path. We have $c \leqq r$: $c = r$ if, and only if, no branch is selected more than once.

Deleting the cutset of c branches divides the network into two parts, M_1 connected to V_1 and M_2 connected to V_v. M_1, which contains b_1 branches and v_1 nodes, is a tree, since it is connected and since all of its nodes are of weight 1, so that there is only one path from V_1 to each node. Thus $b_1 = v_1 - 1$, as for any tree.

M_2 is connected to V_v and thus includes at least one tree. Let one of the trees included in M_2 have b_2 branches and v_2 nodes, with $b_2 = v_2 - 1$. Then there are two possible situations. i) M_2 is a tree. In that case $r = b - v + 2$. Or ii) M_2 is larger than a tree, and includes b_3 branches beyond the b_2 branches in a tree which it includes. In that case $r > b - v + 2$. We will prove the labelled statements.

i) If M_2 is a tree, then $b = c + b_1' + b_2 = c + (v_1 - 1) + (v_2 - 1) = c + v - 2$, or $c = b - v + 2$. Since each branch in the cutset connects two trees, it completes just one path, so the number of paths $r = c$, and $r = b - v + 2$.

Q.E.D.$_i$

ii) If M_2 contains b_3 branches beyond those contained in a tree, then $b = c + b_1 + b_2 + b_3 = c + (v_1 - 1) + (v_2 - 1) + b_3 = c + v - 2 + b_3$, or

$$c = b - v + 2 - b_3. \tag{41}$$

Now each branch among the b_3 has weight $\geqq 2$ by construction, so it lies on at least two paths. Without these b_3 branches, V_v has weight at least c, since the c branches in the cutset have weight 1 each and are connected to V_v. Adding each of the b_3 additional branches adds a weight $\geqq 2$ to V_v, since each of them is connected to V_v through the tree included in M_2. Thus the total weight r of V_v is $r \geqq c + 2b_3$. Combining this with (41) gives

$$r \geqq c + 2b_3 = b - v + 2 + b_3 > b - v + 2. \tag{42}$$

Q.E.D.$_{ii}$

For a network M which is in D but not in D_0, $r > b - v + 2$; and it is impossible to independently vary the path transmissions. For $b - v + 2$ is the cyclomatic number of the graph M' obtained from M by adding a branch B_{b+1} directed from V_v to V_1, and is thus the maximum number of linearly independent cycles in a graph-theoretic sense. Thus the set of r cycles in M', each of which consists of a path R_i from V_1 to V_v followed by the branch B_{b+1} from V_v to V_1 are linearly dependent in the graph-theory sense. Therefore, so is the set of the paths themselves in M.

The linear dependence of the R_i implies, by taking logarithms in (17), one or more linear relations between the logarithms of the path transmissions $\log T_i$, leading to constraints of the form

$$\log T_i + \log T_j = \log T_m + \log T_n, \quad \text{or} \quad T_iT_j = T_mT_n \tag{43}$$

and no selection of values for the branch transmissions t_k can provide independent control of all path transmissions.

It may still be possible to achieve equality in (19) and (20) for a network in D which is not in D_0, however, if the optimizing values of the path transmissions happen to satisfy the additional constraints of the form (43) imposed by the network topology. This happens in particular for the networks which are in D_1 but not in D_0.

Lemma 3

Given a network M in D, and a network M' in D_0. Let M'' be constructed by replacing branch B_j in M by the network M'. Then the value of the parameter N''_{opt} of M'' will be the same as the value of the parameter N_{opt} of M if the latter is evaluated using the parameter value N'_{opt} of M' for B_j. The path transmissions obtained in computing N''_{opt} will lead to the same set of transmissions for the subnetwork M' as are obtained directly in the computation of N'_{opt}.

The network M'' is equivalent to the network M with some value of the parameter N_j for branch B_j by the argument following (2), i.e., the subnetwork M' is equivalent to *some* noisy branch B_j, and the only question is what its parameter value is. The optimum set of path transmissions for M'' must lead to the same transmissions inside M' as does the direct optimization of M'. Any other choice would give a larger value to the parameter of M'' by Lemma 1.

Lemma 3 completes the proof of the theorem. Lemma 2 covers networks in D_0 and Lemma 3 justifies the extension of the results to networks in D_1. More practically, it permits the solution of network problems of large order by local reductions—the combining of series or parallel branches, etc.—which greatly reduces the computation. Unfortunately the other tool used for the local reduction of resistive networks—the star-mesh transformation—cannot be used for Gaussian channels, since it leads to transformed branches which have correlated generators. This takes us outside of our present model. Networks with correlated noise present problems which are discussed briefly in a later section.

Proof of Earlier Results

The result of (5) follows from the theorem by noting that for a series network $r = 1$, and $||G_{ij}|| = ||G_{ii}||$. Thus,

$$G_{ii} = \prod_{i=1}^{b} (1 + N_i) = 1/G_{ii}^{-1}. \tag{44}$$

Equation (6) follows by noting that for a parallel network, $r = b$ and $||G_{ij} - 1||$ is diagonal with elements $G_{ii} = N_i$, so that

$$[G_{11} - 1]^{-1} = S_i, \qquad \sum_{i=1}^{b}\sum_{j=1}^{b} [G_{ij} - 1]^{-1} = \sum_{i=1}^{b} S_i. \tag{45}$$

Equation (11) follows from the evaluation of (20) for the network of Fig. 1. Equation (9) follows by letting S_2 approach infinity in (11), for $k = 2$. For larger k, the first three branches are combined into an equivalent forward branch of capacity $C_1 + C_2$ and it is combined with the next noisy forward branch and the next noiseless feedback branch in the same way, etc.

Equations (12) and (13) have already been justified. Equation (14) follows by throwing away all but the linear terms, i.e., terms having a single N_i as a factor, in (33). By (25) this reduces the right side and provides a lower bound to N_{opt} or an upper bound to S_{opt}. The resulting equations are those for a set of resistors—the noisy branches—with resistance $= N_i$, all in parallel—both the forward and the feedback branches—with the noiseless branches acting as short circuits at the two ends and the conductances $S_i = 1/N_i$ adding.

Reduction of Another Problem to the Above

A more general problem concerning networks of Gaussian channels can be reduced to the previous results. Consider the class of two-terminal networks as in D (mentioned previously), but in which each node may supply a different linear combination of the voltages on its incoming branches to each outgoing branch. This model still leaves the operation at the node simple and linear, and provides an increased number of independently controllable path transmissions. Thus it enlarges the class of networks for which explicit solution is possible and for which equality holds in (19) and (20).

As an example, the network shown in Fig. 4 consisting of five vertices connected by four branches forming a directed path from V_1 to V_2 to V_3 to V_4 to V_5, with three additional branches from V_1 to V_3, V_3 to V_5, and V_2 to V_4 has $b = 7$, $v = 5$, and $r = 5$, and is thus not in D_0: it has no two-terminal subnetworks, and is thus not in D_1.

The reduction to the former case replaces each node V_j which has $I_j > 1$ incoming branches and $O_j > 1$ outgoing branches by I_j nodes at each of which one of the incoming branches arrives and O_j nodes from each of which one of the outgoing branches leaves, together with I_jO_j noiseless branches connecting each of the I_j arrival nodes to each of the O_j departure nodes. The added noiseless branches permit the formation of the desired different linear combinations of input branch voltages for each output branch. In the case of the five-node network already described, replacing V_3 by 4 nodes and 4 branches, as shown in Fig. 5, adds 3 nodes, 4 branches and no paths. Thus $b - v + 2 = 7 + 4 - (5 + 3) + 2 = 5 = r$, and the resulting net is in D_0.

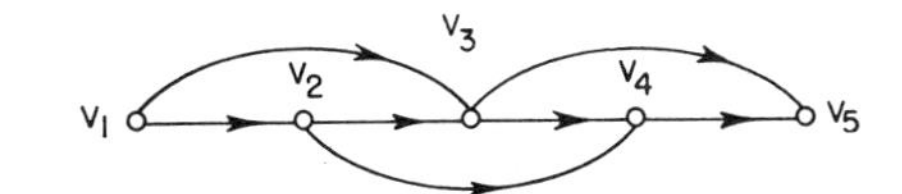

Fig. 4. A network not in Class D_1.

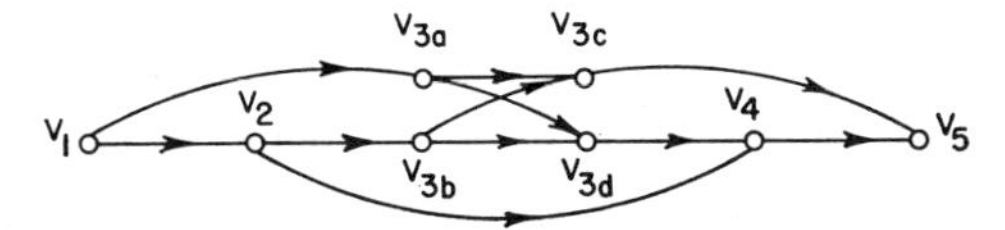

Fig. 5. A reduction of the network of Fig. 4 to a network in Class D_1.

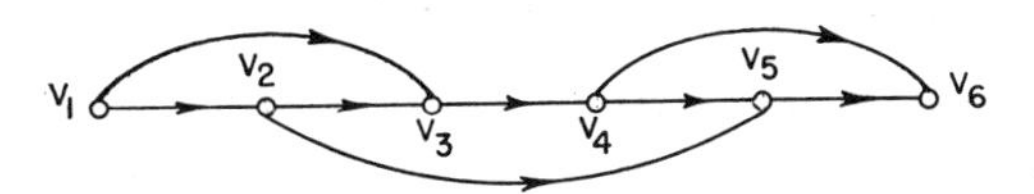

Fig. 6. A network not in Class D_1 which cannot be reduced.

The simplest network which cannot be expanded by the above substitution, has no two-terminal subnetworks, and is not in D_0, is shown in Fig. 6. It consists of six nodes V_1 to V_6 connected in order by five branches, with three additional branches from V_1 to V_3, V_4 to V_6, and V_2 to V_5.

Unfortunately the additional control provided by the change in rules provides no help for networks in F, which remain outside D_0 for $k > 2$.

Networks with Correlated Noise

One can consider networks in which $\overline{n_i n_j} \neq 0$, although the noise and the signal remain uncorrelated. For parallel branches, if we take $\overline{n_i n_j} = G_{ij}$ and T_i as the branch transmission subject to the constraint of (18), then minimizing the mean-square value of the sum

$$\overline{[\sum_i T_i(e_0 + n_i)]^2} \tag{46}$$

leads to precisely the result of the theorem, with equality in (19) and (20), by precisely (35) to (40). In fact, the proof of the theorem may be taken as a proof that the voltages transmitted to the output node V_v by the different paths R_i have the average product matrix $||G_{ij}||$.

For series branches the situation is different, however. Correlated series branches do not commute unless their parameter values are equal. Even the validity of the branch model breaks down. The definition in (1) of the added branch noise power $\overline{n_i^2} = P_i N_i$ is valid as a model of a physical channel so long as the channel is always used at the maximum possible input power. This is always advantageous when branch noises are uncorrelated, so the restriction is not felt in the optimization problem of the theorem. However a more realistic model of a physical channel has an input power limit P_i, and adds a noise of power $N_i P_i$ to any input signal whose power is $\leqq P_i$. With correlated noise in series branches it will sometimes be advantageous to use less than the maximum input power to a branch. No analog to Lemma 1 holds.

As an example consider two identical channels in series. Each accepts inputs of power $\leqq 2$ and adds to them the *same* noise voltage n, of power $\overline{n^2} = 1$. If we apply only 1 unit of signal power to the first channel, invert its output using a branch transmission of -1, and apply the result to the second channel, the output of the second channel has no noise voltage, and therefore an infinite signal-noise ratio. If, however, we apply 2 units of signal power to the first channel, and scale its output voltage by $-\sqrt{2/3}$ to provide 2 units of input power to the second channel, we cannot get a signal-noise power ratio at the output of the second channel which is better than $4/(5 - 2\sqrt{6}) \cong 40$.

Further Bounds on Networks in F

The open questions of greatest interest for applications concern networks in F with $k > 2$ and with noisy feedback. In a feedback system it is reasonable to assume that the transmitter has a limited amount of signal-to-noise ratio S_{odd} available, and that the receiver has a limited amount S_{even}, given by

$$S_{\text{odd}} = \sum_{j=1}^{k} S_{2j-1}$$
$$S_{\text{even}} = \sum_{j=1}^{k-1} S_{2j}, \tag{47}$$

and that they are free to allocate their limited resources between the different forward and feedback channels in the way which maximizes the resulting S_{opt} of F. This freedom may even extend to deciding how large k should be, if the available forward and feedback channels have infinite bandwidth.

In the case of noiseless feedback $k = \infty$ is best and gives the result of (10). When the feedback is noisy, evaluating what S_{opt} the best division of limited power gives and how S_{opt} depends on k involves a great deal of numerical solutions of linear equations subject to constraints of the form of (43). Even evaluating the upper bound to S_{opt} of (20) is not easy. Lower bounds to S_{opt} which are more meaningful than that of (12) can be computed, however, by making use of iteration of networks for which $k = 2$, as shown in Fig. 7.

For the first level network, we assume that the two forward branches have equal signal-to-noise ratio, since this maximizes S_{opt} in (11), for fixed S_{odd}. Denoting their common signal-to-noise ratio as S_1, the feedback branch as S_2, and the resulting S_{opt} as S_3, we have from (11)

$$S_3 = 2S_1 + \frac{S_1^2 S_2}{(1 + S_1)^2 + S_2}. \tag{48}$$

We now consider the second-level network to consist of two forward branches of ratio S_3 and a feedback branch of ratio S_4. The resulting S_{opt} is denoted by S_5, and we have for the kth level

$$S_{2k+1} = 2S_{2k-1} + \frac{S_{2k-1}^2 S_{2k}}{(1 + S_{2k-1})^2 + S_{2k}}$$
$$S_{\text{odd}} = 2^k S_1 \tag{49}$$
$$S_{\text{even}} = S_{2k} + 2S_{2(k-1)} + \cdots + 2^{k-1} S_2.$$

For this network the optimum allocation of S_{odd} among the 2^k forward branches has already been made, and each

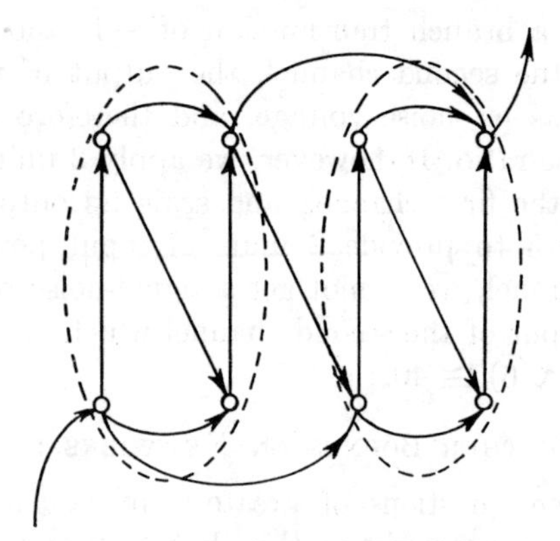

Fig. 7. An iteration of networks in Class F for which $k = 2$.

receives an equal amount. S_{even} is divided unequally, however, with more for higher-numbered branches, in the optimum case. The optimum allocation can be determined by solving (49) for S_{2k}:

$$S_{2k} = \frac{S_{2k+1} - 2S_{2k-1}}{1 - \dfrac{1 + S_{2k+1}}{(1 + S_{2k-1})^2}}. \tag{50}$$

Now differentiating $S_{2k} + 2S_{2k-2}$ with respect to S_{2k-1} for fixed S_{2k+1} and S_{2k-3} and setting the result equal to zero gives

$$\frac{S_{2k-3}^2(1 + S_{2k-3})^2}{[(1 + S_{2k-3})^2 - (1 + S_{2k-1})]^2} = (1 + S_{2k-1}) \frac{(1 + S_{2k-1})^3 - (1 + S_{2k+1})(2 - 3S_{2k-1}) + (1 + S_{2k+1})^2}{[(1 + S_{2k-1})^2 - (1 + S_{2k+1})]^2}. \tag{51}$$

For given S_{2k-3} and S_{2k-1}, this equation is quadratic in S_{2k+1}, and solving it enables us to start with a desired S_1 and S_3 and to generate S_{2k+1} for any k. Alternatively, we may fix S_{2k+1} and S_{2k-1} and solve for S_{2k-3}. Taking the positive square root of each side of (51) gives a quadratic in S_{2k-3} and we can proceed from given values of S_{2k+1} and S_{2k-1} down to S_1. In either case the resulting set of values is optimum in the sense that by keeping the end points fixed, and fixing k, any other division of S_{odd} will take more of it. Choosing all combinations of values for, e.g., S_1 and $S_3 > 2S_1$, generates the full set of optimum curves.

The result, unfortunately, must be displayed as a set of curves rather than an equation. A much weaker lower bound to S_{opt} can be given as an equation. Although it is not the best strategy, we may pick a division of S_{even} which gives us a fixed c such that

$$1 + S_j = c(1 + S_{j-2})^2, \quad \text{odd } j. \tag{52}$$

Then from (50),

$$S_{2k} = \frac{S_{2k+1} - 2S_{2k-1}}{1 - c} \tag{53}$$

and from (49) and (52),

$$S_{\text{even}} = \frac{S_{2k+1} - 2^k S_1}{1 - c} \leqq \frac{S_{\text{opt}} - S_{\text{odd}}}{1 - c}, \tag{54}$$

since $S_{2k+1} \leqq S_{\text{opt}}$. We also have from repeated application of (52)

$$c(1 + S_{\text{opt}}) \geqq c(1 + S_{2k+1}) = c^{2^k}(1 + 2^{-k}S_{\text{odd}})^{2^k}. \tag{55}$$

Together, (54) and (55) provide a useful analytic lower bound to S_{opt}.

References

[1] G. B. Dantzig and D. R. Fulkerson, "On the max-flow min-cut theorem of networks," in *Linear Equalities*. Princeton, N. J.: Ann. Math Studies 38, 1956.

[2] L. R. Ford and D. R. Fulkerson, "Maximal flow through a network," *Canad. J. Math.*, vol. 8, pp. 399–404, 1956.

[3] P. Elias, A. Feinstein, and C. E. Shannon, "A note on the maximal flow through a network," *IRE Trans. Information Theory*, vol. IT-2, pp. 117–119, December 1956.

[4] L. R. Kahn, "Ratio squarer," *Proc. IRE* (*Correspondence*), vol. 42, p. 1704, November 1954.

[5] D. G. Brennan, "On the maximum signal-to-noise ratio realizable from several noisy signals," *Proc. IRE* (*Correspondence*), vol. 43, p. 1530, October 1955.

[6] P. Elias, "Channel capacity without coding," M.I.T. Research Lab. of Elect., Quarterly Progress Rept., pp. 90–93, October 15, 1956.

[7] ——, "Channel capacity without coding," in *Lectures on Communication System Theory*, E. J. Baghdady, Ed. New York: McGraw-Hill, 1961, pp. 363–368.

[8] J. P. M. Schalkwijk and T. Kailath, "A coding scheme for additive noise channels with feedback, Part I: No barrier constraint," *IEEE Trans. Information Theory*, vol. IT-12, pp. 172–182, April 1966.

[9] J. P. M. Schalkwijk, "A coding scheme for additive noise channels with feedback, Part II: Band-limited signals," *IEEE Trans. Information Theory*, vol. IT-12, pp. 183–189, April 1966.

[10] ——, "Center of gravity information feedback," Appl. Research Lab., Sylvania, Waltham, Mass., Research Rept. 501.

[11] B. M. Oliver, J. R. Pierce, and C. E. Shannon, "The philosophy of PCM," *Proc. IRE*, vol. 36, pp. 1324–1331, November 1948.

[12] V. A. Kotelnikov, *The Theory of Optimum Noise Immunity*. New York: McGraw-Hill, 1959.

[13] J. M. Wozencraft and I. M. Jacobs, *Principles of Communication Engineering*. New York: Wiley, 1965.

[14] G. L. Turin, "Signal design for sequential systems with feedback," *IEEE Trans. Information Theory*, vol. IT-11, pp. 401–408, July 1965.

E. R. BERLEKAMP

Block Coding for the Binary Symmetric Channel with Noiseless, Delayless Feedback

ABSTRACT

This paper considers the problem of transmitting messages reliably across a noisy binary symmetric channel which is accompanied by a noiseless, delayless feedback channel. Upper bounds are derived for the error correction capability of arbitrary block coding strategies, and explicit strategies are constructed which achieve these bounds at many rates. At certain rates, these strategies are asymptotically close-packed.

1. Background and Summary of Results

Led by Schalkwijk and Kailath (1966), communication theorists have recently displayed increasing interest in the problem of transmitting information across a noisy channel which is accompanied by a noiseless, delayless feedback channel of large capacity. Although Shannon (1956) showed that the capacity of the channel with feedback cannot exceed the capacity of the same one-way channel, the feedback channel nevertheless proves useful at all rates below capacity. Phenomenal reductions in the probability of error can be obtained when feedback is used with continuous channels which have an average power constraint at the transmitter, and substantial reductions can be obtained when feedback is used with discrete channels or continuous channels with peak-power constraints. Some of the feedback coding schemes which have been suggested for discrete channels [Horstein (1963)] employ a variable length stopping rule, according to which the transmitter continues sending digits until the receiver is "sufficiently sure" which message is intended. Although such schemes attain a low probability of error, they may occasionally require a relatively large number of digits to be transmitted before the receiver is able to reach a decision. For this reason, Shannon and Gallager suggested that it might be interesting to obtain error bounds for the best feedback coding strategy in which the number of digits to be transmitted is fixed in advance, i.e., the best block coding strategy. Several such bounds were obtained by Berlekamp (1964). The present paper extends some of those results which apply to the binary symmetric channel.

The present paper is devoted to the problem of determining n, the block length of the shortest feedback coding strategy which always enables the receiver to decide correctly among $M = 2^{Rn}$ possible transmitted messages, assuming that the channel alters no more than $e = fn$ of the transmitted bits. For very large n, the relationship between the rate R and the error-correction fraction f is shown in Figures 1 and 2.

The asymptotic relationship between f and R for the best one-way binary codes lies somewhere within the shaded region of Figure 1. The best known asymptotic lower bound on f as a function of R is obtained by a nonconstructive argument due to Gilbert (1952); the best upper bound is obtained by an argument due to Elias. Figure 1 also includes a weaker upper bound due to Hamming (1950). Proofs and generalizations of all of these bounds for one-way codes may be found in Chapter 13 of Berlekamp (1968).

The asymptotic relationship between f and R for the best binary feedback block coding strategies is shown in Figure 2. These upper bounds on $f(R)$ are derived in Part 3 of this paper; these lower bounds, in Part 4. At rates in the interval $.2965 < R < 1$, the best asymptotic upper bound on f is the volume bound, which is a generalization of Hamming's bound for one-way codes. For lower rates, $0 < R < .2965$, the best upper bound on $f(R)$ is a straight line which emanates from the point $R = 0$, $f = 1/3$, and joins the volume bound tangentially at $R = .2965$. One of the constructions presented in Part 4 of this paper yields a range of coding strategies which attain the rate and error correction fraction of any point on this straight line. Other constructions yield ranges of coding strategies which attain the rates and error-correction fractions of any point on the straight lines emanating from the point $R = 0$, $f = 1/t$, where t is any integer. These constructions yield asymptotically "close-packed" codes at any of the rates at which one of these straight lines touches the volume bound. Many compelling heuristic arguments suggest that it is actually possible to obtain asymptotically close-packed feedback coding strategies at all rates $\geq .2965$. We give a construction which apparently yields codes along the straight line emanating from $R = 0$, $f = 2/7$ and joining the volume bound tangentially.

A comparison of Figures 1 and 2 reveals that the error-correction capability of the best binary feedback block coding strategies is asymptotically superior to the error-correction capability of the best one-way binary block codes, at all rates $0 \leq R < 1$. It is also worth noting that the lower bounds of Figure 2 are obtained by evaluating the

Figure 1. Asymptotic error-correction fraction of one-way BSC

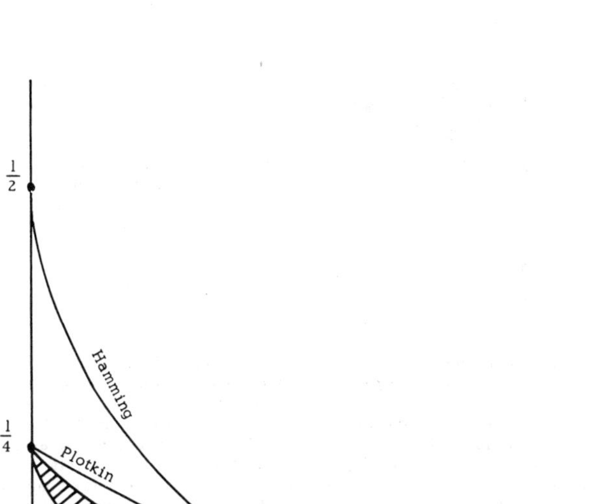

Figure 2. Asymptotic error-correction fraction of BSC with feedback

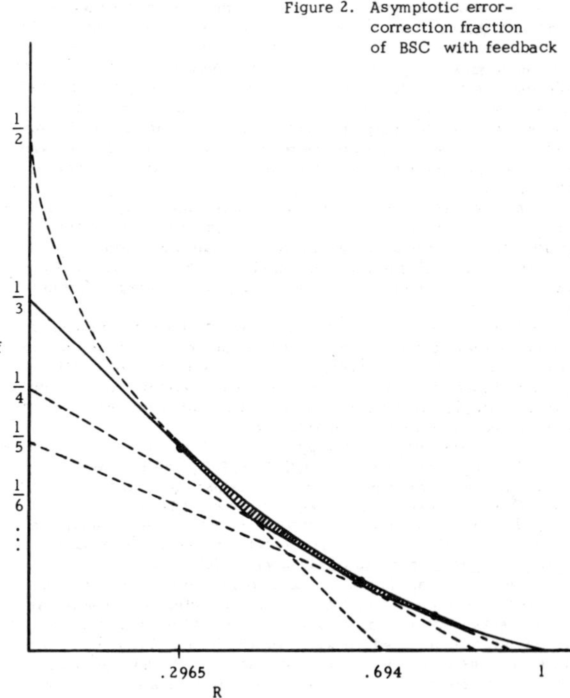

Reprinted with permission from *Error Correcting Codes*, Wiley, pp. 61-85, 1968. Copyright © 1968 by John Wiley & Sons, Inc.

performance of certain explicitly stated strategies, whereas no known "explicit" constructions for one-way codes attain an asymptotically positive f at any positive R .

Before presenting the asymptotic upper and lower bounds on f(R) in Parts 3 and 4 respectively, we introduce some preliminary notions in Part 2.

2. Coder vs Nature; States and Partitions

We now consider the problem of coding for the binary symmetric channel (Figure 3) with noiseless, delayless feedback.

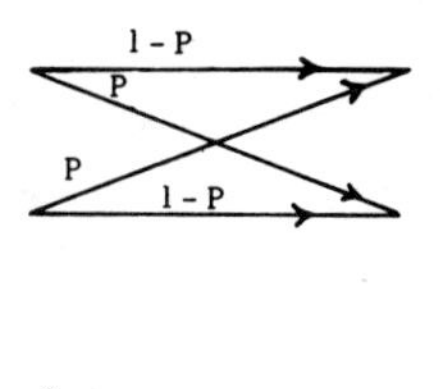

Figure 3. Binary symmetric channel with noiseless delayless feedback

We will often refer to the coding process as a game between two hostile opponents: Coder, a partnership including the transmitter and the receiver, and Nature, who controls the channel transitions.

The game of transmitting one block of information on this channel is played as follows: The source first selects one message from an ensemble of M equiprobable words. He attempts to convey this choice to the receiver by transmitting n bits across the noisy channel. Some of these transmissions may depend on information the source received from the feedback channel as well as on the selected message word.

We adopt the point of view that just prior to each forward transmission the receiver asks the source a yes-no question: "Is the correct message among the set S_i ?" (S_i is a subset of the M possible messages.) The question is transmitted back to the source over the noiseless feedback link, and the source's answer is then sent to the receiver via the noisy channel. The source receives a noisy answer, and then asks another question. At each stage the questioned set, S_i, may depend on the entire past history of the game.

It may first appear that this viewpoint necessitates an unusually large amount of feedback, since at each stage the receiver transmits back a subset S_i, which may be any subset of M possible message words. This transmission seems to require $\log_2 M$ bits of noiseless feedback. Actually, however, only one bit of feedback is required for each bit transmitted, because the transmitter may be endowed with the same deterministic subset-selecting machine as the receiver. The only inputs into this subset-selector are the results of previous questions, i.e., the received sequence of bits. Thus the evolution of the questioning process is determined only by the received sequence of answers. If the feedback channel can accommodate one noiseless bit for each noisy bit sent down the forward channel, the source can be kept informed of the received sequence. He then knows as much as the receiver, and additional feedback cannot be of any additional help.

Any strategy may be viewed as a quiet-question, noisy-answer process of the type just described. One need only consider the set of possible selected messages which would cause the source to transmit a one next, and call this the questioned set. The question-answer viewpoint involves no restriction on the types of strategies permissible.

The receiver may regard each answer he receives as a vote against a certain subset of words. As the process proceeds, different words acquire different numbers of unfavorable votes. After all n transmitted bits are received, the receiver must decide which word was transmitted. He obviously does best to select that word which has received the fewest unfavorable votes.

As an example, let us suppose that M = 8, n = 11. We denote the 8 possible messages by A, B, C, D, E, F, G, and H . We start with all 8 codewords having no votes against them, and 11 questions remaining. The game might proceed as shown in Figure 4.

For example, when there were 5 questions remaining, the receiver asked the question: "Is the selected message among the set FEGH?" The reply that was received was "No".

In this game, Nature caused at least three channel errors. If it caused only three errors, than A was the selected message, and Nature's errors occurred at questions 10, 9, and 4. It is also possible that D was the selected message, in which case Nature caused four errors, at questions 7, 6, 3, and 1; or that F is the message, in which case the four errors occurred at questions 11, 5, 2, and 1. If any of the other codewords was the message, then Nature committed six or more errors.

We now turn our attention to the problem of selecting a questioning strategy which guarantees that after all n questions have been asked, the receiver will be able to deduce correctly which word was transmitted unless the channel had made more than e errors. As

Votes Against	Questions Remaining: 11	10	9	8	7	6	5	4	3	2	1	0
0	ABCDEFGH	ADEH	DE	D	D							
1		BCFG	ACFH	CEF	F	DF	F					
2			BG	AGH	ACE	A	AD	ADF	DF	F		
3				B	BGH	BCEG	EG	-	A	AD	ADF	A
4						H	BC	BCEG	EG	-	-	DF
5							H	-	BC	BCEG	EG	-
6								H	-		BC	EGBC
7									-		-	-
8									H		-	-
9											H	H

Figure 4. A Sample Game

before, after each question we tally the number of negative votes against each possible message word. However, we may now throw away words which accumulate more than e unfavorable votes.

After each question, we record the number of words which have 0 negative votes, the number which have 1 negative vote, ... the number which have e negative votes. We write these numbers as components of a column vector, and call this vector the state of the game. If there are n questions remaining, this vector is called an n-state. The topmost components of this vector are often zeros. For this reason, we index the components from the bottom up and omit any zeros above the highest nonzero component:

$$\begin{matrix} \cdots \\ c_4 \\ c_3 \\ c_2 \\ c_1 \\ c_0 \end{matrix} = \underline{c}$$

The component c_i denotes the number of words which have received $e-i$ negative votes.

With e = 3, the states which occurred in the game of Figure 4 are shown in Figure 5.

Unused Votes Against	Questions Remaining: 11	10	9	8	7	6	5	4	3	2	1	0
3	8	4	2	1	1	0	0	0	0	0	0	0
2		4	4	3	1	2	1	0	0	0	0	0
1			2	3	3	1	2	3	2	1	0	0
0				1	3	4	2	0	1	2	3	1

Figure 5. States of the game of Figure 4

At each question, the receiver partitions the present state of the game into two substates, and asks the source which substate contains the message. The (noisy) answer constitutes a vote against one substate or the other. The next state of the game is then a new list of numbers of words having received various numbers of negative votes. The general situation is depicted below:

n-state	partition	resulting (n-1)-state if answer favors left	resulting (n-1)-state if answer favors right
c_4	$a_4\ b_4$	a_4	b_4
c_3	$a_3\ b_3$	$a_3 + b_4$	$a_4 + b_3$
c_2	$a_2\ b_2$	$a_2 + b_3$	$a_3 + b_2$
c_1	$a_1\ b_1$	$a_1 + b_2$	$a_2 + b_1$
c_0	$a_0\ b_0$	$a_0 + b_1$	$a_1 + b_0$
	$a_i + b_i = c_i$	$c'_i = a_i + b_{i+1}$	$c'_i = a_{i+1} + b_i$

It is frequently more convenient to discuss only the current state and the pair of states which may result from it, without being too concerned with the details of the partition which brings this about. This nonchalantness is justified by the following theorem:

Partitioning Theorem 2.1: There exists a partition which reduces the state $\underline{x}$ into the states $\underline{y}$ and $\underline{z}$ iff

(1) $\underline{x} \geq 0;\ \underline{y} \geq 0;\ \underline{z} \geq 0$ (all components are nonnegative)

(2) $x_{i+1} + x_i = y_i + z_i$ for all i, $0 \leq i \leq e$

(3) For all I, $0 \leq I \leq e$

$$\sum_{i=I}^{e} y_{2i+2} \leq \sum_{i=I}^{e} z_{2i+1} \leq \sum_{i=I}^{e} y_{2i}$$

and

$$\sum_{i=I}^{e} z_{2i+2} \leq \sum_{i+I}^{e} y_{2i+1} \leq \sum_{i=I}^{e} z_{2i}$$

For given $\underline{x}$, $\underline{y}$, and $\underline{z}$, this partition is unique.

Proof: Without (1), the vectors $\underline{x}$, $\underline{y}$, and $\underline{z}$ are not really states and partitioning is meaningless. Among nonnegative vectors, a partition exists if there are two substates $\underline{u}$ and $\underline{v}$ such that

$$\underline{x} = \underline{u} + \underline{v}$$
$$y_i = u_i + v_{i+1};\ z_i = v_i + u_{i+1} \quad \text{for all } i\ .$$

We will show that given $\underline{y}$ and $\underline{z}$, both $\underline{x}$ and the unique partition can be determined subject to conditions (2) and (3). Solving for $\underline{x}$ is most readily accomplished by computing the highest components first and working down.

$$x_e = u_e + v_e = y_e + z_e$$

In general, $x_i + x_{i+1} = u_i + u_{i+1} + v_i + v_{i+1} = y_i + z_i$, and thus x can be computed from the topmost component working down, using the equation $x_i = y_i + z_i - x_{i+1}$.

We may also solve for $\underline{u}$ and $\underline{v}$ in terms of $\underline{y}$ and $\underline{z}$.

$$\sum_{i=I}^{e} y_{2i+1} = \sum_{i=I}^{e} (u_{2i+1} + v_{2i+2});\ \sum_{i=I}^{e} z_{2i} = \sum_{i=I}^{e} (v_{2i} + u_{2i+1})$$

$$v_{2I} = \sum_{i=I}^{e} (v_{2i} - v_{2i+2}) = \sum_{i=I}^{e} (z_{2i} - y_{2i+1})\ .$$

Similar expressions are found for the odd components of v, and for the odd and even components of $\underline{u}$. Condition (3) is the statement that these components be nonnegative. Q.E.D.

For some n-states, it is possible to devise a partitioning strategy for the remaining n questions which ensures that all words but one will eventually receive more than e negative votes; for other n-states no such strategy exists. We call the former winning n-states; the latter, losing n-states. A 0-state is winning if only one word has e or less negative votes. These considerations justify the following definitions:

Definitions 2.2: A 0-state $\underline{x}$ is winning if $\Sigma x_i \leq 1$. A nonzero winning 0-state is called a singlet. An n-state is winning if it can be reduced to two winning (n-1)-states. (The two winning (n-1)-states need not be distinct.)

Any vector which is a winning j-state but a losing (j-1)-state is said to be a borderline winning j-state. Several lemmas follow at once:

Lemma 2.3: Any vector which is a winning n-state is also a winning j-state, for any $j > n$. Singlet states are winning n-states for all n.

Lemma 2.4: The only borderline winning 1-state is

0
...
0
2

Omitting top zeros, this state is written as 2.

If $\Sigma x_i = 2$, $\underline{x}$ is called a doublet. The winning partition of any doublet (if it exists) plays the two words against each other. This consideration leads to the following result:

Lemma 2.5: A doublet $\underline{c}$ is a winning n-state iff $\Sigma i c_i \leq n-1$, with equality in the borderline case.

Lemma 2.6: If $\underline{c} \leq \underline{d}$ (meaning $c_i \leq d_i$ for all i) and $\underline{d}$ is a winning n-state, then $\underline{c}$ is also a winning n-state.

The conclusion of Lemma 2.6 is also valid under slightly weaker hypotheses:

Lemma 2.7: If $\sum_{i=k}^{e} c_i \leq \sum_{i=k}^{e} d_i$ for all k, and $\underline{d}$ is a winning n-state, then $\underline{c}$ is also a winning n-state.

Lemma 2.8: M is a winning n-state if $M \leq 2^n$. The best partition of the state M is one which plays half the words against the other half.

A table of some of the winning n-states for $1 \leq n \leq 9$ is given in Figure 6.

3. Upper Bounds on Error-Correction Capability

An examination of Figure 6 leads us to some more general results. Foremost among these is a volume bound, which is a generalization of Hamming's (1950) bound for one-way codes. The primary difference is that our bound surrounds words at different levels with different sizes of spheres.

The appropriate definition of the volume of an n-state $\underline{x}$ is obtained by surrounding all words at height j by a sphere of radius j:

$$V_n(x) = \sum_{i=0}^{e} x_i \sum_{j=0}^{i} \binom{n}{j}$$

Theorem 3.1 (Conservation of Volume): Let $\underline{x}$ be any particular n-state, and let $\underline{y}$ and $\underline{z}$ be the (n-1)-states which result from it following any given partition. Then $V_n(\underline{x}) = V_{n-1}(\underline{y}) + V_{n-1}(\underline{z})$.

Proof: Let $\underline{x} = \underline{u} + \underline{v}$ be the partition which reduces $\underline{x}$ to $\underline{y}$ and $\underline{z}$. Then the theorem becomes

$$\sum_{i=0}^{e} (u_i + v_i) \sum_{j=0}^{i} \binom{n}{j} = \sum_{i=0}^{e} (u_i + v_i + u_{i+1} + v_{i+1}) \sum_{j=0}^{i} \binom{n-1}{j}\ .$$

SOME WINNING n-STATES, $1 \leq n \leq 9$

i	n:	9	8	7	6	5	4	3	2	1
0		512	256	128	64	32	16	8	4	2
1		50	28	16	8	4	2	2	1	
0		12	4	0	8	8	6	0	1	
2		1	1	1	1	1	1	1		
1		0	0	0	0	0	0	0		
0		456	219	99	42	16	5	1		
2		1	1	1	1	1	1			
1		43	22	11	4	1	1			
0		36	21	11	14	10	0			
.		.	2	2	2	2				
.		.	0	0	0	0				
.		.	182	70	20	0				
		.	2	2						
		.	20	6						
		.	2	22						
		7	3							
		0	0							
		190	145							
		7	3							
		15	14							
		40	19							
		8	4							
		0	0							
		144	108							
		8	4							
		8	8							
		64	36							
		1	1	1	1	1	1			
		4	1	1	0	0	0			
		14	10	0	1	1	0			
		58	36	35	15	0	1			
		.	.	1	1	1				
		.	.	0	0	0				
				5	0	0				
				24	22	6				

Since x_1 is arbitrary, an equivalent theorem is

$$\sum_{j=0}^{i}\binom{n}{j} = \sum_{j=0}^{i}\binom{n-1}{j} + \sum_{j=0}^{i-1}\binom{n-1}{j}$$

This is true because

$$\binom{n}{j} = \binom{n-1}{j} + \binom{n-1}{j-1} \quad \text{Q.E.D.}$$

One immediate application of this result is Theorem 3.2.

Volume Bound Theorem 3.2: If $\underline{x}$ is a winning n-state, then

$$V_n(\underline{x}) \leq 2^n$$

Proof: The theorem is true for $n = 0$, for in fact a singlet state satisfies any volume bound:

$$\sum_{k=0}^{\infty}\binom{j}{k} = \sum_{k=0}^{j}\binom{j}{k} = 2^j$$

For arbitrary n, the theorem follows directly from the conservation of volume theorem by induction. Q.E.D.

In some special cases this bound is the only requirement. Lemma 2.8 showed one such case. Doublet states are another, as is shown by the following restatement of Lemma 2.5.

Lemma 3.3: A doublet state $\underline{c}$ is a winning n-state iff $V_n(\underline{c}) \leq 2^n$. Equality occurs in the borderline case.

Proof: Let the only nonzero components of $\underline{c}$ be $c_i = 1 = c_j$ (where possibly $i = j$). Then the borderline case of Lemma 2.5 becomes $i + j = n - 1$. In this case

$$V_n(\underline{c}) = \sum_{k=0}^{i}\binom{n}{k} + \sum_{k=0}^{j}\binom{n}{k} = \sum_{k=0}^{i}\binom{n}{k} + \sum_{k=n-j}^{n}\binom{n}{n-k} = \sum_{k=0}^{n}\binom{n}{k} = 2^n$$

Q.E.D.

Lemma 3.3 shows that for doublets, the volume bound is the only restriction that must be satisfied. In general, however, there are losing states which still satisfy the volume bound. For example, consider the 4-state $\frac{3}{0}$. Its volume is $3(\binom{4}{0} + \binom{4}{1}) = 15 < 16 = 2^4$, but this is nevertheless a losing state. A generalized version of this limitation is the following.

Translation Bound Theorem 3.4: If $\Sigma x_i \geq 3$, then $\underline{x}$ cannot be a winning n-state unless $T\underline{x}$ is a winning (n-3)-state. $T\underline{x}$ is the translation of $\underline{x}$, defined by $(T\underline{x})_i = x_{i+1}$.

Proof: The basic idea is again an induction on n. We first verify by exhaustion that the theorem is true for small n, as shown in Figure 6. Now suppose that the theorem is true for $n \leq k - 1$, and that $\underline{x}$ is a winning k-state. There must then exist some partition of $\underline{x}$ which reduces it to $\underline{y}$ and $\underline{z}$, which are winning (k-1)-states. When this same partition is applied to $T\underline{x}$, it reduces $T\underline{x}$ to $T\underline{y}$ and $T\underline{z}$. If $\Sigma y_i \geq 3$ and $\Sigma z_i \geq 3$, then the induction hypothesis guarantees that $T\underline{y}$ and $T\underline{z}$ are both winning (k-r)-states. Thus, $T\underline{x}$ is a winning (k-3)-state.

In the exceptional case that $\Sigma y_i \leq 2$, we must resort to special considerations. The translation bound, as stated, does not apply to such states. In fact, from Lemma 2.5, if the doublet $\underline{y}$ is a borderline winning n-state, then $T\underline{y}$ is a borderline winning (n-2)-state. If $\underline{y}$ is not a borderline (k-1)-state, then $T\underline{y}$ is a winning (k-3)-state. Thus only the borderline doublet must be considered.

Define $\underline{x}'$ by $x_0' = 0$; $x_k' = x_k$ for all $k > 0$. Then $\underline{x}' \leq \underline{x}$ and $V_n(\underline{x}') \leq V_n(\underline{x})$. Since $\underline{y}$ is a reduction from $\underline{x}$,

$$\sum_{i=0}^{e} y_i \geq \sum_{i=1}^{e} x_i = \sum_{i=0}^{e} x_i'$$

Thus $\underline{x}'$ is a singlet or a doublet. If it is a singlet, so is $T\underline{x} = T\underline{x}'$, and the proof is completed. If $\underline{x}'$ is a doublet, there are two possibilities. Either $\underline{x}' = \underline{y}$, or $\underline{x}'$ is a borderline winning k-state (because the doublet $\underline{x}$ reduces to $\underline{y}$ and $\underline{y}$ is a borderline winning (k-1)-state).

If $\underline{x}' = \underline{y}$ then $T\underline{x} = T\underline{y}$ and $T\underline{x}$ is a winning (k-3)-state because it is a translate of the winning doublet (k-1)-state.

If instead, $\underline{x}'$ is a borderline winning k-state, then it satisfies the volume bound with equality: $V_k(\underline{x}') = 2^k$. But x is also a winning k-state. $\underline{x} = \underline{x}'$ is the only possibility. In this case $T\underline{x}$ is not a winning (k-3)-state, but $\underline{x}$ is a doublet, so this case lies outside the hypotheses of the theorem.

Having verified the theorem in all the exceptional cases, we have completed the proof. Q.E.D.

Corollary 3.5: If $x_e \geq 3$, then $\underline{x}$ cannot be a winning n-state unless $n \geq 3e + 2$.

Together, the volume bound and the translation bound eliminate most of the losing states. They are not exhaustive, however, for there are a few losing states which satisfy both bounds. For example, $\frac{5\,1}{2}$ is a losing 9-state, even though it satisfies the volume bound and $\frac{5}{}1$ is a winning 6-state. In spite of such isolated cases, however, we shall show in Part 4 that in the asymptotic cases of greatest interest, these two bounds are all inclusive.

The volume and translation bounds apply to all possible n-states. The special n-states of greatest interest are ones in which the initial state $\underline{I}$ contains only one nonzero component: $I_e = M = 2^k$. We wish to find the minimum n for which $\underline{I}$ is a winning n-state.

The information rate is then defined by $R = k/n$; the allowable error fraction is denoted by $f = e/n$. For small values of n, the possible values of R and F can be derived from our table. We note, for example, the occurrence of $\frac{16}{0}$ among our list of winning 7-states. (There is actually a one-way partitioning strategy which accomplishes this win, called the Hamming (7,4) single-error-correcting code.) For $n > 9$, however, the detailed extension of the table involves considerable labor and we must rely instead on asymptotic bounds which we shall now derive. We are interested in the cases when n grows very large while f and R remain fixed.

3.6. The Asymptotic Volume Bound and Its Tangents: Let $f = e/N$. We first consider the volume bound:

$$2^k \sum_{j=0}^{e}\binom{n}{j} \leq 2^n$$

Asymptotically this becomes $H(f) \leq 1 - R$, or $R \leq 1 - H(f)$, where $H(f) = -f \ln f - (1-f)\ln(1-f)$. This is identical to Hamming's bound for one-way codes, as plotted in Figures 1 and 2.

In Part 4 of this paper, the tangents to this curve play an important role. We digress here briefly to derive the equations for these tangent lines.

Consider a straight line which intercepts the f axis at f_0, the R axis at R_0, and which is tangent to the curve $R = 1-H(f)$ at the point $R = R_t$, $f = f_t$. Define $g = 1 - f$. Then the equations giving the point of tangency are

$$R(f_t) = 1 - H(f_t) = R_0(1 - f_t/f_0) \tag{3.7}$$

$$R'(f_t) = \log(f_t/g_t) = -R_0/f_0 \tag{3.8}$$

and

$$f_t + g_t = 1 .$$

These quantities are depicted in Figure 7. Subtracting f_t times Equation 3.8 from Equation 3.7 gives

$$R_0 = 1 + \log g_t$$

Figure 7

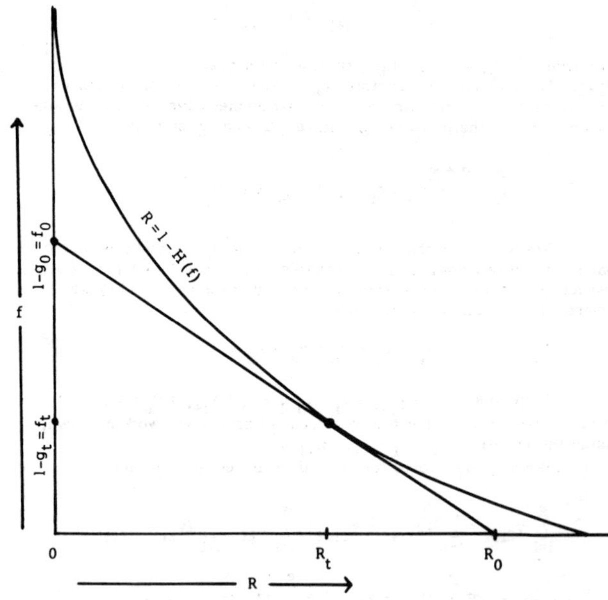

Substituting this expression into Equation 3.8 and exponentiating gives

$$2 f_t^{f_0} g_t^{g_0} = 1 .$$

Introducing the quantity $s = g_t/f_t$, we then have $\log s = R_0/f_0$; $f_t = 1/(1+s)$; $R_t = (f_0 - 1/(1+s))\log s$. These substitutions transform the problem into a single equation in only one unknown:

$$1 = 2 f_t^{f_0} g_t^{g_0} = 2s^{g_0}/(1+s)$$

or

$$2s^{g_0} = 1 + s \tag{3.9}$$

In the special case in which g_0 is a rational number, this equation is algebraic. $s = 1$ is always an extraneous root of this equation; it may be removed by dividing through by $s - 1$. The computation of the coordinates of the tangency point then reduces to the solution of this final algebraic equation.

Example 3.10: $g_0 = 2/3$. In this case the equation is

$$2s^{2/3} = 1 + s$$

$$8s^2 = s^3 + 3s^2 + 3s + 1$$

$$0 = s^3 - 5s^2 + 3s + 1 = (s-1)(s^2 - 4s - 1)$$

$$s = 2 + 5^{\frac{1}{2}} = ((1 + 5^{\frac{1}{2}})/2)^3$$

$$f_t = (3 + 5^{\frac{1}{2}}) = .19095$$

$$R_t = (1/3 - 1/(3 + 5^{\frac{1}{2}})) \log(2 + 5^{\frac{1}{2}}) = .29650$$

$$R_0 = \log((1 + 5^{\frac{1}{2}})/2) = .69425$$

3.11. The Tangential Bound: The best possible asymptotic upper bound on error-correction capability is obtained by the following argument:

We first apply the translation theorem: If the initial state $\underline{I}$ (which has 2^k words at height e) is a winning n-state, then $T^m \underline{I}$ must be a winning $(n - 3m)$-state, for any $0 < m < n/3$. Applying the volume bound to this state gives

$$2^k \binom{n-3m}{e-m} \leq 2^{n-3m}$$

Define $x = n - 3e$; $y = e - m$. The bound becomes

$$8^{-y} \binom{x+3y}{y} \leq 2^{x-k}$$

The validity of this bound is restricted only by the requirement that $0 < y < e$, and it behooves us to choose the best y to obtain the strongest bound. This is accomplished by maximizing the left side of the above inequality. This can be done most readily by setting equal to one the ratio of the value of this expression for y to its value for $y + 1$. For large y, this gives

$$1 = \frac{(x+3y)^3}{8y(x+2y)^2}$$

$$0 = (x+y)(x^2 - 5y^2)$$

$$y = 5^{-\frac{1}{2}} x .$$

Plugging this value into the bound and taking logarithms gives

$$x(1 + 3 \cdot 5^{-\frac{1}{2}}) H(1/(3 + 5^{\frac{1}{2}})) \leq x(1 + 3 \cdot 5^{-\frac{1}{2}}) - k$$

or

$$R(1 + 3 \cdot 5^{-\frac{1}{2}})^{-1} \leq (1 - 3f)(1 - H(1/(3 + 5^{\frac{1}{2}}))) . \quad (3.12,$$

The result is valid in the region $0 < y < e$, which is equivalent to the requirement that

$$(3 + 5^{\frac{1}{2}})^{-1} < f < 1/3 .$$

Comparing these numbers with the tangents to the volume bound computed in Example 3.10, we see that this bound is a straight line which goes from the point $R = 0$, $f = 1/3$ to the volume bound, where it comes in tangentially and then ends. A plot of this bound is given in Figure 2.

This bound is a special case of a more general bound for error exponents given by Shannon, Gallager, and Berlekamp (1967).

In Part 4, we shall show how this bound is actually attainable.

4. The Construction of Asymptotically Optimum Coding Strategies

Having completed proofs of the volume bound, the translation bound, and their asymptotic combination, we are naturally led to investigate the possibility of finding specific winning states which lie on or close to these bounds. We start by an examination of our table of winning states for small n, (Figure 6). We know that in order to find any substantial (≥ 3) number of words at the top component, we are restricted to states for which $n \geq 3e + 2$. Thus if $e = 0$, $n \geq 2$, and we find that 4 is indeed a winning 2-state. If $e = 1$, $n \geq 5$, and we find that $\begin{smallmatrix}4\\8\end{smallmatrix}$ is a winning 5-state. Continuing, we find that $\begin{smallmatrix}4\\8\\36\end{smallmatrix}$ is a winning 8-state. This is quite a bit better than we had bargained for!! We knew that $\begin{smallmatrix}3\\0\end{smallmatrix}$ is a losing 7-state, and were inquiring merely as to whether it is a winning 8-state. We find that not only can we put $4 (> 3)$ words on top, but a sizable number of additional words may be added at the lower levels. If we continue this investigation, we find that $\begin{smallmatrix}4\\8\\36\\152\end{smallmatrix}$ is a winning 11-state. Further extensions of this sequence are found in the first column of the table of Figure 8.

Figure 8

INFINITE SEQUENCE OF BORDERLINE WINNING STATES

Row \ Column:	1	2	3	4	5	6	7	8	9	10	11	12	13	14
1	4	2	1	1	1	1	1	1	1	1	1	1	1	1
2	8	6	4	1	0	0	0	0	0	0	0	0	0	0
3	36	22	14	10	5	1	0	0	0	0	0	0	0	0
4	152	94	58	36	24	15	6	1	0	0	0	0	0	0
5	644	398	246	152	94	60	39	21	7	1	0	0	0	0
6	2728	1686	1042	644	398	246	154	99	60	28	8	1	0	0
7	11556	7142	4414	2728	1686	1042	644	400	253	159	88	36	9	1
⋮														

MOST IMPORTANT PROPERTIES

Let $A_{i,j}$ be the number in the ith row and the jth column:

$\underline{A}_{m,j} = \begin{smallmatrix}A_{1,j}\\A_{2,j}\\\cdots\\A_{m,j}\end{smallmatrix}$ $\underline{A}_{m,j}$ is a borderline winning $(3m-j)$-state. It satisfies volume bound with equality. It can be reduced to $\underline{A}_{m,(j+1)}$ and $\underline{A}_{(m-1),(j-2)}$

$$A_{i,j} + A_{i,(j+1)} = A_{(i+1),(j+2)} \quad \text{(unless } i \leq 2)$$

If $i \geq j$ and $i \geq 3$, then

$$A_{i,j} = 2((1 + 5^{\frac{1}{2}})/2)^{3i-j-2} + 2((1 - 5^{\frac{1}{2}})/2)^{3i-j-2}$$

Figure 9

ANOTHER INFINITE SEQUENCE OF BORDERLINE WINNING STATES

8	4	2	1	1	1	1	1	1	1	1
64	36	20	11	4	1	0	0	0	0	0
744	404	220	120	67	35	16	5	1	0	0
8512	4628	2516	1368	744	407	222	118	22	6	1

We shall now construct this table and derive some of its important properties.

Definition 4.1: The values in the table of Figure 8 are defined recursively as follows: The first two rows are postulated as initial conditions:

$$A_{1,1} = 4; A_{1,2} = 2; A_{1,k} = 1 \quad \text{for} \quad k \geq 3$$

$$A_{2,1} = 8; A_{2,2} = 6; A_{2,3} = 4; A_{2,4} = 1; A_{2,k} = 0 \quad \text{for} \quad k \geq 5 .$$

The remainder of the table is derived recursively by the following rules; applicable only when $i \geq 3$.

For $j \geq 3$, $A_{i,j} = A_{i-1,j-1} + A_{i-1,j-2}$

For $j = 2$, $A_{i,2} = A_{i,3} + A_{i-1,1}$

For $j = 1$, $A_{i,1} = A_{i,2} + A_{i,3}$

Definition 4.2: The state $\underline{A}_{m,j} = \begin{smallmatrix}A_{1,j}\\A_{2,j}\\\cdots\\A_{m,j}\end{smallmatrix}$

Theorem 4.3: For $j \leq 3 \leq i$, $A_{i,j} = 2((1 + 5^{\frac{1}{2}})/2)^{3i-j-2} + 2((1 - 5^{\frac{1}{2}})/2)^{3i-j-2}$

Proof: Notice that the first three columns are defined only in terms of themselves. We introduce the single sequence a_k by the transformation:

$$a_k = A_{(k+3)/3,1} \quad \text{if } k \equiv 0 \bmod 3$$

$$a_k = A_{(k+4)/3,2} \quad \text{if } k \equiv 2 \bmod 3$$

$$a_k = A_{(k+5)/3,3} \quad \text{if } k \equiv 1 \bmod 3$$

The recurrence relations defining $A_{i,j}$ then become

$$a_k = a_{k-1} + a_{k-2}, \quad \text{valid for } k \geq 4$$

The general solution of this equation is of the form

$$a_k = B r_1^k + C r_2^k$$

where B and C are constants determined by the two initial conditions, $a_2 = 6$ and $a_3 = 8$. r_1 and r_2 are the roots of the equation

$$r^2 = r + 1 .$$

Solving gives $r = (1 \pm 5^{\frac{1}{2}})/2$, $B = C = 2$. Transforming back from the a_k to the $A_{i,j}$ gives the desired result. Q.E.D.

Corollary 4.4: Theorem 4.3 also holds in the extended range $i \geq j$, $i \geq 3$.

Proof: These values are obtained by the same recurrence relations as their counterparts in the first three columns, to which they must be equal.

Theorem 4.5: $\underline{A}_{m,j}$ can be reduced to $\underline{A}_{m-1,j-2}$ and $\underline{A}_{m,j+1}$

Proof: We first patch up the exceptional columns on the left boundary, for $j \leq 2$, by defining $A_{m-1,0} = A_{m,3}$; $A_{m-2,-1} = A_{m,2}$. The recurrence relations defining the table are than uniformly stated for all $m \geq 0$. The proof consists of verifying conditions (2) and (3) of the partitioning Theorem 2.1. Condition (2) becomes

$$A_{i,j} + A_{i-1,j} = A_{i,j+1} + A_{i-1,j-2}$$

If $i = 1$ or 2, we observe that this condition is satisfied by the initial conditions. For larger i, we have

$$A_{i,j} = A_{i-1,j-1} + A_{i-1,j-2}$$
$$A_{i,j+1} = A_{i-1,j} + A_{i-1,j-1}$$

Subtraction of these two equations yields condition (2). A sufficient condition for satisfying (3) is

$$y_{2i+2} \leq z_{2i+1} \leq y_{2i} \quad \text{and} \quad z_{2i+2} < y_{2i+1} \leq z_{2i}$$

$$A_{i-2,j-2} \leq A_{i,j+1} \leq A_{i,j+2}$$

The latter inequality follows from the fact that $A_{i,j}$ is a monotonic nonincreasing function of j, for any fixed i. This monotonicity may be established by induction on i and the observation that the monotonicity holds for $j \leq 3$, where $A_{i,j}$ is given by an explicit formula. The former inequality is verified as follows:

$$A_{i,j+1} = A_{i-1,j} + A_{i-1,j-1} \geq A_{i-1,j} = A_{i-2,j-1}$$
$$+ A_{i-2,j-2} \geq A_{i-2,j-2}$$ Q.E.D.

Corollary 4.6: $\underline{A}_{m,j}$ is a winning $(3m-j)$-state. (proof by induction)

Corollary 4.7: $V_{3m-j}(\underline{A}_{m,j}) = 2^{3m-j}$ (proof by induction, using conservation of volume Theorem 3.1).

This concludes our proof of the remarkable properties of the table of Figure 8.

The correct partition of any state shown in column 1 or 2 of Figure 8 divides the words at each level evenly between the two sets. The states shown in the other columns cannot be so partitioned, because they have a single word at the top. To balance the two sets, certain numbers of "compensating words" at various lower levels must be assigned to the other set. It turns out that the correct numbers of compensating words needed to partition the jth column correctly are

$$0$$
$$\binom{0}{j-3} + \binom{1}{j-3}$$
$$\binom{1}{j-4} + \binom{2}{j-4}$$
$$\binom{2}{j-5} + \binom{3}{j-5}$$

For example, the correct partitioning of the 6th column is

1		1 + 0		0 + 0
0		0		0 + 0
1		0		1 + 0
15	=	5	+	5 + 5
60		29		2 + 29
246		123		0 + 123
1042		521		0 + 521
⋮		⋮		⋮

We can use the first column of Table 8 to obtain a lower bound on f and R for large n. The manipulations are simple:

$$2^k \leq 2[(1+5^{\frac{1}{2}})/2]^{3(j-1)} = A_{j,1} \quad \text{(to the nearest integer)}$$

so if

$$k - 1 \leq (j-1)\,3\log[(1+5^{\frac{1}{2}})/2]$$

we have

$$I \leq A_{e+j,1}$$

which is a winning $3(e+j-1)$-state. Thus we may protect 2^k words from e errors by n questions if

$$k - 1 \leq (n-3e)\log((1+5^{\frac{1}{2}})/2)$$

$$R \leq (1-3f)R_0;\ R_0 = \log((1+5^{\frac{1}{2}})/2) \qquad (4.8)$$

Since Equation 4.8 is identical to Equation 3.12, we conclude that for all rates in the region $0 < R < R_t = .30$, (or equivalently, for all error fractions $.19 = f_t < f = e/n \leq f_0 = 1/3$) this straight line gives the best possible asymptotic result. For higher rates (or lower error fractions), however, the bounds differ. The upper bound on R and f departs from the straight line and follows the curve $1 - H(f)$. The lower bound obtained by this simple comparison with the first column of Figure 8 remains on the straight line for small error fractions, f.

We soon decide that in the region of this discrepancy it is the straight line, rather than the volume curve, which is the weak bound. Among other observations, we know from Gilbert's (1952) bound for one-way codes that it is possible to get the rate R up to 1 for sufficiently small error fractions f. We are thus led to search for sequences of winning states which provide better achievable bounds for high rates.

In this region the restraining limits are imposed by the volume bound rather than the translation bound. Intuition suggests that we would do well to construct infinite sequences of winning states which build up more slowly, having fewer nonzero components, but much greater weight. These bottommore components must carry the load for small e/n. These desires can be fulfilled by requiring that the translate of a borderline winning n-state be a borderline winning $(n-4)$-state instead of an $(n-3)$-state. The attempt to construct a table with this property proves successful. The result is shown in Figure 9.

In fact, for any integer $t \geq 3$, we can compute a similar table of winning states. The first two lines of this table are defined as

$$2^{t-1} \ldots 2^{t-u} \quad \ldots 1 \quad \ldots \quad 1 \ldots 11111 \ldots$$
$$2^{t-1}(2^t-2t)\ldots 2^{t-u}(2^t-2t\tfrac{1}{2}u-1)\ldots 2^t-t-1\ldots 2^v-v-1\ldots 11\;4\;1\;0\;0 \ldots$$

The remainder of the table is recursively defined to fulfill the relation

$$A_{i,j} + A_{i-1,j} = A_{i,j+1} + A_{i-1,j-(t-1)}\,.$$

This definition cinches condition (2) of the partitioning Theorem 2.1. Condition (3) of the partitioning theorem is readily verified by methods similar to those elucidated for the special case $t = 3$. This then proves that the constructed table has the basic property:

$$\underline{A}_{m,j} \text{ reduces to } \underline{A}_{m,j+1} \text{ and } \underline{A}_{m-1,j-(t-1)}\,.$$

To compute the rate of exponential growth of the components of the first column of this table, we can again change variables to a_k. The first t columns then become

$$a_t \; a_{t-1} \cdots a_1$$
$$a_{2t} \; a_{2t-1} \cdots a_{t+1}$$
$$a_{3t} \; a_{3t-1} \cdots a_{2t+1}$$

In terms of the a_k, the recursion relation becomes

$$a_k + a_{k-1} = 2a_{k-1}$$

from which we get an algebraic equation for the growth rate r

$$r^t + 1 = 2r^{t-1}$$

The growth rate of the first column, $s = r^t$, satisfies the equation

$$s + 1 = 2s^{(t-1)/t} \qquad (4.9)$$

In terms of s, the achievable asymptote becomes

$$s^k \leq B s^j;\ j + e \leq nt$$
$$k \leq (nt-e)\log s$$
$$R \leq (t-f)\log s$$

Comparing Equation (4.9) with Equation (3.9), we note that if $g_0 = (t-1)/t$, the equations are identical!! This bound is a straight line, emanating from the point $f_0 = 1/t$ (or equivalently, $g_0 = (t-1)/t$) and proceeding up to a point where it touches the volume bound tangentially, and then continues on to the R axis. For lower and lower rates, the best bounds result from higher and higher values of the integer t.

There is considerable heuristic evidence that the correct asymptotic relationship between f and R is given by the volume bound for all rates $R \geq .2965\ldots$. However, the only known lower bounds on f and R are based on tables whose constructions require that the number t in Equation (4.9) be an integer. It is conjectured that similar tables may be constructed in which t is any rational number ≥ 3. An unproved construction with $t = 7/2$ is tentatively presented in Tables 10 and 11.

Binary Communication Over the Gaussian Channel Using Feedback With a Peak Energy Constraint

LAWRENCE A. SHEPP, JACK K. WOLF, MEMBER, IEEE, AARON D. WYNER, MEMBER, IEEE, AND JACOB ZIV, ASSOCIATE MEMBER, IEEE

Abstract—**We consider binary communication over the additive white Gaussian noise channel with no bandwidth constraint on the channel input signals, assuming the availability of a noiseless delayless feedback link. Although the signals at time t can depend on the noise at times $\tau < t$ and are therefore random functions, we require that the signal energy never exceed a fixed level. We show that the optimal probability of error is attainable without the use of the feedback channel by using antipodal signals.**

I. Introduction

We SHALL consider binary transmission over an additive white Gaussian noise channel with the use of a noiseless delayless feedback link. No bandwidth constraint is assumed on the input signals.

By using the feedback channel the signals can be modified according to the channel noise; thus the signal energy per transmission can be a random variable. We assume here that the signal energy is always limited (as distinct from limiting the statistical average of the energy as in [1], [2]). We will show that with this type of energy constraint, no improvement over the attainable error probability for the one-way channel is possible.

Following standard procedures (see [3], p. 266, and [4]) we can assume the following time-discrete model for our communication system. Denote the message to be transmitted by 0 or 1, and assume that the messages are a priori equally likely. Corresponding to message i ($i = 0$ or 1) the channel input is a sequence $x_{i1}, x_{i2}, x_{i3}, \cdots, x_{in}$. The "energy" of the input is limited to S, i.e.,

$$\sum_{k=1}^{n} x_{ik}^2 \leq S, \qquad i = 0, 1. \quad (1)$$

The channel output is a sequence $y_1, y_2, y_3, \cdots, y_n$ given by

$$y_k = x_{ik} + z_k, \qquad k = 1, 2, \cdots, n,$$

where the "noise" sequence $\{z_k\}$ are independent Gaussian variates with zero mean and unit variance. The parameter S in (1) is therefore the signal-to-noise ratio. Since the transmitter has the use of a noiseless feedback link, the kth channel input x_{ik} can depend on the values $y_1, y_2, \cdots, y_{k-1}$, of the channel output on the $(k - 1)$ previous transmissions. Constraint (1) is assumed to hold for all values of the noise sequence $\{z_k\}$.

Manuscript received November 22, 1968.

The authors are with Bell Telephone Laboratories, Inc., Murray Hill, N. J. 07974. J. K. Wolf is currently on leave of absence from the Department of Electrical Engineering, Polytechnic Institute of Brooklyn, Brooklyn, N. Y. 11201. J. Ziv is currently on leave of absence from the Scientific Department, Israel Ministry of Defence, Haifa, Israel.

After n transmissions the receiver examines the received sequence $(y_1, y_2, \cdots, y_n)$ and attempts to decide which message was actually transmitted. Let P_e denote the probability that the receiver makes an error. Our main result states that for all n

$$P_e \geq \Phi(-S^{1/2}), \quad (2)$$

where

$$\Phi(u) = (2\pi)^{-1/2} \int_{-\infty}^{u} e^{-\alpha^2/2} \, d\alpha,$$

is the error function. The right member of (2) is the error probability attainable when $x_{01} = +S^{1/2}$, $x_{11} = -S^{1/2}$, and $x_{ik} = 0$ $(k > 1)$. In this case the feedback channel is not used. For M-ary communication $(M > 2)$ use of the feedback channel can improve the error probability.[1]

II. Formal Statement of Results and Proofs

In this section we give a formal and precise statement of the problem and results. We begin with some definitions. Let $f = (f_1, f_2, \cdots, f_n)$ be a vector-valued function which takes Euclidean n space into itself. We say that f is "nonanticipatory" if for $\boldsymbol{u} = (u_1, u_2, \cdots, u_n)$

$$f_1(\boldsymbol{u}) = f_1, \qquad \text{independent of } \boldsymbol{u} \quad (3a)$$

$$f_k(\boldsymbol{u}) = f_k(u_1, u_2, \cdots, u_{k-1}), \qquad k = 2, \cdots, n. \quad (3b)$$

A (binary) *code* of dimension n is a pair of nonanticipatory functions $\boldsymbol{x}_0$ and $\boldsymbol{x}_1$ which map n space into n space. A *decoder* is a mapping D of n space to the set $\{0, 1\}$.

Let the "noise" be a sequence $\boldsymbol{z} = (z_1, z_2, \cdots, z_n)$ of independent Gaussian variates with zero mean and unit variance. For a given code $(\boldsymbol{x}_0, \boldsymbol{x}_1)$ the random n vectors $\boldsymbol{y}_0 = \boldsymbol{x}_0(\boldsymbol{y}_0) + \boldsymbol{z}$ and $\boldsymbol{y}_1 = \boldsymbol{x}_1(\boldsymbol{y}_1) + \boldsymbol{z}$ are well defined, since the $\boldsymbol{x}_i$ are nonanticipatory. For a given decoder D, the "error probability" is defined by

$$P_e = \tfrac{1}{2} \Pr \{D(\boldsymbol{y}_0) = 1\} + \tfrac{1}{2} \Pr \{D(\boldsymbol{y}_1) = 0\}. \quad (4)$$

The correspondence of the code $(\boldsymbol{x}_0, \boldsymbol{x}_1)$, the noise $\boldsymbol{z}$, the decoder, and the error probability P_e to the corresponding quantities in the communication system of Section I should be clear. We postpone for the moment consideration of the energy constraint.

Now for a given code $(\boldsymbol{x}_0, \boldsymbol{x}_1)$ it is readily shown (paralleling the usual derivation for the one-way channel [3],

[1] E. R. Berlekamp, private communication.

Reprinted from *IEEE Trans. Inform. Theory*, vol. IT-15, pp. 476-478, July 1969.

p. 217, and [4]) that the decoder which minimizes P_e takes $D(y)$ to be that $i = 0, 1$ which minimizes

$$\|x_i(y) - y\|,$$

where "$\| \ \|$" denotes Euclidean norm.

Now consider the random n vector y_0, which corresponds to the received vector when message zero is transmitted. $D(y_0) = 1$, corresponding to an error, only if

$$\|x_0(y_0) - y_0\| \geq \|x_1(y_0) - y_0\|,$$

or since

$$z = y_0 - x_0(y_0),$$

$$\|z\| \geq \|z - [x_1(y_0) - x_0(y_0)]\|,$$

or

$$(z, x_1(y_0) - x_0(y_0)) \geq \tfrac{1}{2} \|x_1(y_0) - x_0(y_0)\|^2, \quad (5a)$$

where "(,)" denotes inner product. Similarly $D(y_1) = 0$ only if

$$(z, x_0(y_1) - x_1(y_1)) \geq \tfrac{1}{2} \|x_0(y_1) - x_1(y_1)\|^2. \quad (5b)$$

Let us define $\mathcal{C}_1(n)$ as the set of n-dimensional codes such that for all y,

$$\|x_i(y)\|^2 \leq S, \qquad i = 0, 1. \quad (6)$$

We want to determine

$$P_e^* = \inf_n \inf_{\mathcal{C}_1(n)} P_e, \quad (7)$$

where the first infimum in (7) is taken over all codes in $\mathcal{C}_1(n)$.

Let us note that the code defined by

$$x_{01} = +\sqrt{S},$$

$$x_{11} = -\sqrt{S},$$

$$x_{ik} = 0, \qquad 2 \leq k \leq n, \quad i = 0, 1,$$

belongs to $\mathcal{C}_1(n)$, $n \geq 1$. For this code $(z, x_1 - x_0) = -2z_1\sqrt{S}$ and $(z, x_0 - x_1) = +2z_1\sqrt{S}$, so that from (5),

$$D(y_0) = 1 \qquad \text{only if} \quad z_1 < -\sqrt{S}$$

and

$$D(y_1) = 0 \qquad \text{only if} \quad z_1 > \sqrt{S}.$$

Thus $P_e = \Phi(-\sqrt{S})$ and

$$P_e^* \leq \Phi(-\sqrt{S}). \quad (8)$$

We will now show that this is in fact an equality.

Define $\mathcal{C}_2(n)$ as the set of n-dimensional codes such that for all y, $\|x_1(y) - x_0(y)\|^2 \leq 4S$. From the triangle inequality it follows that $\mathcal{C}_1(n) \subseteq \mathcal{C}_2(n)$, so that

$$\inf_{\mathcal{C}_1(n)} P_e \geq \inf_{\mathcal{C}_2(n)} P_e. \quad (9)$$

Next define $\mathcal{C}_3(n)$ as the set of n-dimensional codes such that for all y, $\|x_1(y) - x_0(y)\|^2 = 4S$. If a code belongs to $\mathcal{C}_2(n)$ we can, by adding an $(n + 1)$th coordinate (say,

$$x_{1,n+1}(y) = \left[4S - \sum_{k=1}^{n} [x_{1k}(y) - x_{0k}(y)]^2\right]^{1/2},$$

and $x_{0,n+1}(y) = 0$), obtain a code in $\mathcal{C}_3(n + 1)$. Since the decoder can ignore this last coordinate, the error probability is not increased. Thus

$$\inf_{\mathcal{C}_2(n)} P_e \geq \inf_{\mathcal{C}_3(n+1)} P_e. \quad (10)$$

Now consider a code (x_0, x_1) in $\mathcal{C}_3(n)$. Since $\|x_0 - x_1\|^2 \equiv 4S$, we have from (5) that $D(y_0) = 1$ only if

$$\psi_0 = (z, x_1(y_0) - x_0(y_0)) \geq 2S,$$

and $D(y_1) = 0$ only if

$$\psi_1 = (z, x_1(y_1) - x_0(y_1)) \leq 2S.$$

We will show below that regardless of the code (x_0, x_1), the inner products ψ_0 and ψ_1 are Gaussian random variables with mean zero and variance $4S$. Thus for any code in $\mathcal{C}_3(n)$

$$P_e = \tfrac{1}{2} \Pr\{\psi_0 \geq 2S\} + \tfrac{1}{2} \Pr\{\psi_1 \leq 2S\}$$
$$= \Phi(-\sqrt{S}).$$

It then follows from (8)–(10) that

$$P_e^* = \inf_n \inf_{\mathcal{C}_1(n)} P_e = \Phi(-\sqrt{S}),$$

which is our main result.

Our proof will be complete when we prove the assertion about the distributions of ψ_0 and ψ_1. This will follow easily from the following.

Lemma

Let f be a nonanticipatory function which takes n space into n space. Further assume that

$$\|f\|^2 \equiv \lambda^2.$$

Let $z = (z_1, z_2, \cdots, z_n)$ be a sequence of independent Gaussian variates with zero mean and unit variance. Then the random variable

$$\alpha = (z, f(z)) = \sum_{k=1}^{n} z_k f_k(z_1, \cdots, z_{k-1})$$

is Gaussian with zero mean and variance λ^2.

Proof: The proof is by induction on n. If $n = 1$, $\alpha = f_1 z_1 = \pm\lambda z_1$ and the Lemma holds. Assume that the Lemma holds for $n = N$, and consider $n = N + 1$. Write

$$\alpha = z_1 f_1 + \sum_{k=2}^{N+1} z_k f_k(z_1 \cdots z_{k-1}).$$

If we are given that $z_1 = \xi$, then

$$\alpha' = \sum_{k=2}^{N+1} z_k f_k(\xi, z_2, \cdots, z_{k-1})$$

satisfies the hypotheses of the Lemma with λ^2 replaced by $\lambda^2 - f_1^2$. Thus by the induction hypothesis α' is Gaussian with zero mean and variance $\lambda^2 - f_1^2$. Hence given $z_1 = \xi$, the random variable $\alpha = \xi f_1 + \alpha'$ is Gaussian with mean ξf_1 and variance $\lambda^2 - f_1^2$. Thus the conditional density of α given $z_1 = \xi$ is

$$p(\alpha \mid z_1 = \xi) = [2\pi(\lambda^2 - f_1^2)]^{-1/2} \cdot \exp[-\tfrac{1}{2}(\alpha - \xi f_1)^2/(\lambda - f_1^2)].$$

Since the density for z_1 is

$$p_{z_1}(\xi) = (2\pi)^{-1/2}e^{-\xi^2/2},$$

the density for α is by direct computation

$$p(\alpha) = \int_{-\infty}^{\infty} p(\alpha \mid z_1 = \xi)p_{z_1}(\xi)\, d\xi = (2\pi)^{-1/2}\lambda^{-1}e^{-\alpha^2/2\lambda^2},$$

which proves the Lemma.

Let us return to ψ_0 defined by

$$\psi_0 = (\boldsymbol{z}, \boldsymbol{x}_1(\boldsymbol{y}_0) - \boldsymbol{x}_0(\boldsymbol{y}_0)).$$

Since the value of $\boldsymbol{z} = (z_1, \cdots, z_n)$ determines the value of $\boldsymbol{y}_0 = \boldsymbol{x}_0(\boldsymbol{y}_0) + \boldsymbol{z}$ in such a way that y_k depends only on $z_1, z_2, \cdots, z_k$ $(k = 1, 2, \cdots, n)$, the functions $\boldsymbol{x}_1$ and $\boldsymbol{x}_0$ when considered as functions of $\boldsymbol{z}$ are nonanticipatory. Further, since the code $(\boldsymbol{x}_0, \boldsymbol{x}_1)$ belongs to $\mathcal{C}_3(n)$, $||\boldsymbol{x}_1 - \boldsymbol{x}_0||^2 \equiv 4S$. Thus the hypotheses of the above Lemma are satisfied for $\boldsymbol{f} = \boldsymbol{x}_1 - \boldsymbol{x}_0$ and $\lambda^2 = 4S$. Hence ψ_0 is Gaussian with mean zero and variance $4S$. The identical result holds for ψ_1, and the assertion (11) is proved.

Additional Comment

It has come to the attention of the authors that a recent paper by Pinsker [5] considers the asymptotic behavior (with n) of the system treated in this paper. Pinsker proves that the optimal *error exponent* is attainable without the use of feedback. Furthermore he proves that for any finite M the optimal error exponent for an M-ary system with feedback is identical to the exponent for two signals.

References

[1] A. J. Kramer, "Improving communication reliability by use of an intermittant feedback channel," *IEEE Trans. Information Theory*, vol. IT–15, pp. 52–60, January 1969.
[2] J. P. M. Schalkwijk and T. Kailath, "A coding scheme for additive channels with feedback—Part I: No bandwidth constraint," *IEEE Trans. Information Theory*, vol. IT–12, pp. 172–182, April 1966.
[3] J. Wozencraft and I. Jacobs, *Principles of Communication Engineering*. New York: Wiley, 1965.
[4] A. D. Wyner, "On the Schalkwijk–Kailath coding scheme with a peak energy constraint," *IEEE Trans. Information Theory*, vol. IT–14, pp. 129–134, January 1968.
[5] M. S. Pinsker, "Error probability for block transmission on a Gaussian memoryless channel with feedback," *Problemy Peredachi Informatsii*, vol. 4, pp. 3–19, 1968.

Rate Distortion Over Band-Limited Feedback Channels

Abstract—**Although linear feedback is by itself sufficient to achieve capacity of an additive Gaussian white noise (AGWN) channel, it can not, in general, achieve the theoretical minimum mean-squared error for analog Gaussian data. This correspondence gives the necessary and sufficient conditions under which this optimum performance can be achieved.**

Introduction

The use of linear feedback to transmit analog data from a Gaussian source over the additive Gaussian white noise (AGWN) channel was first considered by Elias [1] in 1956. Since then many other articles have appeared considering various aspects of linear feedback communication. Initially, Elias showed that noiseless linear feedback can achieve Shannon's [6] rate-distortion bound (RDB) on the mean-squared error (MSE) for any band-limited Gaussian source with a flat spectrum over the infinite-bandwidth AGWN channel. Later, Kailath [2] and Schalkwijk and Bluestein [3] extended this result to band-limited AGWN channels whose bandwidth is an integral number times the source bandwidth. Recently, Cruise [4] and Fang [5] have shown that the RDB on the MSE can be achieved for any band-limited source with an arbitrarily colored spectrum over the infinite-bandwidth AGWN channel. However, it has not been shown whether noiseless linear feedback is by itself a sufficient technique for achieving the RDB for an arbitrarily colored source over the band-limited AGWN channel. The purpose of this correspondence is to determine the circumstances (if any) under which noiseless linear feedback is by itself sufficient to achieve Shannon's RDB on the MSE for a colored Gaussian source over the band-limited AGWN channel. In addition, some comments about the connection between Fang's [5] and Cruise's [4] infinite-bandwidth schemes, which appear to be different, will be made.

Problem

Without loss of generality, we shall assume that analog data from a Gaussian source is the arbitrary sequence $\{\theta_i\}$ of independent zero-mean Gaussian random variables with a spectrum of variances $\{\lambda_i\}$, respectively, such that

$$E_\theta = \sum_{i=1}^{\infty} \lambda_i < \infty. \tag{1}$$

It is well known that a representation such as this can be obtained for any zero-mean Gaussian process $\theta(t)$ via the Karhunen–Loève series expansion [8]. The problem is to transmit the Gaussian data over an additive Gaussian white noise channel by means of noiseless linear feedback, and to achieve the minimum MSE between $\{\theta_i\}$ and its received facsimile $\{\hat{\theta}_i\}$.

Shannon's RDB on the MSE for a Gaussian source over a Gaussian channel, as given by Kolmogorov [7], is

$$\epsilon_{\min}(H) = \min_{\phi} \sum_{i=1}^{\infty} \min(\phi, \lambda_i), \tag{2}$$

where ϕ satisfies the constraint

$$2H = \sum_{i=1}^{\infty} \max[\ln(\lambda_i/\phi), 0], \tag{3}$$

Manuscript received January 20, 1970; revised July 21, 1970. This work was sponsored by the National Aeronautics and Space Administration under Contract NAS 7-100.

and $H = CT$ is the maximum amount of information transmitted in T seconds over the AGWN channel of capacity

$$C = W \ln(1 + P/N_0 W) \text{ nats/s}, \tag{4}$$

where P in watts is the average transmitted power, W in hertz is the bandwidth of the channel, and N_0 in watts per hertz is the one-sided spectral density of the noise. The number of degrees of freedom or dimensions is denoted by $N = 2TW$, where N is an integer. The signal-to-noise ratio per degree of freedom is $\rho = P/N_0W = 2E/N_0N$, where E is the total average transmitted energy. Thus, we also write

$$C = \tfrac{1}{2} \ln(1 + \rho) \text{ nats/degree of freedom}. \tag{5}$$

The main point of this correspondence is in the following theorem.

Theorem: Noiseless linear feedback is by itself a sufficient technique for transmitting a Gaussian source over the AGWN channel to achieve Shannon's rate-distortion bound on the mean-squared error if and only if the source spectrum is of the form

$$\lambda_i = \phi(1 + \rho)^{-N_i} \tag{6}$$

for all $i \in I \triangleq \{i | \lambda_i > \phi\}$, and N_i is a positive integer satisfying

$$\sum_{i \in I} N_i = N, \tag{7}$$

where the constant ϕ and the index set $I = \{i | \lambda_i > \phi\}$ are determined by (2) and (3). They are, therefore, fixed quantities for the given source-channel pair.

Proof: The proof is for the most part a reorganization of well-known facts. For instance, implicit in the minimization of (2) subject to the constraint (3) is the fact that Shannon's RDB on the MSE can be achieved if and only if the total $H = CT$ nats of information are allocated per θ_i according to

$$H_i = \begin{cases} \tfrac{1}{2} \ln(\lambda_i/\phi) & \text{if } \lambda_i > \phi \\ 0 & \text{otherwise}. \end{cases} \tag{8}$$

Thus, it is necessary only to find a capacity-achieving coding scheme for the AGWN channel that can be used to send $H = CT$ nats of information in amounts of H_i per θ_i. By sending $H_i = C T_i$ nats about θ_i, we would receive a facsimile $\hat{\theta}_i$ with a MSE not more than ϕ and thus achieve the RDB. Alternatively, we could subdivide the channel into subchannels of capacity C_i in such a way that $H_i = C_i T$.

However, although linear noiseless feedback achieves channel capacity, it can not by itself provide such an arbitrary division of information when the channel bandwidth is finite. The reason is that with linear feedback an integral number of iterations must be used per θ_i.

Now, if we transmit θ_i using N_i iterations, or degrees of freedom, and E_i joules of average energy, the MSE that can be achieved in this way is

$$E_i = \lambda_i (1 + \rho_i)^{-N_i} \tag{9}$$

where $\rho_i = 2E_i/N_0N_i$ is the signal-to-noise ratio per iteration. We note that the information transmitted per θ_i when using this approach is

$$H_i = \tfrac{1}{2} N_i \ln(1 + \rho_i). \tag{10}$$

However, although linear noiseless feedback achieves channel capacity, it can not by itself provide such an arbitrary division of because $\epsilon_{\min}(H)$ as given by (2) and (3) is strictly a monotone decreasing function of H. Thus, before feedback is to be used, it is necessary to make sure that $H = CT$ nats will be transmitted. Since

$$\sum N_i \ln(1 + \rho_i) \le N \ln(1 + \rho) \tag{11}$$

Reprinted from *IEEE Trans. Inform. Theory*, vol. IT-17, pp. 110–112, Jan. 1971.

with equality if and only if $\rho_i = \rho$ when the constraints

$$\sum N_i \leq N \tag{12}$$

and

$$\sum E_i = \tfrac{1}{2}N_0 \sum N_i \rho_i \tag{13}$$

are imposed, it follows that

$$H_i = \tfrac{1}{2}N_i \ln(1 + \rho) \tag{14}$$

is adjustable only in steps of $\frac{1}{2}\ln(1 + \rho)$ since N_i must be integer valued. Therefore, noiseless linear feedback can not by itself provide the optimum allocation of information for an arbitrary source spectrum. The optimum allocation is possible if and only if the spectrum satisfies

$$\tfrac{1}{2}\ln(\lambda_i/\phi) = \tfrac{1}{2}N_i \ln(1 + \rho)$$

where N_i is an integer for all $\lambda_i > \phi$, which is the condition (6) we set out to prove.

Corollary: The RDB can be achieved for arbitrary source spectra when the bandwidth of the AGWN is infinite. This has already been shown by Cruise [4] and Fang [5]. However, we shall deduce it again for the sake of completeness. Since N approaches infinity as W becomes infinite, we can allow each N_i to grow without bound. Therefore,

$$H_i = \lim_{N_i \to \infty} \tfrac{1}{2}N_i \ln\left(1 + \frac{2E_i}{N_0 N_i}\right) = E_i/N_0,$$

and since the remaining constraint $\sum E_i = E$ does not prevent us from adjusting the E_i in a continuous manner, it is clear that we can allocate the energies according to

$$E_i = \frac{N_0}{2}\ln(\lambda_i/\phi) \text{ per } \theta_i \qquad i \in I$$

and thus achieve the RDB for all choices of $\{\lambda_i\}$ when the bandwidth is infinite.

Remarks

If we can not meet condition (6), we can still use feedback to reduce the MSE to a small value. However, it will not be as small as $\epsilon_{\min}$. In particular, if we retain the equal-signal-to-noise-ratio degree-of-freedom allocation, it is possible to achieve a MSE of

$$\epsilon = \phi \sum_{i \in I} (1 + \rho)^{\beta_i} + \sum_{i \notin I} \lambda_i > \epsilon_{\min} \tag{15}$$

such that $|\beta_i| < 1$ and

$$\sum_{i \in I} \beta_i = 0 \tag{16}$$

where $\{\beta_i\}$ is obtained as follows. We first take N_i equal to the smallest integer in $\ln(\lambda_i/\phi)/\ln(1 + \rho)$. As a result, there will be an integral number

$$L = N - \sum_{i \in I} N_i < \sum_{i \in I} 1 \tag{17}$$

degrees of freedom or feedback iterations left unused. These extra iterations are not enough to make an additional transmission for each $\theta_i, i \in I$, but we can make one more iteration for some of the θ. The best that can be done is to make one more iteration for those $L\theta$ for which the difference

$$\alpha_i = [\ln(\lambda_i/\phi)/(1 + \rho)] - N_i < 1 \tag{18}$$

is largest. Then

$$\beta_i = \begin{cases} \alpha_i & \text{if } L \text{ or more } \alpha \text{ are larger than } \alpha \\ \alpha_{i-1} & \text{otherwise.} \end{cases} \tag{19}$$

If we discard the requirement that $\rho_i = \rho$ when (6) can not be satisfied, it will not be possible to transmit the full $H = N \ln(1 + \rho)$ nats of information and thus to achieve $\epsilon_{\min}$; however, this does not exclude the possibility of achieving a MSE smaller than (15). The general problem that must be solved is

$$\epsilon = \min_{\{N_i\}\{\rho_i\}} \sum_{i \in I} \lambda_i (1 + \rho_i)^{-N_i} + \sum_{i \notin I} \lambda_i \tag{20}$$

subject to the constraints

$$\sum_{i \in I} N_i \rho_i = 2E/N_0, \tag{21}$$

$$\sum_{i \in I} N_i = N, \tag{22}$$

$$N_i = \text{integer}. \tag{23}$$

It is only when (23) either is ignored or will anyway be satisfied as when (6) is true that (20)–(22) will yield the RDB.

Conclusion

We should like to conclude by considering the apparent difference between Cruise's [4] and Fang's [5] schemes. These schemes "look" different in the sense that they use different pre-emphasis filters. In Fang's scheme, the source is passed through a pre-emphasis filter with weighting coefficients L_i given by

$$L_i^2 = N_0 \ln(\lambda_i/\phi)/2\lambda_i \qquad i \in I, \tag{24}$$

while Cruise uses a filter with weighting coefficients Q_i where

$$Q_i^2 = \frac{N_0}{2}\left(\frac{1}{\phi} - \frac{1}{\lambda_i}\right) \qquad i \in I, \tag{25}$$

which were determined originally by Boardman and Van Trees [10] in a different context.

Actually, Fang's scheme is essentially the same as the scheme referred to by Cruise. To see this, we note that it has been shown in many places that the forward channel signals of a system of the type referred to by Fang have the general form

$$s_k = g_k(\theta - \hat{\theta}_{k-1}) \qquad k = 1, \cdots, N, \tag{26}$$

where θ is some Gaussian message of variance λ, $\hat{\theta}_k$ is its MSE estimate fed back after k transmissions, and $N = 2TW$ is the number of transmissions allocated per θ. Moreover, the MSE after k transmissions is (see, for instance, Butman [9])

$$\boldsymbol{E}[(\theta - \hat{\theta}_k)^2] = \lambda \Big/ \left[1 + (2\lambda/N_0)\sum_{j=1}^{k} g_j^2\right] \tag{27}$$

where $\boldsymbol{E}$ is the expectation operator.

In particular, the MSE for $k = N$ must be ϕ, hence

$$\|\boldsymbol{g}\|^2 \triangleq \sum_{k=1}^{N} g_k^2 = \tfrac{1}{2}N_0\left(\frac{1}{\phi} - \frac{1}{\lambda}\right) = Q^2. \tag{28}$$

Thus, the magnitude of the pre-emphasis filter alluded to by Cruise is the Euclidean norm of the N-dimensional vector $\boldsymbol{g} = \text{col}(g_1, \cdots, g_N)$.

Fang's suggestion of filtering the source amounts to replacing θ above by $\chi = L\theta$. However, the information sent about χ must be the same as about θ. Thus, the transmitted signals will be the same since the system will merely employ a vector $\boldsymbol{g}' = \boldsymbol{g}/L$. This can be done with any nonzero choice of L_i per θ_i for $i \in I$. Thus, the pre-emphasis filter (24) proposed by Fang is not necessary, and it does not determine the system.

We observe that (28) holds for both finite and infinite bandwidth channels and that it specifies only the norm $\|\boldsymbol{g}\|^2$, but not the relative values of the individual elements of the vector. When the bandwidth is finite, these elements are uniquely specified by

$$g_k^2 = (\rho/\lambda)(1+\rho)^{k-1} \qquad k = 1, \cdots, N, \tag{29}$$

for each θ. The above formula also specifies the infinite bandwidth pre-emphasis filter $g(t)$ provided we define $g_k^2 = g^2(t)\,dt$, $k = 2Wt = t/dt$, where $dt = 1/(2W)$, goes to zero as W goes to infinity, and $\rho = P/N_0W = 2P\,dt/N_0$. Then (29) yields

$$g^2(t) = (2P/N_0\lambda)\exp(2Pt/N_0) \qquad 0 \le t \le T,$$

and

$$\int_0^T g^2(t)\,dt = Q^2.$$

However, the above choice is not unique; the choice $g^2(t) = Q^2/T$ will also achieve the RDB in the wide-band case. In fact, we can take

$$g^2(t) = \frac{d}{dt}F(t) \qquad F(t) \ge 0,$$

as long as $F(T) = Q^2$ and $F(0) = 0$, because the average transmitted energy per θ will be $(N_0/2)\ln(1 + F(T)) = (N_0/2)\ln(\lambda/\phi)$ as required.

S. Butman
Jet Propulsion Lab.
Pasadena, Calif. 91103

References

[1] a. P. Elias, "Channel capacity without coding," Res. Lab. of Elec., Massachusetts Institute of Technology, Cambridge, Quarterly Progress Rep., pp. 90–93, October 1956; also in *Lectures on Communication System Theory*, E. Baghdady, Ed. New York: McGraw-Hill, 1961, pp. 363–366.
b. ——, "Networks of Gaussian channels with applications to feedback systems," *IEEE Trans. Inform. Theory*, vol. IT-13, pp. 493–501, July 1967.
[2] T. Kailath, "An application of Shannon's rate-distortion theory to analog communication over feedback channels," presented at the 1st Ann. Princeton Conf. on Inform. Sci. and Syst., March 1967; also *Proc. IEEE* (Letters), vol. 55, pp. 1102–1103, June 1967.
[3] J. P. M. Schalkwijk and L. I. Bluestein, "Transmission of analog waveforms through channels with feedback," *IEEE Trans. Inform. Theory*, vol. IT-13, pp. 617–619, October 1967.
[4] T. J. Cruise, "Achievement of rate-distortion bound over additive white noise channel utilizing a noiseless feedback link," *Proc. IEEE* (Letters), vol. 55, pp. 583–584, April 1967.
[5] R. J. F. Fang, "Optimum linear feedback scheme for matching colored Gaussian sources with white Gaussian channels," *IEEE Trans. Inform. Theory* (Correspondence), vol. IT-16, pp. 789–791, November 1970.
[6] C. E. Shannon, "A mathematical theory of communication," *Bell Syst. Tech. J.*, vol. 27, pp. 379–623, 1948.
[7] A. N. Kolmogorov, "On the Shannon theory of information transmission in the case of continuous signals," *IRE Trans. Inform. Theory*, vol. IT-2, pp. 102–108, December 1956.
[8] W. B. Davenport, Jr., and W. L. Root, *An Introduction to the Theory of Random Signals and Noise*. New York: McGraw-Hill, 1958, pp. 96.
[9] S. Butman, "A general formulation of linear feedback communication systems with solutions," *IEEE Trans. Inform. Theory*, vol. IT-15, pp. 392–400, May 1969.
[10] C. J. Boardman and H. L. Van Trees, "Optimum angle modulation," *IEEE Trans. Commun. Tech. nol.*, vol. COM-13, pp. 452–469, December 1965.

Broadcast Channels

THOMAS M. COVER, MEMBER IEEE

***Abstract*—We introduce the problem of a single source attempting to communicate information simultaneously to several receivers. The intent is to model the situation of a broadcaster with multiple receivers or a lecturer with many listeners. Thus several different channels with a common input alphabet are specified. We shall determine the families of simultaneously achievable transmission rates for many extreme classes of channels. Upper and lower bounds on the capacity region will be found, and it will be shown that the family of theoretically achievable rates dominates the family of rates achievable by previously known time-sharing and maximin procedures. This improvement is gained by superimposing high-rate information on low-rate information. All of these results lead to a new approach to the compound channels problem.**

I. Introduction

THIS PAPER attempts to develop some intuition on the general topic of the simultaneous communication of information from one source to several receivers. Examples of simultaneous communication include broadcasting information to a crowd, or broadcasting TV information from a transmitter to multiple receivers in the area, or giving a lecture to a group of disparate backgrounds and aptitudes.

We will find that our proposed model will also be applicable to the situation of compound channels, where the transmitter does not know the true channel characteristics but wishes to transmit at an interesting rate to the receiver.

The general broadcast channel with k receivers is depicted in Fig. 1. Details of this formulation are made precise in Section III. The basic problem is to find the set of simultaneously achievable transmission rates $(R_1,R_2,\cdots,R_k)$.

Suppose that the transmission channels to the receiver have respective channel capacities $C_1,C_2,\cdots,C_k$ bits per second. The first approach that suggests itself is the maximin approach—send at rate $C_{\min} = \min \{C_1,C_2,\cdots,C_k\}$. Even this modest goal is only possible when the channels are compatible in some sense (see Section IX for the general expression). If the channels are compatible, each receiver will understand perfectly at the rate $R = C_{\min}$ bits/s. Here the transmission rate is limited by the worst channel. At the other extreme, information could be sent at rate $R = C_{\max}$, with resulting rates $R_i = 0$, $i = 1, 2, \cdots, k-1$, for all but the best channel, and rate $R_k = C_{\max}$ for the best channel.

Manuscript received March 23, 1971; revised July 30, 1971. This work was supported by Contract F44620-69-C-0101 and Contract N-00014-67-A-0112-0044. Portions of this work were performed at Bell Telephone Laboratories, Murray Hill, N.J. This paper is based on talks presented at the IEEE International Symposium on Information Theory, Noordwijk, the Netherlands, June 1970, and subsequently at the 6th Berkeley Symposium on Probability and Statistics, Berkeley, Calif., July 1970.

The author is with the Department of Electrical Engineering and the Department of Statistics, Stanford University, Stanford, Calif. 94305. He is currently on sabbatical leave at the Department of Electrical Engineering, Massachusetts Institute of Technology, Cambridge, Mass., and at Harvard University, Cambridge, Mass.

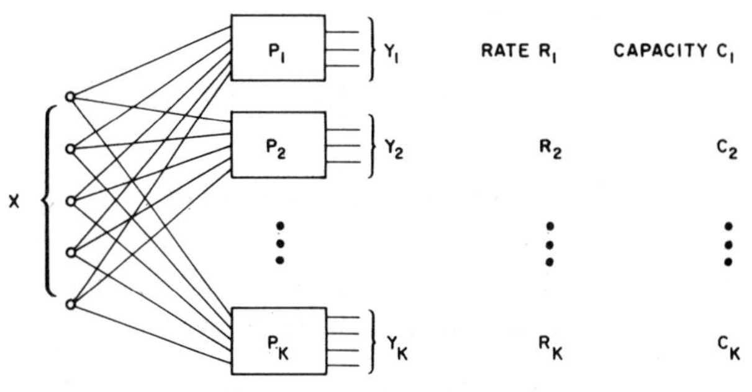

Fig. 1. Broadcast channel.

The next idea is that of time sharing. Allocate proportions of time $\lambda_1,\lambda_2,\cdots,\lambda_k$, $\lambda_i \geq 0$, $\sum \lambda_i = 1$, to sending at rates $C_1,C_2,\cdots,C_k$. Assuming compatibility of the channels and assuming $C_1 \leq C_2 \leq \cdots \leq C_k$, we find that the rate of transmission of information through the ith channel is given by

$$R_i = \sum_{j \leq i} \lambda_j C_j, \qquad i = 1,2,\cdots,k.$$

To our knowledge, no other schemes have been discussed in the literature, nor has the problem of the broadcast channel been formulated.

In this paper, we shall show that even this family of rates can be exceeded. In particular, it will be shown that for a slight degradation in the rate for the worst channel, an incrementally larger increase in the rate of transmission can be made for the better channels. The heuristic that will result from our discussion will be that one should not transmit simultaneously to several channels at the rate of the worst channel, nor should one attempt to transmit information by a time-sharing or time-multiplexing method, but rather one should distribute the high-rate information across the low-rate message.

Examples of good encodings for a family of binary symmetric channels and for a family of Gaussian channels will be presented. Also, the extreme case of orthogonal channels, in which it does not matter that one is trying to send two messages at once to two different people, will be considered, as well as the other extreme of incompatible channels, in which the transmission of information to one receiver precludes the transmission of information to the other.

II. Two Binary Symmetric Channels

Before proceeding with the precise formulation of a broadcast channel in Section III, let us pursue the case of two binary symmetric channels in heuristic detail. Unfamiliar terminology may be found in Ash [1] and Section III.

Let the input alphabet be $X = \{1,2\}$ and the output

Reprinted from *IEEE Trans. Inform. Theory*, vol. IT-18, pp. 2-14, Jan. 1972.

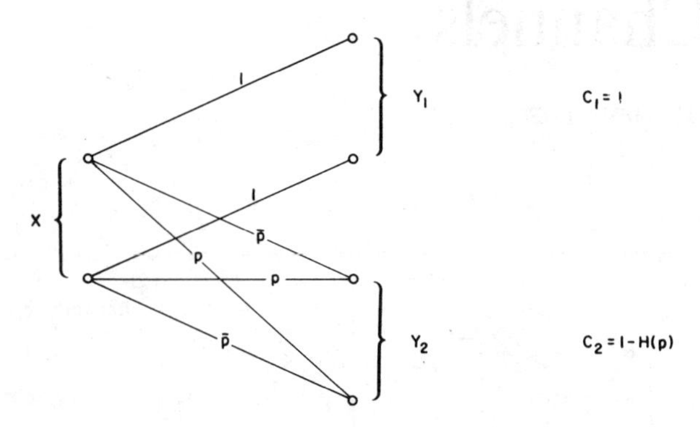

Fig. 2. Two binary symmetric channels.

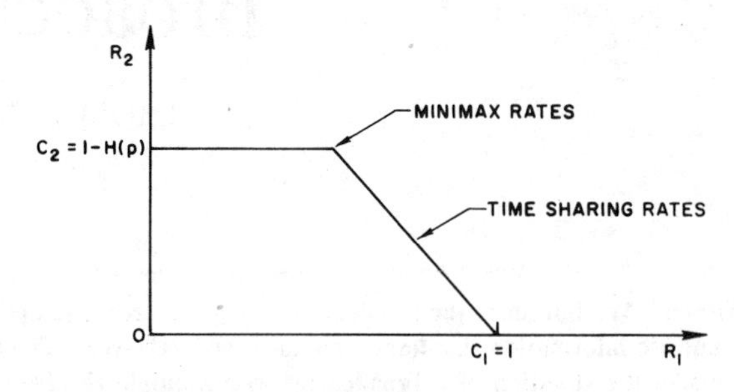

Fig. 3. Some achievable rates for the BSC.

alphabets for receivers 1 and 2 be $Y_1 = \{1,2\}$ and $Y_2 = \{1,2\}$. Let the channel matrices be given by

$$P_1 = \begin{bmatrix} 1 & 0 \\ 0 & 1 \end{bmatrix} \qquad P_2 = \begin{bmatrix} \bar{p} & p \\ p & \bar{p} \end{bmatrix} \tag{1}$$

as depicted in Fig. 2.

Thus channel 1 is noiseless and channel 2 is a binary symmetric channel (BSC) with error probability p. The corresponding channel capacities are $C_1 = 1$ bit per transmission and $C_2 = C(p) = 1 - H(p)$ bits per transmission.

The maximin approach would have us transmit at rates $(R_1,R_2) = (C_2,C_2)$ as shown in Fig. 3. (The maximin points are loosely called the minimax points in the figures. Although not generally equal, minimax and maximin are equal in all examples depicted in the figures.) These rates can indeed be simultaneously achieved by using a standard $(2^{n(C_2-\varepsilon)},n,\lambda_n)$ code for channel P_2 (see Wolfowitz [2]).

At the other extreme, we may send at rate $R_1 = 1$ with zero probability of error to receiver 1, with a resulting rate $R_2 = 0$ for channel 2. Then, by allocating a proportion of time λ to sending at rate (C_2,C_2) and a proportion of time $1 - \lambda$ to sending at rate (1,0), we obtain the family of rates shown by the straight line in Fig. 3. This we shall call the time-sharing lower bound of the set of achievable rates.

Now let us see how to do better. We know, from the random coding proof, that a good $(2^{n(C_2-\varepsilon)},n,\lambda_n)$ code can be generated by choosing at random a subset S of $2^{n(C_2-\varepsilon)}$ elements from the set of 2^n binary n-sequences $X^n = \{1,2\}^n$, and using the decoding rule that assigns the received vector $\boldsymbol{y} = (y_1,y_2,\cdots,y_n)$ to the element of S that is within Hamming distance $n(p + \varepsilon)$ of $\boldsymbol{y}$.

Let us choose a code of this form designed for a somewhat noisier channel; namely, the cascade of a BSC of parameter p and a BSC of parameter α, resulting in a BSC of parameter $\alpha\bar{p} + \bar{\alpha}p$, where $\bar{\alpha} = 1 - \alpha$. Thus there will be only $2^{n(C(\alpha\bar{p}+\bar{\alpha}p)-\varepsilon)}$ codewords in this set, but a larger noise of size $n(\alpha\bar{p} + \bar{\alpha}p)$ will be tolerated.

We now take advantage of this tolerance by packing in some extra message information intended solely for the perfect receiver Y_1.

With each codeword $\boldsymbol{x}$ in $S \subseteq X^n = \{1,2\}^n$, we will associate the set of all codewords at Hamming distance

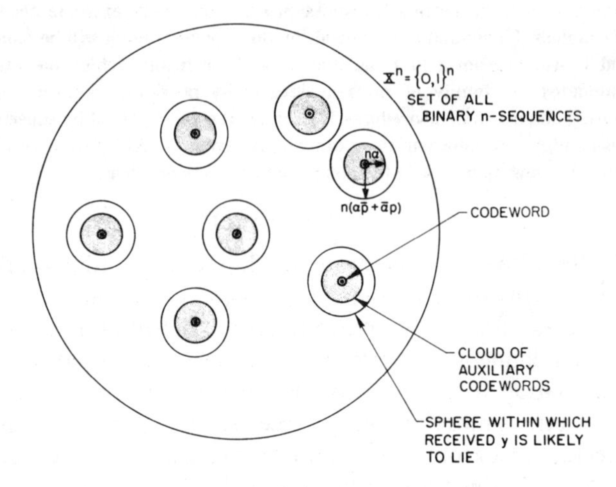

Fig. 4. Space of codewords for BSC.

equal to $[\alpha n]$, as suggested by the clouds of points shown in Fig. 4.

This code structure allows the transmission of an arbitrary integer $r \in \{1,2,\cdots,2^{n(C(\alpha\bar{p}+\bar{\alpha}p)-\varepsilon)}\}$ to both receivers 1 and 2 and an arbitrary integer

$$s \in \left\{1,2,\cdots,\binom{n}{[\alpha n]}\right\}$$

to receiver 1. (See Section III for further elucidation of these ideas.) The message (r,s) is sent in the following manner. The integer r designates the cloud, and the integer s designates the point $\boldsymbol{x} \in \{1,2\}^n$ within the cloud. This n-sequence $\boldsymbol{x}$ is then transmitted. The perfect channel receives $\boldsymbol{y}_1 = \boldsymbol{x}$ and thus correctly decodes both r and s. Since there are

$$\binom{n}{[\alpha n]} \approx 2^{nH(\alpha)}$$

points per cloud, we see that the transmission rate for channel 1 is

$$R_1 \triangleq \frac{1}{n} \log 2^{nH(\alpha)} 2^{n(C(\alpha\bar{p}+\bar{\alpha}p)-\varepsilon)} = C(\alpha\bar{p} + \bar{\alpha}p) + H(\alpha) - \varepsilon. \tag{2}$$

Channel 2 perceives the cloud center as if it had been sent through an additional BSC of parameter α (due to the choice of s). However, since the cloud centers were chosen to be distinguishable over a BSC of parameter $\alpha\bar{p} + \bar{\alpha}p$, we see that r is correctly decoded by receiver Y_2. Thus

$$R_2 = C(\alpha\bar{p} + \bar{\alpha}p) - \varepsilon. \tag{3}$$

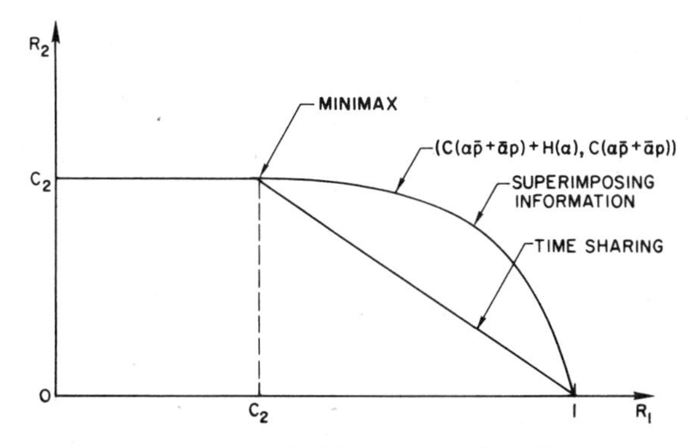

Fig. 5. Set of achievable rates for BSC.

This argument suggests that (R_1,R_2) is jointly achievable. The Appendix contains the proof. Letting α range from 0 to 1 generates the new achievable set of rates shown in Fig. 5. We strongly believe that Fig. 5 exhibits the optimal region of achievable rates.

This curve dominates the time-sharing curve. We note also that near the minimax point, the slope is zero. Thus an infinitesimal degradation in the rate for the poor channel will allow an infinitesimally infinite increase in the rate for the good channel. Consequently, at least for two BSC, superposition of information dominates time sharing.

This example naturally leads to a conjecture concerning the evaluation of the capacity region for a special class of broadcast channels in which one channel is a degraded version of the other.

Definition: Let P and Q be channel matrices of size $|X| \times |Y_1|$ and $|X| \times |Y_2|$, respectively. Q will be said to be a *degraded version* of P if there exists a stochastic matrix M such that $P = QM$. Shannon[6] has shown that the capacity of channel Q is not greater than that of channel P.

Conjecture 1: Let S be an arbitrary $|X| \times |X|$ channel matrix corresponding to the channel density $p(x \mid s)$ and let $p(s)$ be an arbitrary probability distribution on X. Let $p(s)$ induce the joint distribution $p(y_1,y_2,s,x) = p(s)\,p(x \mid s)$ $p(y_1,y_2 \mid x)$ on (y_1,y_2,s,x). Let P_2 be a degraded version of P_1. Then the set of achievable (R_1,R_2) pairs for the broadcast channel $(X, p(y_1,y_2 \mid x), Y_1 \times Y_2)$ is given by $(I(S;Y_2) + I(X;Y_1 \mid S), I(S;Y_2))$; generated by all channels S and probability distributions $p(s)$.

The two-BSC example in this section is a special case of this conjecture. The code that achieves (R_1, R_2) is constructed in an analogous manner. At the time of this writing, P. Bergmans at Stanford has made some progress on the proof of this conjecture. In fact, Bergmans' considerations have allowed me to modify the conjecture from an initially more ambitious version involving a larger class of channels mentioned in [6]. I no longer have any basis for belief in the more ambitious conjecture.

III. Definitions and Notation

We shall define a two-receiver memoryless *broadcast channel*, denoted by $(X,\ p(y_1,y_2 \mid x),\ Y_1 \times Y_2)$ or by $p(y_1,y_2 \mid x)$, to consist of three finite sets X,Y_1,Y_2 and a collection of probability distributions $p(\cdot,\cdot \mid x)$ on $Y_1 \times Y_2$, one for each $x \in X$. The interpretation is that x is an input to the channel and y_1 and y_2 are the respective outputs at receiver terminals 1 and 2 as shown in Fig. 6. The problem is to communicate simultaneously with receivers 1 and 2 as efficiently as possible.

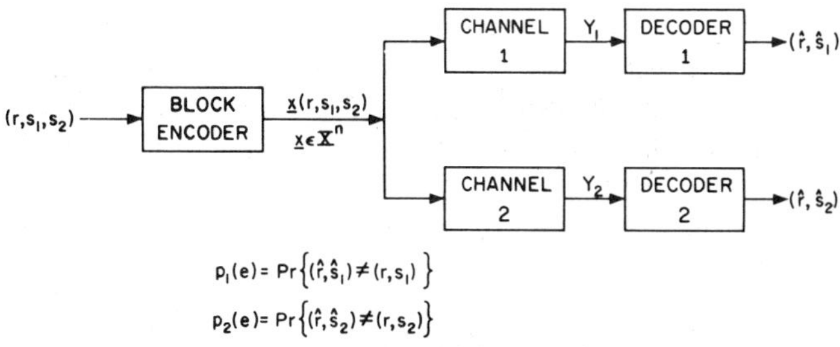

Fig. 6. Encoder and decoder for broadcast channel.

For the development of this paper we shall need knowledge only of the marginal distributions

$$p_1(y_1 \mid x) = \sum_{y_2 \in Y_2} p(y_1, y_2 \mid x)$$

$$p_2(y_2 \mid x) = \sum_{y_1 \in Y_1} p(y_1, y_2 \mid x), \tag{4}$$

which we have designated in the examples by channel matrices P_1 and P_2 of sizes $|X| \times |Y_1|$ and $|X| \times |Y_2|$, respectively. The possible dependence or independence of Y_1 and Y_2 given X is irrelevant, given the constraint that the decoding at the two receivers must be done independently.

The nth extension for a broadcast channel is the broadcast channel

$$(X^n,\ p(\mathbf{y}_1,\mathbf{y}_2 \mid \mathbf{x}),\ Y_1{}^n \times Y_2{}^n), \tag{5}$$

where $p(\mathbf{y}_1,\mathbf{y}_2 \mid \mathbf{x}) = \prod_{j=1}^n p(y_{1j},y_{2j} \mid x_j)$, for $\mathbf{x} \in X^n$, $\mathbf{y}_1 \in Y_1{}^n$, $\mathbf{y}_2 \in Y_2{}^n$.

An $((M_1,M_2,M_{12}),n)$ code for a broadcast channel consists of three sets of integers

$$R = \{1,2,\cdots,M_{12}\}$$

$$S_1 = \{1,2,\cdots,M_1\}$$

$$S_2 = \{1,2,\cdots,M_2\},$$

an encoding function

$$\mathbf{x}\colon R \times S_1 \times S_2 \to X^n,$$

and two decoding functions

$$g_1\colon Y_1{}^n \to R \times S_1;\ g_1(\mathbf{y}_1) = (\hat{r},\hat{s}_1)$$

$$g_2\colon Y_2{}^n \to R \times S_2;\ g_2(\mathbf{y}_2) = (\hat{r},\hat{s}_2).$$

The set $\{\mathbf{x}(r,s_1,s_2) \mid (r,s_1,s_2) \in R \times S_1 \times S_2\}$ is called the set of codewords. As illustrated in Fig. 6, we think of integers s_1 and s_2 as being arbitrarily chosen by the transmitter to be sent to receivers 1 and 2, respectively. The integer r is also chosen by the transmitter and is intended to be received by both receivers. Thus r is the "common" part of the message and s_1 and s_2 are the "independent" parts of the message.

An error is made by the ith receiver if $g_i(y_i) \neq (r,s_i)$. If the message (r,s_1,s_2) is sent, let

$$\lambda_i(r,s_1,s_2) = \Pr\{g_i(y_i) \neq (r,s_i)\}, \qquad i = 1,2, \tag{6}$$

denote the probabilities of error for the two channels, where we note that y_1,y_2 are the only chance variables in the above expression.

We denote the (arithmetic average) probability of error in decoding (r,s_1) averaged over all choices of s_2 by

$$\bar{\lambda}_1(r,s_1) = \frac{1}{M_2}\sum_{s_2=1}^{M_2} \lambda_1(r,s_1,s_2). \tag{7}$$

Similarly, for channel 2 we define

$$\bar{\lambda}_2(r,s_2) = \frac{1}{M_1}\sum_{s_1=1}^{M_1} \lambda_2(r,s_1,s_2). \tag{8}$$

Finally, we define the overall *arithmetic average probabilities of error* of the code for channels 1 and 2 as

$$\bar{p}_1(e) = \frac{1}{M_1M_{12}}\sum_{r,s_1} \bar{\lambda}_1(r,s_1) = \frac{1}{M}\sum_{r,s_1,s_2} \lambda_1(r,s_1,s_2) \tag{9}$$

$$\bar{p}_2(e) = \frac{1}{M_2M_{12}}\sum_{r,s_2} \bar{\lambda}_2(r,s_2) = \frac{1}{M}\sum_{r,s_1,s_2} \lambda_2(r,s_1,s_2), \tag{10}$$

where

$$M = M_1M_2M_{12}. \tag{11}$$

The overbar on $\bar{p}_i(e)$ will serve as a reminder that this probability of error is calculated under a special distribution; namely, the uniform distribution over the codewords.

We shall also be interested in the *maximal probabilities of error*

$$\lambda_i = \max_{r,s_1,s_2} \Pr\{g_i(y_i) \neq (r,s_i) \mid (r,s_1,s_2)\}, \qquad i = 1,2, \tag{12}$$

corresponding to the worst codeword with respect to each channel. Note that $\lambda_i \geq \bar{p}_i(e)$.

We shall define the *rate* (R_1,R_2,R_{12}) of an $((M_1,M_2,M_{12}),n)$ code by

$$R_1 = \frac{1}{n}\log M_1M_{12}$$

$$R_2 = \frac{1}{n}\log M_2M_{12}$$

$$R_{12} = \frac{1}{n}\log M_{12} \tag{13}$$

all defined in bits/transmission. Thus R_i is the total rate of transmission of information to receiver i, $i = 1, 2$, and R_{12} is the portion of the information common to both receivers.

Comment: When λ_i and $\bar{p}_i(e)$ refer to the nth extension of a broadcast channel, we will often designate this explicitly by $\lambda_i^{(n)}$, $\bar{p}_i^{(n)}(e)$.

Definition: The rate (R_1,R_2,R_{12}) is said to be *achievable* by a broadcast channel if, for any $\varepsilon > 0$ and for all n sufficiently large, there exists an $((M_1,M_2,M_{12}),n)$ code with

$$M_1M_{12} \geq 2^{nR_1}$$

$$M_2M_{12} \geq 2^{nR_2}$$

$$M_{12} \leq 2^{nR_{12}} \tag{14}$$

such that $\bar{p}_1^{(n)}(e) < \varepsilon$, $\bar{p}_2^{(n)}(e) < \varepsilon$.

Comment: Note that the total number $M = M_1M_2M_{12}$ of codewords for a code satisfying (14) must exceed $2^{n(R_1+R_2-R_{12})}$.

Definition: The *capacity region* $\mathfrak{R}^*$ for a broadcast channel is the set of all achievable rates (R_1,R_2,R_{12}).

The goal of this paper is to determine $\mathfrak{R}^*$ for as large a class of channels as possible.

Comment: We shall sometimes let $\mathfrak{R}^*$ also denote the set of achievable (R_1,R_2) pairs. However, at this stage in our understanding, it seems that sole concern with (R_1,R_2), with the exclusion of concern with R_{12}, would result in a coarsened and cumbersome theoretical development.

Comment: The extension of the definition of the broadcast channel from two receivers to k receivers is notationally cumbersome but straightforward, given the following comment. The index sets R,S_1,S_2 should be replaced by $2^k - 1$ index sets $I(\theta)$, $\theta \in \{0,1,\}^k$, $\theta \neq \mathbf{0}$, with the interpretation that the integer $i(\theta)$ selected in index set $I(\theta) = \{1,2,\cdots,M(\theta)\}$ is intended (by the proper code selection) to be received correctly by every receiver j for which $\theta_j = 1$ in $\theta = (\theta_1,\theta_2,\cdots,\theta_k)$. Then, for example, the rate of transmission over the nth extension of a broadcast channel to the ith receiver will be given by

$$R_i = \frac{1}{n}\log \prod_{\substack{\theta \in \{0,1\}^k \\ \theta_i = 1}} M(\theta) = \frac{1}{n}\sum_{\theta_i=1} \log M(\theta). \tag{15}$$

In the two-receiver broadcast channel, the corresponding sets in the new notation are $R = I(1,1)$, $S_1 = I(1,0)$, $S_2 = I(0,1)$.

Section IV treats the best two-channel situation and Section V treats the worst.

IV. Orthogonal Channels

In this section we shall investigate a broadcast channel in which efficient communication to one receiver in no way interferes with communication to the other. A movie designed to be shown simultaneously to a blind person and a deaf person would be such an example.

Consider the broadcast channel with $X = \{1,2,3,4\}$, $Y_1 = \{1,2\}$, $Y_2 = \{1,2\}$, with

$$P_1 = \begin{bmatrix} 1 & 0 \\ 1 & 0 \\ 0 & 1 \\ 0 & 1 \end{bmatrix} \qquad P_2 = \begin{bmatrix} 1 & 0 \\ 0 & 1 \\ 1 & 0 \\ 0 & 1 \end{bmatrix} \tag{16}$$

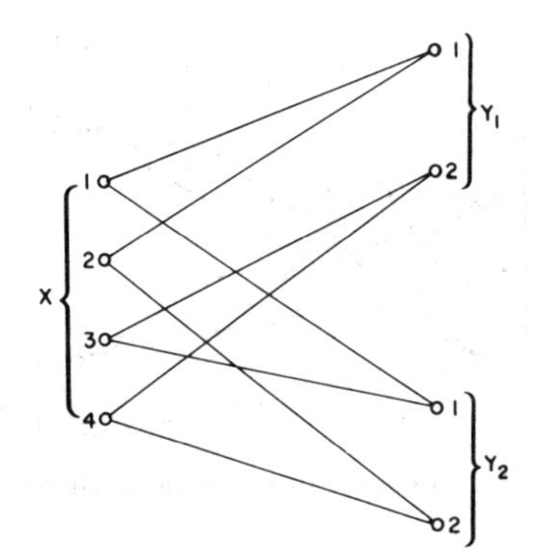

Fig. 7. Orthogonal channel.

as depicted in Fig. 7. As before,

$$(P_k)_{ij} = \Pr\{Y_k = j \mid x = i\},\ k = 1,2;\ j = 1,2;\ i = 1,2,3,4.$$

We easily calculate $C_1 = C_2 = 1$ bit/transmission. Clearly, from the standpoint of receiver y_1, inputs $x = 1$ and $x = 2$ both result in $y_1 = 1$ with probability 1 and can therefore be merged. Proceeding with this analysis, we find that Y_1 can determine only $x \in \{1,2\}$ versus $x \in \{3,4\}$, while Y_2 can determine only $x \in \{1,3\}$ versus $x \in \{2,4\}$.

For this example, $C_1 = 1$ and $C_2 = 1$ and are, respectively, attained for $\Pr\{x = 1\} + \Pr\{x = 2\} = \frac{1}{2}$ and $\Pr\{x = 1\} + \Pr\{x = 3\} = \frac{1}{2}$. Solving these simultaneous equations, we find $(I(X \mid Y_1), I(X \mid Y_2)) = (1,1)$ can be achieved by $\Pr\{x = i\} = \frac{1}{4}$, $i = 1,2,3,4$. This in itself does not guarantee that (C_1,C_2) can be simultaneously achieved. However, there does exist a coding theorem for this channel. Let $u_1 \in \{1,2\}$, $u_2 \in \{1,2\}$ denote the message bits that we wish to transmit to Y_1 and Y_2, respectively.

Make the association from pairs of u to input symbols

$$\begin{aligned}(u_1,u_2) &= (1,1) \mapsto 1\\ (u_1,u_2) &= (1,2) \mapsto 2\\ (u_1,u_2) &= (2,1) \mapsto 3\\ (u_1,u_2) &= (2,2) \mapsto 4\end{aligned} \tag{17}$$

and send the appropriate input symbol x. Then $y_1 = u_1$ and $y_2 = u_2$, and capacities C_1 and C_2 are simultaneously achieved. Since u_1 and u_2 may be chosen independently, we may also achieve $R_{12} = 1$ by this scheme. Fig. 8 shows the set of achievable rates. The upper bound theorem of Section VIII establishes this region as optimal.

The noiselessness of the channels is not crucial. This broadcast channel remains orthogonal in the sense that $(R_1,R_2) = (C_1,C_2)$ may be achieved even if we define the new channels

$$P_1 = \begin{bmatrix} r_1 & \bar{r}_1\\ r_1 & \bar{r}_1\\ \bar{r}_1 & r_1\\ \bar{r}_1 & r_1 \end{bmatrix} \qquad P_2 = \begin{bmatrix} r_2 & \bar{r}_2\\ \bar{r}_2 & r_2\\ r_2 & \bar{r}_2\\ \bar{r}_2 & r_2 \end{bmatrix} \tag{18}$$

In this case, $C_1 = 1 - H(r_1)$ and $C_2 = 1 - H(r_2)$. C_1 and C_2 may be simultaneously achieved by selecting sequences of

$$(2^{n(C_1-\varepsilon)}, n, \lambda_1^{(n)}),\ (2^{n(C_2-\varepsilon)}, n, \lambda_2^{(n)})$$

codes with words in $\{0,1\}^n$ such that $\lambda_1^{(n)} \to 0$, $\lambda_2^{(n)} \to 0$,

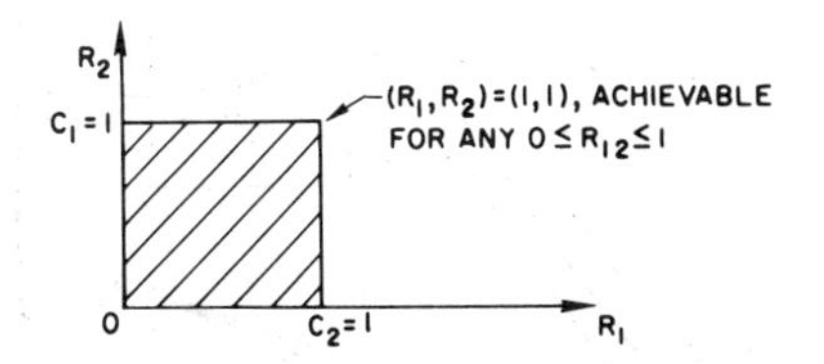

Fig. 8. Achievable rates for the orthogonal channel.

as $n \to \infty$, and selecting $x_i \in \{1,2\}$ or $x_i \in \{3,4\}$ according to the value of the ith bit of the codeword chosen to be sent from the first code and selecting $x_i \in \{1,3\}$ or $x_i \in \{2,3\}$ according to the value of the ith bit of the codeword selected from the second code. Here, any R_{12} such that $0 \le R_{12} \le \min\{C_1,C_2\}$ may also be achieved. Nothing more could be expected, and each channel performs no worse in the presence of the other than it would alone.

V. Incompatible Broadcast Channels

In a search to find the worst case of incompatibility in simultaneous communication we turn to the following practical example which, for obvious reasons, we term the switch-to-talk channel.

Example 1—Switch-to-Talk: Let

$$\begin{aligned} X &= X_1 \cup X_2\\ Y_1 &= \tilde{Y}_1 \cup \{\phi_1\}\\ Y_2 &= \tilde{Y}_2 \cup \{\phi_2\}\end{aligned}$$

and

$$P_1 = \begin{array}{c} \\ X_1 \\ \\ \\ X_2 \\ \\ \end{array}\left[\begin{array}{ccccc|c} \multicolumn{5}{c|}{\tilde{Y}_1} & \phi_1 \\ x & x & x & \cdots & x & 0\\ & & & & & 0\\ x & & & \cdots & x & 0\\ \hline 0 & 0 & & \cdots & 0 & 1\\ & & & & & 1\\ 0 & 0 & & \cdots & 0 & 1 \end{array}\right]$$

$$P_2 = \begin{array}{c} \\ X_1 \\ \\ \\ X_2 \\ \\ \end{array}\left[\begin{array}{ccccc|c} \multicolumn{5}{c|}{\tilde{Y}_2} & \phi_2 \\ 0 & 0 & & \cdots & 0 & 1\\ & & & & & 1\\ 0 & 0 & & \cdots & 0 & 1\\ \hline x & x & & \cdots & x & 0\\ & & & & & 0\\ x & & & \cdots & x & 0 \end{array}\right] \tag{19}$$

as shown in Fig. 9.

Each receiver has an indicator that lights when the sender is communicating with the other receiver. The idea is that when the sender wishes to communicate with Y_1 he uses $x \in X_1$, resulting in $y_2 = \phi_2$, indicating to receiver 2 that the sender is communicating with Y_1. Similarly, to communicate with Y_2, the sender uses $x \in X_2$, resulting in $y_1 = \phi_1$. This might correspond to the situation, for example, where a speaker fluent in Spanish and Dutch must speak simultaneously to two listeners, one of whom understands only Dutch and the other only Spanish.

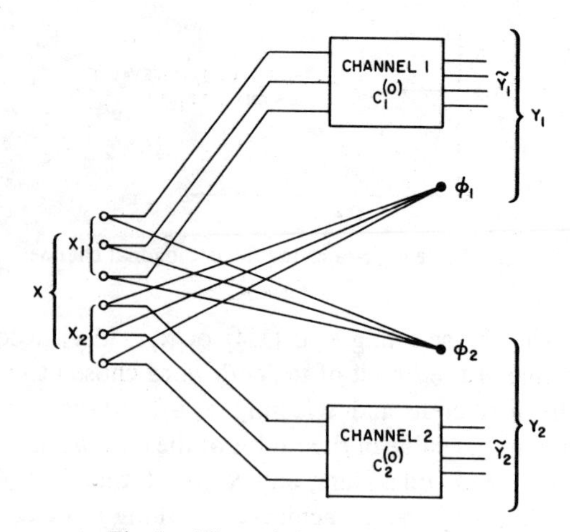

Fig. 9. The switch-to-talk channel.

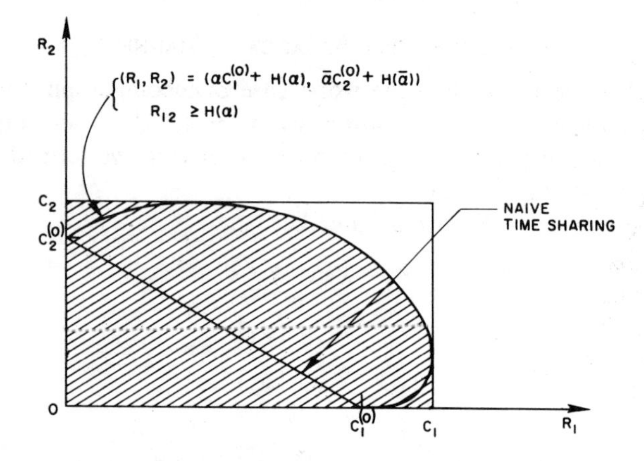

Fig. 10. Achievable rates for switch-to-talk channels.

Let channel 1 have capacity $C_1^{(0)}$ and channel 2 have capacity $C_2^{(0)}$. Using the known result for sum channels (see Shannon [3]) we find

$$C_1 = \log(1 + 2^{C_1(0)})$$

and

$$C_2 = \log(1 + 2^{C_2(0)}).$$

We shall discuss this example informally. Certainly $(R_1,R_2) = (C_1,0)$ is achievable and $(R_1,R_2) = (0,C_2)$ is achievable, and hence, by time sharing, any pair of rates $(R_1,R_2) = (\lambda C_1,\bar{\lambda} C_2)$, $0 \leq \lambda \leq 1$, is achievable. However, additional information is contained in the knowledge of ϕ; and proper encoding of the transmission times to Y_1 and Y_2 can be used to send extra information to both channels. If channel 1 is used proportion α of the time, $\alpha C_1^{(0)}$ bits/transmission are received by Y_1. However, $H(\alpha)$ additional bits/transmission are achieved by choosing which channel to send through independently at each instant by flipping a coin with bias α. In other words, modulation of the switch-to-talk button, subject to the time-proportion contraint α, allows the perfect transmission of one of $2^{nH(\alpha)}$ additional messages to both receivers Y_1 and Y_2.

Thus all (R_1,R_2) of the form $(R_1,R_2) = (\alpha C_1^{(0)} + H(\alpha),\bar{\alpha} C_2^{(0)} + H(\alpha))$ can be achieved by choosing the subset of n transmissions devoted to the use of channel 1 in one of the $2^{nH(\alpha)}$ possible ways. This bound cannot be achieved unless the information rate R_{12} common to both channels satisfies $R_{12} \geq H(\alpha)$. The results are summarized in Fig. 10.

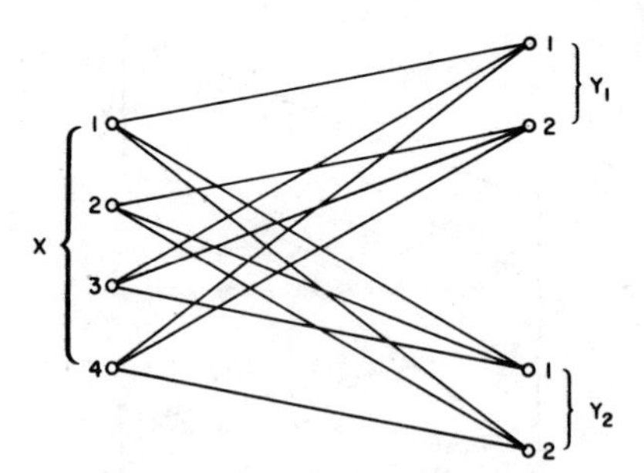

Fig. 11. Incompatible broadcast channels.

It is an easy consequence of Section VIII that Fig. 10 corresponds to the capacity region for this channel, and therefore that this encoding scheme is optimal for the switch-to-talk channel.

The following example illustrates the worst case that may arise in simultaneous communications.

Example 2—Incompatible Case: Let

$$X = \{1,2,3,4\},\ Y_1 = \{1,2\},\ Y_2 = \{1,2\}$$

and let

$$P_1 = \begin{bmatrix} 1 & 0 \\ 0 & 1 \\ \frac{1}{2} & \frac{1}{2} \\ \frac{1}{2} & \frac{1}{2} \end{bmatrix} \qquad P_2 = \begin{bmatrix} \frac{1}{2} & \frac{1}{2} \\ \frac{1}{2} & \frac{1}{2} \\ 1 & 0 \\ 0 & 1 \end{bmatrix} \tag{20}$$

as shown in Fig. 11. Thus if X wishes to communicate with Y_1 over the perfect channel $x \in \{1,2\} \to Y_1$, he must send pure noise to Y_2, i.e., $\Pr\{y_2 = 1 \mid x \in \{1,2\}\} = \frac{1}{2}$. A similar statement holds for X communicating with Y_2.

In Section VIII we shall establish an upper bound on the capacity region by finding the set of all achievable $(I(X \mid Y_1), I(X \mid Y_2))$ pairs. Anticipating these results, we shall make this calculation for this example. Let $\Pr\{x = i\} = p_i$, $i = 1,2,3,4$. Define $\alpha = p_1 + p_2$, $\bar{\alpha} = p_3 + p_4$. Then $H(Y_1) = H(p_1 + \bar{\alpha}/2)$ and $H(Y_1 \mid X) = \bar{\alpha}$, yielding $I(X \mid Y_1) = H(p_1 + \bar{\alpha}/2) - \bar{\alpha}$. Similarly, $I(X \mid Y_2) = I(X \mid Y_2) = H(p_4 + \alpha/2) - \alpha$.

First, fixing α, $\bar{\alpha}$ and maximizing over $0 \leq p_1 \leq \alpha$, $0 \leq p_4 \leq \bar{\alpha}$, we find the maximum values

$$I(X \mid Y_1) = 1 - \bar{\alpha} = \alpha$$

$$I(X \mid Y_2) = 1 - \alpha = \bar{\alpha} \tag{21}$$

achieved by $p_1 = p_2 = \alpha/2$ and $p_3 = p_4 = \bar{\alpha}/2$. This is the upper boundary of achievable $(I(X \mid Y_1), I(X \mid Y_2))$ pairs.

It may also be verified that, for any $\alpha \in [0,1]$, there exist p_1,p_2,p_3,p_4 achieving any $(I(X \mid Y_1),I(X \mid Y_2))$ dominated by $(\alpha, 1 - \alpha)$. Thus we have the set I of achievable $(I(X \mid Y_1), I(X \mid Y_2))$ pairs depicted in Fig. 12.

In Section VIII it will be shown that this region of jointly achievable $(I(X \mid Y_1),I(X \mid Y_2))$ pairs is an upper bound on the capacity region. However, we can trivially achieve any pair of rates (R_1,R_2) on the upper boundary of $\mathscr{R}$ by simply

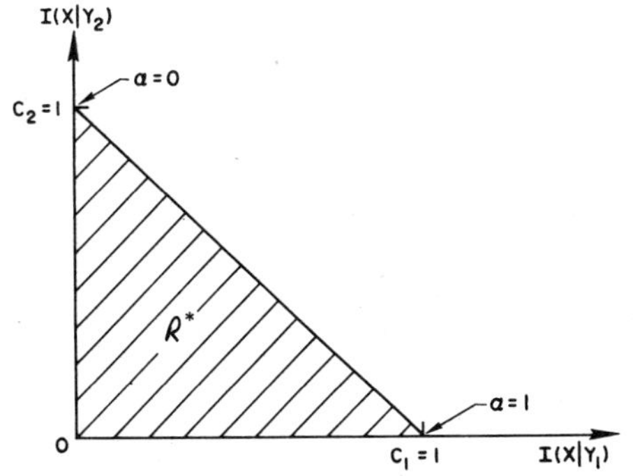

Fig. 12. Capacity region for incompatible channels.

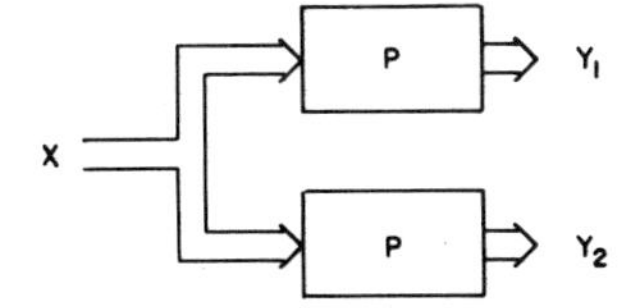

Fig. 13. The bottleneck channel.

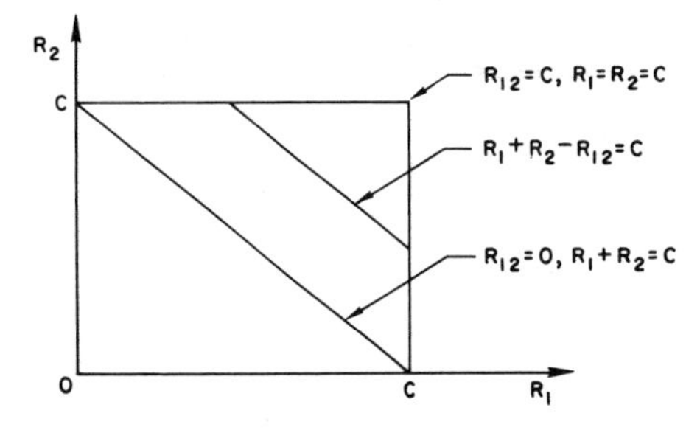

Fig. 14. Achievable rates for bottleneck channel.

time-sharing the two noiseless channels $x \in \{1,2\} \to Y_1$ and $x \in \{3,4\} \to Y_2$. If $x \in \{1,2\}$ is used a proportion α of the time, then rates $R_1 = \alpha$ and $R_2 = \bar{\alpha} = 1 - \alpha$ may be achieved without any additional coding. Thus the upper bound can be achieved with trivial coding procedures, and Fig. 12 therefore corresponds to the capacity region.

Here, then, is an example in which the two channels are so incompatible that one can do no better than time sharing —i.e., using one channel efficiently part of the time and the other channel the remainder. Fortunately, for those wishing to get something for nothing, this is the exception rather than the rule.

VI. The Bottleneck Channel

Consider the broadcast channel in which the two channels have the same structure, i.e.,

$$p_1(y_1 \mid x) = p_2(y_2 \mid x), \forall x \in X, \forall y_1,y_2 \in Y_1 = Y_2 = Y$$

as shown in Fig. 13. We shall term this the bottleneck channel.

Here, we note that any code for receiver Y_1 is also a code with the same error properties for receiver Y_2. Thus Y_1 and Y_2 both perceive correctly the transmitted sequence x with low probability of error.

Let the capacity of channel P be denoted by $C_1 = C_2 = C$ bits per transmission. Now, since both receivers receive the same information about X, it follows that both receivers 1 and 2 will be able to correctly recover r, s_1 and s_2 if and only if (R_1,R_2,R_{12}) is an achievable rate. Counting the number of messages per unit time necessary to transmit (r,s_1,s_2) correctly yields the following proposition [see comment following (14)].

Proposition: (R_1,R_2,R_{12}) is an achievable rate for the broadcast bottleneck channel of capacity C if and only if

$$R_1 + R_2 - R_{12} \le C$$

$$0 \le R_1 \le C$$

$$0 \le R_2 \le C$$

$$0 \le R_{12} \le C. \tag{22}$$

As an important application of these ideas, suppose that we wish to send a random process $U = \{U_n : n = 1,2,\cdots\}$ to receiver 1 and a random process $V = \{V_n : n = 1,2,\cdots\}$ to receiver 2 through the bottleneck channel P with arbitrarily small probability of error. (See Fig. 15).

Assume that $U = \{U_n\}$ and $V = \{V_n\}$ are jointly ergodic processes taking values in finite alphabets. By jointly ergodic, we mean that the process $Z_n = (U_n,V_n)$ is ergodic. We recall that the definition of the entropy of an ergodic process $\{Z_n\}$ is defined by

$$H(Z) = \lim_{n\to\infty} n^{-1}H(Z_1,Z_2,\cdots,Z_n). \tag{23}$$

We assert the following.

Fact: Asymptotically error free transmission of $\{U_1,U_2,\cdots,U_n\} \to \{\hat{U}_1,\hat{U}_2,\cdots,\hat{U}_n\}$ and $\{V_1,V_2,\cdots,V_n\} \to \{\hat{V}_1,\hat{V}_2,\cdots,\hat{V}_n\}$ over the bottleneck channel of capacity C can be accomplished if and only if

$$H(U,V) < C. \tag{24}$$

Proof: The well-known idea of the encoding is to enumerate the $2^{n(H(U,V)+\varepsilon)}$ ε-typical sequences and send the index of the actually occurring sequence $(z_1,z_2,\cdots,z_n)$ over the channel. If $H(U,V) + \varepsilon < C$, then this index will be correctly transmitted with probability of error $\varepsilon/2$ for sufficiently large n. Since the probability that a random $(z_1,z_2,\cdots,z_n)$ will be typical can be made $\ge 1 - \varepsilon/2$ for sufficiently large n, the overall probability of error can be made less than ε. The converse follows the standard argument for a single channel.

The generalization of this result to arbitrary broadcast channels is unknown.

Let us now compare the orthogonal channel with the bottleneck channel. The orthogonal channel of Section IV achieves $(R_1,R_2) = (1,1)$ with *arbitrary* joint rate $0 \le R_{12} \le 1$. Thus fully independent messages $(R_{11} = 0)$ or maximally dependent messages $(R_{11} = 1)$ can be sent simultaneously to receivers 1 and 2.

At the other extreme, in the case of the bottleneck channel with capacity $C = 1$, we can simultaneously achieve $R_1 = 1$, $R_2 = 1$. Here however, it may be seen that achieving $(R_1,R_2) = (1,1)$ implies $R_{12} = 1$. Thus the messages sent to 1 and 2 must be maximally dependent, and in fact equal.

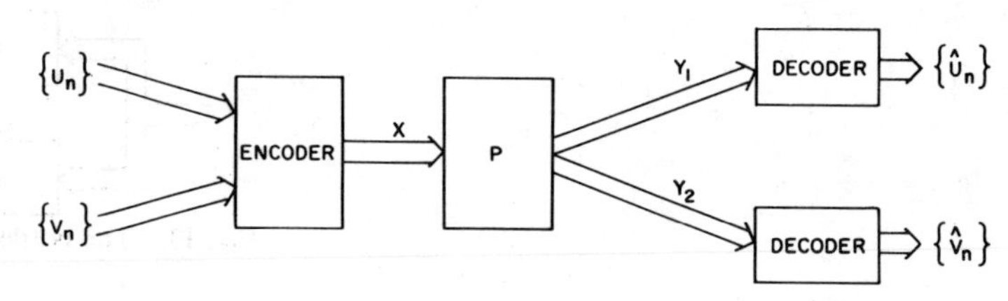

Fig. 15. Sending two random processes over the same channel.

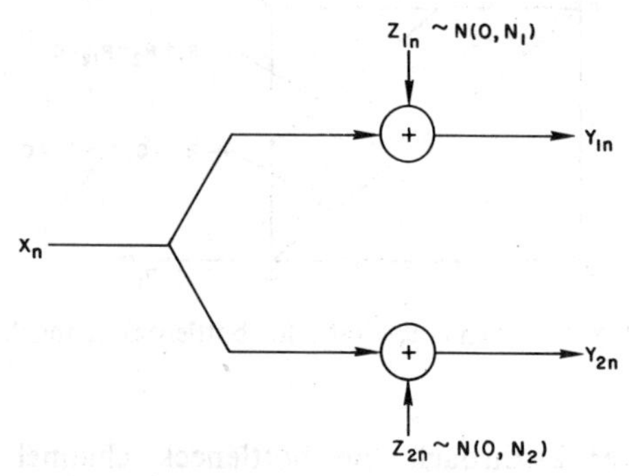

Fig. 16. Gaussian broadcast channel.

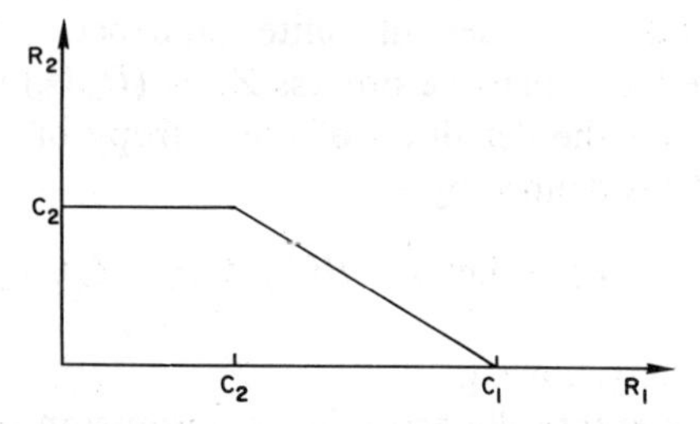

Fig. 17. Time sharing rates for the Gaussian broadcast channel.

VII. Gaussian Channels

Consider the time-discrete Gaussian broadcast channel with two receivers depicted in Fig. 16.

Let $z_1 = (z_{11}, z_{12}, \cdots, z_{1n}, \cdots)$ be a sequence of independently identically distributed (i.i.d.) normal random variables (RV) with mean zero and variance N_1, and let $z_2 = (z_{21}, z_{22}, \cdots, z_{2n}, \cdots)$ be i.i.d. normal RV with mean zero and variance N_2. Let $N_1 < N_2$. At the ith transmission the real number x_i is sent and $y_{1i} = x_i + z_{1i}$, $y_{2i} = x_i + z_{2i}$ are received. In our analysis it is irrelevant whether z_{1i} and z_{2i} are correlated or not (although in the feedback case it may make a difference). Let there be a power constraint on the transmitted power, given for any n by

$$\frac{1}{n} \sum_{i=1}^{n} x_i^2 \leq S \tag{25}$$

for any signal $x = (x_1, x_2, \cdots, x_n)$ of block length n.

It is well known that the individual capacities are $C_1 = \frac{1}{2} \log (1 + S/N_1)$ and $C_2 = \frac{1}{2} \log (1 + S/N_2)$ bits/transmission, where all logarithms are to the base 2.

Time sharing will achieve any convex combination of (C_2, C_2) and $(C_1, 0)$, as shown in Fig. 17.

Now let us see how we can improve on this performance. Think of the signal s_2 (intended for the high noise receiver Y_2) as a sequence of i.i.d. $N(0, \bar{\alpha}S)$ RV. Superimposed on this sequence will be a sequence s_1 that may be considered

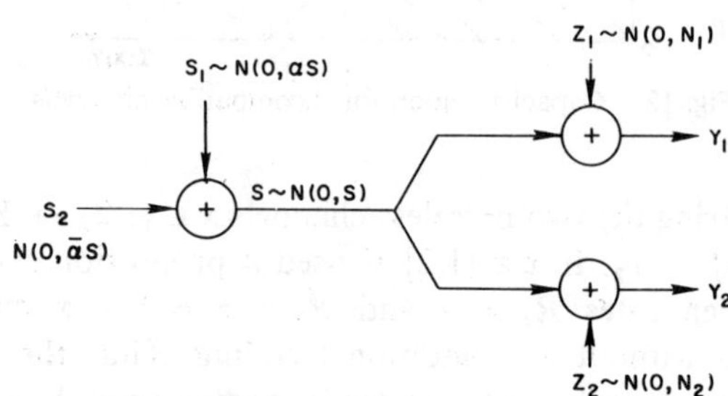

Fig. 18. Decomposition of the signal.

as a sequence of i.i.d. $N(0, \alpha S)$ RV. Here $0 \leq \alpha \leq 1$ and $\bar{\alpha} = 1 - \alpha$. Thus the sequence $s = s_1 + s_2$ will be a sequence of i.i.d. $N(0, S)$ RV. The received sequences $y_1 = s_1 + s_2 + z_1$ and $y_2 = s_1 + s_2 + z_2$ are depicted in Fig. 18.

Now s_1 and z_2 are considered to be noise by receiver 2. We see that $s_{1i} + z_{2i}$ are i.i.d. $N(0, \alpha S + N_2)$ RV. Therefore, messages may be sent at rates less than

$$\frac{1}{2} \log \left(1 + \frac{\bar{\alpha}S}{\alpha S + N_2}\right) \triangleq C_2(\alpha)$$

to receiver Y_2 with probability of error near zero for sufficiently large block length n. That is, there exists a sequence of $(2^{n(C_2(\alpha) - \varepsilon)}, n)$ codes with average power constraint $\bar{\alpha}S$ and probability of error $\bar{p}_2^{(n)}(e) \to 0$.

Now, since $N_1 < N_2$, receiver Y_1 may also correctly determine the transmitted sequence s_2 with arbitrarily low probability of error. Upon decoding of s_2, given y_1, receiver Y_1 then subtracts s_2 from y_1, yielding $\tilde{y}_1 = y_1 - s_2 = s_1 + z_1$. At this stage channel 1 may be considered to be a Gaussian channel with input power constraint αS and additive zero mean Gaussian noise with variance N_1. The capacity of this channel is $\frac{1}{2} \log [1 + (\alpha S/N_1)] = \tilde{C}_1(\alpha)$ bits/transmission and is achieved, roughly speaking, by choosing $2^{n\tilde{C}_1(\alpha)}$ independent n-sequences of i.i.d. $N(0, \alpha S)$ RV as the code set for the possible sequences s_1. Thus receiver Y_1 correctly receives both s_1 and s_2.

This informal argument indicates that rates

$$R_1 = \frac{1}{2} \log \left(1 + \frac{\bar{\alpha}S}{\alpha S + N_2}\right) + \frac{1}{2} \log \left(1 + \frac{\alpha S}{N_1}\right)$$

$$R_2 = \frac{1}{2} \log \left(1 + \frac{\bar{\alpha}S}{\alpha S + N_2}\right) \tag{26}$$

may simultaneously be ε-achieved, for any $0 \leq \alpha \leq 1$. These rate pairs, shown in Fig. 19, dominate the time-sharing rates.

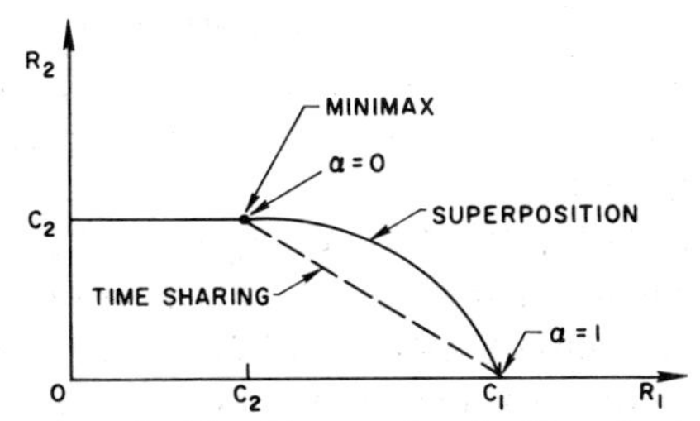

Fig. 19. Set of achievable rates for the Gaussian broadcast channel.

Summarizing the argument, we select a set of $2^{n(C_2(\alpha)-\varepsilon)}$ random n-sequences of i.i.d. $N(0,\alpha S)$ RV, and a set of $2^{n(\tilde{C}_1(\alpha)-\varepsilon)}$ random n-sequences of i.i.d. $N(0,\bar{\alpha}S)$ RV. Now $2^{n(\tilde{C}_1(\alpha)+C_2(\alpha)-2\varepsilon)}$ n-sequences are formed by adding together pairs of sequences, in which the first sequence is chosen from the first set and the second sequence is chosen from the second set, and the pairs are chosen in all possible ways. A message

$$(r,s_1),\ r \in \{1,2,\cdots,2^{n(C_2(\alpha)-\varepsilon)}\},\ s_1 \in \{1,2,\cdots,2^{n(\tilde{C}_1(\alpha)-\varepsilon)}\}$$

is transmitted by selecting the n-sequence corresponding to the sum of the rth sequence in the first set and the s_1th sequence in the second set. Receiver 1 is intended to decode (r,s_1) correctly and receiver 2 is intended to decode r correctly, thus simultaneously achieving rates

$$R_1 = \tilde{C}_1(\alpha) + C_2(\alpha) - 2\varepsilon$$
$$R_2 = C_2(\alpha) - \varepsilon \tag{27}$$

as given in (26).

A full discussion of the Gaussian channel would lead far afield. A direct simple proof of the achievability of the rates given in (27) has been found but will not be presented here.

We shall conclude this section with one observation. If $N_1 = 0$, and channel 1 is therefore perfect, we have $C_1 = \infty$ and $C_2 = \frac{1}{2}\log(1 + S/N_2)$. A compound channel or maximin approach would have us send at rates $(R_1,R_2) = (C_2,C_2)$. However, an arbitrarily small decrement in the rate for channel 2, corresponding to $0 < \alpha \ll 1$ in (26), yields $(R_1,R_2) = (\infty, C_2 - \varepsilon)$ as a pair of achievable rates. Although this rate pair does not dominate (C_2,C_2), it seems vastly preferable.

VIII. An Upper Bound on Achievable Rates (R_1,R_2)

Suppose that $p(x)$, a probability distribution on X, generates the pair of mutual informations $(I(X \mid Y_1), I(X \mid Y_2))$, where, for $i = 1,2$,

$$I(X \mid Y_i) = \sum_{x\in X}\sum_{y\in Y_i} p(x)p_i(y \mid x)\log\frac{p_i(y \mid x)}{p_i(y)}. \tag{28}$$

Given the intuitive properties of mutual information, it is natural to assume that rates $R_1 = I(X \mid Y_1)$, $R_2 = I(X \mid Y_2)$ are therefore simultaneously achievable. This turns out not to be the case. (Close inspection of the example of two BSC in Section II, with $\Pr\{x = 1\} = \frac{1}{2}$ and $I(X \mid Y_1) = 1$, $I(X \mid Y_2) = C_2$, will yield a counterexample.) However, the

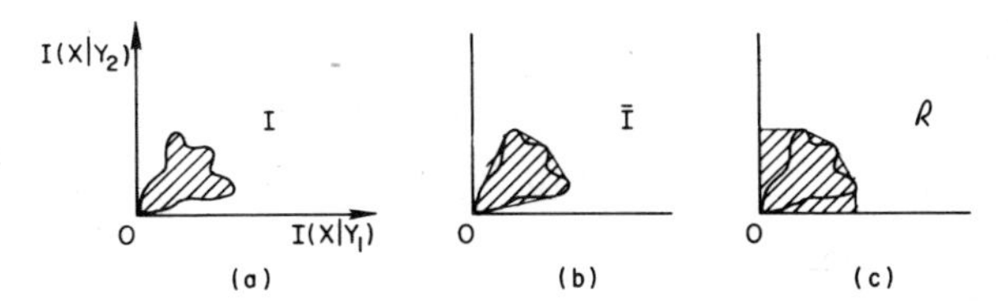

Fig. 20. Upper bound $\mathscr{R}$ on capacity region.

set of jointly achievable mutual-information pairs, properly modified to take into account the possibility of time-sharing and throwing information away, does yield an upper bound $\mathfrak{R}$ on the capacity region $\mathfrak{R}^*$. This upper bound is actually achieved by the orthogonal-channel, switch-to-talk-channel, and incompatible-channel examples.

Thus we proceed to define $\mathfrak{R}$ and establish $\mathfrak{R}$ as an upper bound. Let

$$I = \{(I(X \mid Y_1), I(X \mid Y_2)) \mid p(x) \geq 0, \textstyle\sum p(x) = 1\} \tag{29}$$

denote the set of all pairs $(I(X \mid Y_1), I(X \mid Y_2))$ generated by $p(x)$ as $p(\cdot)$ ranges over the simplex of possible probability distributions on X. Define $\bar{I}$ to be the convex hull of I. Thus $\bar{I}$ may be interpreted as the average joint mutual information achievable by varying $p(\cdot)$ with time. Let

$$\mathfrak{R} = \{(R_1,R_2) \in E_2 \mid R_1 \leq I_1,\ R_2 \leq I_2, \text{ for some } (I_1,I_2) \in \bar{I}\}. \tag{30}$$

Thus $\mathfrak{R}$ intuitively corresponds to the joint mutual information achievable from $\bar{I}$ by throwing information away. These sets are depicted in Fig. 20. We now show $\mathfrak{R}^* \subseteq \mathfrak{R}$.

Lemma 1: Given an arbitrary $((M_1,M_2,M_{12});n)$ code for the nth extension of a broadcast channel, consisting of words $x(r,s_1,s_2) \in X^n$, $r \in R$, $s_1 \in S_1$, $s_2 \in S_2$, $|R| = M_{12}$, $|S_1| = M_1$, $|S_2| = M_2$, $M = M_{12}M_1M_2$; let (r,s_1,s_2) be a random variable with range $R \times S_1 \times S_2$. Let $(y_1,y_2) \in Y_1^n \times Y_2^n$ be the corresponding random output n-sequences received by 1 and 2, generated by sending $x(r,s_1,s_2)$ over the channel. If $p_1(e) = \Pr\{(\hat{r}_1,\hat{s}_1) \neq (r_1,s_1)\}$ and $p_2(e) = \Pr\{(\hat{r}_1,\hat{s}_2) \neq (r_1,s_2)\}$ are the receiver probabilities of error of the code, then,

$$H(X \mid Y_1) \leq 1 + \log M_2 + p_1(e)\log M_{12}M_1 \tag{31}$$

$$H(X \mid Y_2) \leq 1 + \log M_1 + p_2(e)\log M_{12}M_2. \tag{32}$$

Proof: Let the decoding rules corresponding to the code be

$$g_1: Y_1^n \to R \times S_1$$
$$g_2: Y_2^n \to R \times S_2 \tag{33}$$

written

$$g_k(y_k) = (g_{k1}(y_k), g_{k2}(y_k)), \qquad k = 1,2.$$

Thus, given a random message (r,s_1,s_2) and sequence $y_k \in Y_k^n$, receiver k will make an error if and only if

$$g_1(y_1) \neq (r,s_1), \qquad k = 1$$
$$g_2(y_2) \neq (r,s_2), \qquad k = 2. \tag{34}$$

Thus

$$p_1(e) = \Pr\{g_1(y_1) \neq (r,s_1)\}$$
$$p_2(e) = \Pr\{g_2(y_2) \neq (r,s_2)\}. \quad (35)$$

We note that

$$H(X \mid y_1) \leq H(p_1(e \mid y_1), 1 - p_1(e \mid y_1)) + (1 - p_1(e \mid y_1)) \log M_2 + p_1(e \mid y_1) \log (M - M_2), \quad (36)$$

where we have used the inequality

$$H(a_1,a_2,\cdots,a_m) \leq \log m, \quad (37)$$

and a basic composition relation (see Ash [1, p. 8]). We have, of course, conditioned on the events $g_1(y_1) = (r,s_1)$ and $g_1(y_1) \neq (r,s_1)$. Taking the expectation over $Y_1{}^n$, and using the convexity of $H(p, 1 - p)$ in p, we have

$$H(X \mid Y_1) \leq H(p_1(e), 1 - p_1(e)) + (1 - p_1(e)) \log M_2 + p_1(e) \log (M - M_2). \quad (38)$$

Finally, since $H(p, 1 - p) \leq 1$ and $M = M_{12}M_1M_2$ we have

$$H(X \mid Y_1) \leq 1 + \log M_2 + p_1(e) \log (M - M_2)/M_2 \leq 1 + \log M_2 + p_1(e) \log M_{12}M_1. \quad (39)$$

The corresponding argument for $H(X \mid Y_2)$ completes the proof.

We shall need the following lemma Ash [1,p. 81].

Lemma 2: Let $X_1,\cdots,X_n$ be a sequence of input random variables to the (discrete memoryless) broadcast channel and $Y_{11},\cdots,Y_{1n},Y_{21},\cdots,Y_{2n}$ the corresponding received output random variables for 1 and 2, respectively. Then

$$I(X_1,\cdots,X_n \mid Y_{k1},\cdots,Y_{kn}) \leq \sum_{i=1}^{n} I(X_i \mid Y_{ki}), \qquad k = 1,2,$$

with equality iff $Y_{k1},Y_{k2},\cdots,Y_{kn}$ are independent.

Proof:

$$H(Y_{k1},Y_{k2},\cdots,Y_{kn} \mid X_1,\cdots,X_n) \triangleq -\sum p_k(x,y_k) \log p_k(y_k \mid x),$$

but, because the channel k is memoryless, $p_k(y_k \mid x)$ factors into a product $\prod p_k(y_{ki} \mid x_i)$, yielding

$$H(Y_{k1},\cdots,Y_{kn} \mid X_1,\cdots,X_n) = \sum_{x,y_k} p_k(x,y_k) \sum_{i=1}^{n} \log p_k(y_{ki} \mid x_i) = \sum_{i=1}^{n} H(Y_{ki} \mid X_i).$$

Also, by a basic inequality

$$H(Y_{k1},\cdots,Y_{kn}) \leq \sum_{i=1}^{n} H(Y_{ki}),$$

with equality iff Y_{ki} are independent for $i = 1,2,\cdots,n$. Since $I(X \mid Y_k) = H(Y_k) - H(Y_k \mid X)$, the lemma follows.

We now wish to show that $\bar{p}_1{}^{(n)}(e)$, $\bar{p}_2{}^{(n)}(e)$ cannot simultaneously tend to zero for rates $(R_1,R_2) \notin \mathfrak{R}$. This will establish $\mathfrak{R}$ as an upper bound on the capacity region for a broadcast channel.

Let $R_1 = 1/n \log M_1M_{12}$ and $R_2 = 1/n \log M_2M_{12}$ be the rates of communication in bits/transmission for receivers Y_1 and Y_2, respectively. (We recall that $R_{12} = \log M_{12}$ is the transmission rate for information common to both channels.) The proof closely resembles that used by Shannon [4] for the two-way channel.

Theorem: For any sequence of $[(2^{nR_1},2^{nR_2},2^{nR_{12}}),n]$ codes, $(R_1,R_2) \notin \mathfrak{R}$ implies that

$$(\bar{p}_1{}^{(n)}(e),\bar{p}_2{}^{(n)}(e)) \rightarrow (0,0),\ (\lambda_1{}^{(n)},\lambda_2{}^{(n)}) \nrightarrow (0,0), \qquad n \nrightarrow \infty.$$

Thus $\mathfrak{R}$ is an upper bound on the capacity region for the broadcast channel.

Proof: Given an arbitrary $[(M_1,M_2,M_{12}),n]$ code for the nth extension of the broadcast channel, choose a codeword $x(r,s_1,s_2)$ at random according to a uniform distribution $\Pr\{r,s_1,s_2\} = 1/M$, $(r,s_1,s_2) \in R \times S_1 \times S_2$, where $M = |R||S_1||S_2|$. If the codewords $x(r,s_1,s_2) \in X^n$ are not distinct, a simple modification of the proof below will prove the theorem. Thus treating the case where the $x(r,s_1,s_2)$ are distinct, we have $H(X) = \log M$ and $I(X \mid Y_1) = \log M - H(X \mid Y_1)$, under the given uniform distribution on the codewords. As in Section III, let $\bar{p}_1{}^{(n)}(e)$ and $\bar{p}_2{}^{(n)}(e)$ designate the probabilities of error of the code under this distribution. By Lemma 2,

$$I(X \mid Y_1) \leq \sum_{i=1}^{n} I(X_i \mid Y_{1i}). \quad (41)$$

Thus

$$I(X \mid Y_1) = \log M - H(X \mid Y_1) \leq \sum_{i=1}^{n} I(X_i \mid Y_{1i}). \quad (42)$$

Finally, since (31) in Lemma 1 holds for any distribution on the codewords, substitution in (42) yields

$$\log M - 1 - \log M_2 - \bar{p}_1{}^{(n)}(e) \log M_{12}M_1 \leq \sum_{i=1}^{n} I(X_i \mid Y_{1i}), \quad (43)$$

which becomes the basic inequality

$$R_1 \triangleq \frac{1}{n} \log M_{12}M_1 \leq \frac{(1/n) + (1/n) \sum_{i=1}^{n} I(X_i \mid Y_{1i})}{1 - \bar{p}_1{}^{(n)}(e)}. \quad (44a)$$

Similarly, we find

$$R_2 \triangleq \frac{1}{n} \log M_{12}M_2 \leq \frac{(1/n) + (1/n) \sum_{i=1}^{n} I(X_i \mid Y_{2i})}{1 - \bar{p}_2{}^{(n)}(e)}. \quad (44b)$$

Summarizing, an arbitrary code for the nth extension of a broadcast channel must have rates (R_1,R_2) satisfying (44a) and (44b), where

$$\bar{p}_i{}^{(n)}(e) = \frac{1}{M} \sum_{r,s_1,s_2} \lambda_i(r,s_1,s_2), \qquad i = 1,2. \quad (45)$$

Now suppose $(R_1,R_2) \notin \mathfrak{R}$, $R_1 \geq 0$, $R_2 \geq 0$ as in Fig. 21.

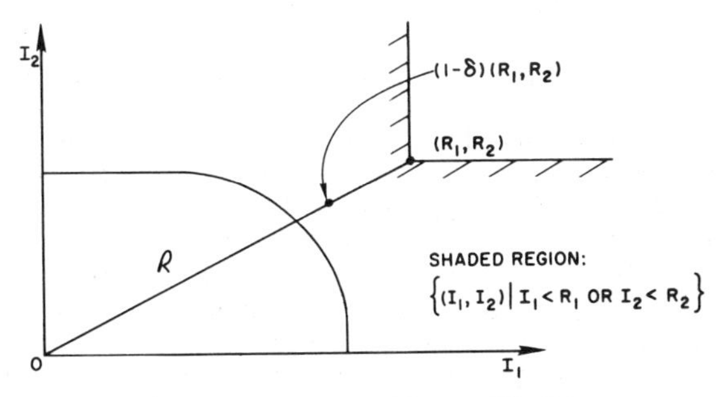

Fig. 21. Unachievable rate (R_1,R_2).

We shall show that $\bar{p}_i^{(n)}(e)$, $i = 1,2$ cannot simultaneously be small.

By the convexity of $\mathfrak{R}$ and $I \subseteq \mathfrak{R}$, we have

$$\left(\frac{1}{n}\sum_{i=1}^{n} I(X \mid Y_{1i}), \frac{1}{n}\sum_{i=1}^{n} I(X \mid Y_{2i})\right) \in \mathfrak{R},$$

for all $p(x)$. Consequently, as illustrated in Fig. 21, either

$$\frac{1}{n}\sum_{i=1}^{n} I(X \mid Y_{1i}) < R_1(1-\delta) \tag{46a}$$

or

$$\frac{1}{n}\sum_{i=1}^{n} I(X \mid Y_{2i}) < R_2(1-\delta), \tag{46b}$$

where $\delta > 0$ is any nonnegative real number such that $(1-\delta)(R_1,R_2) \notin \mathfrak{R}$.

But (44) implies for $i = 1,2$ that

$$\bar{p}_i^{(n)}(e) \geq 1 - \frac{1}{nR_i} - \frac{(1/n)\sum_{j=1}^{n} I(X \mid Y_{ij})}{R_i}. \tag{47}$$

The second term on the right-hand side of (47) tends to zero with n, but the third term must be less than $(1-\delta)$ for either $i = 1$ or $i = 2$, or both. Thus

$$\lim_{n\to\infty} \max \{\bar{p}_1^{(n)}(e),\bar{p}_2^{(n)}(e)\} \geq \delta > 0, \tag{48}$$

and therefore $\bar{p}_1^{(n)}(e)$, $\bar{p}_2^{(n)}(e)$ may not simultaneously be near zero. Also, since the probability of error $\lambda_i^{(n)}$ of the worst codeword for each channel obeys $\lambda_i^{(n)} \geq \bar{p}_i^{(n)}(e)$, $i = 1, 2$, we conclude that if $(R_1,R_2) \notin \mathfrak{R}$, then there exists no sequence of $((2^{nR_1},2^{nR_2},2^{nR_{12}}),n)$ codes for a broadcast channel such that $(\lambda_1^{(n)},\lambda_2^{(n)}) \to (0,0)$.

IX. An Approach to Compound Channels

Let $P_\beta(y \mid x)$, $\beta \in \mathfrak{B}$ be a perhaps infinite collection of channel transmission functions. An index β will be chosen by nature and a sequence of n transmissions $x_1,x_2,\cdots,x_n$ will be sent to the receiver over the discrete memoryless channel $P_\beta(y \mid x)$. The index β is unknown to the sender but may, without loss of generality, be assumed known to the receiver. (Simply sending $\sqrt{n}$ prearranged symbols in n transmissions will allow the receiver to determine β with arbitrarily low probability of error, for finite $\mathfrak{B}$, without affecting the achievable rate R.) Wolfowitz [2] and Blackwell *et al.* [5] have defined the capacity C of the compound channel to be

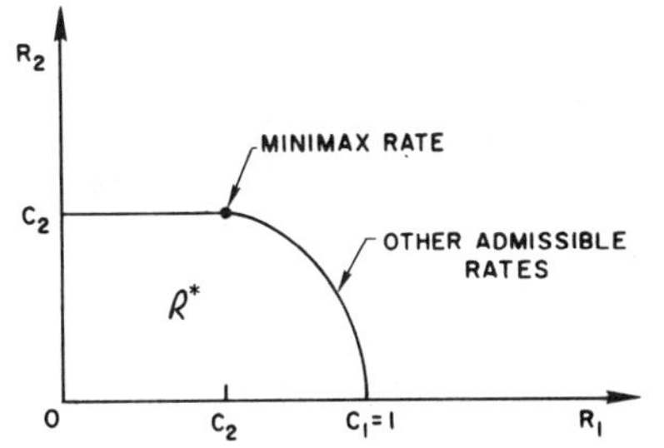

Fig. 22. Set of achievable rates for compound channel.

$$C = C_{\text{maximin}} = \sup_{p(x)} \inf_{\beta} I_\beta(X \mid Y). \tag{49}$$

This rate C is achieved for finite $\mathfrak{B}$ by designing the code for the channel β^* such that

$$C = \max_{p(x)} I_{\beta^*}(X \mid Y). \tag{50}$$

The maximin rate C is then achieved independently of the β chosen by nature.

Now consider a communication link in which it is unknown whether the link is a perfect binary symmetric channel or a binary symmetric channel of parameter p. Thus the channel descriptions $P_\beta(y \mid x)$, $\beta = 1,2$, are given by

$$P_1 = \begin{bmatrix} 1 & 0 \\ 0 & 1 \end{bmatrix} \qquad P_2 = \begin{bmatrix} \bar{p} & p \\ p & \bar{p} \end{bmatrix}. \tag{51}$$

For this compound channel we find

$$C = 1 - H(p). \tag{52}$$

The point of view of this paper suggests instead that we determine the set $\mathfrak{R}^*$ of all achievable rate pairs (R_1,R_2) for the two given channels. See Fig. 22. This yields the entire spectrum of achievable rates under the different contingencies selected by nature.

Thus, for example, if it is known that

$$\Pr\{\beta = 1\} = \pi_1 = 1 - \Pr\{\beta = 2\}, \tag{53}$$

then we may find the maximum expected rate

$$R(\pi_1) = \max_{(R_1,R_2)\in \mathbf{R}^*} (\pi_1 R_1 + \pi_2 R_2). \tag{54}$$

The interpretation is that by using the superimposed codes of Section II we can achieve average rates

$$R(\pi_1) = \max_{0 \leq \alpha \leq 1} [C(\alpha\bar{p} + \bar{\alpha}p) + \pi_1 H(\alpha)], \tag{55}$$

corresponding to points on the boundary of $\mathfrak{R}^*$. These average rates are strictly greater than average rates achievable by time sharing (except for the degenerate prior $\pi_1 = 0$ or 1). Finally, a submessage of rate $C(\alpha\bar{p} + \bar{\alpha}p)$ is sure to be received, regardless of which channel is the true state of nature.

These considerations suggest that the compound channels problem can be reinvestigated from this broadcasting point of view by interpreting the probability distribution on the

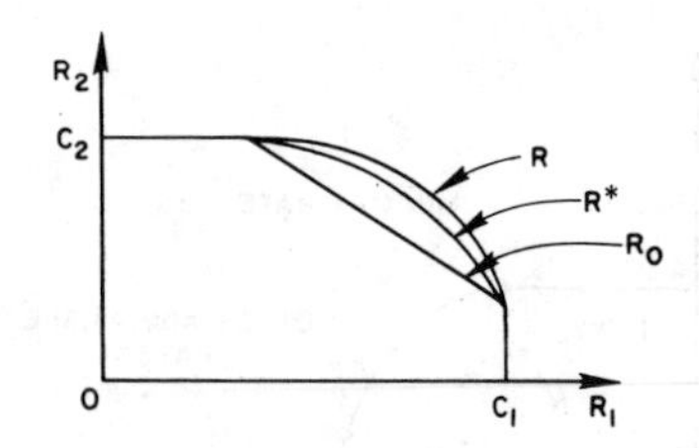

Fig. 23. Bounds on capacity region $\mathscr{R}^*$.

channel parameter β as a probability distribution on the receiver chosen in the multiple receiver broadcast channel formulation. Inspection of the capacity region $\mathfrak{R}^*$ would then yield all achievable probability distributions on rates for the compound channel. The most desirable distribution could then be picked.

X. Conclusions

As before, let the capacity region $\mathfrak{R}^*$ be the set of all achievable joint rates (R_1,R_2) for a given broadcast channel with two receivers. We now know the following. There is a certain information-theoretically defined region $\mathfrak{R}$ generated by $(I(X \mid Y_1),I(X \mid Y_2))$, given in Section VIII, which upper-bounds $\mathfrak{R}^*$. Also, by simple time sharing there is an inner bound $\mathfrak{R}_0$, say, to $\mathfrak{R}^*$, as shown in Fig. 23.

Sometimes these bounds coincide, as they do for the incompatible channel. Here $\mathfrak{R} = \mathfrak{R}_0 = \mathfrak{R}^*$. In other examples, such as the orthogonal channel, in which the bounds do not coincide, there is a simple demonstration that the upper bound can be achieved and therefore that $\mathfrak{R} = \mathfrak{R}^*$. In many of the intermediate cases (for example, the two BSC of section II) we can be reasonably well assured that our *ad hoc* codes achieve $\mathfrak{R}^*$, although proofs of converses appear to be difficult.

The analysis of this problem is made worthwhile by the fact that it is almost always the case that proper coding will achieve rates $\mathfrak{R}^*$ strictly greater than those achievable by simple time-sharing.

The primary heuristic that we garner from these investigations is that high joint rates of transmission are best achieved by superimposing high-rate and low-rate information rather than by using time-sharing. Novels written with many levels of symbolism provide just one example of a mode of communication that may be perceived at many different levels by different people.[1]

Acknowledgment

I wish to thank D. Sagalowicz and C. Keilers for many helpful discussions of the ideas presented in this paper. D. Sagalowicz has helped improve the proof of the upper bound and C. Keilers has helped with some of the examples. I have also benefited from discussions with P. Bergmans and A. D. Wyner.

[1] I am soliciting double- and triple-meaning quotes that illustrate this idea. Consider, for example, the reaction of three different people to the following donated story. Buck and Harry led a beautiful maiden into the clearing by a rope tied around her ankle. "Let's make her fast," said Buck, "while we have breakfast." The anonymity of the authors will be protected.

Appendix

In this section we prove the main result of Section II. Let $C(p) = 1 - H(p)$.

Theorem: For the broadcast channel of Section II, with BSC with parameters $p_1 = 0$ and $p_2 = p$, respectively, $(R_1,R_2) = (C(\alpha\bar{p} + \bar{\alpha}p) + H(\alpha),\ C(\alpha\bar{p} + \bar{\alpha}p))$ is achievable for any $0 \leq \alpha \leq 1$.

Proof: Let $M_{12} = 2^{nR_{12}}$, $M_1 = 2^{n(R_1 - R_{12})}$ be integers and let $R_2 = R_{12}$, $M_2 = 2^{nR_{12}}$. Consider the following random code. Let $x(r)$, $r \in R = \{1,2,\ldots,M_{12}\}$, be i.i.d. n-sequences in $X^n = \{0,1\}^n$, where $x(r)$ is drawn according to a uniform distribution on X^n. Let $\alpha < \frac{1}{2}$, αn be an integer, and let $z(s)$, $s \in S = \{1,2,\ldots,M_1\}$, be an enumeration of all the n-sequences $z \in \{0,1\}^n$ such that

$$\sum_{i=1}^{n} z_i = \alpha n.$$

There are

$$\binom{n}{\alpha n} = 2^{n(H(\alpha) + O(\ln n/n))}$$

such sequences. Define $x(r,s) = x(r) \oplus z(s)$, where the vector addition is termwise modulo 2. Without loss of generality let $p < \frac{1}{2}$.

The decoding rule $g_1: Y_1^n \to R \times S$ for the nth extension for receiver 1 will be to choose the value of $\hat{r} \in R$, $\hat{s} \in S$ such that $y_1 = x(\hat{r},\hat{s})$. We shall declare an error if there is more than one choice of $(\hat{r},\hat{s})$ such that this is true. (Since channel 1 is noiseless, the possibility that no such $(\hat{r},\hat{s})$ exists will not arise.)

The decoding rule $g_2: Y_2^n \to R$ for channel 2 will decide the value of $\hat{r} \in R$ such that $d(y_2,x(\hat{r})) \leq n(\alpha\bar{p} + \bar{\alpha}p) + n\varepsilon$, for a given $\varepsilon > 0$, where d is the Hamming distance. An error for channel 2 will be declared if there are more than one or if there are no such values of $\hat{r} \in R$.

Let us now pick a message (r,s) with probability $1/M_1M_{12}$ and evaluate the expected sum of the probabilities of error $E\{\bar{p}_1(e) + \bar{p}_2(e)\}$ [see (9), (10)] where the expectation is over the random code, drawn as described.

Since channel 1 has perfect transmission (i.e., $y_1 = x(r,s)$), the only possibility of a decoding error for channel 1 is if the (random) code itself has assigned some other index (r',s') to the same n-sequence as (r,s).

By the symmetry of the code generation process, we may fix attention on the transmission of $x(1,1)$. Thus

$$E\bar{p}_1(e) = \Pr\{x(r,s) = x(1,1), \quad \text{for some } (r,s) \neq (1,1)\}, \tag{57}$$

where the probability is defined over the random code assignment.

Now $x(1,1) = x(1,s)$ implies $z(1) = z(s)$, which is impossible for any $s \neq 1$, by the construction of $z(s)$. Thus the only possibility of error is $x(1,1) = x(r,s)$, $r \neq 1$, $s \in S$. But $r \neq 1$ implies $x(r,s)$ and $x(1,1)$ are independent uniformly distributed n-sequences over $\{0,1\}^n$. Thus, for $r \neq 1$,

$$\Pr\{x(1,1) = x(r,s)\} = 2^{-n}. \tag{58}$$

Putting this together with the union of events inequality yields

$$E\bar{p}_1(e) \leq \sum_{(r,s)\neq(1,1)} \Pr\{x(1,1) = x(r,s)\} \tag{59}$$

$$\leq M_1M_{12}\,2^{-n} = 2^{-n(1-R_1)} \to 0, \qquad R_1 < 1.$$

Thus $E\{\bar{p}_1(e)\} \to 0$, as $n \to \infty$, if $R_1 < 1$, where the construction implies

$$R_1 - R_{12} \triangleq (\log M_1)/n = H(\alpha) - 0(\ln n/n). \tag{60}$$

Now consider channel 2. Let $e = (e_1,e_2,\ldots,e_n)$ be a binary n vector of i.i.d. Bernoulli RV with parameter p. Thus we can write $y_2 = x(r,s) \oplus e$ and

$$y_2 = x(r) \oplus z(s) \oplus e. \tag{61}$$

A decoding error can be made in one of two ways. E_1: the true $r = 1$ does not satisfy

$$d(y_2,x(1)) \leq n(\alpha\bar{p} + \bar{\alpha}p) + n\varepsilon, \tag{62}$$

and E_2: there exists an index $r \neq 1$, $r \in R$, such that

$$d(y_2,x(r)) \leq n(\alpha\bar{p} + \bar{\alpha}p) + n\varepsilon.$$

Thus

$$E\{\bar{p}_2(e)\} \leq \Pr\{E_1\} + \Pr\{E_2\}, \tag{63}$$

where here the probability is understood to range over the random choice of code as well as the selection of (r,s). From (61),

$$\Pr\{E_1\} = \Pr\{d(y_2,x(1)) > n(\alpha\bar{p} + \bar{\alpha}p + \varepsilon)\}$$

$$= \Pr\left\{\frac{1}{n}\sum_{i=1}^{n} z(s)_i \oplus e_i > \alpha\bar{p} + \bar{\alpha}p + \varepsilon\right\}. \tag{64}$$

We find the expected value (over e and s)

$$\frac{1}{n}\sum_{i=1}^{n} z(s)_i \oplus e_i = \frac{1}{n}\sum_{i=1}^{n} \Pr\{(z(s)_i,e_i) = (1,0) \text{ or } (0,1)\}$$

$$= \frac{1}{n}\sum_{i=1}^{n} (\alpha\bar{p} + \bar{\alpha}p) = \alpha\bar{p} + \bar{\alpha}p. \tag{65}$$

Also, after some calculation

$$\operatorname{var}\frac{1}{n}\sum_{i=1}^{n} z(s)_i \oplus e_i \leq \frac{p\bar{p}}{n}. \tag{66}$$

It follows that $d(y,x(r)) \to \alpha\bar{p} + \bar{\alpha}p$ in probability and therefore $\Pr\{E_1\} \to 0$ as $n \to \infty$.

We are left with the evaluation of $\Pr\{E_2\}$. We write

$$\Pr\{E_2\} \leq \Pr\{d(x(r),y_2) \leq n(\alpha\bar{p} + \bar{\alpha}p + \varepsilon), \text{ for some } r \neq 1 | x(1) \text{ transmitted}\}$$

$$\leq 2^{nR_{12}} \Pr\{d(x(2),y_2) \leq n(\alpha\bar{p} + \bar{\alpha}p + \varepsilon)\}. \tag{67}$$

But

$$d(x(2),y_2) = wt(x(2) \oplus x(1) \oplus z(s) \oplus e), \tag{68}$$

where wt denotes the number of 1's in the binary n-tuple, and $x(2)$ and $x(1)$ are independent Bernoulli n-sequences with parameter $\frac{1}{2}$. Thus, for any $\varepsilon > 0$,

$$\Pr\{E_2\} \leq 2^{nR_{12}} 2^{n(H(\alpha\bar{p}+\bar{\alpha}p)+O(\ln n/n)+\varepsilon')} 2^{-n}, \tag{69}$$

where $2^{n(H(\alpha\bar{p}+\bar{\alpha}p)+O(\ln n/n)+\varepsilon')}$ denotes the number of points

$$\sum_{i=0}^{n(\alpha\bar{p}+\bar{\alpha}p+\varepsilon)} \binom{n}{i}$$

in the decoding sphere centered at y_2. Consequently, if

$$R_{12} < 1 - H(\alpha\bar{p} + \bar{\alpha}p) - \varepsilon', \tag{70}$$

then $\Pr\{E_2\} \to 0$, as $n \to \infty$. Collecting the constraints of (60) and (70), we see that if

$$R_2 = R_{12} < 1 - H(\alpha\bar{p} + \bar{\alpha}p) \tag{71}$$

$$R_1 < H(\alpha) + R_2 = 1 - H(\alpha\bar{p} + \bar{\alpha}p) + H(\alpha),$$

then

$$E\{\bar{p}_1^{(n)}(e) + \bar{p}_2^{(n)}(e)\} = E\{\bar{p}_1^{(n)}(e)\} + E\{\bar{p}_2^{(n)}(e)\} \to 0. \tag{72}$$

Since the best code behaves better than the average, there must exist a sequence of $[(2^{nR_1},2^{nR_2},2^{nR_{12}}), n]$ codes for $n = 1,2,\ldots$, with

$$R_1 = C(\alpha\bar{p} + \bar{\alpha}p) + H(\alpha) - \varepsilon$$

$$R_2 = C(\alpha\bar{p} + \bar{\alpha}p) - \varepsilon \tag{73}$$

such that

$$\bar{p}_1^{(n)}(e) + \bar{p}_2^{(n)}(e) \to 0, \tag{74}$$

and thus $\bar{p}_1^{(n)}(e) \to 0$, $\bar{p}_2^{(n)}(e) \to 0$.

Taking the limit of (R_1,R_2) as $\varepsilon \to 0$ proves the theorem.

References

[1] R. B. Ash, *Information Theory*. New York: Interscience, 1965.
[2] J. Wolfowitz, *Coding Theorems of Information Theory*, 2nd ed. Berlin: Springer, and Englewood Cliffs, N.J.: Prentice-Hall, 1964.
[3] C. E. Shannon, "A mathematical theory of communication," *Bell Syst. Tech. J.*, vol. 27, 1948, pp. 379–423 (pt. 1) and pp. 623–656 (pt. 2); also Urbana, Ill.: Univ. Illinois Press, 1949.
[4] C. E. Shannon, "Two-way communication channels," in *Proc. 4th Berkeley Symp. Probability and Statistics*, vol. 1. Berkeley, Calif.: Univ. California Press, 1961, pp. 611–644.
[5] D. Blackwell, L. Breiman, and A. Thomasian, "The capacity of a class of channels," *Ann. Math. Statist.*, vol. 30, 1959, pp. 1229–1241.
[6] C. E. Shannon, "A note on a partial ordering for communication channels," *Inform. Contr.*, vol. 1, 1958, pp. 390–397.

Noiseless Coding of Correlated Information Sources

DAVID SLEPIAN AND JACK K. WOLF

***Abstract*—Correlated information sequences $\cdots, X_{-1}, X_0, X_1, \cdots$ and $\cdots, Y_{-1}, Y_0, Y_1, \cdots$ are generated by repeated independent drawings of a pair of discrete random variables X, Y from a given bivariate distribution $p_{XY}(x,y)$. We determine the minimum number of bits per character R_X and R_Y needed to encode these sequences so that they can be faithfully reproduced under a variety of assumptions regarding the encoders and decoders. The results, some of which are not at all obvious, are presented as an admissible rate region $\mathscr{R}$ in the R_X–R_Y plane. They generalize a similar and well-known result for a single information sequence, namely $R_X \geq H(X)$ for faithful reproduction.**

I. Introduction

Notation and Problem Statement

IN THIS PAPER, we generalize, to the case of two correlated sources, certain well-known results on the noiseless coding of a single discrete information source. Typical of the situations considered is that depicted in Fig. 1. Here the two correlated information sequences $\cdots, X_{-1}, X_0, X_1, \cdots$ and $\cdots, Y_{-1}, Y_0, Y_1, \cdots$ are obtained by repeated independent drawings from a discrete bivariate distribution $p(x,y)$. The encoder of each source is constrained to operate without knowledge of the other source, while the decoder has available both encoded binary message streams. We determine the minimum number of bits per source character required for the two encoded message streams in order to ensure accurate reconstruction by the decoder of the outputs of both information sources. The results are presented as an allowed two-dimensional rate region $\mathscr{R}$ for the two encoded message streams as shown in Fig. 2. Note that in $\mathscr{R}$ for this case we can have both $R_X < H(X)$ and $R_Y < H(Y)$ al-

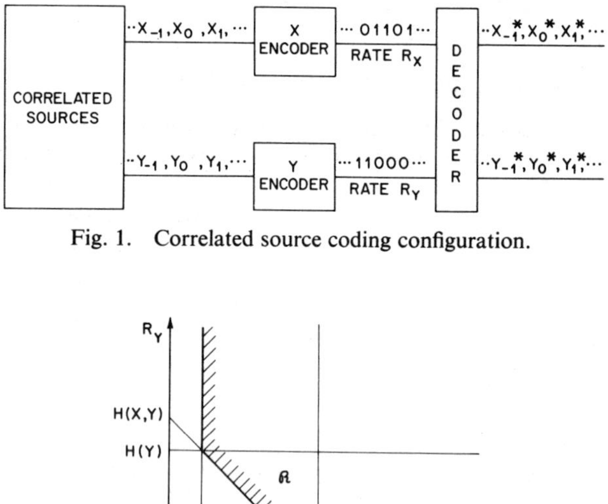

Fig. 1. Correlated source coding configuration.

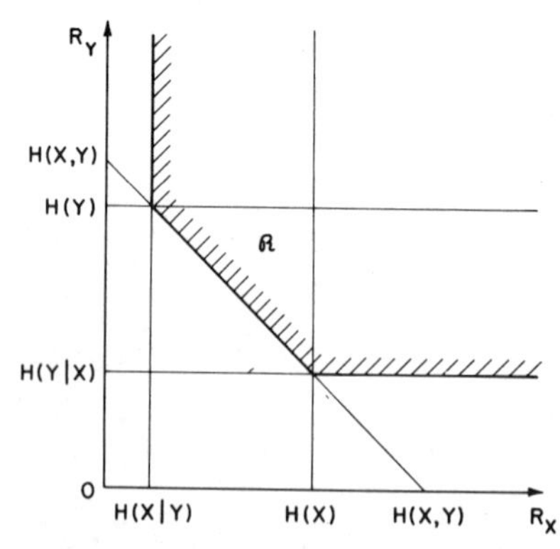

Fig. 2. Admissible rate region $\mathscr{R}$ corresponding to Fig. 1.

Manuscript received August 25, 1972; revised December 28, 1972.
D. Slepian is with the University of Hawaii, Honolulu, Hawaii, and the Bell Laboratories, Murray Hill, N.J. 07974.
J. K. Wolf was with the University of Hawaii, Honolulu, Hawaii, on leave from the Polytechnic Institute of Brooklyn, Brooklyn, N.Y.

Reprinted from *IEEE Trans. Inform. Theory*, vol. IT-19, pp. 471–480, July 1973.

though each encoder sees only its own source. All these notions will be made precise in the following discussion.

First, however, let us review briefly some results for a single source that have long been known. Let X be a discrete random variable taking values in the set $\mathscr{A} = \{1,2,\cdots,A\}$. Denote the probability distribution of X by $p_X(x) = \Pr[X = x]$, $x \in \mathscr{A}$. Now let $\boldsymbol{X} = (X_1,X_2,\cdots,X_n)$ be a sequence of n independent realizations of X so that the probability distribution for the random n-vector $\boldsymbol{X}$ is given by

$$P_{\boldsymbol{X}}(\boldsymbol{x}) = \Pr[\boldsymbol{X} = \boldsymbol{x}] = \prod_{i=1}^{n} p_X(x_i) \tag{1}$$

$$\boldsymbol{x} = (x_1,x_2,\cdots,x_n) \in \mathscr{A}^n, \qquad x_i \in \mathscr{A}, \quad i = 1,2,\cdots,n.$$

Here, we introduce the symbol $\mathscr{A}^n$ to stand for the collection of the A^n different n-vectors $\boldsymbol{x}$, each of whose components is a member of $\mathscr{A}$. We regard $\boldsymbol{X}$ as a block of n successive characters from the output of an information source producing characters independently with letter distribution $p_X(x)$.

While little can be said about individual letters produced by this information source, for large n the composition of blocks of n letters tends to be fixed. In a typical long block, one can expect about $np_X(1)$ occurrences of letter 1, about $np_X(2)$ occurrences of letter 2, etc. The probability of such a typical long sequence is, therefore,

$$\begin{aligned} p_T &= p_X(1)^{np_X(1)} p_X(2)^{np_X(2)} \cdots p_X(A)^{np_X(A)} \\ &= \exp[np_X(1) \log p_X(1)] \cdots \exp[np_X(A) \log p_X(A)] \\ &= \exp[-nH(X)] \end{aligned}$$

where

$$H(X) \equiv -\sum_{1}^{A} p_X(i) \log p_X(i) \tag{2}$$

is called the *entropy* of the source or of the random variable X. These simple observations lead to the useful, though imprecise, statement characterizing the long blocks of such a source: there are only $N_T = \exp[nH(X)]$ likely blocks of length n; each has probability $\exp[-nH(X)]$. This in turn suggests that we can accurately transmit the output of the information source using only $R = (1/n) \log N_T = H(X)$ natural units (nats) of information per character and that at least this rate is required of any transmission scheme that allows accurate recovery of the source output.

These intuitive coding notions can be made precise as follows. An encoder $\mathscr{E}(n,M)$ is any single-valued function $i = f(\boldsymbol{x})$ from the n-vectors $\boldsymbol{x}$ of $\mathscr{A}^n$ to the integers of the set $\mathscr{M} \equiv (1,2,\cdots,M)$. A decoder $\mathscr{D}(n,M)$ is a single-valued function $\boldsymbol{x} = \boldsymbol{g}(i)$ from the integers $i \in \mathscr{M}$ to the vectors $\boldsymbol{x} \in \mathscr{A}^n$. Associated with a source and a particular encoder and decoder pair are the rate R of the encoded messages defined by $R = (1/n) \log M$, and the two random variables $I \equiv f(\boldsymbol{X})$ and $\boldsymbol{X}^* \equiv \boldsymbol{g}(I)$ called, respectively, the encoded message number and the decoded block. We think of the encoder as producing the integer I after observing the n source characters $\boldsymbol{X}$. Then R units of information per source character suffice to communicate the value of I to the decoder. The decoder then produces the message block $\boldsymbol{X}^*$ as its estimate of $\boldsymbol{X}$.

A rate R is said to be *admissible* if for every $\varepsilon > 0$ there exists for some $n = n(\varepsilon)$ an encoder $\mathscr{E}(n,\lfloor\exp(nR)\rfloor)$ and a decoder $\mathscr{D}(n,\lfloor\exp(nR)\rfloor)$ such that $\Pr[\boldsymbol{X}^* \neq \boldsymbol{X}] < \varepsilon$. Otherwise R is called *inadmissible*. Here the symbol $\lfloor x \rfloor$ denotes the largest integer not greater than x. We shall make frequent use of the following well-known theorem. (See, for example, [1, p. 43] or [3, p. 45] for equivalent results.)

Theorem 1: If $R > H(X)$, R is admissible. If $R < H(X)$, R is inadmissible. In this latter case there exists a $\delta > 0$ independent of n such that for every encoder–decoder pair $\mathscr{E}(n,\lfloor\exp(nR)\rfloor)$, $\mathscr{D}(n,\lfloor\exp(nR)\rfloor)$, $\Pr[\boldsymbol{X}^* \neq \boldsymbol{X}] > \delta > 0$. Stated in less formal terms the theorem asserts that for $\eta > 0$ one can achieve arbitrarily small decoding error probability with block codes transmitting at a rate $R = H(X) + \eta$; block codes using a rate $R = H(X) - \eta$ cannot have arbitrarily small probability of error.

We now seek to generalize these notions to correlated sources. Let X and Y be discrete random variables taking values in the sets $\mathscr{A}_X = \{1,2,\cdots,A_X\}$ and $\mathscr{A}_Y = \{1,2,\cdots,A_Y\}$, respectively. Denote their joint probability distribution by

$$p_{XY}(x,y) = \Pr[X = x \text{ and } Y = y], \qquad x \in \mathscr{A}_X, \quad y \in \mathscr{A}_Y. \tag{1}$$

Next let $(X_1,Y_1), (X_2,Y_2),\cdots,(X_n,Y_n)$ be a sequence of n independent realizations of the pair of random variables (X,Y). Denote by $\boldsymbol{X}$ the sequence $X_1,X_2,\cdots,X_n$ and by $\boldsymbol{Y}$ the sequence $Y_1,Y_2,\cdots,Y_n$. The probability distribution for this correlated pair of vectors is

$$P_{\boldsymbol{XY}}(\boldsymbol{x},\boldsymbol{y}) = \Pr[\boldsymbol{X} = \boldsymbol{x}, \boldsymbol{Y} = \boldsymbol{y}] = \prod_{i=1}^{n} p_{XY}(x_i,y_i) \tag{2}$$

$$\boldsymbol{x} = (x_1,x_2,\cdots,x_n) \in \mathscr{A}_X{}^n$$

$$\boldsymbol{y} = (y_1,y_2,\cdots,y_n) \in \mathscr{A}_Y{}^n$$

where $\mathscr{A}_X{}^n$ is the set of $A_X{}^n$ distinct n-vectors whose components are in $\mathscr{A}_X$ and $\mathscr{A}_Y{}^n$ is defined analogously. We regard $\boldsymbol{X}$ as a block of n-characters produced by one of two correlated information sources. $\boldsymbol{Y}$ is the corresponding block produced by the other source.

When it comes to encoding the outputs of these correlated sources a number of possibilities of interest present themselves depending upon the information available to the encoders and decoders. Sixteen cases that we shall consider are shown in Fig. 3. Each setting of the switches S_1,S_2,S_3,S_4 yields a new case. It is convenient to associate with switch S_i a state variable s_i taking the value 0 if the switch is open and the value 1 if the switch is closed, $i = 1,2,3,4$. The quadruple $s_1s_2s_3s_4$, always listed in that order, will be used to specify the setting of the switches. Thus 0101 means that switches S_1 and S_3 are open while S_2 and S_4 are closed. The setting 0011 corresponds to Fig. 1.

An X-encoder $\mathscr{E}_X(n,s_2,M_X)$ is a single-valued function from $\mathscr{A}_X{}^n \times \mathscr{A}_Y{}^n$ to the set of integers $\mathscr{M}_X = \{1,2,\cdots,M_X\}$ of the form $i_X = f_X(\boldsymbol{x},s_2\boldsymbol{y})$. Similarly a Y-encoder $\mathscr{E}_Y(n,s_1,$

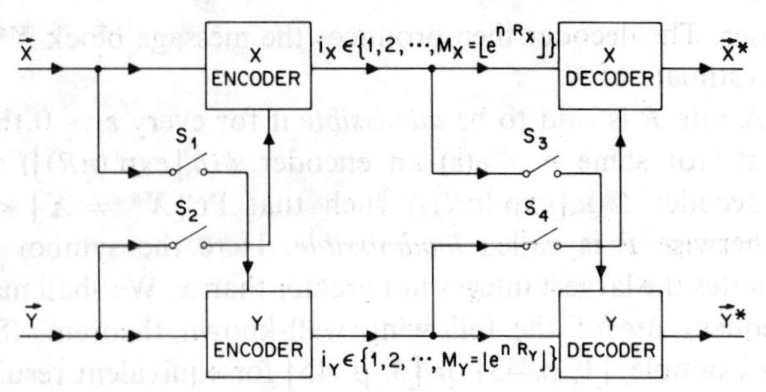

Fig. 3. Sixteen cases of correlated source coding.

M_Y) is a single-valued function of the form $i_Y = f_Y(s_1x,y)$ from $\mathscr{A}_X{}^n \times \mathscr{A}_Y{}^n$ to the set $\mathscr{M}_Y = \{1,2,\cdots,M_Y\}$. Decoders are defined analogously: $\mathscr{D}_X(n,s_4,M_X,M_Y)$ is any single-valued function of the form $x^* = g_X(i_X,s_4i_Y)$ from $\mathscr{M}_X \times \mathscr{M}_Y$ to $\mathscr{A}_X{}^n$ while a Y-decoder $\mathscr{D}_Y(n,s_3,M_X,M_Y)$ is a single-valued function of the form $y^* = g_Y(s_3i_X,i_Y)$ from $\mathscr{M}_X \times \mathscr{M}_Y$ to $\mathscr{A}_Y{}^n$.

Associated with these coders and decoders are rates $R_X = (1/n)\log M_X$, $R_Y = (1/n)\log M_Y$, and random variables $I_X \equiv f_X(X,s_2Y)$, $I_Y \equiv f_Y(s_1X,Y)$, $X^* = g_X(I_X,s_4I_Y)$ and $Y^* = g_Y(s_3I_X,I_Y)$. I_X and I_Y are called the *encoded X-message number* and the *encoded Y-message number*, respectively, while X^* and Y^* are the corresponding decoded blocks. We think of the two encoders as producing the integers I_X and I_Y after n correlated source pairs X,Y have been generated. R_X units of information per source character suffice to transmit I_X to the X-decoder: R_Y units suffice to transmit I_Y to the Y-decoder. The decoders then produce the estimates X^* and Y^* of the input sequences X and Y. The role played by the switches S_1,S_2,S_3,S_4 has been incorporated here into the arguments of various coding functions f_X, f_Y, g_X, and g_Y. Thus, for example, if S_2 is open, $s_2 = 0$ and then $I_X = f_X(X,0)$ depends only on the X-sequence.

We are at last in a position to define our problem. A pair of nonnegative numbers R_X,R_Y is said to be an *admissible rate point* for the case $s_1s_2s_3s_4$ if for every $\varepsilon > 0$ there exists for some $n = n(\varepsilon)$ encoders $\mathscr{C}_X(n,s_2,M_X)$, $\mathscr{C}_Y(n,s_1,M_Y)$, and decoders $\mathscr{D}_X(n,s_4,M_X,M_Y)$, $\mathscr{D}_Y(n,s_3,M_X,M_Y)$ with $M_X = \lfloor \exp(nR_X) \rfloor$, $M_Y = \lfloor \exp(nR_Y) \rfloor$, such that, $\Pr[\{X^* \neq X\} \cup \{Y^* \neq Y\}] < \varepsilon$. Otherwise, the pair (R_X,R_Y) is called an *inadmissible rate point*. We denote by $\mathscr{R}$ the closure of the set of admissible rate points. In this paper, we determine the admissible rate region $\mathscr{R}$ for the 16 cases of Fig. 3 for the correlated source described.

II. Discussion of Results

To describe the admissible rate region $\mathscr{R}$ for the various cases of Fig. 3, we must first introduce the marginal and conditional distributions of X and Y, namely,

$$p_X(x) = \sum_y p_{XY}(x,y)$$

$$p_Y(y) = \sum_x p_{XY}(x,y)$$

$$p_{X\mid Y}(x \mid y) = p_{XY}(x,y)/p_Y(y), \qquad p_Y(y) \neq 0$$

$$p_{Y\mid X}(y \mid x) = p_{XY}(x,y)/p_X(x), \qquad p_X(x) \neq 0 \tag{3}$$

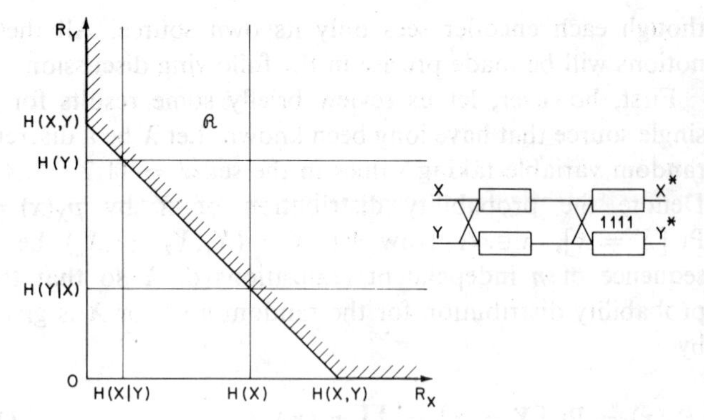

Fig. 4. Admissible rate region.

and the usual associated information-theoretic numbers

$$H(X,Y) = -\sum_x \sum_y p_{XY}(x,y) \log p_{XY}(x,y)$$

$$H(X) = -\sum_x p_X(x) \log p_X(x)$$

$$H(Y) = -\sum_y p_Y(y) \log p_Y(y)$$

$$H(Y \mid X) = -\sum_x p_X(x) \sum_y p_{Y\mid X}(y \mid x) \log p_{Y\mid X}(y \mid x)$$

$$H(X \mid Y) = -\sum_y p_Y(y) \sum_x p_{X\mid Y}(x \mid y) \log p_{X\mid Y}(x \mid y). \tag{4}$$

The regions are described in terms of these quantities.

The 16 cases are covered by Figs. 4–9. Each figure shows a region $\mathscr{R}$ and certain switching configurations that have $\mathscr{R}$ as region of admissible rates. Figs. 5–7 each serve as well for the switch settings shown at the right of the drawing when every X in the figure is replaced by Y and every Y is replaced by X, including those on the small block diagrams.

The cases vary in novelty and interest. For instance, the case 1111 shown in Fig. 4 contains nothing new. To obtain the results shown there, we have only to regard the pair (X,Y) as a new discrete random variable taking on A_XA_Y values and apply Theorem 1. This will be explained in full below.

The case 0011 shown in Fig. 8 is by far the most interesting and novel of our results. Consider for a moment a point near the corner of $\mathscr{R}$ given by $R_Y = H(Y) + \varepsilon$, $R_X = H(X \mid Y) + \varepsilon$, where $\varepsilon > 0$ is thought of as very small. By Theorem 1, a Y-encoder transmitting at this rate R_Y and a Y-decoder exist that allow the Y-source outputs to be recovered with arbitrarily small error probability. We can suppose then that the joint X–Y decoder shown has available the Y outputs. In view of the normal interpretation of $H(X \mid Y)$ as the "uncertainty of X given Y," it is most satisfying then that the X-encoder need only produce a message stream with information rate $R_X = H(X \mid Y) + \varepsilon$. But how can this be done? What properties of X alone must the X-encoder examine and transmit (for it cannot observe the Y source) at the rate $H(X \mid Y) < H(X)$ to allow reconstruction of the X sequence when Y is at last seen at the decoder? The answer is not clear. We obtain our results by a random coding argument which somewhat generalizes that used in the usual noisy channel coding theorem. Since

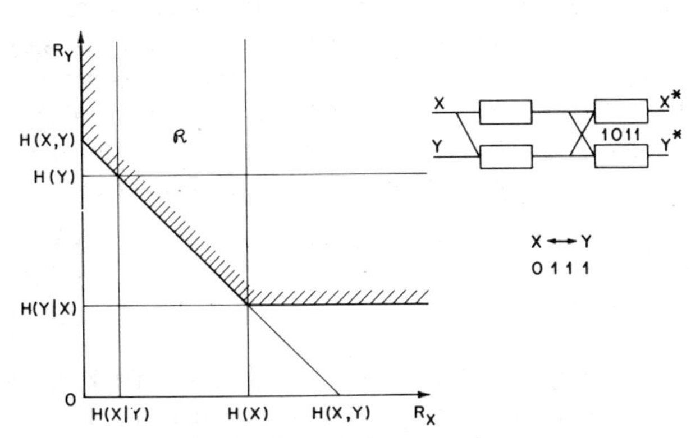

Fig. 5. Admissible rate region.

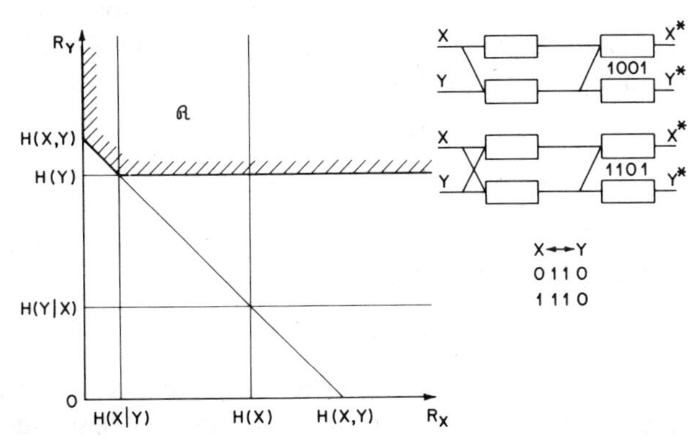

Fig. 6. Admissible rate region.

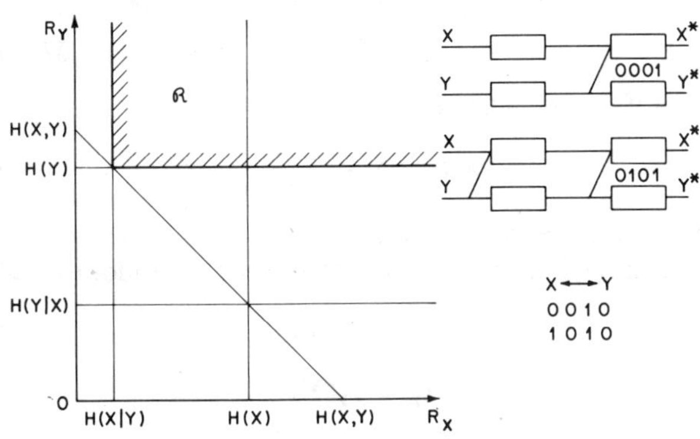

Fig. 7. Admissible rate region.

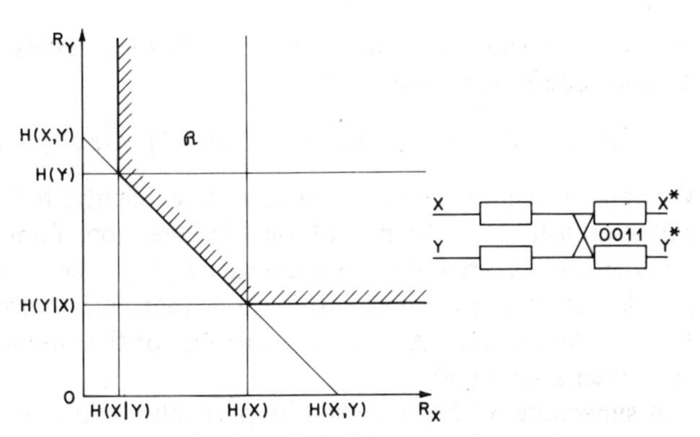

Fig. 8. Admissible rate region.

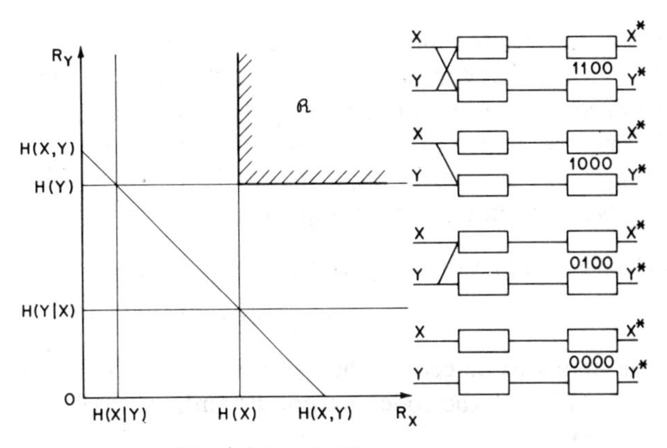

Fig. 9. Admissible rate region.

this is one of the principal contributions of the paper, we turn now to treat this case in some detail, then later proceed to establish more general machinery that allows treatment of the remaining 15 cases with more dispatch.

III. The Case 0011

In this section we prove the following.

Theorem 2:

$$\begin{aligned} R_X &= H(X \mid Y) + \varepsilon_X, \qquad \varepsilon_X > 0 \\ R_Y &= H(Y) + \varepsilon_Y, \qquad \varepsilon_Y > 0 \end{aligned} \tag{5}$$

is an admissible rate point for the case 0011.

In the course of the proof, we shall need the quantities

$$P_{Y|X}(\boldsymbol{y} \mid \boldsymbol{x}) = \Pr[\boldsymbol{Y} = \boldsymbol{y} \mid \boldsymbol{X} = \boldsymbol{x}] = \prod_{i=1}^{n} p_{Y|X}(y_i \mid x_i) \tag{6}$$

with notation as in (1)–(3). We shall also use

$$\begin{aligned} I(X,Y) &\equiv H(X) + H(Y) - H(X,Y) = H(X) - H(X \mid Y) \\ &= H(Y) - H(Y \mid X) \end{aligned} \tag{7}$$

with the H given by (4).

The concepts behind the formal proof that follows are these. By Theorem 1, we know that R_Y is large enough to allow nearly error-free transmissions of the Y-sequences. We shall accordingly think of the n-vector $\boldsymbol{Y}$ as known as the decoder.

Now, from the fact that $p_{XY}(x,y) = p_{Y|X}(y \mid x)p_X(x)$, we can think of the Y-sequences of the correlated source as being generated by applying successive characters of the X-sequence as inputs to a discrete memoryless channel with transition probabilities $p_{Y|X}(y \mid x)$. The coding theorem for such a channel tells us that for large n and any $\varepsilon > 0$, there exists a decoder and a code $\mathscr{X}_1$ composed of $M = \lfloor \exp[n(I(X,Y) - \varepsilon)] \rfloor$ n-vectors $\boldsymbol{x}_{11}, \boldsymbol{x}_{12}, \cdots, \boldsymbol{x}_{1M}$ that can be used as inputs to this channel and decoded with little error probability when the output $\boldsymbol{Y}$ is seen. Now it turns out that we can find many other codes for this channel, say $\mathscr{X}_2, \mathscr{X}_3, \cdots, \mathscr{X}_{M'}$, each with its own decoder, each of the same size as $\mathscr{X}_1$, and each enjoying the same small probability of error. Our scheme then for encoding the X-sequences of our correlated sources X and Y is to transmit to the decoder the index of the first code in the series $\mathscr{X}_1, \mathscr{X}_2,\cdots,\mathscr{X}_M$, that contains X. The X–Y decoder can then use the decoder appropriate for that code $\mathscr{X}_i$ of the $p_{Y|X}(y \mid x)$ channel to determine X. There are $\exp\{n[H(X) + \eta]\}$ highly likely X sequences, so that if the codes $\mathscr{X}_1,\mathscr{X}_2,\cdots, \mathscr{X}_M$, were disjoint, it would require

$$M' = \exp\{n[H(X) + \eta]\}/\exp\{n[I(X,Y) - \varepsilon]\} = \exp\{n[H(X \mid Y) + \varepsilon_X]\}$$

code books to be certain that X was contained in one of them. Although the codes are not disjoint, we shall use just this many.

Let us turn now to the formal proof. Let n be a positive integer, later to be chosen very large, and set

$$M_X = \lfloor \exp(nR_X) \rfloor \qquad M_Y = \lfloor \exp(nR_Y) \rfloor \tag{8}$$

with the R', given as in (5). X- and Y-encoders for the case at hand are functions $i_X = f_X(x)$ from $\mathscr{A}_X{}^n$ to $\mathscr{M}_X = \{1,2,\cdots,M_X\}$ and $i_Y = f_Y(y)$ from $\mathscr{A}_Y{}^n$ to $\mathscr{M}_Y = \{1,2,\cdots, M_Y\}$. A decoder is a pair of functions $x^* = g_X(i_X,i_Y)$ from $\mathscr{M}_X \times \mathscr{M}_Y$ to $\mathscr{A}_X{}^n$ and $y^* = g_Y(i_X,i_Y)$ from $\mathscr{M}_X \times \mathscr{M}_Y$ to $\mathscr{A}_Y{}^n$. In the proof of Theorem 2, it suffices to restrict our attention to coding and decoding functions of a very special form.

To describe the Y-encoder, we must first define the list $\mathscr{T}(\varepsilon,n)$ of typical Y-sequences of length n. Here $\varepsilon > 0$ is a parameter that will stay fixed throughout the rest of this section. Let k be the smallest integer greater than $\sqrt{A_Y/\varepsilon}$ where as before the Y alphabet is $\mathscr{A}_Y = \{1,2,\cdots,A_Y\}$. Let $f_i(y)$ be the number of occurrences of the integer i among the list of components $y_1,y_2,\cdots,y_n$ of y. A Y-sequence y is contained in $\mathscr{T}(\varepsilon,n)$ if

$$|f_i(y) - np_Y(i)| < k\sqrt{np_Y(i)[1 - p_Y(i)]}, \qquad i = 1,2,\cdots,A_Y. \tag{9}$$

If (9) is violated for any i, y is called *atypical* and is not a member of $\mathscr{T}(\varepsilon,n)$. The following facts about typical sequences are well-known. (See, for example, [2, pp. 14–16] for a very readable account.)

Theorem 3: 1) $\Pr[Y \in \mathscr{T}(\varepsilon,n)] > 1 - \varepsilon$. 2) There exists an $A > 0$ such that for every $y \in \mathscr{T}(\varepsilon,n)$

$$\exp[-nH(Y) - A\sqrt{n/\varepsilon}] < P_Y(y) < \exp[-nH(Y) + A\sqrt{n/\varepsilon}].$$

Here A is independent of n and ε. 3) The number N of members of $\mathscr{T}(\varepsilon,n)$ is $e^{n[H(Y)+\delta_n(\varepsilon)]}$, where

$$\lim_{n\to\infty} \delta_n(\varepsilon) = 0.$$

We now assume that n is chosen sufficiently large so that $\delta_n(\varepsilon) < \varepsilon_Y$ of (5). Then the number of sequences y in $\mathscr{T}(\varepsilon,n)$ satisfies

$$N \leq M_Y \tag{10}$$

with M_Y given by (5) and (8).

The Y-encoder for the correlated sources X and Y is constructed as follows. Number the vectors of $\mathscr{T}(\varepsilon,n)$ to obtain the list $y_1,y_2,\cdots,y_N$. Adjoin to this sequence any $M_Y - N$ other vectors of $\mathscr{A}_Y{}^n$ (not necessarily distinct), labeling them $y_{N+1},y_{N+2},\cdots,y_{M_Y}$. We denote the list $y_1,y_2,\cdots,y_{M_Y}$ by $\mathscr{L}$. Now define the Y-encoder by

$$f_Y(y) = \begin{cases} \text{smallest index } i \text{ such that } y = y_i, & \text{if } y \in \mathscr{L} \\ 1, & \text{if } y \notin \mathscr{L}. \end{cases} \tag{11}$$

The mapping is from $\mathscr{A}_Y{}^n$ to $\mathscr{M}_Y$ as required of a Y-encoder.

The X-encoders are of a very special form. Let $\mathscr{X}_i = \{x_{i1},x_{i2},\cdots,x_{iM}\}$ be a list of M vectors of $\mathscr{A}_X{}^n$, $i = 1,2,\cdots, M_X$. The vectors in these lists need not be distinct. We call each list $\mathscr{X}_i$ an *X-code* and we call the collection $\mathscr{X}$ of M_X X-codes an *X-supercode*. We shall specify how M is to be chosen later. The X-encoders we consider are of the form

$$f_X(x) = \begin{cases} 1, & \text{if } x \notin \mathscr{X} \\ \text{smallest index } i \text{ such that } x \in \mathscr{X}_i, & \text{if } x \in \mathscr{X}. \end{cases} \tag{12}$$

To define the decoding functions we set

$$g_Y(i_X,i_Y) = y_{i_Y} \tag{13}$$

for all $(i_X,i_Y) \in \mathscr{M}_X \times \mathscr{M}_Y$. The X-decoder is somewhat more complicated. Denote by $j(i_X,i_Y)$ the smallest index j such that

$$P_{Y|X}(y_{i_Y} \mid x_{i_Xj}) \geq P_{Y|X}(y_{i_Y} \mid x_{i_Xk}), \qquad k = 1,2,\cdots,M. \tag{14}$$

Then the X-decoder is given by

$$g_X(i_X,i_Y) = x_{i_X,j(i_X,i_Y)}, \tag{15}$$

for all $(i_X,i_Y) \in \mathscr{M}_X \times \mathscr{M}_Y$.

As in the introduction, we introduce the random variables

$$I_X = f_X(X)$$
$$I_Y = f_Y(Y)$$
$$X^* = g_X(I_X,I_Y)$$
$$Y^* = g_Y(I_X,I_Y).$$

We wish to show that for every $\varepsilon' > 0$ there exists an X supercode $\mathscr{X}$ such that

$$P_e(\mathscr{X}) \equiv \Pr[\{X^* \neq X\} \cup \{Y^* \neq Y\}] < \varepsilon'. \tag{16}$$

We cannot exhibit such an X-supercode explicitly, but we will establish the existence of one by the now familiar random coding argument. We average $P_e(\mathscr{X})$ over an ensembles $\mathscr{E}$ of X-supercodes and show that this average, $\overline{P_e(\mathscr{X})}$, is less than ε'. At least one member of the ensemble must then satisfy (16).

A supercode of $\mathscr{E}$ is specified by particular values of the M_XM random vectors X_{ij}, $i = 1,\cdots,M_X$, $j = 1,\cdots,M$. The values lie in $\mathscr{A}_X{}^n$. The probability structure of $\mathscr{E}$ is specified by

$$\Pr[X_{ij} = x_{ij}, i = 1,\cdots,M_X, j = 1,\cdots,M] = \prod_{i=1}^{M_X}\prod_{j=1}^{M} P_X(x_{ij}) \tag{17}$$

where

$$P_X(x) = \prod_1^n p_X(x_i) \tag{18}$$

in the notation of (2) and (3). Stated otherwise, the vectors of the supercode are drawn component by component independently from the marginal $p_X(x)$ of the given joint distribution $p_{XY}(x,y)$.

Suppose now all supercodes are enumerated and listed $-\mathscr{X}^{(1)}, \mathscr{X}^{(2)},\cdots$. The average error probability we seek is

$$\begin{aligned}\overline{P_e(\mathscr{X})} &= \sum_j \Pr(\mathscr{X} = \mathscr{X}^{(j)}) \\ &\quad \cdot \Pr[\{X^* \neq X\} \cup \{Y^* \neq Y\} \mid \mathscr{X} = \mathscr{X}^{(j)}] \\ &= \sum_j \Pr[\{X^* \neq X\} \cup \{Y^* \neq Y\} \text{ and } \mathscr{X} = \mathscr{X}^{(j)}].\end{aligned}$$

This last sum can be interpreted as the probability P_e that $(X^*,Y^*) \neq (X,Y)$ in the joint experiment of choosing an X-supercode $\mathscr{X}$ from $\mathscr{E}$ and independently choosing an X and Y from $p_{XY}(x,y)$ to use with that supercode. We proceed to upperbound this quantity.

Let

$$\begin{aligned}P_1 &= \Pr[Y \notin \mathscr{L}] \\ P_2 &= \Pr[X \notin \mathscr{X}] \\ P_3 &= \Pr[Y \in \mathscr{L}, X \in \mathscr{X}, X^* \neq X].\end{aligned} \tag{19}$$

Then clearly

$$\overline{P_e(\mathscr{X})} \leq P_1 + P_2 + P_3. \tag{20}$$

That

$$P_1 < \varepsilon \tag{21}$$

follows directly from statement 1) of Theorem 3 about typical sequences and from the fact that $\mathscr{L}$ includes $\mathscr{T}(\varepsilon,n)$.

We now show that if n is large enough and

$$M = \lfloor \exp\{n[I(X,Y) - \tfrac{1}{2}\varepsilon_X]\}\rfloor, \tag{22}$$

where we assume $I - \frac{1}{2}\varepsilon_X > 0$, then

$$P_2 < 2\varepsilon. \tag{23}$$

We first note that

$$\begin{aligned}P_2 = \Pr[X \notin \mathscr{X}] &= \sum_x \Pr[X = x]\Pr[X \notin \mathscr{X} \mid X = x] \\ &= \sum_x P_X(x)[1 - P_X(x)]^{MM_X}\end{aligned} \tag{24}$$

with $P_X(x)$ given by (18). From Theorem 3, the set of X-sequences of length n can be divided into two parts, one $\mathscr{T}_X{}^c(\varepsilon,n)$ of probability $< \varepsilon$ and a disjoint part $\mathscr{T}_X(\varepsilon,n)$ such that

$$\exp\left\{-n\left[H(X) + \frac{A'}{\sqrt{n\varepsilon}}\right]\right\} \leq P_X(x), \tag{25}$$

for all $x \in \mathscr{T}_X(\varepsilon,n)$. Here $A' > 0$ is independent of n and ε. We write (24) as

$$\begin{aligned}P_2 &= \sum_{x \in \mathscr{T}_X} P_X(x)[1 - P_X(x)]^{MM_X} \\ &\quad + \sum_{x \in \mathscr{T}_X^c} P_X(x)[1 - P_X(x)]^{MM_X} \\ &\leq \left[1 - \exp\left\{-n\left[H(X) + \frac{A'}{\sqrt{n\varepsilon}}\right]\right\}\right]^{MM_X} \\ &\quad \sum_{x \in \mathscr{T}_X} P_X(x) + \sum_{x \in \mathscr{T}_X^c} P_X(x) \\ &\leq \left[1 - \exp\left\{-n\left[H(X) + \frac{A'}{\sqrt{n\varepsilon}}\right]\right\}\right]^{MM_X} + \varepsilon \\ &= Z + \varepsilon.\end{aligned} \tag{26}$$

Now

$$\begin{aligned}\log Z &= MM_X \log\left[1 - \exp\left\{-n\left[H(X) + \frac{A'}{\sqrt{n\varepsilon}}\right]\right\}\right] \\ &\leq -\exp(n[I(X,Y) - \tfrac{1}{2}\varepsilon_X])\exp\{n[H(X \mid Y) + \varepsilon_X]\} \\ &\quad \cdot \exp\left\{-n\left[H(X) + \frac{A'}{\sqrt{n\varepsilon}}\right]\right\} \\ &= -\exp\left\{n\left[\frac{1}{2}\varepsilon_X - \frac{A'}{\sqrt{n\varepsilon}}\right]\right\}\end{aligned}$$

on using (5), (7), (8), and (22). Thus, for any fixed ε, Z approaches zero as n becomes large. We can therefore choose n sufficiently large to make $Z < \varepsilon$ and then, from (26), (23) is true.

For the remaining term in (20) we have

$$\begin{aligned}P_3 &= \Pr[Y \in \mathscr{L}, X \in \mathscr{X}, X^* \neq X] \\ &\leq \Pr[X \in \mathscr{X}, X^* \neq X] \\ &= \sum_{xy} P_{XY}(x,y)\Pr[X \in \mathscr{X}, X^* \neq X \mid X = x, Y = y] \\ &= \sum_{xy} P_{XY}(x,y)A(x,y).\end{aligned} \tag{27}$$

Here

$$A(x,y) = \sum_{i=1}^{M_X} \Pr[X \notin \mathscr{X}_1, X \notin \mathscr{X}_2,\cdots,X \notin \mathscr{X}_{i-1}, X \in \mathscr{X}_i, X^* \neq X \mid X = x, Y = y]$$

and the term $i = 1$ is to be interpreted as $\Pr[X \in \mathscr{X}_1, X^* \neq X \mid X = x, Y = y]$. Now

$$A(x,y) = \sum_{i=1}^{M_X} [1 - P_X(x)]^{(i-1)M} \sum_{j=1}^{M} B_{ij}(x,y) \tag{28}$$

with

$$B_{ij}(x,y) = \Pr[X_{i1} \neq x, X_{i2} \neq x,\cdots,X_{ij-1} \neq x, X_{ij} = x, X^* \neq X \mid X = x, Y = y].$$

The terms B_{i1} are to be interpreted as $\Pr[X_{i1} = x, X^* \neq X \mid X = x, Y = y]$. But

$$B_{ij}(x,y) = [1 - P_X(x)]^{j-1}P_X(x)C_{ij}(x,y) \tag{29}$$

with $C_{ij}(x,y) \le \Pr[P_{Y|X}(y \mid X_{i\alpha}) \ge P_{Y|X}(y \mid x)$, for some $\alpha \ne j \mid \mathscr{D}]$ where $\mathscr{D} \equiv \{X_{i1} \ne x, \cdots, X_{ij-1} \ne x, X_{ij} = x, X = x, Y = y\}$.

$$C_{ij}(x,y) \le \Pr\left[\bigcup_{\alpha \ne j} \{P_{Y|X}(y \mid X_{i\alpha}) \ge P_{Y|X}(y \mid x)\} \mid \mathscr{D}\right]$$

$$\le \left[\sum_{\alpha \ne j} \Pr[P_{Y|X}(y|X_{i\alpha}) \ge P_{Y|X}(y|x) \mid \mathscr{D}]\right]^{\rho} \quad (30)$$

where $0 \le \rho \le 1$. Here we have used the now common union bound, (see [1, p. 136]),

$$\Pr\left[\bigcup_i \{\mathscr{A}_i\}\right] \le \left[\sum_i \Pr[\mathscr{A}_i]\right]^{\rho}.$$

But if $\alpha > j$

$$\Pr[P_{Y|X}(y \mid X_{i\alpha}) \ge P_{Y|X}(y \mid x) \mid \mathscr{D}]$$

$$= \Pr[P_{Y|X}(y \mid X_{i\alpha}) \ge P_{Y|X}(y \mid x) \mid X = x, Y = y] \quad (31)$$

while if $\alpha < j$

$$\Pr[P_{Y|X}(y \mid X_{i\alpha}) \ge P_{Y|X}(y \mid x) \mid \mathscr{D}]$$

$$= \Pr[P_{Y|X}(y \mid X_{i\alpha}) \ge P_{Y|X}(y \mid x) \mid X_{i\alpha} \ne x, X = x, Y = y]$$

$$= \frac{\Pr[P_{Y|X}(y \mid X_{i\alpha}) \ge P_{Y|X}(y \mid x), X_{i\alpha} \ne x \mid X = x, Y = y]}{\Pr[X_{i\alpha} \ne x \mid X = x, Y = y]}$$

$$\le a^{-1}\Pr[P_{Y|X}(y \mid X_{i\alpha}) \ge P_{Y|X}(y \mid x) \mid X = x, Y = y] \quad (32)$$

where

$$a \equiv \Pr[X_{i\alpha} \ne x \mid X = x, Y = y] = [1 - P_X(x)] \le 1. \quad (33)$$

Thus, (30)–(32) give

$$C_{ij}(x,y) \le [\{(j-1)a^{-1} + M - j\} \Pr[P_{Y|X}(y \mid X_{i\alpha}) \ge P_{Y|X}(y \mid x) \mid X = x, Y = y]]^{\rho}$$

$$= [(j-1)a^{-1} + M - j]^{\rho}\left[\sum_{x_{i\alpha}}{}' P_X(x_{i\alpha})\right]^{\rho}$$

where the sum is over the set of values of $x_{i\alpha}$ for which $P_{Y|X}(y \mid x_{i\alpha}) \ge P_{Y|X}(y \mid x)$. Since on this set $P_{Y|X}(y \mid x_{i\alpha})/P_{Y|X}(y \mid x) \ge 1$, we have

$$C_{ij}(x,y) \le [(j-1)a^{-1} + M - j]^{\rho} \cdot \left[\sum_{x'} P_X(x')\left[\frac{P_{Y|X}(y \mid x')}{P_{Y|X}(y \mid x)}\right]^{s}\right]^{\rho}$$

for any $s \ge 0$.

Let us now assemble these results. Equations (27)–(29) and (34) give

$$P_3 < \sum_{xy} P_{XY}(x,y) \sum_{i=1}^{M_X} a^{(i-1)M} \cdot \sum_{j=1}^{M} a^{j-1} P_X(x)[(j-1)a^{-1} + M - j]^{\rho} \cdot \left[\sum_{x'} P_X(x')\left[\frac{P_{Y|X}(y \mid x')}{P_{Y|X}(y \mid x)}\right]^{s}\right]^{\rho}. \quad (34)$$

In the Appendix it is shown that

$$\sum_{i=1}^{M_X} a^{(i-1)M} \sum_{j=1}^{M} a^{j-1}[(j-1)a^{-1} + M - j]^{\rho} \le \frac{M^{\rho}}{P_X(x)}. \quad (35)$$

Insert this bound for the sums into (34) and write $P_{XY}(x,y) = P_X(x)P_{Y|X}(y \mid x)$. There results

$$P_3 \le \sum_{xy} P_X(x)P_{Y|X}(y \mid x)^{1-\rho s} M^{\rho} \cdot \left[\sum_{x'} P_X(x')P_{Y|X}(y \mid x')^{s}\right]^{\rho}.$$

Choose $s = 1/(1+\rho)$ to obtain

$$P_3 \le T \equiv M^{\rho} \sum_y \left[\sum_x P_X(x)P_{Y|X}(y \mid x)^{1/(1+\rho)}\right]^{1+\rho}. \quad (36)$$

The quantity T is well known to information theorists. It plays a central role in the Gallager bound for error probability of a noisy memoryless channel. Recalling (6), (18), and (22), we have

$$T = \left[\exp\left\{n\rho\left[I(X,Y) - \frac{\varepsilon_X}{2}\right]\right\}\right] \cdot \left[\sum_y \left[\sum_x p_X(x)p_{Y|X}(y \mid x)^{1/(1+\rho)}\right]^{1+\rho}\right]^{n}$$

$$\le \exp\left\{-n\rho\left[\frac{1}{\rho}V(\rho) - I(X,Y) + \frac{\varepsilon_X}{2}\right]\right\}.$$

Here

$$V(\rho) = -\log \sum_y \left[\sum_x p_X(x)p_{Y|X}(y \mid x)^{1/(1+\rho)}\right]^{1+\rho}$$

is seen to be analytic in the neighborhood of $\rho = 0$, and indeed $V(0) = 0$. The function $E(\rho) = V(\rho)/\rho$ is also analytic in this neighborhood and by L'Hôpital's rule and a straightforward calculation one finds

$$E(0) = \left.\frac{dV}{d\rho}\right|_{\rho=0} = -\sum_y p_Y(y)\log p_Y(y) + \sum_{xy} p_{XY}(x,y)\log p_{Y|X}(y \mid x) = I(X,Y)$$

by (4) and (7).

Since $E(\rho)$ is analytic at $\rho = 0$, there exists a $\rho_0 > 0$ such that $|E(\rho_0) - I(X,Y)| < \varepsilon_X/4$ whence (36) becomes

$$P_3 \le \exp(-n\rho_0\varepsilon_X/4).$$

It is now seen that by choosing n sufficiently large, $P_3 < \varepsilon$. Combined with (23), (21), and (20), this shows that $P_e \le 4\varepsilon$. Now choose $\varepsilon = \varepsilon'/4$. We have then shown that $P_e(\mathscr{X}) < \varepsilon'$. There must therefore exist a code in the ensemble for which (16) holds. Thus Theorem 2 is proved.

IV. Determination of the Regions $\mathscr{R}$

Table I lists twelve theorems whose applications in connection with Fig. 10 give immediately the admissible rate region $\mathscr{R}$ for the 16 cases. In Table I, the symbol x for the state of a switch means that the theorem holds both when

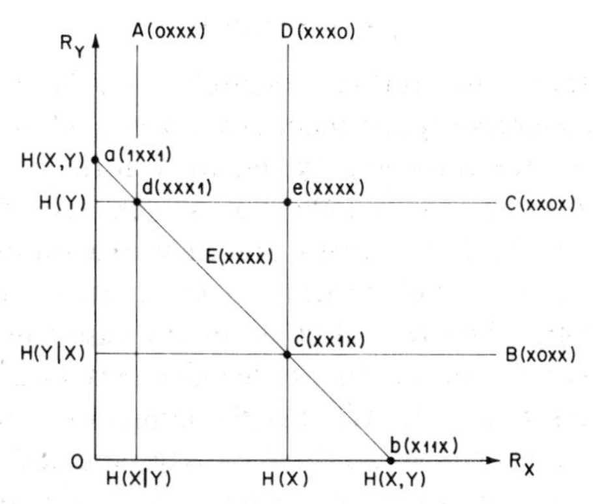

Fig. 10. Lines and points of Table I.

TABLE I
THEOREMS TO DETERMINE $\mathscr{R}$

$s_1s_2s_3s_4$	Theorem Name	Theorem	
		It is necessary that:	
0xxx	A	$R_X \geq H(X \mid Y)$	
x0xx	B	$R_Y \geq H(Y \mid X)$	
xx0x	C	$R_Y \geq H(Y)$	
xxx0	D	$R_X \geq H(X)$	
xxxx	E	$R_X + R_Y \geq H(X,Y)$	
		It is sufficient that:	
1xx1	a	$R_X = 0$	$R_Y = H(X,Y) + \varepsilon_{XY}$
x11x	b	$R_X = H(X,Y) + \varepsilon_{XY}$	$R_Y = 0$
xx1x	c	$R_X = H(X) + \varepsilon_X$	$R_Y = H(Y \mid X) + \varepsilon_Y$
xxx1	d	$R_X = H(X \mid Y) + \varepsilon_X$	$R_Y = H(Y) + \varepsilon_Y$
xxxx	e	$R_X = H(X) + \varepsilon_X$	$R_Y = H(Y) + \varepsilon_Y$
		$\varepsilon_X,\varepsilon_Y,\varepsilon_{XY} > 0$	
		Bit stuffing:	
xxxx	f	$(R_X,R_Y) \in \mathscr{R} \Rightarrow (R_X + \delta_X,\ R_Y + \delta_Y) \in \mathscr{R}$ $\delta_X,\ \delta_Y \geq 0$	
		Limited time sharing:	
xxxx	g	If $(R_X,R_Y) \in \mathscr{R}$, $(R_X',R_Y') \in \mathscr{R}$ and $R_X + R_Y = H(X,Y)$ and $R_X' + R_Y' = H(X,Y)$, then $(R_X'',R_Y'') \in \mathscr{R}$, $R_X'' = \lambda R_X + (1 - \lambda)R_X'$ $R_Y'' = \lambda R_Y + (1 - \lambda)R_Y'$, $0 \leq \lambda \leq 1$	

the switch is open and when the switch is closed. For example, Theorem A asserts that if switch S_1 is open the $\mathscr{R}_X$ coordinate of an admissible rate point must be at least as large as $H(X \mid Y)$. Theorem E asserts that the coordinates of all admissible rate points, independent of the switch settings, must satisfy $R_X + R_Y \geq H(X,Y)$. Theorem a says that if switches S_1 and S_4 are closed, then $R_X = 0$, $R_Y = H(X,Y) + \varepsilon_{XY}$ is an admissible point for every $\varepsilon_{XY} > 0$. Since $\mathscr{R}$ is defined as the closure of the set of admissible rate points, the theorem also asserts that $R_X = 0$, $R_Y = H(X,Y)$ is in $\mathscr{R}$.

On Fig. 10 certain lines and points are labeled with the names of theorems of Table I. The corresponding switching states are affixed. These points and lines can be used along with Theorems f and g to determine immediately the boundary of $\mathscr{R}$ for any of the 16 cases. The bit stuffing Theorem f then shows that the interior of the boundary can be filled in to obtain $\mathscr{R}$. We give several illustrations.

1) For the switch setting 1011, we see at once from Table I, scanning the columns of switching states, that Theorems B E a c d e apply. The first two show that $\mathscr{R}$ cannot extend below the line labeled B on Fig. 10 nor below the line labeled E there. The next four applicable theorems show that the points a, c, d, and e all lie in $\mathscr{R}$. Theorem f then shows that points above a on the R_Y-axis are in $\mathscr{R}$ as well as all points on B to the right of c. Theorem g shows that the line segment $\overline{ac}$ is in $\mathscr{R}$. The region as given in Fig. 5 is thus established.

2) For the setting 0001, Table I shows that A B C E d e all apply. Locating the lines ABCE on Fig. 10, we see that $\mathscr{R}$ can neither extend to the left of line A nor below line C. The point d is in $\mathscr{R}$, and by Theorem f so is every point to the right of it on line C and every point above it on line A. The region $\mathscr{R}$ of Fig. 7 is thus established.

Many of the theorems of Table I are trivial and we do not belabor them.

Theorem E: The pair of random variables X,Y can be regarded as a single random variable Z taking A_XA_Y values. The entropy of this variable is $H(Z) = H(X,Y)$. Any encoding of the pair (X,Y) as described by Fig. 3 can also be regarded as an encoding of Z by indexing the M_XM_Y possible pairs of values (i_X,i_Y) for I_X and I_Y and taking this index of (I_X,I_Y) as the value of I_Z. If (R_X,R_Y) were admissible and $R_X + R_Y < H(X,Y)$, say $R_X + R_Y = H(X,Y) - \delta$, the construction just mentioned would show the existence of Z codes with $M_Z = M_XM_Y = \lfloor \exp(nR_X) \rfloor \lfloor \exp(nR_Y) \rfloor \leq \lfloor \exp[n(R_X + R_Y)] \rfloor = \lfloor \exp[n(H(X,Y) - \delta)] \rfloor$ values for the channel symbols that had error $< \varepsilon$. But this contradicts Theorem 1 as applied to the variable Z. Q.E.D.

Theorem A: Let switch S_1 be open and suppose that $R_X = H(X \mid Y) - \delta$ and $R_Y = R_2$ is an admissible rate point. We first show that this implies that $R_X = H(X \mid Y) - \delta$, $R_Y = H(Y) + \delta/2$ is also an admissible rate point.

Let $\mathscr{C}_X,\mathscr{C}_Y,\mathscr{D}_X,\mathscr{D}_Y$ be encoders and decoders that employ coded message rates $R_X = H(X \mid Y) - \delta$ and $R_Y = R_2$ and that achieve error probability ε. We replace $\mathscr{C}_Y$ by an encoder $\mathscr{C}_Y'$ that produces coded messages at a rate $R_2 = H(Y) + \delta/2$ by using a list of typical Y sequences. We know by Theorem 1 that for large enough n such an encoder and a corresponding decoder $\mathscr{D}_Y'$ exist, ones that reproduce the Y sequence with arbitrary accuracy. We now consider a new decoder $\mathscr{D}_Y''$ consisting of $\mathscr{D}_Y'$ followed by $\mathscr{C}_Y$ and $\mathscr{D}_Y$. The scheme $\mathscr{C}_X,\mathscr{C}_Y',\mathscr{D}_X,\mathscr{D}_Y''$ signals at rates $R_X = H(X \mid Y) - \delta$ and $R_Y = H(Y) + \delta/2$ with error probability $< \varepsilon$. But $R_X + R_Y = H(X \mid Y) + H(Y) - \delta/2 = H(X,Y) - \delta/2$, contrary to Theorem E. Therefore, we must have $R_X \geq H(X \mid Y)$ for an admissible point.

Theorem B: Theorem A with X and Y interchanged.

Theorems C and D: Follow directly from Theorem 1.

Theorem a: Theorem 1 applied to $Z = (X,Y)$.

Theorem b: Theorem a with X and Y interchanged.

Theorem c: Theorem 2.

Theorem d: Theorem 2 with X and Y interchanged.

Theorem e: Follows from Theorem 1.

Theorem f: This theorem follows from the simple observation that for any encoder, say $\mathscr{C}_X(n,s_2,M_X)$, mapping elements of $\mathscr{A}_X^n \times \mathscr{A}_Y^n$ onto integers of the set $\mathscr{M}_X = \{1,2,\cdots,M_X\}$ we can always trivially increase the range of the mapping by increasing M_X (and hence R_X). The new values of the enlarged set $\mathscr{M}_X'$ never occur as values of I_X in this new mapping and so the decoder can be defined arbitrarily for these values. The error probability remains unchanged.

Theorem g: A result somewhat stronger than Theorem 1 is the following. If $R > H(X)$, then for every $\varepsilon > 0$ there exists an $n_0(\varepsilon)$ and an encoder $\mathscr{C}(n_0,\lfloor \exp(n_0 R)\rfloor)$ and a decoder $\mathscr{D}(n_0,\lfloor \exp(n_0 R)\rfloor)$ such that $\Pr[X^* \neq X] < \varepsilon$. Furthermore, for each integer $n > n_0$ there exists an encoder $\mathscr{C}(n,\lfloor \exp(nR)\rfloor)$ and a decoder $\mathscr{D}(n,\lfloor \exp(nR)\rfloor)$ such $\Pr[X^* \neq X] < \varepsilon$. This result is implicit in proofs of Theorem 1 that compute explicit bounds for error probability such as the one given by Jelinek [3, sec. 5.2, p. 86]. Examination of Section III then shows that a correspondingly strengthened statement of Theorem 2 is also possible: if $R_X = H(X \mid Y) + \varepsilon_X$, $R_Y = H(Y) + \varepsilon_Y$ then for every $\varepsilon > 0$ there exist coders and decoders for every n greater than some $n_0(\varepsilon)$ for which $\Pr[X^* \neq X \text{ and } Y^* \neq Y] < \varepsilon$. We call a rate point (R_X,R_Y) *strongly admissible* if for every $\varepsilon > 0$ and all sufficiently large n there exist encoders and decoders using block length n for which $\Pr[X^* \neq X$ and $Y^* \neq Y] < \varepsilon$. When points a, b, c, d, or e are in $\mathscr{R}$, i.e., when the switch settings are appropriate, they are strongly admissible rate points.

For strongly admissible rate points (R_X,R_Y) and (R_X',R_Y'), Theorem g is shown as follows. There are encoders and decoders for all block lengths n greater than $n_0(\varepsilon/2)$ that use encoded message rates R_X and R_Y with error $<\varepsilon/2$. Similarly there are encoders and decoders for all block lengths n' greater than $n_0'(\varepsilon/2)$ that use encoded message rates R_X' and R_Y' with error $<\varepsilon/2$. Let λ, $0 \leq \lambda \leq 1$, be rational and choose n'' so large that $n = \lambda n''$ and $n' = (1-\lambda)n''$ are both integers and $n \geq n_0$ and $n' \geq n_0'$. Now encode $X - Y$ sequences by alternately using first block length n and the code with rate (R_X,R_Y) and then block length n' with the code of rate (R_X',R_Y'). This operation can be regarded as the use of a single block code of length $n'' = n + n'$. For this new code, $M_X'' = M_X M_X' = \lfloor \exp(\lambda n'' R_X)\rfloor \lfloor \exp[(1-\lambda)n'' R_X']\rfloor \leq \lfloor \exp(n'' R_X'')\rfloor$. As in the proof of Theorem f, we can artificially increase M_X'' so that $M_X'' = \lfloor \exp(n'' R_X'')\rfloor$. A similar calculation holds for M_Y''. The error probability for this n'' code is less than $1 - (1-\varepsilon/2)^2 = \varepsilon - (\varepsilon/4)^2 \leq \varepsilon$. This establishes Theorem g for rational λ. But $\mathscr{R}$ was defined as the closure of all admissible points and since the rationals are dense in the reals, Theorem g is established.

A stronger form of Theorem g is indeed true, for examination of Figs. 4–9 shows that $\mathscr{R}$ is convex for all 16 cases. Thus we have the following theorem.

Theorem h: If $(R_X,R_Y) \in \mathscr{R}$ and $(R_X',R_Y') \in \mathscr{R}$, then for every $\lambda, 0 \leq \lambda \leq 1, (R_X'',R_Y'') \in \mathscr{R}$ where $R_X'' = \lambda R_X + (1-\lambda)R_X', R_Y'' = \lambda R_Y + (1-\lambda)R_Y'$.

V. Commentary

Many topics for further research on joint coding of correlated sources suggest themselves. We mention a few.

How does the foregoing extend to N correlated sources instead of two? The number of switch settings grows rapidly with N. Many cases are easy extensions of our results for $N = 2$, but basically new situations arise too. For example, when $N = 3$ consider the case where the X decoder sees I_X and I_Y, the Y decoder sees I_Y and the Z decoder sees I_Y and I_Z. What is the admissible rate region then if the encoder sees all three message sources?

How does the foregoing extend to a rate-distortion theory of correlated sources? The probability-of-error criterion is then replaced by more general measures of decoder output fidelity. A rate-distortion theory would permit a generalization from discrete-valued to continuous-valued random variables.

The design of block codes of given length n to have small error probability is a more difficult and more interesting problem for correlated sources than for a single source, where the problem is solved by providing a list of typical sequences. Here, in cases such as 0011 one wants to take advantage of the known correlation of the sources. Are there better methods than timesharing to achieve rates along the line E between c and d?

What is the theory of variable-length encodings for correlated sources? How does one generalize the Huffman code, say, in the case 0011 to encode X sequences of length n with fewest bits on the average when $R_Y = H(Y)$?

How does the theory extend for correlated sources that are not independent drawings of pairs of correlated variables?

These are but a few of the many interesting problems to be solved in this area.

Acknowledgment

The authors gratefully acknowledge many valuable discussions with Prof. N. T. Gaarder on this research.

Appendix

Here we establish the inequality (35). Jensen's theorem (see, for example, [1, (4.4.4) and (4.4.5), p. 85]) asserts that if $g(x)$ is convex up for $a \leq x \leq b$, i.e., if $g''(x) \leq 0$, for $a \leq x \leq b$, then

$$\sum_1^M p_j g(x_j) \leq g\left(\sum_1^M p_j x_j\right)$$

where $a \leq x_1 \leq x_2 \cdots \leq x_M \leq b$ and

$$\sum_1^M p_j = 1$$

with $p_j \geq 0, j = 1,\cdots,M$. We apply this theorem to the following function

$$g(x) = x^\rho, \qquad 0 \leq \rho \leq 1$$

which is convex up for $x \geq 0$, taking

$$p_j = \frac{1-a}{1-a^M} a^{j-1}$$

$$x_j = [(j-1)a^{-1} + M - j] \qquad j = 1,2,\cdots,M$$

with a given by (33), so that $0 < a < 1$. We find

$$\frac{1-a}{1-a^M}\sum_{j=1}^{M} a^{j-1}[(j-1)a^{-1}+M-j]^\rho$$

$$\leq \left[\frac{1-a}{1-a^M}\sum_{j=1}^{M} a^{j-1}[(j-1)a^{-1}+M-j]\right]^\rho$$

$$= \left[\frac{1-a}{1-a^M}\left\{(M-a^{-1})\sum_{j=1}^{M} a^{j-1} + (a^{-1}-1)\sum_{j=1}^{M} ja^{j-1}\right\}\right]^\rho$$

$$= \left[\frac{1-a}{1-a^M}\left\{M\frac{1-a^{M-1}}{1-a}\right\}\right]^\rho.$$

Here we have evaluated the sums using the two formulas

$$\sum_{j=1}^{M} a^{j-1} = \frac{1-a^M}{1-a}$$

$$\sum_{j=1}^{M} ja^{j-1} = \frac{d}{da}\left(\frac{1-a^{M+1}}{1-a}\right) = \frac{1-(M+1)a^M + Ma^{M+1}}{(1-a)^2} \qquad \text{(A-1)}$$

and performed some algebraic simplification. It follows then that

$$\sum_{j=1}^{M} a^{j-1}[(j-1)a^{-1}+M-j]^\rho \leq \frac{1-a^M}{1-a} M^\rho \left(\frac{1-a^{M-1}}{1-a^M}\right)^\rho$$

$$\leq \frac{1-a^M}{1-a} M^\rho \qquad \text{(A-2)}$$

since if $0 < a < 1$, $[1-a^{M-1}]/[1-a^M] < 1$.

Returning to (35), from (A-1) we have, on replacing M by M_X and a by a^M,

$$\sum_{i=1}^{M_X} a^{M(i-1)} = \frac{1-a^{MM_X}}{1-a^M}.$$

Combining this with (A-2) gives

$$\sum_{i=1}^{M_X} a^{M(i-1)} \sum_{j=1}^{M} a^{j-1}[(j-1)a^{-1}+M-j]^\rho$$

$$\leq \frac{1-a^{MM_X}}{1-a^M}\,\frac{1-a^M}{1-a} M^\rho$$

$$\leq \frac{1}{1-a} M^\rho = \frac{M^\rho}{P_{\mathbf{X}}(\mathbf{x})} \qquad (35)$$

on using the definition (33) and the fact that $1 - a^{MM_X} \leq 1$.

Q.E.D.

References

[1] R. G. Gallager, *Information Theory and Reliable Communication.* New York: Wiley, 1968.

[2] R. Ash, *Information Theory.* New York: Interscience, 1965.

[3] F. Jelinek, *Probabilistic Information Theory.* New York: McGraw-Hill, 1968.

Author Index